Springer Proceedings in Complexity

Springer Proceedings in Complexity publishes proceedings from scholarly meetings on all topics relating to the interdisciplinary studies of complex systems science. Springer welcomes book ideas from authors. The series is indexed in Scopus.

Proposals must include the following:

- name, place and date of the scientific meeting
- a link to the committees (local organization, international advisors etc.)
- scientific description of the meeting
- list of invited/plenary speakers
- an estimate of the planned proceedings book parameters (number of pages/articles, requested number of bulk copies, submission deadline)

Submit your proposals to: christoph.baumann@springer.com

More information about this series at http://www.springer.com/series/11637

Anna Visvizi · Miltiadis D. Lytras ·
Naif R. Aljohani
Editors

Research and Innovation Forum 2020

Disruptive Technologies in Times of Change

Springer

Editors
Anna Visvizi
Research & Innovation Institute (Rii)
Warsaw, Poland

Miltiadis D. Lytras
Effat University
Jeddah, Saudi Arabia

Naif R. Aljohani
Faculty of Computing and Information Technology
King Abdulaziz University
Jeddah, Saudi Arabia

ISSN 2213-8684 ISSN 2213-8692 (electronic)
Springer Proceedings in Complexity
ISBN 978-3-030-62065-3 ISBN 978-3-030-62066-0 (eBook)
https://doi.org/10.1007/978-3-030-62066-0

This Springer imprint is published by the registered company Springer Nature Switzerland AG
The registered company address is: Gewerbestrasse 11, 6330 Cham, Switzerland

Organization

Chairs

Anna Visvizi, Ph.D., Research & Innovation Institute, Poland
Miltiadis D. Lytras, Ph.D., King Abdul-Aziz University

Steering Committee Members

Anna Visvizi, Ph.D., Research & Innovation Institute, Poland and SGH Warsaw School of Economics, Poland
Miltiadis D. Lytras, Ph.D., King Abdul-Aziz University, Saudi Arabia
Naif R. Aljohani, Ph.D., King Abdul-Aziz University, Saudi Arabia
Claudia Toriz Ramos, Ph.D., University of Fernando Pessoa, Porto; CEPESE Research Centre, University of Porto, Portugal
Enric Serradell-Lopez, Ph.D., Open University of Catalonia, Spain

Program Committee

Mu-Yen Chen, Ph.D., National Taichung University of Science and Technology, Taichung, Taiwan
Małgorzata Dziembała, Ph.D., University of Economics in Katowice, Poland
Jari Jussila, Ph.D., Häme University of Applied Sciences, Finland
Eugenijus Kurilovas, Ph.D., Vilnius Gediminas Technical University, Lithuania
Higinio Mora, Ph.D., University of Alicante, Spain,
Marco Moreno Ibarra, Ph.D., Instituto Politécnico Nacional, Mexico
Jennifer Moss Breen, Ph.D., Creighton University, USA
Placido Rogerio Pinheiro, University of Fortaleza, Brazil
Lucie Rohlíková, Ph.D., University of West Bohemia, Czechia

Akila Sarirete, Ph.D., Effat University, Saudi Arabia
Filippo Sciarrone, Ph.D., Roma Tre University, Italy
Chien-wen Shen, Ph.D., National Central University, Taiwan
Marco Spruit, Ph.D., Utrecht University, the Netherlands
Miguel Torres Ruiz, Ph.D., Instituto Politécnico Nacional, Mexico
Orlando Troisi, Ph.D., University of Salerno, Italy
Yves Wautelet, Ph.D., KU Leuven, Belgium
Yenchun Jim Wu, Ph.D., National Taiwan Normal University, Taiwan

rii forum ©

Under the auspices of:
HELLENIC REPUBLIC
Ministry of Digital Governance

Athens, April 15-17, 2020

Research & Innovation Forum

Advancing research on technology, innovation, education and their social impact

https://rii-forum.org

Foreword by the keynote speaker

Rii Forum 2020, COVID-19 and the Society: Towards the New Normal

In my keynote speech at Rii Forum 2019, I wrote a short essay on the responsibility of being an academic in modern times and the search for its true meaning. One of the main arguments of my speech was that soft skills and markedly cognitive ones can help us avoid the drawbacks of automation and overreliance on technology in our daily life. Those innately human characteristics, such as ethics or creativity, help us minimize the harmful effects of globalisation and automation.

Academics and professionals must be alert to signs of the labour market, trends, and technologies to overcome the tsunami of the Fourth Industrial Revolution. I think that in 2019, everyone agreed with that. We also decided to be very attentive to any change in trend and technology to adapt to it. Flexibility was our mantra. A year later, on 15 January 2020, the World Economic Forum published its annual report on the threats to the planet, *The Global Risks Report 2020*. It listed the top 10 risks in the next 10 years (defined as long-term risks). Among them are climate action failure, weapons of mass destruction, biodiversity loss, extreme weather, water crises, information infrastructure breakdown, natural disasters, cyberattacks, human-caused environmental disasters and, closing the list, infectious diseases.

When the experts who participated in the report assessed the probability of the risks of infectious diseases, it was left off the list of *probable* catastrophes. Other ones, such as those related to the environment, were seen as much more likely. The top five global risks in terms of likelihood were extreme weather conditions, climate action failure, natural disasters, biodiversity loss, and human-caused environmental disasters, while in the previous decades, economic and financial crises were considered the most dangerous. Almost as soon as the report was presented, COVID-19 appeared, first affecting China before spreading rapidly throughout most of the countries of the world with similar effect. We have witnessed with our own eyes how entire sectors of the economy ceased to function and pandemics, which we thought were distant in time, relegated to the Middle Ages, became a reality in the twenty-first century. Sectors such as mobility, transport, commerce, tourism, industry, and services have suffered terribly in the current crisis. Here are some specific examples of well-known companies:

- **Primark** went from sales of £650m a month to zero when the last of its stores closed on March 22. Its direct-sales strategy was gutted by COVID-19. The company, which was founded in 1969, did not have an online store to sell its products.
- Disrupting the disruptors: **Airbnb** created a new industry based on idea of bringing the world closer together. Now, with travel suspended and people restricted to near their homes, more than 7 million Airbnb listings spread across 100,000 cities worldwide are mostly unused. This company and other short-term rental companies are suffering from this unexpected whirlwind. Hosts report a dramatic drop in guest numbers, but the neighbourhood may benefit from cheaper prices for rentals.

Who would have told us a few months ago that giants like Primark or Airbnb would be forced to suspend most or all their activity (even temporarily)? No one would have predicted it and meant it, and in the case of Primark, we can certainly say that there was no contingency plan for a situation like this. I noted at the beginning that analysis and critical thinking are strengths of researchers in the process of automation, and it is at times like this that we must demonstrate that human nature can coexist with processes as traumatic as the pandemic and to propose solutions and above all a light of hope for the future.

One of the presentation points of *The Global Risks Report* referred to 'Turbulence: The new normal'. It is curious how this *new normal* has vividly expressed that the changes suffered during this period will in many cases be permanent. The International Monetary Fund has relayed very pessimistic forecasts for the world economy and foresees high levels of unemployment in the short and medium terms. Economic sectors related to mobility, such as tourism and air transport, are suffering sharply and especially affect countries where tourism is a vital part of the economy, such as Mediterranean states.

At the moment, there are three powerful human entities in the world attempting to counter the perverse effects of the pandemic—the US Federal Reserve, which stands ready to print unlimited money, the International Monetary Fund, which will lend large amounts of money to countries that need it, and the European Central Bank, which will facilitate the availability of money to all countries of the European Union, even at the cost of greatly increasing public debt. What impact does this have on our world? The truth is that the COVID-19 pandemic is accelerating the division of the world into two economic spheres—countries in China's orbit and the Western world, nominally led by the USA. But the global supply system is now also being questioned. Before the pandemic, the production of a standard car required between 70,000 and 90,000 parts, many of them originally manufactured in different countries, and an element as simple as a bicycle can have components from as many as 20 different countries. In the post-COVID world, the entire supply chain must be rethought, so the need arises about how to rethink basic elements, such as whether a surgical mask requires a trip of thousands of kilometres from the production centre to the user. The same is applicable to the consumption of raw materials and other basic products.

The **Research & Innovation Forum (Rii Forum)** provides a unique platform for discussion about these issues, and this **Rii Forum 2020 *Proceedings*** book in which I have had the pleasure to participate is a great example of this. The world has turned to digital and electronic commerce to replace face-to-face sales, while video conferencing and the companies that market this type of software have grown exponentially, replacing—perhaps forever—many in-person trips. The economic and social relations of our environment have changed, maybe forever.

Like the companies mentioned above, **Rii Forum 2020** also faced the unforeseen consequences of the pandemic and had to convert into a virtual conference. Even so, it was worthwhile and below, I highlight the conference tracks:

- Smart Cities: Issues and Challenges
- Technology-enhanced Teaching and Learning
- ICT, the Business Sector and Policymaking
- ICT, Politics and the Society.

All of these topics are essential elements to consider in the coming years—I would even say vital. How will technology evolve and adapt to the new normal? What will happen to cities, migration, and the public health system? What will we say about diversity and the need for transformative and adaptive leadership? And finally, what can we say about education? Higher education surely has accelerated more in three months than it would have in years. Universities and educational institutions in general have discovered the online world, and for those of us working on these issues, we consider it a challenge. It is no longer enough to use an online methodology; we must fight to position ourselves and distinguish ourselves, ensuring superior quality and excellence in everything we do. And that is applicable to all sectors and companies.

Perhaps one of the few positive elements of this period has been verifying that humanity has been able to agree on a common goal: to find a cure in the form of a vaccine or treatment. Every day, hundreds of academic and research papers are published openly on all possible perspectives related to that common goal. It has never happened in the history of humanity. When Kennedy in the 1960s proposed that in 10 years the USA would put a person on the Moon that objective materialized as a 'Moonshot' programme, and today, we affirm that we are in a similar situation. Hundreds of research teams around the world are working and sharing information collaboratively to find a solution—and 7.5 billion people are counting on it.

I am sure that in our next meeting—**Rii Forum 2021**—we will be able to give a good account of the new communications and research into the reality of globalization, discuss the effects of this period and reflect on what we could do better, and, above all, what we can do to prevent it.

Good luck and #staysafe.

Enric Serradell-Lopez
Universitat Oberta de Catalunya
Barcelona, Spain
e-mail: eserradell@uoc.edu

Preface by the Rii Forum 2020 Chairs

The Covid-19 pandemic redefines our societies. Amidst the chase for ways of navigating the implications of the pandemic, information and communication technology (ICT) has been reified across issues and domains. Faced with the pandemic, tangible applications of research are sought after in such fields as, to mention just a few, health care, education, business management, communication, governance, public administration, and so on. Today, the dialectical relationship between technology and the society has been brought to the surface of the popular debate and academic research. In this way, the necessity of a thorough discussion engaging policymakers, civil society, and the research community has been recognized. For the second year in a row now, **Research and Innovation Forum** (**Rii Forum**) sought to encourage this kind of debate among researchers, policymakers and the civil society.

Research and Innovation Forum (**Rii Forum**) is an annual conference that brings together researchers, academics, and practitioners to engage in a conceptually sound, inter- and multidisciplinary, empirically driven debate on key issues influencing the dynamics of social interaction today. The role of advances in sophisticated technology stands at the heart of discussions held during **Rii Forum**. Held annually, **Rii Forum** features in-depth cutting-edge research on the most current and the emerging issues that unfold at the intersection of technology and society. Normally, the format of **Rii Forum** would consist of traditional and flipped presentations, interactive workshops, and featured roundtables that would create ample space for exchange of ideas and networking. The Covid-19 pandemic, the travel restrictions and eventually lockdown gradually introduced across the European Union and around the world, made it necessary for the conference to be held virtually.

Originally scheduled to take place in Athens, 17–19 April 2020, **Rii Forum 2020** took place virtually, i.e. via a specialized online conference platform. The **Rii Forum 2020** delegates represented countries from nearly all continents, including North and South America, Asia, the Arab Peninsula, Europe, and sub-Saharan Africa. The conference was held under the Aegis of the Ministry of Digital Governance of the Hellenic Republic.

The cutting-edge quality of scholarly discussion during the **Rii Forum 2020** was possible due to an arduous review, selection, and double-blind peer review process. Specifically, nearly 150 extended paper proposals from all over the world were submitted in response to the call for papers. The latter was organized in line with the following discussion panels:

- Smart Cities: Issues and Challenges
- Technology-enhanced Teaching and Learning
- ICT, the Business Sector, and Policymaking
- ICT, Politics, and the Society.

The paper proposals were reviewed by the **Rii Forum Programme Committee**. Out of the initial 150 paper proposals, 75 paper proposals were accepted to be presented at the conference, and 51—following a rigorous double-blind peer review process are presented—are featured in this volume. The structure of the volume mirrors the key panels as outlined above.

The papers included in these **Rii Forum 2020 proceedings** serve as a proof of the research community's steady engagement with issues and developments shaping today's world. The content of the papers included in this collection suggests as well that the research community seeks to offer ways of bypassing societal needs as they emerge. A careful reading of the content of this volume will allow the reader to get a good grasp of nascent and already present issues that our society must face. Simultaneously, the content of the papers included in this volume serves as a useful primer for all those who seek to deal with diverse implications of the Covid-19 induced lockdown as experienced in the fields of education, health care, politics, and business management.

We remain grateful to the **Rii Forum Steering Committee** and the **Rii Forum Programme Committee** for their commitment, sound judgment, and hard work in the process of organizing the **Rii Forum 2020** and then successfully moving the conference to the virtual space. We would like to say 'thank you' to all contributing authors for their hard work and their patience in subsequent rounds of the 'revise and resubmit' process. This would not be possible, of course, without the reviewers who devoted countless hours to evaluate papers submitted to this volume. Finally, we would like to express our gratitude to the entire Springer team and the Editors of Complexity for their continued support and guidance.

Hopeful that the COVID-19 pandemic will be contained soon and sustainably, we take this opportunity to invite you all to join the **Rii Forum 2021** which will take place in Athens, Greece, in April 2020. Please, check the **Rii Forum** website (https://rii-forum.org) for updates.

Sincerely,

Warsaw, Poland
Jeddah, Saudi Arabia

Anna Visvizi
Miltiadis D. Lytras
Chairs, Rii Forum 2020

Contents

Part IV ICT, Politics, Business and the Society

Contributors

Rabeeh Abbasi CS Department, Quaid-i-Azam University, Islamabad, Pakistan

Fairouz Aldhmour Innovation and Technology Management Department, Arabian Gulf University, Manama, Bahrain

Fahad Alfarsi Department of Electronics and Electrical Engineering, University of Jeddah, Jeddah, Saudi Arabia

Walaa Alhabashi FCIT, IS Department, King AbdulAziz University, Jeddah, Saudi Arabia

Osama Alharbi Department of Electronics and Electrical Engineering, University of Jeddah, Jeddah, Saudi Arabia

Soltan A. Alharbi Department of Computer and Network Engineering, University of Jeddah, Jeddah, Saudi Arabia

Odeh Al-Jayyousi Innovation and Technology Management Department, Arabian Gulf University, Manama, Bahrain

Naif Aljohani FCIT, IS Department, King AbdulAziz University, Jeddah, Saudi Arabia

May Al-Khalifa Innovation and Technology Management Department, Arabian Gulf University, Manama, Bahrain

Faisal Alsalmi Department of Electronics and Electrical Engineering, University of Jeddah, Jeddah, Saudi Arabia

Sachi Arafat FCIT, IS Department, King AbdulAziz University, Jeddah, Saudi Arabia

Jorge Azorín-López Department of Computer Technology, University of Alicante, Alicante, Spain

Zain Balfagih Computer Science Department, Effat University, Jeddah, Saudi Arabia

Mohammed Balfaqih Department of Computer and Network Engineering, University of Jeddah, Jeddah, Saudi Arabia

Benaouda Bensaid Istanbul Sabahattin Zaim University, Küçükçekmece/İstanbul, Turkey

Alessio Maria Braccini Department of Economics and Entrepreneurship, Tuscia University, Viterbo, Italy

Tayeb Brahimi Energy and Technology Research Center, Natural Science, Mathematics and Technology Unit, College of Engineering, Effat University, Jeddah, Saudi Arabia

Antonella Carbonaro Department of Computer Science and Engineering – DISI, University of Bologna, Cesena, Italy

Mario Casillo DIIN, University of Salerno, Fisciano, SA, Italy

Amandine Chagniot Université de Namur, Namur, Belgium;
Haute Ecole Léonard de Vinci, Woluwe-Saint-Lambert, Belgium

Jeng-Chung Chen National Taiwan Normal University, Taipei, Taiwan

Fabio Clarizia DIIN, University of Salerno, Fisciano, SA, Italy

Francesco Colace DIIN, University of Salerno, Fisciano, SA, Italy

Ricardo Colomo-Palacios Østfold University College, Halden, Norway

Rafael Comin-Nunes University Center September 7, Fortaleza, Brazil

Margarida Neves de Abreu School of Nursing of Porto, Porto, Portugal

Wilson Correia de Abreu School of Nursing of Porto, Porto, Portugal

Carlo De Medio Engineering Department, Roma Tre University, Rome, Italy

Massimo De Santo DIIN, University of Salerno, Fisciano, SA, Italy

Niels de Vries Department of Information and Computing Sciences, Utrecht University, Utrecht, The Netherlands

Václav Duffek Centre of Biology, Geoscience and Environmental Education, University of West Bohemia, Plzen, Czech Republic

Małgorzata Dziembała Department of International Economic Relations, University of Economics in Katowice, Katowice, Poland

Aldrin Espín-León Faculty of Sociology and Social Work, Central University of Ecuador, Quito, Ecuador

Raquel Ferreras-Garcia Universitat Oberta de Catalunya, Barcelona, Spain

Jan Fiala Department of Computer Science and Educational Technology, University of West Bohemia, Plzen, Czech Republic

Francesc Font Cot UOC—Universitat Oberta de Catalunya, Barcelona, Spain

Giacomo Frisoni Department of Computer Science and Engineering – DISI, University of Bologna, Cesena, Italy

Fabio Gasparetti Engineering Department, Roma Tre University, Rome, Italy

V. Gilart-Iglesias Department of Computer Science Technology and Computation, University of Alicante, Alicante, Spain

Mara Grimaldi University of Salerno, Fisciano, SA, Italy

Yasmine Ben Hamadi KU Leuven, Leuven, Belgium

Samedi Heng HEC Liège, Université de Liège, Liège, Belgium

Ana Beatriz Hernández-Lara Department of Business Management, Universitat Rovira i Virgili, Reus, Spain

Petr Hořejší Department of Industrial Engineering and Management, University of West Bohemia, Plzen, Czech Republic

Anne-Mari Järvenpää Häme University of Applied Sciences, Hämeenlinna, Finland

Antonio Jimeno-Morenilla Department of Computer Technology, University of Alicante, Alicante, Spain

Akinul Islam Jony Universitat Oberta de Catalunya, Barcelona, Spain

Jari Jussila Häme University of Applied Sciences, Hämeenlinna, Finland

Oduor Kevin Stowelink, Nairobi, Kenya

Nelly Kfir Center for International Migration and Integration (CIMI), Jerusalem, Israel

Krzysztof Kozłowski SGH Warsaw School of Economics, Warsaw, Poland

Eugenijus Kurilovas Vilnius Gediminas Technical University, Vilnius, Lithuania

Pablo Lara Navarra UOC—Universitat Oberta de Catalunya, Barcelona, Spain

Carla Limongelli Department of Engineering, Roma Tre University, Rome, Italy

J. M. Llaurado-Fons Department of Computer Technology, University of Alicante, San Vicente del Raspeig, Spain

Francesca Loia University of Rome "La Sapienza", Rome, Italy

Marco Lombardi DIIN, University of Salerno, Fisciano, SA, Italy

Thai-Ha Luong Department of Business Administration, National Central University, Zhongli District, Taoyuan City, Taiwan

Rustom Mamlook Innovation and Technology Management Department, Arabian Gulf University, Manama, Bahrain

Ana Martinez Department of Computer Technology, University of Alicante, San Vicente del Raspeig, Spain

Anna Masłoń-Oracz SGH Warsaw School of Economics, Warsaw, Poland

Alonso Medina Cortes Centro de Investigación en Computación, Instituto Politécnico Nacional, Nueva Industrial Vallejo, Gustavo A. Madero, México

Julio C. Mendoza-Tello School of Computer Engineering, Faculty of Engineering, Physics and Mathematics, Central University of Ecuador, Quito, Ecuador

Tatiana Mendoza-Tello Insurance Consultant, Quito, Ecuador

Pavel Mentlík Centre of Biology, Geoscience and Environmental Education, University of West Bohemia, Plzen, Czech Republic

Hector Migallon-Gomis Department of Computer Engineering, Miguel Hernandez University, Elche, Spain

Rafa Mollá-Sirvent Department of Computer Science Technology and Computation, University of Alicante, Alicante, Spain

Anna Molnár National University of Public Service, Budapest, Hungary

Higinio Mora Department of Computer Technology, University of Alicante, San Vicente del Raspeig, Spain

Mario R. Morales School of Computer Engineering, Central University of Ecuador, Quito, Ecuador

Marco Moreno Ibarra Centro de Investigación en Computación, Instituto Politécnico Nacional, Nueva Industrial Vallejo, Gustavo A. Madero, México

Gianluca Moro Department of Computer Science and Engineering – DISI, University of Bologna, Cesena, Italy

Rosalba Mosca DIIN, University of Salerno, Fisciano, SA, Italy

Ali Mohamad Mouazen Management and International Management Department, Lebanese International University, Beirut, Lebanon

Bernardo Pereira Nunes Australian National University, Canberra, ACT, Australia

Luciano C. Nunes University of Fortaleza, Fortaleza, Brazil;
University Center September 7, Fortaleza, Brazil

Josephine Ojiambo Peter Ojiambo Foundation, London, UK

Maria L. D. Pereira Ceara State University, Fortaleza, Brazil

Raquel Pérez del Hoyo Department of Building Sciences and Urbanism, University of Alicante, Alicante, Spain

María Luisa Pertegal-Felices Developmental and Educational Psychology Department, University of Alicante, Alicante, Spain

Tuan Pham University of Technology and Education of Danang, Hai Chau District, Danang City, Vietnam

Luana I. C. C. Pinheiro National Confederation of Medical Cooperatives, Fortaleza, Brazil;
Ceara State University, Fortaleza, Brazil

Mirian Calíope D. Pinheiro University of Fortaleza, Fortaleza, Brazil

Pedro G. C. D. Pinheiro University of Fortaleza, Fortaleza, Brazil

Placido Rogerio Pinheiro University of Fortaleza, Fortaleza, Brazil;
Ceara State University, Fortaleza, Brazil

Jiří Polcar Division of Lifelong and Distance Learning, University of West Bohemia, Plzen, Czech Republic

Consuelo O. Pontes National Confederation of Medical Cooperatives, Fortaleza, Brazil

Tomáš Průcha Department of Computer Science and Educational Technology, University of West Bohemia, Plzen, Czech Republic

Francisco A. Pujol-López Department of Computer Technology and Computation, University of Alicante, Alicante, Spain

Jukka Raitanen Häme University of Applied Sciences, Hämeenlinna, Finland

Cláudia Toriz Ramos Universidade Fernando Pessoa, CEPESE-Centro de Estudos da População, Economia e Sociedade, Porto, Portugal

Hector Rico-Garcia Department of Computer Technology, University of Alicante, San Vicente del Raspeig, Spain

Lucie Rohlíková Department of Computer Science and Educational Technology, University of West Bohemia, Plzen, Czech Republic

Karla M. C. Rolim University of Fortaleza, Fortaleza, Brazil

Kawther Saeedi FCIT, IS Department, King AbdulAziz University, Jeddah, Saudi Arabia

Magdalena Saldaña Pérez Centro de Investigación en Computación, Instituto Politécnico Nacional, Nueva Industrial Vallejo, Gustavo A. Madero, México

Jordi Sales-Zaguirre School of Tourism and Hospitality Management Sant Ignasi, Universitat Ramon Llull, Barcelona, Spain

Jose-Luis Sanchez-Romero Department of Computer Technology, University of Alicante, San Vicente del Raspeig, Spain

Domenico Santaniello DIIN, University of Salerno, Fisciano, SA, Italy

Filippo Sciarrone Engineering Department, Roma Tre University, Rome, Italy

Enric Serradell-López Universitat Oberta de Catalunya, Barcelona, Spain

Chien-wen Shen Department of Business Administration, National Central University, Zhongli District, Taoyuan City, Taiwan

Sean W. M. Siqueira Federal University of the State of Rio de Janeiro, Rio de Janeiro, Brazil

Asamaporn Sitthi Department of Geography, Faculty of Social Science, Srinakharinwirot University, Bangkok, Thailand

Humberto Sossa Azuela Centro de Investigación en Computación, Instituto Politécnico Nacional, Nueva Industrial Vallejo, Gustavo A. Madero, México

Marco Spruit Department of Information and Computing Sciences, Utrecht University, Utrecht, The Netherlands;
Department of Public Health and Primary Care, Leiden University Medical Centre, Leiden, The Netherlands

Andrea Sterbini Sapienza University, Rome, Italy

Anu Helena Suominen Tampere University, Tampere, Finland

Aleksander Surdej Department of European Studies and Economic Integration, Cracow University of Economics, Krakow, Poland

Marco Temperini Department of Computer, Control, and Management Engineering, Sapienza University of Rome, Rome, Italy

Marcelo Tibau Federal University of the State of Rio de Janeiro, Rio de Janeiro, Brazil

Miguel Torres Ruiz Centro de Investigación en Computación, Instituto Politécnico Nacional, Nueva Industrial Vallejo, Gustavo A. Madero, México

Orlando Troisi University of Salerno, Fisciano, SA, Italy

Kune-Muh Tsai National Kaohsiung University of Science and Technology, Kaohsiung, Taiwan

Konstantinos Tsilionis KU Leuven, Belgium

Anna Urbanovics National University of Public Service, Budapest, Hungary

Erick Varela-Guzmán Department of Electronics and Informatics, Central American University, La Libertad, El Salvador

Anna Visvizi SGH Warsaw School of Economics, Warsaw, Poland and Effat University, Jeddah, Saudi Arabia

Jun Wang Graduate School of Chinese Academy of Social Sciences, Beijing, China

Wei Wang College of Business Administration, Huaqiao University, Quanzhou City, Fujian Province, China

Yves Wautelet KU Leuven, Belgium

Shiann Ming Wu College of Business Administration, National Huaqiao University, Quanzhou, China

Yenchun Jim Wu Graduate Institute of Global Business and Strategy, National Taiwan Normal University, Taipei, Taiwan

Yung Chang Wu College of Business Administration, National Huaqiao University, Quanzhou, China

Part I
Smart Cities: Issues and Challenges

Chapter 1
Introduction to a Support Model Based on Cryptocurrencies for Social Inclusion Projects

Higinio Mora, Erick Varela-Guzmán, Anna Visvizi, and Rafa Mollá-Sirvent

Abstract Eradication of poverty and social exclusion has been one of the main goals of the international community for several years now. With the implementation of specific policies, frequently directly aligned with the Sustainable Development Goals (SDGs), governments and international organizations adopted measures to assist vulnerable groups worldwide. As data suggests, while existing problems and challenges are far from being solved, new challenges arise. Against this backdrop, in this work, by focusing on specific issues and challenges, and by showcasing developments in the European Union (EU), we propose a model based on cryptocurrency technologies to suggest how certain of these challenges can be addressed. In this view, the findings of research presented here add to the discussion on how to attain SDG 10, i.e. "Reduce inequalities within and among countries".

Keywords Smart government · Cryptocurrency · Smart society · SDGs—Obj.10

Introduction

Since its inception, the fight against poverty and social exclusion has been one of the main objectives of the European Union (EU). Primary and secondary sources of EU law offer a rich background to examine the EU's stance toward poverty and social

H. Mora (✉)
Department of Computer Technology, University of Alicante, San Vicente del Raspeig, Spain
e-mail: hmora@ua.es

E. Varela-Guzmán
Department of Electronics and Informatics, Central American University, La Libertad, El Salvador

A. Visvizi
SGH Warsaw School of Economics, Warsaw, Poland and Effat University, Jeddah, Saudi Arabia

R. Mollá-Sirvent
Department of Computer Science Technology and Computation, University of Alicante, 03690 Alicante, Spain

A. Visvizi et al. (eds.), *Research and Innovation Forum 2020*,
Springer Proceedings in Complexity,
https://doi.org/10.1007/978-3-030-62066-0_1

exclusion. With the implementation of some policies, governments have adopted measures, offering legal protection and creating aid programs to benefit vulnerable groups.

The Article 19 of the Treaty on the Functioning of the European Union [1] establishes that the Council "may take appropriate action to combat discrimination based on sex, racial or ethnic origin, religion or belief, disability, age or sexual orientation". Furthermore, The Council Directive 2000/43/EC: the principle of equal treatment between persons irrespective of racial or ethnic origin [2], among its considerations it maintains that: "The right to equality before the law and protection against discrimination for all persons constitutes a universal right recognized by the Universal Declaration of Human Rights".

All these policies that have allowed improving the way that governments and institutions work against vulnerable groups discrimination, being the main beneficiaries of these programs, people in social exclusion situations and refugees affected by conflicts, persecutions or natural disasters [3]. But this does not mean that the problem is near to be resolved. Achieve a greater integration and development of groups that have a high vulnerability index is a real challenge for today's society [4].

According to the Living conditions in Europe 2018 edition report [5], in 2016, approximately 188 million people, i.e., an estimated 23.5% of the EU member states population, was at high risk of poverty or social exclusion. Also, the Third Overview of Housing Exclusion in Europe 2018 [6] report, calculates that at least 11 million households depend on social resources, third-party accommodation or even live on the street. Furthermore, indicates that the homelessness phenomenon is increasing in Europe, even in countries that are experiencing rapid economic growth.

Furthermore, the presence of immigrants and refugees in the EU exceeds 82 million people, which represents 11% of the total population [7]. According to data from the European Parliament, in recent years, Europe has faced one of the greatest migration challenges since World War II [8, 9]. In 2015, the EU received more than 1.25 million asylum applications; This figure was reduced to 881,000 applications in 2018. With respect to the entries in irregular administrative status, in 2015 more than one million entries were registered by sea, likewise, this figure was reduced to 116,647 entries in 2018 [10].

People in a situation of social exclusion often depend on institutions such as the International Federation of Red Cross and Red Crescent Societies (IFRC), the UN Refugee Agency (UNHCR) and the Urgent Relief & Development Association (URDA). These institutions, governmental or non-governmental, use different programs, most of them, financed by governments, donors and volunteer work and try to offer good conditions and opportunities of a dignified life to vulnerable groups [11]. Despite their important work, these institutions currently face many challenges, one of them is how to win the trust of donors, as they often do not trust the management of funds carried out by organizations that provide social aid [12]. Also, there has been no gradual application of new technologies and the development of new mechanisms to provide significant value to how social aid is offered to vulnerable groups. The NGO sector recognizes that it is a challenge to implement new strategies to promote greater impact, efficiency, and responsibility with vulnerable sectors [13].

Based on the described current situation, in this work, we analyze the feasibility of a model-oriented to smart cities [14] as a new action strategy to support organizations that offer social aid, enabling organizations to improve the way they support people in vulnerable situations while promoting trust among stakeholders. For this, a review of blockchain technology is carried out [15–17]. Due to its characteristics, Blockchain allows the implementation on a completely decentralized platform, independent of any entity and with the ability to trace all its operations [17], making actions such as fraud or counterfeiting impossible [18].

Defining the Problem

The work of organizations that provide social aid is not easy, but this is compounded by the fact that these organizations are experiencing a crisis of reliability. In recent years, skepticism and mistrust towards these institutions have increased [12]. Added to this, the wrongdoing of some of its members affects seriously the reputation of the organization and its ability to supply social assistance, often at irreparable levels, e.g., the case of the NGO Anesvad was affected by a judicial process because its ex-head was accused of embezzlement of 7.5 million euros [19], later, the NGO admitted that it lost 40% of its partners due to this problem. An empirical study found that at least one in four institutions involved in fraud does not survive more than three years beyond the publication of the fraud [20]. Also, not only the organization that has committed fraud is affected but this type of action affects the credibility of the entire sector [21].

Donors are commonly the primary source of funding for a charity [14], capturing the attention of new donors and retaining them is one of the biggest challenges for organizations [12]. Potential donors face questions like "Where is my money going?" or "What assures me that my donation has contributed to a cause?". Cases like NGO Oxfam who was accused of covering up the use of prostitutes by aid workers during Haiti's earthquake emergency in 2010 [22], further weaken the general confidence of citizens.

The European Fundraising Association in its analysis of the 2019 Edelman Trust Barometer [23], indicates: "While the overarching trend is that trust is on the rise, the Barometer reveals that four out of the six markets identified as distrusting NGOs are in Europe, namely the UK, The Netherlands, Italy, and Germany. At the bottom of the scale, only 44% of Germans said they trusted NGOs, although this is a 7% increase on the previous year. European countries with a neutral rating (between 50–59%) were Ireland, Spain, and France. Trust in NGOs fell in two European countries: Italy, by 2%, and Spain, which had the biggest year on year decline among all markets surveyed, with an 8% drop".

Due to all these difficulties and aware that donors need information, organizations have insisted on the need to be transparent [20], so it is usual that they have tools that help to provoke or increase the trust of citizens, offering communication channels focused on facilitating the information they need [24], e.g., the use of transparency portals in their official websites, sharing detailed information on their annual budgets, programs, executed projects and external audits [20]. In the case of North American organizations, it has become a common practice the use of SOX model requirements in NGOs governance structures [25], this model requires that there be a complaint process within the organization and the presence of an independent audit committee with financial experience. A strategy currently implemented to evaluate the correct actions of organizations is the creation of supervisory entities like the international association of national charity monitoring organizations (ICFO) [26]. A worldwide network of organizations that ensure that fundraising for charitable purposes is performed appropriately and that the administrations of the collected funds are trustworthy.

Another challenge that organizations currently have is how to control that beneficiaries use the aid that they receive appropriately. A common practice to have a certain degree of control is the direct delivery of kits with essential items instead of money [27], this strategy is used in the Food Aid Operational Program of the Fund for European Aid to the Most Deprived (FEAD) [28]. In Spain, the FEAD National Program distributes food to the most disadvantaged people, co-financed by the FEAD in 85% and by 15% by the General State Administration, which includes the purchase of food purchased in the local market and delivering them through institutions intermediaries to people in a vulnerable state [27].

After a bibliographic review, it was not possible to find strategies based on information technologies under the concept of a smart society, that promotes solving the problems faced by an organization that provides social aid, such as increasing confidence in donors and controlling the goods offered to people in a vulnerable state. We observe the need to achieve greater integration of new technologies to "empower and promote the social, economic and political inclusion of all, irrespective of age, sex, disability, race, ethnicity, origin, religion or economic or other status" as stated in goal 10 of the sustainable development goals [29], producing positive effects on people's lives and above all, including and benefiting groups in special situations of vulnerability.

Proposed Model

Based on the described current situation, we propose to develop a smart society model based on disruptive technology Blockchain [15, 16, 17] that allows government entities and different organizations to manage social aids towards vulnerable groups and advance in the achievement of the Sustainable Development Goals (SDGs), contributing specifically to the compliance with the goal 10: Reduce inequality within and among countries [29], and at the same time, promote trust among stakeholders.

The research novelty comes from the use of cryptocurrency technologies to address social problems that affect the daily life of citizens. The proposed solution allows progress towards the creation of a smart society through the use of the blockchain and the smart contracts technologies [17, 30–32] as formidable tools capable of empowering and promoting the social, economic and political inclusion of all citizens, solving many of the problems that exist in the current way of providing social assistance and as well as taking advantage of the benefits offered by these technologies.

Considering the characteristics with which blockchain technology was conceived, we plan to develop a support model that promotes transparency and control throughout the process [15] carried out by institutions that provide social assistance, from the entry of assets to the entity to the form in which they invest and the use that people in a vulnerable state gives to the goods they receive.

Among the benefits that can be listed from a blockchain-based model are the following:

- *Audit capacity and traceability of operations.* It is possible to develop a payment system capable of making a complete and detailed trace of the carried-out operations [17]. A blockchain-based model ensures the trust of two parts without the need for a third-party entity to verify the accuracy and integrity of the transactions [32, 33]. So, it is possible to implement a model that logs the transactions carried out by people who receive social assistance, it is even possible to control and specify in which goods and services the aid can be used [18]. With this feature, citizens and entities that provide funds for social assistance, i.e., the stakeholders, have the possibility of evaluating a reliable history of the transactions, being able to verify how the money collected has been used.
- *Counterfeiting prevention.* With a blockchain-based model, counterfeiting or duplication of goods is impossible, this benefit is achieved because Cryptocurrencies are based on a series of complex crypto techniques [17]. Furthermore, accurate identification of the person involved in the transaction is possible, being more reliable than a system based on signatures and stamps [31].
- *Promoting the right of free movement of persons.* After the gradual phasing-out of internal borders under the Schengen agreements and the adoption of Directive 2004/38/EC [30, 34]. With the current model, there are important obstacles for vulnerable groups to exercise the right of free movement, because the tools used to offer social aid are useless if the person that receives the aid decides to move to another region or country. A Cryptocurrency is not based on the legislation of a specific country [35], this versatility allows that there are no limitations in its architecture for the implementation of a social aid system based on a cryptocurrency model in an international context. It is necessary to clarify that the model is not intended to implement an element that is a substitute for money, but rather a tool that can provide levels of control and auditing that fosters trust in the different stakeholders.

Of course, cryptocurrencies have also drawbacks and limitations [36]. Further research is needed in order to minimize the potential issues of cryptocurrencies applied to social inclusion projects. However, we think that to achieve the benefits described in this work worthwhile to take the risk.

Blockchain and Aid Management

Blockchain can be defined as a distributed database through a point-to-point network. Each node has a complete copy of the database. To consider a new transaction as valid, it is necessary that each node authorize and update its database, thus ensuring reliability [15–17].

Figure 1.1 illustrates a flow chart related to the management of social assistance based on the proposed model in this paper. The organization provides social assistance to the person in a vulnerable state, but before the beneficiary uses the aid, internally and automatically, there is data processing characteristic of a blockchain network. Initially, each node verifies and validates the transaction; if the transaction is correct, it becomes part of a transaction block. After some time, the block is

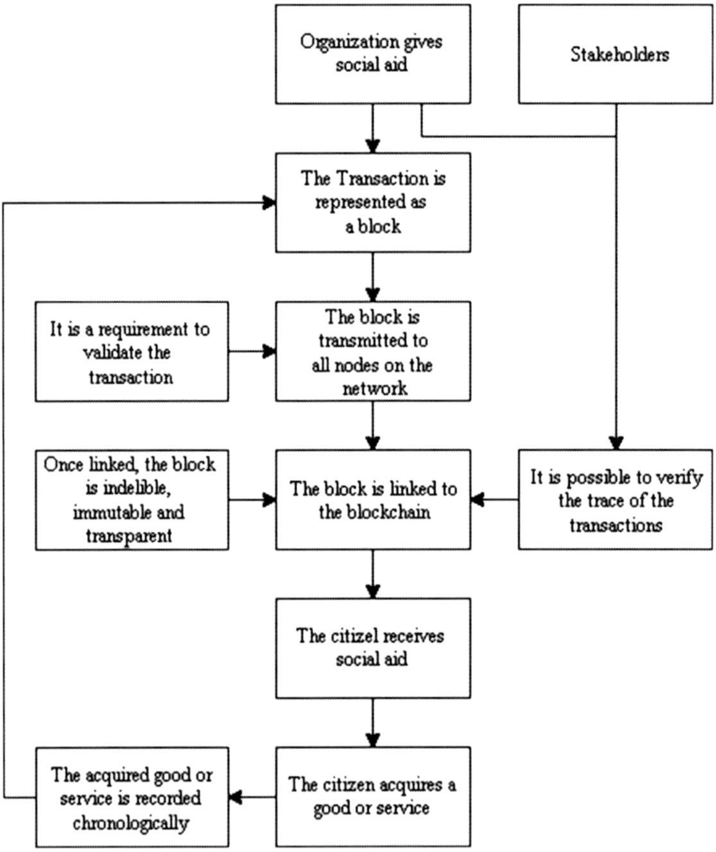

Fig. 1.1 Flow chart of the proposed model

signed with complex encryption techniques and becomes part of the blockchain, that is, it is linked to the previous block, now, this block cannot be modified or removed without modifying all the blocks that are linked to it [17]; it is almost impossible to manipulate, in this way any fraud or counterfeiting could be easily detected [17, 31, 32]. The blockchain has another important characteristic, it is possible to control the traceability of transactions [31], allowing stakeholders (donors, NGOs, government institutions, supervisory entities, et al.) to verify all the paths that a cryptocurrency has followed. Once all this process has been completed, the beneficiary who has received the social aid can make use of it, acquiring the goods and services agreed upon by the institutions. When a person makes use of the aid granted, the verification and validation process of the transaction is performed again.

Conclusions

The key value added of blockchain, seen as a disruptive technology, is to be associated with the most complex problem of interaction among people, i.e. lack of trust. In general, the way to solve this problem has been the use of an intermediary entity that provides trust between two parties when it is necessary to make an important transaction. Blockchain technology provides a method to change the way of working, providing confidence to carry out transactions without the need for intermediary entities.

In this work, a review of key issues requiring attention in context of the debate on poverty and social exclusions was presented. The key problems encountered by various stakeholders involved in providing social aid to groups in a special state of vulnerability were discussed. In this context the value added of the blockchain technology was discussed and its potential to mitigate many of these problems was elaborated. In this context the notion of citizens' trust towards institutions provide social aid was stressed. In the same vein, the prospects of transparent and efficient delivery of assistance by these institutions was also stressed.

As a methodology used in this work, a review of scientific literature, various investigations, and current legislation in the EU have been carried out, dealing with the problems and challenges faced by organizations that provide social aid.

This is only an introductory work that acts as a first contact with the raised issues. The authors of this work aim to continue exploring this line of research in future works and propose solutions based on disruptive technologies for a problem that is difficult to solve with current policies and ways of working and allows progress towards the creation of a smart society using virtual currency systems, blockchain, and smart contracts.

Acknowledgements

This work has been funded by the Participation, Transparency, Cooperation and Democratic Quality Consellería of the Generalitat Valenciana, Spain. The content of this publication is the exclusive responsibility of the University of Alicante and does not necessarily reflect the opinion of the Generalitat Valenciana.

References

1. Publications Office of the EU. The Treaty on the Functioning of the European Union. https://eur-lex.europa.eu/legal-content/EN/TXT/PDF/?uri=CELEX:12012E/TXT. Accessed 25 Mar 2020
2. Publications Office of the EU. Council Directive 2000/43/EC of 29 June 2000 implementing the principle of equal treatment between persons irrespective of racial or ethnic origin. https://eur-lex.europa.eu/legal-content/EN/TXT/PDF/?uri=CELEX:32000L0043. Accessed 25 Mar 2020
3. International Federation of Red Cross and Red Crescent Societies, Vulnerable groups and individuals (2004). https://www.ifrc.org/en/news-and-media/opinions-and-positions/speeches/2004/vulnerable-groups-and-individuals/. Accessed 10 May 2020
4. A. Kennedy, La lucha contra la pobreza, la exclusión social y la discriminación, Fichas técnicas sobre la Unión Europea—2020 (in spanish), European Parliament (2019). https://www.europarl.europa.eu/ftu/pdf/es/FTU_2.3.9.pdf. Accessed 26 Apr 2020
5. Publications Office of the European Union, European Commission: Living conditions in Europe, 2018 edn. Luxembourg (2018). https://doi.org/10.2785/39876
6. European Federation of National Organisations Working with the Homeless, Third Overview of Housing Exclusion in Europe 2018 (2018). https://www.feantsa.org/download/full-report-en1029873431323901915.pdf. Accessed 14 Mar 2020
7. United Nations Population Division. International Migrant Stock (2019). https://www.un.org/en/development/desa/population/migration/data/estimates2/estimates19.asp. Accessed 14 Mar 2020
8. A. Visvizi, M. Pachocka, M.D. Lytras, Managing international migration: rethinking transaction costs red tape, and their impact. Int. Migrat. **57**(5), 271–279 (2019). https://doi.org/10.1111/imig.12648
9. A. Visvizi, On the fringe: Greece, migration, and populism, Roczniki Nauk Społecznych. Ann. Soc. Sci. **46**(3), 63–75, ISSN 0137–4176/eISSN 2544–5812 (2018)
10. European Parliament News, Migration in Europe (2018). https://www.europarl.europa.eu/news/en/headlines/society/20170629STO78632/migration-in-europe. Accessed 07 Mar 2020
11. Red Cross EU Office, Poverty and Social Exclusion (2020) https://redcross.eu/themes/poverty-and-social-exclusion. Accessed 10 May 2020
12. V. Keating, E. Thrandardottir, NGOs, trust, and the accountability agenda. Br. J. Polit. Int. Relat. **19**(1), 134–51. https://doi.org/10.1177/1369148116682655
13. B. Bruce, The change imperative: creating the next generation NGO. Develop. Pract. **21**(6), 905–907 (2011). https://doi.org/10.1080/09614524.2011.582091
14. Mora et al., Virtual currencies in modern societies: challenges and opportunities. Polit. Technol. Post-Truth Era 171–185. https://doi.org/10.1108/978-1-78756-983-620191012

15. H. Mora, R. Pérez-delHoyo, R.M. Sirvent, V. Gilart-Iglesias, Management city model based on blockchain and smart contracts technology. Research and innovation forum 2019. RIIFORUM 2019 Springer Proceedings in Complexity Springer, Cham. https://doi.org/10.1007/978-3-030-30809-4_28
16. M. Sicilia, A. Visvizi, Blockchain and OECD data repositories: opportunities and policymaking implications. Library Hi Tech **37**(1), 30–42 (2019). https://doi.org/10.1108/LHT-12-2017-0276
17. S. Nakamoto, Bitcoin: A Peer-to-Peer Electronic Cash System (2008). https://bitcoin.org/bitcoin.pdf. Accessed 01 Apr 2020
18. J.C. Mendoza-Tello, H. Mora, F.A. Pujol-López, et al., Disruptive innovation of cryptocurrencies in consumer acceptance and trust. Inf. Syst. E-Bus Manage. **17**, 195–222 (2019).https://doi.org/10.1007/s10257-019-00415-w
19. El Pais, Six years for ex-NGO head who embezzled 7.5 million Euros (2012). https://english.elpais.com/elpais/2012/04/25/inenglish/1335380847_182977.html. Accessed 28 Mar 2020
20. D.S. Archambeault, S. Webber, Fraud survival in nonprofit organizations: Empirical evidence. Non-profit Manage. Leadersh. **29**, 29–46 (2018). https://doi.org/10.1002/nml.21313
21. A. Schneiker, Why trust you? Security cooperation within humanitarian NGO networks. Disasters **44**, 25–43 (2020). https://doi.org/10.1111/disa.12363
22. S. O'Neill, Minister orders Oxfam to hand over files on Haiti prostitute scandal (2018). https://www.thetimes.co.uk/edition/news/top-oxfam-staff-paid-haiti-quake-survivors-for-sex-mhm6mpmgw. Accessed 28 Mar 2020
23. European Fundraising Association, Edelman Barometer indicates that trust in NGOs is on the rise (2019). https://efa-net.eu/news/edelman-barometer-shows-trust-in-ngos-remains-lower-across-european-markets. Accessed 28 Mar 2020
24. C.D.B. Burt, Communication: delivering information to donors, in *Managing the Public's Trust in Non-profit Organizations*. SpringerBriefs in Psychology. Springer, New York, NY (2014). https://doi.org/10.1007/978-1-4939-0560-7_4
25. Office of the Legislative Counsel, Sarbanes-Oxley Act Of 2002 (2019). https://legcounsel.house.gov/Comps/Sarbanes-oxley%20Act%20Of%202002.pdf. Accessed 04 May 2020
26. International Association of National Charity Monitoring Organizations. https://www.icfo.org/. Accessed 28 Mar 2020
27. Programa Operativo del Fondo de Ayuda Europea Para las Personas Más Desfavorecidas (in Spanish) (2014). https://www.mscbs.gob.es/ssi/familiasInfancia/inclusionSocial/fead/programaOperativoFead2014.pdf. Accessed 29 Mar 2020
28. European Commission, Fund for European Aid to the Most Deprived (FEAD) (2020). https://ec.europa.eu/social/main.jsp?catId=1089. Accessed 07 Mar 2020
29. United Nations, Goal 10: Reduce inequality within and among countries (2020). https://www.un.org/sustainabledevelopment/inequality/. Accessed 01 Apr 2020
30. H. Mora, R. Mollá-Sirvent, Analysis of virtual currencies as driver of business marketing. Research and Innovation Forum 2019. RIIFORUM 2019 Springer Proceedings in Complexity Springer, Cham (2019).https://doi.org/10.1007/978-3-030-30809-4_48
31. K. Mori, H. Miwa, Digital University admission application system with study documents using smart contracts on blockchain, in Advances in Intelligent Network-ing and Collaborative Systems. INCoS2019 Advances in Intelligent Systems and Computing, vol. 1035 Springer, Cham (2020).https://doi.org/10.1007/978-3-030-29035-1_17
32. J.C. Mendoza-Tello, H. Mora, F.A. Pujol-López, M.D. Lytras (2018) Social commerce as a driver to enhance trust and intention to use cryptocurrencies for electronic payments. IEEE Access **6**, 50737–50751. https://doi.org/10.1109/ACCESS.2018.2869359
33. R. Moro, Blockchain valuation: internet of value and smart transactions, in *The Valuation of Digital Intangibles*. Palgrave Macmillan, Cham (2020). https://doi.org/10.1007/978-3-030-36918-7_16
34. Publications Office of the EU. Directive 2004/38/EC of the European Parliament and of the Council of 29 April 2004. https://eur-lex.europa.eu/legal-content/EN/TXT/PDF/?uri=CELEX:32004L0038. Accessed 29 Mar 2020

35. E. Teyyare, K. Ayyıldırım, The size and taxation of cryptocurrency: an assessment for emerging economies, in U. Hacioglu (ed.), *Blockchain Economics and Financial Market Innovation. Contributions to Economics*. Springer, Cham (2019). https://doi.org/10.1007/978-3-030-25275-5_18
36. F. Brezo, P.G. Bringas, Issues and Risks Associated with Cryptocurrencies such as Bitcoin, in *The Second International Conference on Social Eco-Informatics* (pp. 20–26) (2012)

Chapter 2
Characterization of the Accessibility Information of a City

Rafa Mollá-Sirvent, Higinio Mora, V. Gilart-Iglesias, and Raquel Pérez del Hoyo

Abstract The development of Information Society and Communications Technologies provide new opportunities to monitor and improve the accessibility of modern cities. The concepts of smart and ubiquitous cities make use of processing, sensorization and communications technologies to provide knowledge and intelligence to the city. Knowing this information as up to date as possible is a very helpful tool to improve the lives of citizens, since it allows us to locate accessibility barriers, which helps improve citizen mobility. In this context, this work carries out the task of comprehensively characterizing the information on accessibility available in cities. To do this, we will catalog the information sources that we have available so far from previous works of this research group. The final objective is to find the most suitable format for representing accessibility information, as well as to define what the data to be stored will be: points, routes and areas. For this, the most used systems will be analyzed: GPX, GML and GeoJSON and will be compared to obtain a conclusion.

Keywords Accessibility · Sustainability · Smart city · Urban planning

Introduction

The development of Information Society and Communication Technologies provide new opportunities to monitor and improve the accessibility of modern cities. The concepts of smart and ubiquitous cities make use of processing, sensing and communication technologies to provide knowledge and intelligence to the city [1–3] while offering connectivity, electricity supply and interoperability resources [4–6].

R. Mollá-Sirvent · V. Gilart-Iglesias
Department of Computer Science Technology and Computation, University of Alicante, 03690 Alicante, Spain

H. Mora (✉)
Department of Computer Technology, University of Alicante, San Vicente del Raspeig, Spain
e-mail: hmora@ua.es

R. Pérez del Hoyo
Department of Building Sciences and Urbanism, University of Alicante, 03690 Alicante, Spain

A. Visvizi et al. (eds.), *Research and Innovation Forum 2020*,
Springer Proceedings in Complexity,
https://doi.org/10.1007/978-3-030-62066-0_2

These conditions facilitate the deployment of interconnected intelligent elements that offer services to citizens for efficient decision-making and make better use of resources [7]. For example, interactive maps showing elements of interest to the citizen in real time, automatic identification and management of traffic in situations of congestion or danger, etc. This is an increasingly frequent trend in the representation of geographic information to the citizen, in which, in addition to the map of the territory itself, other information of interest is provided, such as places of leisure, shops, etc.

Along these lines, the generation of elaborated information on the functioning of the city is a great help tool for citizens and administrators in making the correct decisions. Specifically, the elaboration of indicators or indices represents a compact measure easily understandable by society that corresponds to a quantification or quality measure of the specific issue to which they refer. These indices are generally used as an indicator of the current state of affairs and as an instrument to develop strategies [8] (for example: the human development index, schooling index, etc.). In this line, there is a need for developing data ecosystems to meet the social and public good in order to not only improve the quality of life of citizens, but also to protect their digital rights in cities and regions [9].

In the case of urban sustainability, there is growth in the definition of specific indicators of the local urban context. The adoption of adequate indicators is essential to implement sustainable development at the local level, since they provide tools for the analysis and evaluation of the current situation, support the decision-making process and help communication between citizens and society in general. Along these lines, the local and urban dimensions of sustainability are becoming increasingly frequent in international literature [10–12].

The development of indicators consists of a collection of information potentially relevant to the objectives of a use case of the question to be analyzed [13]. This information normally comes from a combination of objective sources (when it comes to quantifiable and measurable information about reality), and subjective sources (when it is based on perceptions of this reality made by individuals) [14]. It is in the construction of indicators that technology plays a fundamental role in collaborating in the relevant and objective collection of information that allows the elaboration of an indicator that reduces subjectivity and is capable of holistically characterizing the appropriate dimension of urban sustainability.

Currently, smart city concept is helpful to address many issues of management of urban areas. In this context, urban computing offers a conceptual and methodological framework that integrates the plethora of increasingly sophisticated technologies and their applications in city/urban space [15, 16]. However, despite the expansion of smart cities, there are still many problems to solve. An important point to improve is the accessibility of cities. Accessibility is a right of all citizens, an essential element of modern society, a determining factor of the habitability of cities and an element of quality of life with universal interest.

However, in most cases, cities and their public spaces were and are designed without taking into account the difficulties that people with physical or sensory limitations (of understanding, communication or perception) might encounter when traveling there [17].

A key aspect to approach evaluation and maintenance tasks, as well as to detect and manage different accessibility problems, is to have a precise awareness of the current state of urban accessibility. This real and up-to-date knowledge about urban accessibility will help improve the mobility and livability of cities and, therefore, the quality of life and well-being of all citizens [18].

For the aforementioned reasons and continuing with the work carried out in previous articles of this research group, in this work we define an essential part of the system for the identification of accessibility problems in cities. This important part is the characterization of the accessibility data that we are able to generate with our own sources or that we consume from others. Knowing what compose our data is going to be crucial when choosing the document format in which we are going to store them, so that later these data can be represented on a map, thus offering a valuable tool for citizens and administrations. In addition, it will be possible to define indices to measure the degree of accessibility of a city or urban area and will allow us to concisely visualize how accessible a city is and its progression over time. In short, the final objectives of this work are to find the most suitable format to represent accessibility information and define the characteristics of the data to be stored: points, routes and areas.

This document is organized in the following sections: Sect. 2.2 shows which are the sources of accessibility data available; In Sect. 2.3, the characterization of accessibility data is developed; In Sect. 2.4, the types of storage format are described and the selection of the best one is made and finally the conclusions are presented in Sect. 2.5.

Urban Accessibility Data Sources

Our research group developed a system for managing accessibility problems found in cities. This system is made up of a mobile application from which citizens register the accessibility barriers they encounter when traveling through the city and a web console for local governments to manage the problems registered from the app [19, 20].

The creation of a new registry by a citizen implies that the user describes the problem they have encountered and grants the application access to their location through the GPS integrated in their smartphone. To facilitate the administration's task in identifying the exact location of the problem and the veracity of it without the need to first move to the location to verify it, the user is given the option to add images to the complaint. A verification process ensures that all recorded issues are correct and prevents inappropriate use of the system. This verification process will be carried out from the web application as indicated below. The registered points will store a history of the states they are passing through in order to follow their time evolution. The possible states of a claim are as follows:

- *Not verified*: These will be the issues that have not yet been validated.

- *Verified*: When the claim has been verified, it is classified into three different states: *Waiting*. It is the first step once the administration reviews the veracity of a claim and plans the solution process. The point needs action to make it stop being conflictive. This state is where decisions are made regarding the priority of one claim over another; *In progress*. At this moment, the administration has already begun working on the solution of the problem; *Solved*. It is the last step of the process. The work has finished and the point has ceased to be conflictive.

The fact of storing a history of the states through which a conflictive point is passing allows administrations to verify that the solutions applied continue to be effective over time and that a problem that has been solved and is affected by other factors may revert to be marked as conflictive to carry out the pertinent actions and solve it.

Another functionality of the application is to save routes from disabled citizens. The user will be the one to indicate when they want to start recording a route, allowing the application to locate it using GPS. When you arrive at your destination, you will stop the recording and can indicate your type of disability and add a description. Thanks to these routes, it will be possible to carry out a follow-up similar to that carried out with the points, maintaining a state of the route compared to the "current" routes obtained through mapping applications such as Google Maps. We are developing this method of comparing routes in parallel to this work and it has been recently presented at an international congress. The routes may have the following states:

- *Not verified*: The route has been registered but has not yet been compared to know whether or not it is accessible.
- *Inaccessible*: When comparing the recorded route with the route that a person without disabilities would take, it has been determined that there is an accessibility problem in it.
- *Correct*: No accessibility issues found.

There are other types of data that we must integrate into our system, such as the information provided by local governments in specific situations in some area of the city. It could happen that a street is blocked due to works and in that case people with disabilities have difficulties when crossing it. It is necessary to integrate these data into the system since they are of great relevance to citizens and for this the collaboration of governments is needed since they are the ones who know them first hand. Our job is to provide them with the necessary tools so that they can do it easily and thus make cities more accessible to everyone.

There exist some open data sources on urban accessibility around the world (https://data.world/datasets/accessibility) but each of a different nature and not standardized. This is why we believe in the importance of this work, which will serve to characterize the data related to accessibility in urban areas and that the data collected by cities is comparable to each other. Pursuing the goal of making cities a better place to live by applying technology for it.

Urban Accessibility Data Characterization

A fundamental aspect when developing an information system is to determine the components that will form the data in order to know which document format is more appropriate to store it with potential to register accessibility information [21]. We must define a structure that allows the integration of the data provided by the different sources mentioned above, taking into account the possible incorporation of other data sources in the future.

In short, knowing the data sources that we have, we need to basically store points, routes and areas. To these three types of elementary data of which our system is formed, we must add certain aggregate data that characterize them as data on urban accessibility.

Below we detail the types of data that will initially form our system.

Points

These are the locations that have been registered through the app developed by our research group. They are characterized by the following data:

- Status: As mentioned when explaining the system for acquiring urban accessibility issues, as time progresses, the issues can change status. Each of the states through which a point passes are stored, relating the type of state to the moment in which it changes.
- Position: formed by the coordinates that locate the point on a map.
- Timestamp: the time tag in which the accessibility issue was recorded.
- Description: detail of the problem found.
- Images: images that have been recorded of the problem found to facilitate the identification and verification task. These images will be stored in an external file that will be hosted on a server and can be accessed through a URL.

Routes

These are the routes that have been registered with the app developed by our research group and that have been recorded by people with a motor disability. They are characterized by the following data:

- Status: following the indications marked in Sect. 2.2.
- Positions: the points that define the route.
- Description: detail of the route that the user wants to indicate.
- Initial Timestamp: the moment at which the user begins to record the route.
- Final Timestamp: the moment when the route ends.
- Disability type: the type of disability of the user who has recorded the route.

Areas

The areas of the city that are in a situation that may interfere with the traffic of disabled people. They could be areas in which actions are being carried out by the administration or areas affected by a meteorological issue, among others. They are characterized by the following data:

- Description: detail of the situation that affects urban mobility.
- Zone: the area that is affected, defined by a polygon formed by geographic coordinates.
- Initial Timestamp - The time at which the problem was reported.
- Final Timestamp: the moment when the problem ceases.

Urban Accessibility Data Storage Formats

The data we want to store is of a geospatial nature to which we add other types of attributes. There exists a lot of Geographic Information Systems (GIS) formats to store geographical data [22]. In this section, we will compare some of the most commons to determine which one is the best for our data. The geographical data format that we need has to accomplish several requirements. It has to be open access and free to use and it has to be possible to transform other formats in it. In addition, it should be possible to define your own labels to be able to add the data related to accessibility and to improve the integration of possible new data sources. Taking in account those requirements we will compare these formats: GPX, GML and GeoJSON.

GPX

GPX (GPS Exchange Format) is an XML schema designed to be the standard format for interchange GPS data between applications [23].

GPX is compatible with most localization devices, hundreds of applications and large number of web pages. The format is open and free to be use, without the need to pay license fees.

It can be used to describe waypoints, tracks, and routes. Latitude and longitude are expressed in decimal degrees, and elevation in meters. Dates and times are not local time, but instead are Coordinated Universal Time (UTC).

It is designed so that developers can create their own extensions where their own objects and attributes are defined. In addition, these extensions could be added to the standard once reviewed by the GPX Developers Forum.

GML

GML (Geography Markup Language) is an XML schema defined by the Open Geospatial Consortium (OGC) to encode geographic data [24, 25].

Is a modeling language for GIS as well as an interchange format for geographical data. The format is open and free to be use, without the need to pay license fees.

GML defines "geometry objects" and "features". The geometry objects are "Point", "LineString" and "Polygon" and are used for describing more complex entities like, for example, roads or buildings. Features are application objects that represents physical entities, e.g. river. A feature may or may not have geometric aspects. A feature can have various geometry properties that describe their geometric aspect or characteristics.

A geometry object defines a location or region instead of a physical entity, and hence it is important to distinct from a feature.

Developers can define specific application schemas that are specialized extensions of GML. These schemas describe types of objects whose data is interesting for the community. For example, an application for tourism can define types of objects such as museums and monuments.

GeoJSON

GeoJSON is a format designed by an internet working group of developers based on JavaScript Object Notation (JSON) for geospatial data interchange [26, 27]. Nowadays is maintained by the Internet Engineering Task Force (IETF).

Is a standard format designed for representing geographical features and their non-spatial attributes. The format is open and free to be use, without the need to pay license fees.

A GeoJSON object may represent a geometry, a feature, or a collection of features. GeoJSON defines several geometry types: "Point", "LineString", "Polygon", "MultiPoint", "MultiLineString", "MultiPolygon", and "GeometryCollection". Features in GeoJSON contain a geometry object and additional properties. It uses a geographic coordinate reference system, World Geodetic System 1984, and units of decimal degrees.

A complete GeoJSON data structure is always an object (in JSON terms). In GeoJSON, an object is a collection of members (name/value pairs). The name of each member is always a string and the value can be a string, number, object, array or one of the literals: true, false, and null. An array consists of elements where each element is a value.

Once the characteristics of each format are known, we have decided that GeoJSON is the most suitable for our project due to the flexibility it offers and the ease of use from our software. This format allows us to store all the accessibility data that we

Fig. 2.1 Example of urban accessibility point represented in the GeoJSON format

```json
{
  "type": "Feature",
  "properties": {
    "id": "point1",
    "description": "no ramp",
    "reportTimestamp": 1583064000,
    "status": "{
      "0":[
      "notVerified",
      1583064000],
      "1":[
      "waiting"
      ,1583078400]},
    "urlImages": {
      "0": "https://..."}
  },
  "geometry": {
    "type": "Point",
    "coordinates": [
      -0.5122750997543334,
      38.387187521214976
    ]
  }
}
```

have so far. In addition, there are many standard tools compatible with the format in which they could already be represented.

Figure 2.1 shows an example of the representation of an urban accessibility issue in the GeoJSON format.

Conclusions

In this work, the characterization of the available urban accessibility data is carried out. As a result of this characterization, it has been possible to make the important decision in which format the data will be stored, taking into account the integration with the Geographic Information Systems and offering the possibility of developing own applications for the exploitation of this data. We have concluded that although all the analyzed formats could be valid to store accessibility data, the one that best suits our needs is GeoJSON. In addition, the possible expansion of accessibility data sources in the future has been considered. With this work we try to standardize the storage of urban accessibility data so that different works, investigations and new data sources can be easily integrated.

References

1. C. Martin, Smart-sustainability: a new urban fix? Sustain. Cities Soc. **45**, 640–648 (2019)
2. M. Alverti, D. Hadjimitsis, P. Kyriakidis, K. Serraos, Smart city planning from a bottom-up approach: local communities' intervention for a smarter urban environment, in *Fourth International Conference on Remote Sensing and Geoinformation of the Environment* (2016)

3. J. Macke, J.A.R. Sarate, S.A. Moschen, Smart sustainable cities evaluation and sense of community. J. Clean. Prod. **239**, 118103 (2019)
4. D. Oktaria, Suhardi, N.B. Kurniawan, Smart city services: a systematic literature review, in *International Conference on Information Technology Systems and Innovation (ICITSI)* (2017)
5. M. Eremia, L. Toma, M. Sanduleac, The smart city concept in the 21st century. Procedia Eng. **181**, 12–19 (2017)
6. E. Ismagilova, Smart cities: advances in research—an information systems perspective. Int. J. Inf. Manage. **47**, 88–100 (2019)
7. D. Bruneo, An IoT service ecosystem for smart cities: the #SmartME project. Internet Things **5**, 12–33 (2019)
8. J. Hou, P.P. Walsh, J. Zhang, The dynamics of human development index. Soc. Sci. J. **52**(3), 331–347 (2015)
9. I. Calzada, E. Almirall, Data ecosystems for protecting European citizens' digital rights. Transforming Government: People, Process and Policy, Vol. ahead-of-print No. ahead-of-print (2020). https://doi.org/10.1108/TG-03-2020-0047
10. E. Feleki, C. Vlachokostas, N. Moussiopoulos, Characterisation of sustainability in urban areas: an analysis of assessment tools with emphasis on European cities. Sustain. Cities Soc. **43**, 563–577 (2018)
11. M. Anisurrahman, H.M. Alshuwaikhat, Determining sustainability assessment indicators for the Holy City of Makkah, Saudi Arabia. Arab. J. Sci. Eng. **44**(5), 5165–5178 (2019)
12. Lu: Assessment of urban environmental change using multi-source remote sensing time series (2000–2016): A comparative analysis in selected megacities in Eurasia. Science of The Total Environment, Vol. 684, 567–577 (2019)
13. V. Bukovszki, D. Apró, A. Khoja, N. Essig, A. Reith, From assessment to implementation: design considerations for scalable decision-support solutions in sustainable urban development, in *IOP Conference Series: Earth and Environmental Science*, Vol. 290, conference 1 (2019)
14. P. Verma, A.S. Raghubanshi, Urban sustainability indicators: challenges and opportunities. Ecol. Ind. **93**, 282–291 (2018)
15. M.D. Lytras, A. Visvizi, M. Torres-Ruiz, E. Damiani, P. Jin, (2020) ieee access special section editorial: urban computing and well-being in smart cities: services, applications, policymaking considerations . IEEE Access **8**, 72340–72346 (2020). https://doi.org/10.1109/ACCESS.2020.2988125
16. E.H. Alkmanash, J.J. Jussila, M.D. Lytras, A. Visvizi, Annotation of smart cities twitter micro-contents for enhanced citizen's engagement. IEEE Access **99**, 1–1 (2019). https://doi.org/10.1109/ACCESS.2019.2935186
17. H. Mora, V. Gilart-Iglesias, R. Pérez-del Hoyo, M.D. Andújar-Montoya, A comprehensive system for monitoring urban accessibility in smart cities. Sensors **17**(8) (2017)
18. R. Pérez-del Hoyo, C. García-Mayor, H. Mora, V. Gilart-Iglesias, M.D. Andújar-Montoya, Improving urban accessibility: a methodology for urban dynamics analysis in smart, sustainable and inclusive cities. Int. J. Sustain. Develop. Plann. **12**, 357–367 (2017)
19. H. Mora, V. Gilart-Iglesias, R. Pérez-delHoyo, M.D. Andújar-Montoya, H.J. Compañ-Gabucio, Interactive cloud system for the analysis of accessibility in smart cities. Int. J. Des. Nat. Ecodyn. **11**(3), 447–458 (2016)
20. R. Pérez-delHoyo, M.D. Andújar-Montoya, H. Mora, V. Gilart-Iglesias, Urban and building accessibility diagnosis using 'accessibility app' in smart cities, in *Proceedings of the 6th International Conference on Smart Cities and Green ICT Systems (SMARTGREENS 2017)*, pp. 99–108 (2017)
21. A. Sirvent-Llamas, H. Mora, V. Gilart-Iglesias, M.D. Andújar-Montoya, R. Pérez-del Hoyo, A. de Ramón-Fernandez, Data Structures Modelling for Citizen Tracking Based Applications in Smart Cities, UCAmI 2017, LNCS 10586 (2017). https://doi.org/10.1007/978-3-319-67585-5_21
22. FileInfo, GIS Files. https://fileinfo.com/filetypes/gis. Last accessed 2020/02/13
23. Topografix, GPX: the GPS Exchange Format. https://www.topografix.com/gpx.asp. Last accessed 2020/02/14

24. OGC, Geography Markup Language. https://www.opengeospatial.org/standards/gml. Last accessed 2020/02/18
25. Sustainability for Digital Formats: Planning for Library of Congress Collection, Geography Markup Language. https://www.loc.gov/preservation/digital/formats/fdd/fdd000296.shtml. Last accessed 2020/02/18
26. The GeoJSON Format. https://tools.ietf.org/html/rfc7946. Last accessed 2020/02/25
27. Technical Overview: GeoJSON. https://web.archive.org/web/20090221220320, http://www.geowebguru.com/articles/97-technical-overview-geojson. Last accessed 2020/02/25

Chapter 3
An Application of Deep Neural Network for Robbery Evidence Using Face Recognition Approach

Alonso Medina Cortes, Magdalena Saldaña Pérez, Humberto Sossa Azuela, Miguel Torres Ruiz, and Marco Moreno Ibarra

Abstract Video vigilance has been implemented at many public places such as streets or buildings to prevent robberies and to obtain proofs from crimes. This information is not always publicly available since has a direct relation with people's privacy. Another commonly used approach to monitor the security of a city, used by smart cities is the Volunteered Personal Information (VPI) where people provide information through different applications and computational solutions focused in solving urban problems. How to use the above mentioned tools for increase the smart cities security? One solution could be the use of video and volunteered personal information of their citizens. The present approach arises from the people's necessity to have proofs when unfortunately a robbery occurs. In the present approach a deep neural network is implemented in order to identify people and faces inside images that could be analyzed when a robbery occurs, the deep neural network is trained and proved with volunteered personal information (VPI) open source datasets and compared with some other recognition algorithms in order to determine its precision and usability. The deep neural network makes use of bounding boxes to identify people and faces from images. This project is part of a bigger one where the deep neural network would be implemented on a wearable device able to take images in real time when a robbery is happening to its user, hence the need for developing a deep neural network dedicated to the people and faces recognition.

A. Medina Cortes (✉) · M. Saldaña Pérez · H. Sossa Azuela · M. Torres Ruiz · M. Moreno Ibarra
Centro de Investigación en Computación, Instituto Politécnico Nacional, Nueva Industrial Vallejo, Gustavo A. Madero 07738, México
e-mail: a190757@cic.ipn.mx

M. Saldaña Pérez
e-mail: amagdasaldana@cic.ipn.mx

H. Sossa Azuela
e-mail: hsossa@cic.ipn.mx

M. Torres Ruiz
e-mail: mtorres@cic.ipn.mx

M. Moreno Ibarra
e-mail: marcomoreno@cic.ipn.mx

A. Visvizi et al. (eds.), *Research and Innovation Forum 2020*,
Springer Proceedings in Complexity,
https://doi.org/10.1007/978-3-030-62066-0_3

Keywords Volunteer personal information · Deep neural networks · Image classification · Face recognition · Single shot detection

Introduction

Several factors increase the number of robberies and crimes in the cities; focusing on urban areas, we can mention factors such as economic inequality, unemployment and inequality of resources distribution. Despite government initiatives and the institutional response, robberies and crimes increase every day, one method implemented to prevent this situation has been the creation of social programs to help people in economic disadvantage and the placement of video vigilance cameras on public spaces.

Vision based detection of people inside images and the ability to detect people in populated and not controlled scenarios is a challenging task. Previous researches have been treating this task from a scientific perspective [1–4], and some others have tackled the subject from an automotive perspective [5, 6], however most of the results from this investigations have been implemented for its private use [7–10].

When it is necessary to identify people inside images, the standard benchmark are the Pattern Analysis Statistical Modeling and Computational Learning (PASCAL) [11], the Common Objects in Context (COCO) [12] and the National Institute for Research in Computer Science and Automation (INRIA) [13] datasets, these datasets contain limited range of scale, location, illumination, pose variation and occlusion.

Our method is based on a deep light convolutional network designed to work with data of pedestrians contained in Caltech pedestrian's dataset [14]. This dataset contains video sequences annotated with the coordinates of the people that appear in their videos. The video obtained in this dataset was recorded with a mounted camera in an automobile through a specific path, people appearing in the data frames are subject to different scales, occlusion and other kind of image noise, making it suitable to training a deep neural network. From this video sequences we obtained series of images to feed the deep neural network so it can extract features from them.

Although there are networks that detect multiple objects with good precision, such as Faster Region Convolutional Neural Network (Faster R-CNN) [15] or You Only Look Once (YOLO) [16], they need high-performance processors to function. The main objective of the present approach is to implement a deep neural network able to identify people and faces from images with a low processing cost and a good precision, in order to have a light detection model useful for electronic devices with low processing power and economic architecture. The first case of study where this deep neural network would be implemented is a wearable device with low cost-integrated vision application, able to take images when a robbery occurs to its user, and able to identify the face and silhouette of the offender. The data sets used to train and test the deep neural network characteristic is that provide images at noisy scenarios with sun light.

Background

During the last years, many researches have been focused on analyzing images in order to detect particular objects, one factor that augmented the interest on image processing was the increased use of artificial neural networks and the huge amount of data available to be processed [16–18]. Nevertheless, not all the available data is public, regarding to images such as photographs of people and pedestrians Google and Facebook own most of the global repositories [19]. Most of the times the available public images repositories belong to researching groups and universities interested on their study.

The first approach from neural network architectures that made a change in the object detection benchmark is AlexNet [20], champion of the 2012 Image Large Scale Visual Recognition Challenge (ILSVRC), after winning the task using 650,000 neurons trained with 60 million parameters.

The architecture training took a week using two NVIDIA GPU. After this achievement, the object detection has been improved to work with many different areas such as medicine, security systems, geoprocessing, among others, proving to have a fast and accurate response and a multiclass variability. This type of architecture needs processing power to carry out its task, so it is not feasible for edge computing [21–23].

Before AlexNet release, some algorithms such as the Histogram of Oriented Gradient, the Local Binary Patterns, Speed-Up Robust Features, and the Scale-Invariant Feature Transform [24] were used to detect objects, highlight textures and contours, and to obtain image characteristics. The problem with these algorithms were the long periods of execution needed to obtain positive results when used as descriptors with machine learning algorithms [25], such as support vector machine (SVM), decision tree algorithm, and K Nearest Neighbours (KNN) classifiers.

To identify specific elements in images and video streaming, architectures of deep learning networks are commonly used; their main variations are related to the number of layers or the different activation functions [26].

One of the most relevant deep learning works is the You Only Look Once v3 model (YOLO) that can run in 22 ms at 28.2 mAP (mean Average Precision) [16], working with the Common Objects in Context (COCO) dataset they obtained 57.9 in the AP50 metric, despite the good results obtained this models needs a lot of computational power. Another relevant model is the Single Shoot Detector (SSD) that uses a different deep learning architecture based on a group of bounding boxes with different ratios and scales allowing the network to be adjust according to the dimensionality of the boxes, with a view to identify the object shape. When working with the Common Objects in Context (COCO) dataset the algorithm got 26.8 mAP (mean Average Precision) [27], and when working with the AP50 (Average Precision) metric it obtained a result of 46.5 mAP (mean Average Precision). Compared versus the You Only Look Once model, the Single Shoot Detector (SSD) requires less computational power; this makes it perfect to be executed on mobile devices, but needs a precision tradeoff [7]. The architectures previously described are products of investigation groups, however nowadays, governments and private companies are interested in complete image processing challenges and researches applied to security applications [28].

To identify specific objects in images (like faces in our approach), many benchmarks have been structured such as the Labeled Faces in the Wild (LFW) and the You Tube Faces (YTF). The above-mentioned benchmarks are used to teach a deep neural network the human face characteristics [29, 30] and to train them making them able to identify faces inside images with one or more persons. These benchmarks have a good performance when working with deep neural networks, however, when testing these models in everyday environments the performance of networks decreases reducing their reliability and usability.

Object detection has recently focused in real time applications like pedestrian and vehicles detection with the purpose of developing autonomous systems for drones and cars. In the pursuit of this task a few datasets were structured such as the Caltech pedestrian dataset [14], and other datasets which contain images from people in different positions and with different image variables such as scales and occlusions. These datasets are suitable for training a neural network, showing an environment outside a laboratory with real ambient conditions.

Research Model

When robberies happen, most of the times the evidence of images where the alleged guilty person appears is useful, what is more, the evidence of the alleged guilty can be analyzed by authorities to support the case; following this idea, in the present approach we detect both, pedestrians inside the images and their faces.

Benchmarks like the You Tube Faces (YTF) and the Labeled Faces in the Wild (LFW) work over the face detection challenge; such datasets contain images of people acting at controlled environments. Since the benchmarks are useful for probing the neural network model implemented on the present approach, the fact of controlled environment affects its performance. On the other hand, datasets such as the Common Objects in Context (COCO) or the Pattern Analysis Statistical Modeling and Computational Learning (PASCAL) (benchmarks structured for object detection algorithms) does not contain images of people with variations that could produce noise in the analysis such as light variation, occlusion, among others, making this datasets suitable only for not noisy scenarios.

Pedestrian and face detection are two different problems since the characteristics of peoples size and body are quite different to the faces characteristics. When working with pedestrian images the data from the person could get corrupted due to the occlusion, pose or illumination noise [14], on the other hand, when working with faces data there are more sources of noise such as scale, pose, occlusion, expression, makeup and illumination [26]. Deep neural networks are able to generalize and detect specific targets despite the processed images different noise levels, such as change of scale and occlusion, as long as the artificial neural network is fed with the sufficient examples.

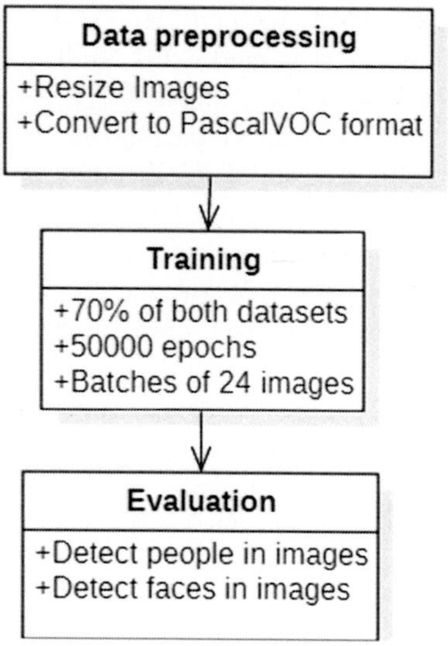

Fig. 3.1 Methodology: the block diagram shows the methodology process and some important aspects to be considered at each stage

In this work, we analyze the pedestrian Caltech dataset with a view to detect persons inside images, and the WIDER face dataset [31] for detecting faces in images. The data set of images used in the deep neural network is composed of two parts, the images that show individual persons, and their respective labels with their anchor boxes contained in a separated data file. An anchor box represents the contour of a referred object through four coordinates (x1, y1, x2, y2). The data will be fed into a Single Shoot Detector (SSD), architecture that consist of several convolutional depth wise layers, a batch norm regularization, rectified linear unit (ReLu) nonlinearity and a softmax layer implemented for the data classification.

The methodology of the approach has been divided into 3 stages as shown in Fig. 3.1. The first stage is the data preprocessing, the second stage consists in the training of the Single Shoot Detector (SSD) artificial neural network, and finally, the third stage consist on implementing the trained neural network and probe its performance when detecting people and faces from different images.

Data Preparation

In the first stage the datasets are processed in order to make them useful for the neural network. Related to the data used on the approach, the WIDER dataset is conformed to 32,203 images that contain 393,703 coordinates of unique faces represented with bounding boxes, the coordinates are clustered in a single file in Pattern Analysis Statistical Modeling and Computational Learning (PASCAL) format. The Caltech Pedestrian dataset is conformed from 10 h of an annotated video sequence, from which we obtained 250,000 images with a total of 350,000 coordinates of pedestrians represented with bounding boxes stored in an xml file. The Pattern Analysis Statistical Modeling and Computational Learning (PASCAL) refers to the way that the coordinates of the bounding box are represented, in this case we need to have the coordinates of two points of the box, the lower left coordinate (xmin, ymin) and the upper right coordinate (xmax, ymax). For training and validating the performance of the neural network the WIDER and Caltech pedestrian datasets are divided into two sets each one, 70% of the samples of each corpus is the considered as a training set, and the remaining 30% of samples at each corpus are considered as the test set.

Our Single Shoot Detector (SSD) architecture has an input image size requirement of 224 $\times$ 224 x 3, that, we have resized the images in order to make them suitable for the Single Shoot Detector (SSD) process. To resizing the images we have developed a Python script that uses Pillow library filters to downgrade the resolution of the image but without losing any information.

Training

In order to train the neural network, it was fed with the images from the training dataset, and the coordinates of the bounding boxes or ground truth from the location of people and faces; by this way the neural network learns how a person and a human face looks like.

The training process is the next, first the neural network is fed with the training corpus from the WIDER data set in order to learn the features of human faces, and after the neural network algorithm is ready it is fed with the Caltech training set, in order to learn the human body features. After each training process the neural network is tested with the test corpus of each dataset (WIDER and Caltech) in order to verify its functionality, as shown in Fig. 3.2.

We fed the neural network with batches of 24 images, after all images went through the neural network architecture described in Table 3.1; the neural network identifies people and faces in the new images. After the prediction we measure the models precision using the Intersection over Union metric (IoU) shown in Eq. (3.1), that consist on dividing the samples Area of Overlap between the Area of Union.

$$IoU = \frac{Area\ of\ Overlap}{Area\ of\ Union} \tag{3.1}$$

The process described above is done in order to upgrade the values of the weights and bias of the network for 50,000 epochs. After the training the neural network predict several bounding boxes in an image, then a score is assigned to each bounding box according to a classification label proposed by the classification network, then their positions are updated slightly using the regression process and a Non Maximum Suppression (NMS) algorithm [32], as a result the location of the person is found. In Fig. 3.3, the process of the NMS algorithm is described.

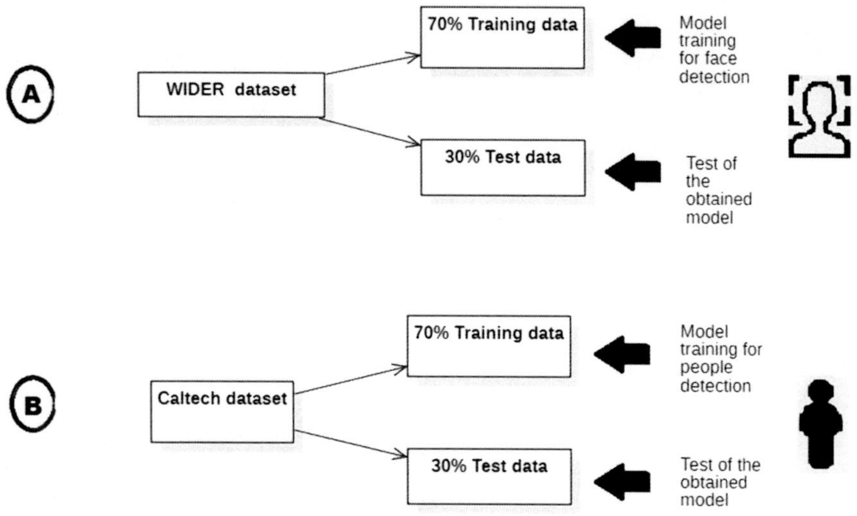

Fig. 3.2 Neural network model training and testing to learn to identify faces and people from images

Evaluation

For evaluating the performance of the neural network we used the precision metric, where the true positive results (True Positives) were divided between the false positive results (False Positives), obtained from the face identification of the deep neural network performance. Also, the precision metric for the body identification was computed applying the same principle, dividing the true positive results between the false positive results obtained from the deep neural network performance. Equation (3.2) shows the described method for each case.

$$Precision = \frac{True\ Positives}{True\ Positives + False\ Positives} \tag{3.2}$$

The neural network obtained a precision of 75.43 mean mAP (mean Average Precision) when working with the Caltech dataset, and a precision of 81.69 mAP (mean Average Precision) when working with the WIDER dataset.

For ensure that the neural network learned from the data and did not try to memorize them, we fed the model with images that the network have never processed before, gathered from volunteers private images (VPI) that were preprocessed and then applied to the model to make inferences. The volunteer's images were obtained from a group of Facebook users who were asked for permission through the Geospatial Laboratory Facebook account.

Table 3.1 MobileNetBody architecture. Layer type, shape and input size (table extracted from [7])

Type/Stride	Filter shape	Input size
Conv/s2	3 × 3 × 3 × 32	224 × 224 × 3
Conv dw/s1	3 × 3 × 32 dw	112 × 112 × 32
Conv/s1	1 × 1 × 32 × 64	112 × 112 × 32
Conv dw/s2	3 × 3 × 64 dw	112 × 112 × 64
Conv/s1	1 × 1 × 64 × 128	56 × 56 × 64
Conv dw/s1	3 × 3 × 128 dw	56 × 56 × 128
Conv/s1	1 × 1 × 128 × 128	56 × 56 × 128
Conv dw/s2	3 × 3 × 128 dw	56 × 56 × 128
Conv/s1	1 × 1 × 128 × 256	28 × 28 × 128
Conv dw/s1	3 × 3 × 256 dw	28 × 28 × 256
Conv/s1	1 × 1 × 256 × 256	28 × 28 × 256
Conv dw/s2	3 × 3 × 256 dw	28 × 28 × 256
Conv/s1	1 × 1 × 256 × 512	14 × 14 × 256
5 × Conv dw/s1 5 × Con/s1	3 × 3 × 512 dw 1 × 1 × 512 × 512	14 × 14 × 512 14 × 14 × 512
Conv dw/s2	3 × 3 × 512 dw	14 × 14 × 512
Con/s1	1 × 1 × 512 × 1024	7 × 7 × 512
Conv dw/s2	3 × 3 × 1024 dw	7 × 7 × 1024
Con/s1	1 × 1 × 1024 × 1024	7 × 7 × 1024
Avg pool/s1	Pool 7 × 7	7 × 7 × 1024
FC/s1	1024 × 1000	1 × 1 × 1024
Softmax/s1	classifier	1 × 1 × 1000

Findings

The deep neural network implemented on this approach is able to learn from the Caltech and WIDER face datasets and is trained to work with images taken on the street at noisy or not controlled scenarios, our model identifies pedestrians with a mean average precision of 75.43 and faces with a mean average precision of 81.69, Fig. 3.4 shows the plots created from the precision computing.

The implemented solution is suitable for working while the image is obtained when there is natural light, according to our investigation of previous works there are no available data sets with night light which limits the deep neural network to work with images taken at environments with light.

After validating the trained neural network with volunteered personal information (VPI) images we observed that the neural network can identify people in images as shown in the downside of Fig. 3.5, also it shows that it has trouble detecting individuals when people are close to each other or doing movements out of common as dancing. The trained neural network showed no trouble detecting multiple faces in images as shown in the upside of Fig. 3.5.

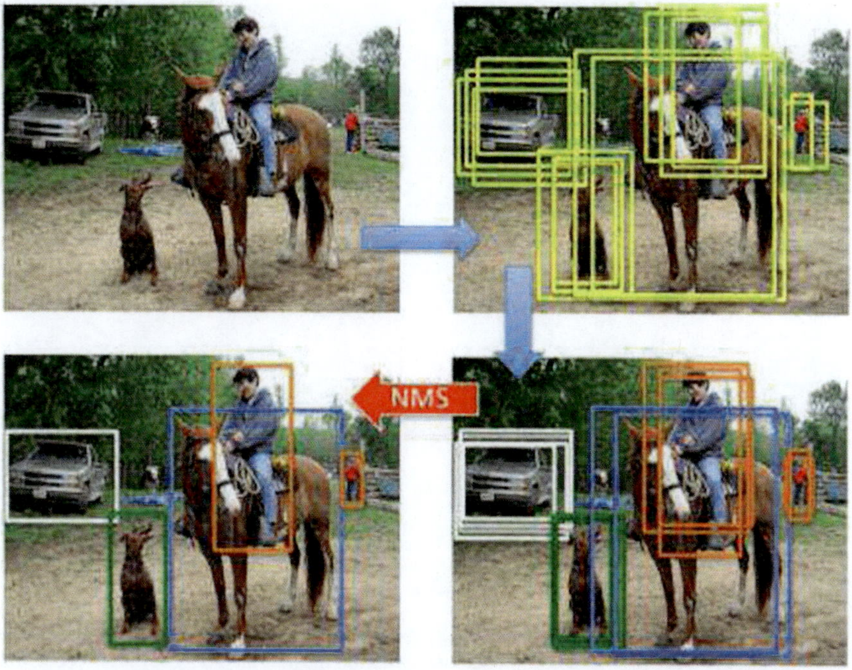

Fig. 3.3 Independent region proposals of the neural network and non-maximum suppression

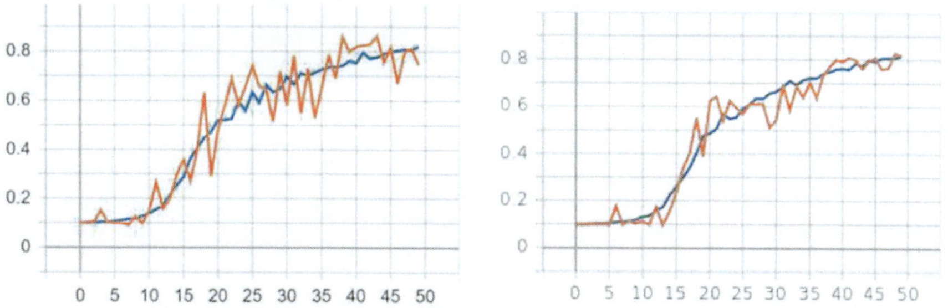

Fig. 3.4 Left: precision graph from Caltech dataset training obtaining a result of 75.43 mean average precision after 50 epochs. Right: precision graph from WIDER dataset training obtaining a result of 81.69 mean average precision after 50 epochs

The proposed approach implements a neural network with a light architecture and low computational cost in exchange for a reduction in precision because it is planned to be implemented on a wearable to take images of robberies in real time (as part of a more complex project). If a greater precision metric was desired, a more complex architecture such as Region Convolutional Neural Network (Faster R-CNN) [33] or Inception should be used. However, the results obtained are higher than expected, since the metric of the average of the average precision resemble the results reported

Fig. 3.5 Object detection in street photos. Up: face detection on VPI data. Down: people detection on VPI data. As it is possible to see, on the face detection most of the image faces are recognized, on the other hand is more difficult for the algorithms to detect people because of their movements as in the girl's picture, or because of the proximity between persons in the same image, as in the musician's picture

in the work of Mobilenet [7] and it is possible to detect people and faces in the images using the trained model.

General Discussions

After the analysis of the results it is possible to see that the proposed model obtains the expected small, medium, and large mean mAP (mean Average Precision) values, when its performance is evaluated using the Common Objects in Context (COCO) dataset. The proposed methodology for the images detection was positive when identifying faces with the WIDER and Caltech dataset.

Since the methodology has been structured to be part of a complex wearable designed to take images and identify faces and people in real time, its performance was probed to identify persons and faces from pictures taken at no controlled environments with natural light variations in the study area of Mexico City, the deep neural network implemented recognition process got a 90% of precision when the experiments were applied. On Fig. 3.6 are shown two graphs, the one on the left show the loss of the test and train procedures applied over the Caltech data set, meanwhile

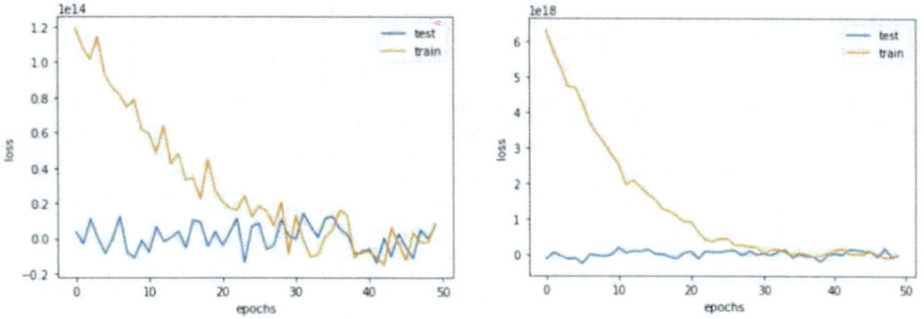

Fig. 3.6 Left: loss graph from Caltech dataset training obtaining a final loss of 0.145 after 50 epochs. Right: loss graph from WIDER dataset training obtaining a final loss of 0.059 after 50 epochs

the graph on the right shows the test and train loss when working with the WIDER data set.

The obtained results from the loss in Fig. 3.6 shows that the face and people detection on images procedure implemented on the present approach is robust and useful to analyze images with different natural light variations taken on different scenarios as in the street.

With these results we can see that the neural network learns to find faces and people in images and that it is possible to carry out the next stage in the development of a larger work where the application against robberies is addressed. Because the robberies are carried out by people directly it is necessary to detect the people who carried out these acts. With the work of this network we can discriminate photos that contain important information of these people from those that do not and thus optimize the storage of information and achieve one of the purposes of edge computing.

The general idea of the total project results from the studied necessity of citizens to generate volunteer private information, and in a way to give people a tool to have physical evidence when a robbery occurs, such project and the implemented procedure for face recognition open a new door for people's interest in their community and their respective participation in the city improvement. The face recognition procedure implemented will be helpful for the raising of citizen complaints.

As a future step the trained network will be uploaded to a portable device that will make new inferences to images provided from volunteers taken by the electronic device. The network implementation is suitable for working with low power devices like ARM microprocessors or smartphones because most of them contain a small graphic processor unit (GPU). In this way any person could use a neural network to detect people and faces in images that could be useful when robberies occur.

In parallel the new images fed to the network will be used as resource to build a new dataset made from volunteered personal information (VPI).

Conclusions

The proposed approach presents an application of computing vision using deep neural networks applied to edge computing. The union of artificial intelligence with volunteer private information as the presented on this approach creates a new feature for smart cities, which results attractive for people that can provide data and images about what happens in their surroundings, and organizations that can process the volunteered personal information (VPI) information to create solutions to urban problems such as insecurity or the increase of efficiently raising of citizens' complaints.

The project has been planned to work directly with users to capture images of robberies and generate citizen complaints. It is decentralized from the authorities so it is possible to strengthen the system with low resources as long as citizens support their usage and developing.

With the results obtained in the training using the datasets and the inference results in the volunteered personal information (VPI) data we guarantee the correct performance of the system with volunteered personal information (VPI), since after meticulously probes applied over the implemented model we obtained good results validating the Neural Network, for example on its performance with noisy data we obtained 75.43 of mAP (mean Average Precision) when detecting people in images, and 81.69 of mAP (mean Average Precision) when detecting faces. These values are comparable good with state of the art architectures like You Only Look Once (YOLO) v2 [34] that obtained 90.9% AP in the Caltech pedestrian dataset.

As future work, the face recognition process will be implemented on a 32 bits microprocessor with a graphic process unit of low energy consume in order to create a wearable device that can be used by people when walking or doing their activities on the street, by this way when they perceive or even worst are victims of a robbery the would be able of take pictures of their aggressors, and after the deep neural network face recognition process, obtain probes for the raising of their complaint with the authorities.

Another idea that comes with the approach is to generate an open database of images from the cities where the approach is used that could be used to ameliorate the training stage of the deep neural network and that could be used on further approaches; also the database could contain images taken with artificial light at night in order to implement the procedure functioning for images without natural light.

Acknowledgements This work was partially sponsored by the Instituto Politécnico Nacional (IPN), the Consejo Nacional de Ciencia y Tecnología (CONACYT) under grant 961713 and the Secretaría de Investigación y Posgrado (SIP) grants number 20201897, 20201863, 20190007, 20200630. Additionally, we are thankful to the reviewers for their invaluable and constructive feedback that helped improve the quality of the paper.

References

1. N. Dalal, B. Triggs, C. Schmid, Human detection using oriented histograms of flow and appearance, pp. 428–441. Springer (2006)
2. W. Ouyang, X. Wang, Joint deep learning for pedestrian detection, in *Proceedings of the IEEE International Conference on Computer Vision*, pp. 2056–2063 (2013)
3. O. Tuzel, F. Porikli, P. Meer, Human detection via classification on riemannian manifolds, in *IEEE Conference on Computer Vision and Pattern Recognition*, pp. 1–8 (2007)
4. N. Dalal, Finding people in images and videos. Doctoral dissertation (2006)
5. A. Ess, B. Leibe, K. Schindler, L. Van Gool, A mobile vision system for robust multi-person tracking, in *IEEE Conference on Computer Vision and Pattern Recognition*, pp. 1–8 (2008)
6. D.M. Gavrila, S. Munder, Multi-cue pedestrian detection and tracking from a moving vehicle. Int. J. Comput. Vision 41–59 (2007)
7. A.G. Howard, M. Zhu, B. Chen, D. Kalenichenko, W. Wang, Weyand, Mobilenets: efficient convolutional neural networks for mobile vision applications. Google Inc. pp. 1–9 (2017)
8. T. Ebrahimi, F. Dufaux, Smart Video Surveillance System Ensuring Privacy. U.S. Patent Application No. 11/631, 806 (2007)
9. LANG, Steve; SISKIND, Barry. Digital video surveillance system. U.S. Patent Application No 11/891, 381 (14 Feb 2008)
10. TIAN, Ying-li, et al., IBM smart surveillance system (S3): event based video surveillance system with an open and extensible framework. Mach. Vision Appl. 315–327 (2008)
11. M. Everingham, et al., The pascal visual object classes challenge: a retrospective. Int. J. Comput. Vision 98–136 (2015)
12. T.Y. Lin, et al., Microsoft coco: common objects in context, in *European conference on computer vision*. Springer, Cham (2014)
13. Dalal, Navneet, and Bill Triggs. "INRIA person dataset (2005)
14. P. Dollár, C. Wojek, B. Schiele, P. Perona, Pedestrian detection: a benchmark, in *IEEE Conference on Computer Vision and Pattern Recognition*, pp. 304–311 (2009)
15. H. Jiang, E. Learned-Miller, Face detection with the faster R-CNN, in *IEEE International Conference on Automatic Face and Gesture Recognition*, pp. 650–657 (2017)
16. J. Redmon, S. Divvala, R. Girshick, A. Farhadi, You only look once: unified, real-time object detection, in *IEEE conference on computer vision and pattern recognition*, pp. 779–788 (2016)
17. P. Viola, M. Jones, Robust real-time object detection. Int. J. Comput. Vision 34–47 (2001)
18. S. Ren, K. He, R. Girshick, J. Sun, Faster r-cnn: towards real-time object detection with region proposal networks. Adv. Neural Inf. Process. Syst. 91–99 (2015)
19. O.M. Parkhi, A. Vedaldi, A. Zisserman, Deep Face Recognition, pp. 1–12. University of Oxford (2015)
20. M.Z. Alom, T.M. Taha, C. Yakopcic, S. Westberg, P. Sidike, Nasrin: the history began from alexnet: a comprehensive survey on deep learning approaches. arXiv, pp. 1–39 (2018)
21. P. Mach, B. Zdenek, Mobile edge computing: a survey on architecture and computation offloading. IEEE Commun. Surveys Tutorials 19.3 (2017)
22. Y. Mao, et al., A survey on mobile edge computing: the communication perspective. IEEE Commun. Surveys Tutorials 19.4 (2017)
23. W. Shi, et al., Edge computing: vision and challenges. IEEE Internet Things J 3.5 (2016)
24. N. Dalal, B. Triggs, Histograms of oriented gradients for human detection, in *IEEE computer society conference on computer vision and pattern recognition*, pp. 886–893 (2005)
25. M.A. Turk, A.P. Pentland, Face recognition using eigenfaces, in *IEEE Computer Society Conference on Computer Vision and Pattern Recognition*, pp. 586–687 (1991)
26. H. Li, Z. Lin, X. Shen, J. Brandt, G. Hua, A convolutional neural network cascade for face detection, in *IEEE conference on computer vision and pattern recognition*, pp. 5325–5334 (2015)
27. W. Liu, D. Anguelov, D. Erhan, C. Szegedy, S. Reed, C.Y. Fu, A.C. Berg, Ssd: single shot multibox detector, in *European conference on computer vision*, pp. 21–37 (2016)

28. T. Jameel, R. Ali, S. Ali, Security in modern smart cities, in *International Conference on Communication, Computing and Digital systems*. IEEE, pp. 293–298 (2019)
29. S. Yang, P. Luo, C.C. Loy, X. Tang, From facial parts responses to face detection: a deep learning approach, in *IEEE International Conference on Computer Vision*, pp. 3676–3684 (2015)
30. F. Schroff, D. Kalenichenko, J. Philbin, Facenet: a unified embedding for face recognition and clustering, in *IEEE conference on computer vision and pattern recognition*, pp. 815–823 (2015)
31. S. Yang, et al.,Wider face: a face detection benchmark, in *Proceedings of the IEEE conference on computer vision and pattern recognition* (2016)
32. N. Bodla, B. Singh, R. Chellappa, L.S. Davis, Soft-NMS—improving object detection with one line of code, in *IEEE international conference on computer vision*, pp. 5561–5569 (2017)
33. R. Girshick, Fast r-cnn, in *IEEE international conference on computer vision*, pp. 1440–1448 (2015)
34. Z. Liu, Z. Chen, Z. Li, W. Hu, An efficient pedestrian detection method based on YOLOv2. Math. Problems Eng. 1–10 (2018)

Chapter 4
A Parallel Metaheuristic Approach to Reduce Vehicle Travel Time for Smart Cities Sustainability

Hector Rico-Garcia, Jose-Luis Sanchez-Romero, Antonio Jimeno-Morenilla, and Hector Migallon-Gomis

Abstract The development of the smart city concept and the inhabitants' need to reduce travel time, as well as society's awareness of the reduction of fuel consumption and respect for the environment, lead to a new approach to the classic problem of the Travelling Salesman Problem (TSP) applied to urban environments. This problem can be formulated as "Given a list of geographic points and the distances between each pair of points, what is the shortest possible route that visits each point and returns to the departure point?" Nowadays, with the development of IoT devices and the high sensoring capabilities, a large amount of data and measurements are available, allowing researchers to model accurately the routes to choose. In this work, the purpose is to give solution to the TSP in smart city environments using a modified version of the metaheuristic optimization algorithm TLBO (Teacher Learner Based Optimization). In addition, to improve performance, the solution is implemented using a parallel GPU architecture, specifically a CUDA implementation.

Keywords Smart cities · Sustainability · Travelling salesman problem · Optimization · TLBO · Parallelism · GPU

H. Rico-Garcia · J.-L. Sanchez-Romero (✉)
Department of Computer Technology, University of Alicante, 03690 San Vicente del Raspeig, Spain
e-mail: sanchez@dtic.ua.es

H. Rico-Garcia
e-mail: hector.rico@gmail.com

A. Jimeno-Morenilla
Department of Computer Technology, University of Alicante, Alicante, Spain
e-mail: jimeno@dtic.ua.es

H. Migallon-Gomis
Department of Computer Engineering, Miguel Hernandez University, 03202 Elche, Spain
e-mail: hmigallon@umh.es

A. Visvizi et al. (eds.), *Research and Innovation Forum 2020*,
Springer Proceedings in Complexity,
https://doi.org/10.1007/978-3-030-62066-0_4

Introduction

The Smart City concept implies smart infrastructure, technology and actions so as to improve the quality of life of inhabitants. One of the main aspects with a relevant impact on this aim involves managing to people and goods movements [1]. Improving traffic management reduces travel times of people, which in turn decreases the stress levels of drivers and improves the arrival at their workplaces. It also entails better use of existing infrastructure and reduces pollution levels, a growing issue given people's awareness of environmental care and health impact [2, 3].

Traffic management in cities becomes more relevant in terms of the movements that transporters and couriers must make to take the different goods from the headquarters to the delivery points. The problem will affect autonomous vehicle fleets in the future, where the driver assumes a passive role and the vehicle itself takes the decisions on the route to be followed to reach its destination in the shortest possible time.

The current development of IoT and other sensorization capabilities of the devices allow a large amount of data and measurements to be available so that researchers can accurately model the different routes within an urban environment [4]. With the information available, the routes are determined not only by the distance between the different geographical points, but by other aspects such as traffic lights and their crossing times, pedestrian crossings, possible traffic jams, etc.

The TSP (*Traveling Salesman Problem*) is a classic combinatorial problem that can be matched with optimizing car movements in urban environments. This problem can be formulated informally as follows: "Given a list of geographical points and the distances between each pair of points, what is the shortest possible route that passes each point only once and returns to the starting point?" A great number of research works have been developed so as to cope with TSP, including various algorithms and heuristics which try to reduce the computational time when providing a solution.

This paper proposes an approach to address TSP applied to smart city environments, using a modified version of the metaheuristic optimization algorithm TLBO (*Teacher Learner Based Optimization*). To improve performance, the solution has been developed using a CUDA (*Compute Unified Device Architecture*) parallel architecture implemented on a GPU (*Graphics Processing Unit*) platform. The paper is organized as follows: Sect. "The Problem of Traffic Management in Smart Cities" provides a review of the traffic congestion management in smart cities. Section "The Travelling Salesman Problem" describes the TSP and its analogy with the management of traffic in urban environments. Section "The TLBO Optimization Method" describes the TLBO method in both original and discrete versions. Section "Parallel TLBO Implementation on GPU" explains the implementation of TLBO on a parallel CUDA architecture. Section "Experimentation" shows the results of the parallel TLBO when applied to several cases from a well-known benchmark. In Sect. "Conclusions", conclusions of the work are synthesized and future lines of research are proposed.

The Problem of Traffic Management in Smart Cities

Vehicle traffic management is a major problem in smart cities. Freight traffic is almost inevitably operated by trucks and vans, while people movement throughout a city is still mostly operated by private vehicles [5]. In 2018, 45% of American people had no access to public transportation [6]. Since the main objective in designing a smart city is to improve "personal satisfaction by utilizing innovation to enhance the proficiency of administrations and address occupants issues" [4], the movements of citizens through a city should be optimized so that travel time is reduced, but concerns about pollution and welfare must also be addressed. Freight transporters must also deliver their cargo in an efficient way, trying to reduce delivery times, pollution rates, and stress and tiredness of drivers.

Several research papers can be found that propose to monitor traffic in urban environments using different technologies and methods. With regard to technology applied to traffic management in smart cities, several studies can be pointed out. In [4], a low cost real-time traffic management smart service is proposed by activating traffic indicators to update traffic data instantly; low-cost vehicle detection sensors are installed on the roads every 500 or 1000 m; IoT is used to acquire and send traffic data; real time streaming data is sent for Big Data analytics. In [7], a system is developed so as to track the movement of people and vehicles monitoring WiFi and Bluetooth signals emitted by smartphones or on-board devices. In [8], a LoRa-based (*Long-Range*) sensor network to monitor pollution levels is developed; the system consist of geo-located nodes embedded in vehicles and fixed places to monitor temperature, relative humidity and CO_2 concentration in urban environments; the data collected allows the system to generate the required commands to the traffic signs and panels that control the vehicles movement. In [9], a method based on an IoT platform is proposed that collects and manages data flows from multifunctional IoT devices and data sources; the data are analyzed so as to identify traffic noise phenomena; if the noise level on a given road exceeds a specified threshold, the traffic light crossing required to redirect traffic to alternative roads is enabled. In [10], a deep learning model is proposed for analyzing IoT data; the model uses Long Short Term Memory (LSTM) networks to predict future values of air quality. In [11], an IoT application is proposed which uses vehicular networks to detect traffic condition of the roads and includes diverse techniques to control the different lanes of the highways according to the level of congestion of these lanes. In [12], an approach of a traffic management system is proposed, including 5G communication, RFID-based parking space monitoring and cloud services for supervisory control and machine learning. In [13], a system that uses IoT devices for city data collection and Big Data analytics is proposed; a smart digital city architecture is proposed which can cope with huge amount of city data and provide guidelines to the local authorities. In [14], a study on the effect of path selection in emission-oriented urban vehicle routing is presented, which is based on a road network including speed and acceleration frequency data; an algorithm to determine all emission-minimal paths for an origin–destination pair given a particular vehicle is tested. In [15], two data-based approaches are proposed to determine time-dependent emission minimization pathways in urban areas and compare their performance with respect to calculation efficiency and solution quality.

With regard to the application of methods so as to manage traffic for reducing pollution, several studies should be highlighted. In [16], research is focused on the pollution routing problem (PRP); a practical PRP model is proposed which uses a minimal subset of the essential factors affecting fuel consumption and pollutant emission of trucks; a solution for this model is developed. In [17], research focuses on minimizing CO_2 emissions from a fleet of vehicles in urban areas; a local search procedure, called tabu search heuristic, is adapted to solve the problem; the research uses instances derived from a set of real data from the road network and 230 million speed observations. In [18], a matheuristic approach is proposed so as to solve PRP; this method integrates Iterated Local Search (ILS) with a Set Partitioning (SP) procedure and a Speed Optimization Algorithm (SOA); the proposed method performs combined route and speed optimization within several local search and integer programming components.

In summary, a wide range of research work is being carried out with regard to improving the quality of life in smart cities in terms of traffic management and thereby reducing fuel consumption, street noise and pollution.

The Travelling Salesman Problem

The TSP has given rise to a wide variety of research work, since it is a complex problem at the time of obtaining a solution [19]. Using graph theory terminology, the TSP is defined as a graph $G = (V, A)$, where $V = \{v_1,\ldots,v_n\}$ is a set of n vertices (nodes) and $A = \{(v_i, v_j)/v_i, v_j \in V, i \neq j\}$ is a set of edges with an associated non-negative cost (distance) matrix $D = (d_{ij})$. The problem is symmetric if $d_{ij} = d_{ji}$ for any pair of nodes $(v_i, v_j) \in A$, and is asymmetric (ATSP) in any other case. Thus, the aim is obtaining the optimal ordering of vertices such that: (i) every one of them is visited once and only once; and (ii) the sum of distances travelled is minimal.

An analogy between the TSP and the problem of managing traffic in smart cities can be established. The set of nodes can be related to the different city locations where a truck must deliver its cargo from a depot. The meaning of *distance* in case of urban environments is different from the mere concept of geographical distance, since it can incorporate traffic lights, pedestrian crossings, speed limits, and even dynamic factor such as temporary traffic jams, road maintenance works, and so on. Data collected on fuel consumption and pollution related to different roads could be included.

Given a set of n nodes, finding the best solution by exhaustive search would involve a complexity of the order of $(n - 1)!$ A large amount list of research works related to TSP has given different methods that reduce this complexity. Much attention has been paid to heuristic methods applied to TSP among researchers. In [20], a survey of swarm intelligence applied to graph search problems is shown, but the paper is almost exclusively focused on Ant Colony Optimization (ACO) and Bee Colony Optimization (BCO). In [21], two different modifications of the Artificial Bee Colony (ABC) method are proposed for TSP; the performance of these two algorithms is compared with eight different versions of Genetic Algorithm (GA), and also with Ant Colony System (ACS) and Bee Colony Optimization methods. In fact, most of the works related to solve TSP by using metaheuristic optimization methods deal with Ant Colony Optimization [22–27], sometimes in a hybrid version with other methods [27].

The TLBO Optimization Method

Original TLBO Formulation

TLBO is a metaheuristic optimization method developed by R. V. Rao [28]. It is based on the behaviour of a *teacher* and a set of students (*learners*) when performing an iterative teaching–learning process. A population of individuals is initially created, each one of them being assigned a random value for each one of the design variables defined in function f, which is the function to be optimized. After this initialization, the method iterates through two stages until a termination criterion is fulfilled:

Teacher stage. In this stage, each individual i evaluates the function f with its own parameters, $f(X(i))$, and the one with the optimal value is labeled as the *Teacher* of the population. Each individual updates its design variables according to the *Teacher* parameters and the mean parameters of the whole population, according to the following expression:

$$X_{new}(i, j) = X(i, j) + rand(0, 1)(X_{best}(j) - TFactor \cdot X_m(j)) \tag{4.1}$$

In (4.1), the design variable j of individual i, $X(i, j)$, is modified by using the value of variable j from the *Teacher*, $X_{best}(j)$, the variable mean from the whole population, $X_m(j)$, and a *TFactor*. This *TFactor* adopts the integer value 1 or 2 according to the following expression:

$$TFactor = round(1 + rand(0, 1)) \tag{4.2}$$

Once the parameters of individual i have been modified, it is evaluated. If the evaluation of $f(X_{new}(i))$ provides a better result than that of the original individual, the new values of the parameters replace the old ones in individual i.

Learner stage. In the next stage, each individual is compared with a random contestant from the population. The individual with a better evaluation is labelled as the *BestLearner*, and the other is labelled as the *WorstLearner*. They are used to create a new individual by using the following expression:

$$X_{new}(i, j) = X(i, j) + rand(0, 1) \cdot (BestLearner(j) - WorstLearner(j)) \tag{4.3}$$

When *Xnew*(*i*) is generated, this new individual is evaluated and compared with the original individual $X(i)$. If the new individual is better than the original one, the parameters of the old one are replaced by the new calculated parameters.

The main advantage of TLBO over other population-based heuristic algorithms is that it has no algorithm-specific tuning parameters, since only population size and generations should be considered. Different research works developed in recent years can be found which demonstrate the better efficiency of TLBO when compared to other optimization methods [29–32]. This is why TLBO is continually being studied, improved and applied in a wide variety of scientific and engineering fields.

Discrete TLBO

TLBO is originally oriented to continuous problems. However, TSP is a combinatorial problem. This sort of problems involves finding a grouping, ordering, or assignment of a discrete, finite set of objects that satisfies a set of conditions. Therefore, an intensive modification must be made on TLBO so as to adapt it to TSP and other combinatorial problems. In the current paper, an implementation of the DTLBO based on the work developed in [33] has been carried out, but several relevant modifications have been performed.

Representation of individuals. Each individual is a sequence of city or urban locations to be visited. As an example, if the problem includes 8 urban locations ($v_0, v_1 \dots v_7$) to be visited, an individual could be represented as shown in Fig. 4.1. In Fig. 4.1, the solution represented consists of starting from urban location v_3 and finishing in location v_4 before returning to the starting point v_3, visiting the different locations in the order $v_3 \rightarrow v_5 \rightarrow v_0 \rightarrow v_1 \rightarrow v_2 \rightarrow v_7 \rightarrow v_6 \rightarrow v_4$.

Initially, each individual is assigned a random permutation of the urban locations to be visited. The global population is divided into subpopulations in order to avoid becoming trapped in local minima. Four subpopulations are considered in [31].

DTLBO Teacher Stage. The best individual of each subpopulation is labeled as the *PartialTeacher* of that subpopulation. It will be used so as to update learners in the subpopulation. A *Mean* individual containing the mean values within each subpopulation is created and used for updating each learner. Moreover, the best individual within the whole population is labelled as the global *Teacher*.

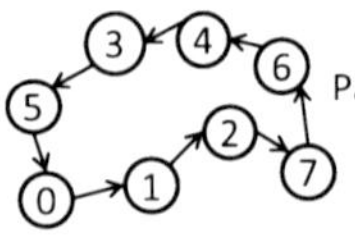

Vector index	0	1	2	3	4	5	6	7
Path sequence	3	5	0	1	2	7	6	4

Fig. 4.1 Representation of an individual solution

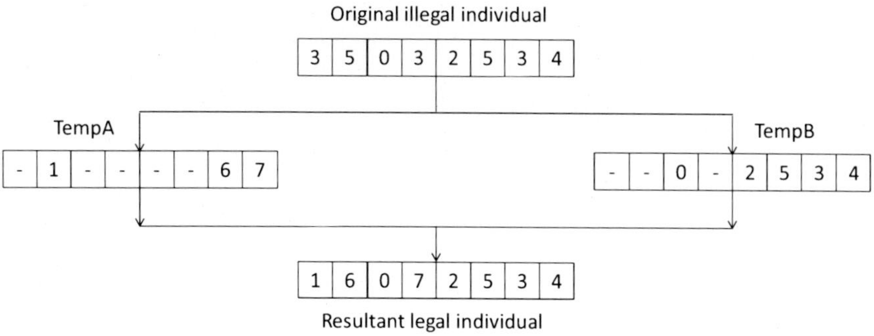

Fig. 4.2 An example of the legality operation

It must be taken into account that the *Mean* individual could be an infeasible solution, since some locations could be repeated and others could not appear. Therefore, a legality operation must be performed on this *Mean* individual. This legality operation shown in Fig. 4.2 works as follows: given an illegal individual, a new vector *TempA* is created where the cities that do not appear are written into its corresponding cell; in vector *TempB*, each city that does appear is written in the cell it appears; if a city appears more than once, it is written in the last cell where it appears; finally, the operation iterates through *TempA* and each city in this array occupies the first free cell found in *TempB*, obtaining thus a legal individual.

Crossover operation. The crossover operation is indicated by symbol ⊗ . There are four different crossover operations available so as to create a new individual:

$$
\begin{aligned}
Xnew(i) &= X(i)\text{Ä}Teacher \\
Xnew(i) &= X(i)\text{Ä}PartialTeacher(i) \\
Xnew(i) &= X(i)\text{Ä}Mean(i) \\
Xnew(i) &= PartialTeacher(i)\text{Ä}Mean(i)
\end{aligned}
\tag{4.4}
$$

In the current work, a random selection of each one of the above crossover operations is performed each iteration and for each individual. In this way, a wide variety of individuals is achieved and, therefore, an additional strategy is included to prevent a subpopulation from being trapped in a local minimum.

Given two individuals, *A* and *B*, the crossover operation that generates a new individual A_c works in the following way: a starting and an ending position in the order of visiting the urban locations are randomly selected; individual *A* replaces its components by the ones of individual *B*. Figure 4.3 shows an example of this operation.

After a new individual is generated, the legality operation must be performed on it.

Mutation operation. Once the new individual is created by means of the crossover operation, a mutation operation is applied on it. Mutation is performed as follows:

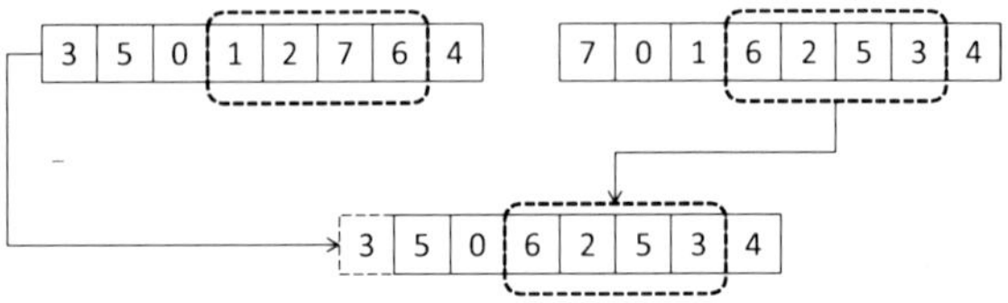

Fig. 4.3 An example of the crossover operation

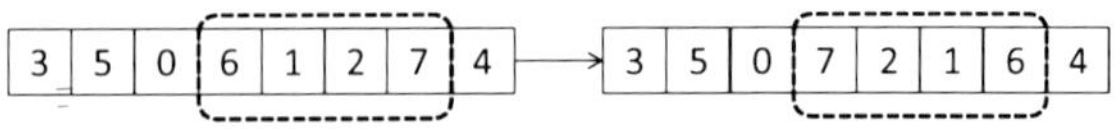

Fig. 4.4 An example of mutation operation

a starting and an ending position in the order of visiting the urban locations are selected; the elements in the range are flipped so as to generate a mutated individual.

In Fig. 4.4, a starting position 3 and an ending position 6 in the order of visiting the urban locations are randomly selected within individual A_c to give A_{cm}.

DTLBO Learner stage. After the teacher stage, the learner stage of DTLBO is carried out. As in the original TLBO, for each individual *i* a contestant *k* is randomly selected within its subpopulation. The learner *Xi* is updated by a crossover operation, working in the same way as the crossover operation in the teacher stage.

Once the crossover operation is carried out, legality of the new individual is checked. Then the mutation operation is carried out just the same way as in the teacher stage.

In [31], DTLBO is compared to ACO, ABC, GA, and Particle Swarm Optimization (PSO) with different TSP instances; it is concluded that DTLBO is better than the above mentioned algorithms in most cases; only ACO and ABC achieve similar or slightly better results in very few instances; conclusions also remark that DTLBO performance decreases with a high number or locations to be visited; therefore, it is suggested that efforts should be made to improve DTLBO performance for large-scale TSP problems.

Parallel TLBO Implementation on GPU

The modifications made on the TLBO to adapt it to the TSP make the iterations more costly in terms of computation time. To reduce the impact of these changes on DTLBO performance, a parallel version of the algorithm has been implemented following a CUDA architecture and using GPU.

Design of the Memory Organization

The first and essential step when proceeding with the parallelization of the algorithm is to create a correct design of the memory structure and the execution flow in order to minimize the global thread locks. GPU memory is a resource that, if mismanaged, can highly increase execution times because transfer operations between memory levels within the GPU are very time-consuming.

Global memory. The information stored in the global memory is: (i) an array with all the necessary points for the generation of the individuals of the population; (ii) an array with the pre-calculated distances between points; and (iii) the best individual of all the populations (global *Teacher*) and its evaluation each iteration, since it is used by the whole population. Most global information is read-only; just the best evaluation (global *Teacher*) will be modified if necessary each iteration.

Block memory. The information shared by a subpopulation is stored in the block memory. A matrix is stored where each row represents an individual and each column stores the index of an urban location; an additional column stores the individual's solution to avoid repeating the evaluation.

Thread memory. Some variables are stored in the thread local memory. These variables are used for the required calculations within each phase of the algorithm and are updated each iteration. This memory is private to each thread and is not shared with the rest of the population.

Execution Flow

An execution flow has been designed with the aim of minimizing the blockages of the threads when synchronizing during the execution of the algorithm. The different threads that run in parallel must synchronize to obtain some information common to the whole population, such as the Mean individual and the Teacher. In addition, the reduction technique will be applied to minimize the iterations required to obtain these values from the population. Each independent thread will be used as an individual of the population; only one thread of the subpopulation (first individual) communicates with the memory so as to avoid problems in accessing this resource.

Experimentation

Experiments are addressed to compare the speed of execution when solving four TSP problems from the TSPLIB library [34]. The comparison is made between a sequential implementation and a parallel GPU implementation using CUDA. The hardware used for the experimentation is a Pentium i7 processor at 3.2 GHz with an NVIDIA GeForce 1060p graphics card, 6 GB GPU RAM and 32 GB DDR4 RAM. The CUDA 9.2 platform was used. Different scenarios are defined for each problem with regard to population size and iterations performed to run the algorithm. The cities from TSPLIB used are Berlin52, Att48, Eil76, and Ch130, with 52, 48, 76, and 130 urban nodes respectively. They represent real urban scenarios taken from Europe, USA, and China.

Table 4.1 Experiments results in terms of CPU and GPU performance

Population	Iterations	City	CPU	GPU	Speedup
64	5,000	Berlin52	0.3652	0.1214	3.01
64	10,000	Berlin52	0.7621	0.2351	3.24
128	5,000	Berlin52	0.7313	0.1309	5.59
128	10,000	Berlin52	1.5281	0.2532	6.04
64	5,000	Att48	0.3543	0.1485	2.39
64	10,000	Att48	0.7023	0.2913	2.41
128	5,000	Att48	0.7199	0.1583	4.55
128	10,000	Att48	1.4395	0.3180	4.53
64	5,000	Eil76	0.5248	0.1808	2.90
64	10,000	Eil76	1.0387	0.3566	2.91
128	5,000	Eil76	1.0513	0.1928	5.45
128	10,000	Eil76	1.9791	0.3766	5.26
64	5000	Ch130	0.9111	0.2791	3.26
64	10,000	Ch130	1.7513	0.5715	3.06
128	5000	Ch130	1.2341	0.2135	5.78
128	10,000	Ch130	2.2981	0.4176	5.50

Table 4.1 shows the results from the average time in ms of 10 blocks of 1,000 runs. It can be observed that the parallel implementation of DTLBO improves performance up to 6× in case of a large number of individuals and iterations, being the average speedup of the whole set of experiments 4.12x. DTLBO reached the optimal in Att48 and just a difference of 2 in Berlin52. Figure 4.5 shows the path obtained for the Att48 TSP problem.

Since the GPU is capable of processing a large number of threads per block, it can be seen that the increase in population does not have as much time penalty as the CPU, affecting only the number of iterations to the GPU times. Also, given the GPU block architecture, different subpopulations can be placed in different blocks of the GPU and, therefore, run in parallel.

Conclusions

Discrete Teaching–Learning-Based Optimization (DTLBO) was proven to be a good choice for solving TSP problems and, therefore, it can be used in smart cities and urban environments to help drivers find the optimal path in terms of distance, fuel consumption, or pollution reduction. In this paper, a parallel implementation of DTLBO using a GPU approach to improve performance is proposed. The developed parallel implementation has substantially improved the speed of the algorithm,

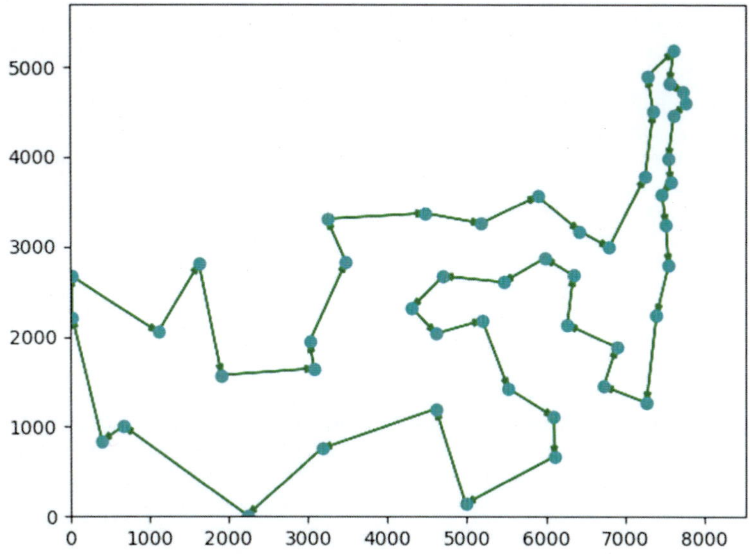

Fig. 4.5 DLTBO solution for Att48

obtaining speedups in the order of 3.86 $\times$ on average, and in case of a large number of individuals and iterations the speedup reaches 6x.

The implementation could still be improved in some parts trying to emphasize the thread blocks organization to optimize the use of GPU resources. In addition, although in this research work parallel implementations of DTLBO have been performed on a desktop computer, they could be run in embedded systems inside cars, since there are already some of them with NVIDIA chipsets and support for CUDA, which are used for image recognition within smart driving. This way, smart driving could be improved by performing route enhancement using checkpoints within the GPU processing without penalizing the CPU. This is an important issue since, in embedded systems in cars, the CPU is a highly demanded resource and over-sharing may be problematic since it must manage more vehicle subsystems.

References

1. C. Benevolo, R.P. Damer, B. D'Auria, Smart mobility in smart city, in T. Torre, A. Braccini, R. Spinelli (eds.), *Empowering Organizations. Lecture Notes in Information Systems and Organisation* (p. 11). Springer, Cham (2016)
2. A. Visvizi, M. Lytras, Rescaling and refocusing smart cities research: from mega cities to smart villages. J. Sci. Technol. Policy Manage. (2018)
3. A. Visvizi, M. Lytras, (eds.), Smart Cities: Issues and Challenges: Mapping Political, Social and Economic Risks and Threats. Elsevier (2019)
4. P. Rizwan, K. Suresh, M.R. Babu, Real-time smart traffic management system for smart cities by using Internet of Things and big data, in *2016 International Conference on Emerging Technological Trends (ICETT)*. IEEE (2016)

5. Transport Statistics Great Britain: 2019 summary. Department for Transport. https://assets.publishing.service.gov.uk/government/uploads/system/uploads/attachment_data/file/870647/tsgb-2019.pdf. Last accessed 4 Aug 2020
6. American Public Transportation Association. Public Transportation Facts. https://www.apta.com/news-publications/public-transportation-facts. Last accessed 4 Aug 2020
7. A. Fernandez-Ares et al., Studying real traffic and mobility scenarios for a smart city using a new monitoring and tracking system. Future Generat. Comput. Syst. **76**, 163–179 (2017)
8. S. Sendra, et al., Collaborative LoRa-based sensor network for pollution monitoring in smart cities, in *2019 Fourth International Conference on Fog and Mobile Edge Computing (FMEC)*. IEEE (2019)
9. A. Kazmi, E. Tragos, M. Serrano, Underpinning IoT for road traffic noise management in smart cities, in *2018 IEEE International Conference on Pervasive Computing and Communications Workshops (PerCom Workshops)*. IEEE (2018)
10. İ Kök, M.U. Şimşek, S. Özdemir, A deep learning model for air quality prediction in smart cities, in *2017 IEEE International Conference on Big Data (Big Data)*. IEEE (2017)
11. M.R. Jabbarpour, A. Nabaei, H. Zarrabi, Intelligent Guardrails: an IoT application for vehicle traffic congestion reduction in smart city, in *2016 IEEE international conference on Internet of Things (Ithings) and IEEE Green computing and communications (Greencom) and IEEE cyber, physical and social computing (cpscom) and IEEE smart data (smartdata)*. IEEE (2016)
12. B. Pawłowicz, M. Salach, B. Trybus, Smart city traffic monitoring system based on 5G cellular network, RFID and machine learning, in *KKIO Software Engineering Conference* (pp. 151–165). Springer, Cham (2018)
13. M.M. Rathore et al., Exploiting IoT and big data analytics: defining smart digital city using real-time urban data. Sustain. Cities Soc. **40**, 600–610 (2018)
14. M. Behnke, T. Kirschstein, The impact of path selection on GHG emissions in city logistics. Transport. Res. Part E: Logist. Transport. Rev. **106**, 320–336 (2017)
15. J.F. Ehmke, A.M. Campbell, B.W. Thomas, Data-driven approaches for emissions-minimized paths in urban areas. Comput. Oper. Res. **67**, 34–47 (2016)
16. Y. Suzuki, A dual-objective metaheuristic approach to solve practical pollution routing problem. Int. J. Prod. Econ. **176**, 143–153 (2016)
17. J.F. Ehmke, A.M. Campbell, B.W. Thomas, Vehicle routing to minimize time-dependent emissions in urban areas. Eur. J. Oper. Res. **251**(2), 478–494 (2016)
18. R. Kramer et al., A matheuristic approach for the pollution-routing problem. Eur. J. Oper. Res. **243**(2), 523–539 (2015)
19. C. Rego et al., Traveling salesman problem heuristics: leading methods, implementations and latest advances. Eur. J. Oper. Res. **211**(3), 426–441 (2011)
20. S.V. Ilie, Survey on distributed approaches to swarm intelligence for graph search problems. Ann. Univ. Craiova-Math. Comput. Sci. Series **41**(2), 251–270 (2014)
21. D. Karaboga, B. Gorkemli, Solving traveling salesman problem by using combinatorial artificial bee colony algorithms. Int. J. Artif. Intell. Tools **28**(01), 1950004 (2019)
22. E. Jabir, V.V. Panicker, R. Sridharan, Design and development of a hybrid ant colony-variable neighbourhood search algorithm for a multi-depot green vehicle routing problem. Transport. Res. Part D: Trans. Environ. **57**, 422–457 (2017)
23. R. Gan et al., Improved ant colony optimization algorithm for the traveling salesman problems. J. Syst. Eng. Electron. **21**(2), 329–333 (2010)
24. M. Shokouhifar, S. Sabet, PMACO: A pheromone-mutation based ant colony optimization for traveling salesman problem, in *2012 International Symposium on Innovations in Intelligent Systems and Applications*. IEEE (2012)
25. J. Bai et al., A model induced max-min ant colony optimization for asymmetric traveling salesman problem. Appl. Soft Comput. **13**(3), 1365–1375 (2013)
26. M. Gła̧bowski, et al., Shortest path problem solving based on ant colony optimization metaheuristic. Image Process. Commun. **17**(1–2), 7–17 (2012)

27. M. Mahi, O.K. Baykan, H. Kodaz, A new hybrid method based on particle swarm optimization, ant colony optimization and 3-opt algorithms for traveling salesman problem. Appl. Soft Comput. **30**, 484–490 (2015)
28. R.V. Rao, V. Savsani, D.P. Vakharia, Teaching–Learning-Based Optimization: a novel method for constrained mechanical design optimization problems. Comput. Aided Des. **43**(3), 303–315 (2011)
29. R.V. Rao, V. Patel, An elitist teaching-learning-based optimization algorithm for solving complex constrained optimization problems. Int. J. Ind. Eng. Comput. **3**, 535–560 (2012)
30. R.V. Rao, V. Patel, Comparative performance of an elitist teaching-learning-based optimization algorithm for solving unconstrained optimization problems. Int. J. Ind. Eng. Comput. **4**, 29–50 (2013)
31. M. Ebraheem, T.R. Jyothsna, Comparative performance evaluation of teaching learning based optimization against genetic algorithm on benchmark functions, in *2015 Power, Communication and Information Technology Conference (PCITC)*, 327–331 (2015)
32. S.R. Shah, S.B. Takmare, A Review of methodologies of TLBO algorithm to test the performance of benchmark functions. Programm. Dev. Circuits Syst. **9**(7), 141–145 (2017)
33. L. Wu, F. Zoua, D. Chen, Discrete teaching-learning-based optimization algorithm for traveling salesman problems, in *MATEC Web of Conferences* **128**, 02022. EDP Sciences (2017)
34. Universität Heidelberg. Institut für Informatik. TSPLIB. https://comopt.ifi.uni-heidelberg.de/software/TSPLIB95/. Last accessed 4 Aug 2020

Chapter 5
An Architecture for Human Action Recognition in Smart Cities Video Surveillance Systems

J. M. Llaurado-Fons, Ana Martinez, Francisco A. Pujol-López, and Higinio Mora

Abstract Facial recognition systems are part of our daily lives. The face is already used to unlock mobile phones, withdraw cash at ATMs, pay at establishments, perform checks at airports or identify suspects at large events such as football matches or concerts. In Smart Cities there is a large amount of information about the environment, so it is very interesting to apply techniques to characterize the different domains and the detection of events and certain situations. In this work we propose a system for access control through facial recognition and data extraction of the individual through an identification card and deep learning. The architecture will be easily scalable and adaptable to new situations and fields of application. Therefore, our system will identify human features in video surveillance sequences using deep learning techniques to support video monitoring systems. The results obtained with different databases provide accuracies over 90%, which proves the validity of our proposal.

Keywords Face recognition · Features recognition · Smart cities · Deep learning · OCR

J. M. Llaurado-Fons · A. Martinez · H. Mora
Department of Computer Technology, University of Alicante, 03690 San Vicente del Raspeig, Spain
e-mail: josemiguel.llaurado.fons@gmail.com

A. Martinez
e-mail: amlopez.ana@gmail.com

H. Mora
e-mail: hmora@ua.es

F. A. Pujol-López (✉)
Department of Computer Technology and Computation, University of Alicante, Alicante, Spain
e-mail: fpujol@ua.es

A. Visvizi et al. (eds.), *Research and Innovation Forum 2020*,
Springer Proceedings in Complexity,
https://doi.org/10.1007/978-3-030-62066-0_5

Introduction

Nowadays, security and automation systems are becoming more and more present and necessary. In the design of Smart Cities, the optimization of spaces and resources, as well as systems that guarantee the security of citizens and infrastructures are a primary objective of development. For this reason, in the field of artificial intelligence and machine learning, efforts are being joined in the design and implementation of systems capable of solving many of these problems in order to guarantee the sustainable development of intelligent infrastructures that help to improve the coexistence and well-being of the population.

Automatic driving systems [1] or traffic management systems in cities, or applied to the early detection of illness in the health system [2, 3] are having a boost due to the great utility they can have to improve the quality of life of people in the near future. In our case, the focus is citizen safety and how to improve it so that our streets are increasingly safe so that there is an improvement in people's lives.

This work has been inspired by some studies already carried out and that have focused on the use of surveillance cameras for facial recognition in order to check how useful these algorithms can be applied to a real situation.

Thus, the SCface study [4] was carried out with 5 surveillance cameras taking different images of several individuals at different distances, in order to check the level of veracity and usefulness of applying facial recognition algorithms to the data collected by these cameras. This study did not obtain great results because when it was carried out, machine learning and deep learning techniques were not as advanced as they are today. However, facial recognition systems are now implemented in many applications, such as smartphones [5]. This new technology is a clear example of how in a short time these algorithms can be very useful and help make people's lives easier.

For this reason, our main goal is the recognition and validation of the identity of individuals using deep learning and convolutional neural networks. To do this, we first implement a system of recognition and verification of individuals with an identity card from which information can be extracted and compared with a database that allows us to verify whether the user's face matches the one in the ID card.

The objectives of this paper are:

- Develop a recognition system by using the data in an ID card.
- Verify a person's facial features and compare them with others stored in a database.
- Implement a system that can be used in access control to ensure the security of critical infrastructure in a Smart City.

To sum up, this work has focused on the extrapolation of a technology already working in a private environment such as mobile phones, to a professional, wider area such as access control to verify individual's identity.

This paper is organized as follows: the implementation of our system is described in Section "Implementation", the results obtained will be presented in Section "Experiments" and finally some brief conclusions will be presented in Section "Conclusions".

Implementation

With the aim of bringing our work closer to a real practical case, the aim was to develop a system that allows the people identification by their face through software that not only tries to recognize if that individual who intends to identify himself is him, but also checks the credentials of an ID card and thus compare the facial data from a database related to that ID and those that the system is able to obtain from the individual who presents it.

To do this have been use of Deep Learning and computer vision tools such as OpenCV and Tensorflow under the Python language, which have helped to extract information from the data captured by the system and implement preliminary software to improve the identification of people.

Method

The development has been divided into two parts, the first one is the obtaining of information from the identification card [6], in which have been applied masks of positions in where the regions of interest (ROI) are located. The information is read using a neural model capable of recognizing the different characters and implementing an OCR engine.

Once it gets this identity data from the card, it will check the ID number data in the database and consult the facial data that have on this person related to this ID number. Then, it obtains the data from the database and by using a webcam obtains the information of the person who tries to cross the security checkpoint by means of a pre-trained model of facial detection (res10 ssd [7]). With the face located it extracts the vector of characteristics of the same one thanks to the model OpenFace [8] that is another model already pretrained in the facial recognition.

To extract the data from the document a series of computer vision libraries will be used using OpenCV as the main support and following the procedure of Fig. 5.1 will be extracted to information.

With the document cropped, the regions of interest (ROI) will be extracted using a system of templates [6] adapted to the ID cards you are working with. With the separated data, the OCR template will be applied to read the data.

With the extracted ID number, the partner is searched in the database to find the characteristic vector. The Euclidean Distance [9] between the database vector and the one obtained is checked.

To improve the reliability of the problem it would be convenient to try to find a threshold of acceptance to see from which distance value it is acceptable to say that a person is who he says he is or on the other hand to reject her/him.

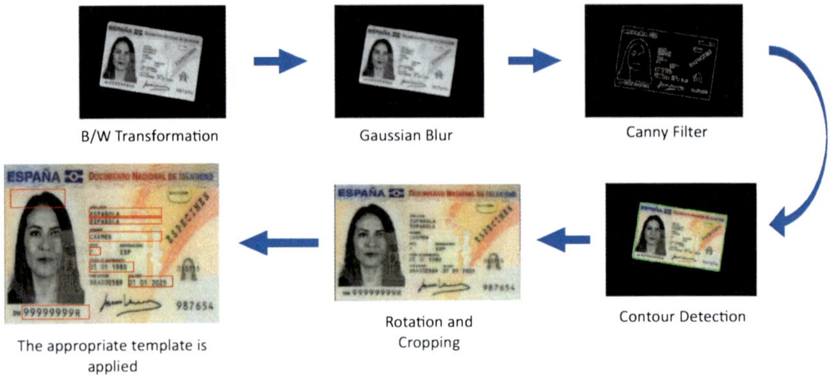

Fig. 5.1 Diagram of the method

Experiments

In the development of this system, two separate tasks have been faced, the first of which is the recognition of document data and the second is the collection of data from individuals for recognition.

For character recognition a neural model has been designed (Fig. 5.2), consisting of a simple convolutional network formed by 3 convolutional layers and 3 pooling layers plus a final layer fully connected to achieve differentiation between the 36 different classes of characters we have.

The Chars74k database [10] has been used for training. It consists of a total of 74,000 images of which only 16,200 have finally been used. To improve the final result, three different models are trained, a model for numbers, another one for letters and finally a model for both letters and numbers. The results are shown in Table 5.1.

As mentioned in the implementation for facial recognition it have been used the OpenFace network [8] which is a pre-trained model for facial recognition, and which gives very good results. To validate our system the Georgia Tech Face database [11] and Faces 95 [12] were used. It provides a total of 2.190 images with a total of 70 different classes.

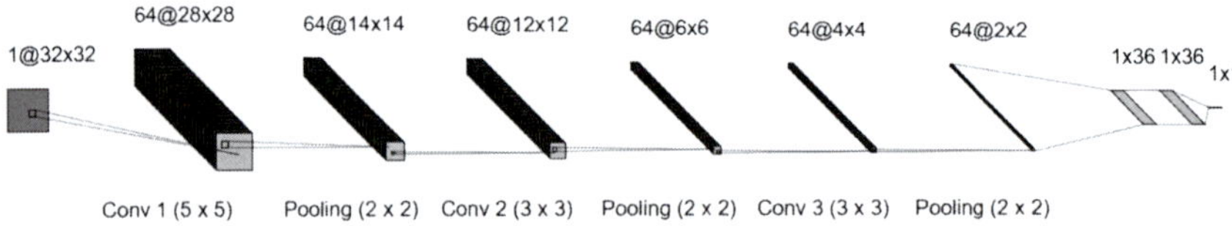

Fig. 5.2 Neural model architecture

Table 5.1 OCR results

Method	Accuracy (%)	Neural layers
Our method	**94**	8
Alexnet	92.6	8
DL-MCD	93.3	9

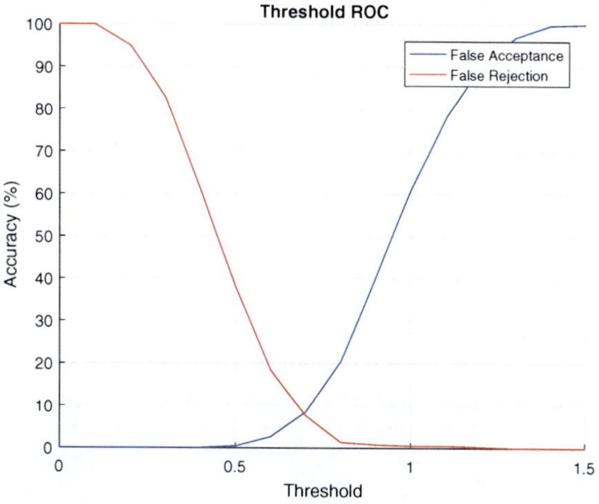

Fig. 5.3 Threshold ROC

To make a correct adjustment of the facial recognition block, a threshold in the Euclidean Distance value has to be estimated to separate positive and negative cases. Therefore, a sweep of the entire image dataset was performed in order to obtain the ROC curves and, as a result, develop a reliable recognition block. After the analysis of each of the possible values, Fig. 5.3 shows that using a threshold value of 0.69 yields an error rate of 8%. This guarantees approximately a 92% success rate in the facial recognition process.

Finally arguably that both blocks have been successfully completed. An OCR system has been achieved with a 94% success rate which equals and even exceeds other similar methods [13] and a facial verification system with a 92% success rate a considerably good level compared to current studies for Smart Cities [14].

This system has shortcomings that can be solved in future work, since the OCR block needs certain boundary conditions in terms of light and homogeneous background to work and the facial recognition block needs a light environment to obtain good results.

Conclusions

During this work a recognition system has been developed by using both facial and character recognition. First it reads the data of an ID card to identify the person and then it compares the stored facial data of that person with the ones in the ID card to verify its identity with a rate of success of 92%.

This system can be very useful in the area of the Smart City since it allows to have an automated control of people at key security points. Furthermore, this work could be the starting point for new projects that could use our proposal to new applications.

References

1. E. Talpes, A. Gorti, G. Sachdev, D. Sarma, G. Venkataramanan, P. Bannon, B. McGee, B. Floering, A. Jalote, C. Hsiong, S. Arora, Compute solution for tesla's full self driving computer. IEEE Micro pp. 1 (2020)
2. C.-C. Tseng, R. Stanciauskas, P. Zhang, D. Woo, K. Wu, K. Kelly, P. Gill, M. Yu, F. Pinaud, A. Lee, GRP78 regulates CD44v membrane homeostasis and cell spreading in tamoxifen-resistant breast cancer. Life Sci. Alliance **2**, e201900377 (2019)
3. K.J. Geras, S. Wolfson, Y. Shen, N. Wu, S. Gene Kim, E. Kim, L. Heacock, U. Parikh, L. Moy, K. Cho, High-resolution breast cancer screening with multi-view deep convolutional neural networks (2017)
4. M. Grgic, K. Delac, S. Grgic, Scface-Surveillance cameras face database. Multimedia Tools Appl. **51**, 863–879 (2011)
5. Apple Inc. Face id security guide. Apple (2017)
6. J.-H. Chiang, P. Gader, Recognition of handprinted numerals in visa® card application forms. Mach. Vis. Appl. **10**, 144–149 (1997)
7. OpenCV Team. Opencv: Github repository, dnn module. https://github.com/opencv/opencv.git (2019)
8. T. Baltrušaitis, P. Robinson, L. Morency, Openface: An open source facial behavior analysis toolkit, in *2016 IEEE Winter Conference on Applications of Computer Vision (WACV)*, pp. 1–10 (2016)
9. J. Vaijayanthimala, P. Theagarajan, Multi-modal biometric authentication system based on face and signature using legion feature estimation technique. Multimedia Tools Appl. (2019)
10. T.E. de Campos, B.R. Babu, M. Varma, Character recognition in natural images, in *Proceedings of the International Conference on Computer Vision Theory and Applications*, Lisbon, Portugal (February, 2009)
11. S. Natu, P. Natu, T. Sarode, H. Kekre, Performance comparison of face recognition using dct against face recognition using vector quantization algorithms LBG, KPE, KMCG, KFCG. Int. J. Image Process. **4** (2010)
12. University of Essex. Collection of facial images: Faces95. https://cswww.essex.ac.uk/mv/allfaces/faces95.html (2019)
13. Y. Weng, C. Xia, A new deep learning-based handwritten character recognition system on mobile computing devices. Mobile Networks Appl. **03** (2019)
14. A. Zam, M. Khayyambashi, A. Bohlooli, Energy-efficient face detection and recognition scheme for wireless visual sensor network. Appl. Soft Comput. **89**, 106014 (2020)

Chapter 6
Disruptive Technologies for Enabling Smart Government

Higinio Mora, Francisco A. Pujol-López, Mario R. Morales, and Rafa Mollá-Sirvent

Abstract This paper presents a review of Blockchain technology and computer protocols called Smart Contracts as a virtual mechanism to expedite the execution of contractual commitments without the intervention of regulatory or verifying entities. Blockchain is considered one of the most important disruptive technologies of recent years and its application can generate profound changes in business models and society in general, even in the public sphere providing potential benefits to citizens and the conformation of a modern e-government. The main contribution of this work is to draw a set of basic considerations that support new e-government processes based on blockchain and smart contracts, and also schematizes an application use case. This is intended to streamline public processes and reduce the problems associated with corruption, lack of transparency and cost overruns, among others; even striving to meet the stated objective of the United Nations regarding the e-government of providing services and information to citizens in a transparent way through technological mechanisms and the use of the Internet.

Keywords Blockchain · Smart contract · E-government · Business model · Tendering · Smart city

H. Mora (✉)
Department of Computer Technology, University of Alicante, San Vicente del Raspeig, Spain
e-mail: hmora@dtic.ua.es

F. A. Pujol-López · R. Mollá-Sirvent
Department of Computer Technology and Computation, University of Alicante, Campus de San Vicente del Raspeig, 03690 Alicante, Spain
e-mail: fpujol@dtic.ua.es

R. Mollá-Sirvent
e-mail: rmolla@dtic.ua.es

M. R. Morales
School of Computer Engineering, Central University of Ecuador, 170521 Quito, Ecuador
e-mail: mmoralesm@uce.edu.ec

A. Visvizi et al. (eds.), *Research and Innovation Forum 2020*,
Springer Proceedings in Complexity,
https://doi.org/10.1007/978-3-030-62066-0_6

Introduction

Since the appearance of white paper "Bitcoin—A Peer-to-Peer Electronic Cash System" [1] Blockchain technology (BCT) has led to the emergence of a new and innovative ecosystem of applications, services and business opportunities beyond the financial sector [2], being considered as one of the most important disruptive technologies of the last decade and that will imply profound changes in business and society [3], playing a fundamental role in the sustainable development of the global economy [4] and whose main characteristic is that it operates in a decentralized architecture without the need of a certifying entity and with a high level of certainty [5]. Among other additional features of BCT is that it provides transparency, robustness, auditability, security, authentication and fault tolerance [6].

Even in the business world, Gartner places it in the top five technology trends in 2018 through its Hype Cycle for Emerging Technologies analysis, offering opportunities for a new value creation [7]. The way in which BCT works implies that it provides a decentralized digital transactional database [7] and there is a complete copy of the information in different nodes [8] in which a transaction it can only be added through consensus methods between such nodes. In addition, previous transactions cannot be modified or altered; in this way enabling all nodes to track the history. The Peer-To-Peer (P2P) architecture helps make transactions recorded in the BCT immutable and secure, and together with the distributed consensus protocol also ensures its integrity [9].

BCT can have various forms of application in public processes such as construction licenses, property registration, birth or marriage certificates, business licenses, vehicle registration, education title registration, voting registration, among others; providing potential benefits to society and the conformation of a modern e-government [10]. In this sense, the study carried out in [11] recommends that governments should update part of their systems with these new technological platforms.

The e-government has had a significant growth in recent years promoting efficiency, accountability and transparency; and still driving citizen engagement in the decision-making processes of governments [12]. According to the United Nations, a fundamental objective of e-government is to provide services and information to citizens in a transparent manner through the use of the Internet [13].

The rest of this work is as follows: chapter two explains the constitutive characteristics of BCT, Smart Contracts and e-government. Chapter three introduces the main considerations for structuring new public processes based on such technologies and with that foundation present a specific case of application. Finally, chapter four presents the most relevant conclusions of the study.

Blockchain, Smart Contracts and E-Government: An Overview

Within a modern society focused on providing better services to the citizen, BCT and SC are becoming fundamental pieces as enablers for the co-creation of innovation. In the same way, the Administration must focus on more efficient, automated and transparent processes, which waste less resources since they are paid by the citizens. Currently, much of it is supported by technology. Consequently, this section reviews the characteristics of each of the components for a better understanding.

Blockchain Technology

BCT is the technology behind so-called virtual currencies (VC), with Bitcoin as the main exponent. One of the objectives of the VCs has been to extend their use as a means of payment, a purpose that has been difficult to achieve; however, according to the [14], in countries with economies in crisis their acceptance could be boosted and could even be considered a weak safe haven.

BCT concept can be self-explanatory: each block contains a different number of transactions and this is added to the chain through a validation scheme called a consensus method that involves the resolution of a complex mathematical algorithm. When the block is added to the blockchain system, it cannot longer be changed or modified in any way. The next block to be added will have a reference to the previous block, which makes Block 2 bounded with Block 1 [15].

Blockchain is a public and auditable record that preserves the anonymity and privacy of its participants [16]. It is also characterized as a distributed consensus ledger in which its records are shared and distributed, so that a reconciliation between the parties is not required [17]. One could say that among the basic characteristics of BCT are: (i) they are mutual, BCT are shared across organizations without a particular predominance of any of the parties, (ii) distributed, maintain a complete copy in each node, providing resilience and robustness to the system, and, (iii) immutable, BCT is a ledger that cannot be modified, adding integrity to the system [18].

Basically, BCT provides a feature of non-repudiation of events that have occurred over time through a group of nodes, usually geographically and proprietarily dispersed, where the added blocks depend on cryptographic keys and consensus methods [16, 19]. BCT is a technology that is constantly evolving and has surpassed its domain of origin related to cryptocurrencies, even more driven by the role played by smart contracts (SC) understood as automated transactions that execute an agreement between parties, which enables it for much more diverse and complex applications, thus being a motivator of innovations that will continue to emerge based on this technology [5, 20].

According to [5, 9] there are several domains that underlie BCT such as: Financial, Business, Privacy and Security, Education, Governance, IoT, Data Management,

Healthcare, Energy, Supply Chain and Logistics, Music Industry and Social Media, Smart Cities, among others, which gives guidance of the interdisciplinary potential of this technology.

The study by [17] states that BCT is very useful when the data needs to be structured, shared and audited among multiple parties without an established trust between them. This could also add value by disintermediating operations and enabling interaction between transactions through SCs.

Etherum and Smart Contracts

The Ethereum born came when cryptocurrency applications began to boom and the Bitcoin model was being recognized. This platform shares common characteristics with other BCTs, especially Bitcoin: a peer-to-peer network, the use of cryptographic primitives, a Byzantine fault-tolerant consensus algorithm and a digital currency (ether); on the other hand, it differs in that its primary objective is not to be a digital currency payment network, since the ether is intended to be a utility currency necessary to pay for the use of the platform [21].

Ethereum White Paper was proposed in 2014 [22] and the project was launched in June 2015 in which its main innovation was to be a computer platform that allowed the construction of smart contracts [23]. Ethereum can be recognized as a transaction-based state machine, in which its status is updated after each transaction occurs. Basically, the Ethereum SCs consist of coins (Ether), executable code and persistent storage, where coin transfers can be made in the form of a normal transaction like in Bitcoin, or through invocation of contracts (see Fig. 6.1) [24]. The ether has the function of restricting the cost of resources in the execution of an SC [21], in other words they are the main internal crypto-fuel used to pay transaction fees [25].

At present, Smart Contracts can be implemented on several platforms, but Ethereum is a second generation BCT that supports a Turing-completeness language supported by a programming language called Solidity that allows to develop complicated processes [15] and build applications distributed complexes [26].

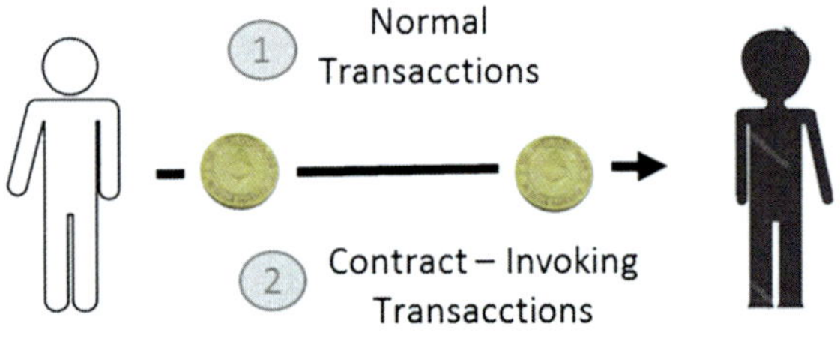

Fig. 6.1 Coin transfer forms on Ethereum platform

E-Government

One of the first definitions of e-government proposed the simplification and improvement of aspects related to democracy, government and business through the use of electronic means that allowed interaction between the government with citizens, businesses and even within the public institutions themselves [12]. At present, the advancement of ICTs has revolutionized the concept of e-government to promote multi-channel delivery of e-services [27]. The maturity models associated with e-government have evolved following the flow of IS technology models; even one of the most relevant academic studies in e-government is associated with the use of web-based technology and Internet applications to reinforce the delivery of citizen information and services from the government [28].

The requirement for governments to provide relevant services and information proactively is increasing even without the need for citizens to require such services but instead be delivered when a life event occurs [29]. A study provided by [30] notes that there are technological enablers that promote proactivity in the services provided by an e-government, such enablers are: (i) technologic components such as database consolidation, data standardization, shared information and security levels, (ii) mobile technology that allows interacting with citizens in a ubiquitous way, (iii) Big data analytics capabilities to adapt e-government services to the needs of citizens. However, the implementation of e-government based on BCT still holds a high degree of uncertainty, so a suggested route would be experimentation in small-scale projects that allow for the assessment of potentialities and limitations and to better understand the requirements of e-government processes [9].

In the evolution of these projects, BCT could be incorporated for storage and transactional processes within the e-government, improving protection, privacy and security. The new scenarios requires citizen-oriented governance through greater e-participation in government decision-making processes [12]; in turn, the government's own responsibility, both fiduciary, legal and to the taxpayer, motivate to ensure value transfers between the relevant stakeholders [17].

As example use case, it is related to urban planning and the necessary licenses to execute projects in this area. These activities require compliance with a set of requirements, burdens and obligations within legislation. The model proposed in this work aims to simplify and automate such a process through the use of BCT and SC in which the requirements are specified and coded for automatically check, then are publicly accessible so that they can be validated without relying on authorities, to while they are decentralized, immutable and transparent [31].

Convergence: Use Case of an Application that Integrates BCT and SC in the Optimization of an E-government Process

Interaction of BCT, SC and the Public Sector

Many governments are already exploring or are planning to do projects related to BCT since their safety, efficiency and speed benefits can be applicable to public organizations, these include reducing the need for intermediary organizations, accelerating transactions, reducing costs, facilitating compliance with regulations [17]. Even BCT could cover a general problem in the public sector that lies in the management and authentication of documents, at the same time that it would be a cost-efficient, scalable solution and enable better citizen services [32]. Like any potential user, governments should analyze weaknesses in their business processes and readjust them based on the advantages brought by the SC and BCT [11]. Even regulatory problems can be seen more as an opportunity than as an obstacle as they could promote greater transparency, improve access to information and simplify regulatory reports [18]; in addition, based on the possibility granted by BCT, government control entities could freely access the chain in real time and verify the authenticity of the information [33], since unlike traditional databases, BCT makes it possible that immutable records can be stored [34].

Special cases in which SC can provide benefits in society are when a series of participants interact among them as in construction of public civil works or in mass events: in these can participate the town hall, police, firefighters, health organizations, all of which they must agree and approve different process conditions. Property titles can be complicated to manage in many countries, in this case SC can register each transaction and prevent the manipulation and/or loss of information, in addition to considering the necessary conditions for a transfer, such as the signature of the legal owner, the validation that there are no mortgages on the asset or that it has a prohibition of disposal, the transfer of money from the buyer to the seller, so that the property is properly transferred and avoid fraud. In all these cases, BCT and SC serve as a mechanism to fight corruption, reduce fraud and improve transparency. For these benefits to be realized, close collaboration between experts and policy-makers is necessary, so that a governance based on BCT can be developed and thus ensure that public values and social needs are taken into account [9].

Although practical applications in e-Government processes are still growing, it is important to review some relevant cases. In [32], a use case for the storage of academic certificates in a blockchain is presented, which could be extended and applied to public sector organizations that require issuance certificates, licenses, etc. and could thus benefit from this technology. The use case in [35] applies BCT for the registration of property titles in Georgia, a country considered a leader in reforms by the World Bank. In this case, the only entity that had control over ownership documents was the National Agency of Public Registry (NAPR). After the implementation of BCT, any interested Sparty can access the hash of the property title document and monitor the validity of the document. The designed process has the flow presented in Fig. 6.2.

Fig. 6.2 Registration of property titles in Georgia using BCT

This scheme enables any person to validate a property title document without the possibility of altering the result of the registration process, making it a more efficient, secure, transparent and accessible process.

Another use case of interest is the e-Residency initiative implemented by Estonia that allows anyone in the world to have a government-backed digital identity, without depending on their citizenship or physical presence in the country. The BCT-based service seeks to attract individuals who wish to participate in the country's economic and business activity, thus capturing investments and innovation while facilitating public and private procedures [19].

Finally, another use case presented in this section is that of electronic voting. Traditional electronic voting solutions have had the difficulty of not being easily auditable or sufficiently transparent. The method proposed in [36] is based on BCT and involves Merkle trees for verification of the voter list and a block explorer for the vote count check. Furthermore, the privacy of the vote is ensured through the inclusion of hashes with multisignature addresses. The model proposed in this study has the procedure summarized in Fig. 6.3.

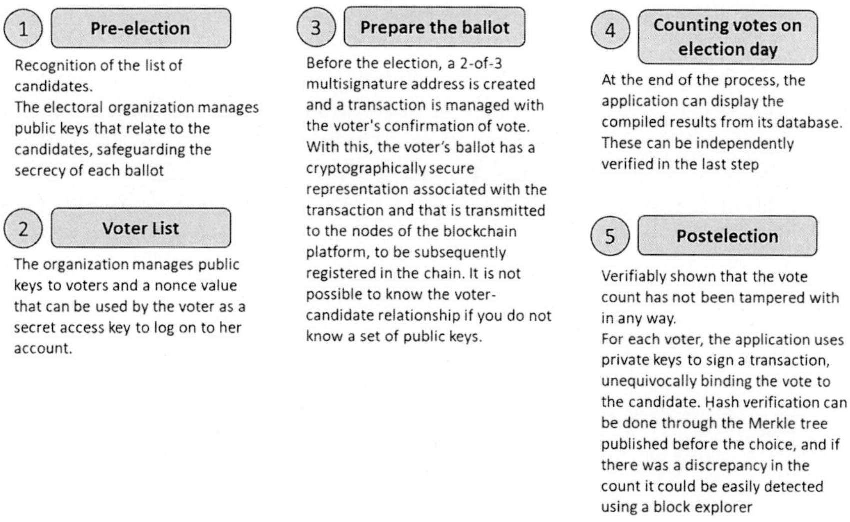

Fig. 6.3 Voting process using BCT

In all these cases, the characteristics provided by BCT are recognized, such as transparency, the independence of a trusted third party, the proper registration and management of documents, so it is relevant to extend the application to other government processes such as a tender.

Bases for the Approach of an Automated Execution Model for Public Processes. Post-Bid Use Case

With the new concept of Internet of Agreements (IoA) proposed in World Government Summit in Dubai in 2017 [11], the bridge linking the Internet with the deals, contracts and regulations that support the daily work is being built, enabling an evolutionary leap in digital commerce. Applications that are built with BCT, SC, IoA will facilitate trade between people, companies and government. The possibility of an evidence-based and more transparent approach will be opened to design and link the policies promoted by the Administration [34]. These systems will adopt verifiable methods to store evidence of transactions and operations carried out between individuals and organizations, and will provide efficient dispute resolution schemes (for example based on proof-of-existence (PoE), proof-of-ownership (PoO), proof-of-integrity (PoI)), even without the participation of official institutions [5].

Ethereum opens up huge possibilities on the horizon, both for the government and for private institutions through applications based on smart contracts. This platform provides tools to create innovative solutions even to old problems, providing them with deep flexibility and functionality [22]. New business process management systems could be maintained on BCT and workflows would be executed through SCs, which would generate a rationalization and automation in internal processes, as well as a cost reduction [37]. The SCs being decentralized and automatically executed do not require trust between the parties for the fulfillment of the obligations, nor of a third party verifier as a government agency to enforce the rules of the contract [6].

The use case proposed in Fig. 6.4 summarizes the different macro-processes that occur in the execution of a tender in a traditional way, from the authorization of the contract through the allocation of guarantees to the settlement once the work is completed and/or accepted the service.

The public sector can absorb probable benefits through the use of SC and BCT [20], in principle as a mechanism that significantly reduces the possibility of fraud and

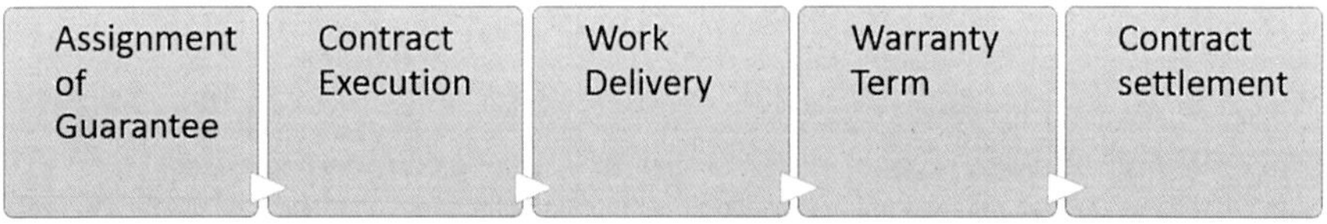

Fig. 6.4 Macro-process flow in the execution of a public tender

corruption in its asset management, but with potentialities much larger ranging from digital identity, mechanism to attract investment, economic growth and promotion of innovation. The evolution of society requires a greater drive for transparency in government activities and systems, and BCT precisely offers a way to reconcile this need with the verifiability of data-based decision-making and the actions that result [34]. This model seeks to contribute to the structuring of an e-government based on BCT.

Based on what is described in this study, a set of general considerations should be met by government organizations as a previous step in structuring a model; these are indicated in the Table 6.1.

With the previous considerations taken into account, the proposed scheme is observed in Fig. 6.5 where automation based on BCT and SC constitutes the fundamental element:

This scheme may still require further study mainly related to existing regulatory frameworks, however it can be taken as a starting point to improve and optimize e-government processes based on disruptive technologies such as BCT and SC. In this way, the Administration could take advantage of the scalable and decentralized nature of BCT, which, added to the functionality provided by SCs, would allow increasing the number of citizen services and extending them to other use cases [41]. Furthermore, it can be seen in the presented concept that bureaucratic processes would be alleviated by simplifying their procedures, thus improving the efficiency and quality of services.

Conclusions

The BCT as a platform and fundamentally the SC as automation mechanisms that allow executing agreements between parties have the potential to bring greater innovations, some of which are initiatives that are already under development. The government and the Administration can be positively impacted with these transformational tools and to the extent that their advantages are exploited, they will respond to the pressing demands that citizens now have: transparency, eradication of corruption, agility and efficiency.

The versatility of these disruptive technologies should be exploited in various domains of government management; further it can be the basis for the provision of proactive and transparent services and information. Within this context, citizens can receive information proactively based on predetermined events, without the need to make an explicit request for them. This would configure a higher level of maturity within the evolution of e-government and a boost from the smart city to make easier the life of citizens.

The definition of projects and the digital transformation supported by BCT and SC should not have a big-bang approach; on the contrary its structuring should be gradual. In that sense, smaller-scale projects must be defined, the objectives and benefits that are expected to be clarified, the process requirements must be understood

Table 6.1 General baseline considerations for implementing a model based on BCT and SC

Consideration	Description
1. Definition of smaller-scale projects	Taking into account that BCT is still a technology that requires maturation and must solve several challenges such as scalability, security, etc., it is advisable to define smaller-scale projects that allow BCT's technological characteristics to fit with the specific requirements of e-government processes [9]. In that sense, start with proof-of-concept to understand the benefits [11]
2. Standardization and integration	One factor that can facilitate the flow of information between the parties is the collaborative use of open standards such as Industry Foundation Classes (IFC), Constructions Operations Building Information Exchange (COBie), among others [38]. Furthermore, it must be considered that these implementations will require both forward flexibility and backward compatibility [39]
3. Infrastructure design	The general options can be three: public, consortium or private. In any of these categories you can find advantages and disadvantages related to the level of security, consensus methods or participation of the nodes in the network [40]. Governments could adopt any of these schemes and take advantage of this technology [11]
4. Configuration of governance rules	The organization should design and configure BCT governance rules, such as who invokes transactions, what is the procedure for accepting transactions, among others. The organization develops and maintains the software for distributed ledger [9]
5. Change in the solution approach: from a tech-driven perspective to a need-driven	While some technical aspects play an important role such as the physical location of nodes and servers due to the law that could be applied; environments with greater maturity move towards a need-driven approach that requires that society's problems and public values be incorporated into e-government applications [9]
6. Requirements specifications	The requirements should be described or documented in detail in machine-readable rule sets. Usually the lack of clarity in the requirements of public contracts makes execution difficult, as a result this may lead to delays, claims and higher prices [38]

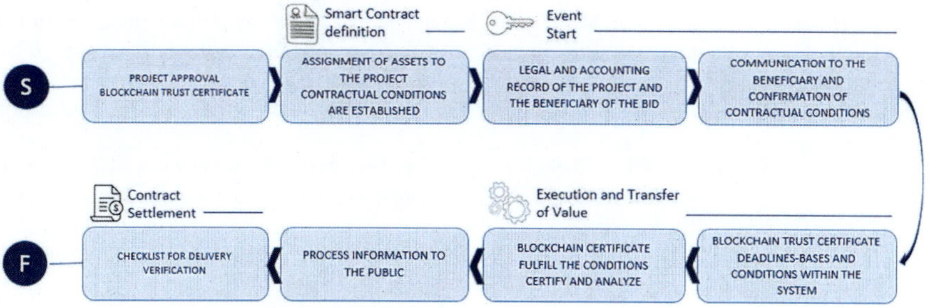

Fig. 6.5 Scheme of automation of an e-government process based on BCT and SC

and defined in greater detail, experts in technology, processes and policy-makers must be gathered, all this to ensure that social needs are included in the project and minimize the uncertainties that these implementations contain.

Finally, there are challenges that must be addressed as the regulatory frameworks that support the incursion of these technologies in the public sphere. There are also technical issues not yet fully resolved such as data and systems integrity, adoption of standards, security and scalability that could currently bring limitations, so it is necessary to deepen the research.

References

1. S. Nakamoto, Bitcoin: A Peer-to-Peer Electronic Cash System (2008)
2. D. Tapscott, A. Tapscott, *"El Protocolo Fiable", in Blockchain Revolution* (Deusto, New York, 2016), pp. 23–51
3. W. Ying, S. Jia, W. Du, Digital enablement of blockchain: evidence from HNA group. Int. J. Inf. Manage. **39**, pp. 1–4 (no. October 2017) (2018)
4. Q.K. Nguyen, Blockchain—a financial technology for future sustainable development, in *3rd International Conference on Green Technology and Sustainable Development (GTSD)*, pp. 51–54 (2016)
5. F. Casino, T.K. Dasaklis, C. Patsakis, A systematic literature review of blockchain-based applications: current status, classification and open issues. Telemat. Informatics **36**, pp. 55–81 (no. May 2018) (2019)
6. I. Makhdoom, M. Abolhasan, H. Abbas, W. Ni, Blockchain ' s adoption in IoT: the challenges, and a way forward. J. Netw. Comput. Appl. **125**(2018), 251–279 (2019)
7. V.J. Morkunas, J. Paschen, E. Boon, How blockchain technologies impact your business model. Bus. Horiz. **2018**, 2019 (2018)
8. M. Muzammal, Q. Qu, B. Nasrulin, Renovating blockchain with distributed databases: an open source system. Futur. Gener. Comput. Syst. **90**, 105–117 (2019)
9. S. Ølnes, J. Ubacht, M. Janssen, Blockchain in government: benefits and implications of distributed ledger technology for information sharing. Gov. Inf. Q. **34**(3), 355–364 (2017)
10. J. Palfreyman, Blockchain for Government: Building trust, demolishing bureaucracy (2017). Available: https://www.ibm.com/blogs/blockchain/2017/02/blockchain-government-building-trust-demolishing-bureaucracy/. Accessed 28 Feb 2020
11. V. Gupta, L.L.C Consensys, *Building the hyperconnected future on blockchains*. World Government Summit, Dubai (2017)

12. N. Bindu, C.P. Sankar, K.S. Kumar, From conventional governance to e-democracy: tracing the evolution of e-governance research trends using network analysis tools. Gov. Inf. Q. 1–15 (no. January) (2019)
13. United Nations, *Global E-Government Readiness Report 2005* (United Nations Publications, New York, 2005)
14. I. Musialkowska, A. Kliber, K. Świerczyńska, P. Marszałek, Looking for a safe-haven in a crisis-driven Venezuela: the Caracas stock exchange vs gold, oil and bitcoin. Transform. Gov. People, Process Policy (2020)
15. G. Gürkaynak, İ Yılmaz, B. Yeşilaltay, B. Bengi, Intellectual property law and practice in the blockchain realm. Comput. Law Secur. Rev. **34**(4), 847–862 (2018)
16. J.C. Mendoza-Tello, H. Mora, F.A. Pujol-López, M.D. Lytras, Disruptive innovation of cryptocurrencies in consumer acceptance and trust. Inf. Syst. E-bus. Manag. 1–28 (2019)
17. J. Killmeyer, M. White, B. Chew, Will blockchain transform the public sector? Deloitte University Press (2017)
18. World EnergyCouncil, PricewaterhouseCoopers, The Developing Role of Blockchain. London (2017)
19. C. Sullivan, E. Burger, E-residency and blockchain. Comput. Law Secur. Rev. **33**(4), 470–481 (2017)
20. H. Mora, F. Pujol, J.C. Mendoza-Tello, M.R. Morales, Virtual currencies in modern societies: challenges and opportunities, in *Politics and Technology in the Post-Truth Era*, Emerald Publishing Limited, pp. 171–185 (2019)
21. A.M. Antonopoulos, G. Wood, *Mastering Ethereum: building smart contracts and dapps*. O'Reilly Media (2018)
22. C. Dannen, *Introducing Ethereum and Solidity* (Apress, New York, 2017)
23. FinTech Network, Smart Contracts—From Ethereum to Potential Banking Use Cases, pp. 1–10 (2016)
24. L. Luu, D.-H. Chu, H. Olickel, P. Saxena, A. Hobor, Making smart contracts smarter, in *Proceedings of the 2016 ACM SIGSAC conference on computer and communications security*, pp. 254–269 (2016)
25. V. Buterin, A next generation smart contract & decentralized application platform. *Ethereum White Paper* (2014). Available: https://blockchainlab.com/pdf/Ethereum_white_paper-a_next_generation_smart_contract_and_decentralized_application_platform-vitalik-buterin.pdf. Accessed on 15 Mar 2020
26. M. Alharby, A. van Moorsel, Blockchain-based Smart Contracts: a systematic mapping study, in *Conference: 3rd International Conference on Artificial Intelligence and Soft Computing*, pp. 125–140 (2017)
27. N.P. Rana, Y.K. Dwivedi, M.D. Williams, V. Weerakkody, Adoption of online public grievance redressal system in India: toward developing a unified view. Comput. Human Behav. **59**, 265–282 (2016)
28. F. Debrí, F. Bannister, e-government stage models : a contextual critique, in *48th Hawaii International Conference on System Sciences*, pp. 2222–2231 (2015)
29. H. Scholta, W. Mertens, M. Kowalkiewicz, J. Becker, From one-stop shop to no-stop shop: an e-government stage model. Gov. Inf. Q. **36**(1), 11–26 (2019)
30. D. Linders, C.Z. Liao, C. Wang, Proactive e-Governance: flipping the service delivery model from pull to push in Taiwan. Gov. Inf. Q. **35**, S68–S76 (2018)
31. H. Mora, R. Pérez-delHoyo, R.M. Sirvent, V. Gilart-Iglesias, Management city model based on blockchain and smart contracts technology, in *Research and Innovation Forum, RIIFORUM 2019* (2019)
32. S. Ølnes, Beyond bitcoin enabling smart government using blockchain technology, in *International conference on electronic government*, pp. 253–264 (2016)
33. J.C. Mendoza-Tello, H. Mora, F.A. Pujol-López, M.D. Lytras, Social commerce as a driver to enhance trust and intention to use cryptocurrencies for electronic payments. IEEE Access **6**, 50737–50751 (2018)

34. M.A. Sicilia, A. Visvizi, Blockchain and OECD data repositories: opportunities and policy-making implications. Libr. Hi Tech **37**(1), 30–42 (2019)
35. N. Goderdzishvili, E. Gordadze, N. Gagnidze, Georgia's blockchain-powered property registration: never blocked, always secured—ownership data kept best!, in *Proceedings of the 11th International Conference on Theory and Practice of Electronic Governance*, pp. 673–675 (2018)
36. P. Noizat, Blockchain electronic vote, in *Handbook of Digital Currency: Bitcoin, Innovation, Financial Instruments, and Big Data*, A. Press, Ed. Elsevier Inc., pp. 453–461 (2015)
37. C. Prybila, S. Schulte, C. Hochreiner, I. Weber, Runtime verification for business processes utilizing the Bitcoin blockchain. Futur. Gener. Comput. Syst. (2017)
38. A. Ciribini, M. Bolpagni, E. Oliveri, An innovative approach to e-public tendering based on Model Checking. Procedia Econ. Financ. **21**, 32–39 (2015)
39. S. Ølnes, A. Jansen, Blockchain technology as s support infrastructure in e-government, in *International Conference on Electronic Government*, pp. 215–227 (2017)
40. V. Buterin, On public and private blockchains, *Ethereum Blog* (2015). Available: https://blog.ethereum.org/2015/08/07/on-public-and-private-blockchains/. Accessed on 05 Mar 2020
41. J.A. Jaoude, R.G. Saade, Blockchain applications—usage in different domains. IEEE Access **7**, 45360–45381 (2019)

Chapter 7
Development Opportunities of Taiwan's Smart Cities from the Viewpoint of Smart Manufacturing

Yung Chang Wu, Yenchun Jim Wu, and Shiann Ming Wu

Abstract Smart manufacturing is an important part in sustainable development of smart cities. The industrial revolution under smart manufacturing evolves the traditional automatic modes with mass production into the efficiency of small, diverse and rapidly customized manufacturing and collaborative manufacturing, and can solve increasingly serious shortage of work force resulting from decline in working-age populations. IoT and Internet services can be used to develop smart factories, so as to enable all mechanical units to have the ability to communicate with each other, through smart integration of sensing and control systems and other technologies. In this paper, the development of Taiwan's smart cities is discussed from the perspective of smart manufacturing development, the concept and the current international situation are introduced, and then smart factories' intellectual technologies, IoT, cloud computing, mass data analysis, artificial intelligence, cyber-Physical systems and Cyber security are introduced in order, and execution status of the smart manufacturing in Taiwan and its integrated manufacturing capacities are finally discussed. Under the vision of smart machine industrialization and industrial wisdom mechanization, readers are more easily to understand Taiwan's role in the competition of sustainable development of smart cities and are expect to cultivate outstanding competitive talents.

Keywords Small and medium enterprise (SMEs) · Information communication technologies (ICT) · Cyber-physical systems · Cyber security · Artificial intelligence

---

Y. C. Wu · S. M. Wu
College of Business Administration, National Huaqiao University, Quanzhou, China
e-mail: Jasonwu988@gmail.com

S. M. Wu
e-mail: wuming.hqu@gmail.com

Y. J. Wu (✉)
Graduate Institute of Global Business and Strategy, National Taiwan Normal University, Taipei 10645, Taiwan
e-mail: wuyenchun@gmail.com

A. Visvizi et al. (eds.), *Research and Innovation Forum 2020*,
Springer Proceedings in Complexity,
https://doi.org/10.1007/978-3-030-62066-0_7

Introduction

Introduction to Intelligent Manufacturing

The smart city is a cross-field concept, which involves the use of the Internet of Things (IoT), cloud computing, big data, mobile Internet, artificial intelligence, and other ICT technologies, to implement urban construction in an information security management environment. Among them, smart manufacturing is also a part of a smart city, which focuses on manufacturing in small quantities, diversity, and rapid manufacturing rate, in order to improve the efficiency of the vertical production chain and horizontal coordinated manufacturing, and solve the increasingly serious problem of labor shortages.

Industry 4.0 is called Productivity 4.0 in Taiwan. Industry 4.0 was first proposed in the 2011 Hannover Industrial Fair in Germany. The concept is to establish smart factories by virtue of the Internet of Things and Internet services through the intelligent integration of sensory control systems and other information communication technologies (ICT), thus, providing each mechanical element with the communication ability. The smart factories are able to monitor the real-time environment, and identify and eliminate problems. Moreover, they can reform the production process and make the operations more flexible in response to different market needs [1]. Intelligent manufacturing technology drives the digital and intelligent development of the industrial chain, and can resolve the labor shortage caused by population aging and the declining birth rate [2].

Since 2011, countries around the world have successively proposed their industrial upgrading programs, which take Industry 4.0 as the core; for example, Germany has incorporated Industry 4.0 into the Top 10 Future Plans under the High-Tech Strategy 2020 Action Plan, and made massive investments in accelerating the computerization, digitization, and intelligence of German manufacturing, thereby, upgrading its manufacturing industry. In the same year, the United States also proposed the Advanced Manufacturing Partnership (AMP), which effectively promoted manufacturing and overseas investment to return to the United States, thereby, enhancing the research and development capabilities of emerging technologies and revitalizing the depressed domestic manufacturing industry. In 2013, Japan proposed the "Japan Industrial Revitalization Plan" to revitalize its manufacturing industry by promoting investments in equipment and R&D. Similarly, China announced "Made in China 2025" in 2015, which focuses on innovation drive, quality priority, green development, structural optimization, and talent orientation, which will transform China from "a big manufacturer" to "a strong manufacturer" by 2025.

In order to accelerate the industrial transformation and upgrade its abilities, Taiwan has created a new economic model, which takes "innovation, employment, and distribution" as the core. In 2015, it proposed the 5 + 2 industrial innovation plan, which consists of "intelligent machinery", "Asian Silicon Valley", "Green Energy Technology", "Biomedical Industry", "National Defense Industry", "New Agriculture", and "Circular Economy". Specifically, "intelligent machinery" is an important policy, which is based on the integration of the concept of Industry 4.0 into the current industrial state [3]. It promotes precision machinery and the ICT industry, and integrates the Internet of Things, big data, CPS, precision management, robotics, 3D printing, and sensors. It aims to upgrade the precision machinery in Taiwan to intelligent machinery, realize intelligent production, and establish a networked service system through the cloud, the Internet, and quick consumer links, thus, making Taiwan a key player in the global manufacturing supply chain.

Connotation of Industry 4.0

Industry 4.0 is to enhance the productivity and quality of the entire manufacturing value chain through the application of automated robots, Internet of Things, supply chain Internet, sales and production, and big data analysis, as based on human–computer collaboration. In the progress of the Internet of Things, Industry 4.0 integrates information with the physical world, creates Cyber-Physical Systems, realizes the growth of technological processes and production business processes in a new era, and creates a manufacturing environment that can be interconnected and communicate intelligently. By taking the core of Cyber-Physical Systems (CPS), Industry 4.0 constructs a comprehensively intelligent industrial system [4]. In addition, Industry 4.0 conducts R&D around the aspects of people, technology, and organization, implements broadband infrastructure construction, work organization design, information security, resource efficiency, personnel training, and sustainable development under the standard and regulatory framework, thereby, achieving the goal of smart factories [5].

Smart Factories

The Industrial Technology Research Institute (2018) concluded the five characteristics of an intelligent manufacturing system: system autonomy, visualization of the overall system into technical practices, coordination, reorganization, and expansion [6]. Each group of computers in the system can learn autonomously and maintain themselves according to the work tasks. Humans co-exist with computers and complement each other at different levels [7]. The Institute for Information Industry also defined smart factories according to the above five characteristics: smart factories are based on the application of ICT hardware, software and system integration technologies, IoT for factory production behaviors, Big Data, artificial intelligence, virtual-real integration, and human–computer collaborative operation.

According to the philosophy and important concepts of German Industry 4.0 [8] and the US Industrial Internet Program [9], the resulting smart factories will be able to control the entire value creation network in real time from order to delivery, and integrate the product and its production system life-cycle engineering; to this

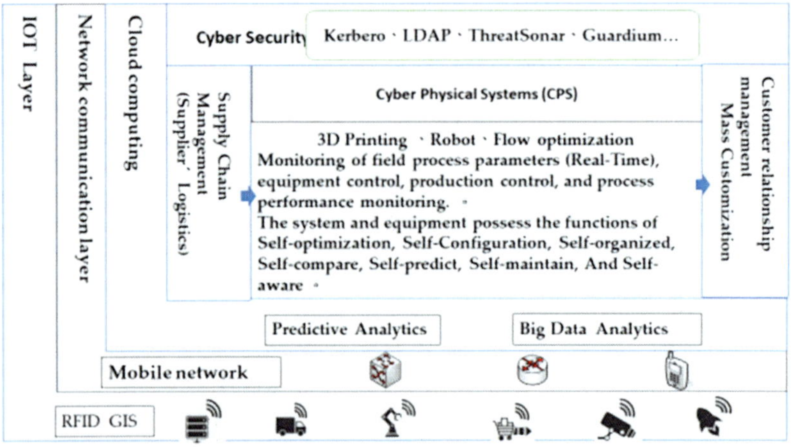

Fig. 7.1 Conceptual architecture of smart factories

end, the factories and internal management processes of the manufacturers must be networked to avoid unnecessary waste, and further reduce inventory and shortened the delivery of customized products. The structure is shown in Fig. 7.1.

Figure 7.1 shows the coordination of ICT and CPS in smart factories. The ICT includes the Internet of Things, Internet access, cloud computing, big data analysis, artificial intelligence, virtual-reality systems, and information security [1]. Under the influence of the supply chain and massive customization, CPS technology includes 3D printing, robotics, and flow optimization, and realizes autonomous field processes in terms of time and space with human control. The essence of CPS is the fusion calculation of people, computers, and objects. Unlike traditional embedded systems, a complete CPS is designed as an interactive network consisting of physical devices instead of just independently-operating devices. This concept is similar to robotic networks and wireless sensing networks.

ICT

In 1998, OECD defined ICT as the combination of manufacturing and service industries for electronic capture, transmission, data display, and electronic information. The Industrial Economics and Knowledge Center (IEK) believed that the main axis of the ICT industry in Taiwan in 2017 was "The AI X Era Begins". It was expected that in the post-Internet of Things era, AI smart devices will gradually enter our lives, and the industrial focus will be extended from the Internet of Things to artificial intelligence [10].

IOT

The IoT plays one of the core roles in Industry 4.0. IEEE P2413 defined IoT as a three-layer architecture, including a Perception layer, Networking and Data Communications layer, and Application Layer [11].

- Perception layer: The perception layer is also called the recognition layer [12]. The perception layer is the lowest layer of the conventional architecture of IoT. Components with sensing or recognition capability are embedded into various real objects, and these sensing components include temperature, humidity, pressure, brightness, flow, distance, acceleration, pressure, gas detection, and other sensors, thus, the objects have the ability to sense environmental changes. In addition, each object is provided with its own identity by a barcode, RFID, or network name. Through identity recognition and feedback, the receiving device knows the change of each object or instrument.
- Networking and Data Communications layer. This layer is the brain of the traditional IoT architecture [13]. Its main responsibility is to help and ensure data transmission between the application layer and the perception layer in the IoT architecture [11]. Based on various transmission technologies, the data from different sensors is uploaded to the cloud server by various extremely low-power-consumption wired and wireless network specification protocols. This layer is the new network of software and hardware component manufacturers, netcom equipment suppliers, telecom operators, Internet Service Providers (ISPs), and cloud service providers that provide various wireless network protocols.
- Application Layer. The application layer is considered the top layer of the traditional IoT architecture, and provides personalized services according to the relevant needs of users [12].

From the perspective of technology and society, IoT covers many fields [5]. By virtue of the available technologies and system architecture, IoT effectively integrates the related signals, networks, and protocols from the underlying sensors and micro-components. Then, the operating system, middleware, cloud solution, application programming, data management, and big data solutions in the software architecture process the signals transmitted by the perception layer; thereby, serving the services and application systems of Industry 4.0. Regarding data management within the IoT architecture, most devices should have self-management and self-optimization functions [14, 15].

$$\text{IoT is expressed as IoT} = \text{Services} + \text{Data} + \text{Networks} \\ + \text{Sensors}$$

The interconnections of IoT and Smart devices improve cloud technology, provide friendly human–computer interfaces, and enhance user experience. IoT integrates various media contents, such as broadcast TV, video, audio, text, images, and data, and incorporates key applications, thus, providing multimedia services with high service quality, satisfying user experience, high security, high interactivity, and high reliability, and inspiring the purchasing intentions of consumers [16]. As IoT serves as the bridge between people and computers, and between computers and computers [17], it is one of the largest data sources for big data. When massive data requires high-frequency processing, transmission, and analysis, IoT transmits and stores big data. "Data security and privacy will play an important role in IoT deployments. As IoT systems will produce and deal with personally identifiable information, data security and privacy will be critical from the very beginning" [18, 19]. Objects transmit data to the cloud or a database through "perception" and "networking"; thereby, developing various applications and services.

Cloud Computing

Cloud computing is a concept, rather than a new technology or a technology [20]. The cloud represents the Internet, and replaces the software originally installed on individual computers or the hard disk space of individual computers through the computing power of the Internet. Instead, various operations are performed by the Internet services, while data is stored in a huge virtual space.

The International Standard for Standardization (ISO) released ISO/IEC 17,788 and ISO/IEC 17,789 in October 2014, which clearly define cloud computing and its reference architectures. In addition, they describe the cloud computing roles, cloud computing activities, cloud operations, and their relationship. Cloud computing is provided for customers through the Internet in the form of services.

In October 2015, the Executive Yuan launched the cloud computing development program [21], which planned to take advantage of the existing ICT industry in Taiwan, consider application service development and the industrial needs of Taiwan, and procure and apply the existing reliable cloud computing solutions, thus, assisting the industry to establish economies of scale. In the meanwhile, it planned to establish the cloud computing application infrastructure of related industries to accelerate the development of open data, big data, IoT, and Industry 4.0, and formulated five promotion strategies: promote the perceptive applications of the public, expand the application development energy, resolve software development technology, implement the cloud infrastructure, and develop green energy efficiency.

Big Data

The Internet oversees and captures everything. Every mouse and button record is stored in the data warehouse, which is used by data exploration to analyze behaviors. Big data applications are the core of Industry 4.0. Cities can realize real intellectual status. We must utilize cloud computing and data mining analysis to efficiently store and manage big data, quickly retrieve and process the data information, mine information and knowledge from big data, and make full use of its value [22].

Big Data Management

Data can be generally divided into three categories: structured data, semi-structured data, and unstructured data. Unstructured data is pre-processed, converted into structured data and stored, which is very important for subsequent exploration analysis. To access the true value of big data, effective management is required for subsequent exploration and cross analysis. The Distributed File System (DFS) is a file system that allows files to be shared across multiple hosts in a network, which allows multiple users on multiple computers to share files and storage space. The industry often uses the NIST SP800-1500 big data specifications as the template for operational management. The Hadoop Distributed File System (HDFS) is widely used in the industry. Through such different designs, big data can be efficiently stored, managed, and used for subsequent data exploration and analysis.

The construction of data-oriented Industry 4.0 requires the strong support of data mining technology, as well as the integration of data in different formats. Data application is the key to data mining research in Industry 4.0. Data visualization technology can present complex urban data to users in a simple and orderly manner, which can fill the understanding gap between technology and users.

Big Data Analysis and Application

Data exploration of business intelligence is to find and identify unknown and hidden relationships and rules within big data by virtue of statistics, artificial intelligence, or other analytical techniques; thereby, achieving classification, estimation, prediction, association grouping, and homogeneity grouping, which are then presented to users in the form of reports or websites (Web Portals). The extraction, retrieval, storage, management, exploration, and analysis methods of big data may encounter different problems depending on the context in which they are used [23]. Big data analysis has diversified applications in Industry 4.0; for example, factories that spend a lot on cutting tools may extend the service life of cutting tools by data analysis; B2C companies with diversified product lines, high marketing, and inventory costs can link customer data, community website data, regional income growth, and population distribution for data analysis and optimization, thus, realizing precise marketing, real-time supply, and reducing inventory [24].

Artificial Intelligence and Robots

Artificial intelligence is the technological science that researches and develops the theories, methods, techniques, and application systems for the simulation, extension, and expansion of human intelligence. It is the core drive of IoT; its logic includes reasoning, knowledge, and learning. Artificial intelligence is an activity dedicated to making computers smart, while intelligence is the ability to make an object implement functionality in a visionary and proper manner in its environment [25]. The artificial intelligence theories and technologies are becoming increasingly sophisticated, such as the rapid development of big data and cloud computing, which expand the field of artificial intelligence applications. The technological products brought by artificial intelligence have entered our lives in recent years as commodities, and this trend is driving a new round of research in artificial intelligence.

Machine learning is a channel to realize artificial intelligence, and refers to training massive data based on algorithms, which can solve problems in artificial intelligence by means of machine learning. After training, a model will be generated, which can be used to make predictions. In view of the fact that traditional machine learning focuses on pattern mining, reinforcement learning shifts the focus to decision-making. This technology will help promote artificial intelligence to relevant research and practice fields in the real world.

Deep learning, which simulates the operation of human neural networks, is the fastest growing field of artificial intelligence. Common deep learning architectures include Multilayer Perceptron, Deep Neural Network (DNN), Convolutional Neural Network (CNN), and Recurrent Neural Network (RNN). Deep learning is popularly used in visual recognition, speech recognition, natural language processing, and the biomedical field, and has generated remarkable results.

Robots are the result of artificial intelligence, and deep learning affects the flexibility of robots. The progresses in reliable machine perception, computing vision, force and tactile perception are mostly driven by machine learning, and they will remain as the key to advancing robot capabilities. According to research by Stanford University, the eight fields in which AI is the most prominent are: transportation, service robots, health care, education, low-resource communities, public safety and security, employment and the workplace, and entertainment [26]. At present, robots are equipped with the capabilities of data retrieval, data response, OCR natural speech recognition, multi-interface input and archiving, automation planning, report list output, work allocation, and automated notification to employees and customers.

Cyber-Physical Systems

Cyber-Physical Systems (CPS) were proposed by the National Science Foundation (NSF) in 2006 and promoted by the German Industry 4.0 in 2011. The development purpose of CPS is that the "objects" enable effective communication and coordination between different "systems" by virtue of "networks", which shorten the distance between the virtual world and the real environment after fully utilizing "computing", "control", and "communication" technologies [27]. Then, CPS can quickly respond to the needs of the environment and faithfully express the information generated in the real environment, and then, grant the systems and equipment with self-optimization, self-configuration, self-organized, self-compare, self-predict, self-maintain, and self-aware functions [28].

The manufacturing process of CPS in smart factories can be divided into the product manufacturing practice stage (four work steps) and the design stage (two work steps) [27]. The four work steps in the product manufacturing practice stage are "precision production", "mechanical control", "production monitoring", and "production management", while the two work steps in the design state are "product design" and "production line simulation" [19].

- Precision production: It mainly includes various sensors for detecting production sites, motors, actuators and robot arms.
- Mechanical control: It mainly includes programmable digital control and integration equipment, such as PLC/CNC and DCS.
- Production monitoring: SCADA/HMI is used to collect the factory environment and equipment information, in order that factory personnel can use the SCADA/HMI system to make real-time response to the factory equipment and manufacturing process. Production management: In the production management process, MES is used to manage the scheduling and maintenance of production contents, such as work documents, raw materials management, and product quality management. Product design: It mainly includes computer-aided design and engineering software, such as CAD/CAE, which needs to cooperate with PDM and physical simulation software for product design related testing.
- Production line simulation: It targets at simulating online production in the future, and thus, avoiding the risks of process or security issues during the deployment of the actual production line.

CPS is a composite overall solution. To improve the CPS in the factories, the affected levels and involved roles are quite complex. Even when CPS is actually introduced, new equipment or new process management will bring a new wave of organizational revolution to factories [19].

Cyber Security

In the digital era, information security is the lifeblood of automated operating systems. The 2017 Global Risk Report of the World Economic Forum (WEF) showed that data fraud or theft ranked fifth in the global risk rankings, while cyberattacks ranked sixth, thus, it can be seen that information security risks have deeply affected human life [29]. The 2018 Global Information Security Survey Report of PWC also pointed out that, more than 40% of global enterprises lacked comprehensive information security strategies [30].

When the risks burst out through complex systems, they increase the possibility of runaway collapse, or lead to a sudden transition to a new and sub-optimal state [29]. As Industry 4.0 is of vital importance to improve people's livelihood, it is necessary to improve the information security concept. To increase the breadth, depth, and speed of information security, the government combines big data analysis and artificial intelligence to build a multiple information security defense system for government agencies, key infrastructure, and local government regional governance, which can predict the trend of security attacks and increase the response speed to information security incidents [31].

The security goals of IoT devices include confidentiality, integrity, and availability (CIA) [17]. These new technologies are used to guarantee end-to-end security and privacy for future IoT system management, and have a significant impact on performance assessment techniques [26]. In addition, the security requirements vary according to different applications [32], and refer to the data security that protects user privacy and access control.

Information security has been predicted as an obstacle to sustainable computing in the future [33]. While the development of cloud computing has brought more convenience to the lives of ordinary people, it must be carefully evaluated when information security and personal data confidentiality are still cloud computing services.

The contextual integrity principle, as proposed by Helen Nissenbaum [34], provides a meaningful idea or framework for solving privacy problems in the era of big data [20]. In data analysis technology, supported by cloud computing, the Internet, and other ICT technologies, the final computing results are obtained through data input and processing calculation analysis [21]. The Expanded Top 10 Big Data Security and Privacy Challenges, as released by the Big Data Working Group of the Cloud Security Alliance and Big Data Analytics for Security Intelligence [22, 24], summarizes the information security into four categories: infrastructure security, data privacy, data management, integrity and response security [31].

Industry 4.0 is composed of different information technology objects, and helps people perceive, make decisions, or actions through integrating a set of systematic software and hardware to meet the needs of various application fields and contexts. Traditional information security technology is still an important component, and has an irreplaceable role in the construction process of Industry 4.0. Traditional grading, domain, and key protection strategies are still applicable to the construction of the Information Security Technology System of Industry 4.0, and the Information Assurance Technical Framework (IATF) is still the top guiding principle [35].

Taiwan Productivity 4.0 in Taiwan

In Taiwan's industrial circle, the concept of Industry 4.0 (such as Germany's Industry 4.0) is called Productivity 4.0, which refers to intelligent manufacturing with intelligent management using IoT, cloud computing, big data, artificial intelligence, and other ICT technologies. Productivity 4.0 requires information communication and construction technology, and Taiwan is one of the regions that have mastered ICT and production, thus, it has excellent conditions to become an important partner in the development of Industry 4.0 worldwide. In the era of IoT and big data, in response to the shortened product life cycle, diversified, and small-sized production, as well as the massive customization demands of the global manufacturing industry, Germany, the United States, Japan, and other developed countries are actively developing intelligent manufacturing technology. Germany Industry 4.0 is a leading model in international intelligent manufacturing.

Enterprises in Taiwan

Taiwan is facing the challenges of shrinking employment and increasing international competition, thus, the government must effectively adjust the industrial structure and break away from the traditional thinking of labor-intensive hardware manufacturing, meaning it should integrate service value and promote industrial innovation and transformation. Taiwan Productivity 4.0 in Taiwan is mainly combined with machinery manufacturing, electronic motors, and communications, which are Taiwan's advantages from the past.

99% of the industrial structure in Taiwan is occupied by SMEs, which provide 60% of the employment market, and account for 31.8% of the total revenue and 41.4% of total investment [36]. Therefore, how to make SMEs willing to practice Industry 4.0 and how to practice Industry 4.0 become an important issue for the government. It will also affect the future development of Industry 4.0 in Taiwan.

In 2015, Taiwan released the development plan of Productivity 4.0 to promote the development of intelligent industries, which mainly focused on intelligent machinery. The plan was piloted in the aerospace industry and semiconductor industry. The promotion structure of Taiwan is based on customer value, and is combined with the two major policy characteristics of Germany and the United States, including CPS from Germany and the value-added service focus of the United States. Based on the integration of the above two characteristics, Taiwan has promoted the networked manufacturing systems consisting of intelligent manufacturing and intelligent services, and applied IoT, intelligent machinery/robots, and big data technologies to guide the transformation of enterprises in Taiwan. The key industries include electronic information, metal transportation tools, machinery equipment, food, and textiles to enhance the new growth drive of industrial manufacturing (Fig. 7.2). Figure 7.2 shows the industrial policy plan, and the government hopes

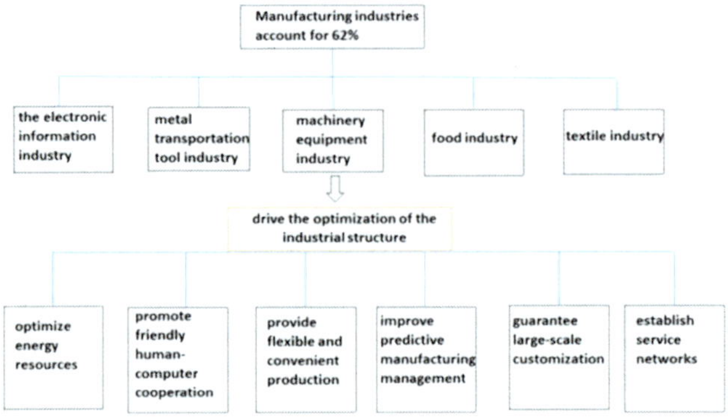

Fig. 7.2 Schematic for the optimization of five major industries in Taiwan's manufacturing. *Source* Wu [38]

that the machinery industry in Taiwan will drive the optimization of the industrial structure, and then, spread to upgrading all relevant manufacturing industries [37].

In Fig. 7.2, Taiwan's progress in intelligent manufacturing is based on the intelligent machinery industry. According to the information released by the Ministry of Economic Affairs, the government will build an intelligent machinery industry ecosystem, and achieve intelligent machinery industrialization and intelligent industrial mechanization [37]. The definitions and scopes are shown in Table 7.1.

The promotional strategy and method of the government is to: link local and promote regional advantageous industries; link future constructs and guide industrial environments; link the world and promote industrial international cooperation. Considering that most manufacturers in Taiwan are SMEs, the Ministry of Economic Affairs put forward the upgrade emphasis on "broadness" and "height" to cater to the SMEs and meet the needs of manufacturers of all scales. Regarding broadness, as the SME manufacturers in Taiwan generally lack the basic digital infrastructure, the government introduced digital systems. After some production management processes are digitized, the production capacity can be greatly increased to meet the needs of industries. Regarding height, the government has assisted large-sized manufacturers, which have already reached the digitization of their production lines to accelerate their entry into intelligent manufacturing, and accelerated industrial diffusion through the establishment and handling of demonstration fields.

Statistics on Taiwan's Manufacturing Industries

According to the Ministry of Economic Affairs, an annual calibration survey was carried out on the 92,946 registered factories (including more than 7,000 temporarily and added registered factories) in June 2018. According to the calibration results, the

Table 7.1 Definitions and scopes of the intelligent machinery industry ecosystem constructed by the government

Item	Definition	Scope
Intelligent machinery industrialization	Intelligent machinery integrates various intelligent technical elements, provides them with intelligent functions, such as fault prediction, accuracy compensation, automatic parameter setting, and automatic scheduling, offers overall solutions, and establishes differentiated competitive advantages	Equipment, robots, IoTs, big data, cps, sensors, and other industries
Intelligent machinery industrialization	Introduce intelligent machinery into the industry, build intelligent production lines, provide a large number of customized products through cloud and online quick links, and form a networked manufacturing service system	Aerospace, semiconductor, electronic information, metal transport, machinery, food, textile, retail, logistics, agricultural products, and other industries

Source Wu [37]

number of operating factories was 87,149 and the number of SMEs was 85,213, which accounted for 97.8% of the total number of factories (Table 7.2 Factory statistics) [39]. The operating income of SMEs accounted for 40.3% (Table 7.3 shows Taiwan's operating income statistics); the number of R&D institutions and the R&D funds in 2017 are shown in Table 7.4: R&D fund statistics. Other relevant data of all the factories for the last five year is included in Table 7.5: Comprehensive comparison of factory operations for the last five years.

Development Energy and Pointer of Taiwan's Manufacturing Industries

Taiwan's manufacturing industries provide considerably support for the economic development and employment market in Taiwan, and possess certain competitiveness in the international market. With reference to the various economic indicators, as released by the Ministry of Economic Affairs and Executive Yuan [40], according to the international competitiveness rankings published by international institutions [41–45], as shown in Table 7.6, Taiwan's manufacturing industries possess quite excellent development energy.

Table 7.2 Factory statistics

Factory type (Quantity)	2014 Quantity	2015 Quantity	2017 Quantity	2017 Proportion	Growth of 2017 compared to 201%
Total number of operating factories	81,986	83,532	87,149	100	2.1
Small-sized factories (with fewer than 20 employees)*	57,361	58,430	61,027	70	2.2
Medium-sized factories (with 20–199 employees)*	22,763	23,225	24,186	27.8	2.0
Large-sized factories (with more than 199 employees)	1862	1877	1936	2.2	1.6

Source Ministry of Economic Affairs [39]
Note* **SMEs refer to the companies that handle company registration or business registration in accordance with the laws, and comply with the following standards: (1) In manufacturing, construction, mining, and earth-rock exploration industries, pay-in capital is below NT$ 80 million, or the number of regular employees is below 200. (1) In other industries, the operating income in the previous year is below NT$ 100 million, or the number of regular employees is below 100

Table 7.3 Taiwan operating income statistics. Unit: NT$ (100 million)

Operating income	2014	2015	2017	Growth of 2017 compared to 2015 (%)
Total operating income of Taiwan	178,953	163,606	166,916	1.0
Small-sized enterprises	12,728	12,311	12,741	1.7
Medium-sized enterprises	57,847	54,984	54,587	−0.4
Large-sized enterprises	108,379	96,311	99,588	1.7

Source Ministry of Economic Affairs [39]

Overview of the Factory Operation in Taiwan's Manufacturing Industries

The number of operating factories, the number of employees at the end of the year, the total annual operating income, the total annual operating profit, the number of R&D institutions, and the R&D funds (such as Table 7.7), as published by the Ministry of Economic Affairs, show slight growth since 2013 [39]. Fortunately, the number of

Table 7.4 R&D fund statistics. Unit: NT$ (100 million)

Year	# R&D institutions	R&D funds	Cost category		Top four industries in R&D funds			
			Regular cost	Capital cost	Electronic component industry	Computer electronics and optical industry	Mechanical equipment industry	Power equipment industry
2014	6847	4591	4214	378	2386	1229	163	132
2015	6944	4870	4450	419	2600	1294	167	138
2017	7056	5302	4729	572	2961	1304	169	130
Average annual growth (%)	0.8	4.3	3.1	16.8	6.7	0.4	0.7	−3.0

Source NDC (National Development Committee) [40]

According to Table 7.4, among the top 4 industries in R&D funds, the electronic components industry ranks first, and the computer electronics and optical industry ranks second, followed by the mechanical equipment industry and the power equipment industry.

Table 7.5 Comprehensive comparison of factory operations for the last five years

Operation overview	2012	2013	2014	2015	2017
Number of operating factories	79,439	81,064	81,986	83,532	87,149
Number of employees at the end of the year (1000 employees)	2538	2577	2660	2701	2815
Total annual operating income (NT$ 100 million)	168,759	170,971	178,953	163,606	166,916
Total annual operating cost (NT$ 100 million)	162,961	161,370	168,286	152,285	153,091
Total annual operating profit (NT$ 100 million)	5798	9601	10,668	11,321	13,825
Total annual fixed-asset investment (NT$ 100 million)	10,788	10,595	11,908	12,494	13,526
Number of R&D institutions	6912	6822	6847	6944	7056
R&D funds (NT$ 100 million)	4134	4133	4591	4870	5302
Number of technology procurers	1721	1675	1603	1844	2036
Technology procurement amount (NT$ 100 million)	1576	1572	1708	1661	903
Number of technology sellers	558	594	593	620	747
Technology sales amount (NT$ 100 million)	313	340	383	412	487

Source Ministry of Economic Affairs [39]

Table 7.6 Development energy and pointers of Taiwan's manufacturing industries

Project/pointer	Development status/evaluation					
	2013	2014	2015	2016	2017	2018
Proportion of Taiwan's manufacturing industries in nominal GDP	28.75	29.99	30.3%	30.68	31.02	30.75
UNIDO (Competitive Industrial Performance index) Latest data is on 201	12th	12th	13th	–	–	–
WEF (state of cluster development)	First	–	5 h	–	13th	–
Proportion of product output of Taiwan's manufacturing industries in the product output of global manufacturing industries	–	2.1%	2.0%	1.8%	1.8%	–

Source Ministry of Economic Affairs [39], Executive Yuan [40], UNIDO [41], WEF [29], IMD [43], Lin [44]; Compiled by this stud

R&D institutions and the R&D funds show considerable growth from 2013–2017, indicating that more and more companies are willing to invest in R&D in Taiwan, and the factory transformation is worth looking forward to (see Fig. 7.3).

Conclusions

The sustainable development of a city is mainly based on residents' spatial perceptions and needs, and the global development of smart cities is already in full swing. The establishment of various evaluation indices aim to explore the uniqueness and service capacity of each city, verify the promotion effect of a smart city, and guide the development of a smart city. There is no absolute measure of smart cities; if a smart city with good performance fails to make good use of the development of ICT technology, and continuously expand its new vision and create new services, its functions may be relatively weakened.

Taiwan is renowned for its focus on the IT industry, and possesses an excellent investment environment. Its high-tech output ranks among the top, and it has sufficient ICT, technology industry, and mechanical industry talents (1). The implementation of Industry 4.0 improves the R&D efficiency and technology of innovation engineering, and reduces the costs (such as salary, benefit costs and raw material costs), which digitalizes the industry, and upgrades the technology integration, innovative production methods, R&D field, and job opportunities, and ultimately, product quality and productivity.

Table 7.7 Overview of the factory operations in Taiwan's manufacturing industries

Operation overview	2013	2014	2015	2017	Remark
Number of operating factories Change compared with previous year (%)	81,064 2.0	81,986 1.1	83,532 1.9	87,149 4.3	The proportion of small-sized factories was the highest, 70%
Number of employees at the end of the year change compared with previous year (%)	2577 1.5	2660 3.2	2701 1.5	2815 4.2	The proportion of medium-sized factories was the highest, 44%
Total annual operating income change (NT$ 100 million) compared with previous year (%)	170,971 1.3	178,953 4.7	163,606 −8.6	166,916 2.0	The proportion of electronic component industry was the highest, 24%
Total annual operating profit change (NT$ 100 million) compared with previous year (%)	9601 65.6	10,668 11.1	11,321 6.1	13,825 22.1	The proportion of electronic component industry was the highest, 51%
Number of R&D institutions change compared with previous year (%)	6822 −1.3	6847 0.4	6944 1.4	7056 1.6	The proportion of electronic component industry was the highest, 23%
R&D funds change (NT$ 100 million) compared with previous year (%)	4133 0	4591 11.1	4870 6.1	5302 8.9	The proportion of electronic component industry was the highest, 53%

Source Ministry of Economic Affairs [39]

Taiwan is featured by population ageing, energy limitations, and a closed market. Under the concept of Industry/Productivity 4.0, driving the industrial upgrade and solving the current substantive problems in Taiwan is a goal that can be explored in the future. At present, the global industry is facing many difficulties, such as improper use of energy resources, environmental damage, shortage of manpower, and the slowing and stagnation of global economic growth. Taiwan, as a member of the international community, should work hard to solve these problems.

When promoting smart cities, cities in Taiwan should refer to smart city evaluation indicators, and then, evaluate Taiwan's local advantages and demands, integrate existing resources, give play to local characteristics, plan the sustainable operations of the city, and facilitate residents' lives. Only in this way can we create a smart city that truly meets the needs of the people, including better quality of life and user friendly and sustainable living environments.

To sum up, Industry 4.0 designs the industrial-appropriate solutions based on the Internet, personalized services, data analysis and decision-making, and energy efficiency to achieve high-value manufacturing and services. Cloud, IoT, and big data are the key technologies to strengthen this connotation. The key promotion fields

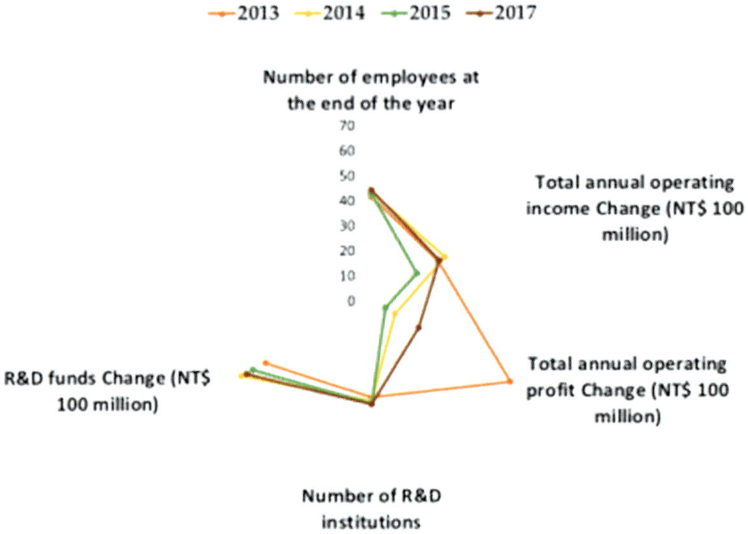

Fig. 7.3 Overview of the factory operations in Taiwan's manufacturing industries (2013–2017)

include equipment intelligence, system virtualization, and industrial intelligence. Replacing old-fashioned equipment investment with knowledge investment is a new trend in intelligent manufacturing for the future; knowledge is the key to the fourth industrial revolution in the future. Only by integrating knowledge into high-value equipment can Taiwan improve its core competitiveness. It is necessary to deepen the advantages of the ICT industry in Taiwan to achieve long-term progress, and to consolidate the manufacturing competitiveness of industrial clusters in Taiwan.

Under the influence of the new development trend of smart cities, the adjustment of traditional laws, regulations, rules, norms, and implementation standards, as well as the transformation of corresponding planning methods and other issues, depend on the cooperation of governmental departments with large enterprises to take the lead in planning and implementation. It is necessary to make good use of the innovative smart city solutions brought by digital technology, such as smart energy, smart manufacturing, and smart environments, in order to achieve the sustainable goals of smart cities, including low carbon, ecological environments, green energy, recycling, and livable environments.

Acknowledgements The authors would like to thank the Ministry of Science and Technology, Taiwan for their support under grant no. 108-2511-H-003-034-MY2.

References

1. W. Mingji, 2016 Taiwan German Productivity/Industry 4.0 Forum. Achievement Report (International Economic Cooperation Association (CIECA), 2017)
2. J. Wang, Y. Ma, The international development trend of Industry 4.0 and Taiwan's response (Central Bank, International Financial Reference, Taipei, 2016)
3. Z. Wu, *Research on Talent Cultivation and Enrollment Strategy in Innovation Industry* (NDC (National Development Committee), Taipei, 2017)
4. A. Botthof, E.A. Hartmann, *Zukunft der Arbeit in Industrie 4.0* (Springer Vieweg Berlin, 2015)
5. E.A. Hartmann, M. Bovenschulte, Paper presented at the using technology foresights for identifying future skills needs, SKOLKOVO Moscow, 2013
6. K. Henning, Recommendations for implementing the strategic initiative INDUSTRIE 4.0 (National academy of science and engineering, Frankfurt, 2013)
7. P.C. Evans, M. Annunziata, Industrial internet: pushing the boundaries of Minds and Machines (Boston, 2012)
8. M. Rüßmann et al., Industry 4.0: the future of productivity and growth in manufacturing industries. Boston Consult. Group **9**(1), 14 (2015).
9. Y. Tan, S. Goddard, L.C. Pérez, A prototype architecture for cyber-physical systems. ACM Sigbed Rev. **5**(1), 26, 2 (2008)
10. I.E.A.K. Artificial intelligence industry accelerated this year (2017)
11. L. Oleg, Standard for an architectural framework for the internet of things (IoT)—IEEE p2413, 22 (IEEE Standards association, 2015)
12. H. Suo, J. Wan, C. Zou, J. Liu, Paper presented at the 2012 international conference on computer science and electronics engineering, Hangzhou, China, 2012
13. A.-F. Ala, M. Guizani, M. Mohammadi, M. Aledhari, M. Ayyash, Internet of Things: a survey on enabling technologies, protocols, and applications. IEEE Commun. Surveys Tutorials **17**, 2347–2376 (2015)
14. B.N. Silva, M. Khan, K. Han, Internet of Things: a comprehensive review of enabling technologies, architecture, and challenges. IETE Tech. Rev. **35**, 15 (2017)
15. L. Atzori, A. Lera, G. Morabito, Understanding the Internet of Things: definition, potentials, and societal role of a fast evolving paradigm. Ad Hoc Netw. **56**, 122–140 (2017)
16. B. Muhammad, A review of internet of things architecture, technologies and analysis smartphone-based attacks against 3D printers. arxiv (2017)
17. P.P. Jayaraman, X. Yang, A. Yavari, D. Georgakopoulos, X. Yi, Privacy preserving Internet of Things: from privacy techniques to a blueprint architecture and efficient implementation. Future Gener. Comput. Syst. **76**, 540–549 (2017)
18. A.H.H. Ngu, M. Gutierrez, V. Metsis, S. Nepal, M.Z. Sheng, IoT middleware: a survey on issues and enabling technologies. IEEE Internet Things J. 1–1 (2016)
19. M.U. Farooq, M. Waseem, A. Khairi, S. Mazhar, A critical analysis on the security concerns of Internet of Things (IoT). Int. J. Comput. Appl. **111**(7), 6 (2015)
20. P. Mell, T. Grance, The NIST definition of cloud computing. NIST SP **800–145**, 7 (2011)
21. *2016*–2018 Cloud computing development programs (2015). https://www.bost.ey.gov.tw/cp.aspx?n=B56ED9F993B2EFA5.
22. S. Wu, T.-C. Chen, Y. Wu, M. Lytras, Smart cities in Taiwan: a perspective on big data applications. Sustainability **10**, 1–14 (2018)
23. A. Akusok, K.-M. Björk, Y. Miche, A. Lendasse, High-performance extreme learning machines: a complete toolbox for big data applications. IEEE Access **3**, 1011–1025 (2015)
24. L. Kang, H.L. Du, H. Zhang, W.L. Ma, Systematic research on the application of steel slag resources under the background of big data. Complexity **2018**, 1–12 (2018)
25. N.J. Nilsson, *The Quest for Artificial Intelligence* (Cambridge University Press, Stanford University, 2009), pp. 707
26. M.-H. Huang, R.T. Rust, Artificial intelligence in service. J. Service Res. **21**, 155–172 (2018)
27. R. Baheti, H. Gill, Cyber-physical systems. Impact Control Technol. **12**(1), 6 (2011)

28. J. Lee, B. Bagheri, H.-A. Kao, A Cyber-physical systems architecture for Industry 4.0-based manufacturing systems. Manuf. Lett. **3**, 23 (2015)
29. A. Collins, *WEF the Global Risk Report 2018* (WEF, Switzerland, 2018)
30. C. Christopher, *Revitalizing Privacy and Trust in a Data-Driven World-Key Findings from the Global State of Information Security Survey 2018* (Cybersecurity and Privacy, PWC Web, 2018)
31. S. Wu, D. Guo, Y. Wu, Y. Wu, Future development of Taiwan's smart cities from an information security perspective. Sustainability **10**, 18 (2018)
32. J. Lee, B. Bagheri, H.-A. Kao, A cyber-physical systems architecture for industry 4.0-based manufacturing systems. Manuf. Lett. **3**, 18–23 (2015)
33. S.M. Wu, T.-C. Chen, Y.J. Wu, M. Lytras, Smart cities in Taiwan: a perspective on big data applications. Sustainability **10**, 106 (2018)
34. H. Nissenbaum, Privacy as contextual integrity. Wash. Law Rev. **79**(1), 119–157 (2004)
35. M. Korotka, L.R. Yin, C. Basu, Information assurance technical framework and end user information ownership: a critical analysis. J. Inf. Privacy Secur. **1**(1), 17 (2005)
36. 2017 SME White Paper (2017). https://www.google.com.tw/search?source=hp&ei=Pe2AXJajBcnO8wXt07d4&q=2017%E4%B8%AD%E5%B0%8F%E4%BC%81%E6%A5%AD%E7%99%BD%E7%9A%AE%E6%9B%B8&oq=2017%E4%B8%AD%E5%B0%8F&gs_l=psy-ab.1.0.0.2486.14339..17309...0.0..0.86.664.13......0....1..gws-wiz.....0..35i39j0i131.j-KmZX2YXK8.
37. M. Wu, *Five Major Industrial Innovation Research and Development Plans, Smart Machinery Industry Promotion Program* (MOEA, Taipei, 2015)
38. M. Wu, 2016 Taiwan German Productivity/Industry 4.0 Forum Results Report—overview of Taiwan productivity 4.0 development (CIECA (Chinese International Economic Cooperation Association), Taipei, 2017)
39. MOEA, Analysis of preliminary statistical summary of factory calibration and operation survey in 2018 (MOEA, Taipei, 2018)
40. NDC, Important statistics manual, (NDC (National Development Committee), Taipei, 2017)
41. UNIDO, Competitive Industrial Performance Report 2016 (UNIDO, Vienna, 2017)
42. UNIDO, Demand for Manufacturing: Driving Inclusive and Sustainable Industrial Development, Industrial Development Report 20*18*. (UNIDO, Vienna, 2018)
43. IMD, imd_world_digital_competitiveness_ranking_2018 (IMD World Competitiveness Center, Lausanne, Switzerland, 2018)
44. H. Lin, Opportunities and challenges for Taiwan's economic growth, (TRA (Taiwan Research Institute) Taipei, 2017)
45. S. Klaus, *The Global Competitiveness Report 2017–2018* (WEF (World Economic Forum), Geneva Switzerland, 2018)

Chapter 8
Sustainable Tourism: Crowdsourced Data for Natural Scene and Tag Mining

Asamaporn Sitthi

Abstract In recent years, the concept of sustainable tourism has emerged at the intersection of debates on visiting somewhere as a tourist and trying to make a positive impact on the environment, society, and economy. By leveraging the power of online infrastructures, we demonstrate that crowdsourced generated data, by the tourists, encode a vast amount of information, such as the physical properties from the photo and description from textual information. Using these online platforms, such as Flickr, users generate crowdsourced geotagged information containing an immense amount of human behavior tracking on scenic views. In this paper, geotagged Flickr data is used for automatic natural scenes classification using an image, and textual features obtained from the crowdsourced data. The proposed method uses the data mining technique with descriptors. The results show that the geotagged Flickr data can imply Urban City interaction with an encouraging accuracy of 90.20% and that the proposed approach improves natural scene classification efficiency if a sufficient spatial distribution of crowdsourced data exists. Hence, social sensing mining the attractiveness of human interaction is an interesting or tourism area using image processing and text mining method with geotagged mobility of users can provide accurate information that challenges for developing sustainable tourism management.

Keywords Sustainable tourism · Mobility · Flicker · Machine learning

Background

Recently, the development of information technology for communication is playing a vital role in the modern society. By leveraging the power of the Internet, users now share and exchange their events or activities in a variety of perspectives through image, text video with location-based service. The range of social sensing data

A. Sitthi (✉)
Department of Geography, Faculty of Social Science, Srinakharinwirot University, Bangkok 10110, Thailand
e-mail: as.sitthi@gmail.com

A. Visvizi et al. (eds.), *Research and Innovation Forum 2020*,
Springer Proceedings in Complexity,
https://doi.org/10.1007/978-3-030-62066-0_8

sources contains a wealth of geotagged and visual information, for example, image sharing services such as Flickr, Facebook, and Panoramio [1].

It has been argued by a number of studies that linking crowdsourced data from social media in geographical terms can certainly improve the performance of existing land cover classification techniques [2, 3]. Since, the social sensing information categories can reflex human behavior, which involves image characteristics or natural scenes and tags opinion from geotagged metadata such as urban area and landscape places. However, the technical challenge raises questions on how to convert the visual feature and tags from huge Flickr data into sematic concepts for the sustainable tourism development.

A number of studies argue that the sustainable tourism is strongly linked to the sustainable consumption of energy, transport and mobility, buildings, living conditions, and management, including trips to urban tourism destinations [4–7]. In Geoinformatics studies, the appropriate data mining technique is essential for detecting spatial patterns in a complex dataset. To understand user behavior, mining techniques are implemented which relevant to tourism and destination management research [3]. There are various studies that highlight the significance of smart city development in an area or particular place [8]. Although some researchers have already investigated the issues related to social media involved smart city and sustainable development, including spatial interaction from human behavior, environmental, and readiness.

In this research, we argue that geotagged information from crowdsources data have a valuable feature such as image propertied, image locations, and tags information that can be leveraged to estimate human interaction in a particular area. In this lie, we implement data mining techniques suitable for the handling of a huge amount of data yet not sensitive to irrelevant features. Furthermore, we mine geotagged social sensing information used for automatic extraction of visual and textural natural scene descriptions. We believe that this integration will be benefited to better understand where there need for further development of sustainable tourism implementations for an increased living standard.

Flicker Geo-tagged Data

For this study, Flickr geotagged data is acquired from the Flickr, focusing on the main Urban area of Tokyo in Japan. The metadata was queried using the Flickr Application Programming Interface (API). Note that Tokyo city covers a total length of 622 km^2. Flickr data contain a large volume of images distributed over this area. The based-map and Flickr geotagged information is shown in Fig. 8.1. Note that Flickr metadata provides some of the fundamental information about image posted, including the username, title, tag, GPS location, and date and time of when the Flickr was obtained and posted. From the available Flickr dataset, between 1 January and 31 December 2018 were obtained about 601,838 data points. The Flickr metadata, such as tags information and location were used to produce a query list of information.

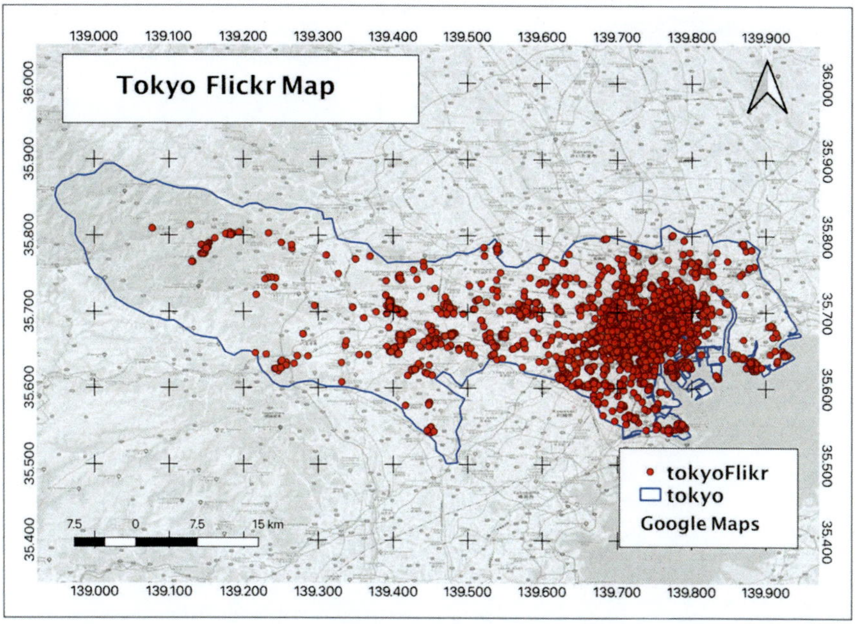

Fig. 8.1 Location of Flickr images in Tokyo city

We plan to use this data to investigate Flickr metadata distribution and the relation of visual feature based natural classification to each geotagged information, which can be used to refer to as urban or city scene. It is implemented on low-level image descriptions and tags features by mining the geotagged information using color, edge, and vegetation index histogram descriptors and a bag of words. We show Flicker meta-data mapping in Fig. 8.2.

Natural Image Features

We leverage image analysis that is a mathematical process to extract, characterize, and interpret information from the image cell. The basic idea of semantic modeling is to extract image properties into the semantic concept of natural scenes, including Water Bodies, Agricultural, Urban Areas, Forest, and Grassland. We use classifiers for the semantic classification of the image between urban/city and non-urban. We employ low-level feature description approaches especially edge detection and vegetation indexes technique for scene categories to classify our data into development (urban area) vs. non-develop or city versus landscape [9, 10].

For the feature extraction from images, we use edge histogram to separate landscape and human-made scenes [11]. As the second kind of feature, we use edge direction histogram, which is computed by grouping the edge pixels. We apply the

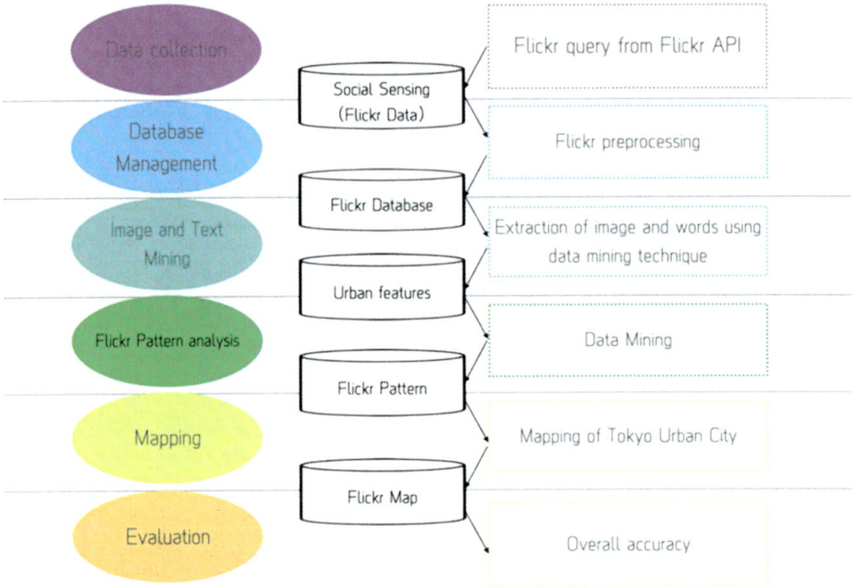

Fig. 8.2 Flickr meta-data mapping

edge histogram method with a set of four 3×3 linear filters operation to detect edges contrasting at horizontal, vertical, 45° diagonal, 135° diagonal orientation. We extract the geo-referenced images feature using edge histogram descriptors, which identify the distribution of edges at different orientations [11]. An appropriate threshold is applied to the outputs of these edges' histogram. As a result of edge orientation (See Fig. 8.3), we find that image of urban scenes has a much higher proportion of horizontal and vertical edges than non-urban [12].

Tags Mining

Text processing is the process of preparing data for mining analysis. To obtain the result, the processes that are most appropriate and minimized distortion of information are implemented as follow;

Firstly, we select the data i.e., a basic method for removing unwanted information in order to reduce the uncertainty in the analysis. The information such as duplicated words, emoji, other language characters, etc. in the texts are regarded as redundancy information. Note that these are some of the key parameters which negatively influence the accuracy and efficiency of word token, text processing, and model training.

Then, we employ tokenization i.e., a process of delimiting the tags into the words (tokens). Note that space is a natural word breaker in languages for English. The

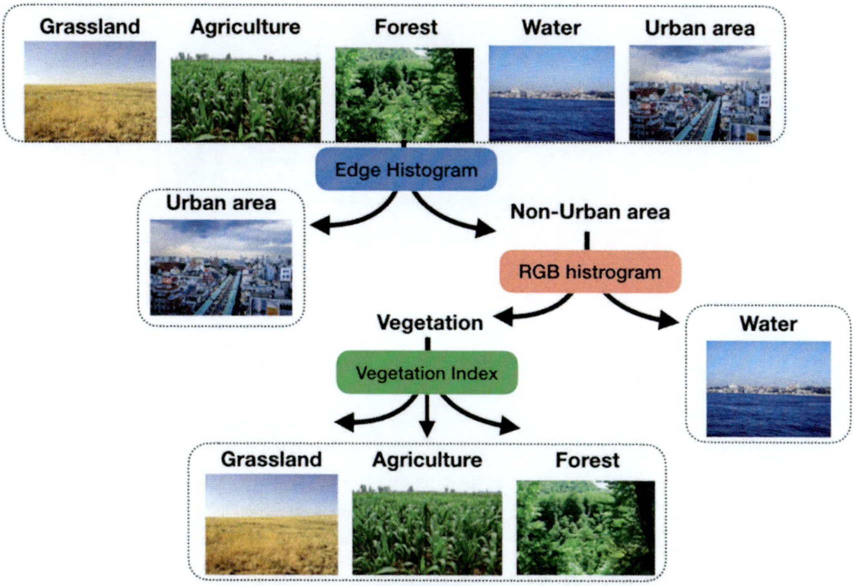

Fig. 8.3 Feature extraction for the land cover classification

Language Technology Platform (LTP) was deployed to segment words, remove punctuation, and stop words [13]. In this study, the Natural Language Processing (NLP) library was applied for text processing, which is a dictionary-based approach. Therefore, the dictionary is important for text tokenization in order to classify words for a specific meaning.

Significant words were token from the tokenization method can used for information extraction to refer to a relationship with human interaction such as city or area name, tourist place, environment, location of interest, human events, and activities (hiking, trekking nightlife, and culture).

After tokenization processing, analyzing the frequency of words that are performed to find the frequency of reference words in Flickr. The frequency of the word's appearance is directly explained in the form of probability in a particular area of interest. To find relevant relationships about human interaction and spatial patterns, words frequency display in the form of a tag cloud and the frequency of words (see Fig. 8.4).

Association Rule Mining

Social sensing data mining has become an interesting research area. In general, it can be categorized into the following four approaches, such as clustering, Association Rules Mining (ARM), path recommendation, and tag analysis [14–17]. In order to

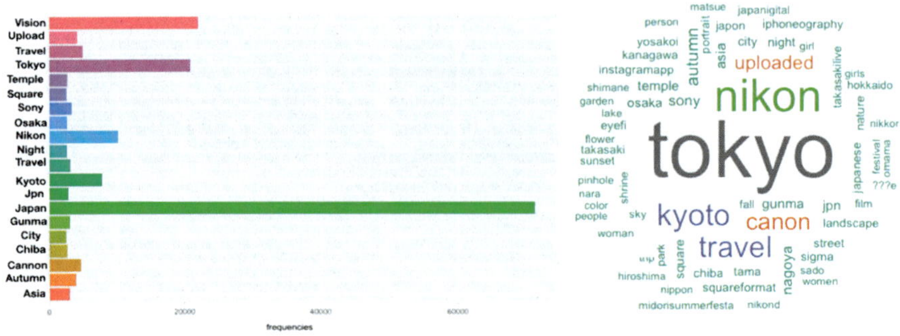

Fig. 8.4 Feature extraction for the land cover classification

understand travelers' photo-taking travel pattern and behavior, we leverage ARM that is the process of finding the relationship between words of location of interest (LoI)—by applying the rules for finding the relationship of data from an individual word by specifying frequency of visitors as main keywords (K) from Flickr data as area name and activities [4]. These words have a set of the collection in the form of word index Ki = {k1, k2, …, kn} when n is the number of sentences in the document. The relationship between the various wording indexes (ki) that are related to the main keyword index (K). In the case that ki has a very good relationship with K (ki $\subseteq$ K), it can be assumed that the ki word index is often found with the main K keyword index or frequently being a group of words [6] (See Table 8.1). As a result, we found

Table 8.1 Example of spatial (area) and activities association words from ARM analysis

Keywords (K)	Occurrence words
Spatial (popular area)	
Harajuku	Street art, shrine, shopping
Shibuya	Night, metro, crossing
Shinjuku	Matro, night, Building, Urban, Town
Roppongi	Sculpture, Lake, coffee bar cafe pub hill
Chiyoda	Tokyo Tower, Lake
Activities	
Night	Cityscape, street, building, traffic, town
Temple	Buddhist, shrines, meijijingu, asakusa
Travel	Cherryblossoms, city, Disney, disney
Hotel	Backpacking, hostel, japanesegarden
Autumn	Fall, flower, October, redleaves

that the keywords of Spatial (area) and activities are highly associated. These can be implying that the tags information that user posted via Flickr can reflect that spatial interaction, mobilities and activities.

Experimental Results

This section presents the experimental results of geotagged classification data and spatial distribution analysis in the following sub-sections.

Geotagged Data Classification from Image and Tags Feature

The main aim of Flickr data mining is to extract human interaction and spatial patterns information such as area, location, and events. This information will be used to classify using the data mining technique. The Flickr geotagged data were classified into two categories urban/city related and normal events.

Naïve Bayes classifier was implemented. The feature description and word frequencies from bag-of-words of two class, including urban/city scenes events and non-urban scenes, have computed the probability of each category, as well as the conditional probability of each word given a category.

The principle of probabilistic minimum risk classification is used to place each input in a category. Each word is represented by the previously described features or bag of words extracted from the Flickr information. The simple but effective naïve Bayes classifier was used because it is fast to train and not sensitive to irrelevant features [12, 18]. Figure 8.5. Shows Mapping of Flickr information that is classified as urban or city area.

Model evaluation was performed using the kappa coefficient and overall accuracy. The availability of classification accuracy depended on the training dataset size, which obtain from Global Land Cover information: an increase in the training dataset size increased the accuracy [18]. Thus, more training data may further improve classification. We obtained an overall accuracy, kappa coefficient of 90.2%, and 0.7183, respectively (see Table 8.2).

Spatial Distribution Analysis

Analysis of the distribution of the Flickr location can determine the pattern density using kernel density analysis. At the thickness (heatmap) or hotspot from Flickr usage with a radius from the point of interest can imply frequency of location that user check-in or travel to these particular places.

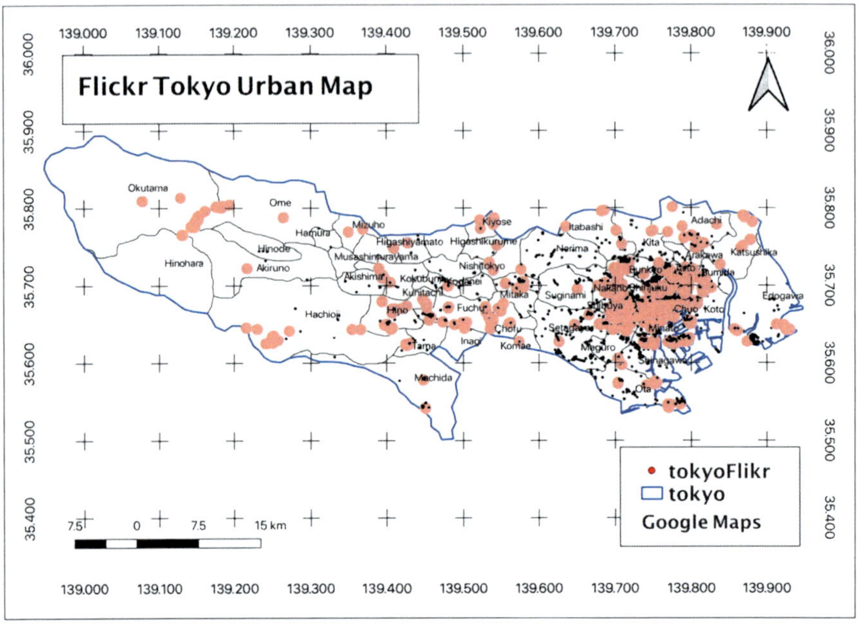

Fig. 8.5 Urban/city area classification map

Table 8.2 A model evaluation from Flickr classification

Predicted class	Observed class		
	Class of urban/city	Non-class of urban/city	Class precision
Class of urban/city	70.32	3.32	96.88%
Non-class of Urban/city	29.68	96.68	100
Class recall	100%	90.91%	90.20%

After classifying individual Flickr post and kernel density analysis, the tags messages from Flickr's user that related to the developed pattern are distributed in the city, as shown in Fig. 8.6. However, it can be assumed that the urban areas have the highest distribution of crowdsourced Flickr heat maps due to rush hours or commercial areas and residential areas. This agreement could be the result due to the density and distribution of available geotagged Flickr data.

The result shows that the highest density of Flickr information in Tokyo is located at popularity tourist places or areas such as Shinjuku, Shibuya, Harajuku, and Ropongi. Therefore, it can be confirmed that the spatial distribution of crowdsourcing has a significant relationship with image and opinion that users posted in social media.

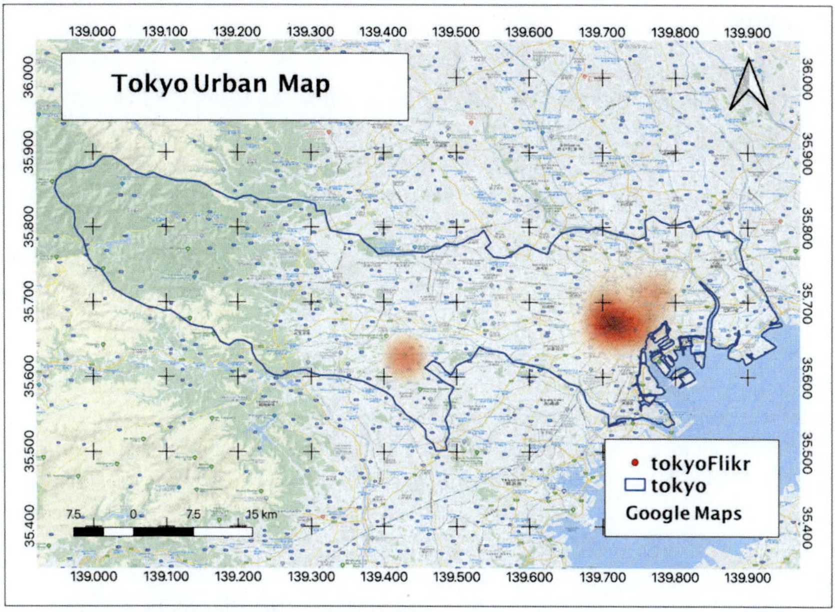

(a)

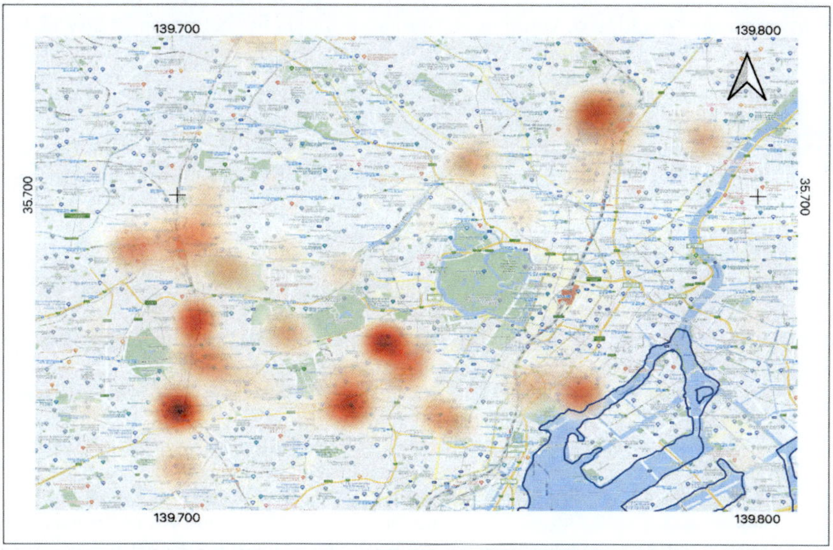

(b)

Fig. 8.6 **a** The spatial distribution of urban Flickr information, **b** zoom area of interest was the most clustering in Tokyo

Conclusion

This paper aims to implement urban/city area estimation using Flicker's images. We employ data mining techniques to discover image descriptions and tags information, which are relevant to natural scene and opinion feature extraction. The salient points of this study can be summarized as follows: Over a hundred thousand geotagged images of Flickr have been queried and download according to their geographic and temporal information via Flickr API. We show that these datasets play a significant source in terms of geographical explorer. We contribute to extract images and text contents for smart cities and sustainable development concepts. We build a framework based on natural scenes classification using social sensing image descriptions and bag of words from tags information by leveraging low-level image analysis techniques including color detection, edge orientation, and vegetation index.

Overall accuracy and evaluation criteria show that aggregate image features and text descriptions provide the best performance. However, we plan to use advanced deep learning techniques [19], temporal mining sequential trajectory patterns [15], and multi-lingual crowdsourcing data in order to improve the efficiency of the classification task to overcome the data limitation issue of this study. Finally, our implementation provides a novel direction to investigate that social sensing dataset can provide valuable knowledge to integrate social sensing datasets for sustainable tourism planning and to develop development policies for better smart city [20–22].

References

1. J. Wang, D.J. Crandall, Observing the natural world with Flickr, in *The IEEE International Conference on Computer Vision (ICCV) Workshops* (2013), pp. 452–459
2. Y. Liu, X. Liu, S. Gao, L. Gong, C. Kang, Y. Zhi, G. Chi, L. Shi, social sensing: a new approach to understanding our socioeconomic environments. Ann. Assoc. Am. Geogr. **105**(3), 512–530 (2015)
3. H. Chourabi, T. Nam, S. Walker, J.R. Gil-Garcia, S. Mellouli, K. Nahon, T.A. Pardo, H.J. Scholl, Under-standing smart cities: an integrative framework, in *Proceedings of the 45th Hawaii International Conference on System Sciences, IEEE Computer Society*, Washington, DC (2012), pp. 2289–2297
4. T. Arreeras, M. Arimura, T. Asada, S. Arreeras, Association rule mining tourist-attractive destinations for the sustainable development of a large tourism area in Hokkaido using Wi-Fi tracking data. Sustainability **11**(14), 3967 (2019)
5. K. Sugimoto, K. Ota, S. Suzuki, Visitor mobility and spatial structure in a local urban tourism destination: GPS tracking and network analysis. Sustainability **11**, 919 (2019)
6. D.M. Tank, Improved Apriori algorithm for mining association rules. Int. J. Inf. Technol. Comput. Sci. **6**, 15–23 (2014)
7. K.J. Fietkiewicz, W.G. Stock, How "smart" are Japanese cities? An empirical investigation of infrastructures and governmental programs in Tokyo, Yokohama, Osaka, and Kyoto, in *The 48th Hawaii International Conference on System Sciences*, Kauai, HI, pp. 2345–2354 (2015)
8. E.H. Alkhammash, J. Jussila, M.D. Lytras, A. Visvizi, Annotation of smart cities twitter micro-contents for enhanced citizen's engagement. IEEE Access **7**, 116267–116276 (2019)
9. A. Bosch, X. Muñoz, R. Martí, Which is the best way to organize/classify images by content? Image Vis. Comput. **25**, 778–791 (2007)

10. J. Xiao, J. Hays, B.C. Russell, G. Patterson, K. Ehinger, A. Torralba, A. Oliva, Basic level scene understanding: categories, attributes and structures. Front. Psychol. **4**, 506 (2013)
11. A. Sitthi, Tags mining analysis using geotagged online social media data. J. Soc. Sci. Srinakharinwirot Univ. **21**(1), 304–319 (2018)
12. J. Xiao et al., Basic level scene understanding: categories, attributes and structures. Front. Psychol. **4**, 1–10 (2013)
13. H. Mahgoub et al., A text mining technique using association rules extraction. Int. J. Comput. Electr. Autom. Control Inf. Eng. **2**(6), 2044–2051 (2008)
14. G. Cai, K. Lee, I. Lee, Mining semantic trajectory patterns from geo-tagged data. J. Comput. Sci. Technol. **33**(4), 849–862 (2018)
15. G. Cai, C. Hio, L. Bermingham, K. Lee, I. Lee, Mining frequent trajectory patterns and regions-of-interest from Flickr photos, in *2014 47th Hawaii International Conference on System Sciences* (2014)
16. I. Lee, G. Cai, K. Lee, Mining points-of-interest association rules from geo-tagged photos, in *46th Hawaii International Conference on System Sciences* (2013)
17. P. Nancy, R.G. Ramani, S.G. Jacob, Mining of association patterns in social network data (Face Book 100 Universities) through data mining techniques and methods, in *Advances in Intelligent Systems and Computing* (2013), pp. 107–117
18. A. Sitthi et al., Exploring land use and land cover of geotagged social-sensing images using naive Bayes classifier. Sustainability **8**(9), 921 (2016)
19. N.N. Samany, Automatic landmark extraction from geo-tagged social media photos using deep neural network. Cities **93**, 1–12 (2019). https://doi.org/10.1016/j.cities.2019.04.012
20. A. Visvizi, M.D. Lytras, Sustainable smart cities and smart villages research: rethinking security, safety, well-being, and happiness. Sustainability **12**, 215 (2020)
21. A. Visvizi, M.D. Lytras (eds.), *Smart Cities: Issues and Challenges: Mapping Political, Social and Economic Risks and Threats* (Elsevier, 2019). ISBN: 9780128166390. https://www.elsevier.com/books/smart-cities-issues-and-challenges/lytras/978-0-12-816639-0
22. M.D. Lytras, A. Visvizi, Who uses smart city services and what to make of it: toward interdisciplinary smart cities research. Sustainability **10**(6), 1998 (2018)

Chapter 9
Salerno Telling: An Approach for Cultural Heritage Enhancing

Fabio Clarizia, Francesco Colace, Massimo De Santo, Marco Lombardi, and Domenico Santaniello

Abstract The cultural tourism market is evolving towards a dimension of complete fulfilment of the tourist's needs. On the one hand, it is enhancing the centrality of the cultural aspect in a 360° travel experience and on the other is paying attention to the choice of one's journey, completed by, according to each context, all needed services. The information overload, the substantial changes of the traveller's decisions and the enlargement of the tools of Knowledge available to each connected user are today's main levers of change and require the use of Context-Aware Systems. In this scenario, a context-aware framework for the fruition of e-Tourism contents and services will be discussed. The proposed system is able of supporting not only tourists but also Public Institutions and professionals, through the dynamic design of core and services for tourism promotion. In order to test the proposed methodology, a prototype has been developed, obtaining promising results.

Keywords Context-aware computing · e-tourism · Cultural heritage applications

Introduction

Tourism and cultural heritage are crucial resources for the enhancement of a territory. Modern tourism should be able to meet the needs of tourists and, at the same

F. Clarizia · F. Colace (✉) · M. De Santo · M. Lombardi · D. Santaniello
DIIN, University of Salerno, Fisciano, SA, Italy
e-mail: fcolace@unisa.it

F. Clarizia
e-mail: fclarizia@unisa.it

M. De Santo
e-mail: desanto@unisa.it

M. Lombardi
e-mail: malombardi@unisa.it

D. Santaniello
e-mail: dsantaniello@unisa.it

A. Visvizi et al. (eds.), *Research and Innovation Forum 2020*,
Springer Proceedings in Complexity,
https://doi.org/10.1007/978-3-030-62066-0_9

time, the host regions by protecting and improving opportunities for the future. All resources should be managed efficiently by satisfying economic and social needs while respecting the territory [1]. Many countries rich in cultural heritage require this type of tourism aimed at sustainable development. In particular, Italy is a country rich in numerous archaeological, cultural, artistic and architectural sites spread throughout its territory. Thus, adequate planning of the cultural services offers, that can make goods easily accessible is necessary [2–4]. Firstly, this implies making museums, archaeological sites and libraries accessible; secondly, upgrading the service network that may promote fruition, information, communication and tourism receptiveness [5, 6].

A way of adding value to these services is making them highly customizable and capable of providing, using new technologies, more immersive and stimulating fruition of information [7–9]. Because of that, it is necessary to use systems capable of gathering and interpreting information from context to adapt to both the user (ex. to his preferences) and the system use situation (location and occupation of the user or available resources close by) [10]. In this scenario, the context may be used to create applications [11, 12] that can adapt their graphical interface, filter data that is relevant, upgrade data gathering precision, discover services, predicting some of the user's choices or create a smart environment [13–15].

Regarding cultural heritage, for example, a context-aware mobile app provided to a museum visitor can use context [16, 17] to adapt the structure and aspect of the graphical interface to the user's preferences, whether he is an elder or a child [18]; it can provide different info according to the visitor's interests or location [19]; it can learn from previous choices made by the user his preferred information [20]; it can suggest certain services like the purchase of tickets for a temporary exhibition or the reservation for a show about dinosaurs that is about to begin in the adjacent room [21]; it can calculate the user's location based on specific sensors that monitor the environment [22]; it can provide up to date content on what is happening in a specific context [23].

Moreover, the user's experience, thanks to the mobility and the territorial and thematic localization, can renew itself using the physical environment like background and plot on which basis structuring and defining stories that wind along with the urban topography. In particular, people (tourists and citizens) and objects (buildings, rooms and sculptures) equipped with appropriate devices (GPS, smartphone, video camera, temperature/humidity sensors) constitute a particular social network in which all the mentioned entities can communicate.

In the following section, a framework will be discussed, based on context rendering models, for the provision of advanced services in the tourism and cultural heritage fields. The intention is of bettering the touristic path projecting him inside the experience, through a system that can not only perceive the entire context but also react to it while providing accurate feedback to the user (regarding services he may need). Provided, for example, that the weather is terrible for the entire duration of the trip, the system will automatically rule out any outdoor activity and will suggest indoor services.

A Context-Aware Architectural Framework

The purpose of this work is to present a framework that shows a high level of context-awareness. It should be understood as a set of technical features capable of providing added value to services in different cultural scenarios [24, 25].

As shown in Fig. 9.1, the proposed framework is composed of different blocks that perform the following main functions:

- Data gathering and service establishment.
- Context rendering and use of said context for the promotion of contents and services.
- Introduction of selected contents/services.

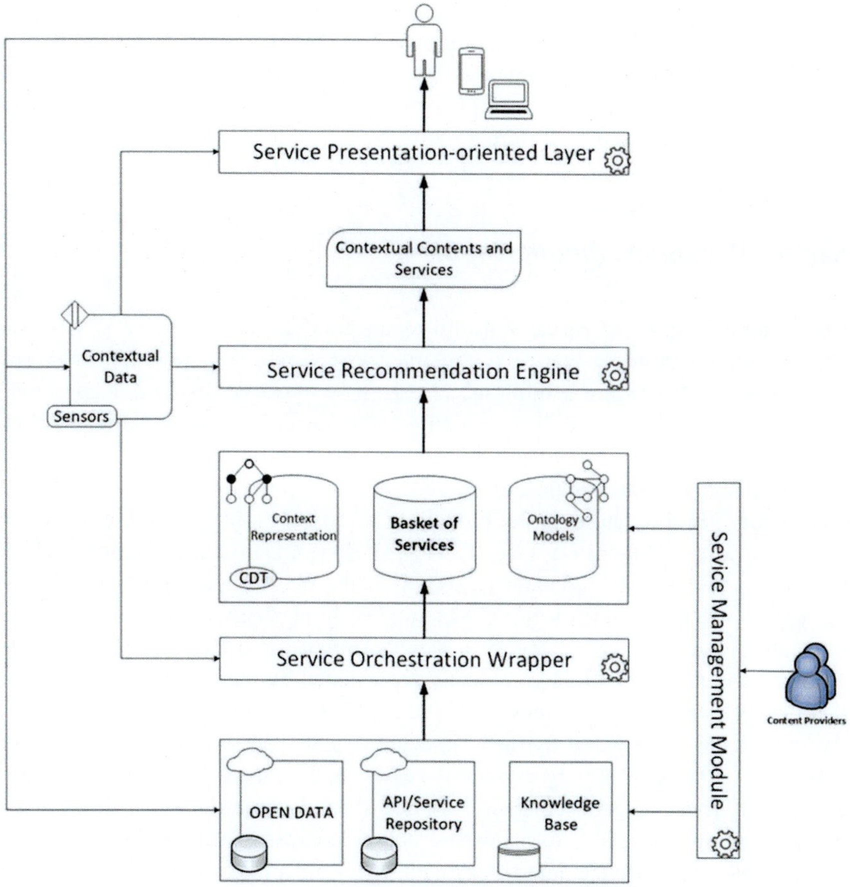

Fig. 9.1 A Context-Aware Framework for e-Tourism

Service Orchestration Wrapper

The service wrapper uses context together with mash-up techniques capable of recovering information from different sources and overcoming content limitations and "pre-packaged" services [26], whose structures must, instead, vary based on the situation during which the fruition happens. The main objective is the capability of identifying appropriate sources, selected according to their suitability to the user's needs, integrating data and services through a single interface.

An example of the benefits is the integration of main content and services to support services for the tourist (Consumer Mashup), like maps or weather info, and/or business services through API (Business Mashup), like booking or purchasing a museum ticket, allowing to upgrade the user experience and the accessibility of content and services. Another example is the use of more than one source for the presentation of the "same" information. For example, data on cultural heritage acquired from the Base Knowledge or through Open Data (Data Mashup): the information is combined to obtain a complete picture of the entity (such as work of art or site of interest) analysed by different tools [27–30].

Service Recommendation Engine

By defining a series of services useful to the tourist (service basket), it is made necessary to integrate the context to the process of recommendation to provide valid suggestions under certain conditions. That is the purpose of the service recommendation engine. For example, if it is raining and the user is walking, he would prefer to receive recommendations of sites of interest that are not too far away.

A deep understanding of the context problem is essential. For this reason, the main contribution, in this phase, is the modelling of context though graphic formalism, like the Context Dimension Tree (CDT) [31, 32]. CDT is a tree composed of a triad <r; N; A> where r indicates its root, N is the set of nodes of which it is made of, and A is the set of arcs joining these nodes. In particular, nodes present within CDT are divided into two categories, namely dimension nodes and concept nodes. A dimension node, which is graphically represented by the colour black, is a node that describes a possible dimension of the application domain. A concept node, on the other hand, is depicted by the colour white and represents one of the possible values that a dimension may assume. In essence, a context is defined as an "and" among different context elements. This tree-shaped model, along with the use of the field ontologies, is able to both represent the context in its main dimensions and interrogate each reference database efficiently to select the right services.

The proposed methodology is divided into three main phases. The first phase deals with the design of the Context Tree, where it is possible to identify relevant context elements for the application considered. The second phase, which deals with the definition of partial views, where it is possible to match each one to a different set of services. The third phase of the composition of global views, where it is possible to elaborate the final responses to the queries. Thus, it is possible, for example, to create a custom itinerary of the visit according to the context and the user through the selection of appropriate services for:

- Planning the itinerary of a trip, customizable in a few steps, and facilitating transportation; during this phase the client can get a taste of the different steps of the itinerary, providing detailed information on the available activities.
- Organising visits to museums or art location, preferably without having to queue and with custom discounts.
- Re-planning dynamically the itinerary basing it on context and user behaviour or emergencies (automatic "travel assistant" services that can suggest actions and modifications to the itinerary).
- Promoting the discovery of unusual places marked by incredible beauty, high cultural value and enogastronomical heritage, the leading example of Made in Italy excellence.

Service Presentation-Oriented Layer

The presentation-oriented layer uses context to adapt the presentation of content and services according to the user and the different channels and devices. Mainly, this form combines context recognition and digital storytelling techniques [33] that can provide, through a custom narrative, the information related to the location.

The aim is to make results available like a modern tourist guide that, thanks to a "storytelling engine" [34], can provide dynamically different narrative contents, in the form of services, accordingly integrated, not necessarily predetermined and adherent to needs and behaviours of the users [35].

The "digital narration" includes:

- Information on the location of the visit (main characteristics and historical notions): the story narrated in the first person by the host (memories, autobiography, family traditions) and stories lived or set in the locations where we are located (novels, legends, songs, movies, historical events);
- Points of interest for the user with all related services, filtered by category and accompanied by in-depth multimedia information;
- Experiences lived by other users, like authentic testimonies of the destination. The site of cultural interest users could be involved in the making of new digital resources (stories/comments, images and videos) that, stimulated, gathered and framed in the best way possible, will contribute to the enrichment and development of new compelling and personal stories.

Experimental Results

An applicative prototype has been developed based on the proposed framework. It is a hybrid mobile app, designed and implemented, together with a server-side component, following the previous description. The App has been created to support tourists that are visiting Salerno, a city of the Campania region in Italy.

In this experimental phase, the main potentially useful API services and contents for the tourists were identified. The App collects information also from social environments adapting the proposed itinerary taking into account the user's communities and interests.

The 70 participants are between 18 and 52 years old and do not know the subject of the study. They are all subscribed to the social network Facebook and own a mobile device. The App, which requires the login by Facebook credentials, has been installed by all participants. After having interacted with the App, participants have answered, according to the Likert scale, to five sentences, summarized here:

1. The system provides correct information on the location of the visit, based on personal preference and current context
2. The system has successfully adapted to the context changes
3. The suggested services are adequate and satisfy the tourist's needs
4. The information was presented adequately, in the form of a modern tourist guide
5. The experience of friends and other users has been useful.

To every sentence, five possible answers have been associated: I totally disagree—TD, I disagree—D, Undecided—U, I agree—A, I totally agree—TA.

Table 9.1 presents a synthesis of the answers of the participants to each sentence.

As shown in this table, of the 70 participants who have interacted with the application, many agree and/or strongly agree that the system gives appropriate contextual information and services about the place and its points of interests, it satisfies the tourist's needs, and the experiences of friends are interesting.

Figure 9.2 shows, in a graphic form, the results obtained from the proposed questionnaire and the positivity of the answers concerning the total. The users show great appreciation for the App: in general, they have appreciated the proposed contents and services.

Table 9.1 Experimental results

Statement	Answer				
	TD	D	U	A	TA
1	1	3	6	38	22
2	0	4	4	32	30
3	1	8	5	31	25
4	2	1	12	46	9
5	3	2	5	44	16

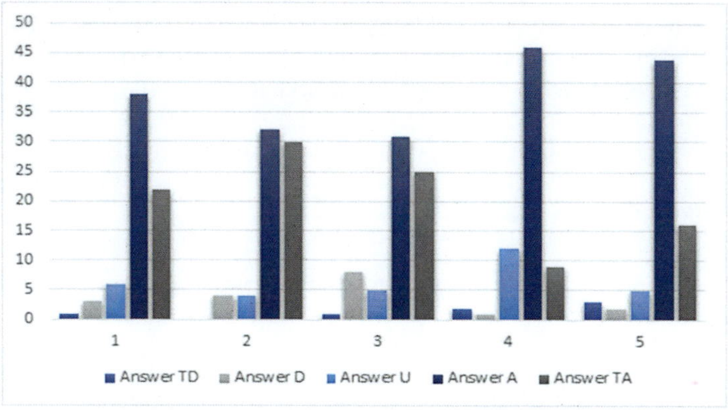

Fig. 9.2 Analysis of results

All the assertions were assessed positively, however, in the case of sentence 4, although satisfied users have not expressed complete satisfaction with the adequate presentation of the content. This type of reaction could be due to the poor presentation of the content. However, the proposed system, being a prototype, has not been particularly careful with the visual appearance and presentation of the content. In order to improve the proposed application, it is necessary to make improvements in the presentation, better integrating the contents and transforming them better into a modern tourist guide. Besides, another aspect to consider could be a higher integration with Open Data and API.

Conclusions

In this document, it was presented a framework capable of favouring logic of integration and interoperability among platforms (existing or new) in order to allow an automatic construction that adapts to the highly customizable context of services, that overcomes the informative phase, facilitating the tourist during each step of his experience.

In this scenario, the proposed framework can be declined in different mobile applications that can behave like a modern tourist guide and follow the user in every step, creating custom paths and content based on variables or contingencies that may occur during the trip.

First experimental results demonstrate the ability of the framework to be effective. However, the results show that the prototype could be better implemented mainly in the presentation of information. Furthermore, an improvement in the integration with Open Data and API services could be useful. Future work activities include the improvement of the developed prototype, especially in the presentation of information and integration with many other services. A more robust experimental campaign that could be conducted in different cities is needed.

References

1. C. K. de Viron, G. Mudri, Integrated approach to sustainable EU smart villages policies, in *Smart Villages in the EU and Beyond* (Emerald Publishing Limited, 2019), pp. 13–27. https://doi.org/10.1108/978-1-78769-845-120191003
2. F. Colace, M. De Santo, S. Lemma, M. Lombardi, A. Rossi, A. Santoriello, A. Terribile, M. Vigorito, How to describe cultural heritage resources in the web 2.0 era? in *Proceedings—11th International Conference on Signal-Image Technology and Internet-Based Systems, SITIS 2015*, 2016. https://doi.org/10.1109/SITIS.2015.50
3. F. Colace, S. Lemma, M. Lombardi, F. Pascale, A context aware approach for promoting tourism events: the case of artist's lights in Salerno, 2017. https://doi.org/10.5220/0006370007520759
4. A. Visvizi, M. Lytras, It's not a fad: smart cities and smart villages research in European and global contexts. Sustainability **10**(8), 2727 (2018). https://doi.org/10.3390/su10082727
5. A. Augello, A. Santangelo, S. Sorce, G. Pilato, A. Gentile, A. Genco, S. Gaglio, A multimodal interaction guide for pervasive services access, in *2007 IEEE International Conference on Pervasive Services, ICPS*, 2007. https://doi.org/10.1109/PERSER.2007.4283923
6. S. Hartman, B. Wielenga, J.H. Heslinga, The future of tourism destination management: building productive coalitions of actor networks for complex destination development. J. Tour. Futur. (2020). https://doi.org/10.1108/JTF-11-2019-0123
7. P. Lops, M. De Gemmis, G. Semeraro, C. Musto, F. Narducci, M. Bux, A semantic content-based recommender system integrating folksonomies for personalized access. Stud. Comput. Intell. (2009). https://doi.org/10.1007/978-3-642-02794-9_2
8. D. Capriglione, M. Carratù, M. Ferro, A. Pietrosanto, P. Sommella, Estimating the outdoor PM10 concentration through wireless sensor network for smart metering (Springer, Cham, 2019) pp. 399–404. https://doi.org/10.1007/978-3-030-04324-7_49.
9. A. Visvizi, J. Jussila, M.D. Lytras, M. Ijäs, Tweeting and mining OECD-related microcontent in the post-truth era: a cloud-based app. Comput. Human Behav. **107**, 105958 (2020). https://doi.org/10.1016/j.chb.2019.03.022
10. J. Zhuang, T. Mei, S. C. H. Hoi, Y. Q. Xu, S. Li, When recommendation meets mobile: Contextual and personalized recommendation on the go, in *UbiComp'11—Proceedings of the 2011 ACM Conference on Ubiquitous Computing*, 2011. https://doi.org/10.1145/2030112.2030134
11. A.K. Dey, Understanding and using context. Pers. Ubiquitous Comput. (2001). https://doi.org/10.1007/s007790170019
12. M. Raento, A. Oulasvirta, R. Petit, H. Toivonen, ContextPhone: a prototyping platform for context-aware mobile applications. IEEE Pervasive Comput. (2005). https://doi.org/10.1109/MPRV.2005.29
13. F. Colace, M. Lombardi, F. Pascale, D. Santaniello, A. Tucker, P. Villani, MuG : a multilevel graph representation for big data interpretation, in *IEEE 20th International Conference on High Performance Computing and Communications; IEEE 16th International Conference on Smart City; IEEE 4th International Conference on Data Science and Systems*, 2018, pp. 1410–1415. https://doi.org/10.1109/HPCC/SmartCity/DSS.2018.00233
14. J. Jin, J. Gubbi, S. Marusic, M. Palaniswami, An information framework for creating a smart city through internet of things. IEEE Internet Things J. (2014). https://doi.org/10.1109/JIOT.2013.2296516
15. F. Colace, M. De Santo, M. Lombardi, F. Pascale, D. Santaniello, A. Tucker, A multilevel graph approach for predicting bicycle usage in London area, in *Fourth International Congress on*

Information and Communication Technology. Advances in Intelligent Systems and Computing, vol. 1027, ed. by X.S. Yang, S. Sherratt, N. Dey, A. Joshi (Springer, Singapore, 2020), pp. 353–362. https://doi.org/10.1007/978-981-32-9343-4_28
16. A. Dix, T. Rodden, N. Davies, J. Trevor, A. Friday, K. Palfreyman, Exploiting space and location as a design framework for interactive mobile systems. ACM Trans. Comput. Interact. (2000). https://doi.org/10.1145/355324.355325
17. F. Colace, M. De Santo, M. Lombardi, R. Mosca, D. Santaniello, A multilayer approach for recommending contextual learning paths. J. Internet Serv. Inf. Secur. **2**, 91–102 (2020). https://doi.org/10.22667/JISIS.2020.05.31.091
18. J. Fink, A. Kobsa, User modeling for personalized city tours. Artif. Intell. Rev. (2002). https://doi.org/10.1023/A:1016383418977
19. L. Ardissono, A. Goy, G. Petrone, M. Segnan, P. Torasso, Intrigue: personalized recommendation of tourist attractions for desktop and hand held devices. Appl. Artif. Intell. (2003). https://doi.org/10.1080/713827254
20. Q. Yang, A novel recommendation system based on semantics and context awareness. Computing (2018). https://doi.org/10.1007/s00607-018-0627-4
21. J. Budzik, K.J. Hammond, User interactions with everyday applications as context for just-in-time information access, in *International Conference on Intelligent User Interfaces, Proceedings IUI*, 2000. https://doi.org/10.1145/325737.325776
22. G. Ghiani, F. Paternò, C. Santoro, L.D. Spano, UbiCicero: a location-aware, multi-device museum guide. Interact. Comput. (2009). https://doi.org/10.1016/j.intcom.2009.06.001
23. D. Gavalas, M. Kenteris, A web-based pervasive recommendation system for mobile tourist guides. Pers. Ubiquitous Comput. **15**(7), 759–770 (2011). https://doi.org/10.1007/s00779-011-0389-x
24. M. Casillo, F. Clarizia, F. Colace, M. Lombardi, F. Pascale, D. Santaniello, An approach for recommending contextualized services in e-Tourism. Information **10**(5), 180 (2019). https://doi.org/10.3390/info10050180
25. F. Colace, M. De Santo, M. Lombardi, D. Santaniello, CHARS: a cultural heritage adaptive recommender system, in *Proceedings of the 1st ACM International Workshop on Technology Enablers and Innovative Applications for Smart Cities and Communities—TESCA'19*, 2019, pp. 58–61. https://doi.org/10.1145/3364544.3364830
26. V. Cassani, S. Gianelli, M. Matera, R. Medana, E. Quintarelli, L. Tanca, V. Zaccaria, On the role of context in the design of mobile mashups, in *Communications in Computer and Information Science*, 2017. https://doi.org/10.1007/978-3-319-53174-8_7
27. S. Heo, S. Woo, J. Im, D. Kim, IoT-MAP: IoT mashup application platform for the flexible IoT ecosystem, in *2015 5th International Conference on the Internet of Things (IOT)*, 2015, pp. 163–170. https://doi.org/10.1109/IOT.2015.7356561
28. G. D'Aniello, M. Gaeta, V. Loia, F. Orciuoli, S. Tomasiello, A dialogue-based approach enhanced with situation awareness and reinforcement learning for ubiquitous access to linked data, in *2014 International Conference on Intelligent Networking and Collaborative Systems*, 2014, pp. 249–256. https://doi.org/10.1109/INCoS.2014.73
29. S. Tomasiello, M. Gaeta, V. Loia, Quasi–consensus in second–order multi–agent systems with sampled data through fuzzy transform. J. Uncertain Syst. 2016
30. F. Amato, A. Mazzeo, A. Penta, A. Picariello, Knowledge representation and management for E-government documents, in *E-Government ICT Professionalism and Competences Service Science* (Springer, Boston, MA, US) pp. 31–40. https://doi.org/10.1007/978-0-387-09712-1_4
31. C. Bolchini, C.A. Curino, E. Quintarelli, F.A. Schreiber, L. Tanca, Context information for knowledge reshaping. Int. J. Web Eng. Technol. (2009). https://doi.org/10.1504/IJWET.2009.025015
32. F. Colace, M. Lombardi, F. Pascale, D. Santaniello, A multilevel graph representation for big data interpretation in real scenarios, in *Proceedings—2018 3rd International Conference on System Reliability and Safety, ICSRS 2018*, 2019. https://doi.org/10.1109/ICSRS.2018.8688834

33. F. Pittarello, A simple story: using an agents' based context-aware architecture for storytelling (Springer, Berlin, 2006), pp. 25–36. https://doi.org/10.1007/11944577_3
34. M. Casillo, F. Clarizia, G. D'Aniello, M. De Santo, M. Lombardi, D. Santaniello, CHAT-Bot: a cultural heritage aware teller-bot for supporting touristic experiences. Pattern Recognit. Lett. (2020). https://doi.org/10.1016/j.patrec.2020.01.003
35. Y. Ioannidis, K. El Raheb, E. Toli, A. Katifori, M. Boile, M. Mazura, One object many stories: introducing ICT in museums and collections through digital storytelling, in *Proceedings of the DigitalHeritage 2013—Federating the 19th Int'l VSMM, 10th Eurographics GCH, and 2nd UNESCO Memory of the World Conferences, Plus Special Sessions fromCAA, Arqueologica 2.0 et al.* 2013. https://doi.org/10.1109/DigitalHeritage.2013.6743772

Chapter 10
A Recommender System for Enhancing Coastal Tourism

Mario Casillo, Francesco Colace, Massimo De Santo, Marco Lombardi, Rosalba Mosca, and Domenico Santaniello

Abstract Nowadays, the coastal and maritime tourism sector is playing an increasingly important role thanks to the extraordinary beauty, variety and cultural richness of the coastlines. These represent distinctive places, which allows planning unique pathways able to attract different types of users. In this scenario, the Amalfi coast contains cities with an important touristic-cultural value. However, at the same time, not having the same investments in the tourism sector as in big cities, it needs to be enhanced through innovative methodologies. This paper aims to propose a proper methodology with a high-degree Context-Awareness able offering tailored services through a Recommender System for enhancing coastal and maritime tourism. The main innovation features concern the information content available to end-users (itineraries, points of interest, etc.), which present three different aspects: context representation, data management, inferential engines.

Keywords Context-awareness · e-tourism · Recommender systems

M. Casillo · F. Colace · M. De Santo · M. Lombardi · R. Mosca · D. Santaniello (✉)
DIIN, University of Salerno, Fisciano, SA, Italy
e-mail: dsantaniello@unisa.it

M. Casillo
e-mail: mcasillo@unisa.it

F. Colace
e-mail: fcolace@unisa.it

M. De Santo
e-mail: desanto@unisa.it

M. Lombardi
e-mail: malombardi@unisa.it

R. Mosca
e-mail: rmosca@unisa.it

A. Visvizi et al. (eds.), *Research and Innovation Forum 2020*,
Springer Proceedings in Complexity,
https://doi.org/10.1007/978-3-030-62066-0_10

Introduction

The tourism sector has a significant impact on the economic and social aspects of modern society, and the positive effects of this phenomenon are diverse. Tourism development brings economic benefits and advantages for entire geographical areas. To have well-developed and organised tourist areas available means that both residents and guests can enjoy the services [1]. In this field, the coastal and maritime tourism sector plays an increasingly essential role thanks to the extraordinary attractiveness, variety and cultural richness that the coasts are able to offer. The coasts attract tourists in search of reserved places, with a wide variety of culture and allowing them to design unique and tailored itineraries. Seas and coasts strongly characterise the Italian tourist scenario. The coastal development of the peninsula and its islands is about 7500 km long and presents very different sceneries and landscapes: low and sandy coasts, high and rocky coasts, jagged and gulfs where an infinite variety of landscapes and sceneries follow one another. It is not surprising, therefore, that coastal tourism represents a natural consequence of Italian geomorphology. The coastline has always had a strong evocative and symbolic power. The presence of the sea defines daily rhythms, seasons, social and cultural life of a city. In many cases, the coastal cities have become touristic cities. An example of such a reality is the Amalfi Coast. The Amalfi Coast is one of the 55 Italian sites included by UNESCO in the World Heritage List. A unique environment protected by UNESCO as a perfect example of Mediterranean landscape with a scenery of great cultural and natural value due to its topographical features and its historical evolution. The entire area extends between the Gulf of Naples and the Gulf of Salerno and includes several municipalities in the province of Salerno. The cities and towns of the Coast are rich in variety; each one has its traditions and peculiarities that make it unique and are all characterised by architectural monuments of historical and cultural value. Due to the presence of variety and the number of monuments and cultural heritage present, an adequate management of the offers for cultural services is necessary that would make them easy to use [2–4]. To add value to such services, it may be necessary to make them highly customizable, allowing users to live tailor-made, highly cultural but unique experiences. This can be achieved by exploiting new paradigms such as Big Data [5] and the Internet of Things (IoT) [6]. We live in the Big Data era, a vast amount of data that is processed for different purposes [7, 8]. This data is generated by mobile and intelligent IoT devices that can be used for various applications [9–11]. Not forgetting the security issues that such devices may pose [12–14].

These devices and this information could be used to process and interpret personal and contextual information. For this purpose, context can be used to create applications [15, 16] that can filter relevant data by providing the right information at the right time and continuously update [17, 18]. As far as coastal tourism is concerned, for example, a mobile application capable of context analysis can guide a user towards personalised experiences, both according to his interests and other contextual factors, such as weather and location. The app, in addition to personal interests, can adapt to the user's profile [19], differentiating for example between a child and an older person, can learn from previous choices and can provide real-time updates concerning the context [20, 21]. All this would allow the user, through a thematic localisation, to benefit from the physical environment, exploiting personalised paths to which stories about the local culture can be linked. In particular, both the facilities, equipped with appropriate sensors, and the citizens, equipped with appropriate devices, are part of an environment in which it is possible to unite interests, recommending tailor-made paths to be followed according to interests and needs. In the following section, we will describe a model, based on methodologies with context-adaptive and predictive ability, which aim at providing advanced services in the coastal tourism sector. This paper aims to improve the touristic experience through a recommender system that would not only be able to meet the users' needs but also to react to context changing according to weather conditions, timetables and day of the week or time available.

This paper is organized as follows. In section two, we describe the System Architecture of the proposed approach. In section three, we assessed the effectiveness of our methodology through a preliminary test activity. Finally, in section four, we draw conclusions and future research directions.

System Architecture

This work aims to introduce a recommendation framework capable of handling contextual information. This approach aims at sharing different methodologies able to enhance different cultural sectors [22, 23], and in particular, in this case, coastal tourism. As shown in Fig. 10.1, the proposed architecture is composed of three main blocks:

- Data Layer
- Inference Layer
- Information Layer.

Data Layer

The data layer collects various types of information from various sources in order to enable the system to provide more meaningful responses than other services [24]. One of the main objectives of this layer is to identify and exploit appropriate sources, considering their Truthfulness and Value [25]. The acquired data are then harmonized through an orchestration algorithm and passed to the next layer.

The main purpose of this layer is to collect and harmonize information from different sources. Therefore, information related to the main tourist support services such as maps, weather conditions or road traffic are combined. In addition, company service information is integrated via API such as booking and ticket purchase. In addition, open data describing the cultural heritage, or the characteristics of places is also considered open data. All this data is added to the user's preferences and

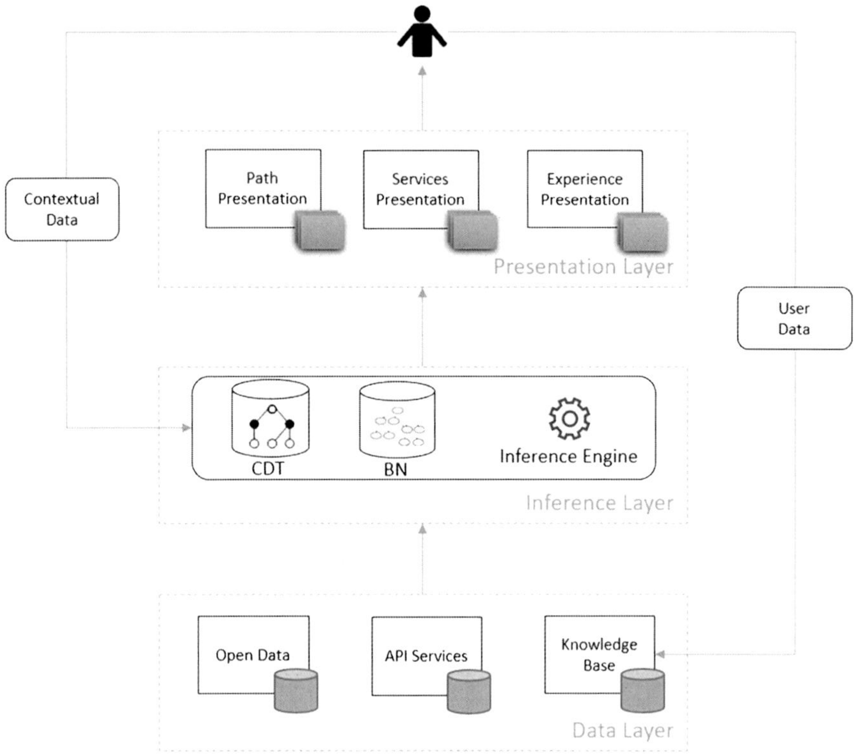

Fig. 10.1 System architecture

processed in order to obtain a complete picture of the points of interest in the territory that can be analyzed by different tools [26, 27].

Inference Layer

In order to define a valid offer to the tourist it is necessary to combine the data obtained from the data layer with the context. The main purpose of the inference layer is to add context to the provided data in order to analyze it and make forecasting operations to recommend the right services to users at the right time. For example, if it is too hot, no user would want to spend an entire afternoon walking in the blazing sun.

One of the main objectives of this phase is to model the context. This is done through the Context Dimension Tree (CDT) [28]. This approach allows to manage the context through a decision tree composed by a triad <r; N; A> where r indicates its root, N is the set of nodes of which it is composed, and A is the set of arcs that join these nodes. In particular, the nodes within CDT are divided into two categories, namely dimensional nodes and conceptual nodes. The dimension nodes describe the possible dimensions of the application domain, the concept nodes collect all the possible values that a dimension can assume. This model is able to represent all possible contexts and query a database efficiently in order to select the right services.

Another main element of this phase is forecasting. Prediction is made through the use of Bayesian networks (BN). BNs are graphical-probabilistic models that represent a set of stochastic variables with their conditional dependencies through the use of a direct acyclic graph (DAG) [29]. Such models exploit the Bayes theorem and are able to predict the probability of occurrence of a given event. Moreover, thanks to their structure, Bayesian networks are able to interface adequately with the CDT [30].

The proposed methodology is divided into three main phases. The first phase is the design of the CDT, which allows us to identify possible application contexts. The second phase deals with the construction of the Bayesian network, which is built through experts in the field and machine learning algorithms, this structure is able to provide us with the conditional probabilities of the nodes that are useful to predict future scenarios. The last phase concerns the interaction of the networks that provide the final answers to users' questions. Through this process it is possible to create custom itineraries that adapt to the context and are periodically updated providing the user with different services:

- Planning of customized routes in a few simple steps. The planning of routes starts in a simplified way, allowing the user to choose among alternatives more suitable to the characteristics of his profile.
- Organization of excursions, museum visits or activities to be carried out according to the time available and avoiding queues, with the possibility of access to customized discounts.
- Dynamic re-programming of the itinerary according to the user's behaviour, context or possible emergencies.
- Adventure itineraries, through which the true local culture of unusual places is enhanced, promoting the coastal and food and wine heritage of local excellence.

Presentation Layer

This layer is designed to present the recommendations of the inferential engine to the user. This information is presented to the user through a simple and intuitive interface with detailed information about the routes and choices made. The aim is to present the customized results as a modern interactive guide combining images and text that make it easy and intuitive to memorize the critical stages of the tourist itinerary.

The interactive guide includes:

- Information about the places of the visit, main features, historical notions;
- Points of interest with any related services;

- Information and testimonials about other users' experiences.

Experimental Results

The validation of the proposed methodology was performed by a prototype, which was developed according to the proposed framework. A hybrid mobile app and a server-side component have been developed and implemented. The App has been designed to support tourists visiting the Amalfi Coast in the Campania region of Italy. In this first experimental phase, some of the main services and points of interest have been identified, and various information on users' preferences and interests have been collected in order to provide tailor-made itineraries. Fifty volunteers aged between 18 and 56 were involved, who were unaware of the purpose of the study. Each participant, on their mobile device, was installed the application, participated in the different itineraries proposed, and after that, they were offered a questionnaire covering different sections:

A. Recommendation
B. Presentation
C. Reliability
D. Usability
E. Performance.

To the assertions of the questionnaire, based on the Likert scale, five possible answers were associated: I totally disagree—TD, I disagree—D, undecided—U, I agree—A, I totally agree—TA. The answers have been collected in Table 10.1.

Besides, for a smaller number of ten participants, a further analysis was conducted to appreciate the system's ability to recommend. In practice, participants, different from previous attendees, were allowed to participate in different routes that the system proposed also based on the experience of previous users. New users could also assess whether or not some of the proposed services and activities were relevant to the context and their needs at that time. The results were collected in the form of a confusion matrix in Table 10.2.

Table 10.1 Questionnaire answers

Section	Answer				
	TD	D	U	A	TA
A	5	0	10	78	57
B	7	9	17	75	42
C	0	21	27	66	36
D	6	0	30	63	51
E	7	10	15	72	48

Table 10.2 Confusion matrix

Predictions	References				
	S1	S2	S3	A1	A2
S1	58	8	2	6	0
S2	7	37	9	4	5
S3	0	6	43	5	7
A1	8	5	5	37	1
A2	0	7	3	4	48
Overall Accuracy: 70, 79%					

As can be seen from Table 10.1 of the 50 participants many agree that the system provides contextual information and appropriate services on the site and its points of interest, meets the needs of the tourist, and the experiences. In Fig. 10.2, we can see, in graphic form, the results obtained from the proposed questionnaire and the positivity of the answers related to the total. Therefore, users show an excellent appreciation for the app: in general, they appreciated the contents and services proposed. Moreover, according to what reported by the confusion matrix (Table 10.2) the system has an overall accuracy higher than 70%, which is very encouraging and could improve over time, based on the increase of experimental data available.

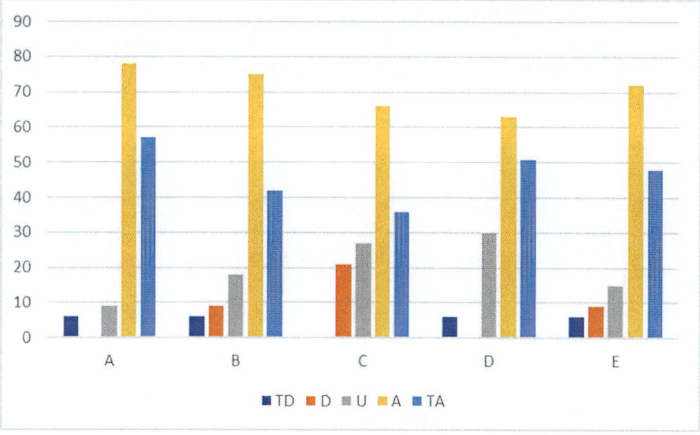

Fig. 10.2 Questionnaire answers trend

Conclusions

In this paper, a framework has been presented to recommend itineraries, services, activities related to coastal tourism tours. This work aims to provide highly customizable and tailored services, making the experience unique and supporting the tourist during each phase of his experience. The innovation of the recommendation system presented lies in the use of a high degree of context-awareness.

The proposed architecture could be declined in different contexts and mobile applications. The experimental results are promising and encouraging, showing that the system is able to recommend appropriately and that the prototype developed is efficient from several points of view such as recommendation and service presentation, reliability, usability and performance shown. Future developments include improvements to the developed prototype and a more consistent and comprehensive experimental campaign.

References

1. A. Visvizi, M. Lytras, It's not a fad: smart cities and smart villages research in European and global contexts. Sustainability **10**(8), 2727 (2018). https://doi.org/10.3390/su10082727
2. M. Casillo, F. Clarizia, G. D'Aniello, M. De Santo, M. Lombardi, D. Santaniello, CHAT-Bot: a cultural heritage aware teller-bot for supporting touristic experiences. Pattern Recognit. Lett. (2020). https://doi.org/10.1016/j.patrec.2020.01.003
3. M. Casillo, F. Clarizia, F. Colace, M. Lombardi, F. Pascale, D. Santaniello, An approach for recommending contextualized services in e-tourism. Information **10**(5), 180 (2019). https://doi.org/10.3390/info10050180
4. O. Wolski, M. Wójcik, Smart villages revisited: conceptual background and new challenges at the local level, in *Smart Villages in the EU and Beyond*, Emerald Publishing Limited, 2019, pp. 29–48. https://doi.org/10.1108/978-1-78769-845-120191004
5. . S. Sagiroglu, D. Sinanc, Big data: a review, in *2013 International Conference on Collaboration Technologies and Systems (CTS)*, 2013, pp. 42–47. https://doi.org/10.1109/CTS.2013.6567202
6. K. Ashton, That 'Internet of Things' thing. RFiD J. (2009). https://doi.org/10.1016/j.amjcard.2013.11.014
7. F. Amato, A. Mazzeo, A. Penta, A. Picariello, Knowledge representation and management for E-government documents, in *E-Government ICT Professionalism and Competences Service Science* (Springer, Boston, MA, US), pp. 31–40. https://doi.org/10.1007/978-0-387-09712-1_4
8. F. Colace, M. Lombardi, F. Pascale, D. Santaniello, A. Tucker, P. Villani, MuG : a multilevel graph representation for big data interpretation, in *IEEE 20th International Conference on High Performance Computing and Communications; IEEE 16th International Conference on Smart City; IEEE 4th International Conference on Data Science and Systems*, 2018, pp. 1410–1415. https://doi.org/10.1109/HPCC/SmartCity/DSS.2018.00233
9. A. Castiglione, F. Palmieri, F. Colace, M. Lombardi, D. Santaniello, Lightweight ciphers in automotive networks : a preliminary approach, in *2019 4th International Conference on System Reliability and Safety*, 2019, pp. 142–147
10. M. Carratu, M. Ferro, A. Pietrosanto, P. Sommella, V. Paciello, A smart wireless sensor network for PM10 measurement, in *2019 IEEE International Symposium on Measurements and Networking, M and N 2019—Proceedings*, 2019. https://doi.org/10.1109/IWMN.2019.8805015

11. F. Amato, V. Moscato, A. Picariello, F. Colace, M. De Santo, F.A. Schreiber, L. Tanca, Big data meets digital cultural heritage: design and implementation of SCRABS, a smart context-aware browsing assistant for cultural environments. J. Comput. Cult. Herit. (2017). https://doi.org/10.1145/3012286
12. A. Castiglione, C. Dambrosio, A. De Santis, A. Castiglione, F. Palmieri, On secure data management in health-care environment, in *Proceedings—7th International Conference on Innovative Mobile and Internet Services in Ubiquitous Computing, IMIS 2013*, 2013. https://doi.org/10.1109/IMIS.2013.120
13. K.T. Chui, P. Vasant, R.W. Liu, Smart city is a safe city: information and communication technology-enhanced urban space monitoring and surveillance systems: the promise and limitations, in *Smart Cities: Issues and Challenges Mapping Political, Social and Economic Risks and Threats* (Elsevier, 2019), pp. 111–124. https://doi.org/10.1016/B978-0-12-816639-0.00007-7
14. A. Castiglione, F. Palmieri, F. Colace, M. Lombardi, D. Santaniello, G. D'Aniello, Securing the internet of vehicles through lightweight block ciphers. Pattern Recognit. Lett. **135**, 264–270 (2020). https://doi.org/10.1016/j.patrec.2020.04.038
15. A.K. Dey, Understanding and using context. Pers. Ubiquitous Comput. (2001). https://doi.org/10.1007/s007790170019
16. M. Raento, A. Oulasvirta, R. Petit, H. Toivonen, ContextPhone: a prototyping platform for context-aware mobile applications. IEEE Pervasive Comput. (2005). https://doi.org/10.1109/MPRV.2005.29
17. F. Colace, M. De Santo, M. Lombardi, F. Pascale, D. Santaniello, A. Tucker, A multilevel graph approach for predicting bicycle usage in London area, in *Fourth International Congress on Information and Communication Technology. Advances in Intelligent Systems and Computing*, vol. 1027, ed. by X.S. Yang, S. Sherratt, N. Dey, A. Joshi (Springer, Singapore, 2020), pp. 353–362. https://doi.org/10.1007/978-981-32-9343-4_28
18. J. Jin, J. Gubbi, S. Marusic, M. Palaniswami, An information framework for creating a smart city through internet of things. IEEE Internet Things J. (2014). https://doi.org/10.1109/JIOT.2013.2296516
19. J. Fink, A. Kobsa, User modeling for personalized city tours. Artif. Intell. Rev. (2002). https://doi.org/10.1023/A:1016383418977
20. G. Ghiani, F. Paternò, C. Santoro, L.D. Spano, UbiCicero: a location-aware, multi-device museum guide. Interact. Comput. (2009). https://doi.org/10.1016/j.intcom.2009.06.001
21. D. Gavalas, M. Kenteris, A web-based pervasive recommendation system for mobile tourist guides. Pers. Ubiquitous Comput. **15**(7), 759–770 (2011). https://doi.org/10.1007/s00779-011-0389-x
22. F. Colace, M. De Santo, M. Lombardi, D. Santaniello, CHARS: a cultural heritage adaptive recommender system, in *Proceedings of the 1st ACM International Workshop on Technology Enablers and Innovative Applications for Smart Cities and Communities—TESCA'19*, 2019, pp. 58–61. https://doi.org/10.1145/3364544.3364830
23. F. Colace, M. De Santo, M. Lombardi, R. Mosca, D. Santaniello, A multilayer approach for recommending contextual learning paths. J. Internet Serv. Inf. Secur. **2**, 91–102 (2020). https://doi.org/10.22667/JISIS.2020.05.31.091
24. V. Cassani, S. Gianelli, M. Matera, R. Medana, E. Quintarelli, L. Tanca, V. Zaccaria, On the role of context in the design of mobile mashups, in *Communications in Computer and Information Science*, 2017 DOI:https://doi.org/10.1007/978-3-319-53174-8_7.
25. M. A. Khan, M. F. Uddin, N. Gupta, Seven V's of big data understanding big data to extract value, in *Proceedings of the 2014 Zone 1 Conference of the American Society for Engineering Education*, 2014, pp. 1–5. https://doi.org/10.1109/ASEEZone1.2014.6820689
26. S. Heo, S. Woo, J. Im, D. Kim, IoT-MAP: IoT mashup application platform for the flexible IoT ecosystem, in *2015 5th International Conference on the Internet of Things (IOT)*, 2015, pp. 163–170. https://doi.org/10.1109/IOT.2015.7356561
27. S. Tomasiello, M. Gaeta, V. Loia, Quasi–consensus in second–order multi–agent systems with sampled data through fuzzy transform. J. Uncertain Syst. (2016)

28. C. Bolchini, C.A. Curino, E. Quintarelli, F.A. Schreiber, L. Tanca, Context information for knowledge reshaping. Int. J. Web Eng. Technol. (2009). https://doi.org/10.1504/IJWET.2009.025015
29. N. Friedman, D. Geiger, M. Goldszmidt, Bayesian network classifiers. Mach. Learn. (1997). https://doi.org/10.1002/9780470400531.eorms0099
30. F. Colace, M. Lombardi, F. Pascale, D. Santaniello, A multilevel graph representation for big data interpretation in real scenarios, in *Proceedings—2018 3rd International Conference on System Reliability and Safety, ICSRS 2018*, 2019. https://doi.org/10.1109/ICSRS.2018.8688834

Chapter 11
A Bibliometric Review of Smart Cities and Migration

Ali Mohamad Mouazen and Ana Beatriz Hernández-Lara

Abstract Among the different topics related to sustainability, developing smart cities became a priority for several countries as far as the aim of those countries is to shift towards sustainability and to improve the quality of human lives. These countries experience internal and external migration from those regions where people are seeking better quality of life, smart services and solutions, better environment and better business activities. However, the research on smart cities and migration is still in its early stage. According to the Web of Science (WOS) and Scopus databases, only 77 papers were published between 2000 and 2019, but it is expected that this field becomes a relevant research topic given the social and economic relevance of both phenomena. The purpose of this paper is to provide a bibliometric review on papers that discuss about the links between smart cities and migration, aiming to determine the evolution of research on this topic. The results confirm the novelty of this research line. The most discussed topics reveal the relevance of planning for smart cities, city development and infrastructure projects, and educated workers migration. Future studies should address how to encourage sustainability and the role of governments in providing sustainable practices.

Keywords Bibliometric review · Smart cities · Migration · Sustainability · Text mining

A. M. Mouazen
Management and International Management Department, Lebanese International University, 146404 Mazraa Mousaytbeh, Beirut, Lebanon
e-mail: ali.mouazen@liu.edu.lb

A. B. Hernández-Lara (✉)
Department of Business Management, Universitat Rovira i Virgili, Avinguda Universitat, 1, 43204 Reus, Spain
e-mail: anabeatriz.hernandez@urv.cat

A. Visvizi et al. (eds.), *Research and Innovation Forum 2020*,
Springer Proceedings in Complexity,
https://doi.org/10.1007/978-3-030-62066-0_11

Introduction

The migration to urban and developed areas started in the nineteenth century and has not stopped growing since then. While almost 70 million people migrate to cities globally each year, it is expected that, by 2050, 70% of the world population will be living in urban areas [1].

The world nowadays is witnessing the emergence of several phenomena like urbanization and digital revolution, that together with sustainability, lead the governments and cities planners to transform big cities to offer their residents and visitors better services and living conditions [2, 3].

In the last decade, this transformation of the urban areas has been evolving from solely focusing on technology development, to new perspectives such as improving service delivery for urban residents and offering better life quality through adopting new technologies, fostering the emergence of the notion of smart cities [3].

Until now there is still little agreement on the definitions of smart cities, since there is no clear description and perception of their meaning [4–9].

Most of the definitions of smart cities included the notion of relying on information and communication technology to improve infrastructure operations; proper and effective communication between business institutions, organizations and individuals; adding to these the environmental issues including waste management and energy production.

For example, Angelidou [9] described smart cities as the utilization of human and technological capitals to conceptualize and develop urban clots or models. Marsal-Llacuna et al. [5] included in the notion of smart cities the relevance of information technologies to enhance urban performance, to monitor and enhance infrastructure, to increase collaboration and cohesiveness between several economic and productive actors, to encourage private and public sectors businesses innovation, and to provide efficient services to city residents. Navarro et al. [4] considered two main dimensions in the notion of smart city. In the first hand, the 'urban future' dimension that is associated with the impact of new technologies advancement on the city development on the long term. On the other hand, the 'knowledge and innovation economy' dimension focused on the knowledge management in the city's framework. Hollands [6] asserted on the role of human capital and citizens rather than believing that only technology can transform normal cities into smart cities.

Several countries reacted to grasp the benefits and opportunities related to smart cities, considered as a rising trend and a source to attract visitors and new residents.

Migration is the movement of people for settling temporarily or permanently in a new country or region. The research on migration and its reasons started since World War II; debates on this research interest included several domain areas including sociology, anthropology, law, international relations, politics, among other fields [10]. Reports provided by the Organization for Economic Co-operation and Development (OECD) indicated that permanent migration flows to OECD rose by 2% in 2018. Those represent almost 5.3 million permanent migrants. In addition, through temporary migration programs, 4.9 million workers migrated to these countries in 2017.

Those developed countries are the ones where smart cities are basically located. In addition, the migrant phenomenon impacts especially on cities and urban areas, that according to the World Bank data concentrate 54% of the world's population, being expected to increase by 50% (6 billion) by 2045 [11]. Yet, despite extensive research since 1963, and the probability of irregular migration continuity, developed countries still evoke continues major challenges resulting from migration [10]. Thus, planning and managing big cities is a must and a major challenge for governments, as far as cities shall be ready to cope with the migrant phenomenon and face its challenges, by providing affordable services, housing, infrastructure and sustainable growth.

The research that relates smart cities and migration is still on its early stages, and there are not bibliometric or text mining researches covering the characteristics and evolution of publications on this topic. Yet, only one publication provided bibliometric analysis on international migration [10]. However, given the current relevance of these phenomena and their emergence and growth as a social and economic issue, studies devoted to their analysis are starting to significantly appear [12–14]; which suggest the opportunity to conduct a bibliometric research to determine the main anchors of this new research line. The present research aims to:

- Analyse the evolution of research and type of publications on this topic.
- Determine the most publishing journals.
- Identify the most cited papers and most productive scholars.
- Identify collaborations among authors, affiliations and countries.
- Identify the most frequent keywords and the most discussed topics.

Smart Cities and Migration

In general, smart and sustainable cities are described as innovative cities that rely on information and communication technologies and other means, to improve the efficiency of urban services and operations, competitiveness, and quality of life, while guaranteeing economic, social and environmental needs for present and future generations [15]. Smart cities have received a considerable attention in the last years for their ability to attract investments, enhance economic growth and improve sustainability. They provide safer places with better services, an innovative environment that facilitate creative solutions, thus creating more jobs and reducing inequality. Consequently, they have become powerful destinations that attract people for living there, looking for their social and economic benefits [16, 17].

For instance, some papers attempted to study how migration flows from rural to urban areas led to cities physical expansion and population growth. Due to this expansion, and to offer better quality of life, the governments launched smart cities mission, which in its terms improved infrastructure and urban services delivery [18, 19].

Also, there are studies that have addressed the migration of youth to smart cities [20, 21]. The first study showed that Romanian youth prefers to migrate to countries where smart cities exists, such as UK, Germany, Netherlands and USA. They prefer places where there are opportunities for long life learning, recycling activities, green spaces, relaxing and recreating sites, low pollution, information technology and management services [20, 21]. While the second study empirically tested the relationship between characteristics of smart cities (i.e. living, environment, mobility, governance, people, and economy) and intention to live in smart cities, whereas this relationship is mediated by perceived usefulness by youth. The results of this study showed a significant relationship between smart city characteristics and intention to live, and this relationship is mediated by youth perceived usefulness [20, 21].

Other scholars have discussed how aggregate characteristics of smart cities attract well educated workers and smart people [22], and the factors that affect their migration [23, 24]. For example, business, employment growth, wage level, access to transportation and airport, crime index, diversity, pollution, taxes, schools, rent and flat size [23, 24].

However, in terms of bibliometric review, only two papers provided a bibliometric review on smart cities [25] and international migration [10]. With respect to the smart cities bibliometric research, the review showed that, environmental studies, urban studies, geography, environmental science, and green sustainable science have developed more attention on this issue than management business and economic sciences. In addition, smart cities, big data, communication technology and urban areas are the most utilized keywords [25]. As for the migration bibliometric review, the publication addressed most productive countries, evolution of different key terms associated to the international migration field, most productive institutions, collaboration and citation network, as well as keyword occurrence [10].

Although, there is a need for researches to understand the perception, facts and results of the relationship between smart cities and migration. Up till now, none of the publications entailed a bibliometric review on smart cities and migration.

This paper contributes to fill this gap providing up to date information to serve the academic and practical perspectives, by systematically reviewing the evolution of this research, type of publications, the most publishing journals, the most cited papers and productive scholars, the collaboration among authors, affiliations and countries, and the most used keywords and discussed topics.

This review will contribute to the state of the art identifying crucial information on this topic that will guide its future research, determining its status and the expected trend, emphasising the positioning of the topic in terms of publishing and examining the more relevant topics and those that will require additional studies.

Methodology

WOS and Scopus platforms were utilized to extract data due to their relevance as an academic bibliographic database. In both platforms, it is possible to classify sources (papers, book) or categories (social sciences, management, engineering etc.) and filter bibliographic information, unlike google scholar.

In order to broaden the search and include the highest number of publications, smart citie(s) and migration keyword solely were utilized instead of relying on titles or keywords. In Scopus platform the Boolean query conducted is: (TITLE-ABS-KEY (smart AND cities) OR TITLE-ABS KEY (smart AND city) AND TITLE-ABS-KEY (migration)), while the subject area was limited to social sciences, business, management and accounting, economics and finances. As for WOS platform, topic query TOPIC: (smart city) AND TOPIC: (migration) was indicated. In WOS, the research categories were limited to educational research, economics, management, social sciences, political science and public administration. Moreover, in both platform that time span was set to less than 2020 since it is still on its early months. Additionally, the subject areas were limited to management and social science categories to analyse publications that discussed smart cities and migration from people and social sciences perspective and not from technological and infrastructure perspectives.

As result of the search, 53 papers from Scopus and 24 papers from WOS were retrieved, being the first paper on the topic published in 2007.

In the first stage, each publication retrieved from both databases was treated separately to identify its main bibliographic information, like authorship, citations, affiliations of the authors and their countries, publishing journals, keywords, research topics and main contents. In the second stage, we aimed to correct discrepancies and standardize entries between the two platforms WOS and Scopus which could affect the results outcome. In the third stage, fine tuning on the data was conducted, 18 Scopus and 3 WOS papers were excluded from the database for not meeting the research indicators criteria (i.e. Smart cities and migration). It was also noticed that 8 papers were commonly included in both WOS and Scopus. Those were named as common papers and treated separately and analysed one time to avoid duplication. After these modifications, the final number of papers subject to study was 48, 13 WOS papers, 27 Scopus papers, and 8 common papers.

Results

Evolution of Research and Type of Publications

The first paper on smart cities and migration was published in 2007 in the "Architektura A Urbanizmus" Journal. From 2007 until 2011 there were not indexed publications on this topic, slightly appearing between 2011 and 2015 when many countries around the world initiated their plans and actions to develop smart cities. Between 2016 and 2018 the research on this topic increased from 2 papers (4.17%) to 9 papers (18.75%), at the beginning in the form of conference papers. In 2019 the publications doubly increased to reach 15 (31.5%), being mostly journal papers.

As for the type of publications, 26 (53.06%) of the publications were journal papers, 10 (22.45%) of the publication were classified as book chapters, 9 (18.7%) were conference papers and 3 (6.12%) publications were classified as conference proceedings.

Most Publishing Journals

Regarding the top journals publishing in this field, "Urban Book Series" appeared in the first rank place with three publications but only have 3 citations. While six other journals and one conference proceeding appeared in the second place with two publications. Among them, "Journal of Regional Science", "Papers in Regional Sciences" and "Journal of Science and Technology Policy Management", received the highest number of citations, with 277, 36 and 35 citations respectively.

Most Cited Papers and Most Productive Scholars

At the top of the list of the most cited papers, it is found that titled "Why Are Smart Cities Growing? Who Moves and Who Stays", published by Winters in 2011. This paper is index in WOS and Scopus platforms and have received 127 citations in the Scopus platform, with an average of 14.11 citation per year, and 122 citation in the WOS platform with 13.55 average citations per year. This paper is published in the "Journal of Regional Science" and discussed the reasons for migrating to smart cities [13]. Receiving 122 citations in both WOS and Scopus platforms in less than two years. The paper published in 2018 by Visvizi and Lytras is ranked the second [26]. It is titled "Rescaling and Refocusing Smart Cities Research: From Mega Cities to Smart Villages" and describes the relationship between mega cities and smart cities on three levels macro, mezzo and micro.

In 2017, Yassine, Singh and Alamri published an paper entitled "Mining Human Activity Patterns from Smart Home Big Data for Health Care Applications" that received 65 citations in two years. This study established that migration is increasing in smart cities, and smart homes are becoming a valuable source of data that could be used to enhance services. It was published in the "IEEE Access" [27]. Additionally, the paper entitled "Irregular Migratory Flows, Towards an ICTs' Enabled Integrated Framework for Resilient Urban Systems", published by Visvizi, Mazzucelli and Lytras, has reserved the fourth place by receiving 63 citations in both Scopus and WOS platforms. This paper navigated the challenges of irregular migration flows for cities and urban systems, and was published in the "Journal of Science and Technology Policy Management" [28].

Visvizi and Winters contributed specifically to the smart cities and migration topic the most, with two publications, included among the top six cited papers (Table 11.1).

Collaborations Among Authors, Affiliations and Countries

The analysis of co-authorship in WOS noticed that the major cluster of authors collaborating in their research on smart cities and migration were formed by 5 scholars.

Table 11.1 Most cited papers

Authors	Year	Number of citations
Winter	2011	249
Visvizi and Lytras	2018	122
Yassine, Singh and Alamri	2017	65
Visvizi, Mazzucelli and Lytras	2017	63

Gonclaves represented the centre of this cluster, and were collaborating mostly with Deggau, Prasath, Neiva, De Amorim, and Osorio de Andrade Guerra. As for the Scopus network, we found a cluster composed by 8 scholars. The centre of this cluster is Kumar, and was connected with Kantakumar, Wagner, Krass, Bharucha, Butch, Schneider and Kroll.

The same analyses were conducted for affiliations. In the WOS database, 4 clusters were identified, the first included Ohio State University and 5 other universities and institutions, the second included University of Barcelona and World Intellectual property organizations, the third was leaded by Hanyang university, while the fourth cluster included Bucharest University and Dree institution. Regarding the Scopus database, three clusters were identified, the first included Bucharest University and other three institutions, the second included two organizations, while the third was leaded by the University of Kiel. It is concluded that most of the affiliations usually cooperate with private or public business organizations to produce researches on smart cities and migration.

As Table 11.2 indicates, the analysis of the countries collaboration based on the affiliation location concluded that affiliations usually prefer collaboration with an institution in the same country. It is the case of India which is considered as the country most engaged in researching this topic. India produced 9 papers, 8 of them occurred in collaboration with institutions within the same country. On the other hand, in USA, out of 5 publications, 4 publications occurred without any external collaboration. In addition, Europe appeared as the first continent researching smart cities and migration with 17 (35.41%) publications, followed by Asia with 16 (33.33%), North America with 9 (18.75%) and South America, Africa and Australia with 2 (4.16%) each.

Most Frequent Keywords and Discussed Topics

The analysis of the occurrences of the most frequent keywords shows that smart cities, migration, productivity, amenities, creative class, United States and sustainable development were the most frequent keywords, that appeared more than twice in the WOS database. Regarding the Scopus dataset, some coincidences appear, being smart cities or smart city the most frequent topics, followed by urban growth and planning, urbanization, climate change, urban population, energy utilization, India, the Internet of things, population growth, urban areas and development, waste management and smart power grids.

Table 11.2 Countries collaboration

Country	N	SCC	%	WC	%	OCC	%
India	9	8	16.00	1	2.00		
Canada	5	3	6.00	1	2.00	1	2.00
US	5	1	2.00	4	8.00		
Germany	4	1	2.00	2	4.00	1	2.00
Greece	3	1	2.00			2	4.00
Australia	2			2	4.00		
Brazil	2	2	4.00				
China	2			2	4.00		
Italy	2	2	4.00				
Spain	2	2	4.00				
Turkey	2	2	4.00				
Bosnia and Herzegovina	1			1	2.00		
Czech Republic	1			1	2.00		
England	1	1	2.00				
Georgia	1	1	2.00				
Luxembourg	1	1	2.00				
Namibia	1	1	2.00				
Romania	1	1	2.00				
Russia	1	1	2.00				
Saudi Arabia	1	1	2.00				
Slovakia	1					1	2.00
South Korea	1	1	2.00				
Tanzania	1	1	2.00				

SCC same country collaboration, *WC* without collaboration, *OCC* other country collaboration

From the most frequent keywords and also from the content of the papers the most relevant discussed topics were extracted, which contemplate mainly planning for smart cities, city development and infrastructure projects, smart cities development opportunities, educated workers migration, electrical efficiency and technology services in smart cities, and waste management.

From the most frequent keywords and discussed topics, it can be concluded the relevance of the plans discussed in the papers for smart cities, their development, issues related to sustainability issues, especially concerned with the environmental dimension of sustainability, the role of infrastructures, as well as the Internet and technologies in smart cities, and population concerns, related to its growth and movement issues, like that of educated workers, as the main migrant source in smart cities.

Conclusion and Recommendations

Although many smart cities are being developed around the world, and the migration towards these cities have increased in the last decade, the research on smart cities and migration is still in its early stages and needs further development to aid government, policy makers and cities developers in their initiatives to promote sustainability, better quality of life, and effective and efficient organizations in these urban areas.

This paper aimed to provide a bibliometric review on smart cities and migration to aid researchers, institutions and practitioners from both, theoretical and practical perspectives. In the first hand this research framed what is done so far by identifying the evolution of research, the main journals, the most cited papers and productive scholars and countries, their collaboration and the most frequent keywords and discussed topics. From the other hand, this research will shed some light on the topics that should be discussed in the future to enrich the research on smart cities and migration.

Reports on internal and external migration indicated that the movement of youth, skilled and educated workers and smart people towards cities is motivated mainly for reason including transportation and good infrastructure, utilities and services, health care, and employment; which are characteristics of smart cities. The analysis of the most discussed topics of this review showed that "planning for smart cities" is addressed the most with 10 (17.5%), followed by "city development and infrastructure projects" and "educated workers migration" with 5 (8.7%) publications each. In addition, other topics were addressed including, smart city and development opportunities, waste management, water management, food security, student and smart workers migration, future smart cities problems, and economic development.

As the migration is expected to increase the upcoming years, several topics on migration still require more attention from the scholars, including world crises, such as and not limited to, Yemen and Venezuela crises [29]. Civil strives, wars, natural and environmental degradations and poverty are also reasons for migration decision [30].

Accordingly, studying and prioritizing problems that affect people, the planet, and organizations' profitability are essential to benchmark and innovate solutions for the future, including those applicable to smart cities among them. This work is not exempt of limitations, mainly derived from the novelty of the issue. This novelty, for example, suggests the need to repeat this kind of review in the future, when more studies are expected on this topic, and postpones further analyses on the research networks that could be developed. Also, the future studies on this aspect should address how to encourage sustainability and sustainable practices, as well as the role of governments in providing guidelines and policies, the role of the active participation of citizens, the need of information technology expertise and training for citizens, and the funds needed to enforce such practices.

The implementation of smart cities requires leadership with a strong future vision, knowledgeable managers in urban designs and information technologies, diversified workforce, and continues training on how to benefit from technologies that can be implemented in daily live government activities and operations. With these requirements and complex social interaction between government, migrants, citizens, policy makers, planners and executives, more studies on this social interaction and collaboration are essential to understand how leaders' vision and style can affect the implementation and development of smart cities, based on the increase of innovation especially when the migrants that these cities attract are mostly students and qualified workers.

All these issues are scarcely analysed and will constitute the future of this research line.

References

1. UNDP. UNDP, Hilton Foundation launch new platform to support smart, inclusive cities (2019)
2. S.M.V. Coutinho, C.C.C. Abilio, M. da Penha Vasconcellos, C.A.A. Netto, Indicadores para cidades inteligentes: a emergência de um novo clichê. Rev Gestão Ambient e Sustentabilidade. Universidade Nove de Julho (UNINOVE), PPGA **8**, 389–405 (2019)
3. S. Riffat, R. Powell, D. Aydin, Future cities and environmental sustainability. Futur Cities Environ [Internet]. Ubiquity Press, Ltd., 2:1 (2016)
4. J.L.A. Navarro, V.R.L. Ruiz, D.N. Peña, The effect of ICT use and capability on knowledge-based cities. Cities (Elsevier) **60**, 272–280 (2017)
5. M.-L. Marsal-Llacuna, J. Colomer-Llinàs, J. Meléndez-Frigola, Lessons in urban monitoring taken from sustainable and livable cities to better address the smart cities initiative. Technol. Forecast Soc. Change (Elsevier) **90**, 611–622 (2015)
6. R.G. Hollands, Will the real smart city please stand up? Intelligent, progressive or entrepreneurial? City (Taylor & Francis) **12**, 303–320
7. A. Caragliu, C. Del Bo, P. Nijkamp, Smart cities in Europe J Urban Technol (Taylor & Francis) **18**, 65–82 (2011)
8. H. Chourabi, T. Nam, S. Walker, J.R. Gil-Garcia, S. Mellouli, K. Nahon, et al., Understanding smart cities: an integrative framework, in *45th Hawaii International Conference System Science* (IEEE, 2012), pp. 2289–2297
9. M. Angelidou, Smart cities: a conjuncture of four forces Cities (Elsevier) **47**, 95–106 (2015)
10. S.-U. Hassan, A. Visvizi, H. Waheed, The 'who' and the 'what' in international migration research: data-driven analysis of Scopus-indexed scientific literature. Behav. Inf. Technol [Internet]. (Taylor and Francis Ltd.) **38**, 924–939 (2019)
11. H. Ritchie, M. Roser, Urbanization. Our World Data [Internet]. 2020; Available from: https://ourworldindata.org/urbanization
12. J.M. Shapiro, Smart cities: quality of life, productivity, and the growth effects of human capital Rev. Econ. Stat. (MIT Press) **88,** 324–335 (2006)
13. J.V. Winters, Why are smart cities growing? Who moves and who stays. J. Reg. Sci. **51**, 253–270 (2011)
14. M. Batty, K.W. Axhausen, F. Giannotti, A. Pozdnoukhov, A. Bazzani, M. Wachowicz et al., Smart cities of the future. Eur. Phys. J. Spec. Top. (Springer) **214**, 481–518 (2012)
15. ITU. Recommendation ITU-T Y.4900/L.1600 overview of key performance indicators in smart sustainable citiesI. Itu-T Sg20. 2016
16. M.E. Kahn, Sustainable and smart cities. (The World Bank, 2014)
17. J.M. Barrionuevo, P. Berrone, J.E. Ricart, Smart cities, sustainable progress. Iese Insight. **14**, 50–57 (2012)
18. A. Alakshendra, City profile: Patna, India. Environ. Urban ASIA. (SAGE Publications Ltd., England) **10**, 374–392 (2019)

19. G. Todd, I. Msuya, F. Levira, I. Moshi, City profile: Dar es Salaam, Tanzania. Environ. Urban ASIA [Internet]. (Sage Publications India Pvt. Ltd.) **10**, 193–215 (2019)
20. M.-C. Suciu, C.A. Florea, An empirical study on the migration among young skilled and creative people. AMFITEATRU Econ. **19**, 727–741 (2017)
21. L.H. Contreras Pinochet, G.F. Romani, C.A. de Souza, G. Rodriguez-Abitia, Intention to live in a smart city based on its characteristics in the perception by the young public. REGE-REVISTA Gest. **26**, 73–92 (2019)
22. M.R. Betz, M.D. Partridge, B. Fallah, Smart cities and attracting knowledge workers : which cities attract highly-educated workers in the 21st century ? Pap. Reg. Sci. **95** (2014)
23. T. Buch, S. Hamann, A. Niebuhr, A. Rossen, How to woo the smart ones? Evaluating the determinants that particularly attract highly qualified people to cities. J. Urban Aff. (Wiley, USA) **39**, 764–782 (2017)
24. L. Zheng, What city amenities matter in attracting smart people? Pap. Reg. Sci. (Wiley Inc., USA) **95**, 309–U378 (2016)
25. M. Li, Visualizing the studies on smart cities in the past two decades: a two-dimensional perspective. Scientometrics [Internet]. (Springer International Publishing) **120**, 683–705 (2019)
26. A. Visvizi, M.D. Lytras, Rescaling and refocusing smart cities research: from mega cities to smart villages. J. Sci. Technol. Policy Manag. [Internet]. (Emerald Group Publishing Ltd.) **9**, 134–145 (2018)
27. A. Yassine, S. Singh, A. Alamri, Mining human activity patterns from smart home big data for health care applications. IEEE Access [Internet]. (Institute of Electrical and Electronics Engineers Inc.) **5**, 13131–13141 (2017)
28. A. Visvizi, C. Mazzucelli, M. Lytras, Irregular migratory flows towards an ICTs' enabled integrated framework for resilient urban systems. J. Sci. Technol. Policy Manag. (Emerland Group Publishing Ltd., England) **8**, 227–242 (2017)
29. A. Visvizi, M.D. Lytras, M. Pachocka, Multiple facets of migration research: Key questions, topics, and avenues yet to be explored. Sustain [Internet]. (MDPI AG) **12** (2020)
30. A. Visvizi, M. Pachocka, M.D. Lytras, managing international migration: rethinking transaction costs, red tape, and their impact. Int. Migr. [Internet]. (Blackwell Publishing Ltd) **57**, 271–279 (2019)

Part II
Technology-Enhanced Teaching and Learning

Chapter 12
Coping with COVID-19: Higher Education in the GCC Countries

Benaouda Bensaid and Tayeb Brahimi

Abstract Today, the COVID-19 pandemic has, among many other things, tested the resilience of higher education institutions, but profoundly reshaped the management of learning, education and skills development, leading to a rethinking of perspectives and approaches within the education business as a whole. This is in addition to prompting transformation systems to better adjust to the devastating effects and implications of this pandemic or other possible potential crises. This study seeks to provide a general descriptive and analytical overview of the various responses of higher learning institutions in the context of Gulf Cooperation Council (GCC) amid seismic changes, with particular emphasis on the ways and means through which respective institutions have adopted to survive in the business of education, salvage the careers of its students' population, all while challenged to protect their community well-being. While events continue to unfold, this study also seeks to highlight the extent to which universities demonstrated their roles in preserving the standards of quality education and social responsibility in general, across the GCC countries during the pandemic of the COVID-19 and to draw attention to responses and coping mechanisms showcasing resilience, creativity, and re-imaginations of future educational business. This study shows that institutions of higher education in the GCC, in their response to COVID-19, have implemented online learning effectively thanks to their already established distance education and digital transformation. Through close coordination with their respective governments, GCC institutions of higher learning demonstrated good governance, showing due attention to the safety and welfare of their students, faculty, and staff, while giving attention to their roles in community service, whether in the form of promotion of awareness campaigns, volunteering, research support, and initiatives. More research should explore the future of higher

B. Bensaid (✉)
Istanbul Sabahattin Zaim University, 34303 Küçükçekmece/İstanbul, Turkey
e-mail: benaouda.bensaid@izu.edu.tr

T. Brahimi
Energy and Technology Research Center, Natural Science, Mathematics and Technology Unit, College of Engineering, Effat University, Jeddah 21478, Saudi Arabia
e-mail: tbrahimi@effatuniversity.edu.sa

A. Visvizi et al. (eds.), *Research and Innovation Forum 2020*,
Springer Proceedings in Complexity,
https://doi.org/10.1007/978-3-030-62066-0_12

institutions in the post coronavirus world as well as the inter-communication and cooperation for better learning and education in the region.

Keywords COVID-19 · Gulf cooperation council · GCC · Higher learning institutions · Coping

Introduction

Coronavirus disease (COVID-19) has unleashed global havoc in 266 countries with 3,256,853 confirmed cases and 233,357 confirmed deaths as of April 30th, 2020 [1]. This pneumonia of unknown cause was first detected in Wuhan, China, and reported to the World Health Organization (WHO) on December 31, 2019 [2]. With the rapid spread of this pandemic however, the WHO, on January 30, 2020, declared the outbreak as a Public Health Emergency of International Concern. The declaration was to put in place strong measures to detect disease, isolate cases, promote social distancing, create networks of experts and scientific research to assess the pandemic globally, and implement a robust risk management plan concerning coronavirus. On May 04, at the global pledging event hosted by the European Commission (EC), over $8 billion was pledged in a digital fundraiser for research and development for the COVID-19 vaccine [3]. Despite all the measures, however, the pandemic spread fast in many countries. In the Gulf Cooperation Council (GCC), the first case of coronavirus infection was reported on January 29 in the UAE. Since then, the number of cases has been growing steadily, reaching a daily record of 1289 new cases on April 17, in Saudi Arabia. As of April 30, Fig. 12.1, three months after

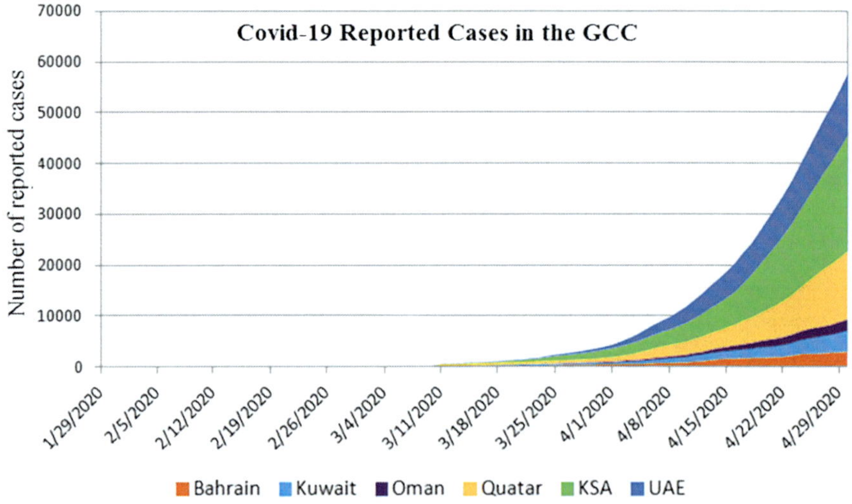

Fig. 12.1 COVID-19 confirmed cases in the GCC countries (as of April 30, 2020)

Table 12.1 Number of reported COVID-19 in the GCC as of April 30, 2020 [4]

Countries	Cases	February	March	April	Total
Bahrain	Confirmed	41	526	2473	3040
	Deaths	0	4	4	8
	Recovered	0	295	1205	1500
Kuwait	Confirmed	45	244	3735	4024
	Deaths	0	0	26	26
	Recovered	0	73	1566	1539
Oman	Confirmed	6	186	2156	2348
	Deaths	0	1	9	10
	Recovered	1	33	461	495
Qatar	Confirmed	1	780	12,628	13,409
	Deaths	0	2	9	11
	Recovered	0	62	1310	1372
Saudi Arabia	Confirmed	0	1563	21,190	22,753
	Deaths	0	10	152	162
	Recovered	0	165	2998	3163
	Confirmed	21	643	11,817	12,481
UAE	Deaths	0	6	99	105
	Recovered	5	56	2368	2429

being declared as a global health emergency, the GCC countries were hit hard by this pandemic, as a total of 58,055 new cases were reported, with a large number 22,753 in Saudi Arabia, 12,481 in the United Arab Emirates, 13,409 in Qatar, and 4024 in Kuwait, while Bahrain and Oman had much less reported cases as shown in Table 12.1 [4]. A comparison between the GCC countries showing the percentages of new cases and recovery cases for each country relative to the overall number recorded in the GCC indicates that Saudi Arabia accounts for almost 40% of the GCC reported case with 30% of recovered cases, followed by the UAE with 23% reported cases and 26% recovered cases as shown in Fig. 12.2. In comparison with many other countries in Europe, Russia, Canada, or all countries in African, the GCC countries quickly imposed restrictions and took all necessary measures to contain the virus. This perhaps helps explain relatively the low numbers of confirmed cases in the GCC countries [5].

To prevent the spread of the disease, the GCC countries took unprecedented steps to contain the virus. Saudi Arabia, for instance, suspended Umrah (voluntary Muslim pilgrimage to Makkah), daily and Friday prayers in about 80,000 mosques, public transportation, and shut down all workplaces except the essential ones [6, 7]. On March 09, in its efforts to control the COVID-19 spread, the Saudi Ministry of Education announced the temporary closure of all schools and universities all across the country [8]. The UAE, on the other hand, imposed even tougher restrictions on

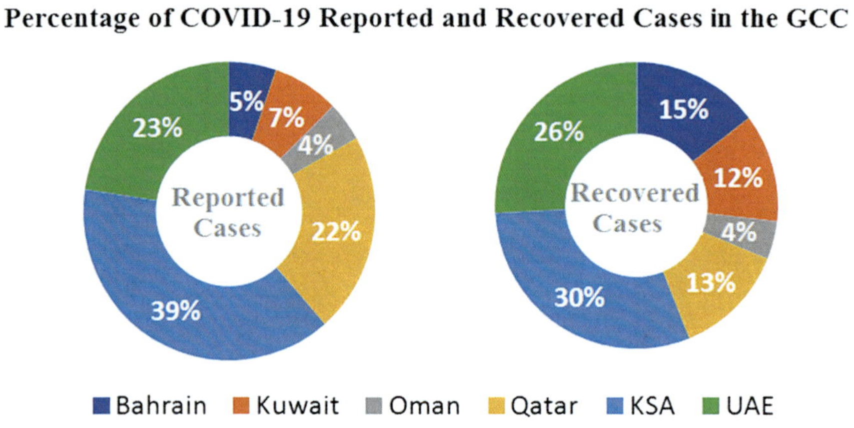

Fig. 12.2 Percentage of confirmed and recovered cases in each of the GCC countries compared to the total number in the GCC (as of April 30, 2020)

movement and travel, public to stay home unless necessary, and individuals entering the country subject to 14-days mandatory quarantine [9]. Following the report of its first COVID-19 case, Kuwait also enforced a nationwide curfew whereby all the public and private schools and higher educational institutions closed, theoretically up to August 04, 2020. Qatar has also banned dining in restaurants and café, along with halted transport across the country. Oman has closed its borders and all of its transportation services until further notice. Bahrain has issued electronic bracelets to citizens who have self-isolated themselves, and it notifies the station when they are 15 m away from their phones. Along with that, all public and private educational institutions have been closed [10].

This present study seeks to explore and investigate the various responses, measures, and coping strategies by the higher educational institutions in the GCC context used to contain the spread of the COVID-19 pandemic. We will first discuss the effect of COVID-19 on Global education with particular emphasis on the GCC institutions of higher learning, followed by some discussion and analysi on the coping strategies and contributions used thereof alongside some implications for the future reimagining of the business of education as a whole.

Effect of COVID-19 on Global Education

The COVID-19 is an unprecedented health crisis that is still growing at an alarming rate. Several countries closed off their public and private educational institutions. The closure started first in China and Mongolia on February 20, where 999,014 enrolled learners were affected [11]. This number increased to 298,368,352 by the end of February, and jumped to 1,600,780,887 by the end of March, then dropped

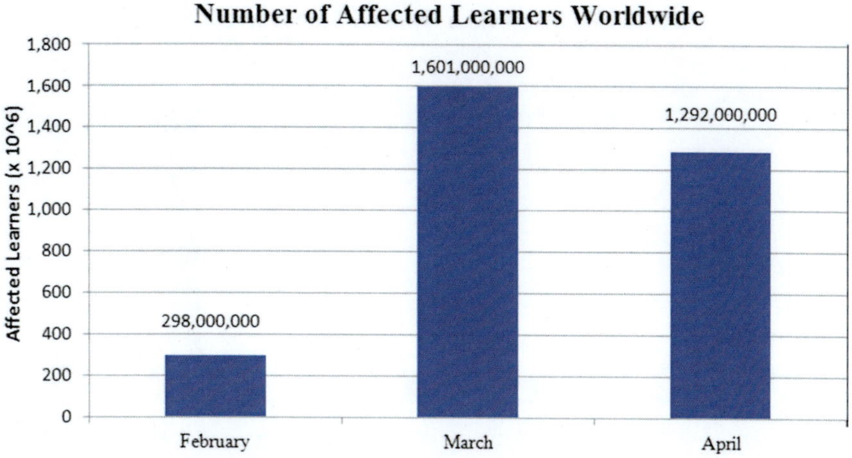

Fig. 12.3 Number of affected learners worldwide due to COVID-19 pandemic

down to 1,292,378,969 by the end of April when some countries started lifting the coronavirus lockdowns, as shown in Fig. 12.3. The closure resulted in about 1.5 billion students out of schools across the globe, which is 87.6% of the world's total enrolled learners [12]. During this period, the number of worldwide COVID-19 confirmed cases hit 86,604 people at the end of February, then increased to 863,184 at the end of March and jumped to 3236 on April 30 with more than 200,000 reported deaths [1]. UNESCO data shows that in the GCC countries, out of a total population of 58,664,098 [13], 12,085,898 learners have been affected because of COVID-19, as shown in Table 12.2 and Fig. 12.4 [11]. This number represents about 21% of the total population in the GCC, among whom 841,026 (males and females) in the KSA alone, and 24% of the total population in the KSA.

The impacts of the COVID-19 pandemic have been daunting as there was a possibility of increased drop-out rates across the world [14]. Along with that, 63 million primary and secondary teachers, several educational support personnel, early childhood educators, technical and vocational trainers, and higher education teachers were drastically impacted as a result of this pandemic [15]. From the outset, the global lockdown of education institutions was expected to cause a major (and likely unequal) interruption in students' learning, disruptions in internal assessments, and

Table 12.2 Number of affected learners in the GCC due to COVID-19 [11]

# of affected learners	Bahrain	Kuwait	Oman	Qatar	KSA	UAE	**Total-GCC**
Males	144,535	376,419	445,430	168,540	4,304,758	712,370	6,152,052
Female	147,894	400,750	454,723	174,984	4,105,506	649,989	5,933,846
Total	292,429	777,169	900,153	343,524	8,410,264	1,362,359	12,085,898

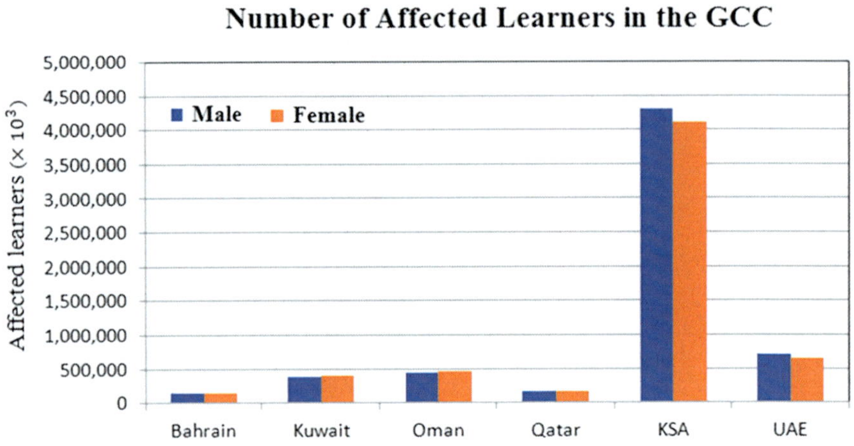

Fig. 12.4 Number of affected learners in the GCC region due to the COVID-19 pandemic

cancellation of public assessments for qualifications or their replacement by an inferior alternative. The education disruption also affected families across the globe as homeschooling is greatly affecting the parents' productivity along with children's learning and social life. Moreover, internal assessments have been canceled, which were the prime source used to evaluate students' progress for both teachers and parents. The external examinations such as GCSEs and A levels have been canceled as well [16]. According to a survey conducted towards the end of March 2020, 30% of respondents residing in G7 countries were very concerned about their children's education because of the (COVID-19). However, 35% of respondents were fairly concerned, meaning nearly 2/3 of respondents harbor concerns over this issue [17]. Open doors, exhibitions, and in-person interviews for student recruitment fairs are not possible anymore, and a digital solution is needed. A study by QS World University Rankings surveying over 400 higher education institutions revealed that 83% said that digital marketing is becoming more important, and about 94% stressed on the importance of one-to-one online meeting [18].

The COVID-19 pandemic has greatly affected the educational sector. There is a strong possibility of deteriorating mental health because of the resulting sense of uncertainty and anxiety among students and faculty members [19]. In these circumstances, students are experiencing major interruptions in learning and teaching, graduations might also get delayed, not to speak about the associated economic challenges [20]. To mitigate the impacts of COVID-19 pandemic, large scale national efforts were employed to use technology to ensure remote and online learning [21]. Seven years ago, the IMF-World Bank Spring Meetings Symposium recognized the role of education as global practice reached by millions of people around the world [22]. More recently, top universities turned towards innovative ways of teaching and learning, such as "Massive Online Open Courses "MOOCs" and Makerpsace [23, 24] which have not only caught the attention of educators in higher education but also the interest of high schools. The International Association of Universities [25] and UNESCO's [26] COVID-19 Education Response compiled a list of initiatives, resources, and tools for Higher Education Institutions around the world to ensure the uninterruptedness of teaching and learning during COVID-19 pandemic. Likewise, to contain the spread of COVID-19 pandemic, UNESCO issued several recommendations to ensure continuous teaching and learning which stress on the readiness, data privacy, program inclusion, psycho-social challenges, course schedule, support to teacher and parent, the limit of the number of platforms/applications, monitor student's learning process, define the duration of the distance learning, and enhance the connection between teachers, parents, and school, and share experience and strategies [27].

Technology-based strategies are being adopted by teachers and educational institutions to provide education. There are many tools available on Blackboard for both teachers and students. These include; Home page, Course & Learning Materials, Announcements, Discussion Boards, Messages, Roster, My Groups, My grades (Grade center), and the assessment tools that include interactive formative exams, assignments, quizzes, and tests. Moreover, many students supporting tools are also available as Contact professors, Student supports, and Professor resources. The virtual classroom is also one of the Blackboard teaching tools. In several countries, radio programs and national televisions are being used to ensure the provision of educational lessons and material, particularly in under-resourced areas that do not have access to technological infrastructure [15]. Reliable and relevant tools were adopted for the provision of online learning to ensure inclusivity. With increased access to technology, it was important to take substantial measures to ensure data privacy and security. Along with addressing educational challenges, it was pertinent to address the psycho-social challenges of students and parents during the quarantine. Before rolling out the digital tool, students, teachers, and parents must be trained about its usage, and most importantly, to assess students learning, there must be some strategies to evaluate and monitor their learning [26].

Coping of GCC Higher Learning Institutions with COVID-19

Higher Education Institutions have been unprecedentedly affected as COVID-19 continues to spread across the world. According to UNESCO, the number of affected learners in the GCC region exceeds 12 million, with 6,152,052 males and 5,933,846 females (see Table 12.2. The pandemic, however, affected not only students and teachers, but also the economic and social consequences, yet opened new avenues for digital learning. As a result of COVID-19, universities across the GCC region have closed their campuses, with most moving to online learning to ensure the uninterrupted delivery of knowledge using the reliable and appropriate tools to facilitate distance teaching [28].

Higher Education Institutions in GCC countries are offering flexible grading systems and providing students the right to accept or decline their cumulative Grade Point Average when unsatisfied with the evaluation system [28]. Universities started offering 24 h of e-library services to support students' and faculty members' research. This e-library contains several electronic databases of journals, e-books, and academic articles [29]. Several numbers of technology initiatives have been currently developed and expanded rapidly to promote distance learning and the internet for learning from home, support countries in their effort for a continuous education through remote learning, and working actively with Ministries of Education in many countries to support remote learning resources concerning coronavirus [26, 30]. Many tools for remote training, web conferencing, presentation, and desktop sharing exist such as Adobe Connect, Google Classroom, Edmodo, Microsoft Teams, Cisco WebEx, Blackboard Collaborate, LearnCube, WhatsApp, Zoom, and many others [11, 31].

In the context of the GCC, internet users are already high, which facilitates the move to distance learning. According to the InternetWold Stats, as of December 31, 2019, the Internet penetration rates in each GCC country exceeds 90% [32]. Fortunately, some of the distance learning platforms already exist, many of which are provided by Ministries of Education such as EduNET in Bahrain, E-learning in Kuwait, MoE in Oman, Ministry of Education Channel in Qatar, iEN portal and Vschool in Saudi Arabia, and Alef Education and EduShare in UAE just to name a few [26]. In preparation for virtual courses, several universities in KSA organized workshops and announced special mechanisms for the final exams [33–35].

As UAE reported its first case of COVID-19, all of the country's universities made the transition to online teaching, which put all universities in a quandary of dealing with digital delivery of education. Zayed University adopted Adobe Connect while the University of Sharjah and United Arab Emirates University is using Blackboard systems, and Heriot-Watt University Dubai is using the Vision learning tool. However, among all universities, Hamdan Bin Muhammad Smart University was the first e-university that has extensive experience of online teaching, and henceforth, it was able to facilitate for the rest of the universities to successfully implement online classes [36]. Khalifa University-trained its faculty members and staff on online delivery tools instruction. University created an online learning ecosystem based on four interconnecting layers. At first, they ensured that all employees and students are on the same online learning page. Secondly, they initiated a Learning Management System to connect faculty with students for sharing course material, conducting online discussions, tests, quizzes, assignments, and administered feedback. At the third level, the university initiated virtual classroom platforms such as Big Blue Button, MS Teams, and BlackBoard Connect and lastly compiled a repertoire of content development and management tools to ensure faculty develop their lessons without hindrances [37].

In the Kingdom of Saudi Arabia, the Saudi Research and Innovation Network and The Integrated Telecom Company collaborated to develop a portal dedicated to sharing resources with teachers, parents, and students to ensure the provision of high-quality virtual classrooms and learning platforms. This free of cost portal consists of apps, websites, and other resources that support teachers, parents, and students in distance learning [38]. For example, King Abdullah University of Science and Technology (KAUST) initiated online teaching and implemented work from home to ensure the safety of employees and students. The fully virtual educational model is being employed to ensure the uninterrupted delivery of instructions [39].

The government of Bahrain also issued directives to close off all public and private universities. Alongside, all training centers and community centers have also been closed [40]. Universities moved to online teaching and online learning systems, which are proving successful in imparting education smoothly [41]. Using Microsoft Teams and Blackboard, Bahrain University started digital classrooms and e-learning to enable students to access instructional content and allow both students and faculty to communicate through chat rooms and live broadcasts [42]. Similarly, in Kuwait, authorities announced on February 26, a two-week suspension of all schools and public and private universities in an attempt to contain the spread of COVID-19 pandemic [43]. The American University of Kuwait rolled out an e-learning platform for students to review course material and also to connect with faculty with plans, not to hold assessments [44]. The University of Kuwait used Blackboard Collaborate to continue the teaching and learning process during pandemic [43].

In Oman, The Ministry of Higher Education instructed all institutions of higher learning to use online teaching methods to complete the current academic year [45]. The Sultan Qaboos University, for instance, initiated an e-learning program in which lectures are delivered in written, audio, or visual format with queries conducted through virtual chat rooms. The alternative options for assessments such as final reports, case studies, projects, or independent research studies are also adopted. However, extensive support is being provided to both faculty members and students to ensure the smooth implementation of the program and follow up conducted to assess the progress of the educational process [46]. Muscat University remained operational in the spring semester using e-learning with assessments conducted using the best alternative strategies to accurately assess the learning [47]. The German University of Technology in Oman moved its classes online but in a pilot program for students in Bachelor of Computer Science to be applied later for other programs [45].

Universities in Qatar decided to hold their face to face learning up until August 01, 2020. The Education City, Student Center, and The Qatar National Library have been closed until further notice [48]. Qatar University Library has compiled all reliable information of organizations and social media accounts regarding COVID-19 situation on their website [49]. They also embarked on distance learning programs using QU distance learning education Blackboard and Blackboard Collaborate Ultra [50]. Qatar University also launched an emergency research grant with a budget of up to QR 150,000 per research track. Along with that, the Ministry of Education and Higher Education Qatar launched the public service portal to obtain the services of the ministry on one platform [51]. Hamad Bin Khalifa University Qatar has continued delivery of learning and education remotely [52] while ensuring the online availability of Books and audiobooks to promote literacy during the pandemic [53].

Also important is the investment of universities in the well-being of its students and faculty and staff members. Counseling and psychological services are also remotely available for students as well [48]. For example, Effat University, in Jeddah, KSA [54], created a Library Contingency Plan on the COVID-19 crisis, stress management, emotional health, and preventing stress, relaxation, and meditation strategies, as well as free access to scholarly materials. Besides that, Effat University Medical Clinic provided services during the outbreak of COVID-19, including Medical Emergency calls, WhatsApp messages, Peer-to-Peer messages on Effat Mobile App, and email consultation. King Abdullah University of Science and Technology has launched the COVID-19 Emotional Support Helpline to provide its students and faculty members with psychological services [55]. In Qatar, Carnegie Mellon University developed an employee assistance program in support of faculty members and staff with resources and information for personal and work-life issues. Muscat University shared students' support contact with all students to help cope with COVID related issues [47]. Khalifa University in the UAE provided remote mental health services and online development courses on leadership, managing stress, and positive thinking [29].

As far as the universities' community service is concerned during this pandemic, a total of 19,533 volunteers from across twenty-seven universities in the Kingdom participated in efforts to combat COVID-19. They worked to provide awareness and education, under educational and community programs for epidemic control [56]. Saudi Universities have also made individual contributions to fight the pandemic. For example, KAUST developed the 'Rapid Research Response Team (R3T) [57] to help to fight against COVID-19 pandemic, with researchers already producing several publications on COVID-19 [58]. More recently, KAUST made the Supercomputer Core Laboratory (KSL) (KSL n.d), including the supercomputer "Shaheen-II" resources available to support projects, research, and innovation to fight COVID-19 pandemic. KAUST also launched the 'Community COVID-19 Innovation Challenge' which includes Education during COVID-19, social connectivity, community management, and other open challenges related to developing new and creative approaches for the control of the coronavirus short and long-term effects. King Abdulaziz University announced the winners of a research initiative on COVID-19 [34] with 42 of 230 proposals selected for further study and also opened a fast-track grant program to develop coronavirus research. King Abdulaziz City for Science and Technology (KACST) opened a fast-track grant for researchers interested in studying and fighting COVID-19 pandemic. By the end of April 29, the KACST received 230 proposals. The King Fahd University of Petroleum and Minerals (KFUPM) [59] established the "Coronavirus Innovation Proof of Grant" to develop innovative products, applications, and services that could help the country better handle the pandemic. The King Abdullah Petroleum Studies and Research Center (KAPSARC) launched a database to monitor COVID-19 updates with data available for research purposes [60]. An educational health program has also been launched by King Saud University—Medical City at King Khalid University Hospital [61]. Besides the above activities, many universities organized workshops on distance learning, online communication tools, and online exam preparation. In research and development, the Islamic University Madinah launched a coronavirus research and study program in a wide range of fields, including legal studies during pandemics, as well as studies highlighting the efforts of Saudi Arabia to combat the coronavirus and its impact on the economy [33]. The King Abdullah Petroleum Studies and Research Center (KAPSARC) also launched a database to monitor COVID-19 updates for research purposes [60]. Similarly, King Saud University-Medical City launched COVID-19 Educational Campaign and a 2-day health educational campaign on CoronaVirus at King Khalid University Hospital [61].

Discussion and Analysis

The inventions of the internet and the World Wide Web have removed all traditional barriers and further reshaped the concept of knowledge sharing [62, 63]. Contemporary online learning platforms not only have the potential of providing rich educational resources but also supporting real-time and asynchronous communication between instructors and learners. This, in spite of the fact that online learning in higher education has been looked down upon for decades [64]. A recent study that compared the online and face-to-face sections of the same course showed that there was no significant difference in the test scores, assignments, participation grades, and final grades of both groups [65]. However, the average online groups were slightly higher. The online course was found either as effective as or more effective than face-to-face courses by 96% of the online students. Several other studies also proved the insignificant differences between learning outcomes achieved through online courses and face-to-face instruction [66]. However, the haste of universities to finish up their online academic year in response to the COVID-19 pandemic could result in unprecedented challenges for faculty, students, and families. This is because the way universities are currently implementing their online programs and courses may create highly uneven and unsatisfying educational learning experiences that can threaten the credibility of distance education as a viable and substantial educational platform [67].

In response to the COVID-19 crisis, Higher Education Institutions (HEIs) across the GCC countries have rushed to carry out their online transition smoothly. This is understandable in view of their keen investment in online learning which already has its roots in the tradition of distance education. In this context, early digital transformation plans in smart education adopted by the GCC countries such as the "Classroom of the Future" including online platforms and ICT content focus [68] played a fundamental role in facilitating the shift from traditional face to face learning to digitally-enabled learning and curricula. Also, as a response to the spread of COVID-19 and the closure of higher education institutions, the GCC countries have taken the following swift actions including administrative measures to limit the spread of coronavirus and establishing policies regarding universities' remote learning, availability of resources including network service, online platforms, and sharing materials, and financial resources as well. All of the GCC countries have been supported by their respective Ministries of Education which helped towards providing several technology platforms needed to promote distance learning while keeping internet penetration rates of more than 90%.

GCC higher learning institutions have also launched many research initiatives and grants on COVID-19 including the educational health programs and workshops and database to monitor the pandemic. Added to that faculty members and staff assistance support program and the well-being of its students and faculty and staff members. The promotion of volunteering work to help fight and slow down the spread of COVID-19.

It is interesting to note, however, that the mass closure of universities to slow down the spread of the COVID-19 pandemic has strongly forced educators across the world to shift quickly their learning and assessments from face-to-face to online and distance learning. However, there was a short time for educators to prepare for this challenging virtual teaching and learning process. The lack of research and studies of similar challenges in the past accompanied with the uncertainty related to when would there be potential treatments and vaccines for this pandemic, have only added an extra challenge to the prediction of future actions of higher education institutions in the event of the cessation of face-to-face activities or the return in the academic year 2020–21.

Now, that this phase is hopefully almost over, at least three conditions must be maintained to return to classes in fall 2020, including the full control of COVID-19, prevention measures and resources still in place, and students' and teachers' safety is guaranteed. Interestingly, the following three scenarios are being currently discussed in the UAE [37]: (1) COVID-19 pandemic goes on, which required maintaining of distance-learning, (2) dissipation of the virus with the gradual return at a rate of 30–50%, and (3) total disappearance of COVID-19 and everything returns to normal, then blended learning at 70% direct education and 30% distance learning.

There is also a critical issue pertaining to the responsibility of accreditation bodies who are to assure and improve the quality standards of higher education institutions. Throughout the history of the accreditation, accreditation bodies have always acted and improved the accreditation process as a result of academic and non-academic pressure [69]. They improve the educational quality of most GCC higher education institutions that are pursuing accreditation by both national and international accreditations bodies [70] for colleges and university programs in applied and natural science, computing, engineering, and engineering technology [71]. The accreditation bodies expect evidence from institutions to measure to which extent the student learning outcomes have been achieved by implementing continuous improvement and quality assurance processes during internal and external reviews. Because of the COVID-19, it is expected that each university produces evidence showing how the processes of student learning outcomes have been accomplished during the pandemic and what students accomplished in terms of student outcomes, assessments, and evaluations. Until those reports become public, it is difficult to analyze students' outcomes, measure the applied evaluation process, or evaluate the extent to which student learning outcomes are being effectively attained as per the requirements of respective accreditation bodies.

Although the solutions to combat the pandemic effects on education are no less than encouraging and innovative, the reality is that many schools and regions are better positioned than others to take the benefits of technological resources and infrastructure, allowing them to respond to educational crises during this pandemic much more effectively. Relevant data indicates that scientific activity is not declining, it is merely shifting to new activities.

In a survey by ResearchGate, it has been found that almost half of the researchers surveyed spent more time reading and studying medical work and scientific research than they had done before the pandemic [72]. The same trend was found in the time spent on writing, publishing, and peer-reviewing scientific articles before and after the pandemic. More research studies are currently required to explore the impact of this emergency immersion into remote learning as well as exploring ways to render selective online practices and innovative approaches solutions for an effective education including problem solving, decision making, adaptability, and flexibility.

Conclusion

As a result of the spread of the COVID-19, Millions of students in the world were affected by the strong measures including the closure of schools and universities, social distancing, and the shift from face-to-face teaching and learning to e-learning. Those changes caused some degree of inconvenience to both students and educators but also gave a glimpse at how education could change rapidly and forced educational policymakers and managers to look at new models of educational creativity and innovation. While most students worldwide were able to find a solution in face-to-face learning, only about 60% [73] were active internet users as of April 2020 which may further increase the gap in education quality worldwide for students living in less developed countries. In response to the COVID-19 crisis, higher education institutions (HEIs) across the GCC countries rushed towards implementing their online learning effectively thanks to their early GCC digital transformation plans to move to smart education and smart government. The key initiatives that helped HEIs maintain the learning cycle while protecting students' learning trajectory were mainly linked to swift administration and policy steps, access to resources, financial support, as well as the help provided in GCC's respective ministries of education. However, the absence of studies on comparable pandemic issues in the past, the uncertainty related to when there would be potential treatments and vaccines, have only contributed to the difficulty to forecast possible future solutions. In the context of the GCC, a more collaborative yet strategic agenda on education and research is needed. The first step in this direction would be for the universities to further boost their collaboration on teaching, research, joint financing, and community service in addition to increased engagement of the industry and the private sector, to ensure effective reverse linkage between universities and the business world. There is also a need to encourage and support students, entrepreneurship training, and development. Quality Assurance systems need to be built to their full strength to create new networks between higher education institutions in the region. Also important is for GCC HEIs to give more attention to the GCC context of job market analyses and trends therein for the sake of guiding students, and the society at large.

Acknowledgements The authors gratefully appreciate the support from Istanbul Sabahattin Zaim University, Istanbul, Turkey, and the College of Engineering at Effat University, Jeddah, Saudi Arabia.

References

1. World Health Organization, Coronavirus disease (COVID-19) Pandemic (2020a). Retrieved April 26, 2020, from World Health Organization (WHO): https://www.who.int/emergencies/diseases/novel-coronavirus-2019
2. World Health Organization, *Novel Coronavirus (2019-nCoV) SITUATION REPORT—1* (2020b, January 01). Retrieved May 13, 2020, from World Health Organization: https://www.who.int/docs/default-source/coronaviruse/situation-reports/20200121-sitrep-1-2019-ncov.pdf?sfvrsn=20a99c10_4.
3. European Commission, *Coronavirus Global Response: €7.4 Billion Raised for Universal Access to Vaccines* (2020). Retrieved May 13, 2020, from European Commission: https://ec.europa.eu/commission/presscorner/detail/en/ip_20_797
4. JHU, Johns Hopkins University and Medicine (2020), Coronavirus Resource Center (2020). (JHU CCSE). Retrieved April 30, 2020. https://data.humdata.org/dataset/novel-coronavirus-2019-ncov-cases
5. T. Hilton, Coronavirus: With 15,000 cases in the Gulf, here is how GCC states are coping (2020). Retrieved April 26, 2020, from Al Arabiya: https://english.alarabiya.net/en/coronavirus/2020/04/14/Coronavirus-With-15-000-cases-in-the-Gulf-here-is-how-GCC-states-are-coping.html
6. M. Alsherebi, *Saudi Arabia's clear response to the coronavirus outbreak is in stark contrast to the West* (2020) Retrieved 29 Apr 2020, from Euro News: https://www.euronews.com/2020/03/27/saudi-arabia-clear-response-to-the-coronavirus-outbreak-is-in-stark-contrast-to-west-view
7. S. Yezli, A. Khan, COVID-19 social distancing in the kingdom of Saudi Arabia: bold measures in the face of political, economic, social and religious challenges. Travel Med. Infectious Dis. 101692 (2020). https://doi.org/10.1016/j.tmaid.2020.101692
8. Godinho, V, *Saudi Arabia to close schools indefinitely from Monday* (2020, March 09). Retrieved May 13, 2020, from Gulf Business: https://gulfbusiness.com/saudi-arabia-close-schools-monday
9. T. Karasik, Lessons the US can learn from the UAE about the decontamination of COVID-19 (2020, April 21). Retrieved April 29, 2020, from Atlantic Council: https://atlanticcouncil.org/blogs/menasource/lessons-the-us-can-learn-from-the-uae-about-the-decontamination-of-covid-19.
10. The National, Coronavirus in the Gulf: everything you need to know about Covid-19 in the GCC (2020, April 12). Retrieved April 26, 2020, from The National: https://www.thenational.ae/world/gcc/coronavirus-in-the-gulf-everything-you-need-to-know-about-covid-19-in-the-gcc-1.994534
11. UNESCO, COVID-19 Educational Disruption and Response (2020a, April 28). Retrieved May 2, 2020, from UNESCO: https://en.unesco.org/covid19/educationresponse.
12. Educational International, Guiding principles on the COVID-19 pandemic (2020, March 27). Retrieved April 27, 2020, from Educational International: https://www.ei-ie.org/en/detail/16701/guiding-principles-on-the-covid-19-pandemic
13. UN, United Nations, Department of Economic and Social Affairs, World Urbanization Prospects—Population Division (2018). Retrieved May 02, 2020, from https://population.un.org/wup/Download.
14. COVID-19 and Education in Emergencies, (2020, March 30). Retrieved April 27, 2020, from: https://www.educationcannotwait.org/covid-19

15. ILO, COVID-19 and the education sector (ILO, 2020). Retrieved from https://www.ilo.org/sector/Resources/publications/WCMS_742025/lang--en/index.htm
16. S. Burgess, H.H. Sievertsen, Schools, skills, and learning: the impact of COVID-19 on education (2020). Retrieved April 2020, 27, from VOX CEPR Policy Portal: https://voxeu.org/article/impact-covid-19-education
17. E. Duffin, *Concerns for education due to COVID-19 in G7 countries March 2020* (2020, March 27). Retrieved May 13, 2020, from Statista: https://www.statista.com/statistics/1107324/covid-19-concerns-education-g7/
18. QS, The New Virtual Student Journey: Changing Dynamics as a Result of the Coronavirus (2020, May 05). Retrieved May 08, 2020, from https://www.qs.com/portfolio-items/virtual-student-journey-changing-dynamics-result-coronavirus-report
19. P. Sahu, Closure of universities due to coronavirus disease 2019 (Covid-19): impact on education and mental health of students and academic staff. Cureus **12**(4) (2020)
20. A. Moopen, Mental health in the time of COVID-19 outbreak (2020, March 29). Retrieved April 30, 2020, from Gulf News: https://gulfnews.com/opinion/op-eds/mental-health-in-the-time-of-covid-19-outbreak-1.70680268
21. J. Miks, J. McIlwaine, Keeping the world's children learning through COVID-19 (2020, April 20). Retrieved April 27, 2020, from UNICEF: https://www.unicef.org/coronavirus/keeping-worlds-children-learning-through-covid-19
22. WorldBank, World Bank Symposium: Assessment for Global Learning. Report (World Bank Headquarters, Washington, DC, 2013)
23. T. Brahimi, S. Khalifa, B. Benaouda, Integrating makerspaces in higher education: constructionism approach to learning, in *Research & Innovation Forum 2019 Springer Proceedings in Complexity* (2019), pp. 65–73. https://doi.org/10.1007/978-3-030-30809-4_7
24. T. Brahimi, A. Sarirete, Learning outside the classroom through MOOCs. Comput. Hum. Behav. **51**, 604–609 (2015). https://doi.org/10.1016/j.chb.2015.03.013
25. IAU, *The impact of COVID-19 on higher education worldwide* (2020, April 20). Retrieved May 13, 2020, from International Association of Universities: https://www.iau-aiu.net/IMG/pdf/covid-19_and_he_resources.pdf
26. UNESCO, *Distance learning solutions* (2020b, May 12). Retrieved May 13, 2020, from https://en.unesco.org/covid19/educationresponse/solutions
27. UNESCO, COVID-19: 10 Recommendations to plan distance learning solutions (2020c, March 06). Retrieved April 27, 2020, from UNESCO: https://en.unesco.org/news/covid-19-10-recommendations-plan-distance-learning-solutions
28. A. Ahmed, COVID-19: University students in the UAE exempted from academic warnings and dismissals (2020). Retrieved 27 Apr 2020, from Gulf News: https://gulfnews.com/uae/covid-19-university-students-in-the-uae-exempted-from-academic-warnings-and-dismissals-1.71090887
29. E. Solomon, Khalifa University's Response to Covid-19 (2020, April 3). Retrieved April 27, 2020, from Khalifa University: https://www.ku.ac.ae/khalifa-universitys-response-to-covid-19
30. World Bank Education and COVID-19, (n.d.). Retrieved from https://www.worldbank.org/en/data/interactive/2020/03/24/world-bank-education-and-covid-19
31. C. Pace, S. Pettit, K. Barker, Best practices in middle level quaranteaching: strategies, tips and resources amidst COVID-19. Becom. J. Georgia Middle School Assoc. **31**(1) (2020). https://doi.org/10.20429/becoming.2020.310102
32. K. Wansink, *Middle East—Mobile Infrastructure and Mobile Broadband* (2019, April 28). Retrieved May 13, 2020, from BuddeComm: https://www.budde.com.au/Research/Middle-East-Mobile-Infrastructure-and-Mobile-Broadband?r=51
33. SPA, *We are all working to serve students, says Saudi education minister* (2020a, April 04). Retrieved May 13, 2020, from Arab News: https://www.arabnews.com/node/1653166/saudi-arabia.
34. SPA, *King Abdulaziz University Announces Winning of 42 Research Proposals on COVID-19* (2020b, April 27). Retrieved May 13, 2020, from Saudi Press Agency: https://www.spa.gov.sa/2079464

35. SPA, *Program launched to support COVID-19 research in Islamic University in Madinah* (2020c, April 11). Retrieved May 13, 2020, from Arab News: https://www.arabnews.com/node/1656761/saudi-arabia
36. J. Crawford, K. Butler-Henderson, J. Rudolph, M. Glowatz COVID-19: 20 Countries' higher education intra-period digital pedagogy responses. J. Appl. Teach. Learn. (JALT) **3**(1) (2020)
37. S. Salama, *Coronavirus UAE: E-learning could continue into next academic year if COVID-19 crisis goes on* (2020, May 09). Retrieved May 13, 2020, from Gulf News: https://gulfnews.com/uae/education/coronavirus-uae-e-learning-could-continue-into-next-academic-year-if-covid-19-crisis-goes-on-1.71416631
38. A. Draycott, Coronavirus drives remote learning's acceptance in the Middle East and beyond (2020, March 22). Retrieved April 27, 2020, from Arab News Pakistan: https://www.arabnews.pk/node/1645091/middle-east
39. KAUST responds to COVID-19 (2020a). Retrieved April 28, 2020, from King Abdullah University of Science and Technology: https://www.kaust.edu.sa/en/covid-1
40. Khaleej Times, Covid-19: Schools, universities closed for two weeks in Bahrain over coronavirus fears (2020, February 26). Retrieved April 29, 2020, from Khaleej Times: https://www.khaleejtimes.com/coronavirus-outbreak/covid-19-schools-universities-closed-for-two-weeks-over-coronavirus-fears
41. K. Gillet, How Bahrain is leading from the front on coronavirus containment (2020, March 24). Retrieved April 28, 2020, from The National: https://www.thenational.ae/world/gcc/how-bahrain-is-leading-from-the-front-on-coronavirus-containment-1.996771
42. Zawya, Universities in Bahrain launch digital classroom (2020, March 21). Retrieved May 03, 2020, from Zawya: https://www.zawya.com/mena/en/life/story/Universities_in_Bahrain_launch_digital_classrooms-SNG_169661507/
43. Kuwait University, Dean of Admission and Registration: The University Council Approves the Proposed Amendments to the Academic Calendar for the Year 2019/2020 and the Academic Calendar for the Year 2020/2021 (2020, April 21). Retrieved May 04, 2020, from Kuwait University: https://kuweb.ku.edu.kw/ku/Announcements/KUW19573
44. American University of Kuwait, AUK Responds to COVID-19 (2020). Retrieved April 28, 2020, from American University of Kuwait: https://www.auk.edu.kw/about-auk/auk-responds-to-covid-19
45. German University of Technology in Oman, Announcements (2020, April 04). Retrieved May 03, 2020, from German University of Technology in Oman: https://www.gutech.edu.om/registration-decision-ss2020-online-teaching/#toggle-id-3
46. Oman Observer, SQU to implement e-learning from April 12 (2020). Retrieved April 28, 2020, from Oman Observer: https://www.omanobserver.om/squ-to-implement-e-learning-from-april-12
47. Muscat University, COVID-19 Update (2020, April 23). Retrieved April 28, 2020, from Muscat University: https://muscatuniversity.edu.om/covid-19-update
48. COVID-19 Information (2020, April 28). Retrieved April 29, 2020, from Carnegie Mellon University in Qatar: https://www.qatar.cmu.edu/alert-covid-update
49. Resources (COVID-19), (2020, April 21). Retrieved May 03, 2020, from The Library Qatar University: https://library.qu.edu.qa/en/covid-19-resources#.Xq72AfZuLIU
50. Remote Learning Tools and Resources, (2020, April 14). Retrieved May 03, 2020, from Qatar University: https://www.qu.edu.qa/offices/cetl/distance-learning
51. Meeddubai, Latest on coronavirus impact on Middle East (2020). Retrieved from https://www.meed.com/latest-news-on-the-pandemics-economic-impact
52. Student Circular: Academic Continuity, (2020, March 12). Retrieved May 03, 2020, from Hamad Bin Khalifa University: https://www.hbku.edu.qa/en/coronavirus/students
53. HBKU, HBKU Press Highlights New Ways to Promote Literacy at Home for Children (2020, March 28). Retrieved May 03, 2020, from Hamad Bin Khalifa University: https://www.hbku.edu.qa/en/news/literacy-home-children
54. Effat University LibGuides, Stress Management: Home (2020). Retrieved May 13, 2020, from https://effat.libguides.com/Stress_Management

55. KAUST Health, Wellness Services (2020). Retrieved April 30, 2020, from KAUST Health: https://kh.kaust.edu.sa/CoronaVirus/Pages/Wellness-services.aspx
56. Umm Al-Qura University, Umm Al-Qura University is leading the voluntary efforts of Saudi Universities to combat Coronavirus (COVID-19). (https://uqu.edu.sa/en/App/News/77947)
57. KAUST, *KAUST Rapid Research Response Team and COVID-19* (2020b). Retrieved May 13, 2020, from King Abdullah University of Science and Technology: https://www.kaust.edu.sa/en/r3t-covid-19
58. T. Jamil, I.S. Alam, T. Gojobori, C. Duarte, No Evidence for Temperature-Dependence of the COVID-19 Epidemic (2020). https://doi.org/10.1101/2020.03.29.20046706.
59. Deanship of Scientific Research, *The Coronavirus Innovation Proof of Grant is Established* (2020, April 11). Retrieved May 13, 2020, from Deanship of Scientific Research: https://dsr.kfupm.edu.sa/about/news/coronavirus
60. Data Source KAPSARC, *Home* (n.d). Retrieved May 13, 2020, from Data Source KAPSARC: https://datasource.kapsarc.org/pages/home
61. KSU, *King Saud University- Medical City Launches CoronaVirus Educational Campaign* (n.d). Retrieved May 13, 2020, from King Saud University: https://medicalcity.ksu.edu.sa/en/news/details/king-saud-university-medical-city-launches-corona-virus-educational-campaig
62. G. Michelsen, P.J. Wells, *A Decade of Progress on Education for Sustainable Development. Reflections from the UNESCO Chairs Programme.* Published by the United Nations Educational, Scientific and Cultural Organization, France (2017), pp. 1–135m from: https://www.iau-hesd.net/sites/default/files/documents/252319e.pdf
63. A. Visvizi, M.D. Lytras, A. Sarirete (eds.), *Management and Administration of Higher Education Institutions in Times of Change.* (Emerald Publishing, Bingley, UK, 2019). ISBN: 9781789736281. https://books.emeraldinsight.com/page/detail/Management-and-Administration-of-Higher-Education-Institutions-in-Times-of-Change/?K=9781789736281
64. M. Shachar, Y. Neumann, Twenty years of research on the academic performance differences between traditional and distance learning: Summative meta-analysis and trend examination. MERLOT J. Online Learn. Teach. **6**(2) (2010)
65. C. Neuhauser, Learning style and effectiveness of online and face-to-face instruction. Am. J. Distance Edu. **16**(2), 99–113 (2002)
66. M. Goyal, R. Krishnamurthy, in Optimizing student engagement in online learning environments advances in educational technologies and instructional design, *Optimizing Student Engagement in Online Learning Environments* (2018), pp. 187–219. https://doi.org/10.4018/978-1-5225-3634-5.ch009
67. W. Durden, Turning the Tide on Online Learning (2020, April 08). Retrieved April 30, 2020, from Inside Higer Ed: https://www.insidehighered.com/views/2020/04/08/online-learning-can-only-be-viable-if-it-offers-certain-connection-points
68. Deloitte, National Transformation in the Middle East A Digital Journey. Report (Deloitte & Touche, M.E., 2017)
69. J.F. Volkwein, L.R. Lattuca, B.J. Harper, R.J. Domingo, Measuring the impact of professional accreditation on student experiences and learning outcomes. Res. High. Educ. **48**(2), 251–282 (2006). https://doi.org/10.1007/s11162-006-9039
70. ABET, *The Accreditation Board for Engineering and Technology Home Page* (n.d). Retrieved 13 May 2020, from Abet: https://www.abet.org/
71. T. Brahimi, A. Sarirete, S. Khalifa, Impact of accreditation on engineering education, Ch5 in *Enhancing Knowledge Discovery And Innovation in the Digital Era*, ed. by M.D. Lytras, L. Daniela, A. Visvizi (Information Science Reference, Hershey, PA, 2018)
72. ResearchGate Reports, How COVID-19 Has Affected the Global Scientific Community (2020, April 14). Retrieved May 13, 2020, from https://statnano.com/news/67570/ResearchGate-Reports-How-COVID-19-Has-Affected-Global-Scientific-Community
73. Statista, *Global digital population as of April 2020* (2020, April 24). Retrieved May 13, 2020, from Statista: https://www.statista.com/statistics/617136/digital-population-worldwide

Chapter 13
Pre-Service Teachers' Immersive Experience in Virtual Classroom

Václav Duffek, Jan Fiala, Petr Hořejší, Pavel Mentlík, Jiří Polcar, Tomáš Průcha, and Lucie Rohlíková

Abstract In this paper, the results of two pilot studies, carried out in a 3D classroom simulator (based on the Unity 3D engine), will be presented. A virtual classroom user uses a virtual reality headset, and teaches virtual learners (didactic-controlled), who report, ask, and answer questions, but also disturb. A total of 10 future geography teachers and 5 future computer science teachers participated in the research. In the virtual classroom, they underwent three-phase training, which was aimed at developing their pedagogical-psychological skills. In the first phase, they briefly tried working in the virtual classroom, so that they could prepare more accurately for the second phase, during which they were asked to explain the selected curriculum, or to assign virtual or individual work to virtual pupils. The performance in the virtual classroom was always followed by an in-depth analysis of the output with the facilitator and other students. A video recording of the whole performance was also made for students' self-reflection. Students had the opportunity to reflect on their output from various points of view, and prepare for the last phase—repetition of the output in the virtual classroom (this time with the inclusion of comments, new approaches to the output, etc.). The research showed that reflective lessons in the virtual classroom have enormous potential for opening specific didactic topics during didactics lessons, both for the development of the study group discussion, and for deep reflection. In the third phase of the training, all of the students observed an improvement in

V. Duffek (✉) · P. Mentlík
Centre of Biology, Geoscience and Environmental Education, University of West Bohemia, Univerzitni 8, Plzen 306 14, Czech Republic
e-mail: duffekv@rek.zcu.cz

J. Fiala · T. Průcha · L. Rohlíková
Department of Computer Science and Educational Technology, University of West Bohemia, Univerzitni 8, Plzen 306 14, Czech Republic

P. Hořejší
Department of Industrial Engineering and Management, University of West Bohemia, Univerzitni 8, Plzen 306 14, Czech Republic

J. Polcar
Division of Lifelong and Distance Learning, University of West Bohemia, Univerzitni 8, Plzen 306 14, Czech Republic

A. Visvizi et al. (eds.), *Research and Innovation Forum 2020*,
Springer Proceedings in Complexity,
https://doi.org/10.1007/978-3-030-62066-0_13

a wide range of aspects of teaching. In the third phase of training, most students also experienced strong immersion—students get used to the virtual classroom, almost to the degree of forgetting where they are in the real world. This is all in spite the fact that the current prototype of the virtual classroom used is only at the beginning of its development, and will be gradually modified, complemented, and developed.

Keywords Virtual reality · Higher education · Pre-Service teachers · Teacher training · Curriculum · Didactics

Introduction

Jaron Lanier [10] considered the virtual environment to be a world without limits as early as the 1980 s, and that is exactly how we can access it today; using all of the possibilities that the current hardware level provides, we can simulate almost everything in virtual reality in any way. As a result, simulation games can be taken to a new level and be used as a learning method. Widely used simulation games with a modelled environment, or a process that a player/student can penetrate or just follow, can motivate the user, and leave a deeper footprint in it, just because they stimulate more senses. Evidence of the popularity of such games is their constantly increasing number in the market. Learning by simulation depends on the environment in which students fully immerse and behave in it as in a real environment. For example, the results of studies summarized by Jensen and Konradsen [9] show that pupils who used HMD (Head-Mounted Display) in learning spent more time on tasks, and acquired better cognitive, psychomotor, and affective skills. These virtual environments are also used in health care, the military, and even in crisis training.

Ferguson et al. [7] looked at other factors that, in addition to the above-mentioned technical factors, affect virtual reality learning and user immersion. It focused on the content of the curriculum, the degree of interaction, and the story it offers to the student, which serves as a motivation for learning. The authors investigated two things: the degree of freedom in passing simulations and movement in the environment, and the structure of the story being simulated.

Depending on the degree of freedom of movement in the simulation, motion can be distinguished between active and passive. Active interaction is one in which the user can freely explore the virtual environment. Such an environment is intended to stimulate creativity, a positive attitude to the subject and immersion. It has been found that an active environment can provide a greater presence to the user. A passive passage, on the other hand, is characterized by a clear leadership of the author of the environment, based on Vygotsky's [20] Zone of Proximal Development, which stimulates cognitive learning processes, and leads the user to the intended goal, without distracting "side issues".

Teachers' performance in a virtual simulation environment was followed by research using the TLE TeachLivE application [3]. The research examined the effect of the use of simulations in a virtual environment on the performance of mathematics teachers. It was attended by 135 teachers from 10 different locations. Each of them expected four sessions of 10 min each in the virtual environment. Moreover, the evaluation of the research was not only in the final self-evaluation of the participants, but in the subsequent observation of the practiced behaviour in the real class. The whole research was based on the hypothesis that the skills (high-leverage practices) that teachers learn in a virtual environment will subsequently be put into practice, and moreover, they will be more effective in teaching.

The authors see the reasons for using simulations as being: a safe environment, the possibility of repeating the simulation, and providing immediate feedback to the supervising teacher. Participants in the simulations can therefore continuously improve the specific skills required. One of the indisputable advantages is also the recording of the simulated training on video, without the consent of legal representatives. SimSchool authors [4] have similar arguments, and talk about the possibility to better prepare future teachers for real situations, because during the simulation, they can train and practise specific skills in a controlled environment.

The individual research itself consisted of observing and evaluating teachers in a virtual classroom TLE TeachLivE—a class of pupils projected on screen, filled with avatars controlled by the supervising teachers. The evaluation showed that almost all of the teachers (80%) agreed that the virtual classroom was very close to the real class, and 90% confirmed that the controlled avatars behaved in the same way as real children. The presence rate was therefore high during the training. Practicing in virtual reality has also led teachers to ask questions more effectively, and provide feedback to pupils. In addition, teachers were able to put the acquired skills into practice. As a recommendation for further research, the extension of simulations to other subjects, and testing of different length and frequency of simulations, are given.

Spencer et al. [18] looked at whether training in a virtual environment had a more beneficial effect on the training of two collaborating teachers in the class ("co-teaching"). A total of 90 teachers (from experienced to non-practitioners) were divided into two groups—one practicing roles ($n = 42$) with each other, and the other ($n = 48$) practicing in a virtual environment where they communicated with a trained actor controlling the avatar of the collaborating teacher. Based on a questionnaire survey, it was found that students who experienced virtual simulation rated their experience as very beneficial and useful, compared to those who played roles with each other. The authors of the research attribute this finding to the fact that in real contact, students were basically shy with their colleagues, and did not fully engage in roles. However, they admit that the results may have been biased because the students divided themselves into the groups studied, and did not go through both variants of the simulation. Despite this, all of the above-mentioned advantages of simulations in virtual reality are stated to be beneficial for practicing the required skills of the teacher. Lorenzo [14] at the University of Alcala examined the same phenomenon. The study was attended by twenty high school teachers and two experts in the learning process. The research results relevant to this work are consistent with the previous one, that simulation in a virtual environment was more effective in teacher education than playing real-world roles.

Virtual Classroom

Virtual classroom is an application to support the teaching of future teachers, created in the Unity development environment. The project originated at the University of West Bohemia. More about the formation of this simulator can be found in Duffek et al. [6].

Training using the simulator takes place in the classroom, which is divided into two parts. In the first part, enough space is created for safe movement of the student in the virtual environment (see Fig. 13.1a), which is monitored by the camera during the study. The camera monitors the student's movements to monitor his/her non-verbal expressions, and the spoken word is very audible in both parts of the classroom. In the second part of the classroom, there is a session for other participants, and a large projection device (Fig. 13.1b), which displays the image from the monitoring web camera, and in particular, the virtual classroom image (Fig. 13.1c) that the student sees in the virtual environment. This ensures that observers can evaluate both the participant's speech in the virtual classroom, and their real movements. The same image can be seen on the monitor by a didactic (supervising teacher) who controls the simulator (see Fig. 13.1d).

Virtual pupils—avatars, with speech from the teacher—didactic who controls the whole scene, or perhaps the classmates of those who are teaching in the virtual classroom speak for them. The didactic can place the student in a situation where students ask questions which are more amusing for the class, divert the teacher's attention from the topic of the lesson, or which check the teacher's professional knowledge. So it is up to the student teacher not only to teach the pupils in the

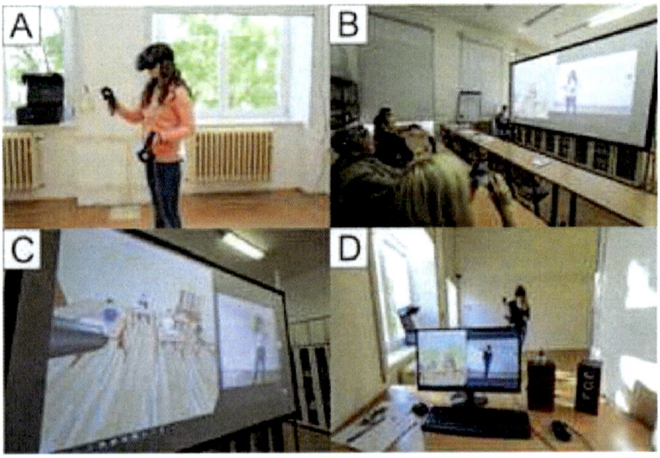

Fig. 13.1 Demonstrations of the environment in which the training is carried out, using the simulator. **a** Space for safe movement in a virtual environment. **b** Space for other study participants with a view of the projection equipment. **c** Detailed view of the projection equipment. **d** View from the side of the didactic who controls the virtual class *Source* author's photo

virtual classroom a set of the curriculum, but also to be able to react appropriately in a problematic situation, to respond adequately to the situation, and to manage the lesson correctly.

Methodology

The simulator suitability test was conducted in three stages (similar to Haynes et al. [1], Dawson and Lignugaris [15]), according to the same principle (see Fig. 13.2) as in the pilot simulator testing presented by Duffek et al. [6].

In the first stage, the participants learned to use the created simulator, and provided feedback on the virtual classroom environment. In the second stage, participants in the virtual classroom already presented the prepared topics. The output of each student was thoroughly analyzed and evaluated by the didacticist and other participants in order to obtain the most detailed and best feedback possible. The third stage of the pilot study was the same as the second. The participants presented the same issue, but with already incorporated comments, which had appeared in the second phase. The whole process was again recorded, and thoroughly analyzed and evaluated. After completing all three levels, the participants filled in the final anonymous questionnaire, in order to get their views and attitudes to the simulator, as well as to the virtual classroom project itself. [6].

A group of 5 future teachers of computer science (three males and two females) aged 25–26 years (one participant 43 years), and a group of 10 future teachers of geography, four males and six females, aged 23–25 years, were selected to verify the suitability of using the virtual classroom in preparing future teachers of computer science. At the time of testing, all 15 students studied the first year of a follow-up Master's degree of teacher training in computer science or geography. Almost all of them had no experience of virtual reality in research, but were aware of it. On the other hand, almost all students (due to their age) had very good teaching experience. Most of them were already employed in primary or secondary schools at the time of the pilot study. All of the participants signed their permission to provide personal data, and to enable recording and photography, or use these records to popularize research.

The feedback from each participant after each stage of the research was transcribed into a text file, and coded with the help of Atlas.ti [8, 19]. Individual answers were assigned codes, according to their positive, neutral, and negative evaluation. For

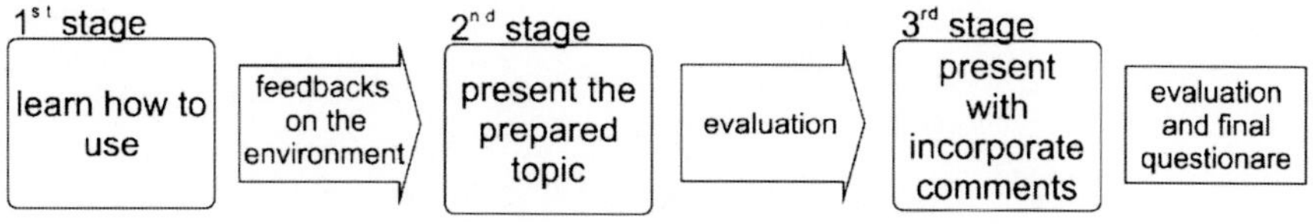

Fig. 13.2 Visualisation of the methodological approach

the integrity of the encoded file, one more code has been created to bring together participants' suggestions to improve the virtual classroom environment.

Possibilities of using the virtual classroom in the preparation of future computer science teachers were determined at the end of the study, by comparing feedback from participants of both groups and their anonymous evaluation in the final questionnaire.

Results

In this chapter, feedback on the virtual simulator environment after each stage of research will be presented, as well as the results of the anonymous evaluation obtained by the final questionnaire. All evaluations will be compared between the two groups, in order to identify any differences. The evaluation was obtained only from those participants who did not suffer from cyber sickness during the performance. Therefore, for the group of future teachers of geography, feedback on the environment and the results of the final questionnaire were collected from eight participants; for computer scientists, from four.

Feedback Comparison

The participants of both groups provided relatively balanced feedback after their outputs (see Table 13.1). Feedback after the first session was not included in the results, as the impetus for technical improvement of the virtual simulator was very prevalent in both groups. These responses were coded with a "response to change" code. As an example, here are some authentic answers that were recorded after the first stage: "*It definitely needs work on pupils' mimicry*", "*I would definitely add a clock so that we can see the time.*", "*There are no visual aids.*", "*Writing on the board is a rather complex matter.*"

In spite of the relatively balanced feedback after the 2nd and 3rd levels, we can discover some partial findings in the evaluation. In the group of pre-service teachers of geography, the amount of positive reactions increased with the time spent in the virtual simulator (grades between 2 and 3). The opposite was true for computer science group. The more inputs a group of computer science teachers in the virtual simulator experienced the more negative their feedback was. Reactions to the change/improvement of the environment appeared only after the third output.

Table 13.1 Comparison of the feedback of the participants of the groups after the second and third stage of the research, with the demonstration of the most representative answers of the participants

Feedback after 2nd level	
Pre-service Geography Teachers	Pre-service Computer Science Teachers
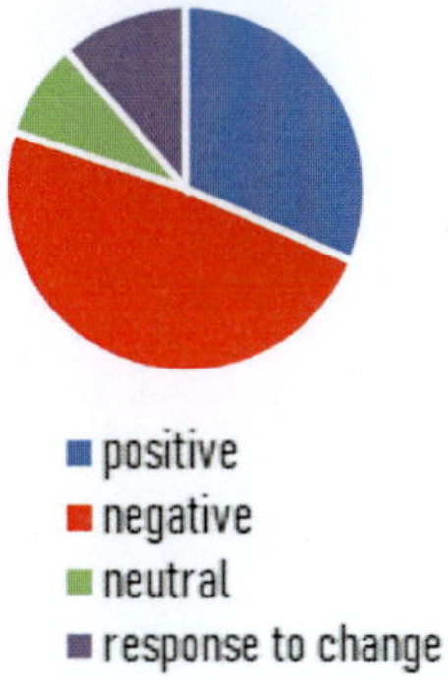	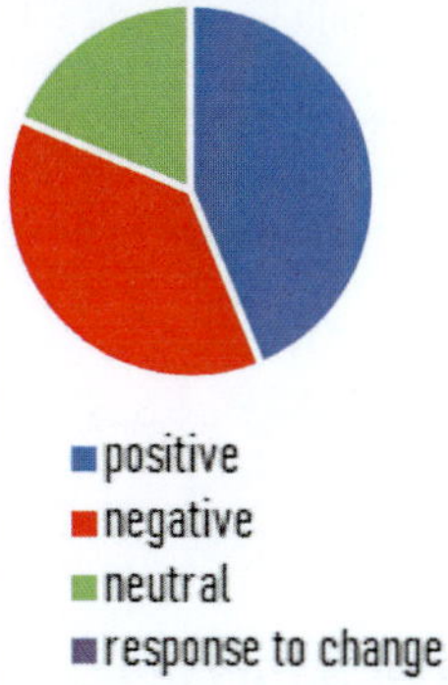
• Terrible • Pleasant • Well written • Pleasant environment • Pupils stare strangely (blank expressions) • I'm afraid of movement (after some time but adaptation) • Overall, it cannot replace the interaction between pupils and teachers • Inability to have support materials	• Good, you need to get your hand on it • Writing should be thoroughly tested • The illusion is accurate enough • Weird, when I teach already • Nice • I was looking forward to AI, but it wasn't AI. • No interaction • I see it more positively than my colleagues • There is silence compared to a normal class • Immersion was possible in a virtual environment
Feedback after 3rd level	
Pre-service Geography Teachers	Pre-service Computer Science Teachers
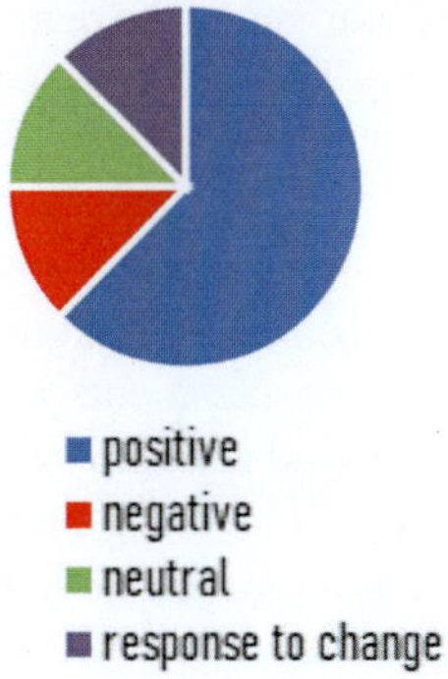	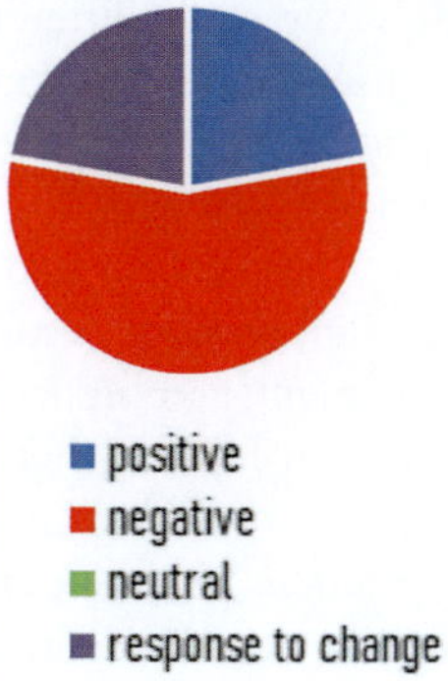

(continued)

Table 13.1 (continued)

Feedback after 2nd level	
Pre-service Geography Teachers	Pre-service Computer Science Teachers
• I felt at home • It isn't pleasant, I can't wipe the board • I felt better • I can already work with HMD—I did it better • Better, I feel more confident • In the beginning, my head was spinning, and later it was OK • The same experience (positive)	• Not enough children • I understand the importance for a person who does not teach • You can't move • I can't have teaching aids • I see no purpose • VR gives me no added benefits • Better than nothing • Good when I have never stood in front of a classroom • No projection

Comparison of Statements in the Final Questionnaire

For comparing the evaluation of the virtual simulator, only those statements that directly aimed at product evaluation were selected from the final questionnaire. Therefore, this chapter lacks a comparison for Statement No. 3, to which participants responded in the same way as Statement No. 2 (In the virtual classroom, I lack more teaching aids and options that I could use in real-life lessons.) I did not experience any stress or stage fright in the virtual classroom.), and No. 9 (I think I did my best pedagogical performance as far as possible.).

The first statement of the questionnaire examined whether participants found the virtual classroom environment to be the same as the real environment. None of the participants stated that they fully agree with the statement "*Virtual classroom represents an environment identical to the real reality.*" Authentic comments that the participants have written on this evaluation can be seen in Table 13.2. Authentic comments make it difficult to determine whether the scores of the two groups differ significantly. However, slight differences can be seen in the comparison of the degree of agreement or disagreement on the Likert [13] scale. Looking at Table 13.2, we can say that geography teachers rated the virtual classroom environment more positively than computer science teachers.

Statement No. 2 focused on finding the perfections/imperfections of the virtual simulator. The participants once again expressed their level of agreement or disagreement with the statement, and supplemented or complemented their assessments with authentic comments in Table 13.3. The participants of both groups did not even agree once with the statement "*This designed virtual simulator is perfect. Nothing needs to be changed or added.*" Both groups presented ideas and necessities that should be incorporated into the environment, according to their own specific focus. One very interesting thing is the incentive to incorporate artificial intelligence into the environment, which was recorded twice by a group of future computer science teachers. Again, from the authentic comments, we cannot claim that the ranking of one group

Table 13.2 Comparison of statements of future teachers of geography and computer science in statement No. 1: Virtual classroom represents an environment identical to real reality

Statement 1: The virtual classroom represents an environment identical to the real reality				
	Totally agree	Rather agree	Rather disagree	Completely disagree
Pre-service Geography Teachers	0	1	5	2
Pre-service Computer Science Teachers	0	0	2	2

Authentic comments	
Pre-service Geography Teachers	Pre-service Computer Science Teachers
• I miss the interaction between teacher and pupil • The relationship between teacher and pupil is missing • You can't move a class like in reality • You can't see if the pupils are bored or excited about the lesson • VR has limited use of teaching aids. • I also feel limited mobility in VR • I have to change my normal behaviour to apply in the real class	• Unfortunately, the virtual classroom lacks basic signs of authenticity as opposed to reality, such as noise • Absence of pupil feedback (impossibility to adjust style and pace) • Inability to use teaching aids • In a virtual classroom, children do not respond to stimuli • Students don't respond authentically, there are a small number of them in the virtual classroom You can't move between students

Table 13.3 Comparison of statements of future teachers of geography and computer science in statement No. 2: The virtual simulator conceived in this way is perfect

Statement 2: The virtual simulator conceived in this way is perfect. Nothing needs to be changed or added.				
	Totally agree	Rather agree	Rather disagree	Completely disagree
Pre-service Geography Teachers	0	2	5	1
Pre-service Computer Science Teachers	0	0	1	3

Authentic comments	
Pre-service Geography Teachers	Pre-service Computer Science Teachers
• Possibility to work with a map, a globe • Improve virtual classroom communication • Expand field of motion • Add time	• I miss the possibility of wiping the board • Parameterization of the number and ages of pupils • Possibility to use the whole classroom area (to walk between desks) • AI connection • To be truly "perfect", the virtual classroom would need an artificial intelligence expansion

is more positive than the other. However, this trend can be observed with a record of the degree of agreement or disagreement with the statement. Looking at the upper half of Table 13.3, similar to the first statement, we can say that geography teachers tend to be more consensual about the statement. Similar to the recorded feedback of participants after their outputs, this statement helped to group important information on possible further development of the virtual classroom.

For the next statement (No. 4), participants were asked to express their agreement with the statement "*I'd rather improve my learning in the virtual classroom than in front of a real class.*" If the participants indicated other options than "*I fully agree*", they had to state why (see the authentic comments at the bottom of Table 13.4). Most of the participants stated that classroom teaching is better, because the classroom is real, the pupils are real, and their answers and responses are authentic. Based on the degree of agreement or disagreement, it could be argued that prospective computer science teachers are more in agreement with geography teachers to improve their pedagogical and didactic skills in the virtual classroom, since half of them tend to agree with statement 4 (see Table 13.4).

The fifth statement was conceived as a YES/NO question. Participants had to indicate whether they would rather make mistakes in front of real pupils than virtual avatars. Looking at Table 13.5, we can say that the results are contradictory. The group of computer science teachers would prefer to make mistakes in front of virtual avatars, while geography teachers would rather make mistakes in front of real pupils.

Table 13.4 Comparison of statements from future teachers of geography and computer science to statement 4: I'd rather improve my learning experience in a virtual classroom than in front of a real class

Statement 4: I'd rather improve my learning experience in a virtual classroom than in front of a real class.				
	Totally agree	Rather agree	Rather disagree	Completely disagree
Pre-service Geography Teachers	0	1	5	2
Pre-service Computer Science Teachers	0	2	2	0
Authentic comments				
Pre-service Geography Teachers	Pre-service Computer Science Teachers			
• A real class is more human, personal • I prefer to see how students react—whether they understand the subject matter, whether they enjoy it • We can talk to real pupils normally (natural situations); • I feel like a person, and not like a computer • Real pupils give me some feedback • Freedom, nature, interaction with a real group of pupils	• It's definitely the classroom environment and climate. Authentic questions from pupils and real reactions • Certainly the reality of the situatio • Students give authentic responses			

Table 13.5 Comparison of statements of future teachers of geography and computer science in statement 5: It is easier for me to make mistakes in front of avatars (virtual pupils) than real pupils in the classroom

Statement 5: It is easier for me to make mistakes in front of avatars (virtual pupils) than real pupils in the classroom.		
	I agree	I disagree
Pre-service Geography Teachers	3	5
Pre-service Computer Science Teachers	2	1

For the 7th statement, the participants evaluated the virtual simulator compared to classical didactic teaching. Half of the geography teachers were totally in agreement with the statement, "*To experience teaching in the virtual classroom is more meaningful for me than to sit and learn about teaching.*" Despite the fact that the evaluation of computer science teachers was not as positive as the evaluation of geography teachers (see the upper part of Table 13.6), we can state that almost all of the participants would prefer to experience teaching in a created simulator, rather than just passively learning about teaching. It should be noted, however, that there

Table 13.6 Comparison of statements of future teachers of geography and computer science in statement 7: Experience teaching in the virtual classroom is more meaningful for me than sitting and learning about teaching

Statement 7: Experience teaching in the virtual classroom is more meaningful for me than sitting and learning about teaching.				
	Totally agree	Rather agree	Rather disagree	Completely disagree
Pre-service Geography Teachers	4	3	1	0
Pre-service Computer Science Teachers	0	3	1	0

Authentic comments

Pre-service Geography Teachers	Pre-service Computer Science Teachers
• We should learn about how to teach, and to do this is a good addition to the virtual classroom • If there was no practical experience in schools, then this would be better than nothing. But I would definitely not replace the classic practice with VR • I am glad I could try the VR, it is variegation, but it certainly will not replace practical experience • If we are learning about some didactic novelties + practical demonstrations for the lessons, I would rather learn	• I think it's definitely better to try teaching this kind of things "rough" than sitting and listening to the same things again and again, how to correctly set the lesson goals, or how to motivate pupils • Practical experience enriches me more than theory • Learning about teaching is just as important as trying teaching in practice, but I don't see the virtual classroom as a full-fledged replacement for practical experience

Table 13.7 Comparison of statements of future teachers of geography and computer science in statement No. 8: The graphic side of the virtual simulator is sufficient

Statement 8: The graphic side of the virtual simulator is sufficient.				
	Totally agree	Rather agree	Rather disagree	Completely disagree
Pre-service Geography Teachers	4	4	0	0
Pre-service Computer Science Teachers	0	1	3	0

Authentic comments	
Pre-service Geography Teachers	Pre-service Computer Science Teachers
• Poor overall graphics and resolution • Poor pupils' faces • Pupils are stiff and follow you every time you move; it's strange • Facial expressions of pupils	• Unfortunately, I don't like pupils' facial expressions, and hands close to their bodies • Unreliable graphic presentation of pupils prevents "being drawn into" VR • The fact that objects fall through the benches is not very beneficial • Huong doesn't have trousers

were concerns about moving their own teaching practices to the virtual classroom (see some comments at the bottom of Table 13.6). The authors reiterate here that this idea is definitely not the goal of the whole project, nor has it ever been considered.

Statement No. 8 focused on finding attitudes to the graphic side of the simulator. Moreover, the statement helped to gain interesting suggestions for possible further improvements of the virtual classroom. According to the authentic comments (lower part of Table 13.7) written by the participants of both groups, we could say that their ranking of the graphic page is very similar. However, looking at the upper half of Table 13.7, we must refute this claim. Geography teachers perceive the graphic side of the simulator as sufficient, while on the other hand, computer science teachers tend to be critical (rather disagree with the statement). Half of the geography teachers agree with the statement "The graphic side of the virtual simulator is sufficient." And the other half rather agree with the statement. In contrast, three computer science teachers tend to disagree with the statement, and only one agrees with the statement.

The last statement of the questionnaire was aimed at finding meaningfulness or uselessness of the created simulator. Again, the statement helped to gather very interesting and beneficial results, on the basis of which the possible inclusion of this simulator in the training of future teachers can be adjusted. Authentic comments from participants of both groups are almost in agreement. The simulator makes sense, but there are also comments that completely criticize the virtual classroom. This assessment is also reflected in the recorded level of agreement or disagreement with the tenth statement "This virtual simulator is absolutely unnecessary. I don't see the point" (see the upper part of Table 13.8). Here we can see slightly more criticism of geography teachers than computer science teachers, because two geography teachers said they rather agree with the statement.

Table 13.8 Comparison of the statements of future teachers of geography and computer science in statement No. 10: This virtual simulator is absolutely unnecessary

Statement 10: This virtual simulator is absolutely unnecessary. I don't see the point.				
	Totally agree	Rather agree	Rather disagree	Completely disagree
Pre-service geography teachers	0	2	3	3
Pre-service Computer Science Teachers	0	0	2	2

Authentic comments	
Pre-service Geography Teachers	Pre-service Computer Science Teachers
• I think this simulator makes sense, especially in showing the mistakes that the teacher makes • It is possible to practice various things on it—summoning pupils, noticing more activity of pupils, where I look after class as I stand out in front of class • It will not replace real reality • I don't know what it's for. It really differs from the real world/reality. Nowhere will you teach for only 5 min • As an alternative way of improving communication, it's good, and I think it has its use	• I think the virtual classroom is good, because it allows you to try how to lead classroom lessons • It only makes sense for non-practice teachers, and provided that it is not possible to gain practical experience in any other way (micro output, real class) • I strongly disagree that the simulator as such would be completely unnecessary, I just think that there are still gaps in the current stage of development • I find the use of such a simulator suitable for completely inexperienced teachers, e.g. before their first practical experience. For students already teaching, I find it rather useless, although the experience is interesting

Summary and Discussion

It should be noted that participants in virtual reality training may suffer from so-called "cyber sickness", which is manifested by unpleasant headaches and nausea. In the pilot study, 2 participants (20%) from the geography teacher group and one participant from the computer science teacher group (20%) were affected by the illness. According to scientific literature, 20–80% of the population suffer from this illness [17], but it is also affected by the specific hardware used for virtual reality (Polcar and Hořejší [16]. Unfortunately, there is nothing to do with the manifestations as such [12], so participants with symptoms immediately stopped their output, and continued the study only as evaluators. However, participation in the individual stages of the pilot study remains beneficial for these students. Participants can improve their assessment, discuss issues, or discuss new didactic or subject topics.

As for Dieker et al. [5], a positive change in the subjective perception of the virtual environment was recorded, depending on the amount of time, respectively, the number of outputs that participants spent in the virtual environment. Conversely, this was the case for the group of computer science teachers who worsened their feedback depending on the number of outputs the participants spent in the simulator.

In both groups, after the first stage, the most suggestions for improvement of the simulator were noted, but during the study the number of statements with inspiration for changing the environment varied significantly. While the geography teachers gradually declined, we did not see any reaction to change after the second level, but again after the third level. We believe that these two phenomena are closely related, and that they are evidence of the higher technological demands of computer science teachers on any didactic technique.

By looking in more detail at the responses to the final questionnaire statements, a difference across groups can be observed when evaluating various issues. When the statements were directed to the graphical or technical side of the simulator (Table 13.7), its authenticity (Table 13.2), or its level of perfection (in terms of application environment) (Table 13.3), geography teachers rated the created simulator more positively than computer science teachers. For the purpose of training didactic skills, geography teachers see the simulator graphically and technically sufficient, while the group of computer science teachers more noticed the graphic and software deficiencies, and evaluated them more negatively. Even computer science teachers pointed out several times that artificial intelligence should be incorporated into the simulator. This apparent contrast in the evaluation of particular aspects of the virtual classroom can be explained by the above-mentioned higher individual demands of computer science teachers on the technical and graphical execution of simulations and games, which can be given by the computer science teachers' focus, and their experience with these environments.

In evaluating the statements concerning the teaching itself (Tables 13.4 and 13.5), the simulator was more positively evaluated by computer science teachers. Geography teachers would prefer to improve their learning in front of a real class, and would also make mistakes in front of the real pupils. Computer science teachers do not have such troubles in front of virtual avatars, and computer science teachers responded more positively than geography teachers to the statement "I'd rather improve my learning in the virtual classroom than in front of a real class". This contrast can be given again by the personalities of computer science teachers, who are probably more used to gaming and simulated environments than geography teachers are.

Both groups agreed on the practicality of the created simulator (Table 13.6). According to almost all participants, training in didactic and subject didactic skills in the virtual classroom is better than attending lectures in general, or subject didactics. Very positive for the simulator authors are reactions to statement No. 10 "This virtual simulator is absolutely unnecessary. I don't see the point." (Table 13.8). None of the participants in both pilot studies fully agreed with this statement, and only 2 geography teachers rather agreed. The rest of the participants do not consider the simulator unnecessary.

Despite the overall positive assessment of both groups, it should be noted that compared to similar studies (e.g. Dieker et al. [11], the created simulator is still perceived more negatively. Everything is obviously due to the low degree of reality of the virtual classroom, which in turn leads to low immersion in virtual reality, as

Lanier (1989) points out. Other authors also assume that the more a user can immerse into a virtual environment, the response should be as if they really were [2].

It should be noted, however, that the virtual classroom is still in the development phase, with the gradual incorporation of specific comments from individual pilot studies in the future. At the same time, we are aware that the number of future teachers who have completed the research is low, and the results can only be considered as a pilot orientation study.

Conclusion

By comparing the assessments of the two groups tested, it can be argued that a virtual classroom, in addition to the training of future teachers of geography, could be a useful complement to the training of future teachers of computer science. However, it turns out that for training in the field of didactics, it is necessary to be able to use subject-specific teaching aids in the virtual classroom. It is therefore essential for the further development of the virtual classroom to extend the library of devices with industry-specific equipment, and to work with other types of classroom layout. Following the pilot studies, we plan to use the virtual classroom in compulsory, optional, and non-compulsory subjects, rather than as a compulsory basis for pre-service teacher training.

References

1. A. Hayes, S. Hardin, C.E. Hughes, Perceived presence's role on learning outcomes in a mixed reality classroom of simulated students. Lecture Notes in Computer Science, 8022, Springer-Verlag, Heidelberg, pp. 142–151 (2013)
2. G. Billingsley, S. Smith, S. Smith, J. Meritt, A systematic literature review of using immersive virtual reality technology in teacher education. J. Interact. Learn. Res. **30**(1), 91–106 (2019)
3. C. Straub, L. Dieker, M. Hynes, C. Hughes, Using virtual rehearsal in TLE TeachLivE™ mixed reality classroom simulator to determine the effects on the performance of mathematics teachers. 2014 TeachLive National Research Project: Year 1 Findings, pp. 1–50 (2014)
4. R. Christensen, G. Knezek, T. Tyler-Wood, D. Gibson, SimSchool: an online dynamic simulator for enhancing teacher preparation. Int. J. Learn. Technol. **6**(1), 201–220 (2011)
5. L.A. Dieker, C.E. Hughes, M.C. Hynes, C. Straub, Using simulated virtual environments to improve teacher performance. Sch. Univ. Partnerships **10**(3), 62–81 (2017)
6. V. Duffek, P. Hořejší, P. Mentlík, J. Polcar, T. Průcha, L. Rohlíková, Pre-service teacher training in the virtual classroom: pilot study, in *DisCo 16 2019: E-learning—unlocking the gate to education around the globe—14th conference reader*, ed. by J. Beseda, L. Rohlíková (Center for Higher Education Studies, Prague, 2019)
7. C. Ferguson, E.L. van den Broek, H. van Oostendorp, On the role of interaction mode and story structure in virtual reality serious games. Comput. Educ. **143**, 103671 (2020)
8. J. Hendl, *Kvalitativní výzkum: základní teorie, metody a aplikace*, 4th edn. (Portál, Prague, 2016)
9. L. Jensen, F. Konradsen, A review of the use of virtual reality head-mounted displays in education and training. Educ. Inform. Technol. **23**(4), 1515–1529 (2018)

10. K. Kelly, A. Heilbrun, B. Stacks, Virtual reality: an interview with Jaron Lanier. Whole Earth Review **64**, 108–120 (1989)
11. L.A. Dieker, J.A. Rodriguez, B. Lignugaris/Kraft, M.C. Hynes, C.E. Hughes, The potential of simulated environments in teacher education: current and future possibilities. Teacher Educ. Special Educ., 37(1), 21–33 (2014)
12. J.J. LaViola Jr., A discussion of cybersickness in virtual environments. ACM Sigchi Bull **32**(1), 47–56 (2000)
13. R. Likert, A technique for the measurement of attitudes. Arch. Psychol. **22**, 5–55 (1932)
14. C.M. Lorenzo, Teacher's skill improvement by role-play and simulations on collaborative educational virtual worlds. J. Educ. Comput. Res. **50**(3), 347–378 (2014)
15. M.R. Dawson, B. Lignugaris/Kraft, Meaningful practice: generalizing foundation teaching skills from TLE TeachLivE™ to the classroom. Teacher Educ. Special Educ. **40**(1), 26–50 (2017)
16. J. Polcar, P. Horejsi, Knowledge acquisition and cyber sickness: a comparison of VR devices in virtual tours. MM Sci. J. **2015**(June), 613–616 (2015)
17. L. Rebenitsch, C. Owen, Review on cybersickness in applications and visual displays. Virtual Reality **20**(2), 101–125 (2016)
18. S. Spencer, T. Drescher, J. Sears, A.F. Scruggs, J. Schreffler, Comparing the efficacy of virtual simulation to traditional classroom role-play. J. Educ. Comput. Res. **57**(7), 1772–1785 (2019)
19. R. Švaříček, K. Šeďová, *Kvalitativní výzkum v pedagogických vědách*, 1st edn. (Portál, Prague, 2007)
20. L.S. Vygotsky, *Mind in society: the development of higher psychological processes*, 1st edn. (Harvard University Press, Cambridge, Mass, 1978)

Chapter 14
A Community of Practice for the Development of Teachers' TEL Skills: A Social Network Analysis Perspective

Carla Limongelli, Filippo Sciarrone, A. Sterbini, and Marco Temperini

Abstract Teacher's, at all levels, are confronted daily with the need to acquire, refine, and apply their skills in Technology Enhanced Learning. It is since many years that professional development initiatives for teachers, and adoption of technology-based tools, have been undertaken to foster new ways of teaching; yet much is still needed, as the recent needs for distance learning, in countries where the schools and universities were closed for social health reasons, has revealed quite dramatically. In this paper we describe an experiment of use of a Community of Practice (CoP), where teachers of different disciplines participated, with the aim to (1) get in contact with technologies of relatively easy adoption, and (2) interact with colleagues, to discuss with and learn from them. The experiment was conducted within the framework of a EU project aiming to bridge the gap between teachers and pupils in the use of the Web technologies. The CoP was designed based on the foundational Wenger's concepts of domain, community, and practice. It supported an educational program on Web2.0 educational technologies for Vocational Education teachers. We present a discussion of the social aspects of the CoP dynamics, using evaluation metrics coming from the Social Network Analysis research area.

Keywords Technology enhanced learning · Community of practice · Vocational education

---

C. Limongelli (✉)
Department of Engineering, Roma Tre University, Rome, Italy
e-mail: limongel@ing.uniroma3.it

F. Sciarrone
Engineering Department, Roma Tre University, Rome, Italy
e-mail: sciarro@ing.uniroma3.it

A. Sterbini
Sapienza University, Rome, Italy
e-mail: marte@diag.uniroma1.it

M. Temperini
Department of Computer, Control, and Management Engineering, Sapienza University of Rome, Rome, Italy
e-mail: sterbini@di.uniroma1.it

A. Visvizi et al. (eds.), *Research and Innovation Forum 2020*,
Springer Proceedings in Complexity,
https://doi.org/10.1007/978-3-030-62066-0_14

Introduction

Teacher's, at all levels, are confronted daily with the need to acquire, refine, and apply, their skills in Technology Enhanced Learning (TEL). It is since many years that professional development initiatives for teachers, and adoption of technology-based tools have been undertaken to foster new ways of teaching; yet much is still needed, as the recent emergency application of distance learning, in countries where the schools and universities were closed for social health reasons, has revealed quite dramatically. In this paper we recall and describe an experiment of use of a Community of Practice (CoP) where teachers of widely different disciplines participated, with the aim to get in contact with technologies of relatively easy adoption, and to interact with colleagues, with which to discuss, and from which to learn.

CoPs can be interpreted as vertical evolutions of Social Networks (SN), where members share a common interest in a particular domain or area, and exchange practical experiences, to increase knowledge and skills related to that specific field [1]. CoPs are not in principle dependent on the use of the Internet, and yet their usefulness and spread is significantly boosted by the use of network technology and Web-based communication tools. In terms of usefulness, CoPs are known to be effective in supporting the learning and development in professional learning communities. It is generally agreed that a significant part of knowledge, protocols, strategies and rules of a professional activity may remain only partially covered, hidden, or just implicit, in traditional educational activities: the mutual engagement fostered by the participation in a CoP can ease this problem, for instance by the construction and sharing of a "collective meaning in the daily routines in the work place" [2]. Other topics in skills development through TEL are broadly discussed in [3, 4].

In this paper we present some evaluation aspects of an experiment carried out within the framework of the European project UnderstandIT.[1] The experiment involved the use of a CoP (UnderstandIT[2]) providing an educational program on Web2.0 Technologies for education, dedicated to Vocational Education and Training (VET) teachers. Technical information and best practices were presented, about how to use some Web 2.0 tools and systems for the development and administering of educational activities pertaining everyday teaching activity. Typical generic tools were discussed (forum, blog, chat and wiki), as the main methods for interaction between participants. Several tools and systems, available through the Web, were also presented and experimented (such as tools to easily develop multimedia resources, systems to manage personal e-portfolio, tools for the self-evaluation of technological skills, web sites on teaching methods and theories). The underlying assumption of the project was that the use of these tools can make teaching more effective, and their knowledge by the "digital immigrants" (the teachers) can let them meet productively the "digital natives" (the students) on the common ground, languages, and services of the modern Web Tools [5]. The UnderstandIT CoP was designed based on the

[1] http://aitel.ist.no/understandit/.

[2] http://understandit.di.uniroma1.it/.

concepts of domain, community and practice [6]. The domain of shared competence addressed here is the teaching activity for VET education; the community members are VET teachers while the practice is the use of the Web 2.0 tools during teaching activities. We used the open source ELGG social network engine[3] as the technological platform.

The evaluation of the UnderstandIT CoP is based on techniques and concepts coming from Social Network Analysis (SNA): we use some centrality measures such as betweeness, centrality, and closeness, in order to gather more information about the network dynamics.

After the description of some related work and background, done in Section "Related Work", we present the UnderstandIT CoP implementation in Section "The CoP Web Application", and its actual use (Section "The UnderstandIT CoP at Work"). Section "Analysis of the UnderstandIT CoP at Work", then, shows the use of SNA means to describe the CoP dynamics. In Section "Discussion and Conclusions" we provide some further discussion and concluding remarks.

Related Work

The concept of CoP has started developing in the world of enterprises, as a mean to effectively engage senior and junior members into professional development, also by the use of mentoring. CoPs are based on the consideration of Learning as a social activity in the first place, to be fostered in the framework of a social network of individuals, where overlapping personal universes of knowledge, values, and practical experience are communicated, compared, shared, and developed by personal and social exchange [1].

Initially Lave and Wenger defined a theory of learning based on practice, where the core concept is the Legitimate Peripheral Participation in CoPs [7]. The novices firstly access the community from the periphery. They acquire experience though the support of more experienced members, and gain reputation also as a consequence of the support that they in turn are able to provide to companions. In this way, they finally achieve full participation and membership. Self-development originates by active participation to the community, and the community develops together with its members. Knowledge is acquired from and applied back to everyday real settings, while discussing it with peers and experts in a rich social system [6].

CoPs have raised a lot of interest among TEL researchers in the last 2 decades. In [8] the authors underline *the growing need to integrate educational research and practice* in order to connect what we know with what we do. The risk of failing to promote personal exploration and responsibility by participating must be countered by the CoP, as an entity, and by its members by encouraging and motivating each person to analyse and constructively criticize, so to support the main accomplishment

[3]https://elgg.org/.

of the network, i.e. allowing mutual complementing of members' experiences and knowledge.

The SEDA project [9] proposes a CoP to support members working in higher education institutions. It is an environment where educational developers can highlight their needs and fruitfully share their experiences. The spirit of the SEDA project is the mutual support provided by the members of the community.

The UnderstandIT project is a Leonardo da Vinci - Transfer of Innovation project, supported by the EU under the Lifelong Learning Programme.[4] Its partners were from Germany, Lithuania, Denmark, Italy, Portugal, and Norway. The project aimed to include ICT in the competences of VET teachers, trainers and tutors. The project activities went through the design, implementation and experimentation of a course dedicated to VET and general High School teachers, and directed to the development of abilities regarding the use of Web2.0 tools and methodologies to attract and motivate students, and to enhance the overall learning experience. Four courses (localized to fur partners' country languages) were provided through dedicated CoPs. Each course was articulated in nine lessons, organized in the structure of a "group" of the UnderstandIT Community of practice. The course activities were designed to provide the teachers with the pedagogic concepts related to the use of technology for education, and with ample opportunities to use tools to implement their course works. This was accompanied by the CoP related activities, related to discussion and exchange of knowledge and opinions. The teachers were also exposed to concepts related to coaching, as they were supposed to spread in their school the proceedings of the course afterwards.

[10] presents a CoP dedicated to early childhood teachers. Besides the support to development of professional skills, and stakeholders' engagement (such as parents), the benefits of the use of a CoP are shown also in terms of how social interaction allowed to support the teachers each other, elicited self-reflection of one's professional identity, and ultimately allowed to meet with challenging professional occurrences at the workplace with greater success.

In [11] the social, collaborative nature of a CoP is stressed as a factor providing the participants with the opportunity of co-creating solutions to practical problems. The paper provides tips about the implementation of a CoP, that can be of general use, although the focus is on CoP's suitability for the collaborative development of best practices in Health professions.

About SNA, it is useful here to recall that it deals with the analysis of social networks in order to trace the relationships holding among members, learn their meanings and apply such findings to better understand the dynamics of members, and groups interactions.

A Social Network can be rendered as a directed or undirected graph, where (1) the actors are represented by the nodes, (2) the relationships among the actors are indicated by the graph's edges, and (3) weights can be assigned to the edges between nodes, to designate different interactions strengths [12–14].

---

[4]Project number: 2010-1-NO1-LEO05-01839, http://aitel.hist.no/understandit.

Because of this representability of SN as graph, SNA borrows many concepts and tools from the graph theory. In Sect. 5 we will analyse the dynamics of the UnderstandIT CoP and visualize them by means of the *Gephi* graph tool,[5] an open source tool useful for SNA 15]. Several metrics and operations are defined to allow for such an analysis [16, 13], such as *Size* and *Density* (related to the amount of interaction), *Clique detection* (which is essential to see how the members group in *sub-communities*), and several declinations of the concept of *centrality*, which represents the importance, and presence, of a member in the network alongside the others. *Degree* centrality measures the number of interactions that a node (member) has with the others, possibly its popularity; *Betweenness* centrality says how a node is important as a connector for other nodes, so fostering exchanges among members; *Closeness* centrality, based on shortest path computations, provides an evaluation of how information can spread more or less quickly from certain nodes towards the others. Besides these, several other useful information can be extracted from the log files of a social network, and of a CoP: They can allow to comprehend the existing ways of interactions, and possibly help acting to make them more fruitful [17].

The CoP Web Application

The design of the system started from the requirement analysis, based on an input coming from a sample of VET teachers to identify a list of useful features to deploy:

- Groups common goals sharing, and shared editing tools;
- Friends, groups, stream of interactions and event calendar and RSS feed import;
- Antispam protection;
- FAQ managing;
- Activities sharing among friends;
- Embedding external contents;
- Tracking changes/hot topics;
- Communication: Sync. (chat) and async. (blog, comment, tweet, forum).

Then we selected the ELGG framework. ELGG is a LAMP (Linux, Apache, MYSQL, Php) web application that provides a basic set of services for a social network system, allowing for a complete interface configuration and extension of the functionalities (by adding plug-ins that are either newly implemented or selected among the many already available from the ELGG development community). Finally, the system was customized and deployed.

Members share and exchange knowledge by operations such as adding a new bookmark, posting a link to a particular video, publishing new pages and so on.

Other members might feel the need to access, comment, modify, spread, the knowledge shared by other members, so, among the several features offered by a social networking platform, the setting of access rights associated to each contributed

[5]http://www.gephi.org.

resource, is relevant. In particular, the author of a contributed resource can specify a group (a class of users) to whom allow either read access or read and write access rights on the resource.

The group is a subset of CoP members. The following group specifications are available: *Everybody*, *Members*, *Group*, *Your friends*, *A subset of your friends*, *Private*. Depending on the write access rights of a post, the contributed contents can be edited asynchronously by different users. Each post in the system can be tagged by the author, i.e. it can be associated with one or more keywords. This allows for searching through the system resources by both full text and tag-based selection.

Another characteristic of the system is that the UnderstandIT CoP was configured in such a way to allow for discussions in several language-specific communities, with a common discussion space in English. This feature was implemented through the concept of *Group* managed in ELGG, where a Group is a subset of all the CoP members that is owned by a member and can be moderated. From an operative viewpoint, each group represents a (sub-)community whose members can interact through:

- Group Blog posts, Group Bookmarks, and files and Group pages;
- Embedding videos/podcasts and RSS feeds in posts, and uploading files (File Tool);
- Embedding external content by reference through the Media Tool;
- Commenting on posted resources (files, media, bookmarks and blogs);
- Sharing interesting resources through the Bookmarks Tool;
- Monitoring (importing) RSS feeds through the Feed Tool;
- Managing and or participating in Forums and Chats.

To increase support to collaboration, a synchronous shared editing plug-in was installed, based on the Etherpad open source project.[6] Through it, Group members can collaborate editing a document, also being supported by a small chat.

Users are kept updated about the CoP activities either through automatic e-mail notifications or through the CoP web interface. Both the former and the latter collect information from many relevant areas of the network, offering notifications such as:

- E-mail, RSS feeds, status tweets;
- Group activities;
- Last changes in (1) the main CoP's page, (2) the user's dashboard, (3) each Group main page, or (4) in each page's history;
- Contextual tag cloud with information about terms used in the CoP and their location.

Members can report occurrences of spamming, eliciting an intervention from the administrator. The discussions can be externally disseminated towards *Twitter*. The interface is in the languages of the UnderstandIT project partners (Lithuanian, Italian, Portuguese, German, Danish, and English). Cooperation among different language groups is supported by a *Google Translate*-based tool.

---

[6]https://etherpad.org/.

The UnderstandIT CoP at Work

The UnderstandIT CoP was used as a container for an online course deployed adopting peer to peer coaching and mentoring strategies. The course was attended by a sample of 77 teachers, coming from different didactic experiences, as shown in Fig. 14.1.

The course was designed based on 16 Learning Outcomes (LO). The comprehensive LO was: *I can choose relevant resources using the CoP approach for any planned learning activity including ICT-tools where these are the most appropriate and I am able to change my teaching style so to use a more coaching oriented approach.*

The course included several tasks such as:

- Practice with Web 2.0 tools for doing and analyzing one's own teaching;
- Analysis of existing CoPs, SNs, blogs to coach sessions with course tutor and peers;
- Final design of a plan to experiment acquired Los in current teaching activity.

Figure 14.2 reports on some data about participation and overall CoP activity in terms of standard metrics. Members have been positively engaged in the CoP life. For instance, about blog activity 90 discussions were managed, with 700 posts.

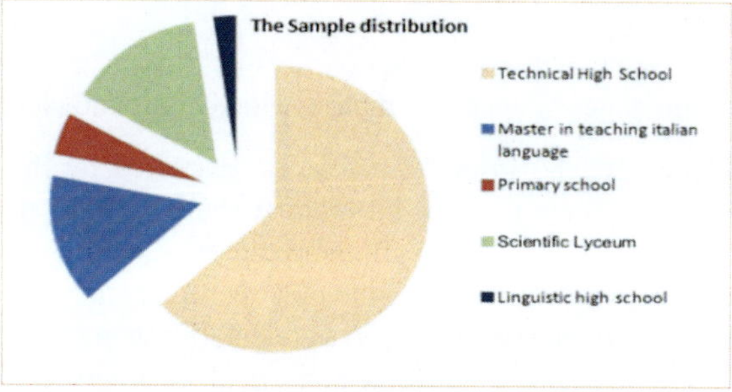

Fig. 14.1 The sample distribution of the didactic work experiences of the CoP members

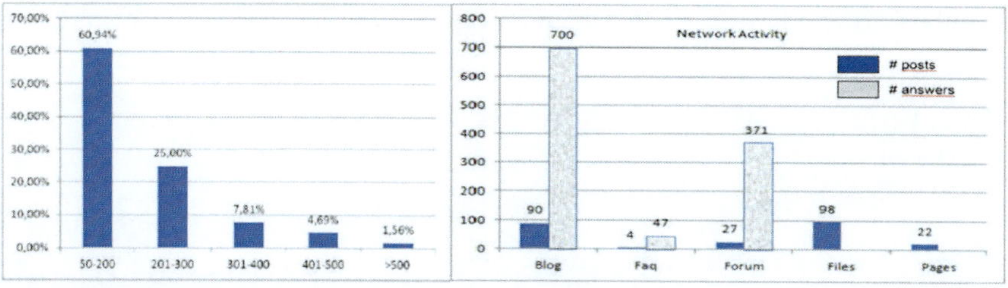

Fig. 14.2 Member's visits to the web site (left) were about 2000, with an average per member of 31.25. About the overall CoP activity (right), blue bars represent number of posts, while the grey bars represent the number of answers originated by those posts in Forum, Blogs and FAQ

Analysis of the UnderstandIT CoP at Work

Among the participants to the experiment, 70% have had already experience in at least one online course, although 71% did not know of CoPs. They had previous experience with some web tools: 79% had been in a Forum, 29% knew of RSS, and blog, chat, podcasting, and wiki, were between the above values. About user satisfaction and perceived success we used a Likert scale, five values from lowest assessment to highest, and asked whether (1) the expectations about the course were met, and (2) would the participant feel prepared to be a coach to colleagues on the acquired skills.

To the first question 62% answers were above the neutral (that was 20%): we interpreted this a good result, considering that the members kept participating while continuing to work in their classrooms. The second question had overall 56% of positive answers, with 30% to neutral: we decided to give this question, and not another more generically measuring the personal acquisition of skills, as being a reference point for colleagues seemed the toughest factor to measure the effectiveness of the course, and we interpreted these figures as a good (if not enthusiastic) result.

Then we evaluated the CoP dynamics by means of three main SNA measures in order to appreciate information about the following relationships:

- Friendship relationships (How the members built their list of friends to interact with)
- Participation to group activities;
- Exchange of messages between members, which elicit knowledge sharing.

Figure 14.3 presents the Friendship relation extracted by the log data of the CoP. The graph helps seeing how each member builds her/his list of "privileged counterparts". It is easy to spot four sub-communities.

In Fig. 14.4 the network representing the participation of the community members to social activities is shown. We have again a clear partition in four sub-communities; however, two sub-communities show a good deal on intersection, meaning that several of their members were active in the same groups, and so showing a remarkable commonality of interests.

Finally, Fig. 14.5 shows the graph representing the exchange of knowledge among members, after a brief period of the online activities. The graph presents the shape

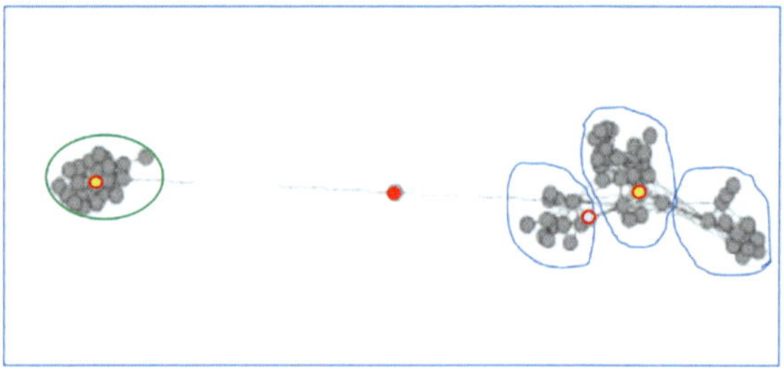

Fig. 14.3 The Friendship Relationships

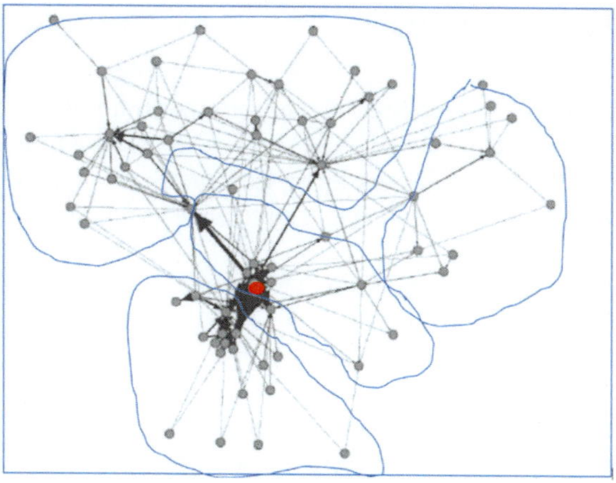

Fig. 14.4 The development of group activities in the CoP

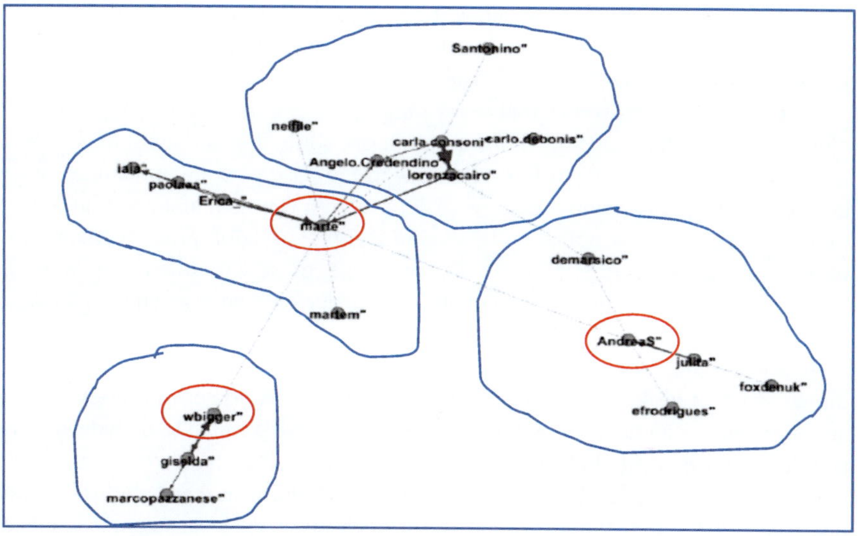

Fig. 14.5 Knowledge exchange among members

of the message exchanges among members; while it is not surprising, in view of the previous figures, that we see again a partition in four sub-communities, here we can also appreciate how "main" (more "central" in SNA parlance) members are operating in each sub-community, and so how they are in turn relevant with respect to certain topics of interest developed in the whole CoP.

Discussion and Conclusions

TEL studies cover a widespread set of application of information and communication technologies, with educational aims, ranging from more formal and traditional applications [18, 19], through less traditional and more social-collaborative instances [20, 4], to learning experiences making a more wide use of the connecting power of social networks [21].

In this paper we have presented the design and implementation of the UnderstandIT CoP, the results of using it in an experimental setting, and the application of SNA methods and techniques to this case study.

Through SNA we performed a preliminary structural analysis of the network. We took into consideration the distribution of relationships among all the members involved in the learning process, focussing on the dynamics of such members' relationships.

The analysis of our findings allowed us to evaluate some basic aspects of the interactions occurring through the network. For example, we discovered different sub-communities showing some integrations among all the members in all the activities carried out. We also saw that there are key figures in the information flows between the members.

Our experience showed us that everything, in these communities, happens in the context of a spontaneous flow of information, so, while it is not possible (say, by an administrator of the CoP) to force changes in the pace of communication, several characteristics in the mentioned flow of information can be maintained, in order to make the members' participation lively and fruitful. An appropriate feeding by the most estimated members (the core of the CoP, i.e. those with higher reputation), a prompt feedback to inquiries, most of all from novice members, and the organization of online events, can maintain the community alive and healthy.

In a worthwhile multi-lingual setting, it would be quite natural that the language dimension should prevail on other aspects, unless appropriate translation services are provided. As a matter of fact, in a future perspective we plan to include keyword translation for multilingual labelling of contents and online translation of pages. On the other hand, a hierarchical inspection of detected clusters and the use of finer measures, or even of the same ones on a restricted set of participants, can help highlighting more covered processes.

References

1. E. Wenger, Communities of practice and social learning systems: the career of a concept. *Social Learning Systems and Communities of Practice* ed. by C. Blackmore, 179–198. The Open University, (2010)
2. A.-C. Wennergren, U. Blossing, Teachers and students together in a professional learning community. Scand. J. Educ. Res. **61**(1), 47–59 (2017)
3. A. Visvizi, L. Daniela, Technology-enhanced learning and the pursuit of sustainability. Sustainability **11**(15), 4022 (2019). https://doi.org/10.3390/su11154022

4. A. Visvizi, L. Daniela, Ch.W. Chen, Beyond the ICT—and sustainability hypes: a case for quality education, Computers in Human Behavior, vol 107, June 2020
5. M. Prensky, Digital natives, Digital Immigrants. University Press, **9**(5), (2001)
6. E. Wenger, Communities of practice: learning as a social system. Syst. Thinker **9**(5), 2–3 (1998)
7. J. Lave, E. Wenger, *Situated learning: legitimate peripheral participation* (Cambridge University Press, Cambridge, England, 1991)
8. V. Buysse, K.L. Sparkman, P.W. Wesley, Communities of practice: connecting what we know with what we do. Except. Chil. **69**(3), 263–278 (2003)
9. S. Nixon, S. Brown, A community of practice in action: seda as a learning community for educational developers in higher education. Innovations Educ. Teach. Int. **50**(4), 357–365 (2013)
10. J. Kirkby, L. Walsh, A. Keary, A case study of the generation and benefits of a community of practice and its impact on the professional identity of early childhood teachers. Prof. Dev. Educ. **45**(2), 264–275 (2018)
11. M.A. de Carvalho-Filho, R.A. Tio, I. Steinert, Twelve tips for implementing a community of practice for faculty development. Med. Teach. **42**(2), 143–149 (2020)
12. S. Wasserman, K. Faust, *Social Network Analysis: Methods and Applica-tions* (Press, Cambridge Un, 1994)
13. M. Zhang, Social network analysis: history, concepts, and research. In *Handbook of Social Network Technologies*, ed. by Borko Furht, Springer, 2010, pp. 3–21
14. A. Marin, B. Wellman, *Social network analysis: an introduction*, ed. by Peter Carrington, John Scott, Handbook of Social Network Analysis, Sage, 2010
15. M. Bastian, S. Heymann, M. Jacomy, Gephi: an open source software for exploring and manipulating networks. Int. AAAI Conference on Weblogs and Social Media, (2009)
16. D.M.A. Hussain, *Investigation of Key-Player Problem in Terrorist Networks using Bayes Conditional Probability*, ed. by Borko Furht, Handbook of Social Network Technologies, Springer, 2010
17. K. Slaninova, J. Martinovic, P. Drazdilova, G. Obadi, V. Snasel, *Analysis of Social Networks Extracted from Log Files*, ed. by Borko Furht, Handbook of Social Network Technologies, Springer, (2010)
18. F. Gasparetti, C. De Medio, C. Limongelli, F. Sciarrone, M. Temperini, Prerequisites between learning objects: Automatic extraction based on a machine learning approach. Telematics Inform. **35**(3), 595–610 (2018)
19. C. Limongelli, G. Mosiello, S. Panzieri, F. Sciarrone, Virtual industrial training: joining innovative interfaces with plant modeling. Proc. Int. Conf. on Information Technology Based Higher Education and Training, ITHET, IEEE
20. M. De Marsico, F. Sciarrone, A. Sterbini, M. Temperini, Supporting mediated peer-evaluation to grade answers to open-ended questions. Eurasia J. Mathe. Sci. Technol. Educ. **13**(4), 1085–1106 (2017)
21. C. Limongelli, F. Sciarrone, M. Temperini, A social network-based teacher model to support course construction. Comput. Hum. Behav. **51**(PartB), 1077–1085, Elsevier, (2015)

Chapter 15
Key Factors that Boost the Effectiveness of Virtual Teamwork in Online Higher Education

Akinul Islam Jony and Enric Serradell-López

Abstract A globalized nature of today's business along with the advancement of Information and Communication Technology (ICT) has made virtual teamwork a significantly important skill to the employer as well as to the future graduates for their professional career success. At the same time, it is believed that higher education is the institutional foundation for a graduate to be prepared for his/her professional career. But, the key factors that affect the effectiveness of virtual teamwork development are still not explored enough in the field of online higher education. Therefore this paper investigates the key factors to develop effective virtual teamwork in online higher education which eventually proposed as a conceptual model. This conceptual model of the key factors of virtual teamwork is derived based on the existing literature. The application of this model can help future graduates from online higher education to gain the required skills, knowledge, and ability to participate in virtual teamwork successfully and foster their early success in their professional careers. It can be used in the evaluation of a virtual team. Also, top managerial personnel of a globalized company can apply this model for the management of virtual teamwork.

Keywords Virtual teamwork · Higher education · E-Learning · Effectiveness · Key factors

Introduction

The growing demand for virtual teamwork to employers in business organizations [55] has rationale the importance of implementing virtual teamwork in higher education. So, graduates from higher education can become skilled in virtual teamwork

A. I. Jony (✉) · E. Serradell-López
Universitat Oberta de Catalunya, Barcelona, Spain
e-mail: ajony@uoc.edu

E. Serradell-López
e-mail: eserradell@uoc.edu

A. Visvizi et al. (eds.), *Research and Innovation Forum 2020*,
Springer Proceedings in Complexity,
https://doi.org/10.1007/978-3-030-62066-0_15

before starting their professional careers, and achieve early success in their professional life. Besides, in online higher education, teamwork only can take place virtually as there is no option for face-to-face interaction. Considering these facts it is become essential to develop effective virtual teamwork in online higher education, where identification of the key factors of virtual teamwork and the relationships among them are the primary concern for developing such a model. So, the purpose of this paper is to develop and propose a conceptual model with a set of key factors of virtual teamwork and the relationships among them that foster the effectiveness of virtual teamwork in online higher education.

In virtual teamwork, *performance* is the main responsible factor like the traditional face-to-face teamwork [52] to evaluate its effectiveness [25, 47]. Performance is defined as the measurable outcome of any task [65] to achieve the desired outputs or the goals [4]. In this study, performance means to measure, evaluate or judge the effectiveness of teamwork [26] in the context of online higher education, which is also supported by many research works (e.g. [4, 15, 18, 47, 48, 71]). However, there are other factors that are critical and affect *performance* directly or indirectly, and eventually, foster the effectiveness of virtual teamwork in online higher education. Content analysis is performed on the existing literature to identify the key factors of virtual teamwork, which has revealed some key factors that boost the effectiveness of virtual teamwork in online higher education (e.g. [15, 18, 36, 47]).

The rest of the paper is structured as follows. How the content analysis is performed is presented in Section "Methodology". The detailed description of the key factors is presented in Section "Key Factors". The relationships between the factors are presented in Section "Relationship & Hypotheses Development" along with the derivation of hypotheses. Finally, the conceptual model along with the set of hypotheses is proposed and presented in Section "The Proposed Conceptual Model", and the concluding remarks are drawn in Section "Conclusion".

Methodology

The plan of this article is to discover and explore more about virtual teamwork in online higher education, and its performance criteria (i.e. effectiveness criteria) that influence the effectiveness of virtual teamwork. Therefore, qualitative research method will be used in this paper for exploring the topic and discovering the key factors of virtual teamwork that affects the effectiveness of virtual teamwork because this method is suitable for discovering and exploring more about a topic in terms of people's experiences and perspectives [31, 45]. Besides, in general, a qualitative research method used for interpreting or exploring information on a topic, whereas, the quantitative research method is used for measuring the data points statistically about the topic [32]. In addition, qualitative methodology is suitable for examining the unit of people in detail on a chosen portion of the range [45]. Finally, the type of exploration of this study topic towards developing a conceptual model is particularly aligned with the qualitative research method, and hence, chosen and applied here.

For developing and proposing a conceptual model based on the existing literature, an extensive systematic review is performed. The systematic review is performed based on the prestigious, popular, and widely used database, Web of Science. However, for not missing any related information and research work, the other common databases (such as Scopus, ACM Digital Library, Springer, IEEE Xplore, and Google Scholar) are also considered for the exploration and content analysis of the topic towards developing the conceptual model of this research work.

Key Factors

The set of selected key factors that affect the effectiveness of virtual teamwork in online higher education, is depicted in Fig. 15.1. Besides these key factors, there are other factors found in the literature which influence the effectiveness of virtual teamwork but those are not critical in the context of online higher education, and hence, are not considered in this study. The selected key factors will eventually constitute the conceptual model that is going to propose at the end of this paper. The description of these factors actually represents the so-called coding scheme of the conceptual model and the justification of being included in the conceptual model.

Performance

In a virtual team, each team member works independently on a given task to complete it [24], where the performance means to get the desired output from the assigned task,

Fig. 15.1 Key factors of virtual teamwork in online higher education

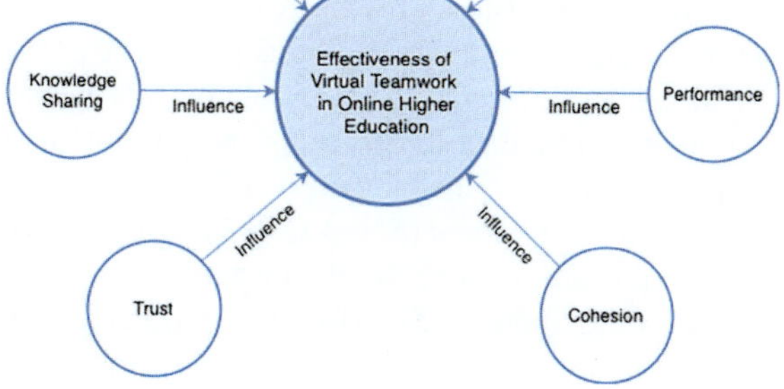

by following the planned objectives and requirements of the team's goal [4]. However, team member performance is boosted if they are given importance, value, encouragement, and appreciation [24]. Basically, each team member is responsible for their own performance, and a good performer is tried hard to improve the performance at the same time correcting their mistakes. On the other hand, team performance means combined work for achieving the team's goals [33, 60] by interacting and collaborating with the members of the team [68]. Thus, team performance measurement should be clearly defined according to the team's goals. In teamwork, if all team members are at the almost same level in terms of communication and involvement, it's easier to achieve higher team performance [60] because it ensures the awareness about the responsibility among the team member that is, who is performing which duties in the team [11, 15] to meet the team's goals.

Besides, a team leader's performance can ensure the performance of teamwork by managing and executing required activities for achieving the team's goals and objectives [48, 71]. Thus, in a virtual environment, a team leader should be skilled in various aspects and competencies [15, 61]. A skilled team leader builds trust among the team members and maintains a neutral atmosphere in the team [61]. Besides, a team leader is responsible for managing all resources for performing teamwork and maintaining a clear plan and schedule for the successful completion of the teamwork according to the team's goals [71]. But, in higher education, having a team leader in a virtual team of students is not a very common scenario, and hence, it is not considered in this study.

In a nutshell, performance will be considered as the target key factor in the conceptual model of this study that evaluates the effectiveness of virtual teamwork in the context of online higher education.

Motivation

In teamwork, motivation is the level of enthusiasm which encourages team member to actively work in a team to achieve the team goal [38, 63]. It influences the performance of the team [14] as a consequence, affect the effectiveness of teamwork in a virtual environment as well. On the other hand, without motivation team member feels discouraged which affects the performance and decreases the effectiveness of the teamwork [38]. Therefore, motivation is considered as one of the key factors in terms of the effectiveness of virtual teamwork [22, 51].

Moreover, a motivated team member always stays active within the team and willingly contributes to achieving the goal of the team [3]. For that reason, it is important to know how a team member get motivated while working in a team in the virtual environment [51]. A team member can be motivated in various ways such as giving incentives or rewards for good performance or showing positive attitudes whatever the situation [51]. However, in the context of online higher education, team members can be motivated when they are challenged by the tasks given, or when they are confident about the learning outcome and future prospect, and/or when it contributes sufficiently to their grades.

Communication

In virtual teamwork, communication refers to the exchange of information and discussion between team members by means of electronic documents, text or voice messages, video conferences, emails, blogs, and forums [15, 47, 56]. Technological development makes the communication process easier for an activity like virtual teamwork. In the context of virtual teamwork, communication is the first thing to be considered seriously for its effectiveness [12, 15, 42, 61, 71] because fruitful communication makes knowledge sharing, periodical meeting, and discussion easier. Therefore, communication between team members has a strong influence on the performance and effectiveness of virtual teamwork [11, 61, 64, 71].

On the other hand, poor communication causes many problems in virtual teamwork [57], and eventually affects the effectiveness of virtual teamwork negatively. For example, a study by Schwalbe mentioned that failure in communication between team members is one of the major causes against the success of virtual teamwork [60]. Besides, the importance of communication is sometimes overlooked and undervalued [15, 56] which leads to failure in virtual teamwork. That's why, team members without having proper communication skills, abilities, and knowledge it is difficult to obtain the effectiveness of virtual teamwork [12]. Thus, communication is considered as one of the key factors that affect the effectiveness of virtual teamwork, and it is advisable to measure the communication factor for effective virtual teamwork development in the context of online higher education.

Knowledge Sharing

A team consists of individuals with the necessary skills and expertise to collaborate on organizational tasks [23]. This type of orientation of a team enables the distribution of knowledge and information among team members and facilitates effectiveness in teamwork and achievement of more effective outcomes [34]. So, the exchange of knowledge, or information, or experiences between team members is called knowledge sharing in the case of virtual teamwork.

However, knowledge sharing is a voluntary activity in a team [18]. So, a team member should have a willingness to share knowledge/information in the team [29] when necessary for the effectiveness of virtual teamwork. By the means of knowledge sharing, team members can acquire new knowledge, information, experiences, know-how, and organizational and business policy which improves their performance [18]. As a consequence, it affects the effectiveness of teamwork [1]. That means knowledge sharing has a positive influence on the effectiveness of virtual teamwork [1, 18].

Moreover, knowledge sharing practices within the team is considered as the key activity of virtual teamwork to enhance the performance and effectiveness of the virtual teams [72]. Therefore, knowledge sharing is also considered as the key factor for the conceptual model of virtual teamwork of this study.

Trust

Trust is referred to as the basis of interpersonal cooperation in organizations [43] and hence considered as the key factor that affects the performance or effectiveness of virtual teamwork [15, 35, 71]. The presence of trust in the team encourages the sharing of knowledge and makes knowledge sharing activity easier and effective [1]. Besides, trusting each other in the team makes easier to accept ideas, share knowledge, and come up with a decision that eventually fosters the performance of teamwork [1, 15, 71, 60], and eventually affects the effectiveness of virtual teamwork. Therefore, trust is considered a key factor and has a significant impact and positive influence on the success and effectiveness of a virtual team [19, 21, 28].

Moreover, Rad and Levin (in [48]), and Wise (in [71]) mentioned that the higher level of trust between team members increases the performance level, and hence, increases the possibility of successful completion of virtual teamwork [15, 16]. That's why trust can be stated as the key factor for assessing between high and low performing virtual teams [36]. Therefore, trust is also considered as one of the key factors of virtual teamwork for the development of the conceptual model of this study.

Cohesion

In teamwork, cohesiveness means to stick together in the team until the team goal is not achieved, and the team members are satisfied with the outcomes of the teamwork [66]. Carron et al. defined team cohesiveness as "a dynamic process that is reflected in the tendency for a group to stick together and remain united in the pursuit of its instrumental objectives and/or for the satisfaction of member affective needs" [6].

The above definitions of cohesion indicate it as one of the key factors for the success and effectiveness of teamwork. Besides, research suggests that cohesion is a critical factor that fosters the effectiveness as well as the performance of the virtual teamwork (e.g. [20, 38, 42]). A study by Cohen and Bailey also stated that cohesion is a critical factor that improves the performance of the team and the team members [8]. Another study by Powell et al. (in [55]), and Maznevski and Chudoba (in [42]) also provided a similar suggestion that cohesion is one of the key factors for the improved performance and effectiveness of virtual teamwork. So, team cohesion helps team members to stay united in order to achieve the team's goal. The higher the cohesiveness in the virtual teams, the better the performance and effectiveness.

Coordination

In a virtual team, coordination is positively linked to the performance or effectiveness of virtual teamwork [30, 42]. Coordination means the effort of virtual team members for managing the collective resources of the team and maintaining a consistent and coherent team activity [54, 55]. In addition, coordination can play a big rule in conflict management in the virtual team and team performance [44]. These suggest that coordination is crucial for the effectiveness of virtual teamwork. Therefore, coordination is also considered as one of the key factors of virtual teamwork for the development of the conceptual model in the context of online higher education.

As a summary, all the key factors are listed in Table 15.1 along with their supporting references which justifies their inclusion in the conceptual model (to be proposed) as key factors for the effective development of virtual teamwork in the context of online higher education.

Relationship & Hypotheses Development

This section presents the relationships between the key factors to construct the conceptual model for effective virtual teamwork development in online higher education. All the relationships are explained, discussed, and presented based on the existing literature and theories. That's why their inclusion in the conceptual model are justified undoubtedly with sufficient supporting references. All these relationships of the key factors are presented and discussed as follows along with their supporting references from the existing literature and theories, and according to the best of the knowledge. In addition, a set of hypotheses is also derived and proposed from the following explained relationships of the key factors.

Table 15.1 Set of key factors with supporting references

Key Factors	Supporting References
Motivation	[3, 22, 51]
Communication	[12, 15, 42, 61, 64]
Knowledge Sharing	[1, 18, 72]
Trust	[16, 19, 21, 28, 71]
Cohesion	[8, 20, 38, 42, 55]
Coordination	[30, 42, 44]
Performance	[4, 33, 60]

Motivation to Communication

Motivation factor stimulates desire and energy in team members to be continually interested and committed to a job or role or a specific topic to make an effort for achieving a team's goal [38, 63]. Since team members in virtual teams do not have traditional face-to-face interaction, virtual collaboration might meet challenges, where motivation factors can play a vital role to overcome the challenges [14]. The reason is that a motivated team member always contributes to teamwork willingly, and stays active and focused within the team for achieving the team's goal [3]. On the contrary, lack of motivation discourages team members which eventually decreased the performance and effectiveness of virtual teamwork [38]. That's why it is necessary to motivate the team members while working in a virtual team for the effectiveness of virtual teamwork [51]. Besides, researchers agreed that communication is the first thing to be considered seriously for the effectiveness of virtual teamwork [12, 15, 42, 61, 71]. These suggest that the *Motivation* factor should have a strong positive impact on the *Communication* factor towards developing effective virtual teamwork in online higher education. Therefore, according to the discussion and explanation of this relationship between *Motivation* and *Communication* factors, the following hypothesis is derived. $\boldsymbol{H}_{MT \rightarrow CM}$: Motivation affects positively to Communication.

Communication to Knowledge Sharing

Effective communication between team members facilitates the sharing of information and knowledge clearly and accurately to achieve the team goal [7, 9, 60, 62]. Besides, communication plays a vital role to articulate team member's opinions, express plans and goals, share ideas or knowledge or information and discuss each other's viewpoints [7, 17, 60]. That's means, *Communication* affects *Knowledge Sharing* towards the effective development of virtual teamwork in online higher education. According to the discussion and explanation of this relationship between *Communication* and *Knowledge Sharing* factors, the following hypothesis is derived. $\boldsymbol{H}_{CM \rightarrow KS}$: Communication affects positively to Knowledge Sharing.

Communication to Trust

Effective communication and participation among the team members ensure the trust in the team which eventually leads to reliable performance and concern for each other in teamwork [11, 71]. The building of trust among the team members and in the team gradually improves over time based on communication [41, 58]. According to Jarvenpaa and Leidner, communication is the key to building trust in virtual teams [28]. That means *Communication* has a significant impact on *Trust* building in the virtual team which enhances the performance and eventually fosters the effectiveness of virtual teamwork. According to the discussion and explanation of this relationship between *Communication* and *Trust* factors, the following hypothesis is derived. $\boldsymbol{H}_{\boldsymbol{CM \rightarrow TR}}$: Communication affects positively to Trust.

Communication to Cohesion

Cohesion is considered as an important factor for the effectiveness of virtual teamwork [67]. However, a significant effort needs to take in the virtual team's communication than in the traditional face-to-face team for building effective cohesion in virtual teams [54]. Besides, in comparison to the face-to-face team, the virtual team generally has weaker interpersonal relationships between team members [70]. These suggest that communication can foster cohesiveness in virtual teams. That's means, *Communication* has a positive impact on *Cohesion* towards developing effective virtual teamwork in online higher education [54, 69, 70]. According to the discussion and explanation of this relationship between *Communication* and *Cohesion* factors, the following hypothesis is derived. $\boldsymbol{H}_{\boldsymbol{CM \rightarrow CH}}$: Communication affects positively to Cohesion.

Knowledge Sharing to Coordination

The exchange of information or knowledge between team members fosters or improves the coordination of virtual teamwork [40, 42]. That means virtual team members should share and distribute knowledge adequately for effective collaboration in virtual teamwork [49, 53, 72]. Otherwise, virtual teamwork will be less efficient and poor in coordination, for example, suffering from relevant information, difficulties in decision-making, miscommunication, etc. [53]. These suggest that *Knowledge Sharing* has a positive impact on *Coordination* towards developing effective virtual teamwork in online higher education. According to the discussion and explanation of this relationship between *Knowledge Sharing* and *Coordination* factors, the following hypothesis is derived. $\boldsymbol{H}_{\boldsymbol{KS \rightarrow CR}}$: Knowledge Sharing affects positively to Coordination.

Trust to Coordination

Trust is considered an essential factor for the coordination process in teamwork which ultimately improves the performance of the virtual team. The act of one team member affects other members of the team [23]. The relationship among individuals in a virtual team has a big influence on the team's action, coordination, resources management, and collaboration [27]. The research reported trust as a key factor in reducing risk management, dealing with complexity and uncertainty, facilitating a positive atmosphere to team members for collaborating within a social system like virtual teamwork (e.g. [5, 28, 50]). Besides, the research found that trust has a direct effect on the team's outcome (e.g. [46, 49, 50]). These suggest that *Trust* has a positive impact on *Coordination* factor towards developing effective virtual teamwork in online higher education. According to the discussion and explanation of this relationship between *Trust* and *Coordination* factors, the following hypothesis is derived. $\boldsymbol{H}_{TR \rightarrow CR}$: Trust affects positively to Coordination.

Cohesion to Coordination

Cohesion plays a vital role in successful coordination in virtual teams [35, 59]. A strong cohesion between team members leads to better coordination in virtual teamwork [10]. These suggest that cohesiveness nature of teams positively affects the coordination in virtual teamwork. That's means, *Cohesion* has a positive impact on *Coordination* towards developing effective virtual teamwork in online higher education [10, 35, 59]. According to the discussion and explanation of this relationship between *Cohesion* and *Coordination* factors, the following hypothesis is derived. $\boldsymbol{H}_{CH \rightarrow CR}$: Cohesion affects positively to Coordination.

Knowledge Sharing to Performance

A team performs better when it contains individuals with relevant knowledge and skills [13, 39, 49]. In other words, knowledge sharing practices within the virtual teams enhance the performance of the team [72]. Besides, Alsharo et al. suggested that knowledge must be shared or exchanged among the team members for the effectiveness of the virtual team [1]. These suggest that *Knowledge Sharing* is positively related to *Performance*, or has a positive impact on the performance of the effectiveness of virtual teamwork. According to the discussion and explanation of this relationship between *Knowledge Sharing* and *Performance* factors, the following hypothesis is derived. $\boldsymbol{H}_{KS \rightarrow PF}$: Knowledge Sharing affects positively to Performance.

Trust to Performance

Team performance can be improved if team members among the team trust and rely on each other, otherwise it leads to failure [4]. Research indicates that trust has positive influences on performance in virtual teamwork (e.g. [2, 58, 60]). The higher level of trust among team members improves the performance levels of the virtual teams [48, 71]. Thus, trust can be used as the evaluator between high and low performing virtual teams [36]. So, *Trust* has a positive impact on the *Performance* of the effectiveness of virtual teamwork. According to the discussion and explanation of this relationship between *Trust* and *Performance* factors, the following hypothesis is derived. $\boldsymbol{H}_{TR \rightarrow PF}$: Trust affects positively to Performance.

Cohesion to Performance

Team cohesiveness is considered as critical for improving the performance of virtual teamwork [20, 38, 42]. Also, different studies by Cohen and Bailey (in [8]), and Powell et al. (in [55]) mentioned that cohesion is a critical factor that influences the performance factor. These suggest that *Cohesion* has a positive impact on the *Performance* of the effectiveness of virtual teamwork. According to the discussion and explanation of this relationship between *Cohesion* and *Performance* factors, the following hypothesis is derived. $\boldsymbol{H}_{CH \rightarrow PF}$: Cohesion affects positively to Performance.

Coordination to Performance

Coordination is the core of virtual teamwork for managing and coordinating it successfully [54, 55]. Besides, it deals with conflict management to foster performance [44]. These suggest that coordination is positively related to the performance and success of virtual teamwork. Moreover, different research studies also outlined that coordination is positively linked to the performance or effectiveness of virtual teamwork (e.g. [30, 37, 42]). Therefore, according to the discussion and explanation of this relationship between *Coordination* and *Performance* factors, the following hypothesis is derived. $\boldsymbol{H}_{CR \rightarrow PF}$: Coordination affects positively to Performance.

As a summary, all the relationships of the key factors are listed in Table 15.2 along with their supporting references which justifies their inclusion in the conceptual model for effective virtual teamwork development in online higher education.

The Proposed Conceptual Model

This section will present and propose the conceptual model for effective virtual teamwork development in online higher education as per the key factors and their relationships presented in the previous two sections (i.e. 3 and 4).

Finally, considering these derived key factors, relationships, and hypotheses, the conceptual model of this study is proposed and presented in Fig. 15.2. As can be seen, it includes the key factors, relationships, and the set of corresponding hypotheses.

Table 15.2 Set of relationships with supporting references

Relationships	Supporting References
Cohesion to Coordination	[10, 35, 59]
Cohesion to Performance	[8, 20, 38, 55]
Communication to Cohesion	[54, 69, 70]
Communication to Knowledge Sharing	[7, 17, 60]
Communication to Trust	[28, 41, 58, 71]
Coordination to Performance	[30, 37, 42, 44]
Knowledge Sharing to Coordination	[40, 42, 49, 53, 72]
Knowledge Sharing to Performance	[1, 13, 39, 49, 72]
Motivation to Communication	[3, 14, 51]
Trust to Coordination	[5, 28, 46, 50]
Trust to Performance	[2, 48, 58, 71]

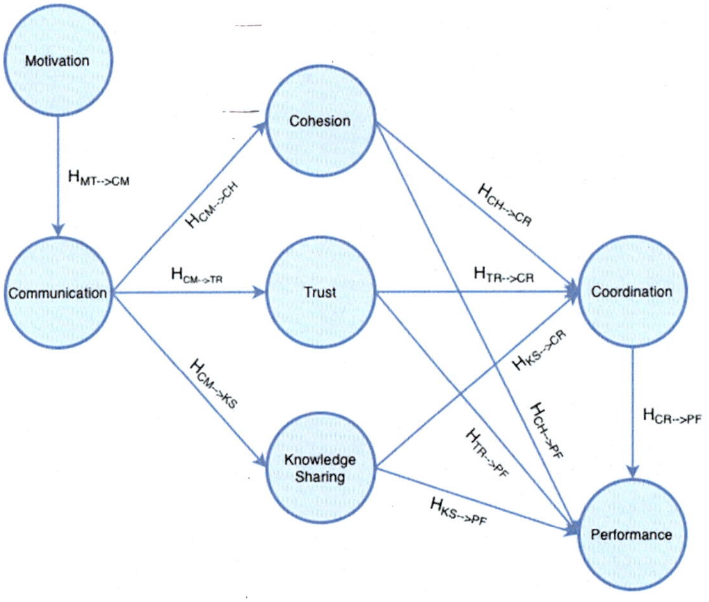

Fig. 15.2 The proposed conceptual model

Conclusion

The set of key factors that boost the effective development of virtual teamwork in the context of online higher education is presented and discussed in this paper along with the relationships among them. A set of hypotheses is derived based on the relationships of the key factors which are justified and supported by the existing literature.

A conceptual model is eventually developed and proposed at the end to develop effective virtual teamwork in online higher education along with the set of derived hypotheses. The key factors and the set of hypotheses of the proposed conceptual model are selected, developed, and justified based on the existing literature.

This model can be implemented in online higher education to promote teamwork skills among future graduates. Besides, a business organization can also be benefited from this model for managing and organizing their virtual teams effectively.

In future research, a measurement model can be developed to assess the latent variables (i.e. constructs) of the conceptual model. Besides, an in-depth empirical study is necessary to conduct in order to evaluate the proposed conceptual model for testing the statistical significance, which is part of future research.

Acknowledgements This work is part of a Doctoral Thesis funded by and conducted at *Universitat Oberta de Catalunya* (Barcelona, Spain).

References

1. M. Alsharo et al., Virtual teamwork effectiveness: The role of knowledge sharing and trust. Inf. Manage. **54**, 479–490 (2017)
2. V. Anantatmula, M. Thomas, Managing global projects: A structured approach for better performance model. Proj. Manage. J. **41**(2), 60–72 (2010)
3. A. Ardichvili et al., Motivation and barriers to participation in virtual knowledge-sharing communities of practice. J. Knowl. Manage. **7**(1), 64–77 (2003)
4. M. Armstrong, *Performance Management: Key Strategies and Practical Guidelines*, 3rd edn. (Kogan Page, London, 2007)
5. M.L. Baba, Dangerous liaisons: Trust, distrust, and information technology in American work organizations. Hum. Organ. **58**(3), 331–346 (1999)
6. Carron, A. V., et al. "The measurement of cohesiveness in sports groups." *Advancements in sports and exercise psychology measurement,* ed. by J.L. Duda, 1998, pp. 213–226
7. J.P. Clements, J. Gido, *Effective Project Management*, 5th edn. (South-Western Cengage Learning, Mason, OH, 2012)
8. S. Cohen, D.E. Bailey, What makes teams work: Group effectiveness research from the shop floor to the executive suite. J. Manage. Dev. **23**(3), 239–290 (1997)
9. D.W. Johnson et al., *Cooperative Learning Methods: A Meta-Analysis*, Minneapolis, MN, University of Minnesota
10. D.R. Deeter-Schmelz et al., Enriching our understanding of student team effectiveness. J. Mark. Educ. **24**(2), 114–124 (2002)
11. L. Duarte, *Mastering Virtual Project Teams: Strategies, Tools, and Techniques that Succeed* (Jossey-Bass, San Francisco, CA, 2002)
12. L. Duarte, N.T. Snyder, *Mastering Virtual Teams: Strategies, Tools, and Techniques that Succeed* (The Jossey-bass business & management series, San Francisco, CA, 2006)
13. H. Gardner et al., Dynamically integrating knowledge in virtual teams: Transforming resources into performance. Aca. Manage. J. **55**, 998–1022 (2012)
14. S. Geister et al., Effects of process feedback on motivation, satisfaction, and performance in virtual teams. Small Group Research **37**(5), 459–489 (2006)
15. R.L. Gordon, W. Curlee, *The Virtual Project Management Office: Best Practice, Proven Methods* (Management concepts inc, Vienna, VA, 2011)
16. R. Greiner, G. Metes, Trust and virtual organization. Harvard Bus. Rev. **73**(3), 40–50 (2005)

17. H. Grutterink et al., Reciprocal expertise affirmation and shared expertise perceptions in work teams: their implications for coordinated action and team performance. Appl. Psychol. **62**(3), 359–381 (2012)
18. S.W. Hahm, 'Information sharing and creativity in a virtual team: Roles of authentic leadership, sharing team climate and psychological empowerment. KSII Trans. Internet Inf. Syst. **11**(8), 4105–4119 (2017)
19. M. Hakonen, J. Lipponen, It takes two to tango: The close interplay between trust and identification in predicting virtual team effectiveness. J. E-Working **3**, 11–23 (2009)
20. S.T. Hannah et al., Leadership in action teams: Team leader and member's authenticity strength, and team outcomes. Pers. Psychol. **64**, 771–802 (2011)
21. K. Henttonen, K. Blomqvist, Managing distance in a global virtual team: the evolution of trust through technology-mediated relational communication. Strateg. Chang. **14**(2), 107–119 (2005)
22. G. Hertel et al., Managing virtual teams: a review of current empirical research. Hum. Resour. Manage. Rev. **15**(1), 69–95 (2005)
23. M. Hoegl, H.G. Gemuenden, Teamwork quality and the success of innovative projects: A theoretical concept and empirical evidence. Organ. Sci. **12**(4), 435–449 (2001)
24. T. Holmes, Ten characteristics of a high performance work team. Handful Handouts **29**, 179–182 (2012)
25. A. Idrus et al., Prioritizing project performance criteria within the client perspective. Res. J. Appl. Sci. **3**(10), 1142–1151 (2011)
26. J. Zigon, "Team performance: a process for creating team performance standards." Manage.: Compensation Benefits Rev. **29**(1), 38–47 (2000)
27. J.S. Coleman, "Social capital in the creation of human capital." Am. J. Sociol. **94**, 1988
28. S. Jarvenpaa, D. Leidner, Communication and trust in global virtual teams. Organ. Sci. **10**(6), 791–815 (1999)
29. R.M.-M. Jessica, A.D. Leslie, Information sharing and team performance: a meta- analysis. IEEE Eng. Manage. Rev. **40**(1), 119–136 (2012)
30. C. Johansson et al., Software engineering across boundaries: student project in distributed collaboration. IEEE Trans. Prof. Comm. **42**(4), 286–296 (1999)
31. R. Johnson, A.J. Onwuegbuzie, Mixed methods research: A research paradigm whose time has come. Educ. Researcher **33**, 14–26 (2004)
32. R. Johnson, J. Waterfield, Making words count: the value of qualitative research. Physiotherapy Res. Int. **9**, 121–131 (2004)
33. B.L. Kirkman, J. Mathieu, The dimensions and antecedents of team virtuality. J. Manage. **3**(5), 700–718 (2007)
34. A. Lam, Tacit knowledge, organizational learning and societal institutions: An integrated framework. Organ. Stud. **21**(3), 487–513 (2016)
35. J. Lipnack, J. Stamps, *Virtual Teams: People Working Across Boundaries with Technology*, 2nd edn. (Wiley, New York, 2000)
36. H. Lippert, V. Dulewicz, A profile of high-performing global virtual teams. Team Perform. Manage.: An Intl. J. **24**, 169–185 (2018)
37. M. Lu et al., Virtuality and team performance: understanding the impact of variety of practices. J. Global Inf. Technol. Manage. **9**(1), 4–23 (2006)
38. J. Lurey, M. Raisinghani, An empirical study of best practices in virtual teams. Inf. Manage. **38**(8), 523–544 (2001)
39. J. Malhotra, A. Majchrzak, Enabling knowledge creation in far-flung teams: best practices for its support and knowledge sharing. J. Knowl. Manage. **8**(4), 75–88 (2004)
40. A.P. Massey, et al., "Synchronizing pace in asynchronous global virtual project teams." *35th Hawaii international conference on system science*, 2002
41. R.C. Mayer et al., An integrative model of organizational trust. Acad. Manage. Rev. **20**(3), 709–734 (1995)
42. M. Maznevski, K. Chudoba, Bridging space over time: Global virtual team dynamics and effectiveness. Organ. Sci. **11**(5), 473–492 (2000)

43. D.J. McAllister, Affect and cognition-based trust as foundations for interpersonal cooperation in organizations. Acad. Manage. J. **38**(1), 24–59 (1995)
44. M. Montoya-Weiss et al., Getting it together: temporal coordination and conflict management in global virtual teams. Acad. Manage. J. **44**(6), 1251–1262 (2001)
45. J. Mouton, *How to Succeed in your Master's and Doctoral Studies: A South African Guide and Resource Book* (Van Schaik Publishers, Pretoria, 2001)
46. M. Muethel et al., When do we really need interpersonal trust in globally dispersed new product development teams? R&D Management **42**(1), 31–46 (2012)
47. P. Ludden, A. Ledwith, "A typology framework for virtual project teams: an empirical investigation." *Pmi research and education conference*, ed. by C. Messikomer (Project Management Institute, Portland, OR, 2014)
48. P.F. Rad, G. Levin, "Achieving project management success using virtual teams." J. Ross, 2006
49. F. Pangil, J.M. Chan, The mediating effect of knowledge sharing on the relationship between trust and virtual team effectiveness. J. Knowl. Management **18**(1), 92–106 (2014)
50. D. Paul, R.R. McDaniel, Effect of interpersonal trust on virtual collaborative relationship performance. MIS Q. **28**(2), 183–227 (2004)
51. T.M. Peterson, Motivation: How to increase project team performance. Proj. Manage. J. **38**, 60–69 (2007)
52. Pinar et al., "The relationships between leadership behaviors team learning and performance among the virtual teams." Int. Bus. Res. **7**(5), 68–79 (2014)
53. P. Pinjani, P. Palvia, Trust and knowledge sharing in diverse global virtual teams. Inf. Manag. **50**(4), 144–153 (2013)
54. A. Pinsonneault, O. Caya, Virtual teams: what we know and what we don't know. Int. J. e-Collab. **1**(3), 1–16 (2005)
55. A. Powell et al., "Virtual teams: a review of current literature and directions for future research." The Data Base for Advances in Info. Systems, **35**(1), 6–35 (2004)
56. Richardson et al., "A process framework for global software engineering teams." Inf. Softw. Technol. **54**(11), 1175–1191 (2012)
57. B. Rosen et al., Training for virtual teams: an investigation of current practices and future needs. Hum. Res. Manage. **45**(2), 229–247 (2006)
58. L.M. Roth, Essentials for great teams: trust, diversity, communication, and joy. J. Am. Board Fam. Med. **25**(2), 146–148 (2012)
59. S. Sarker et al., Using an adapted grounded theory approach for inductive theory building about virtual team development. DataBase Adv. Inf. Syst. **32**(1), 38–56 (2001)
60. K. Schwalbe, *Information Technology Project Management* (Cengage Learning, Boston, 2014)
61. N. Setle-Murphy, *Leading Effective Virtual Teams: Overcoming Time and Distance to Achieve Exceptional Results* (Auerbach Publications, Denver, CO, 2013)
62. Slechta, R. "Teamwork: Components of an effective team." 2007. www.lmi-inc.com/articles/teamwork_components.pdf. Accessed 20 February 2020
63. Sridhar et al., "Analyzing factors that affect the performance of global virtual teams." *International conference on management of globally distributed work*, 2007
64. J. Suchan, G. Hayzek, The communication characteristics of virtual teams: a case study. IEEE Trans. Prof. Commun. **44**(3), 174–186 (2001)
65. Swanepoel et al., *South African Human Resource Management for the Public Sector* (Kenwyn, Juta, 2011)
66. W. Forrester, A. Tashchian, "Modeling the relationship between cohesion and performance in student workgroups." Int. J. Manage. **23**(3), 2006
67. W.D. Salisbury et al., "Cohesion in virtual teams: validating the preceived cohesion scale in distributed settings." ACM SIGMIS Database. **37**, 2006
68. W.L. Bedwell et al., "What is collaboration? a multidisciplinary review." *4th annual conf. of the interdisciplinary network for group research* (Colorado Springs, CO, 2009)
69. J.B. Walther, Interpersonal effects in computer-mediated interaction: a relational perspective. Commun. Res. **19**(1), 52–90 (1992)

70. M.E. Warkentin et al., Virtual team versus face-to-face teams: an exploratory study of web-based conference system. Decis. Sci. **28**(4), 975–996 (1997)
71. T.P. Wise, *Trust in Virtual Teams* (Gower, Surrey, 2013)
72. Y.C. Xiao, Y.H. Jin, The hierarchical linear modeling of shared mental on virtual team effectiveness. Kybernetes **39**(8), 1322–1329 (2010)

Chapter 16
An Evaluation of Virtual Teamwork Model in Online Higher Education

Akinul Islam Jony and Enric Serradell-López

Abstract Virtual teamwork is considered as one of the key skills to an employer in the recruitment process because of the globalization of business and technological advancement. The company needs employees skilled in virtual teamwork to run their globalized business process. So, it becomes important for graduates to have good skills, knowledge, and ability of virtual teamwork through their higher education before starting their professional careers. Therefore, this paper is going to evaluate a virtual teamwork model which was proposed for developing an effective virtual teamwork in the context of online higher education so that graduate can achieve a sound knowledge and skills about virtual teamwork. The model is assessed by applying the partial least squares structural equation modeling (PLS-SEM) statistical technique to validate the model's statistical significance. The evaluation of the model shows that the model has a significant positive impact on the effectiveness of virtual teamwork in online higher education. Hence, it is believed that the application of the model would be useful to enhance the knowledge, skills, and ability of future graduates in online higher education, and can foster their early success in their professional career. Also, any company which practiced virtual teamwork in their business process can be benefited from this model for the better management of their virtual teams.

Keywords Model evaluation · Partial least squares · PLS-SEM · Structural equation modeling · Virtual teamwork · Higher education · E-Learning

A. I. Jony (✉) · E. Serradell-López
Universitat Oberta de Catalunya, Barcelona, Spain
e-mail: ajony@uoc.edu

E. Serradell-López
e-mail: eserradell@uoc.edu

A. Visvizi et al. (eds.), *Research and Innovation Forum 2020*,
Springer Proceedings in Complexity,
https://doi.org/10.1007/978-3-030-62066-0_16

Introduction

Technological advancement and globalization of business have made virtual teamwork as one of the key activities and skills to an employer [1, 15, 19, 21]. It also reflected in the recruitment process where employers are asking for virtual teamwork skills from jobseekers with regular importance [3]. Therefore, it becomes important for the graduates to be skilled in virtual teamwork before starting their professional career. A graduate can achieve this skill through higher education with the facility of building virtual teamwork skills. Considering the importance of developing virtual teamwork in higher education, this paper is going to evaluate a model which is particularly proposed for developing an effective virtual teamwork in online higher education so that graduate can achieve the required knowledge and skills about virtual teamwork. This model of virtual teamwork is proposed as a theoretical or conceptual model by Jony and Serradell-Lopez [18], which needs a complete evaluation for the statistical significance.

Therefore, this research work is mainly aimed to present the systematic evaluation of the model for justifying its statistical supports. The model is assessed and validated by applying the partial least squares structural equation modeling (PLS-SEM) statistical technique. The analysis shows that the model has a substantive statistical significance. Hence, it can be said that the application of the model in online higher education can be useful for developing an effective virtual teamwork. As a consequence, the students (future graduates) from higher education will be able to enhance their knowledge, skills, and ability on virtual teamwork, which eventually help them to foster their early success in their professional career. Also, any globalized company can be benefited from this model for managing and evaluating their virtual teams more effectively.

The rest of the paper is structured as follows. Section "Methodology" describes the methodology of this empirical study. The conceptual model of virtual teamwork [18], which is going to evaluate in this study using the PLS-SEM technique is presented as a PLS path model in Section "The PLS Path Model". The assessment of the measurement model and the structural model are presented in Sections "Evaluation of Measurement Model" and "Evaluation of Structural ModelEvaluation of Structural Model" respectively to illustrate the evaluation of the conceptual model, and the quality assessment of the results. Finally, Section "Evaluation of Structural Model" concludes the paper.

Methodology

The empirical research study of this paper aimed to evaluate the virtual teamwork model in online higher education which is proposed by Jony and Serradell-Lopez [18]. Specifically, it's aimed to measure the relationship between Motivation, Communication, Knowledge Sharing, Trust, Cohesion, Coordination, and Performance constructs of the model, and assess the impact on the effectiveness of virtual teamwork development in online higher education. The existing literature and theories have served as the source of identification of the contents of each of the constructs in the theoretical or conceptual model. Table 16.1 shows the complete set of the questionnaire (indicator variables) for measuring all the constructs of the model. For simplicity, only the set of indicators are listed in this paper but the detailed description of the measurement model development is not included here for limiting the length of this paper. However, the items were created and refined for ensuring their quality. The final questionnaire contained 5-point Likert scale items to record the responses.

The data collection of this empirical study was based on a survey from a virtual teamwork competition held in 2019 [7]. The questionnaire of the survey is answered by the virtual teams of undergraduate students from different countries. These virtual teams were the participants of a competition (Virtual Tournament) where they asked

Table 16.1 Indicators for the constructs of the model

Indicator Name	Description of Indicator
Motivation_1	All the team members had the opportunity to develop knowledge and skills
Motivation_2	As team members, we were able to add value to the teamwork
Motivation_3	In our team, we found that we were challenged by the teamwork
Communication_1	Communication in our team was open and honest
Communication_2	Team members were in contact with each other on a regular basis in order to conduct the teamwork
Communication_3	Team members exchanged information clearly and accurately
Knowledge_Sharing_1	Knowledge and information sharing were understood to be the norm in our team
Knowledge_Sharing_2	Team members exchanged knowledge and information with each other to solve a problem together
Knowledge_Sharing_3	All the team members exchanged their opinion in important decision making
Trust_1	Team members consulted with each other if they needed support
Trust_2	Our team valued individual input from the team members
Trust_3	There was no mutual distrust between team members
Cohesion_1	Our team was a very cohesive unit
Cohesion_2	Our team members experienced a sense of shared goals and objectives
Cohesion_3	Team members had interpersonal connections with each other
Coordination_1	Our team coordinated tasks effectively among each other
Coordination_2	Team members in our team displayed high levels of cooperation
Coordination_3	When disagreement occurred, they were addressed promptly in order to solve them
Performance_1	Our team worked effectively
Performance_2	Our team is satisfied with the outcomes
Performance_3	Our team generally worked on time

to work virtually in their virtual teams for a month to complete the designated task. After the competition, each team has asked to answer the questionnaire voluntarily. Among all the registered virtual teams in the competition, 159 data samples (one sample per virtual team, where all the answers are based on the combined opinions of each team member of the team) are collected of which 150 samples were usable after applying missing value treatment. Data was analyzed and evaluated using the SmartPLS which is one of the widely used and popular software [24] of the PLS-SEM statistical technique.

The PLS Path Model

As per the conceptual model [18], and the measurement items of the constructs in the conceptual model presented in Table 16.1, the following PLS path model (presented in Fig. 16.1) is developed which will be evaluated using the PLS-SEM statistical technique. Before going for the step-by-step evaluation of the model, let's check whether all the relationships of the model are supported statistically or not. Afterward, the model will be assessed with all the standard evaluation criteria of the PLS-SEM technique, separately for both measurement and structural parts of the model.

After running the PLS-SEM algorithm and bootstrapping procedure for the model estimation, there are found 3 relationships, that is *Knowledge Sharing* → *Performance* ($t = 0.460$, $p = 0.645$), *Trust* → *Performance* ($t = 0.515$, $p = 0.607$), and *Cohesion* → *Performance* ($t = 0.089$, $p = 0.929$) are not statistically significant because none of them has p value less than 0.05, and also none of them has t value above the threshold of 1.96 at least for the significance level of 5%. As these 3 relationships are not statistically supported according to t values, and p values, these 3 relationships can be eliminated from the model. In addition, these predecessor constructs *Knowledge Sharing*, *Cohesion*, and *Trust* have already affected positively on the *Performance* construct through the *Coordination* construct. So, it is believed that the elimination of these 3 relationships does not degrade the quality and orientation of the model.

As there are 3 relationships in the PLS path model which are not statistically significant, the following PLS path model (presented in Fig. 16.2) after eliminating those insignificant relationships will be evaluated next thoroughly in this paper to statistically validate the model.

The complete evaluation of the measurement and structural models of the PLS path model are presented in the following two subsequent Sects. 4 and 5 respectively.

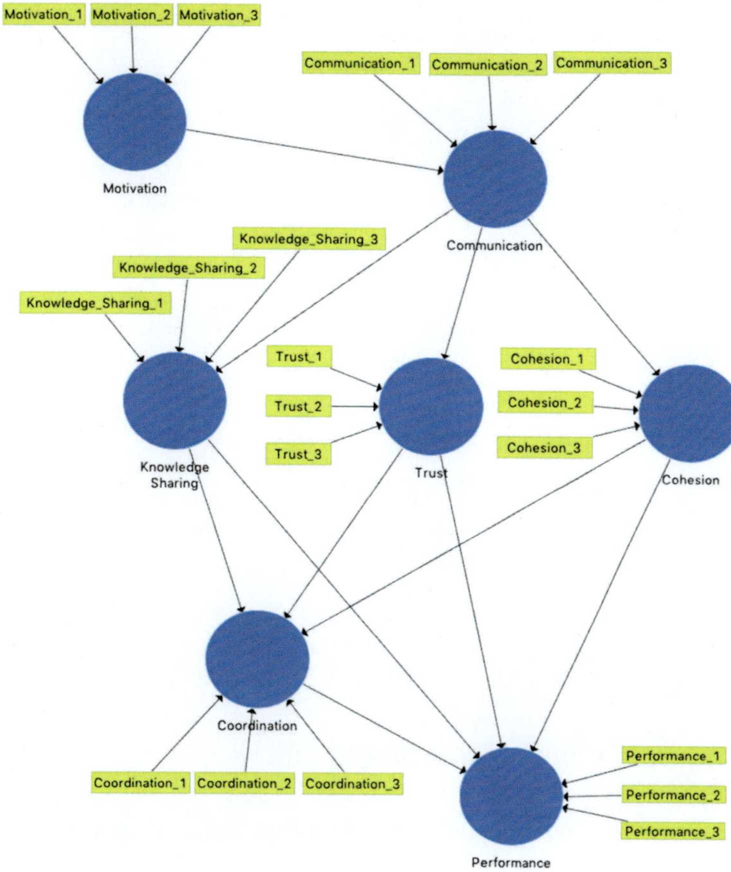

Fig. 16.1 The PLS path model

Evaluation of Measurement Model

This section presents the measurement model evaluation of the PLS path model of applying PLS-SEM. This empirical study has considered using a formative measurement model only. It is to note that the same evaluation criteria used for reflective measurement cannot be applied directly to assess the formative measurement model [5]. Thus, a different set of evaluation criteria as recommended in existing literature (e.g. [2, 12, 14]) are considered to evaluate the formative measurement model. A systematic procedure of such measurement model evaluation criteria is compiled and mentioned in an article [17], which is followed in this study to assess the measurement model.

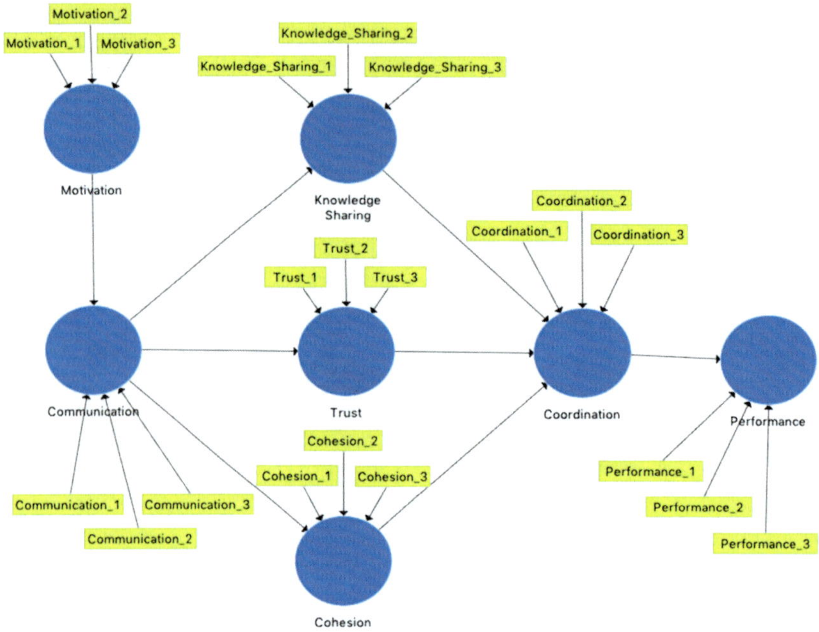

Fig. 16.2 The final PLS path model to be estimated

Assessment of Convergent Validity

In a formative measurement model, the convergent validity assessment is served as the examination of construct validity. It examines whether the formatively measured construct is highly correlated with the same construct measured reflectively to check whether the constructs exhibit convergent validity. It is also called the redundancy analysis [5]. In this assessment, separate redundancy analysis is carried out for each construct in the measurement model. For doing redundancy analysis, a global indicator for each of the constructs needs to identify that summarizes the essence of the formatively measured construct [26]. Table 16.2 presents a set of global single-item measures with generic assessments of the 7 constructs, Motivation, Communication, Knowledge Sharing, Trust, Cohesion, Coordination, and Performance—which will be used as measures of the dependent construct in the redundancy analysis.

In this analysis, the formatively measured construct (as an exogenous construct) is connected with a reflective measure of the same construct (as an endogenous construct) with its corresponding single-item global measure and check the value of path coeffcient (for definition see [22, 36]), and R^2 value (for definition see [10, 31]) for convergent validity of the construct. A desirable minimum path coeffcient value is 0.70, and R^2 value is 0.50 or above. Anything below the threshold means that it does not contribute at a suffcient degree to its intended content [12].

Table 16.2 Set of Global Indicators

Global Indicator Name	Description of Global Indicator
Motivation_Global	We feel valued as team members in our team.
Communication_Global	The methods used to communicate with each other were effective
Knowledge_Sharing_Global	Team members were open to sharing any knowledge and information
Trust_Global	Members of our team trusted each other
Cohesion_Global	Our team members help each other deal with problems or resolve issues
Coordination_Global	There has been coordination among the team members in the team to achieve the goals
Performance_Global	Our team met the team's objective

In the redundancy analysis, the path coeffcients for motivation, communication, knowledge sharing, trust, cohesion, coordination, and performance constructs are 0.773, 0.765, 0.839, 0.748, 0.849, 0.900, and 0.808 respectively. As can be seen, all these values are above the recommended threshold of 0.70. Thus, it indicates that all the formatively measured constructs exhibit convergent validity.

Assessment of Collinearity Issues

In the evaluation of the measurement model, the assessment of collinearity is served as the examination of construct reliability, which concerns the internal consistency of a measurement model [28]. This type of assessment checks whether each formative indicator in the measurement model contributes its intended meaning to it its formative construct [12].

In the context of PLS-SEM, variance inflation factor (VIF) (for definition see [16, 32]) values are the standard approach to statistically assess the presence of critical collinearity levels in the formative measurement model. As a rule of thumb, VIF value is less than 5 is mostly acceptable in the literature (e.g. [12]), anything greater than or equal to 5 indicates the presence of collinearity problem [13]. The VIF values for all the indicators are shown in Table 16.3 to assess the collinearity issues of the measurement model.

As can be seen, the highest VIF value is 3.686 for *Motivation_2* indicator, which indicates that all the VIF values of the indicators are below the threshold of five. Therefore, it can be said that collinearity for any constructs does not reach critical levels and hence, is not an issue for the PLS path model estimation.

Table 16.3 Collinearity statistics (VIF values of indicators)

Indicators	VIF Values
Cohesion_1	1.836
Cohesion_2	1.656
Cohesion_3	1.579
Communication_1	2.402
Communication_2	2.469
Communication_3	2.890
Coordination_1	2.157
Coordination_2	2.419
Coordination_3	1.699
Knowledge_Sharing_1	2.805
Knowledge_Sharing_2	3.641
Knowledge_Sharing_3	2.258
Motivation_1	3.614
Motivation_2	3.686
Motivation_3	1.789
Performance_1	1.784
Performance_2	1.474
Performance_3	1.912
Trust_1	2.795
Trust_2	2.591
Trust_3	1.258

Assessment of Significance and Relevance of Indicators

In the previous step of measurement model evaluation, there was found no collinearity issues so it is time to assess the significance and relevance of the indicators next. The outer weights (for definition see [20]) of the indicators will be analyzed to assess the significance and relevance of the indicators. Later, the outer loadings (for definition see [20]) of the indicators will be analyzed to assess the absolute contribution (or absolute importance) of the indicators to (for) their constructs.

The significance of outer weights can be obtained by means of the bootstrapping procedure, where (recommended) 5000 subsamples and 0.05 Significance level are the parameter settings to execute the bootstrap procedure. The statistics in Table 16.4 exhibits that all the indicators in the measurement model are significant at a 5% level (i.e. $p < 0.05$), except *Performance_2* ($p = 0.099$). However, this non-significant indicator should not delete as per the outer weight only, rather it should consider outer loading as well for examining the indicator's absolute contribution. The reason is that if an indicator's outer weight is non-significant but its outer loading is high (i.e. more than 0.50) then the indicator should be interpreted as absolutely important

Table 16.4 Outer weights of the indicators

Indicators - > Constructs	Original	t Values	p Values
Cohesion_1 - > Cohesion	0.457	6.852	0.000
Cohesion_2 - > Cohesion	0.522	8.888	0.000
Cohesion_3 - > Cohesion	0.189	2.631	0.009
Communication_1 - > Communication	0.386	5.116	0.000
Communication_2 - > Communication	0.376	4.984	0.000
Communication_3 - > Communication	0.347	4.519	0.000
Coordination_1 - > Coordination	0.368	6.631	0.000
Coordination_2 - > Coordination	0.519	7.780	0.000
Coordination_3 - > Coordination	0.246	3.470	0.001
Knowledge_Sharing_1 - > Knowledge Sharing	0.244	2.548	0.011
Knowledge_Sharing_2 - > Knowledge Sharing	0.499	4.457	0.000
Knowledge_Sharing_3 - > Knowledge Sharing	0.353	3.664	0.000
Motivation_1 - > Motivation	0.413	3.334	0.001
Motivation_2 - > Motivation	0.309	2.205	0.028
Motivation_3 - > Motivation	0.395	4.146	0.000
Performance_1 - > Performance	0.560	4.493	0.000
Performance_2 - > Performance	0.138	1.651	0.099
Performance_3 - > Performance	0.450	3.371	0.001
Trust_1 - > Trust	0.439	2.531	0.011
Trust_2 - > Trust	0.444	2.870	0.004
Trust_3 - > Trust	0.297	2.560	0.010

but not as relatively important, and hence, in this situation, the indicator generally should be retained in the measurement model [12].

The outer loadings of the indicators are presented in Table 16.5, which indicates all the indicator' outer loadings are above the threshold of 0.50, and also all the p values of the indicator's outer loadings are clearly below 0.01%. So, it suggests that all outer loadings are significant at a level of 1%. Considering this fact, the only non-significant indicator as per the value of outer weight is *Performance_2* but it has outer loading 0.651, which clearly above the threshold of 0.50, and so, it absolutely contributes to (absolutely important for) its construct. That's why it should be retained in the measurement model. Thus, there is no problem to proceed with the evaluation of the structural model.

Table 16.5 Outer loadings of the indicators

Indicators - > Constructs	Original	t Values	p Values
Cohesion_1 - > Cohesion	0.877	25.104	0.000
Cohesion_2 - > Cohesion	0.890	28.641	0.000
Cohesion_3 - > Cohesion	0.712	8.195	0.000
Communication_1 - > Communication	0.898	33.738	0.000
Communication_2 - > Communication	0.898	25.498	0.000
Communication_3 - > Communication	0.912	30.530	0.000
Coordination_1 - > Coordination	0.878	26.485	0.000
Coordination_2 - > Coordination	0.937	36.903	0.000
Coordination_3 - > Coordination	0.774	11.354	0.000
Knowledge_Sharing_1 - > Knowledge Sharing	0.870	19.221	0.000
Knowledge_Sharing_2 - > Knowledge Sharing	0.956	47.832	0.000
Knowledge_Sharing_3 - > Knowledge Sharing	0.880	20.971	0.000
Motivation_1 - > Motivation	0.923	26.676	0.000
Motivation_2 - > Motivation	0.910	23.240	0.000
Motivation_3 - > Motivation	0.855	16.329	0.000
Performance_1 - > Performance	0.916	14.893	0.000
Performance_2 - > Performance	0.651	7.688	0.000
Performance_3 - > Performance	0.883	18.117	0.000
Trust_1 - > Trust	0.921	18.440	0.000
Trust_2 - > Trust	0.899	21.073	0.000
Trust_3 - > Trust	0.662	5.976	0.000

Evaluation of Structural Model

This section presents the assessment of structural model using different evaluation criteria as recommended in existing literature (e.g. [4, 5, 12, 29]) in order to estimate the model of this research study. A systematic procedure of assessing the structural model is compiled and mentioned in [17], which is followed in this study to assess the measurement model. The structural model assessment builds on the results from the standard PLS-SEM algorithm, the bootstrapping routine, and the blindfolding procedure, which is be presented thoroughly in the following subsections of this section.

Assessment of Collinearity Issues

The first step is to check the structural model for collinearity issues by examining the VIF values (i.e. Inner VIF) of all set of predictor constructs in the structural model. The VIF values of all combinations of endogenous constructs, and corresponding exogenous (i.e., predictor) constructs are presented by the following sets of (predictor) constructs for collinearity issues:

- Motivation as a predictor of Communication (VIF value is 1.000),
- Communication as a predictor of Knowledge Sharing, Trust, and Cohesion (VIF values are 1.000 for all three sets),
- Knowledge Sharing, Trust, and Cohesion as predictors of Coordination (VIF values are 4.314, 2.855, and 3.678 respectively), and
- Coordination as a predictor of Performance (VIF value is 1.000).

The results indicate that all the VIF values are clearly below the threshold of 5. Hence, the collinearity is not a critical issue among the predictor constructs in the structural model.

Assessment of Coeffcient of Determination (R^2 Value)

The next evaluation of the structural model is the coeffcient of determination (R^2 values) (for definition see [10, 31]) of endogenous constructs, which measures the predictive power of the model. That means, it is an in-sample predictive power [23, 25] measurement. In addition, as there might have a possibility of inherent bias towards a complex model. That's why the adjusted coeffcient of determination (R^2 adjusted value) is also considered here as the evaluation criteria to avoid the bias. The R^2 value ranges from 0 to 1, of which higher value indicates the higher levels of predictive accuracy.

So, both R^2 and R^2 adjusted values of endogenous constructs of the model are presented in Table 16.6 to evaluate the predictive power of the model. As can be seen, both values of R^2 and R^2 adjusted of *Cohesion*, and *Coordination* are found substantial (i.e., above the threshold of 0.75), whereas, *Communication*, *Knowledge Sharing*, *Performance*, and *Trust* are found moderate (i.e., above the threshold of 0.50), which altogether statistically proves the predictive power of the model.

Table 16.6 R^2 and R^2 adjusted values of endogenous constructs

Endogenous Constructs	R^2 Values	R^2 Adjusted Values
Cohesion	0.803	0.801
Communication	0.737	0.735
Coordination	0.835	0.831
Knowledge Sharing	0.711	0.709
Performance	0.680	0.678
Trust	0.587	0.585

Assessment of Effect Size F^2

In addition to exploring R^2 values, the changes in R^2 can also be explored in the evaluation of the structural model, which is known as effect size (f^2) (for definition see [8, 33]). This assessment was firstly presented by Cohen, which examines the impact on the dependent construct by the independent construct of the structural model [6].

Following the rules of thumb (for interpreting f^2 values), the f^2 effect sizes of Motivation on Communication (2.799), Communication on Knowledge Sharing (2.464), Trust (1.424), and Cohesion (4.065), Cohesion on Coordination (0.453), and Coordination on Performance (2.125) are found large (i.e., above the threshold of 0.35), whereas, Knowledge Sharing on Coordination (0.091), and Trust on Coordination (0.080) are found small (i.e., above the threshold of 0.02), which statistically proves the effect of exogenous constructs on their corresponding endogenous constructs of the model. That means, there are no exogenous constructs in the model which have no effect on their corresponding endogenous constructs.

Assessment of Significance and Relevance of Relationships

Next to assess the significance and relevance of the structural model relationships which represent the hypothesized relationships among the constructs in the model. This can be examined by the path coeffcients which have standardized values approximately between −1 to +1. The path coeffcients values close to +1 represent strong positive relationships, whereas, path coeffcients values close to −1 are not statistically significant, which means represent weaker relationships.

The path coeffcients of relationships in the structural model are represented graphically in Fig. 16.3. The statistics indicate that all the relationships in the model are positive and strong. Among all the relationships in the model, *Trust* → *Coordination* relationship (0.194) has the lowest values in comparison to the other relationships but still the value is closer to +1 than −1, and hence, it also can be considered as a positive relationship.

Another interesting evaluation is the examination of total effects, which assess the influence of predecessor constructs on the key target constructs (i.e. Performance). For example, among all the predecessor constructs Coordination (0.825) has the strongest effect on the key target construct Performance, followed by Communication (0.688), Motivation (0.590), Cohesion (0.433), Knowledge Sharing (0.210), and Trust (0.160). On the other hand, the Coordination target construct mostly influenced by Communication (0.834), followed by Motivation (0.716), Cohesion (0.525), Knowledge Sharing (0.255), and Trust (0.194) predecessor constructs. Besides, Communication has three direct effects on Knowledge Sharing, Trust, and Cohesion constructs by 0.843, 0.766, and 0.896 respectively, of which the strongest effect

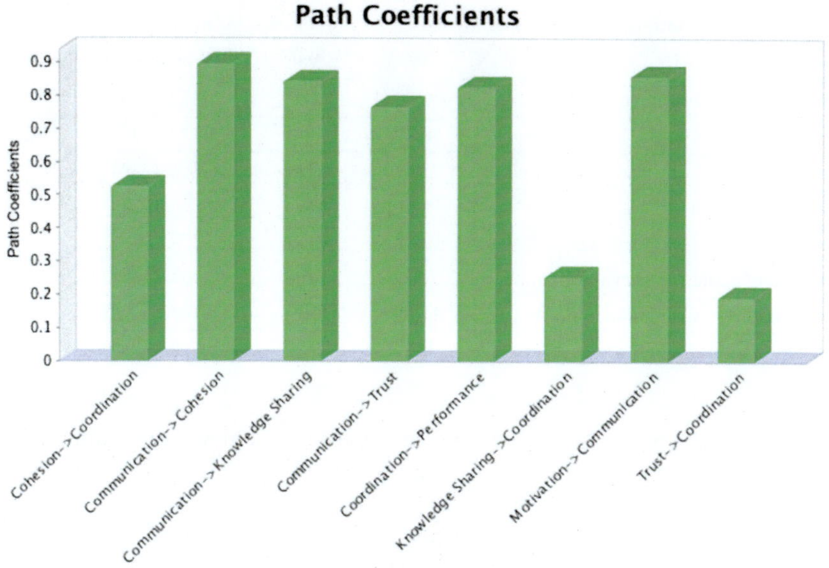

Fig. 16.3 Path coeffcients of relationships in the model

is on Cohesion. So, it is clear that all the predecessor constructs influence the target constructs significantly.

Though the path coeffcients of relationships obtained by the PLS-SEM algorithm are statistically significant that solely not enough to declare it. Therefore, the empirical t values (for definition see [30, 34]) for all path coeffcients of the structural model is needed to prove the statistical significance, and also consolidate the statistical significance so far achieved. The empirical t values can be obtained by means of the bootstrapping procedure. Along with the t values, the empirical p values (for definition see [9, 35]) are also used here to assess the significance levels (in this study, the p value has to be smaller than 0.05 to be considered as a significance level of 5%).

After running the bootstrapping procedure, a detailed overview of the results of path coeffcients for the model is shown in Table 16.7 including t values, p values, and bootstrap confidence intervals. The statistics indicate that all the relationships in the structural model are found statistically significant (i.e., all p values less than 0.05) at 5%.

Besides, Table 16.7 shows the bootstrapping confidence intervals (bias-corrected) which provides additional information on the stability of the estimated path coeffcients of the structural model. Hence, it allows assessing whether a path coeffcient is significantly different from zero, and its true value is somewhere within the range assuming a certain level of confidence (i.e. 95%). If the estimated path coeffcient's confidence interval does not include zero then it can be considered as a significant effect. The statistics of confidence intervals for the path coeffcients of relationships

Table 16.7 Bootstrapping results for path coeffcients of relationships in the model

Relationships	t Values	p Values	95% Confidence Intervals
Cohesion - > Coordination	6.784	0.000	[0.382, 0.691]
Communication - > Cohesion	38.634	0.000	[0.835, 0.930]
Communication - > Knowledge Sharing	25.679	0.000	[0.755, 0.891]
Communication - > Trust	15.951	0.000	[0.630, 0.839]
Coordination - > Performance	19.763	0.000	[0.708, 0.887]
Knowledge Sharing - > Coordination	2.515	0.012	[0.054, 0.455]
Motivation - > Communication	25.550	0.000	[0.758, 0.907]
Trust - > Coordination	2.430	0.015	[0.035, 0.351]

indicate that the true values of the path coeffcients are somewhere within the range with a 95% probability. Also, all the relationships are significant as their confidence intervals do not include zero value.

So far the relationships of the structural model are found statistically significant. Now it is time to assess the relevance of significant relationships (by means of the bootstrapping procedure) because the path coeffcients may be significant but their size may be small. For that reason, the relevance of relationships is examined here by the total effects (i.e. the sum of indirect and direct effects).

The following Table 16.8 shows the total effects of the predecessor constructs on the target constructs by considering both direct and indirect effects. As can be seen, all total effects are significant at a 5% level (i.e. all p values are less than 0.05). That means, it proves the relevance of significant relationships of the structural model.

Besides, Table 16.8 contains the bootstrap confidence intervals (bias-corrected) of total effects for significance testing. The statistics of confidence intervals for the total effects of predecessor constructs on the target constructs indicate that the true values of the total effects are somewhere within the range with a 95% probability. Also, all the total effects are significant as their confidence intervals do not include zero value.

Assessment of Predictive Relevance Q^2

Previously, the R^2 values are used as the evaluation criterion in the structural model for the model's predictive accuracy, which indicates the model's in-sample predictive power. In addition to this evaluation, there has another evaluation criterion to assess the model's predictive relevance, which indicates the model's out-of-sample predictive power. Stone-Geisser's Q^2 value [11, 27] is used for testing the predictive relevance of the structural model, by using a blindfolding procedure with a specified omission distance to generate the Q^2 values [5]. Omission distance value 7 is mostly recommended in the literature [2], providing that the number of observations (used

Table 16.8 Bootstrapping results for total effects

Relationships	Total Effects	t Values	p Values	95% Confidence Intervals
Cohesion - > Coordination	0.525	6.784	0.000	[0.382, 0.691]
Cohesion - > Performance	0.433	6.187	0.000	[0.308, 0.583]
Communication - > Cohesion	0.896	38.634	0.000	[0.835, 0.930]
Communication - > Coordination	0.834	23.826	0.000	[0.732, 0.885]
Communication - > Knowledge Sharing	0.843	25.679	0.000	[0.755, 0.891]
Communication - > Performance	0.688	12.256	0.000	[0.542, 0.775]
Communication - > Trust	0.766	15.951	0.000	[0.630, 0.839]
Coordination - > Performance	0.825	19.763	0.000	[0.708, 0.887]
Knowledge Sharing - > Coordination	0.255	2.515	0.012	[0.054, 0.455]
Knowledge Sharing - > Performance	0.210	2.422	0.015	[0.042, 0.384]
Motivation - > Cohesion	0.769	16.929	0.000	[0.636, 0.836]
Motivation - > Communication	0.858	25.550	0.000	[0.758, 0.907]
Motivation - > Coordination	0.716	13.408	0.000	[0.568, 0.798]
Motivation - > Knowledge Sharing	0.724	14.068	0.000	[0.580, 0.800]
Motivation - > Performance	0.590	9.070	0.000	[0.427, 0.695]
Motivation - > Trust	0.658	10.869	0.000	[0.489, 0.749]
Trust - > Coordination	0.194	2.430	0.015	[0.035, 0.351]
Trust - > Performance	0.160	2.465	0.014	[0.029, 0.287]

in the model estimation) divided by omission distance value should not be an integer [12]. Hence, omission distance value 7 is also used herein the blindfolding procedure to generate the Q^2 values because it does not yield an integer value while dividing the number of observations. In this assessment, if the Q^2 values for the endogenous constructs are greater than 0 then they indicate the presence of predictive relevance of the path model, whereas, zero, or negative values of Q^2 indicate the absence of predictive relevance.

The following Table 16.9 exhibits the blindfolding results, where the cross-validated redundancy approach is used to generate the Q^2 values because this approach includes both structural model and measurement model, and hence perfectly fits in PLS-SEM. The specification of Table 16.9 is given below. Detailed information and definitions about Q^2 (= 1-SSE/SSO) can be found in [11, 12, 20] and [27].

- SSO shows the sum of the squared observations,
- SSE shows the sum of the squared prediction errors, and
- Q^2 is the final value obtained from the calculation of (1 − SSE/SSO) to judge the model's predictive relevance with regard to each endogenous constructs in the structural model.

As can be seen in Table 16.9, all the endogenous constructs in the structural model have Q^2 values clearly above zero. More specifically, *Coordination* has the highest Q^2 value (0.559), followed by *Communication* (0.546), *Knowledge Sharing* (0.536), *Cohesion* (0.514), *Performance* (0.407), and *Trust* (0.380). These results provide clear support for the model's predictive relevance regarding the endogenous constructs in the structural model.

Conclusion

A complete evaluation of the model proposed by Jony and Serradell-Lopez in [18] is presented in this paper to justify its statistical support. As a statistical tool, PLS-SEM is used for assessing and analyzing the model. Based on the recommendation from the existing literature of PLS-SEM, all the required evaluation criteria have been applied to systematically assess the measurement model and structural model towards the statistical estimation of the conceptual model. In this model estimation process, the PLS-SEM algorithm, bootstrapping routine, and blindfolding procedure are applied for testing the significance and relevance of the model. However,

Table 16.9 Q^2 values for endogenous constructs

Endogenous Constructs	SSO	SSE	Q^2 (= 1-SSE/SSO)
Cohesion	450.000	218.904	0.514
Communication	450.000	204.111	0.546
Coordination	450.000	198.555	0.559
Knowledge Sharing	450.000	208.592	0.536
Motivation	450.000	450.000	
Performance	450.000	266.742	0.407
Trust	450.000	279.085	0.380

in general, the bootstrapping procedure is ultimately suffcient to assess the significance and relevance of the model's relationships (e.g., t values, p values, bootstrap confidence intervals, and total effects), and to conclude whether the model or in particular the relationships of the model are statistically significant and relevant. The results of the evaluation indicate clear statistical supports for the model's significance and relevance. So, according to the evaluation of the model, it is believed that the model would be useful for developing an effective virtual teamwork in online higher education.

Acknowledgements This work is part of a Doctoral Thesis funded by and conducted at *Universitat Oberta de Catalunya* (Barcelona, Spain).

References

1. V. Anantatmula, M. Thomas, Managing global projects: a structured approach for better performance model. Proj. Manage. J. **41**(2), 60–72 (2010)
2. P. Andreev et al., "Validating formative partial least squares (PLS) models: methodological review and empirical illustration." ICIS 2009 proceedings, 2009
3. R. Bennett, Employers' demands for personal transferable skills in graduates: a content analysis of 1000 job advertisements and associated empirical study. J. Vocat. Educ. Training **54**(4), 457–476 (2000)
4. Y.M. Chatelin et al., "State-of-art on pls path modeling through the available software." HEC research papers series 764, HEC Paris, 2002
5. W.W. Chin, "The partial least squares approach to structural equation modeling." *Modern methods for business research*. ed. by G.A. Marcoulides (Mahwah, NJ: Lawrence Erlbaum, 1998), pp. 295–358
6. S. Cohen, *Statistical Power Analysis for the Behavioral Sciences*. Lawrence Erlbaum, 1988
7. Company Name, "I torneo acbsp - companygame de simulacion empresarial." *Company Game | Learn Simulating*, 2019, http://www.companygame.com
8. P.D. Ellis, *The Essential Guide to Effect Sizes*. Cambridge University Press, 2010
9. M. Fisz, "Significance Testing." *Probability Theory and Mathematical Statistics*. 3rd edn. (John Wiley and Sons, New York, 1963)
10. B.S. Everitt, A. Skrondal. *The Cambridge Dictionary of Statistics*. 4th edn. (Cambridge University Press, 2010)
11. S. Geisser, A predictive approach to the random effects model. Biometrika **61**, 101–107 (1974)
12. J.F. Hair et al., *A Primer on Partial Least Squares Structural Equation Modeling (pls-sem)*, 2nd edn. (Sage, Los Angeles, 2017)
13. J.F. Hair et al., "Pls-sem: Indeed a silver bullet." J. Mark. Theory Pract., **19**, 139–151 (2011)
14. J.F. Hair et al., "When to use and how to report the results of pls-sem." Eur. Bus. Rev. **3**(1), 2018, https://doi.org/10.1108/ebr-11-2018-0203
15. G. Hertel et al., Managing virtual teams: a review of current empirical research. Hum. Resour. Manage. Rev. **15**(1), 69–95 (2005)
16. G. James et al., *An Introduction to Statistical Learning*, 8th edn. (Springer, New York, 2017)
17. A.I. Jony, E. Serradell-Lopez, "A pls-sem approach in evaluating a virtual teamwork model in online higher education: why and how?", Rii Forum Proceedings of Research & Innovation Forum (rii forum) Annual Conference (Springer, Athens, Greece, 2020)
18. A.I. Jony, E. Serradell-Lopez. "Key factors that positively affects the effectiveness of virtual teamwork in online higher education." Rii forum proceedings of research & innovation forum (rii forum) annual conference (Springer, Athens, Greece, 2020)

19. B.L. Kirkman et al., The impact of team empowerment on virtual team performance: the moderating role of face to face interaction. Acad. Manag. J. **47**(2), 175–192 (2004)
20. H. Latan, R. Noonan, *Partial Least Squares Path Modeling: Basic Concepts, Methodological Issues and Applications*. Springer, 2017
21. P. Ludden, A. Ledwith, "A typology framework for virtual project teams: An empirical investigation." *Pmi research and education conference*, ed. by C. Messikomer (Portland, Project Management Institute, OR, 2014)
22. M.E. Olobatuyi, *A User's Guide to Path Analysis*. UPA, 2006
23. E.E. Rigdon, Rethinking partial least squares path modeling: in praise of simple methods. Long Range Plan. **45**, 341–358 (2012)
24. C.M. Ringle et al., Smartpls3. SmartPLS GmbH. Bönningstedt. 2015, http://www.smartpls.com
25. M. Sarstedt et al., On the emancipation of pls-sem: a commentary on Rigdon (2012). Long Range Plan. **47**, 154–160 (2014)
26. M. Sarstedt et al., "Measuring reputation in global markets: a comparison of reputation measures' convergent and criterion validities." J. World Bus. **48**, 329–339 (2013)
27. M. Stone, Cross-validatory choice and assessment of statistical predictions. J. Royal Stat. Soc. **36**, 111–147 (1974)
28. D. Straub et al., Validating guidelines for is positivist research. Commun. Assoc. Inf. Syst. **13**, 380–427 (2004)
29. M. Tenenhaus et al., Pls path modeling. Comput. Stat. Data Anal. **48**(1), 159–205 (2005)
30. R.E. Walpole, *Probability & Statistics for Engineers & scientists*, 7th edn, Pearson, 2006
31. Wikipedia. "Coefficient of determination." 29 April 2020, https://en.wikipedia.org/wiki/Coefficient_of_determination
32. Wikipedia. "Variance Inflation factor." 8 June 2020, https://en.wikipedia.org/wiki/Variance_inflation_factor
33. Wikipedia. "Effect size." 6 June 2020, https://en.wikipedia.org/wiki/Effect_size
34. Wikipedia. "T-statistic." 11 May 2020, https://en.wikipedia.org/wiki/T-statistic
35. Wikipedia. "p-value." 19 May 2020, https://en.wikipedia.org/wiki/P-value
36. S. Wright, The method of path coefficients. Ann. Math. Stat. **5**, 151–215 (1934)

Chapter 17
A PLS-SEM Approach in Evaluating a Virtual Teamwork Model in Online Higher Education: Why and How?

Akinul Islam Jony and Enric Serradell-López

Abstract The Statistical analysis tool is considered an essential tool for the research study in the social science domain. Of which partial least squares structural equation modeling (PLS-SEM) is a second-generation statistical modeling technique to be used for developing theories in exploratory research. That's why PLS-SEM is considered to be used in the research field of education and ICT to validate a conceptual model of virtual teamwork in the context of online higher education. However, there should have a clear justification for choosing PLS-SEM for particular research. Also, a systematic procedure is needed to apply it to report the analysis and evaluation appropriately. Therefore, this paper mainly aims to present the reasons for choosing PLS-SEM statistical method for evaluating a virtual teamwork model in online higher education, and how to apply this method to evaluate or assess the validity of the model. Though the intent of this paper is to provide a comprehensive guideline about PLS-SEM for evaluating a virtual teamwork model in online higher education it can also be useful for the other research fields. So, the outcome of this paper will help researchers in any field for designing their research who considered applying PLS-SEM in their research study. Also, a new researcher will find it as a comprehensive overview of PLS-SEM, and why and how to apply PLS-SEM in research work.

Keywords Partial least squares · PLS-SEM · Structural equation modeling · Statistical modeling technique · Analysis tool · Virtual teamwork · Higher education · E-Learning

A. I. Jony (✉) · E. Serradell-López
Universitat Oberta de Catalunya, Barcelona, Spain
e-mail: ajony@uoc.edu

E. Serradell-López
e-mail: eserradell@uoc.edu

A. Visvizi et al. (eds.), *Research and Innovation Forum 2020*,
Springer Proceedings in Complexity,
https://doi.org/10.1007/978-3-030-62066-0_17

Introduction

The power of structural analysis is to develop, explore, and confirm research findings that undoubtedly makes it one of the most essential tools for social science researchers or scientists for many years. In which partial least squares structural equation modeling (PLS-SEM) is considered as one of the most emerging second-generation statistical tools for multivariate analysis. Covariance-based structural equation modeling (CB-SEM) is another second-generation statistical technique which was most popular until 2010, but, in resent time PLS-SEM becomes widely used technique in terms of the number of publications [1].

Nowadays, PLS-SEM has been broadly accepted and used in many disciplines, such as human resource management [2], marketing management [3], organizational management [4], operations management [5], strategic management [6], management accounting [7], and management information systems [8], etc. Besides, the PLS-SEM technique is widely applied in a vast range of additional disciplines as well such as economics, engineering, environmental sciences, medicine, political sciences, and psychology [9]. Recently PLS-SEM has been advanced and adopted in a banking and finance discipline [10]. Also, PLS-SEM is addressed, proposed, and illustrated in different forms of publication such as special issues of scholarly journals [11, 9], textbooks [12, 13], and edited volumes [10], which indicate the increasing popularity and applicability of PLS-SEM in a wide range of disciplines and research works.

Therefore, PLS-SEM is chosen as a statistical tool for the research study in the field of education and ICT (information communication technology) and discussed its applicability in this paper. More precisely, the main purpose of this paper is to address and justify what are the reasons of choosing PLS-SEM, and how to apply it in the research of education and ICT for evaluating the conceptual model of virtual teamwork in online higher education (a detailed description of the conceptual model can be found in [14]. Though the discussion about PLS-SEM of this paper is based on a particular research context, however, it can be useful in general all research contexts. Such as, why and when PLS-SEM can be considered in research work, and how to apply PLS-SEM to develop, explore, or analyze any research findings. In addition, it can be used as a guideline to apply PLS-SEM statistical technique in any research study.

The rest of the paper is structured as follows. Section "Background" defines some terms related to the topic in order to give some general foundation. Section "Reason for Choosing PLS-SEM" explains the reasons for choosing PLS-SEM. Section "How to Apply PLS-SEM" presents how to apply PLS-SEM, specifically it compiled three systematic procedures for the evaluation of measurement model, structural model, and model fit respectively. Finally, Section "Conclusion" concludes the paper.

Background

This section briefly describes several terms related to the topic to provide the foundation before starting the main discussion of the paper indented.

SEM in a Nutshell

The researchers of social science usually applied first-generation statistical techniques (such as factor analysis, cluster analysis, logistic regression, multiple regression, multidimensional scaling, etc.) to analyze multiple variables on a regular basis in the past. However, to overcome the weakness of first-generation techniques, nowadays, they are increasingly moving into second-generation techniques, which typically called structural equation modeling (SEM). SEM is a complex statistical technique [15, 16], but it becomes powerful because of its ability in assessing relations between constructs, between latent (unobserved) variables and indicator (observed) variables, and measurement error in observed variables [17]. There are two approaches in an SEM: covariance-based SEM (CB-SEM) and partial least squares SEM (PLS-SEM) [18, 19].

Cb-SEM Versus PLS-SEM

The CB-SEM technique (developed by Jöreskog [20]) uses covariance and considers only common variance to estimate the model parameters [1]. But it is typically very restrictive and based on numerous assumptions [19]. On the other hand, PLS-SEM statistical technique (developed by Wold [21–23]) is a two-step method, of which the first step refers to path estimates of the measurement (outer) model, and the second step refers to path estimates of the structural (inner) model [24]. PLS-SEM method combines principal components analysis with ordinary least squares regressions to assess partial model structures [25].

CB-SEM uses covariance to estimate the model parameters, whereas PLS-SEM uses total variance to estimate the model parameters [1]. Besides, CB-SEM uses constructs as common factors to explain the covariance of indicator variables, on the other hand, PLS-SEM uses proxies for a particular construct in the model [26]. That's why PLS-SEM facilitates a composite-based approach whereas, CB-SEM uses common factors with strong assumptions to explain all the covariance of indicators [27] Rigdon, *Rethinking partial least squares path modeling*). In addition, PLS-SEM used mainly for prediction purposes, whereas, CB-SEM mainly focused on parameter estimation. In terms of sample size, PLS-SEM can estimate with a smaller sample size [28], while CB-SEM always needs a larger sample size for better estimation. In terms of the application of these two methods, CB-SEM is applied to confirm or reject theories, whereas, PLS-SEM is applied to develop theories in exploratory research. Basically, the choice between CB-SEM and PLS-SEM depends on the characteristics and objectives of the research study [3]. When the primary objective of a research study is to predict and explain the target construct, and when the theory is less developed or no prior knowledge then the PLS-SEM approach should consider for the study [29].

So, in a nutshell, it can be said that PLS-SEM and CV-SEM techniques differ in terms of objectives, assumptions, parametersestimates, latent-variable scores, model complexity, and sample size [30].

PLS Path Modeling

A path model is a diagram that is developed to visualize the hypotheses and variable relationships and then examined those hypotheses and relationships by using the SEM technique [19]. A PLS path model can be described from the perspective of two models: the structural model (also called inner model) and the measurement model (also called outer model) [31–33]. An example of a PLS path model is shown in Fig. 17.1.

The structural model contains constructs and relationships between constructs (e.g. in Fig. 17.1, C_1 to C_5 constructs). On the other hand, measurement models contain indicator variables and relationships between indicator variables and constructs (e.g. in Fig. 17.1, I_1 to I_{13} are indicator variables). There are also error terms

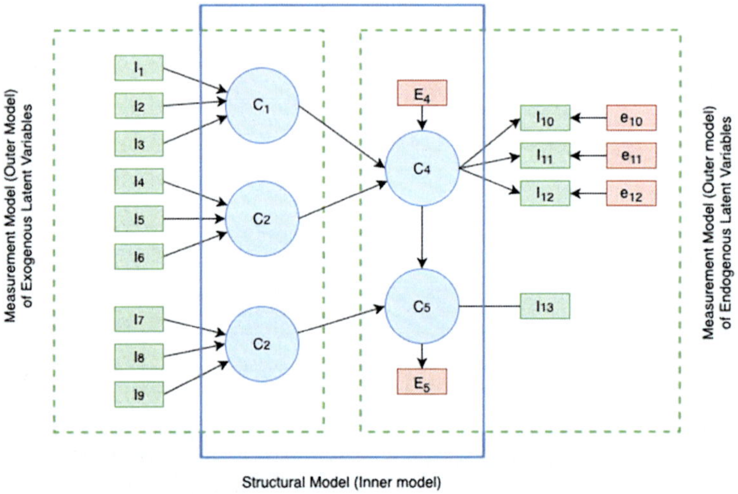

Fig. 17.1 An example of a PLS path model. *Source* Adapted from Hair et al. in [26]

in the path models represented as rectangles, which are connected to the endogenous latent variables, and the reflectively measured indicator variables. For example, in Fig. 17.1, e_{10} to e_{12} are the error terms for the reflectively measured indicators I_{10} to I_{12} respectively (C_5 is a single-item construct, so there is no error term connected to I_{13}). Also, E_4 and E_5 are the error terms for endogenous latent variables C_4 and C_5 respectively.

Moreover, in a PLS path model, a construct is called independent or exogenous latent variable (e.g. in Fig. 17.1, C_1, C_2, and C_3 are independent variables) if it is starting off a sequence in the structural model, and which is at ending or middle of a sequence, is called dependent variable or endogenous latent variable (e.g. in Fig. 17.1, C_4 and C_5 are dependent variables). It is to mention that, as the independent variables explain the dependent variables in the path model, they do not have error terms. Besides, there are two types of measurement model based on the latent variables: (a) measurement model of exogenous latent variables (if the latent variables explain other constructs in the model), and (b) measurement model of endogenous latent variables (if the constructs are explained by other constructs in the model), which can be seen in Fig. 17.1 as a left and right dashed rectangle respectively.

Finally, there are mainly two approaches of measurement model: formative and reflective measurement model. The choice between these two has to be taken cautiously which mainly depends on the particular research study. In a formative measurement model, the arrow directions are pointing from the indicator variables to the constructs, which refers to predicting or casual relationships (e.g. in Fig. 17.1, C_1, C_2, and C_3 are formatively measured constructs). On the other hand, in a reflective measurement model, the arrow directions are pointing from the constructs to the indicator variables, which refers to the assumption relationships (e.g. in Fig. 17.1, C_4 and C_5 are reflectively measured constructs). It is to note that formative measures are assumed to be error-free [31, 34], whereas, in a reflective measure, all the indicator variables have error terms [35].

Virtual Teamwork in Online Higher Education

Nowadays virtual teamwork is a common activity in business organizations considering the technological growth and globalized nature of business [36]. Besides, it is frequently mentioned in a graduate job advertisement which indicates that companies are looking for candidates who are skilled in virtual teamwork [37]. Therefore, higher education should value this fact, and emphasize on building virtual teamwork skills in future graduates. A study by Jony and Serradell-Lopez proposed a conceptual model to develop effective virtual teamwork in online higher education [14]. They identified seven key factors such as motivation, communication, knowledge sharing, trust, cohesion, coordination, and performance that impact on the effectiveness of virtual teamwork, and also outlined relationships among the key factors to develop the model. In order to validate that model and evaluate its statistical significance, PLS-SEM is chosen as a statistical technique. The reasons for choosing PLS-SEM and how it can be applied to test this model are discussed and explained in the rest of this article.

Reason for Choosing PLS-SEM

In general, PLS-SEM is a recommended approach to apply as a statistical tool when (a) primary objective is to explore and predict target constructs, (b) constructs are measured formatively, (c) the model is complex in nature, and (d) the sample size is small due to its small population size [38]. These recommendations are matched for choosing the PLS-SEM in order to evaluate the conceptual model of virtual teamwork development in online higher education. For example, the model of virtual teamwork to be estimated is a complex model with many constructs and indicators variables, the constructs are formatively measured, and the primary objective is to predict and explore the target construct in the model, which clearly indicate PLS-SEM as a suitable technique for the research topic. Besides, the sample size issue is clearly an advantage of using PLS-SEM because it works well with a small sample size due to the small population size, and, it works very well with a large sample size as well. However, a small sample size should not be the sole argument for choosing the PLS-SEM technique in any empirical study but the focus should be on the goal of the study and its empirical analysis [39].

Moreover, Hair et al. (in40) presented an overview of points researchers should consider when choosing PLS-SEM as an appropriate SEM technique for a study, of which many of them are already discussed above recommended by Sarstedt et al. (in [38]). However, apart from these, other recommendations of choosing PLS-SEM are (a) when the analysis is to study a conceptual model for predicting target constructs, (b) When data distributions assumption is absent (i.e. nonnormal or nonparametric) [40].

The research is primarily based on existing literary works for developing a conceptual model, which is exploratory in nature for analyzing the conceptual model of virtual teamwork in online higher education. This argument is also matched in favor of choosing PLS-SEM for the study topic. Another reason for choosing PLS-SEM for this research study is the absence of distributional assumptions of data (i.e. lack of normality) as this technique relies on nonnormal data. This is another big advantage of using PLS-SEM, but should not be the sole argument, rather in combination with other main arguments for choosing the PLS-SEM technique [40].

From the above discussion, a list of points is summarized in Table 17.1 to highlight the main reasons for choosing the PLS-SEM for evaluating the virtual teamwork model in online higher education. These arguments are listed based on the characteristics of the study and indicate that the PLS-SEM is an appropriate SEM technique for applying it in the evaluation of the conceptual model on virtual teamwork development in online higher education.

Table 17.1 Table captions should be placed above the tables

Characteristics	Arguments in favor of PLS-SEM
Nature of study	Exploratory
Study purpose	Estimate a theoretical framework or a conceptual model
Primary objectives	Predict, or explore target constructs, and/or detect important constructs
Model Complexity	Complex with many constructs, indicator, and relationships
Sample size	More than the recommended minimum sample siz
Measurement model	More than one formatively measured constructs
Follow-up-analyses	Latent variable scores
Theoretical framework	Developed based on existing literary works, theories, and logic
Data distributions	Nonnormal or nonparametric

How to Apply PLS-SEM

This section describes the procedure of applying PLS-SEM for evaluating virtual teamwork model in the context of online higher education. In PLS-SEM, a path model is developed from two perspectives: structural model and measurement model [31–33]. So, both the structural and measurement model has to be evaluated separately to validate the PLS path model. Additionally, it is always fair enough to assess the model fit for judging how well the hypothesized model structure fits the empirical data. The following three subsections describe the evaluation criteria considered for assessing the structural model, the measurement model, and the model fit of applying PLS-SEM respectively.

Assess Measurement Model

Among the two types of measurement models (i.e. reflective and formative), this study only considers the formative measurement model, thus, only formative measures evaluation criteria will be discussed only in this paper. In a formative measurement model, measures are not covaried [26], and indicators are error-free [31, 34]. That's why the same evaluation criteria used for reflective measurement cannot be applied directly to the formative measurement model [17]. However, many researchers applied the same evaluation criteria in formative measures used for the reflective measurement model which is incorrect [3, 6,].

Based on the recommendations from existing literature (e.g. [26, 40, 41]), the following set of evaluation criteria is compiled (as can be seen in Fig. 17.2) to evaluate the formative measurement model.

Assess Convergent Validity. The convergent validity assessment can be used for the examination of construct validity for formative constructs. It means to test whether a formatively measured construct is highly correlated with the same construct

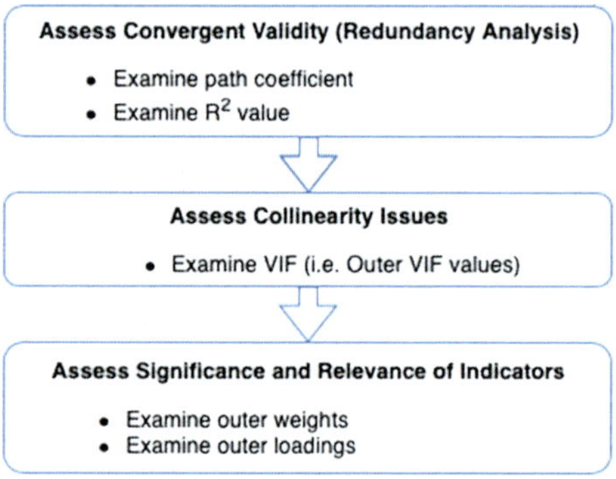

Fig. 17.2 A procedure for assessing the formative measurement model

measured reflectively with multi-item or measured with a global single item. It is also called redundancy analysis because the same construct is included as both formative and reflective construct [17]. In this analysis, a formatively measured construct (as an exogenous construct) is connected with a reflective measure of the same construct (as an endogenous construct) and assesses path coefficient (for definition see [42, 43]) and R^2 value (for definition see [44, 45]) for the convergent validity of the construct. A desirable minimum path coefficient value is 0.70 or above, which means that the acceptable minimum R^2 value is 0.50 or above for the reflectively measured construct. Anything below the threshold value indicates that the formative indicators of the construct do not contribute at a sufficient degree to its intended content [26].

The challenging part of this analysis is to include a number of suitable indicators for the reflectively measured construct in data collection. However, multi-item reflective indicators could increase the survey length [26], thus an alternative approach is to include a global indicator in the survey that summarizes the essence of the formatively measured construct [46], which is much more flexible, less time-consuming, and more importantly does not affect the survey length.

Assess Collinearity Issues. The collinearity assessment is used for testing construct reliability, which concerns the internal consistency of a measurement model [47]. It assesses whether each formative indicator contributes its intended meaning to the formative index because indicator' information could be redundant if it has a high correlation with other indicators of the same construct [26]. The high correlation between indicators of the same construct indicates the presence of collinearity at a critical level which needs to be fixed.

In the context of PLS-SEM, VIF (variance inflation factor) (for definition see [48, 49]) is the standard approach to statistically assess the presence of critical collinearity issues at a critical level. As a rule of thumb, a VIF value of less than 10 indicates the absence of multicollinearity [15]. However, research in PLS-SEM the widely accepted VIF value is less than 5 (e.g. [26]). The VIF value 5 or higher indicates the presence of potential collinearity problems [19]. In the case of collinearity at a critical level, the corresponding indicator of the construct should be removed, providing that the remaining indicators sufficiently captures the content of the construct.

Assess Significance and Relevance of Indicators. Finally, it is very important to examine whether the formative indicators contribute significantly to their corresponding constructs both relatively and absolutely by examining their outer weights and outer loadings (for definitions see [50]) respectively. Outer weights are the statistics of multiple regression [18], which state the relative contribution of each indicator to its corresponding construct, whereas, outer loadings are the statistics of a simple regression, which state the absolute contribution of each indicator to its corresponding construct [26]. The outer loading or absolute contribution of any formative indicator comes to handy when the indicator's outer weight is found non-significant but the indicator's outer loading is found significant. In that case, the indicator can be kept in the measurement model.

The assessment of significance and relevance of indicators is calculated by applying a bootstrapping procedure which is a non-parametric method to examine the stability and significance of various coefficients based on resampling subsamples with replacement from the original sample [26]. PLS-SEM uses a bootstrapping procedure because it does not assume a normal distribution of data, and hence, parametric significance tests cannot be applied for testing the statistical significance of coefficients such as outer weights, and outer loading. Therefore, PLS-SEM relies on a non-parametric bootstrapping procedure to test the significance of coefficients [51]. A big number of subsamples are drawn randomly from the original data sample, where each contains the same number of observations as in the original sample. A widely used recommended number for the bootstrap samples is 5000 [26].

Assess Structural Model

The evaluation criteria of structural model evaluation criteria are discussed and presented in this subsection. The following systemic procedure outlined in Fig. 17.3 is considered for the assessment of the structural model of this study which is chosen based on the suggestions from existing literature (e.g. [17, 26, 32, 33]).

Assess Collinearity Issues. The path coefficients of the structural model might be biased if the estimation indicates the presence of collinearity at a critical level among the predictor constructs. For that reason, it is necessary to assess whether the collinearity reaches a critical level among the predictor constructs by examining VIF values. The VIF values above 5 in the predictor constructs indicate the presence of collinearity at a critical level, and in that case, the corresponding constructs should remove, or merge into a single construct, or create higher-order constructs to treat collinearity issues for a structural model [26].

Assess Significance and Relevance of Relationships. The significance and relevance of the structural model relationships can be examined by the path coefficients

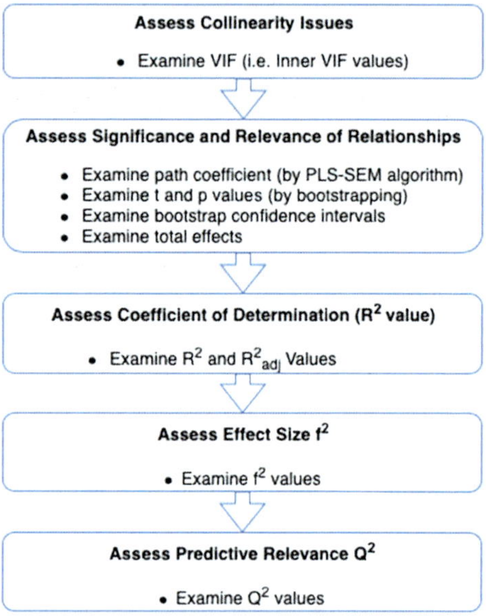

Fig. 17.3 A procedure for assessing the structural model

which have standardized values approximately between −1–+1. The path coefficients values close to +1 are usually statically significant and represent strong positive relationships. On the other hand, path coefficients values close to −1 are usually not statically significant and represent weaker relationships.

However, the path coefficients obtained from running the PLS-SEM algorithm is not enough to say that the path coefficients are statistically significant. In this case, the bootstrapping procedure is needed here to compute the empirical t values (for definition see [52, 53]) and p values (for definition see [54, 55]) for all path coefficients of the structural model in order to prove the ultimate statistical significance. If the empirical t value is larger than the critical value then it indicates that the path coefficient is statistically significant at a certain error probability. The common critical values of t for a two-tailed test that used to examine the significance of path coefficients are 1.65 (significance level = 10%), 1.96 (significance level 5%), and 2.57 (significance level = 1%). The choice of significance level depends on the objective and field of the study. In general, the 10% significance level can be assumed in exploratory research, and a 5% significance level can be assumed in marketing research [26]. However, a 5% significance level is considered for this study which mostly used for testing the significance of path coefficients. The empirical p values can also be used to assess the significance levels. The p value of less than 0.05 is considered as a significance level of 5%, and the p value less than 0.01 is considered as a significance level of 1%.

In addition to these, the bootstrapping confidence intervals provide information on the stability of the estimated path coefficients of the structural model, and hence allows assessing whether a path coefficient is significantly different from zero, and the true values are somewhere withing the range assuming a certain level of confidence (e.g. 95%). If the estimated path coefficient's confidence interval does not include zero then it can be considered as a significant effect [26].

After assessing the significance of structural model relationships, it is important to assess the relevance of significant relationships because the path coefficients may be significant but their size may be small. For that reason, the relevance of relationships is crucial for interpreting and concluding the results even though the relationships are significant [26]. It can be examined by the total effects (i.e. the sum of indirect and direct effects).

Therefore, the bootstrapping procedure is ultimately used to assess the significance and relevance of structural model relationships (such as t values, p values, bootstrap confidence intervals, and total effects) to conclude that whether the relationships of the structural model are statistically significant and relevant.

Assess Coefficient of Determination (R^2 Value). The coefficient of determination (R^2 Value) (for definition see [44, 45]) measures the predictive power of the model. It represents the combined effects of all the linked exogenous constructs on the endogenous construct. That is, it represents a measure of in-sample predictive power [29] ranges from 0 to 1, with higher levels indicating higher levels of predictive accuracy. A common rule of thumb for acceptable R^2 values is difficult to provide because it depends on the model complexity and study field. For example, in marketing research, R^2 Values of 0.75, 0.50, and 0.25 for endogenous constructs are considered as substantial, moderate, and weak respectively [19, 56]. Though these rules of thumb are particularly for marketing research, in general, it is considered as rules of thumb for R^2 values.

The R^2 Values should not be considered solely to understand the predictive power of the model because there might be an inherent bias towards a complex model. As a remedy, the adjusted coefficient of determination (R^2 adjusted value) can be considered to avoid bias towards complex models [26].

Assess Effect Size f^2. In addition to exploring R^2 values, the changes in R^2 can also be explored, which is known as effect size (f^2) (for definition see [57, 58]), which was first presented by Cohen in [59]. The f^2 effect size examines the substantive impact of each independent construct on the dependent construct in order to assess the structural model. More precisely, in this evaluation, a specified construct is omitted from the model to examine the impact of the omitted exogenous construct on the endogenous constructs. The rules of thumb for assessing the f^2 effect size are 0.35, 0.15, and 0.02, which indicate large, medium, and small effects of exogenous construct respectively [59]. There is no effect if the value of the effect size is less than 0.02.

Assess Predictive Relevance Q^2. The model's predictive relevance or out-of-sample predictive power of the model is examined by Stone-Geisser's Q^2 value [60, 61]. That means Stone-Geisser's Q^2 value can be used for testing the predictive relevance of the structural model. A blindfolding procedure with a specified omission distance is used to generate Q^2 values [17].

Blindfolding procedure is a sample reuse technique that runs iteratively until each data point has been omitted and the model has been re-estimated with remaining data points [17, 24, 56]. Where, the omission distance should be an integer ranging from 5 to 10 [3, 17, 62], however, omission distance value 7 is mostly recommended in the literature [41]. The number of observations divided by omission distance should not be an integer, otherwise, it will delete the same set of observations in each round [26]. A Q^2 value greater than 0 indicates the presence of predictive relevance of the path model for a particular dependent variable, whereas the Q^2 value of zero, or below indicates the absence of predictive relevance.

Assess Model Fit

Unlike CB-SEM, the notion of model fit is not fully transferable to PLS-SEM because it works differently than CB-SEM [26]. However, there are several early stages of development model fit measures for PLS-SEM. For example, goodness-of-fit index (GoF), one of the earliest fit indices proposed by Tenenhaus et al. ([32]) for validating the PLS model globally. But Henseler and Sarstedt challenged the usefulness of the GoF both conceptually and empirically and stated that it does not represent a goodness-of-fit criterion for PLS-SEM [63]. That's why it is not considered in this study as well. Another measure of model fit, the root mean square residual covariance (RMS_{theta}) introduced by Lohmöller which relies on covariance [64]. But this criterion has not been explored yet in a PLS-SEM context until recently by the PLS-SEM researchers [26], and hence, it is not considered as well in this study.

Finally, the following model fit measures outlined in Fig. 17.4 are considered to assess the model fit in the case of the PLS-SEM. A brief description of each of the model fit measures is presented in the rest of the section. More information can be found in [65].

Normed Fit Index (NFI). Normed fit index (NFI) is one of the first model fit measures proposed in the SEM literature by Bentler and Bonett, which computes the Chi^2 value of the proposed model and compares it against the Chi-square value of the null model (a meaningful benchmark) [66]. Afterward, the NFI is defined as 1

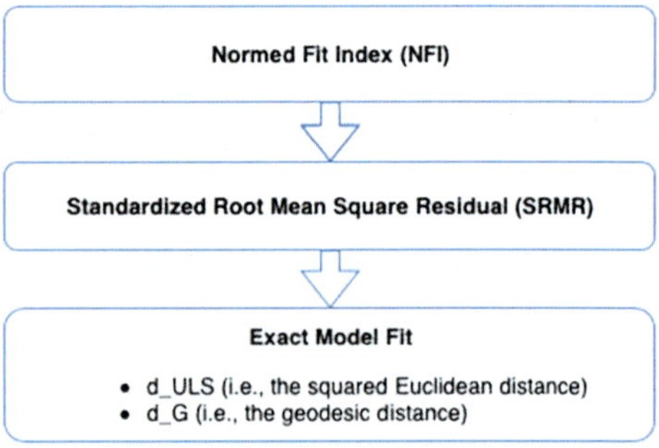

Fig. 17.4 A procedure for assessing model fit

minus the Chi-square value of the proposed model divided by the Chi-square values of the null model. The NFI results in values between 0 and 1. Generally, the NFI value above 0.90 is considered an acceptable fit. However, this explication is quite difficult to achieve. Therefore, in general, the NFI values closer to 1 indicate the better model fit [65].

Standardized Root Mean Square Residual (SRMR). The standardized root mean square residual (SRMR) measure fit is defined as the difference between the observed correlation and the model implied correlation [65], which is originally known for CB-SEM. However, Henseler et al. (*Common beliefs and reality about pls*) assessed the efficacy of the SRMR in a PLS-SEM context and suggested the SRMR as a goodness of fit measure for PLS-SEM (that can be used to avoid model misspecification) as well. The value of less than 0.08 is a more conservative version and generally considered a good model fit when applying CB-SEM, however, this threshold value is likely too low for PLS-SEM, so a value less than 0.10 is also considered a good fit [67].

Exact Model Fit. The exact model fit tests the statistical inference of the discrepancy between the observed correlation (empirical covariance) matrix and the model implied correlation matrix (covariance matrix implied by the composite factor model) [65]. Unlike SRMR, the discrepancy is expressed in terms of distances (i.e. not in the forms of residuals like in SRMR). There are two different approaches to calculate this discrepancy, d_ULS (i.e., the squared Euclidean distance) and d_G (i.e., the geodesic distance), defined by Dijkstra and Henseler [68]. The PLS-SEM bootstrapping procedure provides the confidence intervals of these discrepancy values. The original value of the exact d_ULS and d_G fit criteria have to be compared against the confidence interval created from the sampling distribution, and then the original value should be included in the confidence interval [65]. That means, the upper bound (is at the 95% or 99% point) of the confidence interval should be larger than the original value of the exact d_ULS and d_G fit criteria to indicate that the model has a good fit.

Conclusion

PLS-SEM statistical technique is used in many disciplines of research nowadays, and also applicable in education and ICT research. The content of this paper discussed on PLS-SEM statistical technique to deliver the message about why, and how it can be applied in the research work, like virtual teamwork development in online higher education. This paper mainly summarized and synthesized the key points of choosing PLS-SEM and how it can be applied in the research field like virtual teamwork development in online higher education for empirical study. It is believed that the detailed discussion of this chapter consolidated the foundation and justification of choosing PLS-SEM in the target research topic of education and ICT. Though the content of the paper is presented for a specific topic bear in mind, however, it can be useful in general for other research studies as well and can be used as a guideline of applying the PLS-SEM statistical technique indented.

Acknowledgements This work is part of a Doctoral Thesis funded by and conducted at *Universitat Oberta de Catalunya* (Barcelona, Spain).

References

1. J.F. Hair et al., Mirror, mirror on the wall: a comparative evaluation of composite-based structural equation modeling methods. J. Acad. Mark. Sci. **45**(5), 616–632 (2017)
2. C.M. Ringle et al., "Partial least squares structural equation modeling in HRM research." Int. J. Hum. Res. Manage., (2018)
3. J.F. Hair et al., An assessment of the use of partial least squares structural equation modeling in marketing research. J. Acad. Mark. Sci. **40**, 414–433 (2012)
4. J.J. Sosik et al., Silver bullet or voodoo statistics? a primer for using the partial least squares data analytic technique in group and organization research. Group Org. Manage. **34**(1), 5–36 (2009)
5. D.X. Peng, F. Lai, "Using partial least squares in operations management research: a practical guideline and summary of past research." J. Oper. Manage. **30**(6), pp. 467–480 (2012)
6. J.F. Hair et al., The use of partial least squares structural equation modeling in strategic management research: a review of past practices and recommendations for future applications. Long Range Plan. **45**, 320–340 (2012)
7. C. Nitzl, The use of partial least squares structural equation modelling (pls-sem) in management accounting research: directions for future theory development. J. Acc. Lit. **37**, 19–35 (2016)
8. Ringle, C. M., et al. "A critical look at the use of pls-sem in mis quarterly." MIS Quarterly, **36**(1) (2012)
9. N.F. Richter et al., "European management research using partial least squares structural equation modeling (pls-sem): editorial." Eur. Manage. J. **34**(6), 589–597
10. N.K. Avkiran, C.M. Ringle (eds.), *Partial Least Squares Structural Equation Modeling: Recent Advances in Banking and Finance* (Springer, 2018)
11. W.L. Shiau et al., Editorial: internet research using partial least squares structural equation modeling (pls-sem). Internet Research **29**(3), 398–406 (2019)
12. G.D. Garson, *Partial Least Squares Regression and Structural Equation Models* (Statistical Associates, Asheboro, 2016)
13. T. Ramayah et al., *Partial least squares structural equation modeling (pls-sem) using smartpls 3.0: an updated and practical guide to statistical analysis*, (Pearson, Singapore, 2016)
14. A.I. Jony, E. Serradell-Lopez, "Key factors that positively affects the effectiveness of virtual teamwork in online higher education." Rii forum proceedings of research & innovation forum (rii forum) annual conference, Athens, Greece, Springer, 2020
15. D. Gefen et al., Structural equation modeling and regression: guidelines for research and practice. Commun. Assoc. Inf. Syst. **4**(7), 1–70 (2000)
16. T.F. Golob, Structural equation modeling for travel behavior research. Transp. Res. Part B: Methodol. **37**(1), 1–25 (2003)
17. W.W. Chin, "The partial least squares approach to structural equation modeling." *Modern Methods for Business Research*, ed. by G.A. Marcoulides (Mahwah, NJ: Lawrence Erlbaum, 1998), pp. 295–358
18. J.F. Hair et al., *Multivariate data analysis* (Prentice Hall, Englewood Cliffs, NJ, 2010)
19. J.F. Hair et al., Pls-sem: indeed a silver bullet. J. Mark. Theory Pract. **19**, 139–151 (2011)
20. K.G. Jöreskog, "A general method for estimating a linear structural equation system." *Structural Equation Models in the Social Sciences*, ed. by A.S. Goldberger, O.D. Duncan (Seminar Press, New York, NY, 1973), pp. 255–284

21. H.O.A. Wold, "Partial least squares." *Encyclopedia of statistical sciences*, ed. by S. Kotz, N.L. Johnson (Wiley, New York, NY, 1985), pp. 581–591
22. H.O.A. Wold, "Path models with latent variables: the nipals approach." *Quantitative sociology: international perspectives on mathematical and statistical modeling*, ed. by H.W. Blalock, A. Aganbegian, F.M. Borodkin, 1975, pp. 307–357
23. H.O.A. Wold, "Soft modeling: the basic design and some extensions." *Systems under indirect observations: Part ii*, ed. by K.G. Jöeskog, H.O.A. Wold (Amsterdam, North-Holland, 1982), pp. 1–54
24. M. Tenenhaus, Component-based structural equation modeling. Total Qual. Manage. Bus. Excellence **19**(7), 871–886 (2008)
25. G. Mateos-Aparicio, Partial least squares (pls) methods: Origins, evolution, and application to social sciences. Commun. Stat. **40**(13), 2305–2317 (2011)
26. J.F. Hair et al., *A Primer on Partial Least Squares Structural Equation Modeling (pls-sem)*, 2nd edn. (Sage, Los Angeles, 2017)
27. J. Henseler et al., Common beliefs and reality about partial least squares: comments on rönkkö & evermann (2013). Organ. Res. Methods **17**, 182–209 (2014)
28. M. Sarstedt et al., Estimation issues with PLS and CB-SEM: where the bias lies! J. Bus. Res. **69**, 3998–4010 (2016)
29. E.E. Rigdon, Rethinking partial least squares path modeling: in praise of simple methods. Long Range Plan. **45**, 341–358 (2012)
30. W.W. Chin, P.R. Newsted, *Structural Equation Modeling Analysis with Small Samples Using Partial Least Squares, Thousand Oaks* (Sage Publications, CA, 1999)
31. A. Diamantopoulos, The error term in formative measurement models: Interpretation and modeling implications. J. Model. Manage. **1**, 7–17 (2006)
32. M. Tenenhaus et al., Pls path modeling. Comput. Stat. Data Anal. **48**(1), 159–202 (2005)
33. Y.M. Chatelin et al., "State-of-art on pls path modeling through the available software." *HEC research papers series 764*, HEC Paris, 2002
34. J.R. Edwards, R.P. Bagozzi, On the nature and direction of relationships between con structs and measures. Psychol. Methods **5**, 155–174 (2000)
35. A. Diamantopoulos, Incorporating formative measures into covariance-based structural equation models. MIS Q. **35**, 335–358 (2011)
36. A. Powell et al., Virtual teams: a review of current literature and directions for future research. Data Base for Advances in Information Systems **35**(1), 6–35 (2004)
37. R. Bennett, Employers' demands for personal transferable skills in graduates: a content analysis of 1000 job advertisements and associated empirical study. J. Vocat. Educ. Training **54**(4), 457–476 (2002)
38. M. Sarstedt et al., "Partial least squares structural equation modeling." *Handbook of Market Research*, ed. by C. Homburg, M. Klarmann, A. Vomberg (Springer, Heidelberg, 2017)
39. E.E. Rigdon, Choosing pls path modeling as analytical method in European management research: a realist perspective. Eur. Manage. J. **34**(6), 598–605 (2016)
40. J.F. Hair et al., "When to use and how to report the results of pls-sem." Eur. Bus. Rev. **3**(1), (2018) https://doi.org/10.1108/ebr-11-2018-0203
41. P. Andreev et al., "Validating formative partial least squares (PLS) models: methodological review and empirical illustration." *ICIS 2009 proceedings*, 2009
42. M.E. Olobatuyi, *A User's Guide to Path Analysis*. UPA, 2006
43. S. Wright, The method of path coefficients. Annal. Math. Stat. **5**, 151–215 (1934)
44. B.S. Everitt, A. Skrondal, *The Cambridge Dictionary of Statistics*. 4th edn. Cambridge University Press, 2010
45. Wikipedia. "Coefficient of determination." 29 April 2020, https://en.wikipedia.org/wiki/Coefficient_of_determination
46. M. Sarstedt et al., Measuring reputation in global markets: a comparison of reputation measures' convergent and criterion validities. J. World Bus. **48**, 329–339 (2013)
47. D. Straub et al., Validating guidelines for is positivist research. Commun. Assoc. Inf. Systems **13**, 380–427 (2004)

48. G. James et al., *An Introduction to Statistical Learning*, 8th edn. (Springer, New York, 2017)
49. Wikipedia. "Variance Inflation factor." 8 June 2020, https://en.wikipedia.org/wiki/Variance_inflation_factor
50. H. Latan, R. Noonan, *Partial Least Squares Path Modeling: Basic Concepts, Methodological Issues and Applications*. Springer, 2017
51. B. Efron, R. Tibshirani, Bootstrap methods for standard errors, confidence intervals, and other measures of statistical accuracy. Stat. Sci. **1**, 54–75 (1986)
52. R.E. Walpole, *Probability & Statistics for Engineers & scientists*, 7th edn. (Pearson, 2006)
53. Wikipedia. "T-statistic." 11 May 2020, https://en.wikipedia.org/wiki/T-statistic
54. Fisz, M. "Significance Testing." *Probability theory and mathematical statistics*. 3rd ed. (John Wiley and Sons, New York, 1963)
55. Wikipedia. "p-value." 19 May 2020, https://en.wikipedia.org/wiki/P-value
56. J. Henseler et al., The use of partial least squares path modeling in international marketing. Adv. Int. Mark. **20**, 277–320 (2009)
57. P.D. Ellis, *The Essential Guide to Effect Sizes*. Cambridge University Press, 2010
58. Wikipedia. "Effect size." 6 June 2020, https://en.wikipedia.org/wiki/Effect_size
59. S. Cohen, "*Statistical Power Analysis for the Behavioral Sciences* (Cambridge University Press, Bootstrap methods and their application, Cambridge, UK, 1988)
60. S. Geisser, A predictive approach to the random effects model. Biometrika **61**, 101–107 (1974)
61. M. Stone, Cross-validatory choice and assessment of statistical predictions. J. Royal Stat. Soc. **36**, 111–147 (1974)
62. H. Apel, H.O.A. Wold, "Soft modeling with latent variables in two or more dimensions: Pls estimation and testing for predictive relevance." *Systems Under Indirect Observation*, ed. by K.G. Jöreskog, H.O.A. Wold (Amsterdam, 1982), pp. 209–247
63. J. Henseler, M. Sarstedt, Goodness-of-fit indices for partial least squares path modeling. Comput. Stat. **28**, 565–580 (2013). https://doi.org/10.1007/s00180-012-0317-1
64. J. Lohmöller, *Latent Variable Path Modeling with Partial Least Squares* (Physica, Heidelberg, Germany, 1989)
65. SmartPLS-Documentation. "Model Fit", SmartPLS, 2020, www.smartpls.com/documentation/algorithms-and-techniques/model-fit
66. P.M. Bentler, D.G. Bonett, Significance tests and goodness-of-fit in the analysis of covariance structures. Psychol. Bull. **88**, 588–600 (1980)
67. L.T. Hu, P.M. Bentler, Fit indices in covariance structure modeling: sensitivity to under parameterized model misspecification. Psychol. Methods **3**, 424–453 (1998)
68. T.K. Dijkstra, J. Henseler, Consistent and asymptotically normal pls estimators for linear structural equations. Comput. Stat. Data Anal. **81**(1), 10–23 (2015)

Chapter 18
Sense of Virtual Community in Wikipedia Online Community for Technology-Enhanced Learning in Higher Education

Anu Helena Suominen and Jari Jussila

Abstract Currently, global crises along the digital revolution challenge the organization of higher education (HE) world-wide. Hence, there is an urgent need to better understand how Web 2.0-technology with their online communities can support learning in HE particularly in distance mode. This article investigates technology-enhanced learning (TEL) in HE, taking place in the Wikipedia online community. Particularly the study focuses on the role of sense of virtual community (SOVC) in the social process of learning. Therefore, this study investigates the student perceptions of purpose and benefits of the Wikipedia community, as well as the perceived SOVC while being an active contributing participant in the community. The study was carried out in one university course, where technology intervention was carried out by using Wikipedia in a learning assignment of collaborative writing. The case study results illustrate that mainly students perceive the purpose and benefits of Wikipedia being cognitive, although only less than one-third of the students experienced SOVC as active participants. The article contributes to the field of TEL in HE, more specifically to Wikipedia-enhanced learning and the significance of SOVC in Wikipedia community during learning assignment as active community participants. The presented case study provides insights into the SOVC in online communities for HE-learning during technology interventions.

Keywords Sense of virtual community · Sense of community · Technology-enhanced learning · Wikipedia-enhanced learning · Online community · Higher education

---

A. H. Suominen (✉)
Tampere University, P.O. Box 1001, Tampere FI33014, Finland
e-mail: lncs@springer.com

J. Jussila
Häme University of Applied Sciences, 13100 Hämeenlinna, Finland

A. Visvizi et al. (eds.), *Research and Innovation Forum 2020*,
Springer Proceedings in Complexity,
https://doi.org/10.1007/978-3-030-62066-0_18

Introduction

Pandemics, armed conflicts and other global crises, alongside digital revolution challenge the organization of higher education (HE) world-wide. Hence, there is an urgent need to better understand how 2.0-technology with their online communities [1] can support learning in HE, particularly in distance mode. This article investigates technology-enhanced learning (TEL) [2–4], more specifically Wikipedia-enhanced learning taking place in the Wikipedia online community. Although enhanced via technology, inherently learning is a social process [5, 6]. Accordingly, for the social aspect of learning to occur in the online communities where the students participate for learning, they should experience a sense of virtual community (SOVC). However, SOVC is a complex concept containing an individual's feelings, that has antecedents of community-type action such as helping others, as well as needs to be fulfilled. Although in previous studies SOVC has been detected in online communities both via reading and posting actions, posting actions seemed to enhance SOVC better [7]. Thus, being an active participator instead of lurking in Wikipedia could potentially enhance student learning with the fortification of SOVC. However, the role of SOVC in learning has not been previously studied among TEL with Wikipedia in HE. Therefore, this study investigates the student perceptions of Wikipedia's purpose and benefits, as well as the extent the students perceive SOVC in Wikipedia community in one university course, where technology intervention was carried out by using Wikipedia in a learning assignment of collaborative writing.

The article is structured as follows: First drawing from the literature, technology-enhanced learning, Wikipedia as an online community, and aspects of SOVC in online communities, such as Wikipedia, are presented. Then a case study of technology intervention with Wikipedia in HE institution is described. The case study results illustrate the student experiences from the purpose and benefits of Wikipedia and SOVC in its community while creating knowledge in Wikipedia. Following the results of the case study, theoretical contribution and practical implications are presented and future research directions are proposed.

Theoretical Background

Technology-Enhanced Learning

The term "technology-enhanced learning" (TEL) is used world-wide yet lacks a unanimously accepted concept definition in the educational literature. The use of the TEL term has proliferated, as the concept is extended beyond its original scope to encompass teaching, for example. We delimit our study on TEL that is solely focused on learning [3, 4, 8]. Learning focused TEL has its characteristic practices and underlying principles and deals with how technology supports cognition and metacognition [4]. Rationales behind technology interventions in Higher education (HE) include not only replicating or supplementing existing teaching but transforming teaching and/or learning processes as well ways to transform learning processes include, for instance, redesigning activities to provide active learning opportunities and to support qualitatively richer learning among students. Therefore, the design of TEL intervention studies should extend beyond the technology itself: outlining the intervention goals, designing the evaluation, and interpreting the results [8]. Furthermore, surpassing the technology layer, the versatile contemporary TEL-studies incorporate the social aspect of learning [6], particularly regarding the social media, such as in the study of Twitter-enhanced teaching and learning [9]. Yet, there is a growing need to research the phenomenon of technology-use in education, specifically evaluating learning by using descriptive and explanatory studies [10]. This article answers the call by conducting a descriptive and explanatory study of TEL in the context of Wikipedia online community.

Wikipedia as Part of New Learning and Network Collaboration Paradigm

Wikis are an example of social media that uses Web 2.0-technologies and regarded as part of the new learning and network collaboration paradigm [11, 12]. However, all social media is not equal, and varying classifications of social media have been carried out in the last decade. Kaplan and Haenlein classified Wikipedia among "*Collaborative projects*", distincting it from e.g. "*Social networking sites*", such as Facebook [13]. Furthermore, different types of social media focus on various functional building blocks defined in honeycomb-framework. Honeycomb-framework has seven building blocks of identity, conversations, sharing, presence, relationships, reputation, and groups, that explain the engagement needs of the social media audience [14, 15]. Moreover, according to 5C-model, Wikipedia belongs to the "*Collaborating: collective content creation*"-category, that has a different basic purpose than e.g. "Communicating", which includes discussion forums and "Connecting"-categories, which includes Facebook and LinkedIn, for example [16]. Additionally, although Wikipedia is regarded as an example of a social media platform, it is not a social networking platform, since the primary aim of the the engagement is not at forming relationships with like-minded individuals or groups [17].

Wikipedia is considered the most successful application of wiki technology [18]. It is the world's largest reference website, a collaborative encyclopedia, attracting 1.5 billion unique visitors monthly, with over 6 million articles in English and 38 million registered users reported in March 2020 [19]. Reshaping the creation of collaborative knowledge, the essence of wikis is a generally open practice of content addition and revision in a dynamic process of a large number of individuals, constantly displaying the status quo of collaboration [12, 20, 21]. Furthermore, the Internet and social media are considered vehicles for knowledge acquisition that are actively used in working life settings, but underused in university learning [22]. In reality, Wikipedia has been the de facto standard for a knowledge base in business [11]. From learning viewpoint, the Internet and social media promote learner agency, autonomy, and engagement in online communities and social networks, thus shift control from teacher to learner [23]. Recent studies of Wikipedia use in HE has covered e.g. patterns of use according to student perceptions [24] as well as Wikipedia's influence on the course design and students' learning process [25]. Furthermore, in a recent study, an acculturation process to the virtual community was detected while utilizing Wikipedia community in HE [26] However, the role of the Wikipedia online community [1] or the sense of virtual community for HE students has not been thoroughly researched.

Sense of Virtual Community in Wikipedia Online Community

Online communities have no universally accepted definition, but they are understood e.g. "*as web-based online services with features that enable members to communicate with each other*", which focal challenge is the user participation and thus the vitality of community [27]. Sense of community (SOC), is a shared emotional connection and individuals' feeling of belonging in a group [28], which is also experienced in online social groups as a sense of virtual community (SOVC) [7, 27]. SOVC is defined as "*human experience of a community feeling in a virtual environment*", which can be experienced both with reading and posting in an online community [7]. As a relatively new research domain, SOVC as a concept is not fully developed. However, it is established as a complex, individual-level concept, that incorporates feelings, with antecedents of community-like behaviour and expectations i.e. need fulfilment [7]. Tonteri et al. have operationalized SOVC with five dimensions of individual's feelings: (1) membership and one's rights and obligations in the community, (2) influence in the community and of being influenced by the community, (3) among the individual members of having a distinct identity in the community, (4) having a common social identity and identifying with the community, and (5). a strong emotional connection among the community members. These feelings that compose SOVC is differentiated from the preceding community-level phenomenon of community-like behaviour that takes the form of participation [7].

Online participation is regarded as a dichotomy of passive and active participation. Therefore, becoming an active participant incorporates transformation from consumer i.e. 'lurker' to a creator, and entails a transition in the community from the periphery towards the centre. Lurking, i.e. passive yet non-negative participation is detected particularly among online community newcomers with lessened SOVC and membership [27, 29, 30]. In Wikipedia, active participation, i.e. contribution, is mainly driven for the public good, that is altruism and ideology [31]. SOVC has been explored in different types of virtual communities [7]. However, when Wikipedians start experiencing the SOVC, their perception change: they see it as a community rather than a collection of articles [29].

In their study of SOVC, Tonteri et al. utilized the Nambisan and Baron's categorization of expected benefits [7, 32]. The categorization consists of four types of expected benefits 1. cognitive (i.e. knowledge and learning), 2. social-integrative (i.e. a sense of belonging to a group due to created social ties), 3. personal-integrative (i.e. self-efficacy due to influence), and 4. hedonic and affective (i.e. pleasurable experiences) [32]. We also adopted this view while studying the purpose of using Wikipedia from both general and personal views, as well as benefits for using Wikipedia.

Methodology

Following pragmatism [33] as research philosophy, we chose a case study approach [34] to investigate empirically the virtual sense of community in Wikipedia during university studies. We selected a master level knowledge management course of fall 2019 as our research case. The specific class was chosen due to its idiosyncrasy: it involved students of multiple universities, nationalities, and disciplines, and was offered also with distance learning option. Hence, it can be assumed that the results of the study could be also transferred to different language areas, and disciplines that make use of Wikipedia in learning. We present only the results of those who did not forbid the use of their survey responses in research (n = 56) and accomplished Wikipedia learning assignment, and the demographic information of that group is presented in Table 18.1. Altogether, the study was carried out with three surveys during the course: pre- and post-surveys as well as a survey covering the perceptions of Wikipedia learning assignment. In this article, we present the results of four statements with Likert-scale response options and five open-ended questions regarding Wikipedia's community and the sense of virtual community.

Results

Wikipedia Survey's Statements with Likert-Scale Response Options

In Table 18.2, the survey responses of four mandatory survey statements with numerical 5-point Likert scale response options regarding Wikipedia as a community and its sense of virtual community (SOVC) are presented.

Majority of the students felt that the Wikipedia community does have the capacity to influence other communities (76,8%). However, trust to members of Wikipedia was not that high: only more than one third agreed (35,7%), whereas almost half of the students (46,4%) had a neutral stance on the statement.

The effect of contribution to the SOVC was conflicting: contribution increased the SOVC only for one-third of students (30,4%), whereas over the third (35,7%) disagreed that the contribution increased their SOVC, and over one third (32,1%) expressed neutral stance to the effect of contribution. However, reading other user's

Table 18.1 Demographic information on the students n = 56

Variable	Values					
Sex	*Male*	*Female*				
	66.1%	33.9%				
Age	*18–24*	*25–34*	*35–44*	*45–54*		
	37.5%	41.1%	16.1%	5.4%		
Nationality	*Finnish*	*German*	*French*	*Other*		
	78.6%	14.3%	1.8%	5.4%		
Highest education degree	*Bachelor's degree*	*Master's degree*	*Trade/technical/vocational training*	*Some college credit, no degree*	*High school graduate, diploma or the equivalent*	*Some high school, no diploma*
	78.6%	5.4%	1.8%	5.4%	7.1%	1.8%
University where enrolled	*Course university*	*Uni 2*	*Uni 3*	*Uni 4*	*Uni 5*	
	64.3%	19.6%	7.1%	5.4%	3.6%	
Study	*Full time*		*Full time, but working on the side*		*Part-time*	
	44.6%		35.7%		19.6%	
Employment status	*Employed for wages*	*A student*	*Self-employed*	*Out of work*	*A homemaker*	*No answer*
	42.9%	48.2%	1.8%	1.8%	1.8%	3.6%
Work experience years	*1–2*	*3–5*	*6–10*	*10–15*	*15–20*	*over 20*
	37.5%	25.0%	17.9%	5.4%	7.1%	7.1%

Table 18.2 Results of four statements Wikipedia's community and SOVC

Survey statements	Responses n = 56				
5—Strongly agree, 4—Agree, 3—Neither agree nor disagree, 2—Disagree, 1—Strongly disagree	5%	4%	3%	2%	1%
Wikipedia community can influence other communities	21.4	55.4	21.4	1.8	0.0
I can trust members in Wikipedia community	0.0	35.7	46.4	17.9	0.0
Contributing to Wikipedia increased my SOVC in Wikipedia community	1.8	28.6	33.9	26.8	8.9
Reading other users' messages of the revision history in Wikipedia-pages increased SOVC in Wikipedia community	0.0	39.3	42.9	14.3	3.6

messages did increase the SOVC of 39,3% of the students, yet a little more, 42,9% had a neutral stance on the effect of reading other people's messages to their SOVC, whereas 17,9% disagreed, that reading other people's messages increased their SOVC.

The Purpose of the Wikipedia Community in General and Personally and the Benefits of Participating in Community

Table 18.3 portrays the results of three voluntary open-ended statements in the Wikipedia-survey:

- "*What is the purpose of the Wikipedia community in general?*"
- "*What is the purpose of the Wikipedia community for you personally?*"
- "*What do you perceive as potential benefits in participating in the Wikipedia community?*"

The data is analyzed by the number of mentions in the open-ended responses according to the benefit types of Nambisan and Baron [32]. The results show that the purpose of Wikipedia to students, when viewed through its benefits both in general and personally, are mostly cognitive and particularly knowledge benefits. Similarly, responses to the experienced benefits were cognitive, particularly concerning learning from others and learning to write objective and readable text, paying attention to the references. Personal integrative benefits were not mentioned at all in the survey, but some social integrative and affective benefits could be interpreted from the open-ended questions of the survey.

Table 18.3 Results of three open-ended survey statements regarding Wikipedia's purpose and benefits

	The Purpose of Wikipedia community		***Benefits of participating***
Benefits	*General n=43*	*Perso-nally n= 35*	n=36
Cognitive	**73**	**26**	**27**
Knowledge			
What: *Information source and collection for sharing and creation of knowledge, helping to understand which content is allowed and which not*	31	23	7
To whom: *Accessibility and editability to all worldwide*	16		
How: *Continuously checked and controlled relevant, accurate, good quality knowledge for a quick overview or basic knowledge with free and easy access*	26	1	
Improving learning opportunities			
Learn what others have found and written, develop in the topic, learn to write objectively and text that anyone can comprehend, be more wary of references, improve language skills, learn how to make additions to Wikipedia, observe and check the revision history		2	20
Social integrative: Social ties between participants	**3**	**2**	**5**
Network and discuss valuable knowledge, the more writers on Wikipedia, the faster the information will increase, get in touch with other professionals in the same field, contacts with specialists, get in touch with others who are interested the same subjects, get constructive feedback	3	2	5
Personal integrative: a sense of self-efficacy			**11**
Contribution: making own expertise and important information accessible to the general public to improve general knowledge for common good, to help others			11

(continued)

Table 18.3 (continued)

	The Purpose of Wikipedia community		*Benefits of participating*
Hedonic or affective	**3**	**3**	
Appreciation to people who maintain Wikipedia voluntarily for free use, making a better world, help in the community	3	3	
Conflicting or negative expressions	**2**	**7**	**3**
Unreliability as a source of information or Wikipedia as nothing special or the community behind being invisible	2	7	1
The negative feeling of community as the contribution was deleted right away			2

The Manifestation of SOVC and Methods for Improving SOVC in Wikipedia

In Table 18.4 are presented results of two voluntary open-ended questions:

Table 18.4 Results of two open-ended statements of Post-Wikipedia survey

Manifestation of SOVC	Qnty n=6
Community type behaviour of communication seeking influence *The positive effect of communication and connection, seeking communication via potential feedback, being aware of other people while editing texts, feeling proud of sustainable input*	*4*
Negative manifestations	*2*
Activities or things that potentially increase SOVC	**Qnty n=19**
Community type of behaviour: Own active participation	**6**
Own regular behaviour by participation: contribution, information adding and communication with other users with comments and conversation	
Community type of behaviour: Communication	**6**
Messaging and regular feedback, perhaps even automated after (first) edit or added texts, particularly reasons for deleted or reversed modifications, tips for future, and discussions about the articles, forum	
Community type of behaviour: Co-operation	**4**
Co-operation between authors and transparency of the respective authors of their information and how they received it, verification by peer-review	
Gamification	**1**
Gamifying: rewarding by points	
Community type of behaviour: the behavior of others	**1**
With people being more helpful and not judging instantly	

- *"In case you felt SOVC while using Wikipedia, how did it manifest to you?"*
- *"In your opinion, what activities or things could increase your SOVC in Wikipedia community?"*

The data is analyzed by the number of mentions in the open-ended responses according to the view of Tonteri et al., where SOVC incorporates feelings, with antecedents of community-like behaviour and expectations i.e. need fulfilment [7]. The results show that the manifestation of SOVC is more of antecedent, i.e. community-like behavior of seeking influence via communication and feedback, as well as connection, not feelings per se. Unfortunately, rude behavior was also manifested. However, the number of responses were low, thus limiting the validity of the result.

The activities students expressed potentially increasing SOVC are also antecedents of SOVC, i.e. community-like behavior. This behavior is anticipated from both themselves as well as from the community.

Conclusion

This article contributes to the field of Technology-enhanced learning (TEL) literature, focusing on the Wikipedia-enhanced learning in Higher education (HE) [3, 4, 8, 10]. More specifically, it contributes to the the sense of virtual community (SOVC) emergence [7] in Wikipedia-enhanced learning. In the field of TEL, Wikipedia as a Web 2.0-technology with the online community provides learning opportunities for HE students [6] as both lurkers and active participants [27, 29, 30]. As an online community, there is also the potential for a sense of virtual community (SOVC) to emerge in Wikipedia use that can enhance learning with social interaction between participants. This study investigated the identified purpose and benefits, the degree of perceived SOVC, together with its manifestations, by university students as active participants performing a learning assignment in Wikipedia.

The results regarding students' conceptions of the purpose and benefits of the Wikipedia community were mainly cognitive: purpose in general and personally was information and knowledge and benefits particularly regarding learning. Considering Wikipedia as the world's largest encyclopedia, these results are not surprising. However, some students did mention personal integrative benefits of contribution as well as social integrative benefits of the Wikipedia community, which implies that some students experience Wikipedia additionally from the community's altruism and interaction viewpoints. These results add to the literature of TEL, particularly social media-enhanced learning by evaluating Wikipedia via the model of Nambisan and Barron [32], besides the discussion forums previously evaluated by Tonteri et al. [7].

The results regarding students perceived SOVC in Wikipedia community were conflicting. Contributing to Wikipedia increased a SOVC only for one-third of students. Interestingly, reading other users' messages did increase a SOVC for almost 40% of the students. This implies that even unidirectional interaction seems to affect perceived SOVC for some students. Yet, these results are conflicting compared to previous results, where posting increased SOVC more than reading [7]. Moreover, the manifestations of Wikipedia's SOVC were more regarding the antecedents of SOVC i.e. community-like behaviour aiming at e.g. influence, than the dimensions of an individual's feelings themselves [7]. Reasons for such conflicting results compared to previous ones may be caused by Wikipedia being inherently different type of social media than discussion forum [13–16]. Wikipedia is categorized as "*Collaborative project* "or "*Collaborating: collective content creation*", whereas discussion forums belong to "*Communicating*" types of social media, perhaps even to "*Social Networking sites*" or "*Social networking platforms*" [13, 16, 17]. Therefore, the goal of entering, using and engaging in Wikipedia is fundamentally different, thus most likely affecting also the mechanisms of SOVC emergence. Moreover, the manifestations of Wikipedia's SOVC regarding its the antecedents aiming at e.g. influence via knowledge, are most likely due to the nature of content creation for universal knowledge. After all, knowledge is power and with knowledge you can influence or be influenced with feedback all over the world. Therefore, we suggest that the SOVC emergence in different types of social media should be studied furthered to reveal their social and community affecting aspects.

One limitation of the study was that the learning assignment in Wikipedia was mandatory for students that did not have previous experience of contributing to Wikipedia. Forcing students to participate in the Wikipedia community can introduce a bias concerning SOVC. Secondly, the experience of contributing to Wikipedia during only a short period may limit the manifestation of SOVC that could develop over a longer period, which implies that for students there is an acculturation process while transforming from lurker to active participant in Wikipedia [26].

We argue, Wikipedia-enhanced learning i.e. learning enhancement aimed particularly with technology interventions with mandatory learning activities in Wikipedia online community as active participants, do not necessarily result in the SOVC and those altruistic benefits that voluntary activity in other social media might. However, the purpose and the benefits of Wikipedia are perceived by students cognitive, i.e. knowledge and learning, despite the lack of SOVC. Thus, we conclude that mandatory Wikipedia learning activities do produce perceived learning, although not all the gains of the online community are fully accomplished. This is due to the social media type Wikipedia as collective content creation platform inherently is and to an acculturation process of students to the online community while transforming from lurkers to active Wikipedia participants.

References

1. S. Faraj, G. Von Krogh, E. Monteiro, K.R. Lakhani, E. Monteiro, Online community as space for knowledge flows, 0–17 (2016)
2. L. Price, A. Kirkwood, Using technology for teaching and learning in higher education: A critical review of the role of evidence in informing practice. High. Educ. Res. Dev. **33**, 549–564

(2014). https://doi.org/10.1080/07294360.2013.841643
3. S. Bayne, What's the matter with 'technology-enhanced learning'? Learn. Media Technol. **40**, 5–20 (2015). https://doi.org/10.1080/17439884.2014.915851
4. D. Passey, Technology-enhanced learning: Rethinking the term, the concept and its theoretical background. Br. J. Educ. Technol. **50**, 972–986 (2019). https://doi.org/10.1111/bjet.12783
5. L.S. Vygotsky, *Mind in Society: The Development of Higher Psychological Processes* (Harvard University Press, Cambridge, Mass. 1978)
6. A. Visvizi, L. Daniela, C.W. Chen, Beyond the ICT- and sustainability hypes: A case for quality education. Comput. Human Behav. **107**, 2018–2020 (2020). https://doi.org/10.1016/j.chb.2020.106304
7. L. Tonteri, M. Kosonen, H. K. Ellonen, A. Tarkiainen, Antecedents of an experienced sense of virtual community. Comput. Human Behav. (2011). https://doi.org/10.1016/j.chb.2011.06.018
8. A. Kirkwood, L. Price, Technology-enhanced learning and teaching in higher education: what is 'enhanced' and how do we know? A critical literature review. Learn. Media Technol. **39**, 6–36 (2014). https://doi.org/10.1080/17439884.2013.770404
9. A. Visvizi, J. Jussila, M.D. Lytras, M. Ijäs, Tweeting and mining OECD-related microcontent in the post-truth era: a cloud-based app. Comput. Human Behav. **107**, 105958 (2020). https://doi.org/10.1016/j.chb.2019.03.022
10. J.W.M. Lai, M. Bower, How is the use of technology in education evaluated? A systematic review. Comput. Educ. **133**, 27–42 (2019). https://doi.org/10.1016/j.compedu.2019.01.010
11. J. Albors, J.C. Ramos, J.L. Hervas, New learning network paradigms: communities of objectives, crowdsourcing, wikis and open source. Int. J. Inf. Manage. **28**, 194–202 (2008). https://doi.org/10.1016/j.ijinfomgt.2007.09.006
12. P. Prasarnphanich, C. Wagner, The role of wiki technology and altruism in collaborative knowledge creation. J. Comput. Inf. Syst. **49**, 33–41 (2009)
13. A.M. Kaplan, M. Haenlein, Users of the world, unite! The challenges and opportunities of Social Media. Bus. Horiz. **53**, 59–68 (2010). https://doi.org/10.1016/j.bushor.2009.09.003
14. J.H. Kietzmann, K. Hermkens, I.P. McCarthy, B.S. Silvestre, Social media? Get serious! Understanding the functional building blocks of social media. Bus. Horiz. **54**, 241–251 (2011). https://doi.org/10.1016/j.bushor.2011.01.005
15. J.H. Kietzmann, B.S. Silvestre, I.P. Mccarthy, L.F. Pitt, Unpacking the social media phenomenon: towards a research agenda. J. Public Aff. (2012). https://doi.org/10.1002/pa.1412
16. V. Vuori, J. Jussila, The 5C categorization of social media tools. Acad. 2016 - Proc. 20th Int. Acad. Mindtrek Conf. 26–33 (2016). https://doi.org/10.1145/2994310.2994367
17. U.T. Lama, Understanding social media: functionality and differentiating social media and social networking sites. Int. J. Res. Eng. Appl. Manag. (2018). https://doi.org/10.18231/2454-9150.2018.0576
18. B.W. Carver, R. Davis, R.T. Kelley, J.A. Obar, L.L. Davis, Assigning students to edit Wikipedia: four case studies. E-Learning Digit. Media. **9**, 273–283 (2012). https://doi.org/10.2304/elea.2012.9.3.273
19. Wikipedia. https://en.wikipedia.org/wiki/Wikipedia:About
20. I. Elgort, A.G. Smith, J. Toland, Is wiki an effective platform for group course work? Australas. J. Educ. Technol. **24**, 195–210 (2008). https://doi.org/10.14742/ajet.v24i2.1222
21. S. Evenstein Sigalov, R. Nachmias, Wikipedia as a platform for impactful learning: a new course model in higher education. Educ. Inf. Technol. **22**, 2959– 2979 (2017). https://doi.org/10.1007/s10639-016-9564-z
22. F.J. García-Peñalvo, R. Colomo-Palacios, M.D. Lytras, Informal learning in work environments: training with the Social Web in the workplace. Behav. Inf. Technol. **31**, 753–755 (2012). https://doi.org/10.1080/0144929X.2012.661548
23. C. McLoughlin, M.J.W. Lee, Personalised and self regulated learning in the Web 2.0 era: International exemplars of innovative pedagogy using social software. Australas. J. Educ. Technol. **26**, 28–43 (2010). https://doi.org/10.14742/ajet.1100

24. N. Selwyn, S. Gorard, Students' use of Wikipedia as an academic resource - Patterns of use and perceptions of usefulness. Internet High. Educ. **28**, 28–34 (2016). https://doi.org/10.1016/j.iheduc.2015.08.004
25. Meseguer-Artola, Antoni Rodríguez-Ardura, Inma Ammetller, G., Rimbau-Gilabert, E.: Learning with Wikipedia in Higher Education: Academic Performance and Students' Quality Perception, in *Research & Innovation Forum 2019 Technology, Innovation, Education, and their Social Impact*. pp. 117–124. Springer Proceedings in Complexity (2019)
26. A. Brailas, K. Koskinas, M. Dafermos, G. Alexias, Wikipedia in Education: Acculturation and learning in virtual communities. Learn. Cult. Soc. Interact. **7**, 59–70 (2015). https://doi.org/10.1016/j.lcsi.2015.07.002
27. S. Malinen, Understanding user participation in online communities: A systematic literature review of empirical studies. Comput. Human Behav. **46**, 228–238 (2015). https://doi.org/10.1016/j.chb.2015.01.004
28. A.L. Blanchard, M. Lynne Markus, The Experienced "Sense" of a Virtual Community: Characteristics and Processes. Data Base Adv. Inf. Syst. (2004). https://doi.org/10.1145/968464.968470
29. S.L. Bryant, A. Forte, A. Bruckman, Becoming Wikipedian: Transformation of participation in a collaborative online encyclopedia, in *Proceedings of the International ACM SIGGROUP Conference on Supporting Group Work* (2005)
30. B. Nonnecke, D. Andrews, J. Preece: non-public and public online community participation: Needs, attitudes and behavior. Electron. Commer. Res. (2006). https://doi.org/10.1007/s10660-006-5985-x
31. O. Nov, What motivates Wikipedians? Commun. ACM. (2007)https://doi.org/10.1145/1297797.1297798.
32. S. Nambisan, Baron, A. Robert, Interactions in virtual customer environments: implications fo product support and customer relationship management. J. Interact. Mark. **21**, 42–62 (2007). https://doi.org/10.1002/dir
33. D.L. Morgan, Pragmatism as a paradigm for social research. Qual. Inq. **20**, 1045–1053 (2014). https://doi.org/10.1177/1077800413513733
34. N. Siggelkow, Persuation with Case Studies. Acad. Manag. J. **50**, 20–24 (2007). https://doi.org/10.5465/AMJ.2007.24160882

Chapter 19
Virtual Hackathons—A Novel Approach for University-Industry Collaboration

Jari Jussila, Jukka Raitanen, Anu Helena Suominen, and Anne-Mari Järvenpää

Abstract Pandemics with their lockdowns have proven that radical collocation, collaborating intensively in a same physical space in the same time is not always possible. However, radical collocation is one of the necessary attributes of hackathons, one type of innovation contests. Yet, digital platforms enable virtual collocation, i.e. collaborating at the same time in a virtual place. This paper addresses the virtual hackathon as an innovation contest method for uses in situations, where radical collocation is unfeasible. Specifically, it focuses on virtual hackathon as a method for university-industry collaboration. Although collocation in general plays an integral part in hackathon process, either radical or possible virtual collocation has not yet been the focus of hackathon research. Therefore, this paper presents a case study of a university-industry collaboration involving five organizations in Finland. As a result, the paper reveals benefits, disadvantages and challenges collocation in virtual format causes to the hackathon as an innovation contest event. Presenting conclusions for both academics and industry, the paper contributes to the literature on hackathons used particularly with virtual collocation in the university-industry collaboration.

Keywords Hackathon · Virtual hackathon · University-industry collaboration · Radical collocation · Virtual collocation

---

J. Jussila (✉) · J. Raitanen · A.-M. Järvenpää
Häme University of Applied Sciences, 13100 Hämeenlinna, Finland
e-mail: jari.jussila@hamk.fi

A. H. Suominen
Tampere University, P.O. Box 1001, 33014 Tampere, Finland
e-mail: lncs@springer.com

A. Visvizi et al. (eds.), *Research and Innovation Forum 2020*,
Springer Proceedings in Complexity,
https://doi.org/10.1007/978-3-030-62066-0_19

Introduction

Pandemics and other global crises, alongside advances in ICT, challenge the organizations of higher education (HE) world-wide. For instance, COVID-19 pandemic locked down a significant amount of schools and higher education institutions around the world and enforced remote learning. In parallel, many industries shifted as much as possible to remote work, by the necessity of governments and cities increased regulation for social distancing. The circumstances challenged how university-industry collaboration can be conducted, while people are unable to meet and interact physically. Virtual hackathons provide one alternative for conducting university-industry collaboration, where universities can e.g. contribute by developing ideas and innovations for industry.

This article investigates to what degree virtual hackathons meet the necessary and sufficient features of hackathons, explores the benefits, disadvantages and challenges from virtual collocation compared to radical collocation, and finally outlines benefits that industry can gain by participating in virtual hackathons and giving a challenge to be solved in university-industry collaboration.

Theoretical Background

Innovation in University-Industry Collaboration

Innovation is seldom a straightforward activity. It can be characterized as uncertain, co-constructive, experimental and interactive [1, 2]. University-industry collaboration aims at mutually beneficial knowledge and technology exchange between higher education and industry. Despite the growing interaction between higher education and industry, partners in university-industry collaboration have challenges in utilizing the results of their joint efforts [3, 4]. One root cause for the challenges is that the primary goal of universities is to create open and public knowledge, and provide education [5], whereas industrial partners have strong focus on capturing valuable knowledge that can create competitive advantage that is often directly associated with new product development and innovative functioning of the company [5, 6]. Contradictory objectives, organizational goals and culture have been found to limit the positive effects that can be achieved through university-industry collaboration [7, 8]. Mechanisms and practices for university-industry collaboration include students projects (for example hackathons [9]), thesis projects, tailored degree courses and jointly organized courses [10].

One of the most well-known models of university-industry collaboration is the Triple Helix [11] principle that is based on the institutional triangle of government, business and academia. The entrepreneurial university following Triple Helix principle encompasses a 'third-mission' of economic development in addition to their research and teaching activities [12]. Economic development can, for instance, take the form of developing products and services [13] for business as part of education. Governments can support such activities by, for example, funding research and development projects that involve both business and academia. Hackathons as one type of innovation contest method provides one vehicle how innovations can be developed in university-industry collaboration [14, 15] especially in the front-end of innovation.

Hackathons

Hackathons, as one type of innovation competition [16, 17], originated among information technology (IT) practitioners [18]. The roots of hackathons date back to MIT in the 1960s, where students gathered together to code in self-imposed 24-hour 'marathon bursts' [18]. From coding the use of hackathons spread to other domains and use cases, [18, 19]. For the purpose of this article, we consider hackathon as *"one type of innovation contest, a short time-bounded event with a challenge to be solved creatively in coopetition and with the radical collocation of teams, whose output is recognized in a ceremony at the end of the event."* [14].

Hackathons have also received criticism. The goal achievement of hackathons have been found questionable. The ideas resulting from hackathons are regarded as rarely being effective or adopted in addressing the problems that inspired the hackathons Hackathons have also been found to suffer from a lack of institutional memory [20]. Furthermore, participants have also experienced frustration resulting from expectations about the results of the hackathon. [21] The lack of commercialized results have led [22] to conclude that there is still something missing from the hackathon method.

Virtual hackathons are a new breed of hackathons, where all the activities take place online, and where the participants are gathered together virtually. Virtual hackathons face the same challenges as 'regular' hackathons, however, operating fully in the virtual space provides also new challenges. The virtual space needs to e.g. accommodate all the participants, enable interactions between them and also provide tools how the participants can work together with the problem and the solution.

Collocation: Radical Versus Virtual in Hackathons

Hackathons are described mainly as an event where participants collaborate intensively [20, 23], during which participants come together and form teams [21, 24–28]. Furthermore, hackathon "implies an intense, uninterrupted, period of programming." [22] and [25] have stated that "the teams are usually collocated". Radical collocation is a situation, where team members are together in a physical space for the duration of the project [25, 29, 30]. Collocation is seen as beneficial for technical work, speeding up software development, facilitating enduring relationships [31] and improving productivity [30]. Also known as a 'war room', radical collocation is seen to help coordination, problem-solving and learning [30]. The physical method of collocation has its benefits in e.g. new product design teams, such as improving communication, building relationships, reducing time involved in design reiterations, and improving project, etc. In physical collocation people communicate via face-to-face communication. In addition to physical collocation, the collocation can be virtual. [32].

In virtual collocation, the members and teams are usually integrated via information and communication technology (ICT), such as personalized ICT tools, e.g. instant messaging, or collective ICT tools, e.g. shared cloud documents [33]. However deploying ICT in collaboration may often results negative aspects, such as misunderstandings, unclear communications, team members' status differences, and the challenges caused by task complexity. The basic differences between physical and virtual collocation is: (1) close versus remote proximity, (2) the great amount of work and non-work related information versus minimal informal exchange, which can affect relationship building, (3) increase the opportunity for allocation and sharing both physical and non-physical resources versus access to resources via ICT infrastructure with its restrictions, (4) control and accountability by monitoring of activities and events and ability to respond to requirements, versus 'out of sight out of mind' (5) sharing ideas and dilemmas versus motivational problems of isolation and frustration causing low performance, (6) similar and complementary cultural and educational background versus variation in both, in addition to time zones and expertise, (7) compatibility versus incompatibility of the technological systems. [32].

Methodology

We chose a case study approach [34] to investigate two virtual hackathons aimed at university-industry collaboration. The cases were purposefully selected based on convenience. The research team consisted of three experienced hackathon facilitators and one hackathon jury member. The hackathons were aimed to create solutions and ideas for industry that were viable, desirable and feasible. The criteria was derived from design thinking and business design [35–37].

First hackathon was aimed at developing a solution for city bikes at the city of Riihimäki. Riihimäki Hackathon examined the maturity of Riihimäki area as a platform for urban cycling. Urban cycling is one among the rising trends in creating city environment. Riihimäki is a city with low cycling modality, where there is limited cycling structure. In addition, the lack of cycling structure and investments in cycling facilities were seen a factor leading to challenges in creating new type of cycling culture [38]. In Riihimäki Hackathon the challenge was formulated by city of Riihimäki and Kaakau Oy, which is a city bikes platform company providing the bikes for Riihimäki and several other cities in Finland. Four teams of students from Häme University of Applied Sciences (HAMK) worked for one day to create solutions to the challenge. The hackathon was facilitated by HAMK Design Factory, an interdisciplinary product and service design and learning platform at HAMK.

The second hackathon was aimed at developing a solution to utilizing surplus fabrics and fabrics with minor quality defects for textile company A. Textile company A, located in Forssa develops and procures high quality fabrics having the largest product selection in its market area of workwear, military and outdoor clothing fabrics. The challenge was to understand the regional ecosystem and business ecosystem [39] and create a viable business model utilizing surplus fabrics of textile company A. The solution could involve either an industrial ecosystem or regional ecosystem. Industrial ecosystems create value propositions for specific industry whereas in regional ecosystems the value is created to support the development of certain area [40]. The second hackathon was facilitated by HAMK Design Factory involving student teams from HAMK and Forssa Vocational Institute. In order to understand the benefits that industry can gain by giving a challenge and participating in a hackathon organized in university-industry collaboration interviews of the industry representatives were conducted after the hackathons were completed by email survey. As for increasing understanding on the disadvantages and challenges collocation in virtual format causes to the hackathon observations were performed by two facilitators of the hackathon events and one teacher that participated in the second hackathon. The observation notes and findings were additionally reviewed and discussed with researcher that has also personal experiences of hackathons, but who did not participate in the two hackathons. Investigator triangulation was used to add breadth to the phenomena of interest [41].

Results

First, the results portray virtual hackathons evaluated in terms of necessary and sufficient hackathon features. Then the observed benefits, disadvantages and challenges of radical collocation versus virtual collocation are presented. Finally, the results introduce perceived benefits of virtual hackathons perceived by the industry, based on case study interviews. Virtual hackathons evaluated in terms of necessary and sufficient hackathon features [14] are illustrated on Table 19.1.

Based on the necessary and sufficient features of hackathon defined by Halvari et al. [14], we can observe that the two case studies of virtual hackathons met all other necessary and sufficient features except radical collocation. In Zoom and Teams it is possible to have meetings, where you can see live video feed of each team member thus you can have appearance of everyone being virtually present in the virtual space at the same time. However, due to slow network connections of students, teachers and industry representatives participating from home office during the pandemic, it was not possible to have video feed from every participant, rather only from those presenting one at a time. When everyone is participating from home office it follows that there is no opportunity to enjoy meals or refreshments together with the team members and other participants. As a consequence, virtual co-location is disrupted when individual participants take breaks. In situations, where the participants are muted and video is not display it is not possible to determine are the participants

Table 19.1 Evaluation of virtual hackathon against necessary and sufficient hackathon features

Theme	Attribute	Sufficient feature	Case Riihimäki; Case Forssa
Short time-bounded event	Short duration	t < week, ideally 1–3 days	1 day; 1 day
	Team	n > 1 teams, with 1 > members per team	4 teams, 3–5 members/team; 2 teams, 4–5 members/team
	Challenge	Task	Challenge presented and owned by company in both cases
Coopetition	Creation process	A team formation, creation process	Creating a concept of the solution in both cases
	Ceremony process	An idea pitching jury, possible winner recognition	Both cases had a jury that decided on a winning team
	Collaboration	Individuals and teams, organizers and participants	Interaction between team members and organizers in Zoom and Teams meetings, collaboration using Teams, Office365 and Google Drive
Radical collocation	Co-location	Separated from daily business, food often served	Virtual co-location via Zoom and Teams meetings, everyone working from home office. Food served on self-service basis
	Consistency	Intensive and consistent	In both cases students were dedicated for completing the hackathon for full day. Interruptions possible

actually present unless prompted. Operating from home office opens up also the possibility interruptions that would not similarly occur in radical collocation.

Table 19.2 summarizes the observed benefits, disadvantages and challenges of radical collocation in contrast to virtual collocation in virtual spaces, such as Teams and Zoom used in the case studies.

According to the industry representative of Riihimäki Hackathon the hackathon did not yield much direct benefits. It was found interesting to follow the teams work and the results they achieved. However, the results were perceived rather superficial, which could be turned to beneficial outcomes perhaps after a second hackathon. By the point of view of industry representative of Forssa Hackathon, the benefit of virtual collocation was the need to prepare the presentation more carefully, by using simple language and examples to clarify the case. This kind of preparation gave more

Table 19.2 Benefits, disadvantages and challenges of radical collocation versus virtual collocation

Theme	Attribute	Benefit	Challenge/disadvantage
Radical collocation	Co-location	Physical co-location ensures separation from daily business and family life. Participants can benefit from informal interactions during breaks	It can be time-consuming and costly to organize people to be physically present at the same location. If co-location, e.g. in the case of social distancing, is restricted it may be impossible to organize co-location
	Consistency	Coordination of tasks inside team easy when everyone is present and you see what they are working on. Concentration can be more intense when there are not outside interruptions. Facilitators can control the interruptions and optimize workflow	Getting to know each other and team formation needs to take place physicaly, most often during the event, whereas in virtual environment there are opportunities to interact and comtemplate before the event
Virtual collocation	Virtual co-location	Virtual co-location removes the need to travel to venue, which reduces cost and can save time. In virtual spaces new rooms and dedicated spaces can be created instantly and people moved without the physical delay of movement between locations, buildings, rooms, etc. Preparing of presentations for virtual co-location can develop industry representatives own understanding	Virtual co-location opens up potential interruptions from business and family life, depending on physical location. In virtual spaces participants can be distracted or even absent from working on shared goals, which may not show on virtual space. Lack of interaction between industry representatives and participants brings challenges to ensure that the presentation/challenge is understood
	Consistency	People may find it easier to dedicate to virtual hackathon, e.g. participate in introductory talk, mentor teams, present the challenge and participate as jury member when they do not need to reserve time for the full day or duration of the hackathon, but can easily and quickly do their part	Team members may not be as dedicated to collaboration as there is less social pressure in virtual compared to physical space. Consistency can be disrupted when participants are interrupted or have conflicting activities going on at the same time. Inability to see where, how and on what the team members are working on. Facilitators cannot control the interruptions or optimize workflow as effectively in virtual hackathons

understanding for the industry representative as well. The challenges related to the fact that the industry representative did not see the participants and because of that, the presenter couldn't get any any visual cues and feedback from the participants' facial expressions and gestures. Because of the lack of interaction, it was difficult for the industry representative to know, did the participants get the point or not. The industry representative proposed two related development ideas. First to prepare the participants beforehand to promote discussion. This could be done by sharing the presentation material of the challenge beforehand for the participants and asking them to prepare questions for the industry representative. Second idea related to the need of seeing peoples faces, at least when they are talking—as well as participants and facilitators. This could create more fluent atmosphere. After the Forssa Hackathon, the giver of the challenge continued university-industry collaboration with HAMK in the form of a tailored course, where student teams of bioeconomy engineering continued to develop solutions for the company following design thinking process [37].

Conclusion

This article began with investigation to what degree virtual hackathons meet the necessary and sufficient features of hackathon concept as defined by Halvari et al. [14]. It was found that the necessary feature of radical collation and consistency are not achieved as such in virtual hackathons. This can be interpreted as a need to refine the concept definitions or that virtual hackahtons should be considered as a new type of hackathon concept that is related, but with differences in attributes and design elements.

The exploration of virtual collocation in contrast to radical collocation yielded several insights from facilitators, participants and industry representatives perspective. The results support the findings of Pawar et al. [32] on physical and virtual collaboration on several aspects. From the perspective of both facilitators and teams members it is challenging to monitor the progress of work without physical co-location. Isolation of team members from close proximity can result in motivational problems, reduced performance and also in the case of working from home office provide increased opportunities for interruptions and distractions that would not take place in physical close proximity. Intrestingly, the industry representative also pointed out the lack of seeing facial expressions and bodily gestures of hackathon participants, i.e. lack of visual feedback, as a challenge in communication and collaboration. This insight supports information richness theory [42] and the findings of studies that investigate new media use in collaboration and innovation [43, 44] in this new context of hackathons.

The benefits of short time-bounded events, such as hackathons, were perceived limited by the industry. Hackathons support fast iteration of innovations, mainly yielding new ideas and concepts that can be further developed in slow iteration of innovations. Slow iteration of innovations in university-industry collaboration can be for example thesis projects and research and development projects spanning e.g. 1–3 years. Hackathons, nevertheless, provide universities with the opportunity to increase access and deepen relationships with industry, as is evidenced in the Forssa Hackathon case.

To conclude, virtual hackathons have similarities with the physical hackathons, but they must be planned differently in order to use them efficiently. ICT-tools, selected and prepared virtual environment and facilitation have a key role in making the virtual hackathon successful. The challenges and disadvantages discovered in this study can be used by the practitioners to overcome the most obvious pitfalls of virtual hackathons.

References

1. B. Edvardsson, B. Tronvoll, T. Gruber, Expanding understanding of service exchange and value co-creation: a social construction approach. J. Acad. Mark. Sci. **39**, 327–339 (2011)
2. J. Jussila, J. Kukkamäki, M. Mäntyneva, Open data and open source enabling smart city development: a case study in häme region. Technol. Innov. Manag. Rev. **9**, 25–34 (2019)
3. A.B.L. Pennacchio, University of knowledge and firm innovation: evidence from European countries. J. Technol. Transf. **41**, 730–752 (2016)
4. L. Kunttu, *Learning Practices in Long-Term University-industry Relationships*. https://osuva.uwasa.fi/handle/10024/8180 (2019)
5. K.J. Lee, From interpersonal networks to inter-organizational alliances for university–industry collaborations in Japan: the case of the Tokyo Institute of Technology. R&D Manag. **41**, 190–201 (2011)
6. J. Bruneel, P. D'Este, A. Salter, Investigating the factors that diminish the barriers to university–industry collaboration. Res. Policy **39**, 858–868 (2010)
7. J.F. Gomes, P. Hurmelinna, V. Amaral, K. Blomqvist, Managing relationships of the republic of science and the kingdom of industry. J. Work. Learn. **17**, 88–98 (2005)
8. L. Kunttu, Y. Neuvo, Balancing learning and knowledge protection in university-industry collaborations. Learn. Organ. **26**, 190–204 (2019)
9. J. Porras, A. Happonen, J. Khakurel, A. Knutas, J. Ikonen, A. Herala, Hackathons in software engineering education—lessons learned from a decade of events. in *SEEM'18, May 27-June 3, 2018, Gothenburg, Sweden*. (Association for Computing Machinery, Cothenburg, 2018), pp. 40–47
10. L. Kunttu, Educational involvement in innovative university–industry collaboration. Technol. Innov. Manag. Rev. **7**, (2017)
11. H. Etzkowitz, Innovation in innovation: the triple helix of university-industry-government relations. Soc. Sci. Inf. **42**, 293–337 (2003)
12. H. Etzkowitz, A. Webster, C. Gebhardt, B. Terra, The future of the university and the university of the future: evolution of ivory tower to entrepreneurial paradigm. Res. Policy **29**, 313–330 (2000)
13. I. Kunnari, T. Ho, T. Nguyen, *Rethinking learning towards education 4.0*. HAMK Unltd. J. (2019)
14. S. Halvari, A. Suominen, J. Jussila, V. Jonsson, J. Bäckman, Conceptualization of hackathon for innovation management, in *The ISPIM innovation conference—celebrating innovation: 500 Years Since daVinci*. (ISPIM, 2019)
15. A.H. Suominen, S. Halvari, J. Jussila, World heritage meets smart city in an urban-educational Hackathon in Rauma. Technol. Innov. Manag. Rev. **9**, (2019)

16. S. Hartmann, A. Mainka, W.G. Stock, Innovation contests: how to engage citizens in solving urban problems? in *Civic Engagement and Politics: Concepts, Methodologies, Tools, And Applications*. (IGI Global, 2019), pp. 58–77
17. S. Hartmann, A. Mainka, W.G. Stock, Opportunities and challenges for civic engagement: a global investigation of innovation competitions. in *Civic Engagement and Politics: Concepts, Methodologies, Tools, and Applications*. (2019), pp. 607–623
18. S. Leckart, The hackathon is on: pitching and programming the Next Killer Appl WIRED
19. S. Zukin, M. Papadantonakis, Hackathons as Co-optation ritual: socializing workers and institutionalizing innovation in the "New" Economy. Precarious Work Res. Sociol. Work. **31**, 157–181 (2017)
20. G. Briscoe, C. Mulligan, Digital innovation: the Hackathon Phenomenon. Creat. London. 1–13 (2014)
21. C. Granados, M. Pareja-Eastaway, How do collaborative practices contribute to innovation in large organisations? Case Hackathons. Innovation. **00**, 1–19 (2019)
22. M. Komssi, D. Pichlis, M. Raatikainen, K. Kindstrom, J. Jarvinen, M. Komssi, D. Pichlis, M. Raatikainen, K.K. and J.J., What are Hackathons for?. in *IEEE Software*, (2015), pp. 60–67
23. E. Almirall, M. Lee, A. Majchrzak, Open innovation requires integrated competition-community ecosystems: Lessons learned from civic open innovation. Bus. Horiz. **57**, 391–400 (2014)
24. A.M.M. Angarita, A. Nolte, Does it matter why we hack ?—Exploring the impact of goal alignment in hackathons. in *Proceedings of the 17th European Conference on Computer-Supported Cooperative Work: The International Venue on Practice-centred Computing and the Design of Cooperation Technologies—Exploratory Papers* (Reports of the European Society for Socially Embedd, 2019), pp. 1–15
25. E.P.P. Pe-Than, A. Nolte, A. Filippova, C. Bird, S. Scallen, J. Herbsleb, Designing corporate hackathons with a purpose. IEEE Softw. 15–22 (2019)
26. K. Gama, B. Alencar, F. Calegario, A. Neves, P. Alessio, Hackathon methodology for undergraduate. in *IEEE Fronties in Education Conference* (FIE). (2018), pp. 1–9
27. B. Rosell, S. Kumar, J. Shepherd, Unleashing innovation through internal hackathons. In *Digest of Technical Papers—InnoTek 2014: 2014 IEEE Innovations in Technology Conference* (2014)
28. H. Kienzler, C. Fontanesi, Learning through inquiry: a global health Hackathon. Teach. High. Educ. **22**, 129–142 (2017)
29. E.P.P. Pe-Than, J.D. Herbsleb, Understanding Hackathons for Science: Collaboration, Affordances, and Outcomes. in *Information in Contemporary Society, Proceedings of 14th International Conference, iConference*, eds by N.G. Taylor, C. Christian-Lamb, M.H. Martin, B. Nardi. (Springer, Washington, DC, USA, 2019), pp. 27–37
30. S. Teasley, L. Covi, M.S. Krishnan, J.S. Olson, How does radical collocation help a team succeed?. in *AMC Conference on CSCW 2000*. (Philadelphia, PA, USA, 2000), pp. 339–346
31. E.H. Trainer, A. Kalyanasundaram, C. Chaihirunkarn, J.D. Herbsleb, How to Hackathon: Socio-technical Tradeoffs in Brief, Intensive Collocation. Presented at the February (2016)
32. K.S. Pawar, S. Sharifi, Physical and virtual collaboration in new product development: A comparative analysis. in *2017 International Conference Engineering Technology Innovations Engineering Technology Innovations management Beyond 2020 New Challenges, New Approaches, ICE/ITMC 2017 - Proc. 2018-Janua*, (2018), pp. 997–1002
33. P. Lohikoski, J. Kujala, J. Härkönen, H. Haapasalo, M. Muhos, *Enhancing communication practices in virtual new product development projects* (Int. J. Innov. Digit, Econ, 2015)
34. N. Siggelkow, Persuation with Case Studies. Acad. Manag. J. **50**, 20–24 (2007)
35. H. Plattner, C. Meinel, L. Leifer, *Design thinking: understand—improve—apply* (Springer, New York, 2010)
36. A. Faljic, The ultimate business design guide. d.MBA (2019)
37. J. Jussila, J. Raitanen, A. Partanen, V. Tuomela, V. Siipola, I. Kunnari, Rapid product development in university-industry collaboration: Case study of a smart design project. Technol. Innov. Manag. Rev. 10, (2020)

38. R. Félix, F. Moura, K. Clifton, Maturing urban cycling: Comparing barriers and motivators to bicycle of cyclists and non-cyclists in Lisbon. J. Transp. Heal. 15, (2019)
39. J. Pulkkinen, J. Jussila, A. Partanen, I. Trotskii, A. Laiho, Smart mobility: services, platforms and ecosystems. Technol. Innov. Manag. Rev. **9**, 15–24 (2019)
40. E. Den Ouden, Innovation Design: Creating Value for People, Organizations and Society. (Springer Science & Business Media, Berlin, 2011)
41. N. Denzin, *Sociological methods: A sourcebook* (McGraw-Hill, New York, NY, 1987)
42. R. Lengel, R. Daft, Information richness: a new approach to managerial behavior and organization design. Res. Organ. Behav. **6**, 191–233 (1984)
43. J.J. Jussila, H. Kärkkäinen, M. Leino, Learning from and with customers with social media: A model for social customer learning. Int. J. Manag. Knowl. Learn. **1**, 5–25 (2012)
44. A.R. Dennis, S.T. Kinney, Testing media richness theory in the new media: The effects of cues, feedback, and task equivocality. Inf. Syst. Res. **9**, 256–274 (1998)

Chapter 20
A Curriculum Design Method for New Product Development

Yenchun Jim Wu, Jeng-Chung Chen, and Kune-Muh Tsai

Abstract An entrepreneurship movement is sweeping the world; new product development (NPD) is the core activity of an organization's competitive strategy, which includes concept design and the successful development of new products that can be launched in the market. However, a surprising dearth of research has been conducted on what the components and knowledge should be embedded in the process of new product development. To orchestrate a timely curriculum, this study aims to analyze the published papers on new product development to shed light on comprehensive developments and trends within the field. Thus, the Latent Semantic Analysis (LSA) was applied to extract research tropics based on scientific articles published between 2015 and 2019 and interpreted the results using Bloom's Revised Taxonomy. The study identified ten educational objectives of NPD curriculum from the initial results. Furthermore, the domain knowledge and key components were derived from the corresponding educational objective. Consistent with previous findings and others' assumptions, the study found that the scope of NPD curriculum is still expanding continuously; and the results herein can guide educators to design a more appropriate curriculum and to enhance students' learning performance.

Keywords Curriculum design · Latent semantic analysis · New product development · Text mining

---

Y. J. Wu (✉)
Graduate Institute of Global Business and Strategy, National Taiwan Normal University, Taipei, Taiwan
e-mail: wuyenchun@gmail.com

J.-C. Chen
National Taiwan Normal University, Taipei 106, Taiwan
e-mail: chenjengchung@gmail.com

K.-M. Tsai
National Kaohsiung University of Science and Technology, Kaohsiung, Taiwan
e-mail: kmtsai@nkfust.edu.tw

A. Visvizi et al. (eds.), *Research and Innovation Forum 2020*,
Springer Proceedings in Complexity,
https://doi.org/10.1007/978-3-030-62066-0_20

Introduction

Today's enterprises are facing fierce competition from domestic and foreign organizations [1]. To that end, the development of new products and services is the core activity of an organization's competitive strategy, which includes concept design and the successful development of new products that can be launched in the market. While enterprises come under pressure to develop innovative products, the process of new product development and innovation (NPD) is extremely challenging [2]. Statistics show that the NPD failure rate for the established firm is 40% to 65%, while for startup companies it even exceeds 90% [3, 4]. Scholars estimate that approximately 3000 original ideas are needed to produce a successful new commercial product [5].

In a wave of entrepreneurship, the commercialization of business ideas has become both a core issue and a major challenge. Although existing research has generated a wealth of knowledge in NPD area, prior work generally shares a commonality of being set in the context of the firm level. Compared with other subjects in the field of business management, such as marketing, entrepreneur-ship, and social responsibility, studies on the development and teaching methods of NPD courses are relatively limited [6]. According to a survey on technology management courses by Mallick and Chaudhury [7], scholars and administrators both agree that NPD is one of the key knowledge and skills. The coverage of NPD has been continuously expanding across various fields, including, but not limited to, innovation, product design, finance, teamwork, manufacturing, and marketing. The development of new products and services is a complex process, which is why it is so difficult for scholars to design a curriculum that suits a practical purpose. However, the majority of current NPD textbooks available to the public fail to meet the current requirements of innovation and practical guidance, and therefore, do not satisfy the demands of business society.

To fill this gap, the study aims to disclose key components of NPD curriculum for this emerging issue by considering the research topics of extant journal articles so as to educate students becoming a creative individual who possesses knowledge, skills, and enthusiasm to bring an idea from conceptualization, idea generation, and finally to idea implementation. As NPD related courses gain increased interests in higher education institutions (HEIs), the study intends to provide some guidelines for program and curriculum design in business schools. In particular, we seek to answer the following two questions:

(1) What are the educational objectives should be drawn up in the NPD course?
(2) Which key components should be included in the innovative NPD curriculum?

The rest of the paper is organized as follows: we first review the background related to this study. We then introduce research method. Finally, we discuss the implications of our findings for NPD curriculum design and provide directions for future research.

Background

New Product Development (NPD) for Entrepreneurship

As defined by Fixson [1], new product development consists of nine stages, namely ascertaining opportunity, investigating market and user, creating idea and concept, refining and choosing conception, designing the product, building prototype, testing, evaluating financial capacity, and introducing marketing. The NPD process, hence, is often viewed as a flow that integrates the various phases from conceiving and transforming new ideas to delivering commercialize products [8]. The lifecycle of a consumer product generally encompasses four stages, namely introduction, growth, maturity, and decline. Theoretically, in the final stage of the lifecycle, the sales volume of the product will decrease gradually and it should be terminated when it is unprofitable for the company [9]. To keep pace with the competitors, companies should introduce new products with high quality, and innovativeness with shorter time to market. Also, a previous study found that regularly launching a new product is a crucial strategy for a company that attempts to survive in extremely competitive and fragmented markets [10]. Additionally, the first step of developing a successful product is to comprehend variables of the targeting market, such as consumer needs, economic situation, demography, technology, and political and legal shifts [11].

It cannot be overemphasized that new product development plays an essential role in innovation, which is the engine of growth in today's marketplace [12]. The pressures to innovate are relentless, and highly innovative companies usually make more profit than less innovative companies [1]. Innovation generally involves five states, namely creating the idea, research, and development, building prototype, production and delivery, marketing and sales [13]. From the stake-holders' view, innovation is a value-added process that through developing novel products, services, manufacturing procedures, and solutions. However, without entrepreneurial activities exploiting opportunities when they emerge in organizations, innovation remains little more than enthusiasm, rather than an objective goal [13].

Entrepreneurship is the driving force for promoting innovation and competitiveness in response to globalization and the development of technology. Prior research stressed entrepreneurship not only as a foundation to underpin innovation but also as a vital element for innovation [13]. Entrepreneurship was defined as an effort in promoting innovation in an ambiguous situation. Within the innovative process, the primary roles of entrepreneurship are to challenge extant rules and regulations, to evaluate new opportunities, to allocate and exploit re-sources, and to keep the innovation process moving forward [14]. Therefore, entrepreneurship education (EE) in higher education institutions (HEIs) gains increased interests because exerting EE on learners is beneficial to their pursuit of knowledge and skills as well as increasing intention becoming entrepreneurs [15]. Entrepreneurship education is generally categorized into three approaches [16].

(1) Teaching "about" entrepreneurship means a content-laden and theoretical approach aiming to give a general understanding of the phenomenon. It is the most common approach in higher education institutions.
(2) Teaching "for" entrepreneurship means an occupationally oriented approach aiming at giving budding entrepreneurs the requisite knowledge and skills.
(3) Teaching "through" entrepreneurship means a process-based and often experiential approach where students go through an actual entrepreneurial learning process.

The majority of scholars today now agree that learning through own experience is the best way to learn to become an entrepreneur [16–18]. Entrepreneurs are uncommon learners who acquire knowledge and skill by doing, from the surrounding, and person, such as customers, suppliers, competitors and employees [19]. In addition, Lackéus [20] stated that the learning-by-doing is that one learns by experience in which include coping, experiment, problem-solving and opportunity taking and by learning from making mistakes. Learning-by-doing as an important approach to experiential education, which has emerged as the preferred methodology in higher education [21]. Consistent with this context, the study considers that learning-by-doing is a feasible pedagogic method for designing NPD curriculum.

Bloom's Taxonomy and Educational Framework

Bloom's Taxonomy was initially proposed by Benjamin Bloom in 1956 for guiding educators to design more appropriate curricula and to enhance students' learning performance [22]. In the framework, the educational objectives are classified into three domains, namely affective, psychomotor and cognitive domains. The affective domain narrates the interactions between learners and educators who draw up objectives to induce awareness, attitudes, emotions, and feelings. In the psychomotor domain, the educational objectives are identified and categorized about physical manipulation of tools or instruments. The cognitive domain, which has been widely researched in many disciplines, consists of the hierarchy of sub-levels including knowledge, comprehension, application, analysis, synthesis, and evaluation [23]. Besides, this framework stresses that learning occurs at six levels of cognition: knowledge, comprehension, application, analysis, synthesis, and evaluation.

However, the original Bloom's Taxonomy possessed some shortcomings and given cause for criticism; thus the revised Taxonomy was proposed in 2001 [22]. In Bloom's Revised Taxonomy, the verb and none represent the process of cognitive dimension and the knowledge base respectively which allow the use of more information to define educational objectives and planning activities [23]. As shown in Table 20.1, the educators in HEI can apply Bloom's Revised Taxonomy to draw up educational objectives and to design an appropriate curriculum, as well as the framework facilitates students clarifying their purpose [24].

Table 20.1 Bloom's revised taxonomy

Hierarchy and Cognition domain	Description	Action verbs
Lower order thinking activities		
1. Remember	Retrieving, recognizing, and recalling relevant knowledge from long-term memory	Recognizing, recalling
2. Understand	Constructing meaning from oral, written, and graphic messages through interpreting, exemplifying, classifying, summarizing, inferring, comparing, and explaining	Interpreting, exemplifying, classifying, summarizing, inferring, comparing explaining
3. Apply	Carrying out or using a procedure for executing, or implementing	Executing, implementing
Higher order thinking activities		
4. Analyse	Breaking material into constituent parts, determining how the parts relate to one another and to an overall structure or purpose through differentiating, organizing, and attributing	Differentiating, organizing, attributing
5. Evaluate	Making judgments based on criteria and standards through checking and critiquing	Checking, critiquing
6. Create	Putting elements together to form a coherent or functional whole; reorganizing elements into a new pattern or structure through generating, planning, or producing	Generating, planning, producing

Research Method

Data Collection

The study retrieved the textual data from the Scopus, which is one of the most extensive peer-reviewed research repositories in the Social Sciences, by searching the keywords of "new product development" or "new product design" or "new product innovation", which was determined by relevant experts in the field. Next, the criteria from the previous study [25] were adopted to filter the article source and to select six journals that have the highest usage scores of NPD related articles in Scopus. There is a total of 221 articles, published from 2015 to 2019, were extracted, and their title,

Table 20.2 Distribution of NPD journal articles by subject area

	Subject area				
Source title	Business, management, accounting	Engineering, computer science	Social sciences	Decision science	Total
J. Prod. Innov. Manage.[1]	55	0	0	0	55
	22.4%	0.0%	0.0%	0.0%	22.4%
Ind. Mark. Manage.[2]	31	0	0	0	31
	12.6%	0.0%	0.0%	0.0%	12.6%
IEEE Trans. Eng. Manage.[3]	7	10	0	0	17
	2.8%	4.1%	0.0%	0.0%	6.9%
Technovation	2	6	0	0	8
	0.8%	2.4%	0.0%	0.0%	3.3%
Int. J. Technol. Manage.[4]	1	0	2	0	3
	0.4%	0.0%	0.8%	0.0%	1.2%
Eur. J. Innov. Manag.[5]	15	0	0	0	15
	6.1%	0.0%	0.0%	0.0%	6.1%
R D Manage.[6]	16	0	0	0	16
	6.5%	0.0%	0.0%	0.0%	6.5%
Int. J. Innov. Manag.[7]	41	0	0	0	41
	16.7%	0.0%	0.0%	0.0%	16.7%
Int. J. Prod. Res.[8]	5	12	0	8	25
	2.0%	4.9%	0.0%	3.3%	10.2%
Res.-Technol. Manag.[9]	2	8	0	0	10
	0.8%	3.3%	0.0%	0.0%	4.1%
Total	180	46	1	18	221
	73.2%	18.7%	0.8%	7.3%	100.0%

Note: 1: Journal of Product Innovation Management; 2: Industrial Marketing Management; 3: IEEE Transactions on Engineering Management; 4: International Journal of Technology Management; 5: European Journal of Innovation Management; 6: R And D Management; 7: International Journal of Innovation Management; 8: International Journal of Production Research; 9: Research Technology Management

abstract, and keywords were downloaded for furthering analysis. The outline of data sources and subject areas was summarized in Table 20.2.

Data Analysis

The study conducted the Latent Semantic Analysis (LSA) on a corpus of documents each represented by the title, abstract and keywords of the documents. LSA is especially appropriate for this study because NPD is characterized by many specific vocabularies in a complex manner and possessed latent concepts in this context [26]. However, traditional word-counting text mining approaches, count the word frequency merely, are unsuitable for analyzing this type of latent concept. LSA works in a way that is similar to the human brain works: It extracts the contextual meanings of a concept and identifies the complex structure between a latent concept and its relative words or terms [26].

To extract the key components of NPD curriculum from existing articles, the study utilized the package "lsa" for R [27] and the suggestion of a previous study [28] to performed LSA with the following steps.

Step 1: Text Preprocessing: It is one of the most elaborate steps to convert textual data to an appropriate format by using the following sub-steps.

(1) Text cleanup is to remove unnecessary and meaningless elements, such as white space, stop words, and abbreviations.
(2) Tokenization aims to identify meaningful keywords. Hence, the process includes splitting sentences into words, removing all punctuation marks from textual data, and giving words of text which is called tokens.
(3) Stemming is the process of consolidating the different forms of a word into a standard representation, the stem. For instance, the words: 'developed', 'development', 'developing' could all be condensed to a common representation 'develop'; and plurals to root like women to woman, men to man and horses to horse.

Step 2: Text Transformation: In general, the input of this step is bag-of-word (BOW) and it cannot be applied to the algorithms directly. The BOW should be transformed into the term-document matrix (TDM) which pit 'terms (y-axis)' against 'documents (x-axis)', and the cells give the amounts or frequencies of a term in a document. In addition, there are several methods can be used to find out the term importance.

Step 3: Feature Selection: To decide a subset of important features for use in model creation, the study utilized the Term Frequency inverted document frequency (TF-IDF) for filtering and removing redundant or irrelevant features. The function of TF-IDF is that calculating the relative frequency of words in a particular document compared to the inverse ratio of that word over the entire document corpus. This calculation establishes how relevant a given word to a specific document. Words that are prevalent in a single or a small portion of documents are prone to have higher TF-IDF numbers than common words such as articles and prepositions. Given that a document collection D, a word w, and an individual document $d \in D$, the TF-IDF can be calculated as follows [28].

$$w_d = f_{w,d} \times \log\left(\frac{|D|}{f_{w,D}}\right)$$

where:

$f_{w,d}$ Equals the number of times w appears in d;

$|D|$ Is the size of the corpus; and
$f_{w,D}$ Equals the number of documents in which w appears in D.

Step 4: Singular value decomposition (SVD): In this step, the TF-IDF weighted frequency-by-document matrix was decomposed using SVD, a factorization method. As the statement of prior literature [29], the calculation of SVD is based on a theorem from linear algebra, and a rectangular matrix A can be disassembled and produced three matrices—an orthogonal matrix U, a diagonal matrix S, and the transpose of an orthogonal matrix V:

$$A_{mn} = U_{mm} S_{mn} V_{nn}^T$$

where $U^T U = I$, $V^T V = I$, the columns of V are orthonormal eigenvectors of $A^T A$, the columns of U are orthonormal eigenvectors of AA^T. S is a diagonal m × n matrix containing the square roots of eigenvalues from U or V in descending order.

Step 5: Factor reduction: In the previous step, SVD generated several latent topics. However, these topics are not equally important. Some topics contain more information and some contain less information. To reduce the number of factors and retain most of the information in the original text, one can restrict the matrix S to relatively higher eigenvalues and get a simplified matrix $\hat{S}_{kk}$ by deleting rows and columns from S. For the matrix multiplication to go through, the corresponding row vectors of U and corresponding column vectors of V^T have to be removed and results in another two matrices Û and V̂. The result looks like this:

$$\hat{A}_{mn} = \hat{U}_{mk} \hat{S}_{kk} \hat{V}_{kn}^T$$

where k $< min(m, n)$, $\hat{A}$ is an approximation matrix A.

Choosing a different k value will give different $\hat{A}_{mn}$. This approximates the original matrix A and therefore captures more or less "trivial" topics.

Step 6: Factor Interpretation: The previous step generated several groups of topics with a unique level of prevalence (coherence) rate and a serial highly load terms that can identify specific meaning. Thereby, both authors utilized Bloom's taxonomy action verbs and its synonym (see Table 20.1) to interpret these topics and inferred the key components of NPD curriculum independently. The identified topic labels in question are discussed with the third author to complete topic labels. The interpretation of the identified educational objectives is discussed in the following sections.

Results and Discussion

The dataset of textual data retrieved from ten journals related to the subject matter of the study included a total of 221 articles published between 2015 and 2020. Table 20.2 reported that the subject area of these articles was mostly business, management, and accounting (73.2%); engineering and computer science (18.7%); and decision science (7.3%). In addition, the main source journals based on the number of publications were: Journal of Product Innovation Management (22.4%), International Journal of Innovation Management (16.7%), Industrial Marketing Management (12.6%), International Journal of Production Research (10.2%) (see Table 20.2).

The Latent Semantic Analysis (LSA) provided further insight based on the research trend about the new product development into the topics of curriculum design. The study adopted ten topics, which offered an acceptable compromise between the level of prevalence and coherence. As shown in Table 20.3, the topics are defined by the factor loading of terms rather than by hard clustering a word may appear in more than one topic. The ten identified topics can be interpreted using Bloom's Revised Taxonomy to disclose the implications of educational objectives.

The Latent Semantic Analysis (LSA) provided further insight based on the research trend about the new product development into the topics of curriculum design. The study adopted ten topics, which offered an acceptable compromise between the level of prevalence and coherence. As shown in Table 20.3, the topics are defined by the factor loading of terms rather than by hard clustering a word may appear in more than one topic. The ten identified topics can be interpreted using Bloom's Revised Taxonomy to disclose the implications of educational objectives.

Consistent with previous findings [15], Table 20.3 revealed that the educational objectives of NPD course have been expanding across various fields, including innovation, product design, teamwork, engineering, manufacturing, and marketing. Therefore, the NPD curriculum is expected to facilitate the enhancement of students' ability in innovating new products, collaborating with external stakeholders, and involving in open innovation. Besides, the ability of application of computer-based technology, such as CAD, man-machine interface, and data science, are deemed as major objectives of NPD course.

To infer the key components of NPD curriculum, the highly loaded terms of each topic are synthesized to create the corresponding domain knowledge and key component. Table 20.4 shows that the domain knowledge of NPD was classified into two categories as "field of business and management" and "field of engineering and computer science". The field of business and management contains six domain knowledge, which is the foundation of new product development, including knowledge-based NPD, stakeholder engagement, sustainable development, lean principle, marketing management, and open innovation. On the other hand, the field of engineering and computer science focuses on practical application and the domain knowledge ranges from computer technology, text analytics, social media, and big data in business. The domain knowledge, however, is de-rived from the educational objectives and the key components of each domain knowledge were induced from the highly loaded term within each topic. There-fore, these key components of new product development, such as innovation management, knowledge management, supplier/customer involvement, and computer-aided design and manufacturing, not only align with the essentials of new product development but also facilitate educators and learners comprehending the full picture of the program and achieving the corresponding objectives.

Table 20.3 Summary of LSA results and interpretation

Topic	Prevalence (coherence)	Interpretation (educational objectives)	Highly loaded terms
1	100.608 (−0.003)	• Ability to innovate new products and collaborate with team members	Npd, innov, knowledge, performance, design, collabor, team, supplier, customer, capability
2	4.654 (0.240)	• Ability to adopt and operate CAD and e-assembly service	Cad, service, assembl, design, architecture, models, assembly models, cad assembly, module partition, partition
3	3.688 (0.097)	• Ability to involve in open innovation, and knowledge intensive business services	Lean, slack, organizational slack, kibs, goodyear, market performance, involv, oi, lean principles, time market
4	3.396 (0.290)	• Understanding of knowledge about social media and social networks	Networks, champions, propos, social, project proposals, sustainability, social media, organizational, media, social networks
5	2.594 (0.589)	• Understanding of knowledge about circular economy and alliance	Ce, ce_npd, alliances, circular, circular economy, confucianism, taoism, customer participation, economy, private
6	2.101 (0.044)	• Ability to collaborate with supplier and govern their involvement	Purchasing, collabor, knowledge, supplier, innovation, govern, npd, suppliers, kibs, involvement
7	2.084 (0.298)	• Ability to derive insights from big data analysis for decision	Driv, pricing, collaboration, govern, big data, humanitarian, big, support, decision, behavior
8	2.048 (0.248)	• Ability to adopt man-machine interface for enhancing market capability	Marketing capability, mmi, supplier, architectures, adapt, sustainable, ambiguity, backer, networks, bma
9	1.793 (0.002)	• Understanding of knowledge about text analytics and product life cycle	Knowledge, prototyp, supply chain, chain, supply, text, life, cycle, text analytics, life cycle
10	1.621 (0.041)	• Ability to develop new product with the formal process and open innovation	Standardization, oi, patenting, formal npd, architectures, acceptance, acceptance formal, product architectures, knowledge transfer, transfer

Table 20.4 Key components of NPD curriculum

Domain knowledge	Key elements of NPD curriculum
Field of business and management	
• Knowledge-based NPD	Innovation management, Knowledge management
• Stakeholder engagement	Supplier involvement, Customer involvement
• Sustainable development	Alliance, Circular economy
• Lean principles	Knowledge intensive business service, Lean manufacturing
• Marketing management	Marketing-manufacturing integration
• Open innovation	Standardization, Patent management, Knowledge transfer
Field of engineering and computer science	
• Computer technology	Computer-aided design and manufacturing
• Text analytics	Text mining technique
• Social networks	Social media
• Big data in business	data science

Conclusion

Nowadays, under the fierce competition brought along by the globalization of the world economy, there are competitors all around the organization. For survival, therefore, the development of new products and services has become the essential capability, which encompasses concept design and the successful development of new products that can be launched in the market [4]. The process of new product development is highly challenging and uncertainly [1]. Therefore, a previous study by Mallick and Chaudhury [7] stressed that it is difficult to design an NPD curriculum that suits a practical purpose because the scope of NPD continuously expands across various fields.

Thus, collection and analysis of related literature are crucial to getting insight into the development trend in the future for product innovation and clarify the core components for curriculum design. To disclose the essentials of an NPD curriculum, the study performed the Latent Semantic Analysis to analyze collected literature on the subject matter. The results reveal that there are ten educational objectives must be considered for designing the NPD curriculum (see Table 20.3). Furthermore, as shown in Table 20.4, the domain knowledge corresponding to each educational objective was identified, and a serial key component of the NPD curriculum was derived from the domain knowledge.

In the field of business and management, there are six domain knowledge must be considered and brought into the NPD curriculum. At first, the management skills and knowledge, such as innovation management and knowledge management are deemed as the foundation of entrepreneurship education to enhance students' ability to set up a new venture or develop and grow an existing business. Stakeholder engagement also is a crucial factor for reducing risk during new product development [15]. Therefore, the knowledge and skill of supplier involvement and customer involvement are the second key element of the NPD curriculum. Pursuing sustainable development and resilience is one of today's most challenging tasks for management teams [25]. Thus, educating students with the knowledge of alliance and circular economy can promote their ability to attaining this goal. Finally, launching a new product efficiently is a crucial way to create organizational competition, regardless of its capacity and resource. Thus the knowledge and skill from product design, manufacturing and selling are the necessary components of the NPD curriculum.

In the field of engineering and computer science, the study highlights that the domain knowledge of the application of computer technology, text analytics, social network, and data science are important components of the NPD curriculum. As the statement from previous research [6], computer-aided design and manufacturing (CAD/CAM) can demonstrate and assembles the prototype in virtual space which reduces risks of new product development substantially. In addition, IT has been a powerful tool and widely applied to gather information from the relevant industry sector and customer [30]. Therefore, cultivating the ability about operation and manipulation of data science and social media is necessary for the NPD course.

The majority of existing NPD curricula tend to focus on theories and processes, rather than providing systematic methods and resources for the actual commercialization of ideas. The study presents several unique and important contributions to the education field. First, the study provides further insight into the trend of new product development which is expected to satisfy a growing interest in improving the learning performance of entrepreneurship education. Compared with the current entrepreneurship education program, the results of the study can be applied to design a comprehensive and systemic new product development program. Therefore, the NPD course not only provides students with entrepreneurial knowledge and skill but also encourages them to practice their capability by experiencing the new product development process.

However, the results and findings presented in this paper are based on the textual dataset retrieved from 221 articles that are related to the study. Hence, our future research will utilize this research method to analyze additional literature, and more information can be extracted for facilitating curriculum development and preparation work at school levels.

Acknowledgements The authors would like to thank the Ministry of Science and Technology, Taiwan for their financial support under grant no. 108-2511-H-003-034-MY2.

References

1. S.K. Fixson, Teaching innovation through interdisciplinary courses and programmes in product design and development: an analysis at 16 US schools. Creativity Innov. Manag. **18**(3), 199–208 (2009)
2. K.D. Wowak, C.W. Craighead, D.J. Ketchen, G.T.M. Hult, Toward a "theoretical toolbox" for the supplier-enabled fuzzy front end of the new product development process. J Supply Chain Manage **52**(1), 66–81 (2016). https://doi.org/10.1111/jscm.12084
3. R. Adams, *If You Build It Will They Come?: Three Steps to Test and Validate Any Market Opportunity*. (Wiley, 2010)
4. G. Castellion, S.K. Markham, New product failure rates: Influence of argumentum ad populum and self-interest. J. Prod. Innov. Manag. **30**(5), 976–979 (2013). https://doi.org/10.1111/j.1540-5885.2012.01009.x
5. G. Stevens, J. Burley, 3000 raw ideas equals 1 commercial success! Res. Technol. Manag. **40**(3), 66–81 (1997)
6. M.G. Violante, E. Vezzetti, Guidelines to design engineering education in the twenty-first century for supporting innovative product development. Eur. J. Eng. Educ. 1–21, (2017). https://doi.org/10.1080/03043797.2017.1293616
7. D.N. Mallick, A. Chaudhury, Technology management education in MBA programs: a comparative study of knowledge and skill requirements. J. Eng. Technol. Manag. **17**(2), 153–173. 2000/06/01/ (2000). https://doi.org/10.1016/S0923-4748(00)00019-9
8. D.A. Kolb, *Experiential learning: Experience as the source of learning and development* (PrenticeHall, Englewood Cliffs, NJ, 1984)
9. W.-H. Wu, W.-C. Yan, H.-Y. Kao, W.-Y. Wang, Y.-C. J. Wu, Integration of RPG use and ELC foundation to examine students' learning for practice. Comput. Hum. Behav. **55**(Part B), 1179–1184, 2016). https://doi.org/10.1016/j.chb.2014.10.023
10. R.J. Calantone, C.A. Di Benedetto, An integrative model of the new product development process: an empirical validation. J. Prod. Innov. Manag. Int. Publ. Prod. Dev. Manag. Assoc. **5**(3), 201–215 (1988)
11. F.-H. Leung, D. Martin, H. Batty, A theory-based curriculum design for remediation of residents' communication skills. Med. Teach. **31**(12), e555–e559 (2009)
12. W. Drechsler, M. Natter, P.S. Leeflang, Improving marketing's contribution to new product development. J. Prod. Innov. Manag. **30**(2), 298–315 (2013)
13. A.D. Smith, Competitive approaches to new product development: A comparison of successful organizations in an unstable economic environment. Team Perform. Manag. Int. J. **17**(3/4), 124–145 (2011)
14. D. Francis, J. Bessant, Targeting innovation and implications for capability development. Technovation **25**(3), 171–183 (2005). https://doi.org/10.1016/j.technovation.2004.03.004
15. E. McFadzean, A. O'Loughlin, E. Shaw, Corporate entrepreneurship and innovation part 1: the missing link. Eur. J. Innov. Manag. Rev. **8**(3), 350–372 (2005). https://doi.org/10.1108/14601060510610207
16. O. Ahonen, U.-M. Kinnunen, J. Heinonen, G.-B. Lejonqvist, E. Rajalahti, K. Saranto, Students' competence as eHealth and eWelfare service developers based on the International Medical Informatics Association IMIA's curriculum structure and design thinking (2018)
17. A. Espín-León, A. Jimeno-Morenilla, M.L. Pertegal-Felices, J. Azorín-López, Quantification of cultural identity through artificial intelligence: a case study on the Waorani Amazonian ethnicity. Soft Comput. (2019). https://doi.org/10.1007/s00500-019-04469-3
18. C.-H. Yuan, Y.J. Wu, Mobile instant messaging or face-to-face? Group interactions in cooperative simulations. Comput. Human Behav **113**, 106508, 2020/07/28/ 2020, https://doi.org/10.1016/j.chb.2020.106508
19. S.L. Nielsen, P. Stovang, DesUni: university entrepreneurship education through design thinking. Educ. Training. **57**(8/9) 977–991, (2015)
20. M. Lackéus, An emotion based approach to assessing entrepreneurial education. Int. J. Manag. Educ. **12**(3), 374–396 (2014). https://doi.org/10.1016/j.ijme.2014.06.005

21. S. Dhliwayo, Experiential learning in entrepreneurship education: A prospective model for South African tertiary institutions. Educ. Training **50**(4), 329–340 (2008). https://doi.org/10.1108/00400910810880560
22. A. Huq, D. Gilbert, All the world'sa stage: transforming entrepreneurship education through design thinking. Educ. Training. **59**(2), 155–170 (2017)
23. C. Pietersen, Research as a learning experience: A phenomenological explication. Qual. Rep. **7**(2), pp. 1–14, 2002. [Online]. Available: https://nsuworks.nova.edu/tqr/vol7/iss2/3
24. D.R. Krathwohl, A revision of bloom's taxonomy: An overview. Theory Pract. Rev. **41**(4), 212–218 (2002). https://doi.org/10.1207/s15430421tip4104_2
25. Y.C. Wu, M. Goh, C.H. Yuan, S.H. Huang, Logistics management research collaboration in Asia. Int. J. Logistics Manag. **28**(1), 206–223 (2017). https://doi.org/10.1108/IJLM-09-2013-0104
26. H. Perks, R. Cooper, C. Jones, Characterizing the role of design in new product development: An empirically derived taxonomy. J. Prod. Innov. Manag. **22**(2), 111–127 (2005). https://doi.org/10.1111/j.0737-6782.2005.00109.x
27. N. Bashir, K.N. Papamichail, K. Malik, Use of social media applications for supporting new product development processes in multinational corporations. Technol. Forecast. Soc. Change. **120**(Supplement C), 176–183, 2017/07/01/ (2017). https://doi.org/10.1016/j.techfore.2017.02.028
28. S.V. Gaikwad, A. Chaugule, P. Patil, Text mining methods and techniques. Int. J. Comput. Appl. 85(17), (2014)
29. V. Gupta, G.S. Lehal, A survey of text mining techniques and applications. J. Emerg. Technol. Web Intell. Rev. **1**(1), 60–76 (2009). https://doi.org/10.4304/jetwi.1.1.60-76
30. M.J. Eppler, F. Hoffmann, S. Bresciani, New business models through collaborative idea generation. Int. J. Innov. Manag. **15**(06), 1323–1341 (2011)

Chapter 21
ICT Education for Sustainable Development: Saudi Institutions of Higher Learning, the Case of Effat University

Tayeb Brahimi

Abstract This study explores the crossroads between ICT education and sustainable development in the context of higher learning of Saudi institutions. In particular, this research seeks to examine the ways and means according to which ICT education at Effat University contributes to several critical domains of sustainable development. The study includes bridging the gender equality gap, reducing poverty and marginalization, enhancing social inclusion, promoting active community engagement, and redefining the role of women in socio-economic development and political participation. The current research examines the contributions of the ICT program at Effat University to sustainable development, specifically in ICT education, research and training, ethics of sustainability, community engagement, and industrial collaboration. The study concludes that Effat University is currently making foundational contributions to national growth and sustainable development, while also effectively paving the way to the 2030 transformation agenda of Saudi Arabia. Further comparative research carried out on the crossroads between ICT and specific sustainability targets in the context of higher education institutions would improve findings.

Keywords ICT education · Sustainable development · Effat university · Saudi institutions of higher learning · 2030 saudi vision

Introduction

The understanding of the role of Information Communication Technology (ICT) education in the sustainable development agenda is of critical importance and highlights the position of education as a catalyst to achieving the main objectives of sustainable development. Education for sustainable development (ESD) is the process of providing students with the knowledge, understanding, and skills needed to work and live in a manner that safeguards environmental, social and economic

T. Brahimi (✉)
Energy and Technology Research Center, Natural Science, Mathematics and Technology Unit, College of Engineering, Effat University, Jeddah 22332, Saudi Arabia
e-mail: tbrahimi@effatuniversity.edu.sa

A. Visvizi et al. (eds.), *Research and Innovation Forum 2020*,
Springer Proceedings in Complexity,
https://doi.org/10.1007/978-3-030-62066-0_21

wellbeing, both in the present and future generations. It further seeks to encourage students to the following [1]: *(a) Consider what the concept of global citizenship means in the context of their own discipline and in their future professional and personal lives; (b) consider what the concept of environmental stewardship means in the context of their own discipline and in their future professional and personal lives; (c) think about issues of social justice, ethics and wellbeing, and how these relate to ecological and economic factors; (d) develop a future-facing outlook; learning to think about the consequences of actions, and how systems and societies can be adapted to ensure sustainable future.*

The interaction between ICT and sustainable education needs a strategic reflection if we want to take advantage of the resources and opportunities that it offers [2]. The socio-cultural context of the Middle East, inclusive of its vibrant trends of change and transformation, makes the topic even more interesting, especially given that it brings into play dynamics of societal destinations, religious and cultural re-definitions, and the emergence of new concepts and identities. Prior to discussing the issue at hand, let us first draw on the notion of sustainability and the 2030 Agenda for Sustainable Development.

Sustainable development speaks of "development that meets the needs of the present generation without compromising the ability of future generations to meet their own needs" [3]. It involves a holistic approach to defining solutions that deliver benefits to all while minimizing negative impacts [4]. In September 2015, world leaders ratified a series of goals to eliminate poverty, counter inequality, preserve the earth and strive for prosperity, in fulfillment of the 2030 Sustainable Development Agenda. The present agenda makes a significant departure from traditional conceptions of development, identifying 17 targets to be fulfilled over 15 years. In essence, it's goals are underpinned with the understanding that planetary preservation and eliminating poverty are the synonymous target. To this end, broad-spectrum strategies are needed addressing a broad range of social needs while tackling the environment and climate change [5]. However, the agenda has a stand-alone goal on gender equality and the empowerment of women and girls [6].

Prior to addressing the interconnection between ICT and sustainability in the Saudi context, it is worth highlighting the broader Arab perspective of sustainability to which Saudi Arabia belongs. The Middle East is challenged by rapid urbanization, growing youth demographics, and employment demands, increasing populations, not to mention shortages of basic resources. In light of the Arab Spring, the need for securitization and good governance is now more essential than ever before. With their varying qualities and potentials, development approaches in the region vary significantly [7].

In 2013, the Arab Regional Implementation Meeting (RIM) of the 20th Commission for Sustainable Development, emphasized the Rio Principles, calling for implementation through technology, finance, and capacity development [7]. RIM attendees underlined the necessity for energy diversification, supporting renewable energy technology, countering desertification, ensuring access to water, and ramping job creation to attend to increasing population demand [8]. However, SDGs will need to be more relevant for national policy-making and national development situations.

The approach of "global goals and national targets" would allow countries to set targets most relevant to them, such as reducing particular sub-national disparities or closing gaps in governance [7].

Saudi Perspective on ICT and Sustainable Development

In the following, we shall outline in brief, however, the pledge of the Kingdom of Saudi Arabia towards the SDGs, how it is envisaging impactful effects on sustainability through its recent Vision 2030 and National Transformation Program 2020 ("Vision 2030," n.d.; "National Transformation Program," n.d.), how the vision of the Saudi government is set to fulfill SDGs, and how is it committed to aligning its diverse socio-economic initiatives along those broad lines. Presently, Saudi Arabia is undergoing the diversification of traditional revenues, intending to undertake fiscal measures to allow for targets outlined in both Vision 2030 and the SDG agenda. According to Vision 2030, Saudi has set several initiatives and targets for sustainable development. With regards to the seventeen SDGs, Saudi is committed to achieving them all with specific national socio-economic targets ("National Transformation Program," n.d.).

More specifically, however, is the interest to explore the inherent crossroads between ICT and sustainable development within the Saudi Higher Education Institutions. This is crucial since sustainability is inextricably linked to technology because the sustainability framework is frequently applied to situations that involve technology [9]. ICT helps to achieve SDGs in areas like Smart Water Management, Smart Energy, Smart City Mobility, Smart Building, Smart Agriculture, All Digital Solutions, and Smart Conservation [5]. Through ICT, tangible results are witnessed in energy efficiency, production, and consumption. Smart use of socially-integrated technologies have visible results and will go a long way to furthering SDG realization [5]. By leveraging the potential of current services and industry with ICT capacities, significant energy, and consumption optimization is feasible. However, technology by itself has never been a solution. It must be properly deployed- aimed at social purposes—and extended to the poor and remote regions and must be combined with a will towards the common good which requires harnessing it to the global objectives embodied by the Millennium Development Goals (MDGs) and SDGs [10]. The 2030 Agenda for Sustainable Development recognizes that: "*the spread of information and communications technology and global interconnectedness has great potential to accelerate human progress, to bridge the digital divide and to develop knowledge societies, as does scientific and technological innovation across areas as diverse as medicine and energy.*"

While none of the SDGs specifically concerns ICT, several targets refer to ICT, innovation, and technology, and every objective will be supported by ICT. All three pillars of sustainable development. Namely, economic development, social inclusion, and environmental protection need ICT as a key catalyst; and ICT [10]. Described as a catalyst for achieving the SDGs, ICT will promote gender equality and empowerment, allowing women and girls to access essential information for their efficient, reproductive, and civic roles and involving women in urban planning. Women's sustainable livelihoods can be enhanced through expanded access to markets, education, training, and employment [10].

Interestingly, in Saudi Arabia, the advancement of women is taken as one of the urgent priorities. The Kingdom is currently seeking ways and means to reposition women at the core of the agenda of sustainability. The argument is that women's empowerment yields strong economic returns and maybe the most critical enabler to achieving a wide range of development and other objectives. Moreover, it is a direct investment in communities and society. Particularly, in the field of education, the Kingdom seeks to provide equitable quality education for all, bridge learning divides, enhance the quality of learning, strengthen inclusion, and promote long-life learning opportunities for all. Based on its Vision of 2030, it plans to continue investing in education and training, with a focus on innovation in advanced technologies and entrepreneurship while ensuring that the outcomes of the education system are aligned with the requirements of the labor market needs and increasing women's participation in the workforce from 22 to 30% ("National Transformation Program", n.d.).

SDG seeks to enhance the use of enabling technology, mainly information and communications technology (ICT), to promote women's empowerment. Research indicates that people are better able to cope with risk or invest in a business by engaging in the financial system and funding expenditures [11, 12]. Women are often the household Chief Financial Officers, managing multiple financial streams within the family, including income, expenses, and savings. ICT is a means of reaching vulnerable groups such as disadvantaged women currently excluded from financial services and of helping to fulfill a shared global commitment to ensuring that no-one is left behind [10]. Extensive research on the role of ICT for women's empowerment indicates that women can benefit immensely from ICT led financial services. ICT helps women to access the capital needed to establish and grow their businesses. Women's participation in economically productive activities is constrained by limited access to finance and lack of ICT skills. ICT skills should be provided to help women take control of their finances and leverage ICT for business activities [10].

Currently, in Saudi Arabia, the Communications and Information Technology Commission (CITC) stresses that National Transformational Program 2020 ("National Transformation Program", n.d.) includes strategic aims in the field of communications and information technology, including developing the infrastructure of communications and information technology, broadband, innovation in advanced technologies, investment in the digital economy, national human resources, building an effective generation using the latest technologies, and establishing a knowledge-based economy in line with the Kingdom Vision 2030 through a public-private partnership.

Research Methodology

Prior to addressing the different ICT contributions of Effat University to sustainability, one would first introduce the study sample and the adopted framework of discussion used in this paper. First, Effat University is a private non-profit institution of higher education for women in Saudi Arabia ("Effat University", n.d.), which was established in 1999 as a nonprofit educational institute operating under the umbrella of King Faisal's Charitable Foundation. Effat University is Saudi Arabia's first private university institution for women. It has been named as one of the top 100 universities in the QS Arab Region University Ranking 2018. Research achieved steady growth since 2009 and is aspiring to become one of the leading research universities in the Kingdom.

It is equally important to highlight some of the current sustainability-based frameworks used to understand and assess the implementation of sustainable development, which was developed to determine priority areas in sustainability and strategies to achieve economic, social, and environmental developmental goals. The *Triple Bottom Line* is the first of those models and can be adapted and modified by adding culture, ethics, equality, social responsibility, politics, and economic, social, and environmental factors. The second is the *Natural Step Model,* which seeks to integrate decision-making frameworks to help people make the right choices, with a more focus on society's interactions with the earth. The third is the *Five Capitals Model,* which uses economics as the starting point and maintains that any government or organization has five capitals or stocks to manage: natural, social, human, financial, and manufacturing. This model emphasizes environmental management but plays down social and economic development [13]. However, the Corporate Social Responsibility (CSR) model attempts to apply sustainability to guide the behavior of business concerning both society and the environment as well as its responsibility to stockholders.

However, despite the sophistication of those models, there is still little agreement about how higher education could best contribute to global sustainable development through its teaching and research, institutional management, building, and various operational activities such as purchasing. OECD [13] identifies two opportunities for higher education institutes to engage in sustainable development. Firstly, by forming a link between knowledge generation, transferring the knowledge to society, and into the labor market. Such preparation includes the education of teachers. Secondly by actively contributing to the societal development through outreach and service to society. Cortese [14] seconds this notion, stating the following: "*Higher education institutions bear a profound, moral responsibility to increase the awareness, knowledge, skills, and values needed to create a just and sustainable future. Higher education often plays a critical but often overlooked role in making this vision a reality. It prepares most of the professionals who develop, lead, manage, teach, work in, and influence society's institutions*" (Cortese, p. 17).

Currently, universities and intergovernmental institutions have developed more than 31 Sustainability in Higher Education (SHE) declarations, with more than

1400 universities signing a SHE declaration globally. There are 31 SHE declarations including the Stockholm Declaration in 1972, the Tbilisi Declaration in 1977, the Talloires Declaration in 1990, the Halifax Declaration in 1991, World Declaration on HE in 1998, the Barcelona Declaration, the Graz Declaration in 2005, the Rio Declaration, the Bonn Declaration (2009), the Sapporo Sustainability Declaration (2008), the Lubeck Declaration in 2009, and the Kyoto Declaration among others [15–17]. These initiatives set the foundation for incorporating sustainability in higher education. For instance, the Talloires declaration' ten action points including among many others: Educate for Environmentally Responsible Citizenship; Foster Environmental Literacy for All; Collaborate for Interdisciplinary Approaches; Enhance Capacity of Primary and Secondary Schools; and Broaden Service and Outreach Nationally and Internationally [18].

Different approaches emphasize the moral imperative of universities to strive towards sustainability in society, and what is to be passed down to future generations. Analysis of core constituents of higher education systems (curricula, research, operations, assessment, and dissemination) can be supplemented through university collaboration, ensuring sustainable development is part and parcel of the institutional vision, immersive on-campus engagement; and educator capacity building. The I3E model for embedding education for sustainability (EfS) in undergraduate systems highlights the need to inform the university community, engage different university stakeholders in change; Empower agents to institute changes within their domains, and finally Embed sustainability in existing structures [19]. Alshuwaikhat et al. [20] use specific units of analysis to assess the implementation of sustainability in Saudi state schools, evaluating teaching and curriculum, research and scholarship, campus management, and financial management [20].

Instead of enumerating all of the action items pertaining to the implementation of sustainable development in higher education institutions, it is clear that there is no one-way strategy to actualize sustainability. It is argued that the study would further demonstrate that impactful policies can be developed through an understanding of the university system, particularly re-envisioned teaching, facilities, management, research, curricula, and operational practice [13]. In the following section, we will survey the different contributions of Effat to sustainability, however, according to the idea that one can adopt or combine and apply any of the models to the sustainable management and operation of the University.

The Contribution of Effat University to Sustainability

HEIs have a critical and tangible role in developing the principles, qualities, and awareness not only needed to perpetuate the sustainable development philosophy but to improve upon its delivery [21, 22]. Yet to speak of the role of Effat in the field of ICT and sustainable development, one also needs to remember that the role of universities is much broader than generating patents, licenses, and publications, or education and training. As argued by Lester [23], universities should be seen not only as a source of creators, receptors, and interpreters of innovation and ideas, but also as sources of human capital, and a key component of social infrastructure and social capital. Universities contribute to the growth and coherence of society through the promotion of their core ethical values of justice, openness, and equality. By their engagement with local government, universities can build services that involve the local community more efficiently than the government or the private sector can, while retaining their core educational and research competencies [24]. However, the channels through which these contributions can be enhanced for the benefit of local and regional communities should be further explored [25].

ICT Education, Research, and Training

In 2005, the Decade of Education for Sustainable Development was announced [26]. It aimed to integrate sustainable development principles in all levels of education and learning [13]. This was predicated on the potential for maximizing technology in the achievement of development targets, seeking out force multipliers through effective educational integration with its broad implications for all aspects of societal development and growth. This approach emphasizes the necessity for higher education institutions to actively engage with this campaign, given the potential they hold for mass education. Several means were identified, including their function as sources of innovation and knowledge, and the potential role they could bring about bringing about affordable, accessible green technology for suitable markets.

Rapid developments and advances in information and communication technology (ICT) are revolutionizing the way we communicate, learn, and live. Today, ICT represents a powerful tool that we need to adopt for accelerating the achievement of the 2030 Agenda for Sustainable Development. For instance, using ICT in the health sector to gather, manage, analyze, improve, and exchange information will ensure a healthy life and promote wellbeing for all. In education, the use of ICT ensure inclusive and equitable quality education and promote lifelong learning opportunities. Effat University Research and Consultancy Institute (RCI) works to develop and sustain a dynamic campus research culture that fosters creativity and innovation among faculty, researchers, staff, and students. In terms of attracting best research from local and foreign talent groups that would like to conduct research at Effat University, regional and global goals for research are also the RCI's key priority. These goals were also considered in the development of the four research centers at the University: (1) Smart Buildings Research Center (SBRC) to study and explore the use of innovative smart platform technologies with intelligent building components, (2) Virtual Reality Research Center (VRRC) to explore the use of high-end VR technologies for applications in architecture, construction, healthcare, training, and education, to serve both the global and local research community, (3) E-Arabization Research Center (E-ARC) to investigate the design, development, contextualization, and translation of digital content in the Arabic language, and (4) the Business Innovation Entrepreneurship Research Center (BIERC) to develop tools and concepts for sustainable economic growth.

In the energy sector, Saudi Arabia's Vision 2030 and the National Transformation Program set out an ambitious roadmap. However, to gain expertise in renewable energy, Engineers need to be well trained, competent, and have strong theoretical and practical skills. Many studies called for a transformation of current programs to incorporate education and training programs in renewable energy and sustainability [27–29]. One of the recommendations for the energy education program is to offer a combination of academic and hands-on skills by conducting labs, practical demonstrations, field visits, and installation of the actual renewable energy system.

Aligning its mission with the Saudi Vision 2030, Effat University introduced a Master in Energy Engineering and a Solar Energy training program to help students and faculty alike acquire the knowledge and experience needed to diversify and at the same time capitalize on renewable energy sources. The Master of Science in Energy Engineering program prepares Saudi women to meet the demand for engineers, scientists, and researchers in Saudi Arabia's booming energy sector. The objectives are to produce female graduates who thoroughly understand the energy-engineering theory and recent developments in fundamental research.

Effat University is also committed to supporting sustainable use and exploration of the potential of renewable energy technology to become a high-level expert on renewable energy technology and environmental impact. Students are strongly encouraged to embark on research and hands-on practice. For instance, female seniors and juniors at Effat University participated in the country's first solar energy training program to learn how to install a rooftop solar photovoltaic system created by Saudi female engineers. This project of PV Rooftop is designed to provide engineering and architecture students with an educational and practical opportunity to participate in the National Transformation Program and 2030 vision. Recently high school girls from Jeddah and Makkah have participated at an innovation laboratory for the development of science, technology, engineering, art, and math (STEAM) organized and hosted by Effat University to connect future innovators among high school students. STEAM is a new program organized at Effat 's University and includes a full range of courses such as Cryptography, Robotics, Drama and Theatre, English, Emotional Intelligence, Engineering, Play and Create Games, Creative Arts, Cinematic Arts, Photography, and Physics in Everyday Life. According to the President of the University, the interest of the program is to encourage and develop high school students, from Makkah and Jeddah regions, in subjects that are not usually covered by their school curricula, and provide them with a variety of topics and unique opportunity to highlight their imagination and creativity and apply innovative thinking to their selected subjects [30].

Effat University also hosts three-week Mawhiba enrichment and creativity program for talented students for female students to develop their skills and abilities, which enable them to continue building quality, cumulative experience in Electrical Engineering, Engineering Design, Encryption, Chemistry in Society, and Advanced Physics. During the last Mawhiba 2017 enrichment and creativity program, students visited the King Abdullah University for Science and Technology (KAUST) and the Water Desalination Company in Jeddah as part of the program's main events. According to the President of the University, the Mawhiba targets talented female students to identify their potential and offer the support needed to develop and refine their skills. This program comes within the framework of the University's contribution toward building a promising society characterized by talent and creativity and efforts in developing young, talented Saudi leaders with superior educational outcomes who can play their awaited developmental roles with capability and merit [31].

For the senior projects, all of Effat Engineering programs require completion of a capstone research project for which students would select respective organization or industry as a client to explore a particular research problem and accordingly develop a model and propose a solution. In these capstone projects, ICT represents a powerful means and platform for achieving the SDGs such as accelerating critical services in health, ensuring inclusive education and promoting lifelong learning opportunities, facilitating and ensuring access to energy for all, building resilient infrastructure, and fostering innovation, and enhancing gender equality. Projects challenge students to think beyond the boundaries of the classroom, develop the skills of the 21st-century and solve current and future problems faced by our society in healthcare information systems, remote nutrition care for diabetic patients, neuro-feedback to treat children with ADHD, blood donation application, 3D bio-printing, augmented reality system through wearable technology for sensory impairments applications, smart cities, green smart library station, environment, renewable energy, virtual reality, electronic school directory, web application, railways application, document mining, mobile application, cloud computing, real estate management, internet of things, supermarket application, cybersecurity and fingerprint authentication, civil defense emergency, volunteering management system, emergency and rescue management system, sign language recognition system, Green Smart Library Station. As per Fig. 21.1, the distribution of capstone projects in the last two years (2015–2019) at the College of Engineering (CoE) shows that 36% of the students' projects are related to SDG #9; 30% to SDG #3; 18% to SDG #4, and 16% to SDG#7. It is interesting to see here that a high percentage of projects in engineering is dedicated to good health and wellbeing.

Ethics of Sustainability

Sardar [32] considers the roots of present-day ecological crises as axiomatic, for they lie in the very mindset/belief and value structures that shape the human relationship with nature, with each other and the lifestyles. The current definitions of sustainability highlight responsibility for protecting the quality of human life and natural resources, environment, and imply moral commitment not only to today's generations but to futures ones as well; this makes the building an ethical mindset

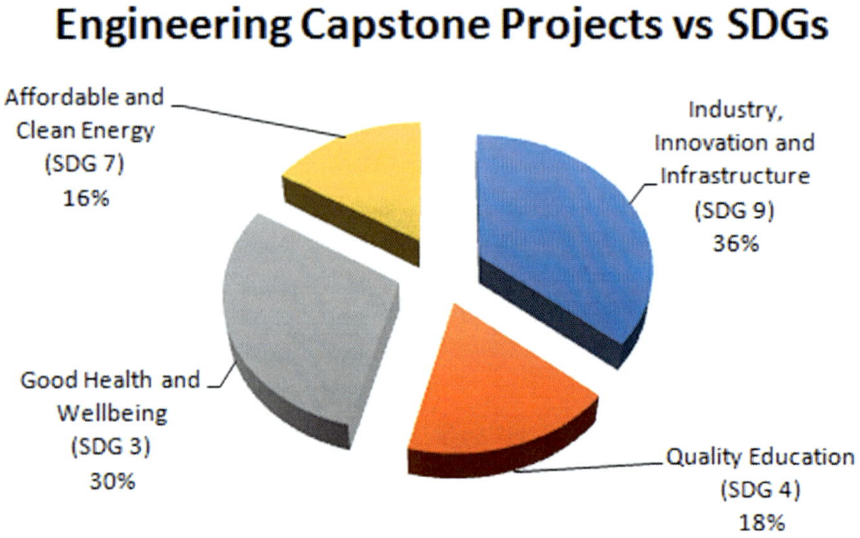

Fig. 21.1 Engineering capstone projects by SDG, Effat University 2015–2019

of sustainability and social responsibility unavoidable. This also implies consideration of native approaches to sustainability and attention to the context of beliefs, values, socio-cultural dynamics. Grine et al. [33] highlighted the need to explore the essence and characteristics of worldviews underlying the faith and practice of religious societies because it helps to understand the Islamic religious and moral values, spirituality, integration of religiosity and social action, and conception of sustainable living.

The moral philosophy of Effat University stands on clear core values referred to as IQRA ("IQRA Core Values", n.d.) which are based on the divine commandment IQRA, 'read,' and embrace life-long research, ethical social and educational values, responsible and creative leadership, and effective communication and reaching out to others. According to its literature, Effat University has produced individuals who have solid values at a time when a rising emphasis on technological advancement has prevented us from concentrating on both spiritual and moral questions (which appear to erode further with each new development). Effat is also committed to preparing students to become effective members of their respective institutions and useful contributors to their society. Effat's mission is to promote ethical behavior in the management of activities in academia, research and community service, and the interaction with stakeholders. Effat has identified and designated eight fundamental elements of its code of conduct called *Tarbawiyyat* Effat which make up the University Code of Ethical Conduct, and include piety, manners, stewardship, cooperation, moderation, modesty, ethics and integrity, and Guidance ("Code of Conduct", n.d.).

The above makes it clear that Effat deeply recognizes the critical role of ethics and morals in the shaping of sustainability and the definition of responsible contribution to societal development. With this idea in mind, Effat provides its students with courses on ethics and corporate social responsibility, and assess the semestrial improvements on set targets of ethical achievements. Similarly, Effat provides an Ambassador Program to enhance its students' commitment to ethics and service and offer them the skills they need to succeed, which aims to develop Effat students' personal, professional, leadership, and academic skills, see Fig. 21.2 ("Effat Ambassadors Program", n.d.). Effat University offers the Effat Ambassador Program as a student development program that will allow every student at Effat University to fulfill the university core values of IQRA through the demonstration of the four Effat University Graduate Characteristics ("Effat Ambassadors Program", n.d.). Achieving ethical, social, and educational values is one such important intended Learning Outcomes of the program which seeks to ensure students' practice of integrity and ethical values; demonstration of understanding and respect to others' culture and values; commitment to health and wellbeing for herself, her community, and the world; reaching out effectively to others; engaging actively in activities to serve the community while demonstrating awareness of the global challenges

Effat also invests in the religious and spiritual capital residing in the community and in imbued in the socio-historical makeup of the community and works carefully

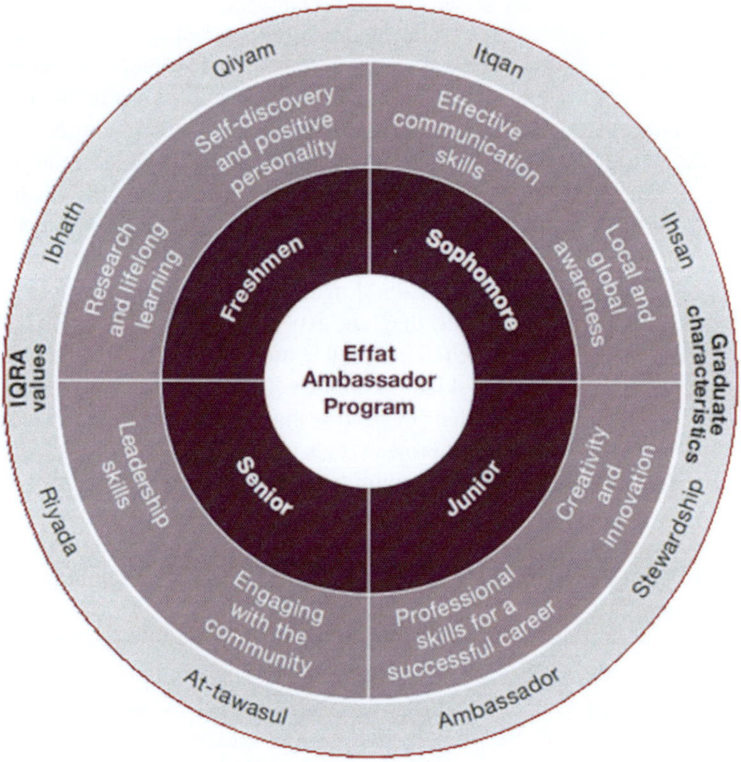

Fig. 21.2 Effat ambassador program

to interweave those ethical and spiritual elements with the nurturing of the graduates' personality. Such a conscious integration not only gives due attention to the ethical background of students but also ensure channeling those profound beliefs towards the responsible attitude to development and ethical stewardship. With proper education and training, it is hoped that those religious and spiritual values help shaped a sound ethical mindset for sustainable development, and ensure highest degrees of synergy between deep-seated religiosity and spirituality with the demands of global sustainability while helping students defining meaning in their daily spiritual rituals and practices.

Community Engagement and Indusial Collaborations

Vidican [25] noted that the universities' role in community engagement had been described as significant in the creation and promotion of principles of sustainability. Therefore, both campus and local communities need to be involved in changing practices and thinking about implementing policies. It is according to the philosophy of stewardship and community service that Effat University dedicates itself to spontaneous volunteering and selfless efforts to serve the community and preparedness to collaborate with government organizations and the private sector to provide an efficient and successful operational framework. Through its programs, the University is closely connected with the community, which promotes continuous development. The idea is that the engagement and ability of Effat to serve the society improve its relations with the community and help meet the need for highly qualified and skilled citizens. They will eventually serve and represent their respective institutions. ("Effat Ambassadors Program," n.d.). Along the same lines, Effat students are required to enroll in the course of Civic Engagement at their sophomore or junior stage. In their final year, however, they are expected to apply the skills learned in the course and other courses such as communication skills, presentation skills, time management skills, leadership skills. In addition to the requirement of completing academic capstone projects in their respective majors, students are also required to fulfil certain amount of hours in projects that involve community service and are also expected to showcase their community projects towards the end of semester ceremony. Along similar lines, Effat faculty are expected to demonstrate degrees of meaningful community service according to which their yearly performance will be evaluated. This is known as the Faculty Portfolio, which assigns community service a percentage of 5%.

Throughout their study, students are encouraged to be engaged in community projects, whether through local clubs on campus or collaborations with the industrial sectors. One such program, The Science and Engineering Fair, was first introduced in 2005 in collaboration with the Ministry of Education and the Administration of Gifted students in Jeddah's High schools. Students compete for the best projects in areas of Engineering and Information Technology. This one-day full program also invites role models women engineers to share their experiences with participants. Experts are invited from different education sectors and universities to judge students' work. Students exhibit their projects and discuss their ideas in front of the judges. Then University and Ministry representatives make their recommendation on what they think is the best project to participate in national and international competitions.

The STEAM Innovation Lab Program is also another exciting program set to help high school students acquire knowledge and skills in STEM and Art related fields. Each student selects two courses (STEM and Art). By the end of this program, students are expected to have participated in hands-on projects, have enjoyed several recreational activities, have learned useful life skills, and have developed experience on a postsecondary education campus, and experience university life.

The Women in Technology Training, known as WIT, is yet another interesting community program that seeks to empower disadvantaged Saudi women and provide them with quality, hands-on training in computer literacy and work-related skills. The program also aims at contributing to ongoing, national efforts to increase the presence of skilled, professionally independent Saudi women in the workforce as part of the country's futuristic Vision 2030. Established in 2006, this program trained a large number of students in the fields of Information Technology and business skills. Throughout the years, Effat University has collaborated with leading industry and government institutions such as Microsoft, Human Resources Development Fund (HRDF), and Gulf International Bank to deliver the program. This year's WIT 10-week training program covered topics like Fundamentals of Computer Applications I & II, basic work skills, and networking. By the end of the program, students were able to use various computer applications, communicate via emails, create web pages, use multimedia forms and spreadsheets, conduct an Internet search, and master other work-related skills. Program instructors also assessed students on several personality skills, such as learning motivation and attitude, behavioral performance in class, creativity, and leadership. Figure 21.3 displays the survey results of their skills before and after joining the WIT Program.

Also important is the current collaboration held between Effat College of Engineering and an established stakeholder like Siemens Company, which involves the organization of the yearly Siemens Challenge under the supervision of Siemens senior Chief Technology Officer. In this program, students are expected to select a topic on projects and innovation associated with sustainable development and to propose their solutions to real problems before the jury. In the academic year of 2016/2017, the challenge pertains to the theme of "Energy Efficiency—and how to address this topic and thereby support the Saudi Vision 2030". Other workshops focused on making the kingdom cleaner and greener. Future Plans for WIT are to follow up with alumnae through the career portal system. A workplace paid internship will be offered for two top performers after conducting, with GIB managers, interviews with all students for future job opportunities. The monetary award will be used as a seed fund for the two top student's startup businesses. An English training component will be also be added to the program to help those in need to improve their English language.

At the institutional level, Effat's commitment to minimizing its environmental impact extends well beyond its environmentally friendly Waste Disposal Program

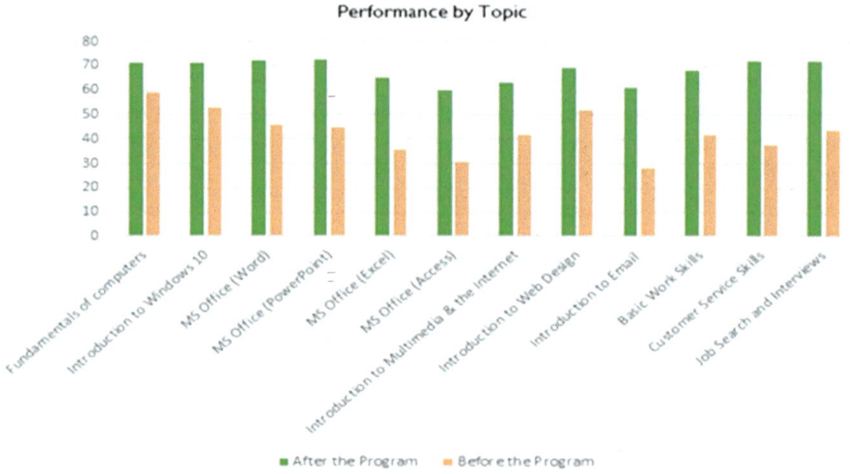

Fig. 21.3 WIT program performance by topic

to an interdisciplinary project, whereby waste materials are recycled for a noble purpose. The University, for example, reduces energy consumption through both active and passive measures, reduce its water consumption whilst maintaining reasonable comfort levels for building occupants. Effat University is also dedicating its efforts to enable mobility, however, without increasing CO_2 emissions, and has regulated its transportation services to limit daily trips to two per day only. As a measure, it divided neighborhoods into zones and cost regions, thereby limiting the number of University vehicles on the road ("Sustainability," n.d.).

Effat continues to forge partnerships with national and institutions and organizations that advocate social inclusion and the increased role of women in society and technology and take an interest in improving women's professional and financial conditions. Those partnerships are set around the core issues of sustainable development. For example, Effat has made many collaborated with institutions like Microsoft, Human Resources Development Fund (HRDF), the Gulf International Bank, Siemens, Schlumberger, Aramco, Unilever, Procter & Gamble, Nestle, Cisco, KAUST, KACST, NCB, and Samba Bank, and has partnered with national and international institutions to host summer training programs for junior students.

Another interesting cooperation, which is set to enhance development is the 'SAP Collaboration' which is to provide students in the fields of Computer Science, Information systems, and Operation and Information Management with high-class curriculum and SAP certifications. In essence, this collaboration seeks to bridge the gap between education and industry, and for this very purpose, Effat Faculty were trained on the SAP curriculum. Currently, they continue to deliver the curriculum to students under the supervision of SAP specialists who visit Effat University regularly and also work closely with the College of Engineering and Business. SAP collaboration not only speaks of effective bridging with industries, but also draws on very viable channels of capacity building and skills transfer, and certainly paving the way to ICT solutions to issues and problems of development and life in general.

However, despite the considerable contributions made to sustainable development, several challenges remain. Higher education institutions remain predominantly focused on environmental and ecological concerns to the detriment of the broader SDG agenda for several reasons. These include relative SDG illiteracy [28, 34]; challenges to academic credibility from entrenched views [35]; oversaturated curricula [36, 37]; the absence of funding or support; perceptions of irrelevance to courses or domains of study; and uncertainty regarding what it takes to adopt and incorporate sustainable development [38]; not to mention restrictive organizational systems [39].

Sustainability continues to suffer from narrow terms of definition for institutions with a strong focus on activities to do with environmental sustainability. Yet, less interest was shown in faculty, community, and student wellbeing, with less attention directed to economic sustainability. Other challenges include the need for increased national coordination and supportive linkages to encourage synergy and cooperation on the bilateral and internal levels. To this end, the need for strategic linkages with external non-HEI parties is recognized as essential to remaining engaged with societal efforts at achieving sustainable development. In this sense, regional cooperation permits deeper participation and adaptability while allowing room for organic knowledge and education. Moreover, there is an urgent need to utilize measurable indexes in implementation, monitoring, and assessment of sustainable development that is compatible and permit interchangeability with national development indicators. In this respect, further research is needed to review practiced measures, management, and impact of ICT on the sustainable development agenda. Besides that, as reported by Visvizi et al. [40], Higher Education Institutions (HEI) need to adapt to the continuously changing environment where advanced technology, politics, economics, and societal change present several opportunities and challenges.

Conclusion

Given the commitment of the sustainable development goals to promoting peaceful and inclusive communities, ensuring access to justice for all and creating open, efficient, and accountable institutions at all levels, in addition to fighting for ending poverty, building economic growth and addresses social needs including education, health, social protection and participation, and job opportunities while tackling climate change and environmental protection, it is clear that Effat University is making conscious contributions to the actualization of the agenda of sustainability in the region. Its evolving partnerships with the community and industry and efforts to help women take its role back in the various professional development is also considered critical to the promotion of women participation and inclusion in the workplace and society in general. Increased partnerships with institutions of higher learning and professional training will undoubtedly speed the fulfillment of SDGs targets and set indexes. However, meeting the goals of SDGs and 2030 Agenda requires continuous and increased efforts from all partners, including the ICT industry, civil society, local communities, research centers, and institutions.

References

1. J. Longhurst, L. Bellingham, D. Cotton, V. Isaac, S. Kemp, S. Martin, D. Tilbury, Education for sustainable development: Guidance for UK higher education providers (2014)
2. A. Visvizi, L. Daniela, Technology-enhanced learning and the pursuit of sustainability. Sustainability **2019**(11), 4022 (2019). https://doi.org/10.3390/su11154022
3. G.H. Brundtland, *World Commission on Environment and Development: Our Common Future* (Oxford University Press, New York, 1987)
4. T. Strange, A. Bayley, Sustainable *Development: Linking Economy, Society, Environment.* (Organisation for Economic Co-operation and Development (OECD), Paris, 2008)
5. T. Ono, K. Iida, S. Yamazaki, Achieving sustainable development goals (SDGs) through ICT Services. Fujitsu Sci. Tech. J. **53**(6), 17–19 (2017)
6. UN Women, Women and Sustainable development goals, east and southern Africa regional office, 3, http://www2.unwomen.org/-/media/field%20office%20africa/attachments/publications/2016/03/analysis%20of%20women%20and%20sustainable%20development%20goals-%20web%20rev.pdf?la=en&vs=1720 (2016)
7. ESCWA, *SDG Priority Conceptual Issues: Towards an Arab Approach for the Sustainable Development Goals.* (Technical paper, United Nations Economic and Social Commission for Western Asia, Arab Regional Consultative Dialogue on the Sustainable Development Goals, Tunisia, 18–19 November 2013), https://sustainabledevelopment.un.org/content/documents/2973E_ESCWA_SDPD_13_TP-8_E.pdf (2013)
8. IISD, Summary of the Asian and pacific regional implementation meeting on Rio 20 outcomes: 22–24 April 2013. (2013, April 26). Retrieved from http://enb.iisd.org/vol05/enb05308e.html
9. C.J. Kibert, L. Thiele, A. Peterson, M. Monroe, *The Ethics of Sustainability* (Wiley, UK, 2011)
10. Earth Institute, & Ericsson., *How Information and Communications Technology Can Achieve The Sustainable Development Goals(Publication).* Earth Institute. (ERIC Document Reproduction Service, 2016)
11. S. Hendriks, The role of financial inclusion in driving Women's economic empowerment. Dev. Pract. **29**(8), 1029–1038 (2019). https://doi.org/10.1080/09614524.2019.1660308
12. S. Singh, Connecting financial inclusion and ICTs to Women's Empowerment. SSRN Electron. J. (2018)
13. OECD, Higher education for sustainable development (2017). Retrieved from http://www.oecd.org/education/innovation-education/centreforeffectivelearningenvironmentscele/45575516.pdf
14. A.D. Cortese, The critical role of higher education in creating a sustainable future. Plan. High. Educ. **31**(3), 15–22 (2003)
15. J.S. Alejandro-Cruz, R.M. Rio-Belver, Y.C. Almanza-Arjona, A. Rodriguez-Andara, Towards a science map on sustainability in higher education. Sustainability **11**(13), 3521 (2019). https://doi.org/10.3390/su11133521
16. T.S. Grindsted, Sustainable universities from declarations on sustainability in higher education to national law. SSRN Electron. J. (2011). https://doi.org/10.2139/ssrn.2697465
17. T. Wright, The evolution of sustainability declarations in higher education. *Higher Education and the Challenge of Sustainability*, 7–19. https://doi.org/10.1007/0-306-48515-x_2 (2004)
18. R. Lozano, R. Lukman, F.J. Lozano, D. Huisingh, W. Lambrechts, Declarations for sustainability in higher education: becoming better leaders, through addressing the university system. J. Clean. Prod. **48**(17–18), 34 (2013)
19. G. Cebrián, The I3E model for embedding education for sustainability within higher education institutions. Environ. Educ. Res. **24**(2), 153–171 (2018)

20. H.M. Alshuwaikhat, Y.A. Adenle, B. Saghir, Sustainability assessment of higher education institutions in Saudi Arabia. Sustainability. **8**, 750 (2016). https://doi.org/10.3390/su8080750
21. N. Blanco-Portela, J. Benayas, L.R. Pertierra, R. Lozano, Towards the integration of sustainability in Higher Education Institutions : a review of drivers of and barriers to organisational change and their comparison against those found of companies. J. Clean. Prod. (2017). https://doi.org/10.1016/j.jclepro.2017.07.252
22. J.W. Yen-Chun, S. Ju-Peng, Higher education for sustainable development: a systematic review. Int. J. Sustain. High. Educ. **17**(5), 633–651 (2016). https://doi.org/10.1108/IJSHE-01-2015-0004
23. R.K. Lester, Universities, innovation, and the competitiveness of local economies. A summary report from the local innovation systems project: Phase I. Massachusetts institute of technology, Industrial Performance Center, Working Paper Series, (Cambridge, 2005), pp. 5–10
24. A. Rappaport, S.H. Creighton, *Degrees That Matter: Climate Change and the University* (MIT Press, Cambridge, 2007)
25. G. Vidican, The role of universities in innovation and sustainable development. Sustain. Dev. Plann. **IV**, 37–38 (2009). https://doi.org/10.2495/sdp090131
26. UNESCO, A *Decade of Education for Sustainable Development.* (2005). Retrieved from http://www.unesco.org/new/en/santiago/education/education-for-sustainable-development/decade-of-education-for-sustainable-development-desd/
27. L. Hancock, S. Nuttman, Engaging higher education institutions in the challenge of sustainability: sustainable transport as a catalyst for action. J. Cleaner Prod. **62**, (2014) https://www.sdg.sa/wps/portal/sdg/sdgnews/CONT-news-14092017%201)
28. F.J. Lozano, R. Lozano, Developing the curriculum for a new Bachelor's degree in engineering for sustainable development. J. Cleaner Prod. **64** (2014)
29. A.E. Wals, Sustainability in higher education in the context of the UN DESD: a review of learning and institutionalization processes. J. Cleaner Prod. **62** (2014)
30. Saudi Gazette, Engaging future innovators (2018, February 01). Retrieved from http://saudigazette.com.sa/article/527458/SAUDI-ARABIA/Engaging-future-innovators
31. Arab News., Effat *University Hosts Participants of Mawhiba Program.* (2017, September 17). Retrieved from http://www.arabnews.com/node/1162856/corporate-news (2017)
32. Z. Sardar, *Islamic Futures.* (Mansell Publishing Limited, New York, 1985). https://www.worldcat.org/title/islamic-futures-the-shape-of-ideas-tocome/oclc/11470033
33. F. Grine, B. Bensaid, R. Mohd Nor, T. Ladjal, Sustainability in multi-religious societies: an Islamic perspective. J. Beliefs Values **34**(1), 72–86 (2013). https://doi.org/10.1080/13617672.2013.759363
34. S.A. Davis, J.H. Edmister, K. Sullivan, C.K. West, Educating sustainable societies for the twenty-first century. Int. J. Sustain. High. Educ. **4**(2), 169–179 (2003)
35. D. Peet, K. Mulder, A. Bijma, Integrating SD into engineering courses at the Delft University of Technology: the individual interaction method. Int. J. Sustain. High. Educ. **5**, 278–288 (2004)
36. S.A. Abdul-Wahab, M.Y. Abdulraheem, M. Hutchinson, The need for inclusion of environmental education in undergraduate engineering curricula. Int. J. Sustain. High. Educ. **4**(2), 126–137 (2003)
37. K.W. Chau, Incorporation of sustainability concepts into a civil engineering curriculum. J. Prof. Issues Eng. Educ. Pract. **133**(3), 188–191 (2007)
38. R. Lozano, Diffusion of sustainable development in universities' curricula: an empirical example from Cardiff University. J. Clean. Prod. **18**(7), 637–644 (2010)
39. L. Velazquez, N. Munguia, M. Sanchez, Deterring sustainability in higher education institutions: an appraisal of the factors which influence sustainability in higher education institutions. Int. J. Sustain. High. Educ. **6**(4), 383–391 (2005)
40. A. Visvizi, M.D. Lytras, A Sarirete. (eds) (2019) Management and administration of higher education institutions in times of change (Emerald Publishing), Bingley, UK, ISBN: 9781789736281, https://books.emeraldinsight.com/page/detail/Management-and-Administration-of-Higher-Education-Institutions-in-Times-of-Change/?K=9781789736281

41. Code of Conduct (TARBAWYYAT). (n.d.), Retrieved from https://www.effatuniversity.edu.sa/English/Student-Life/Why-Choose-Effat/Pages/Code-and-Conduct.aspx
42. Effat Ambassadors Program (n.d.), Retrieved from https://www.effatuniversity.edu.sa/English/Student-Life/Why-Choose-Effat/Effat-ambassadors-program/Pages/default.aspx
43. Effat University. (n.d.), Retrieved from www.effatuniversity.edu.sa
44. IQRA Core Values. (n.d.), Retrieved from https://www.effatuniversity.edu.sa/English/About/Why-Effat/Pages/IQRA_Core_Values.aspx
45. National Transformation Program (n.d.), Retrieved from https://vision2030.gov.sa/en/programs/NTP
46. Sustainability. (n.d.), Retrieved from https://www.effatuniversity.edu.sa/English/About/Impact/Sustainability/Pages/default.aspx
47. Vision 2030 (n.d.), The Kingdom of \Saudi Arabia. Retrieved from http://vision2030.gov.sa/en

Chapter 22
A Social Approach to a *Wiki Course* Building

Carlo De Medio, Fabio Gasparetti, Carla Limongelli, and Filippo Sciarrone

Abstract Wikipedia has become the most used source of didactic materials with its immense number of Wiki pages, freely available to anyone. Wiki Course Builder is a Web platform designed to help teachers building courses composed by Wikipedia pages only, directly taken by the Wikipedia free encyclopedia. One of the interesting and novel features is the *Community* module, that provides a collaborative environment, through a suitable Graphic User Interface, where teachers can interact with all the courses in the platform. The Community is a graphical visualization of the Wikipedia pages used by all the teachers in the systems, grouped in clusters, each representing a particular topic. Another characteristic is the information about the Teaching Styles of those teachers that have used that page in their courses. This feature has a twofold advantage. First, each Wikipedia page is tagged with the Teaching Styles. Second, this social aspect gives useful suggestions for possible extensions or changes in the courses. We present a complete use case showing the usefulness of the social *Community* environment to speed up and improve the course building activity.

Keywords Course building · Social learning · Machine learning

---

C. De Medio · F. Gasparetti · F. Sciarrone (✉)
Engineering Department, Roma Tre University, Via della Vasca Navale, 79, Rome, Italy
e-mail: sciarro@ing.uniroma3.it

C. De Medio
e-mail: demedio@ing.uniroma3.it

F. Gasparetti
e-mail: gaspare@ing.uniroma3.it

C. Limongelli
Department of Engineering, Roma Tre University, Via della Vasca Navale, 79, Rome, Italy
e-mail: limongel@ing.uniroma3.it

A. Visvizi et al. (eds.), *Research and Innovation Forum 2020*,
Springer Proceedings in Complexity,
https://doi.org/10.1007/978-3-030-62066-0_22

Introduction

In recent years, distance learning has made significant progress thanks to the ever-growing availability of meaningful information material, in the form of HTML pages or other formats, including multimedia ones. Moreover, in the last time, due to the pandemic caused by COVID-19, the use of distance learning and web platforms has become of fundamental importance to guarantee the continuity of teaching for students of all ages in the context of smart cities as well [12, 14]. Many Web platforms have been implemented, both to provide distance courses and to offer a service for collecting and filtering learning materials on the Web. For instance, *Merlot* (http://www.merlot.org) is a web educational repository that makes learning objects freely available to teachers and in general to the educational community. *Coursera* (www.coursera.org) is the most used Web educational platform which delivers courses to hundreds of thousands of learners, acting as a Massive Open Online Course platform. Consequently, many teachers and instructional designers enthusiastically are more and more embracing the multiple educational opportunities offered by the Web 2.0. Environments, such as blogs and wikis, have rapidly increased their popularity because of their powerful opportunities to use new educational paradigms such as *Social Constructivism* [15], *Social* and *Collaborative Learning* [1], *Active Learning* [11] and *Communities of Practice* [16]. Wikis are online environments suitable for collaborative projects, having an intrinsically social and collaborative nature. Today many teachers have adopted Wikis, transforming them into a sort of *agora* of their courses since they promote a more spontaneous and flexible interaction between teachers, learners and didactic environments. Wikipedia is the most famous and used online Wiki-encyclopedia with free, collaborative, multilingual and free content, born in 2001, supported and hosted by the *Wikimedia Foundation*, a no-profit organization. Many teachers and students use this resource as a source of learning material to be used in their courses, to teach or to learn, also because their content is not bound by strong constraints on copyright rules being of *Creative Commons* (www.creativecommons.org) type. Moreover, Wikipedia is a large set of HTML pages, with conceptual connections among them, represented by links, easily to be managed by suitable API. *Wiki Course Builder* (WCB), is a Web-platform developed at our department, working with Wikipedia pages and developed to help teachers to build on-the-fly online courses [3, 8]. In this paper, we show a new particular module of the system, the *Community* module, which allows teachers building a new course. Teachers can take advantage of courses already developed by colleagues, by using resources, i.e., Wikipedia pages, already in use, thereby saving time. All this through a suitable graphical interface. Furthermore, each teacher using the system, is modelled through her *Teaching Styles* according to the Grasha model [6]. The interface allows to select teaching materials belonging to other courses through a set of colours that express similar teachers, that is teachers having almost the same teaching styles. Here we present a case study showing the Community process, with the aim to verify, through a sample of teachers, the feasibility of such an approach.

The remainder of the article is organized as follows. Section "Related Work" illustrates some important related works. Section "The WCB System" shows the main features of the WCB system, focussing on its *Community* feature. Section "A First Evaluation" shows an experimentation based on a use case while the conclusions reported in Section "Conclusions and Future Work" close the article.

Related Work

Distance learning literature mainly focuses on student learning, so most of the proposed systems are student-centred. Collaborative e-learning environments often intend the role of teachers as a support and a coordination among groups of students groups during their learning activities [9] and during the peer review phase [2, 13]. WCB is a system that has been developed to help teachers to quickly build a new course composed by Wikipedia pages only. Here we present some comparisons with similar systems where the collaborative aspect in the construction of a new course is present.

MoodleREC [4] is a recommending system that helps teachers configure new courses in the Moodle Learning Management System (www.moodle.org), based on courses already built by the community of teachers using the same platform. This system uses some standard Learning Object Repositories, like Merlot and Ariadne, to retrieve didactic materials. Moreover, it uses a social approach to course building by suggesting a learning object for a new course taking into account other courses where it was already used, in the Amazon recommendations style. Differently, WCB uses teaching styles to recommend a new Wikipedia page, using a graph-oriented environment, thus making the recommendation process simpler.

In [5], a machine learning approach for the identification of relationships between text-based resources is proposed, to build new learning paths on Wikipedia pages. A feature selection methodology allows to consider the most relevant attributes to the predictive modeling problem for the set of topics under consideration. These features are extracted from both the input material and weak-taxonomies available on the web. Natural language annotation processes the input data in such a way that patterns and inferences of interest are more easily found by an automated analysis. Finally, the prerequisite identification is input to a binary statistical classification task. This approach does not take into account the teaching styles of each teacher and learning paths are built based on some machine learning algorithms. Our system presents a personalization aspect in the selection of learning material, thus making the course building process more adapted to the teacher.

In [10], the authors present a method for automatic generation of learning paths for education or self-education. As the knowledge base, their system uses the semantic structure of Wikipedia, leveraging on its broad variety of covered concepts. The system experimental evaluation showed its usefulness. Our system is a complete system that allows to build and manage a complete course, other than suggesting learning paths.

In [4, 7] the authors present a system that gives some support to the operations of retrieving, analyzing, and importing LOs from a set of standard Learning Objects Repositories, acting as a recommending system. In particular, it is designed to support the teacher in the phases of (i) retrieval of LOs, through a keyword-based search mechanism applied to the selected repositories; (ii) analysis of the returned LOs, whose information is enriched by a concept relevance metric, based on both the results of the searching operation and the data related to the previous use of the LOs in the courses managed by the Learning Management System; and (iii) LO importation into the course under construction.

The WCB System

In this section, we briefly present the WCB system, focusing on the *Community* module. For a deep insight into the system, the reader can refer to [3].

The General Architecture of the System

The general architecture of the system, developed in java and based on the *mongo* no-sql database, is shown in Fig. 22.1.

The system is composed of several functional modules:

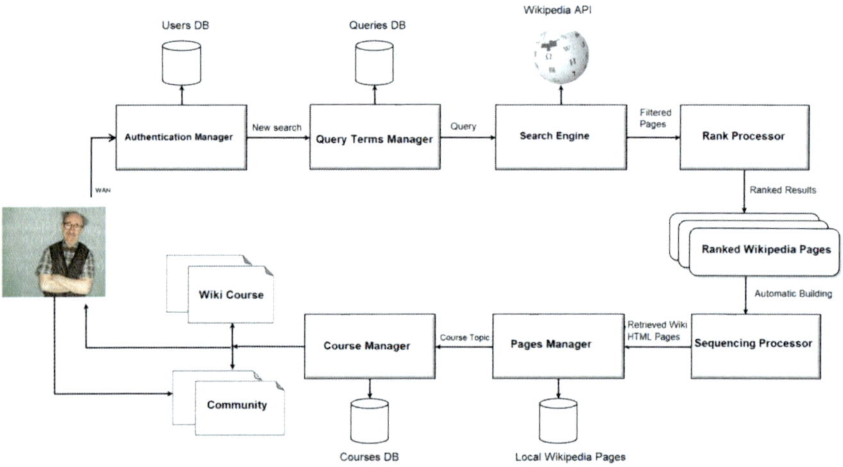

Fig. 22.1 The WCB architecture

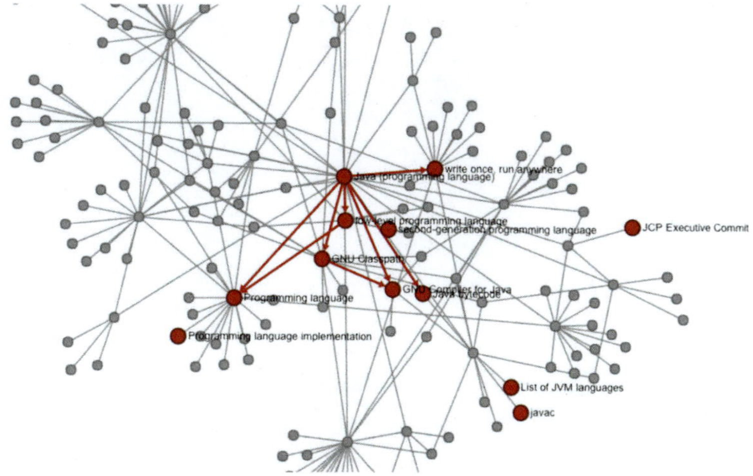

Fig. 22.2 A portion of Wikipedia graph after a query

- The Authentication Manager. The login manager manages the access to the platform, allowing users to sign in to the system, profiling each teacher's learning styles, using the Grasha and Riechmann test;
- The Search Engine. The search engine retrieves the links to probably relevant Wikipedia pages, taking into account different evaluation metrics, among which the one based on the teacher's teaching styles is the most important;
- The Recommendation Processor. The recommendations processor deals with the retrieving and the ranking of the Wikipedia pages. Starting from the page selected by the user during the search phase, this module extracts a set of Wikipedia pages. The pages are selected by exploring and analyzing a portion of the Wikipedia semantic graph, that is the graph built starting from a Wikipedia page and taking all the linked pages. Each node represents a page together with the concept associated with it. The arcs are the links between the pages. The user can view the sub-graph composed by the retrieved pages, as shown in Fig. 22.2 where, after a query, the teacher can analyze the learning path generated by the sequencing of the retrieved relevant pages. Figure 22.2 shows the idea of the graph that can be explored by the teacher instead of navigating the Wikipedia pages directly on the web.

The Community Module

The *Community* module is the relevant feature focus of this work. When launched, this module presents all the courses, where each Wikipedia page is tagged with five colours, each representing the mean of all the teaching styles of the teachers who used it. So, a teacher can explore the graph searching for new pages adapted to her

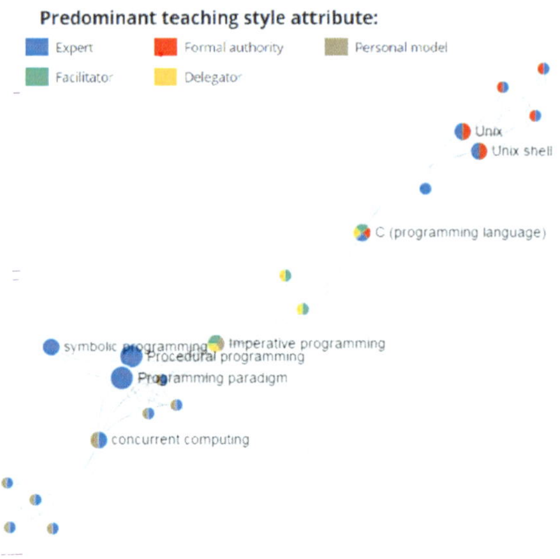

Fig. 22.3 An example of the Community graph with their associated teaching styles

way of teaching other than to her course. An example of a community graph is given in Fig. 22.3 and shows all the courses currently in the system (see the use case in the next section), coloured according to the values of the teacher's teaching styles. The graph can be filtered by topics allowing teachers to use this social tool to enrich their courses with didactic materials belonging to other courses and with similar teaching styles, improving the sharing and reuse of Wiki pages.

A teacher t_i is characterized by a set of five real numbers $TS_i = [ts_i^1, ts_i^2, \ldots, ts_i^5]$, being $ts_i^k \in [1, \ldots, 5]$ indicating her level in that particular category of the teaching styles set, that is ts^1: Formal Authority, ts^2: Expert, ts^3: Personal Model, ts^4: Facilitator, ts^5: Delegator.

Each wiki page used by n teachers in their courses has an associated TS given by the mean of all the TS_i of the teachers that have chosen that page for their courses. So, the algorithm tags each page with the TS of the teachers that used it. The rationale behind this choice is that it is possible to see which kind of teachers (on average) have chosen a given page. In this way, implicitly, we specify that a given page is more appropriate for a teacher with a set of given teaching styles, because (on average) it has been already chosen by teachers with similar teaching styles. In other words, a generic page wp_j, has associated a TS set, after n choices of the same page, that is computed as follows:

$$ts^k(wp_j) = \frac{\sum_{i=1}^{n} ts_{ji}^k}{n} \qquad (22.1)$$

where $ts^k(wp_j)$, with $k = 1...5$ is the value of the k-th teaching style associated to the wp_j page. The Equation 22.1, expresses the updating page rule.

A First Evaluation

In this section, we present a first experimentation of the system, with the aim of verifying our research question.

The Sample

We experimented the *Community* environment by means of a sample of 10 teachers who produced the learning materials shown in Table 22.1. As summarized in the Table, were used 192 Wikipedia pages, created 72 topics and 21 courses on the

Table 22.1 The learning material produced by the sample

Course ID	Title	# Courses	# Pages
1	Computer History	4	12
2	C Language	3	11
3	Artificial Intelligence	5	16
4	The Microprocessor	4	12
5	Ricursion	4	12
6	The Microprocessor	3	9
7	The router	3	9
8	Internet of Things	4	10
9	Files management in C language	3	10
10	Pointers in C Language	4	12
11	Binary arithmetic	3	9
12	Binary Representation	3	10
13	Astract Data Type	4	13
14	Sorting	3	9
15	Binary Trees	3	9
16	Prime Numbers	3	9
17	Image processing	3	9
18	Deep Learning	3	9
19	Object Oriented Programming	3	9
20	The Assembly Language	4	12
21	The Computer Hardware	3	9
		72	190

Table 22.2 The Grasha teachers' model of the sample

Statistics	Delegator	Expert	Facilitator	Formal authority	Personal model
$\bar{x}$	2.9875	3.6625	3.2875	3.7875	3.45
$\bar{s}$	0.487674	0.409896	0.316146	0.586979	0.195138889

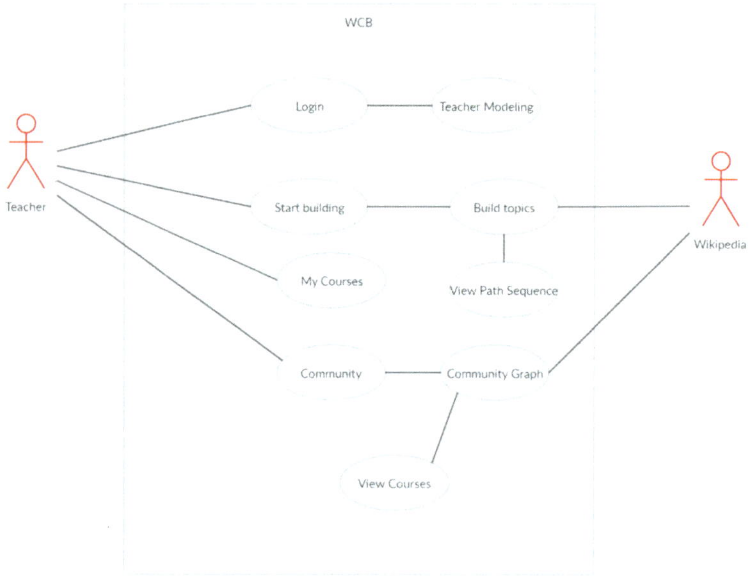

Fig. 22.4 The use case

Computer Science domain. Each course created by the sample was composed by at least 3 topics and each topic by at least 3 Wikipedia pages. All the invited teachers filled in the Grasha questionnaire with the results, shown in Table 22.2 where the values of the standard deviations are much lower than the averages, thus indicating a fairly homogeneous sample.

All the teachers accomplished the requested task to build new courses. After that, each of them used the *Community* feature in order to verify the added value of the proposed social approach. In the following, we show a use case of a user who used the social instrument to improve her course.

A Use Case

Figure 22.4 shows the use case that expresses the user-system interactions in the classic UML use case diagram.

The results of your teaching style survey are as follows:

2.875	3.375	3.375	3.25	2.875
expert	formalauthority	personalmodel	facilitator	delegator
Moderate	High	Moderate	Moderate	High

Fig. 22.5 The teaching styles associated to user 1 by the Grasha-Riechmann Questionnaire

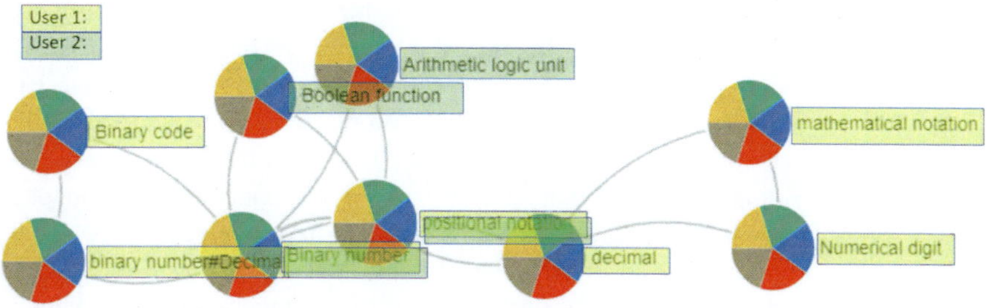

Fig. 22.6 A snapshot of the community environment

Once a teacher (user1) signs in into the system, she has to fill the Grasha Questionnaire. It is composed by 40 questions articulated according to statements, such as: —"My objectives and teaching methods address a variety of learning styles of the students", or —"Take the time to consult the students on how to improve their work in individual projects or group". In Fig. 22.5 the teaching styles of the $user_1$ are shown.

Building a new course Let us assume that $user_1$ is the user of the sample who wanted to prepare a module about *Binary Arithmetic* concerning the general topic *Computer Representation*. She just digits the appropriate keywords and selects four wiki pages: binary code; binary number#digit; binary number and positional notation.

The Community Module Going into the community she finds the situation given in Fig. 22.6.

Namely, $user_1$ observed that someone else in the community has already used the wiki pages "Binary number" and "positional notation", linking these pages to other pages such as "Boolean functions" and "Arithmetic Logic Unit". Since the topic $user_1$ is going to create is oriented towards computer arithmetic, she included in her module these new two pages.

This experimentation is currently in progress because the sample is still using the system. To date, other 5 teachers have used the Community feature to enrich their courses with other pages taken from other courses. This is a good result that encourages us to continue the experimentation.

Conclusions and Future Work

In this article, we introduced a new feature of the WCB system, a system that helps teachers building new courses based on Wikipedia pages. The new feature is based on the possibility of using a social and collaborative environment, typical of communities of practice. Through a graphical user interface, teachers can filter all the other courses built by other teachers through the most appropriate teaching styles associated to Wikipedia pages. In this way they can select teaching materials, i.e., Wikipedia pages, provided by other teachers with similar teaching characteristics. This work presents a first experimentation carried out with 10 teachers, who have expanded their courses with materials semantically related to their teaching goals. The limitation of this approach is that teachers can only navigate on Wikipedia pages. On the other hand, this approach allows a rapid creation of courses and contributes to a first rapid approximation of what will be the final course. As a future development, we are going to allow to integrate didactic resources taken elsewhere on the Web into the courses. A further and more stimulating development is to extend the Teacher Model. It could be enriched with more information, for example, with the "didactic history" of the teachers represented by their courses. This will allow a fine-grained exploration of the community. Teachers could be classified on the basis of their teaching styles and on the basis of the topics taught.

References

1. A. Bandura, *Social Learning Theory* (Prentice Hall, Englewood Cliffs, NJ, 1977)
2. M. De Marsico, F. Sciarrone, A. Sterbini, M. Temperini, Supporting mediated peer-evaluation to grade answers to open-ended questions. Eurasia J. Math. Sci. Technol. Educ. **13**(4), 1085–1106 (2017)
3. C. De Medio, F. Gasparetti, C. Limongelli, F. Sciarrone, Automatic extraction and sequencing of wikipedia pages for smart course building, in *Proceedings of the 21st International Conference Information Visualisation (IV)*, pp. 378–383 (2017)
4. C. De Medio, C. Limongelli, F. Sciarrone, M. Temperini, Moodlerec: A recommendation system for creating courses using the moodle e-learning platform. Comput. Hum. Behav. **104** (2020)
5. F. Gasparetti, C. De Medio, C. Limongelli, F. Sciarrone, M. Temperini, Prerequisites between learning objects: Automatic extraction based on a machine learning approach. Telematics Inform. **35**(3), 595–610 (2018)
6. A.F. Grasha, A matter of style: The teacher as expert, formal authority, personal model, facilitator, and delegator. College Teaching **42**(4), 142–149 (1994)
7. C. Limongelli, F. Sciarrone, M. Temperini, A social network-based teacher model to support course construction. Comput. Human Behavior **51**(PartB), 1077–1085 (2015)
8. C. Limongelli, G. Mosiello, S. Panzieri, F. Sciarrone, Virtual industrial training: Joining innovative interfaces with plant modeling, in *The 11th International Conference on Information Technology Based Higher Education and Training - ITHET 2012* (IEEE, New York, 2012), pp. 1–6
9. N. Maqtary, A. Mohsen, K. Bechkoum, Group formation techniques in computer-supported collaborative learning: A systematic literature review. Technology, Knowledge and Learning (2017)

10. C. Perez-Martinez, G. López-Morteo, M. Reyes, A. Gelbukh, Wikipedia-based learning path generation. Computacion y Sistemas **19**, 589–600 (2015). https://doi.org/10.13053/CyS-19-3-2079
11. J. Piaget, *Science of Education and the Psychology of the Child* (Longman, London, 1971)
12. F. Sciarrone, M. Temperini, K-openanswer: A simulation environment to analyze the dynamics of massive open online courses in smart cities. Soft Comput. (2020). (In Press)
13. A. Sterbini, M. Temperini, Analysis of open answers via mediated peer-assessment, in *Proceedings of the 17th International Conference on System Theory, Control and Computing*, pp. 663–668 (2013)
14. A. Visvizi, L. Daniela, Technology-enhanced learning and the pursuit of sustainability. Sustainability (Switzerland) **11**(15) (2019). https://www.scopus.com/inward/record.uri?eid=2-s2.0-85070416969&doi=10.3390
15. L.S. Vygotsky, *Thought and Language* (MIT Press, Cambridge, MA, 1962)
16. E. Wenger, Communities of practice: A brief introduction (2011)

Chapter 23
Think-Aloud Exploratory Search: Understanding Search Behaviors and Knowledge Flows

Marcelo Tibau, Sean W. M. Siqueira, and Bernardo Pereira Nunes

Abstract This paper describes an experiment that uses Concurrent Think-Aloud protocol (CTA) and person-to-person interviews to map searching behaviors and knowledge flows during search sessions. The findings are: (1) the most used searching strategy during exploratory searches was the "Metacognitive Domain"; and (2) online searching experts have a fair ability to deal with ideas prompted by browsing the search results. The main contributions of this research lie in the understanding of the process in which people find, access, decide what content is useful and apply online data to their different information needs.

Keywords Information searching behavior · Data discovery · Innovation from data · Innovation as data

Introduction

The richness of search engines goes beyond their ability to retrieve data from the Web, forming knowledge from users' search paths. Search paths are composed of queries, information needs and users' intents, and the set of search paths can be defined as a knowledge flow. Knowledge flows can provide hints regarding how search engines are affected by the experience of users and vice-versa.

The current paper describes an experiment that uses Concurrent Think-Aloud protocol (CTA) and person-to-person interviews to map knowledge flows during search sessions. CTA is a research methodology that requires one or more participants

M. Tibau (✉) · S. W. M. Siqueira
Federal University of the State of Rio de Janeiro, Rio de Janeiro, Brazil
e-mail: marcelo.tibau@uniriotec.br

S. W. M. Siqueira
e-mail: sean@uniriotec.br

B. P. Nunes
Australian National University, Canberra, ACT, Australia
e-mail: Bernardo.Nunes@anu.edu.au

A. Visvizi et al. (eds.), *Research and Innovation Forum 2020*,
Springer Proceedings in Complexity,
https://doi.org/10.1007/978-3-030-62066-0_23

to speak aloud her/his thoughts as s/he performs a task or activity [3]. This paper advances the understanding of the search process, focusing on the impact of the user's search expertise to identify useful sources based on her/his information need. A query reformulation taxonomy was applied to this study, in addition to an online information searching strategy framework, to explain users' query reformulation process and search behaviors during exploratory searches. Two Research Questions were the focus of this study. **RQ1:** Which information searching strategies are favored by searching expert users? **RQ2:** How good is the search ability of expert users in dealing with unknown subjects?

This paper is organized as follows: Section "Related Work" reviews closely related literature. Section "Method for Data Collection and Analysis" introduces the method for data collection and analysis, as well as describes the experiment. Section "Results" shows the results. Finally, Section "Discussion and Final Remarks" presents some findings and future works.

Related Work

Query strategies were studied by [7] to address the impact of domain expertise on searching performance and query formulation and reformulation. The results of their work showed that participants with prior knowledge in the domain are more efficient than others and that experts and non-experts used different query strategies. The relationship of cognitive styles and Web searching was investigated by [2], who determined there were differences in Web searching strategies between those who think in mental pictures or images (deemed as *imagers*) and those whose mental process favored words or verbal associations (*verbalizers*). Reference [8] applied an online information searching strategy framework to examine how information searching behaviors, task completion time and rates differ depending on information searching experience and cognitive style. They found out that experience (e.g., users with expertise in online information searching) is more influential on task completion time and task completion rate compared to cognitive styles.

The study presented in this paper extended previous approaches by applying a query reformulation taxonomy in addition to an online information searching strategy framework to provide a method to understand users' search behavior. The query reformulation taxonomy was used to verify the impact of both behaviors (query reformulation and searching strategies) on Online Information Searching Behaviors (OISB), and the way users perceive, remember, and assess the retrieved information. It was also elaborated on OISB of users to try to explain, for instance, why some use advanced search forms and browsing tools.

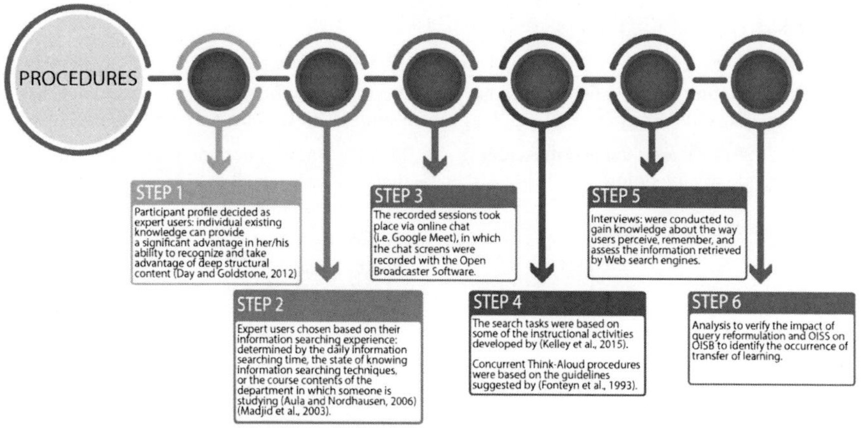

Fig. 23.1 Research workflow

Method for Data Collection and Analysis

This section is composed of three parts. The first part describes the procedures defined for the Think-Aloud Protocol. The second part presents the three search session tasks proposed to the four selected users. Finally, the third part shows the online information searching strategies framework used as a basis to analyze the recorded videos displaying the users implementing each task. Figure 23.1 shows the research steps in a procedural way.

Concurrent Think-Aloud Procedures. Four users were selected from a group of Master and PhD. Computer Science students. The selection was based on the degree specialization of the users, all users hold a higher degree in Information Retrieval. This study requires users with this particular profile to perform three search tasks about subjects not within their knowledge domain. The search tasks were planned with different difficulty levels to provide situations in which the users could apply their skills while performing exploratory searches. They were selected from a Research Group of 22 participants, based on their information searching experience. Information searching experience can be determined by the daily searching time, the state of knowing information searching techniques, and the course contents of the department in which someone is studying [5] [1]. In recorded individual sessions, the users received a sequence of tasks to be performed in three different search sessions. The protocol sessions took place via online chat (Google Meet), in which the screen shared by the users were recorded with the Open Broadcaster Software. Besides recording users' screens, all users were asked to speak aloud her/his thoughts and to explain the search approach used.

The Sequence of Search Sessions. The three search tasks were based on some of the instructional activities developed by [3] and are summarized in Table 23.1. The first two aim to encourage a search with learning intent, and the third one intends to

Table 23.1 Overview of the proposed search tasks

Search session	Task description
Search 1 (20 min): Introduction to weight, mass, volume and density	Describe weight, mass, volume and density. Compare and contrast weight and mass
Search 2 (25 min): Skeletal and muscular systems and movable joints	Identify major bones and muscles in the human body. Identify and locate examples of movable joints in the human body
Search 3 (15 min): Prosthetic limb	Explain the features of movable joints to prosthetic limb

propose a challenging activity that demanded not only a search task but the use of the information retrieved from the previous two searches. The questions used in the interviews were:

- How did you choose the terms used, the search engine, and the most useful retrieved search results?
- How did you evaluate whether the retrieved search results were useful and consistent with your needs?
- What led you to decide to redo the search (or to proceed with a new search)?
- How did you plan your response to the proposed tasks?
- How do you evaluate what you have learned from the searches?

Data Analysis. The domains and strategies undertaken during online searching processes used to identify the searching behaviors were based on [10] framework, showed in Table 23.2. Indicators defined for each strategy are thoroughly described at [8]. A query reformulation taxonomy (ESKiP Taxonomy of Query State, [9]) was used to identify the Query States[1] (see Table 23.3). This taxonomy allows the identification and classification of the presence of different Search Profiles.[2]

Results

The Query States were determined based on the terms used to formulate the initial query on the user's Web search engine of choice—the users were asked to open a Web search engine to initiate the tasks—and the subsequent query reformulations were performed either on the search engine, on a website or a platform embedded searching system. In addition to the terms, all the sequences of actions and spoken descriptions were recorded. The online information searching strategies (OISS) were the same presented by [8] and already shown in Table 23.2.

[1] Set of terms used to reformulate queries within a session, determined by a user's decision-making process regarding whether to include, exclude, modify or keep the set of terms.

[2] A set of implemented user search strategies, which determine a Web search pattern.

Table 23.2 Online information searching strategies [8]

Domain	Description	Strategies
Behavioral Domain	Concerned with basic skills required for manipulating and searching the Web.	– **Control:** skills required for manipulating Web searching applications – **Disorientation:** learner's self-awareness about their searching orientation
Procedural Domain	Concerned with content-general searching approaches and overcoming problems that occur during the searching process	– **Trial and error:** skills in trying different searching approaches – **Problem-solving:** skills and commitment to overcome problems or frustrations resulting from searching
Metacognitive Domain	Concerned with monitoring the searching process, identifying key information, as well as interpreting and evaluating the information retrieved	– **Purposeful thinking:** skills required to self-monitoring the searching process – **Selection of the main ideas:** skills to identify key information concepts from the retrieved batch – **Evaluation:** skills to judge and organize the retrieved information

Table 23.3 ESKiP taxonomy of query state [9]

Query state	Definition
Initial State (IS)	Qi contains at least one term and represents the start of a search
Return State (RS)	Qi contains at least one term and represents the start of a search or a previous search query; Qi+n contains exactly the same term of Qi
Generalization (GE)	Qi and Qi+1 have at least one term in common; Qi+1 contains fewer terms than Qi
Specialization (SC)	Qi and Qi+1 have at least one term in common; Qi+1 contains more terms than Qi
Repeat (RP)	Qi and Qi+1 may contain exactly the same terms, but the format of these terms may be different
Word Substitution (WS)	Qi and Qi+1 have at least one term in common; Qi+1 might have the same length of Qi, but contains some terms that are not in Qi
New (NW)	Qi and Qi+1 do not have any common terms
Related (RE)	Qi and Qi+1 do not have any common terms, but have a similar or related meaning

The option to use the Query State **Specialization (SC)** is connected to the desire to restrict the searched domain (see definition in Table 23.3), but it was found out that it could be achieved by using different OISS. During Task 1, **USER1** employed a Control strategy (C7—using Boolean logic operators for narrowing/widening the search parameters) (Table 23.4), while **USER4** preferred a Purposeful Thinking strategy (PT1—narrowing down the searching field) (Table 23.5). The terms used seem to corroborate a difference in profile, as the first used a logic operator (i.e., "what is density of materials AND physics"), and the later opted for specifying a particular website to "force" a particular result (i.e., "weight, mass, volume and density Wikipedia"), also adding a new strategy, Selection of the main ideas (SMI1—directly opening a website that is known to be relevant to a given search task), although maintaining the same OISS Domain (Metacognitive).

Usually, the Query State **New (NW)** is used to start a new search within the same search session [4, 9]. Unexpectedly, the Query State **Word Substitution (WS)** was used with the same intent. It could be a change in only one word (e.g., **USER1** in Task 1 in Table 23.4), the use of fewer words (e.g., **USER2** in Task 2 in Table 23.6), or even the inclusion of more words to the query or modifying part of the query. In most of these situations, users seem to prefer a systematic approach, using the Procedural Domain through the OISS Trial and error (TE1—modifying the keywords), and also performing several Metacognitive strategies concomitant.

Also, the Query State **Word Substitution (WS)** can be used as a means to reformulate a query from keywords to more direct questioning. It is a way to take advantage of commercial search engines' focus (e.g., Google, Bing, Yandex, and Baidu) on developing support for lookup searches[3] [6]. The use of search terms as a question (e.g., **USER1** in Task 1 in Table 23.4) is accompanied by the OISS Selection of the main ideas (Metacognitive domain), specially SMI2 (typing specific terms about the search task) to maintain the searching thread and SMI6 (looking through the titles on a website) to identify key information. Changing one or two words from the question used as a search term, the user can obtain the effect described in the last paragraph (start a new search).

In a previous study analyzing query reformulation behaviors from transaction logs of an online student-teacher professional community and Yahoo! search engine, [9] found a predilection for repeated terms in consecutive queries at the online community. The authors indicated that "users may want to revisit information or even find or check for new information on topics they have previously explored". Though, it was found out that the **Query State Repeat (RP)** is used as a cross-platform strategy. It means that this reformulation type can be used to apply the same query to commercial search engines and then to in-site search engines (i.e., YouTube's embedded search system) within the same search session (e.g., **USER2** in Task 2 in Table 23.6). The situation (the repeated query) is preceded by the OISS Selection of the main ideas (SMI5—looking through the hyperlinks provided on a website) and prompted by

[3]Lookup searches return discrete and well-structured objects such as numbers, names, short statements, or specific files of text or other media.

the OISS Purposeful Thinking (PT4—doing an in-site search). Note once again the prominence of the Metacognitive Domain.

Discussion and Final Remarks

Although the topics proposed to each task varied in difficulty (from easier to harder), they did not fall within the users' knowledge domain expertise (i.e., computer science). Thus, as the users were experts in online searching, but not in the search topics, it is supposed that the users would apply their past experiences and try different searching approaches until they find the approach that fits best to the tasks at hand. Therefore, there was an expectation of observing searching strategy behaviors more connected to the OISS Procedural Domain (see Table 23.2) as the answer to **RQ1** (which information searching strategies are favored by searching expert users?). Unlike the expectations, the most used strategy was the OISS Metacognitive Domain (67.82%, used 177 times), the second most used was the Behavioral strategy (19.54%, used 51 times) followed by the Procedural (12.64%, 33 times). Figure 23.2 shows the results.

As for the answer of **RQ2** (how good is the search ability of expert users in dealing with unknown subjects?), the results indicate that online searching experts have a fair ability to deal with ideas prompted by browsing the searching results. This abstraction skill aids their assessment and decision-making procedures regardless of the difficulty of the given exploratory search task. Perhaps the Metacognitive domain prominence might be related to the types of the search executed and the information

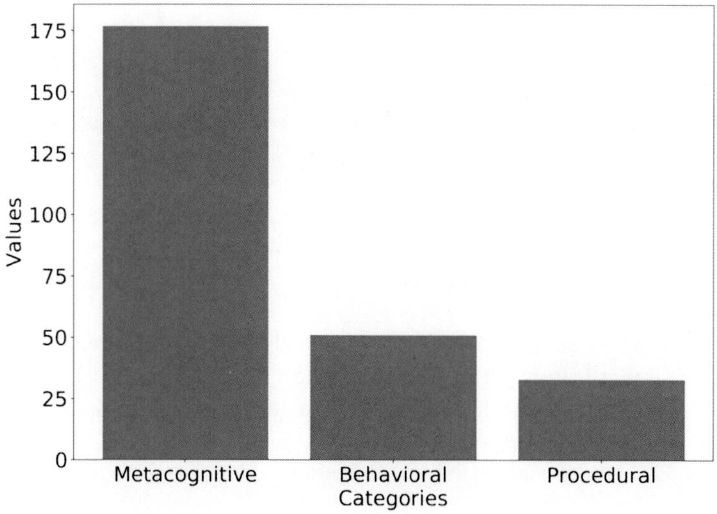

Fig. 23.2 OISS domain used

search task characteristics (e.g., fact-finding, ill or well-structured) or type (e.g., exploratory search) rather than the search task difficulty level or complexity. In this case, this study extends the findings of [8].

The number of users considered in the study can be seen as a limitation of this work. However, it is important to notice the complexity of dealing with the subjective nature of the study, and the number of users was sufficient to achieve the sample saturation.

As future work, it is intended to understand how users evaluated what they learned from the search. Then, the transfer of learning will be included as a means for assessing learning in future analysis. It is also intended to broaden the users' profile to include novices on Web searching and users with different knowledge domain backgrounds.

Innovation from searching could also be investigated within a broader innovation ecosystem. Thus, it is intended to discuss the user's learning through Web searching from the Metacognitive Domain's perspective. It means that when mapping users' behavior while searching the Web with learning intent, known as Searching as Learning (SaL),[4] it is envisioned to measure the Metacognitive strategies as an independent variable, and useful links as a dependent variable (e.g., number of links clicked that were useful) to check the relationship between their use and successful search outcomes.

List of Online Information Searching Strategies' Indicators

Behavioral (Behav)

- **Control**
  - **C1**: Using the most familiar or known search engine in the first place.
  - **C2**: Searching by typing the name of the search engine on the browser.
  - **C3**: Entering the name of the website on the search engine.
  - **C4**: Entering the name of the website on the address bar.
  - **C5**: Using the "home" button to return to the beginning of the search.
  - **C6**: Using the "next" and "previous" buttons of the browser.
  - **C7**: Using Boolean logic operators for narrowing/widening the search parameters.
  - **C8**: Doing a customized search with the help of the images, videos, maps, and other similar features of the search engine.
  - **C9**: Utilizing the advanced search options of images, videos, maps, and other similar features of the search engine.
  - **C10**: Utilizing the advanced search options of the search engine.

[4]Consists of learners using Web search engines as a technology to drive their learning process.

- **Disorientation**
  - **D1**: Giving up in the case of failure to find an answer.
  - **D2**: Using search terms that are not given in the search task.
  - **D3**: Not having any idea about what to do when doing an Internet search.
  - **D4**: Feeling bad in the case of failure to retrieve the desired information.

Procedural (Proced)

- **Trial and Error**
  - **TE1**: Modifying the keywords.
  - **TE2**: Using different search engines.
  - **TE3**: Opening different websites.
- **Problem-Solving**
  - **PS1**: Doing one's best to resolve any problem that occurs during a search.
  - **PS2**: Trying to find out the possible reasons for any problem that occurs during a search.

Metacognitive (Metacog)

- **Purposeful Thinking**
  - **PT1**: Narrowing down the searching field (subject).
  - **PT2**: Accessing additional websites from a main website.
  - **PT3**: Simultaneous information searching from different sources.
  - **PT4**: Doing in-site search.
- **Select Main Ideas**
  - **SMI1**: Directly opening a website that is known to be relevant to a given search task.
  - **SMI2**: Typing specific terms about the search task.
  - **SMI3**: Following the search suggestions of the search engine.
  - **SMI4**: Following the outputs of in-site search.
  - **SMI5**: Looking through the hyperlinks provided on a website.
  - **SMI6**: Looking through the titles on a website.
  - **SMI7**: Keeping an account of the information relevant to the search task found on a website.
  - **SMI8**: Looking for specific words in a website by means of Ctrl + F.

- **Evaluation**
  - **E1**: Evaluating the relationships between the retrieved pieces of information.
  - **E2**: Comparing and contrasting the pieces of information retrieved from different websites.
  - **E3**: Determining whether a piece of information from a website is worth referencing.
  - **E4**: Assessing how to combine and present the data gathered from the Web.

Users' Searching Behavior Tables

See Tables 23.4, 23.5 and 23.6.

Table 23.4 User 1 searching behaviors in Task 1

Participant	Terms used	Query states	Obs.	OISS	Code	OISS Domain
USER1	Mass weight volume density	Initial State (IS)		Control	**C2**	Behav
				Selection of the main ideas	**SMI2**	Metacog
				Selection of the main ideas	**SMI7**	Metacog
	What is density	Word Substitution (WS)	Changed to a question	Trial and error	**TE1**	Proced
				Purposeful Thinking	**PT1**	Metacog
	What is density of materials AND physics	Specializa-tion (SC)		Control	**C7**	Behav
				Selection of the main ideas	**SMI7**	Metacog
				Evaluation	**E1**	Metacog
	What is volume of materials AND physics	Word Substitution (WS)	Changed one word, but this word changed the search's subject	Control	**C7**	Behav
				Selection of the main ideas	**SMI2**	Metacog
				Selection of the main ideas	**SMI7**	Metacog
				Evaluation	**E3**	Metacog

Table 23.5 User 4 searching behaviors in Task 1

Participant	Terms used	Query States	Obs.	OISS	Code	OISS Domain
USER4	Weight, mass, volume and density	Initial State (IS)		Control	**C1**	Behav
				Selection of the main ideas	**SMI2**	Metacog
	Weight, mass, volume and density wikipedia	Specialization (SC)	Specified to get a particular result	Purposeful Thinking	**PT1**	Metacog
				Trial and error	**TE3**	Proced
				Selection of the main ideas	**SMI1**	Metacog
				Evaluation	**E2**	Metacog
				Evaluation	**E3**	Metacog

Table 23.6 User 2 searching behaviors in Task 2

Participant	Terms used	Query states	Obs.	OISS	Code	OISS Domain
USER2	Osseous and muscular system	Initial State (IS)		Control	**C1**	Behav
				Selection of the main ideas	**SMI2**	Metacog
				Purposeful Thinking	**PT1**	Metacog
				Trial and error	**TE3**	Proced
				Evaluation	**E2**	Metacog
				Evaluation	**E3**	Metacog
				Control	**C6**	Behav
				Control	**C8**	Behav
				Evaluation	**E3**	Metacog

(continued)

Table 23.6 (continued)

Participant	Terms used	Query states	Obs.	OISS	Code	OISS Domain
	Particular system	Word Substitution (WS)	Used less words, but it changed the search's subject	Trial and error	**TE1**	Proced
				Selection of the main ideas	**SMI2**	Metacog
				Selection of the main ideas	**SMI5**	Metacog
	Particular system	Repeat (RP)		Purposeful Thinking	**PT4**	Metacog
				Selection of the main ideas	**SMI4**	Metacog
				Selection of the main ideas	**SMI5**	Metacog
				Selection of the main ideas	**SMI6**	Metacog
				Evaluation	**E1**	Metacog
				Evaluation	**E3**	Metacog

References

1. A. Aula, K. Nordhausen, Modeling successful performance in web searching. J. Am. Soc. Inform. Sci. Technol. **57**(12), 1678–1693 (2006). https://doi.org/10.1002/asi.20340
2. N. Hariri, M. Asadi, Y. Mansourian, The impact of users' verbal/imagery cognitive styles on their Web search behavior. Aslib J. Inf. Manage. **66**, 401–423 (2014). https://doi.org/10.1108/AJIM-02-2013-0019
3. T. Kelley, B. Capobianco, K. Kaluf, Concurrent think-aloud protocols to assess elementary design students. Int. J. Technolo. Des. Educ **25**(4), 521–540 (2015)
4. C. Liu, J. Gwizdka, J. Liu, T. Xu, N.J. Belkin, Analysis and evaluation of query reformulations in different task types. Proc. Am. Soc. Inf. Sci. Technol. **47**(1), 1–9 (2010). https://doi.org/10.1002/meet.14504701214
5. I. Madjid, C. Stéphane, M. Daniel, The effect of individual differences on searching the web. Proc. Am. Soc. Inf. Sci. Technol **40**(1), 240–246 (2003). https://doi.org/10.1002/meet.1450400130
6. G. Marchionini, Exploratory search: From finding to understanding. Commun. ACM **49**(4), 41–46 (2006). https://doi.org/10.1145/1121949.1121979
7. S. Monchaux, F. Amadieu, A. Chevalier, C. Mariné, Query strategies during information searching: Effects of prior domain knowledge and complexity of the information problems to be

solved. Inf. Process. Manage. **51**(5), 557–569 (2015). https://doi.org/10.1016/j.ipm.2015.05.004

8. I. Reisoğlu, A. Çebi, T. Bahçekapılı, Online information searching behaviours: examining the impact of task complexity, information searching experience, and cognitive style. Interactive Learning Environments, pp. 1–18 (2019). https://doi.org/10.1080/10494820.2019.1662456
9. M. Tibau, S.W.M. Siqueira, B. Pereira Nunes, T. Nurmikko-Fuller, R.F. Manrique, Using query reformulation to compare learning behaviors in web search engines, in *2019 IEEE 19th International Conference on Advanced Learning Technologies (ICALT)*, vol. 2161-377X, pp. 219–223 (2019). https://doi.org/10.1109/ICALT.2019.00054
10. M.J. Tsai, C.C. Tsai, Information searching strategies in web-based science learning: the role of internet self-efficacy. Innov. Educ. Teaching Int **40**(1), 43–50 (2003). https://doi.org/10.1080/1355800032000038822

Chapter 24
Mining and Classifying Social Network Data: The Case on King Abdul-Aziz University Twitter Accounts

Walaa Alhabashi, Kawther Saeedi, Naif Aljohani, Sachi Arafat, and Rabeeh Abbasi

Abstract Social media, and especially Twitter, in specific domains such as healthcare, education and politics, turned into the key venue of social interaction today. Many higher education institutions (HEIs) seek therefore to utilize the value-added of Twitter to disseminate and to collect information from students in view of improving the quality of education at their institutions. In this view, the ability to mine, classify and interpret the content of Tweets is crucial. By examining the case of the King AbdulAziz University in Saudi Arabia, this paper offers a preliminary insight into what kind of information can be collected and in which ways it can be useful for a given HEI as regards teaching, administration and overall management. To this end, this paper examines the usability of three machine learning models, including Support Vector Machine (SVM), K Nearest Neighbor and, finally, Random Forests (RF). The outcomes of this study that this paper elaborates on suggest that in terms of accuracy, SVM is the best performing classifier. Meanwhile, even if the RF proved to be a strong classifier too, it did not perform as well as the SVM.

Keywords Data mining · Social network mining · Tweet classification · Machine learning algorithms · Case study · Arabic tweets

---

W. Alhabashi · K. Saeedi (✉) · N. Aljohani · S. Arafat
FCIT, IS Department, King AbdulAziz University, Jeddah, Saudi Arabia
e-mail: Ksaeedi@kau.edu.sa

W. Alhabashi
e-mail: Walhebshy@stu.kau.edu.sa

N. Aljohani
e-mail: Nraljohani@kau.edu.sa

S. Arafat
e-mail: Sarafat@kau.edu.sa

R. Abbasi
CS Department, Quaid-i-Azam University, Islamabad, Pakistan
e-mail: Rabbasi@qau.edu.pk

A. Visvizi et al. (eds.), *Research and Innovation Forum 2020*,
Springer Proceedings in Complexity,
https://doi.org/10.1007/978-3-030-62066-0_24

Introduction

Social networking sites and our ability to mine data included therein and turn it into information is one of the greatest advantages of the big data paradigm [1]. Twitter, with more than 500 million messages posted daily, with its potential to reach substantial numbers of stakeholders, represents one of the most interesting cases in point here. For this reason, many HEI begun to deploy Twitter to facilitate exchanges between faculty, administration, and students, to concomitantly improve the quality of service/education and overall management of their institution. This paper elaborates on the case of King Abdul-Aziz University (KAU), Saudi Arabia, which implements a broad strategy aimed at exploiting the potential inherent in Twitter to improve KAU's performance. The specific question that this paper explores is which methodological approach to apply to classify tweets generated by the KAU community and therefore effectively engage with data extraction and information generation. While this paper explores a tiny chunk of the process, the most recent data suggest that the strategy implemented by KAU already yielded substantial positive results in times of the Covid-19 pandemic [2]. The argument in this paper is structured as follows. Following a brief outline of the research design and methodology, a literature review follows. The objective here is to map machine learning techniques and identify the most efficient to classify Arabic tweets generated by the KAU community. Discussion and conclusions follow. The value added of this paper consists of a proposal of framework suitable for conducting large scale classifications of tweets.

Research Design

In general terms, the study elaborated on in this paper consisted of 9 stages: (i) conceptualization of the research problem; (ii) literature review and validation of the relevance of the research problem; (iii) design of the research method and methodology, including identification of a classification model and algorithm considered as performing best in achieving predictive accuracy; (iv) data collection (to create a dataset drawn from the KAU tweeter community, including faculties, research centers, deanships and other carefully selected stakeholders (see Table 24.1 for details); (vi) data examination, through categorization of tweets included in the dataset into broad topic groups; (vii) data interpretation through testing and validation of the previously identified models and algorithms; (viii) identification of the 'best' in terms of predictive accuracy model; (ix) presentation of results and discussion. This paper elaborates in detail on selected parts of this study, as outlined above.

Table 24.1 Categories of tweets

Category	Definition
Admission	Matters pertaining to registration and admission including general enquiries, entry exam results and dates
Exams	Matters pertaining to exams including dates, e-exams and schedules
News	Matters pertaining to student forums, alumni, honoring, celebration and meetings
Faculties	Matters pertaining to faculty-level and faculty staff issues including facilities management, workshops, announcements, and jobs advertisements
Students	Matters pertaining to student enquiries, classes and timetables, lectures and student events, student pastoral and academic services, calendars, scholarships and other support services
Libraries	Matters pertaining to university libraries, club, books, and research
Others	All other miscellaneous matters

Research Methodology

To observe and analyze Twitter account content pertaining to the Deanship of e-Learning and Distance Education at KAU, a framework for classifying Arabic text on Twitter has been developed. Comments and opinions that convey positive, neutral or negative attitudes were grouped, and Support Vector Machine (SVM) and Naïve Bayes (NB) algorithms were applied. The study found that the SVM classifier attained a greater accuracy level compared to the NB classifier [3]. In addition, to analyze user opinion tweets related to a Ministry of Education service account (in Saudi Arabia), specifically the university system recently initiated, a sentiment analysis model developed to categorize the Arabic texts into those that are positive, those neutral and those negative, this model captured tweets that included the hashtag #new_university_system [4].

The tweets were divided into different functions' categories. Using the categorizations outlined in Table 24.1, we split the KAU twitter community was divided into distinct groups—deanships, centers, faculties, and other bodies.

The research is split into distinct stages: (i) dataset description and preparation; (ii) feature engineering; (iii) machine learning model development. Calculations of the f-measure, accuracy, recall and precision of every model were made, and dividing the entire dataset into k-folds and conducting concomitant training and evaluation of k-1 times using the K-fold technique. The Kth fold was deployed for testing these and at each run the K-1 folds were treated as training data [5]. A ten-fold cross validation in addition to measurements of recall, f-measure, accuracy and precision were used to evaluate every model. To understand the models' behavior according to the scale of the data pertaining to features CF, SF and LF across the sets of training data, models were run using four dataset sizes. Figure 24.1 offers a visualization the entire framework.

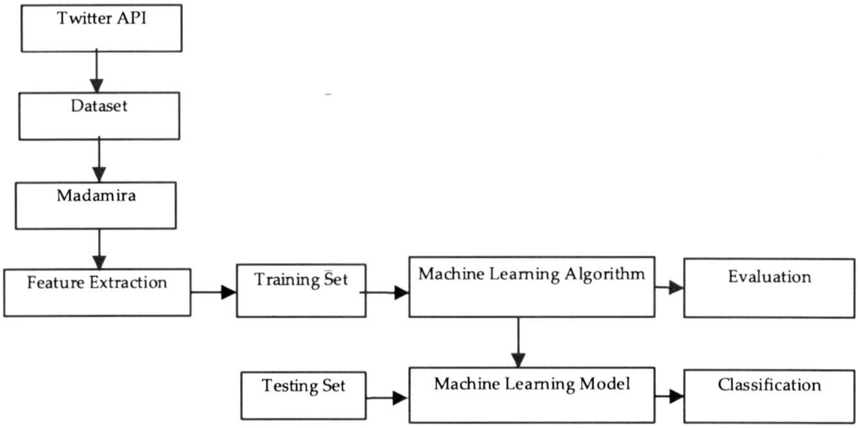

Fig. 24.1 Model framework

Dataset Description and Preparation

In collating and analyzing extensive series of tweets, Twitter gives researchers and developers access to its API services [6]. The present study used the API streaming service to collect the data and conduct the analyses, with 10,000 tweets being taken from the KAU account in the period from 2016 to 2018. Sampling was random. Tweets that are marked as being Arabic by Twitter API responses comprise the only tweets included in the study scope. Through the process of putting the category proceeding each tweet text, 2000 tweets were all labeled manually. Three human coders from different educational backgrounds have labeled the dataset according to the categories as the example in Table 24.2. The categories was selected under consultation with educational experts in the university. Tweets that do not match the categories were labeled as "Other".

A key natural language processing system named MADAMIRA used for Arabic texts was used to process all tweets (see https://camel.abudhabi.nyu.edu/madamira/). Including a part-of-speech tagger and a tokenizer, MADAMIRA constitutes a system that enables morphological and disambiguation analyses of Arabic texts. The system enables the homogenization of certain letters in Arabic of different shapes, referred to as 'normalization' [7].

Machine Learning Algorithms

By testing the accuracy and effectiveness of SVMs, K-nearest neighbor models and random forest models, this study explores and evaluates the validity of common machine learning models when used in a specific educational context. The standard

Table 24.2 Manual tweets labelling

Tweet	Translation	Label
ارغب بتحويل التخصص من ادارة عامة إلى ادارة اعمال	I would like to transfer the major from general management to business administration	Admission
متى سيتم نشر جدول الاختبارات النهائية	When will the final test schedule be published?	Exams
نذكركم بحضور الملتقى العلمي بعنوان: التكامل المعرفي بين العلوم الإنسانية وأثره في تنمية المجتمع	We remind you to attend the scientific forum entitled: Knowledge integration between the humanities and its impact on community development	News
تشمل Safari Books Online على مجموعة من المراجع العلمية والبحثية في التقنية وتحوي آلاف الكتب والمقالات والفيديو من أهم الناشرين	Safari Books Online includes a collection of scientific and research references in technology and contains thousands of books, articles and videos from the most important publishers	Libraries

machine learning function classifies tweets by feeding tweets with known polarity, so that the ensuing system has the capacity to learn and predict with accuracy.

Many regression analyses and classification projects are undertaken using SVMs. In cases of text categorization SVMs are especially helpful. With every value of each feature equivalent to the value of a specific coordinate, SVMs represent each data item as a point in n-dimensional space whereby n is the number of features [8].

As an instance-based algorithm, the K-nearest neighbor (KNN) does not act on assumptions about the distribution of data, instead memorizing training instances and thereafter deploying them as knowledge during prediction. The similarity between all points in the dataset to the unknown observation is calculated when a new data point is assumed by the KNN during classification [9].

Combining a range of decision tree models synonymously using several unique features, the random forest algorithms comprise a collated technique. It is often used and is deployable in classifying as well as for regressions. By combining the results of several separate decision trees the technique is orientated around the notion of collective classification [6].

Machine Learning Model

The machine learning models were built through several phases.

Pre-processing Phase

Processing of the data follows the data collection phase. URLs, @ mentions and hashtags had to be removed. The tweets were thereafter allocated to a category. Cleaning of data was then performed, and the PHP script was generated to read all tweets. By saving tweets as text files named according to their tweet IDs and placing tweets of the same categorical label into collective folders, the data were processed.

Features Engineering Phase

To construct features for the labels, feature engineering was conducted that would be deployed so as to train the machine learning model. Three types of feature engineering were used. After this, the MADAMIRA tool was used to conduct the feature engineering of the Arabic texts. Lexical, character and syntactic feature engineering types were applied. The case study utilized an Arabic text tool developed at Columbia University called MADAMIRA. In this tool, every word that constitutes an input is subjected to the following features: base phrase chunking, named entity recognition, discretization, tokenization, part-of-speech tagging, and lemmatization [7].

PHP scripts stored the various test cases, following the cleaning and removal of noisy data via text analysis libraries in Python. MADAMIRA was utilized alongside Python to extract the character, syntactic and lexical features of the respective categories as well as to develop features-based models for classification. Various open source libraries were used. The study used Anaconda—an environment management system designed to create and load and switch between local computer environments—and Pandas—designed to frame load and manipulate data—and Tweepy—designed to stream tweets from Streaming API onto a local database.

- Character Feature: Character features for every respective category were calculated to consider features that are numeric in form, when used in tweets, such as event dates and so on. The total number of characters in any given tweet, the amount of white space and the uses of special characters are also factored in [10].
- Lexical Feature: Tweets are brief texts, being restricted to 280 characters. For this reason, Lexical Features were conducted to calculate where words have occurred once, twice or many times in the tweet. Such, N-grams—unigrams, bigrams and trigrams of words, form bags of words that we know as lexical features [11].
- Syntactic Feature: Another feature is the ratio of part-of-speech occurrence in a tweet, which MADAMIRA's part-of-speech function enables researchers to identify as well as the ratio of stop words [11].

Model Generation Phase

A free software machine learning library for Python was used and specifically a Scikit-learn library generated three distinct classification models and the training on the training data. To select random tweets and save them, the study used a Python code. Saving them in various directories, random tweets numbered at 500, 1000, 1500 and 2000 were selected.

To comprehend the behavior of all models based on size of data, and to train the model for different data sizes and record accuracy, the training data were prepared for the feature combination CF + LF + SF. Cross validation with ten folds for the different sets of tweets and the storage of ensuing results were performed via the training process for the three classification models. The various sets of training data and the models providing multiple sets of parameters (10-folds) for combined features are shown in Table 24.3. As the table demonstrates, increased training dataset size and increased size of tweets in the modelling correlated with decreased performance across all measures.

Table 24.3 Accuracy comparison of learning model

		No. of tweets	Validation techniques			
			Accuracy	Precision	Recall	F1 measure
Classification approach	SVM	500	0.6454	0.6282	0.6032	0.6076
		1000	0.6425	0.6175	0.6008	0.6031
		1500	0.6308	0.6149	0.5925	0.5983
		2000	0.6180	0.5872	0.5764	0.5777
	KNN	500	0.3810	0.4840	0.3500	0.3516
		1000	0.3712	0.4983	0.3414	0.3473
		1500	0.3585	0.5057	0.3333	0.3407
		2000	0.3497	0.4445	0.3217	0.3180
	Random forest	500	0.6435	0.7028	0.5851	0.5972
		1000	0.6347	0.6820	0.5751	0.5929
		1500	0.6229	0.6838	0.5605	0.5736
		2000	0.6210	0.7009	0.5571	0.5715

Results and Discussion

The findings are clear that the greatest accuracy was achieved by the SVM algorithm. High levels of precision were accrued from the random forest algorithm too. The results were compared via cross validation scores. The present section undertakes an extensive discussion of the results.

Comparison of the Dataset Size

To gain an understanding of the effects of tweets' sizes on the model, the sets of tweets were created randomly. These were grouped as 500, 1000, 1500 and 2000 tweets respectively.

- 500 Tweets: The best accuracy was attained by this group in combining the features.
- 1000 Tweets: The accuracy began to decrease as the group became larger
- 1500 Tweets: As the study progressed to 1500 tweets, accuracy was further decreased
- 2000 Tweets: The least accuracy was achieved when using 2000 tweets.

All this indicates that accuracy is negatively affected by the number of tweets being classified, and this can create challenges for researchers. Algorithms can produce varying accuracies according to training dataset size, thus meaning that the necessary training set size must be modified to suit the specific algorithm being used.

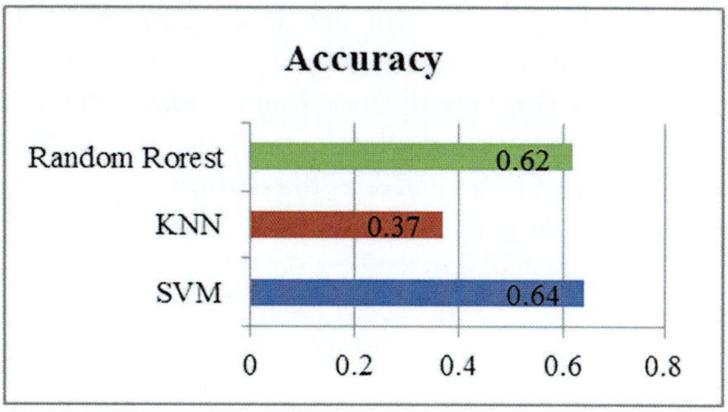

Fig. 24.2 Comparison of algorithms accuracy

Comparison of Classification Techniques

SVM proved to be the best performing classifier, with the random forest emerging as strong but not as well performing, and the KNN as the least well performing in terms of attaining accuracy. As outlined above, different feature types were used in combination for each respective algorithm. The results were significantly affected by the training set size (Fig. 24.2).

Conclusions

This case study evaluated machine learning systems and concomitant feature engineering methods used in HEIs to identify the best performing machine learning algorithms. The thrust of the challenge here was to identify a machine learning system able to classify Arabic social media texts split into groups relating to sub-organizational sector. The features used in training the model significantly affected the cross-validation scores of the algorithms. Tweet sizes and tweet amounts also had a significant effect on the accuracy of the algorithm models, where a random sampling method has been used to select the tweets.

For the purpose of the examination a comparative testing was applied to ascertain which of three machine learning algorithms—KNN, RF and SVM respectively—would perform best in terms of accuracy, using a combined set of character, lexical and syntactic features to produce the results. KNN emerged as the weakest model with the lowest accuracy across all number groups—500, 1000, 1500 and 2000 tweets—whilst SVM emerged as, overall, the strongest. The random forest technique did, however, perform well in the 500 and 2000 datasets. The dataset size correlated negatively with model accuracy, so as the number of tweets in the dataset increased all models performed worse in terms of accuracy.

This area of research looks set to continue to be an area of academic interest. This case study faced challenges in scoping the research focus and in identifying and understanding the parameters affecting classification, and more work to refine the models is needed. The data domain was limited to the Twitter account of KAU, and ongoing research would benefit from widening this scope. Also, they could benefit from the classification framework with more detailed analysis for the number of features tested, and data normalization. In addition, further research should be conducted that uses and tests the performance of other algorithms not explored here, including neural networks, boosted trees and logistic regression.

This paper adds new empirical material and theory to the wider field of social network mining and tweet classification, focusing on higher education institutions and their use of tweets to enable a broader conception of the various machine learning methods used for categorizing tweets into broad topic areas and their effectiveness in large higher educational institutions such as KAU.

References

1. M. Mora-Cantallops, S. Sánchez-Alonso, A. Visvizi, The influence of external political events on social networks: the case of the Brexit Twitter network. J. Ambient Intell. Human Comput. (2019). https://doi.org/10.1007/s12652-019-01273-7
2. A.O. Al-Youbi, A. Al-Hayani, H.J. Bardesi, M. Basheri, M.D. Lytras, N.R. Aljohani, The king Abdulaziz University (KAU) pandemic framework: a methodological approach to leverage social media for the sustainable management of higher education in crisis. Sustainability. **12**, 4367 (2020)
3. H. Al-rubaiee, R. Qiu, K. Alomar, D. Li, Sentiment analysis of Arabic Tweets in e-Learning. J. Comput. Sci. **12**, 553–563 (2016)
4. M. Alsulami, R. Mehmood, Sentiment analysis model for Arabic Tweets to detect users. Opinions about government services in Saudi Arabia: Ministry of education as a case study. in *al Yamamah Information and Communication Technology Forum, Riyadh* (2018), pp 1–8
5. J. Brownlee, A gentle introduction to k-fold Cross-validatio. Mach. Learn. Mastery. **5**, 23 (2018). https://machinelearningmastery.com/k-fold-cross-validation/
6. C. Nagpal, *Application of Machine Learning Techniques for Twitter User Classification* (Indian Institute of Science, Bangalore, 2014)
7. M.A.-B. Arfath Pasha, MADAMIRA: a fast, comprehensive tool for morphological analysis and disambiguation of Arabic. LREC Conf.. (2014)
8. R. Gandhi (2018) Support vector machine — introduction to machine learning algorithms. Towards Data Sci. (2018). https://towardsdatascience.com/support-vector-machine-introduction-to-machine-learning-algorithms-934a444fca47
9. I. Jose, KNN (K-Nearest Neighbors). Towards Data Sci. (2018). https://towardsdatascience.com/knn-k-nearest-neighbors-1-a4707b24bd1d
10. D. Network, Feature selection (data mining). Microsoft (2016). https://msdn.microsoft.com/en-us/library/ms175382.aspx
11. C.V. Bosch, Exploring lexical and syntactic features for language variety identification. Assoc. Comput. Linguist. 190–199 (2017). https://doi.org/10.18653/v1/w17-1224

Chapter 25
On Top-Down Versus Bottom-up Personalisation and Evaluation of Augmented Reality Learning Systems

Eugenijus Kurilovas

Abstract The paper aims to analyse the problem of personalisation and evaluation of quality of Augmented Reality (AR) learning systems. AR is often used in education to enhance students' motivation by visualizing learning content and activities. In the paper, first of all, systematic review of relevant scientific literature on the research topics was conducted. The author's AR learning systems quality personalisation and evaluation frameworks are presented in the paper. Evaluation of quality of AR learning systems should be based on applying both expert-centred (top-down) and user-centred (bottom-up) quality evaluation methods consisting of creating quality models (systems of criteria) and evaluation methods. AR-based learning systems including learning content (i.e. learning objects) and activities should be suitable, acceptable and usable for particular learners. Personalisation of AR learning systems should be based on learners' models/profiles using students' learning styles (bottom-up method), and educational data mining (top-down method). AR personalisation method is aimed to personalise learning by applying well-known learning styles models, educational data mining methods and techniques, and intelligent technologies, and thus to ensure that suitable AR-based learning systems should be selected for particular users to improve their learning motivation and thus—quality and efficiency. The method of identifying students preferring to actively use AR-based learning systems is based on identification of probabilistic suitability indexes to choose the most suitable AR-based learning systems for particular students. Experimental research is also performed, and its results are presented in the paper. The research is multidisciplinary, including computer science, education, operations research, and educational psychology areas.

Keywords Augmented reality · Learning personalisation · Learner model · Evaluation of quality · Educational data mining · Evaluation of suitability · Acceptance and use

---

E. Kurilovas (✉)
Vilnius Gediminas Technical University, Vilnius, Lithuania
e-mail: jevgenij.kurilov@vgtu.lt

A. Visvizi et al. (eds.), *Research and Innovation Forum 2020*,
Springer Proceedings in Complexity,
https://doi.org/10.1007/978-3-030-62066-0_25

Introduction

Augmented Reality (AR) research and application development demonstrate its capacity of AR to change radically many areas experience in the near future. One of those areas is education, where AR applications seem to be very helpful. AR market for education is increasing dramatically during the last several years, and research on optimisation of acquisition of AR-based systems for educational institutions to enhance learning motivation, quality and efficiency becomes very relevant.

Research on AR-based learning systems (incl. learning objects, activities, and environment) becomes more and more demanded in scientific literature. Possibilities of AR application in education are very wide and bring many advantages to students of all ages, although much needs to be done.

However, only several studies directly address personalisation question of AR-based systems in education. Many authors agree that the problem of personalisation of AR-based learning systems and resources is relevant and should be further analysed.

Therefore, original personalisation method of AR-based learning systems, based on applying learners' profiles/models, psychological questionnaires, and educational data mining (EDM), is formulated and presented in the paper. Implementation of this method in real educational practice should optimise acquisition of AR-based systems for educational institutions according to users' (i.e. students') needs. The problem of EDM application to personalise AR to particular students' needs according to their profiles seems to be totally new in scientific literature.

Personalisation of learning became very popular topic in scientific literature during last years [1–3]. Application of learners' profiles/models based on different learning styles models and intelligent (smart) technologies to personalise learning are recognised to be effective in terms of improving learning motivation, quality and efficiency [4–6].

In the paper, AR evaluation method is also proposed based on the same methods: both expert-centred (top-down) and learner-centred (bottom-up).

The reminder of the paper is organised into following interrelated sections. Systematic literature review findings on AR systems application in learning and its personalisation and on EDM application to personalise AR are presented in Section "Systematic Literature Review". Section "Research Methodology" is aimed to present research methodology and describe the method of personalisation and evaluating AR suitability for particular learners according to appropriate probabilistic suitability indexes. Section "Results and Discussion" presents the results of experiment and discussion of results. In Section "Conclusion and Future Work", conclusions are made and future work directions are defined.

Systematic Literature Review

Review on AR Personalisation in Education and Learning

In order to identify existing scientific methods, applications and results on AR-based learning systems and their personalisation, systematic literature review method devised by Kitchenham [7] has been used.

The following research questions have been raised for systematic review:

RQ1 What are the best practices of implementing AR in education?
RQ2 What are the ways to personalise AR learning systems according to learners' needs?

Systematic literature review was performed in Clarivate Analytics Web of Science database. Search protocol is presented in Table 25.1.

In 2002–2017, 2855 published papers were found in Clarivate Analytics Web of Science database according to the topic ("Augmented reality" OR AR) AND (education OR learning), and only 38 papers were found according to the keywords ("Augmented reality" OR AR) AND (personalisation). We searched for articles, proceedings papers, and reviews. The results were combined using OR tool and that gave us a set of 2.881 general results from the Web of Science.

The process of selection of resources included the following procedures: analysis of all the results and excluding resources non-related to the topic of interest (e.g. specific medicine studies, papers that used "AR" abbreviation with different meanings). From the reminder of the results, we selected the most cited papers during the period 2002–2017. The most cited 2002–2013 publication set was enriched with the newest relevant publications in the field published during 2014–2017. The period of 2014–2017 was selected due to the meta-analysis studies we found while analysing the papers published before 2013. On the last stage of [7] systematic review methodology, 18 suitable publications were selected.

The review of those 18 publications revealed that possibilities of AR application in education seem to be endless and bring many advantages to students of all ages. Few are creating content that may be used for educational purposes, with

Table 25.1 Search history in web of science database (AR in education)

Set No.	Search phrase	Research question	Results	Search options
1	TS = [("Augmented reality" OR AR) AND (education OR learning)]	RQ1	2.855	Language: English Document types: (Article OR Proceedings Paper OR Review) Timespan: 2002–2017
2	TS = [("Augmented reality" OR AR) AND personalisation]	RQ2	38	

most advances being made in the entertainment industry, but many understand and realise the future and importance of education applying AR. Many studies report that their analysed AR-based systems were more effective in comparison with traditional ones, especially for situated, inquiry-based, collaborative and self-regulated learning. AR system may bridge the gap between formal and informal education. Teachers and students like learning activities based on AR systems, almost all studies mention improved students' motivation and satisfaction. The results also confirm AR educational systems' positive effect in developing spatial skills and creativity.

On the other hand, although the concept of AR is not new, most applications are still limited, the existing systems have little intelligence in terms of awareness about the current state of the scene and the user's context, and many authors agree that the use of AR systems for learning should be further evaluated (e.g. [8]). Many studies of AR still focus on development, usability, and initial implementation of AR tools (e.g., [9]).

However, out of 18 analysed studies, only one meta-analysis work [10] directly addresses personalisation question of AR-based systems in education (this was one of the systematic review research questions raised by the author). There were only three studies found, published in 2009, 2013 and 2016, addressing some aspects of AR system personalisation. Other studies found under the keyword, containing "personalisation", publish research results with aspects of personalisation from travel industry, manufacturing (e.g. footwear personalisation), etc., and do not relate to personalised learning using AR.

As reported in another meta-analysis work on the topic [11], some studies on learner differences have shown that low and average achieving students showed learning gains through the AR experience, while high achieving students did not receive the same benefits. In fact, the high achieving students showed more learning gains in a traditional classroom where AR was not used.

Another aspect of AR environments negative effects is that students may be cognitively overloaded by the large amount of information they encounter, the multiple technological devices they are required to use, and the complex tasks they have to complete [9, 12]. These are the possible areas for further research on AR learning systems and their personalisation.

The majority of studies evaluate AR learning systems using experiment with a small or medium research samples (e.g. one class of students).

Finally, no studies have proposed technologically and pedagogically sound methods to personalise AR learning systems, models, prototypes or the whole AR-based learning units.

Review on AR Educational Data Mining for Personalisation

According to [13], the student model is a key component of intelligent tutoring systems since enables them to respond to particular needs of students. In the last years, educational systems have widespread in school and industry and they produce data which can be used to know students and to understand and improve the learning process. The student modelling has been improved thanks to educational data mining, which is concerned with discovering novel and potentially useful information from large volumes of data. To build a student model, the authors have used the data log of a virtual reality training system that has been used for several years to train electricians. The authors compared the results of this student model with a student model built by an expert. They rely on Bayesian networks to represent the student models. In [13], the authors present the student models and the results of an initial evaluation (Table 25.2).

Reference [14] realise the essential principles of self-regulated learning model in the library and developed a learning system that utilises the concept of combining mobile AR, indoor navigation and data mining algorithms.

Reference [15] conducted research on sophisticated analytics to mine rich data streams collected on students' devices, using each learner's interactions to help in developing personalised educational experiences. The authors also studied "augmented realities" that infuse virtual data and authentic, simulated experiences into real world settings, facilitating transfer of learning from classrooms to life. However, to realise the full power of ubiquitous learning for educational transformation, educators must overcome numerous challenges related to devices and infrastructure, safety and privacy, digital assets and assessments, and human capital.

According to [16], the immersive 3D Virtual Reality environment allows saving the logs of its users, which gives us a great opportunity to create a database to mine the data about learners' behaviour in the particular educational scenario. It is considered as an important issue of [16] because the analysis of data can be practically used by both teachers and application developers. It is assumed in [16] research work that the data received from the environment would allow teachers to monitor the process of gaining knowledge by students, the educators would be able to correct the scenario and unify it for different learning subjects, and, as a result, the software developers will bring it to life.

We see that educational data mining was almost never analysed for students' profiling and AR personalisation according to students' needs.

Table 25.2 Search history in web of science (AR and educational data mining)

Results	Save history/Create alert open saved history
7	TS = (educational data mining AND Virtual Reality) *Indexes = SCI-EXPANDED, SSCI, A&HCI, CPCI-S, CPCI-SSH, ESCI Timespan = All years*

Research Methodology

AR Personalisation: Top-Down Versus Bottom-up

All learning process (e.g. learning activities or learning units/scenarios) should be personalised according to the main characteristics/needs of the learners/users. Learners' needs and characteristics are usually described in learners' profiles (modules) and include prior knowledge, intellectual level, interests, goals, cognitive traits (working memory capacity, inductive reasoning ability, and associative learning skills), learning behavioural type (according to his/her self-regulation level), and, finally, learning styles.

All those personal students' needs and characteristics could be identified using EDM/learning analytics methods and techniques using top-down (expert-centred) method described in systematic review.

According to [17], the wide range of data about the behaviour of students should be used to generate good quality, real-time predictions about suitable material and activities and success in acquiring knowledge and skills. Students and teachers should be able to plan their work on the basis of reliable tools that can produce detailed and personalised recommendations about what should be done to achieve the best learning outcomes.

The data on real students' behaviours in learning environment obtained by using EDM/learning analytics methods and techniques should be used to correct students' profiles according to the data obtained. Thus, learning analytics software agent should be created to correct students' profiles according to their behaviour in the learning environment (e.g. AR). Students practically use some learning objects and learning activities in real learning practice in a learning environment before personalisation, i.e. identifying appropriate probabilistic suitability indexes [18] and recommending suitable learning units. Here we could hypothesise that students prefer to practically use particular learning activities and objects that fit their learning needs mostly.

Thus, using appropriate EDM/learning analytics methods and techniques, it would be helpful to analyse what particular learning activities and objects were practically used by these students in the learning environment, and to what extent.

Therefore, in order to personalise learning in any learning environment (incl. AR) we could use EDM/learning analytics. In this case, several stages are required:

(1) Learning components (learning objects, learning activities) should be interlinked with students' learning styles.
(2) Students should be analysed to identify their individual learning styles filling in the appropriate learning styles questionnaire;
(3) Probabilistic suitability indexes [18] should be calculated for each analysed student and each AR learning activity to identify which learning objects and activities are the most suitable for particular students, and to what extent;

(4) Using appropriate EDM/learning analytics methods and techniques, we could analyse what particular learning objects and activities were practically used by these students in AR, and to what extent.
(5) The data on the practical use of AR-based learning objects and activities should be compared with students' probabilistic suitability indexes obtained during the second (2) stage. In the case of any noticeable discrepancies between learning units created using students' learning styles identified by the questionnaire only and based on their previous real behaviour identified by learning analytics we can e.g. use personal qualitative survey methods to identify students' learning styles and other data from students' profiles (models) more precisely.

Another way to personalise AR is to apply EDM/learning analytics-based stage (4) not for students' learning styles [after stage (2)] but after stage (3) for the whole personalised learning units.

In order to identify students' learning styles, first of all, the author had prepared Soloman-Felder-based questionnaire [19] for expert evaluation of suitability of Felder-Silverman learning styles [20] and learning objects and activities based on application of AR. The following question was formulated: "What do you think is suitability level of learning systems based on application of AR to Felder-Silverman learning styles (excellent, good, fair, poor, or bad)".

After that, details explaining Felder-Silverman learning styles model (FSLSM) according to [4, 20] were provided for the experts. According to FSLSM, all students could be divided into four dimensions and eight learning styles:

(1) By information type: Sensory (SEN) Versus Intuitive (INT);
(2) By sensory channel: Visual (VIS) Versus Verbal (VER);
(3) By information processing: Active (ACT) Versus Reflective (REF), and
(4) By understanding: Sequential (SEQ) Versus Global (GLO).

After filling in the questionnaire, the author could easily calculate the average values of suitability of learning styles and learning systems applying AR. These values of suitability can be calculated using conversion of linguistic variables into triangular [21] or trapezoidal non-fuzzy values [17], that will be used in this paper: "Excellent"—1.000, "Good"—0.800, "Fair"—0.500, "Poor"—0.200, "Bad"—0.000. The average values of suitability should be easily calculated by division of the total sum of all non-fuzzy values by the number of the experts.

According to [18], an example could be obtained by filling in the questionnaire [19]. If a student answers e.g. 7 questions favourable to the Sensory learning style, and 4 questions favourable to the Intuitive learning style, then $PR_{SEN} = 7/11 = 0.64$, and $PRI_{INT} = 4/11 = 0.36$, and further on to all dimensions of FSLSM. Thus, we could obtain e.g. the following learning styles initially stored in student profile/model (Table 25.3).

According to [18], if we should multiply probabilistic values of particular students' learning styles according to Table 25.3 (PR) by AR-based learning systems and learning styles suitability values according to Table 25.2 (V), we should obtain

Table 25.3 Example of learning style initially stored in student profile/model

Learning styles							
By information type		By sensory channel		By information processing		By understanding	
SEN	INT	VIS	VER	ACT	REF	SEQ	GLO
0.636	0.364	0.818	0.182	0.727	0.273	0.455	0.545

probabilistic values (SI) of suitability of particular AR-based learning systems to particular student according to Formula 25.1:

$$SI_{ACT} = PR_{ACT} \times V_{ACT} \tag{25.1}$$

This is the example of Active learning style of the particular student. In the same way, we could calculate probabilistic suitability indexes of all learning styles of particular student according to Table 25.3.

AR Evaluation: Top-Down Versus Bottom-up

AR evaluation is proposed to perform using the same idea of Top-down (expert-centred) Versus Bottom-up (learner-centred) methods. In this case, Top-down method is based on the quality model (system of criteria) identified by the experts (e.g., [21, 22]) and evaluation methods (e.g., [21, 22]) like Fuzzy, AHP etc. Bottom-up method is based on evaluation of suitability, acceptance, and use of AR by particular students [17, 23].

Results and Discussion

AR Personalisation: Top-Down Versus Bottom-up

3 experts (researchers having solid experience in technology-enhanced learning and personalisation) have filled in the questionnaire by selecting one of the linguistic variables. The results are presented in Table 25.4.

Table 25.4 Evaluation results

Learning style	SEN	INT	VIS	VER	ACT	REF	SEQ	GLO
Average value	0.866	0.933	1.000	0.400	0.700	0.133	0.600	0.700

Table 25.5 Respondents' learning styles (%) according to the questionnaire results

No.	Information type		Sensorial channel		Information processing		Understanding	
	SEN	INT	VIS	VER	ACT	REF	SEQ	GLO
1	54.5	45.5	72.7	27.3	72.7	27.3	54.5	45.5
2	27.3	72.7	100.0	0.0	45.5	54.5	18.2	81.8
3	36.4	63.6	36.4	63.6	72.7	27.3	45.5	54.5
4	63.6	36.4	27.3	72.7	72.7	27.3	45.5	54.5
5	72.7	27.3	54.5	45.5	63.6	36.4	45.5	54.5
6	72.2	27.3	90.9	9.1	27.3	72.2	72.7	27.3

In Table 25.4, the experts have expressed their opinion on suitability of AR-based learning systems to all FSLSM-based learning styles.

If we want to calculate probabilistic indexes of suitability of these learning systems to particular students described by Table 25.3, we should use the methodology of creating probabilistic suitability indexes [18] and calculate these suitability indexes according to Formula 25.1, i.e. to calculate the indexes of particular learning component's (e.g., learning objects/activities/environments) suitability to FSLSM-based learning styles described by Table 25.4.

We'll try to demonstrate the application of the method exploring the group of 6 students of Bachelor study programme of one of Universities analysed in [17].

The respondents have filled out the Questionnaire (44 questions) [19] translated into Lithuanian. The results (Table 25.5) have shown that:

(a) 4 respondents prefer to process information in Active way, and 2 in Reflective;
(b) 4 respondents are mostly Sensory, and 2 Intuitive learners by information type;
(c) 4 respondents are mostly Visuals and 2 Verbal learners by sensorial channel; and
(d) 1 respondent is clear Sensorial and 1 is clear Global learner by understanding.

Here we apply the methodology presented in [18] to evaluate AR learning systems suitability to particular students according to their learning style preferences.

By applying suitability indexes obtained from the experts' evaluation (Table 25.4), we generate AR suitability for each student (Table 25.6).

Suitability indexes, presented in Table 25.6, are: (1) average values taking into account each component of learning style (by Information type, Sensorial channel,

Table 25.6 AR suitability indexes for the respondents

AR suitability for student						
	1	2	3	4	5	6
Average	36.542	37.344	34.084	33.174	34.493	35.457
Max	72.700	100.000	59.339	55.078	62.958	90.900

Information processing, and Understanding) and (2) maximum values, taking into account only one dominant style (e.g. Visual or Active). In practice, student has more than one learning style preference, and all the preferences should be considered. However, maximum value for one dominant preference and most suitable for AR-based learning systems, may help to identify students who could benefit most from using AR learning systems.

In Table 25.6, we see that AR-based learning systems are most suitable for students 1, 2 and 6.

These suitability indexes should be included in recommender system, and all learning components (e.g., learning objects, activities or environments) should be linked to particular students according to those suitability indexes. The higher suitability index the better the learning component fits particular student's needs, and vice versa.

An optimal learning unit/scenario (i.e. learning unit/scenario of the highest quality) for particular student means a methodological sequence of learning components (learning objects to be learnt, learning activities how to learn and learning environment where all the learning process has place) having the highest probabilistic suitability indexes.

The level of students' competences, i.e. knowledge/understanding, skills and attitudes/values directly depends on the level of application of optimal learning units/scenarios in real pedagogical practice.

Evaluation results presented in Table 25.4 have shown that learning systems based on application of Augmented Reality are (1) the most suitable for Visual (value 1.000), Intuitive (value 0.93) and Sensor (value 0.866) learners, and (2) the most unsuitable for Verbal (value 0.400) and Reflective (value 0.133) learners.

The results also show that there is almost no difference in preferences on using AR for learning styles by Understanding dimension (Sequential Versus Global).

AR learning systems are suitable for Activist learners (value 0.700), however this value could be higher if we specified certain types of AR, e.g. AR involving to act within the learning scenario. Therefore, in order to strengthen these students' motivation and improve their learning results, optimal learning scenarios based on active use of AR should be created and used in their learning process.

According to [4, 20], most people of college age and older are visual. Visual learners remember best what they see but not hear. If something is simply said to them they will probably forget it. Thus, AR is extremely suitable for Visual learners while Verbal learners prefer written and spoken explanations. Verbal learners remember and learn well from discussions, prefer verbal explanation to visual demonstration, and learn effectively by explaining things to others.

Therefore, for Visual learners, the optimal learning scenarios should include e.g. visual representations of presented material—pictures, films, diagrams, time lines, flow charts, demonstrations.

On the other hand, according to [20], Active learners learn by trying things out, working with others. They do not learn much from lectures because they require them to receive information passively. They work and learn better in situations that allow for group work and hands on experimentation. They prefer to actively use AR while

Reflective learners learn by thinking things through, working alone. Active learners do not learn much in situations that require them to be passive. An Active learner is someone who feels more comfortable with, or is better at, active experimentation than reflective observation. Active experimentation involves doing something in the external world with the information—discussing it or explaining it or testing it in some way.

On the opposite, according to [20], Reflective learners require situations that provide opportunity to think about the information being presented, and they work well alone and do not requite to actively use AR.

It is obvious that for hypothetic student described by Table 25.2 and student 1 described by Table 25.5 AR activities are very suitable (they are Visual learners), but, as an Activists, they prefer mostly learning units/scenarios based on active use of AR. For the respondents 2 and 6 described by Table 25.5 AR activities are suitable as for Visual learners, while other learning preferences of these students may give more information on what kind of AR activities and sequences the students would prefer. Thus, while student 6 is Visual and Reflective, AR could be useful for the presentation of new material, giving place for student's reflective practice.

Thus, if an educational institution (e.g. University) has a majority of Visual, Intuitive or Sensor students/users, the institution should actively acquire AR-based systems for their learners. If there is a majority of Active learners in the educational institution, "active" (e.g. simulation, not demonstrational) AR systems should be acquired and actively used in real pedagogical practice.

AR Evaluation: Top-Down Versus Bottom-up

Bottom-up (i.e. learner-centred) method to evaluate suitability, acceptance, and use of AR to particular students is presented earlier in [17, 23]. Examples of Top-down evaluation of AR and comparison with the results of Bottom-up evaluation should be the topic of future work.

Conclusion and Future Work

In this paper, original methods of Top-down Versus Bottom-up personalisation and evaluation of AR are presented. Top-down personalisation should be based on educational data mining to identify (or check) the main learners' preferences according to their profiles/models. These are the main items of scientific novelty presented in the paper.

Literature review presented in the paper revealed that there are already some methods and techniques to apply educational data mining in AR learning systems. Those data and methods should be used while implementing Top-down method to personalise AR in conformity with particular students' needs. EDM-based personalisation of such learning systems is an area where much needs to be done.

This approach is more reliable in comparison with purely psychological methods (e.g., questionnaires) or only EDM/learning analytics methods. The approach analyses both data on students' real behaviour in learning environment and students' own opinion on their behaviour obtained by questionnaires.

Research results have shown that AR-based learning systems are the most suitable for Visual, Intuitive and Sensor learners, and most unsuitable for Verbal and Reflective learners.

The presented method is to be used in learning recommendation system.

Research results had shown that the problem of acquisition of AR for educational purposes highly depends on learning institutions' students' learning needs and it should be more effective if corresponding probabilistic suitability indexes and students' data on their real behaviour in learning system should be taken into account.

Future work should include large-scale analysis of students using both Top-down and Bottom-up personalisation and evaluation of particular kinds of AR in real educational practice.

References

1. A. Juskeviciene, E. Jasute, E. Kurilovas, J. Mamcenko, Application of 1:1 mobile learning scenarios in computer engineering education. Int. J. Eng. Educ. **32** (3), 1087–1096 (2016)
2. T. Jevsikova, A. Berniukevičius, E. Kurilovas, Application of resource description framework to personalise learning: systematic review and methodology. Inf. Educ. **16**(1), 61–82 (2017)
3. S.R.I. Yoshioka, L. Ishitani, An adaptive test analysis based on students' motivation. Inf. Educ. **17**(2), 381–404 (2018)
4. E. Kurilovas, S. Kubilinskiene, Lithuanian case study on evaluating suitability, acceptance and use of IT tools by students—An example of applying technology enhanced learning research methods in higher education. Comput. Hum. Behav. **107** (June 2020), Article 106274. https://doi.org/10.1016/j.chb.2020.106274 (2020)
5. A. Visvizi, L. Daniela, Ch. W. Chen, Beyond the ICT- and sustainability hypes: a case for quality education. Comput. Hum. Behav. **107** (June 2020). https://doi.org/10.1016/j.chb.2020.106304 (2020)
6. A.Visvizi, L. Daniela, Technology-enhanced learning and the pursuit of sustainability. Sustain. **11**(15), 4022. https://doi.org/10.3390/su11154022 (2019)
7. B. Kitchenham, Procedures for performing systematic reviews. Joint technical report Software Engineering Group, Keele University, United Kingdom and Empirical Software Engineering, National ICT Australia Ltd., Australia (2004)
8. N. Petersen, D. Stricker, Cognitive augmented reality. Comput. Graph. **53**, 82–91 (2015)
9. H. Wu, S.W. Lee, H. Chang, J. Liang, Current status, opportunities and challenges of augmented reality in education. Comput. Educ. **62**, 41–49 (2013)
10. J. Bacca, R.F. Gesa, S. Graf, Kinshuk, S.M. Navarro, Augmented reality trends in education: A systematic review of research and applications. J. Educ. Technol. Soc. **17** (4), 133–149 (2014)
11. I. Radu, Augmented reality in education: A meta-review and cross-media analysis. Pers. Ubiquit. Comput. **18**(6), 1533–1543 (2014)
12. M. Dunleavy, C. Dede, R. Mitchell, Affordances and limitations of immersive participatory augmented reality simulations for teaching and learning. J. Sci. Educ. Technol. **18**(1), 7–22 (2009)

13. Y. Hernandez, M. Cervantes-Salgado, M. Perez-Ramirez, M. Mejia-Lavalle, Data-driven construction of a student model using Bayesian networks in an electrical domain. Lect. Notes Artif. Intell. **10062**, 481–490 (2016)
14. T.C. Huang, Y. Shu, T.C. Yeh, P.Y. Zeng, Get lost in the library? An innovative application of augmented reality and indoor positioning technologies. Electron. Library **34**(1), 99–115 (2016)
15. C.D. Kloos, D. Gillet, R.M.G. Garcia, F. Wild, M. Wolpers, Emerging technologies, ubiquitous learning, and educational transformation. Lect. Notes Comput. Sci. **6964**, 1–8 (2011)
16. N. Cherenkova, S. Chuprina, Educational scenario testing and its data mining in the immersive 3D virtual reality environment. Proc. 9th Eur. Conf. e-Learning **1**, 733–739 (2010)
17. E. Kurilovas, Advanced machine learning approaches to personalise learning: Learning analytics and decision making. Behav. Inf. Technol. **38**(4), 410–421 (2019)
18. E. Kurilovas, J. Kurilova, T. Andruskevic, On suitability index to create optimal personalised learning packages, in *ICIST 2016, Communications in Computer and Information Science (CCIS)* ed. by G. Dregvaite, R. Damasevicius, vol. 639, pp. 479–490 (2016)
19. B.A. Soloman, R.M. Felder, *Index of Learning Styles Questionnaire*. http://www.engr.ncsu.edu/learningstyles/ilsweb.html
20. R.M. Felder, L.K. Silverman, Learning and teaching styles in engineering education. Eng. Educ. **78**(7), 674–681 (1988)
21. E. Kurilovas, Evaluation and optimisation of e-Learning software packages: Learning object repositories, in *Proceedings of the 4th International Conference on Software Engineering Advances (ICSEA 2009)* (Porto, Portugal, 20–25 Sept 2009), pp. 477–483
22. E. Kurilovas, I. Vinogradova, Improved Fuzzy AHP methodology for evaluating quality of distance learning courses. Int. J. Eng. Educ. **32**(4), 1618–1624 (2016)
23. E. Kurilovas, On data-driven decision-making for quality education. Comput. Hum. Behav. **107**, Article 105774 (June 2020). https://doi.org/10.1016/j.chb.2018.11.003 (2020)

Chapter 26
Business Intelligence for Teaching Analytics: A Case Study

Alessio Maria Braccini, Carla Limongelli, Filippo Sciarrone, and Marco Temperini

Abstract In recent years, there has been a radical change in the world of teaching and training. This is causing numerous schools, universities and companies to adopt the most modern Information and Communication Technologies, mainly based on the Web, for distance education. Moreover, the widespread use of web-based environments is producing a considerable volume of data which could be used to monitor the learning processes to improve them. While Educational Data Mining analyzes this data from a technical point of view, Learning Analytics focuses on educational aspects to optimize online learning opportunities, involving all the stakeholders. In this paper, we present a case study concerning the analysis of data generated by a learning process, in a Learning Management System (LMS). The main goal is to test a particular Business Intelligence platform, the Knime platform, to extract hidden significant educational features from data. The case study strengthens our approach, with interesting pedagogical results.

Keywords Educational data mining · Learning analytics · Machine learning

---

A. M. Braccini
Department of Economics and Entrepreneurship, Tuscia University, Via del Paradiso, 47, Viterbo, Italy
e-mail: abraccini@unitus.it

C. Limongelli
Department of Engineering, Roma Tre University, Via della Vasca navale, 79, Rome, Italy
e-mail: limongel@ing.uniroma3.it

F. Sciarrone (✉)
Engineering Department, Roma Tre University, Via della Vasca navale, 79, Rome, Italy
e-mail: sciarro@ing.uniroma3.it

M. Temperini
Department of Computer, Control, and Management Engineering, Sapienza University of Rome, Viale Ariosto 25, Rome, Italy
e-mail: marte@diag.uniroma1.it

A. Visvizi et al. (eds.), *Research and Innovation Forum 2020*,
Springer Proceedings in Complexity,
https://doi.org/10.1007/978-3-030-62066-0_26

Introduction

In recent years, there has been a radical change in the world of teaching and training which is causing numerous schools, universities and companies to adopt the most modern information technologies, mainly based on the Web, as new knowledge management and sharing tool. This change is prompted by the countless advantages offered by distance learning, first of all for the extreme flexibility of time and space: the learner is no longer tied to the unity of time and place with the teacher and can, instead, freely choose the times and places of fruition [8]. If we add to this the increase in the quality of the learning material, its more flexible management, the possibility of easily measuring the results and the decrease in costs, it is easy to understand why distance learning presents itself as the preferred objective of almost all educational and training environments. Moreover, in the last time, due to the pandemic caused by COVID-19, the use of distance learning has become of fundamental importance to guarantee the continuity of teaching for students of all ages [7, 19]. In this perspective, the e-learning system can reach its maximum potential and effectiveness by using its produced (big) data in a personalized way: personalization of learning is carried out to adapt the learning process to the real needs and skills of the individual learner. Consequently, some disciplines have been proposed and used to exploit such (big) data from different points of view. The first one has been *Educational Data Mining* (EDM), i.e., "a discipline concerned with developing methods for exploring the unique types of data that come from educational settings, and using those methods to better understand students, and the settings which they learn", as reported in the EDM official website(http://www.educationaldatamining.org). EDM represents a technical approach to discovering student learning patterns, precisely the application of Data Mining (DM) and Machine Learning (ML) in general to the large amounts of data produced by students' interactions with the learning [2, 3, 10, 13]. So, another discipline, called *Teaching Analytics* (TA) [5, 16, 18], closely supported by EDM, was born in recent years to optimize online learning opportunities, using the data analyzed precisely through EDM. An example of TA concerns the design of new personalized reporting and visualization methods: easy for students to understand and connected to mechanisms for improving their learning [17]. Based on these new needs to analyze the large amount of data produced by learning processes, Business Intelligence (BI) and Big Data Analytics platforms have been proposed in recent years for the analysis of large quantities of data. Among the most used we cite some freeware platforms: *Knime* (https://www.knime.org), *Weka* (https://www.cs.waikato.ac.nz/ml/weka/) and *R* (https://www.r-project.org/). These platforms provide stakeholders with a dashboard for the control and optimization of those system variables that affect a learning process from different points of view. Here we present a case study of BI applied to TA, through the use of three main components: (i) the BI Knime platform; (ii) the use of the *moodle* e-learning platform; (iii) the use of a standard TA model. Our work aims to illustrate the effectiveness of such a combination of tools and approaches to the study of a learning process, to discover interesting hidden data patterns. The case study concerns the navigational behaviours of a group of 1855 learners during a course, through the analysis of the log files generated by their actions, stored in about 2, 000.000 of navigational log records. To this aim we used the DM *K-means* clustering algorithm [9]. The results strengthen our approach. The remainder of the article is divided as follows. Section "Literature Review" illustrates some important related works. Section "Teaching Analytics" shows the main features of the BI platform. Subsequently, Section "Teaching Analytics" shows the reference TA model we used in our case study. Section "The Case Study" shows the use case while the conclusions reported in Section "Conclusions and Future Work" close the paper.

Literature Review

Our system provides a modular architecture in the sense that the BI platform is external to the learning system, in our case the Moodle platform. This choice allowed us not to write new lines of code but only to integrate the two systems (see next section). In the literature, several systems have been proposed as TA systems integrated into the *Moodle* platform while there are few examples of BI systems applied to the data produced by these platforms. So here we propose three systems: the first of pure EDM, the second a dashboard for course simulation and the last for social interactions representations within a course.

In [4] an experiment similar to ours was presented using the *Weka* platform, where the aim was to determine the Learning Styles of the learners of a course by analyzing the same data that is used in this work. In that case, however, the goal was limited to EDM and not to TA, that is, to apply clustering algorithms to verify any associations and deduce Learning Styles. Our work instead, while working on the same data, presents a wider spectrum of possibilities that go beyond.

In [15], another system is illustrated that can provide the teacher with a simulated environment for the management of *Massive Online Open Courses*. In this case, an analysis environment of a 3D visualization engine is used, the teacher simulates a MOOC having particular statistical characteristics and then possibly switches to a real MOOC. It is a What If dashboard. ML algorithms based on K-nn are used. The TA system that we present has a more ambitious purpose, namely to provide modular learning functions and graphical interfaces that are easily modularized.

SNAPP [1], is a software tool that allows users to visualize the network of interactions resulting from discussion forum posts and replies. The network visualizations of forum interactions provide an opportunity for teachers to rapidly identify patterns of user behaviour—at any stage of course progression. SNAPP has been developed to extract all user interactions from various commercial and open-source learning management systems (LMS) such as BlackBoard (including the former WebCT), Moodle and Sakai. Most of the student data generated from Learning Management Systems (LMS) include reports on the number of sessions (log-ins), dwell time (how long the log-in lasted) and the number of downloads. SNAPP uses information on who posted and replied to whom, and what major discussions were about, and how expansive they were, to analyse the interactions of a forum and display it in a Social Network Diagram. This project aims at representing the social behaviours of learners through a visual instrument that can be applied to different LMSs.

The Knime Platform

In this Section, we briefly show the main characteristics of the Knime BI platform. Figure 26.1 illustrates a screenshot of the working environment proposed by the platform. The workflow shown in the figure is part of that used for the extraction and analysis of the data carried out for our case study.

The platform offers the following main features:

- Data Input/ETL. It is possible to import data, transforming them by harmonizing them with different formats and, finally, exporting or uploading them to other systems. ETL stands for Extract, Transform, Load. ETL processes are really easy and immediate to set up in Knime. It is obtained obtained by connecting functional *nodes*, which are the main components composing a workflow. In fact, each workflow is a set of functional nodes each with a specific goal. Connecting different nodes, we obtain a data flow, like the one shown in Fig. 26.1.
- Machine Learning. Knime includes in its basic version over 200 machine learning algorithms implemented and easily configurable even by non-experts. For such reason we used this platform. It is a good compromise between researchers and instructional designers;

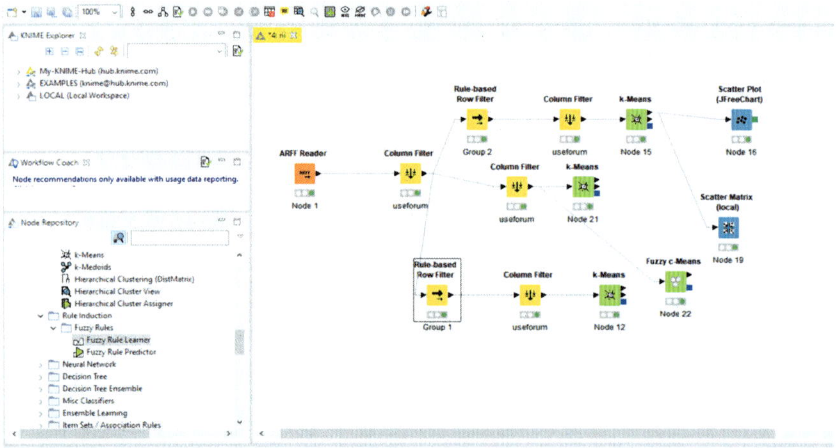

Fig. 26.1 The Knime environment. In the lower left part there is the repository of functional nodes while in the center one can see the work environment with a workflow already set

- Data Visualization. The platform offers several ways to visualize the information, easily configurable. Moreover it is possible to export data directly into Excel files in order to be managed by non-expert people;
- User-friendliness. The Knime user interface is very intuitive, simple to navigate and visually attractive, proposing the *What You You See Is What You Get* paradigm.

The operating mode is very simple. The user can select the functional nodes from a special repository and subsequently build with them a directed graph or workflow, where each node performs a specific processing function on the data given in input.

Teaching Analytics

In this section, we illustrate the essential characteristics of some important TA models, with particular reference to what will be used for the case study. EDM is the discipline that uses the data produced during the process of learning by students and uses the techniques of DM and Machine Learning, in general, to manage such data to discover useful information such as any learning patterns hidden among the data itself [12]. As already said, the goal of TA is to build better pedagogies, enhance active learning, reach students at risk and evaluate the factors that influence student completion and success (for an introductory article to TA models refer to [14]). A definition of TA is [6, 11]:

Teaching Analytics refers to the methods and digital tools to help teachers analyze and improve the educational designs prior to the delivery. More recent developments on Teaching Analytics also support analysis of how the teacher delivers the educational designs .

In particular, among the aforesaid three disciplines, our work can be mainly focused on TA. Our reference TA model is represented by the following *6-steps cycle* [16], shown in Fig. 26.2:

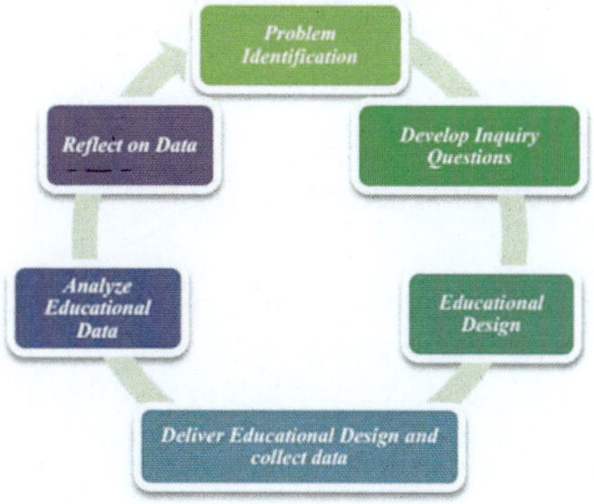

Fig. 26.2 The 6-steps cycle TA model

- *Problem Identification*. The teacher identifies some critical issues to be addressed to improve the learning process;
- *Develop Inquiry Questions*. In this step the teacher identifies what she has to investigate, what data have to be collected by the system and so on;
- *Educational Design*. In this phase the teacher identifies the educational project to put into practice to start the inquiry;
- *Deliver Educational Design and collect data*. In this phase the teacher collects the data produced by the learning process;
- *Analyze educational data*. The teacher analyses educational data to elicit insights to answer the inquiry question;
- *Reflect on Data*. The teacher revises critically the learning process in such a way to improve it.

The Case Study

In this Section, we show the usefulness of the use of the Knime BI platform for the analysis of a learning process, following the TA model shown in Section "Teaching Analytics".

Educational Design

We want to address the following Research Question for our educational design:

Is the BI approach based on the Knime BI platform useful to find out any critical issues hidden in the learning process?

In particular, the goal is to to analyze if there is any differences among learners in the use of some navigational resources: *practical resources* versus *theoretical resources*. The total set of learning resources, available to the sample taking the course is shown in Table 26.1, where in the first column there is the Id of the resource while in the second column its explanation.

Deliver Educational Design and Collect Data

The educational design has been delivered through the moodle platform, where ten disciplinary groups were formed. For example, one group belonged to the IT area while another group belonged to the linguistic area. The learning process lasted about 6 months. We collected data from the log files generated by the learners interactions during the course. The case study concerns the analysis of a (big) data-set composed by about 2,000.000 of records, representing all the interactions between learners and

Table 26.1 The set of resources composing the set of navigational behaviours to analyze

Id	Feature	Significance
a_1	groupid	Identification of the disciplinary group to which the student belonged during the course
a_2	allresourceview	Total access to any type of resource
a_3	resourcepracticalview	Access to tutorials and quizzes
a_4	resourcetheoryview	Access to theoretical resources
a_5	extra	Access to additional material (insights, glossary ..)
a_6	totmoveforum	Total movements of any kind (reading and writing) to all for
a_7	visiteforum	Access to all forums for consultation
a_8	useforum	Posting and editing messages in forums
a_9	totmoveforumdiscipl	Total access to disciplinary forums (discussion workshops)
a_{10}	totmoveforumgeneral	Total access to general forums (Notice Board and Ask the tutor)

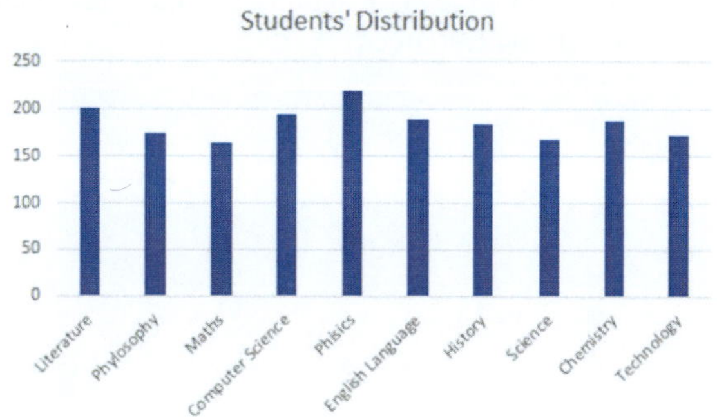

Fig. 26.3 The sample distribution into disciplinary groups

the LMS and by a set of 1855 learners composing the sample. In Fig. 26.3 is shown the distribution of the sample of learners into the disciplinary groups.

Analyzing Data

For the complete data analysis, we designed a workflow, where the interactions with all the resources was taken into account. The workflow begins with an ETL engine, the first node in the left (see Fig. 26.3), which imports data from the log file into a readable table. Subsequently it filters them, for each disciplinary group producing for each of them some clusters built using the *K-means* clustering algorithm [9] with $K = 5$. So, the whole workflow is divided into the following sequential functional nodes, one workflow line for each of the ten groups:

1. Rule-based row-filter node: it produces a table for a specific group;
2. Column-filtering node: it removes the groupId;
3. K-means node: it builds 5 clusters;
4. Excel sheet appender: it saves the cluster into an Excel sheet.

In Fig. 26.4 the use of the nine features by the disciplinary groups is shown. The comparison between the use of two particular resources in cluster 0 for each disciplinary group.

Reflect on Data

In this phase, one can reflect on the data, analyzing all the possible relationships among the use of the resources. In our case study, we note that for each disciplinary group, regardless of the chosen discipline, i.e., the group, the ratio between the total of resources *allresourceview* versus *Resourcetheoryviews*, that is, between the use of practical and theoretical learning resources remains constant among all groups. This result could lead us to think that the disciplinary difference does not affect the use of learning resources.

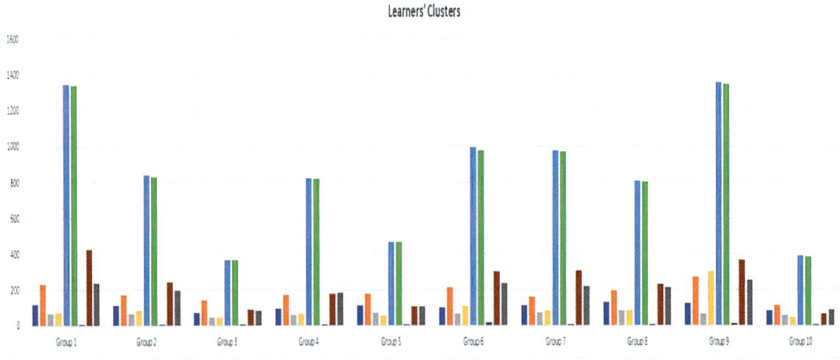

Fig. 26.4 x-axis are represented the ten groups of learners while in the y-axis the frequency of each of the nine features shown in Table 26.1. The groupid feature is not considered

Problem Identification

If we look at Fig. 26.5, we see that there is an imbalance between the use of some resources by all groups. These resources concern the use of the forum. Why is there such an imbalance and was it thought this way? These are questions that the system output suggests to the teacher or in any case to those who follow the learning process.

Develop Inquiry Questions

Based on the foregoing, the teacher must modify the learning process by making some modifications if necessary. For example, from the case study, it appears that students are led to a social approach, as evidenced by the data. Then this aspect could be enhanced by stimulating new discussions.

Conclusions and Future Work

In this article, we have presented a case study to show the effectiveness of the use of a BI platform for Teaching Analytics with the aim of understanding the learning processes that take place in distance learning platforms, like Moodle. Data from a course delivered at our university for the training of high-school teachers was used as the sample for our experiment. We used the Knime platform, a complete BI open-source tool for the analysis of the big educational data produced by our sample. Our experiment showed that the teacher can have feedback by the discovery of hidden behavioural patterns in the learning process. The case study highlighted interesting patterns for a more accurate understanding of the learning process. In particular, some

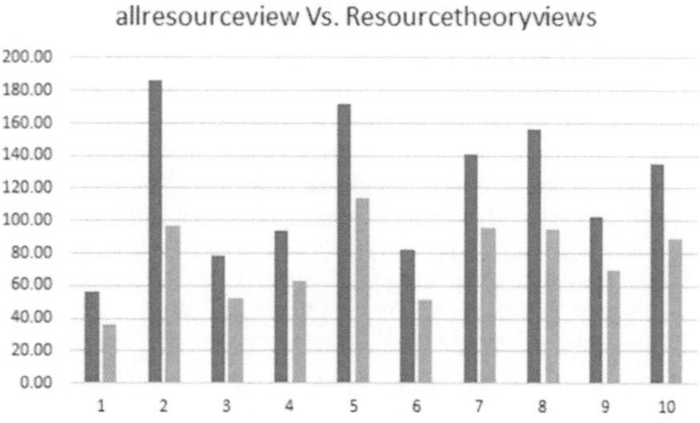

Fig. 26.5 The comparison between two main navigational resources for cluster 0. In the x-axis the ten disciplinary groups

hidden relationships among data showed a strong social behaviour of the sample. As a future work we plan more extensive experimentation together with the use and comparisons of other BI platforms.

References

1. A. Bakharia, S. Dawson, SNAPP: A bird's-eye view of temporal participant interaction, in *Proceedings of the 1st International Conference on Learning Analytics and Knowledge*, pp. 168–173. LAK '11, Association for Computing Machinery, New York, NY, USA (2011). https://doi.org/10.1145/2090116.2090144
2. B. Bakhshinategh, O. Zaïane, S. Elatia, D. Ipperciel, Educational data mining applications and tasks: A survey of the last 10 years. Educ. Inf. Technol. **23**, 537–553 (2018). https://doi.org/10.1007/s10639-017-9616-z
3. M. De Marsico, F. Sciarrone, A. Sterbini, M. Temperini, Supporting mediated peer-evaluation to grade answers to open-ended questions. EURASIA J. Math. Sci. Technol. Educ. **13**(4), 1085–1106 (2017)
4. V. Efrati, C. Limongelli, F. Sciarrone, A data mining approach to the analysis of students' learning styles in an e-learning community: A case study. Lecture Notes in Computer Science (including subseries Lecture Notes in Artificial Intelligence and Lecture Notes in Bioinformatics) **8514 LNCS**(PART 2), 289–300 (2014)
5. F. Gasparetti, C. De Medio, C. Limongelli, F. Sciarrone, M. Temperini, Prerequisites between learning objects: Automatic extraction based on a machine learning approach. Telematics Inf. **35**(3), 595–610 (2018)
6. G. Gauthier, Using teaching analytic to inform assessment practices in technology mediated problem solving tasks, in *Proceeding of Workshop on Teaching Analytics at the 3rd Conference on Learning Analytics and Knowledge LAK 2013*, pp. 1–8 (2013)
7. C. Limongelli, F. Sciarrone, M. Temperini, A social network-based teacher model to support course construction. Comput. Human Behavior **51**(Part B), 1077–1085 (2015)
8. C. Limongelli, G. Mosiello, S. Panzieri, F. Sciarrone, Virtual industrial training: Joining innovative interfaces with plant modeling, in *The 11th International Conference on Information Technology Based Higher Education and Training - ITHET 2012* (IEEE, New York, 2012), pp. 1–6
9. T.M. Mitchell, *Machine Learning*, 1st edn. (David McKay, New York, NY, 1997)
10. M.O. Hegazi, M.A. Abugroon, The state of the art on educational data mining in higher education. Int. J. Comput. Trends Technol. (IJCTT) **31**(4), 46–56 (2016)
11. L.P. Prieto, S. Villagrá-Sobrino, I.M. Jorrín-Abellán, A. Martínez-Monés, Y.A. Dimitriadis, Recurrent routines: Analyzing and supporting orchestration in technology-enhanced primary classrooms. Comput. Educ. **57**, 1214–1227 (2011)
12. C. Romero, S. Ventura, Educational data mining: A review of the state of the art. IEEE Trans. SMC Part C **40**(6), 601–618 (2010)
13. C. Romero, S. Ventura, Educational data mining: A survey from 1995 to 2005. Expert Syst. Appl. **33**(1), 135–146 (2007)
14. F. Sciarrone, M. Temperini, Learning analytics models: A brief review, in *Proceedings of the 23rd International Conference on Information Visualisation (IV-2019)*, vol. 2019-July, pp. 287–291 (2019)
15. F. Sciarrone, M. Temperini, K-openanswer: A simulation environment to analyze the dynamics of massive open online courses in smart cities. Soft Comput. **24**(15), 11121–11134 (2020)
16. S. Sergis, D. Sampson, Teaching and learning analytics to support teacher inquiry: A systematic literature review, in *Learning Analytics. From Research to Practice*, ed. by A. Peña-Ayala (Springer, Berlin, 2017), pp. 25–63

17. A. Sterbini, M. Temperini, Supporting assessment of open answers in a didactic setting, in *2012 IEEE 12th International Conference on Advanced Learning Technologies (ICALT)*, pp. 678–679 (2012)
18. A. Sterbini, M. Temperini, Analysis of open answers via mediated peer-assessment, in *2013 17th International Conference on System Theory, Control and Computing (ICSTCC)*, pp. 663–668 (2013)
19. A. Visvizi, L. Daniela, Technology-enhanced learning and the pursuit of sustainability. Sustainability (Switzerland) **11**(15) (2019). https://www.scopus.com/inward/record.uri?eid=2-s2.0-85070416969&doi=10.3390

Part III
ICT, the Business Sector and Policymaking

Chapter 27
Achieving Business and IT Alignment in Higher Education Institutions Using Conceptual Modeling: A GDPR Implementation Project as Case Study

Konstantinos Tsilionis, Amandine Chagniot, and Yves Wautelet

Abstract Higher education institutions have typically a multifaceted mission. Besides their educational activities, they also aim to support research activities and provide collective services while ensuring that the institution breaks-even financially (as a non profit organization). These varying domains can actually be translated in a set of strategic objectives that need to be achieved in the long term. At the same time, new IT developments and projects can contribute to (or occasionally hamper) the realization of an institution's strategy. Such IT developments suggest indeed new functions that could potentially support the individual achievement of one or many of the institution's long-term strategic objectives. This article depicts a conceptual model evaluating the business and IT alignment of a Belgian university college from the French community with the assessment of the strategic fit of a new GDPR implementation (IT) project serving as case study. The purpose of the case is twofold as it aims to illustrate the generic strategy of this organization and to show secondarily how the alignment between the GDPR implementation project and that strategy can be identified, represented and evaluated. Finally, the case suggests that the realization of specific project objectives contributes positively to the realization of the university college's broadly-based strategic objectives.

Keywords Higher education · Business IT Alignment · Conceptual modeling · Goal-oriented modeling · NFR Framework · Case study

K. Tsilionis · Y. Wautelet (✉)
KU Leuven, Belgium
e-mail: yves.wautelet@kuleuven.be

K. Tsilionis
e-mail: konstantinos.tsilionis@kuleuven.be

A. Chagniot
Université de Namur, Namur, Belgium
e-mail: amandine.chagniot@vinci.be

Haute Ecole Léonard de Vinci, Woluwe-Saint-Lambert, Belgium

A. Visvizi et al. (eds.), *Research and Innovation Forum 2020*,
Springer Proceedings in Complexity,
https://doi.org/10.1007/978-3-030-62066-0_27

Introduction

Higher education institutions have begun employing IT solutions to improve educational and research activities as well as the efficiency of their administrative processes ranging from curriculum (re)design to facility management [5, 26]. However, IT systems need to be continuously evaluated and aligned with the needs of students, academic and administrative staff. Likewise, different systems and platforms used by employees in different departments are not always integrated and when re-engineering and/or new IT projects are considered, the alignment with the institutional objectives should be envisaged to identify business value at a strategic level [3].

Goal-oriented conceptual models can represent the business and IT strategies as a set of (long-term) objectives. The design of information systems can then be seen as a pivot that traces back and forward from the strategic objectives to the physical implementation through code (see for example [15, 28] for an alignment between human organizational behavior and an e-learning system). More precisely, goal-based models can be used to assess the impact of organizational changes on the realization of these strategic objectives and evaluate the state of business and IT alignment [20, 30]. Wautelet [27] proposes MoDrIGo, a model-driven corporate and IT governance framework allowing to evaluate the alignment of an organization with strategic (business and IT) objectives. MoDrIGo suggests supporting governance by modeling the (long-term) business and IT objectives and studying their operational support. The business and IT objectives are determined by interviewing C-level executives and traced with operational execution representations through business IT services to determine added business value of the latter to the former(s). This approach is (at least partially because we do not use the notion of business IT service for synchronization) applied in this paper to align the strategic objectives of an higher education institution with an IT project concerning the GDPR implementation. MoDrIGo has formerly been applied in various sectors and notably in the field of healthcare (see [27, 31]).

Assessing and sustaining the strategic alignment of business and IT objectives in higher education is not an easy task [21], especially after the continuous deployment of new IT developments. This paper introduces firstly a conceptual model depicting the (business) strategy of a higher education institution called *Haute Ecole Léonard de Vinci* (this university college will be hereafter called *HE Vinci*) in terms of business objectives; this serves here as full case study. This artifact is further used to evaluate the university college's strategic objectives' alignment with the goals of a specific IT project. The latter is intended to support GDPR compliance to this strategy. We manage to visualize, delineate and decompose the IT project's goals into sub-goals and operationalized goals while assessing how their realization contributes to the fulfillment of higher education missions expressed through strategic objectives. In particular, our approach:

- Assesses the business and IT alignment in the sector of higher education using the *Model-Driven IT Governance* (*MoDrIGo* [27]) approach and more specifically the

Non Functional Requirements (NFR [8]) framework for visual representation at a strategic level. Business and IT alignment has not received as much attention in higher education institutions compared to traditional market-oriented organizations [7]. For this reason, we provide a graphical structure that layers distinctively the strategic and broadly-based missions of a higher education institution as well as its operationalized project-objectives. Our model-driven approach ensures the traceability between rather stable (long-term) mission objectives and fast-evolving project goals;

- Acts as a common reference for managers, strategists, academics, planners, computer scientists and legal advisers educated to GDPR. In that sense, our approach can be perceived as a first point towards stakeholder-based governance by mapping the needs of various faculty stakeholders and partnering institutions to maximize shared value;
- Creates an archetype for the evaluation and establishment of any newly-introduced IT project in higher education while maintaining uninterrupted the daily mode of operations for such institutions. This is achieved through the use of conceptual modeling acting as an analysis 'blueprint' before starting any actual project-related development and coding;
- Provides a method for the constant refinement and re-evaluation of strategic objectives and IT development-related goals at any given moment through the use of the MoDrIGo approach (with the NFR tree). The latter visualizes a system of inter-dependencies from top to leaf-level strategic objectives while casting a path to them from project operationalizations (goals).

The paper is organized as follows: Section "Background" depicts the background of the research notably with the state-of-the-art regarding the business and IT alignment especially within the sector of higher education. Section "Research Method and Steps" describes the case study. Section "Model-Driven Business and IT Alignment Evaluation of the GDPR Implementation Project" presents the application of the model-driven IT governance framework at HE Vinci for the introduction of a new IT project targeting the institutions' compliance to GDPR. Section "Discussion" further discusses the lessons learned through the case study while Section "Conclusion" concludes the paper.

Background

Business and IT Alignment in Higher Education

Business and IT Alignment (BITA) is being studied extensively for over three decades now [13]. Early research focused solely on aligning business and IT strategies in a top-down manner [2]. However, the Strategic Alignment Model by Henderson and Venkatraman [11] incorporated a cross-domain alignment perspective considering simultaneously the strategic layer and operational integration. Luftman et al. [17]

identified that well-prioritized IT projects as well as IT involvement in (business) strategy development are some of the enablers that influence BITA but Tallon and Kraemer [24] support that organizational alignment will vary according to the strategic value placed within the IT function of each organization. In any case, there is consensus that organizations are bound to perform better when key IT resources (i.e. systemic competences, infrastructural components and IT skills) are aligned with the business strategy [9, 17].

There is nevertheless still a missing roadmap that would help organizational actors operationalize, measure and assess their BITA status [9, 13, 21]. This gap has profound implications in sectors where technological repercussions are neither benign nor transparent and they are always attached to social, cultural, organizational and political reverberations; such is the sector of higher education [23].

Tertiary-level institutions have diverse business goals supporting an academic dimension but also having financial constraints. The former aims at pedagogical excellence while the latter strives for breaking even revenues and costs (as a non-profit organization). The multifaceted mission of such institutions complicates the reconciliation of technological necessities with other institutional priorities and demands of various faculty stakeholders [3, 21, 33]. For instance, higher education institutions are time-pressured to choose and deploy technologies not proven to add value to education. This is further exacerbated by the incompatibility between the monthly (even weekly) cycles of technological advancements and the annual (or biannual) planning cycles in academia. Under these circumstances, Luftman and Kempaiah [16] rank the education sector very low in terms of BITA maturity compared to other industrial sectors while the study of Byungura [6] reports findings indicating a strong misalignment between IT and the provided services in higher education institutions.

The Haute Ecole Léonard de Vinci

HE Vinci is an institution of higher education based in Brussels, Belgium (and Louvain-la-Neuve); it is subsidized by the Belgian French Community. It was formed in 1995 through the grouping of a number of pre-existent institutions and it is currently functioning as a non-profit organization collaborating with other educational institutions in Belgium.[1] HE Vinci offers short cycle (bachelors), long cycle (masters) and specialized tertiary education in fields like Health, Human and Social Sciences and Science/Technology. HE Vinci is considered as a university college meaning that it is a state-funded institution of higher education belonging to one of the three Communities of Belgium, and is not specifically a university.

[1]More information about its structure can be found: http://www.vinci.be/fr-be/Pages/La-haute-%C3%A9cole.aspx.

GDPR Implementation Project at HE Vinci

General Data Protection Regulation (GDPR) came into effect in 2018 aiming to reinforce the rights of EU/EEA's citizens to informational privacy. These regulations require organizations with more than 250 employees to maintain records of personal data processing activities. Accordingly, higher education institutions have to reevaluate and adapt their data privacy procedures considering that they process not only personal data about past, current, and prospective students/employees but several other sensitive data entries related to collaborative projects, research initiatives and educational activities [18]. In essence, higher education institutions are urged to assume the role of *data control centers* overseeing internally the objectives of (personal) data processing activities while reporting them externally to other civic, data safeguarding agencies [10].

Article 30(1)[2] of the GDPR reports that, at minimum, the records of processing activities should include:

- the name and contact details of the data controller and, where applicable, the joint controller, the controller's representative and the data protection officer;
- the purposes of the processing;
- a description of the categories of data subjects and of the categories of personal data;
- the categories of recipients to whom the personal data have been or will be disclosed including recipients in third countries or international organizations;
- where applicable, transfers of personal data to a third country or an international organization, including the identification of that third country or international organization and the documentation of suitable safeguards;
- where possible, the envisaged time limits for erasure of the different categories of data;
- where possible, a general description of the technical and organizational security measures referred to in Article 32(1).[3]

The HE Vinci wishing to comply with this new regulatory obligation, has set up a GDPR working group since September 2017, comprised of computer scientists, data managers and a legal division employing a Data Protection Officer (DPO) and the GDPR project leader. The first objective of this group was to be able to create and support records of (personal) data processing activities. Within this scope, a database (Project 'Médor') was developed to support the HE Vinci with the maintenance of these records. The GDPR working group decided to develop this database internally 'from scratch' as one of the recurrent problems encountered on the market at that particular time was the lack of availability of data inventories adapted to the field

[2]Regulation (EU) 2016/679 of the European Parliament and of the Council of 27 April 2016 on the protection of natural persons with regard to the processing of personal data and on the free movement of such data, and repealing Directive 95/46/EC (General Data Protection Regulation).

[3]More information about the Article 32(1) can be found here: https://eur-lex.europa.eu/eli/reg/2016/679/oj.

of higher education. Even though the database was developed for internal use, the establishment of a partnership with the HE 'Galilée' in 2018 has made the database (as well as its inventories) available to the partnering institution. Since then, other HEs asked to join the database development project. In May 2019, a second partnership was established with Namur University (UNamur)in Belgium. This partnership is based, among other things, on the joint effort to enrich the inventories supporting the records of processing activities.

Aside from rendering the data controllers (HE Vinci at first and the collaborating institutions in a later phase) predominately accountable for any personal data they possess, Project Médor is meant to infuse an end-to-end, privacy-oriented viewpoint. Therefore, the working group had to develop a database that would provide not only a generic overview of the processing of personal data in the context of the HEs' activities, but also an overview of the high-end services, processes, actors, and storage locations involved in the treatment of these records. At the end of January 2018, the working group decided to develop the database internally. The requirements' analysis phase was carried out by the GDPR working group's data managers since they would be the main database users. The order for its development was then placed within the IT department. The working group established that the tool had to be based on ten inventories (developed in advance) whose elements had to meet a defined terminology. These inventories were developed in parallel with the database between February and March 2018. The first version of the Médor database was available in March 2018.

Research Paradigm

Our primary goal is to help alleviate the remissness of BITA in the section of higher education through the use of goal-oriented modeling and the NFR framework in particular. In this perspective, our study follows the paradigm of design science [12] combined with case study research [22].

This research follows the paradigm of design science in the sense that it uses artifacts developed previously in [8, 27] and instantiates them on a real-life case. This allows to bring further validation to these artifacts and study the lessons learned. Based on these instantiations, we suggest improvements for BITA-evaluation notably in higher education.

Case study research provides the opportunity to compare different theories and observations from empirical data [19, 25] and fits particularly well for software engineering research. Case study research has been applied here in the sense that a complete case has been performed and studied in its entirety by the research team.

Research Method and Steps

Data Collection and Validation

In order to understand what personal data is processed within HE Vinci's daily activities, an interview phase was organized in two constituent HE Vinci institutes (the 'Marie Haps' Institute and the 'Parnasse-ISEI'). The data managers prepared a comparative analysis from the results of the two 'pilot' institutes between July and September 2018. The purpose of this analysis was to detect all preliminary differences in processing practices for HE Vinci and gather them in a summarizing document. The first analysis of this document was carried out in January 2019 and it prioritized on the processes relating to academic affairs. The institutional pre-registration and registration processes were the first ones to be addressed, then the validation phase.

Determining the Business Objectives

Since one of our first major goals in this case study research was to represent the generic (business) objectives of the HE Vinci, we had to search and analyze the legal texts that regulate the field of higher education in the Belgian French Community. At that time, we also went through an internal document of the institution called *Priorities of the HE Vinci going towards 2020* describing its strategic objectives as they were established by HE Vinci's top management. The legal texts that were identified[4] mention that higher education in the Belgian French Community is a public service of general interest; they stipulate that one of the missions of higher education institutions is to "transmit, both through the content of the teaching and through other activities organized by the establishment, humanist values, creative and innovative traditions, as well as the artistic, scientific, philosophical and political cultural heritage, the historical foundations of the teaching content, while respecting the specificities of each individual". In short, legal compliance by higher education institutions is more of an obligation than a mission. This legal compliance of any public service is reinforced by the fact that higher education institutions have a mission to transmit philosophical and political values. Our analysis is based primarily on the 'Decree of March 31th 2004' (also known as the "Bologna Process") and secondarily to the 'Decree of November 7th 2013' (also known as the "Paysage Decree") since they are the mandates that dictate the missions in higher education. The former describes the stipulations regarding the integration of higher education systems across Europe into a single European Higher Education Area (EHEA) while the latter document specifies the landscape of higher education and the academic organization of studies within the Wallonia-Brussels Federation in Belgium. We started reviewing, codifying, comparing and identifying the missions of higher education described in

[4]They are presented in Table 27.1.

Table 27.1 Relevant decrees taken as input in the research

Decree 31th March 2004	Decree 7th November 2013
Institutions have to offer high quality initial and continuing education, according to their habilitations, and thus certify the skills and knowledge acquired by their graduates	Institutions have to offer initial and continuing higher education and training courses, corresponding to levels 5 to 8 of the French-speaking framework of qualifications, and to certify the corresponding knowledge and skills acquired, at the end of the study cycles or by valorization of personal, professional and training achievements; Institutions also have to transmit—both though the content of teaching and through other activities—humanistic, cultural, artistic, scientific, philosophical and political values as well as be responsible for creative and innovative traditions
Institutions have to participate in research and/or creative activities in their discipline	Institutions have to participate in individual or collective activities of research, innovation or creation and thus ensure the development, conservation and transmission of knowledge and cultural, artistic and scientific heritage
Institutions have to provide services to the community, in particular through collaboration with the educational, social, economic and cultural world	Institutions have to provide services to the community through their specialized expertise and their duty of independence, listening to societal needs, in collaboration or dialogue with the educational, social, cultural, economic and political communities

these legal texts along with the strategic objectives included in the *Priorities of the HE Vinci going towards 2020*. Table 27.1 presents the main points of both decrees. We determined eventually that they converge in three primal mission objectives: (i) *Provide high-quality education*, (ii) *Provide services to the community* and (iii) *Participate in research activities*.

Determining the Strategic Objectives and the GDPR Implementation Project's Goals

Besides the generic mission objectives that can be instantly accessible to anyone, there are long-term objectives that shape the business strategy of the institution and are internally confined to C-level management. The HE Vinci keeps an internal documentation of these objectives forming the strategic plan of the institution. The elements identified and refined in the mentioned documentation have been the subject of reflection with the top management of HE Vinci. Our final summation of these elements was validated by the director of the institution; so this was done internally at HE Vinci.

As far as the IT project is concerned, we determined its initial broadly-based goals with the exclusive contribution of the top management of HE Vinci and the GDPR working group. At a later stage, we organized interviews and participated in working meetings with various internal and external stakeholders (to HE Vinci) in order to define the requirements of the Médor software. During this phase, members of our research team worked together with various project stakeholders from the partnering institutions of HE Vinci, HE Galilée and UNamur with previous experience in software engineering and data processing inventories. We collected data through semi-structured interviews and working meetings organized with prospective users of the Médor software from the partnering institutions. We focused our attention on members responsible for developing and maintaining the records of processing activities. The information from the working meetings was compared and combined with the resulting material from the organized interviews (specific to the collection and prioritization of requirements). Table 27.2 summarizes the organization of these meetings for the determination of the IT project objectives and the elicitation of the software-specific requirements.

Model-Driven Business and IT Alignment Evaluation of the GDPR Implementation Project

This section introduces the application of the MoDrIGo approach to the HE Vinci's GDPR implementation project case study. Section "General (High Level) Mission of Higher Education Institutions" introduces a generic representation of the strategic objectives of higher education institutions in the French community of Belgium. While the latter representation is valid for all of these institutions, the representation built in Section "Modeling the Strategic Objectives of HE Vinci" depicts the strategic objectives of HE Vinci. These are further decompositions of the general strategic objectives and are specific to HE Vinci. Finally, Section "Business and IT Alignment of the GDPR Project" depicts the GDPR implementation project's specific goals and traces their impact on HE Vinci's strategic objectives for BITA evaluation.

General (High Level) Mission of Higher Education Institutions

This section builds a conceptual model to depict a representation of Belgian French Community's higher education institutions' highest level objectives in the form of a set of strategic objectives. The relevant concepts to understand the representations made in this section are summarized in Fig. 27.1. Our representation can be applied to any higher education institution whose mission is shaped by the 'Decree of March 31th 2004' promoting the integration of higher education systems across Europe

Table 27.2 Performed meetings and interviews

Partnering institution	Participating member	Date and type of data collection
HE Galilée	Lawyer	4th of July 2018 (informal exchange of information)
HE Galilée	Lawyer	30th of July 2018 (meeting discussing the requirements of the Médor software and the data inventory management)
HE Galilée	Computer Scientist	19th of August 2019 (meeting discussing the requirements of the Médor software and the data inventory management)
UNamur	DPO	23rd of November 2018 (meeting regarding the requirements of the Médor software)
UNamur	DPO	29th of January 2019 (meeting regarding the requirements of the Médor software)
UNamur	DPO	14th of May 2019 (meeting regarding the data inventory management)
UNamur	DPO	2nd of August 2019 (semi-structured interview)
HE Vinci	GDPR working group	7th of February 2019 (meeting with the GDPR working group with the exclusive participation of Data Managers)
HE Vinci	GDPR working group	6th of November 2018 (meeting with GDPR working group with the exclusive participation of Data Managers/Project Leader/DPOs)
HE Vinci	GDPR working group	18th of February 2019 (meeting with GDPR working group with the exclusive participation of Data Managers)
HE Vinci	GDPR working group	31st of July 2019 (semi-structured interview with the Data Managers)
HE Vinci	GDPR working group	13th August 2019 (semi-structured interview with the Project Leader)

Fig. 27.1 Relevant concepts and their icons

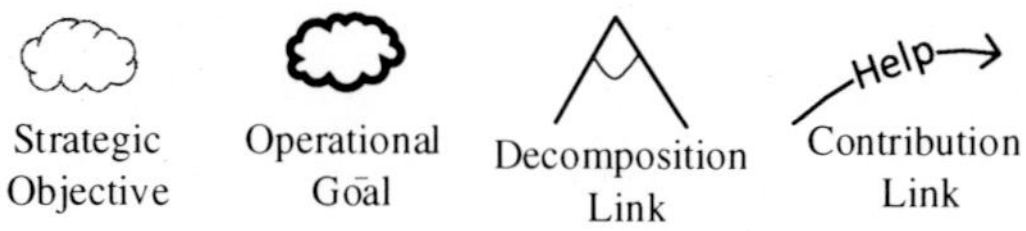

into a single European Higher Education Area (EHEA). Using this mandate as a legislative source, we have prioritized three main higher education mission objectives; (i) *Provide high-quality education*, (ii) *Provide services to the community* and (iii) *Participate in research activities*.

The first mission objective prescribes the pedagogical excellence that any EHEA-incorporated institution is compelled by law to deliver; it can be further decomposed in two sub-objectives:

- *Respect the right to privacy*. All the phases that a prospective student has to undergo during his/her educational journey (i.e. registration to a particular educational curriculum, exam registrations, receiving individual exam grades etc.) must be (it is a legal obligation) organized considering the student's right to privacy. The same right covers all personnel and various faculty stakeholders involved in the provision of educational services;
- *Transmit philosophical and political values*. The provision of teaching or other educational activities within the specialized domain of every higher education institution must include the transmission of political and philosophical values and its historical foundations.

The second mission objective stipulates the provision of services that such institutions aspire to offer though international and intercommunity-level collaborations. It can be further decomposed in the following sub-objectives:

- *Set up collaborations with partnering institutions*. EHEA-institutions have the freedom to establish collaborations and partnerships with other institutions and organizations in order to reinforce and augment their teaching, research, communicative and skill-augmenting abilities of its students and personnel;
- *Manage collaborations with the educational, social, economic and cultural community*. The establishment of collaborations is followed by the responsibility to manage and sustain them in the best possible way.

The third mission objective ensures the fulfillment of the academic identity of higher education institutions through their expectation to participate in high quality research activities. This top-level objective can be further elaborated in two distinct sub-objectives:

- *Participate in research and innovation activities*. A higher education institution has to fundamentally cultivate innovation and engage in high quality research in one or several domains;
- *Ensure knowledge development*. Higher education institutions have to create the structures, processes and mechanisms encouraging the continuous development,

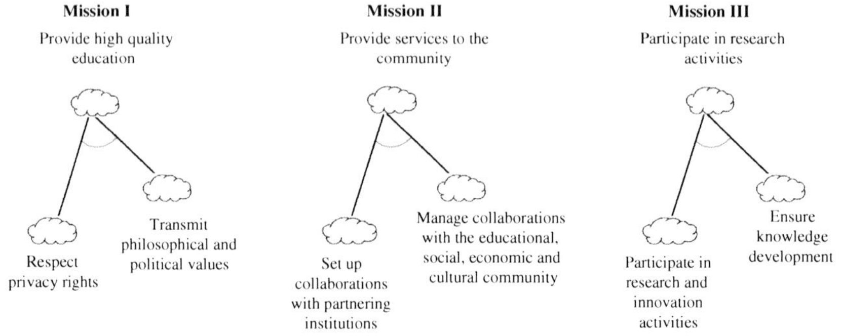

Fig. 27.2 General high level mission of higher education institutions

conservation and dissemination of knowledge for their academic, research and administrative personnel.

Figure 27.2 represents the generic strategic missions of higher education institutions; these need to be further refined and instantiated to specific higher education institutions.

Modeling the Strategic Objectives of HE Vinci

Compliance-wise, HE Vinci has to be aligned primarily with the 'Decree of March 31th 2004'[5] and secondarily with the 'Decree of November 7th 2013'.[6] The latter follows the legislative spirit of the former while specifying the landscape of higher education and the academic organization of studies within the Wallonia-Brussels Federation in Belgium. The overall (business) strategy of HE Vinci is considered complete when the three broadly-based mission objectives are elaborated further to fit the specificities of its strategic objectives.

Strategic objectives are part of the business or IT strategy pursued by an organization; they are perceived as targets that the organization aims to reach within a long time horizon in order to gain and/or sustain a competitive position [27]. In our case, all the strategic elements were extracted from an internal document[7] detailing the strategic plan of HE Vinci towards the year 2020. These strategic objectives were identified and validated by the top management of the institution; their decomposition is represented in the strategic-level diagram of Fig. 27.3.

[5]More info on the 'Bologna Process' can be found on: https://ec.europa.eu/education/policies/higher-education/bologna-process-and-european-higher-education-area_en.

[6]Also known as the 'Paysage Decree'. More information can be found on: https://www.gallilex.cfwb.be/document/pdf/39681_008.pdf.

[7]Internal documentation 'Priorities of the HE Vinci going towards 2020'.

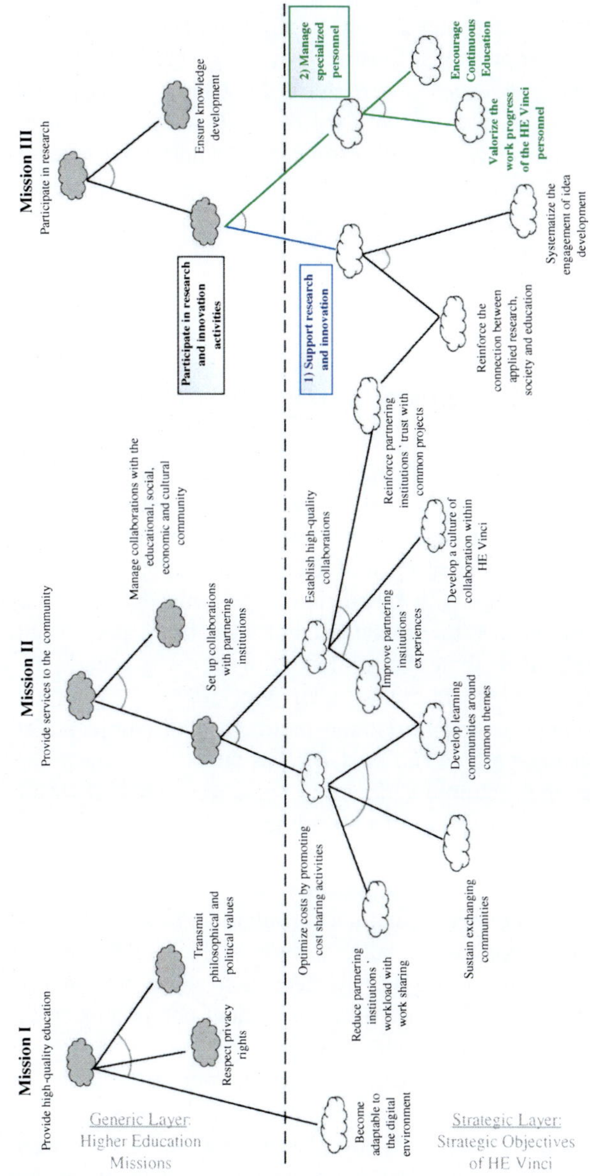

Fig. 27.3 Representation of the overall strategy of HE Vinci

We observe that parent-level elements within the strategic layer are the outcome of higher level decompositions. Illustratively, the generic sub-objective *Participating in research and innovation activities* can be achieved through satisficing the *(1) Support that will be placed into these research and innovation activities* and the *(2) Management of specialized personnel.* Satisficing the latter would entail the strategic *Encouragement of continuous education* as well as strategically *Valorize the work progress of the HE Vinci personnel.* So realizing the participation in research and innovation activities entails the realization of each of the strategic objectives mentioned previously.

Business and IT Alignment of the GDPR Project

This section describes the alignment between the business objectives discussed in the previous section and the IT project that aims to support GDPR compliance at HE Vinci. Figure 27.4 illustrates the newly-introduced project objectives as operationalizations assisting the achievement of the targeted business objectives and examines their inter-dependencies with the previously recognized strategic objectives.

What follows next is the instantiation of some of the project objectives as illustrative examples describing the fulfillment of the three generic missions and their accompanying strategic objectives for HE Vinci.

The working group involved in this particular endeavor identified the *Rendering of the HE Vinci compliant with the GDPR* as the primal project objective. This can be further decomposed into various sub-objectives which describe, among others, the maintenance of a record of processing activities or updating the institution's documentation in terms of the new GDPR legislation. We have to note at this point that are many more project sub-objectives that can be derived from this primal objective; these have been identified during the consultation phase with the top management and the GDPR working group but they have not been added in the formal diagram because of readability issues. We have decided to add significant sub-objectives that can support this design science exercise without compromising the letter and the spirit of the project's implementation goals.

Hence, the above-mentioned primal project objective can be decomposed into sub-objectives, among which we distinguish here: (i) *Adapt the institution's official documents to the new GDPR* and (ii) *Develop and support the record of processing activities for HE Vinci*; the latter yields two direct satisficing objectives:

- *Reduce the number of personal data breach incidents* **(Fulfillment of Mission I)**;
- *Ensure the accuracy within the records of processing activities* **(Fulfillment of Mission I)**.

Even though the project was originally purposed for the institution's internal use, it was decided at a later stage to be shared among various educational institutions[8]

[8]The higher education institution 'Galilée' and University of Namur.

in Belgium with the intention to create a cost-sharing and innovative consortium. Hence, the *Develop and support record of processing activities* sub-objective can be further elaborated in (i) *Set up a collaboration among partnering institutions around the Médor project* **(fulfillment of Mission II)** and (ii) *Propose an innovative solution for the amelioration and support of the records of processing activities* **(Fulfillment of Mission III)**. The latter yields two direct satisficing goals:

- *Improve the Médor software.* An optimized software tool accelerates firstly the identification of personal data utilizing resources and secondly the gradual automation in the encoding of the records of processing activities. The latter enables the records' maintainability in the long run by saving time and effort from the personnel having to develop and maintain them manually. The satisficing of that goal influences directly the fulfillment of the strategic objective *Valorize the work progress of the HE Vinci personnel.* As seen in Section "Modeling the Strategic Objectives of HE Vinci", the latter is elicited indirectly from the *Participation in research and innovation activities*;
- *Improve inventory management.* This goal facilitates the content management of the inventories that support the records of processing activities. This enables long-term time savings and workload reduction for the personnel responsible for the maintenance of these inventories. The satisficing of that project goal influences directly the fulfillment of the strategic objective *Valorize the work progress of the HE Vinci personnel* stemming from the higher education sub-objective calling for *Participation in research and innovation activities.*

The previous illustrations support the alignment between business and IT project-related objectives in distinct stages. We observe that the realization of specific project objectives contributes positively to the realization of broadly-based higher education (business) objectives.

Discussion

The Benefits of Creating a Consortium Around the IT Project

The above delineation of the strategic objectives and IT-project goals allows us to think critically of the initial state of loose collaboration among the partnering institutions in terms of managing the IT project. Under that form of collaboration, HE Vinci had the primal role in developing (and maintaining at a later stage) the software solution while bearing the projects' accompanying costs. The partnering institutions could provide a secondary contribution to the project without pooling their efforts and benefiting from the latest updates in the data inventory management process. It is indicative that at some point, HE Vinci was considering sharing the data inventories with its partners in an Excel format making them practically unmanageable. However, the strategic representation and project evaluation of Fig. 27.4 demonstrates

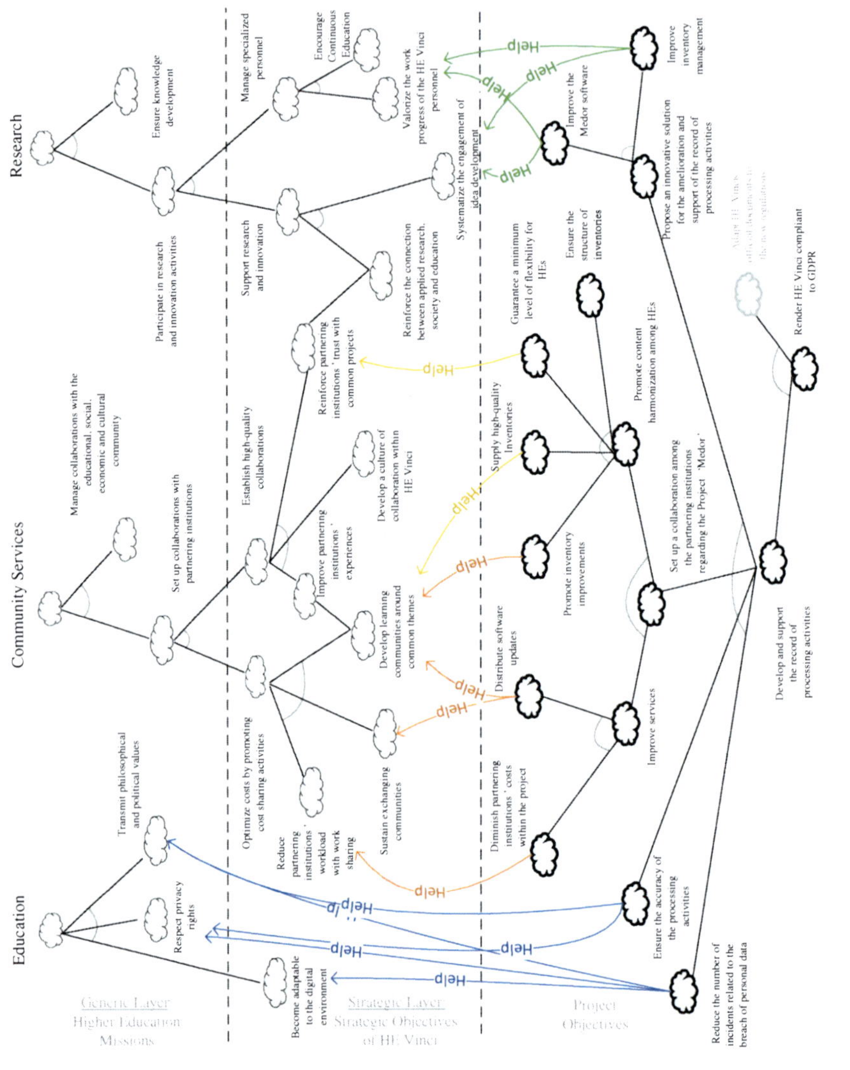

Fig. 27.4 Representation of the alignment between the long term strategic objectives and the GDPR implementation project's operational goals at HE Vinci

that the achievement of the majority of goals relies on the establishment of a collaborative form that entails the exchange of innovative solutions for the development and amelioration of the software tool. This condition is not only in alignment with the second mission objective, but it also facilitates the reconciliation between the IT project objectives and the generic business objectives in the middle layer. In general terms, Barringer and Harrison [4] converge that consortia (as collaborative forms) are able to:

- Enable the pooling of efforts and costs in terms of software development and maintenance;
- Guarantee the neutrality of a specific IT project;
- Allow for an independent mode of governance with its overall IT strategy temporarily oriented towards technology and innovation. In our case, the use of novel technologies such as machine learning and augmented reality could be regarded as opportunities supporting the continuous fulfillment of the projects' objectives.

The above considerations necessitated the creation of a consortium of collaboration among the partnering institutions around the IT project, instigating a signed agreement between HE Vinci, Unamur and HE Galilée. These agreements stipulate that the Médor software is made available to the partnering institutions, free of charge, in exchange for their contributions in terms of inventories and software development and maintenance. They also detail that HE Vinci remains the owner of the software and its accompanying inventories but can still be used by the partnering institutions. When it comes to the procedure of sharing the data inventories among the partnering institutions, a proposal for the development of a contemporary inventory management system has been placed by the data managers of the HE Vinci. The validation of the proposal is still ongoing.

Of course, the establishment of such a collaboration would render it in a constant state of 'coopetition'; this refers to the state of simultaneous cooperation and competition allowing the partnering institutions to increase combined welfare through cooperation while maximizing individual gains through competition [20]. Therefore, the consortium would have to create change management mechanisms to handle the dynamic nature of collective and individual member objectives over time; resolution processes to alleviate mismatch between the consortium members' objectives and the interests of the partnering institutions' stakeholders that do not participate in it or even network members pursuing partly conflicting objectives[32]. Future studies could determine how do these partnering institutions handle this constant state of coopetition.

More Lessons Learned from the Case Study

The case study in higher education has been mostly driven by the business strategy. The IT strategy of such institutions is something that is still not a formal element in this development. The incorporation of a formal IT strategy next to the business one would have created more complex representations that would be harder to deal with. Similarly, we could have pushed the analysis further by restructuring the IT offer in terms of Business IT Services like suggested in [27, 29]. This would nevertheless have introduced unnecessary complexity for the purpose of the study so that we proceeded through project's goal representation rather than a more complex service structuring.

Finally, while performing the case study we understood the importance of having a supporting Computer-Aided Software Engineering (CASE) tool supporting the developments. We consequently extended the Descartes-architect tool [1, 14] for full method support.

Conclusion

This paper provided a case study for BITA evaluation in an higher education institution. Even though we describe an IT project with a specific aim (supporting GDPR compliance), the developed representations can serve as pattern to be customized by any higher education institution (i) sourcing its mission objectives from the Decree of 31th March 2004 (and beyond) or (ii) even more generally higher education institutions having the triple mission of education, research and community services. To this extent, we contribute to BITA research by offering practical directions and prerequisites whose satisfaction guarantees BITA in the section of higher education. We also negotiate ways of further enabling BITA with the creation of the consortium and the prerequisites that it should satisfy.

Our representations also allow to study the dual positioning of higher education institutions as publicly subsidized institutions that need to break-even financially. This is interesting considering that literature often addresses BITA concerns considering one dimension at a time.

Acknowledgements The authors want to thank all the people that participated to this research and in particular Mrs. Valérie Biéva for her active support and effort she provided during this research.

References

1. The Descartes architect case-tool (2020). http://www.isys.ucl.ac.be/descartes/
2. H. Alghamdi, L. Sun, Analysis of the fit of learning and management systems in higher education institutions: A case study from Saudi Arabia. Int. J. Bus. Econ. Affairs **2**(4), 260–267 (2017)
3. H. Alghamdi, L. Sun, Business and it alignment in higher education sector. Integration **10**, 1 (2017)
4. B.R. Barringer, J.S. Harrison, Walking a tightrope: Creating value through interorganizational relationships. J. Manage. **26**(3), 367–403 (2000)
5. I. Brown, I. Motjolopane, Strategic business-it alignment, and factors of influence: A case study in a public tertiary education institution: Reviewed article. South African Comput. J. **2005**(35), 20–28 (2005)
6. J.C. Byungura, Improving IT Integration for Higher Education Institutional Performance: Towards a Contextualised IT-Institutional Alignment Model. Ph.D. thesis, Department of Computer and Systems Sciences, Stockholm University (2019)
7. J.C. Byungura, H. Hansson, Development and validation of a holistic it-institutional alignment model for higher education institutional performance, in *Society for Information Technology &*

Teacher Education International Conference (Association for the Advancement of Computing in Education (AACE), 2019), pp. 1279–1290
8. L. Chung, B. Nixon, E. Yu, J. Mylopoulos, *Non-functional Requirements in Software Engineering* (Kluwer Academic Publishers, 2000)
9. T. Coltman, P.P. Tallon, R. Sharma, M. Queiroz, Strategic IT alignment: Twenty-five years on. JIT **30**(2), 91–100 (2015). https://doi.org/10.1057/jit.2014.35
10. A. Habbabeh, B. Schneider, P.M. Asprion et al., Data privacy assessment: An exemplary case for higher education institutions. Int. J. Manage. Knowl. Learn. **8**(2), 221–241 (2019)
11. J.C. Henderson, H. Venkatraman, Strategic alignment: Leveraging information technology for transforming organizations. IBM Syst. J. **32**(1), 472–484 (1993)
12. A.R. Hevner, S.T. March, J. Park, S. Ram, Design science in information systems research. MIS Q. 28(1), 75–105 (2004). http://misq.org/design-science-in-information-systems-research.html
13. A. Karpovsky, R.D. Galliers, Aligning in practice: From current cases to a new agenda. JIT **30**(2), 136–160 (2015). https://doi.org/10.1057/jit.2014.34
14. M. Kolp, Y. Wautelet, *Descartes architect: Design case tool for agent-oriented repositories, techniques, environments and systems* (Université catholique de Louvain, Louvain-la-Neuve, Belgium, Louvain School of Management, 2007)
15. M. Kolp, Y. Wautelet, Human organizational patterns applied to collaborative learning software systems. Comput. Human Behavior **51**, 742–751 (2015). https://doi.org/10.1016/j.chb.2014.11.094
16. J.N. Luftman, R.M. Kempaiah, An update on business-it alignment: "a line" has been drawn. MIS Q. Executive 6(3) (2007). http://misqe.org/ojs2/index.php/misqe/article/view/135
17. J.N. Luftman, R. Papp, T. Brier, Enablers and inhibitors of business-it alignment. CAIS **1**, 11 (1999). http://aisel.aisnet.org/cais/vol1/iss1/11
18. M.G. Marković, S. Debeljak, N. Kadoić, Preparing students for the era of the general data protection regulation (GDPR). TEM J. **8**(1), 150 (2019)
19. D.M. McCutcheon, J.R. Meredith, Conducting case study research in operations management. J. Oper. Manage. **11**(3), 239–256 (1993)
20. V. Pant, E.S.K. Yu, Modeling simultaneous cooperation and competition among enterprises. Bus. Inf. Syst. Eng. **60**(1), 39–54 (2018). https://doi.org/10.1007/s12599-017-0514-0
21. J.A. Pirani, G. Salaway, Information technology alignment in higher education (2004)
22. P. Runeson, M. Höst, A. Rainer, B. Regnell *Case Study Research in Software Engineering - Guidelines and Examples* (Wiley, 2012). http://eu.wiley.com/WileyCDA/WileyTitle/productCd-1118104358.html
23. I. Snyder, S. Marginson, T. Lewis, 'an alignment of the planets': Mapping the intersections between pedagogy, technology and management in Australian universities. J. Higher Educ. Policy Manage. **29**(2), 187–202 (2007)
24. P.P. Tallon, K.L. Kraemer, Investigating the relationship between strategic alignment and information technology business value: The discovery of a paradox, in *Creating Business Value with Information Technology: Challenges and Solutions* (IGI Global, 2003), pp. 1–22
25. A.H. Van de Ven, Nothing is quite so practical as a good theory. Acad. Manage. Rev. **14**(4), 486–489 (1989)
26. A. Visvizi, M.D. Lytras, A. Sarirete *Management and Administration of Higher Education Institutions at Times of Change* (Emerald Publishing Limited, 2019)
27. Y. Wautelet, A model-driven it governance process based on the strategic impact evaluation of services. J. Syst. Software **149**, 462–475 (2019)
28. Y. Wautelet, S. Heng, M. Kolp, L. Penserini, S. Poelmans, Designing an MOOC as an agent-platform aggregating heterogeneous virtual learning environments. Behaviour & IT **35**(11), 980–997 (2016). https://doi.org/10.1080/0144929X.2016.1212095
29. Y. Wautelet, M. Kolp, Business and model-driven development of BDI multi-agent systems. Neurocomputing **182**, 304–321 (2016)
30. Y. Wautelet, M. Kolp, Conceptual modeling as a tool for corporate governance support: State of the art and research agenda, in *Research & Innovation Forum 2019 - Technology, Innovation,*

Education, and their Social Impact, RIIFORUM 2019, Rome, Italy, April 24–26, 2019 ed. by Visvizi, A., Lytras, M.D. (Springer, Berlin, 2019), pp. 535–548. https://doi.org/10.1007/978-3-030-30809-4_49
31. Y. Wautelet, M. Kolp, S. Heng, S. Poelmans, Developing a multi-agent platform supporting patient hospital stays following a socio-technical approach: Management and governance benefits. Telematics Inf. **35**(4), 854–882 (2018). https://doi.org/10.1016/j.tele.2017.12.013
32. I. Winkler, Network governance between individual and collective goals: Qualitative evidence from six networks. J. Leadership Organ. Stud. **12**(3), 119–134 (2006)
33. R. Yanosky, J.B. Caruso *Process and Politics: It Governance in Higher Education* (ECAR Key Findings, 2008)

Chapter 28
AI Solutions and Productivity of Public Services: Insights from Poland

Aleksander Surdej

Abstract The strategic objective of improving the quality of public services in modern democracies is increasingly being pursued with the use of a broad range of AI tools. Three areas seem to display a potential for efficiency enhancing transformations, namely: energy, public health and transportation. The paper describes the conditions for effective implementation of AI technologies in these areas including public regulation and expected changes in citizens' behavior. Public policy dilemmas will be identified guided by the research question how countries can leapfrog their peers and gain on efficiency by applying AI-enhanced solutions.

Keywords Artificial intelligence · Public services · Public regulations · Trustful AI

Introduction

Public services are key components of the basket of entitlements and rights citizens expect from their governments in modern democracies. Citizens expect that these services are provided for free or guaranteed at highly subsidized prices. Unequal access to them is one of main causes of public discontent [1]. AI tools have potential to enhance the productivity of provision and to improve quality of many public services [2]. Governments, thus, can increasingly rely on AI tools to support the execution of administrative tasks, to deliver services and to do more complex analytical activities. This in turn can contribute to the increase of public sector productivity and in general to quality improvements of public services. But, achieving these goals requires a smart design of public policies to take advantage of applied-technology-enabled innovative tools and it can be achieved only thanks to a holistic approach [3].

How can countries get ready to take advantage of AI tools to improve the productivity of public services [4]? Figure 28.1 presents the approach applied in this exploratory paper which helps to focus on three essential factors: the supply of reliable

A. Surdej (✉)
Department of European Studies and Economic Integration, Cracow University of Economics, Rakowicka 27, 31-510 Krakow, Poland
e-mail: aleksander.surdej@uek.krakow.pl

A. Visvizi et al. (eds.), *Research and Innovation Forum 2020*,
Springer Proceedings in Complexity,
https://doi.org/10.1007/978-3-030-62066-0_28

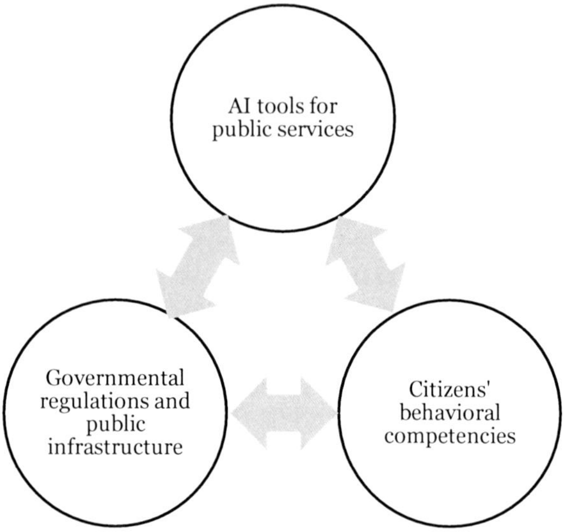

Fig. 28.1 Dimensions of AI potential in public services. *Source* Author

AI tools, the development of efficient regulatory framework and high quality public infrastructure and the capability of citizens to efficiently use available technologies framed by public regulations.

The development of AI tools is driven by continuous technological progress, better and cheaper sensors, powerful computers, cloud computing services and advances in "machine learning" (ML) [5]. Based on these technical achievements there has been observed strong progress in creating applications in areas such as natural language processing (NLP), AVs (autonomous vehicles) or computer vision. Growing investments by venture capital in AI start-ups promise that new commercial and public applications will continue to emerge. The investments in AI tools for services in health, energy and transportation sectors are dominated by private enterprises [6].

In technological progress there is a mysterious component of human creativity [7], and technological advances cannot be planned. But, inventions usually find their ways to commercialized applications [8] thanks to the profit drive of private enterprises.

Yet, in the field of AI the role of governments is especially important as public authorities are asked to fulfill multiple tasks like protecting data privacy [9] and regulating their transfers without which markets will not be efficient. Governments thus need to lay down regulatory frameworks for data collection, storage and usage. Their actions need to complement businesses [10] by taking actions to foster digital transition and reap possible benefits emerging from the development of AI tools especially in their core area of responsibility, namely the provision of public services.

The public policies needed to support AI capacity building include support for AI ecosystems [11], development of digital infrastructure and designing regulatory framework that boosts investment and innovation,[1] improved STEM education [13], and programs for continuous skills upgrading. Governments should also support the diffusion of AI tools in smaller firms as it is likely that adoption of AI tools will be uneven: it will happen faster in large enterprises where the human capital of employees is higher.

It is important to prepare citizens to approach AI tools with an appropriate degree of trust. Individuals usually do not know how much to rely on new technologies and will be thorn between an excessive trust in automated decision-making systems, which may lead to high costs (losing money on investments, getting incorrect medical diagnoses or suffering the costs of software failures) and almost atavistic distrust of them. Yet, distrusting faulty automated instructions leads also to losses as people could benefit from taking advantage of computer systems to get the support in making judgements on complex situations. Progress in ML (machine learning) must be accompanied by improvements in human capacity to make judgements [14].

Following the general approach developed by Naude et al. [15] this paper explores public policy dilemmas for effective implementation of AI tools in three crucial sectors dominated by public regulations, public financing or public delivery, namely public health, energy and public transportation searching to answer how the government can lead the AI developments in the interests of higher productivity and better quality of public services. An exploratory insight into the case of Poland will serve to verify these research questions.

AI-Enabled Solutions and Their Productivity Potential

There is no simple and precise definition of AI.[2] One may rather think of AI technologies as tools to enhance human brain power thanks to faster and more powerful computers, which apply complex machine-learning algorithms[3] to big data sets[4] in order to identify patterns and predict outcomes, recommend actions or even take decisions. Possible applications of AI are broad and mostly complementary to human (natural) intelligence.

---

[1]D. Rodrik suggests to guide technological developments to the benefit of lower skilled employees since (a) they have so far paid high costs of adjustments to new technologies, and (b) it is possible to steer technological developments more to the benefit of traditional middle classes—[12].

[2]As an example of definition one can give this one: AI is "The theory and development of computer systems able to perform tasks that normally require human intelligence" ([16], p. 36).

[3]"An algorithm is a specific set of instructions for carrying out a procedure or solving a problem, usually with the requirement that the procedure terminate at some point. Specific algorithms sometimes also go by the name method, procedure, or technique."—see: https://mathworld.wolfram.com/Algorithm.html.

[4]Mullainathan and Spies write: "The phrase "big data" emphasises a change in the scale of data. But, there has been an equally important change in the *nature* of this data. Machine learning can deal with unconventional data that is too high dimensional for standard estimation methods, including image and language information that we conventionally had not enough thought of as data set we can work with, let alone include in a regression" ([17], p. 99).

Despite recent significant advances AI tools are still in their early stage of development, but the pace of progress is rapid and possible applications may significantly transform many areas of economic activities [18]. At present AI is perceived through its successful, but still at a limited scale, applications in computer vision to track movements of objects, animals or humans, in natural language processing, in virtual assistants, smart robotics, and in autonomous vehicles (AVs). The diffusion of productive AI solutions might be slow since they are dependent on costly infrastructure and expensive skilled labour.

While, being realistic about the expected benefits generated by AI technologies we can envisage the expansion of their applications to improve the efficiency of the delivery of public services. The AI seem to exemplify the productivity paradox of all great innovations: they initially do not generate any significant productivity growth, but later lead to deep transformations in many areas [19]. This phenomenon is additionally amplified by the specificity of public services which are plagued by obstacles to productivity improvements.[5]

AI is a kind of general-purpose technology that has capacity to improve productivity across many areas. Yet, the materialization of this potential takes time and requires significant earlier investments in physical infrastructure, in organizational changes, in managerial and cognitive skills. The measurement of AI's contribution to public service productivity growth encounters the problem of measurement common to all areas where partially inputs and outputs are intangible [20].

AI has the potential to support the increase of productivity in public services since it helps to create and to exploit the economies of scale. AI-based automation is likely to make many routine tasks obsolete and depress the demand for administrative jobs [21]. The provision of public services is defined by laws and procedures, they are to be homogenous and supplied continuously. In principle, it should be possible to automate or eliminate many of them contributing thus to the growth of labor productivity and to the increase of total productivity. This in turn could generate significant budgetary savings.

AI-Enabled Solutions in Public Health

In its basic functions of improving predictions, identifying recommendations and, in certain situations, automatically making decisions, AI tools offer large possibilities in their applications to public health [22]. These applications range from timely and precise surveillance of contagious diseases like coronavirus (Covid-19) to offering individualized recommendations to persons with symptoms that might signal a given disease. AI tools, like NLP, may exploit the variety of sources to early identify a threat to public health: not only an information collected from general practitioners (GPs) that report a high frequency of certain symptoms, but also from other sources like posts on social media. Data scientists, and not only medical doctors, will be needed to

[5]For instance, procedural complexity and stricter employment regulations.

map out the patterns in outbreaks of communicable and non-communicable diseases [23].

AI might become a powerful tool in the medical research speeding up the discovery of new drugs. For instance, a new antibiotic effective against untreatable diseases was discovered by MIT biological engineers who "were using a deep-learning algorithm developed by computer scientist Regina Barzilay, which was trained to analyse the structure of 2500 molecules, including current antibiotics and other natural compounds such as glucose, to determine their anti-bacterial potency" [24].

AI might support the improvements in medical diagnosis. Thus, for instance, algorithms can examine the data from thermal imaging of breasts looking for the asymmetry of their thermal structure, which can be correlated with cancer thus better selecting the patients for further and more costly examinations [25].

Software can help medical doctors to complete medical documentation and thus to diagnose more patients during the same working period [26]. Furthermore, the accuracy of the diagnosis can be improved since having images stored on their computers medical doctors can consult their colleagues in case of doubt and reduce the frequency of medical errors [27].

Increasingly, patients them-selves monitor their health, which leads to improved self-care and well-being. The control of weight and fight against obesity is supported by new applications such as *Obegital*[6] and by the usage of smartphones to track the number of steps or to be reminded about the need of physical activities. Self-monitoring of health parameters like respiratory rate, heart rate and body temperature generates the data that used in machine learning (ML) procedures identify persons who might need medical treatments. Electronic equipment generating medical data allows a medically trained person to significantly increase the number of monitored patients leading to significant productivity improvements.

In short, AI technologies promise and provide methods to reduce waste in health services by supporting prevention, avoiding unnecessary treatments, prioritizing medical interventions to the benefit of medical effectiveness and the general increase of productivity of medical services.

Yet, the applications of AI to health services encounters several serious obstacles. First and foremost, the governments need to introduce regulations that guarantee a high degree of the security of the medical data as the right to privacy of treated persons needs to be respected. Availability of various types of data might be needed to improve medical services. Thus, for instance, researches can work on genetic records not only of living but also of dead persons. These persons and their families expect that such data are protected, stored and made available only for research purposes and in formats that defend them from personal identification [28].

The effectiveness of public regulations is important not only for the protection of privacy. There is also a need for assuring sufficient technical standards for AI tools since if applied to medical diagnoses, they might identify "false positives" generating stress and uneasiness of users. AI medical tools, in their various applications, need to be tested and made reliable and this calls for their oversight by public

[6]Obegital developed by start-up Fedmind—see: https://www.fedmind.com/obegital/.

health safety regulator [29]. The government might need to order specific AI tools monitoring public health issues and helping to nudge citizens choices. Their efficient implementation depends on public regulations and trust.

AI-Enabled Solutions in Energy Sector

Electricity is a special commodity as it needs to be supplied continuously, because it cannot be easily stored. The share of electricity in the total energy consumption is supposed to increase in coming decades [30], while the electricity production needs to be increasingly based on low emission resources, mainly renewables, due to growing environmental constraints.

Renewables, especially solar and wind energy, are characterized by intermittency. Their integration into electricity production and distribution requires reliable technical and organizational solutions as supply and demand for electricity need to be matched in real-time. In addition, the production of renewable energy is dispersed—their integration into the electricity grids need to be locally matched as the losses on long distance electricity transmissions are significant.

AI can support the flexible functioning of electricity grids in two ways [31]. First, it can serve to diminish congestion and improve the capacity to transport electricity where there is an adequate demand. Second, AI can help to identify the patterns of supply and demand in order to pool available electricity resources in response to the changes in electricity consumption. As a result, AI tools can help to territorially optimize the supply and transportation of electricity [32].

As electricity supply increases are limited by upper boundary of deployed power generation capacity, faster and cheaper adjustments are expected from consumers and especially households than from electricity producers. Proper electricity market models allow to price the energy efficiently to encourage individual consumers to use it when it is cheaper and abundant, and save it when it is in short supply and more expensive. If the energy producers apply real time pricing mechanisms, households are exposed to incentive to save electricity [33]. Electricity consumers can thus improve their decisions since they are expected to react to real time information about the electricity price, its consumption and budgetary expenditures on electricity [34]. In addition, the deployment of digital ("smart") electricity meters is expected to generate on average 3% energy saving [35].

Energy markets require a specific type of policy intervention to take advantage of AI solutions: in this case it is more about the access to the data than about their sensitivity. General legal and competitiveness rules should be sufficient to trigger energy savings promised by AI tools.

AI-Enabled Solutions in Public Transportation

The future of transportation is coloured by a vision of the advent of autonomous electric vehicles (AVs), piloted by artificial intelligence and serviced by other "intelligent" machines [36]. Due to complex regulatory problems (safety, liability etc.) this might be a still distant future. In the short- to medium term public policy perspective the application of AI solutions to transportation systems will be differentiated and gradual.

There are different levels and dimensions of the usage of AI in transportation sector. At the individual level AI tools improve the driving security, increase the comfort of driving and may even entirely displace drivers. Many modern cars integrate for instance AEBS (Advanced Emergency Braking Systems), which use AI tools and allow to detect obstacles ahead of a car reducing thus the number of accidents by 50% [37], but a full potential of AI is demonstrated by the development of autonomous vehicles (AVs) in which the automated driving system performs the entire dynamic driving task without any constraints.[7]

AI based tools help to reduce road congestion, reduce driving time and consumed fuel thanks to the availability of real time adjusted road maps signaling to drivers the itineraries to select. Such information is a public good in itself, although it is supplied by private (commercial) agents. As a result AVs and driver supporting AI tools generate positive economic impacts thanks to fewer crashes, less congestion and saved lives.

Public transportation systems would also benefit from the development of AVs and other AI tools. Direct cost savings might come from the reduction in the number of employed drivers (eliminating the need for a human driver) and savings due to reduction in the number of buses and trams needed to serve the same number of passengers.

AVs and other AIs tools need to be taken into account in the planning of the development of public transportation of the future. These new technologies raise however a number of serious regulatory issues including safety, liability (the consequences of the damage liability in case of AI's errors), infrastructure (separating or combining human drivers and AVs), reliability of equipment (especially sensors) and access to data to train and adjust systems. These are complex solutions which need to be first piloted and only next implemented on a larger scale.

Policies need to address the legal issues of liability of AVs operators and regulatory agencies need to monitor technical standards of the equipment. These are crucial factors for the public trust in AVs and other AI solutions in the transportation industry and public transportation specifically.

[7]See: The Society of Automotive Engineers International's—standard J3016. For more information—https://www.sae.org/news/2019/01/sae-updates-j3016-automated-driving-graphic—accessed 5 February 2020.

Table 28.1 Assessing Poland's regulatory framework for AI solutions

Regulatory dimension		Early stage	Advanced stage	Mature stage
AI General Regulations (for instance regarding algorithm accountability)		XXX		
AI relevant civil law regulations (for instance regarding liability or employment rules)			XXX	
Industry specific AI regulations	Health		XXX	
Industry specific AI regulations	Transportation	XXX		
Industry specific AI regulations	Energy			XXX
Data privacy protection			XXX	

Source Author

Assessing Poland's Readiness for AI's Based Productivity Growth in Public Services

The World Bank indicated that to fully benefit from the digital revolution, countries need to improve, sometimes strengthen, but more generally make more efficient, regulations that ensure market competition, invest in workers' skills to adapt them to the digitalized economy and to make public institutions effectively accountable [38].

For any government the initial departure point is a review and improvement of existing legal and regulatory solutions. In Poland the development of AI solutions has become the explicit public policy goal and the government has launched the process of preparing the *Polish AI Strategy for the years 2019–2027* [38] to frame its actions (Table 28.1).

Except for energy markets Poland still needs to work on publication regulations to allow speedy dissemination of AI solutions. This is especially so since in Poland, as in many other EU member states, there has been an acceleration of the developments of AI tools. These developments are business-led [7]: it is expected that until 2025 there will be in Poland approx. 750 companies developing AI applications. In the examined cases of health, energy and transportation services commercially developed applications can serve the public purpose, but there are also solutions (like disease monitoring or health education) that will be developed only if the government inserts them in its policies towards a given sector.

In summary, against the background of other OECD member states Poland is well placed to take advantage of AI solutions: in the Oxford Insights' the Government AI Readiness Index 2019 and, among Central European countries is only second to Estonia (see Graph 28.1).[8]

[8]The Index measures a country readiness to use AI tools in the provision of public services and in particular "public governance", data and infrastructure and education—more see: Oxford Insights, https://www.oxfordinsights.com/ai-readiness2019—accessed on 31 March 2020.

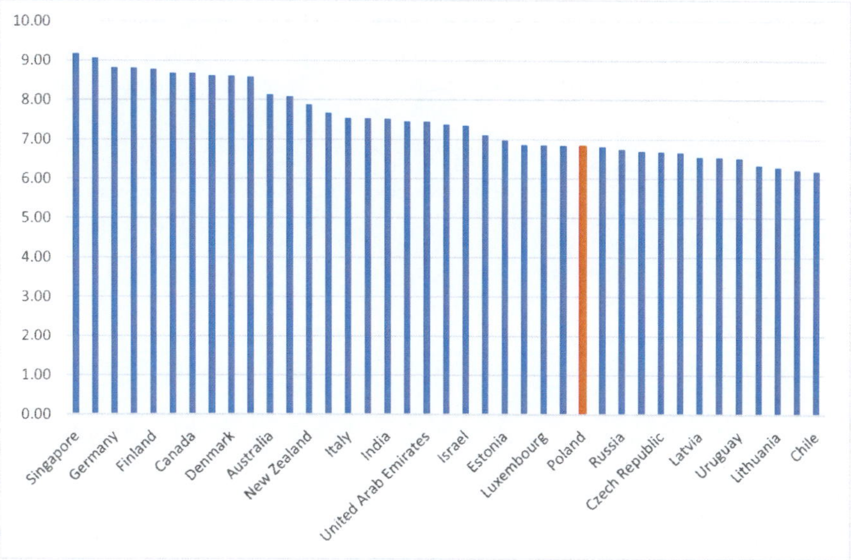

Graph 28.1 Government AI readiness index 2019. *Source* Oxford Insights, https://www.oxfordinsights.com/ai-readiness2019—accessed on 31 March 2020

Yet, the positive effects of public policies directed to enable effective dissemination of AI solutions to public services are cumulative in nature and require time to get materialized -their success depends on public trust in technology and institutions. Subsequent Polish governments have been working on preparing a comprehensive package of actions to make Poland the leader of "Trustworthy AI".[9]

In addition, large public investments are made to strengthen the system of continuous education, which should complement good results of Polish students in 2018 PISA tests [40], to increase the number of scientific doctorates and to improve research conditions in Polish universities in order to prevent the outflow of talents (Table 28.2).

The AI promoting policy tools need be fine-tuned to better target support for relevant research and the transformation of their results into products and solutions that can be exploited for public policy purposes.

In the above mentioned *the Polish AI Strategy* knowledge, skills, financing, infrastructure and data are interlinked and integrated into the concept of AI Ecosystem, whose complexity needs to be taken into account when searching for efficient and effective policy interventions [42]. Poland feels special responsibility to maintain high trust in AI applications—thus it wants to create an efficient regulatory framework for AI, which in turn needs also to be compatible with EU solutions [43].

The Polish government is planning to order AI-enabled products that will be useful to provide various social services: from education to health. Such purchases

[9]Trust in AI tools requires that their reliability and lack of biases is assured—[39].

Table 28.2 Poland and other central European countries: AI-enabling competencies

Country	I-Com Industry 4.0 Index Score 2017 [Highest = best]	Fixed broadband subscriptions per 100 inhabitants, by technology, June 2019 [Highest = best]	IMD Ranking on Digital Competitiveness Ranking, 2019 (of 63) [Lowest = best]	Global Cyber-security Index Score, 2018 [Highest = best]	Density of Industrial Robots in 2016 per 10,000 employees [Highest = best]	The EU Innovation Scoreboard 2019 [Highest = best]
Bulgaria	64	NA	45	0.721	NA	49
Czech Republic	78	30.9	37	0.569	101	89
Hungary	68	32.1	43	0.812	57	81
Lithuania	85	27.2	30	0.908	NA	69
Poland	**66**	**19.6**	**33**	**0.815**	**32**	**61**
Romania	53	NA	46	0.568	15	34
Slovakia	77	28.2	47	0.729	131	69
Slovenia	84	29.3	32	0.701	137	88

Source Author's compilation based on data from IMD Digital Competitiveness Index; Compagnucci et al. [41] and International Federation of Robotics data), http://www.oecd.org/internet; https://ec.europa.eu/growth/industry/policy/innovation/scoreboards_en

will be made according to the rules of public procurement in order to choose the most reliable and effective products. A high priority will be given to the development of the standards for AI medical tools and to the assurance of their quality. Public data will be made accessible to the extent possible in order to help develop reliable products and the execution of *the Polish AI Strategy* will be in the hands of a committee working in direct contact with the office of Prime Minister. Already existing program GovTech Polska serves to help public administration to design and introduce solutions based on AI and modern technologies.

In sum, the governments of Poland are aware of the need to guide the development of AI tools since their effective applications depend on good regulations and readiness of enterprises and citizens to use them. Yet, it seems that the discussed *Polish AI Strategy* is not sufficiently targeting the development of pro-productivity AI applications in public services.

Conclusions

This research supports the thesis that the promise and potential inherent in AI solutions to improve the productivity of public services requires comprehensive responses that would help public authorities, citizens and business stakeholders to undertake and sustain complementary activities [44]. The development and delivery of AI-enabled smart public policies does not assume a top-down centralizing approach, but a framework that enables the collaboration between all stakeholders of a given public policy.

Productivity advances needed, and possible, in public services will be feasible, if governments create right regulatory framework, encourage private investments and induce the cooperation of interested parties Beyond regulating, the strategic objective of improving the quality of public services calls governments to pursue policies to assure the even quality of digital infrastructure, to build safe storages, to guarantee indiscriminate access to data, and high protection of their privacy, to improve AI literacy and STEM education. Only such a comprehensive agenda will create conditions to sustainably exploit the AI potential for productivity growth across the economy and especially in public services.

Literature

1. D. Rodrik, Mettre la technologie au service des emplois déclassés. *Les Echos*, 18 Mar (2020)
2. McKinsey & Company, *Rewolucja AI. Jak sztuczna inteligencja zmieni biznes w Polsce* (AI Revolution. How AI will Change Business in Poland), Warszawa (2017)
3. A. Visvizi, M.D. Lytras (eds.), *Politics and Technology in the Post-Truth Era, Bingley* (Emerald Publishing, UK, 2019). ISBN 978-1-78756-984-3
4. Y.S. Lee, B. Larsen, M. Webb, M. Cuéllar, How Would AI Regulation Change Firms' Behavior? Evidence from Thousands of Managers. SIEPR Working Paper 19-031 (2019)
5. J.F. Stanilko, R. Kroplewski (ed.) *Polityka rozwoju sztucznej inteligencji w Polsce na lata 2019–2027* (The Policy of AI Development in Poland for the years 2019–2027), Warszawa (2019)
6. OECD, *PISA Country Note: Poland*, OECD, available at: https://www.oecd.org/pisa/publications/PISA2018_CN_POL.pdf, accessed: 6 March 2020 (2019b)
7. J. Mokyr, *The Lever of Riches* (Oxford University Press, Technological Creativity and Economic Progress, 1992)
8. A. Koene, C. Clifton, Y. Hatada et al., *A Governance Framework for Algorithmic Accountability and Transparency* (European Parliament, Brussels, 2019)
9. M.S. Bargh, S. Choenni, R. Meijer, On addressing privacy in disseminating judicial data: Towards a methodology. Transforming Gov People, Process Policy **11**(1), 9–41 (2017)
10. A. Rodríguez-Pose, The revenge of the places that don't matter (and what to do about it). Cambridge J. Reg. Econ. Soc. **11**(1), 189–209 (2017). https://doi.org/10.1093/cjres/rsx024
11. J.S. Winter, E. Davidson, Governance of artificial intelligence and personal health information. Digital Policy, Regul. Gov. **21**(3), 280–290 (2019)
12. M. Short, Applications of artificial intelligence in the electricity supply industry. Aslib Proc. **47**(6), 153–160 (1995)
13. L. Lyczko, *Sztuczna inteligencja – kluczowe aspekty prawne* (AI – Key Legal Aspects), PWC Warszawa, available at: https://www.pwc.pl/pl/artykuly/2018/sztuczna-inteligencja-kluczowe-aspekty-prawne.html. Accessed 6 Mar 2020 (2018)
14. F. Liu, Addressing STEM in the context of teacher education. *J. Res. Innov. Teaching & Learn.* Vol. ahead-of-print No. ahead-of-print (2020)
15. OECD, *Artificial Intelligence in Society*, Paris (2019a)
16. Deloitte, *Global Manufacturing Competitiveness Index* (2018)

17. W. Naude, A. Surdej, M. Cameron, Ready for Industry 4.0? The Case of Central and Eastern Europe, in *Industry 4.0 and Engineering for a Sustainable Future*, ed. by M. Dastbaz, P. Cochrane (Springer, Berlin, 2019)
18. Le Monde, Yann Le Cun, Nul Besoin d'avoir peur d'un scenario a la Terminator pour l'IA» *Le Monde*, 9–10 Fevrier (2020)
19. E. Plastino, M. Purdy, Game changing value from Artificial Intelligence: eight strategies. Strategy Leadersh. **46**(1), 16–22 (2018)
20. P. Dunleavy, Public sector productivity: Measurement challenges, performance information and prospects for improvement. *OECD J. Budgeting* 17/1 (2017)
21. G.M. Cortes, N. Jaimovich, C.J. Nekarda, H.E. Siu, The Micro and Macro of Disappearing Routine Jobs: A Flows Approach. NBER Working Paper No. 20307 (2014)
22. World Bank, *World Development Report 2016: Digital Dividends* (2016)
23. Economist (The), New World Curriculum. Briefing Covid-10, *The Economist*, 7 Mar (2020)
24. FT, AI discovers antibiotics to treat drug-resistant diseases, *Financial Times*, 20 Feb 17. IEA (2018) *World Energy Outlook*, Paris (2020)
25. FT, Femtech' adapted to benefit women and girls in poorer countries, *Financial Times*, 29 Nov (2019)
26. M. Cavallone, R. Palumbo, Debunking the myth of industry 4.0 in health care: insights from a systematic literature review. *TQM J.* Vol. ahead-of-print No. ahead-of-print (2020)
27. Accenture, *Artificial Intelligence: Health Care's New Nervous System*, available at: https://www.accenture.com/_acnmedia/PDF-49/Accenture-Health-Artificial-Intelligence.pdf#zoom=50. Accessed: 4 Mar 2020 (2017)
28. J.P. Simon, Artificial intelligence: Scope, players, markets and geography. Digital Policy, Regul. Gov. **21**(3), 208–237 (2019)
29. A. Lauterbach, Artificial intelligence and policy: Quo vadis? Digital Policy, Regul. Gov. **21**(3), 238–263 (2019)
30. ITF, *Safer Roads with Automated Vehicles?* (International Transport Forum, Paris, 2018)
31. A. Visvizi, M.D. Lytras, E. Damiani, H. Mathkour, Policy making for smart cities: Innovation and social inclusive economic growth for sustainability. J. Sci. Technol. Policy Manage. **9**(2), 126–133 (2018)
32. M. Sicilia, A. Visvizi, Blockchain and OECD data repositories: Opportunities and policymaking implications. Library Hi Tech **37**(1), 30–42 (2019)
33. K.T. Chui, M.D. Lytras, A. Visvizi, Energy sustainability in smart cities: Artificial intelligence, smart monitoring, and optimization of energy consumption. Energies **11**, 2869 (2018)
34. H. Allcott, Rethinking real-time electricity pricing. Resour. Energy Econ. **33**(4), 820–842 (2011)
35. EU, Smart Metering deployment in the European Union, available on: https://ses.jrc.ec.europa.eu/smart-metering-deployment-european-union. Accessed on 5 Feb 2020 (2020)
36. A. Blitz, K. Kazi, Mapping technology roadblocks and opportunities in the transportation revolution. Strategy Leadersh **47**(4), 43–46 (2019)
37. U. Lichtenthaler, Extremes of acceptance: employee attitudes toward artificial intelligence. *J. Bus. Strategy* Vol. ahead-of-print No. ahead-of-print (2019)
38. D. Sudharshan, D. Sudharshan, Enabling technologies, in *Marketing in Customer Technology Environments* (Emerald Publishing Limited, 2020), pp. 209–248
39. A. Kuyumdzhieva, Ethics challenges in the digital era: Focus on medical research, in *Ethics and Integrity in Health and Life Sciences Research* (*Advances in Research Ethics and Integrity, Vol. 4*) (Emerald Publishing Limited, 2018), pp. 45–62
40. OECD, Digitalisation and productivity: A story of complementarities, in *OECD Economic Outlook*, vol. 2019 Issue 1 (OECD Publishing, 2019c)
41. S. Compagnucci, G. Berni, G. Massaro, M. Masulli, Thinking the Future of European Industry: Digitalization, Industry 4.0 and the Role of EU and National Policies, EU Study 3.17. iCom, institute for competitiveness (2017)
42. C. Brooks, T. Vorley, C. Gherhes, Entrepreneurial ecosystems in Poland: panacea, paper tiger or Pandora's box? J Entrepreneurship Public Policy **8**(3), 319–338 (2019)

43. Y. Ma, K. Ping, C. Wu, L. Chen, H. Shi, D. Chong, Artificial Intelligence powered Internet of Things and smart public service. Library Hi Tech **38**(1), 165–179 (2019)
44. P. Yaghmaie, W. Vanhaverbeke, Identifying and describing constituents of innovation ecosystems: A systematic review of the literature. *EuroMed J. Bus.* Vol. ahead-of-print No. ahead-of-print (2019)

Chapter 29
Assessing the Ecosystem of Innovation in GCC: Policy Implications and Strategic Directions

May Al-Khalifa, Odeh Al-Jayyousi, Rustom Mamlook, and Fairouz Aldhmour

Abstract The Gulf Cooperation Council (GCC) nations which includes six states; i.e., United Arab Emirates, Bahrain, Saudi Arabia, Oman, Qatar and Kuwait, are embarking on a transformative vision to align their economies with knowledge economy and sustainable development goals (SDGs) for 2030. This transition implies a change in the strategy, business culture, business models, mindset, and innovation policies. The center of gravity for science, technology and innovation (STI) is moving to GCC due to investment in R&D, infrastructure, open innovation networks, and human capital. The aim of this paper is to explore the determinants and enablers for a resilient ecosystem for innovation in GCC to cope with global changes. Besides, the paper will devise a set of policy implications and strategies to foster innovation in GCC and to be in line with the fourth Industrial Revolution and smart society model. The methodology for this paper will be based on document analysis. A framework for assessing innovation ecosystem will be outlined based on literature, which includes actors, technology regimes, and knowledge diffusion, and innovations systems, legal and institutional frameworks. The role of innovation policy in shaping the ecosystem of innovation will be discussed. The paper recommendations to enhance the innovative capabilities of GCC countries to be ranked among the innovative countries corresponding to GDP indicators.

Keywords National innovation system · Innovation policy · Innovation strategy · Sustainability

M. Al-Khalifa · O. Al-Jayyousi · R. Mamlook (✉) · F. Aldhmour
Innovation and Technology Management Department, Arabian Gulf University, Manama, Bahrain
e-mail: rustomm@agu.edu.bh

M. Al-Khalifa
e-mail: sh.maymak@agu.edu.bh

O. Al-Jayyousi
e-mail: odehaj@agu.edu.bh

F. Aldhmour
e-mail: fairouzm@agu.edu.bh

A. Visvizi et al. (eds.), *Research and Innovation Forum 2020*,
Springer Proceedings in Complexity,
https://doi.org/10.1007/978-3-030-62066-0_29

Introduction

The Gulf Cooperation Council (GCC) states are experiencing salient transformations to respond to global changes in energy transitions, disruptive technologies, digital platforms, and environmental challenge [1, 2]. The concept of innovation ecosystem emerged as a strategy to foster organizational capability, resilience, sustainability, and value creation in and digital economy [1]. The ecosystem of innovation is underpinned by access to markets, regulatory support, access to capital, and financial expertise. The imperative for fostering an ecosystem of innovation is shaped by GCC visions to adopt clean energy, diversify economy, and support service sector as part of readiness to post-oil era. SMEs play a vital role in supporting such an ecosystem for fostering and diffusing cleantech and meeting targets for SDGs. This paper outlines the conditions and enablers for an ecosystem of sustainable innovation in GCC.

An ecosystem for innovation includes many actors including governmental entities, venture capital, investors, incubators, R&D institutions, and end users. These actors are sustained and hosted in industrial districts and clusters within science parks or technology parks [3]. The emergence of new forms of ecosystems of innovation is part of supporting knowledge flows within the National Innovation System (NIS), Sectoral Innovation System (SIS), and regional innovation of networks to create value by utilizing Information and Communication Technology (ICT) platforms.

Innovation is the process that entails a non-linear and multi-actors-sectoral interactions to create value [4, 5]. Determinants and enablers for innovation include incentives and organizational culture [6]. The interactions between innovation activities include collaboration and knowledge diffusion. Besides, ecosystems are underpinned by innovation capability, which includes financial support, leadership, and culture.

Beyond the concepts of value chains and supply networks, the notion of ecosystem describes a broad-based definition of interactions and relationships between relevant organizations and actors as articulated in innovation management related literature [5, 7]. It is argued that planning and strategy should shift from firm level to ecosystem [8] which informs new forms of non-linear relationships between actors based on co-evolution and co-creation [7]. Moreover, in a digital era, strategic networks are recognized as a driver for competition [3].

The ecosystem provides new insights into the innovation management of networks of organizations that operate around a focal firm or a platform [5]. Models of networks entail many forms including value networks; value constellations which emphasize the linkage between social domain and economic action and shed light on the nature of innovation that is led by multiple actors [9].

Based on the above discussion, the gap of this research is the absence of an appropriate framework, which identifies the main enablers of innovation and/or determinants of ecosystems in GCC countries. Moreover, a second gap is the absence of a broader approach that addresses the overrating of the technological focus by including value, knowledge exchange and social dimension in GCC countries. Therefore, this paper will address the following questions:

What are the main determinants and enablers for an ecosystem of innovation in GCC countries?

What are the policies and implications for fostering an ecosystem of innovation in GCC countries?

This paper aims to explore the determinants and enablers for a resilient ecosystem for innovation in GCC to cope with global changes. The next sections of the paper will discuss the literature review, methodology, analysis of results, conclusions, and recommendations.

Literature Review

In this section, empirical and theoretical studies addressing ecosystems for innovation will be reviewed. Innovation policy was developed to enhance innovation capability. It is noted that new frameworks for the innovation system were referred to as the national innovation ecosystem [4]. Innovation ecosystem is a dynamic network of interrelated organizations to create a value throughout the innovation process [4].

An innovation ecosystem is a concept adopted from ecology to reflect the holistic and systems approach to innovation. It is a network of socio-technological organizations that help to analyze and sustain innovation through harnessing the knowledge flow within the Triple Helix of business, government and academic institutions [10] which influence the diffusion of innovations [4], value creation [5], co-creation [11], service needs [12], business model [12], and capabilities [6]. Innovation ecosystems is conceptualized at corporate level and open innovation [13], regional and national innovation systems [4, 14]; digital innovation ecosystems; city-based innovation ecosystems [3]; high-tech SMEs centered ecosystems [3], university-based ecosystems. Innovation ecosystems are defined by actors and business models [5].

Innovation ecosystems differ from science parks in terms of the systemic networks open innovation, linkages with actors and digital connectivity [15], niche organizations [16], market forces, emergent and complex systems [13]. Carayannis and Campbell [10] suggest that knowledge systems compete and co-evolve in innovation systems. Hage et al. [17] suggest the connectedness, knowledge and technology are key elements of innovation ecosystems. Mamlook et al. [18] and Xu et al. [19] included the notion of Total Innovation Management (TIM) theory which takes into account non-technological dimensions including culture, strategy and organization [10].

Moreover, Schot and Geels [16] emphasized niches and experimentation in the evolution of technological and socio-technical regimes.

Khorsheed [20] proposed the framework of national innovation ecosystem. Which includes; infrastructure (physical, institutional and digital infrastructure); human capital or intellectual assets, academic, education, training research personnel, training and up skilling courses for business. In addition, entrepreneurship development plans in organizations of higher learning); innovation capability and integrity (transparent procedures, Instilling high-quality governance, finest practices, and Security); inducements (admission to public support and funded research 2–4% of GDP, strategy and policy. To develop source for R&D, economic and non-economic policies to hearten R&D activities, objective policy to improve income and resources, private financial support, impartiality markets to finance technology. In addition, knowledge—concentrated companies and innovation futures market); interface (Communication between all partners, and role of transporting businesses, knowledge hubs, attractive manufacturing clusters and regional industrial clusters). Institutional coordination (universal strategy framework, appropriate ministries and organizations for STI formulation, regulating, investment and management, R&D center, R&D players, technology and knowledge transfer hubs, products. In addition, service suppliers, chambers of business and commerce, Science Parks, associations, and social networks).

In light of a critical literature review, a set of dimensions for assessing the ecosystem of innovation was proposed as outlined in Fig. 29.1. Eight dimensions were integrated with the three main research themes in innovation management as illustrated in Table 29.1.

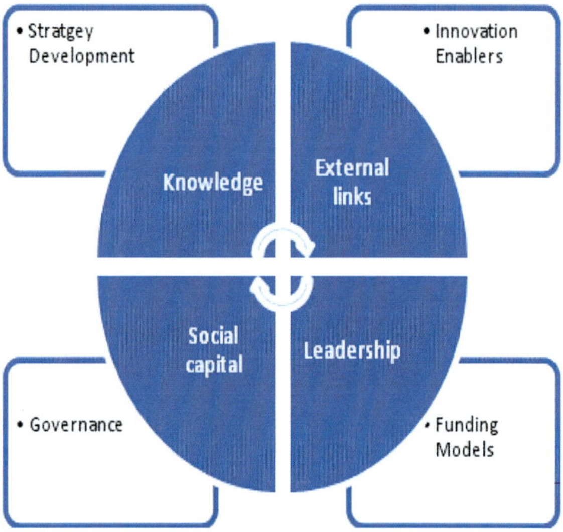

Fig. 29.1 A Framework for assessing innovation ecosystem in GCC. *Source* Adapted by authors

Table 29.1 Dimensions for the assessment of innovation ecosystem in GCC

Dimensions	Value creation	Network embeddedness	Network management
Innovation enablers	Collaboration with global actors in R&D and business	Co-creation of value through joint ventures and partnerships	Focus on organizational capabilities
Funding models	Regional funded industries	Joint ventures and global partnerships	New business model, market-based, crowd funding
Strategy development	Value chains, green value chains	Clusters and science parks	Open sources and crowd sourcing
Governance	Hierarchical	Decentralized	Organic structures
External links	University-industry linkages	National and regional innovation systems	Global networks
Knowledge management	Knowledge sharing	Knowledge creation	Co-creation and emergence
Leadership	Situational	Contextual	Transformational
Social capital	Family business model	SME-large firm linkages and regional alliances	Social enterprise

Source Compiled by authors

Methodology

This study was based on a critical literature review and document analysis (a qualitative analytical method) to explore the determinants and enablers for a resilient ecosystem for innovation in GCC to cope with global changes [21].

Data Collection

Overall, 24 documents, reports, and articles retrieved from the (Ministry of finance and National Economy of Bahrain, Information and e-Government Authority of Bahrain, government e-services, and National Higher Education strategy, information, and communications technology, and gateway to GCC information). Secondary data was reviewed based on document and content analysis during the period January–March 2020. The records collected included official and peer-reviewed journals related to GCC context.

Analysis of Results

The following analysis will shed light on the eight dimensions with reference to specific industries in Bahrain. The first-dimension addresses innovation enablers, which include legal, financial, economic policies, and ICT infrastructure. Specifically, the GCC invested in education, public sector and infrastructure to foster innovation [1, 20].

Moreover, an enabling environment for a sustainable ecosystem of innovation entails a shift in strategic direction to support green policies by adopting green business models as reflected in GPIC and SABIC strategy. Specific enablers for innovation include collaboration with global actors in R&D, co-creation of value through strategic alliances, and harnessing organizational capabilities [1, 2].

The second dimension, *funding models*, entails a variety of modalities including regional funded industries and localized ecosystems like Dhahran Innovation Valley, Al-Jubail Industrial Complex, and KAUST in Saudi Arabia, Al-Waha fund of funds and Global Fintech fund in Bahrain, joint ventures and global partnerships, and new business model, market-based, crowd funding.

Strategy development dimension covers a variety of models including value chains, green value chains, clusters and science parks, and open sources and crowd sourcing. Lundvall [22] argued the determinants of competitive advantage is evident at local level which is in line with Porter [23] notion that competitive advantage is created through localized ecosystems as manifested in clusters.

The *governance* dimension shows the strong relations of organizational structure on the agility of the innovation ecosystem. Among the variety of structures. i.e., hierarchical, decentralized, and organic structures, the latter offers more space for innovative culture. The dimension related to *external links* stresses the significance of university-industry linkages, national and regional innovation systems, and global networks. A regional innovation system (RIS) is crucial for innovation ecosystem and NIS [14]. Porter [23] argued that due to global economic competition national policies play a key role to leapfrog and catchup in technological innovation, which is evident in Bahrain and UAE Fintech implementations.

Knowledge management dimension highlights the processes of knowledge flows including knowledge sharing, knowledge creation, and co-creation and emergence. Edquist [24] outlined a set of innovation activities, which includes competence building, networking, and organizational change. National investments in innovation is a key determinant to a nation's innovation capability. The intent of an innovation system is to create an enabling environment that fosters knowledge flow and systemic linkages among sectors [24].

Leadership dimension sheds light on the linkages of leadership style; i.e., situational, contextual, and transformational and shows that the latter has a positive role on innovation ecosystem. The adoption of vision 2030 in GCC nations embodies a transformational leadership model [2].

Social capital dimension illustrates the benefit of social networks and institutional capital to nurture and sustain businesses. Many models embed different forms of social capital as reflected in family business SMEs, regional alliances, and social enterprise. The environment conducive for innovation include technology companies and networks of venture capital to finance start-ups [25]. The GCC strategic intent is to develop a knowledge-based economy, underpinned by innovation system [1]. Despite the fact that GCC countries invested in education and technology adoption and diffusion, this was not reflected in the innovation capability, which was attributed to cultural factors [26].

Conclusions and Recommendations

The GCC strategy is to transform to a model of sustainability and knowledge-based economy. Despite the fact that GCC countries invested in ICT infrastructure and public services the innovation capability needs to match high innovative nations. The GCC country readiness had improved during the period 2000 and 2012. However, only UAE and Qatar had attained an innovator country status and there is no positive correlation between GDP growth and the innovation index in most GCC countries.

GCC had established a sound innovation ecosystem like the model of Research University like KAUST and technological city, Qatar Foundation, Masdar City, Kuwait Foundation for Advancement of Science (KFAS) and Kuwait Institute for Scientific Research (KISR). Besides, GCC has invested in nurturing innovation ecosystems as illustrated in UAE and Bahrain models by establishing clusters in technology, financial services, media, and education as manifested in Masdar City, Internet City, Media City, Knowledge Village, and Bahrain Fintech.

References

1. O.R. Al-Jayyousi, *Integral innovation: New worldviews* (2017)
2. A.O. El-Kholei, O.R. Al Jayyousi, Technol.-Driven Innov. Municipalities **2018**, 24–48 (2019). https://doi.org/10.4018/978-1-5225-9012-5.ch002
3. A.V. Anttiroiko, *City-as-a-Platform: The Rise of Participatory Innovation Platforms in Finnish Cities* (2016)
4. R.M. Yawson, The ecological system of innovation: A new architectural framework for a functional evidence-based platform for science and innovation policy. SSRN Electron. J. **2**, 48–55 (2009). https://doi.org/10.2139/ssrn.1417676
5. S. Brusoni, A. Prencipe, *The Organization of Innovation in Ecosystems: Problem Framing, Problem Solving, and Patterns of Coupling* (Emerald Group Publishing Limited, 2013)
6. M. Moussa, A. McMurray, N. Muenjohn, A conceptual framework of the factors influencing innovation in public sector organizations. J. Dev. Areas. **52**, 231–240 (2018). https://doi.org/10.1353/jda.2018.0048
7. R. Adner, R. Kapoor, Value creation in innovation ecosystems: how the structure of technological interdependence affects firm performance in new technology generations. Strateg. Manag. J. **31**, 306–333 (2010). https://doi.org/10.1002/smj.821
8. M. Iansiti, R. Levien, Keystones and dominators: Framing operating and technology strategy in a business ecosystem. *Harvard Bus. Sch. Work. Pap.* 3–61 (2004)

9. R. Normann, R. Ramírez, From value chain to value constellation: Designing interactive strategy. Harv. Bus. Rev. **71**, 65–77 (1993)
10. E.G. Carayannis, D.F.J. Campbell, "Mode 3" and "Quadruple Helix": Toward a 21st century fractal innovation ecosystem. Int. J. Technol. Manag. **46**, 201–234 (2009)
11. R. Ramírez, Value co-production: Intellectual origins and implications for practice and research. Strateg. Manag. J. **20**, 49–65 (1999)
12. H. Chesbrough, Business model innovation: Opportunities and barriers. Long Range Plann. **43**, 354–363 (2010). https://doi.org/10.1016/j.lrp.2009.07.010
13. J. Zhang, L. Chen, The review of SMEs open innovation performance. Am. J. Ind. Bus. Manag. **04**, 716–720 (2014). https://doi.org/10.4236/ajibm.2014.412077
14. B.Å. Lundvall, *National Systems of Innovation: Toward a Theory of Innovation and Interactive Learning* (Pinter Publishers, London, 1992)
15. M.L. Fetters, P.G. Greene, M.P. Rice, J.S. Butler, The development of university-based entrepreneurship ecosystems: Global practices. Edward Elgar Publ. (2010)
16. J. Schot, F.W. Geels, Strategic niche management and sustainable innovation journeys: Theory, findings, research agenda, and policy. Technol. Anal. Strateg. Manag. **20**, 537–554 (2008). https://doi.org/10.1080/09537320802292651
17. E. Hage, J.P. Roo, M.A. van Offenbeek, A. Boonstra, Implementation factors and their effect on e-Health service adoption in rural communities: A systematic literature review. BMC Health Serv. Res. **13**, 19 (2013)
18. R. Mamlook, O. Al-Jayyousi, S. Alabyooki, S.A.A. Attallah, *Smart City and Technology Transfer: Towards Sustainable Infrastructure in Bahrain* (2019)
19. Q. Xu, J. Chen, Z. Xie, J. Liu, G. Zheng, Y. Wang, Total innovation management: A novel paradigm of innovation management in the 21st century. J. Technol. Transf. **32**, 9–25 (2007). https://doi.org/10.1007/s10961-006-9007-x
20. M.S. Khorsheed, The National Innovation Ecosystem: A vehicle toward knowledge-based economy. Int. J. Bus. Manag. Study. **2**, 43–45 (2015)
21. A. Strauss, J. Corbin, *Basics of Qualitative Research (3rd ed.): Techniques and Procedures for Developing Grounded Theory* (SAGE Publications, Oak, 2008)
22. Lundvall, B.-Å., *Innovation, Growth and Social Cohesion: The Danish Model* (Edward Elgar Publishing Ltd., 2002)
23. M.E. Porter, Location, competition, and economic development: Local clusters in a global economy. Econ. Dev. Q. **14**, 15–34 (2000)
24. Edquist, C.: *Systems of Innovation: Perspectives and Challenges* (Oxford University Press, 2006)
25. M. Castells, Globalisation, networking, urbanisation: Reflections on the spatial dynamics of the information age. Urban Stud. **47**, 2737–2745 (2010)
26. O.M. Nour, S.S.: science and technology development indicators in the arab region. Sci. Technol. Soc. 10, 249–274 (2005). https://doi.org/10.1177/097172180501000204

Chapter 30
A Chatbot for Training Employees in Industry 4.0

Mario Casillo, Francesco Colace, Massimo De Santo, Marco Lombardi, and Domenico Santaniello

Abstract Within what is called the Fourth Industrial Revolution, one of the problems that companies frequently are experiencing, in order to ensure themselves, their products and services over time, is the need for continuous employee training. If continuing training is a problem, e-learning could represent a natural solution. However, to be effective, e-learning should be both a company's instrument, which allows monitoring and provides reliable results and an easily accessible tool to the end-user. It should provide an agile, simplified and meaningful path on an educational, cognitive and relational level. Chatbots are a tool that is used both in the e-learning sector and in the industry 4.0 paradigm. The chatbots could allow to build individual learning paths and monitoring the learning phase. This paper aims to propose a system capable of providing a constant, reliable and friendly help through a practical and helpful bot, which takes advantage of NLP techniques. In particular, the proposed chatbot acts as a reminder following the user during his company training, ready to provide, when needed, useful teaching material to complete the tailored educational path. A prototype has been developed and tested in the real scenario with promising results.

Keywords Industry 4.0 · E-learning · Chatbot · Context aware computing

M. Casillo · F. Colace · M. De Santo · M. Lombardi (✉) · D. Santaniello
DIIN, University of Salerno, Fisciano, SA, Italy
e-mail: malombardi@unisa.it

M. Casillo
e-mail: mcasillo@unisa.it

F. Colace
e-mail: fcolace@unisa.it

M. De Santo
e-mail: desanto@unisa.it

D. Santaniello
e-mail: dsantaniello@unisa.it

A. Visvizi et al. (eds.), *Research and Innovation Forum 2020*,
Springer Proceedings in Complexity,
https://doi.org/10.1007/978-3-030-62066-0_30

Introduction

The Industry 4.0 term refers to the so-called Fourth Industrial Revolution: a set of enabling technologies, which support the automation processes and are able to exchange a vast amount of data during the production chain [1]. This process is enabled by the advent of the Internet of Things (IoT) paradigm, cognitive computing and cloud computing. All of which can lead to evolution from a traditional industry to a smart industry [2]. The IoT, which nowadays is used in several areas [3–7], represents the paradigm of smart devices able to connect and to communicate with each other through the Internet. This paradigm allows a real revolution within always-connected industries in order to exploit data to optimise their production processes. In particular, the economic accessibility of modern technology has encouraged its increasing use by integrating it into several sectors and production cycles. This advent has provided both large and small companies with the opportunity to experiment with new management, communication and training models among employees [8].

In particular, employee training is crucial to properly teach procedures and company "know-how", which are useful to workers job activities. With the coming of Industry 4.0 and modern low-cost technologies, many companies interested in the growth and updating of the human capital available, push to make their "know-how" and company procedures accessible to employees in a simple, immediate and effective way even outside working hours and on the road. E-learning, which is capable of building accurate and monitorable training plans, could represent an efficient tool to meet the company's training needs. According to this, it is possible to move forward in the design of tailor-made, engaging and meaningful educational experience, which can make use of Virtual Coach. The learning process is based on human-machine interaction, which can create personalised paths, interweaving the experiences of the employee with those of his colleagues through Chatbots.

Chatbots are probably the best tool in personalised training in the workplace that does not involve high costs for the company, and that makes the learning activity pleasant and not dull at the same time. If an employee needs support for a given activity, an e-learning chatbot could effectively guide him towards a specific training path. The user can freely consult the material, and the courses based on their personal needs, or the chatbot will—according to the company's directives—communicate a mandatory training course and the time to complete it. Companies will be able to monitor the status of employee training in real-time. A chatbot can also act as a virtual assistant in a training path. If the user has not completed a course provided by the company, the bot can remind him to proceed indicating the possible terms to do so. This type of function always puts the users in a position to correctly complete their training and help them not to forget the required learning module.

This paper aims to present a framework able to design tailored educational experiences supported by chatbots, which are capable of providing personalised recommendations through friendly approaches. The chatbots also allow to build individual learning paths and provide quick and precise reviews. Taking advantage of NLP techniques, the proposed approach aims to provide a constant, reliable and friendly help through an active and helpful chatbot, which is able to provide useful teaching material and acts as a reminder following the user during his company learning path.

This paper is organized as follows. In section two are presented the related works. In section three we describe the System Architecture of the proposed approach. In section four, we assessed the effectiveness of our methodology through a preliminary test activity. Finally, in section five, we draw conclusions and future research directions.

Related Works

Natural language is widely used in human-computer interaction due to its efficiency, flexibility and user-friendly communication. Currently, many are the applications that concern Natural Language Processing in the paradigm of Industries 4.0, where the aim is an increasing interaction and cooperation between man and machine [9]. Many industries, in different sectors, are using chatbots in order to support workers during production phases and to monitor real-time processes by collecting data that can be analysed by making decisions remotely. In [10], a chatbot has been proposed, equipped with an interface, which allows direct communication to the customer. An interesting example is FinderBuddyis [11], which is a chatbot designed to assist the procurement process in industries. It is able to analyse data and provide support on different features such as checking the availability of various items in the inventory, preparing or cancelling orders, reviewing past orders, and so on. Besides, it is able to monitor bids making it easy to supply materials at affordable prices and better quality. In [12] is proposed a chatbot that is able to coordinate and assist workers in production and which acts as an intelligent assistant.

Also, in the field of e-learning, chatbots have recently been used with encouraging results, becoming a reference for the distance education sector [13]. They can represent a more natural and versatile interface for users by offering tailor-made services and support based on interests and habits [14]. A fascinating application was to assimilate a chatbot to a virtual assistant. In [15], the authors propose AutoTutor an intelligent tutoring system able to have a conversation with the user providing help and support as a real tutor. AutoTutor has been shown to provide users with the same support as a real tutor while maintaining the same educational standards under certain conditions.

As regards the teaching of algebraic analysis, in [16] is introduced Ms. Lindquist, a virtual assistant has been implemented in order to support the teaching paths. It does not offer specific instructions or learning paths but allows students to learn through practice. It has been demonstrated, through an experimental phase, that students who use this system tend to remember the problem-solving phases better than those who do not use it. The system has been able to contribute significantly through its ability to engage in conversation with users by motivating them and keeping their attention alive. Besides, in [17, 18], chatbots have been proposed to offer e-learning, in the field of cultural heritage, tailored to users based on their interests, habits and context.

Therefore, it seems that the main peculiarity of chatbots used in the field of learning is the ability to exploit the communicative aspect able to makes them particularly familiar. The ability of such systems to collect data that can be analysed or exchanged, in case of special needs, with teachers is no less critical. In the field of e-learning, making the best use of natural language could lead users to increase their level of self-assessment and critical capacity, helping them to understand their strengths and fill gaps [19].

The purpose of this work is to combine two areas where chatbots are currently used: Industry 4.0 and E-learning. In this way, it is possible to provide a system that can assist a company's employees in training in a smart way through a virtual assistant. Besides, such a system can be useful for companies to monitor and manage the training processes of their employees.

System Architecture

This section deals with the architecture of the proposed system. This architecture, shown in Fig. 30.1, will presented through the modules of which it is composed, its features and its core, i.e. the approaches and methodologies that exploit the inferential engine to reach the proposed aims.

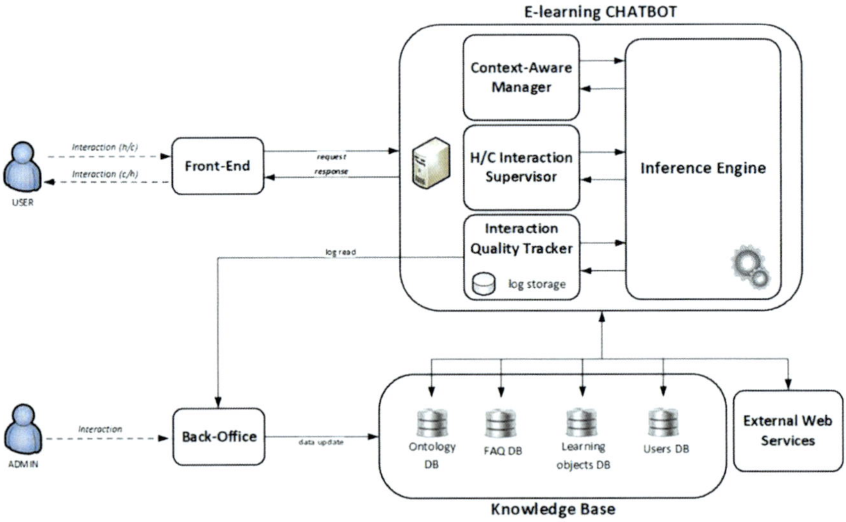

Fig. 30.1 System architecture

Modules

The proposed architecture includes an Inference Engine module, composed of four sub-modules: Text pre-processing, LDA Analyzer, Query classifier, Interaction manager. These modules allow the analysis of the application, the attribution of meaning and the ontology mapping. Besides, there is the formulation of a response (final or intermediate) to the user. They lean on the Knowledge Base, that is the databases related to courses and teaching materials or offered services.

Other modules that make up the architecture are:

- Topics and Dictionary Builder: the module that allows to define key terms to be identified in user questions and the annotated lexicon of equivalent terms.
- Domain Ontology Builder: the module that, through a web-based interface, allows to build and manage the domain ontology.
- Dialogue Ontology Builder: it allows the identification of the topic that characterizes a dialogue, in order to be able to build adequate answers subsequently.
- Interaction Quality Tracker: the module that allows monitoring of the interactions between the user and the chatbot system, producing synthetic quality indicators and highlighting critical aspects of the interactions subject to improvement.
- Human-Computer Interaction Supervisor: the module that has the objective of supervising the dialogue, keeps track of the interaction times, identifies asked questions ambiguously, non-convergent, or too long dialogue sessions.
- Context-aware Information Manager: it allows to guide the dialogue based on the user profile and its environment (including the user's location); it is based on the representation of all possible contexts, through the Context Dimension Tree [20].
- Starting Context Integration: it takes care of transforming the initial context into useful information for the bot, using information such as the navigation path, the current page etc.
- Personalization: in the case of an authenticated user, it takes care of correlating the user with the dialogue and its status, that is "remember" information about the user himself (training courses already followed, etc.).
- Post-chat Integration: it allows to manage the chat output to other systems (e.g.: chat with human-tutor, send mail, etc.) passing all the information collected during the chat.
- Post-chat reply form: the module enables the user to provide feedback on the quality level of the interaction performed.
- Anti-pranks: it detects and limits attempts to divert the bot and generate deliberately ridiculous or embarrassing responses; it manages the exit in case of offensive user behaviour.
- Spell-corrector: it takes care of recognizing and correcting typing errors by users and continuing the dialogue; it is dependent on the language adopted.
- Chat recording and warning: the module that analyzes the conversations carried out, identifies anomalous situations (sudden lowering of the success rate, lengthening of conversations, etc.) and allows to broaden the Knowledge Base or the interaction methods.

Features

The main features are:

- Real-time assistance/Live Support: the inference engine with that of context awareness will be able to process the questions asked by users in real-time, offering precise explanations and advice concerning what is required.
- Efficient training: users will be guided towards the discovery of new teaching materials; in doing so, research costs will be reduced, and the learning process accelerated, which will be improved in terms of efficiency.
- Personalised/Customised and contextualised training: through the context-aware module, the system will be able to analyse user behaviour to determine interests and the context in which it is located. This feature is crucial in order to search for "custom-made" training material and provide it at the most appropriate time.

Core

The core of the architecture is the Inference Engine which includes the analysis of the text and the elaboration of the user's real intentions and needs, obtained through collaboration with other minor modules and, in particular, through the use of the module that allows context extraction, based on the Context Dimension Tree (CDT) [21, 22]. This last module indeed has a crucial role. In particular, it is assumed that the text generated by a chat is a concoction of contextual information and that the use of some words contributes to defining the different context elements useful in the research for the context itself.

In this context, Latent Dirichlet Allocation (LDA) represents the appropriate model for making a series of observations and for explaining the correlation between keywords and topics [23, 24]. In particular, according to the proposed approach, each topic must refer to a specific context element. In this way, the chatbot—first through the analysis of the chat and then through the elaboration of the current context - is able to define the real intention of the user. It is essential in order to satisfy his needs better, that is to recommend properly filtered training content.

Through the textual analysis it is possible to know the employee (factory worker, office worker, manager, etc.), where he is (production department, administrative office, meeting room, etc.), the purpose of his work (assembly, contracts, project management, etc.) and what it needs (technical specification of a machine, legal advice, telephone contact of a customer, etc.). For example, a possible context could describe a user who must process personal data of employees and wants to do so under the directives contained in the GDPR.

The user's interaction with the chatbot is divided into clusters, i.e. short and simple sentences, using keywords with the help of appropriate ontological filters. The proposed method assumes that there is a probability that a specific word W belongs to an N_C concept in the CDT. This probability is directly proportional to the number of times this word has been used in the topic. This model is able to automatically design the communication path with the user without having to specify the semantic value of the words present in the text [25]. In our domain of interest, i.e. the industrial field, through the use of the LDA it is possible to extract the Mixed Graph of Terms (mGT), which is able to provide us real-time contextual constraints useful to design the Context Dimension Tree. Analyzing the chats through the LDA allows to generate the topics, which can be processed as contextual elements when using the Context Dimension Tree [26, 27].

Based on the LDA approach, a distribution of terms i for each topic is represented as a multinomial distribution φ_i drawn from a symmetric Dirichlet distribution with parameter β:

$$p(\phi_i|\beta) = \frac{\Gamma(W\beta)}{[\Gamma\beta]^W} \prod_{v=1}^{W} \emptyset_{iv}^{\beta-1}$$

The topic distribution for a document d is also represented as a multinomial distribution Θ_{d} drawn by a Dirichlet distribution with parameter α:

$$p(\theta_d|\alpha) = \frac{\Gamma\left(\sum_{i=1}^{K} \alpha_i\right)}{\prod_{i=1}^{K} \Gamma(\alpha_i)} \prod_{i=1}^{K} \emptyset_{di}^{\alpha_i-1}$$

In this way, the topic z_{dn} for each index n token can be chosen from the distribution of the document topics as:

$$p(z_{dn} = i|\theta_d) = \theta_{di}$$

Each token w is chosen from a multinomial distribution associated with the selected topic:

$$p(w_{dn} = v|z_{dn} = i, \phi_i) = \emptyset_{iv}$$

LDA aims to find patterns of co-occurrence terms in order to identify consistent topics. If you use LDA to learn a topic i and $p(w = v|z = i)$ is high for a certain term v, then every document d that contains the term v has a high probability for the topic i.

It is possible to state that all the terms co-occurring with the term v are more likely to have been generated by the topic i.

A complex structure like the mGT can allow to capture and represent the contextual information contained in a set of chats that belong to a specific domain (for example, industry). This graph can be extracted automatically and used for the classification of the text, or to label the N_{c} concept nodes and know which of the nodes participate in the definition of the context at a given time. Formally, it can be defined as a graph $g = \langle N, E \rangle$ where:

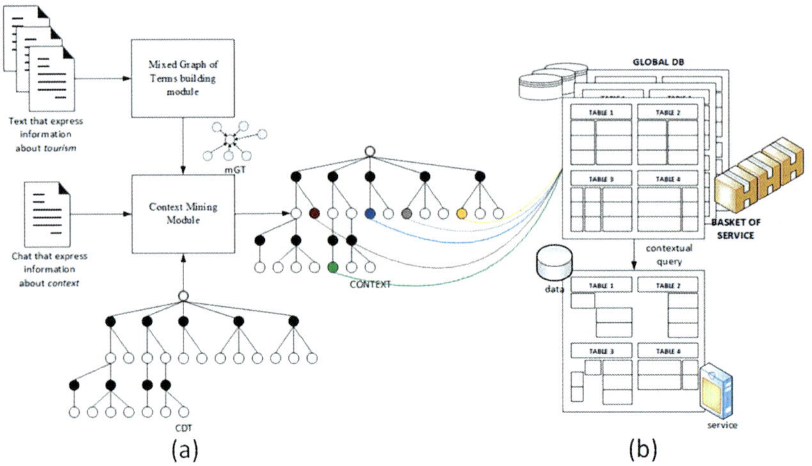

Fig. 30.2 Use of the context defined through mGT and CDT: **(a)** the construction of the Mixed Graph of Terms and a module for extracting the context elements; **(b)** the selection of useful services related to the context identified

- $N = \{R, W\}$ is a finite set of nodes, whose elements can be aggregates or aggregators.
- $E = \{E_{RR}, E_{RW}\}$ is a set of arcs that connect the aggregates and aggregators.

The approach described is based on two modules located inside the inferential engine. As shown in Fig. 30.2a, one module deals with the construction of the Mixed Graph of Terms and the other module deals with the extraction of context elements.

Mixed Graph of Terms module. Through this module, the mGT is created through a series of documents that belong to a specific sector. These documents have been previously labelled with the contextual information contained. Furthermore, the mGT is also useful for the construction of the Context Dimension Tree.

Context Mining Module. This module provides the extraction of the different context elements through the use of the mGT as a contextual filter. The module requires a general chat in input and allows to have the context referred to the chat in output.

The contextual data are allocated in specific sections of the database. Therefore, it is possible to execute contextual queries, automatically defining a global view composed of many associated partial views. This mechanism is used for the tailoring of contextual services, as shown in Fig. 30.2b.

Experimental Results

We developed a prototype based on the proposed system architecture—a company e-learning chatbot designed and implemented with a server-side component. The prototype of the chatbot, at the moment, aims to support the workers of a company in the south of Italy to learn the GDPR (General Data Protection Regulation), in Fig. 30.3 some screenshots of the prototype are shown. For the experimental phase, after the

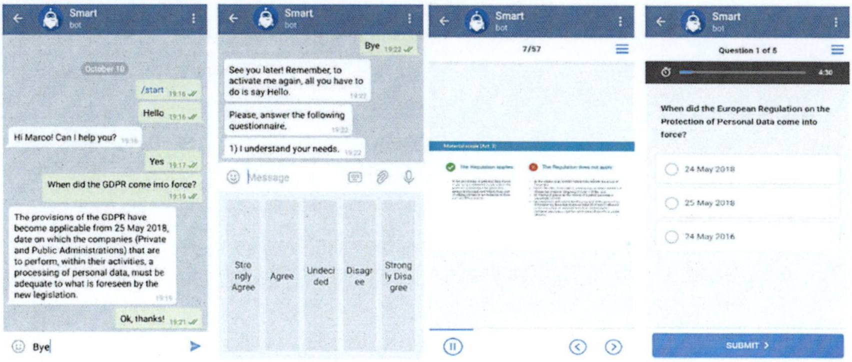

Fig. 30.3 Some screenshots of the chatbot

interaction with the chatbot, employees were asked to perform a five-section questionnaire, which is based on the Likert scale. At every assertion, present in a specific section, have been associated with five possible answers: “I totally disagree”—TD, “I disagree”—D, “Undecided”—U, “I agree”—A, “I totally agree”—TA. A total of 2580 employees from the same company, aged between 21 and 58, were involved, all belonging to different sectors (administration, sales, quality, design, maintenance and production). The employees involved also belong to different types of contracts and levels. In total, 2580 interactions were collected.

Section A: recommendation

1. The proposed contents have satisfied the needs of the user.
2. The system has managed to adapt based on personal preferences and the current context.

Section B: dialogue

1. The dialogue with the chatbot took place smoothly and without unexpected jumps.
2. The system was able to understand the intentions of the user correctly.

Section C: presentation

1. The information provided by the system has been adequately presented.
2. The information provided by the system has been exhaustive.

Section D: usability

1. The chatbot interface is user-friendly.
2. Response times are adequate.

Section E: future development

Table 30.1 Analysis of results

Section	Percentage		
	Negative (%)	Neutral (%)	Positive (%)
A	5.64	2.85	91.51
B	10.54	5.45	84.01
C	10.09	4.92	84.99
D	9.54	4.45	86.01
E	9.17	4.32	86.51

1. It would be useful to include in the chat other users (work colleagues) with whom to share the training experience.
2. It would be interesting to apply the same approach in other scenarios.

In Table 30.1 are reported the results of questionnaire divided in section, which show promising results.

In Fig. 30.4 are shown the results of the questionnaire; the plot shows that the lowest number of positive feedbacks belongs to the conversation section. Several users highlighted difficulties in communication fluidly with the chatbot, which sometimes had problems related to understanding the requests. This issue can be addressed to typos or slang words, which the system is not trained to recognise. Furthermore, section A, which is related to the recommendation module and based on the context, shows promising results. The proposed system is able to understand the needs of the user and dynamically adapting itself to the current context. The results suggest paying particular attention to machine learning in order to improve human-machine interaction.

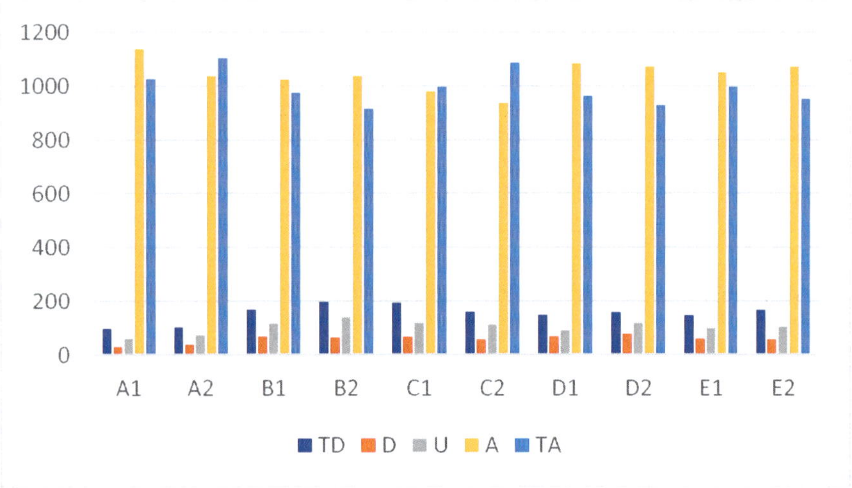

Fig. 30.4 Trend of answers to statements

Conclusions

This work aimed to propose an e-learning system through the use of chatbots that could be used in the 4.0 industry paradigm. Therefore, this work aimed to takes advantage of chatbots which are currently used both for e-learning and within the 4.0 industry paradigm. Through NPL techniques such as the LDA, it has been possible to implement a solution that brings advantages to the simplifying of learning path and in the reduction of training time. The developed tool is suitable both for the training and continuous updating of employees, necessary in case of introduction of new work processes and for the training of new employees, trying to reach, in this way, a further step towards the Industry 4.0 paradigm. The results obtained are promising. However, there are some issues related to the fluidity of conversation with users that need to be further improved. Moreover, future developments will concern a more robust experimental phase, which would involve more companies, more productive sectors and considering different learning pathways. In this way, it would be possible to make an analysis that would consider a grouping of different categories of employees.

References

1. Y. Lu, Industry 4.0: A survey on technologies, applications and open research issues. *J. Ind. Inf. Integr.* (2017). https://doi.org/10.1016/j.jii.2017.04.005
2. L. Da Xu, L. Duan, Big data for cyber physical systems in industry 4.0: a survey. *Enterp. Inf. Syst.* (2019). https://doi.org/10.1080/17517575.2018.1442934
3. M. Gaeta, V. Loia, S. Tomasiello, A generalized functional network for a classifier-quantifiers scheme in a gas-sensing system. Int. J. Intell. Syst. (2013). https://doi.org/10.1002/int.21613
4. F. Colace, M. De Santo, M. Lombardi, F. Pascale, D. Santaniello, A. Tucker, A multilevel graph approach for predicting bicycle usage in London area, in *Fourth International Congress on Information and Communication Technology. Advances in Intelligent Systems and Computing, vol 1027*, ed. by X.S. Yang, S. Sherratt, N. Dey, A. Joshi (Springer, Singapore, 2020), pp. 353–362. https://doi.org/10.1007/978-981-32-9343-4_28
5. F. Colace, M. Lombardi, F. Pascale, D. Santaniello, A. Tucker, P. Villani, MuG: A multilevel graph representation for big data interpretation, in *2018 IEEE 20th International Conference on High Performance Computing and Communications; IEEE 16th International Conference on Smart City; IEEE 4th International Conference on Data Science and Systems (HPCC/SmartCity/DSS)*, pp. 1408–1413, June 2018. https://doi.org/10.1109/hpcc/smartcity/dss.2018.00233
6. F. Abate, M. Carratù, C. Liguori, M. Ferro, V. Paciello, Smart meter for the IoT, in *I2MTC 2018—2018 IEEE International Instrumentation and Measurement Technology Conference: Discovering New Horizons in Instrumentation and Measurement, Proceedings* (2018), pp. 1–6. https://doi.org/10.1109/i2mtc.2018.8409838
7. F. Amato, F. Moscato, V. Moscato, F. Colace, Improving security in cloud by formal modeling of IaaS resources. Futur. Gener. Comput. Syst. (2018). https://doi.org/10.1016/j.future.2017.08.016
8. D. Gorecky, M. Schmitt, M. Loskyll, D. Zühlke, Human-machine-interaction in the industry 4.0 era, in *Proceedings—2014 12th IEEE International Conference on Industrial Informatics, INDIN 2014* (2014). https://doi.org/10.1109/indin.2014.6945523

9. S. Quarteroni, Natural language processing for industry: ELCA's experience. Informatik-Spektrum (2018). https://doi.org/10.1007/s00287-018-1094-1
10. W. Rahane, S. Patil, K. Dhondkar, T. Mate, Artificial intelligence based Solarbot, in *Proceedings of the International Conference on Inventive Communication and Computational Technologies, ICICCT 2018* (2018). https://doi.org/10.1109/icicct.2018.8473172
11. A. Chawla, A. Varshney, M.S. Umar, H. Javed, "ProBot: An online aid to procurement, in *Proceedings of the 2018 International Conference on System Modeling and Advancement in Research Trends, SMART 2018* (2018). https://doi.org/10.1109/sysmart.2018.8746954
12. S. Mantravadi, A.D. Jansson, C. Møller, User-Friendly MES Interfaces: Recommendations for an AI-Based Chatbot Assistance in Industry 4.0 Shop Floors (2020), pp. 189–201. https://doi.org/10.1007/978-3-030-42058-1_16
13. F. Colace, M. De Santo, M. Lombardi, F. Pascale, A. Pietrosanto, S. Lemma, Chatbot for e-learning: A case of study. Int. J. Mech. Eng. Robot. Res. (2018). https://doi.org/10.18178/ijmerr.7.5.528-533
14. F. Amato, A. Mazzeo, V. Moscato, A. Picariello, Semantic Management of Multimedia Documents for E-Government Activity, in *2009 International Conference on Complex, Intelligent and Software Intensive Systems* (2009), pp. 1193–1198. https://doi.org/10.1109/cisis.2009.195
15. A.C. Graesser, P. Chipman, B.C. Haynes, A. Olney, Auto tutor: An intelligent tutoring system with mixed-initiative dialogue. IEEE Trans. Educ. (2005). https://doi.org/10.1109/TE.2005.856149
16. N.T. Heffernan, E.A. Croteau, Web-based evaluations showing differential learning for tutorial strategies employed by the MS. Lindquist tutor. *Lect. Notes Comput. Sci. (including Subser. Lect. Notes Artif. Intell. Lect. Notes Bioinformatics)* (2004). https://doi.org/10.1007/978-3-540-30139-4_46
17. F. Colace, M. De Santo, M. Lombardi, D. Santaniello, CHARS: A cultural heritage adaptive recommender system, in *Proceedings of the 1st ACM International Workshop on Technology Enablers and Innovative Applications for Smart Cities and Communities—TESCA'19* (2019), pp. 58–61. https://doi.org/10.1145/3364544.3364830
18. M. Lombardi, F. Pascale, D. Santaniello, An application for cultural heritage using a Chatbot, in *2019 2nd International Conference on Computer Applications & Information Security (ICCAIS)* (2019), pp. 1–5. https://doi.org/10.1109/cais.2019.8769525
19. A. Kerry, R. Ellis, S. Bull, Conversational agents in e-learning, in *Applications and Innovations in Intelligent Systems XVI—Proceedings of AI 2008, the 28th SGAI International Conference on Innovative Techniques and Applications of Artificial Intelligence* (2009). https://doi.org/10.1007/978-1-84882-215-3-13
20. F. Clarizia, S. Lemma, M. Lombardi, F. Pascale, A mobile context-aware information system to support tourism events, in *Lecture Notes in Computer Science (including subseries Lecture Notes in Artificial Intelligence and Lecture Notes in Bioinformatics)* (2017). https://doi.org/10.1007/978-3-319-57186-7_40
21. F. Daniel, M. Matera, E. Quintarelli, L. Tanca, V. Zaccaria, *Context-Aware Access to Heterogeneous Resources Through On-the-Fly Mashups* (2018), pp. 119–134. https://doi.org/10.1007/978-3-319-91563-0_8
22. F. Colace, M. De Santo, M. Lombardi, R. Mosca, D. Santaniello, A multilayer approach for recommending contextual learning paths. J. Internet Serv. Inf. Secur. **2** (May), 91–102 (2020). https://doi.org/10.22667/jisis.2020.05.31.091
23. R. Krestel, P. Fankhauser, W. Nejdl, Latent Dirichlet allocation for tag recommendation, in *RecSys'09—Proceedings of the 3rd ACM Conference on Recommender Systems* (2009). https://doi.org/10.1145/1639714.1639726
24. M. Casillo, F. Clarizia, G. D'Aniello, M. De Santo, M. Lombardi, D. Santaniello, CHAT-Bot: A cultural heritage aware teller-bot for supporting touristic experiences. Pattern Recognit. Lett. (2020). https://doi.org/10.1016/j.patrec.2020.01.003
25. F. Colace, M. De Santo, L. Greco, F. Amato, V. Moscato, A. Picariello, Terminological ontology learning and population using latent dirichlet allocation. J. Vis. Lang. Comput. (2014). https://doi.org/10.1016/j.jvlc.2014.11.001

26. F. Amato, V. Moscato, A. Picariello, F. Colace, M. De Santo, F.A. Schreiber, L. Tanca, Big data meets digital cultural heritage: Design and implementation of SCRABS, a smart context-aware browsing assistant for cultural environments. J. Comput. Cult. Herit. (2017). https://doi.org/10.1145/3012286
27. M. Casillo, F. Clarizia, F. Colace, M. Lombardi, F. Pascale, D. Santaniello, An approach for recommending contextualized services in e-Tourism. Information **10**(5), 180 (2019). https://doi.org/10.3390/info10050180

Chapter 31
Analysing Competencies by Gender in Internships

Raquel Ferreras-Garcia, Jordi Sales-Zaguirre, and Enric Serradell-López

Abstract This paper aims to analyse whether acquisition of competencies is influenced by students' gender in an internship environment. In order to analyse these effects, we present an introductory and empirical study to know the relationship between gender and competencies in a tourism internship course. A quantitative analysis was conducted on 389 students from the bachelor's degree programme on Tourism and Hospitality Management at the School of Tourism and Hospitality Management Sant Ignasi. The analyses show that women achieve higher levels of competencies than men, except for those related to initiative and entrepreneurial spirit and sensitivity and intercultural awareness. Moreover, significant differences between men and women were found in competencies related to orientation towards achievement and service and planning and organisation. The key finding is the positive effect of being female on competencies, which reveals that in general the level of competencies of men were poorer than those of women when participating in internships.

Keywords Gender · Internships · Competencies · Higher education

R. Ferreras-Garcia · E. Serradell-López (✉)
Universitat Oberta de Catalunya, Barcelona, Spain
e-mail: eserradell@uoc.edu

R. Ferreras-Garcia
e-mail: rferreras@uoc.edu

J. Sales-Zaguirre
School of Tourism and Hospitality Management Sant Ignasi, Universitat Ramon Llull, Barcelona, Spain
e-mail: jordi.sales@htsi.url.edu

A. Visvizi et al. (eds.), *Research and Innovation Forum 2020*,
Springer Proceedings in Complexity,
https://doi.org/10.1007/978-3-030-62066-0_31

Introduction

The European convergence process involves large-scale changes to university education in European countries. This far-reaching overhaul is based on a paradigm shift from a teaching-based to a student-based learning model, stressing the role of competencies that provide a common language to define and express the academic and professional profiles that are considered optimum [1].

One of the main aims of the European Higher Education Area (EHEA) is to orient university studies towards employability by developing the professional skills required for subsequent professional life. Royal Decree 1393/2007 [2], which regulates official education, establishes that the central objectives of programmes of study for students to obtain a university qualification must be the acquisition of competencies.

Based on these indications, the programmes of study White Paper for tourism includes the directives provided by stakeholders in the tourism industry, in which competencies play an essential role in guiding the learning process.

The main objective of tourism management education is to provide industry with highly qualified graduates equipped with appropriate competencies. In that sense, programmes of study should prepare future graduates and give them the competencies and training required to access the labour market with sufficient grounding for their professional development [3].

Aware of its role in developing students' learning, the School of Tourism and Hospitality Management Sant Ignasi centres its learning models on employability. One of its most important actions is to offer students the opportunity of complementing their training with internships in national and international companies during the programme, thus fostering their employability and helping students in the sector to find work quickly. Internships provide a link between education programmes and the job market, by bringing industry, academia and students together.

Despite research showing the importance of internships, there are very few studies on the extent to which competencies developed during internships are acquired and factors affecting their acquisition. Thus, a review of the literature shows a significant gap in the relation between competencies and gender.

The aim of this article is to approach this gap in the literature and provide insight into competencies and gender in the specific context of higher education in tourism and employability. To do this, an analysis was carried out in the context of an internships. Acquisition of competencies by students during their internship was investigated by an exploration of whether gender is a determining characteristic in students' acquisition.

Literature Review

Competencies in Tourism

The tourism industry is highly diverse and the knowledge and skills required by employees differ depending on the subsector and organisation [4].

Competencies provide reference points for teachers in terms of results and learning levels, while also facilitating easy comparison between disciplines [5]. In the Tuning project, competencies represent 'a dynamic combination of attributes, abilities and attitudes. Fostering these competencies is the object of educational programmes. Competencies are formed in various course units and assessed at different stages' [6].

Different researchers identify a considerable variety of competencies, many of which reoccur, but all of them have the common denominator of responding in one way or another to general industry requirements. Competencies identified as especially important for companies who use tourism graduates are communication, applying knowledge in practice and the ability to learn [6, 7]. Moreover, other competencies such as problem-solving, teamwork, foreign languages and entrepreneurship are also identified as important in the tourism sector.

Numerous studies analyse competencies from different points of view. Some authors analyse the importance of competencies among tourism graduates for employers and academics, ranking the most highly valued and necessary competencies for working in the tourism industry [3, 5, 7–10]. There are also studies on how closely students' academic education matches the demands of the labour market, comparing the importance given by supervisors and students to different competencies [11, 12] and on specific skill sets within the competency framework of tourism education [13].

Although most studies analyse the competencies employers in the tourism sector consider important in graduates, there has been little research into the extent these competencies are genuinely acquired and how this relates to other dimensions, such as gender.

Gender and Competencies

In recent years, the profile of students and their inclusion in the job market has changed considerably in favour of women [14]. Furthermore, the role of gender and its relation to competencies has been an academic topic for some time. Previous studies found that gender influences certain competencies. Petrović et al. [15] found gender differences among hotel staff, where procedures were seen as more important for men, while service and customer orientation and emotional control were more important for women. Gender differences are found in behaviour and in interpersonal relations, suggesting that men and women differ in how they participate in social relations [16, 17]. Caruso et al. [18] note that working women have a great ability to understand other people's emotions and are more closely involved in communication with others.

In higher education, differences have also been found between men and women with regard to their learning process and performance [19, 20]. These differences also exist in perceptions, where [12] conclude that male and female interns generally receive favourable assessments from their supervisors, although women are underestimated compared to their male colleagues, possibly due to a lack of confidence in women.

Other studies focus on the tourism and hospitality industry, where the number of women employees has increased in recent years, and show that female managers tend to minimise difficulties and highlight their personality [21]. Yoonjoung Heo et al. [22] include gender in their study and conclude that, for women, trust and commitment influence the quality of relationships with employees.

The diversity of results of previous research shows that a deeper analysis is necessary in order to better understand the role that gender plays in acquisition of competencies by students on curricular internships.

A number of authors [4, 23–26] have shown the effectiveness of internships on the acquisition of knowledge, operational procedures and competencies such as problem-solving, goal achievement and orientation, communication, planning, time management, motivation, confidence, responsibility and interpersonal skills involved in teamwork and contact with customers.

Bearing this in mind, the aim is to establish the level of competency acquisition among men and women on curricular internships and whether competency acquisition is influenced by students' gender.

These issues are expressed in the following research question:

RQ: What is the impact of gender on competency development among intern students?

Methodology

Measurement of Variables

The sample consisted of students on the Tourism and Hospitality Management bachelor's degree programme at the School of Tourism and Hospitality Management Sant Ignasi, who took curricular internships during the 2011–2017 academic years.

Intern students were assessed at the end of their internships by their company supervisors, using a questionnaire with a Likert-type scale (where 1 = strongly disagree, 2 = disagree, 3 = agree and 4 = strongly agree). The questionnaire was divided into four parts in order to assess the level of competency acquisition among students during their internship. Part 1 dealt with general information on students, such as name of the company, department, location and gender; part 2 to personal

Table 31.1 Competencies

Personal cross-disciplinary competencies
G1. A commitment to ethics
G2. Orientation towards achievement
G3. Initiative and entrepreneurial spirit
G4. Self-knowledge and emotional self-control
G5. Flexibility/adaptability and self-confidence
Social and relationship cross-disciplinary competencies
G6. Empathy and interpersonal comprehension
G7. Team work and collaboration
G8. Sensitivity and intercultural awareness
G9. Service orientation
Management cross-disciplinary competencies
G10. Planning and organisation
G11. Development of organisational culture
G12. Personal development
G13. Change management
G14. Leadership

cross-disciplinary competencies; part 3 to social and relationship cross-disciplinary competencies; part 4 to management cross-disciplinary competencies. A total of 389 questionnaires were collected from the supervisors. Table 31.1 shows the variables used.

Data Analysis

The statistics software SPSS version 24 was used to carry out the quantitative analysis. To answer the research question, a descriptive calculation analysis of the mean for the different competencies was carried out, moreover a Student-Fisher *t*-test was carried out on the difference between the means.

Cronbach's alpha coefficient was calculated to check the reliability of the measurement scale [27]. The results show an alpha coefficient of 0.966 for the sample, which means it is highly reliable [28].

Table 31.2 shows the frequencies and percentages for the 'gender' variable.

Table 31.2 Distribution by gender

Gender	Frequency	Percentage
Male	114	29.3
Female	275	70.7
Total	389	100

Table 31.3 Means of the competencies by gender

Gender	G1	G2	G3	G4	G5	G6	G7
Male	3.80	3.59	3.24	3.71	3.31	3.61	3.59
Female	3.82	3.72	3.20	3.74	3.39	3.69	3.63
Gender	G8	G9	G10	G11	G12	G13	G14
Male	3.74	3.51	3.33	3.66	3.55	3.58	3.52
Female	3.71	3.65	3.53	3.68	3.58	3.65	3.54

Results

Data obtained follow a normal distribution. The means of the competencies were calculated for men and women to ascertain whether there are gender differences in competency acquisition among students during their curricular internship. The results show (Table 31.3) that competency acquisition levels are generally higher for women than for men. Only in the competencies of initiative and entrepreneurial spirit and sensitivity and intercultural awareness (G3 and G8) do men score higher.

In addition, competencies with a higher acquisition level segmented by gender relate to ethics, orientation towards achievement, self-knowledge and intercultural awareness (G1, G2, G4 and G8). The least acquired competencies are those related to initiative and entrepreneurial spirit and flexibility/adaptability (G3 and G5).

Thus, it may be said that women are better at acquiring most competencies and therefore gender is a variable that affects students' competency acquisition level.

A Student-Fisher *t*-test comparing the means was conducted to analyse possible differences by gender (Table 31.4). At a significance level of $\alpha = 0.05$, the results show significant differences between men and women for general competencies related to orientation towards achievement and service, and planning and organisation (G2, G9 and G10). In all of them, women showed greater skill than men.

Discussion and Conclusions

Internships in companies are the ideal opportunity to develop and assess students' competencies in a professional environment. Students can put into practice the knowledge they acquire during their training; companies can see whether students obtain the competencies demanded by the sector; and universities can see whether students' education meets the demands of the job market.

The aim of this study is to explore whether gender is a key factor in the acquisition of competencies among university students during their internship.

The results indicate that competency acquisition is generally higher among women than among men. Only competencies related to initiative and entrepreneurial spirit, and intercultural awareness are more successfully acquired by men. Despite generally good competency acquisition levels, the best acquired competencies are those

Table 31.4 Parametric statistical tests of the competencies by gender

Typology		Levene's test for equality of variances		*t*-test for the equal means				
		F	Sig	*t*	df	Sig. (2-sided)	Difference of means	Difference of SE
G2	Equal variances assumed	8.2	0.00	−2.2	387	0.03	−0.12	0.06
	Equal variances not assumed			−2.1	183.4	0.04	−0.12	0.06
G9	Equal variances assumed	4.7	0.03	−1.9	387	0.05	−0.14	0.07
	Equal variances not assumed			−1.8	200.1	0.05	−0.14	0.07
G10	Equal variances assumed	2.9	0.09	−2.47	387	0.01	−0.19	0.08
	Equal variances not assumed			−2.36	191.6	0.02	−0.19	0.08

involving ethics, orientation towards achievement, self-knowledge, and intercultural awareness. The results confirm partially previous studies, such as that of [29], which analyses competency acquisition during business degree internships, and that of [5] who carried out a comparison of competency acquisition, considering students' gender.

In addition, significant differences between men and women were only found in competencies related to achievement and service orientation and planning and organisation. For all other competencies, gender did not affect the level of acquisition during internships. It is worth noting that no previous studies have analysed whether gender differences in competencies were significant, which validates our results compared to previous ones.

The results also suggest that when universities define curricular programmes, they should take into consideration the effect of gender and other individual characteristics to optimise competency development and properly prepare students for the job market.

References

1. A. Mir, Las competencias transversales en la Universidad Pompeu Fabra. La visión de los docentes y estudiantes de segundo ciclo. REDU Rev. docencia univ. (1), 2 (2008)
2. Ministry of Education and Science of Spain, Real Decreto 1393/2007, de 29 de octubre, por el que se establece la ordenación de las enseñanzas universitarias oficiales. Bolet. Ofic. Estado **260**, 44037–44048 (2007)
3. M.C. Dhiman, Employers' perceptions about tourism management employability skills. Anatolia **23**(3), 359–372 (2012)
4. A. Donina, I. Luka, The compliance of tourism education with industry needs in Latvia. Eur. J. Tour. Hosp. Recreat. **5**(3), 91–120 (2014)
5. J.M. López-Bonilla, L.M. López-Bonilla, Holistic competence approach in tourism higher education: an exploratory study in Spain. Curr. Issues Tour. **17**(4), 312–326 (2014)
6. J. González, R. Wagenaar, *Tuning Educational Structures in Europe* (University of Deusto, Bilbao, 2003)
7. A.M. Munar, J.J. Montaño, Generic competences and tourism graduates. J. Hosp. Leis. Sports Tour. Educ. **8**(1), 70–84 (2009)
8. C.R. Mayo, C. Thomas-Haysbert, Essential competencies needed by hospitality and tourism management graduates as determined by industry professionals and hospitality educators. Consortium J. Hosp. Tour. **9**(2), 5–16 (2005)
9. Y.F. Wang, Constructing career competency model of hospitality industry employees for career success. Int. J. Contemp. Hosp. Manag. **25**(7), 994–1016 (2013)
10. Y.F. Wang, C.T. Tsai, Employability of hospitality graduates: student and industry perspectives. J. Hosp. Tour. Educ. **26**(3), 125–135 (2014)
11. M. Mareque, E. De Prada, Evaluación de las competencias profesionales a través de las practicas externas: incidencia de la creatividad. Rev. Investig. Educ. RIE **36**(1), 203–219 (2018)
12. M.B. Marks, J.C. Haug, H. Hu, Investigating undergraduate business internships: do supervisor and self-evaluations differ? J. Educ. Bus. **93**(2), 33–45 (2018)
13. J.V. García Manjón, M.D.C. Pérez López, El grado en turismo: un análisis de las competencias profesionales. Cuad. Tur. **21**, 67–83 (2008)
14. D.G. Schoffstall, A profile of hospitality program characteristics and the gender composition of students and faculty. J. Hosp. Tour. Educ. **27**(2), 69–79 (2015)
15. M.D. Petrović, T. Jovanović, J.J. Marković, T. Armenski, V. Marković, Why should gender differences in hospitality really matter? A study of personnel's service orientation and job satisfaction in hotels. Econ. Res. Ekon. Istraž. **27**(1), 799–817 (2014)
16. A.H. Eagly, *Sex Differences in Social Behavior: A Social-Role Interpretation* (Erlbaum, Hillsdale, NJ, 1987)
17. G.M. Reevy, C. Maslach, Use of social support: gender and personality differences. Sex Roles **44**(7–8), 437–459 (2001)
18. D.R. Caruso, J.D. Mayer, P. Salovey, Relation of an ability measure of emotional intelligence to personality. J. Pers. Assess. **79**, 306–320 (2002)
19. F. Harris, S.R. Harper, Masculinities go to community college: understanding male identity socialization and gender role conflict, in *New Directions for Community Colleges*, vol. 142 (2008), pp. 25–35
20. J.D. Wolfle, M.R. Williams, The impact of developmental mathematics courses and age, gender, and race and ethnicity on persistence and academic performance in Virginia community colleges. Community Coll. J. Res. Pract. **38**(2–3), 144–153 (2014)
21. C. Ng, R. Pine, Women and men in hotel management in Hong Kong: perceptions of gender and career development issues. Int. J. Hosp. Manag. **22**, 85–102 (2003)
22. C. Yoonjoung Heo, S. Kim, B. Kim, Investigating the impact of relationship quality during an internship on millennials' career decisions and gender differences. J. Hosp. Tour. Educ. **30**(2), 71–84 (2018)

23. N.T. Dhevabanchachai, W. Wattanacharoensil, Students' expectations, experiences, and career direction after in-house internship: the pedagogical role of the hotel training center in Thailand. J. Teach. Travel Tour. **17**(2), 118–134 (2017)
24. D. Ruggiero, J.D. Boehm, Project-based learning in a virtual internship program: a study of the interrelated roles between intern, mentor and client. Comput. Educ. **110**, 116–126 (2017)
25. M. Thomas, Community college education abroad and business internship programs cultivation of competency in communicating, collaborating, and critical thinking?, in *International Education at Community Colleges* (Palgrave Macmillan, New York , 2016), pp. 323–335
26. G. Velez, G. Giner, Effects of business internships on students, employers, and higher education institutions: a systematic review. J. Employ. Counsel. **52**(3), 121–130 (2015)
27. L.J. Cronbach, Studies of acquiescence as a factor in the true-false test. J. Educ. Psychol. **33**(6), 401 (1942)
28. R.F. De Vellis, L.S. Dancer, Scale development: theory and Applications. J. Educ. Meas. **31**(1), 79–82 (1991)
29. I. Luka, A. Donina, Challenges of tourism education: conformity of tourism curriculum to business needs. Acad. Tur. **5**(1), 85–101 (2012)

Chapter 32
Redesigning Business Models for Data-Driven Innovation: A Three-Layered Framework

Orlando Troisi, Mara Grimaldi, and Francesca Loia

Abstract The paper investigates how big data analysis and analytics can reshape business model design to encourage the co-development of innovation through the application of a data-driven orientation. The main objectives of the study are: (1) to detect the enabling dimensions of business models for the co-development of innovation; (2) to investigate how the enabling dimensions can be oriented strategically toward the emergence of data-driven innovation. The empirical research is based on a case study performed through qualitative content analysis to explore an Italian agrifood company, Amadori. The results obtained allow the introduction of a three-layered framework that can encourage future studies to: (1) investigate business models enablers for value co-creation and the way in which their combination can produce innovation; (2) detect the different data-oriented and management strategies that redefine business models to develop innovation systematically.

Keywords Business model · Business model innovation · Big data analysis · Value co-creation · Technological innovation · Data-driven innovation

Introduction

Big data analysis allows companies at extracting value from high volumes of heterogeneous sources by generating new value, thus innovation [9], and improving firms' readiness to forecast new strategies [2]. In this scenario, (co-)innovation can be pursued thanks to systematic and systems analysis and sharing of data, which should be turned into insights and ideas thanks to the co-creation with users and stakeholders [22].

O. Troisi · M. Grimaldi (✉)
University of Salerno, Via Giovanni Paolo II, 132, 84084 Fisciano, SA, Italy
e-mail: margrimaldi@unisa.it

F. Loia
University of Rome "La Sapienza", Via del Castro Laurenziano, 9, 00161 Rome, Italy

A. Visvizi et al. (eds.), *Research and Innovation Forum 2020*,
Springer Proceedings in Complexity,
https://doi.org/10.1007/978-3-030-62066-0_32

Despite the recognized impact of the adoption of technology and analytics on the development of innovation, extant research focuses on the identification of the different smart tools available in current markets without exploring deeply how the emergence of innovative insights can be managed [8, 14].

Thus, the use of technologies in business process does not allow per se the development of innovation [18, 27]. There is the need to understand how business models should be redefined [6, 13] and redesigned toward the proper integration of data into corporate strategies and supply chain management.

The most recent theories on business models explore how service design, and the processes of innovation sharing and emergence, can be strengthened thanks to the application of smart technologies. Business model innovation (BMI, [16]) conceptualizes how complex organizations should redefine their business models to assimilate technology [6].

However, despite the recognized need to analyze the digital transformation of business models, extant research does not introduce a framework that defines the different dimensions of innovative business models [15] and that examines how technologies reframe business structure, processes, relationships and value creation.

Therefore, the paper aims at investigating how current business models should be refocused to involve big data analysis and analytics in the process of innovation development [26].

Based on a re-elaboration of extant research of business models and on big data management, it can be revealed the need to re-read value creation and innovation process through the application of an orientation that considers data as a strategic asset for organization refocusing. For this reason, the necessity of a data-driven approach [12, 20] can be highlighted to identify how the growing amount of data collected can be included strategically in business orientation to improve decisions and give birth to innovative actions.

According to a data-driven reinterpretation of traditional business models, the study seeks to address the following research questions:

RQ1: which are the key enabling dimensions of business models for the co-development of innovation?

RQ2: how can these dimensions be oriented strategically toward data-driven innovation?

The empirical research adopts the case study methodology [28] based on a qualitative content analysis that explores an Italian company operating in the Agrifood sector (Amadori). The findings contribute to reveal: (1) (a classification of the) enabling dimensions of business model innovation, (2) the strategies that can foster the co-development of data-driven innovation.

The paper is structured as follows. In the first section, a theoretical background on the main enabling dimensions of business model and of data-driven innovation are presented. Then, a three-layered framework that guides the content analysis is introduced and the research design based on a qualitative approach is described. The findings are debated and some innovation strategies for data-driven companies are revealed. Lastly, conclusion, implications and limitations are discussed.

Theoretical Background

Business models are defined usually as tools for the synthesis, analysis and evaluation of business strategies and processes, which help identifying the ways in which value can be created and captured [23, 31].

In particular, business models conceptualize the content (resources, information, capabilities), the structure (interactions and relationships between actors who share resources) and the governance (actors that manage the content and structure) of business process that can create value to exploit business opportunities [1, 31].

As discussed above, the final aim of the implementation and dynamic combination of the different dimensions of business models is value creation [1].

The broadening of the perspective adopted in the most recent contributions reframes business model strategically as a set of tools that can ensure the fit between business strategic orientation and the strategies executed to give birth to the creation and co-creation of value [5, 10].

The integration between the systems and value-based view on business models and extant research on service can lead to the redefinition of value co-creation as a process that can help reframing business models by encouraging, thanks to the synergy arising from the activation of strategies for the integration of resources, the emergence of new value and, thus, potential innovation.

However, the studies that explore value creation through business models do not define systemically the entire process that marks the transition from value design, to sharing and capture [30].

For this reason, business models should be redefined according to a broader orientation that encompasses strategically all the phases of value creation. This complex value-based process involves the integration of organizational strategies (value propositions), technologies and resources to stimulate the enrichment of users' skills through resource integration, which is managed constantly to guarantee the fit between strategic objectives and strategy execution.

The recognized need to re-read business models through the application of new technologies [29] has been addressed in literature through the proposition of a new orientation underlying business processes: the *data-driven decision-making* (DDDM, [12, 20]). This approach allows at combining constantly research and data collection activities with their proper interpretation leading to transform data into information and, through incremental processes of exchange, into knowledge and value [26].

Thanks to the integration between the main dimensions of data-driven orientation and the key steps of data analysis cycle [4] (Ortiz-Repiso et al. [21]), a process-based view in which the main data-driven strategies can reframe traditional business models [19] is adopted in this study.

Three Layers for the Analysis of Data-Driven Innovation

The analysis of the literature reveals the need to adopt a systems perspective that explores how companies should integrate data and innovation mindset with the design of their business models and value co-creation strategies.

The integration of a systems view on business model with value-based perspective and data-driven orientation allows the definition of three different "layers" that enable the creation of data-driven innovative companies: (1) business models design; (2) value co-creation strategies; (3) data-driven orientation.

The co-existence of the three layers can give birth to data-driven innovation, which can be defined as the complex transformation of data into information, knowledge and new value thanks to the adoption of business models based on coherent value co-creation strategies. Thus, the following assumptions can be hypothesized:

1. *Business models dimensions*: the main enabling dimensions of innovation in business models can start from the definition of a coherent strategy aimed at creating value through the proper selection of resources, the implementation of technology, the adoption of flexible and proactive management skills and the fit alignment between strategic goals and strategies execution;
2. *Value co-creation-based strategies*: the joint development of practices that fosters (co-)innovation development is based on a circular and complex process of value propositions design, diffusion and integration;
3. *Data-driven orientation*: the data-driven redefinition of business orientation can be realized according to a culture based on the inclusion of data as strategic asset, on an integrated technological infrastructure, on the activation of data management skills, on a process management for the transformation of data into knowledge.

The empirical validity of the three layers can be assessed through the analysis of an Italian Agrifood company, Amadori. The case has been selected due to its degree of innovativeness in business models and in the smart use of technologies. The goals of the empirical research are: (1) to define the enabling dimensions of business models and the key co-creation strategies to foster (co-)innovation (RQ1); (2) to explore how these dimensions are integrated with data analysis and analytics to pursue data-driven innovation (RQ2).

Methodology

The study adopts a qualitative exploratory approach based on a case study carried out through content analysis as inquiry [17], which allows to extract from the texts (the analysis unit) fewer categories and to trace some focal points [11] related to the different variables investigated through the application of semantic criteria established by the researcher.

The case study methodology [28] permits to explore in depth the unit of analysis by providing the possibility to investigate the dynamics underlying a single context of investigation through a systems perspective [7, 24]. Therefore, case study can be considered adequate to investigate variables at different levels of analysis in different organizational sub-systems (supply, production, consumption, etc.) connected to each other through different relationship modalities.

Starting from the different macro-areas included in the research questions (business model innovation, value co-creation, data-driven orientation) some categories have been derived to provide researchers with a set of key topics to enable data collection.

The main sources used to collect data are: the official website of Amadori in all its sections, the official websites of the main international innovation events and conferences (Smau, Innovation Days), the websites dedicated to the individual projects carried out by the company and their partners, the main documents available on various websites (such as vision, mission, budgeting, social report, strategic plan), and the official pages of the firm on various social networks (Twitter, YouTube, Facebook).

Therefore, the results are obtained through a semantic interpretation of the texts from which the macro categories identified by the researchers are linked to the main research variables and classified to identify the enabling elements of the business model innovation (RQ1) according to a data-driven orientation that can give birth to the co-development of different innovation practices (RQ2).

Results

RQ1: Business Model Innovation and Value Co-Creation Strategies

Innovation-oriented goals and attitude are included in Amadori's strategic planning and are pursued through relayed business' activities and actions. Open innovation leads to the adoption of collaborative processes for joint co-development of innovative products and services. A new approach to innovation is the key core of strategies and the use of technology is planned strategically in the business strategies.

Partnerships and joint collaborations with start-up are key levers to support the development of new digitalized business models. Thanks to a series of joint projects with other start-ups and organizations in the food industry from suppliers to other producers, the company developed some pilots. Amadori established a partnership with "Foodtech Accelerator", which promotes the relationships between and among Made in Italy companies to innovate agrifood and retail sectors. Moreover, the collaboration with the Israelian Wasteless gives birth to the launch of a set of Amadori products based on dynamic pricing with the aim of reducing food waste.

Regarding technology, the integrated strategies from supplying, production and distribution are turned into strategies for the digitalization of the supply chain and of the relationships with customers (farms, distributors) and final users.

The software house "Sinfo One" from Parma managed the digital transformation of the industrial processes of Amadori by creating an integrated system of smart factory based on Agile platforms, Oracle Product Life management and on updated tools for ERP (*enterprise resource planning*). A real digitalized control centre contributes to improve production, to involve employees at each level, to strengthen corporate culture. The "step-by-step" permits to deliver faster products in the market. Oracle Agile helps the management of processes and information across the entire lifecycle of the products and is a software very common in smart farm.

"Amadori where you want" is an e-commerce platform that allows consumers to order products online and to receive them in physical stores. The aim is to adapt food sector to the global trend of E-commerce, still underdeveloped in Italy. Consumers can order the product online and withdraw it from their favorite stores and can choose from a wide range of products. Stores can take advantage of an online visibility on a national portal, can expand the range of products without increasing warehouse risks and remain the single point of contact for end consumers.

From the analysis of the main activities of the company, the implementation of three kinds of fit which ensure the alignment of strategic objectives with the business activities executed can be revealed: (1) technological fit; (2) informational fit; (3) relational fit.

Technological tools are used to align business objectives with technological practices thanks to the creation of an internal platform for e-commerce, orders and stocks management associated with an external cloud system to store and manage data. Informational fit refers to the enhancement of information ì flows and of the synergy between front-office employees and IT department thanks to the use of smart tools in the point of sales and to the use of analytics to analyse the data gathered from online orders. Lastly, the activation of relational fit derives from the digitalization of supply chain and from the e-commerce platform that help the execution of integrated strategies and tactics to enhance providers' and final customers' loyalty (combination of virtual, physical tools, ICTs tools and the human intervention of front-office employees).

The implementation of open innovative business models promotes an innovation mind-set that envelops strategies, resources and stakeholder selection, technology use, skills and translates into co-innovation strategies leads to the diffusion of a digital culture that introduces changes in data analysis and management.

Thanks to the development of education programmes and sharing of knowledge with start-ups, internal employees and workforce, actors are engaged in the development of new ideas for products and services.

Value integration and sharing is realized: (1) internally, by engaging actual employees; (2) externally, by attracting potential employees (students and young talents). Value co-creation strategies can be applied through the enhancement of internal skills, or through the acquisition of knowledge thanks to outsourcing strategic alliances.

The new values co-created concern: (1) new knowledge, thanks to the sharing of different know-how with start-ups and with IT organizations such as Sinfo One; (2) new IT structures: co-developed with Wasteless, start-ups and Sinfo One that release new services and distribution and communication modalities: (3) new improved processes.

The harmonization and renewal of value are ensured thanks to the enhancement of employee's satisfaction, of users' loyalty to the point of sales (thanks to engagement in the new e-commerce services). The integrated platform for e-commerce can help monitoring final users and retailers' behaviour, by detecting changes in real time, negative evaluations, or new requirements and by improving the quality of relationships and their stability over time.

RQ2: Data-Driven Orientation

The approach to data management seems to be included in business strategic planning thanks to the development of a flexible, proactive and looking-forward orientation (RQ1) that permits to experience different approaches in the co-creation of new products, services, selling modalities and marketing approaches by promoting the adoption of a culture based on open innovation at 360° (RQ2). This orientation all-encompasses data analysis, management, the arrangement of technological infrastructure, the promotion of employees and management technical and analytical skills, the production of continuous improvement.

Amadori creates an in integrated set of analytics (internal and external platforms) to collect and interpret data from multiple sources, stakeholders, point of sales across the different departments of supply chain.

The technological infrastructure realized to collect, analyse and manage data is based on the e-commerce platform and the related cloud computing systems and analytics developed to extract and store data on user's behaviour. The introduction of the infrastructure improves the management of stocks and helps the connection between suppliers, retailers and final customers. The platform is based on a series of tools that collect big data from users both through online e-commerce platforms and in stores, thanks to tablets in the point of sales in which customers should insert personal data and information about their orders.

Moreover, the architecture is supported with Salesforce, a cloud computing system that gathers real time information in the point of sales and improves the digitalization of Continuous Replenishment Planning (CRP). The system stores the purchase orders and analyses the data through dashboards and visualizations. In addition, some surveys are administered to users to assess their satisfaction. Thanks to API (Application Programming Interface), some procedures that simplify the communication between several applications to avoiding redundancies, and web scraping, users' feedback are analyzed through sentiment analysis and other statistics techniques.

The data collected are analyzed and interpreted through the intervention of data analysts with technical and methodological skills.

The firm settles collaborations with external researchers to support the work of data analysts and to optimize the large amount of information obtained from ordering processes, from front-office encounters, from users' evaluations. Thanks to the continuous enhancement of internal skills, the company develops the analytical and technical skills to manage data internally by creating and recruiting new professional figures able to deploy knowledge starting from data visualization and interpretation.

Amadori hires front office employees and organizes meetings with researchers from universities that can give a contribution to data analysis and o the extraction of insights from the order histories and users feedbacks provided face-to-face and on the platform.

The "total" research approach to data collection adopted by the company requires (not only) the technical capabilities to extract and analyse data (programming, machine learning) but also the ability to correlate the results obtained with the strategic objectives (improvement of orders, stocks management, customer's loyalty). The recruitment of big data analysts enhances the synergy in data management and interpretation across the different departments (front-office employees and IT department).

Thanks to data analysis, customers profiles are created and interpreted as "histories" to increase trust and loyalty. Moreover, the company can predict users' behavior by storing their past orders and can offer some discounts, personalized or limited editions of products in some period of the year (Christmas, Easter) based on consumers habits.

Therefore, the main goals pursued through the implementation of a data-driven culture are: (1) the digitalization of processes and sales; (2) the optimization of information flows within the organization, with retailers and the improvement of control power on the distribution; (3) the strengthening of relationships and loyalty with retailers and final customers.

Thus, the data-driven innovation outcomes are related to: (1) technological innovation; (2) process innovation and business-model innovation; (3) new knowledge.

The digitalization of services and processes introduce new standards for production, by reducing risks for storage and costs for stock (*Continuous Replenishment Planning*, CRP), by increasing sales. Moreover, the integrated technological platform (internal for e-commerce and external for cloud) introduces some technological advancements in the supply chain, in business models and relationships management modalities. Thanks to a distributed computing architecture, the firm can reduce costs and times for decision-making, rise competitiveness, increase internal and external communication flows.

Data-driven strategies can generate multiple (economic, knowledge-based and marketing) advantages throughout the entire supply chain and in the enhancement of relationships with customers (CRM and loyalty). The adoption of integrated strategies and tactics for data analysis enhance users' loyalty (thanks to the combination of physical stores, front-office and virtual tools, such as IT and ICTs platforms).

New knowledge is created thanks to the enrichment of data analysts' skills through internal and external training activities and through collaborations with start-ups and young talents that share technological skills and a forwarding-looking attitude.

Table 32.1 synthesizes the main results discussed above, categorized into different sub-dimensions for each of the three layers/macro-areas employed to analyze the data.

Discussion

The results of the case study, obtained through the content analysis performed thanks to the main macro-variables emerged from literature review, allow at detecting the main sub-dimensions within each of the three layers of the framework. Thus, by categorizing the different sub-dimensions revealed through the analysis (Table 32.1), a synthesis framework can be introduced to guide business networks to the understanding of the most appropriate strategies and drivers to encourage co-created innovation through proper data management strategies can be introduced (Fig. 32.1).

As Fig. 32.1 shows, the three (embedded) layers can be intended as recursive "steps" of co-innovation processes that can start from the adoption of innovation-oriented business models (layer 1) that translates into the development of value creation strategies to turn information into knowledge and value (layer 2) that can be turned into innovation thanks to the implementation of data-driven approach (layer 3).

The three layers depicted in the Figure are strictly interconnected. Only through the combination of proper business process modelling (layer 1), based on targeted value co-creation strategies (layer 2), realized through data-driven orientation (layer 3), the synergistic innovation outcomes can be attained.

In the light of the findings discussed above, the core of the process of business model redesign according to the principles of open innovation (*layer 1*) is based on the joint co-creation of new value and innovation (*layer 2*), deriving from: (1) a strategic orientation grounded on a total approach to data and on the promotion of digital culture (*strategy and value design*); (2) selection and management of resources that ensure the acquisition and sharing of know-how from key stakeholders (*resources and sense-making*); (3) the adoption of an integrated technological architecture that encourages the sharing of businesses' values and the resources integration and collects and extract data based on strategic objectives (*technology and value integration and sharing*); (4) agile management to optimize processes, relationships according to the interpretation of data in line with strategic objectives (*process management*); (5) the joint development of innovative outcome ranging from technological and product innovation, to process innovation and sustainable innovation (*co-creation of new value*); (6) the renewal of the value generated over time thanks to the propagation of knowledge and the diffusion of a constant tension to innovation (*continuous improvement and renewal/harmonization*).

Table 32.1 The main findings

Research question	Variables	Main sub-dimensions	Findings
RQ1	Business model	*Strategy*	Open-innovation mind-set for strategy, selection and engagement of stakeholders and skill's enhancement Integrated strategies for the digitalization of supply chain
		Resources	Selection of partners with technological know-how and digital competencies
		Technology	Smart factory integrated systems for ERP Data control centre for Agile production management E-commerce platform
		Human skills	Sharing of a smart approach from start-ups
		Fit alignment	Technological, informational, relational fits
	Value co-creation strategies	*Design*	Knowledge sharing with start-ups Training programs for employees
		Diffusion and sense-making	Teamwork improvement Relationships with young talents
		Integration and sharing	Internal: enhancement of employees' strategic skills External: acquisition of strategic know-how from partners
		Creation and co-creation of value	New knowledge and competencies New IT structures New services Improved processes
		Harmonization/renewal	Monitoring of retailers' behaviour Enhancement of users' loyalty

(continued)

Table 32.1 (continued)

Research question	Variables	Main sub-dimensions	Findings
RQ2	Data-driven orientation	*Data-driven culture*	Flexible, proactive and looking-forward orientation that all-encompasses data analysis and management
		Technological infrastructure	• Cloud computing systems • E-commerce platform and tablets in store • Salesforce (cloud) for CRP • Artificial intelligence based on Reinforcement Learning algorithm
		Skills	Technical and methodological skills
		Process management	Data analysts that correlate results with strategic objectives
		Continuous improvement	Diffusion of a new methodological attitude oriented to constant change
	Innovation outcomes		• Process innovation: digitalization of supply chain and stocks management • Business model innovation: data analysis reframes strategies and tactics in the supply chain (loyalty, CRM)

Moreover, findings confirm the existence of a strategic data-driven approach to big data analysis (*layer 3*) in the firm analysed, which starts from the common adoption of a data-oriented mind-set (*culture*), deriving from the strategy identified in the first layer, that leads to the application of an integrated infrastructure that combines a varied set of applications, technological tools and analytics (*technology adoption*), managed thanks to specialized competencies (*management, technical and methodological skills*) and the constant search for the fit between strategic objectives and the data interpreted (*process management*) in order to achieve co-learning and innovation over time (*continuous improvement and innovation*).

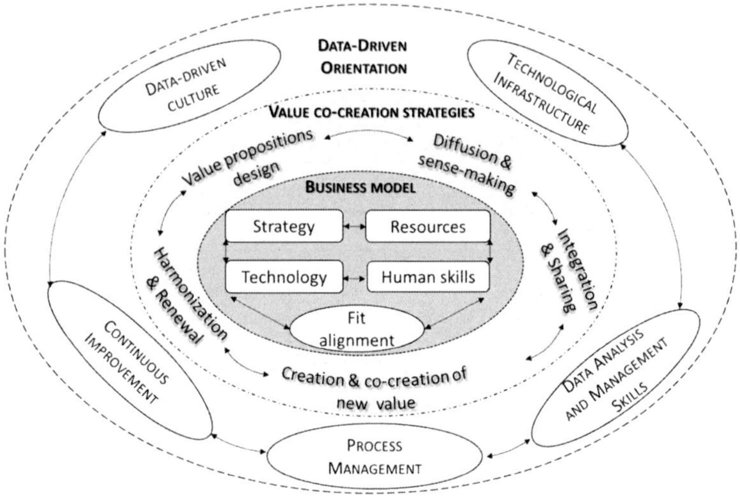

Fig. 32.1 Modelling data-driven companies: a three-layered framework

Limitations

The main limitation of the study concerns the adoption of case study methodology which does not allow at drawing any generalization of results. However, the exploratory research proposed can be considered as a first qualitative step in the investigation of data-driven innovation that can encourage the development of further research. The paper introduces a research agenda that can stimulate the proposition of future studies which: (1) investigate business models enablers for value co-creation and the way in which their combination can produce innovation; (2) detect the different data-oriented and management strategies that redefine business models to develop innovation systematically.

Furthermore, the sample of the study can be broadened to include firms operating in other B2B and B2C markets and to combine qualitative (observation or netnography) and quantitative (regression analysis, structural equation modelling) techniques. The adoption of mixed method, in fact, seems to be the most adequate approach to explore multidimensional concepts and to detect the emergence of complex phenomena such as innovation. For instance, quantitative analysis can explore the statistical relations between data-driven culture, use of analytics and the development of innovation. The sub-dimensions identified in the three-layered framework can be employed as a starting point for the elaboration of measurement items.

Moreover, customer's and users' (students, researchers, citizens) viewpoint can be investigated in future research and comparisons between case studies from different business sectors can be performed to reveal the existence of varied data-driven strategies in different markets or in the implementation of different users' engagement practices [18].

Conclusions

The analysis of the case study can contribute to the identification of the main innovation strategies which data-driven oriented companies can apply to reframe their business models and exploit the possibilities offered from smart technologies and analytics.

The proposition of the three-layered framework reveals the need to focus on some strategic assets grounded on the development of a "total" data approach which can be co-designed with stakeholders and promoted and diffused over time through constant knowledge sharing and continuous improvement as key levers for innovation, relationships and skills management, smart use of technologies. In this way, is reframed as the complex result of virtuous circles of co-creation based on the co-design of joint innovation-oriented projects, which introduce synergistic results superior than the "mere" sum of the knowledge exchanged in the network [3].

From a theoretical point of view, the study reveals the main strategic dimensions and the drivers for achieving data-driven innovation in corporate networks, by highlighting how this complex process cannot be obtained from the "simple" application of technologies to business processes and model, but can be ensured thanks to the diffusion of a cohesive culture, the strengthening of relationships, the exchange of knowledge between the actors and the involvement in proper co-creation strategies. The study categorizes the different enablers of business models and data-driven innovation by identifying some categories of analysis and drivers that can be applied to different contexts (B2B and B2C) and to different segments (from agrifood to automotive markets).

From a managerial point of view, the identification of the key enabling dimensions for the creation of sustainable and joint value can enhance practitioners' and mangers' understanding of innovation management in a data-driven perspective through a classification of the different innovative outcomes ranging from technological, process and environmental dimensions. The classification can contribute to the identification of new strategies for the use of technologies to produce different innovative results based on the different types of stakeholders involved.

References

1. R. Amit, C. Zott, Creating value through business model innovation. MIT Sloan Manag. Rev. **53**(3), 41–49 (2012)
2. D. Antons, C. Breidbach, Big data, big insights? Advancing service innovation and design with machine learning. J. Serv. Res. **21**(1), 17–39 (2018)
3. C. Baccarani, G.M. Golinelli, Le parole dell'innovazione. Sinergie **94**, 9–14 (2014)
4. D. Blazquez, J. Domenech, Big data sources and methods for social and economic analyses. Technol. Forecast. Soc. Chang. **130**, 99–113 (2018)
5. N.M. Bocken, S.W. Short, P. Rana, S. Evans, A literature and practice review to develop sustainable business model archetypes. J. Clean. Prod. **65**, 42–56 (2014)
6. H. Chesbrough, Business model innovation: opportunities and barriers. Long Range Plan. **43**(2–3), 354–363 (2010)
7. K.M. Eisenhardt, Building theories from case study research. Acad. Manag. Rev. **14**, 532–550 (1989)
8. N.J. Foss, Knowledge governance in a dynamic global context: the center for strategic management and globalization at the Copenhagen Business School. Eur. Manag. Rev. **4**(3), 183–191 (2007)
9. P.S. Fosso Wamba, Big data analytics and business process innovation. Bus. Process Manag. J. **23**(3), 470–476 (2017)
10. M. Geissdoerfer, D. Vladimirova, S. Evans, Sustainable business model innovation: a review. J. Clean. Prod. **198**, 401–416 (2018)
11. K. Krippendorff, Reliability in content analysis: some common misconceptions and recommendations. Hum. Commun. Res. **30**(3), 411–433 (2004)
12. S. LaValle, E. Lesser, R. Shockley, M.S. Hopkins, N. Kruschwitz, Big data, analytics and the path from insights to value. MIT Sloan Manag. Rev. **52**(2), 21–32 (2011)
13. S.M. Lee, S. Trimi, Innovation for creating a smart future. J. Innov. Knowl. **3**(1), 1–8 (2018)
14. S. Lee, D. Olson, S. Trimi, Co-innovation: convergence, collaboration, and co-creation for organizational values. Manag. Decis. **50**(5), 817–831 (2012)
15. F. Li, The digital transformation of business models in the creative industries: a holistic framework and emerging trends. Technovation **92–93**, 102012 (2018)
16. Z. Lindgardt, M. Reeves, G. Stalk, M.S. Deimler, *Business Model Innovation. When the Game Gets Tough, Change the Game* (The Boston Consulting Group, Boston, MA, 2009)
17. G. Losito, *L'analisi del contenuto nella ricerca sociale*, vol. 1 (Franco Angeli, Milan, 1996)
18. M. Lytras, A. Visvizi, Who uses smart city services and what to make of it: toward interdisciplinary smart cities research. Sustainability **10**(6), 1998 (2018)
19. M.D. Lytras, V. Raghavan, E. Damiani, Big data and data analytics research: from metaphors to value space for collective wisdom in human decision making and smart machines. Int. J. Semant. Web Inf. Syst. (IJSWIS) **13**(1), 1–10 (2017)
20. A. McAfee, E. Brynjolfsson, Big data: the management revolution. Harv. Bus. Rev. **90**(1), 61–67 (2012)
21. V. Ortiz-Repiso, J. Greenberg, J. Calzada-Prado, A cross-institutional analysis of data-related curricula in information science programmes: a focused look at the iSchools. J. Inf. Sci. **44**(6), 768–784 (2018)
22. V. Ramaswamy, K. Ozcan, *The Co-Creation Paradigm* (Stanford University Press, Stanford, CA, 2014)
23. D. Teece, Business models, business strategy and innovation. Long Range Plan. **43**(2–3), 172–194 (2010)
24. W. Tellis, Application of a case study methodology. Qual. Rep. **3**, 1–19 (1997)
25. D. Trabucchi, L. Talenti, T. Buganza, How do big bang disruptors look like? A business model perspective. Technol. Forecast. Soc. Chang. **141**, 330–340 (2019)
26. O. Troisi, M. Grimaldi, A. Monda, Managing smart service ecosystems through technology: how ICTs enable value cocreation. Tour. Anal. **24**(3), 377–393 (2019)

27. A. Visvizi, M.D. Lytras, E. Damiani, H. Mathkour, Policy making for smart cities: innovation and social inclusive economic growth for sustainability. J. Sci. Technol. Policy Manag. **9**(2), 126–133 (2018)
28. R.K. Yin, *Case Study Research* (Sage, Thousand Oaks, CA, 1984)
29. Y. Zhan, K.H. Tan, R.K. Perrons, A proposed framework for accelerated innovation in data-driven environments. Ind. Manag. Data Syst. **118**(4) (2018)
30. J. Zhang, S.E. Beatty, D. Mothersbaugh, A CIT investigation of other customers' influence in services. J. Serv. Mark. **24**(5), 389–399 (2010)
31. C. Zott, R. Amit, L. Massa, The business model: recent developments and future research. J. Manag. **37**(4), 1019–1042 (2011)

Chapter 33
Boosting ICT Entrepreneurship by Means of Cooperation

Ricardo Colomo-Palacios

Abstract Entrepreneurship is recognized nowadays as the driving force of a market economy and technology-driven entrepreneurship tends to become a standard for early-stage businesses in Europe and worldwide. In fact, Information and Communications Technology (ICT) holds a major part of any business nowadays. In this paper, Computing Innovation for Technology Entrepreneurship (CITE) project is presented. The project addresses the development of education for ICT Entrepreneurship in Romania, aiming to increase the development of new businesses, based on innovation and new technologies. By means of this project, both human capital and entrepreneurship initiatives will be enhanced. Founded by the EEA Grants Programme and developed in cooperation between University Politehnica of Bucharest (Romania) and Østfold University College (Norway), the primary targets are the development of teaching materials at master and bachelor level in order to train students in Management and Engineering and in Computer Science but also the development of entrepreneurial skills among academics. Increasing the quality and quantity of Romanian IT-driven business initiatives addresses directly the first two priority sectors of the EEA grants: Innovation, Research, Education and Competitiveness, as well as Social Inclusion, Youth Employment and Poverty Reduction.

Keywords Entrepreneurship · Lean startup · Collaboration

Introduction

According to Cambridge Dictionary, entrepreneurship can be defined as a "skill in starting new businesses, especially when this involves seeing new opportunities". For the Collins Dictionary, entrepreneurship is "the state of being an entrepreneur, or the activities associated with being an entrepreneur". Finally, for the Oxford English Dictionary, the term is defined as "activity, behaviour, or attitudes that characterize an entrepreneur or entrepreneurs". From these three definitions we can conclude

R. Colomo-Palacios (✉)
Østfold University College, 1757 Halden, Norway
e-mail: ricardo.colomo-palacios@hiof.no

A. Visvizi et al. (eds.), *Research and Innovation Forum 2020*,
Springer Proceedings in Complexity,
https://doi.org/10.1007/978-3-030-62066-0_33

that entrepreneurship is a skill, behaviour or an attitude and, consequently, could be developed.

Scientific literature has covered the topic in a very deep way. According to Eisenmann [1], citing Stevenson, entrepreneurship is the pursuit of opportunity beyond controlled resources. This general and not too specific definition is, however, one of the most accepted ones in the research community. This lack of specificity is connected to the fact that entrepreneurship is difficult to define [2], given this connection with an irrational world [3].

However, entrepreneurship is considered to contribute to societal development [4]. Consequently, entrepreneurship courses have been around since the forties when Harvard Business School offered the first course on the topic [5]. Currently, entrepreneurship education programmes are pervasive [6] and many countries are substantially investing in entrepreneurship education [7] in various levels including universities, high schools and even primary schools. Focusing on the first environment, one of the missions of modern higher education institutions is to encourage the social and economic development of their surroundings through venture creation training and entrepreneurship development [8].

Engineering Entrepreneurship Education [9] and more precisely, Information Technology Entrepreneurship Education [10] is a hot topic, given the key role of technology in economic growth. Regardless of the form of technology and digital entrepreneurship (digital, digital technology or technology entrepreneurship), technology has changed the entrepreneurship panorama from its roots [11] and the corresponding education deserves attention from researchers and policy makers alike.

In this paper, a project called Computing Innovation for Technology Entrepreneurship (CITE) is presented. The project, adopting a Lean Startup approach, is aimed to increase the development of new businesses, based on innovation and new technologies in Romania by means of a collaborative project funded by EEA Grants. The main aims and outcomes of the project will be presented in the last section of the paper.

Lean Startup

The lean startup framework is considered as one of the most popular contributions in the practitioner-oriented entrepreneurship literature [12] to become a worldwide practice nowadays [13]. The approach is inspired by the "Lean philosophy" originated in the manufacturing world [14]. This approach intends to direct production systems towards customers and value generation. Ries [15] adapted this approach in the field of entrepreneurship during his studies at the Steve Blank's customer development class at the University of California, Berkeley. He took several tools from Blank [16, 17], who introduced a set of tools like agile engineering or minimum viable product (MVP). Ries was Blank's student [12].

The lean startup presents five main aspects in its approach, combining a set of tools from previous approaches:

1. Seeking and prioritizing market opportunities
2. Designing business models
3. Validated learning (including customer development)
4. Minimum viable products (MVPs)
5. Learning whether to persist with or pivot from the current course of action.

The whole lean startup concept is directed towards reaching a good product-market fit by treating business ideas as hypotheses to be tested in front of potential customers as quickly and cheaply as possible [18], by means of the build-measure-learn loop. The ultimate goal of the lean startup process is to guide entrepreneurs in finding this fit [19].

The importance of the approach in several fields has been highlighted in the literature and a recent work analyses its impact in business arenas [20]. This influence is also quite important in educational scenarios, thus literature has been reporting cases of lean startup adoption in educational settings in general e.g. [21–23] and in computing in particular, mainly by linking the concept to software startups [24–27]. As a consequence of its popularity, tertiary studies appear on the topic [28], as well as studies on patterns and anti-patterns [29] or global studies [30].

ICT Panorama in Romania

According to Eurostat, Romania presents a population 19,401,658 by 2019. The country has been continuously losing population in the last decade. According to the International Monetary Fund, the GDP per capita (nominal) is $12,483, being the 65% of the EU GDP per capita by the end of 2019.
(https://ec.europa.eu/eurostat/databrowser/view/tec00114/default/table?lang=en).

Interestingly, the digital society is growing, thus ICT contributes 6–7% to Romania's GDP.
(https://ec.europa.eu/romania/sites/romania/files/docs/semestru_european_raport_romania_2018_frank_siebern-thomas.pdf). In the Digital Economy and Society Index (DESI), Romania is ranked in the second last position just before Bulgaria (2019). However, according to a recent report by McKinsey, the potential economic and development benefits of digitization may reach up to €42 billion in additional GDP, by 2025 [31]. One of the digitalization enablers identified by this report is the stimulation of startup ecosystems by means of entrepreneurship. Currently, Romania presents just 28 startups per million citizens, compared to 58 in the Central and Eastern Europe (CEE) region and 215 among Digital Frontrunner markets. This is the gap we aim to bridge through this project.

With regards to the composition of the ICT workforce, in Romania only 2.2% of the employees are ICT specialists in 2018, according to Eurostat https://ec.europa.eu/eurostat/statistics-explained/index.php/ICT_specialists_in_employment, whereas the average in Eu-28 is 3.9%. With regards to the gender distribution of specialists, Romania presents in the market 23.5% of women, just behind Bulgaria and Lithuania as the more gender-balanced country in EU in the

ICT specialist market. However, it is also worth to note that by 2008 this percentage was 33.9%.

The last aspect to analyze is the effect of brain drain. As stated before, Romania is losing population and currently, according to UN (2017) there are 3.6 million Romanians living abroad, 2.7 of them of working age. According to the World Bank, an increasing number of Romanians are enrolled in universities across Europe and the United States, reaching 33.4 thousand in 2016, while the number of students enrolled in the country has been decreasing since 2008.

The Project

In this scenario, there is a need to act in order to maximize the opportunities in the country. In this scenario, the project addresses the development of education for ICT Entrepreneurship in Romania, aiming to increase the development of new businesses, based on innovation and new technologies.

The main objective of the project is the development of materials for the ICT Entrepreneurship discipline in two versions—for master programs in Management and Engineering, as well as for master programs in Computer Science. The transfer of knowledge from the Norwegian side to the Romanian academic staff will help the professors to deal with this challenge in the years to come. Teaching materials like case studies for ICT Entrepreneurship in Romania and in Europe will be provided and developed. Another objective of the project partners is to perform research in IT by means of future projects and related scientific publications. There will be addressed not just the scientific and technical issues related to ICT, but also the methodologies to develop businesses and products. In order to do so, lean startup will be the method to guide these developments.

The target groups for this project are the staff from the Romanian Institution (Polytechnical University of Bucharest) receiving assistance in developing new subjects as well as research projects and the Romanian students facing the possibility to start their own companies and to materialize their ideas. Researchers from Romania and Norway will have to guide them in this endeavor, performing research in IT in order to find out convincing business designs and to validate the student's ideas for start-ups.

A specific aspect the project will tackle is the improvement in the percentage of women entrepreneurs in ICT by means of specific actions directed towards women. The topic of women entrepreneurship in emerging economies still lacks of studies [32], and studies on ICT entrepreneurship and women are scarce [33]. In fact, there are several initiatives in the world (e.g. ActuaUPM from Spain) we can analyze to perform benchmarking and implement measures inspired and backed up with previous results.

The best results will be given by successful IT businesses developed in Romania. Research in IT will be performed together with Norwegian researchers in order to foster collaboration in the broad field of computing.

The project integrates well in the efforts to increase the economic, educational and research capabilities of Romania through this cooperation project in higher education. The objectives of the project, together with the activities which will permit their achievement are presented, as follows:

Obj1. Develop materials for the courses. The activities which will target the objective are:

O1 Develop specific curriculum, including teaching materials for the "Information and Communications Technology Entrepreneurship" discipline to be included in the teaching portfolio.

O2 Prepare teaching materials of the subjects included in the curriculum. They will include case studies and business models in ICTE. Former students will participate in the creation and at the assessment of the materials. Practical results from the concluded projects will be included. This represents a lifelong learning strategy, reinforcing the link between higher education and the working force.

O3 (continuous). Prepare web portal with learning materials, included in the website of the project. Include also success stories. Include specific aspects and teaching materials for women.

Obj2. Strengthen the relationship among partners. The activities to target this objective are:

O4 Perform trainer of trainers events. Personnel from the Romanian institution will visit Norway in order to learn from their experience in the field.

O5 Create jointly research papers as well as proposals and look for ideas that can be transformed in practice.

Obj3. Prepare the Romanian students in an established and competitive environment. The activities to target this objective are:

O6 (continuous). Constant training using the web portal.

O7 Involve the students to create jointly research papers in ICT and look for ideas that may be transformed in practice.

Conclusions and Future Work

Technology has disrupted the business panorama in general and entrepreneurship in particular. In fact, the so-called Information Technology Entrepreneurship has demonstrated significant contributions to the economical and societal development. Given the relevance of the topic, higher education institutions have not only acknowledged the need to focus on this aspect but also have taken action to enhance IT Entrepreneurship education. One of these attempts is the CITE project, which is a collaborative effort between University Politehnica of Bucharest (Romania) and Østfold University College (Norway). This project adopts a Lean Startup approach and aims to enhance the development of education for ICT entrepreneurship and to encourage the flourishment of new innovative businesses in Romania.

The aforementioned objectives will be achieved through activities undertaken in two directions. Firstly, we intend to develop and provide materials for Romanian students enrolled in the master programs of Management and Engineering and Computer Science. Additionally, trainer of trainers events will be organized to transfer the knowledge in the field from Norwegian to Romanian professors. The collaboration continues with jointly research papers between Norwegian and Romanian professors, with the involvement of students. Another interesting aspect that we will address is increasing the number of woman entrepreneurs in ICT. Despite the fact that the literature focused on women and ICT entrepreneurship is scarce, success cases do exist in practice. By replicating the actions undertaken by other countries, we believe the enhancement of this aspect in Romania will be noticeable.

Future work will focus on preparing the web portal with learning materials, which will be included in the website of the project. This portal will be of use for Romanian students, who will be trained and prepared in this competitive environment. Furthermore, it would be interesting to deeply focus on the phenomena of brain drain and analyze the impact of our undertaken actions on this particular aspect.

Acknowledgements The work if partially funded by the project "Computing Innovation for Technology Entrepreneurship" Project Code: 18-COP-0012 funded by The EEA Grants.

References

1. T.R. Eisenmann, Entrepreneurship: a working definition. Harv. Bus. Rev. **10** (2013)
2. A.R. Anderson, M. Starnawska, Research practices in entrepreneurship: problems of definition, description and meaning. Int. J. Entrep. Innov. **9**, 221–230 (2008). https://doi.org/10.5367/000000008786208731
3. A. Sackett, O. Sheldon, On missed boats and sunken ships: asymmetric tolerance for errors in entrepreneurial entry decisions. Acad. Manag. Proc. **2019**, 16074 (2019). https://doi.org/10.5465/AMBPP.2019.16074abstract
4. Z. Su, The co-evolution of institutions and entrepreneurship. Asia Pac. J. Manag. (2020). https://doi.org/10.1007/s10490-019-09703-y
5. K.H. Vesper, W.B. Gartner, Measuring progress in entrepreneurship education. J. Bus. Ventur. **12**, 403–421 (1997). https://doi.org/10.1016/S0883-9026(97)00009-8
6. G. Nabi, F. Liñán, A. Fayolle, N. Krueger, A. Walmsley, The impact of entrepreneurship education in higher education: a systematic review and research agenda. Acad. Manag. Learn. Educ. **16**, 277–299 (2017). https://doi.org/10.5465/amle.2015.0026
7. S.G. Walter, J.H. Block, Outcomes of entrepreneurship education: an institutional perspective. J. Bus. Ventur. **31**, 216–233 (2016). https://doi.org/10.1016/j.jbusvent.2015.10.003
8. V. Barba-Sánchez, C. Atienza-Sahuquillo, Entrepreneurial intention among engineering students: the role of entrepreneurship education. Eur. Res. Manag. Bus. Econ. **24**, 53–61 (2018). https://doi.org/10.1016/j.iedeen.2017.04.001
9. A. Huang-Saad, C. Bodnar, A. Carberry, Examining current practice in engineering entrepreneurship education. Entrep. Educ. Pedagogy **3**, 4–13 (2020). https://doi.org/10.1177/2515127419890828
10. S.-S. Han, J.-J. Kim, The effects of reemployment trainees' IT entrepreneurship education on entrepreneurship willingness. J. Korea Inst. Inf. Electron. Commun. Technol. **11**, 63–69 (2018). https://doi.org/10.17661/jkiiect.2018.11.1.63

11. F. Giones, A. Brem, Digital technology entrepreneurship: a definition and research agenda. Technol. Innov. Manag. Rev. **7**, 44–51 (2017)
12. D.A. Shepherd, M. Gruber, The lean startup framework: closing the academic–practitioner divide. Entrep. Theory Pract. 1042258719899415 (2020). https://doi.org/10.1177/1042258719899415
13. X. Yang, S.L. Sun, X. Zhao, Search and execution: examining the entrepreneurial cognitions behind the lean startup model. Small Bus. Econ. **52**, 667–679 (2019). https://doi.org/10.1007/s11187-017-9978-z
14. J.P. Womack, D.R. Jones, D. Roos, *The Machine That Changed the World* (Macmillan, New York, 1991)
15. E. Ries, *The Lean Startup: How Today's Entrepreneurs Use Continuous Innovation to Create Radically Successful Businesses* (Crown Publishing Group, New York, NY, 2011)
16. S. Blank, Why the lean start-up changes everything. Harv. Bus. Rev. **91**, 63–72 (2013)
17. S. Blank, *The Four Steps to the Epiphany* (K & S Ranch, 2013)
18. E. Ries, *The Startup Way: How Modern Companies Use Entrepreneurial Management to Transform Culture and Drive Long-Term Growth* (Crown Books, 2017)
19. Y. Mansoori, T. Karlsson, M. Lundqvist, The influence of the lean startup methodology on entrepreneur-coach relationships in the context of a startup accelerator. Technovation **84–85**, 37–47 (2019). https://doi.org/10.1016/j.technovation.2019.03.001
20. D.S. Silva, A. Ghezzi, R.B. de Aguiar, M.N. Cortimiglia, C.S. ten Caten, Lean startup, agile methodologies and customer development for business model innovation: a systematic review and research agenda. Int. J. Entrep. Behav. Res. (2019). https://doi.org/10.1108/IJEBR-07-2019-0425
21. R. Harms, Self-regulated learning, team learning and project performance in entrepreneurship education: learning in a lean startup environment. Technol. Forecast. Soc. Change **100**, 21–28 (2015). https://doi.org/10.1016/j.techfore.2015.02.007
22. H.M. Neck, P.G. Greene, Entrepreneurship education: known worlds and new frontiers. J. Small Bus. Manag. **49**, 55–70 (2011). https://doi.org/10.1111/j.1540-627X.2010.00314.x
23. D. Maresch, R. Harms, N. Kailer, B. Wimmer-Wurm, The impact of entrepreneurship education on the entrepreneurial intention of students in science and engineering versus business studies university programs. Technol. Forecast. Soc. Change **104**, 172–179 (2016). https://doi.org/10.1016/j.techfore.2015.11.006
24. A. Järvi, V. Taajamaa, S. Hyrynsalmi, Lean software startup—an experience report from an entrepreneurial software business course, in *Software Business*, ed. by J.M. Fernandes, R.J. Machado, K. Wnuk (Springer International Publishing, Cham, 2015), pp. 230–244. https://doi.org/10.1007/978-3-319-19593-3_21
25. K. Buffardi, C. Robb, D. Rahn, Learning agile with tech startup software engineering projects, in *Proceedings of the 2017 ACM Conference on Innovation and Technology in Computer Science Education* (Association for Computing Machinery, Bologna, Italy, 2017), pp. 28–33. https://doi.org/10.1145/3059009.3059063
26. R. Chanin, A. Sales, L. Pompermaier, R. Prikladnicki, Challenge based startup learning: a framework to teach software startup, in *Proceedings of the 23rd Annual ACM Conference on Innovation and Technology in Computer Science Education* (Association for Computing Machinery, Larnaca, Cyprus, 2018), pp. 266–271. https://doi.org/10.1145/3197091.3197122
27. J. Melegati, R. Chanin, X. Wang, A. Sales, R. Prikladnicki, Perceived benefits and challenges of learning startup methodologies for software engineering students, in *Proceedings of the 50th ACM Technical Symposium on Computer Science Education* (Association for Computing Machinery, Minneapolis, MN, USA, 2019), pp. 204–210. https://doi.org/10.1145/3287324.3287382
28. R. Chanin, A. Sales, L. Pompermaier, R. Prikladnicki, A systematic mapping study on software startups education, in *Proceedings of the 22nd International Conference on Evaluation and Assessment in Software Engineering 2018* (Association for Computing Machinery, Christchurch, New Zealand, 2018), pp. 163–168. https://doi.org/10.1145/3210459.3210478

29. F. Fagerholm, A. Hellas, M. Luukkainen, K. Kyllönen, S. Yaman, H. Mäenpää, Designing and implementing an environment for software start-up education: patterns and anti-patterns. J. Syst. Softw. **146**, 1–13 (2018). https://doi.org/10.1016/j.jss.2018.08.060
30. R. Chanin, D. Khanna, K.-K. Kemell, X. Wang, A. Sales, R. Prikladnicki, P. Abrahamsson, Software startup education around the world: a preliminary analysis, in *CEUR Workshop Proceedings* (RWTH Aachen University, 2018)
31. McKinsey, The rise of digital challengers. How digitization can become the next growth engine for Central and Eastern Europe. https://digitalchallengers.mckinsey.com/. Accessed 2020/02/05
32. S. Rashid, V. Ratten, A systematic literature review on women entrepreneurship in emerging economies while reflecting specifically on SAARC countries, in *Entrepreneurship and Organizational Change: Managing Innovation and Creative Capabilities*, ed. by V. Ratten (Springer International Publishing, Cham, 2020), pp. 37–88. https://doi.org/10.1007/978-3-030-35415-2_4
33. C. Armuña, S. Ramos, J. Juan, C. Feijóo, A. Arenal, From stand-up to start-up: exploring entrepreneurship competences and STEM women's intention. Int. Entrep. Manag. J. (2020). https://doi.org/10.1007/s11365-019-00627-z

Chapter 34
An Integrative Framework for Startups at Early Stage: Promoting Evidence-Based Design and Evaluation in Early Stage Startups

Francesc Font Cot, Pablo Lara Navarra, and Enric Serradell-López

Abstract Startups sector, especially those more technological, but also those who have contributed new business models, has led to and driven innovation and growth in recent economic history. At the same time, it is well recognized that while overall startup contribution is crucial, the high risk and reward strategy followed by these startups has significant failure rates, with mortality rates around 90% before three years and a low proportion of successful startups. Despite this high percentage of failures, the literature tends to focus on successful startups and quantitative studies that seek determinants of success, but the literature is also starting to appear lately, the many lessons behind failure by examining the stories of failed initiatives. Entrepreneurial strategy can be described by different frameworks and different dimensions, and despite all the literature in the startups topic, we find a gap identified in the creation of a model that helps to follow a structured business development strategy in the early stage, learning to prioritize efforts based on right decisions in its strategy to enable the company to survive the early stage. This work aims to fill this gap and contribute to the literature by providing scalable, repeatable methodology that can be applied to databases of both failed and successful startups that passed the early stage to jump into the growth stage.

Keywords Startup · Early stage · Critical success factors · Failure · Operations · Success · Model · Simulator · Methodology

F. Font Cot (✉) · P. Lara Navarra · E. Serradell-López
UOC—Universitat Oberta de Catalunya, Avinguda del Tibidabo, 39, 08045 Barcelona, Spain
e-mail: ffontco@uoc.edu

P. Lara Navarra
e-mail: plara@uoc.edu

E. Serradell-López
e-mail: eserradell@uoc.edu

A. Visvizi et al. (eds.), *Research and Innovation Forum 2020*,
Springer Proceedings in Complexity,
https://doi.org/10.1007/978-3-030-62066-0_34

Introduction

A long-lasting research topic of literature is the startups success in the contemporary economy, studying the impact of startups on value creation, innovation and the economy itself for each region. This belief in entrepreneurship as a potential solution to unemployment, economic growth, regional development and innovation, leads to substantial levels of public support. Despite the almost universally accepted belief outside academia that entrepreneurial activity is a positive driver of the economy, evidence is largely inconclusive [1]. In fact, we can see that a large portion of high-performing companies drive the most innovation, wealth creation and new generation of jobs, while most startups have a marginal impact [2] and a mortality rate of 92% before the age of three (getautopsy.com).

Research in entrepreneurship and startup studies over the last two decades has been quite extensive, with an emphasis on understanding the determinants of failure and success for new companies. There has recently been studied and coded new processes followed by entrepreneurs to create successful startups (the popular Lean Startup or Scalable Startup methods) and researchers have begun to study how entrepreneurs work within these frameworks [3]. Returning to the key factors of failure or success, we see how it is especially relevant for high-tech startups as these companies follow high-reward strategies that often lead to the failure or a great success, a strategy that is well suited to the business model of venture capital but leaves a high mortality rate and poor distribution of value creation.

The success and failure of early stage companies are usually studied with quantitative approaches based on financial data [4] and examining correlations with elements that have a great influence on success or failure, such as entrepreneurial ability, basic company characteristics and the characteristics of the relevant market. The literature explaining these approaches is very extensive. It has been a long time since the first models that [5, 6] proposed using a firm's financial data to predict its probability of failure, with discriminant or regression analysis models. In the 1980s, artificial intelligence models began to be used to predict financial bankruptcy. Financial data-driven approaches had the potential of being applied to a large number of companies, as data could be collected from their annual reports. However, when it comes to passing stages, specially the early stage, there are aspects that are critical such as the capacity of the entrepreneur or his origin, the basic competencies of the company, the market, etc. In this regard, other research studies investigated whether these aspects could also contribute to the success or failure of a business. For example, entrepreneurship analysis examined the influence of gender and ethnicity on the likelihood of success or failure [7]. Logistic regression analysis was also adopted [8] to model the relationship between small business mortality rates and aggregate levels of internal and external risks (e.g., bankruptcy related to interest rates, the shutdown of a business or property, etc.).

Other researchers focused on entrepreneurial attitudes and the failure caused by the mismatch between corporate goals and founders' goals [9]. Others focused on the concept of failure, all analyzed from an entrepreneurial perspective [10], realizing that valuable lessons can be learned from their mistakes at the time the entrepreneurs launch a new venture [11].

We also find literature that focuses on emphasizing how the environment could influence startup success. The factors investigated by these works were mainly related to the differences between the regions and/or the existing industrial structure [12]. We find more recent work that looked at other potential determinants of success, such as deciding to innovate a product [13] or relying on support from a Business angel/venture capital [14].

On the other hand, there is a lot of old literature regarding the entrepreneurial strategy and how this strategy can lead to a company survival. For example, one approach to new venture strategy follows the generic strategy logic of Porter [15]. One dimension of a generic strategy is the decision to serve a broad or narrow market segment. A second is deciding to compete on the basis of low cost or differentiation. A low cost strategy aims for the firm to have lower costs than its rivals; the objective of a differentiation strategy is to offer products or services that have a distinctive difference and perceived advantage over those of the venture's competitors. Typical differentiation strategies emphasize attributes such as quality, speed, customer service, or innovation as methods to achieve competitive advantage [15–17]. All these strategies can be applied to all kind of companies, not only startups, that's why we see recent literature what decisions have been made by entrepreneurs in the early stage and how those decisions lead to failure [18].

Beyond the little research in the field of early stage companies, there is ample evidence by the entire entrepreneurial community that every startup company that focuses primarily on providing products or services to customers who actually they need and want, they will maximize their oversight over time [18]. This perspective is consistent with the entrepreneurship literature, especially one that relies on an unmet need for satisfaction.

All of the above literature tends to work from top to bottom, with researchers hypothesizing causal models for new success or business failure, testing new hypothesis into new data. On the other hand, and despite failure means non-survival, there is a very common result in tech startups researchers' attention, focused on studying success, basically due to better data availability, less attention on the reasons for failure and even less on the critical factors of survival, with very little empirical research.

Proposal

Entrepreneurial strategy can be described by different frameworks and different dimensions. The purpose of this article is to propose an integrative framework that helps fill the gap in our understanding of the effectiveness of early stage startup decisions. Based on the entire literature on startups, we defend the need in this work, for a methodology that analyzes startups in the early stage and allows them to receive inputs from their company, in order to reduce the mortality rate of companies at early stage.

In fact, it is in more recent studies that decisions have been made by entrepreneurs in the early stage and how those decisions lead to failure [18]. Asking many entrepreneurs, they admit that strategic decisions did not fit the stage in which the company was at that stage, which often led to the death of the company. It is here that we aim to study a model that can serve as specific decision-making for companies that are in the early stage.

If we focus on the last 20 years, and on the occasion of digital transformation, many startups are born looking for incremental and/or disruptive innovation [15–17], but they fail to pass the early stage. In this period of time, there is a lack of research that examines the connection between the success and sustainability of those companies that manage to go beyond the 'early stage' to the 'growth stage'. The results show that the uncertain and volatile conditions of early stage companies need to be based on a decision-making model whose primary purpose is survival. Current strategies based on traditional decision-making in companies are not effective in making positive decisions in early stage scenarios that maximize survival and sustainability. That is why it is proposed to research and create a model based on the study of critical success factors (CSF) which are necessary to ensure the success and survival of a startup in early stage [18].

Analysis of the Main Models

In reference to the strategy to be followed for early stage startups, three models have been considered as the basis for the integrative model that seeks to fill the gap when taking early stage company decisions: Sand Cone Model [18], Shell Model for early stage startups [19] and the popular Business Model Canvas [20]. The goal is to be able to integrate these three models into a joint framework that allows entrepreneurs to make decisions, taking into account all possible angles.

Sand Cone Model [18]

This model explains how early stage companies often face conflicting pressures on where and how to prioritize their efforts. This model focuses on the advantages of different competitive priorities, using an operation model sand cone, specifically applied to the context of new initiatives at the outset. The model, based on data generated by an early stage panel of companies, proposes that a competitive priority to serve client's needs is associated with a higher likelihood of survival when companies appear. On the other hand, Stock argues that innovation and traditional marketing activities does not help survival in early stage phase. Looking at the sand cone model,

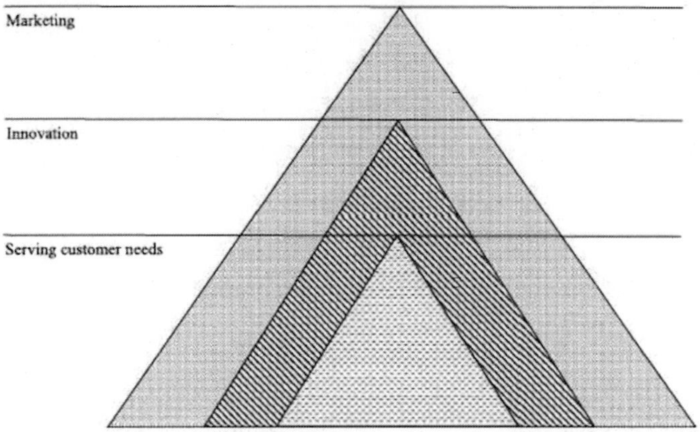

Fig. 34.1 The sand cone model of nascent entrepreneurial strategy

Stock purposes to prioritize the lower triangle, focused to serve the current needs of customers (Fig. 34.1).

The work concludes with implications for professionals and future research, and it is precisely from this point of view that it is proposed that this Sand Cone model can be integrated within a more complete framework that helps early-stage entrepreneurs making decisions to maximize their risk of survival and move to the growth stage.

In this sense, we defend the basis that the presented model is right from the point of view of operations strategy in relation to product/service and where to prioritize investments, ruling out innovations that do not have a short-term impact and avoiding costs of unnecessary marketing in the short term. But despite the success and failure cases described in the extensive literature, the reasons that can lead to non-survival go far beyond a successful operating strategy that can prioritize the important in the early stages. An example is the attitude of the entrepreneur, the alignment with their investors, the context of the market etc. For this reason, it is proposed to integrate this model within a more complete framework for early stage decision making for companies.

Shell Model for Early Stage Startups [19]

This model is based on the premise that there is a gap in the literature that analyzes the main reasons for failure of early stage startups. With this model, it is proposed a scalable, repeatable methodology that can be applied to post-mortem unstructured document databases derived from patterns that lead to early stage failure. The statistics presented at this work show how the lack of a structured business development strategy emerges as the key determinant of startup failure in most cases. In this sense,

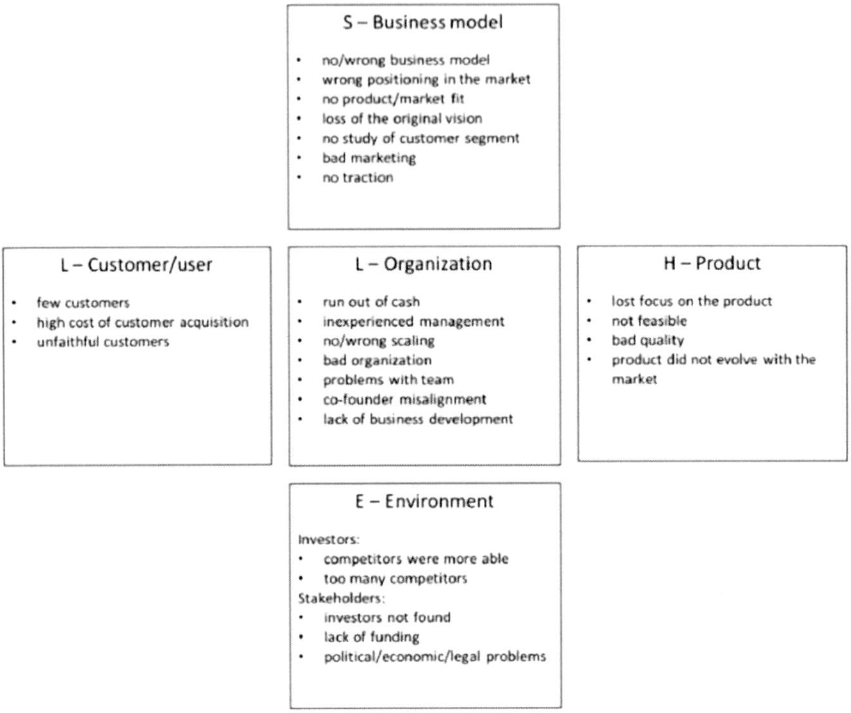

Fig. 34.2 SHELL model adapted for the classification of startups failure

it is considered a more complete model than the sand cone, which focuses only on the prioritization of certain decisions in the company's strategy for survival (Fig. 34.2).

The first major problem of this model is that it focuses only on the causes of failure, ignoring the success factors that are also critical when making decisions that can help. companies to take the leap from the early stage to the growth stage. Obviously, it is critical to avoid the main reasons for failure, but if you do not add the factors that lead to success, the model no longer makes sense for the analysis of companies that are alive and looking for answers on how to guide their strategy at this early stage.

One of the other issues detected in the model is that it generates lots of coincidences that are caused by the fixed architecture, abusing the filling exercise without deepening and correlating how quadrants can affect each other. By representing this model, a company should be able to first understand what where the real reasons for failure or see that they are in a wrong strategy in their business. Only then, companies can evaluate the model and finally consider some adaptations. However, it is a very complete model that to detect potential causes of failure, and that's why we understand that once oriented in a simulator that encompasses the potential causes of failure but also those of success (not considered in this model) and guiding the entrepreneur to an operational strategy that prioritizes survival, can be a very good role model for contributing to the integrative framework that we intend to build.

Business Model Canvas [20]

The Business Model Canvas is a startup strategic and startup management template for developing new or documenting existing business models. It is a visual graph with elements that describe the proposal, the infrastructure, the clients and the finances of a company or a product. As a very visual model, it helps businesses align their activities by illustrating possible deletions. In this sense, it allows prioritizing strategic decisions and helps to focus a business as long as the employer is able to understand and fill in each section so that he can evaluate the model and make adaptations to the initially thought business model (Fig. 34.3).

One of the biggest criticisms of the model is that it is designed by companies with a long history of innovation, but far from startups that have the vital goal of survival.

That is why Lean Canvas, an adaptation of Alexander Osterwalder's Business Model Canvas, was created by Ash Maurya in the spirit of Lean Startup (Fast, Concise and Effective startup). It focuses on problems, solutions, key metrics and competitive advantages. The structure is similar to the Business model canvas, but some sections have been exchanged, making a much more designed model for early stage startups (Fig. 34.4).

Both Alexander Osterwalder and Ash Maurya model are models that help a lot to get to know a business internally, but which neglect fundamental questions regarding

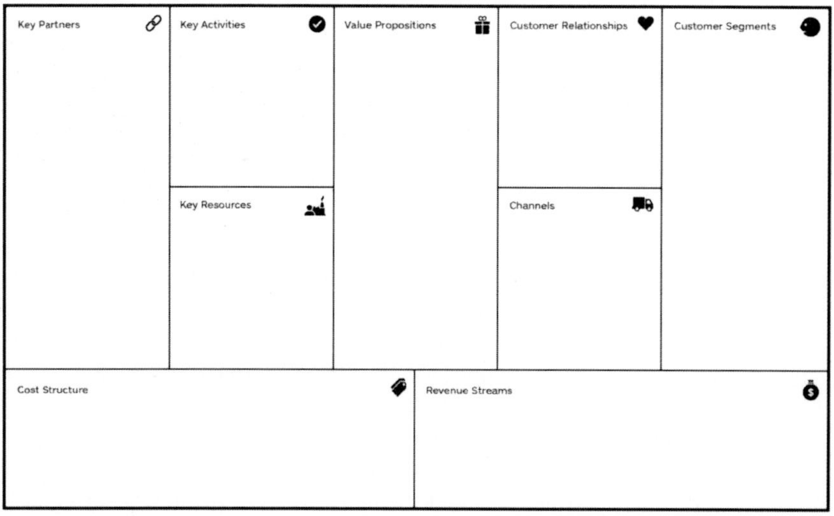

Fig. 34.3 Business model canvas

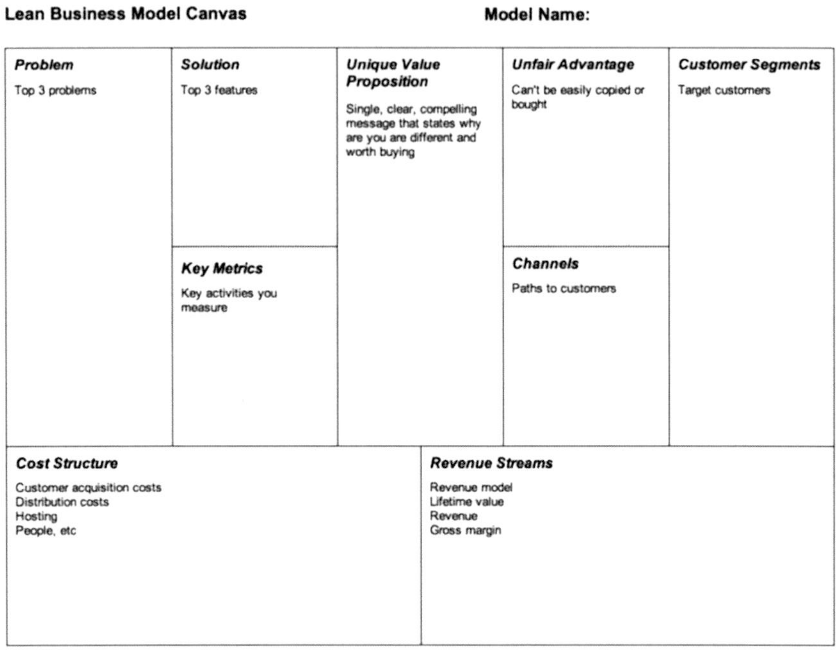

Fig. 34.4 Lean canvas model adapted for startups from the business model canvas

the survival of a new startup relates, in particular to factors outside the company, such as the level of competition, the industrial and investment environment … In this sense, we intend to integrate this model into a more complete and operational model, for early stage decision-making, complementing the Sand Cone Model and the SHELL model analyzed.

Methodology

The methodology will be implemented in a decision-making simulator, based on qualitative and quantitative information on each business project. The simulator will contrast information to extract the factors that must enable startup survival and success beyond the 'early stage' stage. Much of the literature has been focused on value proposition and differentiation [15, 21] and on internal factors in the company, such as training and entrepreneurial experience [22].

Startup performance can be measured by a variety of outcomes, such as survival (e.g., [23–25]), revenue, profitability or success of a sale [26, 27]. In this research project, we will focus on survival, understanding that it is the key factor in the first stage of a startup.

Given the multisectoral and cross-cutting approach to the study, the research project seeks to answer some key issues that may be of interest to those who start up businesses, as well as investors and business professionals in early stage phase:

What is the characterization of a startup stage in the 'early stage' stage?

What are the critical factors that determine success or failure in early stage startups?

How to generate a theoretical model for the validation of critical factors for early stage startup success?

What are the drivers of an 'early stage' value proposition?

How do we develop a co-creation model to build a successful early stage environment simulator?

What factors do we consider that a company must fulfill in order to jump from the early stage to the growth stage?

To solve these questions, we perform three main activities: a literature review, a discussion with knowledgeable actors in the ecosystem, and an exploratory analysis of the data. The proposed research will be organized into different stages with the following structure.

Units of Analysis

In order to build a solid simulator model, a detailed analysis of the literature is needed in order to let us a better understanding of the critical factors of success in startups [28]. This understanding must be conceptually grounded, look at the inputs, processes and results, and be able to approach evaluation beyond the level of the successful professionals and entrepreneurs in the industry from whom we will extract qualitative and quantitative information.

Based on the foregoing considerations, the research will include different units of analysis, as shown in Fig. 34.5.

In more detail, the research project will compile and analyze the frameworks and indicators proposed to successful professionals and entrepreneurs through in-depth interviews and focus groups.

In addition, the research will continue with the collection and analysis of early stage start-ups, of their best practices and success stories in recent years. Subsequently, the description and representation of the data and information collected above will be carried out, with the production of schemes and frameworks for models, indicators and descriptors that will ensure the quality of the model. Finally, it will focus on research, design, testing and definition of a prototype/Simulator model for early stage startup decision making.

Given the above units of analysis, the goal will be to identify assessment models, address inputs, processes and outcomes, and define quality assurance mechanisms

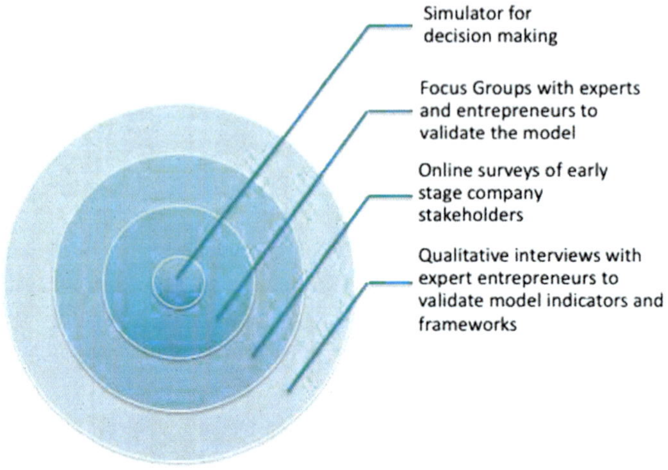

Fig. 34.5 The different units of analysis, up to the simulator. *Source* Self-elaboration

with appropriate indicators and descriptors to support evidence-based evidence in practices related to start-up business and their potential critical success factors.

Data Collection Model

The research will begin with a double data source, as described in Fig. 34.6.

The research will use different qualitative and quantitative methods in a mixed strategy. It will begin with the collection of qualitative data on career guidance policies, strategies and systems that will be illustrated through case studies; the research will use a methodological design of at least 4 case studies:

- one for startups born with less than 24 months of life;
- one for startups that have passed the early stage and are consolidated;
- one dedicated to companies that have participated in accelerators;
- one dedicated to start-ups that have emerged from incubators.

These case studies will assist in the construction of model and the models adopted and used in different birth contexts of companies. Therefore, different frameworks of business birth will be brought together, which will help build the model based on data collection and compare internationally.

Given the difficulty on obtaining all the online indicator data; some of the accelerator and incubator coordinators will be contacted to request that they share their materials and the results of the companies that have been part of them. In addition, it would be important to consider primarily those successful start-ups and "example of good practice."

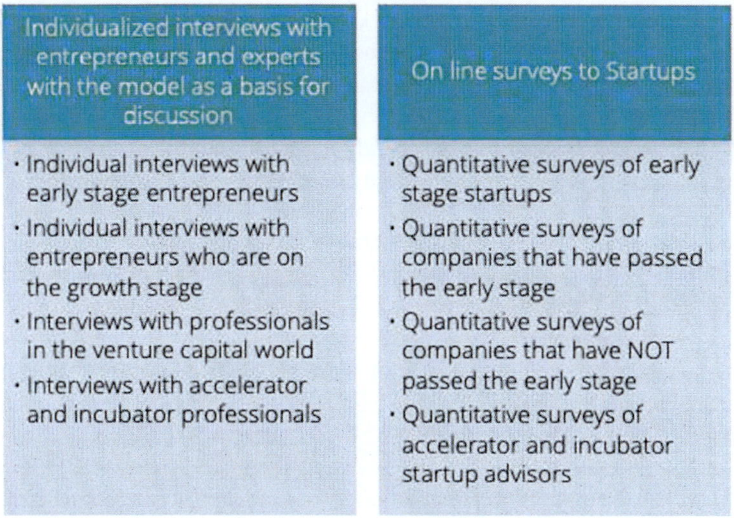

Fig. 34.6 First two different data sources for the research proposal. *Source* Own research

Subsequently, a quantitative data set will be collected during a survey, organized with startup professionals. Case study data will provide questionnaire input material to be conducted by successful professionals and early stage entrepreneur, face-to-face and online web-based questionnaires designed specifically for this research and provide additional information based on model testing, quality assurance indicators and descriptors to guide it, that will enrich the analysis of qualitative data obtained from case studies.

The purpose of collecting this data is to provide you with reliable data on the nationally-represented business orientation process of a sample of start-up entrepreneurs. These data can be complemented by other panels of entrepreneurs [29]. The data will be extracted using a Barcelona startup pool, which will allow us to extract data for the creation of the model and simulator that will be extrapolated internationally. As all studies indicate, Barcelona has become one of the world's leading startup hubs. As you can see in this link, the startup industry is diversified by sector and the community of startups, accelerators and incubators is growing: http://w153.bcn.cat/#/infographics.

The increase in new startups in Barcelona in recent years have been accompanied by the consolidation of startups that have been considered successful and with significant profits or exits to other groups. As you can see in the following link, the volume of investment and new venture capital vehicles has increased significantly in the last decade: https://startupxplore.com/blog/informe-vision-del-ecosistema-inversor-startup-of-Spain-2017/.

The Barcelona startup market is considered a 'mature' market, with successful startups, a large number of new early-stage startups, a consolidated venture capital sector and a large number of accelerators and incubators. All this has also enabled

Barcelona to be a talented attraction market. For this reason, Barcelona is considered to be an excellent region for collecting both quantitative and qualitative data, which can be used for other startup pools internationally.

Qualitative and quantitative data, obtained respectively from case studies and online questionnaires, will be used to define a set of indicators and descriptions for successful decision-making that will ensure the survival of startups, with a focus on building a prototype model of decision simulator. In fact, the research will also include a more experimental phase; Not only will analyze and describe what already exists, but the ultimate goal will also be to propose key new indicators for innovative policies, programs and practices in decision-making that will help nascent startups identify the critical success factors (CSF—Critical Success Factors).

The research is based on a series of indicators and descriptors, already identified in the literature review and in multiple cases of startups (founding education and training, previous experience, financial resources, size market, value proposition ...); it will analyze the policies and practices existing under these indicators, adapt the indicators to make the model valid internationally and the actual conceptual frameworks, and analyze again and compare the results. If the results are positive and meaningful, the idea would be to try to extrapolate and propose indicators and descriptors in various fields to start-ups, in more specific areas.

In summary, the model will be based on several hierarchically organized research strategies as in Fig. 34.7.

Finally, the research will conclude with another set of qualitative data collected through two focus groups organized with around 8–10 entrepreneurship experts (venture capital professionals or business accelerators) and entrepreneurs who have created enduring and successful businesses over time. In these focus groups they will test the decision-making simulator to evaluate their quality and indicators, developed in the previous stages of the research project.

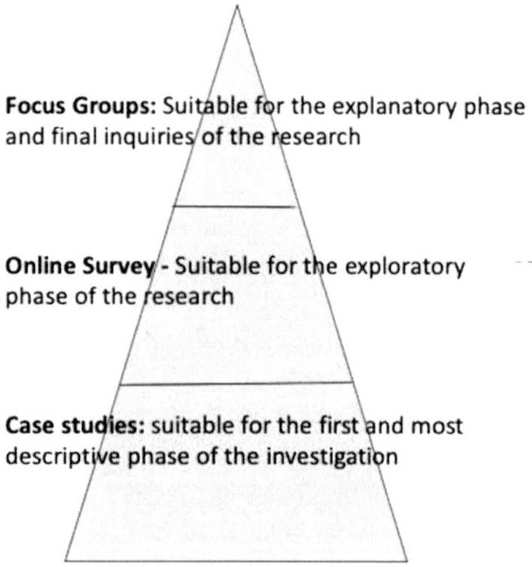

Fig. 34.7 The hierarchically proposed research strategies

Discussion and Conclusion

This article encompasses a large and recent startup research, proposing an integrative framework for designing and evaluating the right strategy, and making decisions that maximize startup survival. This framework proposes a holistic approach to examining the effects that are not usually included in the different models of startup analysis.

In order to produce the evidence needed to build confidence in the need for this integrative framework, there have been analyzed the entire literature on startups that analyze the critical factors that have led to failure and also those that have led to success. In the same way, there have been analyzed several models that intent to help start-ups make decisions, but which do not cover all the information needed to create a strategic framework of operations that maximizes the survival of early stage companies. That's why it is considered to create an integrative framework that fills all the gap.

These quantitative studies through modeling and experimentation allow us to examine the effectiveness of their model [19, 18], together with the strategic management model 'Business Model Canvas' [20] serve as a basis for proposing the creation of a simulator for startups that covers internal factors, external factors, critical success and failure causes.

The integrated system moved in a future simulator, aims to show that each company needs its specific strategy for its initial phase. Unifying the internal factors of the company (financial, human resources, value proposition for the client …) with the external factors (competition, economic situation, market maturity …) we can define a unique model for each startup, in order to maximize its survival and subsequent leap to the growth phase. To do this, we propose a simulator that feeds the data collected and must allow decision-making by companies in the initial phase. The research concludes with implications for professionals and future research in order to improve the model and simulator.

Conflict of Interest Statement The author declares that there is no possible conflict of interest regarding the research, authorship, and/or publication of this article.

Funding The author has not received any financial support for the research, authorship and/or publication of this article.

References

1. D.J. Storey, *The Ten Percenters. Fourth Report: Fast Growing SMEs in Great Britain* (Deloitte & Touche Tohmatsu, London, 1998)
2. P. Nightingale, A. Coad, Muppets and gazelles: political and methodological biases in entrepreneurship research. Ind. Corp. Chang. **23**, 113–143 (2014)
3. D. Frederiksen, A. Brem, How do entrepreneurs think they create value? A scientific reflection of eric ries' lean startup approach. Int. Entrep. Manag. J. **13**, 169–189 (2017)
4. O. Amat, Predicción del éxito o del fracaso de una empresa en base a variables cualitativas y cuantitativas, Tesis doctoral, Universidad Autónoma de Barcelona, 1990

5. E.I. Altman, Financial ratios, discriminant analysis and the prediction of corporate bankruptcy. J. Finance **23**(4), 589–609 (1968)
6. W.H. Beaver, Financial ratios as predictors of failure. J. Account. Res. **4**, 71–111. (1966)
7. L. Arne, K.T.L. Kalleberg, Gender and organizational performance: determinants of small business survival and success. Acad. Manag. J. **34**(1), 136–161 (1991)
8. J. Everett, J. Watson, Small business failure and external risk factors. Small Bus. Econ. **11**, 371–390 (1998)
9. D. Seshadri, Understanding new venture failure due to entrepreneur-organization goal dissonance. Vikalpa **2007**(32), 55–74 (2007)
10. K.M. Hmieleski, D.A. Lerner, The dark triad and nascent entrepreneurship: an examination of unproductive versus productive entrepreneurial motives. J. Small Bus. Manag. **54**, 7–32 (2016)
11. J. Cope, Entrepreneurial learning from failure: an interpretative phenomenological analysis. J. Bus. Ventur. 604–623 (2011)
12. Y. Vaillant, E. Lafuente, Do different institutional frameworks condition the influence of local fear of failure and entrepreneurial examples over entrepreneurial activity? Entrep. Reg. Dev. **2007**(19), 313–337 (2007)
13. A.W. Mackelprang, M. Habermann, M. Swink, How firm innovativeness and unexpected product reliability failures affect profitability. J. Oper. Manag **38**, 71–86 (2015)
14. S. Dutta, T.B. Folta, A comparison of the effect of angels and venture capitalists on innovation and value creation. J. Bus. Ventur. **31**, 39–54 (2016)
15. M. Porter, *Corporate Strategy* (New York, NY, 1980)
16. R.E. Miles, C.C. Snow, *Organizational Strategy, Structure, and Process* (McGraw-Hill, New York, NY, 1978)
17. D. Miller, P.H. Friesen, *Organizations: A Quantum View* (Prentice-Hall, Englewood Cliffs, NJ, 1984)
18. G.N. Stock, C. McDermott, M. McDermott, Early stage venture strategy and the survival of nascent entrepreneurial firms: a sand cone model approach. IEEE Eng. Manage. Rev. **46**(3), 108–116 (2018)
19. M. Cantamessa, V. Gatteschi, G. Perboli, M. Rosano, Startups' roads to failure. Sustainability **10**(7), 2346 (2018)f
20. A. Osterwalder, Y. Pigneur, Business model generation. a handbook for visionaries, game changers and challengers; Wiley: Hoboken, NJ, USA (2010)
21. A.V. Roth, Operations as marketing: a competitive service strategy. J. Oper. Manag. **10**(3), 303–328 (1991)
22. J. Bruderl, P. Preisend orfer, R. Ziegle, Survival chances of newly founded business organizations. Am Sociol Re, **57**(2), 227–242 (1992)
23. J.P.J. De Jong, O. Marsili, Founding a business inspired by close entrepreneurial ties: does it matter for survival? Entrep. Theory Pract. **39**(5), 1005–1025 (2015)
24. E.A. Morse, S.W. Fowler, T.B. Lawrence, The impact of virtual embeddedness on new venture survival: overcoming the liabilities of newness. Entrep. Theory Pract. **31**(2), 139–159 (2007)
25. P. Stenholm, M. Renko, Passionate bricoleurs and new venture survival. J. Bus. Ventur. **31**, 595–611 (2016)
26. C.G. Brush, P.A. Vanderwert, A comparison of methods and sources for obtaining estimates of new venture performance. J. Bus. Ventur. **7**, 157–170 (1992)
27. V. Collewaert, Angel investors' and entrepreneurs' intentions to exit their ventures: a conflict perspective. Entrep. Theory Pract. **36**(4), 753–779 (2012)
28. M. Parra Alviz, G. Rubio Guerrero, Distinctive factors of entrepreneurship that led to success: case studies in entrepreneurs from Ibagué, Tolima. Rev. Cient. Pensam Gestión [Internet] **43**, 89–127 (2017)
29. R. Curtin, *Panel Study of Entrepreneurial Dynamics II Screener þ Wave A—Wave F Codebook* (Survey Research Center, University of Michigan, Ann Arbor, MI, 2011)
30. E. Piva, C. Rossi-Lamastra, Human capital signals and entrepreneurs' success in equity crowdfunding. Small Bus. Econ. **51**(3), 667–686 (2018)

Chapter 35
Exploration of Social Media Opinions on Innovation for Sustainable Development Goals by Topic Modeling and Sentiment Analysis

Chien-wen Shen, Thai-Ha Luong, and Tuan Pham

Abstract This research utilizes social knowledge extracted from user-generated content from Twitter to identify the social media users' concerns on different domains of innovation regarding Sustainable Development Goals (SDGs) during 4 years, whereby understanding the emotional expression of public opinion upon those SDGs innovation dimensions. Topic analysis with latent semantic approach is the most suitable approach for exploring topics of interest from large text corpus; while sentiment analysis using Python-based library Vader is effective to investigate the Twitter users' sentiment underneath those extracted topics. The importance of gender equality and youth empowerment on innovation; and the innovation in sustainable agriculture, education, eco-friendly materials, green energy, and economic development are intensively discussed topics. The component-terms of corresponding topics in different years are further scrutinized to highlight the evolution of Twitter users' concerns on a particular innovation dimension. The result of sentiment analysis suggests the predominance of "very positive" and "positive" sentiment in almost topics during 4 years. The "neutral" sentiment prevails in a certain topic that generally addresses many angles of SDGs innovation in its majority of tweets without focusing on any specific dimensions. The high of "negative" sentiment during 4 years is noticed in the agriculture innovation topic in 2018.

Keywords SDGs · Innovation · Topic analysis · Sentiment analysis · Twitter

---

C. Shen · T.-H. Luong (✉)
Department of Business Administration, National Central University, No. 300 Zhongda Road, Zhongli District, Taoyuan City 32001, Taiwan
e-mail: randyluong1990@gmail.com

C. Shen
e-mail: cwshen@ncu.edu.tw

T. Pham
University of Technology and Education of Danang, No. 48 Cao Thang Road, Hai Chau District, Danang City 550000, Vietnam
e-mail: ptuan@ute.udn.vn

A. Visvizi et al. (eds.), *Research and Innovation Forum 2020*,
Springer Proceedings in Complexity,
https://doi.org/10.1007/978-3-030-62066-0_35

Introduction

The Sustainable Development Goals (SDGs) is a plan of action for people, planet, and prosperity in order to reach peace and freedom for everyone [1, 2]. Including 17 SDGs and 169 associated targets, this universal agenda can be considered as benchmarking for global-scale problem solutions. The existing poverty, social inequality, environmental degradation, political unrest, cultural conflict, peace and justice are hardly improved without applying state-of-the-art innovative approaches [3–6]. Innovation is the key to those problems [7], which require synergy and collective efforts not only from governments and private business sectors but also from the entire societies. Studies have looked into SDGs as a whole, or only investigated specific goals, using various research methods. However, questions on how crucial the innovation contributes to the progress of SDGs being perceived by the global societies, or which innovation dimensions among 17 SDGs should be urgently prioritized in different years under public opinions, are remained unknown. Social media has been the most interactive dialogic communication platform [8, 9]; where information can broadly reaches global audiences, and huge amount of public reactions/opinions on variety of concerns are published in real time [10]. Due to that reason, this research utilizes knowledge extracted from user-generated content from Twitter to explore how the global public uses social media to address their concerns on how innovation should be deployed amid global SDGs implementation. A text mining technique, topic analysis, applied in this paper is suitable for exploring topics of interest from large corpus of text dataset using latent semantic analysis (LSA) algorithm, from which identifying similarities and dissimilarities between topics in different years [11, 12]. Tweets addressing SDGs-related innovation from 2016 to 2019 are retrieved and afterward tokenized into bag-of-words for subsequent frequency and weight values computing; from which documents and terms of high correlation are extracted for the forming of topics of concerns. The comparison of extracted topics in the latter stage sheds light on how SDGs innovation have been implemented and progressed in different years by social media users. In case of resemblance, the component terms of corresponding topics are further scrutinized to highlight the evolution of SDGs innovation within similar domains from one year to another. Sentiment analysis is also conducted on all topics to measure how positively or negatively Twitter users think about a certain topic, and whether or not that sentiment on such topic change in different years. User-generated content from Twitter, together with advanced information and communication technology (ICT) based methods of analysis, including text mining, topic modeling, and sentiment analysis offer invaluable information regarding individual user's perception about particular domains in online environments [13], which strengthen the relevance of this research's topic and methodology.

Literature Review

The definition of SDGs, although varies in different published documents, consistently manifests the triple bottom lines relating to economic development, environmental sustainability and social inclusion. Sustainable growth indicates a repeatable, ethical, responsible growth that can be sustained without having to significantly increase the financial leverage and resource use. The rapid economic growth of emerging countries, which relies on heavy industry, brings serious environmental trade-offs with the increasing carbon-dioxide emission, water contamination and pollution. The focus on economic prosperity should be interconnected with environmental protection and social wellbeing, from which all individuals should be able to have access to safe water and sanitation, electricity, primary health care, and be protected from natural hazards [1]. Inflating food-production scale without concerning about environmental consequences makes more harm than good that result in food supply shortage, and boost food prices in a long run, putting millions of people in food insecurity [14]. Social inclusion of SDGs inclines toward the identical nutrition and intellect care for children from early childhood for the robust physical and mental foundation, and then focuses on education and skills training for young people to ensure the effective from-school-to-work transition for young labor force, with special concerns on girls [15]. However, the poorest and most disadvantaged mothers and children are still being endangered at high health risk due to the inadequate access to proper diagnosis and treatment [16].

Innovation has long been investigated in numerous researches on variety of disciplines. Abundant definitions of innovation can be referred to from past researches, such as the mostly cited "An innovation is the implementation of a new or significantly improved product (good or service), or process, a new marketing method, or a new organizational method in business practices, workplace organization or external relations" [1]. An innovation can be integrated in either product or service, or new materials, new supply sources, new markets, or new approaches applied on existing businesses [17–21]; and can be perceived under different attributes, such as types of innovation, origin of innovation, subject of innovation, level of innovation, source of value creation, classification of innovation. SDGs, with the strong focus on triple bottom lines critically needs new technological breakthrough innovations to enhance living standard of human being and provide better protection of natural ecology [1]. Innovation, especially technological innovation, is believed to be a main driver that helps reduce pollution and waste [22]. Technological innovation is described as a process where company aim to increase its market competitiveness by concentrating large proportion of its resources and capabilities to develop the technical side of the product, service, or process [23]. Investment in renewable and clean energy and better energy allocation are also emphasized as imperative factor toward the sustainability [22]. Meanwhile, business innovation is commonly perceived as either novel business model, or the renovation of an existing business that occurs when the organization utilizes outside ideas as impetus for innovativeness and sustainable growth. Innovation, instead of being a total replacement of R&D, has utility as a paradigm for industrial innovation beyond high tech to more traditional and mature industries; whereby becoming supportive and beneficial for sustainable development in business domain [24]. Innovations in ICT have directly impacted the political landscape and the way in which politics operates today [25]. The incapability in envisioning, designing, and implementing synthetic policies, which enhance the mutual gain of social goals, economic and political interests, may cause the failure of sustainable development achievement. Therefore, multi-purpose policy design should be broadly integrated across different domains of sustainable development where technological innovation largely holds accountable.

Methodology

Tweets associated with the topic "innovation in SDGs" are mined for investigation. Twint, a Python library, is used for the retrieval of real-time Twitter data [26]. Searched terms for tweet collection is set as "SDGs" and "innovat", from which "SDGs" stands for Sustainable Development Goals and "innovat" represents group of vocabulary composed by the prefix "innovat" such as: innovation, innovate, innovative, innovating, etc., 29,492 tweets are retrieved from the beginning of 2016 to the end of 2019, containing both terms "SDGs" and "innovat". The format of retrieved data is converted to CSV format for the convenience of subsequent Python processing stages. Following data-collection stage, data cleansing is performed using Python to remove noises and unnecessary characters within tweets that may cause misinterpretation problems for subsequent topic analysis, and sentiment analysis. Critical steps for data cleansing are described as: converting upper case letters to lower case; removing URLs by filtering regular expressions (http|https|fpt|pic)://[a-zA-Z0-9\\./]+; removing user reference using the regular expression—@(\w+); removing hashtags using the regular expression—#(\w+); removing duplicate tweets from the same users.

Afterward, topic analysis and sentiment analysis on all extracted topics are conducted. Topic analysis identifies the major concerns of social media users on innovation amid SDGs implementation from large textual corpus, using LSA algorithm, to construct term-by-document matrix for topic modeling

$$w_{i,j} = tf_{i,j} * \log\left(\frac{N}{df_i}\right)$$

where $w_{i,j}$ is TF-IDF score or weight value of ith term in jth document, $tf_{i,j}$ is frequency of ith term in jth document, N is the total number of documents in corpus, and df_i is number of documents containing ith term. Meanwhile the sentiment analysis, using Python-library Vader, is conducted to understand the sentiment intensity of expressed by individual tweets, hence identifying the proportion of different emotion of every extracted topic. The standardization of [27] is used to measure the intensity of sentiment of individual tweet

$$y = \frac{x}{\sqrt{x^2 + \alpha}}$$

where x represents the aggregating intensity score of Vader-dictionary registered words that appear in tweets, α is standardization restriction with default value of 15. The higher positive/negative value carried by x results into y to be closer to 1 or -1. The 5 labels applied for sentiment intensity scales are "very negative" (denoted as VN, $-1 \leq y < -0.6$), "negative" ($-0.6 \leq y < 0.05$), "neutral" ($-0.05 \leq y \leq 0.05$), "positive" ($0.05 < y \leq 0.6$), "very positive" (denoted as VP, $0.6 < y \leq 1$). The scale $-0.05 \leq y \leq 0.05$ or $-0.1 < y < 0.1$ is often used for "neutral" in sentiment analysis projects.

Result Discussion

Topic Analysis

The topic analysis results from 2016 to 2019 tweets are presented from Tables 35.1, 35.2, 35.3 and 35.4 respectively, each of which contains 6–8 topics. 5 terms with highest TF-IDF scores of each topic are chosen to represent the topic.

The role of global partnership, public-private partnership, science and technology, young people empowerment as leading influences on SDGs innovations; and innovations applied in education, healthcare, agriculture and economy domains are popularly discussed topics on social media during 2016. Table 35.1 exhibits 6 topics extracted from corpus of tweets in 2016. Topic 1 highlights the partnership and collaboration between governments and political institutes to facilitate the innovation on global scale that relates to SDG 17; while Topic 3, although correspondingly emphasizes the partnership, focuses on the collaboration of public-private sectors

Table 35.1 Topics 2016

Terms/TF-IDF scores	Topic
Political circumstance (0.722), **collaboration** (0.507), **partnership** (0.467), **innovation** (0.326), **depaolis** (0.321)	Topic 1
Science (0.692), **technology** (0.661), **sti** (0.411), **cstd19** (0.389), **unctad14** (0.327)	Topic 2
Partnership (0.652), **helenclarkundp** (0.525), **isdb_dev** (0.513), **innovation** (0.364), **public-private** (0.212)	Topic 3
Education (0.644), **africa** (0.557), **globalcitizen** (0.409), **africanbrains** (0.18), **health** (0.12)	Topic 4
ffa2016 (0.577), **innovation model** (0.417), **agricultural innovation system** (0.411), **debate** (0.403), **forumforag** (0.301)	Topic 5
Business (0.717), **solution** (0.346), **business action** (0.156), **youth** (0.156), **data** (0.142), **spaceshipearth** (0.132)	Topic 6

Table 35.2 Topics 2017

Terms/TF-IDF scores	Topic
Clean energy (0.261), **hunger** (0.259), **inequality** (0.255), **economy** (0.254), **sanitation** (0.253)	Topic 1
Science (0.733), **technology** (0.352), **globalpeoplesummit** (0.199), **summit** (0.198), **innovation** (0.19)	Topic 2
Innovative solution (0.732), **youth** (0.51), **UnleashLab2017** (0.181), **UnleashLab** (0.127), **aidfguide** (0.114)	Topic 3
Data (0.76), **innovative big data** (0.349), **reliefweb** (0.312), **UNdataForum** (0.311), **AI** (0.108)	Topic 4
Business (0.563), **impinv** (0.334), **youth** (0.253), **leadership** (0.223), **hlpf2017** (0.204)	Topic 5
Financing (0.589), **innovative finance** (0.456), **going4goals** (0.318), **impinv** (0.316), **sdgs finance gap** (0.23)	Topic 6

Table 35.3 Topics 2018

Terms/TF-IDF scores	Topic
Seedsandchips (0.416), **foodinnovation** (0.362), **young people** (0.288), **foodtech** (0.263), **SAC18** (0.26)	Topic 1
Energy (0.418), **solarpower** (0.269), **climatechange** (0.209), **SDG7** (0.172), **SDI18** (0.151)	Topic 2
Innovative solution (0.938), **UnleashLab2018** (0.078), **UNDP** (0.064), **business** (0.063), **Singapore** (0.059)	Topic 3
Business (0.694), **CSR** (0.405), **impinv** (0.177), **ESG** (0.167), **socialimpact** (0.136)	Topic 4
Youth (0.714), **entrepreneurship** (0.249), **women** (0.248), **young people** (0.154), **startup** (0.134)	Topic 5
Science (0.73), **technology** (0.482), **STIforum** (0.11), **STI** (0.11), **policy** (0.101)	Topic 6
Economic growth (0.492), **job creation** (0.35), **biotech innovator** (0.336), **greenbiz** (0.333),	Topic 7
Innovative finance (0.301), **UNDP** (0.299), **data** (0.293), **partnership** (0.218), **financing** (0.165)	Topic 8

in bringing up innovations, demonstrated by partnership (0.652) and public-private (0.212). Topic 2, of which STI (0.411), UNCTAD14 (0.389), CSTD19 (0.327) respectively indicate Science Technology Innovation, Commission on Science and Technology for Development, and United Nation Conferences on Trade and Development, relates to SDG 9 that puts the sustainability of industry, innovation and infrastructure development at the central. Topic 4 emphasizes the importance of technological innovation in education and health education, which associates with SDG 3 and SDG 4, in Africa where the majority of citizens living under poverty line and lacking access to education and healthcare. Regarding SDG 2, topic 5 underlines the innovative agricultural system, demonstrated by FFA2016 (0.577) that denotes Forum for the Future of Agriculture 2016, innovation model (0.417), agricultural innovation

Table 35.4 Topics 2019

Terms/TF-IDF scores	Topic
Impactinvestment (0.621), **summit** (0.285), **socialinnovation** (0.282), **ESGinvesting** (0.278), **socialimpact** (0.273)	Topic 1
Horyou (0.523), **davos2019** (0.514), **HYT** (0.509), **WEF19** (0.206), **action-oriented social network** (0.198)	Topic 2
Businessethics (0.312), **humanisticmanagement** (0.284), **socialinnovation** (0.275), **CSR** (0.262), **SocEnt** (0.249)	Topic 3
Sustainable peace (0.43), **STEM** (0.362), **datascience** (0.206), **unique biosphere** (0.195), **education** (0.194)	Topic 4
Technology (0.695), **AI** (0.287), **data** (0.133), **innovation** (0.165), **startup** (0.149)	Topic 5
Youth (0.657), **woman** (0.217), **young people** (0.212), **entrepreneurship** (0.171), **innovative idea** (0.146)	Topic 6
Climatechange (0.467), **environment** (0.231), **energy** (0.207), **zerohunger** (0.202), **economy** (0.185)	Topic 7
Innovative solution (0.347), **UNDP** (0.11), **UnleashLab** (0.089), **UnleashLab2019** (0.085), **lab** (0.074)	Topic 8

system (0.411), forumforag (0.301). Topic 6 identifies the role of young people and data science in creating innovative solutions for business sector that links to SDG 8.

The innovative finance, social impact investment, business ethics, ESG investing, CSR, green business is believed to be stimulus of the sustainable economic growth in 2017. Table 35.2 exhibits 6 topics extracted from corpus of tweets in 2017. Topic 1 includes tweets that generally address 17 SDGs related innovation in the majority of tweets; therefore it is noticeable that clean energy (0.261), hunger (0.259), inequality (0.255), economy (0.254), sanitation (0.253) indicate different innovation dimensions within 17 SDGs. Topic 2 and topic 4 mutually examine the impact of science, technology, artificial intelligence (AI), and big data on innovation, demonstrated by science (0.733), technology (0.352), globalpeoplesummit (0.199), summit (0.198), innovation (0.19) of topic 2 and data (0.76), innovative big data (0.349), reliefweb (0.312), UNdataForum (0.311), AI (0.108) of topic 4. Topic 3 and topic 5 apparently advocate the necessity of youth empowerment on leading economic innovation. Also associated with SDG 8 that considers sustainability in economic thriving as emphasis, topic 6 identifies the critical role of innovative finance, especially the impact investment on sustainable economic development, explained by financing (0.589), innovative finance (0.456), impinv (0.318), going4goals (0.316), sdgs finance gap (0.23).

The innovation in sustainable economy is among the major concerns of Twitter users in 2018 as 4 over 8 topics are associated with SDG 8, including topic 3, topic 4, topic 7, and topic 8 (Table 35.3). While Seedsandchips (0.416), foodinnovation (0.362), SAC18 (0.288), foodtech (0.263), young people (0.26) of topic 1 investigate the innovation and technology in sustainable agriculture that relate to SDG 2; Energy (0.418), solarpower (0.269), climatechange (0.209), SDG 7 (0.172), SDI18 (0.151) of topic 2 apparently address the need to tackle climate change using renewable power and innovative green energy such as solar power that link with SDG 7. Youth empowerment and gender equality are primarily described in topic 5, explaining that the more decision-making power given to young people (0.154) and women (0.248), especially regarding entrepreneurship (0.249) and startup (0.134) domains, would ensure higher returns on SDGs achievement. Topic 6 points out the imperative role of science, technology and innovation in infrastructure establishment that relates to SDG 9. Regarding SDG 8, topic 3 calls for attention on innovative solutions in business; while topic 8 highlights innovative finance (0.301) as an essential catalyst for sustainable economy. Those terms business (0.694), CSR (0.405), impinv (0.177), ESG (0.167), socialimpact (0.136) of topic 4 and economic growth (0.492), job creation (0.35), biotech innovator (0.336), greenbiz (0.333) of topic 7 portray the importance of innovation in impact investment, social responsibility, and environmental responsibility of enterprises on economic growth.

Consistent with the topic analysis result of 2018, concerns on impact investment, business ethic, and social responsibility still emerge among key attention of social media users in 2019, explained by Topic 1 and Topic 3 (Table 35.4). The component terms Horyou (0.523), Davos2019 (0.514), HYT (0.509), WEF19 (0.206), action-oriented social network (0.198) of Topic 2 denote the in-depth discussion on Horyou Token during World Economic Forum (WEF) 2019 in Davos that implies the block chain technology and crypto currency. Another topic that underlines the importance of innovative solutions in business sector is Topic 8. Topic 4 puts forth that the innovative education, which includes STEM (0.362) (Science, Technology, Engineering and Math) and data science (0.206) and links with SDG 4, can be globally deployed regardless countries' wealth status when the sustainable peace (0.43) that relates to SDG 16 is ensured. Topic 5 addresses the prerequisite requirement to apply innovative technology (0.695), AI (0.287) for startups (0.149) to enhance the competitiveness for their newborn businesses. The idea of empowering young people and women, noticed in Topic 5, is again underscored as impetus to gear the SDGs innovation forward. Topic 7 addresses the climate change issue, through which the energy innovation emerges as the effective solution that creates solid foundation for sustainable agriculture and economic growth afterward.

Sentiment Analysis

Topic-based sentiment analysis (Table 35.5) is conducted to understand the sentiment of social media users on various topics related to innovation in SDGs during 2016–2019, presented from Tables 35.1, 35.2, 35.3 and 35.4. The vast majority of social media users express positive sentiment toward all innovation topics in 2016. From 86 to 97% tweets in 2016 are "positive" and "very positive" (VP) aggregated, from which tweets highlighting partnership between governments and public-private sectors account for the highest percentage of positivity referred to Table 35.1 (Topic 1 and Topic 3). Because Topic 1 in 2017 generally indicate the innovation of almost 17 SDGs in every individual tweet without particular emphasis on any single SDG;

Table 35.5 Sentiment analysis

Year	Sentiment	Topic 1 (%)	Topic 2 (%)	Topic 3 (%)	Topic 4 (%)	Topic 5 (%)	Topic 6 (%)	Topic 7 (%)	Topic 8 (%)
2016	VP	7	30	46	27	46	37		
	Positive	90	57	43	60	44	49		
	Not clear	2	10	8	10	7	12		
	Negative	1	3	1	3	2	2		
	VN	0	0	0	1	1	0		
2017	VP	10	33	60	35	38	41		
	Positive	12	53	34	54	49	55		
	Not clear	78	13	6	10	10	4		
	Negative	1	1	0	1	3	1		
	VN	1	0	0	0	0	0		
2018	VP	46	58	75	62	65	62	70	66
	Positive	34	32	19	29	28	29	26	27
	Not clear	6	7	5	7	7	7	3	5
	Negative	14	2	1	2	1	2	0	1
	VN	0	0	0	0	0	0	1	0
2019	VP	30	75	96	58	77	56	72	58
	Positive	30	20	3	27	18	33	21	32
	Not clear	39	4	1	15	4	7	5	7
	Negative	0	1	0	1	0	3	1	2
	VN	1	0	0	0	0	0	0	1

therefore, it is explicable for "not_clear" sentiment to peak to 78%, although tweets that score "positive" (12%) and "very positive" (10%) still outnumber "negative" (1%) and "very negative" (VN) (0%) tweets. The percentage of "not_clear" sentiment of Topic 2 in 2017 is the second highest (13%) among that of other topics in the same year. Compared to the similar topic in 2016, while tweets of the topic Science and Technology Innovation (Topic 2) in 2016 use the particular hashtags CSTD19 and UNTAD14 to underline the importance of scientific and technological innovations in SDGs; tweets of the topic Science and Technology Innovation (Topic 2) in 2017 seem more general with less attentive hashtags used that results in the rise of ambiguity or "not_clear" sentiment of social media users. Except Topic 1 with "not_clear" sentiment accounting for the highest percentage, all other topics in 2017 are prevailingly "very positive" and "positive" with the combined percentage varying from 86 to 96% of total tweets. The percentage of "very positive" sentiment in 2018 and 2019 increases intensively compared to those figures in 2016 and 2017. The highest of "negative" sentiment during 4 year is noticed on the topic innovation in agriculture (Topic 1) in 2018 that hits 14%. However, it may not indicate the negative perception of social media users on the topic since the aggregated percentage of "very positive" and "positive" still features 80% overall. The high of "very positive" sentiment is observed on Topic 3 in 2019, which emphasizes the innovation in CSR and humanistic management. The "not_clear" sentiment of the topic impact management of ESG investment and business innovation (Topic 1) in 2019 that hits 39% is the highest within the year and the second highest among all "not_clear" sentiment during 4 year, although those corresponding topics in previous years including Topic 5 in 2017 and Topic 4 in 2018 score high on the aggregated percentage of "very positive" and "positive" sentiment.

Conclusion

This research explores social media users' concerns on innovation regarding sustainable developments and the sentiments associated with those opinions in different years during 2016–2019, using topic analysis algorithm and sentiment analysis respectively. While social media users consider AI and big data as main drivers for innovation in science and technology domains; the innovation in sustainable economic development is the most discussed topic during 2016–2019. The evolution in individual opinions can be observed within corresponding topics that occurred in different year. While youth empowerment and data science are broadly addressed as innovative drives for business sector in 2016; the innovative finance, social impact investment, business ethics, ESG investing, CSR, green business is believed to be stimulus of the sustainable economic growth in the following years. Agricultural innovation model, food technology, and the global food innovation conference Seeds and Chips 2018 (SAC18) highlight social media users' concerns on agriculture innovation in 2016 and 2018; while education and healthcare innovation topics are consistently identified in 2016, 2017, and 2019. The great amount of tweets on energy innovation are uncovered from 2017 to 2019 with the 2017 tweets generally mentioning about clean energy; while the tweets in 2018 and 2019 particularly pinpoint the severe pollution issue and climate change caused by industrial emission, through which calling for the use of eco-friendly alternatives such as renewable energy, bio energy, solar power. The focus on international partnership and collaboration as a solid base for SDGs innovation in 2016 is shifted to the sustainable peace in 2019.

Sentiment analysis result spotlights the prevalence of "very positive" and "positive" sentiments over almost topics during 4 years. However, it also suggests that "not clear" sentiment increases where a topic indicates many dimensions of SDGs innovation at once; in other words, sentimental ambiguity is noticed in those tweets without particular focus on a specific dimension of SDGs innovation. Considering corresponding topics, a topic with less number of hashtags used in individual tweets generate the higher ambiguous sentiment. The high of "negative" sentiment during 4 year is noticed on the topic innovation in agriculture in 2018, nevertheless, it may not indicate the negative perception of social media users on the topic since the aggregated percentage of "very positive" and "positive" still features 80% overall. The innovation in CSR and humanistic management topic in 2019 has the highest "very positive" sentiment. Although the aggregated percentage of "very positive" and "positive" sentiment on impact investment topic is prevailing in 2017 and 2018 (Topic 5 in 2017, and Topic 4 in 2018); the positivity spectrum of impact investment topic drops drastically in 2019 (Topic 1 in 2019), from which the overwhelming neutral sentiment is observed. Beyond the discussion referred from analysis result, the sentiment analysis reveals several important implicit findings on the progress of SDGs innovation. Although innovative technologies have been increasingly integrated in pursuit of higher agricultural productivity at the reduction of natural resources use, the achievement of agricultural sector so far may not be as expected. Although impact investment and ESG investing is a highly concerned SDGs innovation topic from 2017 to 2019, nevertheless, the shrink in positivity sentiment in the most recent year may implicate the reduced impression of social media users on the achievement of such domain. Novelty, continuous improvement or better achievement on impact investment and ESG investing may again put Twitter users' impression on hype. Regarding social media content management, individuals or organizations can consider to use more specific and relevant hashtags in tweets (as keywords/highlights in official documents) to attract more attention from subscribers or followers.

References

1. J.D. Sachs, From millennium development goals to sustainable development goals. Lancet **379**, 2206–2211 (2012)
2. United Nations, *Transforming Our World: The Agenda for Sustainable Development* (United Nations, New York, NY, 2015)
3. D. Mebratu, Sustainability and sustainable development: historical and conceptual review. Environ. Impact Assess. Rev. **18**, 493–520 (1998)
4. T. Jackson, *Prosperity Without Growth: Economics for a Finite Planet* (Earthscan, London, New York, NY, 2009)
5. World Wide Fund for Nature (WWF), *Living Planet Report 2014: Species and Spaces, People and Places* (World Wide Fund for Nature, Grand, Switzerland, 2014), p. 2014
6. O. Boiral, C. Baron, Environmental leadership and consciousness development: a case study among Canadian SMEs. Bus. Ethics **123**, 363–383 (2014)
7. F. Damanpour, M. Schneider, Phases of the adoption of innovation in organizations: effects of environment, organization and top managers. Br. J. Manag. **17**, 215–236 (2006)
8. E. Bonsón, et al., Local e-government 2.0: social media and corporate transparency in municipalities. Gov. Inf. Q. **29**(2), 123–132 (2012)
9. A.L. Kavanaugh et al., Social media use by government: from the routine to the critical. Gov. Inf. Q. **29**(4), 480–491 (2012)
10. E. Bonsón, S. Royo, M. Ratkai, Citizens' engagement on local governments' Facebook sites. An empirical analysis: the impact of different media and content types in Western Europe. Gov. Inf. Q. **32**(1), 52–62 (2015)
11. H. Kwon, J. Kim, Y. Park, Applying LSA text mining technique in envisioning social impacts of emerging technologies: the case of drone technology. Technovation **60**, 15–28 (2017)
12. D. Valdez, A.C. Pickett, P. Goodson, Topic modeling: latent semantic analysis for the social sciences. Soc. Sci. Q. **99**(5), 1665–1679 (2018)
13. A. Visvizi, et al., Tweeting and mining OECD-related microcontent in the post-truth era: a cloud-based app. Comput. Hum. Behav. **107** (2019)
14. G. Rasul, Managing the food, water, and energy nexus for achieving the sustainable development goals in South Asia. Environ. Dev. **18**, 14–25 (2016)
15. S.Q. Xu, P. Cooper, K. Sin, The "Learning in Regular Classrooms" initiative for inclusive education in China. Int. J. Incl. Educ. (2018)
16. K. Watkins, Longer lives and unfinished agendas on child survival. Lancet **388** (2016)
17. E. Bucherer, *Business Model Innovation—A Guideline for a Structured Approach* (Shaker Verlag, 2010)
18. M. Trapp, *Realizing Business Model Innovation—A Strategic Approach for Business Unit Managers* (Springer, 2014)
19. B. Halecker, *New Perspective and Insights on Business Model Innovation Using Systems Thinking and Action Case Studies* (University of Potsdam, Potsdam, Germany, 2016)
20. G. Stampfl, *The Process of Business Model Innovation—An Empirical Exploration* (Springer, 2016)
21. A.-A. Schaller, R. Vatananan-Thesenvitz, M. Stefania, Business model innovation roadmapping: a structured approach to a new business model, in *Proceedings of PICMET 2018: Technology Management for Interconnected World* (2018), pp. 1–9
22. N.A. Ashford, R.P. Hall, The importance of regulation-induced innovation for sustainable development. Sustainability **3**, 270–292 (2011)
23. A. Osterwalder, Y. Pigneur, *Business Model Generation* (Wiley, Hoboken, 2010)
24. H. Chesbrough, A.K. Crowther, Beyond high tech: early adopters of open innovation in other industries. R&D Manag. **36**, 229–236 (2006)
25. A. Visvizi, M.D. Lytras, *Politics and Technology in the Post-Truth Era* (Emerald Studies in Politics and Technology, 2019)
26. C. Zacharias, F. Poldi, *TWINT—Twitter Intelligence Tool* (GitHub, Florida, 2018). Retrieved from https://github.com/twintproject/twint

27. C.J. Hutto, E.E. Gilbert, VADER: a parsimonious rule based model for sentiment analysis of social media text, in *Proceedings of the Eighth International AAAI Conference on Weblogs and Social Media* (2014)

Chapter 36
Evidence-Based Technological Solutions for Migrant Workers: Is a Technological Information Platform for Thai Migrant Workers in Israel a Solution?

Nelly Kfir

Abstract CIMI is a professional NGO that has partnered with the government of Israel to oversee the arrival of Thai migrants and implement best practices as part of that process. Approximately 25,000 Thai migrants work in remote agricultural areas around Israel, and have little knowledge about life and working conditions in the country. Studies show that increased internet access, specifically mobile information and communications technology (ICT), can lead to a significant improvement in migrants' effective access to information, informed decision-making and work standards. Building on literature that examines how internet usage gaps still exist across social categories despite equal access, the study presented in this paper examines how Thai migrant workers in Israel use mobile ICT, in order to find ways to enhance and utilize it to improve their wellbeing. As a next step and based on a survey conducted among this population for this research, CIMI would aim to apply the findings and operationalize solutions for internet usage which can benefit these migrants by targeting their specific needs and assets. A technological platform to deliver reliable information and utilities can have a significant impact on over 25,000 Thai migrants in Israel. The project also has the potential of being expanded to include other migrant workers' communities in Israel, ranging from over 40,000 Filipino caretakers to over 10,000 Moldovan, Ukrainian and Chinese construction workers.

Keywords Migrant workers · Information · Wellbeing · Decent work · Mobile ICT · Digital divide · Internet usage gap · Inclusion · Community development

Introduction

Researchers suggest that increased internet access, specifically mobile internet and communications technology (ICT) can have a significant positive effect on migrants' wellbeing. ICT, combined with the development of the skills and literacy needed to

N. Kfir (✉)
Center for International Migration and Integration (CIMI), Jerusalem, Israel
e-mail: nellyk@cimi.org.il

A. Visvizi et al. (eds.), *Research and Innovation Forum 2020*,
Springer Proceedings in Complexity,
https://doi.org/10.1007/978-3-030-62066-0_36

effectively access information [1] can encourage migrants to make better informed decisions about the journeys to their destinations and their integration process after arrival [2–5], can help determine their work conditions [6, 7], and can help migrants organize [8] and build their sense of community through networks [9, 10]. The primary goal of the present study is to examine whether and to what extent use of ICT by migrant workers can enhance and improve their wellbeing and expand their understanding of legal rights and the social context of the host society.

Approximately 25,000 Thai agricultural workers are currently working in Israel as part of the Thailand Israel Cooperation (TIC) project. The Center for International Migration and Integration in Israel (CIMI) has partnered with the government of Israel to oversee the project and implement best practices as part of this process. Thai agricultural workers are being placed for work in remote agricultural areas around Israel, with little knowledge about life and working conditions in their regions and in the country. CIMI has found a number of challenges in distributing relevant information and utilities to TIC workers and connecting them after their arrival in Israel. To achieve the goal of the study and in light of the literature on the topic the paper examines the extent to which an existing or a tailored technological platform can create an accessible repository of information and utilities for these migrants.

In order to develop an evidence-based solution tailored to the needs of the target population, CIMI conducted a phone survey among 50 Thai agricultural workers who were sampled according to age and region. The results indicate that there is a significant gap between the wide availability of smartphone technology among Thai workers in Israel, and the migrants' access to information and communications, which was highly inefficient. Thai migrants report an eagerness to access information that can help them during their five-year stays in Israel. This information includes up-to-date knowledge about their legal rights as well as information about other practical issues, such as how to reduce their living expenses. These findings are consistent with literature that suggests that despite having access to internet, that structural social inequality, as well as individual traits, profoundly affect their ability to search for information on the internet. When concerning vulnerable populations, this in turn can have a negative impact on their resources and their agency with respect to improving their situation and mobilizing as a community. This argument is also known as the digital usage divide [11–13].

By better understanding ICT usage among Thai migrant workers, CIMI will be in a position to develop a project that assists these migrants in actively navigating for relevant information that can improve their wellbeing, help them connect with each other, and strengthen their community. A technological platform developed by CIMI in the form of an app tailored to the needs, capabilities of the users, which will effectively deliver reliable information and functionalities can have a significant impact on more than 25,000 Thai agricultural migrant workers in Israel. The project also has the potential to be expanded to include other groups of migrant workers' communities in Israel, ranging from over 40,000 Filipino caretakers to over 10,000 Moldovan and Chinese construction workers.

Migrants' Informed Navigation on the Internet: Tailoring Platforms, Enhancing Skills and Using Community Assets

Studies have examined inequalities in access to internet and ICT technologies, conferring the term "the digital divide" on these inequalities [14]. This literature further identifies differences and inequality in how such access can and is used in practice [12, 15]; and the link between the kind of use and social status and social mobility [13]. These studies trace a series of factors that influence or predict differences in use along socio-demographic attributes [11]. Indeed, differences in use of ICT technologies are influenced by structural social inequality as well as by personal and individual traits. Buchi et al. [11] offer a typology of internet usages, one of which is information seeking. Their results show that the predictors of the ability to seek information on the internet are younger age, academic education, and experience in internet usage.

Much of the literature on the issue of migrant workers' access to ICT draws on comparative studies across countries and across different socioeconomic categories within countries. The importance of access and internet usage in light of the large-scale global migration movements still requires additional attention and empirical case studies. This increased focus is needed because migrant worker populations often have weaker socioeconomic backgrounds and need more means for improving their position and seeking information for that purpose. At the same time, because these populations are transient, they are more dependent on the internet for acquiring and consuming useful information, and accessing low-cost international communications and various other services.

Sharma and Grote [1], who focused on internet usage of internal migrants in Southeast Asia, found that socioeconomic background is not a sufficient predictor of using the internet for purposes beyond that of entertainment, such as information seeking. However, they did find that a far more important factor is developing the ability to engage in informed active navigation through large masses of information. This skill needs to be encouraged, and internet users who are less experienced in internet navigation must be made aware of it. Furthermore, ICT professional experts working with vulnerable populations and developing applications for social good, stress the importance of being user oriented. In the context of digitalization and development work, developing and designing user interface needs to be human-centered and designed using a participatory process that includes the target population and other relevant stakeholders [16, 17]. These professionals stress the importance of tailoring the usage and design to the specific needs and structure of opportunities and constraints of the target populations [18].

Workers, especially migrant workers, have an extensive need for information on employment opportunities and employment rights as well as on guidelines for safety at work and costs of money transfer. The literature shows how workers can make more informed decisions with regard to employment conditions and opportunities using ICT [3, 5]. In addition, according to studies, technology can boost workplace safety by facilitating the transmission of health and safety information to workers in a culturally sensitive way [7]. Finally, additional research has examined digitalization as a way of promoting decent work and work conditions, recommending the use of existing digital and social media platforms by migrant workers to help them access services and connect and organize with their peers [6, 8]. In addition, the ILO [6] supports using specialized apps as a way of meeting the needs of migrant workers.

Buchi et al. [11] call for further empirical and theoretical investigation of the ways in which the sociodemographic usage divide persists even when access gaps are closed. This is precisely the situation identified by CIMI in its experience and engagement with Thai migrants in Israel and is grounded in the survey findings, a situation in which many Thai migrants seem to have technological devices with internet access, but nonetheless exhibit very low levels of utilizing their access to seek reliable information that can potentially improve their situations. This paper responds to the call for further investigation of how internet usage gaps still persist across social categories despite equal access. It presents the empirical case of Thai migrants in Israel in order to lay the foundation for advancing solutions for operationalizing internet usage which can help close these usage gaps and benefit the migrant populations by targeting their specific needs and assets.

Methodology

In order to provide Thai migrant workers in Israel better access to information about their rights and available services by enabling them to better search on existing online platforms (such as Facebook or LINE) or alternatively, to use a tailored mobile app as a solution, CIMI conducted a phone survey among fifty Thai agricultural workers in Israel who had previously contacted CIMI's call center during the month of July, 2019. The phone survey was carried out by two interviewers who speak the Thai language. All respondents were workers who had called the call center before and who agreed to be contacted again for the purpose of future studies. Each interviewer completed twenty-five surveys. The respondents' answers were documented in writing and submitted by the interviewers. Participants were asked a series of twenty-three multiple choice, ranking, and fill-in-the-blank questions covering the following categories: demographics, need for information, devices, access to information, and type of information.

The underlying knowledge from CIMI's work in the call center was that 1. there is a problem of lack of information and lack of awareness but it is unclear how migrants perceive it and what information they feel is missing 2. it is difficult to distribute comprehensive and up-to-date information to Thai migrants through existing platforms because there is no evidence to which extent they are using which platform, and for what purpose 3. many migrants have smartphones of some kind but there was little evidence on exactly how widespread this is, on the exact technical features they have and what is their connectivity. There was no prior knowledge on the extent to which they would want a technological platform through which they can seek and receive information regularly. Accordingly, the questions were aimed at understanding: the need and access respondents currently have for information, their usage of specific technological devices and their access to internet, their level of usage of the available technology to search for information, and finally, what types of information they would most want to have access to and to what extent the access should be through a technological platform.

Possible factors that may have affected the results are differences in communication styles between the two interviewers, and differences in how the interviewers may have explained or presented the questions to the respondents. Another possible selectivity factor could be the fact that because the study respondents consisted only of individuals who had already contacted CIMI's call center for Thai migrant workers, these respondents already had access to both mobile phones and had acquired some information about their basic rights and where they could turn for help. Finally, while respondents were classified according to age groups and geographical region, as seen in Table 36.1, the proportion of the total population aged 40–44, which is 16%, was not reflected in the sample, where this age group amounted to 10% of the study population. Similarly, the sample of those from the north and from the south of the country did not fully reflect their proportion in the population (north: 24% in the sample instead of 31% in the population; south: 50% in the sample instead of 48% in the population), which may have resulted in an overrepresentation from some groups and underrepresentation from others. Lastly, all fifty respondents were males, meaning that the female population of Thai migrant workers in Israel (3% of the total population of Thai migrants), small though it is, was not represented in this survey.

According to the results presented in Table 36.1, all participants were male, in their thirties, early in their stays in Israel, living in the south of the country, and working with a large number of co-workers (almost 10) in their workplaces.

Results: The Availability and Widespread Use of Smartphone Technology Among Thai Migrant Workers in Israel Versus the Ability to Access Information via This Medium

The results displayed in Table 36.2 show that most Thai workers feel that easy access to information for migrant workers is very important to their lives (ranked 4.08 of 5), and that currently they only sometimes search for information regarding their employment in Israel. There were mixed results regarding the question of whether or not they currently have the information they need regarding their employment.

The data reveal that all workers included in the sample have a mobile phone, and almost all of these workers also have internet on their plan. All participants also stated that they use their mobile phone more than any other electronic device. In addition, 86% of the respondents stated that they would prefer to receive information

Table 36.1 Descriptive statistics (definitions, mean/percent and standard deviations [SD]) (N = 50)

Variable	Definition	Mean/proportion	SD
Gender			
	Male	100%	
Age		33.12	4.48
Age groups			
	25–29	24%	
	30–34	36%	
	35–39	30%	
	40–44	10%	
Number of months in Israel		18.34	16.32
Region			
	South	50%	
	Central	26%	
	North	24%	
Number of co-workers		8.84	
Need for information			
	Do you have the information you need regarding your employment in Israel? (1 = not at all, 5 = too much)	2.45	0.58
	How important is easy access to information for migrant workers to you? (1 = not at all, 5 = extremely important)	4.08	0.75
	How often do you search for information regarding your employment in Israel? (1 = never, 5 = always)	3.06	1.04

electronically. These results support the argument that there is a need to develop technology to disseminate information electronically via mobile phones to migrant workers.

The results also show that respondents rarely use LINE, an app for free calls and messages. Indeed, LINE is ranked as the second most difficult way of accessing information after asking employers. While LINE, a communication app originating in Japan, is commonly used in many Asian countries, it appears that Thai workers in Israel have not widely adopted its use for the purpose of finding information. In addition, the respondents only sometimes search for information, using mostly Facebook (82% of those who search online). The medium for accessing information that received the highest ranking (4.16 of 5) was asking other Thai agricultural workers, while websites, mobile phone apps, and asking employers were shown to

Table 36.2 Distribution and prevalence of access to internet, to ICT devices and usage among Thai workers in Israel (mean/percent and standard deviations [SD]) (N = 50)

Variable	Definition		Mean/proportion	SD
Internet access and ICT device usage				
	Own a mobile phone		100%	SD
	What kind of mobile phone do you have?			
		Android	94%	
		iPhone	6%	
	Has an internet plan?		100%	
	What kind of device do you use most/least? (1 = the most out, 3 = the least)			
		Mobile phone	1.00	0.00
		Computer	1.36	1.43
		Tablet	1.78	1.06
	Do you prefer to receive information electronically or physically?			
		Electronically	86%	
		Physically	14%	
Access to information				
	How often do you search for information regarding your employment in Israel using a mobile phone to call an organization or a helpline? (1 = never, 5 = always)		2.78	0.82
	How often do you search for information using the LINE app? (1 = never, 5 = always)		1.60	0.83
	How often do you search for information using Facebook/social media? (1 = never, 5 = always)		3.32	1.35
	Which social media platform do you use?			
		Facebook	82%	
		YouTube	12%	
		Instagram	2%	
	How often do you search for information using websites? (1 = never, 5 = always)		1.92	1.10
	How often do you search for information using a mobile phone app? (1 = never, 5 = always)		1.40	1.11
	Which app?			
		Facebook Messenger	10%	
		WhatsApp	6%	
	How often do you ask your employer for information? (1 = never, 5 = always)		1.94	1.28

(continued)

Table 36.2 (continued)

Variable	Definition		Mean/proportion	SD
	How often do you ask other Thai agricultural workers or other friends for information? (1 = never, 5 = always)		4.16	0.98
	How often do you search for information by contacting or visiting the Thai Embassy? (1 = never, 5 = always)		1.44	0.70
	Which five resources are the easiest ways for you to access information (rank 1 = easiest, 5 = most difficult)			
		Facebook/social media	2.18	0.99
		Mobile calling	2.41	1.27
		LINE	3.95	1.04
		Employer	4.33	1.05
Type of information most important				
	What type of information is the most important and useful for your life during employment period in Israel? (1 = most important, 6 = least important)			
		Employment rights and procedures	1.66	1.12
		Health	3.26	1.29
		Safety	3.34	1.48
		Remittances	3.54	1.53
		Life in Israel (cultural adaptation, words in Hebrew, tips, dos and don'ts)	4.26	1.69
		Transportation	4.94	1.15
	How helpful would a phone App with information regarding your employment be to your life as a worker in Israel? (1 = very unhelpful, 4 = very helpful)		3.93	0.27
		Very helpful	92%	
		Helpful	8%	
		Unhelpful	0%	
		Very unhelpful	0%	

be used only rarely. Turning to employers were rated to be the most difficult way of accessing information (4.3 out of 5).

Thai migrant workers ranked employment rights and procedures as the most important type of information they needed for their lives during their employment in Israel, followed by information about health and safety. Finally, the response to the last question about whether a phone app with information regarding employment would be helpful was overwhelmingly positive, with 92% of participants choosing "very helpful" (ranked 3.93 out of 4).

Discussion

The major goal of the study was to explore ways to overcome the difficulties involved in disseminating useful information to Thai migrants about their rights and life in Israel. CIMI, an NGO that strives to implement best practices with respect to international migration in Israel, has recognized this difficulty while assisting the government of Israel in overseeing the process of migrants' arrivals, pursuant to a bilateral agreement. CIMI's basic question was whether and how an existing or a tailored technological platform could help create an accessible repository of information and services for Thai migrant workers in Israel. Building on previous research investigating how ICT can improve migrants' conditions, CIMI conducted a survey among Thai migrants in order to understand and map their usage patterns and needs.

The results reaffirm what the literature has identified as "the digital usage divide" and demonstrates that there is a gap between the widespread availability of technological devices among Thai migrants, and the migrants' ability to access useful information. A large majority of Thai migrants have a smartphone with internet access, but they all report difficulties in finding important and reliable information online. Asking other Thai agricultural workers is currently the easiest and most popular method of accessing information while in Israel. This channel can be an asset in some respects. However, used as an exclusive source of knowledge, it tends to spread inaccurate information, and is not a reliable source for up-to-date facts on employment in Israel and on workers' rights. Finally, in terms of content, Thai migrant workers ranked employment rights and procedures, health, and safety as the most important categories of information about their lives during their employment in Israel. This type of information is complex and difficult to navigate, and it would be advisable for any future project disseminating information to Thai agricultural workers to give it special attention.

The survey examined the question of whether this gap between easy access to digital devices but difficulties in accessing information could be closed by using existing platforms such as the LINE communication app or Facebook. It appears that LINE is the least used means of seeking information by workers. This is understandable, as LINE is designed for individual or group chats rather than for presenting information and referrals to services. Some official information is already made available to Thai migrants via a Facebook page. However, for the purpose of spreading large volumes of information, Facebook is not an effective medium for presenting and organizing such information thematically. Nor does Facebook adjust to the needs and interests of the user. Rather, Facebook publishes news and updates from time to time. In addition, it is not possible to automatically refer users or create access to various services, including CIMI's call center, through Facebook.

Thus, it seems that while all respondents have smartphones with various apps, the current apps in use are not suited to accessing information that could help answer questions and solve problems migrant workers face during their stays and work in Israel by referring them to relevant sources or services. This deficiency indicates that there is a need for a customized smartphone platform, whether an app, a web-app or a generic platform (e.g., LifeSpots) that can be tailored to a specific need that will allow Thai migrant workers to access information using the technology already available to them.

An operationalization of effective internet usage for the purpose of information seeking would need to include raising awareness, testing and improving the user experience in a participatory approach, utilizing assets of the target community and finding ways to assist Thai migrants in developing their information searching skills. As Sharma and Grote [1] have suggested, ICT, combined with developing skills and literacy, is key to making accessing information more effective.

Acknowledgements I would like to express my special graditude and thanks to Prof. Moshe Semyonov for his valuble guidance and comments. I wish to thank the staff and interns at CIMI for their part in the project of developing technological solutions for migrant workers, for their valuable professional input, and for their diligent assistance in making this study possible. Special thanks are also due to Alon Akerman from CIMI for his important part in the project and to Yoav Roll from Tel Aviv University for his assistance in handling the data.

References

1. R. Sharma, U. Grote, Determinants of internet use among migrants in South-East Asia: a case study of internal migrants in Thailand and Viet Nam. Migr. Res. Ser. IOM **58**, 1–18 (2019). https://publications.iom.int/system/files/pdf/mrs_58.pdf. Accessed 2020/3/8
2. K. Alam, S. Imran, The digital divide and social inclusion among refugee migrants: a case in regional Australia. Inf. Technol. People **28**(2), 344–365 (2015)
3. R. Chinoracký, T. Čorejová, Impact of digital technologies on labor market and the transport sector. Transp. Res. Procedia **40**, 994–1001 (2019)
4. S. Gelb, A. Krishnan, *Technology, Migration and the 2030 Agenda for Sustainable Development* (Swiss Agency for Development and Cooperation, 2018). https://www.odi.org/sites/odi.org.uk/files/resource-documents/12395.pdf. Accessed 2020/3/19
5. R. Hartje, M. Hubler, Smartphones support smart labour. Appl. Econ. Lett. **24**(7), 467–471 (2017)
6. ILO, Digitalization to promote decent work for migrant workers in ASEAN (2019). https://www.ilo.org/wcmsp5/groups/public/---asia/---ro-bangkok/---sro-bangkok/documents/publication/wcms_713546.pdf. Accessed 2020/3/8
7. D. Tutt, A. Dainty, A. Gibb, S. Pink, *Migrant Construction Workers and Health & Safety Communication* (Construction Industry Training Board (CITB)-Construction Skills, Bircham Newton, King's Lynn, Norfolk, UK, 2011)
8. B. Farbenblum, L. Berg, A. Kintominas, *Transformative Technology for Migrant Workers: Opportunities, Challenges, and Risks* (Open Society Foundations, New York, NY, 2018). https://www.opensocietyfoundations.org/uploads/dfe50370-e15a-4a78-99f9-3954c0e73bb3/transformative-technology-for-migrant-workers-20181107.pdf. Accessed 2020/3/11
9. S. Bunmak, Cell phone networks and migrant networks: the case of Nayu migrant workers in Malaysia. Geografia Malays. J. Soc. Space **8**(1) (2017)
10. A. Le Duc, The role of social media in community building for illegal Vietnamese migrant workers in Thailand. JIMS Themat. Art. Commun. Immigr. Commun. **10**(1), 4–21 (2016)
11. M. Buchi, J. Natascha, M. Latzer, Modeling the second-level digital divide: a five country study of social differences in internet use. New Media Soc. 1–20 (2015)
12. A. Van Deursen, J. Van Dijk, The digital divide shifts to differences in usage. New Media Soc. **16**(3), 507–526 (2014)

13. N. Zillien, E. Harggittai, Digital distinction: status-specific types of internet usage. Soc. Sci. Q. **90**(2), 274–291 (2009)
14. E. Harigittai, Y.P. Hsieh, Digital inequality, in *The Oxford Handbook of Internet Studies*. ed. by W.H. Dutton (Oxford University Press, Oxford, 2013), pp. 129–150
15. J.M.E. Ortega, M.R. Menendez, M.V.R. Gonzales, Diffusion and usage patterns of internet services in the European Union. Inf. Res. **12**(2), 1–20 (2007)
16. S. Rippin, User Experience (UX) Design & International Development (2016). https://www.sarahrippin.com/ux-design/ux-design-international-development/. Accessed 2020/3/8
17. T. Lee, What human-centered design can do for international development (2015). https://www.caktusgroup.com/blog/2015/11/20/what-human-centered-design-can-do-international-development/. Accessed 2020/3/8
18. D. Burton, Post-dignity design and how to avoid it in international development UX (2016). https://medium.com/@idiotgrace/post-dignity-design-and-how-to-avoid-it-in-international-development-ux-de483c1b3fc. Accessed 2020/3/8

Dr. Nelly Kfir is the Director of Research, Evaluation and Development at the Center for International Migration and Integration (CIMI) in Jerusalem, Israel. Founded by JDC Israel in 1998, CIMI is an independent non-profit organization dedicated to assisting Israel in meeting its migration challenges. Originally founded to leverage Israel's extensive practical expertise in the fields of immigrant integration and diaspora-homeland partnerships, today, CIMI supports Israel in a wide range of migration management issues and activities, including labor migration, asylum, counter-trafficking, and assisted voluntary return and reintegration. CIMI strives to help Israel meet its obligations under international law and apply international standards and best practices in the field of migration.

Chapter 37
Online Financing Campaigns' Comments: Insights from Crowdfunding Pitches

Wei Wang and Yenchun Jim Wu

Abstract Online financing projects allow users to post comments on projects including various topics in different stages. As a typical user generated content, due to the dynamic generation of comments on online financing projects, user will produce different text topics among various stages. Based on the comment text of online financing projects, this study analyzes the dynamic topics evolution in the online comments, considering both the time and topic factors. Using 5,405,498 comments from Kickstarter as corpus, dynamic topic model is employed, and the online comments are divided into two stages: antecedent stage and posterior stage. The experimental results show that, in the antecedent stage, the topics such as social promotion and content of the project are the mainstream. In the online comments after the financing, namely, posterior stage, there are huge differences in topics. The topics are distributed among the progress inquiry, fund usage, product discussion, etc. This phenomenon may be due to the differences between project schedules and reality. This study shows the dynamic changes of online investor's concerns, which provides a channel for the analysis of online investment opinions.

Keywords Topics extraction · Online financing · Topic model · Kickstarter · Topic over time

Introduction

Among the factors that affect the fundraising outcomes of crowdfunding projects, online review is one of the key factors that is often ignored. The influence of online reviews on subsequent users' purchase intention has been widely proved in other fields [1], for example: hotels, E-commerce, restaurants, films, books, etc. These

W. Wang
College of Business Administration, Huaqiao University, Quanzhou City, Fujian Province, China

Y. J. Wu (✉)
Graduate Institute of Global Business and Strategy, National Taiwan Normal University, Taipei, Taiwan
e-mail: wuyenchun@gmail.com

A. Visvizi et al. (eds.), *Research and Innovation Forum 2020*,
Springer Proceedings in Complexity,
https://doi.org/10.1007/978-3-030-62066-0_37

online reviews are usually analyzed from the perspective of sentiment analysis, that is, if the user's evaluation on the product is positive, it means that the early consumers affirm the product quality and promote the later consumer behavior, and vice versa. However, few studies focus on online reviews of online financing projects. Taking Kickstarter as an example, users can comment on any crowdfunding project, which shows users' evaluation and attitude towards the project. However, it lacks sufficient research on the comments of crowdfunding projects.

Furthermore, the impact of online reviews on online financing projects is different from the way of hotels and E-commerce. The biggest difference is that the evaluation of online E-commerce is based on the feelings after purchasing products, namely, the evaluation of product quality and experience when users actually consume products. But reviews on online crowdfunding projects include posterior comments and antecedent comments. Previous studies are usually based on the posterior comments, that is, the evaluation of products based on experience and clear knowledge of product quality, which represents the real experience on the products. However, crowdfunding projects can only give investors a commitment at the financing stage. Whether they can fulfill their commitment is influenced by many factors. Obviously, the comments on online crowdfunding projects is a kind of pre-comment, they are a kind of predictive evaluation.

Comments on online crowdfunding projects can be divided into several stages, including comments before financing, comments during financing and comments after financing. Only comments before the end of financing have an impact on the fundraising outcomes of online crowdfunding projects, because once the financing is completed, any other comments will not have an impact on the fundraising results. Therefore, the online comment time needs to be analyzed in depth. This kind of online comment with clear timelines is rarely conducted in other fields.

Online reviews can be divided into different topics. Users tend to publish different topics at different stages. Therefore, this study focuses on the review of online crowdfunding platform, analyzes the topics of online reviews and the changes of these topics at different stages.

Literature Review and Research Questions

Literature Review

Crowdfunding is the practice of funding a project or venture by raising small amounts of money from a large number of people, typically via the Internet [2, 3]. Crowdfunding is a form of crowdsourcing and alternative finance [4]. Crowdfunding is widely used in financing by entrepreneurs. Fundraising outcomes are affected by many factors, including social connections, linguistic features [5, 6].

Many literatures show that linguistic features are one of the key factors that affect users' purchase intention, for example, the sentiment of text description [7]. Online reviews have a significant impact on the sales volume of E-commerce products. In a study of online reviews of hotels, it is found that the strategy of managers' response to the online reviews of hotels and the linguistic features are one of the main factors affecting the subsequent consumers' behavior. Therefore, when responding to online reviews, hotel managers should take a strategic approach [8].

Through the text mining and sentiment analysis, Airbnb users tend to evaluate their experience based on the reference framework of hotel accommodation in the past. The three key attributes identified in the data are location, convenience, and host. But price is not a key factor. And the analysis shows that Airbnb users' comments have a positive bias, while the negative sentiment are mainly caused by noise [9].

Information search behavior of hotel consumers is varied, many factors may affect their information search behavior, and the utilization and processing of online reviews, including information overload, information confusion, information processing, information presentation format, information presentation mode, etc. Many factors may affect the information search and information processing of hotel consumers, especially their use pattern of online reviews in the decision-making process [10].

Social factors also have an impact on the financing performance of crowdfunding projects. The early funds collected from private networks and social media networks have a significant impact on the financing performance of projects and the participation intention of subsequent investors. In addition, success drivers related to the number of investors include financing goal, duration of funding, provision of financial information in publicity and positioning of the company's products. On the contrary, a comprehensive assessment of the company in terms of team, market, concept, scalability, stage and transaction terms does not seem to improve predict the success of equity crowdfunding [11].

The influence of financier's identity on financing performance mainly comes from the individual influence of financier, which is influenced by nationality, environment, credit level, education level, occupation and other aspects of financier [12]. The progress of project financing will also greatly affect the willingness of users to participate in investment. When users believe their investment is important to the project, they are more willing to participate in project investment. That is, when the project is about to be the edge of financed successfully, the willingness of investors to participate in investment is the strongest, and the feeling of "go for the goal" is enjoyed by most users. However, once the project reaches the expected goal, the user's willingness to participate will be significantly reduced [13].

Some researches integrate the text and images of project description to predict the financing performance of crowdfunding projects. The dynamic information (comments and updates) makes the project creator and platform difficult to predict the results in time. Some researches use neural network to combine information from different modes to study the influence of complex interaction among text, images and metadata on prediction. Experiments show that images can improve the performance of prediction, especially for items with less text information [14].

Research Gap and Questions

Nowadays, more and more projects attract supporters with interactive text, such as comments, but few scholars mine the comments or evaluate their impact on fundraising outcomes. Throughout the current research, there are deficiencies at least in the following aspects: (1) in the research of online comment, many researches regard online comment as a static variable, and the method similar to LDA is employed to extract the topic of online comment, ignoring the time information of the online comments; (2) we do not know the topics over time of online comments of the online crowdfunding projects.

In view of this, this study puts forward the following research questions: (1) we focus on extracting the time of the comments by employing a large number of online reviews of online crowdfunding projects; (2) identify the topics of online comments on these crowdfunding projects; (3) analyze the time evolution of different topics, that is, the changing rules of users' comments on crowdfunding projects.

Experiment Data and Methods

Experiment Data

The experiment data is from Kickstarter, the world's largest reward-based crowdfunding project platform. The project received 70 online comments. However, given the characteristics of crowdfunding projects, some online comments are posted after financing ends, while some online comments are generated before financing ends.

The process of data collection includes the following seven steps: (1) crawl the successful projects and the ongoing ones on the front page to build an initial project list; (2) crawl the content based on project list (including founders, backers, pledge goal, title, etc.); (3) extract the backer list for each project; (4) crawl the backed projects for each backer (including both successful campaigns and failed ones); (5) select the failed campaigns to build a new project list; (6) repeating the step (2) to step (5) until all projects have been crawled; (7) crawl relevant data (updates, comments, etc.). We collected 5,405,498 comments from Kickstarter as corpus.

Methods

The key issue of text mining is to identify the topic of the comments. In the Kickstarter, users often generate a comment as shown in Fig. 37.1, and the founders can reply to the comment. Python is employed as the main language data extraction and processing, and a Python toolkit—TOT (topic over time)—is used to capture the topic structure changes over time.

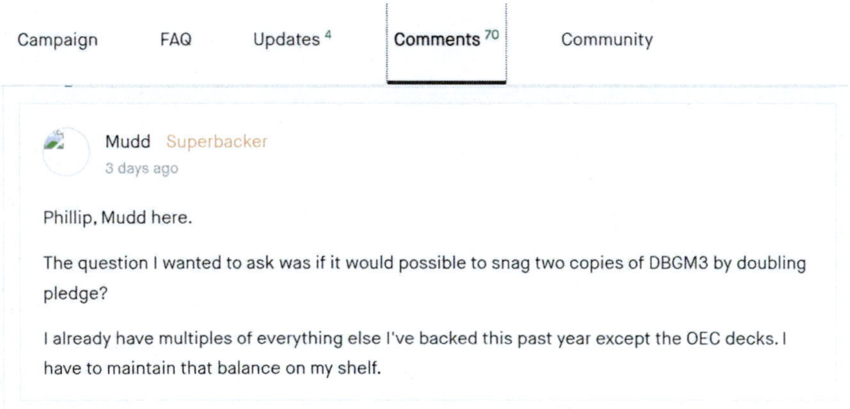

Fig. 37.1 An example of comment

Therefore, we can detect the text topic on the basis of text mining, namely, the supervised classification, and the text pre-process is as follows:

(1) There are many conjunctions which are meaningless for text analysis, we remove the conjunctions in the first step. Table 37.1 shows the stop words list we adopted.
(2) Then we do the following text preprocessing: data standardization, remove the non-frequent words, word stemming and word lemmatization (*PorterStemmer*), document matrix transformation. Word stemming can match the words in various forms, for example: the stemmer word is *ship* for *shipped*, *shipping* and *ship*. With stemmer word, it is easy to identify and contrast in the subsequent process.
(3) Then we use LDA [15] (Latent Dirichlet Allocation) to identify the text topic.
(4) In order to accurately determine the topics over time, TOT (topic over time) is employed to extract dynamic topics. TOT relies on Markov assumptions or discretization of time, each topic is associated with a continuous distribution over timestamps, and for each generated document, the mixture distribution over

Table 37.1 The stop words list we adopted

i, me, my, myself, we, our, ours, ourselves, you, your, yours, yourself, yourselves, he, him, his, himself, she, her, hers, herself, it, its, itself, they, them, their, theirs, themselves, what, which, who, whom, this, that, these, those, am, is, are, was, were, be, been, being, have, has, had, having, do, does, did, doing, a, an, the, and, but, if, or, because, as, until, while, of, at, by, for, with, about, against, between, into, through, during, before, after, above, below, to, from, up, down, in, out, on, off, over, under, again, further, then, once, here, there, when, where, why, how, all, any, both, each, few, more, most, other, some, such, no, nor, not, only, own, same, so, than, too, very, s, t, can, will, just, don, should, now

topics is influenced by both word co-occurrences and the document's timestamp. This study captures not only the low-dimensional structure of data, but also how the structure changes over time [16].

Results and Discussion

Results

In LDA, we need to manually specify the number of topics. We start from 20 topics, try to cluster the online comments and output the top 20 keywords of each cluster. Then observe the keywords of each cluster. It means there is an overlap between the clusters if there is an overlap between the output keywords, in this case, we reduce the number of topics, train the model again, and output the top 20 keywords. Repeat the above steps until the output clusters are not overlapped. Finally, the online comments of crowdfunding projects are clustered into six topics. Table 37.2 shows the topic results of online comments. The six topics of online comments include progress comments, shipment comments, product quality evaluation, fund usage evaluation, social networking promotion. These six topics represent the degree of investors' concern for different aspects of the crowdfunding project.

The topic of online comments is time-related. We divide the time of online comment into six stages. The first stage is antecedent stage, that is, the comments generated before the end of project financing. In other words, the antecedent comments are generated in the process of project financing. After the completion of crowdfunding project financing, investors have generated a large number of comments on the project, which are posterior comments. These comments are divided into five stages according to the date, which are noted by period 1, period 2, period 3, period 4 and period 5 respectively. When we consider the relationship between topic and time, the result is shown in Table 37.3. From the results, users pay attention to different topics at different stages.

Table 37.2 Topic clustering results of online comments

Cluster	Topic	Examples of keywords
1	Progress comments	Progress, schedule, delay
2	Shipment comments	Delivery, post, ship
3	Product quality evaluation	Quality, expectation, product
4	Fund usage evaluation	Fund, money, budget
5	Social networking promotion	Mail, twitter, forward
6	Project content discussion	Content, project, campaign

Table 37.3 Relationship between topics and time

Topic	Antecedent	Posterior				
	Funding duration (%)	Period 1 (%)	Period 2 (%)	Period 3 (%)	Period 4 (%)	Period 5 (%)
Progress comments	8.98	11.24	16.75	18.34	22.75	21.94
Shipment comments	6.63	10.45	15.98	20.32	21.67	24.95
Product quality evaluation	11.45	16.34	17.31	17.87	18.54	18.49
Fund usage evaluation	9.12	17.32	18.91	18.27	17.49	18.89
Social networking promotion	50.21	18.19	10.45	8.43	6.43	6.29
Project content discussion	22.82	18.75	17.46	16.02	14.31	10.64
Progress comments	8.98	11.24	16.75	18.34	22.75	21.94

Discussion

In the antecedent stage, users pay heavy attention to the topics of social network promotion (50.21%) and project content discussion (22.82%). As the main purpose of social networking promotion is to advertise and attract more investors to participate in investment. Therefore, it can be considered that the main purpose of antecedent stage is to achieve the success of financing. In addition, there is a high proportion of project content discussion.

In posterior comments, topics related to progress inquiry, shipment inquiry and product quality evaluation show an increasing trend. The farther away from the end of project financing, the more comments on these topics. However, the quantity of fund usage evaluation in each stage is relatively balanced. The number of topics related to social network promotion and project content discussion shows a decreasing trend, that is, as time goes on, users are less concerned about social network promotion and project content.

Conclusion and Future Directions

Conclusion

In this study, we employ 5,405,498 comments from Kickstarter as corpus, dynamic topic model (TOT) is employed, and the online comments are divided into two stages: antecedent stage and posterior stage. And we get the following conclusion. (1) In the antecedent stage, the topics such as social promotion and content of the project are the mainstream. (2) In the posterior stage, the topics are distributed among the progress inquiry, fund usage, product discussion. It indicates that users pay attention to different aspects at different stages.

Future Directions

Although we do a lot of detection on dynamic comments, but it also has some shortcomings and future directions are: (1) we mainly employ text mining to detect the topic of the comments with LDA. However, we do not distinguish the popularity of projects; (2) We do not distinguish the difference between successful projects and failed ones, which is also a direction in the future study.

Acknowledgements This work is partially supported by the NSFC Grant (71601082, 71771177), Natural Science Foundation of Fujian Province (2017J01132), Education and Research from China's Ministry of Education (2019J01012), Huaqiao University's High Level Talent Research Start Project Funding (16SKBS102), and Ministry of Science and Technology, Taiwan (MOST 108-2511-H-003-034 -MY2 & 109-2511-H-003-049-MY3).

References

1. Q. Zhao, C.-D. Chen, J.-L. Wang et al., Determinants of backers' funding intention in crowdfunding: Social exchange theory and regulatory focus. Telematics Inform. **34**(1), 370–384 (2017)
2. J.P. Berns, M. Figueroa-Armijos, V.S.P. Da Motta et al., Dynamics of lending-based prosocial crowdfunding: Using a social responsibility lens. J. Bus. Ethics **161**(1), 169–185 (2020)
3. L.A. Lasrado, A. Lugmayr, Crowdfunding in Finland: a new alternative disruptive funding instrument for businesses. *Proceedings of International Conference on Making Sense of Converging Media* (2013), pp. 194–201
4. M. Schulz, P. Haas, K. Schulthess et al., How Idea Creativity and Hedonic Value Influence Project Success in Crowdfunding. *Wirtschaftsinformatik Proceedings 2015* (2015)
5. K. Menon, H. Karkkainen, J. Jussila et al., Analysing the role of crowdfunding in entrepreneurial ecosystems: a social media event study of two competing product launches. Int. J. Entrepreneurship Small Bus. **33**(4), 575–606 (2018)

6. J. Huhtamäki, L. Lasrado, K. Menon et al., Approach for investigating crowdfunding campaigns with platform data: case Indiegogo. *Proceedings of the 19th International Academic Mindtrek Conference* (2015), pp. 183–190
7. W. Wang, K. Zhu, H. Wang et al., The impact of sentiment orientations on successful crowdfunding campaigns through text analytics. IET Softw. **11**(5), 229–238 (2017)
8. K. Xie, L. Kwok, W. Wang, Monetizing managerial responses on tripadvisor: Performance implications across hotel classes. Cornell Hospitality Q. **58**(3), 240–252 (2017)
9. M. Cheng, X. Jin, What do Airbnb users care about? An analysis of online review comments. Int. J. Hospitality Manage. **76**, 58–70 (2019)
10. D. Gursoy, A critical review of determinants of information search behavior and utilization of online reviews in decision making process (invited paper for 'luminaries' special issue of international journal of hospitality management). Int. J. Hospitality Manage. **76**, 53–60 (2019)
11. A. Lukkarinen, J.E. Teich, H. Wallenius et al., Success drivers of online equity crowdfunding campaigns. Decis. Support Syst. **87**, 26–38 (2016)
12. Y. Zhao, X. Xu, M. Wang, Predicting overall customer satisfaction: Big data evidence from hotel online textual reviews. Int. J. Hospitality Manage. **76**, 111–121 (2019)
13. V. Kuppuswamy, B.L. Bayus, Does my contribution to your crowdfunding project matter? J. Bus. Ventur. **32**(1), 72–89 (2017)
14. C. Cheng, F. Tan, X. Hou et al., Success prediction on crowdfunding with multimodal deep learning. in *Proceedings of the 28th International Joint Conference on Artificial Intelligence* (AAAI Press, 2019), pp. 2158–2164
15. D.M. Blei, A.Y. Ng, M.I. Jordan, Latent dirichlet allocation. J. Mach. Learn. Res. **3**(1), 993–1022 (2003)
16. X. Wang, A. Mccallum, Topics over time: a non-Markov continuous-time model of topical trends. in *Proceedings of the 12th ACM SIGKDD International Conference on Knowledge Discovery and Data Mining* (2006), pp. 424–433

Chapter 38
Using i*-Based Organizational Modeling to Support Blockchain-Oriented Software Engineering: Case Study in Supply Chain Management

Yasmine Ben Hamadi, Samedi Heng, and Yves Wautelet

Abstract Popularized by the rise of cryptocurrencies, blockchain technology has been gaining an increasing exposure. The potential of the distributed ledger lies in decentralization and immutable information. It promises complete transparency and a unique source of truth. In this article, a blockchain project is considered from a business and social perspective rather than from a technical one. More precisely, the focus is on understanding, from a functional analyst perspective, how the technology can fit the adopting organization's supply chain. More specifically, the i* organizational modeling framework is proposed to depict the interactions between involved actors and the roles of the multiple stakeholders of a blockchain. i* allows to identify the relations and goals within the actors of a complex network at early and late stages of the blockchain's adoption. The applicability and advantages/limitations of i* to the organization of blockchain projects is tested through a case study of a food retail group who is adopting the distributed ledger for Supply Chain Management (SCM).

Keywords i* framework · Blockchain · Blockchain-oriented software engineering · Conceptual modeling · Supply chain management · Distributed ledger

Introduction

The 2017 World Economic Forum and its new Center for the Fourth Industrial Revolution presented blockchain technology as an answer to information integrity and

Y. B. Hamadi
KU Leuven, Leuven, Belgium

S. Heng
HEC Liège, Université de Liège, Liège, Belgium
e-mail: samedi.heng@uliege.be

Y. Wautelet (✉)
KU Leuven, Belgium
e-mail: yves.wautelet@kuleuven.be

A. Visvizi et al. (eds.), *Research and Innovation Forum 2020*,
Springer Proceedings in Complexity,
https://doi.org/10.1007/978-3-030-62066-0_38

transparency concerns. "*Blockchain (also called distributed ledger), the technology enabling cryptocurrencies like bitcoin and Ethereum, is pulling us into a new era of openness, decentralization and global inclusion*" [30]. This promise of a unique and unalterable source of truth, has triggered an explosion of interest in this decentralized technology that goes beyond "the hype around crypto currency" [1].

Industrials and scientific researchers strongly believe in the incredible blockchain opportunities. However, the intrinsic attributes of this distributed ledger are not enough for a blockchain project to succeed. In fact, the success of a blockchain project strongly depends on how the stakeholders behave in stewarding its development [30]. Meaning how cooperation and governance are managed to make sure that this extraordinary technology [30] is not sub-optimized. Blockchain intrinsic characteristics provide great opportunities but they also imply greater software engineering challenges which urges for tailored techniques for *Blockchain-Oriented Software Engineering (BOSE)*.

The expansion of blockchain adoption in a significant range of business sectors urges for enhanced BOSE techniques but also for a wider understanding of the distributed ledger basics. The practical goal and theoretical contribution of this paper is to suggest a modeling-based solution to these two challenges. More concretely, the i* organizational modeling framework is applied to blockchain. First, it is an attempt to enhance requirements engineering for blockchain projects at their early and late stages. The aim is to tackle governance issues resulting from the extensive cooperation needed between numerous stakeholders for the implementation of blockchain in a company's information systems (IS) environment. The framework is applied to blockchain project for Supply Chain Management (SCM) of TheBestFood[1] supermarket chain. Besides, the limits of the model will be highlighted and an extension to its conceptual artifacts will be suggested. Second, i* and business process modeling will be generically applied to model blockchain functioning and its main characteristics. This is an attempt to provide a design-based approach to acquire a basic knowledge about the trending technology that is blockchain. Ultimately, the targeted research question is: *to what extent can the i* framework be considered as effective to model blockchain technology-based applications and the organizational requirements that they imply?*

Given the exploratory character of this research, a grounded theory approach [4] is adopted. In fact, taking into account that, to the best of our knowledge, this is the first attempt to apply i* framework to blockchain and that blockchain itself is an immature technology in the business field, prior commitment to any theory was avoided. We mainly proceeded through a literature review and 3 interviews with current notable blockchain field actors in Belgium. The methodology of the literature review consisted in identifying and analyzing, for each stage of BOSE, the relevant existent literature that aims to enhance or evaluate the existent techniques and methods. The interviews of people involved in the blockchain project performed at TheBestFood started with general open questions that became more precise as

[1]The case study is a supermarket chain active in Belgium but for confidentiality reasons its real name has been changed.

the scope of the research narrowed down. Besides, the research is based on a case study in order to get the necessary information for the modeling but also get more insight on how a blockchain is practically integrated within the configuration of an enterprise. The reason for selecting TheBestFood's blockchain project for SCM, is the early stage of the project when the interviews first started on 9 January 2019 that allowed to follow the evolution of the blockchain adoption and the challenges that came up as the project progressed. It is important to note that all of the produced representations have been discussed and validated ex-post with the interviewees.

Background

Emergence and Evolution

We are noticing since late 2015 a great number of researches that study and encourage blockchain possible applications beyond cryptocurrencies [10]. Peaks of peer reviewed papers concerning blockchain applications were observed in 2017 and 2018 [6]. Besides the increasing interest of the research community, there is an increasing number of blockchain projects being implemented and going live in a context of business applications [27]. The reason for this explosion of interest, lies in blockchain opportunity of decentralization that allows to achieve the same functionalities with the same amount of certainty no middle man needed [22]. Additionally, data in a blockchain can never be modified or deleted, it is permanent which ensures integrity of exchanged information on blockchain. This opportunity of transparent, immutable and open source data led Internet to enter a second era based on blockchain [30].

This "second generation of blockchains", blockchain for cryptocurrency being the first one, was made possible by smart contracts [11]. In simple terms, smart contracts are a digital version of the regular paper-based contracts. It binds the contract parties to respect the transaction's conditions by combining protocols with user interfaces [19]. Smart contracts are indeed, from a technical point of view, self-executing lines of formal code that are run on top of a blockchain [8]. The executions of these agreements stored on blockchain are forced by a computer program [7]. Comparing to the traditional contracts, smart contracts minimize the need for a trusted intermediary and the accidental or malicious hazards linked to their execution [29]. They are accessed, stored and managed on a decentralized application like for example Parity Wallet that uses Ether for incentivization. In their paper, Dutchman et al. [8] propose a business process model to understand the functioning of a smart contract without focusing on the technical aspect. It is interesting to show their constructed model as they use, similarly to this paper, a modeling-based approach to explain different aspects of blockchain.

Besides, as different business sectors have been realizing the added value of smart contracts, it is interesting to highlight the output of [19]. In the latter paper, the authors propose a way to democratize the code writing of a smart contract so that technical IT skills or qualified developers become not required anymore.

To conclude, blockchain is a "cryptographically verifiable list of data" and its great potential mainly lies behind its promise of data decentralization and guarantee of data integrity [12]. Since 2008, we have witnessed blockchain 1.0 era when the distributed ledger was only considered as the technology behind cryptocurrency. Now, we are living the blockchain 2.0 era that allowed smart contracts and that is characterized by cryptocurrency being mostly the incentive mechanism behind it.

Types of Blockchain

In this subsection, a technical background is presented. The aim is to explain the mechanisms behind this decentralization that permitted an open access to a unique source of truth no middle man needed. There are mainly 2 types of blockchain related to data verification rights. On one hand, permissioneless blockchains are open source. This means that everyone can add new blocks to the distributed ledger by validating transactions [21]. On the other hand, Ethereum made possible permissioned blockchains that grant these privileges to a certain number of nodes or a central authority [11].

Now, there are 3 types related to data access and readability which are *public*, *private* and *consortium* [21]. Public blockchain can be permissioneless and is mainly used for cryptocurrency transactions. Public permissioneless blockchain is also referred to as "Bitcoin blockchain" [14]. Anyone with internet connection is allowed to participate in the consensus, meaning data verification and adding of new blocks, and to access the data on blockchain [11] which makes it difficult to control [21]. Real and full decentralization and complete transparency of data are essentially made possible by public blockchain [14]. However, it is mainly criticized for being inefficient due to the important number of verifying nodes that slows down the adding of new blocks. "*Consensus protocols are also slow, requiring up to an hour to reasonably confirm a payment*" [14].

Permissioned blockchain can be private or consortium. Private blockchain are managed and examined by "a central locus of decision-making" [14]. In other words, it allows the middle man back in the transactions to ensure trust. In fact, few nodes are allowed to access data on the ledger and a central authority is accountable for the verification. Although private blockchain guarantees privacy to its users thanks to read restrictions, it is criticized for being subject to data tempering due to the lack of transparency caused by absence of decentralization [21]. One notable example of private permissioned blockchain is the one of IBM Food Trust used by many big retailers. Finally, consortium blockchain, also called hybrid blockchain, is a balance between private and public blockchain [25]. However, there is also a risk of tampering due to reduced decentralization [21].

According to Hintzman, the different types of blockchains are "simply different tools that serve different purposes" [14]. Hence it is important for blockchain

Table 38.1 Type of blockchain

Feature	Public	Private	Consortium	References
Access	Open source	Strictly controlled	To authorized nodes	[11, 21]
Verification	By all nodes	By centralized authority	By authorized nodes	[14, 21]
Efficiency of verification	Less efficient	Very high	In between	[14, 21, 22, 31]
Decentralization	Fully decentralized	Presence of middle man	Partially decentralized	[11, 14]
Security	Anonymous 51% attack	High	High	[14, 21]
Privacy	Low	High	In between	[14, 21]
Scalability (number of verifying nodes)	High	Potentially poor	In between	[11, 21, 31] Interview
Transparency	Guaranteed	Poor	In between	[14, 21, 30]
Immutability	Guaranteed	Risk of tampering	Risk of tampering	[21]
Cost	Potentially high volatility (if no tokens)	Subject to price increases	Partially controllable	Interview
Principal drawback	Inefficiency of verification	Limited decentralization	Governance challenges	Interview

projects stewards to choose the best suited blockchain for their use case requirements. Table 38.1 summarizes the characteristics of public permissioneless blockchain and private and consortium permissioned blockchains.

From Software Engineering to Blockchain-Oriented Software Engineering

The study of Chakraborty et al. [3] aims to unveil the existence of significant differences between software engineering for traditional software development and for centric blockchain projects due to the inherent complexity of the distributed ledger technology.

Destefanis et al. [7] also calls for BOSE. The conclusions are based on a single case study approach. They tackle blockchain software engineering practical problems encountered by developers namely the poor source code of smart contracts and the absence of testing framework [7]. Destefanis et al. [7] urge for the deployment of ad-hoc methodologies for coding and solidity testing for blockchain smart contracts namely, a BOSE discipline. They define BOSE as the need for new professional roles, enhanced security and reliability, more adapted modeling and verification frameworks and specialized metrics. The need for BOSE is demonstrated based on the Parity Wallet hack case study. The researchers point out the vulnerabilities of smart contracts by thoroughly analyzing the poor source code lines and inadequate testing tools that led to 513,774.16 Ether being frozen. They finally suggest the best coding and testing practices that could have helped avoiding this hack.

The paper of Kshetri [20] demonstrates that blockchain is a promising use case for supply chain and not only because of the promising return on investment. In fact, it also enhances trust between actors of the supply chain (e.g., farmers, shippers, suppliers, big retailers, customs, end consumers, etc.), and ensures sustainability and waste reduction. Using a grounded theory approach and based on eleven use cases, the paper's major output consists in a table that lists blockchain roles and the mechanisms behind these roles for six supply chain performance dimensions namely cost, speed, dependability, risk reduction, flexibility and sustainability. One of the recurrent blockchain mechanisms behind enhancing SCM performance is the end to end digital integration of data held by business partners and value-added service providers all along the supply chain.

The i Framework*

The i* framework is an agent-oriented and goal-oriented graphical requirement modeling notation [9]. It allows an early requirement engineering analysis in environments where social actors depend on each other for goals to be achieved, tasks to be performed, and resources to be furnished [9]. Previous researches proved the relevance and utility of i* to model organizational requirements of a "multi agent system" [35] facilitating stakeholder's interactions by depicting their dependencies and hence providing a mean for coordination. i* was previously used to model several organizational settings such as online stores [15], hospital beds management [33, 35], health care [9], supply chains and more specifically outbound logistics [32], production support in the steel industry [16, 36] and also for the development of higher education platforms like collaborative learning software [17] and MOOCs [34]. On one hand, by showing the goals, the model becomes a reference point as it comes to the assessment of the project progress. On the other hand, the focus on the actors directly involved in goals' achievement, allows the analysis of agents' "strategic intents" [9]. Hence, disposing of a model that helps to visualize the whole actors' network configuration would allow the functional analyst to better understand the different roles, responsibilities and relations of each node leading to more adequate decision making.

The i* framework is divided in two parts providing each a different level of abstraction: The Strategic Dependency (SD) and the Strategic Rationale (SR) model [9]. Figure 38.1 provides the core concepts and icons of the i* framework. The SD model shows dependencies and the SR model depicts internal intents.

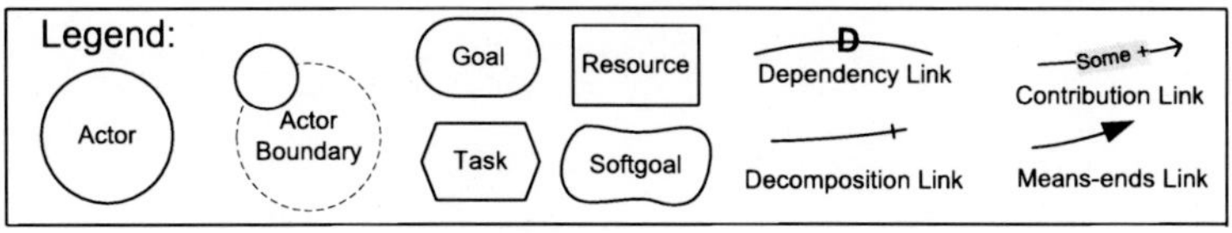

Fig. 38.1 Relevant i* concepts and their icons

As explained in Section "Background", blockchain decentralization and adoption necessitates the cooperation of multiple stakeholders. Hence, we need a model that not only depicts the sequences of processes and activities but also the socio-technical environment of the blockchain.

This paper proposes a first attempt to model a blockchain system with i*. The organizational requirements modeling tool will be both applied to blockchain technology in a general context and to TheBestFood blockchain adoption case study in particular. On one hand, the generic application of the model to blockchain aims to introduce and explain the main aspects to know about the technology. On the other hand, applying i* in the specific context of TheBestFood's blockchain project aims at evaluating to what extent the model can express the organizational requirements stemming from the interactions and roles of the multiple stakeholders in a blockchain system applied to a business context.

Applying Organizational Modeling to Blockchain Technology: Generic Approach

In this section, the framework is used to sketch the functioning and main characteristics of the distributed ledger. The aim is to facilitate the understanding of the decentralized technology thanks to an i*-based software modeling.

The purpose behind Fig. 38.2 is twofold. On the one hand, it depicts in a nutshell the principal interaction between public blockchain actors. On the other hand, it allows to familiarize with the dependencies logic and reading of the framework. As explained in Section "Background", a `Blockchain User` (i.e., the depender) depends on `Miners` (i.e., the dependee) to access `Data verified` (i.e., the dependum). The `Data verified` is the goal. Miners are "*members of the general public using their computers to help validate and timestamp transactions*" [28]. When a blockchain user gets the correct data, i.e. when a new block is added at the end of the chain, the miner receives crypto currency like bitcoin for instance. This mechanism incentivizes the cooperation [37]. This relationship between "intermediate nodes" [13] can be modelled with an SD.

Figure 38.3 is a business process model that shows the flow of activities accomplished by the miner in order to authenticate data. The Business Process Model and Notation (BPMN) [24] representation is here more intuitive to build but also

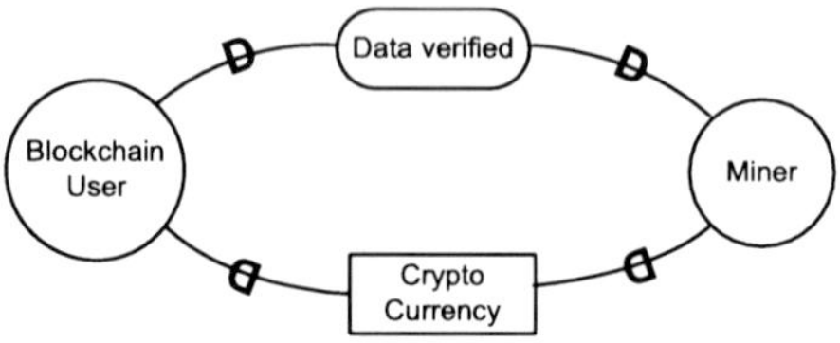

Fig. 38.2 Blockchain incentivized cooperation

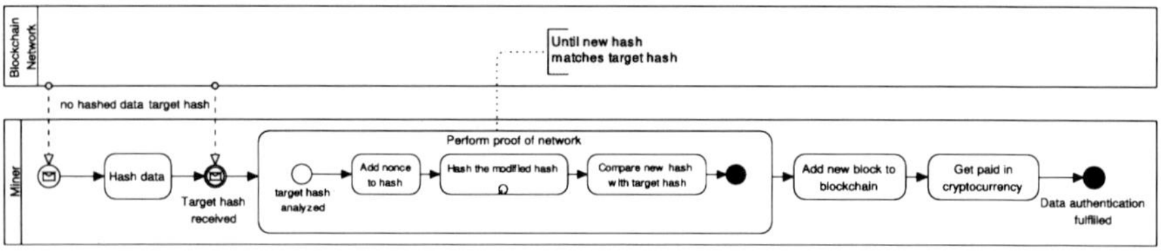

Fig. 38.3 Blockchain data authentication process

a better tool to introduce the verification process to someone who is not familiar with the technology. Figure 38.3 depicts how the miner achieves the goal of `Data verified`. As explained in Section "Types of Blockchain", new data authenticity is verified thanks to a hash comparing process. The miner receives a block of non-hashed data constituted of multiple transactions. Each transaction lead to a specific hash obtained by an algorithm constituted of hashing functions. All data is hashed along a Merkle tree [23]. The latter, also called authentication tree or hash tree, is a series of binary codes organized in a tree where each leaf constitutes a hash [2]. This data structure allows the miner to authenticate data destined for blockchain. Hence, those hashes are structured through a Merkle tree thanks to which one single root hash is obtained and represents the whole block. In order for a miner to be rewarded with cryptocurrency, the miner has to be the most efficient in proof of work. This consists in randomly adding a number between 0 and 256 bits, namely the nonce, to the root hash until the new obtained root hash matches the target hash requirements. Looping through these tasks until finding the right nonce to add to the root hash is referred to as the *proof of work* . The "winner" is the first to find the hash combination that matches the target hash and to get the new block added to the public blockchain.

In Section "Background", the three blockchain types' main characteristics were compared. Among the key early decisions that have to be made in a blockchain software project, is the type of blockchain that best suits the configuration of the enterprise and best serves its interests. The *Non-Functional Requirements* (*NFR*) diagram [5] in Figure 38.4 focuses on the contribution of each blockchain type to the non-functional requirements expected from the technology. This model architecture was used in [15] to guide the selection of an architectural style during IS engineering.

The three blockchain types are represented as "operationalized soft-goals" which are the three clouds in bold at the basis of the model. Verification, access and trust are "AND-decomposed" into their main quality attributes. By the soft-goal "feasible" it

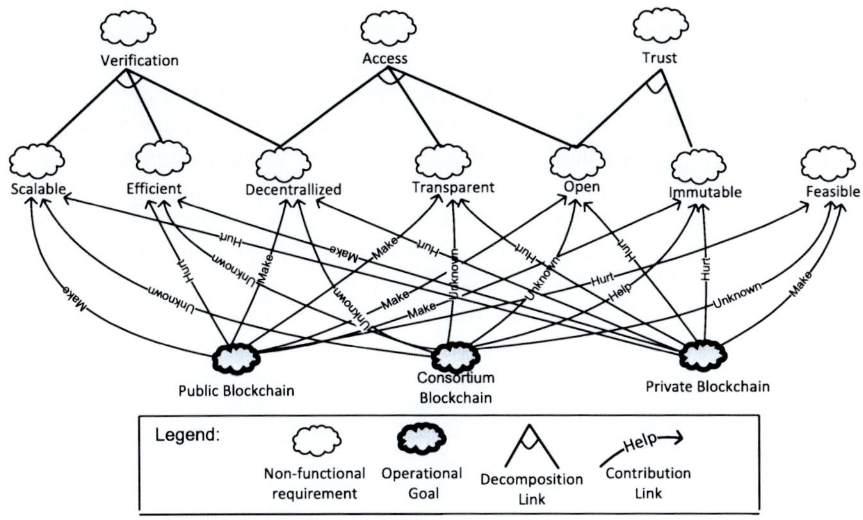

Fig. 38.4 Selecting a blockchain

Fig. 38.5 Selecting a blockchain adoption strategy

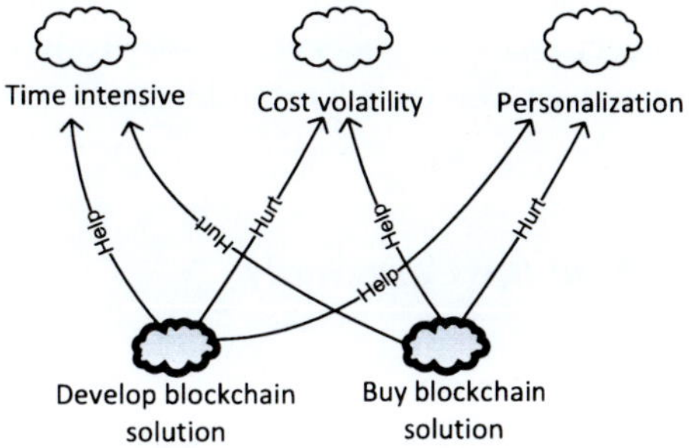

is meant to what extent the blockchain type implications in terms of decentralization, transparency and privacy will be accepted among the stakeholders and how well it will suit the enterprise configuration. In other terms, to what extent it is possible to adopt that particular type of blockchain given the requirements of the internal and external environment of the enterprise aiming to adopt blockchain. In the proposed model it is shown that public blockchain would contribute negatively (i.e., hurt or break) to the feasibility of the project. In fact, the complete decentralization and transparency of a public blockchain implies, particularly in a business context, privacy issues that may repel the stakeholders, who possibly have conflicting interests and confidentiality conditions, from blockchain adoption and hold them off collaborating and integrating their data in the blockchain.

After selecting the blockchain type, the enterprise needs to decide between internally developing the blockchain from scratch or opting for an off-the-shelf solution. Figure 38.5 depicts the adoption strategy using the main non-functional requirements considered when making such a decision.

Choosing a blockchain as a service strategy saves time (Interviewee 1, first interview, 2019) hence the negative contribution of the buy option to the non-functional requirement "time intensive". However, the fee paid to the provider may vary (Interviewee 2, third interview, 2019) which explains the positive contribution of the buy alternative to the "cost volatility". Also, the blockchain solution providers like IBM and Microsoft, deliver more or less the same service to their clients which contributes negatively to the non-functional requirement "personalization". Developing the blockchain solution from scratch by the IT department of the company will lead to a more tailored application. Interviewee 1 even said that to ensure a competitive advantage with the blockchain it has to be custom-developed by TheBestFood's IT department (Interviewee 1, first interview, 2019).

Case Study

This section presents how a blockchain is adopted for SCM through the case study of TheBestFood Belgium. Three interviews were conducted with TheBestFood's innovation officer and blockchain project responsible Interviewee 1. The two first interviews were conducted respectively in January and April and the third and last interview was conducted in July.

About the Case Study

Since its start-up, TheBestFood has been a pioneer in Information Technology (IT) adoption. At TheBestFood, business actors and IT developers have always worked hand in hand from the early stages of IS development. In fact, since the rise of digitization, TheBestFood started to define its business strategy around IT producing an "unusually well-aligned IS strategy". All of the business decisions have always been made at the light of the capabilities and restrictions arising from IT. The big retailer was the first to adopt an end-user-driven IS. In other words, users were always involved along with the IT specialists in the requirement engineering and design phases and not only during the implementation phase. TheBestFood started developing its IS with the software engineering practices as we know them today 20 years before they were described by researchers and IS professionals as the best techniques. TheBestFood has been seizing IT advancements opportunities to enhance its business strategy since its creation. The paper also forecasted that the food retailer will continue to take advantage of the latest IT findings to develop a high alignment between the business needs and its supporting IT.

TheBestFood's Bananas Supply Chain: A High Potential Use Case

It appears that TheBestFood continues its momentum by introducing blockchain technology for the banana's SCM. In this context, blockchain is one of the most prominent technologies to increase cooperation between supply chain members and enhance customer trust [26]. However, in Interviewee 1's words, "*Introducing blockchain for only proving to customers food provenance does not provide a satisfying return on investment*". In fact, the business case behind employing blockchain for TheBestFood's SCM is twofold. First, according to Interviewee 1, "*conflict of interest between stakeholders is among what makes SCM particularly sensitive*". However, by ensuring transparency, blockchain discourages opportunistic behaviors among supply chain actors which reduces relational risks [20]. Hence, one of TheBestFood's main expectations behind introducing blockchain is ensuring a sustainable relationship between all the supply chain actors. Second, TheBestFood's annual bananas waste is significant (Interviewee 1, first interview, 2019). The big retailer aims to unveil the reasons behind this inefficiency in order to prevent waste and get the product in the right condition, in a timely manner and at the lowest possible costs. "*Our biggest argument behind introducing blockchain is to gather insights from every single node involved in the process, to find out why bananas are thrown away*" (Interviewee 1, 2019). Hence, besides the obvious value of ensuring traceability for customers, what makes blockchain employment in bananas supply chain a high potential use case for TheBestFood lies in enhancing trust between all the supply chain actors and decreasing bananas waste. This aligns with Mao et al. [22] claiming that "*consumers are not the main beneficiaries of food traceability systems*".

Selecting a Blockchain Type

Among the actors of the supply chain there are competing businesses which explains the reluctance towards a complete decentralization that involves complete transparency implying privacy issues. "Private channels" [21] are then needed to preserve confidentiality of some transactions and privacy for the parties involved. However, the aim behind this project is to take maximum advantage of the decentralization of blockchain that implies untampered and verified data thanks to complete transparency. Hence, a hybrid blockchain seems to be the most feasible solution in order to maintain good relationship with the stakeholders and get their necessary agreement to collaborate (Interviewee 1, first interview, 2019). Add to that, the use of IoT, like sensors, to collect information concerning bananas temperature, age, etc. requires "*the improved performance, user privacy and data control of a permissioned blockchain*" [21]. The access to and manipulation of data on the blockchain will be bounded to some conditions and requirements defined by a limited number of key actors of the supply chain, namely the consortium. In this context, the representation in Figure 38.4 can be used to evaluate the different alternatives and decide which type of blockchain to adopt.

Organization of TheBestFood's Stakeholders

The different roles of the supply chain nodes are:

- Plantation: harvest the bananas from the trees;
- Packing station: collect the harvested bananas and put them on pallets;
- Harbor: this is where the bananas are stored until all clearance documents are accepted and product shipment is authorized;
- Ripener: responsible for accelerating the ripening process of bananas. It consists in spreading the right amount of ethylene gas to obtain the right color of bananas;
- Carrier: group of individuals responsible for bananas transport from one point to another.

According to the Sustainable Food Trade Association, the sustainable development of food trade involves the participation of a diverse mix of organic producers, processors, manufacturers, distributors and retailers, as well as related vendors, suppliers, aligned organizations and individuals. To ensure this business to business integration, the solution analyst at TheBestFood is collecting the necessary data to integrate in the blockchain from all the IS of the nodes involved in the bananas supply chain. In fact, blockchain answers trust requirements and offers time stamping functionalities but the decentralized technology itself is not a solution for gathering and standardizing data held by the different stakeholders. In Interviewee 1's words, "getting everyone on board" is the biggest challenge faced in this project (Interviewee 1, 2019). This task is even more critical given the length of the bananas supply chain that involves five stakeholders, namely plantation, packing station, harbor, ship and ripener, whose cooperation is needed for the efficiency of the new blockchain system. Figure 38.6 depicts the organizational setting of bananas supply chain through the different stakeholders involved in TheBestFood's bananas supply chain and their dependencies. For confidentiality matters, TheBestFood's partners company's names and locations are not disclosed. As explained in the theoretical background, an i* SD diagram is suited to visualize all the dependencies between the different stakeholders allowing network governance. At the strategic level, the bananas supply chain involves seven main actors depending on each other to get the bananas on TheBestFood's shelves.

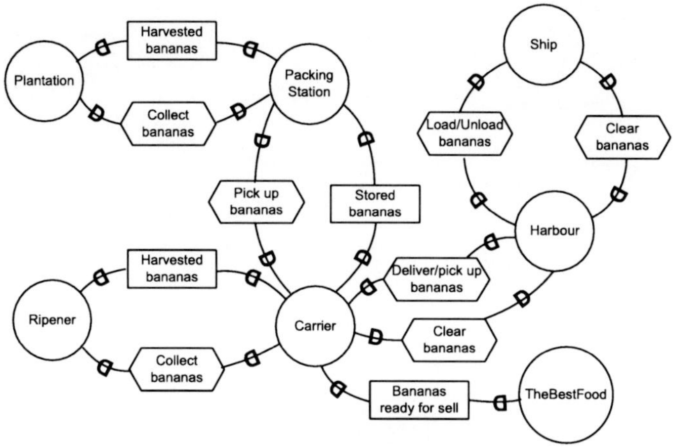

Fig. 38.6 Organizational setting of bananas supply chain

*Modeling the New Blockchain-Based Information System with I**

In this section, the organizational requirements of blockchain adoption for TheBestFood's bananas supply chain are modeled using i*. TheBestFood's blockchain project for bananas is divided into three phases namely, the *descriptive*, *predictive* and *prescriptive* phases (Interviewee 1, 2019). An i* representation, as well as the major milestones, will be presented for each of the three phases. These three phases have not been mentioned in the studied blockchain literature. It is important to specify that all blockchain projects do not necessarily go through the three phases (Interviewee 1, 2019). In fact, it depends on the use case and the defined business case of the project. For instance, if the aim behind blockchain adoption consists only in proving a product's provenance, the project is completed after the descriptive phase (Interviewee 1, 2019).

Descriptive Phase The descriptive phase consists in collecting data concerning the bananas from the major nodes of the bananas supply chain network namely, the plantation, packing station, harbor, ship and ripener. TheBestFood had to get his partners to agree on collaborating by sharing and putting their data on the blockchain. This constitutes the first challenge encountered (Interviewee 1, 2019).

The product attributes to be integrated in the blockchain, like bananas' age, temperature, etc. are collected from IoT devices (e.g., sensors) integrated in the banana's containers. The device data is directly added to the blockchain.

Additionally, TheBestFood needs to decide which data is required from the stakeholders and constitute an input for the blockchain. At the descriptive phase, it is referred to the collected data from the supply chain nodes as "attributes". "Product feedback" concerns the state of the bananas sold to the customers. Product feedback can come from TheBestFood's quality assessors and customers. This phase is quite

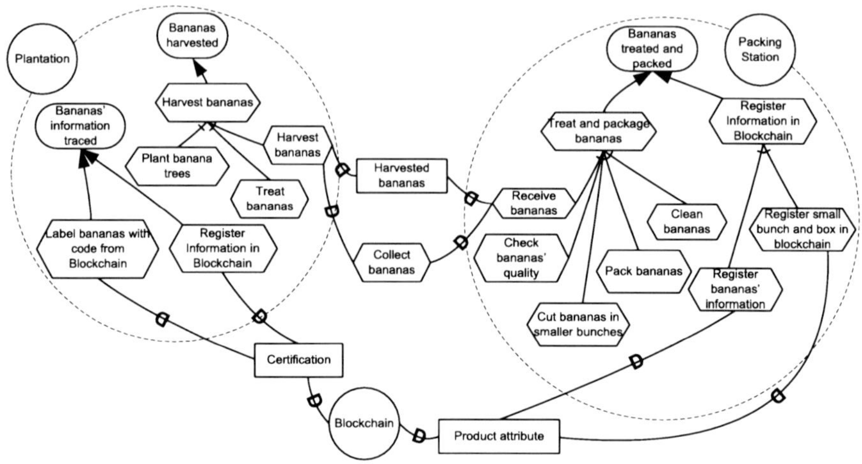

Fig. 38.7 Partners' intentional analysis in descriptive phase

explorative in the sense that the collected data is a list of attributes that can potentially, but not necessarily, impact bananas' quality. After collecting attributes related to a significant volume of bananas, business intelligence solutions have been put in practice in a second phase. Currently, these are some of the data inputs, related to a bananas case, that are being collected from the partners to be integrated in TheBestFood's blockchain: weight, geotagging, box code, identification of organization, temperature, production data, quality report, quality control, humidity levels, ripening schedule, stock data, waste data, price data, etc. Among those attributes some are needed to be furnished by all the nodes of the network and some are specific to particular actors given their role in the chain.

The use of requirements and business modeling languages such as i* and BPMN is beneficial for this phase. It helps in communicating and understanding the end-to-end blockchain application for involved partners, and in deciding which data are required and shared to blockchain. In addition, it allows each partner to reengineer their business processes when adopting blockchain.

Figures 38.7, 38.8 and 38.9 represent the descriptive phase outcome. Figure 38.7 shows an SR diagram for analyzing tasks involved for each partner. Figure 38.8 shows the to-be business process when using blockchain. Figure 38.9 is a SD diagram; it depicts dependencies amongst partners and the end-to-end of the blockchain application. In the latter, the blockchain is represented as an actor and the data needed to be integrated in the blockchain is represented as a resource furnished by the different stakeholders (i.e., actors) of the supply chain. For instance, the blockchain depends on the ripener to furnish product attribute data. Besides the common data he/she has to furnish (e.g., identification of organization, humidity levels, weight, stock data, waste data, price data, etc.), the ripener furnishes data related to its specific role namely, the ripening schedule.

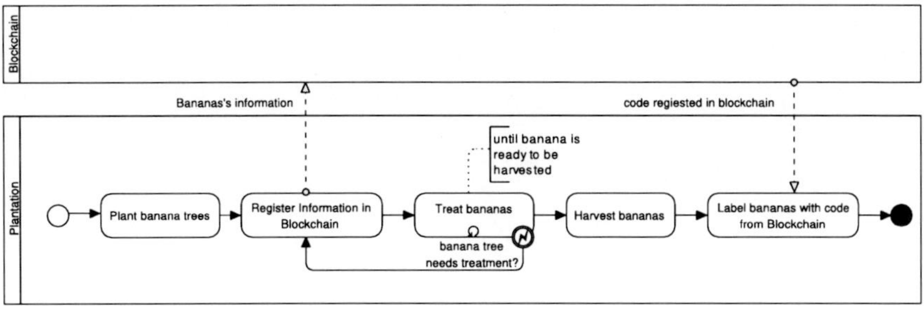

Fig. 38.8 To-be business process in descriptive phase

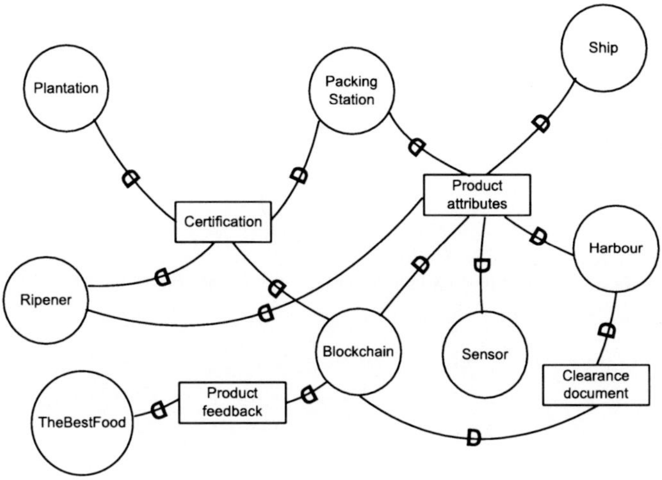

Fig. 38.9 Configuration of the blockchain-based system at descriptive phase

It is possible to have a different configuration of the new system by using smart contracts from the descriptive phase. However, its function will not consist in binding the contract parties to respect the transaction conditions. At this stage the smart contract can play the role of an intermediate container of data (Interviewee 2, 2019). In fact, instead of integrating the collected data directly to the blockchain, that is the main container, data will first be put on the smart contract that will force its sequential organization (Interviewee 2, 2019). This is depicted in Fig. 38.10.

Predictive Phase The data scientist uses the data inputs in the blockchain from the first phase to perform descriptive analytics. The aim is to establish a pattern for bananas' quality. In other words, the data scientist will find a link between the "product attributes" and the actual quality of bananas that arrive at TheBestFood meaning the "product feedback". The main question to be answered is: *what combination of attributes is causing this obtained condition of bananas (i.e. wrong color, ripeness, etc.)?*

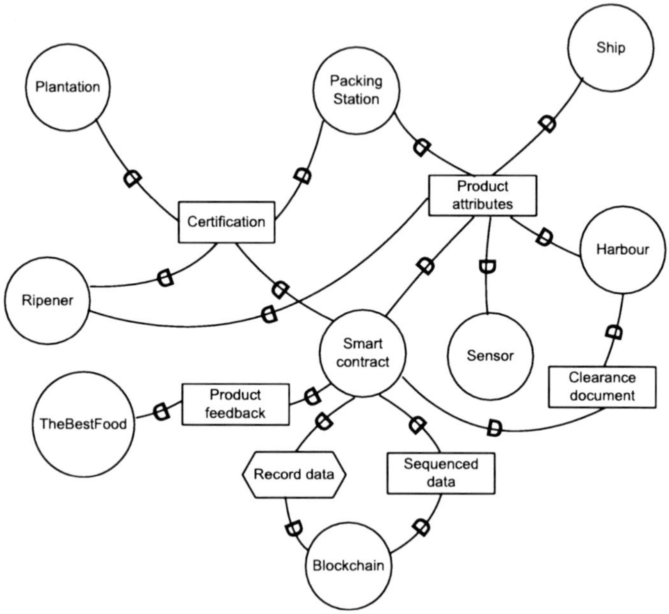

Fig. 38.10 Configuration of the blockchain-based system at descriptive phase: an alternative option using smart contracts

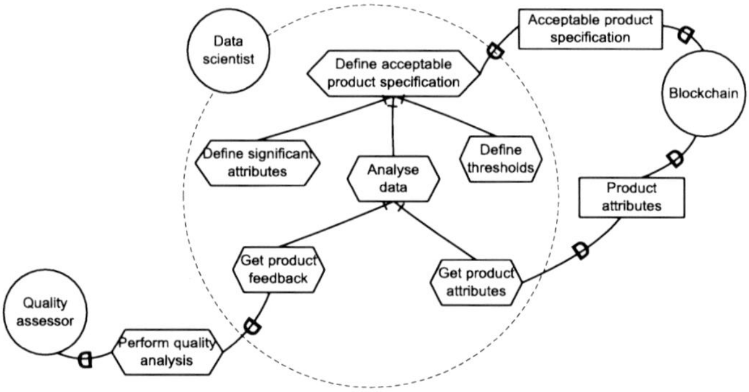

Fig. 38.11 A strategic rationale model for acceptable product specification definition

At the end of this phase, TheBestFood has identified the significant product attributes to be collected from the partners and the threshold for each attribute that need to be strictly respected by all the nodes of the supply chain. The output of this phase is "acceptable product specification" that is the data constituted of the significant attributes and the threshold for each significant attribute. The SR diagram of Fig. 38.11 shows how TheBestFood's data scientist obtains the "acceptable product specification" to be integrated in the blockchain.

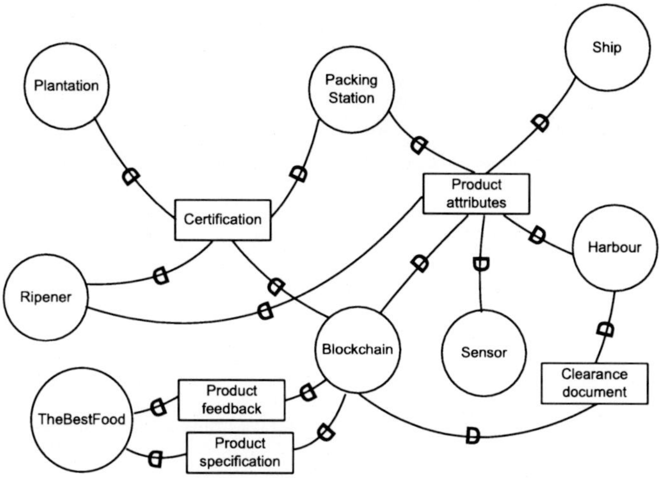

Fig. 38.12 Configuration of the new system at predictive phase

The obtained data will be integrated to the blockchain in order to verify at each node that the product specifications are strictly respected. This will allow to reduce TheBestFood's bananas waste by detecting on time if bananas will not reach TheBestFood's shelves in the required condition. Hence, the bananas that do not match the product specifications inserted in the blockchain will not continue their journey to TheBestFood where they would have been thrown away. Instead, as soon as a specification is not ticked, the bananas are sent back. This will not only reduce waste but also unnecessary costs related to shipment, storage, etc. The output of this second phase is sketched in Fig. 38.12. Configuration of the new system at predictive phase.

Prescriptive Phase At the prescriptive phase, the smart contract is finally used to bind the supply chain nodes to respect the bananas acceptable product specification of TheBestFood. The aim is to oblige the stakeholders to respect the product specifications and hence to ensure that no unit of the product leaves a node if it does not fit all product specifications on the blockchain. This prescriptive character of transactions between the supply chain actors are possible thanks to the smart contract. In Fig. 38.13, smart contracts have two possible execution modes namely, "eager execution" and "lazy execution" [19]. For confidentiality reasons we kept the representation limited at this stage. With the eager execution mode, the transactions are automatically initiated through the code of the smart contract. With the lazy execution mode, transactions are only initiated by one of the parties of the transaction. In the SD diagram depicted of Fig. 38.13, the abstraction "contract party" represents the nodes of the bananas' supply chain.

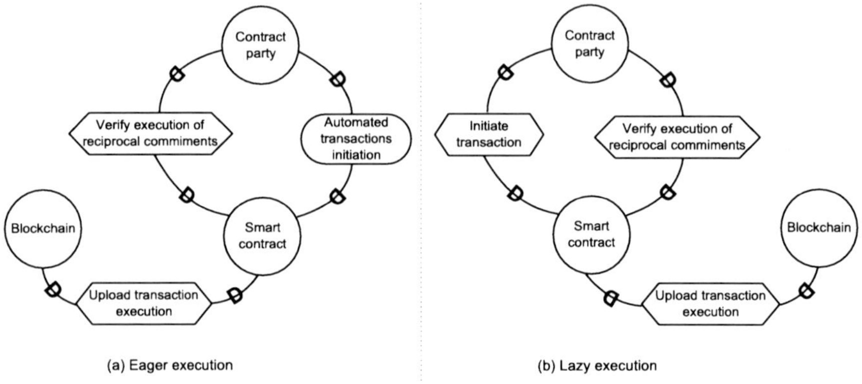

Fig. 38.13 Configuration of blockchain system with two different smart contract execution modes

Discussion

As shown in the previous section, the i* framework is suited for an early analysis of the blockchain actors and their relationships/dependencies. One of the elements highlighted as important in the blockchain context but where the framework was seen as limited is its ability to represent privacy concerns. Privacy is, in this context, a much more important concern than security.

Refinements could thus be proposed to the i* **Resource** elements to model privacy concerns accurately. More specifically we distinguish the need to represent the following characteristics:

- **Access Control**: the access to the bloc content is restricted to a certain number of nodes and/or, to nodes that check specific conditions. Access to data can be restricted due to, for example, confidentiality or anonymity matters;
- **Confidentiality**: the owner of the data prefers to hide the data content. Hence, the access can be restricted.

We also identify at the level of the i* **Actor** element, the need to represent the following characteristics:

- **Privacy accountability**: if the data that obeys to some privacy requirements is manipulated by a third party, the latter is accountable to respect those requirements. This third party has to be monitored to ensure the respect of the role-based restrictions;
- **Anonymity**. The actor who delivered the data may wish to partially or completely hide his personal information to the rest of the network.

Refinements to the meta-model of i* and a formal specification of these refinements is nevertheless left for future work; the Computer-Aided Software Engineering (CASE) Tool Descartes Architect [18] will also be modified to include those refinements.

Conclusion

This paper is an effort to overcome the hype over blockchain technology by presenting its concrete opportunities through gathering information based on a study case and literature review, tackling specific challenges stemming from the complexity of the distributed ledger and proposing a concrete modeling-based solution. The main contribution of this paper is the application of the i* framework to blockchain for organizational representation to better deal with BOSE. The aim behind the proposed representations is indeed twofold. On one hand a proposition to easily understand the technology functioning and characteristics. On another hand, it is an attempt to model the organizational requirements in a blockchain project that comes with governance and cooperation challenges and to grasp the configuration of a system that uses the distributed ledger to manage and access data.

The i* representations based on TheBestFood case study, depicted the organizational requirements in a blockchain project and showed the different stages of the project, the configuration of the system when deploying the distributed ledger. The level of detail of these models can be criticized as low. This is due to the exploratory character of the research. Privacy issues constitute the main reason behind the reluctance to engage in blockchain adoption. The proposed extension of the i* framework, is a non-extensive first attempt to model privacy requirements in a blockchain project starting at its early stages. For the conceptual representations to be more complete, accurate and practical further research needs to be conducted focusing on analyzing and modeling privacy implications of blockchain from an organizational point of view. This could be a possible next step for the software development research community who is getting increasingly involved in the rather ad-hoc and informal world of BOSE.

References

1. G. Blossey, J. Eisenhardt, G. Hahn, Blockchain technology in supply chain management: An application perspective, in *Proceedings of the 52nd Hawaii International Conference on System Sciences* (2019)
2. B. Carminati, Merkle trees, in *Encyclopedia of Database Systems*, ed. by L. Liy, M.T. ÖZsu (Springer US, 2009), pp. 1714–1715
3. P. Chakraborty, R. Shahriyar, A. Iqbal, A. Bosu, Understanding the software development practices of blockchain projects: A survey, in *Proceedings of the 12th ACM/IEEE International Symposium on Empirical Software Engineering and Measurement*, pp. 1–10 (2018)
4. K. Charmaz *Constructing Grounded Theory: A Practical Guide Through Qualitative Analysis* (Sage, 2006)
5. L. Chung, B.A. Nixon, E. Yu, J. Mylopoulos, *Non-Functional Requirements in Software Engineering, International Series in Software Engineering*, vol. 5 (Springer, Berlin, 2000). https://doi.org/10.1007/978-1-4615-5269-7
6. M. Dabbagh, M. Sookhak, N.S. Safa, The evolution of blockchain: A bibliometric study. IEEE Access **7**, 19212–19221 (2019)

7. G. Destefanis, M. Marchesi, M. Ortu, R. Tonelli, A. Bracciali, R. Hierons, Smart contracts vulnerabilities: A call for blockchain software engineering? in *2018 International Workshop on Blockchain Oriented Software Engineering (IWBOSE)* (IEEE, New York, 2018), pp. 19–25
8. F. Duchmann, A. Koschmider, Validation of smart contracts using process mining. ZEUS. CEUR Workshop Proc. **2339**, 13–16 (2019)
9. S. Eric, P. Giorgini, N. Maiden, J. Mylopoulos *Social Modeling for Requirements Engineering* (MIT Press, 2011)
10. P. Forte, D. Romano, G. Schmid, Beyond bitcoin-part I: A critical look at blockchain-based systems. IACR Cryptol. ePrint Archive **2015**, 1164 (2015)
11. M. Garriga, M. Arias, A. De Renzis, Blockchain and cryptocurrency: A comparative framework of the main architectural drivers. arXiv preprint arXiv:1812.08806 (2018)
12. H. Halpin, M. Piekarska, Introduction to security and privacy on the blockchain, in *2017 IEEE European Symposium on Security and Privacy Workshops (EuroS&PW)* (IEEE, New York, 2017), pp. 1–3
13. Y. He, H. Li, X. Cheng, Y. Liu, C. Yang, L. Sun, A blockchain based truthful incentive mechanism for distributed p2p applications. IEEE Access **6**, 27324–27335 (2018)
14. Z. Hintzman, Comparing blockchain implementations-NCTA technical papers, in *2017 Fall Technical Forum SCTE-ISBE* (NCTA, CABLELABS, 2017)
15. M. Kolp, Y. Wautelet, S. Faulkner, Sociocentric design of multi-agent architectures, in *Social Modeling for Requirements Engineering*, ed. by Yu, E., Giorgini, P., Maiden, N., Mylopoulos, J. (MIT Press, 2011)
16. M. Kolp, S. Faulkner, Y. Wautelet, Social structure based design patterns for agent-oriented software engineering. IJIIT **4**(2), 1–23 (2008). https://doi.org/10.4018/jiit.2008040101
17. M. Kolp, Y. Wautelet, Human organizational patterns applied to collaborative learning software systems. Comput. Human Behavior **51**, 742–751 (2015)
18. M. Kolp, Y. Wautelet, DesCARTES architect: design CASE tool for agent-oriented repositories, techniques, environments and systems. Louvain School of Management, Université catholique de Louvain, Louvain-la-Neuve, Belgium. Descartes Case Tool (2020). https://www.isys.ucl.ac.be/descartes/
19. J. de Kruijff, H. Weigand, An introduction to commitment based smart contracts using reaction ruleML, in *VMBO*, pp. 149–157 (2018)
20. N. Kshetri, 1 blockchain's roles in meeting key supply chain management objectives. Int. J. Inf. Manage. **39**, 80–89 (2018)
21. I. Makhdoom, M. Abolhasan, H. Abbas, W. Ni, Blockchain's adoption in IoT: The challenges, and a way forward. J. Network Comput. Appl. **125**, 251–279 (2019)
22. D. Mao, Z. Hao, F. Wang, H. Li, Innovative blockchain-based approach for sustainable and credible environment in food trade: A case study in Shandong Province. China. Sustainability **10**(9), 3149 (2018)
23. R.C. Merkle, A digital signature based on a conventional encryption function, in *Advances in Cryptology - CRYPTO '87*, ed. by C. Pomerance (Springer, Berlin, 1988), pp. 369–378
24. OMG: Business process model and notation (BPMN). Version 2.0.1. Tech. rep., Object Management Group (2013)
25. M. Pilkington, Blockchain technology: Principles and applications, in *Research Handbook on Digital Transformations* (Edward Elgar Publishing, 2016)
26. M.M. Queiroz, S.F. Wamba, Blockchain adoption challenges in supply chain: An empirical investigation of the main drivers in India and the USA. Int. J. Inf. Manage. **46**, 70–82 (2019)
27. F. Renkes, C. Sommer, Blockchain in the context of business applications and enterprise databases. BTW **2019**, (2019)
28. J. van Rossum, The blockchain and its potential for science and academic publishing. Inf. Services & Use **38**(1–2), 95–98 (2018)
29. N. Szabo, Smart contracts: Building blocks for digital markets, extropy. Transhumanist Thought (1996)
30. D. Tapscott, A. Tapscott, Realizing the potential of blockchain. A multistakeholder approach to the stewardship of blockchain and cryptocurrencies, in *World Economic Forum* (2017)

31. M. Vukolić, Rethinking permissioned blockchains, in *Proceedings of the ACM Workshop on Blockchain, Cryptocurrencies and Contracts*, pp. 3–7 (2017)
32. Y. Wautelet, Representing, modeling and engineering a collaborative supply chain management platform. Int. J. Inf. Syst. Supply Chain Manage. (IJISSCM) **5**(3), 1–23 (2012)
33. Y. Wautelet, A model-driven IT governance process based on the strategic impact evaluation of services. J. Syst. Softw. **149**, 462–475 (2019). https://doi.org/10.1016/j.jss.2018.12.024
34. Y. Wautelet, S. Heng, M. Kolp, L. Penserini, S. Poelmans, Designing an MOOC as an agent-platform aggregating heterogeneous virtual learning environments. Behaviour & Inf. Technol. **35**(11), 980–997 (2016)
35. Y. Wautelet, M. Kolp, S. Heng, S. Poelmans, Developing a multi-agent platform supporting patient hospital stays following a socio-technical approach: Management and governance benefits. Telematics Informatics **35**(4), 854–882 (2018)
36. Y. Wautelet, M. Kolp, L. Penserini, Service-driven iterative software project management with I-Tropos. J. UCS **24**(7), 975–1011 (2018)
37. G. Wood et al., Ethereum: A secure decentralised generalised transaction ledger. Ethereum Project Yellow Paper **151**(2014), 1–32 (2014)

Chapter 39
Self-Service Data Science for Adverse Event Prediction in Electronic Healthcare Records

Marco Spruit and Niels de Vries

Abstract Healthcare is a data intensive industry in which data mining has a great potential for improving the wellbeing of patients. However, a multitude of barriers impedes the application of machine learning. This work focuses on medical adverse event prediction by domain experts. In this research we present AutoCrisp as a self-service data science prototype for multivariate sequential classification on electronic healthcare records to facilitate self-service data science by domain experts, without requiring any sophisticated data mining knowledge. We performed an empirical case study with the objective to predict bleedings with the use of AutoCrisp. Our results show that multivariate sequential classification for medical adverse event prediction can indeed be made accessible to healthcare professionals by providing appropriate tooling support.

Keywords Clinical decision support system · Self-service data science · Data mining · CRISP-DM · Multivariate sequential classification · Adverse event prediction

Introduction: Problem Investigation

Never in time has so much data been collected and analyzed as we do nowadays. With the advent of the digitalization revolution, millions of patient records containing medical data exist today in large data warehouses. Whereas formerly only a select group were in a position to collect large quantities of data, we have now reached a situation in which almost all medical researchers have access to technologies which enables them to generate large datasets [1]. The amount of healthcare

M. Spruit (✉) · N. de Vries
Department of Information and Computing Sciences, Utrecht University, Princetonplein 5, 3584 CC Utrecht, The Netherlands
e-mail: m.r.spruit@uu.nl

M. Spruit
Department of Public Health and Primary Care, Leiden University Medical Centre, Albinusdreef 2, 2333 ZA Leiden, The Netherlands

A. Visvizi et al. (eds.), *Research and Innovation Forum 2020*,
Springer Proceedings in Complexity,
https://doi.org/10.1007/978-3-030-62066-0_39

data is overwhelming and medical researchers are increasingly drowning in their own data. Data forms the fundamental base of knowledge, however, deriving usable and relevant knowledge from this data remains a major challenge in healthcare [2]. Different studies have already proven the great potential of data mining in healthcare [3, 4,5]. This includes detecting diseases at earlier stages more easily and effectively, managing specific individual and population health, and detecting healthcare fraud more quickly and effectively [6]. Proper data analytics could result in saving lives. Despite this great potential, Neff claims that in general the application of data analytics in the healthcare industry is lagging behind [7]. Omta points out this is due to the lack of data mining tools and methods available for non-experts [1, 8]. As indicated in [9], the best data scientists have skills from across a range of fields including mathematics, computer science and economics, which are not the foremost set of skills you typically find in medical experts. This fact combined with the lack of methods and tools provides an explanation as to why medical researchers are not making full use of the data available to them.

Technological advancements in collection and storage of healthcare data have resulted in data warehouses, containing complex temporal datasets, characterized by time series of multiple variables. These temporal datasets are also known as electronic health record (EHR) data and encompass lab values, medication records, ECG data and more [10]. Previous researches have demonstrated their potential for medical event detection and prediction [11–13]. Prediction of medical events could reduce costs, improve quality care and save lives, which makes it a highly relevant topic. Still, due to the dynamic characteristic of EHR, building classification models capable for this task belongs to the most challenging topics in data mining research [14]. Its applicability covers a broad spectrum in healthcare, ranging from early predicting of cardiovascular complications to neurologic complications in the intensive care unit [15]. A shortage of data scientists together with their lack of medical knowledge leads to great challenges. Tools such as WEKA have been developed for making machine learning techniques available to non-experts which have been used in varying fields, proving that machine learning techniques can be made available for use by non-experts [16]. EHR data for event prediction can be summarized as temporal, heterogeneous, sparsely sampled data, which unfortunately is not suitable for processing by WEKA. In addition, WEKA focuses specifically on the modeling phase, whereas we believe that phases prior to modeling (e.g. data understanding and preparation) are equally crucial for the success of data mining projects in healthcare. Finally, we note the increasingly better performing automated machine learning techniques to support self-service data science, even in healthcare [17, 18].

In this study, our objective is to develop a self-service data science prototype that allows healthcare professionals to perform a full data mining project aimed at adverse event prediction by analysing structured EHR data. By following the design science cycle [19], we first investigate the problem thoroughly in the problem investigation phase, followed by the treatment design phase in which we develop our artifact, and lastly, the treatment validation phase in which we perform an extensive case study in the Cardiology department of a Dutch academic hospital to evaluate the artifact.

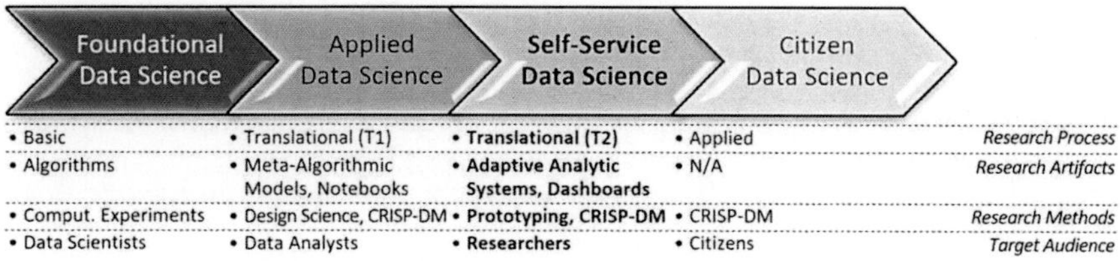

Fig. 39.1 Self-service data science research in context of related data science disciplines

We define Self-Service Data Science by integrating the concepts of Applied Data Science, Meta-Algorithmic Modelling and Self-Service Capability as follows [20, 21, 22, 23]: "*Self-service data science is the knowledge discovery process in which analytic systems are designed and evaluated to empower domain professionals to perform their own data analyses on their own data sources without coding in a reliable, usable and transparent manner within their own daily practices*". Figure 39.1 visualizes its relationships with adjacent data science disciplines, together with their distinctive attributes. One highlight is its recognition of the adjacent data science disciplines of Applied Data Science and Self-Service Data Science as the two main stages (T1 and T2) as defined in Translational Data Science [24].

Problem Statement

There is a growing body of literature that recognizes the importance of event prediction in healthcare [6]. To avoid ambiguity, we define an event prediction project in technical terms as *multivariate sequential classification* (MSC), which refers to the classification task on multivariate time series of multiple instances. To illustrate such a classification task, assume the following. Let $D = \{x_i, y_i\}$ denote an EHR dataset such that $x_i \in X$ is a multivariate temporal variable up to some timepoint t_i for patient p_i and $y_i \in Y$ is the class label associated with x_i at timepoint t_i for patient p_i. In MSC our goal is to correctly classify class label y_i that denotes whether an event (e.g. bleeding) is present for a p_i at t_i or ultimately whether an event is likely to occur for p_i at t_i.

Figure 39.2 presents an example of simulated EHR data, that shows that EHR time series can be distinguished from other time series due to the following characteristics:

1. *Multivariate*: multiple time series variables measured per patient.
2. *Heterogeneous*: the type of data varies per variable, ranging from numeric values for lab values, to categorical values (diseases) up to time durations (medication).
3. *Irregular in time*: time intervals vary both per variable as per patient.
4. *Sparse*: different treatment tracks per patient resulting in missing/unknown values.

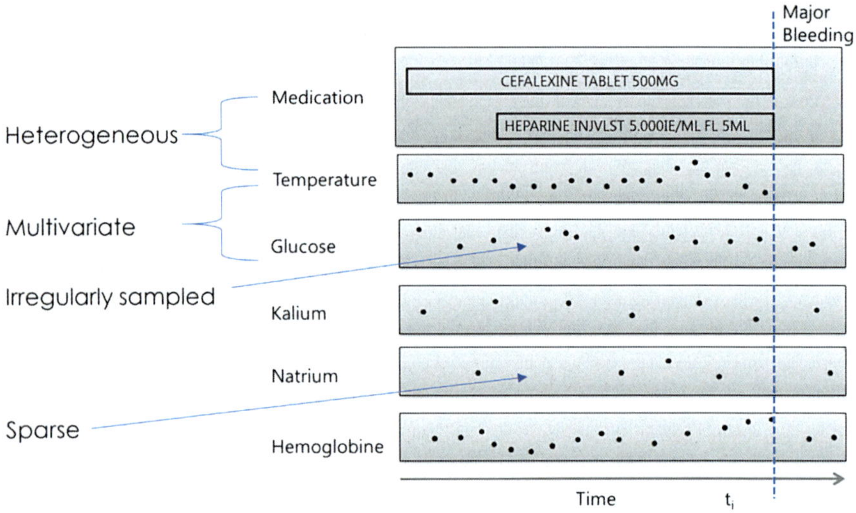

Fig. 39.2 Simulated EHR data illustration

Our objective in this study is to make multivariate sequential classification accessible to domain experts, without the need for coding. Making machine learning available to domain experts has already been proved by the development of tools such as WEKA and Orange [16, 25]. However, due to the previously mentioned characteristics of healthcare data, these tools are not capable of processing raw EHR and performing MSC. Therefore, we address the following problem in this research: *Healthcare professionals including medical researchers do not have the resources for exploring and preparing healthcare data, performing multivariate sequential classification and evaluating these models for the purpose of predicting adverse events.*

Research Approach

The cross-industry standard process for data-mining (CRISP-DM) is one of most commonly used process models for the practice of data mining [26]. The method consists of one iterative life cycle that is broken down into six phases; business understanding, data understanding, data preparation, modeling, evaluation and deployment. In the previous section we have demonstrated that EHR brings certain challenges when performing MSC. In our opinion the core problem that is caused by these challenges is related to technical phases of the CRISP-DM, namely; data understanding, data preparation and modeling. To overcome these barriers, we followed a two-step approach to develop our analytic system prototype, which integrates CRISP-DM with the Design Science Research (DSR) approach as described in [22]. Firstly, all tasks specified within the selected phases of the CRISP-DM were evaluated and modifications were made. Secondly, the AutoCrisp tool was developed to support healthcare professionals in executing the phases without requiring sophisticated data mining knowledge and coding skills. Our tool attempts to automate the most challenging CRISP-DM tasks and is therefore called AutoCrisp. The source code based on R Shiny is available on GitHub: https://github.com/marcospruit/AutoCrisp.

Methods: Treatment Design

In this section the components that have been developed are described and the rationale as to why they were included. AutoCrisp is chronologically built up following the selected phases of CRISP-DM.

Data Understanding

According to the CRISP-DM, the data understanding phase has a number of goals: becoming familiar with the data, verifying the data quality and gaining initial insights. The data understanding phase is split up into four tasks; collecting data, describing data, exploring data and lastly, verifying the data quality. The data to be collected is dependent upon each user and the data gathering process itself. In this study we assume that users have exported non-cleaned but structured tables from healthcare databases. The second step consists of describing the data followed by the third step of exploring this data. CRISP-DM suggests ending the data understanding phase by verifying the data quality. From our point of view valuable insights into the data can only be obtained if the data quality is sufficient. Therefore we suggest a small modification to the process steps in which the data quality verification task takes place before exploring the data. The data understanding phase is supported in AutoCrisp by two main tabs, in which the first tab allows users to insert and specify their datasets (data collection task). The second tab contains three pages which supports the last three data understanding tasks of describing data, exploring data and verifying data quality by automatically creating three reports.

Data collection. EHR data captures medical profiles of patients, containing heterogeneous data originating from multiple sources. According to [27], this heterogeneity forms one of the challenges in data processing. To overcome this barrier, users can upload four different types of datasets into AutoCrisp, as we distinguish medical data in time point data, time interval data, baseline data and event data. Time point data contains data about a dynamic variable at one time point (e.g. temperature measurement), whereas time interval data contains data about a dynamic variable during a specific period of time (e.g. medicine X from March to June). Baseline data concerns data about a static variable where there is no temporal aspect included (e.g. gender).

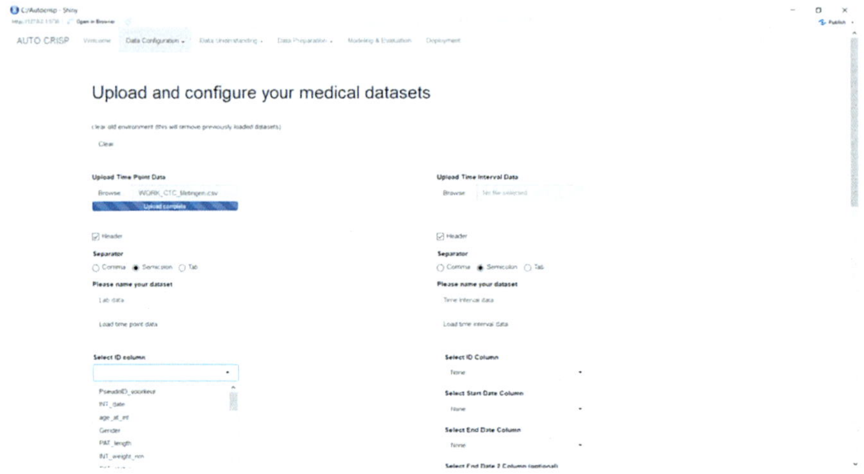

Fig. 39.3 User interface of AutoCrisp for configuring the data sources

Lastly, event data is time point data which contains the events to be predicted, also known as class labels (e.g. mortality). AutoCrisp assumes datasets are structured in a long format and requires users to specify their columns such PatientID and date. Figure 39.3 shows the AutoCrisp user interface, which allows users to insert their EHR datasets.

Data description. The second task concerns the development of a data description report. Based on the user-specified and loaded data, AutoCrisp creates a basic data description report containing statistics such as the volume of datasets (amount of time points, time intervals, patients, complications), event information (patients and percentage targeted) and baseline descriptions. The goal of our data description report is allowing users to become familiar with the data they are working with and check whether the data is correctly processed by AutoCrisp, as shown in Fig. 39.4.

Data quality verification. Data quality in healthcare has gained far more attention and is recognized by multiple studies as one of the biggest barriers of data analytics in healthcare [6, 28, 29]. To assess the data quality and pinpoint data components that might require an extra check, AutoCrisp automatically outputs an interactive data quality report. Inconsistency in EHR datasets originating from multiple sources was designated by [30] as one of challenges in data integration and therefore a similarity plot is presented that measures the consistency between datasets [10] describes the case in which patients with the same disease could undergo different examinations. Linegraphs and 3D plots were therefore included to locate all missing data. EHR data originate from healthcare systems with different sources, such as laboratories, pharmacies, doctors and nurses. Modifications in the collection process affect the data and might introduce noise into models. To make these apparent, we developed our own special kind of heatmap that plots the complete lifespan of all patients in

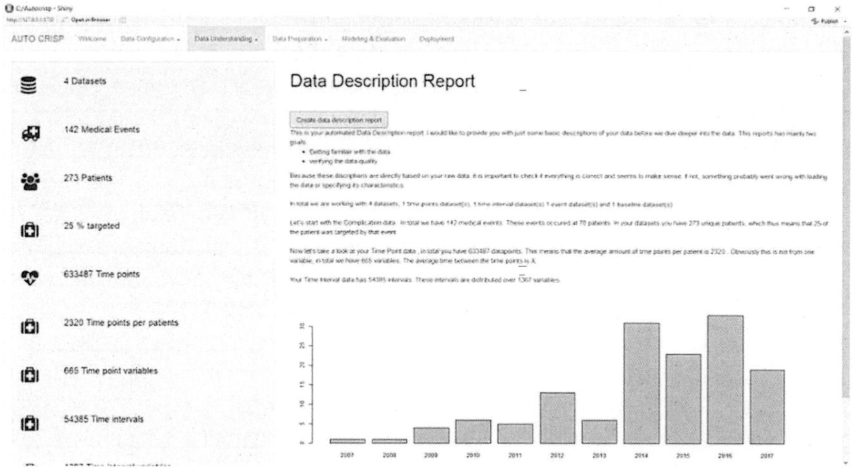

Fig. 39.4 User interface of AutoCrisp for the data description report

the cohort and colors the use of a user selected time interval (e.g. medication) and events.

Data exploration. The data quality report provides information about the quality of the data but does not contain any information about the content of this data. Therefore AutoCrisp outputs a data exploration report that should help users to become more familiar with the data by retrieving first insights into the data, which might even directly address the data mining goals. The first section of the report contains event descriptions, such as Kaplan–Meier curves and descriptive statistics about variables at the day of event (such age, gender, lab values etc.). The disadvantage of these statistics is that they do not address the sequential aspect. In this study we try to support healthcare professionals in event prediction and therefore an interactive line chart is created, that depicts the sequential development of both absolute values as the moving average of the time point variables before an event. The line is surrounded with an area representing either one standard error from the mean or the standard deviation. By creating an interactive interface, users can select multiple time series to hypothesize about relationships between time point variables and the class label. In addition, a correlation matrix is added to support users in discovering relationships between variables.

Data Preparation

As pointed out in the problem investigation chapter, EHR is generally perceived as highly dynamic data, which brings the challenge of preparing the data in a format that enables it to be analyzed. Garbage in results in garbage out. Without high quality input, the desired results will not be achieved and thus proper data preparation is required. Most classification algorithms require a design matrix as input consisting of a fixed set of feature vectors stacked in a feature space. The main objective of data preparation is to create such a feature space that represents the data in such a way that classification algorithms can learn from it in the most effective way and therefore most accurately predict the class labels of a test set.

According to [31] two approaches for data preparation are applicable to MSC. Instance-based classification relates to a dimensionality reduction approach where distance measurements are used between time series. Distance measurements capture information about the shapes and development of time series. To correctly identify these shapes, time series are assumed to be regularly sampled, which is often not the case in healthcare. Therefore we focused in this research on the second approach, feature-based classification, of which a vast number of features are extracted from time series using rolling window frames. In order to construct a design matrix from which a classifier can most effectively learn, based on the CRISP-DM, we determined that the following tasks should be performed; data selection, data cleaning and feature extraction. AutoCrisp supports the data preparation tasks in two pages, wherein the first page allows users to select the data to include and the second page requires users to specify their data imputation method and feature extraction method.

Data selection. The data selection section provides users the possibility to select variables, merge multiple variables and specify the class label. Models can be trained for detection but also for early classification. To do this, the user is required to specify the amount of days before the event which should be considered as the positive label. As data collected closely after an event could potentially include noise into the model AutoCrisp enables users to specify the amount of days after the occurrence of the event that should be discarded from the dataset. As EHR data is irregularly sampled, sliding the window on the regular interval would result in missing a majority of the data. Instead of time dependence, we suggest to slide the window with the occurrence of a new time point. Users can select on which time point variable they would like to create window frames.

Data cleaning. Once the task of data selection have been completed, the data cleaning can be performed. Four data imputation methods are included for handling missing data by replacing them with substituted values, namely Mean replacement, MICE, Amelia and missForest. Mean replacement simply replaces each missing value with the mean of that variable whereas MICE (Multivariate Imputation via Chained Equations) uses regression models to predict the missing value [32]. Amelia [33] and missForest [34] are also both multivariate imputation methods where Amelia uses a bootstrapping and expectation–maximization algorithm and missForest uses a random forest algorithm to predict missing values.

Feature extraction. Feature extraction concerns the process that transforms high dimensional data into a lower-dimensional feature space through the application of mapping. AutoCrisp provides five methods of performing this task. The simplest method is last values, in which the last three available measurements of a window frame are selected. The second method, proposed by [35], is where the mean, standard deviation, skewness and kurtosis for first and second order features are computed. First order features base their calculation on the actual values of the time series whereas second order features are based on the difference of nearby values. The third method included in AutoCrisp is proposed by [12] in which a fixed set of temporal features for continuous values are used such as last values differences, slopes, nadir, apex and nadir differences. The fourth method, proposed by [36], presents a novel presentation of time series combining trend-based and value-approximations of a time series. A SAX algorithm was used for creating value-approximations whereas a least squares line was fitted to generate a trend-approximations. Lastly, the fifth and most complex method is based on a research of [14] that uses association rule mining to determine the features. This is done via first creating abstractions of all attributes, secondly performing a pattern mining algorithm to mine the most predictive and frequently occurring sequential patterns between all attributes which are then used to create a binary vector for each window stating whether each pattern is apparent or not.

Modeling

The objective of the modeling phase is to develop a model based on prepared data that best reflects the truth. The predictive performance depends on the entered data, the algorithm used and the parameters configuration. In this section we separate the modeling phase into the tasks of feature selection, data sampling and algorithm selection which are implemented in the Modeling page of AutoCrisp.

Feature Selection. Feature selection is the process of identifying and removing redundant features from the dataset that do not contribute to the predictive performance. A simpler model is preferred because it is more rapid and more easily understood and potentially improves the performance as redundant features potentially include noise into models. Feature selection techniques are categorized in filter and wrapper methods. Filter methods use univariate statistical measurements to rank variables on importance whereas wrapper methods iteratively generate subsets by constructing and testing new models. Based on inferences drawn from previous models, it decides to add or remove features. AutoCrisp provides one filter and two wrapper methods for selecting important features. The filter method creates a subset by measuring the variable importance by computing the association between class labels and dependent features. The first wrapper method used is recursive feature elimination (RFE) which is a greedy optimization algorithm where for each iteration, the best or worst performing feature is excluded from the model. It creates a new model with the remaining features and ranks all features in the order of elimination. The second wrapper method implemented in AutoCrisp is developed by [37] called Boruta. Boruta is built around a random forest algorithm and uses randomly created features to determine which features are redundant by excluding the ones that perform worse than the random features.

Data Sampling. EHR datasets in MSC data mining projects are usually very unbalanced; the classes are not equally distributed. To combat this issue some researches argue data sampling needs to be performed to make datasets more balanced and from which classifiers can learn more effectively [38]. AutoCrisp therefore allows users three sampling options; the majority class can be randomly undersampled, the minority class randomly oversampled or running the ROSE algorithm. ROSE is a bootstrapped based oversampling method, which synthetically generates data that provides a better estimate of the original data [39].

Classification algorithm. After constructing a design matrix and optionally performing feature selection and data sampling, AutoCrisp offers a subset of classification algorithms in order to build a predictive model. The classifiers implemented are decision trees, naïve bayes, random forest, support vector machines and logistic regression. AutoCrisp uses a five-fold cross validation as a test design from which models can be evaluated. The default in cross-validation assumes that partitioning is done randomly, so that the train and test design best represent the population. As explained in the previous section, sliding overlapping windows are used as the main structure in our prepared data. Because windows overlap each other, randomly partitioning the data would be incorrect as parts of the training data will also be included in test data. To address this problem, instead of randomly dividing the datasets, AutoCrisp splits the dataset in a sequential order. When users have determined their optimal settings, they can build a final model on the complete dataset. In addition new cohorts can be uploaded and applied to the model.

Results: Treatment Validation

Experimental Setup

In cases of severe heart failures, a left ventricular assist device (LVAD) can act as a solution to increase survival rates and improve quality of life. An LVAD is a mechanical pump that takes over the function of the left heart chamber that is responsible for providing blood. Frequently occurring adverse events are bleedings which could be life risking. In this case study, structured data was inserted in AutoCrisp in order to build a model that is capable of predicting bleedings. Based on expert opinions we defined five relevant settings for event prediction. In this research we will evaluate all techniques for all of the following four settings:

Setting 1: classify whether a bleeding is present at time point t_i.
Setting 2: classify whether a bleeding will occur in the following three days.
Setting 3: classify whether a bleeding will occur in the following week (7 days).
Setting 4: classify whether a bleeding will occur in the following month.

A standard set of the most straightforward approaches were selected for all design choices, which are used as the default. For all design choices, every option is evaluated while using the default settings for all remaining design choices. This entails the following: the first seven days after the event are removed, mean replacement is selected as default data imputation technique, last values for feature extraction and no feature selection and data sampling method is set as default for modeling. In addition, a random forest classifier is selected as default classifier because it is known as a stable classifier and therefore possibly best handles unbalanced datasets.

Data Understanding

In this case study, four datasets were uploaded in AutoCrisp. The time point dataset consisted of irregularly measured lab values of all LVAD patients. The time interval dataset consisted of medication data that specified for each medication from when someone started using the medication until the date last used. Baseline data consisted of basic data such gender and age and lastly, an event dataset was uploaded about all the complications that occurred in the cohort.

Data description report. Based on the datasets inputted, AutoCrisp automatically created three data understanding reports. The description report contained basic descriptive statistics such as 273 patients were found in cohort and 142 bleedings were selected which occurred in 25% of the patients. In total 633.487 time points were inserted and each patient had on average 2.320 lab values available related to 665 lab variables. The time interval dataset consisted of 54.385 medication records and thus an average of 1.367 medications were registered per patient.

Data quality report. The data quality report provided a comprehensive report containing multiple plots and descriptions. Firstly, a data similarity plot was presented which showed that patients were missing in both the medication and event dataset. Line charts and 3D plots showed many variables had only a limited amount of measurements and thus only a small amount of variables were regularly measured. In addition to selecting a required medication, the heatmap showed that the first patients who were included did not use these medications which suggested that the medication registration in the first years was insufficient.

Data exploration report. The third report produced by AutoCrisp concerned the data exploration report from which insights were gained from the data. A Kaplan–Meier curve showed that the events were relatively equally distributed over time. Surprisingly, some event statistics showed that bleeding events occurred significantly more in males than females. Some interactive time series plots presented the sequential development of lab values. Although most did not reveal any surprising patterns, one lab value, lactate dehydrogenase, showed a remarkable upwards trend in the 20 days prior to an event. However, it also showed a grey area representing the standard deviation which suggested that some patients had extremely high measurements before a bleeding. A more detailed event plot where only single events were presented confirmed the suggestion that most patients had normal values while other patients had extremely high values. The report ends with a correlation matrix that unfortunately did not reveal any surprising significant correlations. Nonetheless, hemoglobin was indicated to have the strongest correlation with the bleeding event which was in accordance with the expectations of domain experts.

Data Preparation

Data selection. AutoCrisp divides the data preparation phase into a data selection section and a data engineering section. In the data selection section, in addition to the baseline data, 19 lab values and 8 different medication records were selected. Furthermore, the class label was specified according to the previously described settings. As AutoCrisp suggests to discard the first seven days after an event occurrence, we performed a test in which we first ignored this setting and secondly specified to remove the first seven days. As depicted in Table 39.1, the suggestion of excluding data points after an event, improved the predictive performance in all cases. This indicates that the week after an event occurred does introduce noise into the model and therefore it is recommended to discard them.

Data engineering. In the treatment design four different data imputation methods were presented which are implemented in AutoCrisp. The mean replacement is by far the simplest method implemented in AutoCrisp and was suspected to perform significantly worse than all remaining data imputation methods used in this research.

Table 39.1 Results of cleaned dataset test after discarding the first seven days after an event

Setting	Full window	Cleaned dataset
1: 0	0.799	**0.824**
2: 1–3	0.779	**0.792**
3: 1–7	0.779	**0.788**
4: 1–30	0.769	**0.776**

Table 39.2 Results of data imputation methods

Setting	Mean replacement	MICE	Amelia	MissForest
1: 0	0.824	0.790	**0.827**	0.794
2: 1–3	0.792	**0.797**	0.786	0.778
3: 1–7	**0.788**	0.787	0.785	0.787
4: 1–30	0.776	**0.779**	0.766	0.776

Table 39.3 Results of feature extraction methods

Setting	Last Value	Statistics	Summary	TVA	PatternMining
1: 0	0.824	0.817	**0.830**	0.820	0.820
2: 1–3	0.792	0.785	0.780	0.787	**0.828**
3: 1–7	0.788	**0.799**	0.796	0.777	0.794
4: 1–30	0.776	**0.803**	0.796	0.780	0.721

However, as can be seen in Table 39.2, none of the data imputation methods extensively improved the performance.

Feature extraction. Feature extraction methods determine the data structure from which a classifier is able to learn. AutoCrisp provides five different methods that could be used to execute this task. The frequent pattern mining approach is by far the most complex approach and was therefore expected to outperform the others. However, although as depicted in Table 39.3, it shows competitive results, it only outperformed the remainder in the second setting. The summary feature extraction technique achieved the overall highest AUC whereas the statistics feature extraction technique outperformed the remainders in the last two settings. The statistics feature extraction technique does not include any absolute values but only statistical measurements over the complete time series and was therefore expected to predict more reliably over the longer time series whereas the last values method only includes the most recent absolute values and was therefore expected to predict particularly well in the first settings.

Modeling

Feature selection. The goal of feature selection is to determine a subset of features from which the model can most effectively learn. Three different methods were presented in the previous chapter using different underlying techniques. The first observation when using the feature selection methods, was that Boruta and RFE required far more time than expected. In this case study where we created approximately 6.000 objects and 74 features, it required both methods almost an hour while applying fivefold cross-validation. Regardless of the fact that these methods required far more time to process, they did not produce significantly different results in this case study. Moreover, Boruta did not exclude any features from the model in setting 3 and 4 and therefore resulted in the same AUC score (Table 39.4).

Data sampling. Data sampling was suggested by several researches to be very useful when working with very unbalanced datasets [38, 40]. In our datasets the majority of 98% was related to the negative class label (no bleeding) and 2% positive. As discussed in the treatment design, we included three different sampling techniques which are all used for training.

Table 39.4 Results of feature selection methods

Setting	None	Boruta	RFE	Filter
1: 0	0.824	**0.826**	0.800	0.785
2: 1–3	0.792	0.784	0.779	**0.797**
3: 1–7	**0.788**	**0.788**	0.767	0.755
4: 1–30	**0.759**	**0.759**	0.742	0.731

Table 39.5 Results of data sampling methods

Setting	None	Over	Under	Rose
1: 0	0.824	**0.843**	0.799	0.718
2: 1–3	**0.792**	0.731	0.812	0.760
3: 1–7	**0.788**	0.739	0.789	0.734
4: 1–30	**0.776**	0.748	0.751	0.702

Table 39.5 shows the results which in our opinion were less promising than anticipated. Only oversampling in the first setting showed a surprisingly high AUC. However, as explained in the treatment design, a random forest consists of multiple trees by inputting samples via a bootstrapping method and is therefore expected to be more stable. Therefore, we performed an extra test in which we used a single decision tree classifier instead. As can be seen in Table 39.6, oversampling and undersampling significantly improved the classifiers performance. However, the Rose algorithm which was expected to provide the best performance, performed even worse.

Classification Algorithm. In the previous sections, the random forest classifier was used as the default classifier in the tests. However, AutoCrisp provides different classifiers. As shown in Table 39.7, the random forest classifier scored best in all

Table 39.6 Results of data sampling using a decision tree

Setting	*None*	*Over*	*Under*	*Rose*
3: 1–7	0.534	0.648	**0.689**	0.528

Table 39.7 Results of classification algorithms

Setting	Random forest	Naïve bayes	Decision tree	Logistic regression	Support vector machines
1: 0	**0.824**	0.690	0.501	0.793	0.772
2: 1–3	**0.792**	0.773	0.494	0.787	0.701
3: 1–7	**0.788**	0.773	0.524	0.754	0.657
4: 1–30	**0.776**	0.770	0.490	0.705	0.708
Baseline	**0.594**	0.542	0.512	0.568	0.518

cases, indicating that this is a very suitable method for our case study. However, logistic regression which is a generalized linear model showed in most settings competitive results, while still being a transparent model, as will be explained in the following section. In accordance with our expectations the decision tree scored the worst in all cases. As discussed in the previous section and in accordance with the results shown in Table 39.7, it seems that decision trees are indeed very sensitive to class imbalance.

Transparency. All AUC scores presented in the previous section were calculated while performing a fivefold cross-validation. When users have found their ideal settings for all design choices, they can build their final model. In the previous section, we have shown that the selected type of classifier greatly affects the predictive performance. In Table 39.7, it has been shown that the decision tree did not perform very well, nevertheless the added value of decision trees are transparency. Therefore, AutoCrisp automatically visualizes the decision tree model as shown in Fig. 39.5. This tree has been built while using undersampling and setting 3. In line with our findings during the data exploration, it creates the first split on hemoglobin, indicating that this is important predictor. Interestingly, the right leaf nodes of the tree made several splits on LD (lactate dehydrogenase), indicating that the tree also observed the remarkable pattern that we found during the data exploration report. In addition, the plot mostly splits on first and second measurements (most recent), which implies that these are more important for the tree than third measurement.

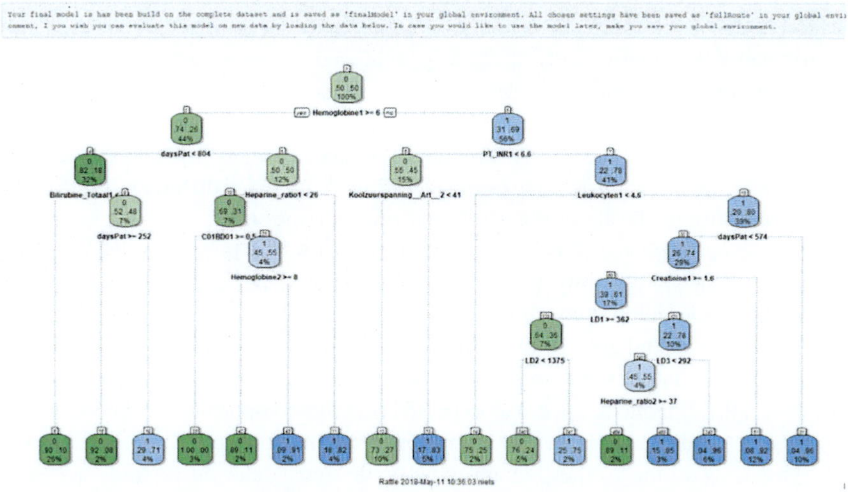

Fig. 39.5 The decision tree as provided by AutoCrisp to support transparant decision making

Conclusion

Early adverse event prediction is a difficult task for healthcare professionals making it nearly impossible to perform a full data mining project aimed at predicting medical events at an early stage. Healthcare professionals are busy practitioners who lack the required data mining knowledge, skills and time for performing a complete knowledge discovery process using multivariate sequential classification (MSC) effectively. This is not surprising considering the statement that MSC projects belong to the most difficult tasks in data mining due to the challenging EHR data in combination with the data mining techniques to be applied [14].

Over the last decade new tools have arisen that provided the evidence that data mining can be made accessible to domain experts. Unfortunately, due to their characteristics, MSC projects cannot adequately be performed by these tools. In this study we have demonstrated with AutoCrisp that this gap can be bridged.

Three reports were automatically created by AutoCrisp for performing the data understanding phase. Our generic tool demonstrated that it is feasible to gain specific medical insights whilst only providing a limited amount of information. The automated data quality report depicted crucial existing data points that did not line up with the truth. In addition, it created relevant plots showing where most data was missing, from which healthcare professionals were able to provide possible causations. In the data exploratory report, insights from the data were given that were recognized by healthcare professionals based on their medical knowledge. The goals of data understanding are becoming familiar with the data, verifying the data quality and gaining initial insights. Although these goals are highly ambiguous, in the perspective of the participants in our case-study, these were achieved.

Our main goal of the data preparation and modeling section in AutoCrisp was to provide a self-service data science service to enable, and thereby empower, healthcare professionals to apply machine learning classifiers on their raw datasets that most reliably predict medical adverse events. For this we included different design choices that according to the literature are based on assumptions that should improve the predictive performance. First, a suggestion of AutoCrisp to remove the data points after an event significantly improved the classifiers' performance in all settings. According to domain experts this was very reasonable due to a disruption of lab values when being hospitalized. In our problem statement, we concluded that EHR data was characterized by sparsely sampled data, resulting in many missing data points. Based on this conclusion, we had expected that more sophisticated data imputation methods would greatly affect the overall performance but surprisingly, this was not the case. According to the domain experts, the most relevant features are usually measured because in their opinion they provide the most information. Features with most missing data could possibly be the less predictive ones, and likewise the impact of data imputation methods could be limited in this context.

In this research, the feature extraction process was the most labor-intensive one. In all settings one of the more complex techniques outperformed the simplest last values method. The frequent sequential pattern mining approach as proposed by

[14] achieved the best performance in the second setting. As the technique is able to capture interactions between different features at different sequential moments, it was also expected to outperform the others in the early classification of the event. However, it might be that the full potential of this method is not recognized in this case study due to the sparsely sampled data.

A possible explanation for not improving the classifier while using data imputation techniques for missing data was given due to none relevant features. This explanation increased our expectations in the improvement of applying feature selection techniques as they remove the redundant features. Although the application of these techniques required a considerable amount of time, it surprisingly did not pay off in improving the predictive performance. As most techniques are widely researched and have proven their performance and we only performed one case study, we cannot make solid conclusions about their performance on EHR data.

As our datasets were very unbalanced, we included three different sampling techniques to create balanced datasets. The first test did not show any promising results but a second test revealed a different perspective. From this, we can conclude that the impact of the sampling method is dependent on the classifier used. Although random forest is based on decision trees, it showed that decision trees are more sensitive to class imbalance.

Although not all our expectations of improving the classifier by introducing different techniques were fulfilled, we have been able to demonstrate the potential for the concept of self-service data science: providing an appropriate analytic system tool can enable and empower healthcare professionals to perform a data mining project aimed at event prediction themselves without requiring exhaustive data mining knowledge.

References

1. W.A. Omta et al., HC StratoMineR: A web-based tool for the rapid analysis of high-content datasets. Assay Drug Dev. Technol. **14**(8), 439–452 (2016). https://doi.org/10.1089/adt.2016.726
2. D. Tomar, S. Agarwal, A survey on data mining approaches for healthcare. Int. J. Bio-Sci. Bio-Technol. **5**(5), 241–266 (2013). https://doi.org/10.14257/ijbsbt.2013.5.5.25
3. K. Srinivas, B. Rani, A. Govrdhan 'Applications of data mining techniques in healthcare and prediction of heart attacks'. *International Journal on Computer Science and Engineering* (02 Jan 2010), pp. 250–255. 10.1.1.163.4924
4. M. Durairaj, V. Ranjani, Data mining applications in healthcare sector: A study. Int. J. Sci. Technol. **2**(10), 29–35 (2013)
5. V. Marx, Biology: The big challenges of big data. Nature **498**(7453), 255–260 (2013). https://doi.org/10.1038/498255a
6. W. Raghupathi, V. Raghupathi, Big data analytics in healthcare: Promise and potential. Health Inform. Sci. Syst. **2**(1), 3 (2014). https://doi.org/10.1186/2047-2501-2-3
7. G. Neff, Why big data won't cure us. Big Data **1**(3), 117–123 (2013). https://doi.org/10.1089/big.2013.0029
8. W.A. Omta et al., PurifyR: An R package for highly automated, reproducible variable extraction and standardization. Syst. Med. **3**(1), 1–7 (2020). https://doi.org/10.1089/sysm.2019.0007

9. T.H. Davenport, D.J. Patil, 'Data scientist: The sexiest job of the 21st century: Meet the people who can coax treasure out of messy, unstructured data', *Harvard Business Review*, (Oct 2012), p. 9. https://doi.org/10.1007/978-1-4302-4873-6_9
10. I. Yoo et al., Data mining in healthcare and biomedicine: A survey of the literature. J. Med. Syst. **36**(4), 2431–2448 (2012). https://doi.org/10.1007/s10916-011-9710-5
11. M.F. Ghalwash, Z. Obradovic, Early classification of multivariate temporal observations by extraction of interpretable shapelets. BMC Bioinform. **13**(1), 195 (2012). https://doi.org/10.1186/1471-2105-13-195
12. M. Hauskrecht et al., Conditional outlier detection for clinical alerting, AMIA ... annual symposium proceedings/AMIA symposium. AMIA Symp. **2010**, 286–290 (2010)
13. D. Kagen, C. Theobald, M. Freeman 'CLINICIAN'S CORNER Risk prediction models for hospital readmission a systematic review'. **306** (15) (2015)
14. I. Batal et al., 'Mining Recent Temporal Patterns for Event Detection in Multivariate Time Series Data', *Proceedings of the 18th ACM SIGKDD international conference on Knowledge discovery and data mining* (2012) pp. 280–288 https://doi.org/10.1002/oby.21042.Prevalence
15. C. Rubinos, S. Ruland, 'Neurologic complications in the intensive care unit'. Curr. Neurol. Neurosci. Rep. **16**(6). (2016) https://doi.org/10.1007/s11910-016-0651-8.
16. M. Hall et al., '*The WEKA Data Mining Software: An Update the WEKA Data Mining Software: An Update*' (Nov 2008)
17. K. Chauhan et al., 'Automated machine learning: The new wave of machine learning', in *2020 2nd International Conference on Innovative Mechanisms for Industry Applications (ICIMIA)*. (IEEE, 2020) pp. 205–212.
18. R. Ooms, M. Spruit, '*Self-Service Data Science in Healthcare with Automated Machine Learning*' (2020), pp. 1–18. https://doi.org/10.3390/app10092992
19. R. Wieringa, 'Design Science as Nested Problem Solving', *International Conference on Design Science Research in Information Systems and Technology*, (2009) pp. 1–12. https://doi.org/10.1145/1555619.1555630
20. D.M. Rubio et al., 'Defining translational research: implications for training', Academic medicine: Journal of the association of American medical colleges. NIH Public Access **85**(3), 470 (2010)
21. M. Spruit, R. Jagesar, 'Power to the People!—Meta-algorithmic modelling in applied data science', *Proceedings of the 8th International Joint Conference on Knowledge Discovery, Knowledge Engineering and Knowledge Management*, 1(Ic3k) (2016), pp. 400–406. https://doi.org/10.5220/0006081604000406
22. M. Spruit, M. Lytras, 'Applied data science in patient-centric healthcare: Adaptive analytic systems for empowering physicians and patients', *Telematics and Informatics* (2018), pp. 643–653. https://doi.org/10.1016/j.tele.2018.04.002
23. M.R. Spruit, T. Dedding, D. Vijlbrief, 'Self-service data science for healthcare professionals: A data preparation approach', in *Proceedings of the 13th International Joint Conference on Biomedical Engineering Systems and Technologies (BIOSTEC 2020)—Volume 5: HEALTHINF*. (Valetta: ScitePress, 2020), pp. 724–734
24. C. Baru et al., *Report of the First Translational Data Science (TDS) Workshop* (Illinois, Chicago, 2017)
25. J. Demšar et al., Orange: Data mining toolbox in python. J. Mach. Learn. Res. **14**, 23492353 (2013)
26. R. Wirth, J. Hipp, 'CRISP-DM : Towards a Standard Process Model for Data Mining', *Proceedings of the Fourth International Conference on the Practical Application of Knowledge Discovery and Data Mining*, (24959), (2000) pp. 29–39. https://doi.org/10.1.1.198.5133
27. J. Sun, K.R. Chandan, '*Big Data Analytics for Healthcare*', *Kdd*. (2013)
28. C.S. Kruse et al., Challenges and opportunities of big data in health care: A systematic review. JMIR Med. Inform. **4**(4), e38 (2016). https://doi.org/10.2196/medinform.5359
29. S.R. Sukumar, N. Ramachandran, R.K. Ferrell, 'Data Quality Challenges in Healthcare Claims Data: Experiences and Remedies' (April 2014), (2016)

30. H.V.V. Jagadish et al., Big data and its technical challenges. Commun. ACM **57**(7), 86–94 (2014). https://doi.org/10.1145/2611567
31. B.D. Fulcher, N.S. Jones, Highly comparative feature-based time-series classification. IEEE Trans. Knowl. Data Eng. **26**(12), 3026–3037 (2014). https://doi.org/10.1109/TKDE.2014.2316504
32. S. Van Buuren, K. Groothuis-Oudshoorn, 'MICE: Multivariate imputation by chained equations in R'. J. Stat. Softw. **VV**(II), 1–68 (2010)
33. J. Honaker, G. King, M. Blackwell 'Amelia II: A program for missing data, R package version 1.5., 2012', Available at https://gking.harvard.edu/amelia/, (2012) pp. 1–116
34. D.J. Stekhoven, P. Bühlmann, Missforest-Non-parametric missing value imputation for mixed-type data. Bioinformatics **28**(1), 112–118 (2012). https://doi.org/10.1093/bioinformatics/btr597
35. A. Nanopoulos, R.O.B. Alcock, Y. Manolopoulos, Feature-based classi cation of time-series data. Int. J. Comput. Res. **10**(3) (2001)
36. B. Esmael et al., *A Statistical Feature-Based Approach for Operations Recognition in Drilling Time Series*, vol. 5 (2013) pp. 454–461
37. M.B. Kursa, W.R. Rudnicki, Feature selection with the Boruta package. J. Stat. Softw. **36**(11), 1–13 (2010)
38. R. Blagus, L. Lusa, Joint use of over-and under-sampling techniques and cross-validation for the development and assessment of prediction models. BMC Bioinform. **16**(1), 1–10 (2015). https://doi.org/10.1186/s12859-015-0784-9
39. N. Lunardon, G. Menardi, N. Torelli, ROSE: A package for binary imbalanced learning. R J. **6**(June), 79–89 (2014)
40. M. Bekkar, H.K. Djemaa, T.A. Alitouche, Evaluation measures for models assessment over imbalanced data sets. J. Inform. Eng. Appl. **3**(10), 27–38 (2013)

Chapter 40
An Efficient Method for Rapid Fabrication Using Low Cost 3D Printer

Mohammed Balfaqih, Soltan A. Alharbi, Fahad Alfarsi, Osama Alharbi, and Faisal Alsalmi

Abstract Additive manufacturing has gained significant attentions from industrial and academia. Personal design and fabrication have been enabled by different 3D modeling software. The design phase typically takes several iterations. Printing these iterations consumes a lot of time and materials. Although several works have been proposed to speed up 3D printers for quick design iterations, the cost of 3D printer depreciation has not been considered. This paper aims to minimize the iteration cost and time of printing a 3D objects preview, using a low-cost printer. The printer is designed based on Fused Deposition Modeling (FDM) technique with total cost <100$. Moreover, the compatibility of two wireframe systems with the developed printer to facilitate printing intermediate low-fidelity previews have been examined. The printer slices the 3D model along its vertical axis into horizontal slices, extracts the contours, and finally supports the developed printer to fill the space between slices with a zigzag pattern. The results show that printing time using WirePrint system is less than using WeaveMesh system. The proposed fabrication solution saves more than 75% of the total cost of the conventional printing methods.

Keywords Additive manufacturing · Fast prototyping · 3D printing · Low-fidelity · Design iteration

Introduction

Personal design and fabrication became more attainable due to the advanced development of both hardware and software technologies. The popularity increment of 3D modeling and personal fabrication tools motivates researchers to work towards seamless merging of functional design and personal fabrication [1–3]. 3D printing is

M. Balfaqih (✉) · S. A. Alharbi
Department of Computer and Network Engineering, University of Jeddah, Jeddah, Saudi Arabia
e-mail: mabalfaqih@uj.edu.sa

F. Alfarsi · O. Alharbi · F. Alsalmi
Department of Electronics and Electrical Engineering, University of Jeddah, Jeddah, Saudi Arabia

A. Visvizi et al. (eds.), *Research and Innovation Forum 2020*,
Springer Proceedings in Complexity,
https://doi.org/10.1007/978-3-030-62066-0_40

one of the technologies that enables people to fabricate their own 3D models design. Although, 3D printing is classified as a rapid prototyping tool, it is still slow where a reasonable sized object might require printing overnight. The design process will also slow down to around a single iteration per day.

Several approaches have been proposed to decrease 3D printing time such as using multiple heads to massively parallelize the printing process [4] or assembling objects layer-wise from prefabricated voxels of equal size [5]. Differently, to speed up the design process, low-fidelity techniques have been presented to iterate quickly by giving priority to speed over functionality. Such techniques will be applied in the early stages of design to explore quickly several design versions without consuming much resources [6]. Therefore, an abstract representation of 3D models was introduced to reduce the printing time of all intermediate versions. For example, faBrickation [7] has limited the 3D printing to regions that require high-resolution only and utilized standard building blocks everywhere else.

Unfortunately, the fabrication of each version still cost material waste, and Inevitably the irreversible process slowed design iterations. Moreover, the cost of 3D printer depreciation has not been considered in all the aforementioned works. Therefore, this paper aims to minimize the iteration cost and time of printing a 3D objects preview, using a low-cost printer. Instead of using the main printer to print the iterations of targeted model. A low-cost printer <100$ is developed. Different printing software and models have been examined to determine the most software compatible with the designed printer that achieves the minimize the iteration printing cost. The results show that printing time using WirePrint system is less than using WeaveMesh system. The proposed fabrication solution saves more than 75% of the total cost of the conventional printing methods.

The main contributions of the paper are (i) a low-cost 3D printer where most of the components are recycled materials, such as water plastic pipes, or PLA plastic filament materials. and (ii) a comparative evaluation between WirePrint and WeaveMesh models to determine the model that reduce material waste more. The rest of the paper is organized as follows; Sect. 40.2 explains briefly the research background, specifically, low-fidelity printing concept and types. The proposed fabrication method is introduced in Sect. 40.3, and the results are discussed in Sect. 40.4. Finally, conclusion and future work are presented in Sect. 40.5.

Low-Fidelity Printing

3D printers are slow which may slow designers down to a single iteration per day. Consequently, the iteration process could be extended easily a week even though designing process does not actually take longer than few hours. Thus, 3D printer is considered a bottleneck of fabrication workflow. To address the bottleneck, low-fidelity fabrication speeding up design iteration by printing intermediate low-fidelity previews. The final version is fabricated only as a full 3D print. The workflow of traditional and low-fidelity fabrication is shown in Fig. 40.1. Low-fidelity fabrication

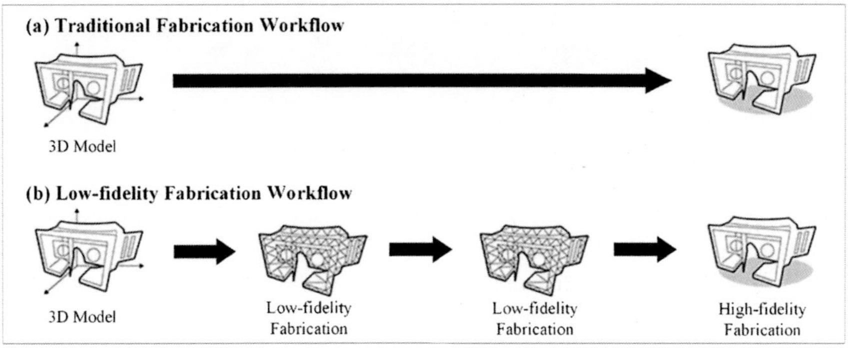

Fig. 40.1 The Workflow of **a** Traditional Fabrication and **b** Low-Fidelity Fabrication [6]

have different ways of implementation based on use case and what is tested, such as faBrickator [7], WirePrint [8], and Platener [9].

FatBrickator system [7] proposed to substitute automatically sub-volumes of an object with standard building blocks and print sub-volumes. It saves printing time by converting firstly a 3D model to Lego and then users can indicate parts of the 3D model to be 3D printed. FaBrickators allows the users to iterate the prototype by reprinting only the elements that changed.

Platener system [9] speeds up the printing by substituting straight and curved plates from 3D model with laser cut parts of the same size and thickness. The 3D printer prints only the regions that are of relevance to the current design in full-detail. These parts are connected by inserting joints automatically. Platener system enable the users to customize the substitution by specifying certain parameters.

In wireframe system [5], the surfaces of a 3D print are replaced with a wireframe mesh, which reduce the printing time significantly. The wireframe represents the overall shape of the object where it allows users to quickly verify the main aspects of 3D design, such as the ergonomic fit. Several research works optimized the wireframe system [10–12]. In [8], WirePrint maximizes the speed by instructing 3D printers to extrude filament in 3D space instead of layer-by-layer. 3D printer creates the edges of the wireframe model one stroke at a time, where it is faster 10 times than the traditional 3d printing. In [10], WeaveMesh is proposed to balance between time and material time by filling a model according to UV mesh map. UV mapping algorithm allows to limit the restoration of big complex differences in a surface.

Proposed Fabrication Method

The traditional fabrication and supply chain could be enhanced with 3D printing by decreasing manufacturing steps, enabling decentralized manufacturing (e.g. at sale centers), printing items on demand only and consequently reducing the need for

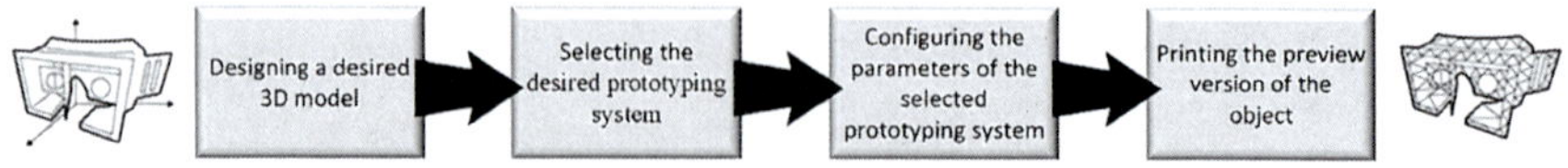

Fig. 40.2 Printing 3D objects preview phases

warehousing, packaging, and transportation [13, 14]. The new community-driven concept (e.g. Shapways) offers online service that enables users to upload their 3D printable file [15]. The designer has to test the products before finalize the design and send it. To enable testing the design with low cost, the proposed fabrication method employs a low-cost printing that minimizes 3D printer depreciation, iteration cost and time of printing a 3D objects preview. The idea of low-cost printing has emerged from early 3D printers that fabricated most of their components [16–19].

The proposed fabrication method to overcome the barriers to design and fabrication for quickly prototyping. In the proposed fabrication method, the phases of printing 3D objects preview, as shown in Fig. 40.2, are: (1) Designing a desired 3D model to utilize the low-cost 3D printer, (2) Selecting the wireframe printing system (i.e. this work considers WirePrint and WeaveMesh system only), (3) Configuring the parameters of the selected system, (4) Printing the preview version of the object. Users then can verify the design of the object and proceed with final fabrication with more complex printers that detail the object appearance. The two main components of the proposed method, developed low-cost 3D printer and the utilized 3D models, are discussed in the following sub-sections.

Low-Cost 3D Printer

The developed printer is a desktop 3D printer that utilizes Fused Deposition Modelling (FDM) which is one of the most widely used additive technologies. Figure 40.3 shows the building of the low-cost 3D printer. To minimize the cost of the printer, the printer body and some components are recycled from electronic waste. The main components of the designed printer include; 3D Printer Arduino

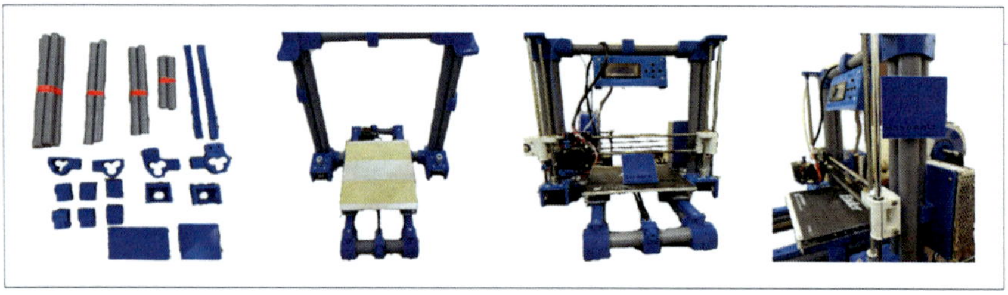

Fig. 40.3 Low-cost 3D printer building

Table 40.1 Detailed cost of the 3D printer

Component	Price ($)	Component	Price ($)
3D Printer Arduino kit Hot end	21	Extruder	20
Stepper motor	31	12-V 10-Amp power supply	9
5 × 8 shaft coupler	4	Hot bed	8
Total cost: 93 $			

kit Hot end, extruder, stepper motor, 12-V 10-Amp power supply, and 5 × 8 shaft coupler. The total cost of the printer is 93$. The detail bill of electrical schematics and materials are shown in Table 40.1.

Printing System

Two wireframe modeling systems are used in the proposed fabrication method; WirePrint [8] and WeaveMesh [10]. WirePrint system uses contour-plus-zigzag approach to convert a 3D object into a wirframe. The 3D model is sliced into horizontal slices, and the contours are extracted. Then, printing a contour and create one layer on top of the contour in a zigzag pattern by moving the printer head up and down. The parameters of a WirePrint model are specified based on the width and length of the object's bounding box at a certain height using a Boolean intersect operation. On the other hand, WeaveMesh system uses an edge contraction-based mesh simplification approach to design the wireframe model. The main aim of this approach is to support self-supportability of the object. WeaveMesh takes into account geometric errors, changes of normal vectors, and self-supportability of edges. The parameters of a WeaveMesh model are quantity and size of the UV strips as well as points.

Results and Discussion

To evaluate the developed low-cost 3D printer, two wirframe systems have been used; WirePrint and WeaveMesh. The default parameters that have been considered in the experiment are shown in Table 40.2. A Rabbit model is printed as shown in Fig. 40.4 with X: 30 mm, Y: 42 mm, and Z: 80 mm dimensions.

Figure 40.5 shows the impact of different layer thickness from 0.1 mm to 0.5 mm on printing time. There is no effect of thickness layer on WirePrint and WeaveMesh systems. The edges of the object are printed in a zigzag pattern by moving the printer head up and down, thus, the printing duration is the same for all thickness layers. On the other hand, printing duration decreases gradually upon decreasing the thickness layers because the whole object is printed layer by layer.

Table 40.2 Default experiment parameters

Parameter	Value	Parameter	Value
Layer thickness	0.2 mm	Print speed	60 mm/s
Nozzle temperature	210°	Target number of edges	1500 edges
Infill percentage	20%	Threshold angle γ	160°

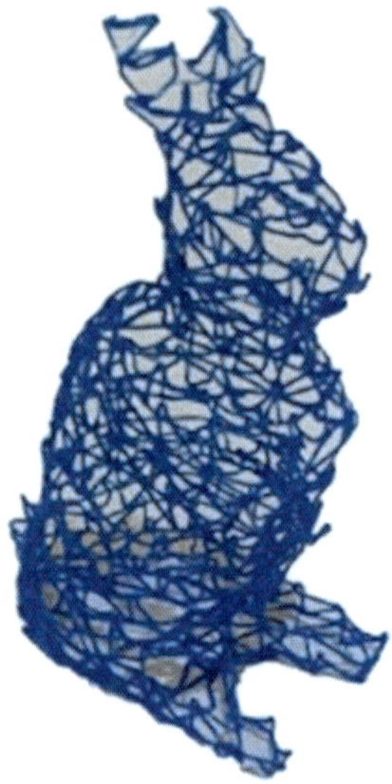

Fig. 40.4 Printing preview of an object using low-cost 3D printer

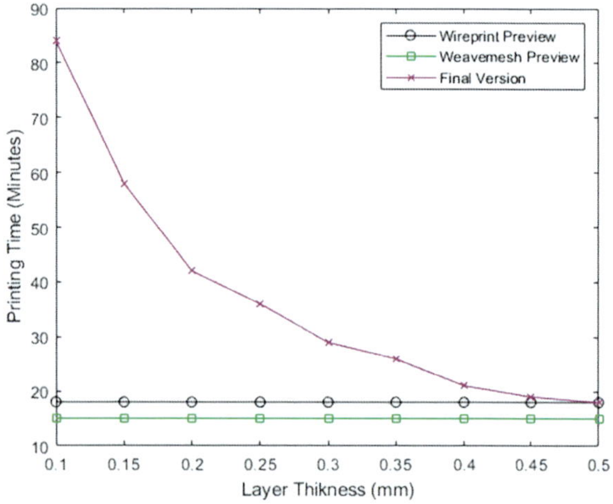

Fig. 40.5 Printing time of the developed printer with different layer thickness

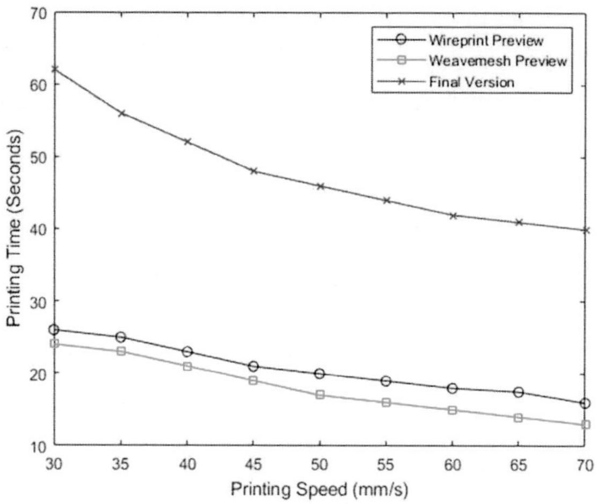

Fig. 40.6 Printing time of the developed printer with different printing speeds

In Fig. 40.6, the printing speed has been changed between 30 and 70 mm/s to study its impact on printing time on printing systems. As expected, increasing the printing speed decreases the printing time gradually. At the minimum examined speed 30 mm/s, the 3D printer completed the printing in 26, 24 and 62 s using WirePrint, Weavemsh and solid printing systems, respectively. On the other hand, the printing time at the maximum speed 70 mm/s was 16, 13 and 40 s, respectively.

Conclusion and Future Work

Different technologies have been developed to enable personal design and fabrication using additive manufacturing. 3D printing is considered one of the fastest fabrication technologies. However, the design phase typically takes several iterations. Printing these iterations consumes a lot of time and materials. Although several works have been proposed to speed up 3D printers for quick design iterations, the cost of 3D printer depreciation has not been considered. Thus, this paper aims to minimize the iteration cost and time of printing a 3D objects preview, using a low-cost printer. The printer is designed based on Fused Deposition Modeling (FDM) technique with total cost 93$. Moreover, a printing software compatible with the designed printer is developed to facilitate printing intermediate low-fidelity pre-views. The results showed that the developed 3D printer is faster during printing preview versions using WeaveMesh system. For future work, more advanced printing systems will be tested. In addition, a new model that aims to increase the accuracy and self-supportability of 3D models will be developed.

References

1. B. Buxton. *Sketching User Experiences: Getting the Design Right and the Right Design*. Morgan kaufmann. (2010)
2. M. Davia-Aracil et al., 3D printing of functional anatomical insoles. Comput. Indus. **95**, 38–53 (2018)
3. M. Davia-Aracil et al., A new methodological approach for shoe sole design and validation. Int. J. Adv. Manufact. Technol. **86**(9–12), 3495–3516 (2016)
4. J. Hiller, H. Lipson, Methods of parallel voxel manipulation for 3D digital printing. 2007 Int. Solid Freeform Fabrication Symp. (2007)
5. N.A. Gershenfeld, *Fab: The Coming Revolution on Your Desktop—From Personal Computers to Personal Fabrication*. Basic Books (AZ), (2005)
6. J. McCrae et al., FlatFitFab: Interactive modeling with planar sections. *Proceedings of The 27th Annual ACM Symposium on User Interface Software and Technology*. (2014)
7. S. Mueller et al., faBrickation: Fast 3D printing of functional objects by integrating construction kit building blocks. *Proceedings of The SIGCHI Conference on Human Factors in Computing Systems*. (2014)
8. S. Mueller et al., WirePrint: 3D Printed previews for fast prototyping. *Proceedings of The 27th Annual ACM Symposium on User Interface Software and Technology*. (2014)
9. D. Beyer et al., Platener: Low-fidelity fabrication of 3D objects by substituting 3D print with laser-cut plates. *Proceedings of The 33rd Annual ACM Conference on Human Factors in Computing Systems*. (2015)
10. Y. Tao et al., Weavemesh: A low-fidelity and low-cost prototyping approach for 3d models created by flexible assembly. *Proceedings of The 2017 CHI Conference on Human Factors in Computing Systems*. (2017)
11. X. Liu et al., Generating sparse self-supporting wireframe models for 3D printing using mesh simplification. Graph. Models **98**, 14–23 (2018)
12. M. Liu et al., WireFab: Mix-Dimensional Modeling and Fabrication for 3D Mesh Models. *Proceedings of The 2017 CHI Conference on Human Factors in Computing Systems*. (2017)
13. T. Campbell et al., Could 3D printing change the world. *Technologies, Potential, and Implications of Additive Manufacturing, Atlantic Council, Washington, DC*, 3 (2011)
14. S.H. Huang et al., Additive manufacturing and its societal impact: a literature review. Int. J. Advan. Manufact. Technol. **67**(5–8), 1191–1203 (2013)
15. www.shapeways.com/
16. R. Jones et al., RepRap–the replicating rapid prototyper. Robotica **29**(1), 177–191 (2011)
17. Ed. Sells et al., RepRap: The replicating rapid prototyper: Maximizing customizability by breeding the means of production. *Handbook of Research in Mass Customization and Personalization: (In 2 Volumes)*. (2010). pp. 568–580
18. C. Baechler et al., Distributed recycling of waste polymer into RepRap feedstock. Rapid Prototyping J. (2013)
19. E.J. Hunt et al., Polymer recycling codes for distributed manufacturing with 3-D printers. Resour. Conserv. Recycl. **97**, 24–30 (2015)

Chapter 41
Blockchain as a Healthcare Insurance Fraud Detection Tool

Julio C. Mendoza-Tello, Tatiana Mendoza-Tello, and Higinio Mora

Abstract Healthcare insurance is intended to help pay for the insured's medical expenses by paying a policy premium. For this, the industry needs the collaboration of some entities, such as: doctors, health care centers, brokers, insurers, reinsurers. In this context, gathering the information necessary to assess and process claims is a major problem. As a consequence, these inconveniences are exploited by fraudsters and scammers. Faced with these challenges, blockchain can help solve them. This paper defines blockchain and investigates how its inherent characteristics can contribute to detecting healthcare insurance fraud. Then, a layered overview and model using smart contracts are defined. Finally, conclusions and recommendations are issued to address its implementation in the insurance market.

Keywords Blockchain · Smart contract · Healthcare insurance · Fraud

Introduction

Fraud is a malicious activity faced by the management of healthcare insurance incidents and claims. This is the branch of the insurance market with the greatest losses through fraud. Currently, the exact value per scam is unknown and may reach tens of billions of dollars per year [1]. The consequences of this abuse in health care claims are reflected in the increase in costs for the use of equipment and medical service

J. C. Mendoza-Tello
School of Computer Engineering, Faculty of Engineering, Physics and Mathematics, Central University of Ecuador, Ciudadela Universitaria, 170129 Quito, Ecuador
e-mail: jcmendoza@uce.edu.ec

T. Mendoza-Tello
Insurance Consultant, Avenida Cristóbal Colón E3-12, 170524 Quito, Ecuador
e-mail: tatymendoza06@yahoo.es

H. Mora (✉)
Department of Computer Technology, University of Alicante, San Vicente del Raspeig, Spain
e-mail: hmora@ua.es

A. Visvizi et al. (eds.), *Research and Innovation Forum 2020*,
Springer Proceedings in Complexity,
https://doi.org/10.1007/978-3-030-62066-0_41

fees, which are demanded by health care providers. In addition, these scams directly affect a loss of cover for patients and a higher payment of tax [2].

For fraud detection, several techniques were developed. However, most detection approaches are static and depend on the control by third parties. In this sense, these techniques can be implemented using blockchain technology.

Blockchain is a distributed ledger with the ability to record and keep transactional events unchanged over time, without the need for a middleman. This technology provides trust through a consensus mechanism based on cryptographic tests, which are executed and provided by network participants. In addition, this innovation has the versatility of transporting smart contracts, which encode events according to strict and predefined rules. Thus, integrity is guaranteed and the supervision of trusted third parties is eliminated [3].

The sections of this document are structured as follows. Section two, notions of healthcare insurance are described. In addition, inherent characteristics of blockchain to combat insurance fraud are explained. With these considerations, section three describes and defines a layered overview and model. Finally, conclusions and future research are presented in section four.

Background

Healthcare Insurance and Fraud

An insured party transfers part of his risk to an insurer by signing up to an insurance policy. In this line, a policy is a contractual agreement between an insured party and an insurer. This contract details the obligations of the parties, conditions, coverage, limitations and exclusions of a set of insured contingencies or events [4]. For this, an insured party pays a fee (called a premium) to an insurer, which undertakes to economically cover an event stipulated in the policy. Within the insurance market, the healthcare line stands out.

A medical insurance policy provides coverage to an insured party in two ways. Firstly, medical expenses as a result of an illness or injury. Secondly, the use of a health care service, as well as the payment of medicines and other medical treatments clearly specified in the policy. When an incident occurs, the insured party sends a claim to the insurer so that the expenses can be covered. An exhaustive analysis of the claim involves the collaboration of various entities: doctors (through hospitals and health care centers attached to the insurance plan), providers of medical supplies (e.g. pharmacies), insurers and the insured patient. In this context, the collection of this data to evaluate and process claims is a major problem. In some cases, the guarantee of integrity and consistency of information depends on human intervention, which is prone to failures and errors. These inconveniences are taken advantage of by fakers, who wish to obtain dishonest benefits and illegitimate rights.

Throughout health insurance activities, various forms of fraud have been identified based on claim anomalies and identification of dishonest behaviors. Thus, the most common types of healthcare insurance fraud are: phantom claims, duplicate claims, bill padding, upcoding, unbundling, kickbacks, excessive numbers of small bills, billing for services not provided, price manipulation of services, claims in short time, overuse of medical services, amongst others [5, 6].

Some investigations to combat these frauds were carried out. Clustering and regression analysis procedures are useful to discriminate the use of medical care and geographic billing regions. Based on the daily distances a patient travels to receive medical care, atypical values can be identified in terms of use and billing [7]. In this sense, overutilization behavior patterns are obtained, when comparing dates of medical care and medication supply of the same class obtained from various pharmacies [2]. Due to the large amount of data generated in this fraud detection process, the use of a data mining technique is necessary. This allows the extraction and feedback of information that compares the behavior of insured parties and medical service providers [8]. In addition, it is important to note that fraud detection procedures should also identify the suspicious behavior of other entities, such as pharmacies and other medical supply providers [9].

Blockchain Characteristics Against Insurance Fraud

Fraud detection audits and identifies anomalies, and unfavorable behaviors in insurance claims [10]. In this line, Table 41.1 summarizes how blockchain features can support this activity.

Blockchain-Based Model to Detect Healthcare Insurance Fraud

In this section, a model to detect healthcare insurance fraud is designed. In this context, Fig. 41.1, an overview describes four layers. Firstly, P2P mining pool layer. This is the base layer; it defines, maintains and executes the protocol to ensure the integrity and reliability of system operations. That is, the guarantee of the consensus process.

Secondly, blockchain layer. This is the layer where nested transaction blocks are implemented sequentially over time. Each transaction can carry digital tokens such as: cryptocurrencies (as a money exchange token), and smart contracts (such as triggers that store and execute previously defined procedures). The correct organization of these tokens depends on the P2P mining pool layer.

Table 41.1 How can blockchain support the detection of healthcare insurance fraud?

Characteristic	Support for insurance fraud detection
Automation using smart contract	• Definition and verification of contractual terms of an insurance policy • Linking various data sources through smart contracts • Coding of anomalous behavior patterns
Digital authentication and access rules	• Keys are stored in digital wallets • Access rules and digital authentication procedure incorporated into smart contracts
Shared intelligence	• P2P Platforms enables having several points of collaboration on the subject of: claims, fraud, capacity and compliance of payments of the insured party, control of hospitalizations and frequent prescriptions
Transparency and immutability of data	• A healthcare record cannot be manipulated or deleted • The medication provided by pharmacists is controlled • Strengthening of the Know Your Customer (KYC) and Anti-Money Laundering processes (AML)
Decentralization	• Decentralized storage through P2P networks • Availability, reliability and fault tolerance • Support to reputation score systems

Thirdly, a smart contract layer. Functional specifications (within smart contracts) are necessary so as to autonomously detect fraud. For this, seven smart contracts are used:

- Authorization rules and authentication. This is the access interface and the gateway from the end user to the system. In this context, the user's digital identity is validated and the correct access to data is guaranteed, according to previously granted privileges. Each entity may update its authorization rule as it deems appropriate.
- Data access audit logs. The access to the blockchain scheme is registered. Each access is encrypted and linked to the user's entity. In turn, each access is registered in an external referential database.
- Healthcare insurance fraud cases. Fraudulent claims made are recorded. Each fraud is linked to the faker's entity.
- Healthcare insurance fraud patterns. To prevent fraud, communication between several partners is necessary. In this line, the required information can be stored in various heterogeneous databases. For this reason, the implementation of a blockchain-based smart contract is necessary in order to solve interoperability difficulties. Thus, a smart contract is defined allowing the registration of data pointers. In this way, the integrity of the information from internal and external sources can be verified and guaranteed. In this context, a decentralized service counter fraud is defined so as to implement and link from external sources, approaches, types and patterns of malicious behavior based on expert recommendations (Fig. 3). To do this, these descriptions can be encoded inside or outside

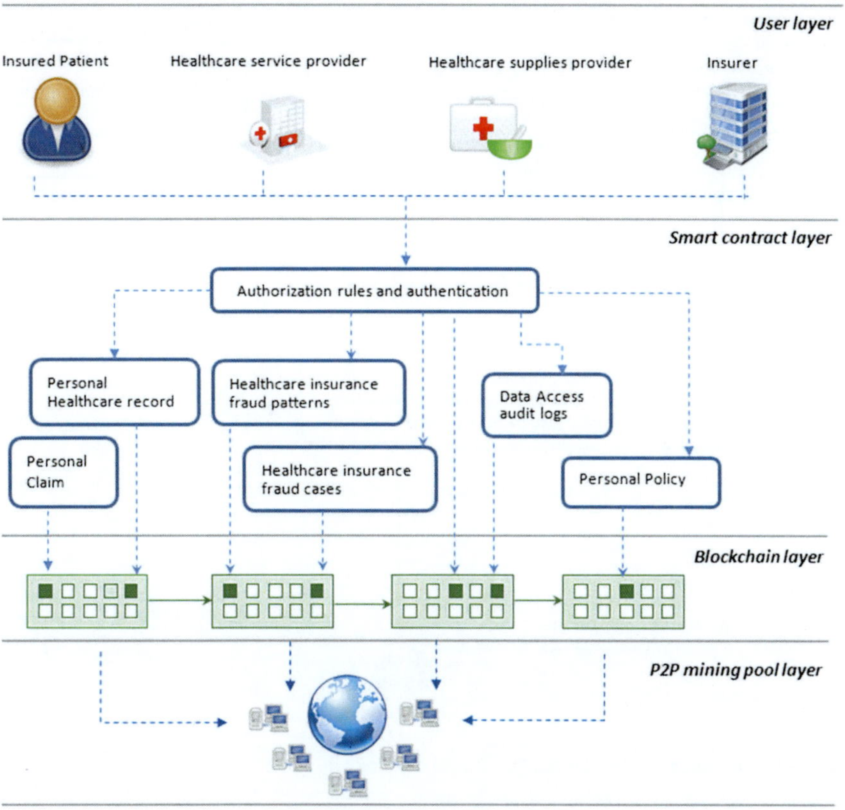

Fig. 41.1 A layered overview to detect healthcare insurance fraud using blockchain

blockchain, using clustering algorithms, regression analysis [7], data mining [8] and the development of decision trees [2].

- Personal Healthcare record. Blockchain records the medical prescriptions and diseases detected for each insured patient. Each token is linked to the identity of the insured party. The insured party (by way of a token) can authorize the display or exchange of their personal healthcare record between entities. This information can also be stored outside blockchain. In this case, each token is a pointer or reference to each of the records stored in other external repositories. In this way, interoperability between different systems could be resolved.
- Personal policy. The contractual commitment between the insured party and the insurer is automated and recorded. The following are specified: benefits, validity of the policy contract, limitations and exclusions of contingencies or insured events.
- Personal claim. Each claim within the blockchain is recorded in encrypted form. A claim is a token linked to the following entities: the insured party, the insurer and the medical service provider. Updated information is provided on the status of

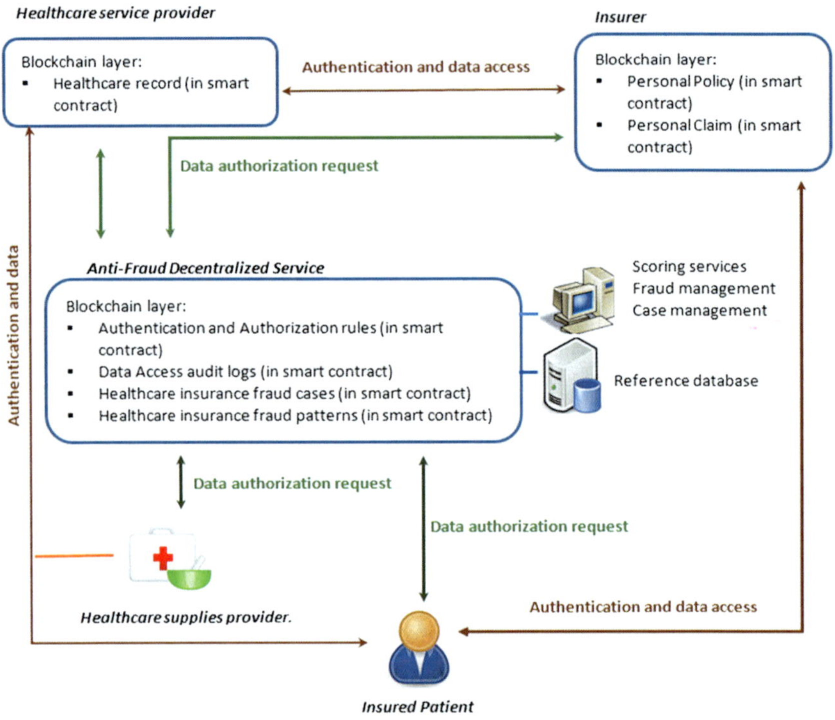

Fig. 41.2 Blockchain-based model to detect healthcare insurance fraud

the claim. This smart contract performs automatic settlement between the entities involved.

Fourthly, a user layer. This represents the five final entities of the scheme. In Fig. 41.2, a functional description of the model is shown.

- Patient. This is the insured entity and it gives permission for its healthcare record to be read. This access is provided by a healthcare service provider through a counter fraud service.
- Insurer. This entity registers and updates the insurance policies of the insured. Consequently, it provides and restricts access to this information through authorization and authentication guidelines configured within an anti-fraud service.
- Healthcare service provider. This entity records and updates healthcare records of patients (through a diagnosis or medical prescription). In addition, it provides access to these records through authorization and authentication options configured within an anti-fraud service.
- Healthcare supplies provider. This is represented by pharmacies and other medical supplies entities. Its function is the sale of medicines based on a medical prescription (which is registered in the healthcare record).

- Anti-fraud decentralized service. This service acts in blockchain layer and has two main functions through smart contracts. Firstly, authentication and authentication rules. Its function is to provide all access guidelines to blockchain records, such as: healthcare records, insurance policies, fraud cases, fraud patterns, and personal claim. Secondly, the rules of collaboration with other external services. Through a smart contract, you can register points that validate information from external sources, as well as the settlement of claims. In this context, collaboration with other fraud management services and scoring services is important. Thus, previous investigations, cases, patterns and techniques to detect healthcare insurance fraud can be linked to the scheme proposed in Fig. 41.2.

Conclusions

Contributions

This research highlights the blockchain characteristics that contribute to the detection of medical insurance fraud, namely: automation using smart contracts, digital authentication and access rules, shared intelligence, transparency and immutability, and decentralization. Based on this versatility, a model is defined from two points of view. Firstly, a referential model that abstracts the connectivity of the components in four layers: P2P mining pool, blockchain, smart contract, and user. This bottom-up scheme provides an overview of the functionality of each component. Secondly, an anti-fraud service is described and defined as a collaborative entity between the various partners that can be integrated into the system. By owning a distributed ledger, claims, policies, healthcare records and medical prescriptions are more visible and integrity is guaranteed. In addition, access will be controlled and fraudulent behaviors will be linked to an entity. Thus, a large number of scam attempts can be quickly detected and blocked before they happen.

Limitations and Future Research

According to these considerations, the benefits of the blockchain-based model to reduce fraud are evident. However, there are situations that need to be addressed because they can evade the controls defined in smart contracts. For example, medical prescriptions depend on the honesty of the person who provides them. Fake certificates issued by a doctor also lead to a billing for services not provided or overused. In both examples, it is clear that the insurer will assume the expenses because it relies on the ethics of the doctor. These fraudulent events in the initial stage of patient treatment cannot be detected automatically due to a lack of information availability. In this sense, it is essential that insurance companies make collaborative efforts to link a large amount of historical information under a legal scheme. Thus, the coordination challenge to minimize the validation time of a claim can be minimized and the benefits of technology are exploited through the value chain of the insurance market.

References

1. Coalition Against Insurance Fraud: By the numbers fraud statistics, https://www.insurancefraud.org/statistics.htm. Accessed 12 Aug 2019
2. A.T. Mailloux, S.W. Cummings, M. Mugdh, A decision support tool for identifying abuse of controlled substances by ForwardHealth medicaid members. J. Hosp. Mark. Public Relations. 34–55 (2010). https://doi.org/10.1080/15390940903450982
3. J.C. Mendoza-Tello, H. Mora, F. Pujol, M.D. Lytras, Disruptive innovation of cryptocurrencies in consumer acceptance and trust. Inf. Syst. E-bus. Manag. 195–222 (2019). https://doi.org/10.1007/s10257-019-00415-w
4. A. Marotta, F. Martinelli, S. Nanni, A. Orlando, A. Yautsiukhin, Cyber-insurance survey. Comput. Sci. Rev. **24**, 35–61 (2017). https://doi.org/10.1016/j.cosrev.2017.01.001
5. A. Abdallah, M.A. Maarof, A. Zainal, Fraud detection system: A survey. J. Netw. Comput. Appl. **68**, 90–113 (2016). https://doi.org/10.1016/j.jnca.2016.04.007
6. D. Thornton, M. Brinkhuis, C. Amrit, R. Aly, Categorizing and describing the types of fraud in healthcare. Procedia Comput. Sci. **64**, 713–720 (2015). https://doi.org/10.1016/j.procs.2015.08.594
7. R.M. Musal, Two models to investigate Medicare fraud within unsupervised databases. Expert Syst. Appl. **37**, 8628–8633 (2010). https://doi.org/10.1016/j.eswa.2010.06.095
8. P.A. Ortega, G.A. Ruz, A medical claim fraud/abuse detection system based on data mining: A case study in Chile. in *Proceedings of the 2006 International Conference on Data Mining, DMIN 2006*. (Las Vegas, Nevada, USA, 2006). pp. 224–231
9. Konijn, R.M., Kowalczyk, W.: Finding Fraud in Health Insurance Data with Two-Layer Outlier Detection Approach. in *Data Warehousing and Knowledge Discovery. DaWaK 2011. Lecture Notes in Computer Science*, ed. by A. Cuzzocrea, U. Dayal (Springer Berlin Heidelberg, 2011). pp. 394–405 https://doi.org/10.1007/978-3-642-23544-3_30
10. Y. Li, C. Yan, W. Liu, M. Li, A principle component analysis-based random forest with the potential nearest neighbor method for automobile insurance fraud identification. Appl. Soft Comput. **70**, 1000–1009 (2018). https://doi.org/10.1016/j.asoc.2017.07.027

Part IV
ICT, Politics, Business and the Society

Chapter 42
The Internet as Public Space: A Challenge to Democracies

Cláudia Toriz Ramos

Abstract Freedom of speech, freedom of information, freedom of association and pluralism are cornerstones of democratic regimes and thus part of the characteristics without which the threshold of democracy cannot even be reached. It would therefore seem that the recent and very substantial changes in information and communication processes, as introduced by new information and communication technologies (ICT), could only foster those freedoms and reinforce pluralism. However, the nature and amount of information, diversity of the actors involved and lack of regulatory framework raise many conundrums that challenge democracy itself. The article departs from a debate on the role of freedom for democracies. It then discusses both the opportunities and the constraints emerging from the use of new ICT. With reference to published data, internet freedom is addressed, in order to grasp information on how political power is reacting to this new and powerful tool of communication with and between the citizens, and both inside and across borders. In an era of debates on 'populism and demagogy', on 'illiberal democracy', and on 'digital authoritarianism', scrutinising the 'virtuous' and the 'vicious' uses of the internet, with reference to the integrity and the quality of democracy, is a necessary academic discussion, in order to understand what may lie ahead for democratic regimes.

Keywords Democracy · Internet freedom · Digital authoritarianism

Introduction

The panel in which this paper is inserted discusses the relation between democracy and the new technologies of information and communication, with the aim of disentangling the benefits of their use from the hindrances. In line with that debate, this paper addresses the internet as public space and its many challenges to the conventional public sphere.

C. T. Ramos (✉)
Universidade Fernando Pessoa, CEPESE-Centro de Estudos da População, Economia e Sociedade, Porto, Portugal
e-mail: cramos@ufp.edu.pt

A. Visvizi et al. (eds.), *Research and Innovation Forum 2020*,
Springer Proceedings in Complexity,
https://doi.org/10.1007/978-3-030-62066-0_42

In the first part the relation between freedom and democracy is addressed, in order to explain how both became mutually constitutive in the context of Western (liberal) democracy. The outcome of this debate is the recognition that pluralism is a fundamental pillar for a regime that carries free choice in its core. The second part debates the impacts of the internet in the public space and the challenges it introduces to the creation of a democratic virtual public sphere. The text ends with a brief summary on the constraints and opportunities thus created for democracy.

The Public Space and Democracy

Freedom of speech, freedom of information, freedom of association and pluralism are cornerstones of democratic regimes and thus part of the characteristics without which the threshold of democracy cannot even be reached. Furthermore, for consolidated democracies, these freedoms must be kept inalienable and unlimited and pluralism must be a permanent underpinning of societal order.

Freedom is a foundational principle of Western, liberal, democracy [1]. Freedom may of course seem a slightly different thing seen from the perspective of philosophers or otherwise from the contemporary legal and political perspective. However, I would rather address freedom as historical construction, under the contingencies of space–time evolution.

Freedom, in the singular, emerged as an inalienable characteristic of the individual in the seventeenth century philosophy of natural law. It entered legal coda with the wave of 'liberal' revolutions of the late eighteenth and early nineteenth century and, ever since, was reified as part of a progressively growing number of citizens' concrete rights, later to be declared a 'universal' right [2].

An indirect indicator of a consolidated (liberal) democracy is therefore the fact that individual freedoms are taken for granted among the majority of the population of a given political community. Freedom and political authority are difficult to balance though. The individual freedoms of an isolated person can be theoretically unlimited but the human condition plunges each and every individual into social interaction and social integration and there, among the others, rules are required for social liberty to emerge while individual liberty is preserved. The dilemmas of individualism vs. communitarianism are well known and widely debated in political theory [3]. Democracy as a political regime does not necessarily equate with one or the other; it rather combines characteristics of both, with substantial space–time variation and thus, at times tending more to one or the other side, depending on historical circumstances and political choices.

What is conventionally designated as liberal democracy considers individual freedom as a pillar and therefore promotes a type of social and political organisation that maximises individual choice, notably for direct political decision-making or for the selection of representatives (the one person one vote rule). Some types of democratic theory will even consider that the political structure can be minimised to its tiniest expression, not to mention liberal anarchists, the theories of which rule out democracy, since they deny the very necessity of the state (or other) political structure [4]. Even if the type of democracy applied is closer to egalitarian democracy, thus giving further consideration to issues of balance (to adopt a general designation) in between citizens, and therefore promoting political action capable of producing that equilibrium, the decision-making mechanisms of a democracy (based on self-government and the rule of law) are the outcome of citizens' free direct or indirect choice. Choice, free choice, hence is at the core of democracy.

In turn, being able to choose implies an array of options and this leads to pluralism. Democratic societies are adverse to monism, they require diversity and divergence, but also discussion and informed, well-argued decision-making. Individual freedom per se will not result in common grounds for solving shared problems. But arguing and negotiating can. A further step is required, that of meeting the others, in order to dialogue. The respect of individual freedom imposes the respect for equal statutory conditions in the dialogue. Choosing implies a vast array of possibilities and the capacity to assess them and to make informed decisions. Therefore it requires an ample and plural public space where opinions can be put forward, information exchanged, and debates held. This is, in democracies, the 'public sphere', a space of mediation, communication and decision-making that makes possible the effective connection between the governed and the governors [5].

Freedom of thought and of opinion is in this context not only an individual right but also a positive input to democracy, because individuals bring forward their perspectives on relevant topics of public discussion, in a political environment that is not repressive and actually values and makes use of the citizens' inputs [2]. Freedom of expression is therefore necessary, in order to feed the discussions with qualified information and abundance of standpoints. Individuals may argue independently, or instead decide to gather in groups, in order to support common ideas, in the public debate and for the purposes of political decision, a behaviour protected under the freedom of association right [2].

Democracies in practice may seem less liberal than theoretical liberalism, because they normally produce a regulated polity, where a collectively agreed form of government rules. Once established, the regime also requires democratic control, notably accountability and transparency, which in turn demand for information and communication tools. Seen from these points of view the emergence of new information and communication technologies and the creation of the internet as a powerful instrument of local, national, regional, global and transnational interconnectedness should be acclaimed as a precious toolbox for fostering democracy [6].

This is not always the case though and often we see established political powers, but also civil-society entities and individuals using these tools in a way such that hinders both common interest and individual rights. It is true that given the strength of those new tools it is risky to underestimate their impacts, which are not merely instrumental, since they touch the very substance of democratic communication. They are a revolution, but they can be a revolution 'for good', that very much depending on the way they are used.

What Are the Impacts of the Internet on the Public Space?

The internet as global electronic communication network enables a dense and complex pattern of communication across the world never seen before in the history of democracy. As such it substantially changes the conditions of the traditional public sphere by creating a complementary or alternative sphere, a 'virtual agora' [7] in which the exchange of information and communication among diverse and even unexpected actors is possible.

Howard and Hussain [8] address the new 'digital media' under three main dimensions: a new information infrastructure; a new type of content; and a new type of users. This tripartite analysis helps in understanding how the internet and democracy can be related, and thus is adopted below. A fourth topic on governance is added [9].

The Infrastructure

In spite of the fact that the material infrastructures required by the internet might seem too sophisticated to guarantee a worldwide coverage, the network of networks expanded rapidly and efficiently. According to ITU [10], "the global penetration rate increased from nearly 17% in 2005 to over 53% in 2019". There is of course a certain divide between the wealthier and the poorer countries, since its availability requires material equipments, markets and financial resources, internet providers, know how, but also legal and administrative provisions, hence a political authority, in order to set it to function. According to ITU [10], 86.6% of individuals are online, in the developed countries, while that percentage is only 19.1 in the least developed countries. As for ICT skills, and although data availability is not wide, ITU [10] shows that less than 50 per cent of the population possesses 'basic' computer skills in 40 out of the 84 countries for which data are available; and less than 50 per cent out of 60 countries, if 'standard' skills are considered.

As a network, the internet is transnational, multilateral and decentralised and thus makes several challenges to conventional political power as further addressed below. Nevertheless, authoritarian governments are sometimes resorting to internet shutdowns, or other physical or legal blockage systems, in order to interrupt access to information, which runs counter the previous idea on worldwide access [11, 12]. Therefore, unevenness, both due to economic and political reasons has to be considered.

The Contents

Contents are abundant, diversified and easy to access, hence from the point of view of the availability of information the internet has very much changed conventional information circuits, has democratised access and has immensely broadened the diffusion of information. This creates a broad public sphere in which the participants can resort to a multiplicity of sources in order to construct their own opinions and to make choices. As such pluralism is reinforced. A problem, however, results from the huge amount of available information, since the users need to sort it, by establishing criteria on interest, relevance, reliability and freshness, among others. Public opinion formation and political choices to be made would need to be grounded on reliable information, which may have become more difficult to sort than in the past.

Furthermore, information may also be deliberately inaccurate. The issue of 'fake news' has invaded the debates on democracy and the internet, because those are creating substantial turmoil and hindering, instead of reinforcing, the virtual agora. Besides, depending on the origin and intention of faking, it can be part of broader strategies of counter-information, political instability and sabotage of democracy, as documented in Freedom House's index on internet freedom. Among other, the index measures manipulation and restriction of contents in terms of: the filtering and blocking of websites; censorship and self-censorship; manipulation of content; and the usage of digital media for social and political activism [11]. The V-Dem Institute [12] also presents relevant information on how governments and foreign governments disseminate false information on the internet.

The fact that information travels fast and is frequently updated can also weight on the positive side of the balance. Updated versions of previous pieces of information are regularly available. The adoption of a chronological criterion for sorting this type of information is also a simple process, which many people are able to adopt. However, because information is rapidly produced and rapidly consumed, it is quite often ephemeral and superficial. On-line mass media, for instance, may rush to provide their readers with the latest news without properly checking their sources and contents. Even worse, the flood of informal information exchanged across the network has no guaranteed accuracy. Furthermore, the dividing line between facts and opinions is often blurred, which may reinforce unfiltered adhesion or rejection of ideas, political standpoints and policy options [13, 14].

The Users

One of the major changes introduced in the public space by the internet is a shift in the origin of the information. No longer is this about reading or discussing together information originating from the conventional mass media, within their conventional mediation role between the people and political power. The overall layout of political communication has become a lot more complex.

First, the citizen, traditionally the receiver, moved central stage as also a sender and a producer of information himself. The success of social networks, or of blogging, for instance, testifies to this fact. Authorship is therefore much broader than in the past, but is also, often, rather undetermined, because there is a lot of word-of-mouth and storytelling running on the internet, with poorly or not even checked sources. Deliberate and unintended mistakes also occur in this process. As a positive feature for the public sphere, the whole process has broadened public debate and has democratised opportunities of participation as never before. Yet, it remains for the citizen to progressively develop the necessary skills for assessing the quality of the information he reaches and diffuses, something that implies digital literacy but goes beyond it, into the fields of political culture and overall educational levels [15]. This idea is conceptually developed under the 'information literacy' label [16].

Second, the conventional mass media are also challenged by this major shift in the paradigm of political communication. They have had to adapt from the point of view of infrastructures (e.g. moving from paper to electronic supports), contents and style. Unlike the common citizens, they are under professional deontology rules, which urge them to check sources and to provide reliable information. Yet, they are also in the market of mass communication, where competition mounted in recent years, namely due to the aforementioned outsiders, a herd of amateurs that suddenly entered the media world in an informal but pervasive way. Major adaptation has therefore come along with the new technological environment and requires from the media enhanced capacities at the technical, professional and financial levels [13, 14].

Third, political institutional actors also discovered a non-mediated (or less-mediated) way to address the citizens. Amongst pressures for transparency in and accountability of the political processes, in political campaigns leading to elections or in public discourse backing major political decisions, politicians now have at their disposal the same tools that the citizens keep handy at home, or in their pockets. The process is tempting because it enables 'direct' communication, the stepping down from formal communication, press conferences and affected speeches, to 'tweets' and 'posts', in a familiar, abridged and often emotional language. The challenge is also enormous for advisory communication offices having to react ex-post facto to political utterances [17].

Political institutions also benefit a lot from these new tools of communication for purposes of transparency and accountability. Meetings (parliamentary, working groups, commissions, etcetera) broadcast live and directly made public via the internet, documents (official journals, reports, working documents, procurement procedures) published online, surveys and consultation procedures available on websites, all became common practice of governments and other organisations.

However, the same infrastructures of communication also serve populist purposes, in that they create the conditions to amplify demagogic discourses, which can in turn be fed by a more accurate perception of on-going online popular debates. From digital interference in elections to the manipulation of social media, many politicians and regimes are being accused of foul play on the internet, and a drift towards 'illiberalism' and 'digital authoritarianism' is pointed out by international research [11, 12].

Last but not the least, the interaction in between these three elements—citizens, government and the media—is more intense and more intricate than ever. Communication is not unidirectional it is multilateral and complex, but has substantially fostered the possibilities for intense interaction between the represented and the representatives. Furthermore, patterns of participatory and deliberative democracy are reinforced with reference to the possibility of using this vast array of new tools, and contribute to relaunching debates on types of democracy and the quality of democracy. Horizontal and transnational communication is also substantially reinforced. Peer to peer communication between citizens and in spite of borders continuously happens, thus creating conditions for a transnational public sphere to emerge gradually. As a whole, the layout of communication is so complex that it requires innovation in governance too.

Governance

Internet governance for the purposes of managing a virtual public space on to which many democratic processes were transferred is a complex issue, given the intricate nature of its structure. Unlike conventional state-society relations where the vertical metaphor of political power applies and thus upwards and downwards movements can be identified, the internet structure is largely horizontal, transnational, decentralised and even non-hierarchical, hence escaping conventional power structures.

States naturally attempt to create management tools, from controlling the provision of infrastructures to managing contents and regulating the use. The regulatory practices are not necessarily undemocratic, quite on the contrary, they may relate to attempts to broaden access, or to protect rights threatened by illegal uses of the internet or to fight criminal behaviour (eg. privacy laws; fighting child pornography). However, a substantial number of provisions in several countries in the world were deliberately created for imposing limits to internet freedom [11]. A virtual and unbound 'territory' as the internet easily escapes conventional judicial control, and thus calls for further action, also at the international level, as expressed in the many international and regional instruments in place [18].

The international legal framework and actual capacity for managing the internet, with a view to protecting citizens' rights and democratic principles is nevertheless insufficient and flawed, as much research provides evidence of. Hindrances to basic freedoms emerge from these flaws. Freedom House, for instance regularly identifies violations of users rights such as restrictions on online activity, surveillance, legal prosecution, imprisonment, physical attacks and other forms of harassment [11]. UNESCO [19] also developed a comprehensive set of indicators on 'Internet universality', based on the following four 'R-O-A-M' principles, in which R stands for 'Rights', O for 'Open', A for 'Accessible to all' and M for 'Multistakeholder participation'. Under the Rights category, UNESCO measures: the overall policy, legal and regulatory framework for human rights and their relation to the internet; freedom of expression; the right to access information; freedom of association; rights to participate in public life; the right to privacy; and economic, social and cultural rights [19].

Freedom on the internet thus became yet another dimension of freedom and bears a major connection with the democratic functioning of political regimes. The fact itself is evidence of how the internet became so pervasive an element of the public sphere.

Final Remarks

Despite all difficulties emerging from the use of the internet as a public space—and some of them quite challenging, as are fake news, electoral interference or censorship—the fact that a powerful tool of information and communication entered the setting of democracies cannot be ignored, in its potential for fostering democracy.

Major challenges lie with the citizens, whose digital literacy needs to go beyond technical skills, into deep and critical analysis of the contents they access. Major challenges lie also with the mass media, which need to keep up to their task of informing, with quality and in an ethical manner. Last but not the least, a major challenge is for the politicians to be able to protect basic freedoms—among which internet freedom has won a place—in order to guarantee that misuses of the internet do not hinder its immense democratic potential. Internationally, it is also for organisations other than the states to stand for regulatory frameworks that foster the safe and ethical use of the internet. This way, it can become a plus for consolidated democracies and an opportunity for the democratisation of regimes in transition. The qualitative change the internet brought to democracies is basically the opening up of the public space into an expanded virtual agora. Critical citizens and reliable politicians are not an outcome of the internet, because the internet is only their instrument, not their master.

References

1. R. Dahl, *Democracy and Its Critics* (Yale University Press, New Haven, 1989)
2. UDHR-Universal Declaration of Human Rights, https://www.un.org/en/universal-declaration-human-rights/. Accessed 25 Mar 2020
3. P. Pettit, *A Theory of Freedom: From the Psychology to the Politics of Agency* (Polity, Cambridge, 2001)
4. D. Held, *Models of Democracy*, 3rd edn. (Stanford University Press, Stanford, 2006)
5. J. Habermas, *The Structural Transformation of the Public Sphere. An Inquiry into a Category of Bourgeois Society* (MIT Press, Cambridge MA, 1991)
6. Z. Papacharissi, The virtual sphere 2.0: the internet, the public sphere, and beyond. in *Routledge Handbook of Internet Politics*, ed. by A. Chadwick, P.N. Howard (Routledge, London, New York, 2009), pp. 230–245
7. A. Visvizi, M.D. Lytras (eds.), *Politics and Technology in the Post Truth Era* (Emerald Publishing, Bingley, UK, 2019)
8. P.N. Howard, M.M. Hussain, *Democracy's Fourth Wave?* (Digital Media and the Arab Spring. Oxford University Press, Oxford, 2013)

9. C.T. Ramos, From the freedom of the press to the freedom of the internet: A new public sphere in the making?, in *Politics and Technology in the Post Truth Era*. ed. by A. Visvizi, M.D. Lytras (Emerald Publishing, Bingley, UK, 2019), pp. 9–22
10. ITU-International Telecommunication Union: Measuring digital development. Facts and figures 2019. ITU, Geneva (2019)
11. Freedom House. Freedom on the net 2019, https://freedomhouse.org/report/freedom-net/2019/crisis-social-media. Accessed 28 Mar 2020 (2019)
12. V-Dem Institute: Democracy Facing Global Challenges. V-Dem Annual Democracy Report 2019. V-Dem Institute, Gothenburg (2019)
13. L. Kung, R. Picard, R. Towse, *The Internet and the Mass Media* (Sage, London, 2008)
14. E. Siapera, *Understanding New Media*, 2nd edn. (Sage, London, 2018)
15. UNESCO-United Nations Educational, Scientific and Cultural Organization: A Global Framework of Reference on Digital Literacy Skills for Indicator 4.4.2. Information Paper Nº 51 June 2018 UIS/2018/ICT/IP/51. https://uis.unesco.org/sites/default/files/documents/ip51-global-framework-reference-digital-literacy-skills-2018-en.pdf. Accessed 28 Mar 2020 (2018)
16. UNESCO-United Nations Educational, Scientific and Cultural Organization: Information Literacy. Retrieved from https://www.unesco.org/new/en/communication-and-information/access-to-knowledge/information-literacy/. Accessed 28 Mar 2020 (2017)
17. R.M. Perloff, *The Dynamics of Political Communication: Media and Politics in a Digital Age* (Routledge, New York, NY, 2014)
18. UNESCO-United Nations Educational, Scientific and Cultural Organization: International and regional instruments. https://en.unesco.org/international_and_regional_instruments. Accessed 28 Mar 2020. (2019a)
19. UNESCO-United Nations Educational, Scientific and Cultural Organization: UNESCO's Internet universality indicators: a framework for assessing Internet development. UNESCO, Paris, https://unesdoc.unesco.org/ark:/48223/pf0000367617/PDF/367617eng.pdf.multi, Accessed 28 Mar 2020 (2019b)

Chapter 43
Simulating and Epidemic Prediction of COVID-19 Transmission in Universities Considering Different Interventions

Zain Balfagih and Mohammed Balfaqih

Abstract Pandemic diseases are fought through efficient intervention procedures including non-pharmaceutical interventions such as social distancing, and closure of events, schools and workplaces. These procedures are ruled by different decisions-makers in local communities, and national authorities. Different procedures are conducted based on the community affected by the pandemic disease to reduce the severity in various social life. During the Coronavirus (COVID-19) pandemic most of universities worldwide closed their campuses and continue the study through online learning. To open back the campuses, universities have to analyze and investigate such decision impact on the outbreaks. This paper develops a simulation model of COVID-19 investigating realistic intervention procedures that can be conducted to control and mitigate outbreaks. The model is constructed to simulate Effat University environment. Our model predicts average COVID-19 attack rates for various intervention scenarios, such as partial closure and social distancing, cleaning and disinfection, COVID-19 symptoms monitoring system are combined with isolate and treat confirmed COVID-19 cases. The findings indicate that combining all the proposed interventions can be substantially more effective than isolate and treat confirmed COVID-19 cases alone from epidemiological standpoint.

Keywords COVID-19 · Epidemic · School opening delay · School closures

Z. Balfagih (✉)
Computer Science Department, Effat University, Jeddah, Saudi Arabia
e-mail: zbalfagih@effatuniversity.edu.sa

M. Balfaqih
Department of Computer and Network Engineering, University of Jeddah, Jeddah, Saudi Arabia

A. Visvizi et al. (eds.), *Research and Innovation Forum 2020*,
Springer Proceedings in Complexity,
https://doi.org/10.1007/978-3-030-62066-0_43

Introduction

Corona virus disease COVID-19, as reported by the World Health Organization (WHO) in March 2020, is threatening people worldwide as a global pandemic due to the fact that this virus is spreading from a person to another by different means. With the fast spread of COVID-19, the increment of death cases caused by COVID-19, in addition to unavailability of an effective treatment and vaccine, several governments applied strict actions to control the pandemic. These actions include community lock down, school closure, travel restrictions, and cancellation all kinds of gatherings [1]. Countries which implemented these interventions at early stage observed a reduce in the spread of the disease and were able to lower down their epidemic curves [2, 3]. However, these actions effected the countries' economy and the social well-being of the population.

Kingdom of Saudi Arabia (KSA) started the implementation of several interventions such as schools closing, travel suspension, controlling gatherings, and many others as shown in Fig. 43.1. Anyway, implementing the social distancing was a challenging task due to its social traditions, religious activities, and hosting two of the major Muslims' gathering which are Hajj and Omrah.

Universities around KSA was locked down and moved to online education since 9th March 2020. Despite the efficiency of the online education, many problems were faced such as system failures, and lack of internet for some of the students [5]. So, it is the responsibility of the University administration to take critical decisions such as classes cancellation, closing research facilities, and identifying the university populations that could enable them to resume in-campus education. Effat university is a private university in Jeddah-KSA that has around 2000 female students and four

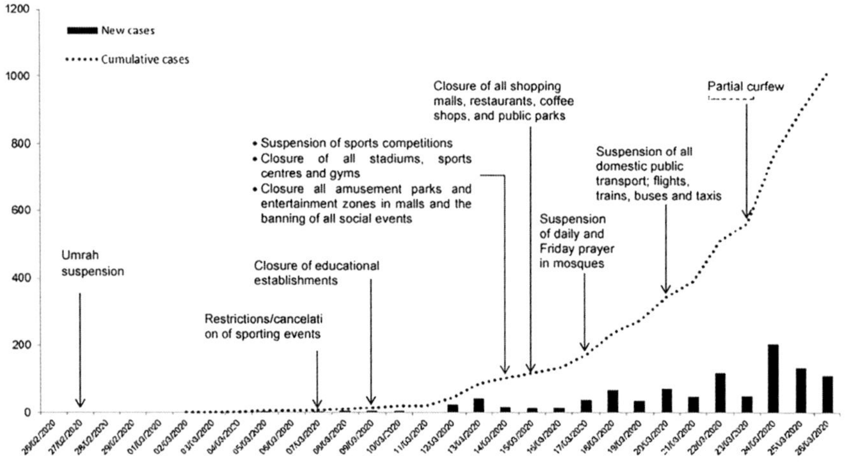

Fig. 43.1 Cumulative and daily cases of COVID-19 and key social distancing interventions in the kingdom of Saudi Arabia as of the 26th of March 2020 [4]

different colleges. The University's goal is to provide the practical procedures and appropriate policies to reduce/prevent the spread of the COIVD-19 while providing the main services in the same time.

In this paper, a simulation model that simulates the effectiveness of implementing different policies during COVID-19 pandemic is presented to test the impact of the intervention strategies that could be used by one of the private universities in the Kingdom of Saudi Arabia, Effat University. This model is designed and developed using Python to measure the effectiveness of the implementation of different policies during the pandemic. The model was calibrated to published illness attack rates and basic reproductive number estimates of COVID-19 as reported by Saudi Ministry of Health (MoH), and is constructed to represent a typical urban community in Jeddah city [6]. The model considers the medical and non-medical strategies that could be taken by the university to control the virus spread. The proposed model predicts average illness attack rates under various scenarios of intervention strategies. We find that certain combination strategies can be more effective than others and can help the university to resume the study in campus.

In the following sections, we will present different research efforts in the literature that tries to mitigate the effects of COVID-19. Then, we will present the proposed model structure and input parameters. Different intervention strategies considered in this model are presented. Finally, we will present and discuss the output analysis with some recommendations.

Related Works

Numerous countries have taken serious actions to control the spread of the virus by imposing the social contact, such as locking down schools, restaurants, and shops, and stopping all public events and imposing studying/working from home. These procedures are known as "social distancing", and are especially efficient for diseases which are transmitted from a human to human within a small range [7]. Several countries (e.g., Germany, China, KSA) have applied social distancing by enforcing lock-downs in the whole country or at least in certain states, while other countries (e.g., Taiwan, Netherlands, Sweden) have taken less strict social distancing procedures. However, there is no clear clue that shows for how long this social distancing will be imposed. Moreover, based on the level of the virus spread, countries might need to impose social distancing in the near future [8].

Social distancing procedures have notable impact on people's daily and routine activities. Unfortunately, many people are working/studying from home, temporarily working as part time or lost their jobs, and most unessential activities and gatherings are cancelled. These actions had a clear effect economically where flights number strongly decreased to almost zero, car traffic went down, and universities' campuses were closed [9–11]. Anyway, the good thing is that this is not a permanent situation where it is expected that out of home activities and flights demand will increase when the social distancing interventions are cancelled although it is not clear how long the adopted interventions will continue.

The mathematical models can help to evaluate when the intervention procedures adopted by the different countries are the optimum to control the disease and to measure how they could affect the virus spread. In [12], authors have proposed a mathematical model representing a quarantine level and the different intervention options that the Indian government adopted to control COVID-19 transmission within the community. They found that reducing the direct contact of humans who could transmit the virus is the most critical factor to be able to control the virus spread and minimize the number of infected people in all times. They have integrated the intervention policies adopted by the Indian government in their mathematical model as a time-dependent control parameter [12].

In another study, authors developed a mathematical model to predict the COVID-19 attack rates in Guangdong province in China [13]. Their model found that the COVID-19 will have another explosion attack when the input population is highly increased. However, the current governmental interventions procedures remarkably decrease the values of infections to around zero [14], proposed Patient Information Based Algorithm (PIBA), for predicting the possible death ratio of COVID-19 in real-time using real-world collected data in the last couple of months. They found the death ratio based on available data from the city which COVID-19 started from (Wuhan) and then in some Chinese cities and it was close from the actual reports in these cities [14]. In another research, the effect of weather conditions on the level of COVID-19 spread in different US states was investigated and aligned with the different weather conditions daily. Based on the obtained results, the research predicted the states in India that have higher attack rates risk, which could highly have an outbreak attacks of COVID-19 within few months from now [15].

Proposed Modeling of COVID-19 Transmission

The proposed simulation modeling is discussed including COVID-19 transmission route, and the simulation input considerations. It simulates the spread of COVID-19 in universities with and without applying interventions. Effat University campus is the considered scenario in this study.

Transmission Route Model

The transmission model is designed based on the Susceptible Infectious Removed (SIR) compartmental model. SIR is an epidemiological model for airborne infectious disease in a small area [16]. The model consists of 4 states; susceptible, infected but not infectious, infectious, and removed (i.e., dead, hospitalized or recovered with immunity). Algorithm 1 shows SIR compartmental model.

Algorithm 1 Transmission route model

```
track the infection status of each person from the university's members each day;
If susceptible university's member is infected then
    proceed to latent period (the period between being infected and becoming infectious);
    proceed an infected person to an infectious period;
    proceed to the removed state (the infected person is dead, hospitalized or recovered with immunity);
else if susceptible university's member is not infected then
    add the susceptible university's member to contact group;
end if;
```

At the first day of in campus study, a random infected university's members are assumed to be infected while the rest are susceptible. The susceptible is the university's member who has the possibility to get infected by the contact groups which represent set of people might be contacted in a day. In a university, the contact group of each university's member mostly includes household, neighborhood, and community of the university staff, colleges, hostels, university's daycare and workgroup. According to the world health organization, anyone with chronic diseases or pulmonology and elderly of ages 65 and higher are at higher risk of severe COVID-19 [17]. This group will be prohibited from attending to the campus. Thus, in our model, everyone has contacts with people in their household, neighborhood, community, colleges, and hostels. Children of ages 0–5 may also visit a daycare center. Workers contact all of the persons in their workgroups.

All data about university members are taken from Student Affairs and Human Resource Departments for Spring 2020. The total of allowed staff and students to attend to the university are approximately 2000 persons. As an intervention for Covid-19 outbreaks, the university will open only the essential facilities including 4 academic buildings, 2 hostels, 1 daycare, and deanship of admission and registration. We formed workgroups of size 20 (i.e., 15 students and 5 faculty staff) to represent the typical number of people with whom a person has close contact during a day.

The model simulates the infection status of each university's member in a day. Accordingly, the simulation system records the infection attack rates in all university's members and different age groups. Daily, the simulation system updates the record and determines the probability of infection for each university's member. The probability, for a university's member, is calculated as one minus the multiplication of the probability of not being infected by infectious contacts in each group of that member. For instance, the probability of infection on a specific day for a lecturer who lives at the hostel is

$$PI_{Lecturer} = 1 - \left[PNI_{col} \times PNI_{hos} \times PNI_{nig} \times PNI_{com}\right] \quad (43.1)$$

where $PI_{Leturer}$ is the infection probability of a lecturer, PNI_{col}, PNI_{hos}, PNI_{nig}, and PNI_{com} represent the probability of not being infected from the contacts at college, hostel, motherhood and community, respectively. In addition, in a particular contact group, the infection probability of a susceptible university's member is determined based on infectious contacts in that group. For instance, the probability that a lecturer being infected in college contact group on specific day is calculated as;

$$PI_{Lecturer\_col} = 1 - [(PNI_{col})^X \times (PNI_{col})^Y] \tag{43.2}$$

In a specific contact group, X is the number of infectious childes and Y is the number of infectious adults. As long as a university's member is not infected, he/she is classified as a susceptible and continues the daily routine till the check at the end of day. If a member is infected with COVID-19, the model state is turned into latent which is the period between being infected and infectious. The latent period according to WHO is lasting 1–3 days [18]. Subsequently, the infected university's member starts an infectious period weather the symptoms are developed or not. Based on WHO reports, the probability that the symptoms are exhibited is 0.2 [18]. Lastly, the removed state is entered either recovered (with immunity for COVID-19), hospitalized, or dead.

Input Considerations

The inputs of the simulation system are different characteristics of university's members, transmission probabilities, and durations of the latent and infectious periods. The university member's contact structure follows the structure in [19]. The transmission probabilities within households is set based on the results obtained in [20, 21].

The transmission probabilities have been calibrated to match the real-world COVID-19 attack rates and R_0 values introduced by [22], in which $R_0 = 2$. Moreover, for the durations of the latent and infectious periods, the empirical cumulative distributions presented in [20, 21] shown in Table 43.1 are used. The duration means of latent and infectious periods are 1.9 and 8.2 days, respectively [20, 21].

Table 43.1 Empirical distributions of latent and infectious periods for COVID-19 [18, 19]

Latent period		Infectious period	
Duration (days)	Cumulative probability	Duration (days)	Cumulative probability
1	0.3	3–5	0.3
2	0.8	6–8	0.7
3	1.0	9–11	0.9
		12–17	1.0

Algorithm 2 The attack rates calibration

```
insert the transmission probability to SIR Compartmental Model;
process SIR Compartmental Model;
if SIR Compartmental Model output does not match with the real-world observed attack rates then
    determine the calibrated attack rates;
    insert the transmission probability to SIR Compartmental Model;
    process SIR Compartmental Model;
end if;
```

Algorithm 2 illustrates the calibrating of transmission probabilities. After setting the input with appropriate transmission probabilities of different age groups, the obtained output is compared with the real-world attack rates and R_0 of each age group. The real-world values are obtained from statistics introduced by Saudi Ministry of Health. The input is tweaked iteratively till the best match is matched. The values of calibrated attack rates and R_0 are close to the real-world values. The identified calibrated attack rates were 11% and 85% for children and adults, respectively. Table 43.2 illustrates the obtained transmission probabilities used in the simulation.

Interventions

Our simulation models the baseline without any intervention, and with different types of intervention. Four interventions have been considered including; partial closure and social distancing, cleaning and disinfection, COVID-19 symptoms monitoring system as well as isolate and treat confirmed COVID-19 cases. The interventions are considered when the overall attack rates of COVID-19 reach to 0.01%.

Table 43.2 Per-contact COVID-19 infection transmission probabilities within contact groups

Contact group	Transmission probability
Household	
Child to child	0.6
Child to adult	0.2
Adult to child	0.2
Adult to adult	0.24
Community	
Children	0.000004
Adult (ages 18–55)	0.000005
Adult (ages 56+)	0.000005
Daycares	0.028
Workgroups	0.075

Isolate and treat confirmed COVID-19 cases. One of significant interventions is reporting of suspected cases and rapid response by transferring a confirmed COVID-19 cases to a designated hospital in coordination with command and control center. The university will ensure providing an intensive supportive care with treatment of symptoms to manage the infection in people. The treatment includes pain relievers, cough syrup or medication, rest, and fluid intake. According to Grifoni et al. [23], 100% of COVID-19 recovered cases made antibodies. Thus, university's members who recovered from COVID-19 will be allowed to attend the university.

Partial closure and social distancing. To control outbreak of COVID-19, educational activities on campus have been suspended at most of the universities worldwide. Although the educational process is currently performed online, it is not suitable for all courses especially practical science. To open back education institutes, the most critical question that must be answered is how the disease transmissibility on campus will be and how could be controlled through. Hence, certain university's members and students will be allowed to present at the university. University's members will be categorized into essential and non-essential. The members that the educational process cannot continue without their attending are classified as essential members while the rest are non-essential. On the other hand, the students are categorized also based on residential status and health situation into essential and non-essential. Students living outside Jeddah city (i.e., Effat University City) will not be allowed to attend the universities.

Cleaning and disinfection. Cleaning and disinfection of university's members and facilities is one of the considered interventions. Wearing face mask will be compulsory for all members at no cost to them. Routine cleaning of surfaces using appropriate cleaning and disinfection methods will be conducted specially to frequently touched surfaces and objects, for example, doorknobs, light switches, handrails, desks, chairs, phones, printers, keyboards and computer mice. In addition, classes, venues and offices will be sterilized daily.

COVID-19 symptoms monitoring system. The suspected COVID-19 case is a patient with at least one of the following symptoms; fever or recent history of fever, cough, sore throat or shortness of breath [17]. Thus, a high temperature monitoring system will be used with real time detection and Artificial Intelligence (AI) based analytics at the entrance of university buildings. The monitoring system is contactless thermal and optical screening detecting high body temperature in crowded areas with 10–15 simultaneous readings. The system includes mobile application that can be used by university's members to enable contact tracing and fast information distribution.

Results and Discussion

The epidemiological results of our simulation are analyzed to investigate the effectiveness of the utilized interventions. For validation, the simulation results found that real-world data are approximately similar. From the calibrated simulation, the average overall COVID-19 attack rates without any interventions is 25.3%. The strategy of partial closure and social distancing reduces the average overall COVID-19 attack rate to 9.8%. The strategies partial closure and social distancing combined with cleaning and disinfection reduce the average overall COVID-19 attack rate to 3.2%. If COVID-19 symptoms monitoring system is also added to the previous interventions, then the average overall COVID-19 attack rate can be reduced to 2.4%. The strategies partial closure and social distancing combined with COVID-19 symptoms monitoring system reduce the overall COVID-19 attack rate to 1.4%. The strategies partial closure and social distancing combined with COVID-19 symptoms monitoring system reduce the overall COVID-19 attack rate to 0.8% if combined with isolate and treat confirmed COVID-19 cases. In general, it is noted that social distancing practices are the most effective intervention in reducing that attack rates. It is expected that several infected individuals will be in campus every day. However, many of them do not show any symptoms but they can transmit the virus to others. Figure 43.2 shows the number of infected people in the campus daily during the first month with and without interventions. It is noted that without applying any intervention, 75% of the university population will get infected in 7 to 10 days. However, with applying all the suggested interventions, only 8% of the population on campus are infected in the same time at maximum. Applying some of the interventions have different impacts on the infection rate.

Applying these interventions does not only reduce the attack rates, it also reduces number of people who need to be hospitalized. Figure 43.3 shows the rate of people who are admitted to hospital with and without applying the proposed interventions.

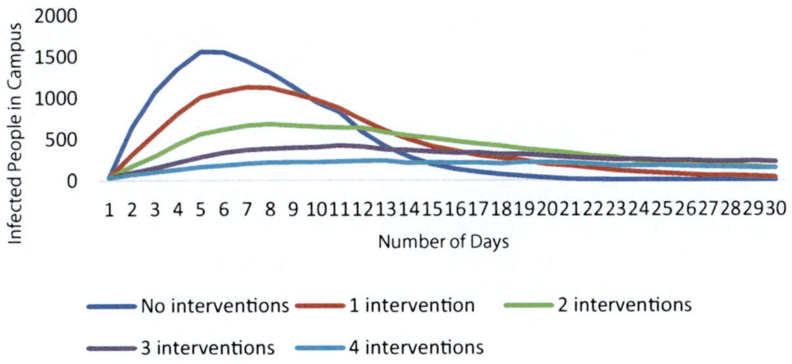

Fig. 43.2 Infected people in campus

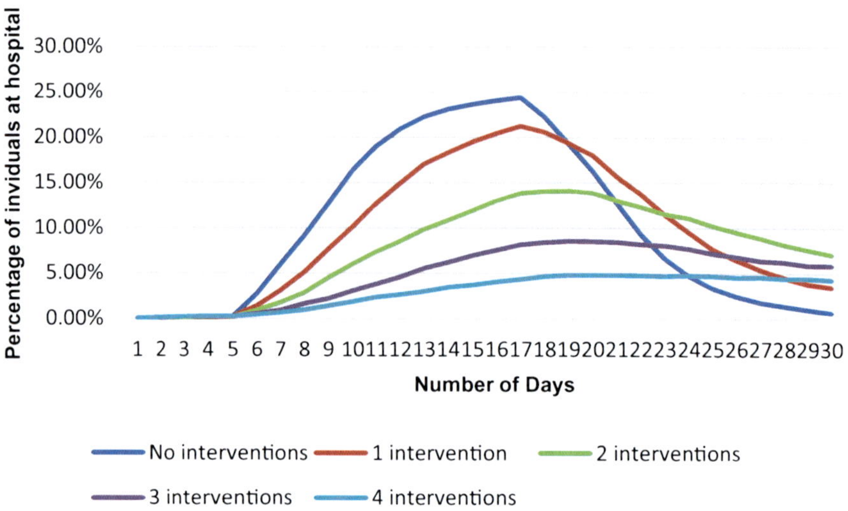

Fig. 43.3 Admission rate in the hospital with and without interventions

Applying all interventions will reduce the rate of individuals who need to be hospitalized to less than 5% at all times. Finally, the death rate without applying the intervention rate was 0.2% and it was almost zero with the interventions. This low rate reflects the real-world data where the death rate in KSA is 0.7% and most of them are either elderly or with chronic diseases. As the whole population has no elderly people or with chronic diseases, the death rate is highly minimized.

Conclusions

This paper develops a simulation model of COVID-19 investigating realistic intervention procedures that can be conducted to control and mitigate outbreaks. The model is constructed to simulate Effat University environment. Our simulation model can represent a typical mid-sized university and predicts average COVID-19 attack rates under various intervention scenarios; specifically, partial closure and social distancing, cleaning and disinfection, COVID-19 symptoms monitoring system as well as isolate and treat confirmed COVID-19 cases. The simulation findings indicate that combining all the proposed interventions is more effective than isolate and treat confirmed COVID-19 cases alone. The interventions decrease attack rates significantly.

References

1. S.H. Ebrahim et al., *Covid-19 and Community Mitigation Strategies in a Pandemic* (2020)
2. D. Fisher, A. Wilder-Smith, The global community needs to swiftly ramp up the response to contain COVID-19. The Lancet **395**(10230), 1109–1110 (2020)
3. V.J. Lee, C.J. Chiew, W.X. Khong, Interrupting transmission of COVID-19: lessons from containment efforts in Singapore. J. Travel Med. **27.3**, taaa039 (2020)
4. S. Yezli, A. Khan, COVID-19 social distancing in the Kingdom of Saudi Arabia: bold measures in the face of political, economic, social and religious challenges. Travel Med. Infect. Dis., 101692 (2020)
5. A. Niwaz, Q.W. Ahmed, S. Kamran, An exploration of issues and challenges faced by students in distance learning environment. Glob. Soc. Sci. Rev. (GSSR) **IV**(IV), 77–83 (2019)
6. Saudi Ministry of Health (2020), https://covid19.moh.gov.sa/ [21 May 2020]
7. A. Wilder-Smith, D.O. Freedman, Isolation, quarantine, social distancing and community containment: pivotal role for old-style public health measures in the novel coronavirus (2019-nCoV) outbreak. J. Travel Med. **27.2**, taaa020 (2020)
8. J.T. Wu, K. Leung, G.M. Leung, Nowcasting and forecasting the potential domestic and international spread of the 2019-nCoV outbreak originating in Wuhan, China: a modelling study. The Lancet **395**(10225), 689–697 (2020)
9. C. Goldbaum, Subway service is cut by a quarter because of coronavirus. The New York Times **24** (2020), https://www.nytimes.com/2020/03/24/nyregion/coronavirusnyc-mta-cuts-.html?searchResultPosition=6 [10 Mar 2020]
10. N. Ferguson et al., *Report 9: Impact of Non-pharmaceutical Interventions (NPIs) to Reduce COVID19 Mortality and Healthcare Demand* (2020)
11. L.K. Gee, M.J. Schreck, A. Singh, From lab to field: social distance and charitable giving in teams. Econ. Lett., 109128 (2020)
12. A. Tomar, N. Gupta, Prediction for the spread of COVID-19 in India and effectiveness of preventive measures.Sci. Total Environ., 138762 (2020)
13. Z. Hu et al., Evaluation and prediction of the COVID-19 variations at different input population and quarantine strategies, a case study in Guangdong province, China. Int. J. Infect. Dis. (2020)
14. L. Wang et al., Real-time estimation and prediction of mortality caused by COVID-19 with patient information based algorithm. Sci. Total Environ., 138394 (2020)
15. S. Gupta, G.S. Raghuwanshi, A. Chanda, Effect of weather on COVID-19 spread in the US: a prediction model for India in 2020. Sci. Total Environ., 138860 (2020)
16. E. Beretta, Y. Takeuchi, Global stability of an SIR epidemic model with time delays. J. Math. Biol. **33**(3), 250–260 (1995)
17. Saudi Center of Disease prevention and control, Coronavirus disease 19 (COVID-19) Guidelines V1.1, Saudi Ministry of Health, February 2020. Available at https://www.moh.gov.sa/CCC/healthp/regulations/Documents/Coronavirus%20Disease%202019%20Guidelines%20v1.1.pdf [13 May 2020]
18. World Health Organization, *Coronavirus Disease 2019 (COVID-19): Situation Report*, 80 (2020). https://www.who.int/emergencies/diseases/novel-coronavirus-2019/situation-reports [21 May 2020]
19. I.M. Longini Jr. et al., Containing pandemic influenza with antiviral agents. Am. J. Epidemiol. **159.7**, 623–633 (2004)
20. Y. Ling et al.,Persistence and clearance of viral RNA in 2019 novel coronavirus disease rehabilitation patients. Chin. Med. J. (2020)
21. K.K.-W. To et al., Temporal profiles of viral load in posterior oropharyngeal saliva samples and serum antibody responses during infection by SARS-CoV-2: an observational cohort study. The Lancet Inf. Dis. (2020)
22. G. Viceconte, N. Petrosillo, COVID-19 R0: Magic number or conundrum? Inf. Dis. Rep. **12.1** (2020)
23. A. Grifoni et al., Targets of T cell responses to SARS-CoV-2 coronavirus in humans with COVID-19 disease and unexposed individuals. Cell (2020)

Chapter 44
Towards Rare Disease Knowledge Graph Learning from Social Posts of Patients

Giacomo Frisoni, Gianluca Moro, and Antonella Carbonaro

Abstract Rare diseases pose particular challenges to patients, families, caregivers, clinicians and researchers. Due to the scarce availability of information and their disintegration, in recent years we are witnessing a strong growth of patient communities on social platforms such as Facebook. Although the data generated in this context are of high value, the currently existing ontologies and resources tend to ignore them. The work presented in this paper studies how to extract knowledge from the large availability of unstructured text generated by the users over time, in order to represent it in an organized way and to make logical reasoning above. Starting from the awareness of the need to integrate different methodologies in complex domains, the research shows a combined use of Text Mining and Semantic Web techniques. In particular, we describe the basis of a novel approach for Knowledge Graph Learning with the aim of introducing a patient-centered vision into the world of Linked Open Data. By identifying and representing correlations between concepts of interest, we show how it is possible to answer patients' questions and provide them with an additional tool for decision making. The outlined contribute minimizes costs through automatic data retrieval and increases the productivity of investigators.

Keywords Text mining · Semantic web · Knowledge graphs · Rare diseases

G. Frisoni (✉) · G. Moro · A. Carbonaro
Department of Computer Science and Engineering – DISI, University of Bologna, Via dell'Università 50, 47522 Cesena, Italy
e-mail: giacomo.frisoni@studio.unibo.it

G. Moro
e-mail: gianluca.moro@unibo.it

A. Carbonaro
e-mail: antonella.carbonaro@unibo.it

A. Visvizi et al. (eds.), *Research and Innovation Forum 2020*,
Springer Proceedings in Complexity,
https://doi.org/10.1007/978-3-030-62066-0_44

Introduction

When reflecting on the concept of rarity in the health sector, we tend to think of a remote event reserved for a few individuals. The moment this phenomenon breaks into a person's daily life, it is realized how the word "rare" is actually frequently used improperly.

A disease is defined as rare when its prevalence, understood as the number of cases present in a given population, does not exceed a set threshold, codified by the legislation of each individual country. Rare diseases (shortened as *RDs*) therefore pose particular challenges to patients, families, caregivers, clinicians and researchers. After being neglected for many years (coming to be called "health orphans"), today they are an important public-health issue.

Due to the scarce availability of information and their disintegration, in recent years we are witnessing a strong growth of patient communities on the web (in correlation with the result of a survey carried out by EURORDIS, which sees rare disease patients more likely to share their health data than the general population [28]). The need to communicate with other people with the same problem is often a natural consequence of the desire to break down the isolation barriers in which a patient is locked up. Social contexts (such as *Facebook* groups) therefore become the environment through which experiences are shared, opinions are requested and information of undoubted relevance is exchanged throughout the whole rare patient path (from symptoms to diagnosis, from therapeutic treatments to specialized centers, from reference doctors to lifestyle impact). Sharing health data to advance scientific research and improve clinical benefits are of particular importance in the field of rare diseases where knowledge and expertise are limited, and patient populations are geographically dispersed.

For what concerns Artificial Intelligence and Knowledge Extraction, *explainability* today is a very important and requested property. Highly advanced methods such as deep neural networks are often not allowed in medical domains, where the ability to explain the learned knowledge is considered essential [26]. On the other hand, *knowledge graphs* are gaining popularity and are looking forward to being one of the most promising technologies in the coming years [12]. The growing complexity of the problems leads many experts to assert how the future is centered on a fusion of quantitative and qualitative solutions.

Patients knowledge extraction is a new research subject that cannot yet boast the presence of numerous contributions. The text mining developments that have taken place in these last months have made it possible the achievement of goals deemed insurmountable until recently. Several ontologies, controlled vocabularies and semantic web resources have been released both in the medical field and — more specifically—in that of rare diseases [16, 19, 32, 33]. However these are primarily intended for nosological use by doctors and researchers. Several applications of interest have been implemented above them [14, 21, 25], but opinions coming directly from patients still tend to be ignored.

This paper arises from an understanding of the value of such data accumulated over time and from the desire to make them easily available. We propose the use of text mining and semantic web techniques to extract the knowledge contained in the high amount of unstructured text generated by users through social interactions, in order to represent it in an organized way and enable deductive logical reasoning above it.

The described framework outlines the basis of a project capable of allowing rapid access to many high-value information (in topics such as epidemiology, symptomatology, diagnosis, treatments, drugs, nutrition), responding to patients' questions and providing them with an additional tool for decision making where insights and multiple points of view are required. All this by minimizing costs thanks to automatic data retrieval and increasing the productivity of researchers. The potential of the data collected by individual patients, aggregated anonymously and discussed on such large scale, create a unique opportunity to combine the real context with the world of scientific research, contributing to the digital health revolution.

The paper is organized as follows. Section "Understanding Text" illustrates the recent trend to adopt Artificial Intelligence solutions which integrate quantitative and qualitative approaches. By analyzing the most interesting works in literature for the objectives of the research, we outline the properties necessary for the extraction and representation of patient knowledge. Section "Semantic Web for Health" discusses the resources and the projects currently existing in the field of rare diseases, with a focus on semantic web ones. Section "Knowledge Graphs" delves into the growing importance of knowledge graphs. Section "Extracting and Representing Patient Knowledge" introduces a new approach for the automatic learning of knowledge graphs starting from the patient's social posts. Finally, Section "Conclusions and Future Works" presents conclusions and future works.

Understanding Text

With the aim of automatically extracting and representing the knowledge contained in the patients' posts, it is necessary to deepen the current approaches for the Natural Language Understanding (NLU) and the trends foreseen for the immediate future.

Towards a Mixed AI

There is a precondition for enabling machines to understand natural language: the textual content must be read and interpreted correctly by them. To do this there are mainly two distinct approaches.

- **Quantitative (Statistical AI)**
  Methods that try to understand or to learn the textual content from its usage. They convey the semantics in an implicit way, through statistical models, Machine

Learning (ML), Deep Learning (DL) and other Natural Language Processing (NLP) techniques [7–9].

- **Qualitative (Symbolic AI)**
  Methods that add explicit semantic information to documents. The Semantic Web is the main example of this category, where the natural language web content is annotated with semantic metadata. The goal is to encode the meaning of the content itself and allow the correct interpretation by machines.

Quantitative approaches get better with more data, are largely unexplainable in their latest evolutions [31], and are good with uncertainty. Qualitative solutions often have mirrored strengths.

Today most of the works fall into one of these two categories. The consideration of mixed solutions has started to develop only recently, and it is one of the key concepts of the research described in this document, with the aim of carefully combining the advantages of both techniques.

From the articles published in the literature, it is easy to deduce how the union of Statistical AI and Symbolic AI constitutes a winning approach and how one strengthens the other. Research is currently being done in both directions.

- **Statistical AI → Symbolic AI**
  Statistical AI techniques can be used to extract knowledge from unstructured text, facilitating the construction and population of ontologies or knowledge graphs capable of representing it (enabling Symbolic AI). So, ontology learning (as well as knowledge graph learning) belongs to this area [2].
- **Statistical AI ← Symbolic AI**
  Integrated and linked data can be used as a more expressive semantic data model to feed the ML algorithms (rather than single and isolated input datasets). Recently a team of researchers at Free University of Amsterdam has published a paper on this topic, where the Knowledge Graph is used as default data model for Machine Learning [35]. Alternatively, as is best known, ontologies can be useful for overcoming several NLP tasks [25].

Knowledge Extraction Using Text Mining

The importance of text mining techniques has increased because of the high number of web-enabled applications which lead to the creation of such unstructured data (without pre-defined models that can describe them).

Learning from text and natural language is a great challenge of Artificial Intelligence and Machine Learning. Any substantial progress in this domain has strong impact on many applications. One of the fundamental problems is to learn the meaning and usage of words in a data-driven way (possibly without further linguistic prior knowledge). Understanding human language is not only understanding words, but also concepts and how they are linked together to create meaning. Ambiguity is what makes NLU (and NLP in general) a difficult problem for computers to master.

Text mining identifies facts, relationships and assertions that would otherwise remain buried in the mass of textual big data. Once extracted, this information is converted into a structured form that can be further analyzed, graphically presented or integrated into databases, data warehouses or business intelligence dashboards. Text mining includes tasks like text classification, text clustering, document summarization, sentiment analysis and any other technique that performs text processing to bring out valuable information.

The concept of *language model* is at the basis of the scientific progress that has taken place in this discipline. Language models go beyond the representation of the text in a machine-understandable form. They are techniques of transformation and dimensionality reduction that allow the construction of a semantic vector space from which perform knowledge extraction, such as the detection of latent similarities between terms and documents. Specifically, language models can be constructed in three ways:

- algebraically (e.g., Latent Semantic Analysis + Singular Value Decomposition [23]);
- probabilistically (e.g., Probabilistic Latent Semantic Analysis [15], Latent Dirichlet Allocation [3]);
- neural network based (e.g., word embeddings [34]).

Formally representing or defining word meaning is a complex problem. The main language models (like Latent Semantic Analysis and word embeddings) are based on relational approaches. As stated in [13], in this case the meaning of a word is expressed on the basis of the relationships it has with other words. Network approaches are also used by symbolic projects (e.g., WordNet database), which formalize knowledge of word meaning within models where the lexicon is seen as a structured system of entries interconnected by sense relations such as synonymy, antonymy, and meronymy.

Latent Semantic Analysis (LSA) is based on the idea that the aggregate of all contexts in which a given word does and does not appear (i.e., local co-occurrence data) provides a set of mutual constraints that largely determines the similarity between words (and groups of words) based on their meaning. Probabilistic LSA (P-LSA) addresses similar goals but applies a statistical model based on multinomial distributions. Latent Dirichlet Allocation (LDA) is a related model that uses a hierarchical Bayesian approach in which Dirichlet priors are placed on the underlying multinomial distributions.

More recently, it has been observed that neural language modeling techniques can perform significantly better than LSA for preserving relationships among words [36]. Word embeddings, like the aforementioned language models, represent terms and documents as dense vectors of real numbers, focusing on a linguistic research area called *distributional semantics*. The underlying idea was popularized by Firth in 1957 and is as follows: *"a word is characterized by the company it keeps"*. Word embeddings therefore consider two words semantically similar if they have a similar context, where the context is a window of other words. Modern contextual word embedding algorithms (e.g., RoBERTa [24], ALBERT [22]) solve one of the main problems of LSA and bag-of-words models where, unless explicitly handled, different terms have different representations, regardless of how they are used. So, an important challenge in word meaning representations concerns context awareness (i.e., multi-sense embeddings, or "one word, different embeddings"). In addition to having the ability to discover advanced relationships between terms by means of vector arithmetic (e.g., slow - slower + shorter = short, man + king - woman = queen), the most advanced word embeddings are also able to take into account the order of words (without losing positioning information). Khattak et al. propose a survey about word embeddings for clinical text [18].

Considering the ultimate goal of understanding the knowledge contained in the patients' posts and using it to answer questions, an interesting text mining task is Machine Reading Comprehension (MRC). Word embeddings can be used for this purpose. However, most state-of-the-art MRC systems are built on *supervised training data*, containing not only the articles but also manually labeled questions about articles and corresponding answers. From this derives a common criticism according to which such tasks are based on the manipulation of representations without any kind of understanding, and adding simple adversarial content easily deceives the models [17].

By examining the proposed overview, it is possible to deduce how research is firm in understanding words. MRC models are currently limited to *extractive Question Answering* on *individual paragraphs*, regarding generic articles [1] or domains taken into account during training. Although the current state-of-the-art models are promising, much research is still needed to be able to answer complex natural language questions concerning the knowledge contained in thousands unlabeled social posts on the vertical domain of a RD.

Another major problem of these solutions concerns *explainability*. Neural word embeddings and highly advanced methods such as deep neural networks are not yet able to explain what they have learned in a human-understandable form, limiting their effectiveness [31]. As black-box algorithms begin making decisions previously entrusted to humans (increasing in complexity and demonstrating spoofability in certain cases), it becomes necessary for these mechanisms to explain themselves (especially in medical domain [26]).

In conclusion, we argue that the extraction of knowledge from documents written and shared by patients should focus on *unsupervised* and *globally explainable* descriptive text mining techniques, requiring *no labeled data*.

Knowledge Representation Using Semantic Web

The Semantic Web (SW) is an emerging research area. Its evolution has moved Knowledge Representation (KR) techniques into the spotlight, aiming at bringing human understanding of the meaning of data to the world of machines. In Semantic Web, knowledge is represented by directed and labeled graph, where nodes are resources and edges are relationships between them.

One of the most powerful features of SW technologies is *reasoning*, which is the process of extracting new knowledge (inferring facts that have not been explicitly stated) from an ontology and its instance base. Semantics is a prerequisite for reasoning support. Reasoners can be used to validate an ontology, making sure that it does not contain any *inconsistencies* among its term definitions.

Since First Order Logic (FOL) has intractability and semi-decidability characteristics, the logical formalism for ontologies is usually provided by *Description Logics (DLs)*, which realize a compromise between expressive power and reasoning complexity. The use of the OWL standard enables more complex queries, which are difficult or impossible to accomplish with relational databases [20].

An important feature linked to the SW refers to the world of *Linked Open Data*. Making use of publicly available interconnected data sources avoids costs, reduces the complexity of data management, can eliminate the need for scraping, and provides constantly up-to-date data without the need for specialized APIs.

Considering also how the graph organization of knowledge is much more natural and *expressive* in certain contexts than other approaches, we believe that Semantic Web technologies are ideal for KR referred to RDs patients.

Semantic Web for Health

Since the birth of this research area, many SW applications have been destined for the healthcare sector. It is interesting to note how the use of the Semantic Web to facilitate the discovery of treatments for diseases was already discussed in the Q&A by Tim Berners-Lee in 2007 [27]. This is also related to the ability of SW technology to open up the boundaries between the silos, allowing scientists to explore hypothesis and looking at how things connect in new combinations never dreamed of before.

Rare Disease Ontologies and Resources

The number of ontologies, vocabularies and databases in the health sector is really very high (more than 300 biomedical ontologies embrace basic, translational and clinical science). Many of them can be found in BioPortal[1] web page.

Orphanet[2] [29] is one of the most important references for rare diseases domain, and is supported by a multilingual relational database. It provides different types of information, such as RDs classification, ORPHA Number nomenclature, relationships with genetic cause, epidemiological data, expert centres and so on.

[1] https://bioportal.bioontology.org/.

[2] https://www.orpha.net.

At the same time, the Orphanet Rare Disease Ontology (ORDO) [33] and the Human Phenotype Ontology (HPO) [19] are considered the most relevant ontologies (OBO/OWL) in the RDs research. While ORDO is derived from Orphanet through periodic extractions, HPO is used for describing the clinical phenotypes observed in patients. Specifically, ORDO is recommended as the primary disease nomenclature ontology, often playing a central role (similarly to DBpedia in the LOD graph). Many HPO annotations have been deduced by carrying out analyzes with text mining techniques on a PubMed corpus, as result of the overwhelming majority of clinical descriptions in the medical literature.

The HPO & ORDO Ontological Module (HOOM) [16] is a recent integration of ORDO and HPO, available from 2018 and based on OBAN format.[3]

Interesting Applications

In the literature there are numerous examples related to the use of these ontologies or of particular interest for the objectives of the paper. Some of the most representative applications are briefly described below.

In [14] is illustrated the use of HPO for the semantic unification of common RDs. A graph is constructed for modeling phenotypic similarities between diseases. The synonyms available in HPO are used to make Concept Recognition within PubMed abstracts. TF-IDF term weighting is used to identify the phenotypes of greatest interest for each disease. In [21] HPO is instead used for a differential diagnosis process with semantic similarity searches. The p-value calculation is used for the construction of a ranking of candidate diseases, starting from the specification of a patient's symptoms.

An ontology-based system for the eHealth domain is presented in [6], where SW technologies are used to provide semantic interoperability among heterogeneous IoT devices. In [25] an ontology containing words considered to be of epidemiological relevance is used for disease name extraction from Twitter messages in order to boost a neural architecture. In [30] ontologies are used as background knowledge through their transformation into Prolog terms and rules.

Considerations

As stated in a recent survey about the state-of-the-art of Semantic Web for Healthcare [37], there are certain limitations and challenges on the current use of Semantic Web technologies in this sector (which continues to produce large volumes of heterogeneous data).

[3]Model used to express the frequency and the provenance of associations, in an ontological design context.

The efforts to ontology development in this domain are quite fragmented and non-standardized. If it is true that a single ontology is not enough to describe the various data, it is equally true that several competitive ontologies exist to describe concepts from the same domain. So, there is also an overlapping problem that does not facilitate the Linked Data objective.

Mapping from proprietary formats to ontological concepts is a difficult and intensive task. Maintaining ontologies and datasets is another challenge and there are not enough tools sufficiently advanced for this purpose.

Despite the large number of ontologies for modeling medical data, there is still lack of resources considering also a non-clinician perspective. To the best of the research carried out for this work, there are no ontologies modeled directly from patients' opinions. Given the incredible impact that such a solution could have on those who live with a rare disease (and not only), the investigation of an approach patient-centered is one of the main contributions of this paper.

Knowledge Graphs

Originally introduced by Google in May 2012 [4], Knowledge Graph (KG) does not yet have a well-established definition. It combine characteristics of several data management paradigms and can be understood as database (for structured queries support), graph (for network data structure) and knowledge base (for formal semantics representation explicitly bound with meaning).

According to the authors of [10], knowledge graphs can be considered ontologies and more. In the community many claim that the real divide between ontologies and knowledge graphs has nothing to do with size or semantics, but rather concerns the very nature of the data. KGs are fact-oriented, while ontologies are schema-oriented. So, not every RDF graph is a knowledge graph. By supporting both the definition of classes and instances (as well as properties), OWL can be used to represent both ontologies and knowledge graphs.

The automatic representation of patient knowledge does not concern smaller and hand-curated collections of assertions, and therefore requires properties for which the use of KG is more suitable.

Knowledge Graph Learning studies the mechanisms to transform KG creation and maintenance from text into a semi or complete automatic process. The evolution of knowledge graph and ontology learning therefore goes hand in hand with that of text mining, and more specifically with that of NLP.

Extracting and Representing Patient Knowledge

We live in a present of exclusion, where among the lost and forgotten data there are those coming directly from the daily personal experience of patients and their families. An important step in the direction of an inclusive future is given by the extraction and representation in an aggregate and anonymous way of the knowledge expressed by patients within social posts, making it accessible and enabling reasoning on it.

One possible way to introduce the vision of patients in the world of Linked Open Data is to extend the nodes related to rare diseases already present within existing and central ontologies, such as ORDO.

In achieving this goal, it is important to focus on the representation of a form of knowledge capable of giving rise to a KG as useful as possible in practical applications. A kind of knowledge directly extractable from the text contained in social data is given by the correlations (i.e., adjacency links) between non-flat terms (or sets of terms) that are highly significant in the context of RDs.

Posts are full of references to entities of various kinds. Patients and caregivers frequently mention concepts such as treatments, symptoms, diagnostic tests, drugs, foods, and places. Modern Named Entity Recognition (NER) systems are able to accurately identify these entities within the text, labeling them with extensive detailed typing. The adoption of a NER system with support for hierarchical types also allows the simple construction of a taxonomy for the concepts identified in the documents ("is_a" relations). From concept extraction, it is also possible to create interlinking towards external references [5].

The sufficiently confident latent semantic correlations between terms can be extracted and represented automatically starting from the application of a language model (possibly explainable in its process). By doing so, a user can identify the concepts most related to one he intends to investigate on the basis of what has been expressed by patients over the years. All with the possibility of rely on the results, being aware of how the correlations are identified.

Figure 44.1 shows a basic example of KG Learning for Esophageal Achalasia.

A KG thus obtained can be used for numerous purposes. For instance, the proposed model can allow a patient, given a medical treatment, to discover the most

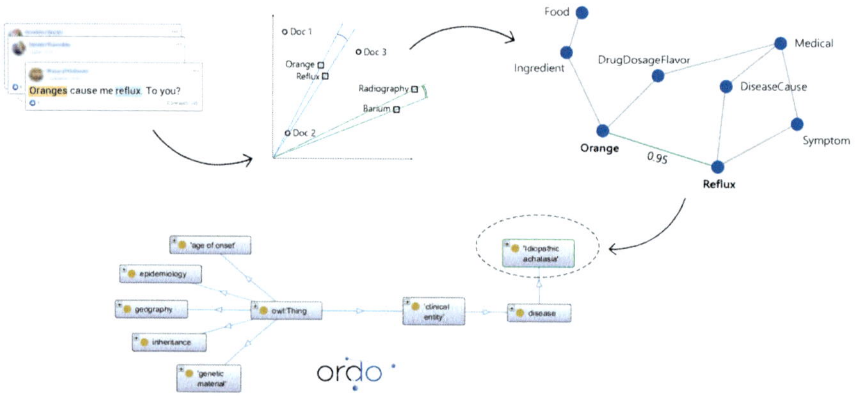

Fig. 44.1 Patient knowledge graph learning example for Esophageal Achalasia

related centers and doctors (and with what confidence). Or, given a symptom, a user can recognize the potentially associated foods or activities (giving the possibility to deepen these entities by means of references to other ontologies, and avoiding the manual reading of posts). Moreover, the knowledge extracted from a certain community con be compared to scientific correlations or to that obtainable from other sources (e.g., groups from other countries).

Conclusions and Future Works

In this paper we hypothesized the combined use of several techniques in the research areas of Text Mining and Semantic Web to automatically extract and represent patient knowledge from social data generated over the years.

Starting from a complete overview of the currently existing approaches for textual content understanding, we identified those most suitable for the domain of rare diseases. Considering the works published in literature and the most recent trends, we recognized globally-explainable unsupervised descriptive text mining and knowledge graphs as two key elements.

The paper also described the most relevant standard ontologies and datasets in the medical field, with a specific focus on RDs. At the end of this evaluation, a lack of resources focused on patients' experience and opinions was found.

Finally, we have outlined the bases for a new methodology capable of solving the problem, showing what can be done on an ontological level.

As a future perspective, we are studying new text mining methods having the properties identified in this research, detecting medical correlations and accompanying the results with their statistical significance [11].

References

1. T. Anusua, T.W. yong, Transfer Learning for Text using Deep Learning Virtual Machine (DLVM). https://bit.ly/2YzR1uO (2018). Accessed 29 Mar 2020
2. M.N. Asim, M. Wasim, M.U.G. Khan, W. Mahmood et al., A survey of ontology learning techniques and applications. Database **2018**, (2018)
3. D.M. Blei, A.Y. Ng, M.I. Jordan, Latent Dirichlet allocation. J. Mach. Learn. Res. **3**(Jan), 993–1022 (2003)
4. G.O. Blog, Introducing the knowledge graph: Thing, not strings. Introducing the knowledge graph: Thing, not strings (2012)
5. A. Carbonaro, Interlinking e-learning resources and the web of data for improving student experience. J. e-Learn. Knowl. Soc. **8**(2), 33–44 (2012)
6. A. Carbonaro, F. Piccinini, R. Reda, Integrating heterogeneous data of healthcare devices to enable domain data management. J. e-Learning Knowl. Soc. **14**(1), (2018)
7. G. Domeniconi, M. Masseroli, G. Moro, P. Pinoli, Discovering new gene functionalities from random perturbations of known gene ontological annotations, in *KDIR*, pp. 107–116 (2014)
8. G. Domeniconi, M. Masseroli, G. Moro, P. Pinoli, Cross-organism learning method to discover new gene functionalities. Comput. Methods Programs Biomed. **126**, 20–34 (2016)

9. G. Domeniconi, G. Moro, A. Pagliarani, R. Pasolini, On deep learning in cross-domain sentiment classification, in *KDIR*, pp. 50–60 (2017)
10. L. Ehrlinger, W. Wöß, Towards a definition of knowledge graphs, in *SEMANTiCS (Posters, Demos, SuCCESS)*, 48 (2016)
11. G. Frisoni, G. Moro, A. Carbonaro, Learning interpretable and statistically significant knowledge from unlabeled Corpora of social text messages: A novel methodology of descriptive text mining, in *International Conference on Data Science, Technology and Applications* (2020)
12. Gartner Hype Cycle for Emerging Technologies 2019. https://gtnr.it/3dB37Is. Accessed 29 Mar 2020
13. L. Gasparri, D. Marconi, Word meaning, in *The Stanford Encyclopedia of Philosophy*, ed. by E.N. Zalta (Metaphysics Research Lab, Stanford University, fall 2019 edition, 2019)
14. T. Groza, S. Köhler, D. Moldenhauer, N. Vasilevsky et al., The human phenotype ontology: Semantic unification of common and rare disease. Am. J. Human Genetics **97**(1), 111–124 (2015)
15. T. Hofmann, Probabilistic Latent Semantic Analysis. arXiv preprint arXiv:1301.6705 (2013)
16. What is HOOM (The ORDO-HOOM Ontological Module)? http://www.orphadata.org/cgi-bin/img/PDF/WhatIsHOOM.pdf. Sept 2019
17. R. Jia, P. Liang, Adversarial examples for evaluating reading comprehension systems. arXiv preprint arXiv:1707.07328 (2017)
18. F.K. Khattak, S. Jeblee, C. Pou-Prom, M. Abdalla et al., A survey of word embeddings for clinical text. J. Biomed. Inf. X **4**, 100057 (2019)
19. S. Köhler, N.A. Vasilevsky, M. Engelstad, E. Foster et al., The human phenotype ontology in 2017. Nucleic Acids Res. **45**(D1), D865–D876 (2017)
20. R. Kontchakov, M. Rezk, M. Rodriguez-Muro, G. Xiao et al., Answering SPARQL queries over databases under OWL 2 QL entailment regime, in *International Semantic Web Conference* (Springer, Berlin, 2014), pp. 552–567
21. S. Köhler, M.H. Schulz, P. Krawitz, S. Bauer et al., Clinical diagnostics in human genetics with semantic similarity searches in ontologies. Am. J. Human Genetics **85**(4), 457–464 (2009)
22. Z. Lan, M. Chen, S. Goodman, K. Gimpel, et al., Albert: A lite bert for self-supervised learning of language representations. arXiv preprint arXiv:1909.11942 (2019)
23. T.K. Landauer, S.T. Dumais, A solution to Plato's problem: The latent semantic analysis theory of acquisition, induction, and representation of knowledge. Psychol. Rev. **104**(2), 211 (1997)
24. Y. Liu, M. Ott, N. Goyal, J. Du, et al., Roberta: A robustly optimized bert pretraining approach. arXiv preprint arXiv:1907.11692 (2019)
25. M.A. Magumba, P. Nabende, E. Mwebaze, Ontology boosted deep learning for disease name extraction from Twitter messages. J. Big Data **5**(1), 31 (2018)
26. S.M. Mathews, Explainable artificial intelligence applications in NLP, biomedical, and malware classification: A literature review, in *Intelligent Computing-Proceedings of the Computing Conference* (Springer, Berlin, 2019), pp. 1269–1292
27. Q&A with Tim Berners-Lee. Businessweek. Retrieved from https://www.bloomberg.com/news/articles/2007-04-09/q-and-a-with-tim-berners-leebusinessweek-business-news-stock-market-and-financial-advice (2007)
28. EURORDIS Rare Barometer Graphic Report. Share and protect our health data! Rare disease patients' preferences on data sharing and protection (2020)
29. A. Rath, A. Olry, F. Dhombres, M.M.Brandt et al., Representation of rare diseases in health information systems: The Orphanet approach to serve a wide range of end users. Human Mutation **33**(5), 803–808 (2012)
30. S. Riccucci, A. Carbonaro, G. Casadei, Knowledge acquisition in intelligent tutoring system: A data mining approach, in *Mexican International Conference on Artificial Intelligence* (Springer, Berlin, 2007), pp. 1195–1205
31. W. Samek, T. Wiegand, K.-R. Müller, Explainable artificial intelligence: Understanding, visualizing and interpreting deep learning models. arXiv preprint arXiv:1708.08296 (2017)
32. L.M. Schriml, C. Arze, S. Nadendla, Y.-W.W.Chang et al., Disease Ontology: A backbone for disease semantic integration. Nucleic Acids Res. **40**(D1), D940–D946 (2012)

33. D. Vasant, L. Chanas, J. Malone, M. Hanauer et al., ORDO: An ontology connecting rare disease, epidemiology and genetic data, in *Proceedings of ISMB*, vol. 30 (2014)
34. B. Wang, A. Wang, F. Chen, Y. Wang et al., Evaluating word embedding models: Methods and experimental results. APSIPA Trans. Signal Inf. Process. **8**, e19 (2019)
35. X. Wilcke, P. Bloem, V. de Boer, The knowledge graph as the default data model for learning on heterogeneous knowledge. Data Sci. **1**(1–2), 39–57 (2017)
36. I.H. Witten, E. Frank, M.A. Hall, C.J. Pal, Moving on: Applications and beyond (Chapt. 13), in *Data Mining*, 4th ed., ed. by I.H. Witten, E. Frank, M.A. Hall, C.J. Pal (Morgan Kaufmann, 2017), pp. 503 – 532
37. X. Zenuni, B. Raufi, F. Ismaili, J. Ajdari, State of the art of semantic web for healthcare. Proc.-Soc. Behav. Sci. **195**, 1990–1998 (2015)

Chapter 45
A Hybrid Model to Support Public Obesity Treatment Policies

Placido Rogerio Pinheiro, Luciano C. Nunes, Mirian Calíope D. Pinheiro, Consuelo O. Pontes, Karla M. C. Rolim, Wilson Correia de Abreu, Margarida Neves de Abreu, Luana I. C. C. Pinheiro, Maria L. D. Pereira, Pedro G. C. D. Pinheiro, and Rafael Comin-Nunes

Abstract Characterized by excess adipose tissue in the body, obesity is a significant public health problem in today's world. Among the various causes that lead to the development of obesity, endocrine, genetic, behavioural, psychological, economic-social, emotional, environmental, and cultural factors stand out. Although not considered a mental disorder, obesity may be associated with several psychological disorders, such as Eating Disorders, Anxiety Disorders; Obsessive–Compulsive Disorders; Depressive disorders; Sleep–wake Disorders. Obesity can also be the cause of many other diseases such as diabetes, hypertension, infertility, and strokes

P. R. Pinheiro (✉) · L. C. Nunes · M. C. D. Pinheiro · K. M. C. Rolim · P. G. C. D. Pinheiro
University of Fortaleza, Fortaleza, Brazil
e-mail: placido@unifor.br

e-mail: lcominn@uol.com.br

M. C. D. Pinheiro
e-mail: caliope@unifor.br

K. M. C. Rolim
e-mail: karlarolim@unifor.br

P. G. C. D. Pinheiro
e-mail: pedrogcdpinheiro@hotmail.com

C. O. Pontes · L. I. C. C. Pinheiro
National Confederation of Medical Cooperatives, Fortaleza, Brazil
e-mail: consueloopontes@gmail.com

L. I. C. C. Pinheiro
e-mail: luanaibiapina1@hotmail.com

W. C. de Abreu · M. N. de Abreu
School of Nursing of Porto, Porto, Portugal
e-mail: wjabreu@esenf.pt

M. N. de Abreu
e-mail: mabreu@esenf.pt

P. R. Pinheiro · L. I. C. C. Pinheiro · M. L. D. Pereira
Ceara State University, Fortaleza, Brazil

A. Visvizi et al. (eds.), *Research and Innovation Forum 2020*,
Springer Proceedings in Complexity,
https://doi.org/10.1007/978-3-030-62066-0_45

of the brain and heart diseases. Because of these factors and causes, governments of various countries have been implementing public health programs in their respective populations that include socio-educational and treatment measures. However, despite advances in the knowledge of obesity, treatment solutions do not show significant progress, since there is still a considerable portion of the population, obese. Furthermore, the present study aims to indicate factors that need to be considered in the decision-making process by health professionals in the definition of treatment, as well as by public managers in the implementation of public policies. For this, a hybrid model structured in decision-making methodologies (Multicriteria Decision Analysis) will be used, associated with a specialist system with representations based on knowledge of rules and production probabilities using Artificial Intelligence.

Keywords Obesity · Public policy · Health promotion · Health technology · Expert systems · Multicriteria

Introduction

Obesity is one of the biggest public health problems today, especially in developing countries. Its prevalence has increased considerably in recent decades. According to the World Health Organization (WHO), 1.9 billion adults are overweight. Of this contingent, 650 million are obese, representing 13% of the world population [1]. Corroborating with the WHO statement, in Brazil, obesity has grown 60% in the last ten years, from 11.8 to 18.9% of the population. One aggravating factor is the fact that more than 50% of the people are in the overweight and obesity range [2]. From a biological point of view, obesity is a chronic disease that is characterised by the accumulation of adipose tissue in the body. The WHO considers that an adult individual is in the category "overweight" when his Body Mass Index (BMI) is greater than or equal to 25 kg/m^2. The "obese" person, on the other hand, is the person with a BMI equal to or greater than 30 kg/m^2.

In the Information Society's era, the digital exclusion leads individuals with a lower educational level and low socioeconomic power to have a diet with a high caloric density, which provides them with nutritional status of overweight and obesity. Thus, obesity is an expanding disease and is associated, too, with

M. L. D. Pereira
e-mail: luciad029@gmail.com

L. C. Nunes · R. Comin-Nunes
University Center September 7, Fortaleza, Brazil

R. Comin-Nunes
e-mail: rafael23.uni7@gmail.com

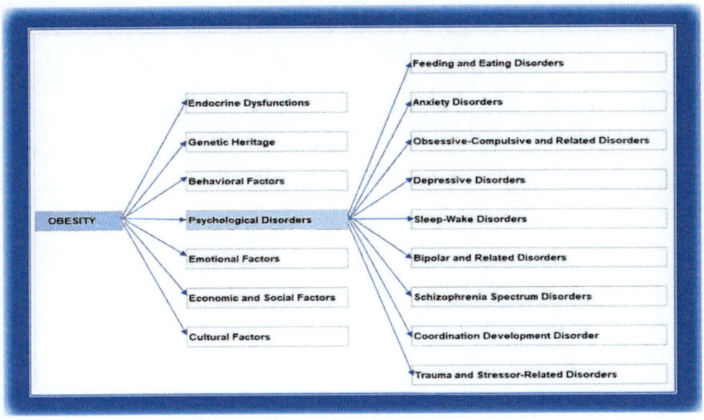

Fig. 45.1 Factors correlated with obesity. *Source* Adapted from the WHO and American Psychiatric Association (APA)'s Diagnostic and Statistical Manual of Mental Disorders (DSM-5)

several pathologies, such as diabetes, hypertension, cardiovascular diseases, infertility, various psychological disorders. Besides this, obese individuals are discriminated against and develop low self-esteem, social isolation, emotional damage with severe consequences, both from a clinical and socioeconomic point of view.

The obesity causes are multifactorial and depend on the interaction of genetic, metabolic, psychological, social, behavioural and cultural factors. It is, therefore, complex multifactorial disease and requires strategies and interventions from the public authorities for its prevention [3]. Figure 45.1 presents a set of factors associated with obesity, showing the complexity of its causes.

Strategies for the control and treatment of obese patients involve changes in behaviour, habits, lifestyle, and nutritional balance. Furthermore, associated with these strategies, there should be a systematic practice of physical activities, as well as the use of technological tools to assist in the effectiveness of weight control and other therapeutic measures to strengthen self-esteem and perseverance in multidisciplinary treatment [3].

Public policies to combat obesity should include, mainly, investment in health management, focusing on Primary Care carried out by a multidisciplinary team composed of: a doctor residing in family and community medicine; nurse working in the management of patients with chronic diseases; as well as psychologist, nutritionist and physical educator. In this context, this article proposes a hybrid model that categorises and classifies, in decreasing order of importance, the multiple factors that cause obesity, to support decision-makers in the development of public policies for the prevention and treatment of obesity. Besides, to categorising the various elements, this model presents, through a specialist system, the diagnosis of obesity of a person under analysis.

A Brief Explanation of the Proposed Model

The proposed model consists of the hybridisation of a multicriteria decision support methodology for categorising and ranking the multiple factors that cause obesity. The result generated by the multicriteria methodology will be associated with control-variables and goal-variables of a specialist system, which is based on intelligent rules, producing a diagnosis of the cause of obesity. It is important to note that hybrid models have been applied to the decision-making process in the health area, to diagnose diseases, and assist in public policies. For example, Castro developed a hybrid model to assist in decision-making towards a neuropsychological diagnosis of Alzheimer's disease [4]. In another study, Menezes presents a proposal for a hybrid model for the early diagnosis of type 2 diabetes, using decision-aid methods [5]. Filho et al. developed a different methodology to support the early diagnosis of gestational diabetes, with data organised by the multicriteria methodology, applying methods of the multi-tax utility theory (MAUT) and structured information in the knowledge base of a specialised system [6]. Nunes et al. present A novel method to Support Decision Making Process in Health and Behavior Factors Analysis for the Composition of IT Projects Teams [7]. Tamanini et al. proposed an evaluation of Alzheimer's disease clinical stages under the optics of hybrid approaches, using verbal decision analysis methods [8]. Nunes et al. developed a hybrid model, Automatic Detection and Diagnosis of Neurologic Diseases [9]. Aragonés et al. proposed a decision support model based on neural networks associated with fuzzy logics for the prognosis of breast cancer relapse [10].

The difference between the models as mentioned earlier and the one presented in this study is in the integrated use of artificial intelligence techniques associated with a multicriteria analysis methodology that beforehand points out which are the main factors causing obesity, ranking them in order of importance to support training in the diagnosis of a specialist system. The integration of technologies aims to increase the accuracy of the best alternatives for the indication of cases of obesity. This research was based on a review of studies on obesity, published by the World Health Organization, associated with what describes the Fifth Edition of the Diagnostic and Statistical Manual of Mental Disorders (DSM-5) of the American Psychiatric Association (APA), also on obesity.

Article's Structure

This article has five sections, including the introductory section, which presents the problems and objectives of the study, in addition to a categorisation of the factors that cause obesity. The second section conceptualises obesity and details the factors causing obesity used in the proposed model. Section three analyses qualitatively and comparatively the factors associated with obesity. This third section further details the proposed model. The fourth section presents a specialist system applied in the diagnosis of the causes of obesity, using the multiple factors described in the second section. The final section offers conclusions and possible future work on the subject.

Obesity Disorder Characteristics

The World Health Organization considers obesity as part of the group of Chronic Non-Communicable Diseases (CNCD). Increased life expectancy, inappropriate environmental factors, a sedentary lifestyle and an inadequate diet are factors responsible for CNCD, which are the leading causes of mortality in today's world [1]. Obesity is a disease characterised by the excessive accumulation of body fat in the individual. The body mass index (BMI) is the parameter commonly used for the diagnosis of obesity in adults. This BMI is calculated by dividing the individual's weight by his height squared. According to the WHO, adult individuals have healthy weight when the result of the BMI calculation is between 18.5 and 24.9. In turn, to be considered obese, the BMI must be above 30. Associated risk factors such as an inactive lifestyle, poor diet, genetics, environment and other comorbidities can decrease the quality and life expectancy of the individual [1]. There are many causes of obesity. Excess weight can be linked to a person's genetic heritage, poor eating habits or, for example, endocrine disorders. The approach to the clinical treatment of obesity requires the adoption of a lifestyle with healthy habits, associated with the prescription of medications, when necessary.

A Scale of Values Among the Multiple Factors that Cause Obesity

The aetiology of obesity is complex, multifactorial, resulting from the interaction of genes, environment, lifestyles and psychosocial factors. There are three primary components in the neuroendocrine system involved with obesity: the afferent system, which involves leptin and other signs of satiety and short-term appetite; the central nervous system processing unit; and the efferent system, a complex of appetite, satiety, autonomic and thermogenic effectors that leads to the energy supply [11]. Figure 45.2 represents the direct psycho-emotional connection of command of the nervous system and its interference at the systemic level.

Fig. 45.2 Representative illustration of the brain and heart connection

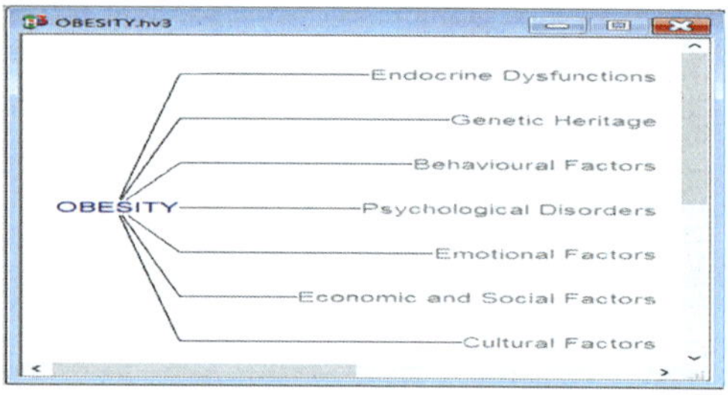

Fig. 45.3 The categorisation of multiple factors that cause obesity

According to the WHO and based on the DSM-5 of the American Psychiatric Association, obesity is a disease with multifactorial causes. These causes may be categorised, according to Fig. 45.3.

Table 45.1 has a description of the main factors that cause obesity.

Each of the multiple factors cited in this study has its respective characteristics, which are called "control events". Thus, taking as an example the factor "Endocrine

Table 45.1 Main causal factors of obesity

Factors	Description
Endocrine dysfunctions	The human endocrine system has as main glands the pituitary, thyroid, parathyroid, pancreas, adrenal, ovaries and testicles. Dysfunctions in this system can compromise an individual's entire health, even generating obesity
Genetic heritage	Studies demonstrate the participation of the genetic component in the incidence of obesity. Estimates indicate that 40–70% of the variation in the phenotype associated with obesity has a hereditary character [12]
Behavioural factors	An individual's eating behaviour is directly associated with that individual's personal history of food. Food consumption, in large quantities and the sequence of stimuli and practices for this attitude, derives from the personal history built since childhood
Psychological disorders	According to the DSM-5, some Psychological Diseases lead individuals to obesity, either because certain psychological disorders induce to the excessive consumption of food as negative compensation, as well as due to hormonal dysfunctions that various drugs used in the treatment of mental disorders may cause to the individual. Figure 45.1, in section one, shows the central Psychological Diseases that cause obesity
Emotional factors	After the occurrence of pleasant or unpleasant events, in the life of an individual, he may react with a consequent exaggerated increase in the consumption of foods, especially rich in carbohydrates
Economic and social factors	An individual's socioeconomic status can provide a disorderly practice of eating. It is observed, however, that there are obese individuals both in the group with high purchasing power and in the group with low purchasing power. Individuals in the first group become obese because they do not follow the standards recommended by specialists and adopt an unruly stance—the individuals in the second group for basically eating foods rich in carbohydrates
Cultural factors	People's cuisine is part of their culture. Depending on the predominant ingredients in cooking, the eating pattern can be a factor causing obesity

0	25	50	75	100
INDIFFERENT	LITTLE INFLUENCE	MODERATE INFLUENCE	MUCH INFLUENCE	DECISIVE INFLUENCE

Fig. 45.4 The scale of the influence of obesity control events

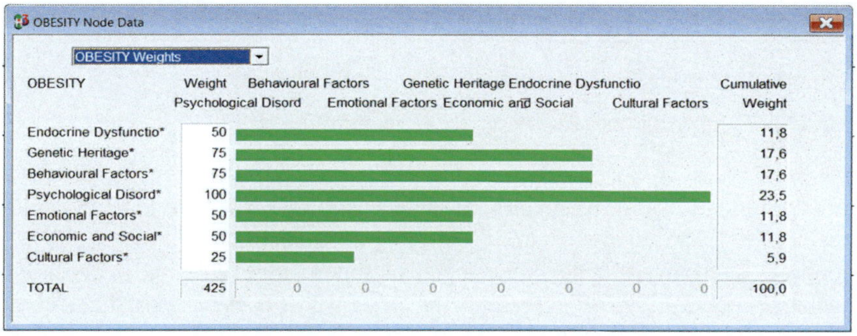

Fig. 45.5 Influence of each factor on obesity

Dysfunctions" that cause obesity, one of its "control events" is monitoring the functioning of the thyroid gland by assessing the "Thyroid Stimulating Hormone (TSH)". It is important to note that individuals with obesity should have their thyroid function investigated, given the prevalence of obesity concomitant with hypothyroidism. Another example may be the factor "Psychological Diseases", more specifically in its form of "Anxiety Disorders". Individuals with anxiety disorders eat foods high in carbohydrates to compensate for their state of anxiety. Such behaviour is characterised, for the hybrid model, as one of the "control events" for anxiety disorder. Therefore, following the line of reasoning of the mentioned examples, for each of the multiple factors of the hybrid model, the mapping of its most relevant characteristics is made, which form the set of "control events" for treatment in the Specialist System, that will be treated in section four. The multiple factors are compared with each other in pairs, aiming to define which one has the most significant influence on the causes of obesity. Therefore, Fig. 45.4 was defined for use in the model, which serves as a scale for measures of influence between factors.

Figure 45.5 shows the influence of each factor on obesity. The numerical values in Fig. 45.5 represent the degree of this influence and come from the scale in Fig. 45.4.

Support Model Proposal in the Diagnosis of Obesity Causes

Developing tools to help people make optimal decisions is the main objective of the research using the Multiple Criteria Decision Analysis methods. Developing such tools requires a combination of interdisciplinary knowledge, such as organisational behaviour, cognitive psychology, and applied mathematics. Methodologies to support

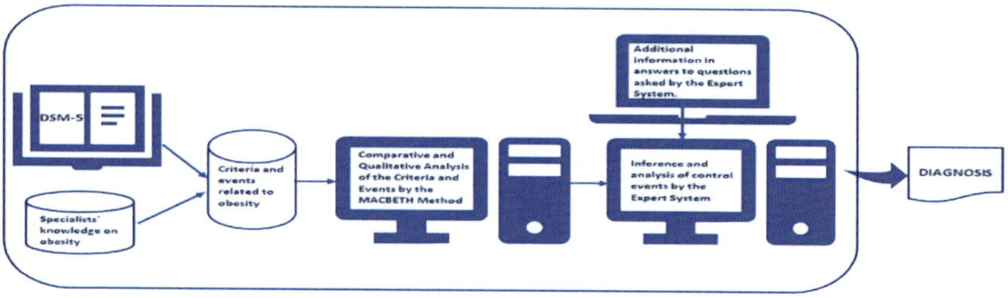

Fig. 45.6 The proposed model to support the diagnosis of causes of obesity

multicriteria decision making provide the decision-maker with techniques and tools that allow structuring the control events, as well as hierarchise these events for the proper classification in order of the degree of importance of each one in the decision-making process. Figure 45.6 presents the process flow of the proposed model in the form of a graph containing the integrations between the technologies used in the present study.

MACBETH Method Application

Technique is a multicriteria decision support approach that allows assigning a value to each alternative through a peer comparison. Given two alternatives, the decision maker should say which is the most attractive and the degree of confidence of this attractiveness in a semantic scale that corresponds to an ordinal scale [13, 15]. In the MACBETH method, the decision-maker makes value judgments about the alternative in each situation due to the attractiveness of this alternative. This task is defined by the construction of a criterion function *vj*, such that [14]:

- For $a, b \in A$, $v(a) > v(b)$, if and only if, for the evaluator, *a* is more attractive than *b (aPb)*;
- Any positive difference, $v(a) > v(b)$, numerically represents the value difference between *a* and *b*, with *aPb* always regarding a fundamental point of view *j(PVFj)* or criterion *j*. Then, for each $a, b, c, d \in A$, with a being more attractive than *b* e *c* being more attractive than *d*, we see that $v(a) - v(b) > v(c) - v(d)$, if and only if, "the difference in attractiveness between *a* and *b* is greater than the difference in attractiveness between *c* and *d*".

The critical question of the MACBETH method is: "Given the impacts *ij*(*a*) and *ij* (*b*) of two potential actions *a* and *b*, from a fundamental point of view *PVFj*, being judged more attractive than *b*, the difference in attractiveness between *a* and *b* is judged to be "null," "very weak," "weak," "moderate," "strong," "very strong," or "extreme."

Figure 45.7 shows the degrees of attractiveness between factors in Table 45.1. The degrees of attractiveness between events form the Current Scale of confidence

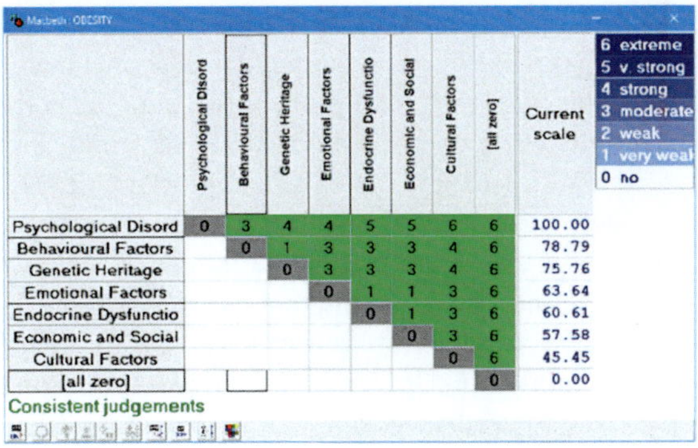

Fig. 45.7 Value judgment matrix and difference in attractiveness between factors causing obesity

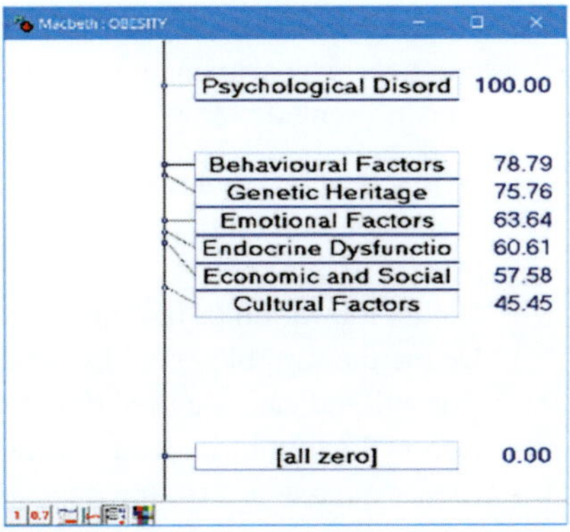

Fig. 45.8 Factors with new values within permissible intervals

factors, indicating "Consistent Judgments" or qualitative consistency in peer-to-peer comparison.

The model, described in Fig. 45.6, proposes that the factors in Table 45.1 come to compose the variables-goal of the Expert System knowledge base described in the fourth section.

Regarding the values of each factor, Fig. 45.8 shows how the decision-maker can adjust these degrees of attractiveness among the factors, pointing in a ruler the limits between the different degrees of the attractiveness of these factors. If the decision-maker modifies these values, the new degrees of attractiveness indicated in the rule are reflected in the current scale of the matrix.

A Specialist System for Diagnosing Causes of Obesity

Specialist systems are associated with the expression of artificial intelligence (AI) and are useful in generating knowledge, which is why they can assist in decision-making processes in the health area. The most common architecture of expert systems involves production rules structured in a set of conditions in the IF … THEN … style, with the possibility of including logical connectives relating attributes in the scope of knowledge and the use of probabilities. For the construction of the Expert System of this study, the ExpertSinta tool was used, which applies AI techniques for automatic generation of expert systems, using a knowledge representation model based on production rules and probabilities. According to Fig. 45.6, the factors described in Table 45.1 and submitted to the influence scale values described in Fig. 45.4 are analysed and compared in pairs by the MACBETH method, generating the values of the attractiveness differences necessary for the judgment of the matrix of values, as exemplified in Fig. 45.7. After this, the information on these factors indirectly feeds the Specialist System as variables-goal. It is important to emphasise that, as the diagnosis of the causes of obesity is predominantly clinical, it requires anamnesis. For this, the Expert System uses information from the factors. Thus, the Specialist System assists in early diagnosis, constituting a useful tool for Primary Health Care Professionals. The sequence of steps for the configuration and use of the Specialist System is presented below.

(a) Define the precedence of the logical operators that will be used by the Inference Machine, indicating one of the following structures: '*(A and B) or C*'; '*A and (B or C)*'.
(b) Define the variables of control (See Fig. 45.9a).
(c) Define the variables-goal from the factors submitted to the MACBETH method that will indicate the final diagnosis (See Fig. 45.9b).
(d) Define the minimum value for the confidence factor for the Expert System.
(e) Create a password for the database.

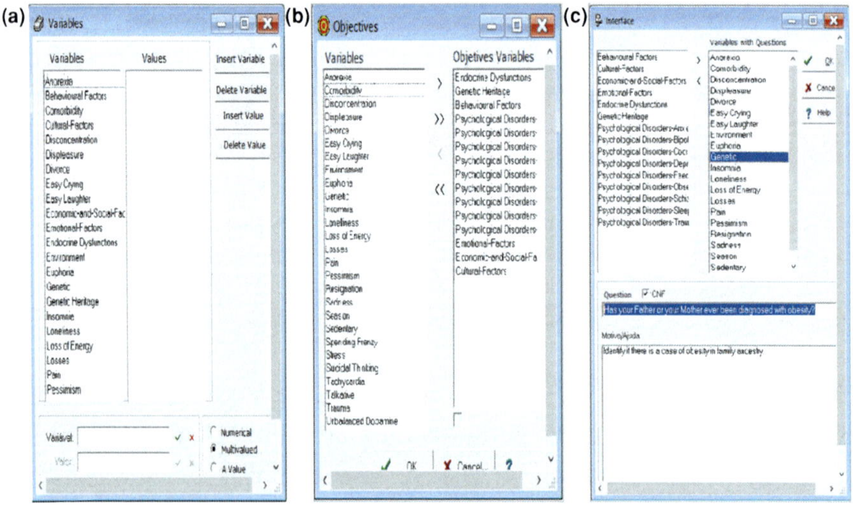

Fig. 45.9 **a** Variables definition, **b**. Objectives definition, **c**. Interfaces definition

(f) Create an interface for user/system interaction. Figure 45.9c illustrates how the questions that are presented to the user are created. Furthermore, it is indicated if, together with the response, the user must inform the confidence factor (field 'CNF' in Fig. 45.9c) of the occurrence of the event that is the subject of the question.
(g) In addition to the variables, objectives and interfaces, the Expert System requires logical rules, from which it will indicate the diagnosis. While executing the expert system, the user interacts through graphic interfaces, as shown in Fig. 45.10. The interaction through these interfaces enables the collection of values that feed the variables and trust factors used by the expert system.
(h) At the end of its execution, the expert system issues a diagnosis for the possible causes of obesity, as shown in Fig. 45.11. The expert system also offers a research tree that consists of the path of logical reasoning conducted by the specialist. This tree assists in the analysis of results after obtaining the diagnosis.

A summary of the model's functioning may be viewed in the algorithm described in Fig. 45.12.

As shown in Fig. 45.12, the decision-maker can build a base of public policies for the treatment of obesity based on the diagnoses generated by the model now proposed, which indicates at the beginning which is the most impacting factor in the disease.

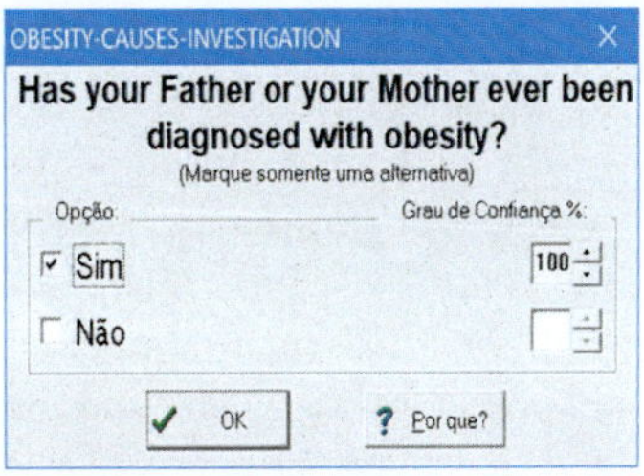

Fig. 45.10 Data entry user interface

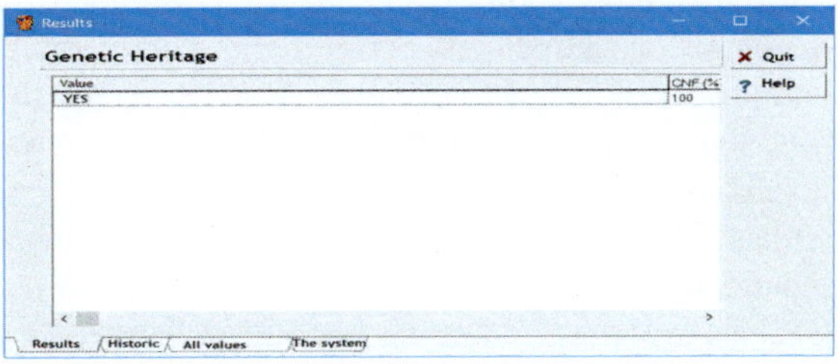

Fig. 45.11 Diagnosis of the genetic heritage

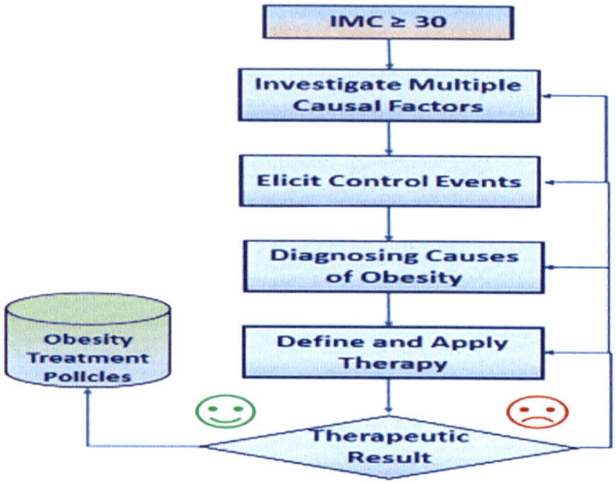

Fig. 45.12 Summary of the model's functioning

Conclusion

Public policies in the health area increasingly require innovative Information Technology solutions to automate the decision-making process, especially when there is a multicriteria analysis associated with incomplete knowledge. This study presented some of the main factors that cause obesity, categorising them and analysing them in a multicriteria methodology, so that they could serve as input as variables-goal as diagnoses of causes of obesity, in a specialist system. Furthermore, these variables-goals, the Expert System had as inputs, data from the DSM-5 and of health experts, with information on each of the factors that cause obesity. This information generated the control variables that enabled the creation of the interfaces for the collection of additional information necessary for the diagnosis of the cause of obesity, making a knowledge base. The automation of the hybrid model, involving the methodology and the expert system is oriented towards future work. A significant advance in the future for this research is the insertion of technologies based on neural networks in the hybrid model proposed in this study, aiming to diagnose patterns of behaviour in a population sample with obesity. This study's research process implies that related entities should encourage more researches on the subject, resulting in more detailed findings. Also, the research on hybrid models is feasible, and organisation interest in using specialised system solutions is increasing. Further studies should be done to enhance the model. We propose enhancements on the user interface, formatting and generation of generic questionnaires, as well as implementation of functionalities such as exporting and importing of data files.

References

1. WHO. World Health Organization. Available at https://www.who.int/news-room/fact-sheets/detail/obesity-and-overweight. Accessed on 03 Dec 2020
2. BRAZIL, Ministry of Health. National Supplementary Health Agency. *Manual of Guidelines for Coping with Obesity in Brazilian Supplementary Health* (ANS, Rio de Janeiro, 2017)
3. ABESO. Brazilian Association for the Study of Obesity and Metabolic Syndrome. *Brazilian Obesity Guidelines*, 4th edn. (ABESO, São Paulo, 2016). Available at https://www.abeso.org.br/uploads/downloads/92/57fccc403e5da.pdf. Accessed on 05 Dec 2020
4. A.K.A. De Castro, P.R. Pinheiro, M.C.D. Pinheiro, A hybrid model for aiding in decision making for the neuropsychological diagnosis of Alzheimer's disease, in *Lecture Notes in Computer Science.* Rough Sets and Current Trends in Computing, vol. 5306 (2008), pp. 495–504
5. A.C. Menezes, P.R. Pinheiro, M.C.D. Pinheiro, T.P. Cavalcante, Towards the applied hybrid model in decision making: support the early diagnosis of type 2 diabetes, in *3rd International Conference on Information Computing and Applications (ICICA)*, LNCS, vol. 7473 (2012), pp. 648–655
6. E.G. Filho, P.R. Pinheiro, M.C.D. Pinheiro, L.C. Nunes, L.B.G. Gomes, Heterogeneous methodology to support the early diagnosis of gestational diabetes. IEEE Access **1**, 1.
7. L.C. Nunes, P.R. Pinheiro, M.C.D. Pinheiro, M.S. Filho, R.E.C. Nunes, Toward a novel method to support decision-making process in health and behavioural factors analysis for the composition of IT projects teams. Neural Comput. Appl. **2018**, 1–22 (2018)
8. I. Tamanini, P.R. Pinheiro, M.C.D. Pinheiro, V.H.C. de Albuquerque, Evaluation of the Alzheimer's disease clinical stages under the optics of hybrid approaches in verbal decision analysis. *Telematics and Informatics*, vol. 35(4) (2018), pp. 776–789; vol. 6401 (Elsevier, 2010c), pp. 427–432
9. L.C. Nunes, , P.R. Pinheiro, M.C.D. Pinheiro, M.S. Filho, R.E.C. Nunes, P.G.C.D. Pinheiro, Automatic detection and diagnosis of neurologic diseases. IEEE Access **1**, 1 (2019)
10. J.M.J. Aragonés, J.I.P. Sánchez, J.M. Doña, E. Alba, A neuro-fuzzy decision model for prognosis of breast cancer relapse, in *Lecture Notes in Computer Science, Current Topics in Artificial Intelligence*, vol. 3040 (Springer, Berlin/Heidelberg (2004), pp. 638–645
11. AMB, Brazilian Medical Association and CFM, Federal Council of Medicine. Right-Wing Project. Obesity (AMB/CFM, São Paulo, 2005). Available at https://diretrizes.amb.org.br/_BibliotecaAntiga/obesidade-etiologia.pdf. It was accessed on 03 Dec 2020
12. I. Marques-Lopes, A. Marti, M.J. Moreno-Aliaga, A. Martínez, Genetic aspects of obesity. Nutr. Mag. **17**(3), 327–338 (2004). https://doi.org/10.1590/S1415-52732004000300006
13. C.A. B.E. Costa, J.M.D. Corte, J.C. Vansnick, MACBETH. LSE-OR Working Paper, 56 (2003)
14. P.R. Pinheiro, G.G.C. de Souza, A.K.A. de. Castro, A Estruturação do Modelo Multicritério para Produção de Jornal. Pesquisa Operacional **28**, 203–216 (2008)
15. L.C. Nunes, P. Pinheiro, M.C.D. Pinheiro, M. Pompeu, M.S. Filho, R. Comin-Nunes, P.G.C.D. Pinheiro, A hybrid model to guide the consultation of children With autism spectrum disorder. Research & Innovation Forum, in *Springer Proceedings in Complexity*, 1st edn, ed. by A. Visvizi, M.D. Lytras (Org.) (Springer International Publishing, 2019), pp. 419–431

Chapter 46
Digital Economy as a Driver of Sustainable and Inclusive Growth in Africa—Case Study

Anna Masłoń-Oracz, Josephine Ojiambo, and Oduor Kevin

Abstract Information and communication technologies (ICTs) can play a critical role in improving health care for individuals and communities. By providing new and more efficient ways of accessing, communicating, and storing information, ICTs can help bridge the information divide that has emerged in the health sector in developing countries—between health professionals and the communities they serve and between the researchers in the health sector and the practitioners who need the research finding (Digital Transformation in Healthcare in 2020: 7 Key Trend. Retrieved from https://www.digitalauthority.me/resources/state-of-digital-transformation-healthcare/). Through the development of databases and other applications, ICTs provide the capacity to improve health system efficiencies and prevent medical errors (Digital economy report 2019, value creation and capture: implications for developing countries, https://www.unctad/der/2019). The aim of this paper is to justify/examine how and what could be done in a sustainable and inclusive manner through the use of innovation to achieve the sustainable development goals through ICTs. This paper examines, the enormous potential for leveraging the digital economy as a driver of sustainable and inclusive growth in Africa. The authors of this paper conclude that further research on the digital economy and increased awareness of e-health coupled with pilot activities would help in better implementing SDG#3 and other related SDGs to enhance the impact of innovation in health on sustainable development and growth. **Methodology**: This paper relied heavily on secondary data from other previously conducted research on digital economy. By the time this paper was written, COVID-19 was already taking toll. Archival study proved suitable since it offered a great opportunity to review numerous researches (State Georgia, University. "GSU Library Research Guides: *Archival Research: Why Archival Research?"

A. Masłoń-Oracz (✉)
SGH Warsaw School of Economics, Warsaw, Poland
e-mail: amaslon@sgh.waw.pl; annamaslonoracz@gmail.com

J. Ojiambo
Peter Ojiambo Foundation, London, UK

O. Kevin
Stowelink, Nairobi, Kenya

A. Visvizi et al. (eds.), *Research and Innovation Forum 2020*,
Springer Proceedings in Complexity,
https://doi.org/10.1007/978-3-030-62066-0_46

Why Archival Research?—*Archival Research—GSU Library Research Guides at Georgia State University, 2016, research.library.gsu.edu/archivalresearch) done on digital economy and make a case on its application as a driver for sustainable and inclusive growth in Africa.

Keywords Digital economy · Africa · Disruptive innovation

Introduction

Digital development and transformation will have implications on virtually all the Sustainable Development Goals (SDG's), and will affect all countries, sectors and stakeholders. However, relevant data on how the use of digital technologies affect for example people's experiences of mental health or their social lives is not collected frequently and if so, in a harmonized manner [4]. This paper was written as the COVID-19 pandemic had begun to destroy the gains made, to some extent, by and in our global health system. Nearly every nation, even those that seem to have resilient health systems have been greatly affected by the COVID-19 pandemic [5]. Even so, we are observing how digitalization of our world is helping us to carry on with our daily life activities. So, eHealth has become a fact of life and it is the only way for the time being to seek the help of a doctor or to receive a prescription. According to WHO, e-health can be defined as "the cost-effective and secure use of ICT in support of health and health-related fields, including health-care services, health surveillance, health literature, health education, knowledge and research." [6]. The digital revolution will transform, and is transforming, our lives and those of our societies on profound scale. At the same time, digital health systems are delivering immense opportunities [7] and responding to daunting challenges. New technologies can make significant contributions to realizing the Sustainable Development Goals. However, we cannot take the positive outcomes delivered by digital health for granted.

Conceptualization of Digitalization of Health and Digital Economy as a Whole

The Digital economy refers to all activities that can be undertaken through the use of ICT tools. The Asian Development Bank Institute defines the digital economy as the usage of the internet, cloud computing, big data, fintech, and other new digital technologies to collect, store, analyze, and share information digitally and transform social interactions [8]. The Digital Economy is a high-policy concern in developing countries due to the very important role it plays in ensuring security for the nation, employment creation and trade revenues. The Digital economy is also the answer to enhanced sustainability and creating jobs in order to ensure the livelihoods of societies in rural settings especially in the developing world. Access to medical care in many rural areas is still very limited. According to Healthy People 2020, access to healthcare is important for [9]:

- Overall physical, social, and mental health status
- Disease prevention
- Detection, diagnosis, and treatment of illness
- Quality of life
- Preventable death
- Life expectancy.

This listing has significant implications for the implementation of the 2030 Agenda for Sustainable Development, presenting major opportunities as well as challenges for developing countries. As outlined in Sustainable Development Goal #3, ehealth is crucial in ensuring healthy lives and promoting the well-being [10] of the world's peoples at all ages by 2030. Achieving healthy lives and promoting well-being is not possible if other SDGs are not tackled as well. The gains in SDG#3 can be frustrated if health is not managed as a cross-cutting development concern. For example, leveraging the digital economy to tackle poverty by creating jobs and thereby lifting standards of living of communities can really help in promoting health. There is overwhelming evidence that poverty and poor health around the world are inextricably linked. Further, poor health is greatly impacted by the political, social and economic injustices, all which can be addressed through the digital economy. World Health Organization estimates point out to around 1.2 billion people who are living in abject poverty in deplorable conditions [11] that make them sick, without shelter and adequate sanitation.

There is no doubt that quality education is linked to good health. Quality education is a green card towards better health choices and these choices have a positive impact on the health outcomes. Quality education is an entry point to health and well-being as an educated population will have a working knowledge of disease prevention. UNESCO's Global Education Monitoring Report revealed that there is considerable reduction in infant mortality and childhood sickness among mothers who attained higher levels of education [12]. 2020 online Education Trends Report indicates that their increase in the volume and traffic of user friendly online educational content [13], making education accessible to many people. This access is brought about by the digitalization of educational content, curriculum and modules.

In the contemporary healthcare industry, eHealth is evolving and developing from decentralized client-provider forms of health care provisioning based on facilities that are improving health information systems from ubiquitous to state of the art ICT systems, and in step with the increasing mobility of populations and resources. eHealth enhances the accessibility of clinical data for decision making by health professionals at all levels and facilitates the visibility of information as feedback for continuous health improvement in communities, by individuals, and in the health systems. In a similar fashion, eHealth augments the existing resources available in healthcare. Gaps, in health manpower resources have been bridged through automation of the health sector [14].

Implication of Digital Economy in Other Life Paradigm on Health

Since health is inextricably linked to other facets of life, there is greater sense that rolling out digital economy to address poverty, political and social-economic injustices or promote access to education [15] is good for health too. This paper has highlighted the interconnectedness between SDG#3 and other SDGs. It is instructive for professionals working in the digital economy to comprehend how leveraging the digital economy to address poverty or any other SDG is of importance to promoting health in a big way. With quality education being accessible in the clouds, it provides a long-term resource as people become more responsible and accountable for their health. Using the digital economy to create jobs, to lift people out of poverty and empower them greatly promotes health and wellbeing. The potential increase in household and individual income related to the digital economy enables people and the communities to afford healthcare [16].

Regulation of ICT in Health Systems

Health systems involve a diverse set of public and private data collection systems, including health surveys, administrative enrolment and billing records, and medical records. These systems are used by diverse entities, including hospitals, physicians, and health plans. While a range of health and health care entities collect data, the data does not flow among these entities in a cohesive or standardized way. Nevertheless, some entities face limitations in health information technology (Health IT), these constraints include internal inertia and frank resistance to its use [17].

As the use of ICT in health is steadily gaining traction, it is becoming equally imperative that regulatory and legal frameworks be put in place to prevent possible misuse of electronic health records. There is an increasing need for stronger standards and guidelines for eHealth to safeguard the role of mobility in the sector and to encourage interoperability between the public and private healthcare providers [18]. The Ministries of health and of ICT have a joint responsibility to put in place interoperable frameworks around the exchange of data and security for effective management of personal health records. What is more, organizations, hospitals or any other healthcare facilities must put in place their own regulation of electronic data management. Unauthorized access must be strictly regulated for patients' data is to remain private and confidential.

Even as these regulations are being put in place, it should not escape the mind that public participation is important. The guidelines and the standards are not developed with the intention to punish offenders but to safeguard patients' rights to privacy and confidentiality-which are supreme. With this in mind, it is prudent that regulation be collectively developed by involving all concerned parties. The public and the private sector must work closely during the development of these regulations and guidelines. This will enhance the acceptability of the regulatory framework.

Why Patients Need Access to Their Data

Patients have the right to access their health data. Even as healthcare is being digitized, patients' access to their data should not be comprised. There are number of reasons why patients should have access to their data [19].

1. Without health data access, patients cannot fully review their personal health information and cannot be fully engaged in their own care.
2. It is important that providers empower patients with access to their own health data and use that to foster patient engagement.
3. Patients are empowered by access to their own health records. In this manner, they can better understand their treatment plans. When they have a thorough understanding of the treatment plans, they are more likely to adhere to these treatment plans.
4. Providers have much going on during a patient visit between current patient needs, remembering their past medical history, and the long to-do list physicians often have. Empowered patients who own their data greatly improve the provider-patient relationship.

Possible Solutions

Patient portals have become a buzzword throughout the healthcare industry, with industry leaders regularly promoting their use. Providers or hospital administrators can send patients their health data, either in hard copy or electronic form. When sending a hard copy, providers must arrange a convenient time for patients to pick up the data, or rather mail the data to the patient. Providers can also send electronic copies of health data to patients. As of 2014, over 50% of patients in the developed world could view, download, or transmit their health data [20]. While patients' portals have been used mostly in the developing countries, there is every reason to believe that Africa can adapt this too and transform patient-physician interaction. Internet penetration in Africa is rising steadily and according to GSMA's **Mobile** Economy Report Series, 84% of the population (1 billion people in Sub-Saharan Africa) will have access to a smartphone connection by 2025 [21]. It is meaningful that patient's portal can be used in Africa too.

The use of social media in health care is also gaining a lot of traction due to the potential of using it to conduct reviews and rate hospitals and doctors. This is particularly important at this time when there is a growing community of thought on quality of health care. Social media reviews and ratings have been predominant in the hospitality industry and has been relied on by customers when making choices on hotels etc. Successful use of social media to rate and review hotels mean that patients can also rate their physicians/health care providers against set metrics to establish the quality of care received. And this can happen is both high and low resource settings. In an article on 10 ways in which digital technologies can transform healthcare, David Roberts, EY Global Health leader submits that there is overwhelming evidence that health care providers and the top leadership in the sector are leveraging patients reviews retrieved from social media and digital survey [22] to establish possible hitches in the delivery of care with a view to amending protocols of care and to improve service delivery.

Similarly, the surge in Africa's population is only an indication of the looming burden on the healthcare. As the global population keeps rising, pressure is piling on healthcare services and systems rendering them unable to effectively meet the growing health care demands. The high cost of training doctors and physicians makes it difficult for the healthcare services to match the growing population. This begs the question of how Africa will be able to decongest hospitals and provide health care without the need for patients to visit the hospital. Telemedicine, which has been tried in some African countries holds the true promise of the digital age [23]. With telemedicine, it is easier for doctors to conduct cloud diagnosis and deliver results without the need to physically meet the patient. The fact that this has been tested and is working is a true testament of its applicability in Africa. This modality has great utility in the management and containment of COVID-19.

On the Flipside

Even with the many documented benefits, ICT can be a major pathway for manipulation of patients' data without their knowledge and consent. This is a threat to human privacy. Unless the ICT is thoroughly regulated using a relevant regulatory toolkit, it remains a risk to feed patient data into the internet. The ICT sector must conform to international best practices as far as health ICT is concerned [24].

What is more, even as we reach the epitome of ICT, there is also an increased threat of cyber hacking. In an unregulated setting, patient information has been accessed by ICT gurus who have constantly introduced algorithms that can manipulate the file stores [25] rendering the commitment privacy and confidentiality fruitless. In some circumstances, files have been deleted and obliterated altogether. This serves to sound a warning bell, that even as we approach the era of ICT and Health, special regulations and securities must be put in place to guarantee security and thoroughly block non-authorized access. Increased vigilance is therefore warranted even as the benefits of the digital economy and eHealth become clearer. Africa must also invest heavily in ICT, train more computer scientists and hire experts to lay strong grounds for the establishment of the technologies.

Sustainable Use of ICT in Health

Review of studies conducted on the impacts of the digital age reveals how digital transformation presents a number of insurmountable impacts [26]. This paper is in agreement with the findings. The automation of jobs and tasks has left many people jobless. But this should not the focus now as we have seen in this paper how leveraging the digital economy holds enormous potential. The question that should be on the table is, "How can ICT be used Sustainably?" How can digital transformation help turn life around instead of jeopardizing the gains? While these questions linger in the mind, this paper offers that health should not fall prey of the deleterious effects of digital transformation already witnessed in other industries.

Even as health is going digital, organizations or health facilities must have a coherent strategy that includes a plan to reskill workers [27]. Whereas the previous technological advancement played over a long period of time, the speed of digital transformation in the current age means that health care industry must adjust so quickly. The digital transformation must be used as an enabler/complement and not as alternative. And when this happens, the workforce must realign their skills for this health care digital revolution. For governments, the challenge is equally pressing. The potential inequality and wage deflation or even social unrest requires urgent action to prepare the workforce for a digital future.

Case Studies on Digitalization of Health

Through the development of databases and other applications, ICTs also provide the capacity to improve health system efficiencies and prevent medical errors. A physician in a remote rural hospital may initially be unable to diagnose a patient with a complex array of symptoms. However, using his MEDLINE search training and the hospital's internet connection, he is soon able to diagnose and successfully treat the patient in theory for a disease that the patient picked up while travelling abroad. Another physician looks at her hospital's prescription trends using the newly created electronic health record system and finds that other physicians are not using the post-surgical antibiotic that is shown to be most effective according to the current international guidelines. She speaks to the administration about advocating a switch in antibiotics that will improve patient recovery outcomes and thereby save the hospital money [28].

A neonatologist, who transmits CT-scans and other medical images by e-mail to his network of personal contacts around the world to help in diagnosing and treating premature new-borns, estimates that teleconsultations have helped him to save numerous lives during the past year. A woman, too embarrassed to ask her physician about reproductive health issues and the risks of sexually transmitted infections, anonymously contacts a physician at a woman's health clinic. Here they have set up e-mail accounts for staff in order to support these types of physician-patient interactions. This woman does not have to present herself physically to the hospital but can take advantage of the digital communication channels to ask questions pertaining her reproductive health [29]. In Table 46.1 we see examples of contemporary ICT usage in the digitalization of health in the East Africa region.

Table 46.1 Examples of usage ICT in digitalization of health in East Africa region [30]

Zipline Inc.	Zipline Inc.—a California-based robotics firm to build the drone port in Shyogwe Sector, Muhanga District—Southern Province, Tanzania in a bid to improve accessibility to blood and emergency medical supplies to remote parts of the country
Asknivi	Asknivi is a digital health platform that bridges the information and access gap that women and girls face in the health field. It is an initiative of Nivi which was launched in 2016 out of the idea that family planning should be simple to understand, access, and use
M-TIBA	M-TIBA is a service on a mobile phone that allows anyone to send, save and spend funds specifically for medical treatment. Money stored in M-TIBA can be used to pay for treatment and medication at partner clinics and hospitals. M-TIBA uses the internationally recognized 'Safe Care' standards to monitor the quality of care available at these facilities. The M-TIBA app is a product of Safaricom.
CareAi	This is an AI-powered computing system anchored on blockchain that can diagnose infectious diseases, such as malaria, typhoid fever, and tuberculosis, within seconds. The platform is engineered to serve the invisible demographic of migrants, ethnic minorities, and those unregistered within traditional healthcare systems
Pharm access	Pharm Access uses blockchain to solve maternal health problems
The Chain of Trust (CoT)	The Chain of Trust (CoT) initiative, which has been using the blockchain to monitor digitally guided pregnancies Considering the major challenges that the healthcare sector has faced, lack of proper records or medical history and the unavailability of funds to mention a few of these challenges, this partnership contributes to solving these gaps by using blockchain technology
Kinect Hub	Kinect Hub is a digital healthcare system aimed at developing world settings such as in nations in Africa. By using blockchain technology in healthcare, it can track the medicine supply chain, personal medical records and donations to avoid corruption and the potential loss of vital patient information. Set up by investment banker Toby Carroll, it also incentivizes patients to attend health appointments by paying them for visits
KWHCoin	KWHCoin, has started working with Nurse in Hand Emergency Response, which provides emergency response services on Kenyan roads. KWHCoin will help the emergency response organization to establish "renewable energy resources for medical stations … and develop a structure for installing energy efficient mobile units throughout Kenya's road network

Source Research papers of authors, listed in the bibliography

The examples above serve to clearly explain how information and communications technologies (ICTs) play a very integral role in improving health care for individuals and communities.

Conclusion

In conclusion, there is no doubt that the digital economy is rapidly unfolding potential around the world. Countries especially in Europe have realized how the digital economy holds huge potential for small and medium sized enterprises and they are leveraging this to stimulate their economies especially in the post COVID scenario. However, as Africa embraces the digital economy to augment health care service delivery, it is becoming imperative that appropriate policies must be developed so as to eliminate the impediments standing in the way of fully leveraging the digital economy. What is more, the examples of successful eHealth programs in Africa provide a window of opportunity to learn how to fully integrate the digital economy in health care while optimizing the advantages and minimizing the risks involved. The time is ripe for health care to go digital and the gains Africa will sustain in the long term are hugely dependent in the commitments that are made today. It will require strong stewardship, creativity and talent on the part of government and the private sector to realize the benefits brought about by the digital age. As internet penetration continues to rise steadily, it is becoming increasingly apparent that Africa will solve its unprecedented challenges in access to health care with investments in digital health.

Appendix

42 African Country National eHealth Strategies, **Below are all the national digital health strategies and architectures that we could find for 54 African countries**.

We believe this is the most comprehensive list of national eHealth strategies for African countries as of March 2020.

- Angola: Estrategico do Sistema de Informação Sanitária 2010
- Benin: Stratégie Nationale de Cybersanté 2018–2022
- Botswana: Health Information and Communication Technology Strategy 2010
- Burkina Faso: Cyberstratégie Sectorielle eSanté 2016–2020
- Burundi: Plan National de Développement de l'Informatique Sanitaire 2011–2015
- Cabo Verde: Política Nacional de Saúde—2020
- Chad: Plan National de Developpement Sanitaire II 2013–2015
- Cameroon: National Digital Health Strategic Plan 2020–2024
- Comoros: Stratégie Nationale de CyberSanté 2017–2021

- Congo: Democratic Republic: Plan National de Développement de l'Informatique de la Santé (PNDIS) 2014
- Congo, Brazzaville: Plan de Développement Sanitaire 2007–2011
- Cote d'Ivoire: Cybersanté en Côte d'Ivoire 2011
- Egypt: National ICT Strategy for 2012–2017
- Eswatini: eHealth Strategy 2016–2020
- Ethiopia: Information Revolution Strategic Plan 2018–2025
- Gabon: Schéma Directeur Stratégique du Système d'Information 2017–2022
- Gambia: Gambian ICT4D-2012 Plan
- Ghana: Ghana e-Health Strategy 2010
- Guinea: Plan National de Développement Sanitaire (PNDS) 2015–2024
- Kenya: National eHealth Policy 2016–2030
- Lesotho: ICT Policy 2005
- Liberia: Health Information System & ICT Strategic Plan 2016–2021
- Madagascar: Plan de Développement du Secteur Santé (PDSS) 2015–2019
- Malawi: Monitoring, Evaluation and Health Information Systems Strategy (MEHIS) 2017–2022
- Mali: Politique Nationale Cybersanté 2013
- Mauritania: Project National de Telemedicine 2005
- Mauritius: National eHealth Strategy 2010–2015
- Mozambique: Plano Estratégico do Sistema de Informação para a Saúde (SIS) 2009–2014
- Namibia: National Health Policy Framework 2010–2020
- Niger: Stratégie Nationale E-Santé 2019–2023
- Nigeria: National Health ICT Strategic Framework 2015–2020
- Rwanda: National Digital Health Strategic Plan 2018–2023
- Senegal: Plan Stratégique du Système d'Information Sanitaire 2018–2023
- Sierra Leone: National Health Sector Strategic Plan 2018–2023
- South Africa: National eHealth Strategy 2019–2024
- South Sudan: Health Sector Development Plan 2011–2015
- Sudan: National eHealth Policy 2005
- Tanzania: Tanzania National Digital Health Strategy 2019–2024
- Togo: Plan Strategique de Development de la Cybersante 2013–2015
- Uganda: Uganda National eHealth Strategy 2017–2021
- Zambia: National eHealth Strategy 2017–2021
- Zimbabwe: National eHealth Strategy 2012–2017

Bibliography

1. *Digital Transformation in Healthcare in 2020: 7 Key Trend*. Retrieved from https://www.digitalauthority.me/resources/state-of-digital-transformation-healthcare/
2. Digital economy report 2019, value creation and capture: implications for developing countries, unctad/der/2019

3. State Georgia, University, GSU Library Research Guides: *Archival Research: Why Archival Research? *Why Archival Research?—*Archival Research—GSU Library Research Guides at Georgia State University* (2016), research.library.gsu.edu/archivalresearch
4. OECD, A measurement roadmap for the future, in *Measuring the Digital Transformation: A Roadmap for the Future* (OECD Publishing, Paris, 2019), www.oecd.org/going-digital/measurement-roadmap.pdf
5. I. Priyadarshini, A survey on some of the global effects of the COVID-19 pandemic. Lancet Glob Health (2020). https://doi.org/10.21203/rs.3.rs-20842/v1
6. World Health Organisation: WHA58.28 eHealth, in *eHealth Resolutions to the 58th Meeting of the World Health Assembly* (2004), pp. 121–123
7. J.A.G.M. Van Dijk, A theory of the digital divide, in *The Digital Divide: The Internet and Social Inequality in International Perspective*, ed by M. Ragnedda, G.W. Muschert. Routledge Advances in Sociology, vol. 73, No. 73. (Routledge, 2013), pp. 29–51
8. Asian Development Bank, Trade in the Digital Economy. *Asian Development Bank*, Asian Development Bank, 2 Mar 2018, www.adb.org/news/events/trade-digital-economy
9. Healthy People (2020). Reterived from https://www.ruralhealthinfo.org/topics/healthcare-access. Access on 1 Mar 2020
10. WHO, Poverty and Health. *World Health Organization*, World Health Organization, 1 Dec 2010, www.who.int/hdp/poverty/en/
11. WHO, SDG 3: Ensure healthy lives and promote wellbeing for all at all ages. World Health Organization, 3 Feb 2017, www.who.int/sdg/targets/en/
12. Writers, Staff. 2020 Online Education Trends Report. *BestColleges.com*, BestColleges.com, 24 Mar 2020, www.bestcolleges.com/research/annual-trends-in-online-education/
13. UNESCO, Global Education Monitoring Report. 2019 GEM Report (2016) gem-report-2019.unesco.org/
14. O. Poppe, B. Jolliffe, D.L. Adaletey, J. Braa, A.S. Manya, Cloud computing for health information in Africa? Comparing the case of Ghana to Kenya. J. Health Inf. Africa **1**(1) (2013). https://doi.org/10.12856/JHIA-2013-v1-i1-45
15. P.E. Brentlinger, Health and poverty. *Understanding Global Poverty* (2016), pp. 72–99. https://doi.org/10.9774/gleaf.9781315316840_5
16. OECD, *Going Digital: Shaping Policies, Improving Lives* (OECD Publishing, Paris, 2019), https://doi.org/10.1787/9789264312012-en
17. J. Collier, What are the Pros and Cons of MHealth? *Medical News Today* (MediLexicon International, 2018), www.medicalnewstoday.com/articles/322865.
18. C. Monson, A. Ros, Radio spectrum management.*ICT Regulation Toolkit* (2015), www.ictregulationtoolkit.org/toolkit/5
19. C. Hung, The right data, the right decisions: why patients need access to their health information. *Hcldr*, 8 Nov 2019, hcldr.wordpress.com/2019/11/08/the-right-data-the-right-decisions-why-patients-need-access-to-their-health-information/
20. GSMA, The Mobile Economy Sub-Saharan Africa 2019. *The Mobile Economy* (2019), www.gsma.com/mobileeconomy/sub-saharan-africa/
21. S. Heath, How patient portals improve patient engagement. *PatientEngagementHIT*, 6 May 2016, patientengagementhit.com/features/how-patient-portals-improve-patient-engagement
22. D. Roberts, *10 Ways Digital Could Transform Healthcare. EY*, EY, 25 Apr 2018, www.ey.com/en_gl/digital/10-ways-digital-could-transform-healthcare
23. M. Mars, Telemedicine in Sub-Saharan Africa. *Telehealth in the Developing World* (2019), pp. 232–241. https://doi.org/10.1201/9781315272917-23
24. I.V. Sika et al., Download research report: ICT and governance in East Africa (IHUB, 2015), ihub.co.ke/ict4gov
25. S. Mowatt, Impacts of the digital economy. Digit Econ (2015), pp. 136–153. https://doi.org/10.4018/978-1-59140-363-0.ch007
26. Aconf, 2018 International Conference on Cyber Security and Protection of Digital Services (Cyber Security), in *2019 International Conference on Cyber Security and Protection of Digital Services (Cyber Security)* (2019), https://doi.org/10.1109/cybersecpods.2018.8560669

27. C. Rudder, Digital transformation strategy: 7 key pieces. *The Enterprisers Project* (2019), enterprisersproject.com/article/2019/8/digital-transformation-strategy-7-pieces
28. Walden, University, Evidence-based practice research: MEDLINE search help. *Academic Guides* (Walden University Library, 2016), academicguides.waldenu.edu/library/healthevidence/medlinesearchhelp
29. V.r. Gruessne, How digital health devices benefit diagnostic services. *MHealthIntelligence*, 2 July 2015, mhealthintelligence.com/news/how-digital-health-devices-benefit-diagnostic-services
30. S. Lee et al., Mapping MHealth (Mobile Health) and mobile penetrations in Sub-Saharan Africa for strategic regional collaboration in MHealth scale-up: an application of exploratory spatial data analysis. Glob Health, BioMed Central, 22 Aug 2017, www.ncbi.nlm.nih.gov/pmc/articles/PMC5568212/

Chapter 47
Belt and Road Initiative (BRI): Mapping the Field and Identifying New Research Avenues

Anna Visvizi, Jun Wang, and Krzysztof Kozłowski

Abstract Since its inception, the Belt and Road Initiative (BRI) sparked considerable attention of research community, on the one hand, reviving memories of the ancient Silk Road and its implications for culture, society, and economy, and on the other hand, triggering concerns about investment patterns, debt issues, and environmental sustainability, to mention just a few. BRI is an evolving policy framework and it is only gradually that it can be examined at full length, with the necessary academic rigor and insight. A look at the existing literature on BRI reveals a burst of interest and corresponding publications on BRI over the past few years. It is too early to claim though that a clearly defined research agenda related to BRI has crystallized. In this research note, factors that may have contributed to that are identified and the frames of a possible future consolidated BRI research agenda are outlined.

Keywords Belt & Road Initiative (BRI) · Conceptual approaches · Research agenda

Introduction

Since its inception, the Belt and Road Initiative (BRI) sparked considerable attention of research community, on the one hand, reviving memories of the ancient Silk Road and its implications for culture, society, and economy, and on the other hand, triggering concerns about nascent investment patterns, debt issues, and environment sustainability, to mention just a few [1–3]. BRI is an evolving policy framework. The logic and the mechanisms, by means of which BRI is implemented, crystallize gradually. Doubtless, inasmuch as—for many reasons—BRI represents one of the

Build the road first if you want to get rich.

A. Visvizi (✉) · K. Kozłowski
SGH Warsaw School of Economics, Warsaw, Poland
e-mail: avisvizi@gmail.com

J. Wang
Graduate School of Chinese Academy of Social Sciences, Beijing, China

A. Visvizi et al. (eds.), *Research and Innovation Forum 2020*,
Springer Proceedings in Complexity,
https://doi.org/10.1007/978-3-030-62066-0_47

most ambitious political and economic collaboration initiatives, it is also one of the most unorthodox ways of expanding a country's economic and political influence. In this view, whilst BRI may seem difficult to delineate, it is also driven by non-linear dynamics. Both factors add to the challenge of conceptualizing and accounting of BRI. Indeed, as BRI gradually marks its presence in the business, economics, and political discourses, it also takes time for the world of academia find ways of examining BRI with the necessary academic rigor and method.

An insight into the existing literature on BRI reveals a burst of interest and corresponding publications on BRI over the past few years [4–6]. It is too early to claim though that a clearly defined research agenda related to BRI crystallized. This calls for an early reflection on the key approaches adopted to the study of BRI so far and on the key issues examined in academic debate on BRI. The objective of this research note is to do just that. The argument is structured as follows. First, the key features of BRI are discussed briefly. Then, a typology of the key analytical lenses that can be applied to the study of BRI is developed. Using the latter as the organizing framework, a stylized literature review follows. Against this backdrop, the drawbacks and limitations of the existing debate are highlighted, and new avenues of research are pointed to.

BRI: A Brief Insight into Its Key Features

For several reasons, the launch of BRI did not automatically trigger academic debate on its diverse facets. For instance, China, its growth model, and the transformation of its international role, including China's role in world trade or China's engagement with Africa, constitute an entire ecosystem of topics and research questions. Implicitly, BRI was seen as a part of these developments, even if, as it shall be argued later, the important causal relations among these may have been overlooked in research. From a different angle, one could argue, the launch of BRI, and the discussion preceding it, coincided with the severe implications of the 2008 global financial crisis and the multidimensional crisis in the euro area, including the crisis in Greece [7, 8]. Finally, a considerable degree of indefiniteness specific to BRI, infused perhaps also with a disbelief that an initiative of this kind could be conceived of in the first place, made the world of academia … catch the breath and wait. By now, the situation changed. BRI established itself as one of the key artefacts of research and policymaking today. This section offers a brief insight into the development of BRI and its key defining features.

The Genealogy of BRI

The conceptual idea of BRI was first introduced by President Xi during his visit to Kazakhstan in September 2013. At that time, the Silk Road Economic Belt (the Land

Silk Road) was proposed. The maritime Silk Road was proposed one month later, in Xi's speech in the Parliament of Indonesia. At that time, Xi unveiled his vision of a Maritime Silk Road that would strengthen the connections within the Association of Southeast Asian Nations (ASEAN) and China, and thus would promote maritime cooperation. To facilitate the implementation of projects that this would require, decision on the establishment of the Asian Infrastructure Investment Bank (AIIB) was taken. Headquartered in Beijing, AIIB begun its operations in 2016.

On March 28, 2015, as joint release of three Chinese ministries, the first official document, a roadmap, specifying Belt and Road's vision and implementation strategy was published, so called BRI White Book. The document is interesting insofar as, on the one hand, it introduces the term Belt and Road Initiative (BRI), and on the other hand, places BRI in a broader historical perspective. BRI in this context is described by reference to co-construction principles, framework ideas, cooperation priorities and cooperation mechanisms. BRI continues evolving.

While initially it was referred to as the One Belt (denoting the Land Silk Road) and One Road (denoting the Maritime Silk Road), in September 2015, the National Development and Reform Commission [9] released a document specifying and regulating the English translation of "一带一路" to "Belt and Road Initiative" (BRI). In this way former translation "One Belt and One Road" (OBOR) was replaced. This seemingly tiny semantic change allowed to bypass the problem of an unnecessary and confusing distinction between 'belt' and 'road'. It also enabled the Chinese authorities to present BRI as a project aimed at connecting diverse stakeholders and fostering regional collaboration, rather than solely as a China-centred initiative.

Issues, Tools, and Approaches

With the aim of reaching the connection in policy, facility, trade, finance and people to people exchange, and along with various forms of BRI, the following dimensions and mechanisms of collaboration have been inscribed in BRI: four (4) routes (land, sea, air, ice), five (5) "connections" (policy, facilities, trade, finance, and people to people bond), six (6) economic corridors, eighteen (18) Free Trade Zones (FTZ), and several major infrastructures projects, including highways and railway connections leading from China through Asia to Europe. Indeed, since 2015 aviation has been a part of BRI. Notably, in June 2017, the Air Silk Road Initiative was officially proposed during Xi's meeting with the Prime Minister of Luxembourg. In July 2017, the establishment of the Arctic Silk Road was discussed by China and Russia [10, 11].

The interesting feature of BRI is that it is not a top-down initiative calling to set up a series of new regulations or standards. Instead, BRI builds on the already existing institutional frameworks and infrastructure, and where necessary adds new pieces of infrastructure. In line with the BRI White Book, the objective of BRI is to foster collaboration and improve coordination among diverse collaboration schemes.

BRI's focus on infrastructure, and so considerable investment effort in this field, reflects the typical Chinese logic, best captured by the old Chinese saying: "If you want to get rich, build the road first". Therefore, the outlook in the 2015 BRI White Book embedded China's unique development theory based on its past 70 years experiences. Additionally, BRI proposed to work closely with countries around instead of merely working with developed countries also reflected the idea of "A good neighbor is better than a brother far off", which is embedded in traditional Chinese works.

Clearly, BRI is a product of Chinese model of growth and development and for that matter also a reflection of China's worldview. Specifically, the logic and mechanisms underpinning BRI can all be found in China's own development path, including bilateral economic parks, resembling successful cases of such parks inside China; special economic zone, like that of e.g. Shenzhen, etc. For this, as well as for other reasons, the success of BRI is very much a function of other countries' ability and/or willingness to follow those and related rules, principles and objectives. Ultimately, BRI is an initiative that focuses on exploring the development path for regional prosperity and emphasizes the value added of spill-over effects of regional collaboration. It remains an open question, whether this development path will meet the demands and expectations of the larger region and thus whether the possible spillover effects will pragmatize.

Querying BRI: Key Approaches and Analytical Lenses

Even if still evolving, BRI, lends itself to a variety of conceptual and empirical approaches derived from a variety of fields and disciplines. It is not easy to build a coherent typology of approaches and analytical lenses applied to the study of BRI. Even if, arguably, inter- and multi-disciplinary approaches are needed to capture the complexity and dynamics involved in BRI and its implementation, very few papers expressly recognize and validate the issue. Clearly, the plethora of approaches and conceptual lenses that BRI enables is vast. Correspondingly, the set of methodologies and techniques that are required to study specific aspects of BRI is expanding. The implications of BRI, short-, mid-term and long-term, require a very specific methodological toolkit to apply if the latter are to be explored. Due to design of BRI, the basic distinction and, so academic approaches to BRI, are dictated by the focus on maritime BRI and land BRI. That distinction predefines the topics of engagements, analytical focus, and issues considered. But the same applies to each aspect of BRI. Table 47.1 offers an insight into the resulting research options and opportunities.

There are several, frequently overlapping, dimensions of the discussion and research on BRI. In this research note, we propose to divide the research opportunities into four broad categories depending (i) on the academic discipline in which a given research items is located; (ii) on the specific issue that is examined in detail; (iii) on the time-frame and time orientation applied in research; (iv) on the level of analysis applied.

(i) It will be a truism to say that it is possible to distinguish approaches clearly associated with a particular discipline of science, e.g. economics [12, 13], and

Table 47.1 BRI: research options and opportunities

	Conceptual approaches	Methodology	Techniques
Discipline of science (social science, engineering, computer science)			
Issue orientation (security, resources, etc.)			
Time frame (short-, mid-, long-term; past-present)			
Perspective/level of analysis (macro-, mezzo-, micro-)			

Source The Authors

research related to key issues that economists are interested in, e.g. capital flow, investment patterns, public finance, including debt issues [14–16]. From a different angle, political scientists look at BRI from a quite different perspective, seeking to figure out what BRI actually is and how to conceive of its implications [17];

(ii) Considering the specific issue that is examined, the variety of topics that is discussed includes, security [18], environment and sustainability [19, 20], infrastructure development [21, 22], energy [11, 23];

(iii) Considering the timeframe [1–3, 24, 25];

(iv) Considering the level of analysis, it is possible to distinguish between macro-, mezzo-, micro- and nano-approaches. In this view, macro-level approaches to BRI would seek to offer the big picture of BRI, possibly most closely related to globalization, global governance, multilateralism, but also geopolitics [6, 26–28], and so on.

The objective of this brief and sketchy insight into current research on BRI is to merely highlight that—as we speak—BRI research agenda consolidates. It is thus important to address the question of which factors and issues drive the emerging academic debate on BRI. It is equally important to ponder which issues in this context seem to have been omitted in the analysis. The following section sheds some light on these questions.

Discussion and Conclusions

BRI remains a largely under-researched and under-conceptualized political and economic collaboration initiative. The expansionist vision that seems to underpin BRI works as one of the key sources of either confusion or bias that can be traced in scholarly and non-academic writing on BRI. Estimates suggest that over the period 2017–2027 BRI investment projects will generate more that USD 1 trillion of outward funding for foreign infrastructure [29].

Instinctively still, rather than based on sound academic proof, a claim can be made that BRI cannot be studied as a yet another case of international collaboration [30]. BRI defies the established conceptual frameworks employed in international relations and political science or economics. And yet, BRI draws from and can only be pursued because of the existing structures of international collaboration, including in particular the regulatory framework underpinning the multilateral trading system. BRI employs, brings together and exploits synergies emerging among traditional forms of regional integration (e.g. corridors), regional collaboration (preferential trade agreements, free trade areas, special economic zones) and novel forms of funding (e.g. multilateral approach to investment; public-private partnerships).

For these reasons, it is challenging to study BRI as a coherent whole. It does not fit the established conceptual frameworks and toolkits associated with the study of globalization, global governance, multilateralism, liberalization of trade, international collaboration. It is all of it, and none of it at the same time. BRI remains an initiative that lends itself to inter- and multi-disciplinary approaches, whereby it is the level of analysis that will be decisive when embarking on a specific research question and a corresponding conceptual framework adopted for the analysis. The challenge is, however, that in this manner our ability to understand and explain what BRI represents and how it works, will remain fragmented. As a result, the collection of insights into specific aspects of BRI will not necessarily enable us to understand BRI. For instance, several issues reflective of a missing BRI research agenda, such as economic migration that BRI implicates, forced migration and displacement, and from a different angle, human resource management, are missing in current debate. Moreover, even if 'greening' of BRI research is under way, research that would capture cross-cutting issues as they evolve along BRI is largely absent.

To conclude, while more research is needed to examine BRI and its intricacies, the key imperative is to set well-defined disciplinary boundaries for the debate. Importantly, these boundaries are needed not to constrain research but to allow conscious attempts of inter-disciplinary research. There is an urgent need to reflect on several categories and dimensions of research that BRI requires. These include (i) the level of analysis, (ii) approaches to and frameworks of analysis, (iii) availability, reliability and quality of data, and above all (iv) conscious effort to match the former with corresponding ontological and epistemological choices. The objective of our forthcoming research is to address this void and its implications for research on BRI.

References

1. T. Ladjal, T. Brahimi, The spirit of silk road: the impact of Medieval Chinese–Arab relations on the contemporary bilateral Sino–Arab relations, in *The New Silk Road Leads through the Arab Peninsula: Mastering Global Business and Innovation*, ed by A. Visvizi, M.D. Lytras, W. Alhalabi, X. Zhang (Emerald Publishing, Bingley, UK, 2019)
2. G. Hawke, Economic historical development and leadership underlying the BRI strategy, in China's Belt and Road Initiative. Understanding the Dynamics of a Global Transformation, ed by Y.W. Chay, T. Menkhoff, L. Low (World Scientific, 2019), pp. 19–43.

10.1142/9789811203275_0003
3. D.T. Van, The rise of China's past in the "belt and road initiative" (from Historical Perspectives), in *Silk Road to Belt Road*, ed by M. Islam (Springer, Singapore, 2019), https://link.springer.com/chapter/10.1007/978-981-13-2998-2_2
4. Y.W. Chay, T. Menkhoff, L. Low (eds), China's belt and road initiative. Understanding the dynamics of a global transformation. World Sci. (2019)
5. M. Islam (ed), *Silk Road to Belt Road* (Springer, Singapore, 2019). https://doi.org/10.1007/978-981-13-2998-2
6. A. Visvizi, M.D. Lytras, W. Alhalabi, X. Zhang (eds), *The New Silk Road Leads Through the Arab Peninsula: Mastering Global Business and Innovation* (Emerald Publishing, Bingley, UK, 2019)
7. A. Visvizi, The crisis in Greece and the EU-IMF rescue package: determinants and pitfalls. Acta Oeconomica **62**(1), 15–39 (2012)
8. Ch. Schweiger, A. Visvizi (eds), *Central and Eastern Europe in the EU: Challenges and Perspectives Under Crisis Conditions* (Routledge, London & New York (Routledge Studies in the European Economy). ISBN: 978-1-13-871402-1 (2018)
9. NDRC, Tuidong gongjian sichouzhilu jingji dai he ershiyishiji haishang sichou zhilu de yuanjing yu xingdong 推动共建丝绸之路经济带和21世纪海上丝绸之路的愿景与行动 [Vision and Actions on Jointly Building Silk Road Economic Belt and 21st-Century Maritime Silk Road]. National Development and Reform Commission, Ministry of Foreign Affairs, and Ministry of Commerce of the People's Republic of China with State Council authorization 经国务院授权, 国家发展改革委、外交部、商务部联合发布 28 Mar 2015 (2015), https://www.fmprc.gov.cn/mfa_eng/zxxx_662805/t1249618.shtml
10. PRC MoFA, China's Arctic Policy, Ministry of Foreign Affairs of the People's Republic of China, fmprc.gov.cn, 26 Jan 2018
11. F. Manta, China's policies on Polar Silk Road and impacts on trade and energy use in Arctic region. Chin. J. Population Resour. Environ. **17**(4), 322–328 (2019). https://doi.org/10.1080/10042857.2019.1685360
12. Z. Hu et al. 胡再勇, "Yidaiyilu yanxian guojia jichu sheshi de guoji maoyi xiaoying yanjiu" "一带一路"沿线国家基础设施的国际贸易效应研究 [Research on Infrastructures' International Trade Effect in Countries along the 'Belt and Road' Initiative]. J. Quant. Econ. 数量经济与技术经济研究 **36**(02), 24–44 (2019)
13. C. Pei, 裴长洪, B. Liu 刘斌 (2020) Zhongguo Kaifangxing Jingjixue: goujian chanshi zhongguo kaifangchengjiu de jingji lilun. 中国开放型经济学: 构建阐释中国开放成就的经济理论 [China's Open Economy: Constructing the Economic Theory of Interpreting China's Open Achievements]. China Soc. Sci. 中国社会科学, 2020(02), 46–69+205
14. W.T. Selmier II., The belt and road initiative and the influence of Islamic economies. Econ. Polit. Stud. **6**, 257–277 (2018)
15. Y. Li, H.J. Schmerer, Trade and the new silk road: opportunities, challenges, and solutions. J. Chin. Econ. Bus. Stud. **15**, 205–213 (2017)
16. G. Jin, 金刚, K. Shen 沈坤荣, Zhongguo qiye dui Yidaiyilu yanxian guojia de jiaotong touzi xiaoying: fazhan haishi zhaiwu xianjing 中国企业对"一带一路"沿线国家的交通投资效应: 发展效应还是债务陷阱 [Chinese Enterprises' Transportation Investment Effect in Belt and Road Countries: Development Effect or Debt Trap]. Zhongguo Gongye Jingji 中国工业经济 **9** (2019)
17. L. Xue, 薛澜, L. Weng 翁凌飞, Guanyu zhongguo yidaiyilu changyi tuidong lianheguo 2030 nian kechixu fazhan yicheng de sikao 关于中国 "一带一路"倡议推动联合国《2030年可持续发展议程》的思考 [Reflections on China's 'Belt and Road' Initiative to promote the UN 2030 Agenda for Sustainable Development]. Bull. Chin. Acad. Sci. 中国科学院院刊 2008, **33**(01), 40–47 (2018)
18. K. Iqbal, Significance and security of CPEC: A Pakistani perspective. China Int'l Stud. **2017**, 66 (2017)
19. Y. Li, X. Zhu, The 2030 Agenda for sustainable development and China's belt and road initiative in Latin America and the Caribbean. Sustainability **11**, 2297 (2019)

20. H.C. Teo, A.M. Lechner, G.W. Walton, F.K.S. Chan, A. Cheshmehzangi, M. Tan-Mullins, H.K. Chan, T. Sternberg, A. Campos-Arceiz, Environmental impacts of infrastructure development under the belt and road initiative. Environments **6**, 72 (2019)
21. K. Kozłowski, BRI and its digital dimension: twists and turns. J. Sci. Technol. Policy Manag., https://doi.org/10.1108/JSTPM-06-2018-0062 (2018)
22. X. Wei, Y. Wei, P. Chen, C. Fan, H. Luo, Q. Zhao, Y. Kong, The analysis of online silk road from the perspective of big data, in *The New Silk Road Leads through the Arab Peninsula: Mastering Global Business and Innovation*, ed. by A. Visvizi, M.D. Lytras, W. Alhalabi, X. Zhang (Emerald Publishing, Bingley, UK, 2019)
23. F. Shaikh, Q. Ji, Y. Fan, Prospects of Pakistan–China energy and economic corridor. Renew. Sustain. Energy Rev. **59**, 253–263 (2016), https://doi.org/10.1016/j.rser.2015.12.361
24. K. Kozłowski, The past, the present, and the future of the new silk road: China as a leader or a free-rider in international relations, in *The New Silk Road Leads Through the Arab Peninsula: Mastering Global Business and Innovation*, ed by A. Visvizi, M.D. Lytras, W. Alhalabi, X. Zhang (Emerald Publishing, Bingley, UK, 2019)
25. F.M. De Soyres, R. Marie, A. Mulabdic, M. Ruta, Common transport infrastructure: a quantitative model and estimates from the belt and road initiative (English). Policy Research working paper; no. WPS 8801 (World Bank Group, Washington, D.C., 2019), https://documents.worldbank.org/curated/en/879031554144957551/Common-Transport-Infrastructure-A-Quantitative-Model-and-Estimates-from-the-Belt-and-Road-Initiative
26. M.K. Khan, I.A. Sandano, C.B. Pratt, T. Farid, China's belt and road initiative: a global model for an evolving approach to sustainable regional development. Sustainability **10**, 4234 (2018)
27. Y. Wang, Offensive for defensive: the belt and road initiative and China's new grand strategy. Pac. Rev. **29**, 455–463 (2016)
28. W. Liu, M. Dunford, Inclusive globalization: unpacking China's belt and road initiative. Area Devel. Policy **1**(3), 323–340 (2016), https://doi.org/10.1080/23792949.2016.1232598
29. OECD, *China's Belt and Road Initiative in the Global Trade, Investment and Finance Landscape, OECD BUSINESS AND FINANCE OUTLOOK 2018* (OECD Publishing, Paris, 2018), p. 3
30. A. Surdej, Multilateralism and international governmental organizations: principles and instruments. Transforming Gov. People, Process Policy **14**(3) (2020), https://doi.org/10.1108/TG-11-2019-0107

Chapter 48
The Role of China's Policy in Transforming Its Position on the Global Innovation Scene

Małgorzata Dziembała

Abstract The main aim of the paper is to discuss the innovation potential of China against the US and the EU and to analyze the growing role of China in the implementation of the innovation policies, including the Made in China 2025 (MIC 2025) plan. It is claimed that the weight of this innovation strategy is not only to enhance the position of China on the global innovation scene but it has also a wider impact all over the world. MIC 2025 promotes the development of strategic sectors being the foundation of the new economic model of China, and there are specific activities that are strongly supported by the government. These actions aim to create China's leading position on the international innovation scene. China increases its innovation performance which is also reflected in international collaboration, including the EU and the US. However, there were many points in disputes between the US and China as well as the EU and China. All these parties have to rethink their relations in the field of science and technology to promote cooperation.

Keywords Innovation · Research & development · Made in China 2025 (MIC 2025) · Innovation policy

Introduction

Promoting innovation becomes a priority for most of the countries in the world. However, for decades, highly developed countries, namely the US and some EU countries have dominated the global innovation scene. Their innovation policies concentrate on creating their potential by implementing specific activities and incurring significant expenditures on research and development (R&D) and innovation. The innovation potential of a country can be seen from two perspectives, namely as an innovation ability i.e. the degree to which a country is able both, to create and to commercialise new ideas, and as an innovation position, which should be

M. Dziembała (✉)
Department of International Economic Relations, University of Economics in Katowice, 1 Maja 50, 40-287 Katowice, Poland
e-mail: malgorzata.dziembala@ue.katowice.pl

A. Visvizi et al. (eds.), *Research and Innovation Forum 2020*,
Springer Proceedings in Complexity,
https://doi.org/10.1007/978-3-030-62066-0_48

deemed as the outcome of the efforts undertaken concerning promoting innovation [1, p. 32]. Thus, a solid foundation for achieving a strong position should be created and reflected in the well-developed innovation systems of countries. The continuous effort concerning innovation potential is being made in the countries characterised by the lower level of economic development, particularly in Asian countries. Among them is China, which since the late 70s has underlined the important role of science and technology in the socio-economic development of the country and then in the mid-2000s making China a world innovating country became a goal [2, pp. 656–659]. At the moment China is regarded as a developing economy [3] for which innovation is very important.

Therefore, science, technology and innovation (STI) policies were developed in different stages of China's development: pre-industrialization, catch-up, and post-catch-up-period since 2012 when innovation and cutting-edge technologies have been priority issues [4, pp. 7–14]. This country transforms its economy in terms of innovation and significant progress has been made within a relatively short time. At present, China appears to be a strong innovator with the share of R&D spending in relation to GDP constantly rising, which amounted to 0.56% in 1996, while in 2017 it amounted to 2.15% GDP [5]. This is also the result of specific activities within the framework of the innovation policy conducted by China in the twenty-first century. The policies promoting STI have evolved over the years concerning their tools and mechanisms. Nowadays they focus on the development of domestic innovation potential through implementing the strategies (plans) including the national Made in China 2025 (MIC 2025) plan that aims to enhance innovativeness of this country, and specific tools and mechanisms within this plan are applied. However, the innovation policy undertaken by China through these strategies (plans) influences the innovation environment of the EU and the US.

The literature discusses changes in the innovation of China and assesses the evolution of its innovation policy, including factors that contribute to this transformation [2, 6–16]. There are some studies related to the innovation strategy of the new plan Made in China 2025, its tools and mechanism which are implemented [17–19]. However, the rapid growth of the Chinese market and the weight of its innovation strategies not only for China itself but for the whole world have not been broadly analysed [20]. Therefore the paper might fill the existing gap.

The main aim of the paper is to discuss the innovation potential of China against the US and the EU and to analyse the emerging growing role of China in the implementation of the innovation policies, including the Made in China 2025 plan. It is claimed that the weight of this innovation strategy is not only to enhance the position of China on the global innovation scene but it has also a wider impact all over the world.

The following research questions were formulated:

Q1—is the implementation of the innovation policy strategies effective (including Made in China 2025 which uses specific innovation tools and policy mechanism) and does it have a major impact on the innovation of China?

Q2—should the significance of the innovation strategies of China particularly Made in China 2025 be perceived in the context of the entire world?

The paper consists of the following parts: in the first section the innovation capacity of the US, the EU, and China has been discussed, and the growing role of China in terms of innovation is underlined. The second section presents the tools and mechanisms of the innovation policy implemented in China within the MIC 2025 plan. Then, some implications of China's innovation policy executed through the aforementioned plan on the global innovation scene based on the example of the US and the EU have been discussed. Conclusions will follow.

The Position of China Against the EU and the US in Terms of Innovation

There are some changes on the innovation scene that have been observed over the last decades. In general, global R&D expenditures are rising. However, not only high-income economies account for those expenditures. The significance of middle-income economies has also increased. In 1996, the share of high-income economies in the global R&D expenditures amounted to 87%, nevertheless it dropped to 64% in 2017 [in absolute figures, in 1986, the share increased from USD 568 million (in PPP) to USD 1079 million (in PPP)]. However, in 2017, the share of upper-middle-income economies (including China) in the global R&D expenditures amounted to 31% (compared to 10% in 1996). The importance of some Asian economies such as China, Japan, the Republic of Korea, and India has increased. Their share in the global R&D in 2017 stood at 40%, while in 1996 it amounted to 22%. The most significant change concerned China that increased its R&D expenditures from 2.6% in 1996 to 24% of the world's R&D in 2017 [21, pp. 3–4]. The role of this country is significant as far as private expenditures are concerned because private R&D expenditures incurred by China accounted for 27% of the world's total expenditures in 2017, compared to the year 1996 when those outlays accounted for only 2% [21, p. 3].

Two groups of indicators measure innovativeness at the country level: input indicators like expenditures on R&D, and the output (outcome) indicators which measure the direct results related to innovative efforts e.g. product, marketing, process, and organizational innovation. However, innovation outcomes are associated with the consequences related to the introduction of the innovation, thus with their impact on the company, the economy, and society. The innovation outcomes could be reflected in the structural change towards more knowledge-intensive industries and the structural upgrading by the different performance of companies within the industry, moving to more knowledge-segments (changes in the intra-sector composition); thus translated into economic effects [22; 23, pp. 3–14].

China has recorded gradual progress in terms of the value of some input indicators against the main competitors which are the US and the EU. China has been gradually increasing its expenditures on R&D. In 2007 this kind of spending amounted to 1.37% of GDP, while in 2017 it reached 2.15%, exceeding this kind of expenditure in the

EU-28 (Fig. 48.1). However, the EU countries could not be treated as the homogenous group because there are countries like Germany which are characterized by a high level of expenditures on R&D as % of GDP (3.07%). In Sweden it amounted to 3.37% while in Romania 0.5% in 2017 [24].

However the distinction should be made between the public and private expenditures (Table 48.1).

What is observed in China in the analysed period, that is increasing contribution of the business sector to GERD, which in 2007 amounted to 0.99% in relation to GDP, while in 2017 it stood at 1.66%. This is reflected in the high share of the business sector in the total R&D expenditure that amounted to 76.5% in 2017.

It is worth mentioning that progress was observed in China not only concerning enormous spending on R&D, but also because this country developed a pool of

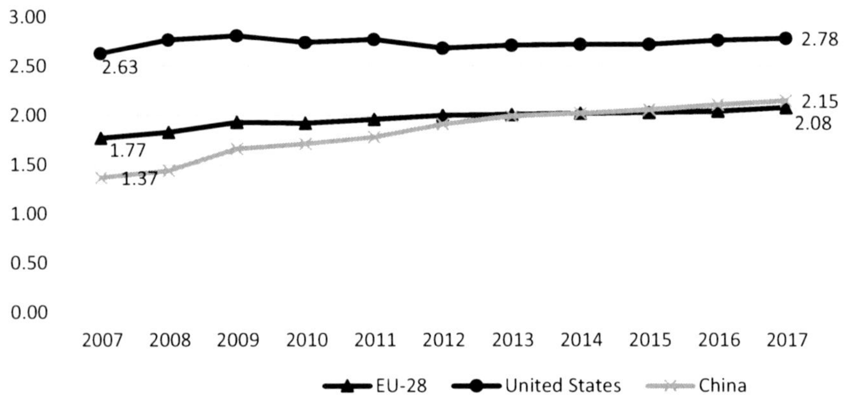

Fig. 48.1 R&D expenditures incurred in EU-28, the US and China in 2007–2017, expressed as % of GDP. *China without Honkong. *Source* Own elaboration based on the [24]

Table 48.1 Expenditures on R&D in China, the US and EU-28 in 2007 and 2017

	Expenditures of business enterprise sector as a percentage of GDP		Expenditures of government sector as a percentage of GDP		Expenditures of higher education sector		Share of business sector in the total expenditures
	2007	2017	2007	2017	2007	2017	2017, in %
China (without Honkong)	0.99	1.66	0.26	0.33	0.12	0.15	76.5
US	1.86	2.03	0.31	0.27	0.35	0.36	63.6
EU-28	1.12	1.37	0.23	0.23	0.4	0.46	58.2

Source Own compilation based on [24], [20.05.2020]

S&T talent taking into consideration the increasing number of graduates, including students of science and engineering and the number of researchers. Significant investments were made in technology and equipment, particularly in ICTs, energy, and environmental protection, which allowed to reduce the gap between China and the developed economies in terms of innovation [25, pp. 623–625]. Some programmes were initiated by the central government in China to prevent brain drain and apply brain gain, brain circulation to attract Chinese citizens educated abroad who have the necessary skills of particular importance to China. Among the several programmes were The Hundred of Talents programme, The Chunhui Scholar Program, The Program of Introducing Discipline-Based Talents to universities, initiated in 2008 the Thousand Talents Plan to attract about 2000 leading researchers (with professorships or equivalent), as well as entrepreneurs, managing staff. There was also The Project of Thousand Youth Talents launched in 2011 for young talented people with Ph.D. degrees obtained from world-class universities [26, pp. 2–3, 23–26; 27].

China's significant position on the innovation scene is reflected in the patent applications filed under the Patent Cooperation Treaty (PCT) and "triadic patent families" (Tables 48.2 and 48.3). The number of China's patent applications under the PCT amounted to more than 48,000 in 2017, which is comparable with the figures achieved by the US and the EU respectively. It also marked the effectiveness of the Chinese innovation system in terms of its output.

However, each country is characterised by the specific patent system. The US legal system offers a relatively quick patenting procedure, with the adoption of the system "first-inventor-to-file" being a component of the American Invents Act (AIA) in 2013. It is relatively an inexpensive process compared to that existing in particular European countries [29, 30].

There are some synthesis indicators that capture different aspects of innovativeness. One of them is the Global Innovation Index (GII). According to GII 2019, out of 129 countries analysed, the following 15 economies are regarded as the most innovative: Switzerland, Sweden, the United States, the Netherlands, the United Kingdom, Finland, Denmark, Singapore, Germany, Israel, South Korea, Ireland, Hong Kong, China, and Japan. In the aforementioned ranking, China took the 14th place [21, p. XXXIV]. When taking into account the most innovative economies by region, in Europe there were Switzerland, Sweden, the Netherlands, the UK, and Finland. In Northern America the leaders were the US, Canada, and in South East Asia and Oceania the front runners were Singapore, the Republic of Korea, Hong Kong, China and Japan [21, p. XXXIV]. When analysing innovation by an income group, the results were as follows: China became the leader in the upper-middle-income group whereas the high-income group included Switzerland, Sweden and the US [21, pp. 10, 16]. The positions of the US, China, and the EU countries in light of the GII 2019 and GII 2008–2009 are presented in Table 48.4.

The US and the EU countries take high positions in the ranking. However, a detailed analysis of the indices shows that sometimes the innovation input do not give spectacular results, but not for China. This country has improved its position over the last ten years, taking into consideration that in the GII 2008–2009 it got into the 37th position while according to the GII 2019, China took the 14th place. The

Table 48.2 Patent applications filed under the PCT in the US, China and selected the EU-28 countries in 2007–2017 (total number of patents)

Country	2007	2010	2015	2016	2017
Austria	1325.1	1411.8	1540.5	1542.0	1539.8
Belgium	1160.1	1255.9	1227.1	1354.8	1199.9
Czech Republic	219.2	153.4	258.1	224.3	198.8
Denmark	1351.8	1143.0	1268.8	1306.7	1318.5
Estonia	46.4	50.7	27.0	38.4	47.6
Finland	1598.0	1568.5	1295.4	1357.2	1379.1
France	6817.3	7215.0	8117.3	7661.8	7267.6
Germany	18,740.4	18,501.7	18,077.6	18,753.6	18,7811
Greece	118.1	90.9	108.4	121.4	117.2
Hungary	248.5	243.6	277.4	262.8	220.9
Ireland	439.2	342.6	442.8	479.2	440.4
Italy	3361.2	3146.2	3643.0	3649.8	3530.2
Latvia	23.7	13.4	31.1	22.6	30.0
Lithuania	17.7	20.8	25.1	35.6	33.1
Luxembourg	40.1	56.5	77.6	95.1	78.0
Netherlands	3580.5	2941.8	3711.2	3512.5	3086.8
Poland	167.0	278.3	543.7	399.8	345.3
Portugal	114.3	134.7	222.3	212.9	198.7
Slovak Republic	48.7	48.9	64.2	75.0	59.4
Slovenia	119.5	134.5	84.6	93.1	125.7
Spain	1540.7	1905.5	1793.8	1755.0	1632.2
Sweden	3164.7	2848.4	3222.1	3396.4	3191.9
United Kingdom	6416.9	5759.5	6330.2	6259.3	5952.0
United States	50,040.7	45,224.0	53,590.5	54,113.3	52,704.4
European Union-28	50,802.4	49,398.1	52,579.6	52,818.8	50,961.3
Bulgaria	29.3	29.5	56.8	46.8	52.9
China	5995.7	13,465.2	33,161.5	42,828.5	48,041.4
Croatia	60.8	47.9	31.1	49.2	34.6
Cyprus	6.2	6.1	10.4	11.6	12.1
Malta	8.5	6.5	11.5	21.0	18.2
Romania	38.4	42.6	80.5	809	69.4

Source Own compilation [28]

Table 48.3 Statistics referring to patents in the EU, the US and China, 2017

	Number of "triadic" patent families	Share of countries in "triadic" patent families
EU	13,106.09	25.05
US	12,454.36	22.85
China	4152.15	7.62

Source Own compilation, 21.05.2020 [28]

more detailed analysis shows that in terms of output, China took the 5th position while in terms of input it got into the 26th place in the GII 2019. The pillars of GII 2019 concerning China are presented in Fig. 48.2.

An improvement in the share of China in the manufacturing Global Value Chains was observed. The share of China in Global Value Chains of manufacturing increased from 6% in 2000 to 19% in 2014, primarily as a result of an increase in its competitiveness. But these were mostly traditional engineering sectors and textiles. Over the analysed period 2000–2014 the following sectors observed a big increase in medium-high tech industries: electrical, machinery, motor vehicles, transport, high-tech electronics, and also low-tech textile industry as far as the Global Value Chains are concerned [32].

China's manufacturing industry is still gaining some advantages and is facing challenges. The advantages include a complete industrial system covering almost all the sectors, the world's largest consumption market important for transformation and industry upgrading, the largest industrial labour resources in the world and visibly increased input for basic R&D. However, despite the changes in the manufacturing industry, per capita increase is still low and technological innovation is not sufficient. There is also a lack of core technologies and there is high energy consumption and large environmental pollution [33, pp. 6–8].

Thus, some changes are necessary for the transformation of the economic model of this country. It is no longer regarded as "the world factory", because its manufacturing sector is not very competitive, also because of rising costs as well as the necessity to transform it into the model based on technology and innovation [25, p. 621]. Therefore, some strategies and plans to improve its innovation potential have been implemented in China.

China's Innovation Policy Within the Framework of the *Made in China 2025* Plan

China makes the effort to become the leader and the prominent competitor on the innovation scene. It is due to, among other things, the coherent policy of the Chinese government to upgrade its position on the international markets.

"The National Medium and Long Term Program for Science and Technology Development (2006–2020) specifies the goals the attainment of which allows

Table 48.4 Position of the US, China and some EU countries according to GII 2008–2009 and GII 2019

	GII 2008–2009			GII 2019
Country	Position in the ranking GII 2008–2009	Position: Input sub-index	Positon: Output sub-index	Position in the ranking GII 2019
Sweden	3	3	4	2
US	1	2	1	3
Netherlands	10	12	8	4
UK	4	4	5	5
Finland	13	11	15	6
Denmark	8	1	21	7
Germany	2	10	2	9
Ireland	21	22	22	12
China	**37**	**47**	**29**	**14**
France	19	23	14	16
Luxembourg	17	24	9	18
Austria	15	15	17	21
Belgium	18	21	19	23
Estonia	29	26	48	24
Czech Republic	33	35	32	26
Malta	38	30	47	27
Cyprus	45	42	43	28
Spain	28	28	31	29
Italy	31	46	25	30
Slovenia	36	34	35	31
Portugal	40	33	54	32
Hungary	47	45	45	33
Latvia	60	50	69	34
Slovakia	35	36	33	37
Lithuania	42	37	46	39
Poland	56	59	56	39
Bulgaria	74	69	78	40
Greece	54	53	57	41
Croatia	62	57	66	44
Romania	69	71	65	50

Source [21, pp. XXXIV, XXXVI, XXXVIII; 31, pp. 9–12]

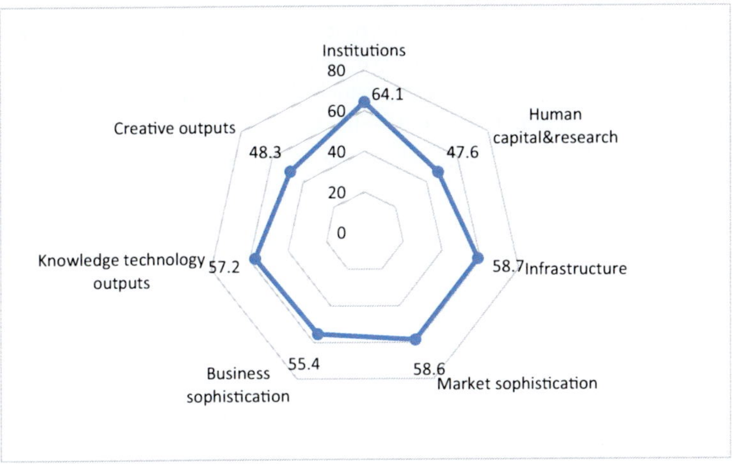

Fig. 48.2 GII 2019 pillars for China and scores. (*Source* Own elaboration based on the available data [21, p. 14])

becoming "a world S&T power by the mid twenty-first century" [34]. Different initiatives were undertaken, including the 2006 "Strategic Emerging Industries (SEI)" Initiative, focusing on technologies to promote advanced technologies, and to get the position in the strategic emerging industries including renewable and alternative fuels. The idea behind this initiative was to get access to foreign intellectual property, but at the same time, access to the market in China was granted [35, pp. 2–4]. A few years later, a new strategy, namely 'the MIC 2025' plan, was announced in order to overcome the middle-income trap, and thus to foster the development through "latent comparative advantage", i.e. investing in industries which have the potential to create this advantage but which, at the same time, need significant investments. The assumption was to invest in industries and identify the ones that could be the source of the comparative advantage. There were many other motives behind this strategy [20, pp. 2–3]. The MIC 2025 initiative follows the "13th five-year plan" encompassing social, economic and political goals as well as some policies, initiatives and regional plans for 2015–2020, and becomes its integral component. As stated in this plan, among others, the key issues will be to make the Chinese organisation leaders in innovation go green and open up by promoting foreign direct investments (FDI) in some sectors, or by promoting the Chinese products and technology abroad [36, p. 6].

The adopted broad strategy MIC 2025 now focuses primarily on the manufacturing process, not only on technical innovations, but also reduces the reliance of China on foreign technology. Therefore, the aforementioned strategy aims at promoting domestic companies that develop R&D and innovation. The 10-year plan MIC 2025 was launched by Prime Minister Li Keqiang, which includes and prioritises the development of high-tech industries. It also takes into account Germany's "Industry 4.0" adopted in Germany. The MIC 2025 plan identifies the key strategic sectors that are regarded as essential for the development in China and should be modernised [26, 32, p. 9; 37]. The aim is to improve the innovation capacity of this country, based on the domestic technological capacity which should be developed. The assumption behind the MIC 2025 plan is to promote the Chinese companies that could compete internationally and no longer depend on imports of foreign technology. The focus is on smart manufacturing and an increase in production, not only in the components but also in final products. In the end, China should move up the value-added chain [32, pp. 9, 48; 38, pp. 1–3].

However, it is important to analyse major tools and mechanisms applied to the achieved goals indicated in the MIC 2025.

There are the following guiding principles which should be implemented: innovation-driven development, quality first, green development of structure optimization and talent-oriented development. The basic principles were also formulated [37, pp. 5–7]. However, the strategic 9 tasks of MIC 2025 include: promoting the manufacturing innovation capability through focusing on research core technologies, innovation design capability, industrialization of scientific and technological achievements, supporting the manufacturing innovation system, reforming the standard system and strengthening intellectual property rights. The second task refers to promoting the integration of IT next generation and industrialization to prioritize "intelligent manufacturing". The third task covers supporting fundamental industrial capabilities, i.e. "four foundations" of the industry: spare parts and components, advanced techniques, also key materials as well as industrial technology. As it was indicated, the target is that 40% of the essential spare parts, also key materials will originate from domestic sources by 2020. By 2025 this share will increase by 70%. The next task will focus on improving the quality of products and building brand-name products not only to enhance the company value but also to improve the image of Chinese brands. The fifth task is related to the green production which is reflected in accelerating green manufacturing, supporting the efficient use of resources and recycling, and strengthening the green manufacturing system and projects. Emission intensity is predicted to be reduced by 20% in major industries. Strategic industries were also identified where special resources will be spent to make breakthroughs. Then another task is to make structure adjustment in manufacturing by promoting technology upgrading of enterprises to reduce the excess production capacity and coordinate the development between large and SMEs etc. Finally, supporting service-oriented manufacturing and manufacturing services will be promoted. The internationalization of Chinese manufacturing is the task that will help certain actions to improve international competitiveness [37, pp. 10–31].

The following implementation mechanisms were established, which included the institutional reform, creating a fair market environment, policies supporting the financing of the projects, including supporting export-import Bank of China in terms of its services, promoting the China Development Bank to provide a greater number of loans for manufacturing enterprises, promoting financial institutions to offer their services for manufacturing industries and developing capital and insurance market to this end, etc., and using fiscal and taxation policy. It envisages the system which promotes multi-level cultivation and developing a policy that supports SMEs. One of the mechanisms refers to manufacturing openness including some foreign investment reforms. The establishment of the national leading group for "rejuvenating Chinese manufacturing" will be also one of the keys mechanisms [37, pp. 32–38; 39]. Thus, the key tasks, as can be seen above, will provide the foundation for developing an industrial base with key sectors. They will support policies, including financial support, which will make this plan feasible. This will modernise the Chinese industry to withstand competition and to effectively be internationalized.

Three steps to achieve the manufacturing goals were identified, first being "Strive to turn China into a major manufacturing power in ten years" [37, p. 7]. The MIC 2025 plan is the first part of the strategy focusing on creating the foundations for the transformation of the Chinese economy through relying on high-tech industries, and thus upgrading the manufacturing sector in this country. However, the position of China in manufacturing will be strengthened in terms of the quality of smart manufacturing technologies. It will be based on the development of world-class enterprises in this country to be achieved by 2025 and they will also become leaders in the value chain. It is also planned to take the second step which is to be achieved by 2035, and which is related to obtaining the level of mid-ranking manufacturing. The leading position in key high-end manufacturing sectors to be reached by 2049 is the third task [32, p. 13; 38, p. 8].

However, the issue begs the question of how these policies are going, to effectively be implemented in terms of mechanisms and tools applied. Some of them are presented below.

The implementation plan was elaborated according to which the guiding documents referring to each of the key industries have been prepared ("X + 1", where X means specific industries), with goals of companies to be attained. Some regulatory measures have been introduced to ensure that the companies follow some standards. Due to the plan, the industrial alliances have been created and the central government is responsible for its creation, e.g. The Alliance of Industrial Internet. The MIC 2025 Action Plan also envisaged the creation of financial incentives for companies to be engaged in this plan, and establishing the innovation centres at the national and regional levels [36, p. 12].

The following nationwide initiatives are to be implemented and related to the establishment of 15 R&D and Innovation Centres (national manufacturing innovation centres) by 2020 and 40 centres by 2025. There will be also implemented smart manufacturing projects in Chinese companies. Green manufacturing projects in the following areas are also envisaged: energy efficiency, environmental protection, re-manufacturing, low-carbon technologies (objectives concerning, among others, building 1000 green factories and 100 green industrial parks by 2020). The last initiative related to Industrial Bases will cover the establishment of four new research centres focusing on developing core industrial components and materials, production technology, also related to techniques and promoting the so-called high-end equipment manufacturing projects in specific sectors [38, p. 12].

There are different sources of funding for the activities within this plan. Among others, they are financed through central government funds supporting high-tech firms, advanced technology, and acquiring foreign technology (e.g. MIIT and China

Development Bank granted USD 45 billion, Special Constructive Fund granted USD 270 billion, local governments for attracting high-tech manufacturing in certain localities). However, the aforementioned funds are oriented towards specific activities set by the government. They function as venture capital or private equity, and capital is raised from state-own enterprises and local governments when implementing some of the projects [20, pp. 3–5].

To sum up, the key role in this plan implementation is played by centralized policy planning, public policies, and other activities. To achieve the formulated goals they serve the formulated explicit targets. Firms, both public and private are stimulated to focus on their decisions on these goals. Some direct subsidies are offered including state funding, low-interest loans, tax breaks, etc. Chinese companies are stimulated to invest in foreign companies to get access to advanced technology. The Chinese acquisition is also in force including state-backed companies, but even some of the private high tech market leaders are supported by the state [40]. The mechanism and the tools applied by the Chinese government have a direct and indirect impact on businesses operating in the countries which are the innovation leaders, i.e. the European Union countries and the US.

China's Innovation Policy and Its Implication on the US and the EU

China and the US

The MIC 2025 plan has significant implications for the relations with other leading competitors in terms of innovation.

The US and China have had close trade and economic ties which have expanded. China is the largest US trading partner in terms of goods, and the third largest partner when it comes to exports (120 USD billion). As far as imports are concerned China was the largest partner (540 billion USD) in 2018, and the trade deficit was significant. Due to the increased tariffs in 2018 the US merchandise exports to China fell by 16% while imports fell by 13%. FDI from China decreased and in 2016 amounted to 45.5 billion USD while in 2018 4.8 billion USD [41, p. 1].

It should be emphasized that mutual relationships between the US and China concerning their cooperation in science and technology have been developed over the years. It is so because US enterprises entered the Chinese market transferring some parts of the value chain (both middle and end chains). It also allowed expanding the activity of the Chinese companies on a global scale incorporating their activity into the industrial value chains. However, the Chinese scientific and technological potential was constantly increasing. The rapid rise was recorded especially in high tech sectors, including ICT, primarily after the financial crisis of 2008 when other economies recorded a slowdown. Therefore some countries, including the US, found the Chinese rise in the cutting-edge technology and science a serious challenge [42, pp. 198–199]. China is no longer regarded as the "cheap factory" and increases its position due to some political and economic initiatives, including Belt and Road Initiative or the MIC 2025 plan. The economic relations with the US have become more competitive [43]. Some measures were taken to tighten China's access to the US market in some sectors, including high technology. This process has intensified since Donald Trump came to power [42, pp. 198–199].

The key points in disputes between both parties were raised by the Trump administration to justify their preventive actions. They are related to US access to China's market which is subject to differentiation, the Intellectual Property (IP) system promoting Chinese companies in China as well as some patent, copyright, and trademark infringement [20, pp. 6–16]. According to some critics specifying the percentage of the essential "domestic content" in the selected products in China, it is the import substitution policy that could negatively influence foreign suppliers specialising in high technology [44]. The problem of recruiting scientists by China was raised as well as the acquisition of US firms that are targeted [45, p. 4; 40]. However, others point out the importance of both economic and political arguments that China's progress in science and technology could undermine the US global hegemony in this field [42, pp. 199–200].

In 2018, some actions against the Chinese policy were announced by the US [44]. The aforementioned issues gave rise to the so-called trade war between the US and China, which finally ended by reaching an agreement between the two parties. The agreement was signed as the so-called "Phase One" trade agreement on the 15th of January 2020. It underlines "the desirability of resolving existing and any future trade and investment concerns as constructively and expeditiously as possible" [10 (Preamble)]. It covers the provisions referring to intellectual property, technology transfer, trade in food and agricultural products, and financial services. There are also provisions related to macroeconomic policies and exchange rates as well as expanding trade. Under the agreement, from 1 January 2020 till the end of 2021 China will purchase and import goods from the US worth more than 200 billion USD. These goods will include manufactured and agricultural goods, energy products, and services (Art. 6.2.1) [8]. Further on, under the agreement the period when any kind of products that will be imported to China from the US is subject to import also during the period from 2022 till 2025 (Art. 6.2.3) [46]. At the same time, it is worth mentioning that access to the Chinese market has been granted to American companies as well as the rules governing intellectual property protection have been applied [25, 47]. But the tense relationship between the US and China affected the whole chain in the technology sector on the global scale, being the disruptive force [43]. In the long run the consequences have multidimensional impact and there is a risk that two sphere of influence could appear, one around China and the second around the US [43]. To prevent the existence of such a scenario in the future both parties should strongly cooperate to create new platforms for cooperation in science and technology. A wider perspective should be taken into account because "easing frictions between the two countries in not only conducive, but also necessary for the sustainable development of global economy and strategic stability" [42, p. 208].

China and the EU Countries

China increases its innovation performance which is also reflected in international collaboration, including the EU. It shows in the high number of co-publications in engineering, physics, astronomy as well as material science and chemistry, which is Chinese specialisation. China has an interest in supporting such collaboration. Chinese programmes are predominantly funding sources of the publications with EU researchers, or funding by particular member states. The internationalisation is seen in the scientific mobility, e.g. the number of Chinese researchers who have moved to the EU or returned, the scale of returns, and the rising number of Chinese STEM graduates [32, pp. 67–70]. 86 R&D centres out of 235 outside China are located or acquired in the EU. On the other hand, some of the R&D laboratories from multinational corporations are from the EU and located in China (349 out of 850), but some of the Chinese companies located their laboratories abroad [32, p. 70].

Bilateral trade increased over the period 2000–2014, including a trade deficit of the EU with China. China's exports to the EU and its imports from the EU increased in terms of the added value. But as far as manufacturing industries are concerned, the decreasing share of the high-tech manufacturing sectors in the EU exports to China was observed in 2014 compared to 2000 (54.8–57.4%) in terms of the added value. The Chinese exports of those products to the EU increased from 38.7 to 50.6% in 2014 in terms of the added value. The same tendency for exports of goods and services for final use were observed whereas EU intermediate exports to China remained stable. As far as services are concerned, the reverse trend was observed. An increase in bilateral trade relations are also reflected in employment related to the industries involved directly or indirectly in exports. Thus, more jobs which amounted to 21.5 million in 2014 were created in China related to exports to the EU while in the EU exports to China created 2.6 million jobs in 2014. However, the EU labour productivity in exports is higher than in Chinese exports to the EU [32, pp. 25–28].

The growing significance of China on the international scene raised the concern of the EU about equal opportunities for businesses from Europe. It is put on the agenda that EU companies penetrate the Chinese market at a slower phase when compared to Chinese entities in Europe. It concerns the obstacles upon entry for example, or EU companies enter the Chinese market through forming joint ventures or investing with minority share rather than mergers & acquisitions (M&A) [32, pp. 49–51]. There has been an increase in the cross-border M&A between China and the EU. Chinese companies are predominantly involved in M&A in the largest EU economies such as Germany, the UK, France, Italy, and the Netherlands. A tendency to move toward manufacturing in foreign-controlled firms was observed with the focus on their investment in high-tech manufacturing. R&D related companies observed an increase, nowadays accounting for 18% of Chinese M&A which have been completed recently. The sectors included in MIC 2025 constitute one-third of China's investments in Europe, mostly in IT next generation, transportation, numerical controls and machinery, robotics, and new materials. They have been made mostly in Germany [32, pp. 33–35].

It should be taken into consideration that China's economic policy implemented within the adopted strategies which in fact strengthen the position of China has the implications for Europe's innovation environment. The implementation of MIC 2025 causes some tension between the EU and China and it "creates uneven playing field" [48], but as it was presented in China's official statement this "initiative brings equal opportunities to foreign and domestic enterprises and will strengthen the role of the market" [48].

Conclusions

Until recently, the US and some EU countries have become leaders on the innovation scene. However, China has been gradually transforming the model of its economic development and has created its strong position in terms of innovation in the world economy. It is reflected not only in the R&D expenditures, but also in the outcome indicators related to patents, for example. It is due to the economic policies implemented through specific strategies that stress the importance of the policy focusing on creating domestic innovation capacity. The MIC 2025 strategy has a significant role in this matter. It promotes the development of strategic sectors being the foundation of the new economic model for China. There are specific activities that are strongly supported by the government, banking sector, and many other units and entities. These actions aim to strengthen and create the leading position of China on the international innovation scene.

China increases its innovation performance which is also reflected in international collaboration, including the US and the EU. Nevertheless, China's innovation policy to strengthen its position on the innovation scene causes a great deal of tension in the relations with other leading innovation world powers, such as the US and the EU. All the parties have to rethink their relations and create new platforms for cooperation in science and technology in the context of the changes in the international environment. Instead of competing, they could create specific alliance which should be mutually beneficial. Nonetheless, it requires a joint effort by all the parties involved.

References

1. M.A. Weresa Narodowy system innowacji i polityka innowacyjna Niemiec – wnioski dla krajów Europy Środkowej i Wschodniej, pp. 65–98. In: I. Bill, K. Cisz, B. Brocka-Palacz, M. Gomułka, F. Kamiński, M.A. Weresa, Gospodarka Niemiec a kraje Europy Środkowej i Wschodniej, SGH, Warszawa, after: M.A. Weresa (2012) *Systemy innowacyjne we współczesnej gospodarce światowej* (PWN, Warszawa, 2006)
2. X. Liu, S.S. Schwaag, U. Tagscherer, A.Y. Chang, Beyond catch-up—can a new innovation policy help China overcome the middle income trap? Sci Public Policy **44**(5), 656–669 (2017). https://doi.org/10.1093/scipol/scw092
3. https://unctadstat.unctad.org/en/classifications.html, 20.05.2020
4. Evolution of science, technology and innovation policies for sustainable development: the experiences of China, Japan, the Republic of Korea and Singapore. United Nations (2018), https://www.unescap.org/sites/default/files/publications/UN_STI_Policy_Report_2018.pdf
5. World Bank data, https://data.worldbank.org/indicator
6. R.P. Appelbaum, C. Cao, X. Han, R. Parker, D. Simon, Innovation in China. *Challenging the Global Science and Technology System* (Polity Press, Cambridge, Medford, 2018)
7. *China's Quest for Innovation*, ed by S. Dai, M. Taube (Routledge, Abingdon, New York, 2020)
8. X. Ding, J. Li, Incentives for Innovation in China. *Building an Innovative Economy* (Routledge, Abingdon, New York, 2015)
9. Evolution of China's Innovation Performance 2000–2013 (Publications Office of the European Union, Luxembourg, 2015)
10. Foreign business in China and opportunities for technological innovation and sustainable economics, in *IGI Global*, ed by A. Visvizi, M. Lytras, X. Zhang, J. Zhao (Hershey, 2019)
11. X. Fu, *China's Path to Innovation* (Cambridge University Press, Cambridge, 2015)
12. X. Fu, W.T. Woo, J. Hou, *Technological Innovation Policy in China: The Lessons, and the Necessary Changes Ahead* (2016), 29 February 2016. https://papers.ssrn.com/sol3/papers.cfm?abstract_id=2798806
13. Innovative China. *New drivers of growth, World Bank Group, Development Research Center of the State Council* (World Bank, The People's Republic of China, Washington, DC, 2019). https://doi.org/10.1596/978-1-4648-1335-1
14. F.-c. Liu, D.F. Simon, Y.-t. Sun, C. Cao, China's innovation policies: Evolution, institutional structure, and trajectory. Res. Policy **40**(7, September), 917–931 (2011)
15. OECD Reviews of Innovation Policy: China 2008. OECD (2008). https://doi.org/10.1787/978 9264039827-en
16. Y. Yu, J. Yu, X. Pan, R. Stough, The Rise of China's Innovation Economy: "Opening Up" Policy to Manufacturing Maturity, and on to Innovation Based Economic Growth and Labour Market Dynamics? August (2016). https://papers.ssrn.com/sol3/papers.cfm?abstract_id=282 0864
17. L. Li, China's manufacturing locus in 2025: with a comparison of "Made-in-China 2025" and "Industry 4.0". *Technological Forecasting and Social Change*, **135**, 66–74 (2018)
18. J. Wübbeke, M. Meissner, M.J. Zenglein, J. Ives, B. Conrad, Made in China 2025. *The Making of a High—Tech Superpower and Consequences for Industrial Countries*, December 2016, No. 2. Mercator Institute for China Studies, Paper on China No. 2, December (2016)
19. M.J. Zenglein, A. Holzmann, Evolving Made in China 2025. China's industrial policy in the quest for global tech leadership, MERICS, Papers on China, No. 8 (2019)
20. A. Malkin Made in China 2025 as a challenge in global trade governance. Analysis and recommendations, CIGI Papers No. 183-August 2018. Centre for International Governance Innovation
21. The Global Innovation Index 2019, *Creating Healthy Lives—The Future of Medical Innovation*, 12th edn, ed. by S. Dutta, B. Lavin, S. Wunsch-Vincent (Cornell University, INSEAD, WIPO, Ithaca, Fontainebleau and Geneva, 2019)

22. Industrial Research and Innovation Indicators, in *Report of a Workshop*, ed by R.S. Cooper, S.A. Merrill (National Academy Press, Washington, D.C.) Chapter four (1997). https://www.nap.edu/read/5976/chapter/5#16
23. J. Janger, T. Schubert, A. Andries, C. Rammer, M. Hoskens, *The EU 2020 Innovation Indicator: A Step Forward in Measuring Innovation Outputs and Outcomes?* Discussion Paper No. 16–072. Centre for European Economic Research, October (2016). https://madoc.bib.uni-mannheim.de/41558/1/dp16072.pdf?rs=true&
24. Eurostat database, https://ec.europa.eu/eurostat/data/database
25. C. Cao, China, Chapter 23, in *UNESCO Science Report Towards* 2030 (UNESCO Publishing, Paris, 2015)
26. Q. Chen, *Globalization and Transnational Academic Mobility: The Experiences of Chinese Academic Returnees* (Springer, 2017)
27. Y. Sharma, China's Effort to Recruit Top Academic Talent Faces Hurdles, May 28, 2013 (2013), https://www.chronicle.com/article/Chinas-Effort-To-Recruit-Top/139485
28. OECD, OECD database https://stats.oecd.org/
29. First inventor to file is here: learn how it works, https://www.uspto.gov/learning-and-resources/newsletter/inventors-eye/first-inventor-file-here-learn-how-it-works, 21.05.2020
30. S. Key, Don't file that patent yet. File a provisional patent application first, Forbes, 2018, January 8 (2018). https://www.forbes.com/sites/stephenkey/2018/01/08/dont-file-that-patent-yet-file-a-provisional-patent-application-first/
31. The Global Innovation Index 2008–2009 (INSEAD, 2009)
32. N. Preziosi et al., China, Challenges and Prospects from an Industrial and Innovation Powerhouse (Publications Office, Luxembourg, 2019). 10.2760/445820, JRC116516
33. H. Ma, X. Wu, L. Yan, H. Huang, H. Wub, J. Xiongb, J. Zhanga (with H. Ma), Strategic Plan of Made in China 2025 and Its Implementations, in *Analysing the Impacts of Industry 4.0 in Modern Business Environments* (IGI Global, 2018), pp. 1–23
34. The National Medium and Long Term Program for Science and Technology Development (2006–2020). https://www.itu.int/en/ITU-D/Cybersecurity/Documents/National_Strategies_Repository/China_2006.pdf
35. Made in China 2025, Backgrounder—June 2018 Institute for Security & Development Policy. https://isdp.eu/content/uploads/2018/06/Made-in-China-Backgrounder.pdf
36. P. Dupont, Made in China 2025. China's industrial vision and New Zealand, 6th July 2018, New Zealand Contemporary China Research Centre, Victoria University of Wellington, Wellington, New Zealand (2018), https://www.victoria.ac.nz/__data/assets/pdf_file/0005/1660811/Made-in-China-2025-Chinas-Industrial-Vision-and-New-Zealand.pdf
37. Made in China 2025, State Council, July 7, 2015. https://www.cittadellascienza.it/cina/wp-content/uploads/2017/02/IoT-ONE-Made-in-China-2025.pdf
38. Made in China 2025. China Manufacturing in the 21st century—opportunities for UK-China Partnership, UK Trade & Investment, China-Britain Business Council. https://www.mta.org.uk/system/files/resource/downloads/Made%20in%20China%202025%20Booklet%20One.pdf
39. China Manufacturing 2025, Putting industrial policy ahead of market forces, *European Union Chamber of Commerce in China* (2017). https://docs.dpaq.de/12007-european_chamber_cm2025-en.pdf
40. J. McBride, A. Chatzky, Is 'Made in China 2025' a threat to global trade? 13 May 2019, Council on Foreign Relations. https://www.cfr.org/backgrounder/made-china-2025-threat-global-trade
41. *U.S.—China Trade and Economic Relations: Overview*. Congressional research service, updated November 19, 2019
42. S. Haiyong, U.S.—China Tech War. Impacts and prospects. 'China Quart. Int. Strat. Stud. **5**(2), 197–2012. https://doi.org/10.1142/S237774001950012X (2019)
43. L. Colback, How to Navigate the US-China Trade War, February 28, 2020. *Financial Times*. https://www.ft.com/content/6124beb8-5724-11ea-abe5-8e03987b7b20
44. W.M. Morrison, *The Made in China 2025 Initiative: Economic Implications for the United States*, Congressional Research Service, In Focus, April 12 (2019), https://fas.org/sgp/crs/row/IF10964.pdf

45. How China's economic aggression threatens the technologies and intellectual property of the United States and the world. White House office of Trade and Manufacturing Policy, June 2018. https://www.whitehouse.gov/wp-content/uploads/2018/06/FINAL-China-Technology-Report-6.18.18-PDF.pdf
46. Economic and Trade Agreement between the United States of America and the People's Republic of China. Phase One, January 15, 2020, https://www.washingtonpost.com/context/u-s-china-trade-agreement/34832562-c807-43f0-b368-580dfa8b2d87/?itid=lk_interstitial_manual_4
47. H. Long, Was Trump's China trade war worth it? The Washington Post **25**(01), 2020 (2020)
48. China daily, 2017–03–13. https://cn.chinadaily.com.cn/2017zgzz2025/2017-03/13/content_28928364.htm

Chapter 49
Effect of Migration Flows on the Ethnic Identity of the Indigenous Peoples of the Amazonian Communities

Aldrin Espín-León, Antonio Jimeno-Morenilla, María Luisa Pertegal-Felices, and Jorge Azorín-López

Abstract The indigenous communities in the Amazonia are in an accelerated process of acculturation. In this context, one of the main elements in order to determine the cause and effect of migration would be to measure the degree of ethnic identity of each indigenous individual or an entire community as a whole. In this research, a study is carried out that quantitatively determines the ethnic identity that the indigenous people of the Waorani Amazon community have in relation to the inhabitants of the cities that have a western identity. For this purpose, a function is provided to measure the distance of cultural identity between two individuals. The proposed method is able to determine the ethnic distances between individuals. Specifically, this article presents a study of the identity distances of three populations with different cultures (the indigenous Waorani community, Tena and Quito), also determining those indigenous Waorani inhabitants who can be considered ethnically closer to their own culture. The results show that ethnic identity is being modified in several ways. Firstly, it affects the relationship that indigenous people have with their habitat and customs. On the other hand, it affects migration to towns and cities, mainly those with western culture.

Keywords Migration flows · Cultural identity measurement · Amazonian communities · Artificial intelligence

---

A. Espín-León
Faculty of Sociology and Social Work, Central University of Ecuador, Quito, Ecuador

A. Jimeno-Morenilla (✉) · J. Azorín-López
Department of Computer Technology, University of Alicante, Alicante, Spain
e-mail: jimeno@dtic.ua.es

M. L. Pertegal-Felices
Developmental and Educational Psychology Department, University of Alicante, Alicante, Spain

A. Visvizi et al. (eds.), *Research and Innovation Forum 2020*,
Springer Proceedings in Complexity,
https://doi.org/10.1007/978-3-030-62066-0_49

Introduction

In Latin America, the migratory movements of indigenous people are slightly less than the migratory movements of non-indigenous people. However, beyond their number, theoretical social models try to differentiate their origin and provide explanations for the phenomenon and its consequences. According to Valdés [28] indigenous migration does not show any explanatory difference with non-indigenous migration, he attributes this to the fact that the instruments available are not capable of detecting differences, especially since these populations integrate historical and cultural variables.

Indigenous migration, mainly labour migration, increased from the 1960s onwards. This indigenous migration injected economic remittances into their villages and led to a social reordering, altering their traditional agricultural production, consumption patterns and ways of living together [21]. While in the case of some indigenous communities in the Amazon, which were not directly subject to the conquest by the West, their colonization began only in the mid-twentieth century. People who would be suffering a strong process of migration and immigration and whose main consequence would be the impact of their own identity [23].

Taking into account this point of view and as mentioned by Moltedo [17], an indicator of social affectation and cultural identity of indigenous populations would be migration. The migratory phenomenon would be affecting some more than others because their populations come from different origins and territories, leading to different degrees of fusion of their cultural identity with other populations. In this sense, Quijano [22] points out that, as a consequence of colonialism and immigration by dominant groups from the West, indigenous communities and nationalities in Latin America have experienced fusions and cultural ethnic changes in an accelerated manner.

Cultural fusion can occur when one culture dominates the other or simply when two cultures come together for various reasons: wars, conquests, colonization, migration, climate change, economic interests, globalization and telecommunications [8]. For Berry [6] acculturation in some indigenous peoples occurs without their will, receiving in their own territory by imposition another culture, while in the case of migrants voluntarily chose new territories. Nevertheless, for Bello [5] some migratory phenomena characteristic of indigenous peoples will continue to be maintained, following their own cultural rules that are little explained by traditional definitions and Western concepts. All these social phenomena, as mentioned by Carrasco-Arroyo [8] produce a territorial heterogeneity and different cultural expressions that make it difficult to have a system that measures and analyzes cultural identity. Visvizi et al. [33, 34] propose the use of technological tools for the challenge of refugee and migrant crises as cities and urban areas revealed that they are not yet prepared, mention that migration also implies moral obligations, costs and bureaucratic procedures for the implementation of policies.

In this context, the authors of this research consider it necessary and as a first instance to quantify, through the use of AI tools, the aspects and cultural distances in the identity of indigenous peoples with respect to other peoples to which they are subject to migration.

Aim of the Research

Determining and measuring how much of an individual's or a people's ethnic identity is compared to the identities of other peoples is one of the first steps in mitigating the cultural effects of migration. Precisely, this study aims to quantify the distance that separates the ethnic identities of the inhabitants of an Amazon indigenous people from populations with a city identity.

The study uses an instrument developed by the authors that categorized and classified populations through variables and indicators, which made it possible to identify factors of ethnic identity that most influence indigenous identity. Quantifying the distance of ethnic identity will allow to know about an indigenous person, how close or distant his identity is from another individual or population with a culture different from his territory. Ethnic distance, as a result of migration, immigration or human mobility, may or may not be altered. To quantify this distance, the study will be carried out with populations that have a different degree of Western culture using techniques based on artificial intelligence.

Structure of the Document

This document has the following structure: Sect. "Background", presents a review of the literature on migration studies and how it influences the identity of indigenous peoples. It also presents studies, which have used methods with Artificial Intelligence tools, related to migration and human mobility. Section "Description of the Indigenous Territory" describes the indigenous territory and the tool that was previously developed by the authors. Section "Description of the Instrument Used to Measure the Distance Between Identities from an Indigenous Approach" analyses the instrument through techniques that use Artificial Intelligence to determine distances of ethnic identity between individuals of the indigenous people and distances to urban individuals. It also analyzes the distances to the individual who has the greatest value in terms of indigenous identity. Section "Conclusions" presents the main conclusions of the study.

Background

The study of migration and the identity of peoples is of social interest, in this sense there are dialogues and debates between academia and decision makers for migration

[14]. The sociologist Beck [3], says that identity is a protagonist because everyone feels threatened by a very powerful rival called globalization.

Migration influences several aspects of identity. Velasco et al. [29] studied in a qualitative way, in a state of Mexico, the impact of migration on a set of cultural aspects such as language, skin colour and ancestral origin in a group of indigenous people, showing that discrimination contributes to a distancing from their own ethnicity, in addition to devaluing their culture.

Other studies analyze ethnic backgrounds and attitudes towards immigration and ethnic groups with a migration background [25]. Carley [7] studied the adaptability of migrants considered as new individuals based on a thesis in which the interaction between individuals leads to shared knowledge, concluded that some groups in the face of the incorporation of new individuals endure more time, others have greater capacity for admission without losing their own character. Meanwhile, in Costa Rica, faced with the accelerated processes of acculturation, migration and globalization, a study of the ethnic identity of the Huetares de Quitirrisí de Mora indigenous people was carried out. The results showed little value in visible cultural elements such as traditional dress, language and customs, yet they continued to be proud of belonging to their ethnic group [18].

Regarding the migration of indigenous people from the Amazon, Frigola [12] studied the work activities that these indigenous people carry out in the city, and determined that the indigenous people prioritize their identity in various cultural areas, while in the work environment their jobs depend on their ancestral knowledge.

Ethnic identification and migration issues are also addressed by ICTs [32] and Artificial Intelligence. Gutta et al. [13] proposed a hybrid classifier based on sets of Radial Basis Functions (RBF), inductive decision trees (DT) and SVM for gender and ethnicity classification. In the ethnic classification they achieved a 92% accuracy rate. Lu et al. [16] proposed recognition of Asian ethnicity based on Linear Discriminant Analysis (LDA). Other studies proposed to classify and identify from public space certain characteristics and attributes of urban identity. Chang et al. [9] used the Principal Component Analysis (PCA) technique and the K-means approach. In this sense and to eliminate noisy and redundant attributes, Tang et al. [26] suggested other techniques, such as dimensional reduction. In migration, AI and biometrics are generally associated with the control and surveillance of individuals, however Pereyra and Estefanía [24] questioned their use in terms of ethnic equality since migratory control by biometrics grants privileges to some and deepens to others, promoting segmented mobility. While for the urban migration study Behnisch and Ultsch [4] proposed the use of AI as a data-ming for the classification of data regarding employment, transport and migration, they used SOM as an unsupervised classifier. Likewise, Abarca-Alvarez et al. [1] studied, through a combination of GIS and AI, the housing occupation of permanent residents and short and long distance migrants in the province of Andalusia.

On the other hand, for the development of smart cities and education, Visvizi et al. [30], Alkhammash et al. [2] studied citizen opinion expressed on websites and social networks. Smart cities must be people-centred and demand-driven not only in urban areas, but also in rural areas, the nested classification model points to a connection between the well-being of individuals, their participation and sustainability [31].

Regarding studies of migration of indigenous peoples who have used AI tools, no evidence was found, however, Espín-León et al. [11] classified and quantified the identity of the Waorani Amazon indigenous people and two cities considered to have a Western culture. They used and compared different classifiers such as: k-Nearest Neighbor (Knn), LDA, Original Self-Organized Map (SOM), the Neural GAS variant (NGAS) and the Support Vector Machines SVM, in addition to a Multiclassiffier (MC), the results were encouraging, in all performance criteria it exceeded 90%, with SVM having the best results with 95%.

The Instrument

The study was carried out with information obtained from indigenous people living in populations of the Amazon, specifically in communities of the Waoranis. Indigenous people who have some contact or influence with western culture and therefore have some kind of migratory flow.

Description of the Indigenous Territory

The Waorani indigenous nationality is located in a territory of about 790,000 ha in the Amazon (South America) between the countries of Ecuador and Peru. In the Ecuadorian territory they are located in the provinces of Pastaza, Napo and Orellana. This indigenous nationality has approximately 13,000 inhabitants, distributed in 22 communities (see Fig. 49.1). Its territory is mostly surrounded by the Yasuní National Park—a park that conserves an exuberant biodiversity—and by an area inhabited by the Tagaeri and Taromenane Indians—an intangible area, peoples who live with their original culture without any contact with western cultures.

Description of the Instrument Used to Measure the Distance Between Identities from an Indigenous Approach

The instrument used in this study was developed by the authors [20]. The instrument was made from a qualitative study of interviews conducted in three Amazonian communities of the Waorani nationality and whose purpose was that the items were obtained from the indigenous territory itself with their individual or community perspective.

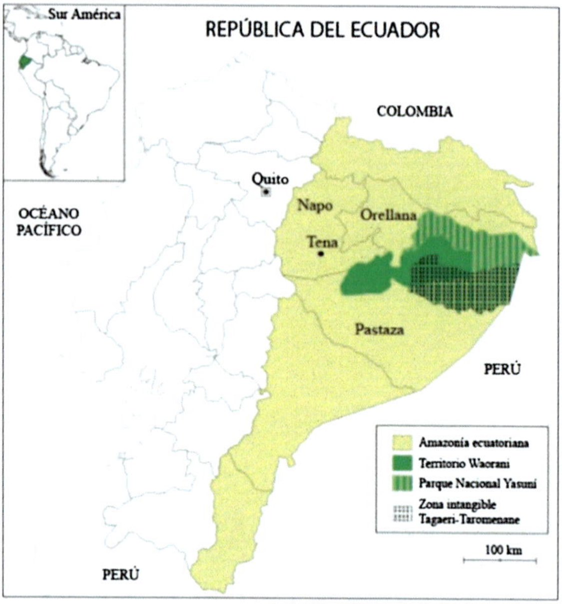

Fig. 49.1 Geographical location of the Waorani territory

Table 49.1 Interviews with the Waorani population

Phase	Interviews	Men	Women	Total
1	Unstructured	4	7	11
2	Semi-structured	6	8	14
	Total	10	15	25

With the authorization of the indigenous leaders, the collection of qualitative data was carried out in two phases (see Table 49.1): (1) The unstructured phase consisted of 11 interviews of free expression of topics that the indigenous person considered of interest and (2) The semi-structured phase with 14 interviews that consisted of themes oriented to ethnic identity.

The 25 interviews were analyzed qualitatively using the Atlasti tool [19]. We obtained 99 items grouped in 30 sub-dimensions and these in 5 dimensions (see Table 49.2).

The instrument was validated by an alternative model called CVR' [15, 27], through the trial of 9 experts: 6 Ecuadorians, 2 Bolivian anthropologists and 1 Colombian political scientist. They also carried out reliability tests (Cronbach's alpha statistics, which showed good results: Alpha General 0.974 in 99 items. While, by dimension the results were: Economy 0.896 in 22 items, Family and reproduction 0.935 in 21 items; Ideology 0.906 in 13 items; Organization 0.602 in 6 items and Social 0.931 in 37 items.

Table 49.2 Dimensions and subdimensiones and items of the Waorani Identity Instrument

Dimensions	Subdimensions	Items
1. Economic (E)	9	22
2. Family and reproduction (F)	6	21
3. Ideological (I)	4	13
4. Organization (O)	3	6
5. Social (S)	8	37
Total	30	99

Experiments

Experimental Setup

In order to establish the distance between the identities of indigenous people and those of city dwellers, the instrument was applied through surveys of 299 people: 88 Waorani indigenous people, 100 inhabitants of the city of Quito and 111 inhabitants of the city of Tena. These three populations represent three types of identities in Ecuador: (1) The Waorani indigenous people who were the subjects of the study, the data were collected in three of the 22 communities that have their population (Konipare, Menipare and Gareno). Quito (the capital of Ecuador) was selected because it is a city with a western culture and (3) The city of Tena is an Amazonian city located geographically between the city of Quito and the Waorani people, it has a western culture, but it receives a migratory flow from the Amazonian indigenous communities, so the identities of its inhabitants would also be located between the Amazonian indigenous people and the inhabitants of the city with a western culture.

With the data of the 99 items grouped in the 30 sub-dimensions and these in the 5 dimensions, the automatic learning process is formed for the identification of the identity of the 299 individuals of the three populations. The learning process was carried out by means of self-organized maps (SOM). Figure 49.2 shows the distances (in red) between the Waorani people (Wao in the figure) and the city of Quito and between the city of Tena and the city of Quito.

In order to determine the ethnic identity distance between inhabitants of the same towns (Waoranis, Tena and Quito), the SOM topographic distance was used. The algorithm proposed is that of minimum roads proposed by Dijkstra [10]. The algorithm is based on exploring the distances between the vertices that connect nodes in a network of nodes and determining the shortest distance between two nodes.

Identity distances were determined for each individual in comparison with each of the other individuals in the three populations (299*299 distances). In addition to determining the distances from a Quaker Indian to a Waorani Indian with the maximum identity score we will call him a "WM" Indian. It also allowed us to determine the identity distance between an indigenous person and his own population or a city with a western culture, see Fig. 49.3.

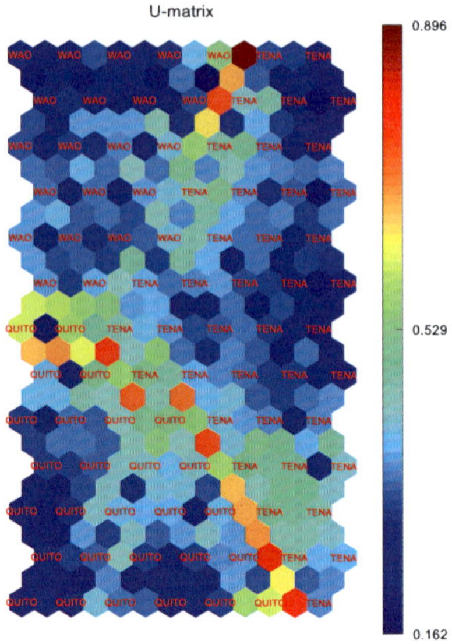

Fig. 49.2 SOM map for the three populations

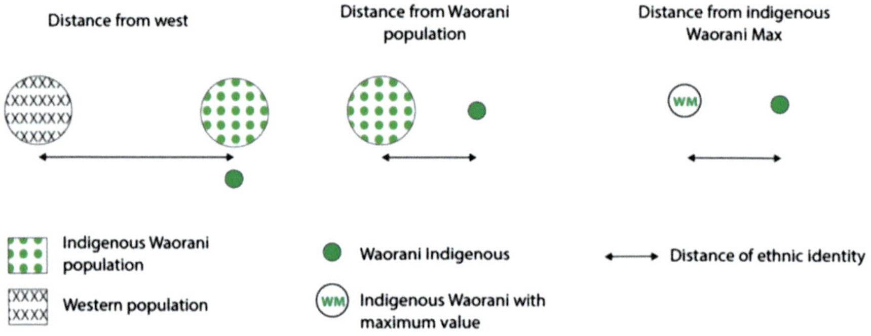

Fig. 49.3 General diagram of identity distances

With each of the 99 items that the instrument has with a value of 4–0, 396 are obtained as the maximum score, a value that would have an inhabitant with an ideal Waorani identity. The result showed that the 37th Indian obtained the highest value with 354 points out of 396 (89.4% of the ideal), corresponding to a woman with 36 years of age.

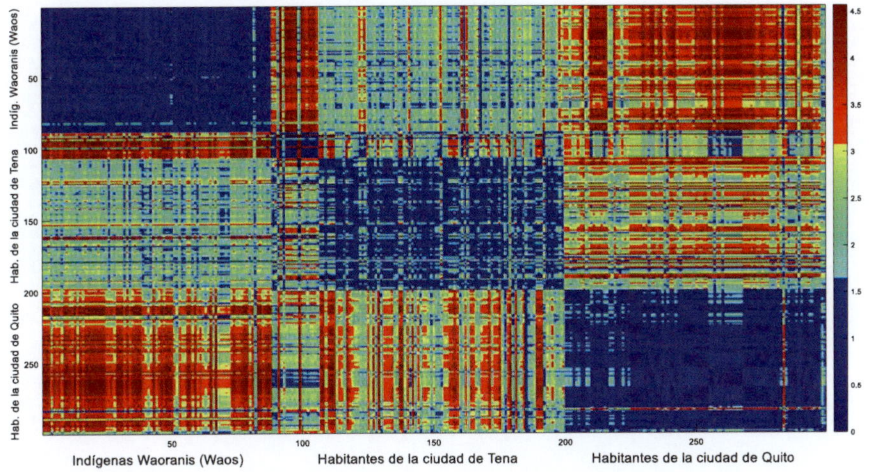

Fig. 49.4 Topographical error: Waorani Indians, city of Tena and city of Quito

Results and Discussion

The topographical distance between individuals per population can be seen in Fig. 49.4 where, blue indicates close identity distances while red indicates distant distances. The data in the graph shows that in the Waorani and Quito populations (in Fig. 49.4 the areas: Waorani-Waorani and Quito-Quito) in addition to their identity distances being close (greater blue color), they also denote greater homogeneity in their distances while, for the inhabitants of the city of Tena (in Fig. 49.4 the Tena-Tena area) their identity distances are not very close and are more heterogeneous (blue and green color). Table 49.3 shows the average distances of individuals belonging to each population and their standard deviation.

A comparison of the distances between the Waorani population and the population of the city of Tena and the city of Quito is shown in Fig. 49.5. It can be seen that the distance separating the Waorani Indians from the inhabitants of the city of Quito is noticeably greater (mostly red) with an average of 3.434 to the distance separating them from the inhabitants of the city of Tena (mostly green) with an average of 2.422 (Table 49.4).

Table 49.3 Statistics on the population distances of each population

Population	Colour predominance	Media	Typical deviation
Waoranis	Blue	0.7823	0.4537
City of Tena	Blue–Green	1.4553	0.9342
City of Quito	Blue	0.9450	0.6170

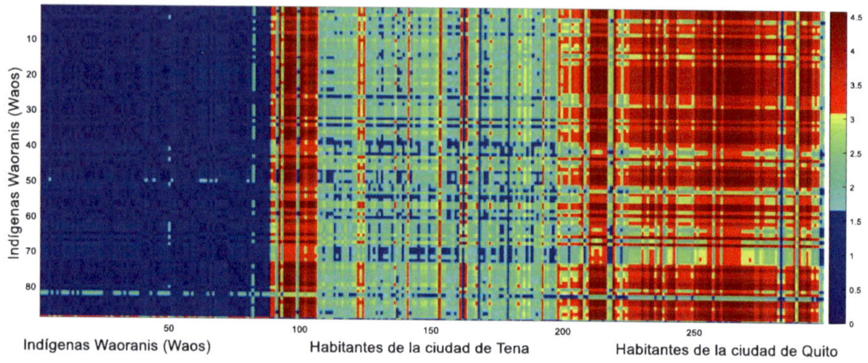

Fig. 49.5 Topographical error of identity of the Waorani indigenous population with respect to the inhabitants of the city of Tena and the inhabitants of the city of Quito

Table 49.4 Distance of identity from the Waorani population

Populations	Colour predominance	Dista. Media
Waoranis–Waoranis	Blue	0.7823
Waoranis–City of Tena	Green	2.4221
Waoranis–City of Quito	Red	3.4349

The topographical distances of the Waorani indigenous inhabitants, the inhabitants of Tena and the inhabitants of Quito were also determined with respect to the Waorani WM. Figure 49.6 shows that the ethnic identity distances of each individual in the Waorani communities are closer to their own member, while the inhabitants of Tena and Quito are more distant. The descriptive statistics of the distances of the three populations with respect to the WM individual can be seen in Table 49.5, evidencing clearly that the average population distance to the WM individual is greater in the inhabitants of Tena and much more in the inhabitants of Quito. The values of the identity distances of all 299 individuals with respect to WM can be seen in the Annex.

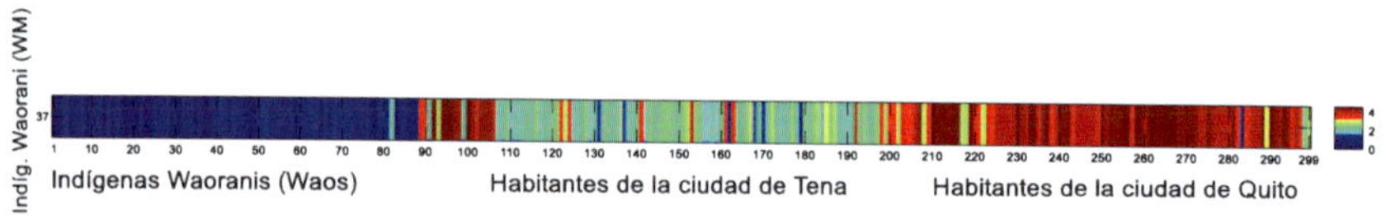

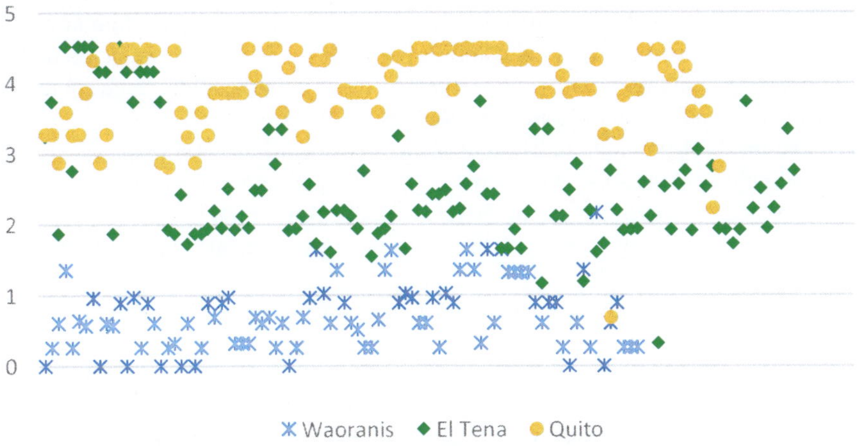

Fig. 49.6 Topographic distance of the Waorani populations, city of Tena and city of Quito from WM, the figure below shows the distribution of the 299 individuals

Table 49.5 Statistics on population distances from the indigenous WM

Population	Media	Medium	Typical deviation
Waorani communities	0.7054	0.6053	0.4787
City of Tena	2.5031	2.2072	0.8378
City of Quito	3.9219	3.9035	0.6152

Conclusions

Ethnic identity is a complex issue that mostly includes variables of a subjective nature. In the Amazonian cultures the process of obtaining variables and applying data collection instruments becomes even more complex due to the limited access to their territory (authorizations of community leaders and chiefs) and the difficulty of interpreting words proper to their language (Wao-terero) that in some cases there is no semantic translation into Western languages such as Spanish or English.

From the review of the scientific literature no research work was found that, from an identity instrument, quantifies or determines topographical distances of cultural identity in Amazonian indigenous communities. On the other hand, the absence of these studies did not allow to make comparisons with the results obtained here.

The indigenous Waorani Amazon people, who are subject to migratory flows as a result of colonization and who mostly come from populations with a western identity, would be subject to processes of acculturation of their people, affecting their own identity.

The application of an instrument that measures identity to three different populations made it possible to distinguish the differences in identity between the individuals in each population. The results obtained showed marked differences in identity between the indigenous Waorani and the city dwellers.

The result of grouping individuals using self-organized SOM maps showed that the individuals of the populations have important separations (considerable distances) mainly the Waorani Indians with the inhabitants of the city of Quito, however, these distances decrease when the comparison is made with the inhabitants of the city of Tena.

The distances of identity between the Indians and between the inhabitants of Quito are both homogeneous and closer to each other, unlike the inhabitants of the city of Tena, where the distances are more heterogeneous and greater. The data, in addition to corroborating with their geographical location (the city of Tena is located between the Amazonian indigenous communities and the city of Quito) the identities of its inhabitants would have greater ethnic-cultural influence on the Waorani and Quito populations.

By considering the distances between the Waorani indigenous people who have the greatest identity value (WM indigenous people with 89.4%) and indigenous people from the same territory, it would be possible to consider in a following measurement whether the migration and immigration flows of individuals from or to indigenous territory would be influencing the identity of an indigenous person in particular or of his or her people in general.

This research could have an impact on political decision-making, since taking into account the distance of cultural identity and the degree of influence by external cultures would make it possible to know whether certain policies aimed at these societies cause undesirable side effects on their own identity, and could reduce the threat that indigenous peoples feel to their identity from people outside their culture.

Annex

Topographical distance of identities from the Waorani (WM)

Waoranis				Tena				Quito			
N	Distan	N	Distan	N	Distan	N	Distan	N	Distan	N	Distan
***1**	**0.0000**	45	0.8867	1	3.2492	56	2.2072	1	3.2873	51	4.3325
2	0.2555	46	0.6054	2	3.7354	57	2.1769	2	3.2873	52	4.1059
3	0.6054	47	0.5157	3	1.8700	58	2.4297	3	2.8844	53	4.3844
4	1.3558	48	0.2555	4	4.5219	59	2.4297	4	3.5923	54	4.3325
5	0.2671	49	0.2555	5	2.7607	60	2.4917	5	3.2729	55	4.3325
6	0.6395	50	0.6537	6	4.5219	61	2.1769	6	3.2873	56	4.5029
7	0.5717	51	1.3558	7	4.5219	62	2.2253	7	3.8727	57	4.5029
8	0.9589	52	1.6338	8	4.5219	63	2.5665	8	4.3325	58	3.4961
***9**	**0.0000**	53	0.8867	9	4.1676	64	2.8267	9	2.8844	59	4.4762
10	0.6054	54	1.0287	10	4.1676	65	3.7417	10	3.2873	60	4.5029
11	0.5717	55	0.9589	11	1.8700	66	2.4297	11	4.5029	61	3.8998
12	0.8867	56	0.6054	12	4.5219	67	2.4297	12	4.3844	62	4.4762
***13**	**0.0000**	57	0.6054	13	4.1676	68	1.6524	13	4.4969	63	4.5029
14	0.9774	58	0.9589	14	3.7417	69	1.6524	14	4.4969	64	4.4762
15	0.2555	59	0.2555	15	4.1676	70	1.9259	15	4.3844	65	4.5029
16	0.8867	60	1.0287	16	4.1676	71	1.6524	16	4.5029	66	4.5029
17	0.6054	61	0.8867	17	4.1676	72	2.1769	17	4.4762	67	4.5029
***18**	**0.0000**	62	1.3558	18	3.7417	73	3.3422	18	2.8844	68	4.5029
19	0.2555	****63**	**1.6371**	19	1.9259	74	1.1682	19	2.8203	69	4.3325
20	0.3267	64	1.3558	20	1.8700	75	3.3422	20	4.4762	70	4.3325
***21**	**0.0000**	65	0.3267	21	2.4297	76	2.1233	21	3.5960	71	4.3325

(continued)

(continued)

Waoranis				Tena				Quito			
N	Distan	N	Distan	N	Distan	N	Distan	N	Distan	N	Distan
22	0.6054	****66**	**1.6371**	22	1.7293	77	2.1233	22	3.2474	72	4.3844
***23**	**0.0000**	67	0.6054	23	1.8700	78	2.4917	23	2.8844	73	4.3325
24	0.2555	****68**	**1.6371**	24	1.8700	79	2.8626	24	3.5923	74	3.8727
25	0.8867	69	1.3153	25	1.9461	80	1.1889	25	3.2729	75	3.8727
26	0.6885	70	1.3153	26	2.2072	81	2.2072	26	3.8727	76	4.3325
27	0.8867	71	1.3044	27	1.9572	82	1.6106	27	3.8727	77	4.1059
28	0.9774	72	1.3153	28	2.5132	83	1.7293	28	3.8727	78	3.8727
29	0.3267	73	0.8867	29	1.9259	84	2.7607	29	3.8727	79	3.9073
30	0.3267	74	0.6054	30	2.1233	85	2.2072	30	3.8727	80	3.8998
31	0.3267	75	0.8867	31	1.9572	86	1.9184	31	4.4969	81	3.9073
32	0.6885	76	0.8867	32	2.4917	87	1.9259	32	4.1059	82	4.3325
33	0.6054	77	0.2671	33	2.4917	88	1.9461	33	3.8998	83	3.2729
34	0.6885	***78**	**0.0000**	34	3.3422	89	2.6004	34	4.4969	84	0.6885
35	0.2671	79	0.6054	35	2.8626	90	2.1233	35	4.4969	85	3.2873
36	0.6054	80	1.3558	36	3.3422	91	0.3267	36	3.5994	86	3.8231
▲ 37	**0.0000**	81	0.2671	37	1.9133	92	2.5326	37	4.2234	87	3.8998
38	0.2555	****82**	**2.1631**	38	1.9461	93	1.9259	38	4.4762	88	3.8998
39	0.6885	***83**	**0.0000**	39	2.1233	94	2.5665	39	3.2474	89	4.4762
40	0.9589	84	0.6054	40	2.5665	95	2.7607	40	3.8231	90	3.0597
****41**	**1.6371**	85	0.8867	41	1.7293	96	1.9133	41	4.3325	91	4.4762
42	1.0287	86	0.2671	42	2.1769	97	3.0597 ara>	42	4.3325	92	4.2234

(continued)

(continued)

Waoranis				Tena				Quito			
N	Distan	N	Distan	N	Distan	N	Distan	N	Distan	N	Distan
43	0.6054	87	0.2555	43	1.6106	98	2.5326	43	4.4762	93	4.1059
44	1.3558	88	0.2671	44	2.2072	99	2.8267	44	3.5923	94	4.4969
				45	2.2072	100	1.9380	45	3.8998	95	4.2433
				46	2.1233	101	1.9259	46	3.8727	96	3.5923
				47	1.9380	102	1.7293	47	3.8727	97	3.8825
				48	2.7607	103	1.9259	48	3.8727	98	3.5923
				49	1.5501	104	3.7354	49	3.8727	99	2.2382
				50	1.8700	105	2.2253	50	3.5994	100	2.8203
				51	1.9380	106	2.5132				
				52	2.1233	107	1.9572				
				53	3.2492	108	2.2366				
				54	1.6524	109	2.5665				
				55	2.5665	110	3.3422				
						111	2.7607				

▲ Waorani individual No. 37 (WM) of maximum value with 354 out of 396 points (89.4%), corresponds to the female gender with 36 years of age
* Waoranis individuals with minimum distance (0.0000) from WM
** Waoranis individuals with maximum distances from WM

References

1. F.J. Abarca-Alvarez, F.S. Campos-Sánchez, R. Reinoso-Bellido, F.J. Abarca-Alvarez, F.S. Campos-Sánchez, R. Reinoso-Bellido (2017) Methodology of decision support through GIS and artificial intelligence: implementation for demographic characterization of Andalusia based on dwelling. Estoa. Revista de la Facultad de Arquitectura y Urbanismo de la Universidad de Cuenca **6**(11), 39–64. https://doi.org/10.18537/est.v006.n011.a03
2. E.H. Alkhammash, J. Jussila, M.D. Lytras, A. Visvizi, Annotation of smart cities Twitter micro-contents for enhanced citizen's engagement. IEEE Access **7**, 116267–116276 (2019)
3. U. Beck, La cuestión de la identidad. El País **11**(11), 2003 (2003)
4. M. Behnisch, A. Ultsch, Urban data-mining: spatiotemporal exploration of multidimensional data. Build. Res. Inf. **37**(5–6), 520–532 (2009)
5. Á. Bello, Viaje de los mapuches de Araucanía a las pampas argentinas: Una aproximación a sus significados socioculturales (siglos XIX y XX). *Disponible desde Internet en* (2000), https://www.mapuche.info/mapuint/bello0000.html
6. J.W. Berry, Immigration, acculturation, and adaptation. Appl. Psychol. **46**(1), 5–34 (1997)
7. K. Carley, A theory of group stability. Am Sociol. Rev., 331–354 (1991)
8. S. Carrasco-Arroyo, Medir la cultura: Una tarea inacabada. *Periférica Internacional. Revista para el análisis de la cultura y el territorio* **7**, 140–168 (2006)
9. M.-C. Chang, P. Bus, G. Schmitt, Feature extraction and k-means clustering approach to explore important features of urban identity. 2018-January (Scopus, 2018), pp. 1139–1144, https://doi.org/10.1109/ICMLA.2017.00015
10. E.W. Dijkstra, A note on two problems in connexion with graphs. Numer. Math. **1**(1), 269–271 (1959)
11. A. Espín-León, A. Jimeno-Morenilla, M.L. Pertegal-Felices, J. Azorín-López, Quantification of cultural identity through artificial intelligence: a case study on the Waorani Amazonian ethnicity. Soft. Comput. **24**, 11045–11057 (2020). https://doi.org/10.1007/s00500-019-04469-3
12. N. Frigola Torrent, Manifestación de la identidad étnica en la construcción de la agencia de migrantes indígenas amazónicos a la ciudad de Lima (2016)
13. S. Gutta, J.R. Huang, P. Jonathon, H. Wechsler, Mixture of experts for classification of gender, ethnic origin, and pose of human faces. IEEE Trans. Neural Netw. **11**(4), 948–960 (2000)
14. S.-U. Hassan, A. Visvizi, H. Waheed, The 'who' and the 'what' in international migration research: data-driven analysis of Scopus-indexed scientific literature. Behav. Inf. Technol. **38**(9), 924–939 (2019). https://doi.org/10.1080/0144929X.2019.1583282
15. C.H. Lawshe, A quantitative approach to content validity. Pers. Psychol. **28**(4), 563–575 (1975)
16. X. Lu, A.K. Jain, Ethnicity identification from face images. Proc. SPIE **5404**, 114–123 (2004)
17. R. Moltedo, Emigración mapuche e identidad étnica. Revista El Canelo **5**(22) (1990)
18. M. Moreno Salas, Identidad étnica en indígenas huetares de Quitirrisí (2008)
19. J. Muñoz Jusicia, M. Sahagun Padilla, Hacer análisis cualitativo con atlas ti. 7. Manual de Uso **135** (2015), https://manualatlas.psicologiasocial.eu/atlasti7.pdf
20. M.L. Pertegal-Felices, A. Espín-León, A. Jimeno-Morenilla, Diseño de un instrumento para medir identidad cultural indígena: Caso de estudio sobre la nacionalidad amazónica Waorani. Revista De Estudios Sociales **71**, 51–73 (2020). https://doi.org/10.7440/res71.2020.05
21. J. Pribilsky, Sustainability of a Life Más Cómodo? Agricultural change, remaking families, and the emerging indigenous middle class in the Ecuadorian Andes. The New Middle Classes, 177–195 (2009), https://doi.org/10.1007/978-1-4020-9938-0_10
22. A. Quijano, Colonialidad y modernidad/racionalidad. Perú indígena **13**(29), 11–20 (1992)
23. D. Racines Izquierdo, La occidentalización de los waoranis, B.S. Thesis, USFQ, Quito (2012)
24. S. Pereyra, S. Estefanía, Biometría y vigilancia social en Sudamérica: Argentina como laboratorio regional de control migratorio. Revista mexicana de ciencias políticas y sociales **63**(232), 247–268 (2018), https://doi.org/10.22201/fcpys.2448492xe.2018.232.56580

25. S.L. Schneider, A.F. Heath, Ethnic and cultural diversity in Europe: validating measures of ethnic and cultural background. J. Ethn Migr. Stud. (Scopus, 2018), https://doi.org/10.1080/1369183X.2018.1550150
26. J. Tang, S. Alelyani, H. Liu, Feature selection for classification: a review. *Data Classification: Algorithms and Applications*, vol. 37 (2014)
27. A. Tristán-López, Modificación al modelo de Lawshe para el dictamen cuantitativo de la validez de contenido de un instrumento objetivo. Avances En Medición **6**(1), 37–48 (2008)
28. M. Valdés, Migración interna indígena y no indígena en América Latina. *Estudios Avanzados, Proyecto Mapunet, Santiago de Chile, septiembre* (2008)
29. L. Velasco Ortiz, D. Rentería, L. Velasco Ortiz, D. Rentería, Diversidad e interculturalidad: La escuela indígena en contextos de migración. Estudios fronterizos, 20 (2019), https://doi.org/10.21670/ref.1901022
30. A. Visvizi, J. Jussila, M.D. Lytras, M. Ijäs, Tweeting and mining OECD-related microcontent in the post-truth era: a cloud-based app. Comput Hum Behav (2019)
31. A. Visvizi, M.D. Lytras, Rescaling and refocusing smart cities research: from mega cities to smart villages. J. Sci. Technol. Policy Manag. **9**(2), 134–145 (2018)
32. A. Visvizi, M.D. Lytras, M. Pachocka, Multiple facets of migration research: key questions, topics, and Avenues yet to be explored. Sustainability **12**(1), 225 (2020). https://doi.org/10.3390/su12010225
33. A. Visvizi, C. Mazzucelli, M. Lytras, Irregular migratory flows: towards an ICTs' enabled integrated framework for resilient urban systems. J Sci Technol Policy Manag **8**(2), 227–242 (2017). https://doi.org/10.1108/JSTPM-05-2017-0020
34. A. Visvizi, M. Pachocka, M.D. Lytras, Managing international migration: rethinking transaction costs, red tape, and their impact. Int Migr **57**(5), 271–279 (2019). https://doi.org/10.1111/imig.12648

Chapter 50
Social Media Communication by Hungarian Anti-establishment Parties in the Context of Italian Movimento Cinque Stelle

Anna Molnár and Anna Urbanovics

Abstract The topic of e-democracy is a steadily growing issue in political science. Political parties exerting a more direct political representation rely on social media use to a great extent. In the present study a comparative analysis of the social media communication is conducted of two Hungarian anti-establishment parties, the Politics Can Be Different (LMP) and Momentum taking as a role model the Italian Movimento Cinque Stelle. The method used for analysis is a mixed approach where quantitative and qualitative aspect are both considered. The scope of the research is 1st March 2018 and 31st December 2019. The analysis is based on contents shared by the examined political parties after identifying their peak periods with respect to the social media use. The studied contents are extracted from the official Twitter profiles of the LMP and Momentum. A filtering process based on the contents' sentiment score and influence score is performed. Research findings show that the two Hungarian parties rely on the social media to communicate their ideologies and political programs, and both have some potential to further develop towards becoming cyber parties. Momentum focuses on European issues, while exhibiting a more direct, and a more aggressive communication against the current administration of Hungary. LMP reflects more on local issues and use digital tools to promote campaigns related to environmental protection issues. Evaluating their social media communication Momentum has more potentials to develop to a cyber party becoming a role model for Hungary in this recently emerging tendency of anti-establishment political parties.

Keywords E-democracy · Social media analysis · LMP · Momentum

A. Molnár (✉) · A. Urbanovics
National University of Public Service, Budapest, Hungary
e-mail: molnar.anna@uni-nke.hu

A. Visvizi et al. (eds.), *Research and Innovation Forum 2020*,
Springer Proceedings in Complexity,
https://doi.org/10.1007/978-3-030-62066-0_50

Introduction

The use of digital tools has become a steadily growing tendency among European anti-establishment political parties. Digital platforms, including mainly social networking sites are suitable forms to communicate alternative ideologies and mobilize citizens for a specific issue. Therefore, the rise of cyber parties can be observed throughout Europe, providing important examples such as the Italian Movimento Cinque Stelle (M5S), the Spanish Podemos or the Greek Syriza. These parties have been successful and competitive role models for cyber parties against the state administrations that has been visible during European Parliamentary elections and national elections as well.

The present study aims to investigate the social media communication of two Hungarian political parties in the context of the Italian M5S. The comparative analysis is based on qualitative methods eyeballing a list of contents from the Hungarian parties—namely the Politics Can Be Different (LMP) and the Momentum. An ideal reference point is identified in the communication of M5S. In our view, these three movements belong to the group of anti-establishment and anti-corruption political parties or movement [1–3] refusing to define themselves on the Left-Right political spectrum. In Italy, the Five Star Movement, which is an anti-establishment political movement, proposed the realisation of direct e-democracy. In Hungary, the LMP and the Momentum have not succeeded in e-democracy, but not even in elections.

Advances in information and communication technology (ICT) created several opportunities for democracy to be exercised also through online communication platforms, including social networking sites (SNS) [4–8].

The present study compares Hungarian political parties based on the example of the M5S being an original contribution of the existing international literature. To the best of the authors' knowledge, this is the first comparative analysis based on social media contents dealing with the LMP and Momentum preparedness for e-democracy and to measure their potentials to become cyber parties. The perspective of conducting on qualitative research further develop the research findings of a quantitative study measuring the general presence and activity of these parties on social media [9].

After the detailed description of methodology used for the analysis, we introduce the role model of M5S based on findings of the existing literature and describe the two Hungarian political parties (LMP and Momentum) from an organizational approach. These are followed by the presentation of research findings and evaluation of social media communication of the examined parties using specific social media contents and mentions extracted from the parties' official Twitter profiles. The scope of the research is restricted to a period between 1st March 2018, the date of the Hungarian national elections, and 31st December 2019.

Methodology

By social listening we analyze the contents, mentions, comments, and shares directly related to the three political parties involved (M5S, LMP or Momentum) in each period. Due to certain simultaneous political events of the two countries, our analysis is restricted within the time framework from 1st February 2018 concerning the Italian party and the 1st March 2018 concerning the two Hungarian parties. We wanted to also study the period just before the parliamentary elections in both countries so for this reason we chose the above-mentioned starting points.

Concerning the platforms used for the research, the official Twitter and Facebook webpages of the parties were used. It was a necessary restriction because otherwise there would have been several irrelevant results where the parties would have been only indirectly mentioned. In this comparative analysis though, our aim was to investigate the social media participation and communication of the parties. The number of total mentions is not comparable and is not significant in our case, as Italy is a much bigger and more developed country than Hungary. However, what is a significant data that must be considered is the share of the two main social networking sites with respect to the raw mentions. Based on data, 42.4% of M5S's raw mentions were generated on Facebook, while for LMP this value is 47.5% and for Momentum is 49.9%. Concerning Twitter raw mentions, 40.8% of M5S's have been originated from there, while in the case of the two Hungarian parties Twitter is not a leading platform. Observing the total reach, 32% for M5S total reach have come from Facebook and 30.8% from Twitter. For LMP 40% is generated on Facebook and for the Momentum this value is 60%.

Considering these numbers Facebook and Twitter have been chosen as the main source for this comparative analysis. Although Facebook and Twitter are both used actively by the political parties, the two social networking sites serve for other purposes. The Twitter is functioning as a platform to share ideas with a broader audience, while the Facebook is built upon personal connections and networks. In the case of the two Hungarian parties the Facebook is more dominant, the Italian M5S uses both sites equally. To gather the relevant contents and mentions we applied several filters as well on the database of raw mentions within the Senti One application. In the following, the result of the filtering process is discussed.

When gathering the content in the case of every party the official Twitter and Facebook profiles have been used. After this, the set of time frame was made which begins one month before the parliamentary elections in both countries and ends at the end of 2019. The time scope is ideal to measure the social media presence of the three parties not just around the elections and during campaigns but between these too. Based on the time frame some peak periods can be determined when the online activity of the parties have been intensive. The comparative analysis is restricted to these peak periods which is summarized in Table 50.1.

The searching results have been further filtered by a preliminary sentiment analysis to gather only the negative or positive sentiment contents. The neutral contents contain technical or informational mentions that do not contain relevant information

Table 50.1 Peak periods of examined political parties by social media communication

LMP	Momentum	M5S
March 2018	Max 2019	February 2018
September 2018	July 2019	May 2018
February 2019	October 2019	August 2018
May 2019		December 2019

with respect to the qualitative analysis. After this, the exported database has been further filtered and put in order in align with the focus points. The results have been filtered by the content author, and only the contents authored by the political parties have remained. The number or mentions through the filtering process is shown in Table 50.2. A larger number of raw mentions can be observed related to the M5S but after the whole process similar number of contents authored by the party itself could have been involved in the analysis. This shows that besides the active online presence of the M5S, their voters and followers are also very active and engage the party in their social media contents.

The order of the results has been constructed based on two indicators. The primary indicator was the sentiment score measuring the sentiment of the content in a negative positive scale with the help of artificial intelligence. The secondary indicator was the influence score measuring the total reach and the number of reactions on the content. These indicators point out the most relevant mentions in the given time periods,

Table 50.2 Peak periods and number of total contents included in the comparative analysis by political parties

		Contents by all authors	Authored by the party
LMP	March 2018	120	76
	September 2018	81	78
	February 2019	101	90
	May 2019	36	4
Momentum	May 2019	94	16
	July 2019	29	2
	October 2019	70	26
M5S	February 2018	1989	11
	May 2018	1698	6
	August 2018	445	19
	December 2019	645	147

indicating the most negative and most positive contents and related issues. Both indicators helped in the extraction of contents for further analysis, on the other hand, only Twitter mentions remained in our database after the filtering process.

Anti-establishment Parties and Digital Opportunities

In recent decades, the rise and electoral success of anti-establishment, anti-corruption and protests parties have been widely discussed in the literature [10–13]. In our research we examine the Italian Five Star Movement (M5S), the Hungarian LMP (Lehet Más a Politika, 'Politics Can Be Different') and Momentum. These political parties or movements belong to the group of anti-establishment or protest parties refusing to define themselves on the Left-Right political spectrum. As a response to the disappointment with traditional parties and to the financial and political crisis these new protest parties were created.

The M5S was founded in 2009. Beppe Grillo created his blog in 2005 as a reference point for his political initiatives. From 2012 this relatively new political player, Beppe Grillo's Five Star Movement, was increasingly challenging traditional party politics and entered national parliament in 2013. Not only the anti-establishment, but also the populist nature of the M5S has been examined by researchers [14].

On the one hand, in Hungary, the two examined parties were not successful at winning the election, but not even at transforming themselves into real anti-elitist digital parties. On the other hand, these two parties are not even accused of using a populist rhetoric. The LMP (Lehet Más a Politika, 'Politics Can Be Different') was founded as an NGO in 2008, and as a party in 2009. During the general elections of 2010, this new green, anti-corruption and anti-elitist party entered the Parliament in 2010 [15]. According to Kovarek and Littvay in recent years the LMP has rebranded 'as an anti-corruption critic of Viktor Orbán's Fidesz government' [16]. However, this new green party did not benefit form opportunities and instruments provided by the digital world and cyber-democracy. In its latest electoral political programs, the LMP promises the launch of digital services in health, the labour market and public administration [17] and a better regulation of tech giants [18]. In its political thematic program, the rule of law, nation politics, climate politics and environmental protection, energy politics, education and urban planning and development are central elements.

The Momentum party as the newest anti-establishment movement in Hungary was founded in 2017. In an interview András Fekete-Győr, founder of the Momentum defined the movement as an anti-establishment party to highlight the centrist position which does not fit the traditional left-right political landscape [19]. As an activist András Fekete-Győr led successfully the NOlimpia referendum campaign to prevent that Budapest host the 2024 Summer Olympics. Despite the fact that the Momentum belongs to the new political generation in Europe, and it is a real youth party, it has not benefitted from digital opportunities. Although the electoral programs of this party mention e-governance, digital economic strategy [20], the digital revolution of the EU [21] or the smart city program [22], elements of cyber democracy as a tool of direct democracy are not listed in the political programs. For the Momentum European issues and European solidarity, the progression and the knowledge society are key areas of activities.

The Italian Five Star Movement has become a digital party which supports e-participation and e-democracy. In order to achieve its political goals, the M5S implemented e-participation platforms in order to strengthen the capacity of ordinary citizens for self-representation and civic engagement [23]. Lindner and Aichholzer summarised the two key concepts of e-democracy and e-participation as follows: e-democracy means "the practice of democracy with the support of digital media in political communication and participation. E-participation encompasses all forms of political participation, making use of digital media, including both formally institutionalised mechanisms and informal civic engagement" [24]. According to Morosini the digital party is the product of technology and ideology. The digital technology covers all those products, without which the digital party could not be born: cheap smartphones and laptops, cheap servers, wireless coverage, internet development. It is no coincidence that M5S was born in 2009, only two years after the iPhone. The ideology that generated the party is digitalism, a way of thinking born in California in the seventies, which sees digital technologies as the tool for the definitive emancipation of humanity. Gianroberto Casaleggio understood the political potential of digital tools [25].

Gianroberto Casaleggio was the main contributor to the movement's institutionalization and communication strategy [26]. According to Hartleb the M5S can be described as an 'anti-elitist cyber party' [27]. Following the electoral success in 2013 Casaleggio, the web guru of the movement emphasised the importance of more citizen-centric governance based on direct democracy made possible by the Web. After the death of Casaleggio this concept was reinforced and continued by his son, Davide Casaleggio [28]. The new anti-establishment party criticizing representative democracy, expressed a strong preference for direct democracy [29–32].

The M5S supports more use of Internet-based direct and deliberative democracy or e-democracy. The Five Star Movement, which was a member of the Europe of Freedom and Direct Democracy (EFDD) group in the European Parliament between 2014 and 2019, utilized an internet-based technological platform, the Rousseau platform to vote on important political decisions internally [33]. The online platform was created in 2016, since then it has been managed by Davide Casaleggio. Through this platform the members are able to comment on and express their opinion proposing amendments on legislative procedures in the Italian Parliament. According to Biorcio and Sampugnaro this project helped the Movement to realise one of its greatest desires: "to transform the ordinary citizen into a potential legislator" [34]. The Rousseau Platform held online votes of people who subscribed to that website on coalition agreements in 2018 and 2019 as well. The movement became the biggest political party in 2018 receiving 32% of votes and started to head the coalition governments. The democratic nature of the Web and especially the use of Rousseau have been debated by researchers in the literature, as it is considered as a new manifestation of populism [35–38].

Discussion and Findings

In the following section the main findings of the qualitative analysis are presented. The empirical comparative analysis deals with the two Hungarian parties based on the theoretical background and role model of digital party provided by the Italian M5S. Another base for comparison is the Salvini—Viktor Orbán tandem in the European communication that the Momentum explicitly criticizes. It is important to note that the results are calculated based on the filtered database and that the time frame clearly determined the most important political events in the political activities of the three parties. The time is also significant in terms of major national and international political events such as the parliamentary and European Parliamentary elections. While in the first case the national and homeland political events have been in forefront, in the second case the European-centric view and the European issues have been pronounced in their communication. In the case of the highly active M5S the division between national and international issues are extremely important because it is difficult to determine the peak points of their social media presence. The party is active on both social networking involved in the analysis as an electronic party engaging their voters through constant social media use. Concerning the two Hungarian parties the peak periods can be determined more easily. These peak times with stronger social media presence are centred around the political campaigns and elections bot hat national and international level. The Hungarian parties use the two platforms mainly to express their political views and inform on their results achieved, however a clear opposition to the governing party is observed as well.

The Case of LMP

The Hungarian green party's social media activity is focused on the election periods but contains two periods between the national and European parliamentary elections discussing ordinary political issues. They use the platforms to express their political goals and achievements to a great extent. Besides these, the engagement in climate politics is observed both at national and international level. Regarding national issues, they deal with specific cases to inform on their official opinion, while related to international issues they communicate about special "world days" climate politics campaigns such as against air or water pollution and the exemption of plastics. In their communication Hungarian political actualities are in the forefront while the European events are not emphasized. Concerning negative sentiment contents, they mainly focus on the critics of the governing party and to emphasize on their anti-establishment nature. The LMP declares itself as an anti-establishment party in its social media communication also. They focus on engaging their voters in political discussions and to communicate the results of traditionally conducted consultations. This motivates an active dialogue between the party and citizens.

In their communication concerning the ruling party, the LMP focuses on specific policy issues and programs. Specific events and atrocities are mentioned instead of ordinary critics. Some examples for this are the followings:

"It is shocking that Hungary is a world leader with respect to the number of patients of colon cancer. It is outrageous that the government is delaying arranging emergency care for the Military Hospital."

"The Hungarian government wants to comply with EU regulations only by 2030, while thousands of Hungarians die every year due to poor air quality!".

Besides these, the issue of the rule of law is significantly present in the party's communication and has been included in their campaign program as well. They demand for greater accountability, transparency, and greater independence for public administrations and public institutions. This can be seen through contents like "The State Audit Office is no longer an independent institution, but an instrument of Fidesz's revenge campaign." The prime minister has been mentioned directly mainly during the run-up of the parliamentary elections in March 2018. It can be interpreted as an open attack aimed at receiving their voters mandate. These voters are mainly citizens disillusioned with the ruling party, so the motives against Fidesz and the motives of the opposition coalition were also significant elements in the 2018 elections. A good example for the direct attack against the prime minister is "Viktor Orbán has lied, is lying and will keep lying."

Turning to the European context, among their European partners, only with the European Greens group they maintain active communication. International issues are not emphasized rather domestic events are reflected. Concerning these issues, besides the government's decisions violating the rule of law, they focus on climate issues and civil programs. The LMP has been actively engaged in programs of environment protections such as movements against air and water pollution in Hungary.

"For our future, the priority is to eliminate unnecessary polluting plastic products that become mere waste after 5–10 min of use."

"The government has not taken any notice of the severity of climate change, and the cabinet is not fighting to improve the quality of life of the Hungarian people in Brussels but is working for the automotive industry."

The LMP maintains active dialogue with citizens about urban development and the development of Budapest and its agglomeration. They focus on specific cases such as "We have launched a petition to solve the asbestos problem in the city park so that the faults can be rectified immediately." Education was also strongly emphasized within the party's campaign before the elections in 2018. This is reflected in their proposal such as "Implementing experience-based education requires a new core curriculum and more autonomy."

When communicating, the party also highlights their own campaign elements and provide information on the public events and activities of their representatives. The climate politics and environmental protection are greatly reflected in their mission and vision as Hungary's green party.

"The future is either green or none! Let's fight climate change together."

"There is no planet B and there is no Plan B. Let's work together for our planet."

The Case of Momentum

In the case of Momentum, a content active social media presence is not observed. The party is more active during the elections; however, the Hungarian parliamentary elections was not a peak period in their communication. They were more present on social media during European Parliamentary elections. Momentum engages in a very interesting dual political communication as it seeks to address domestic and European issues at the same time. Their commitment to the European Union is already reflected in the fact that most of their contents have been written in English. This shows that the party's social media are made for international partner organizations and not exclusively for Hungarian citizens. During the European Parliamentary elections, a significant and actively maintained system of international partnerships could be seen. In their communications, Momentum clearly declares their anti-establishment nature against the ruling party. Concerning this, content primarily mentioning the functioning of the government and the perceived corrupted are listed. However, opposition to government policy decisions is not a prominent element in the party's communication. Momentum was also present during the municipal elections on social networking sites and their political achievements were announced on these platforms. Thus, the 2019 municipal elections are one the peak periods in their online presence. On the other hand, the dialogue and involvement of citizens is not a central element in their political communication as they use social media to share their program and inform on their results.

It is important to note that they accuse the government much more openly in their communications than LMP. The critics do not focus on policy issues, rather the perceived corruption is the core element. Some examples of the criticism on the government and allegations of corruption are the followings:

"Results are coming from all Hungary; we are defeating corruption! Change is here!".

"We can change the corruption political culture in #Hungary… we represent all the people who believe that politics without corruption is possible" says #FeketeGyőrAndrás in front of all our elected officials."

"By not standing against #Borkai, #Orban is showing that he does not respect or stands for the values that #Fidesz preaches. All the speech of Christian values and liberties are just a #populist cover to the true corrupt and power abusing nature of the ruling party!".

Their achievements are not published along the thematic elements of their program rather through the personal successes of their representatives. As a demonstration of their commitment to the European issues and to maintain strong connections with partner organizations, the information related to ALDE party are also shared on Momentum's social networking sites. The commitment to a united Europe and the general European orientation has been a thematic element of the party's program in 2018 elections as well.

"We would like to introduce you to our newly 6 elected mayors…".

"We have a new Vice President from our party in @ALDEParty! Thank you for all the European parties supporting us and for welcoming our vision of a fresh and progressive Europe."

Besides these, another important thematic element in their program is knowledge-based country and support for science at both national and European levels.

"Science Business interviewed Momentum's #MEP candidate Barnabás Kádár, who talked about expanding #Horizon2020 and Europe, instead of spending cohesion funds on R&D, which are more vulnerable to corruption at a national level."

"That's an utter disgrace. Today #Orbán and his party have dismantled one of the cornerstones of #Hungarian science, one of #Hungary's most respected institutions, steeped in tradition. They did it this in a very humiliating way."

In their rhetoric, the dual motives of reform and Europeanism are emphasized, by which the party assume a certain mobilizing role in Hungary. An important element in their communication is the actions against populist and nationalist leaders. With respect to this, a recurring element is the parallel used between Matteo Salvini Italian politician and Viktor Orbán Hungarian prime minister.

"Momentum VP, @donath_anna: Hungary's youth tells us they want us to build a strong Hungary, an alternative to a gvt that drifted towards the fringes of the #EU. They want this strong Hungary to work twd a #strong #Europe, because Europe is their future! #ep2019 #RenewEurope".

"Momentum is and will be in the forefront of the struggle for our common home Europe!".

"I remember the courage of Hungarians fighting the authoritarian #regime in 89. Today we fight together those who want to undermine European #freedom. In #Hungary versus #Orban, in #Italy versus #Salvini. Together, we stand united—says @emmabonino from @Piu_Europa #TeamEurope".

Conclusions

The present study deals with the emerging issue of e-democracy from a qualitative aspect. It is based on social media analysis, presenting empirical findings about the political communication of the LMP and Momentum. Social listening is becoming a popular tool among political scientists providing valuable insights on citizens' reaction, political engagement and the public opinion on current political issues and events. The limitation of the study is found in the restrictive privacy issues, because only freely available public contents are involved. The scope of the research includes the peak periods of the two Hungarian parties between 1st March 2018 and 31st December 2019. The present chapter provides qualitative insights of the two examined Hungarian political parties, Momentum and LMP from the aspect of potentially becoming cyber parties. Given an overview of the M5S representing a role model among European cyber parties, we could successfully position the LMP and Momentum. Main research findings show that both Hungarian parties use social media in their political communication related to traditional political activities, not paying enough attention for engaging and mobilizing their supporters. Their communication strictly follows the key elements of their political programs, in the case of LMP the climate politics, and in the case of Momentum European issues and unity. LMP focuses follows a more well-structured style in their tweets (contents) meaning the structuring of complete sentences, while Momentum uses different hashtags and abbreviations in a shorter, only words form. LMP reflects more on national and local issues, considering the everyday life events and direct campaigns to engage citizens. Momentum uses a broader lens to reflect on issues, mainly considering the most important national issues and European issues. Furthermore, Momentum maintains a more active international network with other European political parties and conducts a more aggressive and demanding way of communication against the current administration led by Fidesz. Momentum relies on its mottos and slogans to a greater extent that are displayed in a hashtag form as well. This way, their main message and ideology is more emphasized and visible for voters. Momentum has better chances to become a cyber party and the first in Hungary of its kind, because they rely more on social media and digital tools, use a more loose language and focus more on showing their results and achievements for citizens.

References

1. D. Vampa, Local representative democracy and protest politics: the case of the five-star movement. Contemp Ital. Politics **7**(3), 232–250 (2015)
2. S. Engler, B. Pytlas, K. Deegan-Krause, Assessing the diversity of anti-establishment and populist politics in Central and Eastern Europe. West Eur. Politics **42**(6), 1310–1336 (2019)
3. B. Fernández-García, Ó.G. Luengo, Electoral scenarios of success for anti-establish-ment political parties in Western Europe: a fuzzy-set qualitative comparative analysis. J. Contemp. Eur. Stud. **27**(1), 77–95 (2019)
4. T.D. Barth, W. Schlegelmilch, The quality of democracy as a key to cyber-democracy, in *Handbook of Cyber-Development, Cyber-Democracy, and Cyber-Defense*, ed. by E.G. Carayannis, D.F.J. Campbell, M.P. Efthymiopoulos (Springer, Cham, 2018)
5. J.C. Bertot, P.T. Jaeger, J.M. Grimes, Promoting transparency and accountability through ICTs, social media, and collaborative e-government. Transf. Gov. People Process Policy **6**(1), 78–91 (2012)
6. K. Axelsson, U. Melin, I. Lindgren, Exploring the importance of citizen participa-tion and involvement in e-government projects: practice, incentives, and organization. Transf. Gov. People Process Policy **4**(4), 299–321 (2010)
7. D.F.J. Campbell, E.G. Carayannis, Epistemic governance and epistemic innovation policy in higher education. Technol. Innov. Educ. **2**(2), 1–15 (2016)
8. A. Visvizi, M.D. Lytras, *Politics and Technology in the Post-Truth Era* (Emerald Publishing, Bingley, 2019)
9. A. Molnár, A. Urbanovics, The role of e-democracy in Italy and Hungary. *Transforming Government: People, Process and Policy Ahead-of-Print* (ahead-of-print) (2020), pp. 1–15
10. S. Ceccanti, S. Curreri, I partiti antisistema nell'esperienza italiana: il Movimento 5 Stelle come partito personale autoescluso. Diritto Pubblico Comparato Ed Europeo **17**(3), 799–832 (2015)
11. A. Batory, Populists in government? Hungary's "system of national cooperation." Democra-Tization **23**(2), 283–303 (2016)

12. B. Verbeek, A. Zaslove, Italy: a case of mutating populism? Democratization **23**(2), 304–323 (2016)
13. Z. Enyedi, Populist polarization and party system institutionalization. The role of party politics in de-democratization. Prob. Post-commun. **63**(4), 210–220 (2016)
14. M. Tarchi, Dieci anni dopo, L'Italia populista e il caso Beppe Grillo. Quaderni di sociologia **65,** 31–49 (2014); L. Manucci, M. Amsler, Where the wind blows: Five Star Movement's populism, direct democracy and ideological flexibility Italian Political Science Review/Rivista Italiana Di Scienza Politica **48**(01), 109–132 (2017); L. Mosca, Beyond left and right: the eclectic populism of the Five Star Movement. West Eur Politics **42**(6), 1258–1283 (2019); M. Caiani, The populist parties and their electoral success: different causes behind different populisms? The case of the Five-star Movement and the League. Contemp. Ital. Politics **11**(3), 236–250 (2019)
15. K. Fábián, Can politics be different? The Hungarian Green Party's entry into Parliament in 2010. Environ. Politics **19**(6), 1006–1011 (2010)
16. D. Kovarek, L. Littvay, Where did all the environmentalism go? 'Politics can be dif-ferent' (LMP) in the 2018 Hungarian parliamentary elections. Environ. Politics **28**(3), 574–582 (2019)
17. LMP Homepage, Új kezdet. Választási program 2018, https://lehetmas.hu/wp-content/up-loads/2018/02/Va%CC%81laszta%CC%81si_program_2018.pdf. Last accessed 27 May 2020
18. LMP Homepage, A Lehet Más a Politika európai parlamenti választási programja (2019), https://lehetmas.hu/wp-content/uploads/2019/03/Ep-program_2019.pdf. Last accessed 27 May 2020
19. Euractiv Homepage, Our time will come'—Hungarian leader says youth will oust Orban, 2022, 31 May, https://www.euractiv.com/section/future-eu/interview/our-time-will-come-hun garian-leader-says-youth-will-oust-orban-in-2022/1242977/. Last accessed 27 May 2020
20. Momentum Homepage, Indítsuk be Magyarországot! (2018). https://program.momen-tum.hu/. Last accessed 27 May 2020
21. Momentum Homepage, Momentum + Europe = Future (2019), https://momentum.hu/wp-con tent/uploads/2019/03/Ne-adjuk-a-j%C3%B6v%C5%91nket-Momentum-EP-Program.pdf. Last accessed 27 May 2020
22. Momentum Homepage, Budapest 2.0 (2019), https://budapest.momentum.hu/wp-content/up-loads/2019/05/budapest_2.0_final-Lite.pdf. Last accessed 27 May 2020
23. M. Deseriis, Digital movement parties: a comparative analysis of the technopolitical cul-tures and the participation platforms of the Movimento 5 Stelle and the Piratenpartei. Inf. Commun. Soc. 1–17 (2019)
24. R. Lindner, G. Aichholzer, E-democracy: conceptual foundations and recent trends. In: L. Hennen, I. Keulen, I. Korthagen, G. Aichholzer, R. Lindner, R.Ø. Nielsen, *European E-Democracy in Practice* (Springer, 2020), pp. 11–45
25. M. Morosini, *Snaturati: Dalla Social-Ecologia al Populismo* (Castelvecchi, Roma, 2019)
26. S. Ventura, Leaders and parties after the Italian elections. Contemp. Ital. Politics **5**(2), 179–195 (2013)
27. F. Hartleb, Anti-elitist cyber parties? J Public Aff. **13**(4), 355–369 (2013)
28. G. Navarria, *The Networked Citizen: Power, Politics, and Resistance in the Internet Age* (Palgrave Macmillan, Springer Nature, 2019)
29. R. D'Alimonte, The Italian elections of February 2013: the end of the Second Republic? Contemp. Ital. Politics **5**(2), 113–129 (2013)
30. P. Corbetta, R. Vignati, Direct democracy and Scapegoats: The Five Star Movement and Europe. Int. Spectator **49**(1), 53–64 (2014)
31. P. Ignazi, Le occasioni mancate. Il Mulino **2**(2), 254–263 (2017)
32. W.L. Bennett, A. Segerberg, C.B. Knüpfer, The democratic interface: technology, political organization, and diverging patterns of electoral representation. Inf. Commun. Soc. **21**(11), 1655–1680 (2018)
33. P. Ceri, F. Veltri, *Il Movimento Nella Rete* (Rosenberg & Sellier, Torino, 2017)
34. B. Biorcio, R. Sampugnaro, Introduction: The Five-Star Movement from the street to local and national institutions. Contemp. Ital. Politics **11**(1), 5–14 (2019)

35. S. Natale, A. Ballatore, The web will kill them all: new media, digital utopia, and political struggle in the Italian 5-Star Movement. Media Cult. Soc. **36**(1), 105–121 (2014)
36. J. Bartllett, The new age of digital populism. New Statesman **147**(5410), 15–16 (2018)
37. J. Bartlett, *The People vs Tech: How the Internet Is Killing Democracy (and How We Save It)* (Ebury Press, London, 2018)
38. C. Stockman, V. Scalia, Democracy on the Five Star Movement's Rousseau platform. Eur Politics Soc, 1–15 (2019)

Author Index

A. Visvizi et al. (eds.), *Research and Innovation Forum 2020*, Springer Proceedings in Complexity,
https://doi.org/10.1007/978-3-030-62066-0

MIX
Papier aus verantwortungsvollen Quellen
Paper from responsible sources
FSC® C105338

Printed by Books on Demand, Germany

The Transition from Foraging to Farming and the Origin of Agriculture in China

The Transition from Foraging to Farming and the Origin of Agriculture in China

Tracey Lie Dan Lu

BAR International Series 774
1999

British Archaeological Reports are published by John and Erica Hedges and by Archaeopress.

This volume has been published by:
John and Erica Hedges
British Archaeological Reports
7 Longworth Road
Oxford OX2 6RA
England
Tel/Fax +44 (0)1865 511560
E-mail: bar.hedges@lineone.net

Enquiries regarding the submission of manuscripts for future publication may be sent to the above address

BAR S774

The Transition from Foraging to Farming and the Origin of Agriculture in China

Volume Editor: John Hedges BSc(Hons), MA, MPhil, FSA, FSAScot, FRAI, MIFA

Printed in England by Biddles Ltd

ISBN 1 84171 100 4

All BAR titles available from:

Hadrian Books
122 Banbury Road
Oxford OX2 7BP
England

The current BAR catalogue with details of all titles in print, prices and means of payment, is available free from Hadrian Books

All volumes are distributed by Hadrian Books Ltd

Once upon a time there was a clever man who taught people to cultivate the five cereals. He was named the God of Agriculture (Shennong).

Chinese legend.

Preface

I am very grateful that this thesis is to be published by BAR. Another year has gone by since I finished this work in the Australian National University. During 1998, details of some archaeological discoveries have been revealed, and I have updated this work with relevant information I can gather up to this date. The three examiners of my thesis have made their comments and have pointed out some errors, which I have also corrected in this version.

The transition from foraging to farming and the origin of agriculture in China is a big issue that requires contributions from many, many scholars of various disciplines. Our knowledge on this issue today has been contributed by a great number of Chines scholars and scholars from other countries. However, we still have a very long way to go, as there are currently more remaining questions on this issue than conclusions. This is probably an issue that requires a life-long commitment, but certainly a worthwhile one.

Tracey Lie-Dan Lu
12/03/1999 in Hong Kong

Acknowledgment

This study could not have been completed without the generous assistance from many people. First and foremost, I am deeply indebted to my supervisor, Dr. Peter Bellwood, who in the past three years not only has given me valuable academic advice, but also endured support for various administrative issues relating to my attempt to carry out this study. My advisers Professor Matthew Spriggs, Dr. Michael Smith, and Mr. Ian Farrington also gave me their valuable advice, and I thank them. I also thank Professor Francesca Merlan for her unfailing support, and Dr. Nicholas Peterson for providing information on the economic activities of Australian Aborigines. Further, I must acknowledge the continuous assistance I have received from the administrative staff of the department of Archaeology and Anthropology during my course. In addition, my special thanks are delegated to Dr. Gail Craswell of the Study Skills Centre in ANU, for her painstaking effort to improve my written English.

The major proportion of the database of this thesis was collected during my field work from October 1995 to October 1996 in mainland China and Hong Kong, which was funded by the Faculty of Arts ANU. During my field work in China, I was allowed to observe collections in numerous museums, archaeological sites and institutes; I received academic advice and updated information from many Chinese scholars and experts. I am deeply indebted to Professor Yan Wenming, Professor Lu Zun'e, Associate Professor Huang Yunpin and Dr. Wang Youpin of Beijing University; Professor Shang Zhitan and Zeng Qi of Zhongshan University; Mr Fu Xianguo, Dr. Chen Xingqan, Mr. Jiao Tianlong, Mr. Lian Zhonghe and Mr. Wu Jia'an of the Archaeology Institute CASS; Mr. Li Ji of the National Bureau for Antiquity Management; Mr. Fu Renyi of Liaolin Provincial Museum; Mr. Xie Fei of Hebei Archaeology Institute; Mr. Zhang Juzhong of Henan Archaeology Institute; Professor Liu Jun of Zhejiang Archaeology Institute; Professor Peng Shifan of Jiangxi Archaeology Institute; Mr. Zhou Houben and Mr. Cheng Jiang of Nanjing Museum; Mr. Yuan Jiarong and Pei Anpin of Hunan Archaeology Institute; Mr. Cheng Jian of the Zhenjiang museum; Dr. Tang Chung, Mr. Steven Ng and Miss Wang Wenchung of the Archaeology Centre in the Chinese University of Hong Kong; and Dr. Li Xiuguo of the Polytechnical University of Hong Kong. All gave generous help during my field work. They either gave me most valuable advice, support and/or information, or granted me access to their unpublished discoveries. Without their support, the database for this thesis could not be completed.

My gratitude is also due to informants in South China, and to the staff of the archaeology station in Qufu City, Shandong province. Without their help my observations on wild rice and millet could not be conducted. And to those anonymous, ordinary Chinese people who helped me one way or the other during my journey of thousands of kilometres, I also thank them.

This thesis has been kindly read in whole or in part by Dr. Peter Bellwood, Professor Matthew Spriggs, and Dr. Doreen Bowdery, and I am most grateful for their valuable comments and their time spent on correcting my written English. Dr. Colin Groves has kindly checked the faunal Latin names. The three examiners have kindly pointed out some errors. I am also very thankful for their kindness and comments. However, I am solely responsible for the arguments and errors that may occur.

Finally, I would also like to express my gratitude towards Dr. Ofer Bar-Yosef of Peabody Museum of Harvard University, Dr. Andrew Garrard of the Institute of Archaeology, UCL, Dr. Michael Smith of the Australian Museum in Canberra, and Dr. Peter White of the Oceania Publications for their kind permissions for reproducing the illustrations. The revision of this thesis was completed when I was a visiting scholar from January to May 1999 in the department of Anthropology, the Chinese University of Hong Kong. I would also like to thank Professor Tan Chee Beng, the head of the department, and other academic staff for their support during my stay there.

Abstract

The origins of agriculture in China is one of the most poorly understood questions in prehistory, primarily due to a lack of data in archaeology and botany, and a lack of co-operation among archaeologists, botanists and agronomists. However, a considerable database has been accumulated since the first rice remain was found in Central China by Anderson in the 1920s. In addition, several important discoveries have been made since the 1970s in the Yellow and the Yangzi River valleys, providing important information on the origins of agriculture in China.

This thesis is an attempt to address the question of agricultural origin in China by summarising, analysing and integrating data from various disciplines, and by theoretical analysis and comparative studies. The database includes archaeological data from the terminal Pleistocene to the early Holocene found in the Yellow and Yangzi River valleys, botanical data on wild millet and wild rice, data on palaeoclimates and palaeoenvironments, and comparative ethnographic data from Australia and North America.

It is argued that there is evidence of cultural continuity in both the Yellow and the Yangzi River valleys from the terminal Pleistocene to the early Holocene, as indicated by lithic implements. The archaeological discoveries also illustrate that wild cereal exploitation was practised prior to the emergence of agriculture, although this is better documented in the Yangzi valley than in the Yellow valley. Agriculture in China emerged from the exploitation of wild millet and wild rice in the Yellow and Yangzi valleys respectively.

It is also argued that the emergence of agriculture in China was a development with various causes. The dry and cold palaeoclimate during the terminal Pleistocene caused significant floral and faunal change, forcing prehistoric people to search for new sources of food. Intensive cereal exploitation might have resulted in a relatively sedentary way of life for foragers, who thereby increased their population size. The imbalance of human and natural resources triggered attempts to increase the productivity of food, such as reseeding wild cereals. This eventually led to the emergence of agriculture.

There are also other important factors that might have accounted for the emergence of agriculture. Pottery in China seems to be present before the emergence of agriculture. It is argued that the invention of pottery was triggered by the need to cook cereal grains by boiling. Once pottery was invented, it would have encouraged more intensive cereal exploitation. Consequently, cereals became staple foods, and demands for increased production eventually triggered cereal cultivation and domestication.

CONTENTS.

List of Tables.

List of Figures.

PART I. INTRODUCTION.

Chapter 1. The Background to this Study.

The origin of agriculture has been one of the most popular topics in world archaeology since the beginning of this century, and there are numerous propositions, hypotheses and theoretical models created by scholars over the world. Given the length limit of this thesis, a comprehensive literature review on this topic is not feasible. However, the majority of these theoretical works are well known, particularly in the English literature. Therefore, a very brief outline of these works is considered sufficient to refresh some points of argument. On the other hand, works by Chinese scholars are much less known to the western academic world. To compensate this imbalance, the major proportion of this literature review will be on the Chinese literature.

1 A. Theories and models on the origin of agriculture in the English literature.

Research on the origin of agriculture was first conducted by botanists, such as De Candolle, at the end of the last century. Generally speaking, they used botanical, particularly taxonomic methods to distinguish between cultivated and wild plants, and to identify the native land of these plants, in order to determine the centres of origin of cultivated plants. The best representative of these botanists, was N. I. Vavilov, whose main work was carried out from the 1920s to the 1930s (Vavilov 1926). Vavilov established several geographical centres of origin of important cultivated plants by using taxonomic-geographical methods. He claimed that Southwestern and Southeastern Asia, the Mediterranean area, Abyssinia and Egypt, and the mountain areas of South America and Mexico were 'the five basic centres with respect to the most important field crops, vegetables and garden plants' (Vavilov 1926:126).

Another important scholar on this issue was V. Gordon Childe. He was 'responsible for bringing a specific focus to the Levant, ... as a likely area for the emergence of agriculture' (Henry 1989: 6). Based upon data from the Levant he created his 'oasis' hypothesis: at the end of Pleistocene, the deterioration of palaeoclimate forced human, animals and plants to congregate into 'oasis' zones, where human beings exploited plants and animals, and this exploitation eventually led to the origin of agriculture (Childe 1951). He claimed that the origin of agriculture was an abrupt event, and referred it as the 'Neolithic Revolution' (Childe 1951), a concept which is still under debate today.

There have been, of course, numerous scholars from various disciplines who have made great contributions to the origin of agriculture debate. We know that Robert Braidwood is the first scholar to bring scientists from various disciplines into the study of the origin of agriculture. His multi-disciplinary approach has been appraised and followed worldwide. Today our knowledge on this topic has significantly advanced. It now seems clear that agriculture, one of the greatest adaptations and inventions made by human beings, independently occurred in both the Old World and the New World; in both the northern and the southern hemispheres. In the Old World, the Fertile Crescent in Southwest Asia, China in East Asia, and arguably the Sahara Savannah in Africa, are now considered as centres of the origin of agriculture, each with its own domesticated plants and animals (Smith 1995). In the New World, agricultural centres have been identified in Middle and South America, as well as in eastern North America (Smith 1995). It is also claimed that evidence of indigenous agriculture is found in highland Papua New Guinea, represented by tuber cultivation (Golson 1989), although this is still a question under debate.

For the last one and a half centuries scholars have been attempting to provide answers to the questions of **where, when, why** and **how** agriculture originated. To date, the answer for the question of **where** seems quite clear as several centres have been identified. The question of **when** is also less puzzling now, mainly due to the invention and application of various absolute dating techniques, especially AMS C^{14} dating. We now know that plant domestication occurred by 10,000 BP in the Fertile Crescent, by 9000-8000 BP in China and probably in New Guinea, and perhaps around 5000 BP elsewhere (Smith 1995). Present data seem to divide these agricultural centres into two chronological groups: the first one within the temporal range of 10,000-8000 BP, consisting of the Fertile Crescent, China and perhaps New Guinea; and the second one at around 5000 BP, formed by centres in America, probably Africa (Smith 1995). Given the argument that they are all independent centres for the origin of agriculture, why should there exist this apparent chronological patterning, with the second phase 4000 to 5000 years later than the first? What implication(s) can this division reveal? This question is not clear yet.

As regarding the **how** question of agricultural origination, our knowledge is more abundant in some areas than in others. In the Fertile Crescent, where research has been undertaken since the last century, cultural sequences from hunting and gathering to cultivating have been built up; numerous theoretical models have been created based upon rich archaeological data (Gebauer and Price 1992, Price and Gebauer 1995). The significance of this great 'Neolithic Revolution' in the Middle East and its expansion to Europe has been discussed by many scholars (e.g. Bar-Yosef and Belfer-Cohen 1989; Braidwood 1975; Smith 1995; Whittle 1985). Early agriculture in the Middle East was characterised by the domestication of einkorn, emmer, barley, goat and sheep, and was the base for the development of civilisation in the western Old World (Braidwood 1975: 95).

In the New World, archaeological plant remains have been discovered since the end of the last century, and extensive research has been carried out since the 1960s (Smith 1995). Today, the transitions from foraging to farming in American centres are quite well documented. It is clear that maize, beans and squash were first cultivated in Middle and South America, while goosefoot (*Chenopodium berlandieri)* and marsh elder were first cultivated in the Eastern woodlands of North America (Smith 1995).

The transitions in Africa, New Guinea and China are less clear, particularly in China, the subject of this thesis. Although it is certain that millets (green foxtail and broomcorn millet) and rice were first domesticated in the Yellow and Yangzi Rivers respectively, little is known about **how** and **when** this domestication occurred. We now know that advanced agricultural societies emerged by 8400 BP in both the Yellow and the Yangzi River valleys, but the initial stages of agriculture have not been discovered so

far. In addition, few theoretical analyses have been undertaken; and much less international attention has been given to this region.

A great proportion of academic attention, however, has been focused on the question of **why** agriculture occurred. Numerous models and propositions have been created on this issue, the majority of them founded on data from the Fertile Crescent and the Americas.

Gebauer and Price listed 37 different explanations in the origins of agriculture in a book they edited in 1992 (Gebauer and Price 1992:2). This alphabetic list is probably the most comprehensive brief list of the ideas on this topic. Very briefly and generally, the models and propositions for the origin of agriculture in the English literature can be classified into a few groups.

The first group has been mainly ecological and environmentally oriented. This group argues that various natural and/or ecological conditions were the cause and/or foundation of early agriculture. Well-known hypotheses in this group include the famous 'oasis' hypothesis proposed by Childe (1951), the current 'climate change' hypotheses such as those by Byrne (1987) and Wright (1993). Although the arguments of these scholars vary in details, they more or less argue that the climatic changes which occurred at the end of Pleistocene, and the associated floral and faunal changes, were vital causes for the origins of agriculture.

The second group has been demography-oriented. This group focuses on the equilibrium between humans and natural resources. The general argument of this group is that once population increased to such a level that local resources were not sufficient, or natural resources decreased significantly, then agriculture was invented in order to provide sufficient food for the increased population. This opinion is represented by Binford (1968), Flannery (1973) and Cohen (1977); but in modern theory it always overlaps with the environmental change hypotheses listed above, since it can often be environmental changes which trigger a population imbalance.

However, the arguments on environmental, climatic changes and population pressure as causal factors of the origins of agriculture differ. Recently Price and Gebauer have argued that 'agriculture is adopted in areas of relatively abundant resources' (Price and Gebauer 1995:7). They also argue that increased population 'does not appear to be a major force in changing subsistence strategies' (Price and Gebauer 1995:7). These arguments are quite different from previous propositions as mentioned above.

The third group, represented by David Rindos, has focused on an evolutionary perspective (Rindos 1984). Rindos proposes a Darwinian selectionist model to explain the origin of agriculture, arguing that humans unconsciously selected suitable plants for themselves, and agriculture is the outgrowth of an evolutionary potential 'which may develop whenever an animal consistently feeds upon any set of food plants' (Rindos 1980: 751). In this view, cultural evolution is an unconscious outcome of natural selection.

In contrast to Rindos' natural evolutionary view, another group has emphasised human social consciousness in the origin of agriculture. For example, Caldwell (1977) argues that cultural complexity in terms of toolkits, storage facilities, dwellings and systematic burial practices emerged prior to the emergence of agriculture. He claims that agriculture was a natural outgrowth of cultural evolution (Caldwell 1977). Bender emphasised the effects of social complexity in the origin of agriculture (Bender 1978). This group claims that the accumulation of surplus food for transformation into more valued items, or competitive feasting aspects of Big Man rivalry, were major impetuses for the origin of agriculture (Gebauer and Price 1992).

There are, of course, other scholars who have held various views on the origin of agriculture. Many argue that the emergence of agriculture was not caused by one simple cause, nor can one model explain this event in all places (e.g. Flannery 1973; Smith 1995). Although it is argued that a general explanation on the origins of agriculture in a global context should be sought (Price and Gebauer 1995: 5), it is also argued that a better approach is to establish propositions firmly based upon data from one specific centre, instead of trying to find a general model for all the centres (Smith 1995).

It is not the intention of this thesis to comment further on the above propositions. But one thing seems clear. After more than one hundred years of research and the accumulation of a vast database from various areas of the world, the origin of agriculture still elude easy explanation. We now know that agriculture emerged in both subtropical and tropical areas; in dry lands such as the Chinese loess, or in wet lands such as the Yangzi valley. In the Levant this event occurred just at the beginning of the Holocene, while in America it happened four to five thousand years later. In these different agricultural centres, there were hunter-gatherers who used different toolkits, lived in different environments, exploited different resources, and finally, cultivated different local plants and domesticated different animal species. On the other hand, we see contemporary foragers who exploit similar wild cereals as the pre-agriculture foragers, but who have not become farmers. For example, some Australian Aborigines collect wild broomcorn millet (*Panicum* sp.) and wild rice (*Oryza* spp) (Jones and Meehan 1989), just as the prehistoric foragers in the Yellow and Yangzi River valleys did or might have done, but the latter developed agriculture while those in Australia remain collectors. Given all these variations, a single reason for the emergence of agriculture, or a single model for the global context, would probably be oversimplified. The intention of this thesis is to focus on the database from one area - China, and build an argument for the origin of agriculture here on this local database.

1 B. A brief literature review on the origin of agriculture in China.

The origin of agriculture in China has been a topic for Chinese scholars in various disciplines since the early 1950s, and the achievements of these study are valuable. In addition, several important discoveries have been made since the 1980s in the Yellow and Yangzi valleys, providing important data for the emergence of agriculture. But, probably due to language barriers, research carried out in China is much less well known in the western academic world. On the other hand, models and theoretical analyses created by western scholars have gained little attention among Chinese scholars. This lack of communication also contributes to the insufficiency of study on prehistoric agriculture in China.

In the Chinese literature, the origin of agriculture was a topic of discussion long before the emergence of modern archaeology. In historical documents such as Guanzi (written in 645 BC) and Mengzi (written in the fourth century BC), the proposition of 'Big men' who taught people to cultivate cereals was raised (cited by Liang and Qi 1980). What is more, historic documents also mentioned the reason(s) for the origin of agriculture, such as a lack of food, or an unsuitable environment for people to live on natural resources, so they had to cultivate cereals instead (Xu 1994).
At the beginning of the 20th century, modern archaeology

was introduced into China by a few western scholars. In 1921, J. G. Anderson discovered the famous 'Yangshao' culture in central China. Among the potsherds he discovered a grain impression which was later identified as a rice husk (Anderson 1929). This was the first direct evidence for prehistoric agriculture in China to be discovered by archaeological excavation.

However, this discovery did not gain attention among archaeologists, and the stratum from which the husk came was not clear. In fact, Anderson's description of the stratigraphy in Yangshao village and his conclusion were both incorrect, as proved by numerous discoveries since. It seems that Anderson failed to distinguish the differences between Yangshao, Longshan and an Eastern Zhou assemblage, all unearthed in the Yangshao site (Huang 1982). This inaccuracy reduced the academic value of the discovered rice husk, which was viewed by other scholars as probably mixed into the Neolithic deposit (Huang 1982).

In 1926, an agronomist called Ding Ying found wild rice (*Oryza rufipogon*) stands in a suburb of Guangzhou City. Based upon his botanical study, and a cross-breeding program of wild and domesticated rice (*O. sativa*), he claimed that wild rice was a hydrophilic plant, and that the progenitor of domesticated rice (*O. sativa*) was the wild rice (*O. rufipogon*) found in South China (Ding 1933). From 1933 to 1964 Ding wrote several articles on the origin of rice cultivation. Based on historical records, archaeological evidence, botanical characteristics and the geological distribution of wild rice, as well as linguistic analysis of words for wild rice, he consistently claimed that the hearth of rice cultivation must have been in South China (Ding 1957). He is undoubtedly the pioneer on the origin of rice cultivation in China.

Since the early 1950s evidence for prehistoric agriculture began to accumulate as the result of numerous archaeological discoveries. Consequently, prehistoric agriculture caught the attention of archaeologists, historians, agronomists and botanists. But archaeological data were mainly found in Central China, and the evidence only came from the middle Neolithic assemblages, such as the Yangshao and Dawenkou cultures. Discussion on the origin of agriculture in the 1960s and early 1970s was therefore focused on Central China, and was basically a synthesis from archaeological discoveries and historical documents. A representative of studies was the book written by Ho (1969), in which, he summarised historical documents and archaeological data, and claimed that the loess plain was the hearth of agriculture in China, characterised by millet cultivation (Ho 1969).

The later 1970s witnessed the beginning of a new era of studying the origin of agriculture in China, mainly stimulated by the discoveries of the Cishan and Peiligang cultures in Central China, and the Hemudu culture in the lower valley of the Yangzi River. The former two were dated to the eighth millennium BP, and the latter at 7000 BP. These discoveries provided solid evidence for sedentism, animal domestication, as well as millet and rice cultivation in Central China and in the basin of the Yangzi River respectively. The large amount of foxtail millet found in Cishan, and the thick deposit of rice in Hemudu illustrated that the agriculture practised by the occupants was far from an initial stage. These discoveries indicated that cereal cultivation must have begun well before the 8th millennium BP. Since then, many archaeologists have attempted to discover evidence for early agriculture in China. Their effort seems to have been rewarded from the early 1990s.

For the last 20 years many scholars have made their contributions towards this topic, including some western and Japanese scholars. In summary, studies have been concentrated on the following issues:

(1) Hearths for the origin of agriculture.

Discussion on this issue focuses on the centres for millet and rice domestication. For millet domestication, it is generally agreed that the hearth is in Central China. But for rice domestication, opinions vary. In summary, there are five schools on this question.

The first school is represented by Yan Wenming of Beijing University. In the late 1980s he proposed that rice domestication originated in the Yangzi valley or/and South China (the Lingnan area) (Yan 1989). This argument is based upon the famous Hemudu culture in the lower Yangzi valley, and the distribution of wild rice in South China. But Yan modified his opinion in 1992 due to the discovery of rice remains in the Pengtoushan culture in the middle Yangzi valley in 1989, which was dated earlier than the Hemudu culture. He now views the middle and lower valley of the Yangzi River as a whole centre of rice domestication, and suggests that South China is not likely to be the centre of rice cultivation due to abundant natural resources and easy access to foodstuffs (Yan 1992a).

The second school argues that rice domestication originated in the middle valley of the Yangzi River. This argument is based upon recent discoveries of rice remains in Pengtoushan and Bashidang, the middle Yangzi valley, dated to the ninth and eighth millennia BP (e.g. Higham 1995; Xiang 1993), and the phytoliths analysis on the Daotonghuan assemblage (Zhao 1998). Recent excavation in Hunan province yielded more rice remains, said to be earlier than the Pengtoushan ones (Zhang and Pei 1997). These new data seem to further strengthen the argument of this school. It is also proposed that rice cultivation expanded from here southwards, and reached Southeast Asia by 3000 years BC (Bellwood 1995, 1996; Higham 1995).

The third school on this topic claims that rice domestication must have been invented first in Southwest China, mainly in Yunnan province; then it diffused along the Yangzi valley to Hemudu and to Southeast Asia (Xu 1998). This opinion is supported not only by some Chinese archaeologists, linguists and botanists, but also by a few Japanese scholars (Chen 1989; Xu 1998). The major argument is based on the widest diversity and dense distribution of wild rice in Yunnan, as well as linguistic analysis (e.g. You 1983).

In the second International Academic Conference on Agricultural Archaeology (or IACAA), another comprise on the origin of rice cultivation has been proposed. Based upon rice remains found in both the middle Yangzi valley and the upper Huai River valley, the latter being in between the Yellow and the Yangzi valleys, it is argued that rice cultivation originated in this area and expanded to other areas (Wang *et al.* 1998).

While the above opinions all agree that rice domestication originated in the Yangzi River and/or further south, there is a hypothesis that rice domestication might have originated from the Yellow valley. Only one scholar holds this view (Chen 1989). This hypothesis is mainly based on interpretation of the pronunciation of certain words in ancient Chinese, and seems to have very little support from the data of archaeology and other disciplines.

The last school is a compromise, arguing that rice could have been domesticated anywhere provided that three basic conditions were met: the existence of wild rice, a suitable environment, and the technology for cultivation (Zhang

1994). This proposition is stimulated by the recent discovery of cultivated rice in Jiahu, a site of the Peiligang culture in the Huai River region, an area north of the Yangzi river, dated to the eighth and seventh millennia BP (Zhang 1994). But Zhang seems to have changed his mind and is arguing that the upper Huai valley is also part of a centre for the origin of rice cultivation (Wang *et al.* 1998).

All these propositions have their strengths and weaknesses. Archaeological data seem to suggest the Yangzi valley as a centre for rice domestication, but the modern distributions of wild rice and the linguistic data seem to favour South and Southwest China as centres. However, there are still many problems remaining on this topic. For example, the arguments on the origin of rice cultivation seem to purely based upon archaeological discoveries, but pay little attention to other factors, such as the local palaeoclimates from the late Pleistocene to the early Holocene, and whether wild rice could have been survived and cultivated in a cold and dry climate.

(2) The ecology of the origin of agriculture.

A few scholars have made propositions on this topic. Tong (1989) has argued that agriculture was likely to originate only in areas where natural resources were not as sufficient as in other ecological zones (i.e. forest areas or lacustrine areas), where hunting and fishing were not able to provide sufficient food (Tong 1989).

Yan Wenming has a similar point of view. He reasons that agriculture could not have been invented in South China due to the richness of natural resources and easy access to food. But in the valley of the Yangzi River, where natural resources were less abundant, where the winter was cold and storable food stuffs were needed, then once people discovered the value of wild rice, they would have started to cultivate this cereal (Yan 1992a).

This hypothesis, which is similar to the 'Marginal Zone' theory proposed by Flannery (1973), is also held by Hung Qixun. He argues that all the species of the Gramineae family are grasses which are not able to grow well in very fertile soil as trees will dominate. It is only in unfertile soil where trees can not grow, that those grasses can dominate and grow into large stands. And as domesticated plants are all grasses, he infers that it must have been in those area where the soil was not very fertile, where wild grasses could grow well, that agriculture originated (Huang 1983).

Another argument on this topic is that agriculture originated in hilly or mountain areas (Li and Lu 1981; Tao 1994). This proposition is mainly based upon ethnographic data. There are quite a few ethnic groups in Southwest China who live in hilly or mountain areas covered by forest, and were still practising the 'slash and burn' cultivation technique up to the early 1950s (Song 1986). The large amount of ash as the outcome of forest clearance by fire provided plant nutrients, and people planted cereals on the land after fire. According to ethnographic data, at an early stage when land was adequate, the 'slash and burn' method was the most practicable one as it required only simple technology, and people could easily shift to other land for cultivation as they needed; but as deforestation progressed, virgin land became less and less, and people had to move down from mountain to river valleys to utilise the land along rivers (Tao 1994). Eventually, the 'slash and burn' method could not be practised due to deforestation, as no trees were left to burn for fertilising ash. The cultivation was then carried out by using hoes and other tools, to deeply dig into the soil, and other fertilisers had to be used, such as ash of grasses (Li and Lu 1981). According to this hypothesis, the initial stage of cultivation must have been 'slash and burn', and agriculture originated from hilly or mountain areas (Li and Lu 1981; Tao 1994).

While all the above-mentioned propositions have their points, it seems that there are many weaknesses and biases in these arguments. First, the richness of natural resources was not necessarily a hindrance to the origin of agriculture. Comparing to South China, the Yellow and the Yangzi valleys might not have been an ecological zone with very abundant natural resources, particularly during the Last Glacial Maximum. On the other hand, it was in the Fertile Crescent of the Middle East, where natural resources were abundant, that agriculture originated. Price and Gebauer propose that agriculture originated from 'areas of relatively abundant resources' (Price and Gebauer 1995:7). The question here is how, and based on what criteria that an area can be defined as with 'relatively abundant resources' (for more detailed discussion see Lu 1999 in processing)? This is an important point to be solved. Second, it is also questionable whether the hilly-mountain regions were ever the centres for the emergence of agriculture. While it is possible for millets to survive in mountains, it is hard to image how rice could be initially domesticated in dry and hilly areas, as botanists have demonstrated that this cereal was originally a hydrophilic plant (e.g. Ding 1933; White 1995).

(3) The transition from foraging to farming and the impetus for the origin of agriculture.

The transition from foraging to farming in China is still not clear, mainly due to the inadequacy of archaeological data. It is proposed that there was a period of extensive hunting and gathering prior to the origin of agriculture based upon archaeological data found in Xiachuan, an assemblage of 18,000 to 13,000 BP in the middle Yellow valley (Shi 1989). The germ of the origin of agriculture has been suggested to be rooted in the Xiachuan assemblage, in which proto-axes and grinding slabs were discovered (Fan 1993). But how this transition occurred has not been discussed in detail.

For the impetus of the origin of agriculture, Yan mentioned that the need for food in winter would have driven prehistoric people to cultivate rice, as this cereal can be stored for a relatively long period of time (Yan 1992a).

Another proposition is that the origin of agriculture was due to the climatic change from the late Pleistocene to the early Holocene (Xu 1994; Zhao 1998). Xu infers that the Last Glacial Maximum caused the extinction of many animals and plants, so humans had to turn to wild cereals as new sources of food. In this way they gained knowledge about cereals: seeds could be stored in the winter; they could be reproduced; etc. After 12,000 BP, as the climate became warmer, food was easier to find and there was a surplus of stored cereals each winter from the previous year. This surplus could be used in the experiment of cultivation, thus came the beginning of agriculture (Xu 1994). But this proposition failed to explain why people would have experimented with cereal cultivation if they already had access to abundant food.

A similar view is also held by Liu (1991). After analysing records in ancient historical documents, Liu proposes that in the Last Glacial Maximum, as resources of animals and plants were reduced, and migration was limited by the possibility of conflict with other groups, prehistoric people were forced to search for new resources of food. Thus they gained a knowledge of wild cereals, and gradually transformed their economy from hunting and gathering to agriculture (Liu 1991).

In fact, discussion on this issue is not easy to take further as archaeological data for the transition in China are far from sufficient, particularly for the period from 18,000 to around

10,000 BP, both in the valleys of the Yellow and the Yangzi Rivers. And without detailed data on the transition, it is even harder to discuss the issue of the impetus for the origin of agriculture, which, even in the Middle East with more adequate archaeological data, has not reached a conclusion agreed by everyone. However, achievement can still be made, step by step, by carefully analysing and comparing the ecological, archaeological and botanical data as well as data from other disciplines, by rebuilding the framework of cultural sequence in each region, also by comparing with studies in the Middle East. This is going to be one of the objectives of this thesis.

(4) The earliest cultigen(s).

In Central China the question of the earliest cultigen seems never to have been debated. As millets were found in the early Neolithic cultures, many scholars agree that the earliest cultivated plants must have been foxtail (*Setaria italica*) and broomcorn millets (*Panicum miliacecum*) (e.g. IA CASS 1984). Most scholars also agree that these millets were indigenously domesticated in Central China (e.g. de Wet *et al.* 1979; Smith 1995). A counter-argument is from Fogg, who proposes that foxtail millet might have been domesticated by Hoabinhian groups in Southeast Asia and dispersed to China (Fogg 1983).

South of the Yellow River basin there is uncertainty about the earliest cultigens. Archaeological data have documented that rice was cultivated at least in the ninth millennium BP (the Pengtoushan culture) in the middle Yangzi valley. But was rice the earliest cultigen in the whole region south of the Yellow River? Tong (1989) proposed that the initial stage of cultivation could have been one of vegetative propagation, and the cultivated plants could have been aroids and yams. This argument is based upon ethnographic data in Southwest and South China, where people still cultivated yams until the early 1950s. Historical documents were cited as evidence of yam cultivation in the fourth century (Tong 1989). After comparing prehistory in South China and Southeast Asia, he argued that there must have been a yam-cultivating period prior to the rice-cultivation stage. Archaeological data can neither support nor deny the existence of such a period, so this proposition remains a question open for debate.

Based upon archaeological discoveries and ethnographic data, recently Pei proposes that the early cultigens in the middle Yangzi valley might have been more than just rice (Pei 1998b). Pei suggests that other plants such as water caltrop and Chinese lotus might have also been cultivated in the vicinity of the Bashidang some 8000 BP as supplements of rice (Pei 1998b).

(5) The origin, evolution and sub-species history of domesticated rice.

In 1926, Ding produced a hybridised variety of wild and domesticated rice, and claimed that there was a close genetic relationship between them, particularly for the *japonica* sub-species of *O. sativa*. He also argued that the perennial wild rice (*Oryza rufipogon*) was the ancestor of the domesticated species (*O. sativa*) (Ding 1933, 1957). This conclusion has since been supported by a comparative isozyme study of wild and domesticated rice in South China (Wu and Chen 1986).

Since then, many scholars have studied the botanical relationship between wild and domesticated rice in Asia (e.g. Chang 1989; Oka 1988; Sun *et al.* 1998; *Wang et al.* 1998). It is now argued that wild rice (*O. rufipogon*) is an annual-perennial complex (Oka 1988). Whether the annual or the perennial type of wild rice was the immediate progenitor of domesticated rice is still a question for debate (for a review, see Ahn 1993). But many botanists tend to perceive the perennial one as the progenitor of the domesticated rice, with the annual one normally being a weed (i.e. see Pang *et al.* 1995; The Institute of Agriculture 1986).

In 1984 a national survey was carried out by scientists in more than ten provinces in China in order to investigate the current distribution of wild rice. During this survey, various species of wild rice were discovered, including annual and perennial types. The result shows that the centre of wild rice distribution today is in South and Southwest China, and only a few small stands were found in the middle and lower valley of the Yangzi River [Fig. 6 (8)] (The National Survey Group of Wild Rice 1984). As it is in the valley of the Yangzi River where the earliest rice cultivation has been discovered, the question arises as whether there was more wild rice seven thousand years ago in the Yangzi River valley due to the warmer climate of the early Holocene.

Further studies on the two sub-species of domesticated rice: *Oryza sativa japonica* (or *O. sativa sinica*) and *Oryza sativa indica,* has also been carried out by many scholars. These studies include the botanical characteristics (Liang 1984), which species is genetically closer to wild rice (Pang *et al.* 1995), whether there is a trend of differentiation towards *japonica* and *indica* within the wild rice (Sun *et al.* 1998) etc. In the 1980s, over 90 sub-species of wild rice from mainland South China and Hainan island were sown and cultivated in Guangzhou by agronomists, and their life cycles in different environments have been studied (Qi *et al.* 1984).

In the 1980s, phytolith analysis was introduced into China, and this method is now applied to study of the origin of agriculture. The recent discoveries in Yuchan Cave in Hunan province, and the Sino-American project carried out in Jiangxi province, have also benefited from phytolith research (Zhao 1998). It is now suspected that rice cultivation might have started as early as 10,000 BP (Peng 1996; Yuan 1996).

While these results are informative, there are still questions. Was the distribution of wild rice the same at some 7,000 years ago as it is now? Further, how long was it required to complete the evolution from wild to domesticated species? According to cultivation experiments using wild wheat and barley from the Middle East, it would have probably taken less than 200 years to complete the process of domestication (Hillman and Davies 1992). Was the evolution from wild to domesticated rice completed within a similar period of time? If the approximate speed of rice domestication can be estimated, then the question of **when** rice was domesticated may be better understood.

(6) Correlationship between sedentism, pottery, rice cultivation and animal husbandry.

Discussion on the correlation between sedentism, pottery, rice cultivation and animal husbandry is far from adequate. Profound evidence for sedentism has been found in many Neolithic sites in China, i.e. the large villages and storage pits found in the Cishan, Peiligang, Pengtoushan and Hemudu cultures; but the importance of sedentism and its significance for the origin of agriculture seem to have caught little attention. Pottery shapes and decoration have been the traditional foci of Neolithic study, for the purpose of typological analysis. It has been proposed that a presence of pottery should indicate an existence of agriculture, since rice and millets were normally cooked by boiling (Feng 1992; Xu 1994). But current archaeological discoveries in China seem to suggest that pottery was invented prior to the

emergence of agriculture, as in Jomon Japan.

Animal husbandry and its relationship to agriculture is only mentioned in a few articles and discussion on this topic has also been quite vague, although identification of animal remains have been carried out wherever they were found. The existence of domesticated animal(s) has mainly been identified by anatomical and statistical analyses of ages at death. For example, the majority of the pigs found at Zengpiyan in South China were killed relatively young, so it is suggested that they must have been husbanded (Li and Han 1978). Remains of domesticated pig, dog and possibly domesticated chicken were also found at Cishan in Central China (Zhou 1981), associated with a large amount of foxtail millet. But in Zengpiyan there is no solid evidence for cereal cultivation. It has been argued that pig domestication was not possible without cereal cultivation, based upon comparison with the prehistory of the Middle East and current agricultural practice in China; thus the practice of rice cultivation in Zengpiyan is assumed (Li 1990). Whether this assumption is valid, however, remains to be investigated.

Obviously, there are many problems regarding the relationship between sedentism, animal husbandry and cereal domestication in China. The agricultural societies, such as the large Cishan and Peiligang villages in the Yellow valley and the Pengtoushan culture in the Yangzi valley, were undoubtedly sedentary. But what happened prior to these advanced agricultural societies? In the Yellow valley so far only one waterlogged site called Nanzhuangtou has been found in Hebei province, dated to about 10,000 BP (Yuan *et al.* 1991). Unfortunately, archaeological data from this site are very limited, and cannot provide information on the life style of the occupants. In the Yangzi valley, a few caves were excavated in the 1990s in Hunan and Jiangxi provinces. But the sedentary pattern for these cave occupants is not clear yet (MacNeish and Libby 1995; Peng 1996; Yuan 1996).

Animal husbandry is another problem. The identification of domesticated animals based upon smaller body size is perceived as not very convincing, because the evidence for size reduction as a result of domestication is not clearly documented (Smith 1995). Therefore, the argument for the presence of domesticated species in archaeological assemblages in China is not very well founded.

(7) Agricultural technology.

The technology for agriculture has drawn the attention of not only archaeologists but also agriculturists. A few books and many articles have been published on this topic, all of which seem to focus on tools discovered in archaeological assemblages, and assumed to have been used in prehistoric agriculture. A study has been carried out on grinding implements found in China in terms of their shapes and functions. It has been proposed that grinding implements could have been used to process not only wild grains, but also nuts, even ochre; that there was a decline in the quantity of grinding implements from early to late Neolithic in central and north China; that typical tools for grain processing from the middle Neolithic onwards must have been the mortar and pestle; that the existence or absence of grinding implements need not imply the existence or absence of agriculture (Wu 1986).

Other tools for cultivation have also been discussed. In Central China, the Neolithic ground axes are suggested as tools for forest clearance, and stone sickles are suggested to have been used for harvesting (Wu 1989). In South China, perforated pebbles are proposed to be 'weight' stones for digging sticks (Xiang 1991). The harvesting tool for rice is not clear, but knives made from shells were probably used as cutting tools (Xiang 1991). It has been suggested that bone and wooden tools might have been used in the initial stage of agriculture, as such tools were discovered in Hemudu (Lin 1992) and recently in Bashidang (Pei 1998a). Further, based upon archaeological and ethnographic data, three phases of technological development for prehistoric agriculture have been proposed: first, a 'slash and burn' phase; then a 'cultivating with hoe' phase, and finally the advanced 'cultivating with hoe' or 'plough' phase (Zhang 1984).

Despite the above-mentioned discussions, one important aspect of agricultural technology, particularly for the emergence of cultivation, seems to have been ignored. Use-ware and starch residue analyses on stone implements presumably used for cereal exploitation and cultivation, and wild cereal harvesting and cultivation, have not been conducted so far in China. The lack of such studies is problematic.

There are currently three groups of archaeological assemblages regarding the origin of agriculture in China. Group one consists of the archaeological assemblages with tools that can be proposed as agricultural tools, and with archaeobotanical remains. The question of whether they were agricultural societies seems positive for this group, yet the actual agricultural techniques remain unclear due to the lack of use-wear and residue analysis on tools and other items (for example see Wu 1989). The second group consists of those with cereal remains but without specialised tools for agriculture, such as the Pengtoushan assemblage. This group is often considered as agricultural societies, but again, the question of agricultural technology remains completely unanswered for the same reason as mentioned above. The last group includes those with tools that can be proposed as agricultural tools, but **without** archaeobotanical remains, such as the Chahai and Xinglongwa assemblages found in northeast China (Xin 1995). For the last group, whether they were agricultural societies are very sceptical. This scepticism might have been at least partially addressed by use-wear and residue analyses.

(8) The significance of agriculture.

The significance of agriculture for Chinese civilization has been discussed in detail. It is without much doubt that the two centres of agriculture in China, one for rice and one for the millets, were the centres of two ancient cultural clusters created by two groups of people. The question is then, how did these cultures contribute to the later development of civilization in ancient China? (Editors of the Archaeology Journal 1989; Yan 1992b). It is commonly agreed that the civilization of Central China was derived mainly from the zone of millet cultivation. The earliest city in the Yangzi valley is claimed to be the Chengtoushan site in Hunan province, dated to around 6000 BP (Jiang 1997b). But no evidence of writing or/and the existence of a state has been reported. According to historical documents, the rice cultivators in the Yangzi River were under strong cultural influence from Central China and learnt writing from there.

The spread of rice and millet cultivation beyond the present boundaries of China has also been discussed by many scholars. There seems little doubt that millet cultivation spread to Korea by 5000 BP (Nelson 1993), although the rout of this cultural diffusion with or without migration is still a question (Li 1982). Based on archaeological and linguistic evidence, it has been also argued that rice cultivation was dispersed from South China to Southeast

Asia as a result of agricultural expansion from 5000 BP onwards (e.g. Bellwood 1995, 1996; Higham 1995).

However, little attention has been paid to the consequences of the emergence of agriculture in China. Did agriculture cause deforestation and/or other environmental changes in prehistoric China? Was there any immediate social change(s) after the emergence of agriculture? Was there any improvement or deterioration of nutrition in farming societies? etc. These questions still await further study.

In summary, considerable achievements on the origin of agriculture in China have been made in the last fifty years. Apart from the above-mentioned results, various scientific methods have also been applied. Phytolith studies have been used to identify cultigens in archaeological remains (Zhang 1994; Zhao 1998). Pollen analysis has been used to reconstruct palaeoenvironments. Flotation is now also used in excavations. Studies on the botanical characteristics of wild rice have been carried out by botanists and agronomists. All these researches provide important data for the origin of agriculture in China. But, despite all these, questions still remain.

1C. Significant questions on the origin of agriculture in China and the objectives of this thesis.

The first question is **how** did agriculture originate? In the Middle East, there was a stage of wild cereal collection prior to the emergence of agriculture. Was there a similar stage in China? Theoretically, there must have been. But this needs to be confirmed by archaeological and botanical evidence.

According to research conducted in the Middle East, sedentism was a vital factor in the origin of agriculture (Miller 1992). What is the significance of sedentism in China? Did sedentism happen spontaneously, prior to, or after the emergence of agriculture in China?

The above questions are in fact part of the question of **how** hunters and gatherers transformed into cultivators in China. Though archaeological data for the transitional period from the end of the Pleistocene to the early Holocene are far from sufficient, there are still clues in terms of cultural continuity as indicated by lithic traditions and other cultural components. It is possible to shed light on this transition by analysing questions of cultural continuity. This will be one of the major tasks of this thesis.

The second question is **where** did agriculture originate? Although archaeological data seem to indicate that this invention most likely occurred in the valleys of the Yellow and Yangzi Rivers, the ecological condition(s) under which agriculture emerged are still not clear. It has been argued that agriculture could not have been the invention of starving people living in an area with poor natural resources, as they would not have been able to afford the risk of experimenting with cultivation (Gebauer and Price 1992). But Yan argues that in regions of rich natural resources such as South China, there would not have been any impetus for people to make the extra effort to exploit new ways of producing (Yan 1992a). Whether agriculture originated in regions with rich or stressed natural resources is still a question needing resolution.

Although archaeological data to date are not adequate to supply a satisfactory answer for the **where** question, archaeobotanical remains of wild and domesticated cereal can provide vital data. By integrating cereal remains, toolkits, and other archaeological data, it is possibly to reconstruct the subsistence strategies of various groups in different regions, and to trace the likely hearth of the origin of agriculture.

Further, there is a question of **why** agriculture was ever developed. In the 1960s and 1970s it was argued, based upon ethnographic studies, that modern hunters and gatherers were able to live comfortably in even very marginal areas (i.e. the Bushman in Africa); that agriculture, in contrast to the general belief, did not lead to a more comfortable life and more leisure time (Cohen 1977). However, more recent ethnographic studies seem to challenge this point of view. The life of hunter-gatherers in Australia was probably not that rosy (Cane 1984; Layton *et al.* 1991), and proto-agriculture practices such as irrigation might have originated from a desire to obtain more food (e.g. Keeley 1992).

In China, the two hearths of agriculture found to date are both landscapes with reasonably abundant natural resources, particularly the Yangzi valley. Then, what made people make the effort to cultivate cereals ? Was population pressure, suggested by Cohen (1977) to be a driving force for the origin of agriculture in the Middle East, also a driving force in China? Or were there other reason(s)? A hypothesis based upon ethnographic and archaeological data, as well as information from other disciplines in China, will be proposed in this thesis.

Finally, although study of the botanical characteristics of wild rice has been undertaken since the 1930s, similar study on the wild millets is virtually non-existent. As demonstrated by research in the Middle East, botanical studies on both wild and domesticated cereals, and experimental studies on the effects of different harvesting methods as part of the domestication process via human selection, are important. Clearly, the lack of similar study on wild progenitors in China is a drawback, which will be partially compensated in this research.

The objective of this thesis is to draw a picture, though a very uncompleted and tentative one, of **how, when** and **where** the cultivation of rice and foxtail millet originated; **how** did the hunters and gatherers of the late Pleistocene transform into farmers of the early Holocene in China. In addition, a hypothesis on the question of **why** agriculture originated will also be proposed. To achieve this objective, the following studies will be undertaken:

- A reconstruction of Chinese palaeoenvironments from the Last Maximum Glacial to the early Holocene as a natural background of the origin of agriculture. Effects of environmental changes, if any, on the emergence of agriculture will be discussed.
- A reconstruction of the cultural sequences in the valleys of the Yellow and Yangzi Rivers.
- Surveys and data collection on the botanical characteristics of both wild and domesticated foxtail millet and rice. An experiment in the harvesting of these cereals is undertaken in order to found a database for analysis of the yield ratio of wild and domesticated cereals, and the efficiency and effects of various harvesting methods which may be considered as factors in the domestication progression.
- An analysis of different stone tool assemblages as representing different cultural groups in the Yellow and the Yangzi valleys from the terminal Pleistocene to the early Holocene .
- A pottery comparison of early Neolithic cultures in China, in order to demonstrate inter-cultural relationships at an early stage of agriculture and the possibility of cultural diffusion.

Based on the above data, some analyses and comparisons will be conducted, focusing on the following topics:

- The characteristics of the archaeological sequence from the terminal Pleistocene into the early Holocene in each region.
- Whether cultural continuity or hiatus is documented.

- If there was cultural continuity, how did the transition occur? What were the characteristics of the transitional period in terms of toolkits, subsistence strategies and settlement pattern? Why did the transition happen? If there was a cultural hiatus or/and discontinuity, were there influences from other cultures? What were the results of these influences? Do we have sufficient data to address these problems?
- What is the best model for a transition to agriculture in China? There are numerous models and hypotheses created by western scholars based upon archaeological data from the Middle East and Mesoamerica on the origin of agriculture. What are the differences and similarities between the Middle East and China in terms of the origin of agriculture? Is there any model which will best fit the situation in China? Or will a new model and/or hypothesis need to be created based on Chinese data?

This thesis will be informative, theoretical and comparative. Recent archaeological data will be revealed and summarised to make these data available to scholars who are not able to access them due to language barriers; the survey and experiments on wild cereals will provide new information. Hypotheses and models created by western scholars will be tested against Chinese data, and propositions made. It is hoped that by this study some light will be shed into the unsolved question of the origin of agriculture in China.

Chapter 2 Terminology and Definitions.

Prior to discussing the origins of agriculture, it is necessary to clarify some key terms which will be frequently used in this thesis.

2A. General definitions.

Agriculture In general, this term covers both plant domestication and animal husbandry. But in this thesis, the term is focused on plant domestication. Animal husbandry in China is not discussed in detail.

Complex. Refers to a group of archaeological assemblages sharing similarities in terms of lithic implements, including identical lithic composition and manufacturing technology. This is primarily a technical term, and a complex might have been produced by one or more groups of people.

Cultivation. It is defined as the process from planting or broadcasting seeds into the ground to harvesting by human beings. The seeds being sown can be either domesticated, or wild. It has been argued that cultivation was the essential step towards domestication, as cultivation will stimulate the mutation of the phenotypes among wild cereals which were beneficial for human exploitation (Hillman and Davies 1992).

Culture This term has been used in various senses. In this thesis, 'culture' refers to a collection of archaeologically observable specific traits, shared among a cluster of archaeological assemblages (Champion 1980).

Domestication An evolutionary process of selecting and modifying favourable genetic forms of some plants by human beings. This process led to the emergence of sub-species which bore botanical characteristics beneficial for human exploitation. Some domesticated plants can only survive under human care.

Farmer This term is reserved for people whose major economic activity is (was) farming. However, hunting, gathering and fishing continued to be supplementary economic activities for farmers. The use of this term does not exclude them.

Forager In this thesis, 'forager' is used to generally refer to people whose primary subsistence strategy was based upon hunting, gathering and fishing, or a combination of any two or three of the above. The special notions of 'forager' and 'collector' as defined by Binford (1980) are generally not applied in this thesis.

2B. Temporal and geographic definitions.

BP and bp. First and foremost, the presentation of 'years before present' needs to be clarified. In this thesis, **BP** refers to calibrated absolute dates before present, or a general statement of the time before present; **bp** refers to an uncalibrated radiocarbon and other absolute date.

In China, radiocarbon dates are often computed using a half-life of 5730 years. All the radiocarbon dates listed in the tables of this thesis are converted to half-life 5568, and calibrated using the CALIB.3 (1993) program designed by Minze Stuiver and Paula J. Reimer of the University of Washington. The results of this calibration may be slightly different from the calibrated dates published by the original laboratories in China.

The time span of this thesis is from 20,000 to about 7,000 BP. The Last Glacial Maximum (LGM) occurred from approximately 18,000 to 15, 000 BP in China (An *et al.* 1990), although the beginning of the coldest climate phase might have started at around 27,000 to 23,000 BP in North and Northeast China (Zhou *et al.* 1984). The LGM had different effects on northern and southern China. After approximately 13,000 BP, the LGM weakened, and the climate gradually became warmer and moister. However, there were climatic fluctuations from 13,000 to 10,000 BP, with sharp deterioration comparable to that of the Younger Dryas in Europe between 12,000 and 10,000 BP. It was at around 10,000 BP the post-glacial epoch of the Holocene in mainland China began.

The early Holocene is set from 10,000 years to 7000 BP in this thesis. Again, this definition is mainly based upon analysis of palaeoclimates and marine transgression and regression. According to pollen analysis, the climate from 10,000 BP onwards was quite warm and moist, and even warmer and moister than that of the present from 9000 to 7000 BP (Li 1987), particularly in the Yangzi valley and South China. It is estimated that the yearly average temperature then was 2° -3° C higher than the present (Dai 1985).

This reconstruction of palaeoclimate is generally in accordance with the evidence for marine transgressions and regressions. The sea level had descended to more than 100 meters below the present level during the LGM in the eastern coast in China; then it rose to 40 meters below the present level at about 10,000 BP (Geng 1981), and kept rising since then with slight fluctuations. The sea level reached the present level at about 7000 years ago. Thus, 7000 BP is used as the boundary between the early and the middle Holocene.

South China This is a term with a broad and a narrow meaning. In its broader meaning, it refers to the vast area south of the Yellow valley, approximately from latitude 32°N. southward. In its narrow meaning, it refers to the area south of the South Mountain Range (Nanling), approximately along latitude 25°N., and consisting of the present administrative areas of Guangdong, Guangxi, and Hainan provinces. In this thesis, the narrow meaning is applied.

2C. Definitions for tools and other stone implements.

As modern Chinese archaeology was established by a few scholars trained in western institutes, many European-derived archaeological terms have been used in Chinese archaeology. Yet these European terms used in Chinese archaeology may have different meanings. For the purpose of avoiding confusion, it is absolutely necessary to clarify some key terms used in this thesis.

Adze The adze is defined by an asymmetrically beveled blade. The earliest flaked adzes in China discovered so far are from the Xiachuan complex [Fig. 4(3):9] (Wang *et al.* 1978). From the early Holocene onwards this tool became quite common in polished form, and from the mid-Holocene there were various types, such as shouldered and stepped adzes (Fu 1989).

Axe The axe is defined by a symmetrically beveled blade, with or without shoulders. Chipped axes were found in the Xiachuan assemblage [Fig. 4(5):2] (IA, CASS and CPAM of Shanxi Province 1978). Fully polished axes occurred by no later than the ninth millennium BP in Central China (IA, Henan Province 1989).

Chisel The chisel has an asymmetrically beveled blade, similar to an adze but narrower and smaller in size. Chisels

seem to be present from the early Holocene onwards. It is proposed that the function of this tool was similar to that of a modern chisel - for woodworking [Fig. 4(18):5].

Debitage This term refers to all the by-products of lithic tool manufacture and retouch, including broken semi- or completed products.

Grinding slab A grinding tool made on a stone block usually with a concave surface produced by wear. The shapes of grinding slabs in China vary, but the majority are rectangular or oval. The earliest grinding slabs are found in Xiachuan, dated from 23,000 to 13,000 BP [Fig. 4 (3):11] (Wang *et al.* 1978). This tool is often associated with a small handstone [Fig. 4(3):10]. It is quite certain that such tools were used to grind ochre, as indicated by remaining colour on the surface; but they might also have been used for grinding or pounding nuts and wild grains, as suggested by archaeological and ethnographic data in Australia and America (Cane 1984; Carter 1980; Kroeber 1925; Veth *et al.* 1997).

Grinding slabs were very common in the early Neolithic in the Yellow valley. They were more finely made, often with four little legs, such as those found in the Peiligang culture [Fig 4(22): 9]. Usually associated with a roller and present in Central China by the ninth millennium BP, the major function of this tool is claimed to have been grinding millet (IA, CASS 1984). This tool becomes increasingly rare after the 7th millennium BP, and is said to have been replaced by mortars and pestles (Wu 1989).

Pick A roughly pointed lithic implement made on a flat and long flake, retouched along the two parallel long sides and the tip. Picks of this type occur in the Middle East (Bar-Yosef and Belfer-Cohen 1989) [Fig. 9(1):23]. Morphologically similar implements occur in the Xiachuan assemblage in China [Fig. 4(5):3], although the Chinese ones appear shorter and wider. According to the excavators of the Xiachuan assemblage, this implement was used to dig and pick (IA, CASS and CPAM of Shanxi Province 1978), but use-wear analysis is needed to validate this proposition.

Point This term in the Chinese context is ambiguous (Lin 1993). However, in this thesis, 'point' is defined as an artefact with a purposely-retouched tip. Whereas the term 'pick' refers to a specific long and flat point with two carefully retouched paralleled sides [Fig. 4(5):3], a point may be with or without parallel retouched sides; the cross-section of points can be triangular, round or irregular; the sizes of points vary. The point is believed to be a multi-functional tool.

Heavy-duty (stone) tools. A lithic industry often found in Central China from the end of the Pleistocene into the early Holocene. It is characterised by the following:

- Tools were made on large pebbles or flakes, normally over 6 cm in maximum dimension. Deep flaking scars on tool surfaces indicate that direct percussion was the major method for flaking and retouch.
- Typical implements of this lithic industry include choppers and chopping tools, grinding slabs and handstones, chipped axes and adzes. The latter four seem to be the forerunners of their Neolithic counterparts.

The time span for some implements in this heavy-duty tool tradition was quite short. Although choppers and chopping tools were made from the early Palaeolithic onwards, grinding tools, flaked axes and adzes are mainly found from the terminal Pleistocene into the early Holocene. After 7000 BP these implements became rare, being replaced by mortars and pestles, polished axes and adzes (Wu 1989). But these tools are the hallmarks of the transition from a chipped stone tool assemblage into a ground tool one, and therefore, they are very important items for examining the transition from the Pleistocene to the Holocene.

The microblades (or 'The Microblade tradition'). This term is used to replace the simpler term 'microlithic' in the Chinese context, since the latter has been quite ambiguously used in the Chinese archaeological literature (for details see Lu 1998). The 'microblade tradition' in China is defined as a lithic tradition which emerged and flourished in the terminal Pleistocene in Central China, prior to the emergence of agriculture there. This lithic tradition is characterised by the following features:

- The majority of cores were prepared prior to blade detachment.
- Indirect percussion is proposed to have been the major method for flaking and retouch. Pressure flaking might also have been used in retouch.
- Typical cores of this lithic tradition are of wedge shaped, prismatic, conical and semi-conical, boat shaped and funnel shaped (Jia 1978). Microblades are under 5 cm in length, 1 cm in maximum width, and usually triangular or trapezoid in cross section (Gai 1985:227).
- According to archaeological discoveries, the major function of this lithic tradition was to provide cutting or piercing edges or points for composite tools, although some microblades could also have been directly used in hand (for detailed discussion see Lu 1998).

The flake tradition (or the flake industry). This was the major lithic tradition in Central China from the early Palaeolithic onward, characterized by the following:

- Major raw materials are quartz, quartzite, chert and agate.
- The major flaking method was direct percussion, including simple direct percussion, block-on-block, and throwing. Direct flaking was used in retouch (Zhang 1989).
- Typical implements of this tradition consist of various scrapers, points and burins.
- Measurements of artefacts vary, but they are usually under 6 cm in length. Some implements, such as the thumbnail scraper, can be as small as 1 cm around (Zhang 1989).

The above three lithic traditions (heavy-duty tools, microblades and the flake tradition) have mainly been discovered in Central China.

Small flakes are also found in the Yangzi valley and South China from the terminal Pleistocene to the early Holocene, distinguished from the above flake tradition by the following:

- Chert is the major raw material, followed by quartz and quartzite.
- Detachment is mainly conducted by direct percussion. Secondary retouch is not commonly applied. Judged from the use-damage on the edges, some of the flakes were used directly after detachment (author's observation 1996).
- The flakes are often irregular in shapes, and the function of them is not clear.
- The sizes of these flakes are very small, often under 4 cm in dimension (author's observation 1996; IA, Hunan Province and Lixian Museum 1990). Some of them are as thin as 0.5 cm or less (author's observation 1996).

The Pebble tool tradition. A dominant lithic industry in the Yangzi valley and South China from the middle Pleistocene to the early Holocene, with the following features:

- Tools were made on pebbles with only a few flakes removed around the edges. Large proportions of the cortex were preserved. The manufacturing technique for this tradition seems quite crude and simple.
- The deep and sharp flake scars indicate that direct percussion was the major technique. Retouch is seldom applied.
- Implements of this tradition are quite simple. Choppers are the major ones. Scrapers and points are present but in much lower frequency. Perforated pebbles appeared towards the end of the Pleistocene, and the quantities of them are said to have increased in the early Holocene in some areas (Jiang and Peng 1990).
- The majority of these artefacts are over 6 cm long. According to published data, many artefacts measure 10 to 18 cm in maximum dimension.

The pebble tool tradition in China is considered to belong to the 'Hoabinhian technocomplex' (Bellwood 1978).

Sickle A cutting tool used for harvesting in the Yellow valley Neolithic, but seldom found in other areas. The typical form in China is a curved stone flake with a denticulated cutting edge [Fig 4(22):3]. This tool was present by the ninth millennium BP in Central China (IA, Henan Province 1989). The so-called 'lithic saw' found in the Xiachuan culture 23,000 to 13,000 years ago might have been a forerunner of these sickles in the Neolithic [Fig. 4(4):8] (Shi 1989), but this remains to be proved.

Spade A flat stone tool effective for soil-digging. Usually longer and flatter than axes and adzes, this tool type first occurred in the early Neolithic in Central China. It is proposed as a tool for cereal cultivation (Tong 1989) [Fig. 4(9):1-3].

Bone Spade A unique tool found in the Yangzi valley. Made from an ungulate scapula, the spade was trimmed into a symmetrical shape [Fig. 7(18):5]. This tool is suggested to have been used for cultivating wet rice fields in Hemudu (Liu and Yao 1993).

2D. Definitions for Pottery.

Pottery was present prior to the emergence of agriculture in China, and was one of the most important components of Neolithic cultures. In China, the earliest pottery, dated to around 12,000 BP, was found in both the Yellow and the Yangzi valleys (see chapters 4 and 7).

The early pottery found in China was coarse, mixed with tempering agents ranging from crushed mica, shell, charcoal and plant parts. The sand is usually coarse and varying in sizes at early stages, as found in Xianrendong (Hill 1995). Finer sand was used as temper by around 8000 BP in both the Yellow and the Yangzi valleys.

Fine pottery is a development appearing later than the coarse pottery in China's context. It is made from sieved fine clay, with very small particles, usually not clearly visible to the naked eye. Fine pottery was present in both the Yellow and the Yangzi valleys by about 7500 BP.

Pottery vessels in China are very diversified. Following are the most commonly found items during the early Holocene in China, some of them with specific names:

Barrel Urn A flat-based vessel with a smooth and slightly curving profile, measuring over 15 cm in maximum diameter and 20 cm in height. The rim diameter can be greater than the maximum diameter of the body [Fig. 4(10): Cishan pottery No. 9].
Bo. A round-based bowl with a smooth and curving profile, the maximum diameter of *bo* is on the mouth. This vessel is used for serving food. The earliest *bo* was found by 8400 BP in Central China. Some of these *bo* have tripods, such as those found in the Cishan, Peiligang and Dadiwan cultures [Fig. 4(10)].

Ding A most commonly found vessel in Neolithic China, specialized for cooking. Evidence from the Houli culture of the lower Yellow valley illustrates that the *ding* was derived from the combination of a *fu* and three stands [Fig. 10(1)]. It was a tripod vessel, the shape of both the legs and the vessel body varying temporally and spatially. The earliest *ding* has been found in Jiahu, a Peiligang culture site in the Huai valley dated to approximately 8400 BP. Some were present at around 7500 BP in phase II of the Houli assemblage, in the lower Yellow valley, and the type was widely spread in the whole valley by 6000 BP. It also occurs in the Yangzi valley by about 5500 BP. In Bronze Age China, the *ding* was a symbol of the state and ruling power. The mysterious and sacred perception of this vessel in Chinese tradition suggests its antiquity and importance in prehistory.

Dou Another vessel widely found in Neolithic China. It is usually a combination of a round-based plate on a high ring foot [Fig. 7(7), Fenshanbao No. 9]. This type of vessel is usually the most decorated item in Neolithic assemblages, either painted, incised, and/or perforated. The earliest *dou* to date has been found in a Fenshanbao assemblage at around 7500 BP in the middle Yangzi valley (The Archaeology Team of Yueyang City and CPAM of Qianlianghu Farm 1994), and occurs widely in both the Yellow and the Yangzi valleys after then. It was used to serve food. In Bronze Age China, the *dou* was an essential vessel for ritual ceremonies.

Fu The earliest type of pottery vessel in China found to date. It has a round body, deep belly and round base. The diameter of the mouth is usually equal to, or even greater than, the maximum diameter of the body. It is sometimes just called 'pot' in some archaeological reports (e.g. Hill 1995). Various types of *fu* have been found in Neolithic China, such as a 'shouldered *fu*' in the lower Yangzi valley, with a 'collar' formed around the rim [Fig 7(18), Hemudu No. 1-2]. The over-cooked rice remains found in one *fu* in the Hemudu assemblage indicate that this vessel was used for rice cooking (Liu and Yao 1993).

Globular Jar with two lugs This is a key vessel found in the early Neolithic cultures from Central China to the middle Yangzi valley. It is a bipartite vessel with a round base and a neck, usually with two lugs or two side-ears on the shoulder [e.g. Fig. 10(2), Pengtoushan No. 4, 5].

Slip A thin layer covering the pottery surface, usually made of a clay wash. In Neolithic China, this slip is usually of a red colour. But a white slip was also used as a background for painting, first appearing in the Fenshanbao assemblage in the middle Yangzi valley (Zhang 1992), and later in Hemudu in the lower Yangzi valley (Liu and Yao 1993).

Yu A special type of vessel mainly found in the Cishan assemblage in Central China. It is a flat-based vessel, with an upright and straight profile, and a straight or slightly everted rim [Fig. 4(10), Cishan No. 2]. Usually around 20 cm in both height and maximum diameter, this vessel is often found with tripod supports, and was probably used for cooking (Jin 1995).

Part II. The Transition from Foraging to Farming in Central China

Chapter 3. Palaeoenvironments and palaeoclimates from the terminal Pleistocene to the early Holocene in Central China

The term 'Central China' here refers to the area approximately between latitude 32.5° and 41° N, and longitude 106° to 123° E, roughly within the present vegetation zone of deciduous forest [Fig. 3(2)] and the middle and lower valley of the Yellow River. The north border of this region is roughly along the Yan Mountains, the south border being the Huai River. The western part of this region reaches the Wei River valley, and the eastern part is the sea [Fig. 3(1)].

There are several mountain ranges in Central China. To the north of the Yellow River lie the Lu-liang and the Tai-hang Mountains, both of north-south direction and with basins in between. To the south of the Yellow River there are the Qin Mountains of west-east direction [Fig. 3(1)], which block part of the summer monsoon from the ocean, and the winter monsoon from the continent. The climate and vegetation on the north and south of the Qin Mountains often vary due to this geographical setting.

The Yellow River is the largest river in Central China. The Huai River to the south is another major river in this region. In the western mountain/plateau area there are the Fen and the Wei rivers, both being tributaries of the Yellow River [Fig. 3(1)]. Apart from these major rivers, there are many other small rivers widely distributed over the region [Fig. 3(1)], forming a water-supply system for human beings.

The current environment of Central China is basically a temperate zone, except for the northwest corner occupied by the Gobi [Fig. 3(2)]. The Tai-hang mountains serve as a border of two geographically different areas: on the west side lies the Loess plateau, which includes the Lu-liang and Tai-hang mountains; while on the eastern side, the North China Plain stretches from the foothills of the Tai-hang mountains eastwards to the sea [Fig. 3(1)]. Current climate in the Loess plateau/mountain area is quite hot in summer and cold in winter. Yearly average temperature in the Loess Plateau is 8 - 12° C, the hottest in July with a mean temperature of 24-26° C, and the coldest in January with a mean temperature of -8 to - 4° C (The Editing Committee of the Physiography of China 1984) [Fig. 3(3)] [Table 11(3)].

The climate in the Loess plateau and the mountains is also quite dry. Yearly rainfall is about 500 to 600 mm in the north, and 700 to 1000 mm in the south around the Wei River [Fig. 3(4)]. But with a local annual evaporation of 800 to 950 mm, it is obvious that not much water can be retained in the ground except in the mountains where moisture increases at higher elevations. Current vegetation in the basins are cultivars and second-growth plants such as shrubs and herbs, as this region has been disturbed by human activity for thousands of years; but the Lu-liang and Tai-hang mountains are still covered by a mixed conifer-deciduous forest (Wu *et al.* 1991).

Modern climate in the North China Plain is moister than in the mountain/plateau, as there are fewer mountain ranges to block the summer monsoon winds from the southern ocean. Precipitation on the North China Plain varies from 400 mm in the north to 900 mm in the south, as the influence of the summer monsoon weakens northwards. The average annual temperature of the North China Plain ranges from 10 to 14° C from north to south. In summer, when the warm and wet summer monsoon dominates this region, the weather is hot with heavy rainfall. July is the hottest month with an average monthly temperature of over 20° C degree. In winter, as the winter monsoon from the northern continent sweeps in, it becomes dry and cold here, with January the coldest with an average monthly temperature below 0° C [Fig. 3(3)&(4)] (The Editing Committee of the Physiography of China 1984).

One of the climatic characteristics of the North China Plain is the fluctuation of precipitation, which can vary from 406 mm in one year to 987 mm in another (Xiong and Xi 1965). Another characteristic is the unequal annual rainfall distribution with 50% to 75% occurring in the summer months (June, July and August), while spring and winter are normally dry. Affected by the monsoon winds from the deserts of inner Mongolia, spring in this region is also windy and dusty (Gu 1991).

Today the vegetation of the North China Plain is basically an agricultural landscape, as this area has been cultivated for thousands of years. Due to the dense population, natural vegetation can hardly survive except in the mountain areas, such as the Yan mountains to the north of Beijing, where some deciduous and coniferous trees can still be found. There are, of course, grassland and shrubs in this region, as second-growth vegetation resulting from human disturbance (Wang *et al.* 1995).

The climatic and ecological pattern in Central China at present, however, is different from the palaeoclimate was and palaeoenvironment was from the end of Pleistocene to the early Holocene. These are reconstructed in the following section.

3A. Palaeoenvironments and palaeoclimates of Central China.

The geotectonic structure of Central China at about 20,000 BP was much the same as it is today. But there are some differences in terms of landforms and vegetation. During the Last Glacial Maximum (LGM), the sea level was about 130-155 meters below the present. As a result, the eastern part of the continental shelf was exposed, and the North China Plain was larger than it is today (Shi and Wang 1982) [Fig. 3(5)] [Fig. 3(12)].

Just as the physiographical appearance of Central China changed quite significantly during the period from the end of the Pleistocene into the early Holocene, so did the palaeoclimate. The palaeoclimatic changes are illustrated by data from pollen profiles, fossil assemblages, palaeosol complexes and sequences, traces of the advances and retreats of snowline and permafrost, remains of marine transgressions and regressions, and stratigraphical sequences reflecting the expansion and shrinkage of inland lakes. Of all these studies, pollen analysis is of great importance as it provides fundamental data for palaeo-vegetation, which is the primary database for the reconstruction of palaeoclimate.

Pollen analysis has been undertaken in this region since the 1960s. In Central China, tens of sites and pollen boreholes have been drilled and pollen sequences from the early to the late Pleistocene have been studied. In relation to the time span of this thesis, there are 27 sites and boreholes with well-presented pollen sequences, dating from the end of the Pleistocene to the early Holocene. These sites and boreholes are densest in the areas around Beijing, the

northern part of the Hebei plain and the Wei river valley [Fig. 3(1)].

There are 7 sites and boreholes containing Quaternary sediments with well-preserved pollen data around present Beijing [Fig. 3(1)]. But only Fenzhuang (site No. 1) has revealed a pollen profile from the end of the Pleistocene to the beginning of the Holocene [Fig. 3(6)], while Gaolizhang (site No. 2) [Fig. 3(7)] yielded a pollen sequence from the early Holocene.

The pollen profile from Fenzhuang showed a prevalence of herbs and grasses from 17,000 to 13,000 BP (Zhang *et al.* 1981) [Fig. 3(6)]. Within the herbs and shrubs, drought-resistant plants such as *Artemisia* and Chenopodiaceae prevailed. As for trees, the majority of the arboreal pollens were from conifers such as fir tree (*Abies*) and pines (Zhang *et al.* 1981). Obviously, this pollen profile represents a landscape of steppe with a few conifers, indicating a very cold and dry climate from 17,000 to 13,000 BP.

The Fenzhuang pollen profile also illustrates a change from about 13,000 to 10,000 BP. During this period, the proportion of arboreal pollen increased, while the pollen of herbs and shrubs decreased. Within the trees, fir (*Abies)*, pine (*Pinus)* and spruce (*Picea)* were major components; but deciduous trees such as elm (*Ulmus*), lime trees (*Tilia*), and oak (*Quercus*) began to appear or to increase in quantity. Amongst the herbs, *Artemisia* was reduced quite significantly, while hygrophilous plants such as *Typha* increased considerably. Symbolising a mixed conifer - deciduous and steppe vegetation, the pollen assemblage of this period indicates that the palaeoclimate was still cool, but slightly warmer and moister than before (Zhang *et al.* 1981) [Fig. 3(6)]. This pollen profile allows an estimation of mean annual temperature of 4.1-4.5° C from 17,000 to 13,000 BP, about 7.5 to 8.5° C lower than the present; and of 10-11° C from 13,000 to 10,000 BP, about 1-2° C lower than the present (Zhou *et al.* 1978; Zhang *et al.* 1981; Zhao 1987) [Fig. 3(6)]. This was an increase of approximately 6-6.5° C over 7000 years.

It should be noticed that there was a sudden decrease of arboreal pollen and an abrupt increase of drought-resistant herbs in the period between approximately 12,250 to 11,850 BP [Fig. 3(6)]. This retreat of forest and expansion of steppe indicates an increase of drought, and the prevailing *Abies* within the arboreal pollen represents a sudden cold change during that period of time. This may be comparable with the Younger Dryas in Europe.

After this cold phenomenon, the trend of warming up continued into the early Holocene. This trend is illustrated by the pollen profiles of Fenzhuang and Gaolizhang (sites Nos.1 and 2). Both show an increase of arboreal pollens, although proportions of the trees and herbs vary. Within the arboreal pollens, there was a significant decline or absence of cryophilic trees such as fir trees and spruce; while the deciduous oak, lime trees and *Carpinus* (see appendix I for common names of plants) increased substantially. Pine was still the major tree; birch was also present in considerable quantity. As for the herbs, although drought-resistant plants still prevailed, the presence and increase of thermophilous and hygrophilous plants, i.e. Polypodiaceae, Cyperaceae and *Typha,* were notable (Zhang *et al.* 1981) [Figs. 3(6) &(7)].

Such a climatic transition from the cold glacial period at the end of the Pleistocene to a warm early Holocene not only occurred in the Beijing region but also in other parts of Central China as well. In the northern part of the North China Plain, pollen profiles from ten sites show a similar sequence. Details of these pollen profiles are not available, but a few summary studies on the palaeoenvironment and palaeogeography of this region have been conducted based upon the pollen sequences of these boreholes. It has been inferred that, at the end of the Pleistocene, the vegetation here was taiga-steppe or coniferous-steppe mixture, as illustrated by the dominance of *Pinus, Picea* and *Abies* within the arboreal pollen assemblage, and *Artemisia* - Chenopodiaceae within the herbs (Tong *et al.* 1983). Apparently this vegetation indicates a cold and dry climate during that time. It has been estimated that the mean annual temperature was about 4-5° C, some 7-9° C lower than the present, based upon the current climate and environment for *Abies* and *Picea* (Tong *et al.* 1983).

Again, there was a change towards a moister and warmer climate after the Pleistocene, as indicated by the presence and increase of deciduous trees such as *Quercus,* and the hygrophilous *Typha* within the herbs. Data from a few boreholes also show an increase of Gramineae in the early Holocene (Tong *et al.* 1983). As the palaeoclimate in the early Holocene was warm and moist, the increase of grasses at some boreholes might have reflected human activities (e.g. deliberately lit fire in order to increase the growth of some grasses, or for land clearance). But further study is needed for this question.

In the central inland area of the Hebei plain, there are four sites and boreholes, all revealing pollen profiles from the Pleistocene to the Holocene except for No. 3, which only revealed a Holocene profile. The pollen sequences of these sites have provided information similar to that from other sites in the Hebei plain (Yang *et al.* 1979) (Zhao 1987) [Fig. 3(8)].

An archaeological site, Nanzhuangtou (site No.3), is particularly important as fragments of pottery, bone and stone tools were found in early deposits radiocarbon dated to between 10,500 and 9700 bp (Yuan *et al.* 1991). These artefacts are the earliest Holocene archaeological remains found in Central China published to date. The pollen sequence of this site revealed that herbs dominated the entire pollen assemblage throughout the whole deposit period, making up 60% - 80% of the total pollen; but trees increased at about 10,000 BP, then declined again shortly after that time. Of the herb pollens, *Artemisia* was the major component; other herbs included Gramineae, Cyperaceae and *Typha.* Pine was the main type among the trees; cryophilic *Abies* and *Picea*, and the deciduous *Quercus* were all found (Yuan *et al.* 1991) [Fig. 3(9)]. This profile again indicates a mixed coniferous - deciduous forest and steppe, representing a cool and quite dry environment. There were, however, climatic fluctuations during this period of time. At around 10,000 BP the pollen of Cyperaceae and *Typha* increased, thus indicating a return to a slightly moister climate (Yuan *et al.* 1991) [Fig. 3(9)].

The vast area from the central part of the Hebei plain southwards to the alluvial plain of the Yellow and Huai rivers was the northern centre for Chinese Neolithic culture. Unfortunately, published pollen data from this region are not very detailed. A few summary studies on the Quaternary of the Huai River Plain, based upon data from a few boreholes (details for these boreholes are not available), can provide a general outline. It is claimed that during the LGM, herbs dominated the pollen assemblage, accounting for over 70% of the total pollen aggregation. Within the herbs, *Artemisia* by far exceeded the others, making up 80.4% - 91.7% of the total herbal pollen. As for the trees, pine was the major one; *Abies* and *Picea* were also present. Based upon summarized pollen data and a few radiocarbon dates, Jin and his colleagues propose a landscape of steppe and semi-desert with only a few trees during the LGM, indicating a cold and very dry climate, probably about 6-7° C lower than the present (Jin *et al.* 1990). But it is noticeable that coniferous trees in this

region were fewer than in the Beijing area, and deciduous trees such as *Quercus* and *Salix* were present, although in minor amounts, during the LGM (Yu and Huang 1993). This can be seen as evidence of a less severe climate in the area around latitude 35° N and below.

After about 10,000 BP, climate became warmer and moister in the Huai River plain as in other areas. Pollen profiles of this period show an increase of arboreal pollen to 31%, with pine being the major type. Deciduous trees such as *Quercus* and *Salix* are found in considerable quantities. Herbs decreased to 41-44% of the total pollen, with less *Artemisia*. Spores also increased to 25-27% of the total vegetation assemblage, the major components being hygrophilous plants (Jin *et al.* 1990).

Similar climatic conditions have been reconstructed for pollen samples from the early Neolithic site of Jiahu (No.4), located between the middle Yellow valley and the Huai River [Fig. 3(1)]. The Neolithic culture in this site is dated to between 8500 and 7500 BP [Table 4(7)], with cultivated rice remains (Zhang 1991). Details of this pollen analysis are not available, but it is claimed that prior to the cultural deposit, the pollen profile consisted of 76% of herbs and shrubs, with *Artemisia* alone accounting for 64% of the total; 6.3% of the pollen assemblage was trees and the rest were fern and algae spores. Pine was the major component within the trees; *Salix, Alnus* and *Quercus* were all present. Within the herbs and shrubs, there were also *Tamarix,* Compositae and Chenopodiaceae (Zhang 1991). This pollen assemblage illustrates a vegetation of mixed coniferous - deciduous trees and grassland growing in a temperate climate, while the presence of hydrophilic ferns indicates relatively high rainfall (Zhang 1991).

It is also claimed for Jiahu that during the early phase of the cultural deposit, arboreal pollen increased to 18.3%, dominated by deciduous trees such as oak, walnuts and elm. Herbs and shrubs declined to 43.7%, dominated by *Artemisia.* Ferns accounted for 11.6% and algae accounted for 26.5%. In the later phase of the cultural deposit, arboreal pollen decreased to 6.3%, still dominated by deciduous trees; herbs and shrubs accounted for 28.5%, dominated by *Artemisia*; algae increased to 62.6%, and the rest was fern (Zhang 1991). While the pollen profile at Jiahu shows a trend of increasing temperature and moisture in the early Holocene, the decline of trees after human occupation and the significantly increased algae may reflect the impact of human activities, particularly wet-land rice cultivation.

In the above section I have focused on the reconstruction of palaeoenvironment and palaeoclimate in the eastern part of Central China, where the alluvial soils enabled good preservation of pollen. But in the western mountains/plateau area of Central China, pollen is less well preserved in the loess, and other methods have been used.

There are only six sites with relatively well-preserved pollen sequences for the period from the end of the Pleistocene to the early Holocene in this latter region. Two profiles are from the northern Sanggan river basin (Yan *et al.* 1979; Zhou *et al.* 1983), and four from the southern Wei river basin [Fig. 3(1)]. Profiles Nos. 5 and 6 are associated with archaeological remains. The Xujiayao (site No. 5) archaeological assemblage is uranium series dated to about 100,000 years bp (Chen *et al.* 1982) and is out of the scope of this thesis. The pollen profile from this location is not precisely dated. However, the pollen profile of Xibaimaying (site No. 6) is relatively well-preserved with two absolute dates. A late Pleistocene assemblage was discovered in this site, uranium series dated to 18,000±1000 and 15,000±1000 years bp (IA, Hebei Province 1989).

Eight stratigraphic layers have been identified at Xibaimaying, of which layer 3 is the one containing a lithic assemblage and animal bones. Ten pollen samples were collected from various layers, and relatively abundant pollen was found. It is claimed that prior to the deposit of layer 3 pollen was quite scarce, and dominated by pine and *Artemisia.* However, in layer 3 there was relatively abundant pollen, of which 0.04% is pine and *Picea* ; 87.59% is *Artemisia*; and the rest are Chenopodiaceae, Gramineae, Ranunculaceae, Convolvulaceae and Cruciferae, plus a few *Selaginella* (IA, Hebei Province 1989). It is clear that the drought-resistant herb *Artemisia* absolutely dominated the pollen assemblage. A similar pollen assemblage was also found in layer 4, overlying layer 3 (IA, Hebei Province 1989) [Fig. 3(10)].

The above pollen assemblages represent a vegetation of dry steppe type with only a few coniferous trees, indicating a cold and dry palaeoclimate. As layer 3 was uranium series dated to 18,000 to 15,000 years bp, it is clearly within the last glacial era. All other layers on top of layer 3 contain very little pollen, except for the modern surface.

As for the southern Wei river basin, Liu has (1988) claimed that during the LGM, the spruce-fir forests in this area appeared at an elevation of 490m, 1500m below their current lower limit, according to pollen data from the Beizhuangcun site in Shaanxi province. From this, Liu has inferred a mean annual temperature 8-10° C lower than the present (Liu 1988).

Similar climatic change during the terminal Pleistocene is also illustrated by a pollen profile from Yangguo (site No. 7) in the Wei river basin. According to this pollen sequence, from 21,000 to 13,000 BP the vegetation in this region was mainly an *Artemisia* steppe (Sun *et al.* 1995). After 13,000 BP the local palaeoclimate seems to have become moister although still cold, as indicated by the presence of *Tsuga* and the increase of pine. From approximately 11,000 BP there occurred an increase of conifers, but in association with a significant increase of *Artemisia* [Fig. 3(11)], representing a continuing steppe environment. This shows that the climate was still very dry. However, the sharp increase of hygrophilous plants *Selaginella* and *Liliaceae* illustrates the increase of moisture after 10,000 BP [Fig. 3(11)], thus marking the beginning of the Holocene.

Although pollen analysis is a major approach for reconstructing palaeoclimate, and pollen diagrams usually reflect vegetation composition, surface pollen analyses indicate that there are possible biases (Yao 1989). For example, in one modern test sample in the Zhongtiao Mountains, Shanxi province, pine pollen was present, but there was no pine growing in the sampling area (Yao 1989). The accuracy of pollen in representing local vegetation is probably affected by different conditions of pollen preservation, deposition, delivery methods and distances of different plants, etc. (Yao 1989). Therefore, additional methods are necessary for reconstruction of the palaeoenvironment and palaeoclimate in China.

In 1983, sedimentary research combining heavy mineral and chemical analysis, as well as pH and organic analyses, disclosed a palaeoclimate cycle from the early to the late Pleistocene in the Dingcun site (Site No. 8), which is located in a basin between the Lu-liang and Tai-hang mountains [Fig. 3(1)]. Minerals in Dingcun sediments of the terminal Pleistocene were illite, plus some chlorite and hydromica, indicating a dry and cold environment (Chen 1991). By comparing this mineral group with current environments of similar mineral aggregation, as well as pollen data, it has been argued that, from 23,000 to 13,000 BP, the mean annual temperature was probably 10° C lower

than the present, and the climate was dry. It is also argued that conditions improved and became warmer after 10,000 BP (IA, Shanxi Province 1994).

The climatic shift to higher temperature from the terminal Pleistocene into the early Holocene is also illustrated by phytolith analyses in the plateau/mountain area. A recent study of 71 samples from loess sections illustrates vegetation from 30,000 to 10,000 BP (Wu *et al.* 1994). Based upon modern phytolith assemblages of 162 samples from different regions of China and current climatic data, Wu and colleagues have plotted relationships between phytolith types with temperature and precipitation. They propose that from 23,000 to 14,000 BP the mean annual temperature in the western mountain/plateau area was 7-9° C lower than the present, with a minimum average yearly temperature of about 5° C; and local precipitation was 240 mm less than the present (Wu *et al.* 1994). They argue that there was a sudden warm change at 14,000 to 13,000 BP, with mean yearly temperatures and precipitation being similar to the present. After 13,000 BP the temperature again dropped by 4-5° C, and the precipitation was reduced by 50-120 mm. Then the palaeoclimate gradually became warm and moist until 11,000 BP, when another abrupt decrease of both temperature and precipitation occurred, with temperature being 3° C lower than the present, and mean precipitation being 50-100 mm less than the present. After the beginning of the Holocene a quick climatic warm-up occurred, and the mean yearly temperature rose to 2-3° C higher than the present, with precipitation being 200 mm more (Wu *et al.* 1994).

The climatic and environmental fluctuations from the terminal Pleistocene to early Holocene are also demonstrated by faunal analyses, which reveal that during the last glacial period, animals of frozen zones such as *Mammuthus primigenius* and *Coelodonta antiquitatis* were widely found in Northeast and Central China, with *Mammuthus primigenius* mainly within the region from latitude 39° N northwards, and *Coelodonta antiquitatis* spread over Central China, down to 33° N [Fig. 3(12)]. Both animals were adapted to a cold to very cold climate, but *Coelodonta antiquitatis* adapted well to both tundra and steppe (Zhou 1978). The widespread distribution of *Coelodonta antiquitatis* further illustrates the cold and dry climate during the terminal Pleistocene in Central China. After 11,000 BP both animals became extinct in China (Cui and Xie 1985).

The presence of *Mammuthus primigenius* and *Coelodonta antiquitatis* in Central China is consistent with other palaeoclimate phenomena. There were signs of an expansion of periglacial condition southwards to about latitude 35° N during the LGM (Cui and Xie 1984), and the edge of the permafrost has been found at about latitude 37° N (Yang *et al.* 1989) [Fig. 3(12)]. Compared to the present edge of permafrost along latitude 44-45° N, this is a southward movement of 7-8° of latitude. Currently the mean yearly temperature is 0-4° C along latitude 45° N and 8-14° C along latitude 37° N; therefore, the mean yearly temperature during the LGM was about 6-8° C lower than the present .

There are, of course, possible biases in faunal analysis as well. Prehistoric animal remains were subject to different conditions of preservation, deposition, transportation, as well as various natural and human disturbances after the animal remains were buried. Further, it should be pointed out that Central China is a vast landmass with various geographical sub-regions, so the climate and environment varies in each region. The animal remains found in Dingcun, Shanxi province and the Upper Cave, Zhoukoudian [Fig. 3(1)] indicate that, prior to the LGM, it was quite warm and moist in the Beijing area and the southern part of the western plateau/mountain area, as animals of warm and moist conditions such as deer, rhinoceros and elephant, were present (Chen 1991; Qi 1989). Dingcun was dated to 26,000 years bp (Chen 1991), and the Upper Cave was dated to 33,000 - 23,000 years bp [Table 4(1)]. On the northern edge of the western plateau/mountain area the palaeoclimate seems to have become already cool and dry at this time, as ungulates dominated the fauna [Table 3(1)].

This spatial climatic difference was also present during the LGM. Pollen analysis at the southern edge of the plateau/mountain area indicates that a few broad-leaved trees, such as hazel (*Corylus* sp.) and walnut (*Juglans* sp.), still survived at around 17,000 to 15,000 BP, although at a very low frequency (Sun *et al.* 1995) [Fig. 3(11)]. These species seem absent in other areas of Central China in this period. This indicates that even during the LGM, the palaeoclimate could have been milder in some regions than in others, particularly at lower latitudes.

Much other research has been done to reconstruct conditions from the LGM onwards, based upon analyses of palaeosol, peat accumulation, and changes in river courses and lake shorelines. As loam in the loess area and peat in the alluvial plains were normally formed under moist and temperate environments, while the loess itself was formed in a dry and windy climate, these sediments can be taken as indicators of palaeoclimatic changes. During the LGM, Central China received a strong winter monsoon, with estimated mean precipitation at about 200 mm, which is 100 to 500 mm less than the present (An *et al.* 1990). During this cold and dry period, dust deposition in Central China accelerated, inland lakes shrank, and the desert in the northwestern and northern areas expanded [Fig. 3(13)]. At about 12,000 BP there occurred 'an episodic strengthening of the summer monsoon' (An *et al.* 1991:20), indicated by the expansion of conifers such as *Abies* and pines in the Hebei plain, the rise of inland lake levels and the formation of palaeo-loam on the loess plateau. Then, from 12,000 to 11,500 BP, the climate abruptly became cold and dry again, as demonstrated by the further deposition of loess, the increase of herbs within pollen assemblages, and the shrinking of lakes. After 11,500 BP the palaeoclimate rapidly became warm and moist, on entering the postglacial regime (An *et al.* 1990). This overall pattern is further supported by oxygen isotope analysis from deep-sea cores (An *et al.* 1990; You and Xu 1981).

During the early Holocene, animals of warm and moist environments such as *Elaphurus davidianus* (David's deer) appeared in Central China, from about 8500 BP onwards (Zhang 1991). The retreat of desert, the expansion of lakes and the rising sea levels, all indicate a warm and moist climate (An *et al.* 1991). It has been estimated that during the period from 9000 to 5000 BP, mean precipitation in Central China rose to 600 mm, similar to current values in the northern part of this region (An *et al.* 1990). Estimation of the annual temperature for this period varies, ranging from 2-3° C to 4-5° C higher than at present (Yang and Lin 1993) [Fig. 3(14)].

In summary, the aggregation of data from various disciplines reveals a general picture of the palaeoenvironment and palaeoclimate from the end of the Pleistocene into the early Holocene in Central China. The last glacial seems to have affected this region quite significantly, particularly during the period from 18,000 to 13,000 BP. Generally speaking, the palaeoclimatic cycle during this time was from a very dry and cold period from 18,000 to 13,000 BP, changing to a milder and relatively wet period from 13,000 to 10,000 BP (Yang and Lin 1993) except for a short cold snap around 12,000 BP. Consequently, the environment in Central China from

18,000 to 13,000 BP was a steppe landscape vegetated mainly by herbs and shrubs, with pine and *Abies* being major trees. From 13.000 to 11,000 BP, as humidity increased, the quantity of pine and *Abies* also increased (Yang and Lin 1993). At around 12,000 to 11,500 there was an abrupt change into a cold and dry climate, 'probably lasted only several hundred years, corresponding to the Younger Dryas in Europe' (An *et al.* 1991:20). After 10,000 BP the whole region of Central China, from north to south, came under the control of a warmer and moister climatic cycle of the early Holocene, as indicated by the presence and increase of deciduous trees and thermophilous plants. It is estimated that the mean annual temperature became for a time even higher than at present (Yang and Lin 1993).

It seems that these natural changes during the transition from the end of the Pleistocene into the early Holocene were significant. What were their effects on fauna and flora, and, consequently, how did these changes affect human beings during this period? These issues are addressed in the following sections.

3B. Local natural resources.

As discussed above, the palaeoenvironment and palaeoclimate from the terminal Pleistocene to the early Holocene went through significant changes, through a cycle of severe cold weather at about 18,000 BP to warm and wet conditions at the beginning of the Holocene. The outcome of this change would have seen alternations in fauna and flora in terms of species extinction and movement.

3B(1). Faunal resources.

Since the 1930s, many faunal assemblages dated from the terminal Pleistocene to the early Holocene have been found in Central China, providing data for faunal changes during this period of time. In addition, it is also useful to look at fauna dated prior to the LGM, so that the faunal changes may be identified more clearly.

Ten faunal sites of the terminal Pleistocene are listed in Table 3(1). In brief, sites 1 to 3 and 8 are dated prior to, or at the beginning of, the LGM; site No. 10 from Xiaonanhai cave contains two faunal assemblages - assemblage A from the lower cultural layer, and B from the upper layer (An 1965). Although absolute dating indicates that there was a time gap of some 10,000 years between the upper and the lower layers, most of the animal species from the two deposits are the same [Table 3(1)]. The other faunas are dated to the LGM in Central China [Table 3(1)] [Table 4(1)].

Among these faunas, Zhoukoudian Upper Cave (site No. 8) is the most abundant in terms of quantity and variety. The fauna was discovered in the lower cellar of the Upper Cave of Zhoukoudian in 1934 by Pei Wenzhong and his colleagues. According to a description given by Weidenreich, 'thousands of hares, thousands of complete lower jaws of Sika deer, several hundreds of Roe deer, several hundreds of tigers, and at least 40 complete skeletons of deer, tiger, bear, hyena, bones of ostrich and representatives of other genera of mammals' (Weidenreich 1939: 161) were found, representing a total of 48 species [Table 3 (1)], in association with a large quantity of fish and amphibian fossils (Jia 1980).

Since the late 1970s, seventeen absolute dates have been obtained for the Upper Cave deposit. Among them, seven dates are for this fauna. Methods of dating range from traditional C^{14} and AMS C^{14} dating, uranium series to thermoluminescence determination (Chen *et al.* 1992) [Table 4(1)]. These absolute dates are not all in good order, probably due to disturbance and sample deviation. Nevertheless, the fauna of the Upper Cave seems to be within the range of 33,200 to 23,150 years bp, which is in the interglacial period prior to the beginning of the LGM, consistent with the environmental implication of this fauna.

The fauna of the Upper Cave has a complex composition. Species such as *Myotis* sp. *Paguma larvata, Acinonyx* cf.*jubatus* and *Rhinoceros* sp. belong to sub-tropical or tropical environments, while animals of cold climates such as *Coelodonta antiquitatis* and *Mammuthus primigenius* were not present. But there are also mammals adapted to a dry steppe environment, including *Ochotona daurica, Myospalax armandi, Equus hemionus* and *Gazella przewalskii* [Table 3(1)]. Furthermore, some forest species were also found, including two to three species of wolf and bear, as well as tiger, leopard and lynx, the last one being a cold climate species [Table 3(1)].

The traditional explanation for this mixture of species is that the palaeoenvironment around the Upper Cave encompassed forest and steppe; and the temperature must have fluctuated dramatically from day to night and from summer to winter (e.g. Jia 1980). But as the dating now demonstrates a time span from 33,000 to 23,000 years bp, over 10,000 years, there must also have been climatic fluctuations. It is likely that the fauna of the Upper Cave represents mixed animal species from various temporal phases, not distinguished in the 1930s excavations.

The fauna of the Upper Cave provides information on the variety and abundance of animal resources before the LGM. As Weidenreich described in 1939, hares and deer were piled up in hundreds within the tiny cave of one meter wide and three meters long (Weidenreich 1939:161). However, it should be noted that the Upper Cave fauna does not relate to the artefacts and human remains found in the upper chamber of the same cave. This fauna only reflects the rich faunal resources prior to the LGM in the Zhoukoudian area, and does not necessarily represent animals being exploited by human beings.

The other nine faunas were all found in association with artefacts, providing information of not only on faunal resources during the terminal Pleistocene, but also on the species being exploited by humans. Nos. 1-7 are from the western plateau/mountain area, and Nos. 8-10 are from the eastern plains [Fig. 3(1)]. A few characteristics of these faunas can be summarised as follows:

- Localities include caves, hilly areas and riverside terraces [Table 3(2)]. Besides the Upper Cave, two other faunas are from caves, four from river terraces, one from a riverside rockshelter, and two from hilly open areas. The sites are all close to water resources; a spring occurs at Mengjiaquan, and rivers occur near the other sites (An 1965; Chen 1993; IA, Hebei Province *et al.* 1991). It is not clear whether these archaeological sites were for residential, butchering or hunting purposes, but it seems that the locations were purposely chosen.
- The most commonly found species are onager (*Equus hemionus*), wild horse (*Equus przewalskii*), woolly rhinoceros (*Coelodonta antiquitatis*), and deer, gazelle and sheep [Table 3(1)]. The individual quantities of each species are often not reported, but teeth of at least 120 wild horses were found in Shiyu (site No. 1)(Jia *et al.* 1972). Assuming that the taphonomic conditions of these faunas are similar, the remains seem to indicate that ungulates were the primary game during the terminal Pleistocene in Central China. According to Wang and Tao (1991), elephant, deer, rhinoceros and wild horse were found at Locality 7701, Dingcun, Shanxi province[Fig. 3(1)], dated to 26400±800 years bp (Wang and Tao 1991). Details of this fauna are not

available, but it seems that the faunal resources in the southern area were more diversified than in the northern mountain/plateau areas prior to the LGM.

- Faunas 1, 2 and 8 are dated to about the same age (prior to LGM), yet the variety of animal species was much less in 1 and 2. The proportions of small carnivores and rodents are much lower in Shiyu and Dangcheng compared to the Upper Cave [Table 3(2)]. Ungulates were major species in most sites except 5 and 8, even in those faunas dated prior to the LGM, such as Nos. 1 and 2 [Table 3(2)]. Small carnivore and rodent species were also in low frequencies in the other faunas [Table 3(1)]. Whether this dominance of ungulates is due to human hunting selection, and/or simply reflects the natural ratio of carnivores to ungulates, is not yet clear.
- Remains of fish were only reported in Mengjiaquan (site No. 9), and fresh-water molluscs in Shizitan (site No. 5) and Xibaimaying (site No. 7)[Table 3(1)]. It is not clear whether the fish and mollusc remains were truly absent in the other faunas. But the discovery of fish and molluscs indicates that the food base for these foragers in the terminal Pleistocene was quite diversified.

According to the pollen analyses, many parts of Central China during the LGM were steppe with sparse trees and/or semi-desert. Trees and grassland disappeared; river and lakes shrank or dried up. Consequently, only animals adapted to a dry and cool steppe environment could have survived. Fauna Nos. 3, 4, 7 and 10A have been dated within the LGM epoch, and the dominance of ungulates and reduction of other species may reflect an increasing human hunting focusing on theses species as other resources became scarce. Onager and wild horse, gazelle and deer occur in these faunas. The first three species are adapted to steppe and semi-deserts, while the deer belongs to a more wooded environment. They can be hunted year-round. These species are fast-running, therefore, some special technique would have been needed to catch them in numbers. New tools and more effective hunting techniques might have been required.

The decrease of animal species is particularly visible in the eastern plain and the southern part of the plateau/mountain area. Geographically, Mengjiaquan is very close to the Upper Cave, yet the diversity of this fauna is reduced significantly compared to that of the Upper Cave fauna. There was clearly a decrease of carnivore species in Mengjiaquan, and ungulates became more dominant [Table 3(2)]. In the southern region of the plateau/mountain area, elephant and rhinoceros disappeared after the onset of the LGM, as in Tashuihe (site No. 3). As rhinoceros is a big but slow-moving animal and relatively easy and rewarding game (Nowak 1991), the disappearance of this species from archaeological assemblages may reflect its disappearance because of environmental factors rather than any change in human hunting selection. This change is consistent with the reconstruction of a cold and dry palaeoclimate and a steppe environment during the terminal Pleistocene in Central China.

The late Pleistocene gave way to the warm and moister early Holocene by 10,000 BP, as shown by faunal data in the period from about 10,000 to 7000 BP [Table 3(3)]. The species of steppe and/or semi-desert, such as wild horses and onagers, hyenas, gazelles and camels, all disappeared. Many species of sub-tropical to tropical forest and grasses returned to Central China, including several species of deer and rhesus monkey (*Macaca mulatta*), which now only live in south and southwest China (Zhou 1981). Fish, mollusc, tortoise and even alligator now appear in archaeological sites of the early Holocene, clearly indicating that there must have been abundant water resources in the vicinities of sites [Table 3(3)] (Zhang *et al.* 1994). All of these animal resources were exploited at the time when millet cultivation became well established (Zhang *et al.* 1994; IA CASS 1984).

3B(2). Floral resources.

The floral resources exploited by human beings can be illustrated by archaeological discoveries. Hackberry seeds, walnuts, hazelnuts and plum stones have been found in at least six early Neolithic sites in Central China, particularly in Cishan, where two storage pits containing a layer of hackberry seeds about 20 cm thick was found, as well as carbonised walnuts and hazelnuts (CPAM of Hebei Province and Handan City 1981). This discovery clearly indicates that even when agriculture was quite well developed, gathering was still an important part of the human economic strategy.

No such botanical discoveries have been made for archaeological assemblages during the LGM regime. Since pollen analysis has documented the retreat of forest and the dominance of steppe at this time, it is unlikely that nut-bearing trees would have been very common during the LGM, and the floral resources available to human beings would mainly have been seeds of grass and herbs.

According to pollen profiles from various boreholes in Central China, the major trees growing during the LGM were pine, fur and spruce, while drought-resistant herbs and shrubs dominated the vegetation [Fig. 3(6), 3 (10), 3 (11)]. In the Xibaimaying site, prior to the LGM, there were hazel (*Corylus sp.*) and walnut trees (*Juglans sp.*). But these species disappeared during the LGM, and even the quantity of pine species was reduced [Fig. 3(10)]. On the other hand, many grass and herb species were abundant during that time, bearing edible seeds. According to ethnographic data, seeds of this type were intensively collected by ethnographic foragers in both Australia and the Great Basin of the U.S., as well as nuts from pines in the latter (Cane 1984; Steward 1938).

The toolkit of some archaeological assemblages in Central China during the terminal Pleistocene suggests that cereal exploitation was practised during the LGM, exemplified by blades with glossed edge from the Xiachuan toolkit, as well as grinding tools in Xiachuan and Shizitan. This issue will be discussed in more detail in chapter 4.

Among all the above mentioned trees and herbs, one member of the grass family - the green foxtail (*Setaria viridis*) - is likely to have been collected by prehistoric gatherers in Central China. As foxtail millet is domesticated in this region, the green foxtail as its progenitor merits special attention.

It is not clear when green foxtail became widespread in the Pleistocene vegetation communities. Neither is the precise distribution of this wild cereal known, both in the present and in the prehistoric period. Only general statements of its morphological features and its rough distribution in Central China are available. Regarding the relationship between the wild and the domesticated foxtail millet, both species have nine pairs of chromosomes, and interspecific crosses between green foxtail and foxtail millet (*Setaria italica*) in the 1940s suggested that green foxtail was probably the immediate progenitor of cultivated millet (Li *et al.* 1945). This proposition has been confirmed by a recent isozymic analysis conducted in China. It is claimed that "the zymogram type of foxtail millet was similar to its wild species, the green foxtail, in the main bands. The distribution of zymogram types of esterase isozyme in the landraces from Europe was different from that in the landraces from China" (Gao and Chen 1988:136). Given the

archaeological discoveries in Central China and this isozymic study, the authors infer that foxtail millet was domesticated from green foxtail in the loess plateau in China (Gao and Chen 1988).

This argument is not new. It was proposed first by Vavilov in the 1920s, although not so precisely, as he suggested a much broader centre of millet domestication in Mongolia and China (Vavilov 1926: 61). Since the 1960s, many Chinese archaeologists have also argued for the same proposition (e.g. Ho 1969), based upon archaeological discoveries where large amounts of domesticated millet have been found in early Neolithic sites. But, contrary to the comprehensive studies undertaken on domesticated millet, very few have been conducted on green foxtail. Does green foxtail have similar characteristics to domesticated millet? What are the difference(s) between the two? How has the domestication process changed the botanical characteristics of green foxtail? None of these questions have yet been answered.

For the purpose of addressing at least some of the above issues, observations of green foxtail were conducted in 1996, with emphasis on the sequences of flowering and ripening, as well as undertaking a few harvesting experiments.

The observations were conducted in Qufu, Shandong province, in the lower Yellow valley in July and August 1996, based on green foxtail stands within an area of about two square kilometres. For the purpose of recording flowering and ripening, 30 plants were randomly chosen and numbered, while others were generally observed every day. The recording concentrated on flowering, ripening and shattering sequences of each tiller. An estimation was made of the percentage of shattering, i.e. how much grain had disappeared and how much was still left on the tiller each day. By this method of observation and recording I recovered the following data:

- Green foxtail is a grass highly adaptive to various, even very harsh environments. The soil where green foxtail grows is normally sandy with a loose structure, often quite dry but sometimes moist where located near ditches.
- Annual plants normally germinate in spring; but some plants of green foxtail sprouted as late as early July, and the tiller production can start within ten days to two weeks. Germination was particularly visible after heavy summer rain. For such late-germinating plants their life cycle will be very short, from July to probably late October. This may be an adaptation of green foxtail to the unreliable rainfall of the loess area. Further observation is needed to confirm this phenomenon. If it is confirmed, then there might be wild variants with short life-cycles in green foxtail, from which some types of domesticated millet with growing seasons as short as 56-80 days (Society of Crop Genetic Resources 1994) might have developed.
- The flowering time of green foxtail varies significantly from one plant to another, even from one floret to another on the same plant. It is quite common that in one individual plant, new tiller growth, flowering, ripening and shattering will all happen simultaneously [Fig. 3(16)].
- The ripening and shattering time of green foxtail varies not only among tillers of the same plant, but also on different parts of the same tiller. Thus one part of a tiller can be flowering while another part of the same tiller is ripening, with a few grains even beginning to shatter [Fig. 3(16), the middle tiller].
- Finally, the time span from beginning to completion of shattering also differs among tillers of the same plant, as well as between individual plants. This difference is not only in terms of the numbers of days for the whole process, but also in terms of the season, as ripening and shattering can happen from early July onwards, maybe earlier. It has been stated that green foxtail flowers and ripens from June to October (The Institute of Botany CAS 1972). I was only able to observe for one full month due to various reasons beyond my control. However, when I began my observation in early July some of the green foxtail had already shattered, and a few plants germinated from early to mid July. When I left in early August the majority of the stands were still in the process of flowering and ripening, as well as tiller-growing. Therefore it is very likely that the flowering and ripening season for green foxtail could be as long as four months. Further, information from local farmers also confirms the long maturing process of green foxtail.

The life schedule of green foxtail is summarised in Table 3(4). It is clear that green foxtail has a highly heterochronous growing schedule. As such, the time span from the emergence of the first tiller to the completion of seed-shattering of all tillers of each plant is very long. Of the 30 plants that I recorded, only two of them - Nos. 15 and 23 - were dead by the end of my 30 days observation. Both died prematurely - before the completion of their shattering, not as the natural end of their life cycle. The majority of the plants were growing well in the hot and moist summer climate, and no one plant completed its ripening and shattering within the period of observation.

The long and highly heterochronous life cycle of green foxtail is a salient feature which differs significantly from its domesticated counterpart, the latter always growing, flowering and ripening at the same time and in the same season, as the result of human domestication. We would not expect the present green foxtail to have exactly the same botanical characteristics as green foxtail at the end of the Pleistocene, as species are always evolving or responding to human manipulation. However, as a single botanical species, the basic characteristics should not have changed significantly over time.

In summary, the palaeoclimate during the LGM in Central China was cold and dry, and grasses dominated the vegetation. These changes would have affected human subsistence strategies. It is quite possible that exploitation of wild cereals was practised in China during and after the LGM regime, and green foxtail would have been one of these cereals. This assumption is based upon the proposal that prior to domesticating a cereal, humans would have exploited it for a substantial period of time. By this exploitation they acquired knowledge about the wild plant, and the knowledge was essential for the later domestication. Such a process from exploitation to cultivation/domestication has been documented in the Middle East from Natufian to the Pre- Pottery Neolithic A (PPNA) (Bar-Yosef and Belfer-Cohen 1989). But for China this assumption needs to be supported by archaeological and archaeobotanic data.

Table 3(1). Faunas of the Terminal Pleistocene in Central China

****

Species	English name	The plateau/mountain area.							The plain area.				Habitat and characteristics
Species	**English name**	No. 1	No.2	No.3	No. 4	No.5	No. 6	No. 7	No.8	No. 9	No. 10 A.	B	
Primates	Primate.												
Homo sapiens sapiens		+	+	+					+	+			
Pongo sp.	Orangutan.										+		Tropical or sub-tropical forest.
Insectivora													
Erinaceus sp.	Eurasian hedgehog.	+	+						+			+	Forest, grassland and scrub etc.
Scaptochir us sp.	Short faced mole.								+				Typical species in dry steppe.
Chiroptera													
Myotis sp.(A)	Brown bat.								+				Sub-tropical and tropical.
Myotis sp.(B)	Brown bat.								+				"
Rhinolophus sp.	Horseshoe bat.								+				
Chiroptera fam.gen.et.													
sp.indet.									+				
Lagomorpha	Lagomorph.												
Ochotona daurica	Pika.								+				Steppe, desert and hilly areas.
Lepus europaeus	European hare.		sp.						+	sp.			Various, steppe, hilly or valley.
Rodentia	Rodent.												
Myospalax sp.	Zokor.	+	+			+							
Myospalax armandi									+				Species of dry steppe.
Myospalax fontanieri	Chinese zokor.										+	+	Various, steppe, hilly or valley areas
Muridae gen.et sp.indet.										+			
**Microtus brandtioides*	Vole.					sp.							Dry steppe.
Rhizomys sp.	Bamboo rat.		+										Bamboo at elevation of over 1200m.
Microtus epiratticeps									+				
Cricetulus barabensis													
obscurus	Ratlike hamster.								+				Open dry steppe/border of desert.
**Cricetulus varians*									+				
Sciurus sp. indet.	Tree squirrel.								+				Deciduous, coniferous/tropical forest.

Table 3(1). Faunas of the Terminal Pleistocene in Central China (continued).

****

		The plateau/mountain area.							The plain area.				Habitat and characteristics
Species	**English name**	No. 1	No.2	No.3	No. 4	No.5	No. 6	No. 7	No.8	No. 9	No. 10		
Rodentia (continued)													
Petaurista sulcatus	Giant flying squirrel.								+				Mountain/hilly forest.
Alticola cf.stracheyi									+				Mountain area.
Meriones sp.									+				
Apodemys sylvaticus	Wood mice.								+				Grass, wooded land and forest.
Rattus rattus	Black rat.					+			+			+	Human commensal.
Hystrix sp.	Old world porcupine.								+				Various-forest/steppe.
Marmota sp.	Marmot.		+										Steppe, alpine meadow/forest edge.
Carnivora	Carnivore												
Canis sp.	Wolf.		+					+					
Canis lupus	Grey wolf.						+		+	+		+	
Nyctereutes procyonoid	Raccoon dog.								+				Forest, thick vegetation with water.
Ursus sp.	Bear.		+							+			
Ursus thibetanus	Black bear.								+				Moist deciduous forest/brush.
**Ursus spelaeus*	Cave bear.								+		+		
Vulpes corsac	Corsac fox.		sp.						+				Hilly and/or forest area.
Vulpes vulgaris	Fox.								+				"
Cuon alpinus	Dhole.								+				Various but avoid desert.
Meles leucurus	Badger.		sp.						+	sp.		+	Forest, dense vegetation.
Mustela cf.altaica	Mountain weasel.								+				Open mountain/grassland.
Mustela eversmanni	Steppe polecat.								+				Steppe.
Crocuta sp.	Hyena.	+											"
**Crotuta cf.ultima*	Spotted hyena.		+						+		+	+	Steppe.
Paguma larvata	Masked palm civet.								+				Tropical or subtropical forest & caves.
Felis chinensis	Chinese cat.					sp.			+			sp.	Forest and steppe.
Lynx Lynx	European lynx.								+				Cold area: forest, mountain and steppe.
Lynx sp.									+				
Panthera tigris	Tiger.	+							+	+			Forest & bush.
Panthera pardus	Leopard.								+			+	Various ecotone. Speed over 60km/h.
Acinonyx cf.jubatus	Cheetah.								+				(Sub)tropical forest, semi-desert, grass.

Table 3(1). Faunas of the Terminal Pleistocene in Central China (continued).

****

		The plateau/mountain area.							The plain area.				Habitat and characteristics
Species	**English name**	No. 1	No.2	No.3	No. 4	No.5	No. 6	No. 7	No.8	No. 9	No. 10		
Proboscidea	Proboscidean												
Elephas sp.	Elephant.							+	+				Tropical & subtropical area.
**Palaeoloxodon*	Nauman's straight-									+			Forest and bush.
namadicus	tusked elephant)												
Perissodactyla													
Equus hemionus	Onager	+			+		+	+	+	+	+	+	Dry steppe/desert. Speed 50-70km/h.
**Equus przewalskyi*	Tarpan.	+	+		+		+	+		+			Cold/dry steppe/grassland.
													Speed over 60km/h.
**Coelodonta antiquitat*	Woolly rhinoceros.	+	+	sp.			+	sp.		+	+	+	Steppe and tundra.
Rhinoceros sp.	Rhinoceros.								+				Tropical and subtropical area.
Artiodactyla													
Moschus sp.	Musk deer.									+			Forest/bush land.
Elaphurus sp.	Tufted deer.									+			" ".
Muntiacus sp.	Muntjacs.									+			Forest/dense vegetation with water.
Megaloceros sp.	Giant deer.		+				+						
**Megaloceros cf.*													Steppe and grassland.
ordosianu	Ordos giant deer.	+											
Cervus elaphus	Red deer.	+		+									Forest and bush. The largest deer.
Cervus hortulorum	Spotted deer.								+				
Cervus canadensis	Canadian deer.		+						+	+			Forest and steppe between 32-60°N.
Cervus grayi				+						+			Grassland and edge of forest.
Cervus sp.			+		+	+	+	+			+	+	
Capreolus manchuricus	Roe deer.		+						+	sp.		+	Forest & grassland in cold area.
Capreolus sp.										+			
Sus sp.			+				+	+	+		+		(Sub)tropical forest and hilly areas.
Sus scrofa	Wild boar.						+			+			
Bovidae gen.et sp.indet(A)		+	+				+						
Bovidae gen.et sp.indet(B)		+											
Bos sp.					+		+		+				
**Bos primigenius*	(aurochs)						+	+		+			Grassland, steppe & edge of forest.
**Bubalus cf.*		+					sp.				sp.	sp.	Forest & bush in moist area.
wansjocki	Wansjock's buffalo.												

Table 3(1). Faunas of the Terminal Pleistocene in Central China (continued).

****

Species	English name	The plateau/mountain area.							The plain area.				Habitat and characteristics
		No. 1	No.2	No.3	No. 4	No.5	No. 6	No. 7	No.8	No. 9	No. 10		
Artiodactyla(continued)											A	B	
Gazella sp.			+		+	+							Steppe and desert.
Gazella przewalskyi	Tibetan gazelle.	+	+				+	+	+	+	+	+	Steppe in cold and dry areas.
Gazella cf.subgutturosa	Goitred gazelle.	+	+										Steppe in dry area.
Ovis sp.	Sheep.			+					+				
Capricornis sp.	Serow.										+		Rugged mountain/edge of forest.
Camelus sp.	Camel.									+			Desert.
**Camelus knoblaki*	" "	+	+										
Pseudois cf. nayaur	Blue sheep.		+	+									Open slope/plateau with abundant grass.
Aves	Birds.												
Struthio sp.	Ostrich.	+			+			+					Desert & steppe.
Struthio anderssoni	" "								+		+		
Pisces	Fishes.												
Cypriniformes													
Cyprinus carpio	Carp.									+			Fresh water.
Ctenopharyngodon idellus										+			
Mollusk	Mollusc												
Mullusc sp.													
Corbicula fluminea	Clam.							+					
Gyraulus convediusculus								+					
Noduloria sp.						+							
Gyraulus compressus								+					
Radxi auricularia	Ear pond snail.							+					
Cipangopaludian sp.						+							
Mullusc (continue)	Mollusc												
Fruticicda sp.						+							
Total species per fauna:		16	23	5	6	9	13	14	48	24	12	13	

Table 3(1). Faunas of the Terminal Pleistocene in Central China (continued).

Note:

No. 1: Shiyu, Shanxi province; radiocarbon dated to 32,220±625 (AMS) and 28,945±1370 years bp (Jia *et al* 1972; Yuan 1993b).

No. 2. Dangcheng, Shanxi province; radiocarbon dated to 31,495±1570 and 28,730±1175 bp (Wu and Sun 1989).

No. 3.: Tashuihe, Shanxi province; uranium and radiocarbon dated to 20,000±1000 and 26,165±1005 bp respectively (Chen 1993).

No. 4: Xueguan, Shanxi province; radiocarbon dated to 13,550±150 bp (Wang *et al.* 1983).

No. 5: Shizitan, Shanxi province; suggested to belong to the terminal Pleistocene (Linfen Cultural Bureau 1989).

No.6.: Dafa, Shanxi province; suggested to belong to the terminal Pleistocene (Gao *et al.*1991; Li and Wang 1992; Wu *et al.* 1990).

No. 7: Xibaimaying, Hebei province; uranium dated to 18,000±1000 and 15,000±1000 bp (IA, Hebei Province 1989).

No. 8: The Upper Cave of Zhoukoudian near Beijing; radiocarbon (AMS and traditional) dated from 33,460±850 to 10,470±360 bp (Chen *et al.* 1992; Qi 1989).

No. 9: Menjiaquan, Hebei province; suggested to belong to the terminal Pleistocene (IA, Hebei Province *et al.*1991).

N0.10: Xiaonanhai cave, Henan province; radiocarbon dated to 24,100±500 and 11,000±500 bp. Sub-group A reprents the fauna from lower layer (3,4), while sub-group B represents the fauna from the upper layer (1,2) (Zhou 1965).

*: Extinct species.

Table 3 (2). A brief summary of the terminal Pleistocene faunas in Central China.

Fauna	Location	Locality	Absolute dates (bp) *	Quantity & % of orders within each fauna			Total sp.	Extinct sp.	Extinct %
				Carnivore	Ungulate	Others			
1. Shiyu	112°17'E 39° 25'N.	Hilly open site.	32,220±625, 28,945±1370	2 sp. 12.5%.	11 sp. 68.75%	3 sp. 18.75%	16	4	25%
2. Dangcheng.	113° 48'E 37° 14'N.	Cave.	31,495±1570 28,730±1175	5 sp. 21.74%	13 sp. 56.52%	5 sp. 21.74%	23	3	13%
3. Tashuihe.	113° 13'E. 35° 29'N	River-side rockshelter.	26,165±1005 20,000±1000.		5 sp. 100%		5	2	40%
8. The Upper Cave.	115° 55'E 39° 41'N.	Cave.	from 33,200 to 23,105 bp.	18 sp. 37.5%	9 sp. 18.75%	21 sp. 43.75%	48	4	8.3%

Table 3 (2). A brief summary of the terminal Pleistocene faunas in Central China (continuing).

Fauna	Location	Situation	Absolute dates (bp) *	Quantity & % of orders within each fauna			Total sp.	Extinct sp.	Extinct %
				Carnivore	Ungulate	Others			
7. Xibaimaying.	114° 14'E. 40° 07'N.	River terrace.	18,000±1000; 15,000±1000.	1 sp. 7.1%	7 sp. 50%	6 sp. 42.9%.	14.	2	14.29%
10. Xiaonanhai.	114° 7' E. 36° 7' N.	Cave.	24,100±500;	(A) 2 sp. 16.67%	(A) 7 sp. 58.33%	(A) 3 sp. 25%.	12	3	25%
			11,000±500	(B) 5 sp. 35.71%	(B) 6 sp. 42.86%	(B) 3 sp. 21.43%	14	2	14.29%
4. Xueguan.	111° E. 36° 27' N.	River terrace.	13,550±150		5 sp. 83.33%	1 sp. 16.67%	6	nil.	
5. Shizitan.	no data.	River terrace.	Terminal Pleistocene?	1 sp. 11.11%	2 sp. 22.22%	6 sp. 66.67.%.	9	nil.	
6. Dafa.	112° 50' E, 37° 40' N.	River terrace.	Terminal Pleistocene?	1 sp 7.69%	12 sp. 92.31%	nil.	13	3	23.1%
9. Mengjiaquan.	117° 47' E. 39° 52' N.	Hilly open site.	Terminal Pleistocene?	4 sp. 17.4%	14 sp. 60.87%	5 sp. 21.73%	23	4	17.4%

*For details of dates see Table 4 (1).

(Sources: An 1965; Chen *et al*. 1992; Chen 1993; Xie *et al.* 1991)

Table 3(3). Faunas of the Early Holocene in Central China.

		The eastern plai			Habitat
Species	**English name**	No.1	No.2	No.3	
Primates					
Macaca mulatta	Rhesus monkey.		+		Sub-tropical forest.
Lagomopha					
Lepus tolai	Mongolian hare.		+	sp.	Steppe.
Rodentia					
Myospalax psillurus	Zokor.		+		
Carnivora					
Meles meles	Badger.		+	sp.	Forest.
Paguma larvata	Masked palm civet.		+		Sub-tropical/tropical forest.
Panthera pardus	Leopard.		+		Sub-tropical and tropical forest.
Canis lupus	Grey wolf.	+	sp.		Forest.
Nyctereutes sp.	Racoon dog.			+	Forest, thick vegetation with water.
Felis sp.	Small cat.			+	
Canis familiaris	Domesticated dog.	+?	+	+	
Artiodactyla					
Moschus moschiferus	Musk deer.	+			
Cervus nippon	Sika deer.		+		Grassland and forest.
Cervus elaphus	Red deer.	+	+		Bush and forest.
Cervus sp.			+	+	
Elaphurus davidianus	David's deer.	+	+	+	Swampy, reed-covered marsh lands.
Capreolus capreolus	Roe deer.	+	+		Forest, sparsely wooded valleys, open fields.
Hydropotes inermis	Chinese water deer.		+		Sub-tropical river/lake side or hilly areas.
Muntiacus muntjak	Muntjacs.		+		Forest/dense vegetation with water.
Sus scrofa	Wild boar.		+	+	Hilly/forest.
Sus domestica	Domesticated pig.	+?	+	+	
Bos exiguus	Ox.		+		
Aves	Birds.				
Anster fabalis	Wild goose.		+		
Gallus gallus domesticus	Domesticated chicken.	+?	+		
Grus sp.	Crane.	+			
Reptilia	Reptile				
Emydidae G. et sp.			+		
Emydidae sinensis	Chinese tortoise.			+	
Alligator sinensis	Chinese alligator.			+	Swampy areas.
Pisces	Fishes.				
Cypriniformes					
Ctenopharygodon idellus			+		Freshwater fish.
Mollusca	Mollusc				
Lamportula sp.			+		
Total species		9	23	11	

No. 1. Nanzhuangtou, date at approximately 12,000-8500 BP (Yuan *et al.*1991; Zhou 1992).
No. 2. Cishan, date to approximately 8000-7500 BP (IA, CASS 1991; Zhou 1981).
No. 3. Jiahu, date to around 8500-7600 BP (IA, Henan Province 1989).

Table 3 (4) The life schedule of green foxtail (*Setaria viridis*).

Life phases of green foxtail (per tiller)	Range (days)	Mean (days)	Sample quantity (tillers)
From green tiller to flowering	1-8.	2.64	100
From flowering to ripening	1-14.	6.60	100
From ripening to shattering commencement	1-12.	4.53	100
Shattering from commencement to completion	7-23.	15.42	59

Chapter 4. Archaeological discoveries in Central China.

Archaeological discoveries dating from the end of the Pleistocene to the early Holocene in Central China are far from sufficient. To date, less than twenty sites and find places with absolute dates for the terminal Pleistocene have been found in this region. Only one site has been found dating to the beginning of the Holocene. It is only after the ninth millennium BP that numerous Neolithic assemblages appeared in Central China, namely the Dadiwan, Cishan, Houli and Peiligang cultures [Fig. 4(1)].

The reasons for such a scarcity of archaeological data for the transitional period from the terminal Pleistocene to the Holocene are not known. This scarcity has certainly caused great difficulty in studying the transition from Palaeolithic to Neolithic in Central China, a region which has long been considered a place of origin of millet cultivation, and consequently, the centre of Chinese civilization. To date, the accumulated archaeological data can only provide a very brief frame for the transition from foraging to farming in Central China.

As discussed in chapter 3, the environments in the west and the east of Central China differ. The archaeological assemblages here will be divided into two groups according to their geographic locations, and presented in chronological order.

4A. Archaeological assemblages in the plateau/mountain region.

This region mainly corresponds with the middle valley of the Yellow River, which makes a huge northern curve before entering the gorges of the western foothills of the Lu-liang Mountains and then flows into the alluvial plains. The majority of this area is covered by loess, except for the mountain areas (Wu *et al.* 1987) [Fig. 4(1)].

Apart from the Yellow River itself, there are other big rivers in this region. The Fen River originates from the northern part of the hilly area and runs southwards through the central part of this region, joining the Yellow River at the southwest corner of the Fen-Wei Graben [Fig. 4(1)]. It is at this corner that the Wei River joins the Yellow River from the western loess plateau [Fig. 4(1)]. Up towards the northern part of the plateau/mountain area there is another river - the Sanggan River (previously known as the San Kan Ho) - which runs northeast towards the Lu-liang mountains before pouring into the Yongding River near Beijing [Fig. 4(1)].

To date, archaeological discoveries have been concentrated within the valleys of these rivers. While archaeological assemblages of the late Pleistocene have been mainly found along the Fen River, early Neolithic assemblages occur in the Wei River basin; and cultural remains from the late Pleistocene to the early Holocene have been discovered in the Sanggan River area [Fig. 4(1)]. Over 25 find places and sites assigned to the late Palaeolithic, and 21 sites of the early Neolithic have been discovered in this region (Chen 1993; Shi 1986). However, many are only surface collections. Only those discoveries that have been properly excavated and dated, with substantial materials uncovered, are discussed here. The focus of discussion will be on lithic toolkits as indicators of the economic strategies employed by people during the terminal Pleistocene and the early Holocene.

The Fen and Wei River reaches will be treated as one sub-region in the following discussion. Archaeological discoveries found in the Sanggan valley are different in terms of their lithic assemblage, and are discussed separately.

4A(1). Archaeological assemblages in the Fen-Wei valleys.

The earliest relevant archaeological assemblages of the terminal Pleistocene, and maybe also the best representatives of the lithic industry in this region, belong to the **Xiachuan** complex [Fig. 4(1)], within Shanxi province. The Xiachuan complex consists of artefacts found from 16 locations within the basin of the Fuyi River (Wang *et al.* 1978). Ten sites occur within an area of about 93 km^2, all with similar lithic implements (Wang *et al.* 1978).

Discovered in 1970, two sites (Fuyihe Geliang and Shuijingbei) were first excavated in 1973 (Wang *et al.* 1978) and yielded thousands of lithic implements. The abundance and importance of these discoveries stimulated further survey and excavations: Fuyihe Geliang (locality No. 1) was further surveyed and excavated in 1974-1975 and 1976-78 (Wang 1986). Apart from this site, another seven sites were dug in the excavation seasons of 1976-78: an open site called Shunwangpin; localities 2-4 on the terraces of the Fuyi river, and three caves in a hill called Shandiyan (IA, CASS and CPAM of Shanxi Province 1978) [Fig. 4(2)]. Numerous surveys have also been conducted since the 1970s (Chen 1996). Thousands of lithic implements were found in Localities 1-4 and Shunwangpin during the excavations in 1976-1978, but the deposits in the caves are of much more recent date (IA, CASS and CPAM of Shanxi Province 1978), and will not be discussed in this thesis.

The results of the Xiachuan research, unfortunately, have not yet been fully published. It is not clear if all the sites excavated in the Xiachuan basin had the same stratigraphic sequence, although the excavators perceived the lithic assemblages from various sites as being contemporary (Wang *et al.* 1978). Nor are the relationships among these sites clear. A preliminary report of the first survey and excavation in 1973-74 was published in 1978 (Wang *et al.* 1978). As for the excavations of 1976-78, only an initial draft was circulated among selected Chinese prehistorians (IA, CASS and CPAM of Shanxi Province 1978). In the 20 years since the last excavation, a few excavators have written articles, which provide important although very limited information on these archaeological assemblages. Fortunately, I was permitted to see the lithic implements found in the 1976-1978 excavations during my fieldwork in China in 1995-1996. Although I was not permitted to make drawings or take photographs, I obtained a copy of the initial draft. These reports, as well as my notes, will be the database for discussion of the Xiachuan complex in this thesis.

The major discovery in Xiachuan is the large quantity of lithic implements from the upper cultural layer in some of the excavated sites. Other archaeological remains are rare except for loose charcoal and a few bones and teeth of ox and wild sheep (Chen and Wang 1989). Thus, the functions of these sites are not certain (Chen 1996; Wang 1986). Different lithic assemblages were found in the lower and upper layers of the Xiachuan sites, and the researchers have concluded that there was technological and cultural continuity from a large flake industry, to a blade and microblade industry (Shi, 1989; Chen and Wang 1989).

Eleven radiocarbon dates have been obtained from the Xiachuan sites, with tested samples mainly being charcoal,

but also some peat and soil. All samples were collected during the 1976-78 excavations. According to the original numbers of the samples, it seems that four samples were from Locality 1, one each from Localities 2- 4, and three samples should have been from Shunwangpin [Table 4(1)].

The radiocarbon dates of Locality 1 appear to be in sequential order, as the earliest date of 36,200 bp belonged to the lower stratum containing large, crude flakes and cores, while the dates from 23,900 to 16,400 bp were from the upper cultural layer which contained implements of the flake tradition and microblades (both defined in chapter 2) [Table 4(1)]. Another three dates were obtained respectively for the same upper cultural layer at Localities 2-4, ranging from 20,700 to 18,375 bp [Table 4 (1)]. It seems that the upper cultural layer of Xiachuan is within the temporal span from 23,900 to 16,400 bp in the terrace sites.

Three radiocarbon dates were obtained for the Shunwangpin site, two from the upper cultural layer, and one from the Holocene deposit overlying the upper cultural layer (IA, CASS 1991) [Table 4(1)]. The date of 2800 bp for the Holocene deposit can be ignored as it is out of the scope of this thesis, although the huge temporal gap between layer 3 (2800 bp) and layer 2 (13,900 bp) also seems problematic (this may be due to the severe erosion in this area). Both the other dates carry statements of their strata: sample ZK-634 was from the lower section of the upper cultural layer, and sample ZK-762 from the upper section of the upper cultural layer [Table 4(1)]. It seems that these two dates were in sequential order and in conformity with the strata, placing the Shunwangpin assemblage within the temporal range from 19,600 to 13,900 bp.

The 'Xiachuan complex' mainly refers to the lithic assemblage from the upper cultural layer, where 1800 lithic implements were unearthed by the first survey and excavations of 1973-1975 (Wang *et al.* 1978), and more than ten thousand lithic implements were found in the upper cultural layer in the excavations in 1976-78. Only 3254 pieces were described in the draft report. Whether the rest were debitage or implements is not clear. All the reported lithic implements are listed in Table 4 (2).

According to the two preliminary reports, the majority of these lithic implements bore secondary retouch and/or use-wear, except for some microblades, so they were perceived as tools (Wang *et al.* 1978; IA, CASS and CPAM of Shanxi Province 1978). I have observed some of these reported artefacts found in 1977-1978, and concur with this statement.

The lithic assemblage found in the two excavations consists of three separate lithic industries in a purely technological sense, namely the microblades, the flake tradition and the heavy-duty stone tools. The flake tradition is the dominant component, accounting for 77.74% of the implements [Table 4(2)]. In Xiachuan, this industry consisted, according to the Chinese excavators, of points, arrowheads, scrapers of great variety, burins, backed blades, perforators, and denticulates such as saws and sickles (Wang *et al.* 1978; Shi 1989).

While the scrapers and points are multiple-function tools, the arrowhead is claimed to be a tool of specific usage. Often being delicately chipped at the bottom [Fig. 4(3): 1, 2], this retouch suggests that the tool was to be attached to a shaft. Studies on the Mesolithic in Europe have already demonstrated the large-scale and persistent use of arrowheads as a new hunting armature (Rozoy 1985). Though "arrowheads" are not present in large quantities within the Xiachuan assemblages, the presence of this tool clearly suggests the employment of this new technique for hunting by the Xiachuan people.

The denticulate is another seemingly new tool type in Xiachuan [Fig. 4(5): 8, 9]. They were made on thin and flat flakes, with one or two edges delicately retouched into a denticulate form (IA, CASS and CPAM of Shanxi Province 1978). Judging from the morphological resemblance of these denticulate tools and the sickles found in Neolithic sites in Central China, it has been claimed that the denticulate was used for cereal cutting, (Shi 1989), but no direct evidence has been supplied for this argument.

Among other flake implements of Xiachuan, some of the long and thin flakes labelled as points or scrapers by the excavators might be of importance. These flakes are normally 5 to 10 cm in length, less than 1 cm in thickness and over 1 cm wide, with delicate, often bifacial retouch, forming denticulate edges [Fig. 4(3):7]. These edges often bear use-wear, some of which looks very like a 'sickle gloss' (author's observation in 1995). As demonstrated by experiments, such sickle gloss is considered to have been caused by cutting cereals or reeds in the Middle Eastern context (e.g. Unger-Hamilton 1985). For China, this is an issue that can only be addressed by experiment and high-power microscopic examination.

Another aspect of the Xiachuan assemblage is the microblade industry, which accounts for 19.84% of the lithic implements (Wang *et al.* 1978; IA, CASS and CPAM of Shanxi Province 1978). This industry is exemplified by various types of microcores, namely wedge, boat, prismatic, conical and semi-conical, polyhedral and funnel shaped; and microblades detached from microcores by indirect percussion, as judged from the tiny and flat flutes [Fig. 4(4)]. After detachment, the majority of microblades were utilized without trimming, but were often truncated on both ends. It has been claimed that the purpose of this truncation was to make short "razors", which were then inserted into handles of soft materials to form the cutting edges of composite tools (Jia 1978). Some of the microblades were retouched and were called "micropoints" by the excavators (Wang *et al.* 1978).

The microblade industry was truly a new lithic tradition in Central China in the terminal Pleistocene. The earliest site with this industry to date is **Locality 7701** Dingcun, Shanxi province [Fig. 4(1)]. Two radiocarbon dates have been obtained for this site, one being 26,495±590, another older than 40,000 bp (Li 1993). These two dates are problematic as the time span between them is huge, and the question of their reliability has been raised (An 1983). The excavator of this site agrees that caution should be taken for the older date (Wang 1986), but he argues that 26,495 bp is reasonable, particularly for the microblades which are similar to those found at Xiachuan, but with a smaller variety of microcores. It has been suggested that the lithic assemblage of Locality 7701 could have been ancestral to Xiachuan (Wang and Tao 1993).

Microblade industries of this type were widespread in North and East Asia, Japan and North America from the late Pleistocene to the early Holocene. It has been demonstrated by Neolithic discoveries in China (e.g. the Huanglianliantai site in Gansu province, of about 4500 BP) that microblades were inserted into bone, antler, even stone handles to form the cutting edges of daggers or knives (Jia 1978). A similar function has been assigned to the microblades of the late Pleistocene (Jia 1978). Recently microblades were found to have been inserted as the cutting edge of a bone harpoon in Xinglongwa in Northeast China, dated to about 8400 BP (Neimeng Archaeology Team CASS 1997). These composite tools would have been effective weapons for foragers.

The last but not the least important lithic industry in Xiachuan was that of the heavy-duty stone tools. Although they only account for 2.42% of the lithic implements, the importance of some of these large tools should not be overlooked. Made on large flakes and pebbles, the heavy-duty stone tools include not only choppers and chopping tools, but most significantly, chipped axes, adzes, and grinding tools - hand-stones and slabs (for definitions see chapter 2) [Fig. 4 (3) & (5)]. The latter were new elements of the toolkit at the end of the Pleistocene.

The grinding tools merit special attention. A total of 27 grinding slabs were found in the excavations (IA, CASS and CPAM of Shanxi Province 1978; Wang *et al.* 1978). Four were reported to be made of sandstone and four of quartzite (IA, CASS and CPAM of Shanxi Province 1978; Wang *et al.* 1978). Raw materials for the others are not clear. They were only found in the upper cultural layer (Shi 1996 personal communication). Made from either flat-surfaced stone blocks or pebbles, the size of these grinding slabs range from 8 to 35 cm in length, and 8.5 to 21 cm in width (IA, CASS and CPAM of Shanxi Province 1978). Their shapes vary. Most have been trimmed around the edge (IA, CASS and CPAM of Shanxi Province 1978) [Fig. 4 (3) & (5)]. But three small handstones were utilized without much prior shaping (IA, CASS and CPAM of Shanxi Province 1978).

The use-wear on these grinding implements varies but mainly consists of four patterns. The first is in the form of a shallow oblong depression (author's observation 1996) with longitudinal striations (IA, CASS and CPAM of Shanxi Province 1978) [Fig. 4(5): 4]. Light pecking is said to have been present on the central surfaces of some grinding slabs (IA, CASS and CPAM of Shanxi Province 1978), which would have been the result of rejuvenation when the surface became too smooth (Smith 1985). Worn grinding slabs of this type show a cross-section in the form of a 'saddle shape' [Fig. 4(5):5]. Such use-wear indicates that the grinding action was a linear one, probably moving from one edge to the other and withdrawing backwards. The precise numbers of this type of grinding implement are not reported, but at least seven examples described in the preliminary report exhibit this use-wear (IA, CASS and CPAM of Shanxi Province 1978).

The second use-wear pattern of grinding slabs is in the form of circular depression (IA, CASS *et al.* 1978; Wang *et al.* 1978) [Fig. 4(3): 11]. The depression is shallow, usually over 5 cm in dimension (author's observation 1996), and relatively flat. Circular striations were present on the surface (Wang *et al.* 1978). This use-wear should have been caused by circular grinding movements. It is not clear how many implements bear such use-wear, but one example was described in the preliminary report of the first excavation (Wang *et al.* 1978), and three were described in the draft report of the second excavation (IA, CASS and CPAM of Shanxi Province 1978).

The third use-wear pattern is also in circular shape, but it is deeper and smaller, like a basin. According to the preliminary report of the second excavation, one example measured 3 cm in diameter and 1.5 cm in depth, and was surrounded by pecking marks (IA, CASS and CPAM of Shanxi Province 1978), which was possibly caused by pounding movements. Three specimens with such use-wear were mentioned in the report for the second excavation (IA, CASS and CPAM of Shanxi Province 1978).

The last use-wear pattern is in the form of a shallow, long groove [Fig. 4(5): 5]. So far, only one fragmentary specimen has been described (IA, CASS and CPAM of Shanxi Province 1978). It is very similar to the use-wear pattern on some grindstones found in Australia (Smith 1985) [Fig. 9(2):1]. According to ethnographic data from Australia, such use-wear is caused by wet-grinding of seeds, mainly seeds with soft husks, such as all the seeds from the grass family (Cane 1984; O'Connell *et al.*. 1983). In Australia, these seeds were ground with water into a paste, which then was dumped from the grinding stone into a wooden container which had been placed under the lip of the stone, and cooked (Cane 1984; Tindale 1977). The grinding slab with groove from Xiachuan is only a fragment, and it is not clear whether the groove was extended to the other edge of the slab. However, the presence of this grinding slab with a groove in Xiachuan suggests the possibility of the practice of a similar seed-processing technique in the Yellow River valley during the terminal Pleistocene. This question awaits further investigation.

In Australia, grinding implements have been used for various purposes, from grinding tubers, cracking nuts and fruits, to grinding fern root, as well as seed-grinding (Smith 1985). As no residue analysis has been conducted on the grinding tools found in Xiachuan, it is risky to assume that the tools were specialised for plant/seed grinding. However, it is legitimate to argue that plant/seed grinding would have been one of the functions of these grinding implements. As well as the ethnographic data from Australia, morphologically similar grinding slabs have been found in Neolithic Jiahu, dated to 8500 BP (IA, Henan Province 1989) [Table 4(7)]. Similar grinding tools were also found in various Yangshao agricultural assemblages, dated after 6500 BP (IA, CASS 1984).

Seven grinding handstones were found in Xiachuan (IA, CASS and CPAM of Shanxi Province 1978). It is reported that the handstones bore either circular or straight-line striations on the grinding surfaces, indicating that they were used either for straight or circular movements (IA, CASS and CPAM of Shanxi Province 1978). Their use-wear patterns are consistent with those on the grinding slabs, indicating that they were used in conjunction. Remains of ochre have also been found on the surface of some grinding tools from Xiachuan (Shi 1989).

The last type of lithic implement to merit attention is a type of discoid chopper. Made of quartzite, this tool was purposely chipped into a round shape (IA, CASS and CPAM of Shanxi Province 1978) [Fig. 4(5):1]. Although the function of this chopper type is not clear, similar tools have been found in early to middle Neolithic sites in the Yellow valley. A few such round discoids were found in Cishan, dated to about 7800 BP, with a relatively thick cross-section and polished edges (CPAM of Handan *et al.* 1977). It has been reported that 2230 such discoid implements were found in Miaodigou (Yangshao culture) in Henan, accounting for 85.90% of the total lithic implements (IA, CASS 1959). Similar discoveries have also been made in adjacent areas (IA, CASS 1959). The presence of this discoid is important evidence for cultural links between the Xiachuan assemblage and the Neolithic cultures in the middle Yellow valley.

Recently, a report of the Xiachuan survey conducted from 1990 to 1992 was published. According to this survey, find places in Xiachuan are more widely spread than previously thought (Chen 1996). Although the artefacts found are similar to those found in the 1970s, the discovery of debitage (defined in chapter 2) in some find places is important, since it implies local tool making (Chen 1996). As the find places were quite dense in the Xiachuan basin, all containing similar artefacts, it is likely that the artefacts found over these places belong to one group, and the Xiachuan basin of about 93 km^2 would have been their catchment area.

Other archaeological assemblages similar or comparable to Xiachuan have been found in the Fen River valley. Some of these are surface collections, but a few have been excavated, such as Xueguan and Shizitan, both in Shanxi province.

The **Xueguan** site is located on a loess-covered terrace of the Xin river, which is a small tributary of the Yellow River [Fig. 4(1)]. The site was discovered in 1964 and excavated in 1979 and 1980. According to the excavation report, some faunal remains and a total of 4777 lithic implements were found. Of the lithic implements, 3630 were classified as debitage and 1147 as cores, flakes and stone tools (Wang *et al.* 1983). The assemblage included a microblade industry, a flake industry and a few heavy-duty tools as listed in Table 4(3). The flake industry dominated the Xueguan assemblage, accounting for 82.47% of the total lithic implements. The microblade industry accounted for 17.27%, and the heavy-duty stone tools only accounted for 0.26% of the total lithic assemblage [Table 4(3)].

The lithic assemblage of Xueguan is comparable to that of Xiachuan, as indicated by its tripartite combination of microblades, flakes and heavy-duty tools, though the proportion of flakes is higher than at Xiachuan. A noticeable difference between Xiachuan and Xueguan is the absence of arrowheads and grinding tools in the Xueguan assemblage. Only one chipped axe was found in Xueguan [Fig. 4(6): 9]. Do these differences represent different economic strategies, or just small sample sizes? The question remains unanswered at present.

The Xueguan microblades were found in association with remains of wild horse, onager, gazelle and deer [Table 3(1)], indicating that the Xueguan occupants were hunters. But details of both animal remains and lithic implements are not fully reported. Only one radiocarbon date was obtained, providing a date of 13,550±150 bp for the Xueguan assemblage (Wang *et al.* 1983) [Table 4(1)]. This date, if valid, places the Xueguan assemblage contemporary with the last stage of the Xiachuan complex.

Another site, **Shizitan**, is located at the southern edge of the Lu-liang Mountains, on a bedrock terrace of the Qingshui river, only two kilometres east of the Yellow River [Fig. 4(1)]. The site is situated at the front of the terrace, facing the river, and there is a rockshelter at the rear of the terrace (Linfen Culture Bureau 1989).

Shizitan was discovered and excavated in 1980. Lithic implements and faunal remains were uncovered. According to the excavation report, the faunal remains were quite fragmentary, and they could only be identified to genera. They included house rat, vole, tiger, wild pig, rhinoceros, ox and deer, plus hundreds of gazelle teeth, of which a substantial proportion were burnt; also ostrich and some molluscs [Table 3(1)] (Linfen Culture Bureau 1989).

A major discovery at Shizitan was the lithic assemblage found in the upper layer (layer 4). A total of 1794 lithic implements was described in the excavation report, but whether there was also debitage is not mentioned. The lithic assemblage is listed in Table 4 (4).

Apart from lithic implements, two shell tools were also found from the upper layer; one appears to be a point and another a scraper. Whether these were marine or river shells is not mentioned in the report. Seven pieces of ochre were discovered; a similar colour was also found on the surface of one grinding slab, suggesting that one of the functions of the slabs was ochre grinding (Linfen Culture Bureau 1989).

The lithic implements from the upper layer of Shizitan again exemplified the three lithic industries: the microblades, the flakes and the heavy-duty stone tools. The microblade industry totals 755 specimens and accounts for 42.09% of the lithic assemblage; the flake industry totals 1016 specimens and accounts for 56.64%; and the heavy-duty stone tools account for only 1.27%. Obviously the flake industry dominated in Shizitan. Shizitan differs from Xiachuan in its larger proportion of microblades, which account for more than 42% in Shizitan, but less than 25% in Xiachuan.

The majority of the other new tool types of the Xiachuan assemblage were also found in Shizitan, such as arrowheads, backed blades, and denticulates. The scraper was the prevailing tool within the flake tools, as at Xiachuan. The composition of heavy-duty stone tools in Shizitan also differs from that of Xiachuan. Chipped axes and grinding slabs were found both in Shizitan and Xiachuan, but flaked adzes and discoid implements were not reported in Shizitan [Fig. 4(7)].

Judging from the toolkit of Shizitan, particularly the quite well-made arrowheads, it seems that the occupants were hunters and gatherers, particularly ungulate hunters. This assumption can be further supported by the large amount of gazelle (*Gazella* sp.) teeth discovered in association with the artefacts. Whether cereals were also collected and the grinding slabs were also used for cereal processing at Shizitan remains a question.

It is also reported that two rock-paintings were found in the rockshelter at Shizitan, stated to be painted in ochre (Linfen Culture Bureau 1989). As ochre was found on the surface of one grinding slab, it is suggested that the paintings were contemporary with the assemblage (Linfen Culture Bureau 1989). But the absolute date of these paintings remains to be investigated.

No absolute dates are reported for Shizitan, but a geological date has been proposed. The microblades and flakes were mainly found in the loam deposit of layer 4. As a similar loam deposit in the nearby Luochuan site, Shaanxi Province has been radiocarbon dated to 10,000 BP, it is inferred that the cultural layer in Shizitan should be of about the same date (Linfen Culture Bureau 1989).

While archaeological data of the terminal Pleistocene are relatively rich in the Fen River basin, early Neolithic assemblages are scarce. Only one site (**Zaoyuan**) and two find places have been discovered so far. It is only reported that Zaoyuan is located at the junction of two local small rivers; that the size of the site was about 20,000 square meters, but that it has been severely destroyed by being underneath a modern village. During a rescue dig, one refuse-pit was excavated and pottery as well as some stone tools were recovered (Anonymous 1994).

The toolkit of Zaoyuan consisted of stone knives, rollers and pottery spindle whorls (Anonymous 1994). This toolkit is partially comparable to that of Cishan and Peiligang in the North China Plain. It has been argued that cereal cutting and processing were probably practised, as indicated by the knives and rollers (Anonymous 1994).

The other two find places in the Fen River valley are **Wenxi** and **Shimen**. For Wenxi, it is only reported that archaeological remains similar to those of Zaoyuan have been found (Anonymous 1994). As for Shimen, only one grinding slab and one roller were collected. The original deposit could not be found (Anonymous 1994). But these two implements are identical to the typical grinding slab and roller set of Cishan and Peiligang (below). This discovery seems to suggest the existence of an early Neolithic culture in southern plateau/mountain area.

What might have been the relationship between the archaeological culture of the terminal Pleistocene as represented by Xiachuan, Xueguan and Shizitan, and the culture of the early Holocene as represented by Zaoyuan? Did the latter develop from the former? All these issues need further exploration, yet none of them can be addressed adequately due to the present scarcity of archaeological data. Whether more early Neolithic assemblages in the Fen River reaches once existed but have been destroyed, or have not been found, or if there are other reasons for this data scarcity, remain unknown.

Contrary to the scarcity of archaeological data from the Fen valley, quite rich archaeological assemblages of the early Holocene have been found in the Wei River valley [Fig. 4(1)]. There is an important surface collection from **Shayuan**, Shaanxi province, located in the lower valley of the Wei River [Fig. 4(1)]. Shayuan is on a small sand dune, with marsh-land on its eastern side. According to Chinese researchers, this would have been a good location for hunting and gathering (An and Wu 1957; Ban-po Museum and CPAM of Dali County 1983).

Shayuan has been surveyed four times, first in 1957 (An and Wu 1957) and then in 1973, 1979 and 1981 (Ban-po Museum and CPAM of Dali County 1983). Discoveries from these surveys are disclosed in two reports. A total of over four thousand lithic implements were found on the surface, but the original deposit could not be traced even by trial excavation (An and Wu 1957; Ban-po Museum and CPAM of Dali County 1983). Only some of the artefacts were described in the two reports, and they are listed in Table 4(5).

There seems to be a difference between the assemblages from the two surveys. For example, backed blades, flaked axes and stone spheroids were not found in the first survey. As the two surveys were carried out by different observers, and no original deposits could be found, it is difficult to tell what such a difference signifies. However, all artefacts are perceived as belonging to one lithic assemblage (Ban-po Museum and CPAM of Dali County 1983).

These lithic implements are morphologically similar to other archaeological assemblages of the terminal Pleistocene, particularly the microblade industry (An and Wu 1957) [Fig. 4(8)]. Here the tripartite combination occurred again. New tool types such as backed blades and chipped axes were found in the second survey, but missing were the grinding tools. As the artefacts had been disturbed by the wind and sand movement of thousands of years before they were collected, it is very likely that only part of the original assemblage was found. But even though this is a surface collection, the researchers emphasised that no potsherds were found. They suggest that the date for Shayuan should be from Mesolithic to early Neolithic (An and Wu 1957).

Some well-dated and excavated Neolithic sites have been discovered in the Wei River region, Shaanxi province, represented by Dadiwan and Baijia. All are dated at around 7800 to 7000 BP, roughly contemporary to the Cishan and Peiligang cultures in the North China Plain [Table 4(7)].

Located on the first terrace of the Wei River, **Dadiwan** is the westmost archaeological site of the early Holocene found in China to date. The site contains archaeological assemblages from early to middle Neolithic (the latter being the Yangshao culture in this region), which demonstrates a cultural continuity in the Wei River valley. The lowest cultural layer of Dadiwan is the early Neolithic deposit in this region, which is relevant to this thesis.

Dadiwan was excavated in five seasons from 1978 to 1982. The tools of the early Neolithic layer found in the first two excavation seasons are classified in Table 4(6). Only brief reports were published for the third to the fifth excavation seasons. It is said that another 90-odd tools were discovered (Dadiwan Excavation Group 1981, 1982), but the exact quantity was not disclosed. It is stated that the Dadiwan toolkit was dominated by lithic implements, particularly flaked tools (Dadiwan Excavation Group 1982), while polished stone tools only accounted for about 10% of the total lithic assemblage; and "the majority of stone and bone tools was only polished at the edge" (Dadiwan Excavation Group 1982: 22). Important tools like axes, adzes and knives were found both flaked and polished, but a new tool - the spade (defined in chapter 2), was polished only (Dadiwan Excavation Group 1981, 1982).

Although axes and adzes could have been multiple function tools, it is claimed that spades were used for soil-digging during cultivation (Tong 1989). As broomcorn millet (*Panicum miliaceum*) and seeds of rape (*Brassica* sp.) were found in a large pit (H398) in Dadiwan (Dadiwan Excavation Group 1982), there is little doubt that agriculture was already practised, and the above-mentioned tools are likely to have been used for cultivation activities, such as land clearance, soil digging and harvesting/cutting.

It is notable that the older tool types such as choppers and scrapers were still found in Dadiwan, as well as stone cores and some small blades, the latter being slightly thicker than microblades; but no microcores were reported. The small blades are particularly interesting as their morphological features are comparable to the microblades of the terminal Pleistocene [Fig. 4(9):10-12]. Does this resemblance indicate any cultural link between Xiachuan, Xueguan and Dadiwan? Such a possibility cannot be ruled out.

Found also were stone "pellets", bone arrowheads, and spindle whorls made of potsherds (Dadiwan Excavation Group 1981, 1982). The appearance of spindle whorls shows the emergence of fibre manipulation, and possibly weaving.

Apart from the toolkits, pottery was another important category of the archaeological assemblage. The pottery was made by a slab-building method with fine sand as temper, and the majority of vessels were red with surface spots of grey and black, indicating uneven temperature and atmosphere in pottery firing (Dadiwan Excavation Group 1981, 1982). The pottery assemblage was quite simple, consisting of *bo* (defined in chapter 2), jars, urns and pots, some of the pots with three small legs. Cord-marking was the main type of decoration, but painting was found on the rims of some bowls (Dadiwan Excavation Group 1981, 1982).

Most sherds and tools were found in pits and semisubterranean houses. Thirteen pits and remains of three houses were found during the first two seasons of excavation, as well as 16 burials (Dadiwan Excavation Group 1981, 1982). The presence of houses, refuse pits and cemeteries suggest a sedentary life for the Dadiwan occupants.

Similar archaeological assemblages have also been found in the site of **Baijia**, Laoguantai, the lowest level of Beishouling, Huaxian and Zijing, as well as in another dozen find places, all in the Wei River valley and the Qin Mountains area [Fig. 4(1)]. Among these sites, Baijia is important due to its rich toolkit.

Located on the northern bank of the Wei river, the Baijia site was excavated in 1982 and 1983. A total of 35 pits, 2 house floors, 17 burials and a large amount of pottery and tools made of stone, bone and antler were found (The Sixth Shaanxi Archaeological Team 1984). While the pottery,

pits, houses and cemeteries were identical to those of Dadiwan, the toolkit of Baijia was richer in terms of its variety. Unfortunately, the quantities of many tools were not reported. It is reported that the lithic implements included polished spades of three types, axes and adzes; one stone knife, some stone rollers, and three flake scrapers (The Sixth Shaanxi Archaeological Team 1984). Tools made of organic materials consisted of spear heads, two arrowheads, one saw and one knife, plus needles and rings, all made of bone. Other organic tools include drills made of bone, animal tooth or antler; and one carved antler; as well as nine sickles and knives made of river shells (The Sixth Shaanxi Archaeological Team 1984). It is also claimed that small stone flakes were found in the site (The Sixth Shaanxi Archaeological Team 1984).

In general, there are some common characteristics of these early Neolithic assemblages:

- Tools are made of stone, bone, antler and fired clay. New tool types such as axes, adzes and spades are major implements within the lithic toolkit, as well as grinding rollers. But scrapers made on flakes and small blades are still found. Bone arrowheads replace the stone equivalents of the terminal Pleistocene. Other new tools are also present - needles, spindle whorls etc. Sickles of shell were discovered in Baijia (The Sixth Shaanxi Archaeological Team 1984). The toolkits seem to indicate that the economic strategies of these early Holocene groups were a combination of cultivation, hunting and gathering. The discovery of broomcorn millet and rape seed in Dadiwan suggests that cultivation already existed then.
- The semisubterranean house floors, pits and graves demonstrate a relatively sedentary life, but it is not clear whether this was all year-round or seasonal. The cultural deposits in all these sites were quite shallow, ranging from 20 to 50 cm in depth (Dadiwan Excavation Group 1981; The Sixth Shaanxi Archaeological Team 1984).
- A large proportion of pottery vessels was red colour with sand temper.

.Since Anderson found the famous Yangshao culture in the 1920s (Anderson 1929) in the Wei valley, the accumulation of archaeological data has demonstrated this region as one of the original centres of Chinese Neolithic and ancient civilization. The stratigraphic and pottery sequences of Dadiwan, Beishouling, Beiliu and a few other site illustrate that the Yangshao - Banpo culture was a direct descendent of the Dadiwan and Baijia assemblages (IA, CASS 1984). The questions are: where were the makers of these archaeological assemblages from? were they local people of the Wei River, or were they from somewhere else? and what was the culture preceding these early Neolithic assemblages?

These questions have not been discussed in detail. A few scholars do mention the possibility of Xiachuan representing an ancestral stage for the early Neolithic culture in Central China as a whole (Shi 1986). But the discussion is very brief and not many arguments have been roused. The major problem regarding this issue is the lack of archaeological data from 10,000 to 8000 BP. However, the small blades and flakes found in Dadiwan and Baijia bear some morphological similarities to their counterparts of Xiachuan. Therefore, it is possible that there were cultural links between the Xiachuan/Shizitan Palaeolithic of the southern plateau/mountain area, and the Neolithic in the Wei valley. Shayuan could have been a central location in this cultural transmission and/or diffusion [Fig. 4(1)]. However, more archaeological data are needed to verify this proposition.

4A(2). Archaeological assemblages from the terminal Pleistocene to the early Holocene in the Sanggan River valley.

In the Sanggan River valley, there seem to be two different lithic assemblages, one with microblades, flake and heavy-duty tools, another without microblades. The former is represented by the Youfang and Hutouliang sites, and the latter by the Xibaimaying site.

Youfang is located on the second terrace of the Sanggan River in Hebei province [Fig. 4(11)]. Excavated in 1986, 697 stone tools, 2675 pieces of debitage and a small number of mammal bones were uncovered (Xie and Cheng 1989). The lithic implements are classified in Table 4 (8).

The Youfang lithic assemblage consists of three industries again - the flake, microblade and heavy-duty tools, comparable to those found in the Fen River region. But there are differences. New tools in Xiachuan such as arrowheads, denticulates, digging tools, grinding slabs and axes were not found in Youfang. Only backed blades, as one of the new non-microblade tool forms of the terminal Pleistocene, were found in this site [Fig. 4(12)].

The microblade industry within the Youfang assemblage accounts for 15.36% of the total number of lithic implements. The presence of microblades indicates that composite tools were probably used in subsistence procurement activities, this being a hallmark technique at the end of the Pleistocene in Central China. But apart from this, very little can be said about the economic activities practised by the Youfang occupants. Were the bow and arrow employed in hunting? Were wild cereals being collected? These questions remain unclear.

Realizing the simplicity of the toolkits recovered, the excavators have proposed that Youfang belongs to the microblade and flake lithic complex represented by Xiachuan, but is more simplified (Xie and Cheng 1989). No absolute dates have yet been obtained for Youfang. It is only proposed that the Youfang assemblage would be within the terminal Pleistocene, approximately contemporary with Xiachuan (Xie and Cheng 1989). The ash and fired earth found in association with the lithic implements, as well as the high percentage of debitage, may indicate that this site was a temporary working place for tool manufacturing.

Xibaimaying is located very close to Youfang [Fig. 4(11)]. Discovered and dug in 1985, a total of 1546 lithic implements, 315 bone flakes and a large quantity of vertebrate and mollusc remains were discovered (IA, Hebei Province 1989).

The animal bones found in association with the lithic implements were quite rich in quantity, but very fragmentary. According to the report, all molluscs were of species existing in the area today. Within the vertebrate fauna, 10 species were found, including ostrich, aurochs, wild horse and onager, gazelle, deer, boar, woolly rhinoceros, elephant and carnivore. The majority of them are existing species, except the woolly rhinoceros and aurochs. This fauna is identical to those from other sites of the terminal Pleistocene in this region, such as Hutouliang (below) and Shiyu (IA, Hebei Province 1989).

Pollen samples were collected from Xibaimaying, and the results suggest a steppe environment dominated by *Artemisia* with a few pines and *Picea,* indicating a palaeoclimate colder and drier than the present [Fig. 3(10)] (IA, Hebei Province 1989). It is notable that the quantity of pollen from the grass family was extremely small, as only one grain was found in the cultural layer (IA, Hebei Province 1989). This scarcity may indicate that many important seed-bearing plants such as green foxtail and wild broomcorn millet were not present when human beings were active at Xibaimaying.

The lithic implements found in Xibaimaying are listed in Table 4 (9). This lithic assemblage apparently differs from archaeological assemblages discovered in the Fen River valley. Only the flake industry was present here; both the microblade and heavy-duty stone tools were missing [Table 4(9)] [Fig. 4(13)]. None of the "new" tools such as denticulates or chipped axes and grinding slabs was found here. The faunal remains indicated that the makers of the Xibaimaying toolkits were hunting similar ungulate game as in the sites of the Fen River basin, but whether their hunting technique differed remains unclear. The abundant shell remains found in Xibaimaying indicate that four species of fresh water molluscs were quite intensively collected.

Besides the lithic implements and animal remains, no other archaeological remains were discovered except for a small amount of charcoal. It is notable that over 68% of the lithic implements were debitage, thus the excavators infer that the Xibaimaying assemblage might have been the remains of a site for manufacturing lithic implements. It was noted that the broken edges of some animal bone fragments were rounded, but the lithic implements did not exhibit any marks of water transportation (IA, Hebei Province 1989).

Only two uranium series dates on animal teeth have been obtained for this assemblage, dated to 18,000±1000 and 15,000±1000 bp (IA, Hebei Province 1989) [Table 4(1)]. After analysing the lithic implements, the excavators argue that this simple lithic assemblage dominated by flakes and scrapers is a representative of another Palaeolithic tradition in the Sanggan River valley during the terminal Pleistocene, being different from the microblade tradition represented by Youfang (IA, Hebei Province 1989).

Another important discovery in the Sanggan River basin is the **Hutouliang** assemblage, which consists of nine sites around Hutouliang village. All sites are located on the second terrace of the Sanggan River within an area of less than 10 square kilometres, with the minimum distance between any two sites only being 500 meters (Gai and Wei 1977).

Although this group of sites was discovered and excavated in 1972 and 1974 in two seasons, only a preliminary report has been published so far. Whether the artefacts found in these sites were all *in situ* is not clear. However, three hearths were found in locality No. 73101, containing a large quantity of charcoal, fired animal bones and egg-shells of ostrich, as well as some stone tools (Gai and Wei 1977). Four pebbles were found around hearth No. II5-6, with many broken pieces of animal bone spread over the pebbles. Lithic implements, debitage, stone flakes and cores at various stages of flake production were also found around the three hearths (Gai and Wei 1977). Based upon these discoveries the excavators infer that this site was a living and stone tool manufacturing place (Gai and Wei 1977).

A substantial quantity of animal bones was found in Hutouliang, the majority being fragments of long bones, mandibles and teeth, without crania, vertebrae and ribs (Gai and Wei 1977). Identifiable species include fox, ostrich, various species of mice, wolf, wild horse and onager, deer, ox, three species of gazelle and wild pig (Gai and Wei 1977). All of them were animals of a temperate steppe ecotone, indicating a cool and dry palaeoenvironment.

The total number of lithic implements from Hutouliang is not reported, only those being considered as representatives of the lithic assemblage have been described. They are summarised in Table 4 (10). It seems that the tripartite combination of microblades, flakes and heavy-duty tools was present again in Hutouliang, although it is not clear which lithic implements are from which site. But it seems that the lithic assemblage of each site differs. Hearths and debitage were found at locality 73101, and it is inferred as a place for manufacturing tools and a camp site (Gai and Wei 1977). At another locality, No. 65040, only a dozen retouched tools were found, and this locality has been inferred as an observation point for hunting (Gai and Wei 1977). Locality 65039 was near a spring; the lithic implements found there were mainly points and scrapers, in association with fragments of sheep horns and broken animal bones, so it is interpreted as a butchering site (Gai and Wei 1977). Locality 72117 yielded a large number of cores, debitage and retouched implements, and is inferred as a site for manufacturing stone tools (Gai and Wei 1977). It seems that the nine sites might have belonged to one forager group. Loc. 73101 might have been a residential camp site, and the others being various 'locations' (Binford 1980). However, more information is needed to address the mobility pattern suggested by the Hutouliang assemblage.

It is noticeable that within the flake industry, several of the so-called points (the quantity is not reported) were carefully trimmed on the bottom, either forming a thinner end, or a sharp end, or a concave one [Fig. 4(14)]. These 'points' are possibly hafted arrowheads or spear heads, otherwise such an effort to systematically retouch the bottom part would not have been made. The large quantity of animal bones found in the sites further suggests that the prehistoric Hutouliang occupants were hunters, probably using these 'points' for their hunting activity. But other tools such as grinding slabs and chipped adze/axes are all missing from Hutouliang. Does this absence rule out the existence of cereal collection? This question cannot be answered yet.

Other noticeable artefacts from Hutouliang include ornaments - three pieces of drilled shell, eight flat beads made of ostrich shell, one flat bead made of bird limb bone, and one perforated stone bead (Gai and Wei 1977). Some of these ornaments were polished. Only one radiocarbon date was obtained for Hutouliang, dated to 10,690 ±210 bp [Table 4(1)], thus placing the Hutouliang assemblage at the end of the Pleistocene.

It is reported that a recent excavation in the Hutouliang area disclosed pottery fragments in association with lithic implements similar to those of Hutouliang (Xie 1996). This discovery was made at **Yujiagou** in 1995, very close to Hutouliang village [Fig. 4(11)]. The pottery was 'coarse pottery, reddish brown colour... and was decorated by finger-nail impressions' (Tang 1997a: 5). One potsherd was dated to 11,870±1720 bp by thermoluminescence (Tang 1997a). This date is close to the radiocarbon date for Hutouliang.

According to a brief news report, the excavation at Yujiagou discovered three cultural levels. Lithic implements, animal remains and ornaments were found in the lower level; polished stone tools, pottery and lithic implements identical to Hutouliang were discovered at the middle level, and Neolithic cultural remains were present in the upper level (Xie 1996). It is claimed that this stratigraphic profile indicates a cultural sequence in the Hutouliang area (Xie 1996). Unfortunately, further details are not yet available.

To the east of Yujiagou and only a few kilometres west of the famous Nihewan village, there is an early Holocene site called **Toumafang** [Fig. 4(11)]. Details for this site are not available yet, but this seems to be one of the earliest Holocene sites in the Sanggan River valley reported so far. It is only reported that artefacts were found in a black soil layer overlying the upper Pleistocene loess (Xie 1996). Artefacts include flaked tools made on small flakes, microblades, one flaked adze, fragments of a grinding

slab(s) and a few polished stone tools; and a large number of flaked and polished bone and antler tools (Xie 1991). Two pieces of brownish red pottery were also found, with sand temper (IA, Hebei Province 1993). The microblade industry is said to be a salient feature of the Toumafang assemblage, represented by a large quantity of typical microblades and microcores (Xie 1991). One radiocarbon date has been obtained for this site, of 7530±100 bp (IA, Hebei Province 1993).

Thus archaeological data from Youfang, Hutouliang, Yujiagou and Toumafang in the Sanggan valley disclose a local cultural chronology from 18,000 to 7500 bp, within which the microblade industry seems to be a hallmark of cultural continuity. The development of the Xibaimaying assemblage is not yet clear.

So far, the major archaeological sites in the plateau/mountain area have been summarised. Apparently, archaeological discoveries of the terminal Pleistocene in this region are richer than in other areas in Central China. But it should be pointed out that no solid evidence for agriculture and/or year-round sedentism in the early Holocene prior to Dadiwan has been discovered in this area. Archaeological data for the period from 13,000 to 8000 BP are very scanty here, and more research is needed.

4B. Archaeological assemblages in the North China Plain.

The North China Plain encompasses part of the middle and the whole lower valley of the Yellow River and the northern part of the Huai River delta, roughly between latitudes 40° and 32.5° N, and from longitude 115° E to the sea [Fig. 4(1)]. The Yellow River runs through the middle of this region, dividing the area into north and south divisions, both with archaeological assemblages of the terminal Pleistocene and the early Holocene [Fig. 4(1)].

4B(1). Archaeological assemblages from the terminal Pleistocene to the early Holocene in the northern part of the North China Plain.

Prior to the 1980s, one of the best-known archaeological assemblages of the terminal Pleistocene in the North China Plain was the Upper Cave at Zhoukoudian, although the quantity of lithic implements found there was quite limited. A polished deer antler, one polished bone needle and a few stone and bone beads, as well as remains of at least 8 human individuals, were found (Zhou 1955). Two conventional radiocarbon dates were obtained in the 1970s, dated to 10,470 and 18,340 bp [Table 4(1)].

However, in 1989 and 1992, two batches of samples from the Upper Cave were dated by AMS dating in Oxford, and the result was a surprise. According to these dates, the archaeological materials from the Upper Cave are actually dated from 29,000 to 24,000 bp, and the animal remains dated to 33,200-23,150 bp (Chen *et al.* 1992) [Table 4(1)]. Therefore, the Upper Cave assemblage falls out of the temporal scope of this thesis. The archaeological assemblages of the terminal Pleistocene in the northern part of the North China Plain are now exemplified by Mengjiaquan and Xiaonanhai, while remains of the early Holocene are represented by Nanzhuangtou and Cishan.
Mengjiaquan is located on the southern edge of the alluvial fan of the Yan Mountains. The site used to be a spring, but as the water dried up in the 1980s, artefacts were discovered. Excavated in 1986 and 1990, remains of *Homo sapiens sapiens* and 23 species of animals, as well as more than 23,000 lithic implements, were discovered (IA, Hebei Province *et al.* 1991).

According to stratigraphical data, this site seems to have been disturbed by water. It is possible that the animal bones and lithic implements have been transported by water, but the tools only show light rolling, suggesting that this transportation was not over a long distance.

It has been reported that the Mengjiaquan lithic implements include microcores and microblades, flakes and flake cores, scrapers, points, perforators, backed blades, choppers, flaked adzes, stone hammers and anvils, and debitage (IA, Hebei Province *et al.* 1991). Debitage accounted for 64.7% of the total pieces, retouched flake tools 10.9%, and the rest were microcores, microblades, small flakes and flake cores (IA, Hebei Province *et al.* 1991). Unfortunately, the preliminary report is very brief, and quantities are not given.

However, this lithic assemblage seems to illustrate the tripartite combination again: with the microblade, flake and heavy-duty elements. It has been claimed that some of the flaked adzes were partially polished [Fig. 4(15)] (IA, Hebei Province *et al.* 1991). One radiocarbon date has been obtained for this site, of 17,500±250 bp (IA, Hebei Province 1993), thus placing this assemblage contemporary with the later phase of the Xiachuan complex.

South of Mengjiaquan, in the central part of the North China Plain, lies **Xiaonanhai** cave. Discovered and dug in 1960, the Xiaonanhai assemblage today remains one of the most important assemblages of the terminal Pleistocene in Central China. The cave is located on the eastern flank of Louding Hill, about 500 meters north-west of the Heng River. A trial excavation was conducted close to the opening of the cave, considered to be the original entrance (An 1965). Based upon stratification and archaeological contexts, An (1965) suggests that this cave was first occupied by hyenas in the late Pleistocene, as coprolites of this animal were found in the lowest layer (layer 5). Then human beings inhabited the cave, as illustrated by burnt animal bones and charcoal, found in layers 4 and 3 overlying layer 5. The large quantity of lithic debitage from these layers indicates stone tool manufacturing. The cave was finally abandoned because of a roof-fall (An 1965).

The raw materials for the lithic implements were mainly chert (90%) and quartz (9.61%), plus a few pieces of flint and chalcedony. The artefacts from Xiaonanhai are classified in Table 4 (11). Apparently the flake industry dominates the Xiaonanhai assemblage, in association with only 4 heavy-duty stone tools; the microblade tradition is completely missing here. Within the flake tradition, there were small and narrow flakes with triangular or parallelogram cross-sections [Fig. 4(16)]. These small flakes are morphologically like microblades but are slightly larger and thicker. An (1965) suggests that these flakes were detached by skilful direct percussion, as judged from their irregular shapes in comparison with true microblades (An 1965). Some of the flakes bear use-damage, suggesting that they were directly used without secondary retouch (An 1965).

None of the new tool types found in the plateau/mountain area, such as "arrowheads", flaked axes and adzes, or grinding tools have been found in Xiaonanhai. Three radiocarbon dates place the Xiaonanhai assemblage within the terminal Pleistocene [Table 4(1)]; but without microblades, the toolkit does not resemble other contemporary assemblages, such as the Xiachuan complex. Whether this indicates a chronological, a regional or cultural difference, remains unclear.

A substantial number of animal bones was found in Xiaonanhai in association with the lithic implements, including woolly rhinoceros (*Coelodonta antiquitatis*), onager (*Equus hemionus*), Tibetan gazelle (*Gazella*

przewalskyi), deer (*Cervus* sp.) and hyena (*Hyaena ultima)*. Teeth of orangutan (*Pongo* sp.) and cave bear *(Ursus cf. spelaeus)* were reported in the lower layer (Zhou 1965) [Table 3(1)]. The majority of these animal bones were fragments; some of the onager bones were burnt, possibly by human action (An 1965). The last phase of Xiaonanhai was at the end of the Pleistocene at 11,000 BP.

Archaeological assemblages from the early Holocene are extremely scarce in Central China. In the late 1980s a site called **Nanzhuangtou**, in Xushui county, Hebei province, was discovered and excavated. The site is located on an alluvial plain of the Bao river, less than 20 km east of the Tai-hang mountains (Yuan *et al.* 1991). According to the report, the layers of Nanzhuangtou from 2 to 6 were all marsh sediments. A typical loess deposit was found underneath layer 6 (CPAM of Baoding District *et al.* 1992). It was in this loess deposit that a small ditch was found filled with grey sandy soil, charcoal and shells of small molluscs, as well as three antlers. One of the antlers was in a vertical position when found, perhaps having been purposely buried (CPAM of Baoding District *et al.* 1992). Remains of grass/wood ash and charcoal were also found in the bottom of layer 5 and in layer 6. Carbonised wood and red fired earth, two lithic flakes, as well as one piece of pig bone and one pig tooth were found on top of the ash. It is suggested that this was probably a hearth of a temporary camp (CPAM of Baoding District *et al.* 1992).

Artefacts found in Nanzhuangtou were very limited: only 15 potsherds, one grinding slab and one roller, as well as one piece of bone and one drill made of antler were uncovered (CPAM of Baoding District *et al.* 1992) [Fig. 4(17)]. Also found were two pieces of wood with cutting marks, one like a handle and another a plank. These artefacts were found in layers 5 and 6 (CPAM of Baoding District *et al.* 1992).

In contrast to the scarcity of artefacts, animal bones were quite rich in Nanzhuangtou. A total of 104 bones included wolf (*Canis lupus*), possibly domesticated pig (*Sus scrofa*) and dog (*Canis familiaris*), red deer (*Cervus elaphus*), David's deer (*Elaphurus davidianus),* roe deer (*Capreolus capreolus)* and possibly domesticated fowl (*Gallus* sp.) (Zhou 1992). This fauna indicates a marshland/forest or forest-edge environment.

If the identifications are certain, then the pig and dog are the earliest domesticated animals found in Central China so far, as their discovery in layers 5 and 6 seems to place them within the date range from 10,815$\pm$140 to 8800$\pm$108 bp, according to radiocarbon dating (Yuan *et al.* 1991) [Table 4(7)]. This discovery, if confirmed, is very important for the origin of animal husbandry in China.

As the Nanzhuangtou cultural layers were mainly a lake and marsh deposit, and no remains of dwellings were found, it was probably not a residential site. However, it is premature to draw any conclusions at this stage, as data are so limited.

To date, one of the best dated representatives of early Neolithic culture in the North China Plain is the **Cishan** culture, named after a Neolithic site of about 80,000 square meters at Cishan village in Wu'an county, Hebei province [Fig. 4(1)]. The site was discovered in 1972 and excavated in 1976 and 1978. Over 640 artefacts, and more than 120 pits were found in 1976; and about 2000 artefacts, 474 pits and two house floors were discovered in 1978 (CPAM of Handan District *et al.* 1977; CPAM of Hebei Province and CPAM of Handan City. 1981).

The Cishan site is located on a loess platform north of a Nanming river, surrounded by hills of the Tai-hang Mountains (CPAM of Handan *et al.* 1977; CPAM of Hebei Province and CPAM of Handan City 1981). The fauna found in Cishan included freshwater fish (*Ctenopharyngodon idellus)*, mollusc (*Lamportula* sp.) and tortoise (Emydidae G.et sp. indet); rhesus monkey (*Macaca mulatta)*, badger (*Meles meles*), masked palm civet (*Paguma larvata*), leopard (*Panthera pardus*), wild boar, and several species of deer, with some birds (*Anser jabalis* and *Gallus gallus*, the last one perhaps domesticated) (Zhou 1981) [Table 3(3)]. This fauna clearly indicates a palaeoenvironment of forest and meadow around the site, with a river to the south, providing various food resources for the prehistoric occupants. Walnuts (*Juglans regia*), hazelnuts (*Geetis bunseana*) and hackberry seeds (*Coryliu leterapluylea)* were also found in Cishan (Zhou 1981). Although millet cultivators, the Cishan people also exploited all these natural resources, and the site catchment area must have included the surrounding mountain areas.

The stratification of Cishan is quite simple. Three cultural layers were found under the modern surface: layer 1 was a Western Zhou (1200-770 BC.) deposit, layers 2 and 3 were Neolithic. Layer 3 was called the first phase and layer 2 the second phase of the Cishan culture (CPAM of Hebei Province and CPAM of Handan City 1981).

In layer 3, 186 pits were excavated, their shapes vary from round, oval to oblong (CPAM of Hebei Province and CPAM of Handan City 1981). Some of the pits were probably subterranean houses, as judged from the discovered artefacts. For example, No. H28 is an oval shaped pit, 2.7 m long, 1.8 m wide and 0.8 m deep. There were three steps down on one side, and fragments of fired earth bearing reed impressions were found on the floor. It is proposed that pits of this type were residential dwellings (CPAM of Hebei Province and CPAM of Handan City 1981).

The oblong shaped pits contained carbonised cereal seeds. Samples were taken from a few pits, and were identified as remains of foxtail millet (*Setaria italica*) (CPAM of Hebei Province and CPAM of Handan City 1981). Cereal remains were found in 62 pits. The thickness of carbonised cereal accumulation in each pit varied between 0.3 and 2 meters when discovered (CPAM of Hebei Province and CPAM of Handan City 1981). It has been claimed that if the decomposed remains from all the Cishan storage pits were converted into fresh millet, that would equal about 50,000 kilograms (Yan 1992a). But whether all the storage pits contained millet is a question raised by a few scholars (e.g. Wu 1989).

The toolkit found in Cishan consists of lithic and organic tools. The lithic toolkit from the lower cultural layer 3 is classified in Table 4 (12). A salient feature of this lithic assemblage is the absence of the microblade and flake industries. Flaked axes dominate the stone implements, accounting for over 56% of the total. Spades account for over 11%. The finish of the axes seems to vary from flaked only, to partially polished or fully polished tools [Table 4(12)]. The significance of this difference, however, is not clear.

Besides stone tools, implements made of bone, antler, animal tooth and shell were also found in the lower cultural layer, listing in Table 4 (13). This tool assemblage consists of tools of various functions, some of them particularly used for hunting, such as the arrowheads; others for fishing, such as the harpoons. Cutting tools like knives might have been used both for cereal collection and other functions. It is notable that such presumed tools for hunting and fishing account for 27.84% of the organic tool assemblage, suggesting the importance of hunting and gathering.

Another important category of Cishan artefacts is the pottery. A total of 156 pottery utensils was found in layer 3. Over 90% of the vessels were made with sand temper, the rest was fine pottery. These pottery utensils were mainly made by pinching and ring-building methods (CPAM of Hebei Province and CPAM of Handan City 1981), and over 50% of the utensils were plain. Decoration includes cord marking, comb-impression, and narrow bands in relief, with cord marking being dominant (CPAM of Hebei Province and CPAM of Handan City 1981). The firing temperature for pottery with sand temper was around 880°C, and 930°C for the fine pottery, according to a test result (CPAM of Hebei Province and CPAM of Handan City 1981). The vessel shapes were relatively simple, with the *Yu* and tripod stands being the major utensils, together with jars, pots and tripod *bo* (CPAM of Hebei Province and CPAM of Handan City 1981) [Fig. 4(10)]. This pottery assemblage seems to be distributed only around Cishan and adjacent areas [Fig. 4(1)].

In cultural layer 2, two round semisubterranean house floors were uncovered in 1978, as well as an 'S' shaped pebble surface, considered to be the remains of some sort of construction (CPAM of Hebei Province and CPAM of Handan City 1981). A total of 380 pits have been discovered. In addition to the two house floors found during the second excavation, some of the large, shallow pits found during the first excavation could also have been subterranean house floors, as judged from the contents unearthed (data from CPAM of Handan *et al.* 1977). The number of such pits is not mentioned in the excavation report.

The pits in layer 2 also varied from oblong to round, oval or irregular shapes. Sizes are not fully reported, but one has been described as 1.1 m long, 0.9 m wide and 3.65 m deep (CPAM of Hebei Province and CPAM of Handan City 1981). A total of 188 oblong pits was found in the second excavation in layer 2. Cereal remains were found in 18 pits, and a substantial quantity of hackberry seeds was found in two pits (CPAM of Hebei Province and CPAM of Handan City 1981). This discovery indicates that gathering was still an important economic activity, and that not only millet but also wild seeds were stored to prevent shortage of food.

The toolkit unearthed from layer 2 illustrates the economic strategies of the Cishan occupants. The first excavation report claimed that 317 stone tools were found, consisting of 154 axes, 11 grinding slabs, 16 rollers and 31 spades; the rest being hammers, chisels, and anvils (CPAM of Handan *et al.* 1977). Tools found in the second excavation were described in more detail and are classified in Table 4 (14). Many of these tools are likely to have been used for cereal cultivation.

A total of 253 artefacts made of bone, antler, tooth and shell was also unearthed from layer 2 in the first excavation, consisting of chisels, drills, net shuttles, harpoons, arrowheads, spades, daggers and needles, as well as ornaments and hair-pins (CPAM of Handan *et al.* 1977), but details are not available. Another 96 organic artefacts were uncovered in the second excavation, classified in Table 4 (15).

The organic tool assemblage from layer 2 is similar to the assemblage found in layer 3, except for the antler tool. Again, tools presumably used for hunting, fishing and gathering, such as arrowheads, harpoons, shuttles and shuttle needles for netting, account for 35.42% of the total assemblage, indicating the significance of these economic activities.

Pottery was also found in layer 2, with similar characteristics to the sherds found in layer 3. Differences between layers 2 and 3 are only small changes in utensil shapes and decoration, as well as a slight decrease in the proportion of pottery with a sand temper (from 90% in layer 3 to 80% in layer 2), and an increase of fine pottery (CPAM of Hebei Province and CPAM of Handan City 1981).

Dated to around 8000 to 7700 cal. BP, the Cishan assemblage, particularly the large amount of cultivated millet, demonstrates a quite advanced agricultural society. Many Chinese scholars infer that the first phase of the Cishan culture could not have been the initial stage of agriculture; that there must have been an initial stage prior to Cishan. But this initial stage has not been discovered to date despite constant searching, although the Nanzhuangtou assemblage may provide some clues.

4(B) 2. Archaeological assemblages from the terminal Pleistocene to the early Holocene in the southern part of the North China Plain.

In the southern part of the North China Plain, there are three important discoveries dated to the end of the Pleistocene. From the north to the south, the first one is a surface collection in Lingjing city; the second one a cave site called Xiaokongshan; and the last one an open site called Dagang [Fig. 4(1)].

The **Lingjing** collection was discovered at a reservoir construction site in Henan province in 1965. Artefacts were collected from yellow sand sediments, which had been dug up from about 10 meters under the ground surface (Zhou 1974). In association with the stone implements, bones of *Homo sapiens sapiens* and 18 animal species, two pieces of ochre, one bone scraper, as well as fired earth and charcoal were found. The materials were very well cemented together, suggesting that they came from the same deposit (Zhou 1974).

A total of 1353 flakes and retouched tools was collected, but only 200 lithic implements were described in the report. They are classified in Table 4 (16). Again, the tripartite combination is present: flakes, microblades and heavy-duty tools. But there are no grinding tools. Only chipped axes were present. The microblade industry appears to be similar to those from Xiachuan and Xueguan (Zhou 1974) [Fig. 4(19)]. This resemblance may suggest some cultural link.

No absolute dates have been obtained for this collection, but the fauna provided information on its geological date. Of the 18 species of animals and molluscs, Anderson's ostrich *(Struthio anderssoni)* and woolly rhinoceros (*Coelodonta antiquitatis)* became extinct in the North China Plain by the end of the Pleistocene. As there is convincing evidence that the lithic implements were in association with the fauna, it is suggested that the Lingjing assemblage belongs to the terminal Pleistocene (Zhou 1974).

Southwest of Lingjing is the **Xiaokongshan** site, which was discovered in 1980. This site consists of two limestone caves: one at a lower level and another at a higher level (Xiaokongshan Excavation Team 1988). The upper cave was excavated in 1987. Animal bones and 113 lithic implements were found, summarised in Table 4 (17). The lower cave of Xiaokongshan is 7 meters beneath the upper cave. It was excavated in 1980 and 1987, and the unearthed lithic implements are classified in Table 4 (18).

The lithic implements of the two caves are similar, and have been considered as one archaeological assemblage (Xiaokongshan Excavation Team 1988). One uranium date is said to place this lithic assemblage within the LGM epoch, though the actual figure is not disclosed

(Xiaokongshan Excavation Team 1988). Clearly, the microblade industry was absent. Tools such as "arrowheads" and backed blades were not discovered either; only a few "burins" were present. Choppers seem to have been an important component within the lithic assemblage, whereas no grinding tools or axes were found [Fig. 4(20)].

This toolkit provides little information on the subsistence strategy of the occupants. In the upper cave, a substantial number of stone tools was found in layer 3, associated with only a few animal bones (Xiaokongshan Excavation Team 1988). As faunal remains associated with the stone tools were so scarce, one excavator argued that the occupants of Xiaokongshan must have been mainly gatherers rather than hunters, based upon ethnographic data (Wang 1988). But no details or further evidence have been provided for this argument.

The Xiaokongshan lithic assemblage resembles the pebble industry in the Yangzi River in many ways. For example, pebble surfaces often remain on tools; and choppers dominate. It should be noticed that three discoid choppers were found in Xiaokongshan, similar to those found in Xiachuan; and similar implements have also been found in the middle Yangzi valley (Yuan 1995). It has been pointed out that the Xiaokongshan lithic assemblage bears characteristics of the stone industries of both the Yellow and the Yangzi valleys (Wang 1988). This mixture may be partially due to the geographical situation of this site, which is located in between the two great river valleys [Fig. 4(1)].

Approximately 110 kms east of Xiaokongshan there is the **Dagang** site (calculated by distance on the map), which is almost on the same latitude as the former [Fig. 4(1)]. Dagang is an open site on top of a small hill. Excavations were conducted in two seasons in 1989 and 1990, revealing a quite simple stratigraphy. Immediately under the surface there was a Han cultural horizon (206 BC. – AD. 220); underneath was layer 3 of the early Neolithic Peiligang culture; and at the bottom was layer 4, containing microblades (Zhang and Li 1996). At this stage only the materials from layer 4 have been published. The lithic implements from layer 4 are summarised in Table 4 (19).

Only the microblade and flake industries were found in Dagang. Heavy-duty tools were absent. New tools were also scanty, being only represented by a few backed blades. The flake assemblage lacks variety, providing little information for the subsistence strategy of the makers of these tools [Fig. 4(21)].

According to the excavators, the lithic implements bear no trace of being rolled, indicating that they were *in situ*. This argument is further supported by the discovery of debitage (Zhang and Li 1996). But no animal bones, no remains of fire, or traces of human settlement were found. Even pollen analysis did not yield many pollen grains from layer 4 (personal communication from Zhang 1996). The excavators suggest that this site could have been a place for tool-making only (Zhang and Li 1996).

Although the archaeological materials are quite limited, the Dagang discovery is still important, not only because of the lithic assemblage itself, but also because of the Peiligang culture remains found in layer 3 above. It has been reported that between layers 4 and 3 there was an erosion horizon (Zhang and Li 1996). This indicates that there was a temporal gap between the Peiligang assemblage and the preceding microblade and flake assemblage. But according to a brief description, the Peiligang assemblage also contains a few microblades (Zhang and Li 1996). Therefore, even though the cultural deposits in Dagang are not continuous, there might have been a possible cultural link between the microblade industry and the early Neolithic Peiligang culture.

While the archaeological assemblages of Lingjing and Dagang are located in the middle valley of the Yellow River, microblades and other lithic implements have also been found in the lower Yellow valley, both by excavation and surface collections (Xu 1994). Xu (1994) claims that over one hundred find-places have been located, and a few sites have been excavated (Xu 1994). However, only brief reports of excavations and collections from a few places have been published. According to these reports, some common characteristics can be summarised as follows:

- Most of the lithic implements were found on the surfaces of low hills, or on the second terraces of rivers (Xu 1994). Two of these sites, Heilongtan and Qingfengling, were excavated in 1984 (Han 1985a, 1985b). Details for Qingfengling are not available. For Heilongtan, it was discovered that the stratum containing the microblade industry overlies a stratum containing flaked lithic implements. No pottery or polished stone tools have been found (Han 1985a). In other places, such as Malingshan and Fenghuangling, the original stratigraphy was not clear (Xu 1994).
- Stone implements including microblades, flakes and heavy-duty stone tools were found in the majority of these places (Xu 1994). The only exception is Rizhao, where no microblades were found (You *et al.* 1989). Within the flake industry, various scrapers, points and burins were present. A flaked spear head was found in Rizhao (You *et al.* 1989). Flaked arrowheads were found in Fenghuangling, Qingfengling, Heilongtan, Malingshan and Yishui (CPAM of Linyi District 1983; Han 1985a, 1985b). However, these discoveries may not be all contemporary, as polished arrowheads were discovered in Fenghuangling (CPAM of Linyi District 1983). Within the heavy-duty stone tools, choppers were present at every site; stone balls were found in Yishui (Xu 1994).

The common characteristics of the microblade and the flake implements found in these places are similar to those of Xiachuan, therefore a cultural connection between Xiachuan and the collections in the lower Yellow valley has been proposed (Han 1985b; Xu 1994). It is also suggested that there might have been cultural links between these microblade assemblages in the lower Yellow valley and the microblades found in Japan (Xu 1994).

No absolute dates have been provided for these lithic assemblages so far, but geological dates have been suggested for a few sites. For example, both the Rizhao and the Qingfengling assemblages are suggested to be at about 10,000 BP, based upon stratigraphic and faunal analysis (Han 1985b).

These discoveries illustrate that, from the terminal Pleistocene to the early Holocene, a combination of microblade, flake and heavy-duty tool industries was widely distributed over the lower Yellow valley. As remains of early Neolithic cultures have already been discovered in the same region, the relationships between the microblade and flake collections and the early Neolithic cultures become an important issue for further study.

The lower Yellow valley is one of the centres of Neolithic cultures. The famous Longshan site was discovered in this region in 1928 (IA, CASS 1984: 97). Since the 1950s numerous Neolithic assemblages have been discovered in this area, dated from 7300 to 5000 BP and onwards (Wu 1989). In the 1990s, a new archaeological assemblage named the 'Houli culture' was discovered in the lower Yellow valley. According to the excavators, this Houli culture was the ancestor of the succeeding Neolithic

cultures in this region (Wang 1993).

The Houli culture is exemplified by archaeological assemblages from the Houli site and the Xihe site. Both were located on river terraces. Houli was excavated in four seasons from 1988 to 1990 (Wang 1993), and Xihe was excavated in 1991 (Tong and Wei 1994).

According to the brief excavation reports of Houli and Xihe, the toolkits of the Houli culture include lithic axes, spades, hammers, grinding slabs and rollers, scrapers and points, plus organic perforators, daggers, knives and sickles (Jiqing Archaeology Team 1992, 1994; Tong and Wei 1994). Although no remains of domesticated cereals have been reported, the Houli toolkit is identical to that of the Peiligang culture, suggesting the practice of cereal exploitation and harvesting, possibly cultivation (Tong and Wei 1994).

The makers of the Houli toolkits seem to have been sedentary, as demonstrated by semisubterranean house floors of square shape, pits, cemeteries and kilns for pottery firing found in Xihe and Houli (Jiqing Archaeology Team 1992, 1994; Tong and Wei 1994]. A large quantity of pottery was also found in these sites, the majority being red in colour (Wang 1994). The firing temperature was quite low, and it has been claimed that some of the potsherds could disintegrate in water (Jiqing Archaeology Team 1994). It is argued that the Houli pottery was at a quite primitive stage of craftsmanship (Wang 1994).

The Houli culture resembles other contemporary Neolithic cultures in the Yellow River in many aspects. The shapes of the dwellings, the toolkits, kilns and burials are similar to those of Cishan and Peiligang (below). Pottery is the major distinguishing cultural component. The major common vessels is a type of big *fu* in Houli (defined in chapter 2) [Fig. 4(10)]. While the Houli economic strategy and life style might have been similar to that of the Cishan and Peiligang cultures, its pottery indicates that it is not just a branch of the latter two. Radiocarbon dates for the Houli assemblage range from 8314 to 7773 cal. BP [Table 4 (7)], contemporary with the Cishan and Peiligang cultures.

The Peiligang culture is a prosperous agricultural complex mainly distributed in the middle Yellow valley [Fig. 4(1)]. Over 100 find places and sites belonging to this culture have been discovered so far, within a distribution area of about 135,530 square kilometers (Sun 1992). Several sites have been excavated (Anonymous 1991). The Peiligang culture is dated to between 8500 and 7500 cal. BP [Table 4 (7)], and can be exemplified by discoveries in the Peiligang and Jiahu sites.

The **Peiligang** site is located on a small river terrace, within the alluvial plain of the Yellow River. This site was excavated in three seasons from 1977 to 1979. Remains of house floors, pits, kilns and cemeteries were discovered, as well as numerous tools and pottery vessels (CPAM of Kaifeng District and CPAM of Xinzheng County 1978; CPAM of Kaifeng District *et al.* 1979; The First Henan Archaeology Team CASS 1984). The lithic toolkit found in Peiligang are classified in Table 4 (20).

Clearly the composition of this toolkit resembles that of Cishan (above). But there are a few differences. First, the axes and spades in Peiligang were all polished, whereas in Cishan some of these implements were flaked. Second, there were lithic flake scrapers and points in Peiligang, whereas in Cishan these tools were absent. Similar small flake tools were also found in other Peiligang sites, such as Huawo (CPAM of Anyang District and CPAM of Qi County 1981), Er'gou (Henan Museum and CPAM of Mi County 1981), Shawoli and Zhongshanzha (The First Henan Archaeology Team CASS 1983, 1991).

It is not clear whether the absence of flaked axes and spades indicates a more advanced tradition of stone tool manufacturing in Peiligang than in Cishan. But the discoveries of scrapers and points made of quartzite or chert in Peiligang are notable. These probably signal a cultural link to the local flake industries of the terminal Pleistocene, as represented by Lingjing and Dagang (above).

Bone and antler tools were also found in Peiligang, but quantities are small [Table 4(21)], only accounting for 10.57% of the total toolkit. The low frequency of tools made on other materials is also a phenomenon of other Peiligang sites except for Jiahu. This is quite contrary to the Cishan site, where a large number of organic tools was found in layer 3 (273 organic implements out of the total of 456 tools, above).

The majority of the lithic toolkit in Peiligang appears to be for agricultural activities, such as land clearance, soil-digging, cereal harvesting and processing [Fig. 4(22)]. The proportion of tools for cereal cultivation in Peiligang by far exceeds the proportion of tools for hunting and gathering. Further, reported animal remains in Peiligang sites were much fewer than in Cishan, except for Jiahu. This probably indicates that cultivation was a more primary subsistence strategy in many Peiligang sites than in Cishan. Carbonised foxtail millet has been found in a few sites such as Shawoli, Peiligang and Shuiquan (The First Henan Archaeology Team CASS 1983, 1984; 1992). But gathering was still important as nuts and red-date seeds were found in almost every site, except for Zhongshanzhai (The First Henan Archaeology Team CASS 1983, 1984, 1991, 1992).

The pottery of Peiligang is dominated by a fine red ware, which is contrary to the dominance of coarse pottery at Cishan. But the Peiligang pottery was mainly made by ring-building, similar to Cishan [Table 5(3)]. Utensils include bowls, *bo*, pots and globular jars with two lugs (The First Henan Archaeology Team CASS 1984) [Fig. 4(10)]. Some are morphologically similar to the utensils found in Cishan. The similarities and differences between Cishan and Peiligang have led to a long-standing argument about whether they were parts of a single culture, or two separate cultures. No conclusions have been reached so far (IA, CASS 1984).

Another important site of the Peiligang culture is **Jiahu,** an open site located on a river terrace in the northern part of the Huai valley [Fig. 4(1)]. Excavated from 1983 to 1987 in six seasons, a total of over 300 pits, at least 10 kilns, over 300 burials and more than 30 house floors have been found in this site. Although the toolkit and pottery were similar to other Peiligang assemblages, there were a few important additional discoveries in Jiahu.

The first one is the discovery of cultivated rice. According to the excavator, imprints of rice husks and rice grains were found in Jiahu (Wang *et al.* 1998; Zhang 1994). Remains of foxtail millet were also reported in Jiahu (Yan 1997). By using phytolith and scanning electron microscope analyses, some rice grains were identified as a cultivated species, but wild rice was also reported (Wang *et al.* 1998). As the Jiahu assemblage has been dated to about 8500 cal. BP (IA, Henan Province 1989) [Table 4 (7)], the cultivated rice is the earliest discovered in the Yellow valley so far. This discovery raises many questions. Is the Huai valley also a centre for the origin of rice cultivation (e. g. Wang et al. 1998; Zhang 1994)? Or was the rice cultivation in Jiahu a result of cultural diffusion from the Yangzi valley? And what was the distribution pattern of wild rice at 8000 BP? All these questions require further study.

The second important discovery at Jiahu is tortoise shells bearing symbols and holes. These are considered to be the earliest symbols ever found in China, providing important data for the origins of Chinese characters (IA, Henan Province 1989).

As Jiahu is located in the Huai valley, between the Yellow and the Yangzi basins, its archaeological assemblage seems to have had mixed cultural characters. The toolkit and pottery of Jiahu resemble those of the Peiligang culture, but not without distinctive features. A large number of organic tools for hunting and fishing was found in Jiahu, in association with polished stone tools (IA, Henan Province 1989). This indicates that hunting and fishing were important subsistence strategies, as well as rice and millet cultivation. Substantial quantities of fish and animal bones, and remains of nuts and fruit seeds, further confirm the variety of the economic activities of the Jiahu occupants.

To date rice cultivation has not been found in any other Peiligang site in the Yellow valley, yet there seems no question about Jiahu belonging to the Peiligang culture, as demonstrated by the similar tools and pottery. Whether the Jiahu assemblage was under cultural influence from the Yangzi River, or was an independent centre for rice cultivation, is a question under debate (e. g. Wang *et al.* 1998; Yan 1997).

Recently, another early Holocene site called **Xiaoshankou** has been discovered in the eastern part of the Huai valley in Anhui province [Fig. 4(1)]. Only one pit, some potsherds, stone and organic tools were found in the earliest cultural layer (The Anhui Archaeology Team CASS 1993). According to the preliminary report, flaked scrapers and points, polished axes, adzes and chisels, as well as one grinding slab and two rollers were found from the site (The Anhui Archaeology Team CASS 1993). The toolkit of this site seems comparable to Houli. The presence of grinding tools also indicates the possible existence of cereal processing, although there was no solid evidence for agriculture. Xiaoshankou is dated between 7942-7613 cal. BP [Table 4(7)]. The presence of the Xiaoshankou assemblage demonstrates the spread of early Holocene culture in the Huai valley.

Summary

In the above sections, the cultural sequence from the terminal Pleistocene to the early Holocene in Central China has been reviewed. Apparently this sequence is far from complete. There are many missing links, particularly in the period between 10,000 and 8500 BP. The insufficient archaeological data makes the discussion on the transition to the Neolithic difficult. However, there is some evidence for cultural continuity within the region, as discussed above.

Since foxtail millet (*Setaria italica*) was the major crop in the early Neolithic in Central China, and has been proposed as having been domesticated in this very region, then the cultural relevance of the botanical characteristics of green foxtail (*Setaria viridis*) becomes important. It is possible that the botanical characteristic of green foxtail might have had some effect on human activities, particularly on gathering activities and mobility patterns. These issues will be discussed in the next chapter.

Table 4 (1) Absolute Dates for Archaeological Assemblages of the Terminal Pleistocene in Central China.

Site	Location	Lab.No	Original No.	Sample	Stratum	Uncalibrated C^{14} date (bp) (h.l. 5730)	Uncalibrated C^{14} date (bp) (h.l.5568)	Calibrated date (BP)	Reference
Xiachuan No.1.	112°2'E, 35°27'N.	ZK-638	78IT1(3)	Charcoal	Square 1, layer 3. 1.65m deep.	$36,200^{+3500}_{-2500}$		n./a.	IA, CASS 1991.
	"	ZK-417	76SCXIT8(2)	Charcoal	Square 8, layer 2.	23,900±1000		n./a.	" "
	"	ZK-384	76SCX1T2-6(2)	Soil & charcoal	Square 2-6, layer 2.	21,700±1000		n./a.	" "
		ZK-385	76SCX1T1(2),	Charcoal	Square 1, layer 2, lower section.	16,400±900	15,938±900	19978-17880	" "
Xiachuan Shunwang-pin		ZK-634	78SCST2(3)	Charcoal	Square 2, layer 3, the upper level of lower section.	19,600±600		n./a.	" "
		ZK-762	79SCST3(3)	Charcoal	Square 3, layer 3, upper section.	13,900±300	13,508±300	16569-15744	" "
		ZK-628	78SCST2(4)	Charcoal	Square 2, layer 4. bottom of Q4 loess.	2830±100	2750±100	2949-2757	" "
Xiachuan No. 2.		ZK-393	76SCXIIT2(2)	Charcoal	Square 2, layer 2.	20,700±600		n/a.	" "

Table 4 (1) Absolute Dates for Archaeological Assemblages of the Terminal Pleistocene in Central China (continued).

Site	Location	Lab.No	Original No.	Sample	Stratum.	Uncalibrated C^{14} date (bp) (h.l.5730)	Uncalibrated C^{14}date (bp) (h.l.5568)	Calibrated date (BP).	References
Xiachuan No.3		ZK-494	77SCXIIIT1-2(2)	Soil	Square 1-2, layer 2.	18,375±480	17,855±480	21930-20623	IA, CASS 1991.
Xiachuan No.4.		ZK-497	77SCXIVT101-103	Peat	Square 101-103, layer 2.	18,560±480	18,035±480	22139-20874	" "
Xiachuan Shandiyan		ZK-493	76SCXSD1(3)	Charcoal	Cave No.1, deposit at entrance.	1380±80	1341±80	1305-1175	" "
Xiaonanhai	114° 7'E, 36° 7'N.	ZK-654	D1B(6)	Charcoal	Layer 6.	24,100±500	23,419±500	n/a.	" "
		ZK-170-0	A1	Bone	Mixed fossils of different layers.	13,075±220	12,705±220	15341-14592	An 1983; IA, CASS 1991.
		ZK-665-0(655)	D1B(3);(2)(3).	Bone and Charcoal	Fossils from layer 2 and layer 3.	11,000±500	10,689±500	13100-11929	" "

Table 4 (1) Absolute Dates for Archaeological Assemblages of the Terminal Pleistocene in Central China (continued).

Site	Location	Lab.No	Original No.	Sample	Stratum	Uncalibrated C^{14} date (bp) (h.l.5730)	Uncalibrated C^{14} date (bp) (h.l.5568)	Calibrated date (BP)	Reference
Upper Cave Zhoukoudian.	115°55'E, 39°41' N.	OXA2773	H4:34:36	Bone	The lower cellar of Upper Cave.		33,460±850	n/a.	Chen *et al*. 1992
		OXA2277	H4:34:44	Bone	"		33,200±820	n/a.	" "
		OXA190	F9:34:42	Bone	"		32,600±2000	n/a.	" "
		OXA1246	J6:34:11	Bone	"		26,500±460	(Uranium series. 21,000±4000)	" "
		OXA1248	F7:34:32	Bone	"		23,150±330	n/a.	" "
		ZK 136-0(2)	F9:39:44	Bone	"	18,865±420	18,340±420	22425-21357	
		OXA2276	O9:33:108	Bone			29,100±520	n/a.	" "
		OXA2271	N6:33:90	Bone			28,680±460	n/a.	" "
		OXA2772	S?:33:?	Bone			27,500±380	n/a.	" "

Table 4 (1) Absolute Dates for Archaeological Assemblages of the Terminal Pleistocene in Central China (continued).

Site	Location	Lab.No	Original No.	Sample	Stratum	Uncalibrated C^{14} date (bp) (h.l.5730)	Uncalibrated C^{14} date (bp) (h.l.5568)	Calibrated date (BP)	Reference
Upper Cave (continued)	115°55'E, 39°41' N.	OXA2274	?:33:70	Bone			27,370±410	n/a.	Chen *et al.* 1992.
		OXA2275	J,K7:33:80	Bone			25,700±360	n/a.	" "
		OXA1247	A1:33:60	Bone			23,700±350	n/a.	" "
		OXA391	J11:33:?	Bone			13,200±160	15993-15475	Chen *et al.* 1992
		ZK-136-0(1)	JK7:33:88, H8:33:109	Bone	?	10,770±360	10,470±360	12761-11696	IA, CASS 1991.
Xueguan	111°E, 36°27'N	BK?	?	Charcoal		13,550±150	13,167±150	15935-15435	Wang *et al.* 1983.
Xibaimaying	114°14'E. 40°07'N.	?		Bos. sp. teeth				(Uranium series 18,000±1000)	IA, Hebei Province 1989.
		?		Bos. sp. teeth				(Uranium series 15,000±1000)	" "
Hutouliang	114°9'E, 40°1'N.	PV-156	73108	Bone	II terrace of sandy loess of the site.		10,690±210	12830-12387	Li *et al.* 1987.
		CUHK?	?	Potsherd		(Thermoluminescence dated to 11,870±1720 years BP)			Tang 1997.

Table 4(2). Lithic implements from the Xiachuan sites.

No	Type of lithic implements	Quantity	% of the total lithic implements
1	Microcores	462	9.14%
2	Microblades	541	10.70%
3	Flake cores	280	5.54%
4	Flakes - unretouched?	425	8.41%
5	Flake "points"	280	5.54%
6	Flake "scrapers"	2557	50.59%
7	Backed blades*	112	2.22%
8	Other retouched flake implements	275	5.44%
9	Pebble choppers.	50	1.00%
10	Pebble hammers	2	0.03%
11	Pebble chopping tools	8	0.16%
12	Flaked adzes	7	0.14%
13	Flaked axes	5	0.10%
14	Grinding slabs	27	0.53%
15	Grinding hand-stones	7	0.14%
16	"Dinging and cutting tools" of large flakes.	16	0.32%
	Total	5054	100%

(Data based upon IA, CASS and CPAM of Shanxi Province 1978 and Wang *et al.* 1978).

*No specific core types have been reported for production of the backed blades. As full reports are not available, it is not clear whether all of the "backed blades" are true blades, or some are just blade-like flakes.

Table 4(3). Lithic implements from Xueguan.

No.	Type of lithic implements	Quantity	% of the total lithic implements.
1	Microcores	87	7.59%
2	Microblades	111	9.68%
3	Flake cores	153	13.34%
4	Flakes	570	49.70%
5	Flake "scrapers"	189	16.47%
6	Flake "points"	28	2.44%
7	Backed blades	2	0.17%
8	Other retouched flake implements.	4	0.35%
9	Chipped axe made on large flake.	1	0.09%
10	Hammers	2	0.17%
	Total	1147	100%

(Data based upon Wang *et al.* 1983).

Table 4(4). Lithic implements from Shizitan.

No.	Type of lithic implements	Quantity	% of the total lithic assemblage
1	Microcores	208	11.60%
2	Microblades	547	30.49%
3	Flake cores	13	0.73%
4	Small flakes	565	31.50%
5	Flake "scrapers"	342	19.06%
6	Flake "points"	49	2.73%
7	Backed blades	8	0.45%
8	Other retouched flake implements	39	2.17%
9	Pebble choppers	6	0.33%
10	Grinding slabs	2	0.11%
11	Pebble grinding hand-stone	1	0.05%
12	Chipped axes	3	0.17%
13	Stone hammers	11	0.61%
	Total	1794	100%

(Data based upon Linfen Culture Bureau 1989).

Table 4(5). Lithic implements from Shayuan.

No.	Type of lithic implements	Quantity (1957)	% of the total assemblage.	Quantity (1983)	% of the total assemblage.
1	Microcores	17	3.31%	123	12.23%
2	Microblades	172	33.46%	91	9.05%
3	Micropoints	6	1.17%	10	0.99%
4	Broken cores*	98	19.07%		
5	Flakes- unretouched.	58	11.28%	114	11.23%
6	Flake "scrapers"	92	17.90%	455	45.33%
7	Flake "points"	39	7.58%	68	6.76%
8	Other retouched flake implements	10	1.95%	104	10.34%
10	Backed blades			6	0.60%
11	Core rejuvenation flakes	22	4.28%	30	2.98%
12	Chipped axes			3	0.30%
13	Stone spheroids			2	0.19%
	Total	514	100%	1006	100%

* It is not mentioned in the report that if these broken cores were for microblade or flake production.
(Data based upon An and Wu 1957; Ban-po Museum and CPAM of Dali County 1983).

Table 4(6). Implements from Dadiwan.

No	Type of tools	lithic (chipped)	lithic (polished)	bone	antler	fired clay	Total
1	Axes	1	2				3
2	Adzes	1?					1?
3	Spades		3?				3?
4	Knives	3	2				5
5	Choppers	3?					3?
6	Scrapers	3					3
7	Abrader.	+*					+*
8	Lithic cores.	2?					2?
9	Small flakes	+*					+*
10	Perforators			1	1		2
11	Chisels			2			2
12	Arrowheads			2			2
13	Pellets	1?					1?
14	Spindle whorls					7?	7?
15	Needles			1			1
	Total						35?

+*: present but quantity not reported. '?': the minimum number mentioned in the report. May be more. (Data based upon Dadiwan Excavation Group 1981).

Table 4(7) Radiocarbon Dates for Archaeological Assemblages of the Early Holocene in Central China

Site	Location	Lab.No	Original No.	Sample	Stratum	Uncalibrated C^{14} date (bp) (hl.5730)	Uncalibrated C^{14} date (bp) (hl. 5568)	Calibrated date (BP)	Reference
Nanzhuang-tou	115°36'E. 39°N.	BK-87088	T(3)6	Wood.	Layer 6, 2.20-2.28 meter deep.	10,815±140	10,509±140	12593-12236	Yuan et al.1991.
		BK-87075	T2(6)H	Charcoal	A pit in layer 6.	10,510±110	10,213±110	12215-11634	" "
		BK-87086	T3(5)	Soil	1.80-2.00 m deep.	9980±100	9698±100	10991-10615	" "
		BK-86120	T1(5)-(6)	Wood		9875±160	9596±160	10971-10389	" "
		BK-89064	T1(5)-(6)	Wood		9850±90	9572±90	10921-10473	" "
		BK87093	T1(5)-(6)	Wood		9810±100	9533±100	10899-10386	" "
		BK-87084	T3(4)	Sludge	1.47-1.60 m deep.	9720±120	9445±120	10800-10300	" "
		BK86121	T1(5)-(6)	Wood		9690±95	9416±95	10782-10298	" "
		BK87083	T3(3)	Sludge	1.40-1.47 m deep.	9535±100	9265±100	10361-10044	" "
		ZK-2661		Wood.	Layer 6, 2 meters from ground.	8800±108	8551±108	9540-9440	IA, CASS 1993.
		BK87082	T3(3)	Sludge		8250±160	8017±160	9185-8563	Yuan et al. 1991.
		BK87080	T3(3)	Sludge	0.80-1.00 m deep.	4605±90	4475±90	5295-4876	" "
		BK87077	T3(2)	Sludge	0.20-0.40 m deep.	2440±70	2371±70	2463-2335	" "

Table 4(7) Radiocarbon Dates for Archaeological Assemblages of the Early Holocene in Central China (continued)

Site	Location	Lab.No	Original No.	Sample	Stratum	Uncalibrated C^{14} date (bp) (hl.5730)	Uncalibrated C^{14} date (bp) (hl. 5568)	Calibrated date (BP)	Reference
Dadiwan	105°35'E. 34°58'N.	?	H363	Charcoal.	Pit No. 363.			*7355±165	Gansu Museum 1981.
		BK81021	T303(4)H10	Charcoal.	Pit No. 10.	6770±80	6579±80	7523-7385	IA, CASS 1991.
		BK81022	F371	Charcoal	House No. 371.	6940±80	6744±80	7620-7478	" "
		BK800025	T2H10	Charcoal	Pit No. 10.	7150±90	6947±90	7884-7639	" "
Lijiacun	107°43'E. 3°58'N.	ZK1267	T2(3)H1	Charcoal	Pit No. 1, upper part.	6245±90	6069±90	7018-6799	IA, CASS 1991.
		ZK1268	T2(3)H1	Charcoal.	Pit No. 1, lower part.	6335±90	6180±90	7179-6912	" "
Cishan	114°5'E. 36°32'N.	ZK0439	T45(2)H145	Charcoal	Pit No. 145.	7355±100	7147±100	8052-7826	IA, CASS 1991.
		ZK0440	T20(3)H48	Charcoal	Pit No. 48.	7235±100	7030±100	7918-7691	" "
		BK78029	T99(2)H453	Charcoal	Pit No. 453.	7060±100	6860±100	7722-7550	" "

* Original uncalibrated date is not reported.

Table 4(7) Radiocarbon Dates for Archaeological Assemblages of the Early Holocene in Central China (continued)

Site	Location	Lab.No	Original No.	Sample	Stratum	Uncalibrated C^{14} date (bp) (hl.5730)	Uncalibrated C^{14} date (bp) (hl.5568)	Calibrated date (BP)	Reference
Houli	(No data)	BK90160	T2324 H2600(11)1	Bone.	Lower section of layer 11.	7300±100	7094±100	7949-7773	Chen et al. 1996.
		BK90158	T3124 H3822(11)1	Charcoal.	Refuse pit under layer 11.	7645±70	7429±70	8314-8126	" "
Xihe	(No data)	BK91036	91SF1(2)	Sludge	In a pottery bowl.	7175±70	6972+70	7887-7667	" "
		BK91034	91SF1(2)	Sludge		7410±80	7201+80	8067-7913	" "
		BK91035	91SF1(3)	Sludge		7325±80	7118±80	7951-7818	" "
		BK91037	91ST11(4).	Sludge.		7905±90	7682±90	8501-8365	" "
Peiligang	113°45'E. 34°25' N.	ZK572	T31(1) & T34(1) (2).	Charcoal	Mixed charcoal from two trenches.	9300±1000	9037±1000	11413-8958	IA, CASS 1991.
		ZK571	?	"		7145±300	6943±300	7993-7519	" "
Shawoli	113°40' E. 34°38' N.	ZK1130	T9(II)H17	Charcoal	Refuse pit in layer 2.	7170±105	6967±105	7898-7642	" "

Table 4(7) Radiocarbon Dates for Archaeological Assemblage of the Early Holocene in Central China (continued)

Site	Location	Lab.No	Original No.	Sample	Stratum	Uncalibrated C^{14} date bp (hl.5730)	Uncalibrated C^{14} date (bp) (hl.5568)	Calibrated date (BP)	Reference
Zhongshan-zhai	112°56'E. 34°8'N.	ZK1366	T101(4)H18	Charcoal	Refuse pit of layer 4.	7390±100	7181±100	8066-7844	IA, CASS 1991
		ZK1367	T103(4)H20	"	"	6955±90	6758±90	7635-7478	" "
Shuiquan	113°12'E 33°45'N.	ZK2345	T205A.H80	Charcoal	Refuse pit.	7270±120	7064±120	7943-7708	IA, CASS 1991
		ZK2344	T108H43	"	Refuse pit.	7100±110	6900±110	7798-7576	" "
Mixian er'gou	113°23'E. 34°32'N.	ZK-580	T79H27	Charcoal.	Pit No. 27.	7240±80	7035±80	7912-7718	Henan Museum 1981.
		WB7839	T24	Charcoal.	Trench No. 24.	7265±160	7060±160	7966-7666	" "
		WB7838	H1	Charcoal.	Pit No.1.	6975±100	6778±100	7653-7479	" "

Table 4 (7) Radiocarbon Dates for Archaeology Assemblage of the Early Holocene in Central China (continued)

Site and Spot	Location	Lab.No	Original No.	Sample	Stratum	Uncalibrated C^{14} date (bp) (hl.5730)	Uncalibrated C^{14} date (bp) (hl.5568)	Calibrated date (BP)	Reference
Jiahu	113°30'E. 33°24'N.	WB83-60	T2H1C:1	Charcoal	1st cultural layer.	7920±150	7696±150	8561-8332	IA, Henan Province 1989.
		BK 91007	WBJI7H82	Charcoal	1st cultural layer	7960±60	7735±60	8546-8411	C^{14} Lab. Dep. of Archae. Beijing U. 1996.
		?	WBJI7H82	Peat.	1st cultural layer.	7561±125	7347±125	8303-7964	IA, Henan Province 1989.
		?	JIH39	Peat.	2nd cultural layer.	7137±128	6935±128	7836-7587	‘’ ‘’
		?	JIH29	Peat.	" "	7105±122	6904±122	7877-7572	‘’ ‘’
		?	JIh55	Peat.	3rd cultural layer.	7017±131	6819±131	7715-7529	‘’ ‘’
Xiaoshankou	116°54'E. 33°36'N.	ZK2622	T101(5)	Charcoal	Layer 5.	7225±160	7021±160	7942-7642	" "
		ZK2621	T101H103 (1)	‘‘	Refuse pit.	7125±110	6924±110	7882-7613	" "
									" "

Table 4(8). Lithic implements from Youfang.

No.	Type of lithic implements	Quantity	% of total lithic assemblage.
1	Microcores	13	1.87%
2	Microblades	94	13.49%
3	Flake cores	72	10.33%
4	Flakes - unretouched	475	68.15%
5	Flake "scrapers"	20	2.87%
6	Flake "points"	5	0.71%
7	Backed blades	2	0.29%
8	Other retouched flake implements	6	0.86%
9	Pebble choppers	4	0.57%
10	Hammers	6	0.86%
	Total	697	100%

(Data based upon Xie and Cheng 1989).

Table 4(9). Lithic implements from Xibaimaying.

No.	Type of lithic implement	Quantity	% of the total lithic assemblage.
1	Flake cores	78	5.05%
2	Flakes - unretouched.	184	11.90%
3	Flake "scrapers"	216	13.97%
4	Flake "points"	11	0.70%
5	Other retouched flake implements	2	0.13%
6	Small flake choppers	1	0.07%
7	Debitage	1054	68.18%
	Total	1546	100%

(Data based upon IA, Hebei Province 1989).

Table 4(10). Lithic implements from Hutouliang.

No.	Type of lithic implements	Quantity	% of the total lithic assemblage.
1	Microcores	253	0.62%
2	Microblades	300	0.73%
3	Flake cores	26	0.06%
4	Unretouched flakes and debitage	about 40,000	97.67%
5	Flake "scrapers"	252	0.62%
6	Flake "points"	42	0.10%
7	Other retouched flake implements	37	0.09%
8	Pebble choppers	33	0.08%
9	Stone hammers	5	0.01%
10	Stone blocks	7	0.02%
	Total	about 40,955	100%

(Data based upon Gai and Wei 1977).

Table 4 (11). Lithic implements from Xiaonanhai.

No.	Type of lithic implements	Quantity	% of the total lithic assemblage.
1	Flake cores	397	5.66%
2	Flakes - unretouched	6500	92.67%
3	"Points"	17	0.24%
4	Flake "scrapers"	96	1.37%
5	Pebble chopping tools	4	0.06%
	Total	7014	100%

(Data based upon An 1965).

Table 4(12). Lithic implements from layer 3, Cishan.

No.	Type of lithic implement	Flaked	Edge-ground	Fully polished	Subtotal	% of the total lithic assemblage
1	Axes	48	10	45	103	56.28%
2	Spades	2	2	17	21	11.48%
3	Pebble chopping tools	4			4	2.19%
4	Scrapers on large flakes	2			2	1.09%
5	Chisels		5	3	8	4.37%
6	Adzes			2	2	1.09%
7	Hammers			3	3	1.64%
8	Grinding slabs	4			4	2.19%
9	Rollers			4	4	2.19%
10	Stone balls			27	27	14.75%
11	Bow pellets			4	4	2.19%
12	Cudgel			1	1	0.54%
	Total	56	17	110	183	100%
	% of each column	30.60%	9.29%	60.11%	100%	

(Data based upon CPAM of Hebei Province and CPAM of Handan City 1981).

Table 4(13). Organic implements from layer 3, Cishan.

No.	Type of tools	Bone	Antler	Shell	Teeth	Sub-total	% of the total bone & antler assemblage.
1	Knives	5		1		6	2.20%
2	Chisels	8				8	2.93%
3	Drills/awls	98	7		5	110	40.29%
4	Harpoons	12				12	4.39%
5	Arrowheads	51		1		52	19.05%
6	Shuttles	15	2			17	6.23%
8	Needles	37				37	13.55%
10	Daggers	23				23	8.42%
11	Hooks		4			4	1.47%
12	Spades			4		4	1.47%
	Total	249	13	6	5	273	100%

(Data based upon CPAM of Hebei Province and CPAM of Handan City 1981).

Table 4(14). Lithic implements from layer 2, Cishan.

No.	Type of tool	Flaked	Edge-ground	Fully polished	Sub-total	% of the total lithic assemblage
1	Axes	107	56	167	330	48.82%
2	Spades	10	18	30	58	8.58%
3	Knives	3			3	0.44%
4	Anvils	24			24	3.55%
5	Chisels		7	13	20	2.96%
6	Adzes		2	8	10	1.48%
7	Bow pellets		6		6	0.89%
8	Sickles			6	6	0.89%
9	Stone balls			105	105	15.53%
10	Hammers			10	10	1.48%
11	Grinding slabs	52			52	7.69%
12	Rollers			50	50	7.40%
13	Grinding handstones			2	2	0.29%
	Total	144	89	443	676	100%
	% of each column	21.15%	13.80%	65.05%	100%	

(Data based upon CPAM of Hebei Province and CPAM of Handan City 1981).

Table 4(15). Organic implements from layer 2, Cishan.

No.	Type of tools	Bone	Antler	Tooth	Shell	Sub-total	% of the total organic tool assemblage.
1	Chisels	8				8	8.33%
2	Knives	3				3	3.12%
3	Harpoons	4				4	4.17%
4	Arrowheads	22				22	22.92%
5	Shuttles	2				2	2.08%
6	Needles	11				11	11.46%
7	Daggers	4				4	4.17%
8	Scraper			1		1	1.04%
9	Drills	41			1	41	42.71%
	Total	94		1	1	96	100%

(Data based upon CPAM of Hebei Province and CPAM of Handan City 1981).

Table 4(16). Lithic implements from Lingjing.

No.	Type of lithic implements	Quantity	% of the total lithic assemblage
1	Microcores	6	3.00%
2	Microblades	36	18.00%
3	Flake cores	3	1.50%
4	Flakes	76	38.00%
5	Flake "points"	12	6.00%
6	Flake "scrapers"	56	28.00%
7	Flake "burins"	4	2.00%
8	Pebble choppers	2	1.00%
9	Chipped axes made on thick flakes	2	1.00%
10	Pebble scrapers	3	1.50%
	Total	200	100%

(Data based upon Zhou 1974).

Table 4(17). Lithic implements from the upper cave, Xiaokongshan.

No.	Type of lithic implement	Quantity.	% of the total lithic assemblage.
1	Flake cores.	27	23.89%
2.	Flakes.	50	44.25%
3	Flake "scrapers"	18	15.93%
4	Flake "points"	6	5.31%
5	Flake "burins"	4	3.54%
6	Flake choppers	5	4.42%
7	Pebble chopper	1	0.89%
8	Core choppers	2	1.77%
	Total	113	100%

(Data based upon Xiaokongshan Excavation Team 1988).

Table 4(18). Lithic implements from the lower cave, Xiaokongshan.

No.	Type of lithic implements	Quantity	% of the total lithic assemblage.
1	Flake cores	56	49.56%
2	Flakes	25	22.12%
3	Flake "points"	7	6.19%
4	Flake "scrapers"	12	10.62%
5	Flake "burin"	1	0.89%
6	Pebble stone balls	4	3.54%
7	Stone hammers	2	1.77%
8	Pebble choppers	6	5.31%
	Total	113	100%

(Data based upon Xiaokongshan Excavation Team 1988; CPAM of Nanyang District 1982).

Table 4(19) Lithic implements from Dagang.

No.	Type of lithic implement	Quantity	% of the total lithic assemblage.
1	Microcores	28	8.56%
2	Microblades	14	4.28%
3	Flake cores.	28	8.56%
4	Flakes	86	26.30%
5	Flake "scrapers"	47	14.37%
6	Flake "points"	10	3.06%
7	Backed blades	3	0.92%
8	Edge-polished flake	1	0.31%
9	Debitage	110	33.64%
	Total	327	100%

(Data based on Zhang and Li 1996).

Table 4(20). Lithic implements from Peiligang.

No.	Type of tool	Flaked	Polished	Subtotal	% of the total assemblage
1	Axes		19	19	12.34%
2	Spades		43	43	27.92%
3	Knives		4	4	2.60%
4	Chisel		1	1	0.65%
5	Sickles		20	20	12.99%
6	Grinding slabs	24		24	15.58%
7	Rollers	24		24	15.58%
8	Spearhead.		1	1	0.65%
9	Grinding stones	6		6	3.90%
10	Pellets		2	2	1.30%
11	Flake "scrapers"		8	8	5.19%
12	Flake "points"			2	1.30%
	Total	54	100	154	100%
	% of each column	35.06%	64.94%	100%	

(Data based upon CPAM of Kaifeng District and CPAM of Xinzheng County 1978; CPAM of Kaifeng District *et al.* 1979; The First Henan Archaeology Team CASS 1984).

Table 4(21). Organic implements from Peiligang.

No.	Type of tools	Bone	Antler	Subtotal	% of the toolkit.
1	Drills	6		6	37.50%
2	Arrowheads	4		4	25.00%
3	Dagger	1		1	6.25%
4	Needle	1		1	6.25%
5	Points	2		2	12.50%
6	Rod.		2	2	12.50%
	Total	14	2	16	100%

(Data based upon CPAM of Kaifeng District and CPAM of Xinzheng County 1978; CPAM of Kaifeng District *et al.* 1979; The First Henan Archaeology Team CASS 1984).

The transition from foraging to farming in Central China should have left its mark on the toolkits used for economic activities. Therefore, toolkits are considered to be primary data for understanding this transition. Another important aspect of understanding this transition is to understand the natural resources available to, and exploited by prehistoric people. Remains of plants and animals, including pollen and phytoliths, provide the best data. In the case of Central China, plant remains in archaeological sites are scarce. But as discussed in chapter 3, the presence of grinding tools in Central China may indicate the practice of plant exploitation, and green foxtail would have been a major resource for prehistoric foragers. Therefore, the botanical characteristics of this plant are relevant for the transition from foraging to farming in Central China.

5A. Toolkits as indicators of the transition from foraging to farming.

The archaeological data in Central China document the existence of at least two major toolkit traditions in the terminal Pleistocene: the tripartition of microblade, flakes and heavy-duty stone tools, exemplified by Xiachuan and Xueguan; and a toolkit mainly of flakes with a few heavy-duty tools (without microblades), represented by Xibaimaying and Xiaonanhai (chapter 4). As both traditions are represented by various archaeological assemblages, each of them is named as a lithic complex. The relationships of these two archaeological complexes are not clear due to lack of archaeological data, but it is possible that there are cultural continuities from the two complexes into some of the early Neolithic assemblages in Central China.

A salient feature of the tripartite complex is the emergence of the microblade industry, first occurring at Locality 7701, Dingcun, and dated to about 26,000 bp; and slightly later at Xiachuan, Xueguan and Shizitan, dated from 26,000 to 10,000 bp, all in the Fen valley. Archaeological data show that, from 18,000 to approximately 10,000 BP, the microblade industry also occurred in the Sanggan River valley, in the present North China Plain and Northeast China [Fig. 4(1)] (chapter 4) (Lu 1998).

Whether this expansion of microblade technology was only a cultural diffusion, or an expansion carried by human movement, is not clear. The expansion could have also taken different forms in different regions. Archaeological data are not sufficient to solve this question at this stage. But there are a few points that can be made.

First, the microblade industry emerged in the terminal Pleistocene, flourished between 18,000 and 10,000 BP, and diminished during the early Holocene in Central China. The florescence of this industry was just within the transitional period from the Pleistocene to the Holocene. Thus, microblade technology should be perceived as a hallmark of this transitional period in Central China (Lu 1998).

Second, archaeological data have demonstrated that the microblade industry diminished after the emergence of agriculture in Central China where agriculture was well developed by 8500 BP (e. g. Jiahu). However, the disappearance of microblades does not indicate the cessation of hunting and gathering activities. The presence of animal bones and nuts in early Neolithic sites illustrates the importance of hunting and gathering, but the tools for these activities were mainly made of organic materials, such as bone arrowheads and knives, as found in Cishan and Jiahu. Composite tools were rarely found in early Neolithic farming sites in Central China, and the function of microblades as the cutting edges of composite tools seems to have been terminated there.

Beside the microblade industry, there were two other lithic industries of the tripartite complex: the flakes and the heavy-duty tools. Both can be traced back to the early Palaeolithic in China (Jia and Huang 1985; Huang 1989; Li 1993). But some new tools within these two industries appeared in the terminal Pleistocene. Chinese archaeologists have claimed that arrowheads and denticulates in the flake industry were present in Xiachuan and other sites (e.g. Shi 1989; Wang *et al.* 1978). Chipped axes, adzes and grinding tools were important implements in the heavy-duty tools.

Arrowheads were for hunting, particularly effective for fast-running ungulates such as onager and wild horse. Denticulates could have been used for plant cutting, axes for land and forest clearance, and grinding slabs and hand-stones for seed processing. The presence of these new tools suggests technical development and an improved ability to exploit natural resources. This improved ability would have been particularly vital for human survival during the LGM epoch, when the variety of both fauna and flora decreased due to the severe palaeoclimate. As discussed in chapter 3, the retreat of forest from Central China during the LGM would have caused a decline or even disappearance of edible fruits and nuts. The withdrawal or extinction of some animal species would have forced humans to focus on animals like wild horse, onager and gazelle, all with running speeds of over 50 km/h (Nowak 1991). The necessity of adjusting subsistence strategies would have triggered the manufacturing of new, effective tools.

And effective these new tools were. The discovery of hundreds of teeth of gazelle in Shizitan (chapter 3) suggests that humans were focusing on one major game species, probably because it was the most abundant faunal resource at that time. Gazelle is an alert animal with a fast running speed (Nowak 1991), but the large number of teeth recovered demonstrates the success of the hunting activities, very likely aided by new tools such as arrowheads.

These new tools then became a major part of the Neolithic toolkit in the early Holocene, except that lithic arrowheads were replaced by bone arrowheads, and other lithic implements became polished. Stone spades also joined the toolkit in the early Holocene. The axes and spades, sickles and grinding tools could have been used for land-clearance, soil-digging, plant harvesting and processing - in short, for cultivation activities. The quantity of those new tools in the lithic assemblages of the terminal Pleistocene was often minimal; but their quantities increased substantially in the early Holocene cultures, as exemplified by data from a few important sites [Table 5(1)].

Based upon their morphological resemblance, Shi has claimed that denticulates and grinding slabs found in Xiachuan were the forerunners of the sickles and grinding slabs found in the early Neolithic sites in Central China (Shi 1989). The development of axes and adzes from flaked to polished may also be detected from Xiachuan to Cishan [Fig. 4(5); 4(18)].

In the early Holocene, not only did the quantity of such tools increase, but the quality also improved. The flaked axes and denticulates of the terminal Pleistocene became the polished axes and sickles of the early Holocene. The grinding slabs and hand-stones of the terminal Pleistocene became much better shaped in the early Holocene, and formed a standardised toolset of grinding slab and roller. These tools also became grave goods in the early Holocene.

In a few Peiligang culture sites, axes and spades were commonly found together in some graves, while grinding slabs and rollers were found together in others. These two sets seldom appeared within one grave (Zheng 1993). Interestingly, the identifiable remains of human skeletons with axes and spades were all male, and those with grinding slabs and rollers all female (Zheng 1993). This indicates that different tasks of cultivation/cereal processing might have already been assigned to different sexes.

In contrast to the florescence of new tools, the other implements of the flake industry and the heavy-duty tools diminished in the early Holocene in Central China. Scrapers and points made of chert, flint, quartzite and quartz were a major part of the lithic assemblage in the terminal Pleistocene; but these tools either disappeared, or became residual in the toolkits of the early Holocene. Choppers and chopping tools were rare. Only a few scrapers and points of quartz and quartzite have been found in the early Neolithic sites of the Dadiwan and Peiligang cultures. It seems that these scrapers and points, as well as choppers and chopping tools, were no longer functional.

One puzzling lithic implement is the stone discoid, which appeared in the Xiachuan and Xiaokongshan assemblages in the terminal Pleistocene, the Dadiwan and Cishan assemblages in the early Holocene, and became very popular in the Yangshao culture. Its function is not clear, but its continuous presence is more evidence of cultural links and continuity in the Yellow valley.

As the new tools for cultivation became dominant, it seems that the focus of tool-making consequently shifted from making small or micro-tools to producing large and/or polished tools. In the early Neolithic cultures, when the new heavy-duty tools such as axes and spades became major implements - and at the same time, it had been discovered that tools for hunting and gathering could be made of organic materials at much lower labour cost - the decline and/or elimination of the small/micro-tools was inevitable. This elimination could have been due to the economic consideration of labour cost in tool manufacturing, as a prehistoric society could have only endured a certain amount of labour cost. It was neither feasible nor economic to make small lithic tools of less importance, particularly when these tools could have been more effectively made by using organic materials.

An important point regarding the lithic assemblages should be noted. As mentioned above, the microblade industry first emerged in the Fen River valley in the terminal Pleistocene, and later expanded to other areas. But not all the other new tools had expanded in the same way. During the terminal Pleistocene, "denticulates" and grinding tools were only found in the Fen valley; they did not appear in other parts of Central China until the beginning of the Holocene. This indicates that, even within the tripartite complex in Central China, archaeological assemblages still differed in some aspects. Whether this difference of tools reflects different subsistence strategies remains a question.

The significance of the lithic complex without microblades is not clear at this stage. It seems that there were also cultural differences within this complex, but the classification of this complex is still under debate (e.g. Li 1993; Zhang 1990). The cultural links between this complex and the early Neolithic in Central China seems weak, and therefore, this complex will not be discussed further in this thesis.

In summary, the tripartite complex was an important toolkit during the transitional period from foraging to farming in Central China. The occurrence of new tools such as arrowheads, flaked axes and grinding tools, suggests intensified foraging activities. These tools became major implements in the early Neolithic cultures. When agriculture emerged, the trend of tool-minimisation reversed, microblades diminished, and heavy-duty tools dominated the toolkit. However, scrapers and points of small flakes seem to have survived longer than the microblades. These flake tools resembled their ancestors of the terminal Pleistocene in terms of raw material and manufacturing technique, indicating a cultural continuity from the terminal Pleistocene to the early Holocene in Central China.

5B. The cultural relevance of the botanical characteristics of green foxtail during the transition from foraging to farming in Central China.

The tripartite lithic complex seems to have been the ancestor of the early Neolithic cultures in Central China. The grinding tools and the possible existence of 'sickle gloss' in Xiachuan seem to indicate the practice of wild plant harvesting, although no botanical data are available to indicate the plant species being exploited. But without doubt foxtail and broomcorn millets were major cultivars in the early Holocene in Central China. The progenitors of these cereals would have been exploited prior to their cultivation, on analogy with research in the Middle East. Thus, if the Xiachuan complex was an ancestor of those Neolithic cultures in Central China, it is legitimate to propose that wild millets must have been exploited by the prehistoric Xiachuan people.

Two species of millet have been found in Neolithic Central China: foxtail millet (*Setaria italica*) and broomcorn millet (*Panicum miliaceum*). Both have been cultivated in this region for thousands of years, although today they are no longer important crops. The wild progenitor of foxtail millet is green foxtail (*Setaria viridis*), identified by interspecific crosses and cytological analysis (Li *et al.* 1945), as well as isozymic studies (Gao and Chen 1988). But the wild progenitor for broomcorn millet is not certain, although it has been suggested that a wild broomcorn millet found in Northwest and Central China could have been its progenitor, as demonstrated by the cross-fertility between the wild and domestic species (The Editing Committee of the Agriculture Encyclopaedia of China 1991). Given the uncertainty about the wild progenitor of the broomcorn millet, and its lower frequency in archaeological sites, broomcorn millet will not be discussed in detail.

As discussed in chapter 3B, my observations of green foxtail have uncovered some of its botanical characteristics. A major difference between wild and domesticated millet is the much longer and highly heterochronous life cycle of the wild species. The times of tiller growing, flowering, ripening and shattering vary not only on each plant, but also on each spikelet. As a result, the flowering and ripening season of this species as a whole can be as long as four months. During this period of time, there would always be some plant/spikelets ripening and ready to be collected.

Assuming that green foxtail in the terminal Pleistocene had similar botanical characteristics, and a similar long ripening season, the question now is how the prehistoric gatherers would have collected this plant. A few harvesting experiments were carried out during my field-work in 1995 to find out the effects of different collecting methods. These experiments included spikelet-shaking and hand-picking, cutting the whole plant with a stone flake, and uprooting. The preliminary outcome of these experiments are summarised in the following:

- **Spikelet-shaking.** The experiment was to hold the ripening spikelet in one hand and gently shake it into a container - in my case, a pottery bowl was used. This

method had two advantages: first, it was an easy and effective way of collecting green foxtail. Within a catchment area of 150 square meters and a labour time of half an hour, 2.1 grams of grain were collected, all ripe and ready to be milled. Second, by this method, only ripe grains were collected, while green or flowering seeds could still remain on the spikelet and continue their life cycle. Therefore, the continuing availability of green foxtail during its ripening season was guaranteed.

- **Spikelet-picking.** This method was to simply pick those spikelets which seemed to be mature, if not the whole spikelet, at least the major part of it. The method was also easy and effective. The nodes of green foxtail are quite weak and the spikelet came off easily by hand-picking. Within a catchment area of 1500 square meters and labour time of 40 minutes, 181 spikelets were collected. Another 70 minutes were spent on beating these spikelets on a wooden surface to shatter the ripe grains. Finally 10 grams of grain were gathered and ready to be milled.
- **Up-rooting.** Again, only those plants on which most spikelets were mature were uprooted. After 10 minutes of labour, 12 plants were uprooted with a total quantity of 81 spikelets. But only 21 spikelets produced ripe grains, the others were all flowering and/or green. This means that only 26% of the total collected spikelets were productive, and the rest had to be thrown away. Once uprooted the green and/or flowering spikelets could not continue their life cycle, thus causing a waste of resources.
- **Cutting with a flake.** In this method, a small flake of chert was used to cut the ripened spikelets of green foxtail. Within a catchment area of 7500 square meters and 40 minutes of labour, 124 spikelets were collected. After half an hour of beating, about 3 grams of grains were obtained. Again, there were many green/flowering grains that remained on each spikelet.

The results of each experiment are compared in table 5(2). All the above harvesting methods were selective in varying degrees, as only ripe seeds and/or spikelets were collected. The most selective method was the spikelet-shaking one. By this method, only ripe grains were collected, while green or flowering seeds remained on the spikelet to continue their growing cycle.

This method also maximised the exploitation of *S. viridis* in one year, within one given area. According to my observations, on 10th August, late summer in the lower Yellow valley, the 30 numbered plants bore a total of 668 spikelets, with a mean number of 22.27 spikelets per plant, and a range from 4 to 56. Most of them had around 14-20 spikelets. As the commencement of maturity of each spikelet was marked by the beginning of seed-shattering, the number of mature spikelets on each day could be traced. My records show that, except for 12th and 13th of July, and 10th of August, there were mature spikelets ready to be picked each day. The daily number of mature spikelets ranges from 1 to 21 [Fig. 5(1)].

It must be emphasised that the stands on which this observation was based were not dense at all. In an area with more dense stands of *S. viridis*, a daily supply of seeds would be certain. If a prehistoric foraging group was wandering in such an area, by spikelet-shaking they could have exploited the seed resource from early July onwards. The collecting season could have lasted for probably four months, until the life cycle of *S. viridis* terminated in October. Thus the continuing availability of seeds during the season of the year would have been reliable, and the exploitation of the wild plant would have been maximised.

If *S. viridis* was the staple food of these gatherers, they could have settled down in one given area for about four months, to collect the ripe seeds each day. This might have led to a seasonally sedentary life style. However, a major drawback of the spikelet-shaking method is that the life cycle of the wild plant was not disturbed at all. In other words, human interference was virtually non-existent in terms of a selection for simultaneous ripening and tough rachis, and the species could have remained wild forever if the spikelet-shaking method had been used.

Contrary to the spikelet-shaking method, the up-rooting method is the one that produces strong human interference by harvesting selection. This method was the most effective one in terms of grain collected per hour within a small catchment area [Table 5(2)]. It was also the quickest to harvest the stands of *S. viridis*. The 30 numbered plants could have been pulled up within 10 minutes by the up-rooting method, and the total stands of *S. viridis* around my working station would have been harvested within a few days. The seasonality of *S. viridis* would have been maintained for as long as four months by the spikelet-shaking method, but could have been reduced to as short as one to a few days by up-rooting or whole-plant cutting.

If this method was used by prehistoric foragers, the stands of *S. viridis* within a given area would have been quickly harvested. How quick this harvest might be in one given area would have depended on various factors, such as the density of the stands and the number of gatherers in that area, as well as the variety and availability of other food resource, and the degree to which gatherers would have preferred *S. viridis* over other wild plants. But there is no doubt that the seasonality of *S. viridis* would have been much shorter if the plant was harvested by this method, and the prehistoric foragers would have had to move to other areas if they still wanted to gather green foxtail.

The up-rooting method terminated the life cycle of many spikelets, which were still green and/or flowering. Thus only those seeds that ripe at about the same time (i.e. within a few days of the harvesting) were collected when the plant was up-rooted. Seeds that had either shattered prior to the harvesting, or were still flowering when the plant was pulled up, would not be collected. Although the former were still able to germinate in the following year and continue to reproduce, the life cycle of the latter was terminated before maturity. A continuity of this practice would encourage the species shifting to simultaneous ripening, which is one of the characteristics of domesticated cereals. If the whole plant was not up-rooted but cut, the effect in terms of harvesting selection in favour of simultaneous ripening would have been the same. In addition, the movement caused in up-rooting or cutting simultaneously shook the plant, and some easily-shattered ripe seeds would have been lost during that process. Only those remaining on the spikelets were collected. This harvesting method may have also produced human selection for a tough rachis (Hillman and Davies 1992).

The other two harvesting methods are somewhere in between the spikelet-shaking and up-rooting methods, in terms of both harvesting efficiency and selection effect. The cutting method seemed less efficient in this experiment, which was probably due to the small size of the lithic flake used (only 1.6x2 cm), and my lack of cutting skill. Nevertheless, both methods were similar to the up-rooting method in terms of harvesting selection in favour of simultaneous ripening and tough rachis. But comparatively, cutting the whole plant, like up-rooting, produces stronger selection for simultaneous ripening than spikelet-cutting or picking does, as the life cycle of an individual plant was terminated by the former. On the other hand, stands of *S. viridis* in a given area would have been cleared more quickly by cutting the whole plant, thus giving a possible stimulus to any conscious move towards planting in the next season.

Now we will look at the returns of each harvesting method. According to the observation record on the 30 numbered plants, a total of 294 spikelets reached their ripe stage during the period from 11th July to 9th August. The average weights of seeds produced from one spikelet was under 1 g (based upon 20 samples. Precise weights could not be measured due to lack of equipment in the working station) . If the spikelet-shaking method was used to collect the seeds, less than 300 grams of seeds would have been collected from these 30 plants during the 30 days, with a daily return under 10 grams. But by up-rooting, if the harvesting was conducted on some days in early August when more spikelets began to mature than on other days (e.g. the 9th of August), 21 mature spikelets would have produced approximately 20 grams of seeds, and the daily return would have been much higher than the return by spikelet-shaking.

Of course the samples here are small, and the stands of *S. viridis* were far from dense in this observation. Nevertheless, the above analysis seems to indicate that while spikelet-shaking provides much longer seasonality for foragers than uprooting or cutting, the daily return of the former was much lower than the latter. More importantly, the spikelet-shaking method does not favour simultaneous ripening and a tough rachis. This method is in fact an exploitation of heterochronous ripening and shattering habit, as it allows and encourages spikelet growing, flowering and ripening at varying times. On the contrary, pulling or cutting the whole plant provides strong human selection for simultaneous ripening and a tough rachis. Therefore, up-rooting or cutting the whole plant seems to be the most likely route to eventual domestication.

As discussed above, different harvesting methods would have had different effects on the seasonal availability of green foxtail for prehistoric gatherers in one year within one given area. At first sight, the spikelet-shaking method would seem to have encouraged temporary settlement by prehistoric foragers, while the up-rooting method would seem to have an adverse effect on sedentism by shortening the harvesting time for the plant in one given area. However, there would have been other factors which would have influenced the mobility pattern of prehistoric foragers, such as how dense the stands of green foxtail in an area were, how diversified their subsistence resources were, and the availability and seasonality of their other food resources.

It is not clear which methods were used by the prehistoric hunter-gatherers in the Yellow River. However, judging from the lithic blades found in a few archaeological assemblages in the middle Yellow River valley (i.e. the Xiachuan complex), seed grass cutting seems to have been practised. Archaeological data document that green foxtail was domesticated by 8000 BP in Central China (chapter 4). As the spikelet-shaking does not seem to be the trajectory leading to domestication, it is not likely to have been used as a major method. Given that abundant lithic blades have been found in a few archaeological assemblages in the terminal Pleistocene in the middle Yellow valley, it is possible that cutting the plant would have been the method used by prehistoric hunter-gatherers.

If this method was used, then the stands of *S. viridis* in a given area would have been harvested in a short period of time, probably within a few days or one or two weeks, depending on the density of the stands, the size of the catchment area and the density of the human population. The possible impact of this harvesting method upon hunter-gatherers would then depend upon other factors. The first one is the size of their territory. If the territory was big, the group was able to move on, then they could have harvested other stands. But if the territory was small, if such movement was restricted or prohibited by dense population in a region and the possible conflicts between groups, then a shortage of plant supply would have occurred. The second factor is the degree of reliance on this plant. If *S. viridis* was not a staple food, then they could have collected other food resources, and would not have had to worry about the shortage of green foxtail. But if green foxtail was a staple food, then they would have encountered a serious problem.

To solve the problem there would have been various solutions, and planting or broadcasting some of the seeds was one of the solutions in terms of increasing the production of the stands. One distinctive characteristics of green foxtail, like many grasses, is its high reproductive ability. One seed can produce a whole plant bearing hundreds or even thousands of grains. This advantage would have been known and manipulated by prehistoric foragers. A continuous practice of cutting the whole plant of *S. viridis* and planting or deliberately broadcasting some of the seeds was probably the pathway towards agriculture.

It is certain that green foxtail alone would not have been sufficient for the dietary needs of the whole forager community. In fact, green foxtail seeds are very small (about 0.2x0.1 cm, although not much smaller than domesticated seeds). A few hours of labour can only procure tens of grams, which would hardly have been sufficient for one person's daily food. There must have been other activities such as hunting and collecting other plants. But why then did human beings finally choose green foxtail to domesticate? It is probably because green foxtail is easy to grow, adaptable to a wide variety of environments, very drought-resistant, able to recover from natural damage, including breakage of spikelets (as I observed in one plant), and its life cycle is quite short.

As discussed in chapter 3B, some green foxtail germinates in summer, and these have a life cycle much shorter than those that germinate in the spring. Plants with a shorter life cycle would have been advantageous for humans, particularly when other foods failed, and planting was urgent. This mutation of short life cycle certainly has been manipulated by farmers. Today some sub-species of domesticated millet in China can ripen within 56 days of sowing (Society of Crop Genetic Resource 1994), and they still serve as an emergency crop after natural disasters.

The nutritive value of green foxtail is also quite high. According to one analysis, it contains 14.5% of crude protein, 5.09% of fat, 32.7% of acid-diet fibre, and the gross energy per gram is 4.2 Kcal (Harrold and Nalewaja 1977). The protein and fat values of green foxtail are even higher that in its domesticated counterpart, which contains 9.27% of protein and 3.15% of fat (Anonymous 1976). An experiment of feeding rats with green foxtail indicates that "the seeds of green foxtail have an energy content comparable to that of barley" (Harrold *et al.* 1980: 131). Domesticated foxtail millet is also rich in vitamin A and B, plus various amino acids. Its nutritive value is said to be higher than rice, wheat and maize (Anonymous 1976). I cannot find corresponding values for green foxtail, but it can be assumed that green foxtail has somewhat similar constituents to the domesticated form.

The nutritional value and the environmental flexibility of green foxtail must have been valuable attributes for people living in the dry and cool conditions of the last glacial in Central China. In addition, the seeds can be stored for a long period of time, which is a great advantage, as storage would have helped people survive food shortages, and would have given them motives to increase the production of this plant. It has been claimed that wind-dried domesticated foxtail millet can be stored for twenty years, and still remain edible (Anonymous 1976: 2), outlasting all

the other domesticated grains. Its leaves can be used to feed domestic animals. Although it is not clear how many years green foxtail can be stored, the time span might have been considerable. It is not without good reason that green foxtail was domesticated and became the major crop in Central China.

5C. Pottery as an indicator of cultural relationships between the early Neolithic assemblages in Central China.

Pottery seems to have been present prior to the beginning of crop domestication and villages in Central China, but pottery assemblages became more diversified after the emergence of the latter two. Comprehensive comparisons of these pottery assemblages have demonstrated that there were at least four different cultures in the Yellow valley during the early Holocene: namely Dadiwan/Baijia, Cishan, Peiligang and Houli (e.g. An 1979, 1984; IA, CASS 1984; Li *et al.* 1996) [Fig. 4(10)][Table 5(3)].

Judging from their pottery assemblages, it seems that Cishan, Peiligang and Dadiwan share similar tripod *bo*, and deep pots, the latter from Dadiwan with tripod [Fig. 4(10)]. The deep pot was probably used as a cooking vessel in both the Dadiwan and Peiligang cultures, but the Cishan people used *yu* (defined in chapter 2) with pottery stands to cook their meals (Jin 1995). Houli people preferred the *fu* set on stone stands for cooking (Tong and Wei 1994). The pottery assemblages of these four early Holocene cultures seem to indicate that there were common elements among them, yet each had its own characteristics, particularly Houli. The other three assemblages in the middle Yellow valley seem to have more similar items, such as the tripod *bo* and globular jars. This similarity is probably a result of cultural contact or influence among the three.

The early Neolithic pottery in Central China was predominantly made by the ring-building method. Firing temperatures are quite low, and the decorative techniques are limited, primarily cord-marking and some incision. The inclusions of the early Neolithic pottery in Central China also vary. Houli phase I pottery seems more primitive, as temper apparently was not purposely added (Wang 1994). Sand was the primary temper in the assemblages in the middle Yellow River: Dadiwan, Cishan and Peiligang. But temper materials seem more diversified in Jiahu, where crushed shell and plant charcoal were also used. This is similar to the Xiaoshankou assemblage. Both are located in the Huai River valley, which may indicate cultural contact.

Summary

So far we have discussed the botanical, faunal and archaeological data in Central China from the terminal Pleistocene to the early Holocene. There is no doubt that the LGM did have a strong influence over this region between 18,000 and 11,000 BP. As a result of this climatic change, there occurred an alteration in the availability of natural resources. During the last glacial, humans were forced to exploit the reduced resources available in the dry and cold steppe for survival, including ungulates and wild plants. This exploitation is illustrated by faunal remains and toolkits, particularly the new tools such as "arrowheads", grinding slabs and axes.

According to accumulated archaeological data so far, this intensive exploitation first emerged in the Fen River valley, represented by the Xiachuan complex. It is not clear why it occurred first in this area, but it seems that the influence of Xiachuan spread into the middle and lower Yellow valley probably after 13,000 BP. The morphological resemblance of some tools found in Xiachuan and the early Neolithic cultures in Central China indicate that the former was very likely the ancestor of the latter.

As the practice of millet cultivation has been documented in the early Neolithic in Central China, there must have been a practice of collecting the wild progenitor of millets in the Xiachuan complex if it was in fact the ancestor of the early Neolithic. This was very likely, as indicated by the glossed edges on its denticulates and the presence of grinding tools. Green foxtail has many advantages: it is adapted to harsh environments, it matures rapidly, and is storable for a long period. When green foxtail became one of the staple foods, the stands within the territory of one group would have been quickly cleared. If a movement to other areas was restricted by reasons such as borders between different groups, then cultivation might have been stimulated, which then would have led to the domestication of green foxtail.

The majority of the early Neolithic societies in Central China were millet cultivating, but rice remains were also found in Jiahu, and two other sites: Lijiacun and Hejiacun, Shaanxi province, south of the Qin mountains [Fig. 4(1)]. Details for the latter two sites are not available, but their pottery and toolkits seem similar to those of the Dadiwan culture. Lijiacun is dated to around 7000 BP [Table 4(7)], slightly later than the earliest phase of the Dadiwan culture. Lijiacun and Hejiacun might represent a southbound expansion of the Dadiwan culture from north of the Qin Mountains. Whether this rice cultivation in Central China was influenced by the Neolithic of the Yangzi River, or originated independently from Central China, is a question under debate (Zhang 1994).

Table 5(1). Comparison of selected lithic implement categories, by number and percentage for major sites.

	Xiachuan		Shizitan		Cishan		Peiligang	
Type of tool	Quantity	% of toolkit	Quan-tity	% of toolkit	Quan-tity	% of toolkit	Quan-tity	% of toolkit
"Arrowheads"	38	0.75%	12	0.67%	74*	6.00%	4*	2.33%
"Denticulates"/ "sickles"	31	0.61%	2	0.11%	6	0.49%	20	11.63%
Axes	7	0.14%	3	0.16%	433	35.11%	19	11.05%
Grinding slabs	27	0.53%	2	0.11%	56	4.54%	24	13.95%
Grinding stone/Rollers	7	0.14%	1	0.05%	54	4.38%	24	13.95%

* Arrowhead was made of bone.

Sources: IA, CASS and CPAM of Shanxi Province 1978; Wang *et al.* 1978; Linfen Culture Bureau 1989; CPAM of Hebei Province and CPAM of Handan City 1981; The First Henan Archaeology Team 1984.

Table 5(2). Comparison of harvesting experiments.

Type of experiment.	Catchment (square meters)	Total labour time *	Grains obtained	Collecting ratio: gram/ per hour.*
Spikelet-shaking	150	0.5 hour	2.1 gram	4.2 g/p.h.
Spikelet-picking	1500	1.83 hour	10 gram	5.46 g/p.h.
Uprooting	150	0.58 hour	3.5 gram	6.03 g/p.h.
Cutting.	7500	1.17 hour	3 gram	2.56 g/p.h.

* This time includes the time for selecting, collecting and beating. E.g. the total time spent to get the grains ready to be milled.

Table 5(3). Pottery comparison in Central China.

Pottery assemblage	Clay and temper	Colour	Building method	Firing temperature	Main decoration patterns.
Dadiwan	Over 90% with fine sand.	Greyish red to red.	Hand-moulding.	About 700 C.	Large proportion of vessels were decorated by cord marking and red slip.
Cishan	90% with sand temper. 10% fine pottery.	Brownish red.	Ring-building and hand-moulding.	Sandy: 850-880 C; Chalk: 700-930 C	Over 50% were plain. Decoration patterns: cord marking, comb-impression and narrow-band in relief.
Peiligang	Over 60% was fine pottery; the rest with sand temper.	Mainly red; some grey.	Ring-building.	Sandy: 820-960 C; Chalk: 910 C	The majority of fine ones was plain. Decoration mainly on sandy ones: cord-marking, comb-impression, incision and pin in relief.
Houli	Natural clay with sand.	Brownish red and red.	Ring building and hand-moulding.	Said to be 'very low' (Wang 1994).	The majority was plain. Narrow band in relief, some cord marking decorated the mouth of some vessels.
Xiaoshankou	Sand, ash of grass/wood and shells as temper.	Brownish red.	Hand-moulding.	Very low.	The majority was plain. A few narrow-band and pin in relief.

Sources: Dadiwan Excavation Group 1981, 1982; CPAM of Hebei Province and CPAM of Handan City 1981; The First Henan Archaeology Team CASS 1984; Jiqing Archaeology Team 1992, 1994; The Anhui Archaeology Team CASS 1993.

Part III. The Transition from Foraging to Farming in the Yangzi River Valley.

While the transition from foraging to farming was marked by the emergence of millet cultivation in Central China, in the Yangzi River valley it was marked by the origin of rice domestication. To date, the earliest cultivated rice has been found in the middle Yangzi valley, dated to 9000-8000 BP. Although there are still divided arguments regarding the centre of the origin of rice cultivation in China, at this stage archaeological data seem to suggest the middle Yangzi valley as the most likely place.

The archaeological assemblages in the Yangzi valley are different from those in Central China. The most prominent difference lies in the lithic assemblages. In Central China the flake industry dominated the whole Palaeolithic era, but in the Yangzi valley there was a prevalent pebble tool tradition, related to that of South China.

Despite this difference, the ensuing Neolithic cultures in the Yangzi valley were as advanced as those were in Central China. Pottery was also present in the Yangzi valley by 10,000 bp (Hill 1995; Yuan 1996). Firmly identified cultivated rice is found in the middle Yangzi valley by no later than 8500 BP, associated with remains of houses, burials and refuse pits (He 1995a; Zhang and Pei 1997). These discoveries are contemporary with, or slightly earlier than the earliest Neolithic remains found in Central China to date.

However, many questions remain. How did Palaeolithic cultures enter the Neolithic in the Yangzi River region? Was the transition a local cultural continuity, or was it due to cultural dispersal from other areas? Was there any significant hallmark of the transition, either in lithic assemblages, or in other aspect(s) of the archaeological assemblages in this region? How did the rice cultivation originate, and where, and when? etc. All these questions are open to debate, as archaeological and archaeobotanical discoveries are not sufficient to provide firm conclusions. However, recent discoveries in this region provide important data for the above questions, thus enabling us to discuss these questions in more tangible terms.

It is the task of part III of this thesis to integrate data from various disciplines, to examine the natural background of the transition from Palaeolithic to Neolithic in the Yangzi valley, including palaeoenvironment, palaeoclimate, fauna and flora. This will be followed by a chronological and regional summary of the archaeological assemblages from the terminal Pleistocene to the early Holocene, and a discussion on the transition from foraging to farming. However, the discussion will focus on the middle and lower Yangzi valley. The cultural sequence from the terminal Pleistocene to the Holocene for the upper Yangzi valley is far from clear at this stage due to very limited archaeological data. Its Neolithic cultures are similar to those found in the middle Yangzi valley, and have been considered a result of cultural diffusion from the latter region (He 1995a; IA, CASS 1984).

Chapter 6. The palaeoenvironments and palaeoclimates from the terminal Pleistocene into the early Holocene in the Yangzi Valley.

The English name 'Yangzi River' is in fact only a translation of the Chinese name for the down stream of this river, which is from the Yangzhou city to the sea (roughly from 119°26' E, 32°23' N eastwards) [Fig. 6(1)]. The proper name for the whole river should be "the Long River" (Changjiang). However, as the name 'Yangzi' has been used for so long in the Western literature, this name will be retained.

The Yangzi River is the longest river in China. Originating in the Qinghai-Tibet plateau, it runs 6300 kms from southwestern to eastern China to enter the East Sea near Shanghai City. The basin of this great river generally refers to the area within about 103-123°E and 24-33°N [Fig. 3(2)] (Liu 1988). Its northern border is the Huai River, and the southern border the Nanling Mountains [Fig. 6(1)].

The Yangzi River basin is divided into upper, middle and lower valley. The upper valley is west of Yicheng city, Sichuan province (111°19' E, 30°42' N); the middle valley is from Yicheng to Hukou city, Jiangxi province (116°14' E, 29°45' N.); and eastwards of Hukou is the lower valley (Ren 1985). These three reaches of the Yangzi valley differ physiographically. The upper valley is mainly a plateau and mountain area which includes the famous Three Gorges. The middle and lower valley is alluvial plains with isolated hills and three large lakes, namely the Dongting, Poyang and Tai lakes from the west to the east (The Editing Committee of the Physiography of China 1981) [Fig. 6(1)].

The middle and lower Yangzi valley is defined as the area within longitudes 111°19' - 123° E, and latitudes 32°30'-25°N. Today, the whole area is temperate to subtropical in climate [Fig. 3(2)]. The present mean precipitation ranges from 1000 to 1600 mm, and mean annual temperature ranges from 14 to 16°C (The Editing Committee of the Physiography of China 1984) [Fig. 3(3); 3(4)]. The flora in this region is a mixture of deciduous and evergreen trees, meadow and woods, as well as cultivated crops, with subtropical fauna.

Although the middle and lower Yangzi valley is roughly within the same latitude, and their current environments are similar, the archaeological assemblages from these two areas are not identical. For the purpose of detailed discussion, the middle and lower Yangzi valley will be considered as two sub-regions; their palaeoenvironments, palaeoclimates and cultural sequences will be discussed separately.

6A. The palaeoenvironments and palaeoclimates of the Yangzi valley.

6A(1) The middle Yangzi valley.

Reconstruction of the palaeoenvironments and palaeoclimates from the terminal Pleistocene into the early Holocene in the middle Yangzi valley are mainly based upon geomorphologic research, analysis of geological sediments, and pollen and faunal data. According to geomorphologic research, during the Last Glacial Maximum, rivers were deeply incised due to the lower sea

level. High platforms and terraces were formed by the down cutting of the rivers. Lakes shrank during that time as the result of a dry and cold climate (Wang *et al.* 1995).

A loess sediment called the 'Xiashu loess' has been widely found in the middle Yangzi valley (Yang 1986). Yang reports that a calcareous concretion containing a deer antler within the loess was dated to 16,620±200 bp (Yang 1986), but the precise location is not revealed. He also states that at another site near Wuhu City, Anhui province, peat sediments on top and beneath the loess were dated to 12,300±465 and 21,500±1700 bp respectively (Yang 1986). These absolute dates indicate that the Xiashu loess accumulated approximately between 20,000 and 13,000 BP. This loess, like its counterparts in Central China, was also formed in a cold and dry palaeoclimate (Yang 1986). The current precipitation in the loess plateau of Central China is between 300-600 mm (Wu *et al.* 1991). Therefore, the precipitation of the middle Yangzi valley during the time of loess deposition could have been similar. This would be a decrease of 700-1000 mm from the present precipitation of 1000 to 1600 mm.

Both faunal and pollen remains in the Xiashu loess are very scarce. Yang (1986) reports that only a few species of molluscs, deer and red deer have been found in the loess (Yang 1986) (localities of these discoveries not revealed). A limited quantity of drought-resistant grass pollen, as well as some *Abies* and *Pine* pollen, have been extracted (Yang 1986). This limited floral assemblage represents a vegetation of coniferous and grassy steppe, reflecting a dry and cold palaeoclimate.

Many boreholes have been drilled in the middle Yangzi valley for pollen analysis (The Pollen Laboratory, IA, Hunan Province 1990), but not many reports are available in detail. Only two boreholes drilled in the Longquan lake in the middle Yangzi valley provide details. According to the pollen profile, the vegetation of the Longquan lake area during the LGM consisted of 60% herbs, 30% trees and 10% ferns (Liu 1991). *Artemisia* and Gramineae were major components within the herbs; fir accounted for 30% of the trees, followed by pine (20%) and birch (15%) (Liu 1991) [Fig. 6(2)].

This pollen assemblage indicates a steppe environment with coniferous and deciduous trees, with conifers dominating. Trees and ferns seem more abundant in the Yangzi valley at this time than in Central China. Among the trees in Longquan, fir (*Abies*) is an important indicator. This tree can only grow in an environment where the maximum average monthly temperature is 10-15°C, and minimum precipitation is 500 mm (Xu *et al.* 1989). Today the maximum average monthly temperature is 28°C in this region, and precipitation is 900-1400 mm (Liu 1991). Thus, the maximum average monthly temperature during the LGM epoch could have been 10-13°C lower than at the present.

Besides the Longquan Lake report, there are a few other summary articles on the palaeo-vegetation of the middle Yangzi valley during the terminal Pleistocene. Liu has claimed that the flora of this region was a mixture of coniferous and deciduous trees, with the conifers dominating (Liu 1995). Trees adapted to a cold climate such as fir, spruce and pine were major components, but birch, elm and oak were also found, in association with ferns and grasses (Liu 1995). This flora seems to represent a steppe and forest palaeoenvironment of a temperate to cool climate, colder than that of the present but relatively moist. The date for this flora is proposed to be between 40,000 to 12,000 BP (Liu 1995). But details are not available, so it is impossible to judge the validity of this argument.

However, opinions on this issue differ. Xu (1985) argues that the mixture of coniferous and deciduous forest mainly existed between 15,000 and 10,000 BP, whereas during the LGM of 20,000 to 15,000 BP, the main vegetation in the middle Yangzi valley should have been steppe with sparse trees; and that only after 15,000 BP did the quantity of trees increase as moisture increased (Xu 1985). A similar view is also held by Zhang and his collaborators (1985). They infer that, under the influence of the East Asia monsoon, the palaeoclimate of the Yangzi valley during the LGM was colder and drier (Zhang *et al.* 1985).

Overall, estimates of mean annual temperature during the terminal Pleistocene in the middle Yangzi valley vary from 4 to 10°C lower than at the present (Xu 1985; Zhang *et al.* 1985). As the minimum precipitation for fir and spruce forest is 500 mm, it is inferred that local precipitation in the terminal Pleistocene should not have been much less than this amount, at least in areas where these trees were found (Yang and Chen 1988).

Whereas changes of floral composition in the middle Yangzi valley seem to be quite apparent during the LGM, faunal change is much less so. The major fauna in the Yangzi valley and South China was an *Ailuropoda* (giant panda)-*Stegodon* animal assemblage from the middle Pleistocene into the early Holocene. Chronological divisions of this fauna cannot be decided by extinct species, so absolute dating is often the main method to determine the date of animal assemblages found in the Yangzi valley. Unfortunately, only a few faunas have been absolutely dated.

So far, four faunas of the terminal Pleistocene have been reported in the middle Yangzi valley. The Zhangnao and Yan'er cave faunas were found near the Yangzi River, and the other two - the Daoxian and Yuchan cave faunas were discovered further south [Fig. 6(1)]. The Zhangnao cave fauna is radiocarbon dated to 13,490±150 bp (Huang *et al* 1987). The Yuchan fauna is proposed to be between 15,000 and 12,000 BP (Yuan 1996, personal communication). The other two are estimated to be terminal Pleistocene (Chen 1986; IA, Hunan Province and Shimen Museum 1994).

Some cool climatic species were present in these faunas. The Zhangnao cave fauna contains the Himalayan goat, while in the Yan'er cave group, brown bear was found, which mainly inhabits in deciduous forest, and today is mainly found in Central and Northeast China (Qi 1989). However sub-tropical species such as macaques, giant tapir and Chinese rhinoceros were found in Yan'er cave, while giant tapir and Sumatra gazelle were present in Zhangnao cave [Table 6(1)]. Such species mix indicates that the climate might have been cooler than the present, but not very severe.

The Daoxian and Yuchan cave faunas were both found within a hilly area north of the Nanling Mountains, along the Xiao river, which is a tributary of the Yangzi River. Current precipitation here is 1400-1600 mm, and average yearly temperature around 18°C, with evergreen trees and subtropical rain forest being the major vegetation (Gao and Dou 1981). The palaeoclimate during the terminal Pleistocene at this latitude would have been comparatively warm, exemplified by macaques, bamboo rats and masked palm civets found in both caves, and crested gibbon, giant panda, Asiatic elephant, giant tapir and Chinese rhinoceros in Daoxian [Table 6(1)].

The beginning of the Holocene at about 10,000 BP brought many changes. The pollen profile in the Longquan Lake shows an increase of trees from 30% to about 40%, and a decrease of herbs from 60% to 40% by 10,000 BP (Liu 1991) [Fig. 6(2)]. Ferns also increased from 10% to 20%.

Within trees, the proportion of fir declined significantly, while evergreen trees increased substantially (Liu 1991) [Fig. 6(2)]. The quantity of pollen grain was also more abundant in the early Holocene stratum than that in the Pleistocene (Liu 1991), indicating the increase of vegetation during a warm and moist period.

It is claimed that similar vegetation changes also occurred in other areas in the middle Yangzi valley (Gu 1991). A mixture of fir, pine, willow and oak with grasses and ferns was found in early Holocene strata in the Dongting Lake area (Gu 1991). In the Pengtoushan site, where one of the earliest Neolithic assemblages in China has been uncovered, there are two pollen horizons, both belonging to the Pengtoushan culture. The major component of the early horizon (roughly at 9000-8000 BP according to C^{14} dating) was arboreal, dominated by broad-leafed China fir (*Cunninghamia* sp.). Also present were grasses including rice, herbs, and ferns (The Pollen Laboratory, IA, Hunan Province 1990). In the later horizon (about 7700-7500 BP), arboreal pollen still accounted for over 80% of the total vegetation, but pine became dominant, with broad-leafed China fir second. It is estimated that the temperature at the beginning of the Holocene at Pengtoushan was about 1°C lower than at the present, but was similar to the present after 8000 BP (The Pollen Laboratory, IA, Hunan Province 1990).

As the climate began to warm up, sea level started to rise, precipitation increased, and so did river outflows. Lakes also expanded during this time. It is claimed that from about 8000 to 7000 BP, the palaeoclimate was even 1-2°C warmer than at the present (Wang *et al.* 1995). As a result, many small tributaries of the Yangzi developed, and a more complex water system was formed. The water levels of rivers became higher, the dissection of rivers weakened, and peat sediments were widely deposited, particularly around the present Dongting lake (Liu 1995).

The improvement of palaeoclimate during the early Holocene must have been favourable to animals. Unfortunately not many animal remains of the early Holocene have been found in the middle Yangzi valley so far. Worse still, details of many animal groups are not reported. It is only known that, remains of hares, deer, pig and sheep were found in two early Holocene sites: Hujiawuchang, radiocarbon dated from 7210±110 to 6960±100 bp; and Zaoshi, dated to 6920±200 bp [Table 6(2); 7(2)] (Hunan Museum 1986; IA, Hunan Province 1993). Details of these faunal remains are not reported, but it has been argued that these remains were residues of human consumption (Hunan Museum 1986; IA, Hunan Province 1993).

Geological research also suggests a more prosperous environment in the early Holocene. The deposition of the Xiashu loess terminated, and loam began to accumulate on top of it (Yang 1986). As loam would have only formed in a temperate and moist climate, the presence of this sediment indicates the improvement of the palaeoclimate. The increased outflow of rivers and the forming of lakes provided abundant water for animals and human beings. This warming trend continued from the early to the middle Holocene, reaching a so-called 'Holocene Megathermal' at about 8000 BP in the middle Yangzi valley (Shi 1993).

6A(2). Palaeoclimates and palaeoenvironments in the lower Yangzi valley.

Research on the palaeoclimates and palaeoenvironments in the lower Yangzi valley have been more intensively conducted than in the middle valley. Since the 1970s, analyses on pollen profiles, sea level changes, geological sediment depositions and fauna have been carried out for the period from the late Pleistocene to the early Holocene. Consequently, a reconstruction of the environments for that period is better-founded.

The first data base for this reconstruction is based upon pollen analysis. Numerous boreholes have been drilled in the lower Yangzi valley since the late 1970s. Sediments from these boreholes are often absolutely dated, providing more accurate and precise information. A good example is a group of seven boreholes drilled in the present Nanjing, Zhenjiang and Wuhu areas, revealing pollen sequences from 30,000 to 4000 BP (Xu *et al.* 1987).

According to the pollen profiles from these boreholes, the local vegetation between 21,000 to 18,000 BP was a mixture of deciduous and coniferous trees, with herbs and ferns (Xu *et al* 1987). The pollen profiles of boreholes ZK 13 and ZK 1391 illustrate that trees accounted for 56-78% of the pollen profile, dominated by oak and followed by pine, spruce and elm; herbs accounted for 10-25%, with *Artemisia* and Gramineae being the major components; the rest was ferns [Fig. 6(3)]. This combination seems to indicate a temperate but slightly dry and cool climate as the beginning of the LGM.

The pollen assemblages changed significantly from 18,000 to 15,000 BP. Both arboreal and fern declined while herbs increased [Fig. 6(3)]. It is reported that the quantity of pollen was much less than that of the previous period, illustrating a less abundant vegetation (Xu *et al.* 1987). Drought-resistant herbs dominated the vegetation, accounting for 79% of the pollen count, while deciduous trees and ferns became minor components [Fig. 6(3)]. This pollen assemblage represents a steppe environment with sparse trees, similar to the vegetation of present northern Central China, where the mean annual temperature is 10°C or lower, with a precipitation of only 150-600 mm (Xu *et al.* 1987). As the current mean annual temperature in the lower Yangzi is 15-16°C, and precipitation 1000 mm, the temperature here during the LGM was 5-6°C lower than at the present, with a precipitation at least 400 mm less than at the present.

From 15,000 to 13,700 BP the climate moderated, demonstrated by a pollen profile from the Zhenjiang borehole [Fig. 6(4)]. Within this pollen assemblage, trees increased to 46-47% of the total pollen composition after 15,000 BP, with oak becoming the major component, followed by pine, elm and willow (Xu and Zhu 1984). Herbs accounted for 36-39%, the rest were ferns. Drought-resistant herbs and hygrophilous plants were both present, indicating that the palaeoclimate was becoming moister and warmer than in the previous stage.

After 13,700 BP the palaeoclimate became colder and drier again. Both the quantity and the variety of the pollen profile were substantially reduced. Drought-resistant herbs dominated, indicating a steppe environment with sparse deciduous [Fig. 6(4)]. It is estimated that the temperature of this period was about 6-7°C lower than at the present, with a yearly precipitation of 500-600 mm (Xu *et al.* 1987).

Between 10,850 to 10,000 BP there were climatic fluctuations, and the quantities of arboreal pollen were up and down during this period [Fig. 6(4)]. But, in general, trees during this period were fewer, while ferns increased significantly, accounting for between 33-84% of the pollen count [Fig. 6(4)]. Although the reason for the decline of trees is not yet clear, the dominance of ferns, as well as the peat sediment accumulated during this time, indicates a moist and warm palaeoclimate.

After 10,000 BP, trees became abundant in this region again. Arboreal pollen increased from less than 25% to a

maximum of 93% within the vegetation [Fig. 6(4)]. Deciduous trees dominated and evergreens were also present (Xu and Zhu 1984). Gramineae and *Artemisia* were the major components within the herbs (Xu and Zhu 1984). This pollen profile represents a temperate to sub-tropical environment.

Similar pollen sequences are also reported for other locations in the lower Yangzi valley. The cave of Xianrendong is particularly important because of its archaeological discovery. The cave was excavated twice during the 1960s, and re-excavated in 1993 and 1995 by a Sino-American team. The pollen profiles reported from Xianrendong and nearby Wangdong cave indicate local climatic fluctuations [Table 6(3)].

The chronology of Xianrendong and Wangdong is based on a combination of absolute dating and cross-cultural estimation. Although a few radiocarbon samples have been tested, the dates are not all in sequential order (MacNeish and Taylor 1995). However, the pollen sequence reveals the palaeoenvironment from the terminal Pleistocene to the early Holocene in the Dayuan basin in the lower Yangzi valley.

The vegetation in this basin was dominated by conifer and drought-resistant herbs during the LGM epoch, but pollen for the period between 20,000 and 16,000 BP is extremely rare (Wang *et al.* 1995), which may reflect sparse vegetation. The vegetation became more abundant around 15,000 BP, but declined again afterwards; then increased after 11,000 BP. The pollen profiles represent a cold and dry phase in the terminal Pleistocene, changing into a warm and wet early Holocene (Wang *et al.* 1995).

The climatic change in Xianrendong is generally consistent with the palaeoclimatic reconstruction for other sites in the lower Yangzi valley, such as Gucheng (No. 5) [Fig. 6(1)]. It is notable that in Gucheng, a sharp decline in trees and an increase of herbs occurred between 11,300 to 11,000 BP, indicating a sudden climatic deterioration, which may be comparable to the Younger Dryas in Europe (Yang *et al.* 1996) [Fig. 6(5)].

The cold palaeoclimate during the LGM in the lower Yangzi valley is also illustrated by faunal analyses. Four faunas of the terminal Pleistocene have been found in the lower Yangzi valley. Two of them have been absolutely dated; the other two are estimated to be terminal Pleistocene (Li and Lei 1980; Li *et al.* 1982; MacNeish and Taylor 1995; Zhang *et al.* 1987). Among them, the fauna from a small island called Sanshan in Tai Lake may represent the LGM epoch. Brown bear and European lynx were found in the Sanshan fauna, both belonging to temperate or cold environments (Zhang *et al.* 1987) [Table 6(1)]. As lynx today inhabits coniferous forest of the North to Northeast China, the presence of this species indicates a cold palaeoclimate.

Another fauna from the Spirit Cave (Shenxiandong) in Lishui county, Jiangsu province, is dated to 11,200±100 bp. Short faced mole, ratlike hamster and brown bear were present in this fauna [Table 6(1)] (Li and Lei 1980). The short faced mole and hamster today live in dry steppe and sandy areas, while the brown bear mainly lives in cool climate deciduous forest and bushy areas (Han and Xu 1989). The presence of these species suggests a cooler and drier palaeoclimate at the end of the Pleistocene.

Apart from the above, the majority of the fauna assemblages found in the lower Yangzi valley include animal species of temperate to sub-tropical environments. Monkeys, porcupines, masked palm civets, pandas, Stegodon and giant tapirs are frequently found [Table 6(1)]. Most of these animals are extant except for Stegodon, although some species, such as panda and giant tapir, are no longer extant in the Yangzi valley. These animals are all adapted to forest and bush environments, indicating that the palaeoclimate in the lower Yangzi valley during the terminal Pleistocene was relatively mild and moist.

Climatic fluctuations have also been illustrated by data from other disciplines. According to geologists, the Xiashu loess also accumulated in some parts of the lower Yangzi valley during the LGM, particularly around the present Nanjing city where the loess was originally named. Mineral analysis illustrates that the Xiashu loess was a deposit of a windy, cold and dry climate (Wu *et al.* 1992). As the loess contains scanty pollen and animal fossils, a palaeoenvironment with sparse vegetation is inferred (Wu *et al.* 1992). However, this scarce might have been partially due to the soil itself, as loess is not a favourable sediment for pollen preservation.

It is also claimed that some of the lakes in the present lower Yangzi valley did not form until after the LGM period. The site of Tai Lake, which is now one of the largest lakes in China, was part of an alluvial plain covered by loess-like sediment during the LGM (Chang and Liu 1996). At about 11,000 BP, the river level rose to form a bay there, and the present Tai Lake was only come into existence since 6500-5000 BP (Chang and Liu 1996).

The other two large lakes in the lower Yangzi valley - Dongting and Poyang lakes – also seem to have gone through a similar development. The present average depths of these lakes differ from 2.12 to 4 meters, with the deepest parts ranging from 3.33 m (Tai), 16 m (Poyang) to 30.8 m (Dongting) (Yang 1986). It is argued that the water depths directly relate to Yangzi water levels, and the latter relates to the sea level (Yang 1986). This reconstruction of lakes suggests that the water resource during the LGM epoch must have been much less than at the present.

Sea level movements from the late Pleistocene into the early Holocene played an important role in changing palaeoenvironments in the lower Yangzi valley, particularly in the Yangzi delta. Comprehensive studies on this issue have demonstrated that the sea level was 155 meters lower than at the present from 16,000 to 14,700 BP (Geng 1981) [Fig. 3(5)]. The present continental shelf then was exposed, and the palaeochannel of the lower Yangzi River was about 45 meters deeper than at the present (Yang 1986). After 14,000 BP the sea began to rise with fluctuation, reaching to 56-61 meters lower than at the present at 12,000 BP (Geng 1981). It then continued to rise, but between 11,100 and 10,900 BP it dropped again sharply from -26 to -33 meters, with a descending rate of 35 mm/per year. This is inferred to correspond to a sharp cooling down of the palaeoclimate, comparable to the Younger Dryas in Europe (Yang and Xie 1984).

After the Younger Dryas the palaeoclimate became more temperate at around 10,000 BP. There are numerous data regarding the palaeoclimate of the early Holocene in the lower Yangzi valley, particularly from palynology. Many boreholes have been drilled and some have been absolutely dated, such as the boreholes at Zhenjiang, Tai Lake and Qidong [Fig. 6(1)]. Additionally, pollen analysis has been conducted in a few archaeological sites.

According to the pollen profiles from Zhenjiang, Tai Lake and Qidong, at around 11,000 to 10,000 BP there was a climatic deterioration, indicated by a sudden increase in fir *(Abies)*, spruce (*Picea)* and drought-resistant herbs (Liu and Chang 1996) [Fig. 6(5), 6(6) and 6(7)]. It is argued that this sudden deterioration was comparable to the Younger Dryas (Liu and Chang 1996). After 10,000 BP there was a

significant increase of broad-leafed trees, representing an expansion of subtropical evergreen vegetation (Liu and Chang 1996), which must have been the outcome of an improving palaeoclimate.

The warm and moist palaeoclimate of the early Holocene seems to have provided a better environment for animals in the lower Yangzi valley. A few faunas have been found in this region, among them the famous Hemudu fauna. A total of 61 animal species were found in Hemudu, including 34 species of mammals, eight species of bird, six species of reptile, 10 species of fish, two species of mollusc and one species of crab (Liu and Yao 1993) [Table 6(2)]. Another fauna from the Jinhua cave, Zhejiang province consisted of 46 animal species (Ma and Tang 1992). The Hemudu fauna is dated to around 7000 BP, and the Jinhua fauna to 7815±385 bp [Table 7(2)]. The majority of animals in these two faunas belong to a subtropical to tropical environment, except for mole shrew found in Jinhua, which inhabits cold mountainous areas today (Ma and Tang 1992). These two faunas suggest that the palaeoclimate of the early Holocene was temperate, probably even slightly warmer and moister than at the present, with forest and meadows around the sites.

Sea level changes also indicate a warm-up of the palaeoclimate. At about 10,000 BP the sea level was about 18 meters lower than at the present, then it rose 9 meters within two thousand years, at an average speed of 5.4 mm per year (Yang and Xie 1984). From 8000 to 7000 BP the marine transgression of the Holocene began, and the lower basins and alluvial plains of the lower Yangzi valley were inundated. It has been estimated that the maximum sea level at about 6000 to 5000 BP was two meters higher than at the present. A few early Neolithic sites, such as Hemudu, were then temporarily inundated by the rising sea level (Lang 1987) [Fig. 3(5)].

In summary, research from various disciplines have documented the palaeoclimatic and palaeoenvironmental changes from the late Pleistocene to the early Holocene in the middle and lower Yangzi valley. It is clear that from 18,000 to 15,000 BP, the LGM strongly influenced the Yangzi valley, and the vegetation was basically a steppe/grassland with a mixture of coniferous and deciduous trees. The mean annual temperature was estimated at about 10°C during the LGM, with precipitation of about 600 mm. This is a palaeoclimate similar to the cool-temperate zone of Central China today, but the temperature was about 4-6°C lower than the current average yearly temperature in the Yangzi Valley, and the precipitation was 400-1000 mm less than at the present.

However, as the latitudes of the Yangzi valley are lower, the cold and dry winter monsoon of the LGM must have been weaker in this region compared to that in Central China. During the LGM, deciduous trees were absent in many parts of Central China, but were still present in the Yangzi valley. Faunal changes were more significant in Central China, as many temperate species withdrew or became extinct, but the fauna in the Yangzi valley seems to have changed much less. Species of temperate to sub-tropical environments such as macaque and deer still existed [Table 6(1)]. A distinctive characteristic of the LGM fauna in the Yangzi valley is the continuing presence of some species from North China, such as brown bear and short-face mole.

After 15,000 BP the palaeoclimate improved. There were fluctuations, particularly around 11,000 to 10,300 BP when a sudden cold event occurred. The beginning of the Holocene witnessed the spread of a warmer and moister climate. Consequently, evergreen trees and deciduous trees dominated, large lakes were formed, and the environments became more favourable to animals and humans.

6B. Natural resources in the Yangzi valley.

The fauna and flora were natural resources for humans inhabiting the Yangzi valley from the terminal Pleistocene to the early Holocene. Therefore, it is necessary to investigate the availability and seasonality of these natural resources during that time.

6B(1) Faunal resource.

Eight faunas have been chosen to represent the animal resources in the middle and lower Yangzi valley during the terminal Pleistocene, and six during the early Holocene [Table 6(1), 6(2)]. Some of these faunas have been absolutely dated, and some were found in association with artefacts. Of the terminal Pleistocene faunas, four were from the middle valley, and four from the lower. Except for the Sanshan fauna, all were found in caves. Of the early Holocene faunas, two were from the middle valley and four from the lower. Most species of the Pleistocene fauna belonged to the giant panda-Stegodon fauna, and a substantial proportion of the species were commonly found both in the middle and the lower Yangzi valley, such as macaque, hog badger, bear, masked palm civet, wild boar and deer [Table 6(1)]. The compositions of these faunas are summarised at table 6(4).

It seems that animal species were relatively abundant in terms of variety during the terminal Pleistocene. Remains in the four archaeological sites included teeth and broken long bones; many were burnt. This is interpreted as residues of human consumption (Huang *et al.* 1987; IA, Hunan Province and Shimen Museum 1994; Yuan 1996). Some of the bones found in Zhangnao and Yan'er caves bear traces of human cutting (Huang *et al.* 1987; IA, Hunan Province and Shimen Museum 1994). Further, the majority of the ungulate teeth found in Yan'er were deciduous (IA, Hunan Province and Shimen Museum 1994), indicating that the animals killed were mainly young individuals.

The remains of these faunas illustrate that ungulates and small carnivores were hunted by humans during the terminal Pleistocene. In the Yangzi River region, ungulates such as rhinoceros and giant tapir, were slow moving with big bodies, relatively easy to catch, and once caught a large quantity of meat would have been obtained.

In general, the animals inhabiting the Yangzi valley during the terminal Pleistocene were comparatively easier to hunt than those in Central China were. Fast running ungulates such as wild horses, gazelle and onager seem not to have been present in the Yangzi valley. However, during the cold and dry LGM, animals like deer, muntjac and giant tapir would have had difficulty to survive, as they were species of forest and meadow with abundant water. Therefore, the availability of these species would have been less.

It is notable that the influence of the winter monsoon would have been much weaker in areas of lower latitudes, thus the palaeoclimate would have changed much less during the LGM. Therefore, animals could have moved from cold and dry to less severe areas. For example, in the Lishui fauna, a few species like short faced mole, brown bear and hamster were animals of Central China (Li and Lei 1980). Their presence in the lower Yangzi valley indicates a movement from the north. In the same fauna, giant tapir and muntjac were absent; so were several species of deer adapted to forest and meadow. Dated to 11,200±1000 bp (Li and Lei 1980) and calibrated to between 14,247 and 11,959 cal. BP by CALIB. 3 (Stuiver and Reimer 1993), the Lishui fauna indicates that the palaeoclimate was rather cool. The

absence of animals adapted to subtropical environments indicates that they might have withdrawn to a more southern area. On the other hand, those species from Central China would have been additional resources for the Yangzi occupants.

Nevertheless, there seems no evidence for any sudden extinction horizon of animals in the Yangzi valley. This is a distinctive difference from Central China. Han and Xu (1989) made a detailed table which listed all the Pleistocene species found in China south of the Yellow River basin. It is clear that many animal genera were present from the middle to the end of the Pleistocene in the Yangzi River region, namely various macaques, hog badger, civet, giant panda, Stegodon, giant tapir, several species of rhinoceros, deer and muntjac, although there were changes of species over this long period (Han and Xu 1989).

This comparatively stable fauna continued into the early Holocene. Giant panda was still found in Jinhua cave in Zhejiang province by approximately 8000 BP, associated with Stegodon [Table 6(2)]. Most other species commonly found in the terminal Pleistocene continued in archaeological sites of the early Holocene, as exemplified by the Hemudu fauna [Table 6(2)]. The additional species found in Hemudu are mainly aquatic animals, such as turtles and molluscs, plus domesticated dog and pig.

Generally speaking, the majority of Yangzi animals could have been hunted year-round, probably except for bears in winter hibernation. Remains of more than 10 species of birds have been reported in the Yuchan cave, Hunan province, accounting for over 30% of the total animal remains (Yuan 1996), but details of the species are not available. Wild geese were found in Neolithic Hemudu and Luojiajiao in the Yangzi delta, both dated to around 7000 BP [Table 7(2)]. Wild geese are migrating birds, they fly from the north after the autumn. Their seasonal availability in the Yangzi River valley would have been from autumn to spring next year.

This relatively stable animal resources of the Yangzi might also have had their effects on human behaviour, particularly on toolkits. In the Yangzi valley and South China there were only limited changes in lithic implements during the Pleistocene and into the early Holocene, compared to the changes in Central China. Is it possible that the stability of animal resources was one of the reasons for the toolkit stability? If prehistoric humans always hunted the same game, and species variety and quantity changed little, there probably was no need to change hunting techniques and tools. Of course, this is only a presumption which needs to be ratified or rejected by further study, particularly by stone tool analysis and experiment.

6B(2). Floral resources.

According to pollen data, plant resources in the Yangzi valley were quite rich. The pollen profile from the Longquan lake borehole in the middle Yangzi valley shows that the LGM witnessed a dominance of coniferous forest and herbs (Liu 1991). Among the trees, pines accounted for 20%. Some species of pines produce edible and nutritious nuts. Several genera of deciduous trees were still present (although in a low frequency), particularly walnuts (*Juglans*) [Fig. 6(2)]. Herbs accounted for 60% of the pollen assemblage, dominated by *Artemisia* and Graminea (the Grass family) [Fig. 6(2)]. Although pollen analysis cannot identify exact species, many species of the Grass family bear edible seeds, particularly wild rice (*Oryza rufipogon*). The beginning of the Holocene in the Longquan borehole was marked by a decline of coniferous trees, and a significant increase of oak and other deciduous and broad-leafed trees. But *Artemisia* and Gramineae were still the major components of the herbs [Fig 6(2)] (Liu 1991).

Similar pollen profiles also occur in the lower Yangzi valley, where the vegetation during the LGM was also mainly steppe/grassland with sparse trees. After 15,000 BP there was an increase in deciduous and evergreen broad-leafed trees. Grasses were present in many boreholes, and was the major component in the Qidong borehole from 12,000 to 10, 810 BP (Liu *et al.* 1992), again providing edible seeds for human beings.

Archaeological plant remains of the terminal Pleistocene were found in Yuchan cave, Hunan province. It is reported that over 40 species of plant macro- and microfossils have been found, the most abundant being hackberry seeds, as well as wild rice (Yuan 1996). Further details of this discovery are not available, but the discovery suggests that hackberry seeds and wild rice were exploited by human beings at the terminal Pleistocene there.

Phytolith analysis at Xianrendong also indicates that wild rice was one of the resources available during the terminal Pleistocene (Zhao *et al.* 1995; Zhao 1998). Isotopic analysis of human bone from Xianrendong, dated to 15,180±90 bp, suggests that wild rice was probably consumed (MacNeish *et al.* 1995).

These discoveries suggest that wild rice (*Oryza rufipogon*) was collected in the middle Yangzi valley, probably also in the lower Yangzi valley during the terminal Pleistocene. In addition, remains of domesticated rice (*O. sativa*) have been found in the middle Yangzi valley by 8000 BP in the Bashidang assemblage (Zhang and Pei 1997; Pei 1998b), and in the lower Yangzi valley by 7000 BP in the Hemudu assemblage (Liu and Yao 1993; Luojiajiao Archaeology Team 1981). Wild rice seems to be the most important plant in the Yangzi River valley, and it merits special attention.

Since the 1930s, issues concerning the classification, botanical characteristics and ecology of both wild (*O. rufipogon*) and domesticated rice (*O. sativa)*, as well as their relationships, have attracted attention from numerous scholars (for a review, see Ahn 1993). Many arguments and propositions have been raised, but there are still many questions. First, there are divided arguments regarding the classification of wild rice in Asia. The Asian common wild rice (*Oryza rufipogon*) is considered by many scholars as the progenitor of the Asian domesticated rice (*Oryza sativa*) (e.g. Ding 1957; Oka 1988; Pang *et al.* 1995; Wang 1994). But it is also argued that common wild rice is a complex consisting of perennial and annual types (Oka 1988; Pang *et al.* 1995). The point of argument is whether the annual type is a hybrid weed of wild and domesticated rice, or is a genuine wild type (Oka 1988). Those scholars who argue that the annual wild rice is a true wild plant tend to consider it as the immediate progenitor of the domesticated rice (i.e. Chang 1989; Tsunoda and Takahashi 1984). Others argue that the annual type is only a weed (Pang and Wang 1996), and that the perennial one is the ancestor of domesticated rice (Ding 1957; You 1995). It is also proposed that an intermediate type was possible between the perennial and domesticated form (Oka 1988). But all scholars seem to agree that the ultimate ancestor of all the annual rice was the perennial form (e.g. Chang 1989; Tsunoda and Takahashi 1984). So it seems certain that the perennial common wild rice (*Oryza rufipogon*) is the ultimate progenitor of domesticated rice, with a question mark over its "immediateness".

A second question concerns the contemporary and ancient distributions of common wild rice in China. For the contemporary distribution, a national survey was conducted

from 1978 to 1982. There are currently three species of wild rice in China: common wild rice (*O. rufipogon*), medical wild rice (*O. officinalis*) and a short wild rice (*O. meyeriana*). The last two species do not relate to domesticated rice. The Chinese distribution of common wild rice (*O. rufipogon*) today is within South China, from 100°40' to 117°08' E, and from 18°09' to 28°14' N (The National Survey Group of Wild Rice 1984), with a few stands in the middle and lower Yangzi valley (Jiang *et al.* 1990; Sun 1990) [Fig. 6(8)].

However, this is only the modern distribution of common wild rice. What was the distribution of this plant in prehistory? The most reliable data come from archaeobotanical discoveries. Remains of common wild rice have been found in archaeological sites outside the current distribution zone, not only in the middle and lower Yangzi valley, but also in the Yellow valley. A recent analysis of the rice remains from Hemudu revealed that, while the majority were domesticated, a few grains of wild rice were also present (Chen and Huang 1994). Further north, wild rice has been found in a Longshan culture site called Lilou, in Henan province, dated to approximately 4000 BP (Wu 1994). These discoveries suggest that the distribution of common wild rice from 7000 to 4000 BP was far more expansive than it is now. Given the warmer and moister palaeoclimate in the early and middle Holocene, this broader distribution is possible.

Is it possible that the remains of common wild rice found in Hemudu and Lilou were brought by people from other areas, for example, from the southern area where wild rice is found today? This possibility cannot be completely ruled out, but it is not very likely. The Hemudu culture was quite advanced in rice cultivation, as a large quantity of cultivated rice was found there. Cultivated rice (both *japonica* and *indica)* was also found in Lilou (Wu 1994). In fact, as discussed in chapter 4, the earliest remains of cultivated rice found in Central China to date have been discovered in Jiahu, dated to between 8500 and 7500 cal. BP (Zhang 1994) [Table 4(7)]. Remains of cultivated rice have also been found in another eight archaeological sites (including Lilou) in Central China, ranging from 7100 to 4500 BP (Zhang 1994) [Fig. 6(8)]. In the Yangzi valley, several archaeological sites have also yielded remains of cultivated rice, dated from 9000-8000 BP onwards (below, chapter 7). These discoveries indicate that rice cultivation was widely spread in Central China and the Yangzi valley from the early to the middle Holocene. In this situation, there was no need for prehistoric people to take common wild rice from other areas, as domesticated rice had been cultivated locally. The logical explanation is that there must have been natural stands in these areas at that time. The presence of wild rice mixed with the domesticated counterpart might have resulted from their locations close to fields; or they might still have been collected by prehistoric people who were cultivating the domesticated rice.

Common wild rice is an aquatic plant of a sub-tropical to tropical environment. It needs more water and a warmer climate to grow than the millets. According to reports on contemporary common wild rice, it can survive in an average annual temperature above 16°C, with a precipitation of 1000 mm [Table 6(5)]. In South China, common wild rice grows well in swampy areas with permanent shallow water (under 1 meter deep); it does not grow well in land with a poor and/or discontinuous water supply (The Guangxi Survey Group of Wild Rice 1990). This plant favours fairly neutral soil (pH6-7), although it can also grow in slightly alkaline land (The Guangxi Survey Group of Wild Rice 1990). The cold tolerance of common wild rice varies during its life-history: it can survive down to 2°C during the early seedling stage, 6°C during the late seedling stage, and 15°C during the panicle developing stage (Lin 1990). But a most distinct feature of common wild rice is that, being a perennial plant, it can regenerate itself by both ratooning and seeding, mainly by ratooning (Oka 1988; Pang *et al.* 1995). Consequently, the quantity of seeds produced by common wild rice is quite limited (Oka 1988).

Chinese botanists and agronomists have also conducted observations on common wild rice in various areas. The life-history of common wild rice varies from south to north, as summarised in table 6(5). It seems that the panicle developing stage of common wild rice begins earlier in the northern area than in the south, with stands in Hainan island being the latest. This is probably a natural adaptation to local climate, as ripening has to be completed in the month prior to the cold winter.

An observation on common wild rice was conducted in Guangdong province in late-October 1995. The observed population was very small, as this wild plant is hard to find nowadays. It grows in a swampy habitat with water about 40 mm deep [Fig. 6(9)]. In late-October there were flowering and shattering on a single plant simultaneously [Fig. 6(10)]. Unfortunately due to various reasons, only two days could be spent on observing ripening process. However, three preliminary harvesting experiments were carried out. It seems that the schedule of flowering to ripening and shattering for common wild rice is quite heterochronous, as in green foxtail. Within one panicle, some florets were flowering but others already ripened and shattered [Fig. 6(10)]. A local informant who knew the plant well said that the whole process could last for one month, and that the shattering was normally completed in mid-November. It is also reported that the flowering season of wild rice found in Hunan province was as long as two months, from mid-September to mid-November (Sun 1990).

The grain size of common wild rice is in fact quite large, approximately 6x2x1 mm, similar to cultivated grains [Fig 6(11)]. According to a quality identification of common wild rice and a comparison between common wild rice and several types of domesticated rice, the grain of common wild rice is slightly longer than the domesticated ones; its yield and starch quantity are similar to domesticated rice; and the taste of cooked common wild rice is said to be better than the domesticated one (Qin and Li 1990). No wonder this plant was exploited by prehistoric people.

Another possible floral resource in the Yangzi valley is green foxtail (*Setaria viridis*), which is commonly found from North to South China today. In Hainan island, millet was still cultivated by local ethnic groups in the 1970s (Zhang 1985), as it was in Taiwan (Fogg 1983). Did millet cultivation originate independently in these islands, or was it due to cultural dispersal from Central China? This is a question unsolved, although it has been proposed that millet was introduced from the north (Zhang 1985). Today millet is seldom cultivated in the Yangzi valley and South China. But remains of cultivated millet have been found in Qinglongquan, Hebei province, dated to approximately 4000 BP (Chen 1993), although it is not clear whether the cereal was cultivated locally. Green foxtail can endure a cold and dry climate, and it might have existed in the Yangzi valley during the LGM. Therefore, the possibility of its being exploited by prehistoric human beings cannot be ruled out.

Unlike many animals, most wild grasses have a seasonal availability. The seasonality of common wild rice varies in different areas from south to north [Table 6(5)]. It seems that plants from the northern area develop their panicles earlier than their southern counterparts, and ripen earlier

[Table 6(5)]. On an average, for each latitudinal increase of 1°N, the beginning of panicle development seems to be about 6.8 days earlier. If this botanical characteristic changed little since the terminal Pleistocene, it can be inferred that the ripening period of wild rice during the late Pleistocene to the early Holocene in the Yangzi valley might have been from late August to early September. According to Sun (1990), the flowering and ripening season of modern common wild rice in Hunan province (two stands at 25°05' and 26°50' N respectively) also starts from late August to mid September, and completed from late October to mid November (Sun 1990).

However, the availability of common wild rice in different areas would have been restricted by the palaeoclimate. As mentioned above, contemporary common wild rice can survive in a minimum average annual temperature of 16°C, with a precipitation of 1000 mm. During its seedling period it can survive at 2°C, and at 15°C during its panicle development. In the present areas along the Yangzi River, the average spring temperature is 3-5°C (The Editing Committee of Physiography of China 1984), which is barely enough for the plant to regenerate and survive. **But**, as discussed above, it is estimated that the average annual temperature during the LGM was about 4-10°C lower than the present. Even taking the minimum 4°C as a parameter, the average annual temperature in the Yangzi valley during the LGM would have been about 12°C, and the temperature in spring would have been well below its current level [Table 11(3)], which seems too cold for the common wild rice. If the cold tolerance of this species has not changed significantly in the last 18,000 years, common wild rice would not have been widely available in areas along latitude 30°N during the LGM.

Could the climate in southern area of the Yangzi valley have been warmer? According to the phytolith analysis report for Xianrendong-Wangdong, wild rice was present during the LGM, but the quantity in the Pleistocene was very limited (Zhao *et al.* 1995). Xianrendong cave is located in a small basin at 28°44' N, with a mean yearly temperature of 16°C, and a precipitation of 1600-1800 mm (The Editing Committee of the Physiography of China 1984). The meagre presence of wild rice phytoliths during the LGM probably reflects the sparse stands of wild rice there.

Common wild rice has also been found in Yuchan cave, Hunan province (Yuan 1996). The average yearly temperature here is 18°C, with a precipitation of 1492 mm. More importantly, the local average temperature in spring is 13°C (Gao and Dou 1981). As the latitude of Yuchan is much lower than that of other sites along the Yangzi valley [Fig. 6(1)], the decrease of palaeo-temperature here during the LGM would have been less significant. Therefore, the local temperature in the spring would have been well over 2°C even during the LGM, sufficient for common wild rice to survive and regenerate.

In summary, assuming the botanical characteristics of common wild rice have not changed significantly over the last 18,000 years, it can be inferred that this plant might not have been abundant in areas along latitude 30°N during the LGM, but it could have survived in areas around latitude 25-28°N [Fig. 6(8)]. In other words, common wild rice might have been a rare resource for prehistoric foragers in areas along latitude 30°N during the LGM. After the LGM it seems that common wild rice expanded northwards [Fig. 6(8)], as the palaeoclimate became warmer and moister.

Summary

In this chapter the palaeoenvironments and palaeoclimates from the terminal Pleistocene to the early Holocene in the Yangzi basin have been reviewed, and the local faunal and floral resources which were possibly available to human beings during this time have been discussed. Although the LGM did have quite significant impact in the Yangzi basin, this impact seems less severe here compared to that in Central China. It seems that both faunal and floral resources in this area were richer than those in contemporary Central China. Species of animals changed little at the end of the Pleistocene to early Holocene; there were more trees with edible nuts; wild rice and other seed-bearing grass were available. It would be legitimate to assume that the foraging life of human beings in the Yangzi basin was easier than that of their counterparts in Central China. What would have been the impact of this stability and relatively abundance of natural resources on the prehistoric occupants in the Yangzi basin? Why did they, even in such a relatively favourable environment, take the trouble to plant cereals and eventually to develop agriculture? These are questions for the next chapter.

Table 6(1) Faunas of the terminal Pleistocene in the Yangzi valley.

Species	English name	Z.N.	Y'e.	D.X	Y.C.	D.Y.	S.S.	L.S.	Habitat and characteristics
Mammalia	Mammals.								
Insectivora					+				
Scaptochirus moschatus	Short faced mole.							+	Typical in dry steppe/sandy area.
Primates									
Hylobates cf. concolor	Gibbon.			+					Sub-tropical forest, lowland and hills.
Macaca sp.	Macaque.		+	+	+	+	+	+	" "
Macaca arctoides	Red-faced macaque.		+						" "
Rodentia									
Cricetulus sp.	Ratlike hamster.							+	Dry and cold steppe/border of desert.
Rattus rattus	Black rat.					sp.	+	+	Human commensal.
Hystrix sp.	Old world porcupine.				+	+		+	Various-forest, mountain, steppe. Weight 10-30 kg.
Hystrix cf. subcristata			+	+		+	+		
Rodentia (type A).		+	+		+				
Rodentia (type B).		+							
Rhizomys sp.	Bamboo rat.			+	+				Sub-tropical bamboo stands.
Leporidae								+	Various-steppe, forest.
Lepus sp.	Hare.		+		+				
Carnivora									
Canidae			+					+	
Nyctereutes procyonoides	Raccoon dog.				sp.		+	+	River valley, forest and steppe.
Arctonyx collaris	Hog badger.			+	+	+	+		Forest up to 3500m elevation. Weight 7-14kg.
Ursus arctos	Brown bear.		+		sp.	sp.	+	+	Deciduous forest/bushy areas.
Ursus thibetanus	Black bear.			+			+		Moist deciduous forest.
Mustela sp.	Polecat.				+		+	+	Steppe.
Meles leucurus	Badger.	sp.			+		+	+	Forest and densely vegetated areas.
Paguma larvata	Masked palm civet.			sp.	sp.	+		+	(Sub)tropical forest.
**Crocuta crocuta ultima*	Hyena.			sp.		+	+	+	Steppe. ungulates as major food.

Table 6(1) Faunas of the terminal Pleistocene in the Yangzi valley (continued).

** **

Species	English name.	Z.N.	Y'e.	D.X	Y.C.	D.Y.	S.S.	L.S.	Habitat and characteristics
Carnivora(continued)									
Felis teilhardi	Wild cat.			+	sp.			sp.	Forest.
Felis bengalensis	Small wild cat.								
Ailuropoda melanoleuca	Giant panda.	+		+					Bamboo-forest.
Vulpes sp.	Fox.	+							Hilly/forest areas.
Cuon javanicus	Javanese dhole.		sp.	+					Various but avoid desert.
Viverra zibetha	Oriental civet.			sp.	+				Forest, bush/meadows. Weight 5-11kg.
Viverricula indica	Rasse.			+	+				Forest, bush/meadows. Weight 2-4kg.
Panthera pardus	Leopard.			+					Various.
Lupus lupus	Grey wolf.			+					
Lutra sp.	River otters.			+					
Lynx lynx	European lynx.						+		Cold environment, forest.
Panthera tigris	Tiger.						+		Forest.
Proboscidea									
**Elephas cf. namadicus*	Straight tusked elephant)					+			Forest and bush.
**Stegodon orientalis*	Stegodon.	+	+	+					Forest and bush.
Elephas maximus	Asiatic elephant.			cf.					(Sub)tropical thick jungles/grassy plains.
Perissodactyla									
**Dicerorhinus cf. kirchbergensis*		+							Cool-temperate forest or forest edge.
**Megatapirus augustus*	Giant tapir.	+	+	+		sp.			(Sub)tropical grassy plains.
**Rhinoceros sinensis*	Chinese rhinoceros.		+	+					(Sub)tropical forest/watered areas.
Artiodactyla									
Bovinae					+	+	+	+	
Bubalus sp.	Buffalo.	+		+					Moist bush/forest.
Bos sp.	Ox.		+						
Cervus sp.	Deer.	+	+	+		+	+	+	
Cervus nippon	Sika deer.				+				Forest/grass. Weight 26-33 kg.
Cervus unicolor	Sambar.				+		cf.		Various. Prefer wooded area. Weight 109-260 kg.
Hydropotes inermis	Chinese water deer.					sp.			(Sub)tropical river/lake side or grassy mountains.

Table 6(1) Faunas of the terminal Pleistocene in the Yangzi valley (continued).

Species	English names.	Z.N.	Y'e.	D.X	Y.C.	D.Y.	S.S.	L.S.	Habitat and characteristics
Artiodactyla(continued)									
Muntiacus reevesi	Small muntjac.						+		Forest/dense vegetated areas with water.
Muntiacus muntjak	Muntjac.		sp.	sp.			+		" ".
Ovis sp.	Sheep.		+				+		Upland areas/grassy mountains.
Gezella sp.	Gazelle.		+						Dry open country or bush lands.
Nemorhaedus goral.	Himalayan goral.	+							Rugged, wooded mountains, elevation over 1000m.
Capricornis sumatraensis	Sumantra serow.								Rugged mountains/ridge of forest/bush.
Kanjereus sp.		+							(Sub)tropical bushy area.
Sus scrofa	Wild boar.		sp.	+	sp.	+	+	sp.	Forest.
Total species.		12	17	24	20	14	19	17	

Note:

Z.N.: Zhangnao cave in the middle Yangzi valley. Radiocarbon dated at 13,490+150 years bp (Huang *et al.*1987).

Y'e: Yan'er cave in the middle Yangzi valley. Estimated to be at the end of the Pleistocene (IA, Hunan Province and Shimen Museum 1994).

D.X.: A cave in Daoxian in the middle Yangzi basin. Estimated to be at the end of the Pleistocene (Chen 1986).

Y.C.: Yuchan cave in the middle Yangzi basin. Radiocarbon dated at approximately 12,000-8000 years BP (Yuan 1996; IA, CASS 1997).

D.Y. Danyang cave in the lower Yangzi valley. Estimated to be at the end of the Pleistocene (Li *et al* 1982).

S.S.: Sanshan site in the lower Yangzi valley. Estimated to be at the end of the Pleistocene (Zhang *et al.*1987).

L.S.: Lishui spirit cave in the lower Yangzi valley, radiocarbon dated at 11,200+1000 years bp(Li and Lei 1980).

Table 6(2) Fauna of the early Holocene in the valley of the Yangzi River.

Species	English name	H.J.	Z.S.	X.R.	J.H.	H.M.	L.J.	Habitat and characteristics.
Mollusk	**Molluscs**							Fresh water.
Anodonta sp.	Fresh-water bivalve					+		
Bellamya quadrata						+		
Scylla serrata	Crab.					+		
Pisces	**Fishes**							
Carcharhin us sp.	Gray sharks.					+		Ocean.
Acipenser sp.	Sturgeons					+		Fresh water.
Cyprinus sp.	Carps.					+		Standing or slowly moving fresh water.
Carassius auratus.	Goldfish.					+		Fresh water.
Aristichth ys sp.						+		
Parasilurus asotus	Catfish.					+		Fresh water and ocean.
Pseudobagrus fulvidraco.						+		
Mugil cephalus	Striped mullet					+		Coastal area of tropical to temperate seas.
Gymnocranius griseus.						+		
Ophiocephalus argus.						+		
Reptilia	**Reptile**							
Chelonia								
Emydidae gen. et sp. indet.					+			Lake and rivers.
Chinemys reevessi	Chinese turtle.		+			+		Various habitats with water.
Amyda sinensis						+		" "
Chelonia mydas	Sea turtle.					+		Sea.
Testudo sp.	Terrestrial tortoise.					+		Humid areas.
Cuora flavomarginata	Turtle.					+		Shallow pools on land.
Crocodylia	Crocodile							
Alligator cf. sinensis	Chinese alligator.					+		Swamy area.

Table 6(2) Fauna of the early Holocene in the valley of the Yangzi River (continue).

****

Species	English name.	H.J.	Z.S.	X.R.	J.H.	H.M.	L.J.	Habitat and characteristics.
Aves	**Birds.**							
Pelecanus sp.	Pelican.					+		Various but in fresh or brackish water.
Phalacrocorax sp.	Cormorant.					+		Open water and marine habitats.
Ardea sp.	Heron.					+		Various, wetland/shallow water.
Grus sp.	Crane.					+		Open landscapes.
Aras sp.	Wild duck.					+		
Anser sp.	Wild goose.					+		Various. Migrating from north in winter.
Corvidae sp.	Crow-like birds.					+		Various ecotone.
Accipitridae	Eagles.					+		Forest or open habitats.
Mammalia								
Cetacea								
Cetacea indet	Whale.					+		Ocean.
Insectivora								
Crociduradracula sp.	Shrew.				+			Damp/dry forest, grassland etc.
Crocidura russula	White-toothed shrew.				+			" ".
Anourosorex squamipes	Mole shrew.				+			Cold mountain area.
Mogera sp.	East Asia mole.				+			Various environments.
Manis sp.	Pangolin.					+		Forest, thick bush, open land or savannah.
Chiroptera	Bats.							
Rhinolophus ferrumequinum					+			Cave.
Rhinolophus pearsoni					+			"
Rhinolophus sp.					+			"
Hippsioderos armiger					+			"
Miniopterus cf. schreibersi					+			"
Myotis sp.					+			"
Murina sp.					+			"
Vespertilionidae gen. et sp.indet.					+			"

Table 6(2) Fauna of the early Holocene in the valley of the Yangzi River (continue).

Species	English name.	H.J.	Z.S.	X.R.	J.H.	H.M.	L.J.	Habitat and characteristics.
Primates								
Macaca speciosa	Red-faced monkey.					+		Forest and hilly area.
Macaca cf. mulatta	Rhesus monkey.			sp.	+	+		" "
Macaca cf. thibetana	Stump-tailed macaque.				+			" "
Macaca cf. nemestrina	Pigtail macaque.				+			" "
Phinopithe cus sp.					+			" "
Presbytis sp.	Langurs.				+			" "
Homo sapiens sapiens				+				
Rodentia								
Eothenomys melanogaster	Pratt's vole.				+			Wood/mountain meadow.
Apodemus sylvaticus	Wood/field mouse.				+			Grassy field, forest/cultivated areas.
Niviventer confucianus	White-bellied rat.				+			Forest.
Leopoldamys edwardi	Long tail giant rat.				+			Mountainous forest.
Rattus rattus	Black rat.	sp.	sp.		sp.	+		Human commensal.
Typhlomys cinereus	Chinese pygmy dormouse.				+			Mountainous forest.
Hystrix subcristata	Old world porcupine.		sp.		+	sp.		Various ecotones.
Lepus sp.	Hare.	+		+				
Carnivora								
Nyctereutes procyonoides	Raccoon dog.			+	+	+		Forest, lakes & streams.
Mustela sibirica	Siberian weasel.			sp.		+		Forest, especially along streams.
Arctonyx collaris	Hog badger.			+	+	+		Forest.
Ursus thibetanus	Black bear.				+	+		Moist deciduous forest.
Paguma larvata .	Masked palm civet.			+	+	+		(Sub) tropical forest.
Felis bengalensis	Leopard cat.			+		+		Forest.
Felis sp.	Small cats.			+				Various.
Crocuta ultima	Spotted hyena.				+			Steppe.
Ailuropoda melanoleuca	Giant panda.				+			Bamboo-forest.
Martes flavigula	Yellow-throated marten.					+		Forest.
Lutra lutra	River otter.					+		All inland waterways.
Lutra perspicillata	River otter.					+		" "
Cuons javanicus	Javan dhole.				+	sp.		Various habibats except desert.

Table 6(2) Fauna of the early Holocene in the valley of the Yangzi River (continue).

Species	English name.	H.J.	Z.S.	X.R.	J.H.	H.M.	L.J.	Habitat and characteristics.
Carnivora (continued)								
Viverricula cf.				+				
malaccensis pallida	Fossil civet.							
Viverra zibetha	Oriental civet.				+	+		Forest, bush/grassland.
Viverricula indica	Rasse.					+		" ".
Panthera tigris	Tiger.				+	+		Forest.
Canis familiaris	Dog.			+		+		
Herpestes uvra	Mongoos.					+		Wide variety of habitats.
Proboscidea								
Stegodon orientalis					+			
Elephas maximus	Asian elephant.					+		(Sub)tropical thick jungles /grassy plain.
Perissodactyla								
Megatapirus augustus	Giant tapir.				+			Wooded and grassy habibats with water.
Didermocerus sumatrensis	Sumatran rhinoceros.				sp.	+		(Sub)tropical savannahs, shrubby and forest.
Rhinoceros cf. sondaicus	Javan rhinoceros.				+	+		Dense rainforest with sufficient water/mud.
Artiodactyla								
Sus scrofa	Wild boar.	sp.		+	+	+		Various. Forest and open land.
Sus sp (cf. Xiaozhu).	Small boar.				+			" "
Sus domestica	Pig.	+?				+		
Cervus unicolor	Sambar.	sp.			+	+		Various but prefer wooded area.
Cervus nippon	Sika deer.		+		+	+		Forest, grassland.
Cervus hortulorum				+				" "
Cervus elaphus	Red deer.			+				Various.
Elaphurus davidianus	David's deer.			+		+		Swamy area. Weight 159-214 kg.
Hydropotes inermis	Chinese water deer.			+		+		Tall reeds/rushes along rivers/mountain.
Capricornis sumatraensis	Sumatran serow.				+	+		Rugged mountain/ridge of forest/bush.
Muntiacus reevesi	Small muntjac.		+			cf.		Forest/area of dense vegetation with water.
Muntiacus gigas	Giant muntjac.					+		
Muntiacus muntjak	Muntjac.			sp.	+			" ".
Ovis sp.	Sheep	+	+	+				Upland areas, grassy mountains, cliffs etc.
Gezella sp.	Gazelle.							Dry open country or bushland.

Table 6(2) Fauna of the early Holocene in the valley of the Yangzi River (continue).

Species		H.J.	Z.S.	X.R.	J.H.	H.M.	L.J.	Habitat and characteristics.
Artiodactyla (continue)								
Nemorhaedus sp.	Gorals.				+	+		Rugged & wooded mountains.
Bubalus mephisto	Water buffalo.	sp.	sp.		sp.	+		(Sub)tropical area.
Total		7	6	18	46	61	9	

H. J. The Hujiawuchang fauna in the middle Yangzi valley, dated to about 7000 years BP.
Z.S. The Zaoshi fauna in the middle Yangzi valley; dated to around 7000 years BP.
X. R. The Xianrendong fauna in the lower Yangzi valley. Although the date for Xianrendong is from the terminal Pleistocene, details for the fauna from preceramic deposit are not available. Therefore, the whole fauna is tentatively placed in Holocene.
J. H. The Jinhua fauna in the lower Yangzi valley, dated to 7815±385 BP. No artifacts were found with this fauna.
H. M. The Hemudu fauna in the lower Yangzi valley, dated to about 7000 years BP.
L. J. The Luojiajiao fauna in the lower Yangzi valley, dated to about 7000 years BP.

(Sources: CPAM of Jiangxi Province 1973; Hunan Museum 1985; IA, Hunan Province 1993; Jiangxi Museum 1976; Liu and Yao 1993; Luojiajiao Archaeology Team 1981; Ma and Tang 1992).

Table 6(3): Pollen Sequences of Xianrendong and Wangdong.

Date (years bp)	Pollen assemblage		Climate
	Xianrendong	***Wangdong***	
20,000-17,000	*Pinus-Compositae* with some Pteridaceae	Scanty pollen. Gramineae, *Concentricystes* and spores.	Cool and dry.
15,200		Pine, Polypodiaceae, Gramineae & *Pteridium*.	Moderate and wet.
13,000	*Salix-Artemisia*. a few spores and other woody pollens.		Cool and dry.
11,900		Scanty pollen. Only herbs and spores.	Cold and dry.
11,300		*Quercus, Pteris* and ferns.	Warm and wet.
11,000-10,500	*Pinus-Artemisia,* Apocynaceae and ferns.		Warm and dry.
10,000	Gramineae-*Typha*.		Warm and wet.
9000		Polypodiaceae and ferns.	Warm and wet.

(Based upon data from Wang *et al.* 1995).

Table 6(4). Faunal Species Classification in the Middle and Lower Yangzi Valley.

Fauna	Date (BP)	Species classification				Extinct sp.	% of extinct sp
		Carnivore (% of the fauna)	Ungulates (% of the fauna)	Others (% of the fauna)	Total		
Zhangnao*	15800-15290	3 (25%)	7 (58.3%)	2 (16.7%)	12 (100%)	3	25%.
Yan'er*	T.P.	3 (17.7%)	8 (47%)	6 (35.3%)	17 (100%)	3	17.7%
Daoxian	T.P.	12 (50%)	6 (25%)	6 (25%)	24 (100%)	5	28%
Yuchan*	12,000-8000?	10 (50%)	4 (20%)	6 (30%)	20 (100%)	-	--
Danyang	T.P.	4 (28.6%)	5 (35.7%)	5 (35.7%)	14 (100%)	3	21%
Sanshan	T.P.	9 (47.4%)	7 (36.8%)	3 (15.8%)	19 (100%)	1	5%
Lishui	14,200-12,000.	8 (47.1%)	3 (17.6%)	6 (35.3%)	18 (100%)	1	6%

* Fauna found in association with archaeological assemblages.
T.P. The terminal Pleistocene.

Sources: Huang *et al.* 1987; IA, Hunan Province and Shimen Museum 1994; Chen 1986; Yuan 1996; Li *et al.* 1982; Zhang *et al.* 1987; Li and Lei 1980.

Table 6(5) Comparison of Common Wild Rice in Various Areas.

****

Areas.	**Stand(s) location**	**Average annual temperature**	**Average precipitation**	**Beginning of panicle development.**	**Beginning of ripening.**	**Length of life-cycle.**
Jiangxi	28°4'-28°10'N. 116°36'E.	17.8°C.	1662 mm	Early September.	n/a.	216 days.
Hunan	26°50'N,113°40'E. 25°05'N, 110°02'E.	17.9-18.7°C.	1500mm or more.	Mid-September.	Mid-October.	195-200 days.
Guangxi	20°54'-26°20'N. 104°29'-110°04'E	16-23°C.	1100-2800mm.	Mid. to late September.	Mid- October.	n/a
Guangdong	Widely found in latitudes 25°-20°N.	20-2°C	1400-2000mm.	Late Sep. to early October.	Mid October.	n/a
Hainan Island	20° N and further south.	22-24°C	1000-2000mm.	Early Oct. to early November	n/a	n/a

(Sources: Qi *et al*. 1984; Sun 1990).

Chapter 7. Archaeological discoveries in the Yangzi valley.

Intensive archaeological research in the middle and lower Yangzi valley only began in the 1950s. To date, archaeological data in this region are skewed. While the Neolithic data found here are comparable to those found in Central China in terms of both quantity and quality, upper Palaeolithic data are quite limited. In recent years, several important discoveries have been made in this region, but unfortunately detailed reports are not available.

However, I was allowed to observe some of the recent important discoveries during my field work in 1996. The information I gathered then is part of the database for the discussion in this chapter. The Yangzi River basin is relatively rich in archaeological assemblages dating to the early Neolithic. In the middle Yangzi valley there are the Pengtoushan and Bashidang assemblages, dated from 9000 to 8000 BP. The Neolithic cultures succeeding Pengtoushan and Bashidang are represented by the Fenshanbao and Hujiawuchang assemblages, dated from 7500 to 7000 BP (He 1995b) [Table 7(2)].

Compared with the middle Yangzi valley, new archaeological discoveries in the lower Yangzi valley are fewer. The earliest Holocene culture in the lower valley is now represented by the upper deposit of the Xianrendong-Wangdong caves assemblage in Jiangxi Province. But the immediate successor of this assemblage is not clear. The well-known Hemudu culture, and the less well-known Luojiajiao assemblage, both dated to 7000 years ago, still represent the first prosperous Neolithic societies in the lower Yangzi valley.

An important characteristic of the archaeological cultures in the Yangzi valley is the early presence of pottery. It must be emphasised here that if pottery is taken as an indicator of Neolithic culture, then the onset of the Neolithic may be considered to have occurred approximately 12,000 to 10,000 BP, for pottery has been found in this region probably at this time at Xianrendong and Yuchan caves. However, the beginning of the Neolithic implies more than just the presence of pottery. The vital element is the emergence of agriculture. Firm evidence of rice cultivation comes from several archaeological sites in the middle Yangzi valley, dated from 9000 to 8000 BP, representing the early Neolithic in this region.

7A. Archaeological assemblages in the middle Yangzi valley.

Archaeological assemblages in the middle Yangzi valley can be chronologically classified into the terminal Pleistocene (20,000-10,000 BP in this thesis) and the early Holocene (10,000-7000 BP) phases. However, not many archaeological assemblages of the terminal Pleistocene have been absolutely dated. And for those dated, some of them seem problematic (e.g. Yuchan, below).

Most of the archaeological assemblages of the terminal Pleistocene in the middle Yangzi valley are surface collections, except for a few sites which have been excavated. These excavated sites are all in caves, except for one open site on the northern side of the Yangzi River, called **Jiangling**.

Jiangling is located on top of a small hill on the alluvial plain of the Yangzi River, in Hubei province. Discovered and excavated from 1992 to 1993, two sub-strata were found. The lower stratum contained remains of a living and working floor, with evidence for lithic tool manufacturing and animal butchering (Liu 1993). Pebble tools were the major implements found in this layer, estimated to be of late Pleistocene date based upon geological analysis (Liu 1993).

The upper stratum is a light yellow soil, about 10-25 cm thick (Liu 1993). Approximately 300 lithic implements were found in this layer, the majority made from small flakes (Liu 1993). The raw materials for these stone implements were quartz, chert and quartzite, and the retouched tools were mainly scrapers (Liu 1993). No absolute date has been obtained for this layer, but it has been estimated to be around 20,000 to 10,000 BP, based upon geological observation and strata profile analysis (Liu 1993).

The presence of stone tools made on small flakes in the upper layer at Jiangling is important. As will be discussed in the following section, the toolkit tradition in the Yangzi valley during the Pleistocene was one of pebble tools. Notably, small flake tools appeared only in the terminal Pleistocene in this region, and these small flake tools were the components of the early Neolithic assemblages.

Another important archaeological assemblage in the middle Yangzi valley comes from a cave called **Zhangnaodong** [Fig. 7(1)]. Located in Hubei province, north of the Yangzi River, this cave was discovered and excavated in 1986. A total of over two thousand lithic implements and bones of 12 species of animals were found. Some bone fragments seem to have been burnt or flaked (Huang *et al.* 1987). The faunal remains included giant panda (*Ailuropoda melanoleuca baconi*), stegodon (*Stegodon orientalis*) and Kirchberge's rhinoceros (*Dicerorhinus* cf. *kirchbergensis*) [Table 6(1)], thus belonging to the giant panda-stegodon fauna of the Yangzi valley. The presence of Kirchberge's rhinoceros indicates that the palaeoclimate might have been cooler than at the present (Huang *et al.* 1987).

According to the preliminary excavation report, the lithic implements are made of black silica rock (44%), vein quartz (40%) and sandstone pebbles. The types of implements consisted of cores, flakes, scrapers, points and choppers (Huang *et al.* 1987). Size of flakes in Zhangnaodong varied significantly, with lengths ranging from 10 mm to 100 mm, and thickness from 3 mm to 50 mm (Huang *et al.* 1987). Direct percussion is said to be the major technique for flaking and retouch (Huang *et al.* 1987). A single radiocarbon date of 13,490$\pm$150 bp [Table 7(1)] places this assemblage towards the end of the Pleistocene.

The quantity of lithic implements found in Zhangnaodong is not reported. No drawings are available for the implements. But it is notable that a type of 'discoid chopper' was reported at this site. Usually made on pebbles and flaked into a discoid shape, this type of implement has been widely found in other surface collections of the terminal Pleistocene, and occurs also in the early Neolithic Hujiawuchang and Zaoshi assemblages in the middle Yangzi valley (below). The presence of the discoid chopper in the terminal Pleistocene in Zhangnaodong may indicate cultural continuity from this phase into the Neolithic in the middle Yangzi valley.

Another important site in the middle Yangzi valley is **Yan'er** cave. Consisting of two caves, this site was

ıseum 1992; Xiang 1992; Yuan 1995). the first and second terraces of various Yangzi River. The majority of the lithic pebble tools, similar to those found in d Southeast Asia [Fig. 7(4)].

the characteristics of the lithic assemblages in Yangzi valley during the terminal Pleistocene ɔws:

ıes of sandstone are major raw materials. Lithic ıements are made either by flaking the whole pebble ɲg one or more edges, or by detaching flakes from ıe pebble [Fig. 7(4)]. Flaked implements are present, out the proportions of pebbles and flakes within each lithic assemblage vary. Flakes, and tools made from flakes, only account for 6.4% of the total lithic implements within the Zhangjiatan collection (Lixian Museum 1992), but accounted for 50.7% in the Jingya collection, both from Hunan province (calculation based upon Xiang 1992).
2) Direct percussion by hard hammers is said to have been the only method of flake detachment and secondary retouch (Lixian Museum 1992; Xiang 1992; Yuan 1995). But many flakes seem to have been used directly without retouch, as indicated by use-wear along the edges (Lixian Museum 1992; Xiang 1992; Yuan 1995).
3) The typological variation is quite limited. Cores, flakes, choppers and chopping tools, stone balls, scrapers and points are the major lithic implements (Lixian Museum 1992; Xiang 1992; Yuan 1995). According to Yuan (1995), there are two different lithic complexes in the middle Yangzi valley; one is characterised by big triangular points and stone balls, which are absent in the other complex (Yuan 1995) [Fig. 7 (4)&(5)]. Whether this difference in toolkits indicates different subsistence strategies remains to be investigated.

A salient characteristic of the lithic industry in the Yangzi valley, as well as in South China generally, is the stability of the pebble tool industry. Choppers and chopping tools occur from the middle Pleistocene to the middle Holocene in these regions, and the manufacturing techniques changed little over such a long period of time. It becomes very difficult to classify tools into different temporal phases, if one only considers morphology and manufacturing techniques. To date, the approximate dates for the lithic assemblages of the middle Yangzi valley are mainly based upon associated faunas and geological studies of strata (Yuan 1995), unless absolute dating is feasible.

Although the archaeological discoveries of the terminal Pleistocene in the middle Yangzi basin are far from abundant, these discoveries do provide important information on the transition from the terminal Pleistocene to the early Holocene in this region, the latter phase being represented by rather rich archaeological data.

To date, the earliest archaeological assemblages of the early Holocene, with convincing evidence of rice cultivation, pottery and sedentism, have been found in the middle Yangzi valley. The Neolithic cultural sequence in this region is also quite clear now, as various assemblages have been found, dating from the ninth to the fourth millennia BP (He 1995a).

The first discovered and now well-known site is **Pengtoushan,** located on a small hill in the Liyang alluvial plain of the Yangzi River [Fig. 7(1)]. There is a small stream running along the western and southern sides of the site. Discovered and excavated in 1988, Pengtoushan has yielded house floors, pits, burials, flaked and polished stone tools, pottery, and most importantly, rice husks used as pottery temper (IA, Hunan Province and Lixian Museum 1990). Since its discovery, 24 absolute dates have been obtained [Table 7(2)], convincingly placing the Pengtoushan assemblage at the beginning of the Holocene.

Only lithic tools were found in Pengtoushan, but quantities have not been published. Three categories are present: flaked pebble tools, small flake implements made of black chert, and a few polished implements (IA, Hunan Province and Lixian Museum 1990). It is reported that the flaked pebble toolkit consisted of choppers, scrapers and stone hammers; the small flake implements included scrapers, burins and drills, while the polished implements consisted of a few small axes (IA, Hunan Province and Lixian Museum 1990).

The toolkit of Pengtoushan is thus quite simple. Direct percussion was the major method for both flake detachment and retouch, but secondary retouch seems not commonly applied (author's observation 1996). The flaked pebble tools are similar to those of the terminal Pleistocene found in this area, in terms of both raw material and workmanship. They are made from either whole pebbles, or flakes detached from pebbles, most of these are sandstone. Only the working edges were flaked, and pebble surfaces were often retained [Fig. 7(6)]. The flaked pebble tools are quite large, normally over 8 cm in length, more than 7 cm in width and over 2 cm in thickness (IA, Hunan Province and Lixian Museum 1990).

The small flakes were made of black or grey chert. Judged from the salient bulbs, some of these flakes might have been detached by direct percussion (author's observation 1996). Their shapes vary [Fig. 7(6)]. The use-damage found on some un-retouched flakes indicates that they were used directly after detaching (author's observation 1996). These small flakes are usually under 5 cm in length, less than 3 cm in width and no more than 1.5 cm in thickness (IA, Hunan Province and Lixian Museum 1990), usually without systematic retouch. They certainly differ from the pebble tools in terms of size and raw materials. Although they have been assigned names such as 'scrapers' or 'burins' (IA, Hunan Province and Lixian Museum 1990), their morphological appearances differ from those found in Central China. Whether the Yangzi flakes were really used as scrapers and burins remains to be investigated.

Polished stone implements in Pengtoushan are made from oil shale and mudstone (IA, Hunan Province and Lixian Museum 1990). Both are soft stones, ranging between 3 and 4 on Mohs' hardness scale (IA, Hunan Province and Lixian Museum 1990). According to the brief excavation report, only a few polished axes were found (IA, Hunan Province and Lixian Museum 1990), which, according to the drawing, seems more like an adze [Fig. 7(6)]. They seem too soft and fragile to have served as practical tools. The majority of the polished implements are polished stone pendants and beads (IA, Hunan Province and Lixian Museum 1990) [Fig. 7(6)].

No tools made of organic materials have been found in Pengtoushan. No tools that seem to have been associated with cultivation activities have been found. Had the remains of rice and rice husks not been preserved in the pottery, no one would have known that rice was harvested in Pengtoushan.

The toolkit of Pengtoushan has provided only limited information on economic activities. On the other hand, the discoveries of large and small house floors, pits and graves are more informative. Quantity of the these discoveries are not revealed, but it is reported that remains of a large square house approximately 33.6 square meters were discovered, the floor of it covered by a mixture of sand and clay (IA, Hunan Province and Lixian Museum 1990). The small houses are simpler, being round semi-subterranean

structures with ovens inside, made of a mixture of fire clay and soil (IA, Hunan Province and Lixian Museum 1990). The floors of these houses were levelled, but no layers of any special material were found (IA, Hunan Province and Lixian Museum 1990). In house No.2, a pot was found on top of a hearth when excavated (IA, Hunan Province and Lixian Museum 1990), clearly indicating that the house had been occupied immediately prior to the abandonment of the site.

A total of 15 pits have been found in Pengtoushan, but no cereal remains were reported. Eighteen burials were found in the site, but the skeletons were poorly preserved. Grave goods were pots or lithic implements, mainly the polished stone beads and pendants (IA, Hunan Province and Lixian Museum 1990). Such findings indicate that the prehistoric occupants of Pengtoushan were living a quite sedentary way of life.

The Pengtoushan pottery clay was mixed with a large quantity of rice straw, grains and husks, as well as other organic materials (IA, Hunan Province and Lixian Museum 1990). Most pottery vessels were black, but seem to have been covered externally and internally by a thin layer of red colour (IA, Hunan Province and Lixian Museum 1990). It is not yet clear whether this red colour was a result of oxidisation of the clay during firing, or was deliberately added as a slip (refer to chapter 2 for definition) (IA, Hunan Province and Lixian Museum 1990). Cord-marking was the major kind of surface finish, together with some incision and carving [Fig. 7(7)]. Small items seem to have been built by hand-pinching, large items by slab-modelling, as the cross-sections of some potsherds clearly show slab joins (author's observations 1996). Vessels forms from Pengtoushan include cooking *fu*, pot, globular jar with two lugs, bowl, plate and pottery stands (refer to chapter 2 for definition) [Fig. 7(7)] (IA, Hunan Province and Lixian Museum 1990).

In summary, Pengtoushan illustrates that at about 9000 to 8000 BP, a group of prehistoric people in the middle Yangzi River had settled down, harvested rice, made pottery and buried their dead. The abundance of rice and rice husks found in the pottery indicates that rice harvesting might have been quite intensive. On the other hand, a lack of specific tools for hunting and gathering, plus the absence of any animal bones, is a bit puzzling. Although the soil is quite acid, this still cannot convincingly explain why no organic materials have been preserved apart from the human burials. Does this absence of materials associated with hunting and gathering indicate that the subsistence strategy in Pengtoushan was primarily based upon rice collection or cultivation? Or were there other factors attributing to this absence? The question cannot be answered at this stage.

The rice grains and husks contained in the Pengtoushan pottery cannot be taken out without damage, so it is difficult to identify whether they are remains of domesticated or wild rice (IA, Hunan Province and Lixian Museum 1990). But the size of the rice pollen in Pengtoushan is quite large, similar to that of domesticated rice (The Pollen Laboratory, IA, Hunan Province 1990). Based upon pollen analysis, the abundance of rice grains and husks, as well as houses and other evidence of sedentary life, many Chinese scholars argue that the rice found in Pengtoushan must have been domesticated (e.g. Liu 1994; Shen 1994), although some of them also point out the lack of tools for cultivation in the site (Yan 1990).

Although the rice found in Pengtoushan cannot be identified in detail, this has been more or less compensated for by another important discovery at a site called **Bashidang,** which was excavated from 1993 to 1996 in three seasons. The excavation report has not been published, but a study on the rice remains has been conducted.

According to the excavator, Bashidang is located on top of a small hill about 20 km north of Pengtoushan. Both sites are located on the Liyang Plain, which is one of the alluvial plains of the Yangzi River. Pei (1998b) claims that there were two layers in Bashidang. The upper cultural layer was dated to approximately 8000 BP (Pei 1998b), and the lower layer should have been earlier. According to Pei (1996, personal communication), over 200 post-holes were found within an area of about 100 square meters in the western part of the lower layer. As the ground was a peat deposit, Pei suggests that the post-holes must be remains of pile-dwellings. Protective ditches and walled fortresses have been reported from Bashidang (Pei 1998b).

There is a riverbed near the site where small flakes of black chert and abundant remains of plants were found (Pei 1996, personal communication). Among the plant remains, more than 10,000 rice grains have been discovered (Pei 1998b). The widths and lengths of these grains were measured; and the glume epidermal phytoliths of four grains were examined under the microscope. The results then were compared with local cultivars and wild rice. The glume epidermal phytoliths of the rice grains from Bashidang are similar to those of cultivated Indian rice (*Oryza sativa indica)* (Zhang and Pei 1997). The length of the prehistoric grains is smaller than that of *O. sativa indica*, but close to *O. sativa japonica* (Zhang and Pei 1997). On the other hand, the widths are smaller than both modern cultivars and contemporary wild rice. The ratio of length/width is quite similar to that of *O. sativa indica* (Zhang and Pei 1997).

In general, the size of the prehistoric rice grains is smaller than both of the contemporary cultivars and wild rice. The possible effect of shrinkage has been considered but seems not significant, as the measurements of the glume epidermal phytoliths are similar to those of contemporary samples (Zhang and Pei 1997). Zhang and Pei (1997) argue that the rice remains from Bashidang have mixed characteristics of *O. sativa indica* and *O. sativa japonica*, perhaps closer to *O. sativa indica*. They further suggest that these remains represent an ancient cultivated rice, which was on the way to developing into the two different sub-species (Zhang and Pei 1997).

No absolute dates have been published for the Bashidang assemblage. Based upon pottery comparison, it has been proposed as slightly later than the Pengtoushan assemblage (Pei 1998a), which suggests that rice cultivation was already practised by 8000 BP. Given the similarity of pottery between Bashidang and Pengtoushan, and the proximity of the two sites, Pei has argued that they clearly belong to one culture (Pei 1995)

It seems that the successors of the Pengtoushan culture were quite diverse. Since the early 1990s, several early-to-middle Neolithic assemblages have been discovered, most located in the middle Yangzi valley [Fig. 7(1)]. Important discoveries include the assemblages from **Fenshanbao**, which is east of Pengtoushan; **Zaoshi** to the west; and **Hujiawuchang** to the south [Fig. 7(1)]. To the north of Pengtoushan a few assemblages of early Holocene date have also been found and assigned to the **Chengbeixi** culture. While Chinese scholars agree that Fenshanbao, Zaoshi and Hujiawuchang are successors of Pengtoushan (He 1995a), the relationships between Chengbeixi and Pengtoushan is still a question under debate (Pei 1995).

Fenshanbao is located on a small terrace in the middle of Dongting Lake, about 140 km east of the Pengtoushan assemblage. Discovered in 1989 and excavated in 1990 and

1991, three cultural layers were uncovered, defined as phase I, II and III from the bottom (Zhang 1992). House remains, post-holes, pits, over 50 burials, as well as stone implements, pottery, animal and plant remains were found (Zhang 1992). Imprints of rice husks were found on some pottery (The Archaeology Team of Yueyang City and CPAM of Qianlianghu Farm 1994). Remains of open pottery firing were also discovered (Zhang 1992).

According to the excavation report, flaked pebble tools, small flakes of black chert, and polished axes were unearthed. Many flakes were used without secondary retouch (The Archaeology Team of Yueyang City and CPAM of Qianlianghu Farm 1994). Choppers, scrapers (made from pebbles or black chert flakes), burins (made of black oil shale) and partially polished axes were found in 1990 (The Archaeology Team of Yueyang City and CPAM of Qianlianghu Farm 1994) [Fig. 7(8)], while polished lithic needles and arrowheads were added to the toolkit in the 1991 excavation (Zhang 1992). But quantities and other details of the lithic assemblage are not available.

Judging from available data, the composition of the Fenshanbao toolkit seems more varying than that of Pengtoushan. Choppers and axes are the major implements, probably used for multiple purposes in land clearing and wood and bamboo working. Polished stone arrowheads indicate hunting. Again, no tools made of organic materials have been reported, but animal bones (details of species unreported) were discovered in 1991 (Zhang 1992). The abundant rice remains suggest the possible existence of rice cultivation (Zhang 1992).

Post-holes were found in all three layers. As the bottom layer was a peat deposit, it is inferred that the post-holes in this layer must be remains of pile-dwellings. But living floors of houses were found in the second and third layers. These house floors were about 10 square meters in area, covered by sand and fired clay, with a big post-hole in the centre (Zhang 1992). It has been proposed that these houses would have been built directly on the ground and covered by conical-shaped roofs supported by the central post (Zhang 1992).

Burials were found within the living areas in Fenshanbao. They were all secondary burials, with a few pottery vessels as grave goods (Zhang 1992). The discoveries of houses, graves and a pottery firing area suggest that the prehistoric occupants of Fenshanbao were quite sedentary.

The pottery in Fenshanbao seems more developed than that found in Pengtoushan. Rice husks and wood charcoal were major temper material in cultural Phase I at Fenshanbao. In phases II and III, mixed sand and wood charcoal were used as tempers, but fine pottery became dominant in Phase III (The Archaeology Team of Yueyang City and CPAM of Qianlianghu Farm 1994; Zhang 1992). All vessels were made by slab-modelling. The majority were of red or brownish red colour, but white and painted pottery were also unearthed (The Archaeology Team of Yueyang City and CPAM of Qianlianghu Farm 1994; Zhang 1992). Types of vessels include pots, globular jars with two lugs, shouldered *fu*, bowls, *dou* (defined in chapter 2), stands and lids [Fig. 7(7)]. Most of the pottery has been coated with a red slip, and the motifs of decoration consist of cord marking, incision, stamping, and different combinations of several motifs, but the proportions of these motifs are not reported. Perforating as another decorative method was first present in Phase II [Fig. 7(7)] (Zhang 1992). It is notable that some of the pottery found in Fenshanbao resemble those of the middle Neolithic (6000 years onwards) pottery found in both the middle and lower Yangzi valley, suggesting a possible cultural continuity.

Three radiocarbon dates have been obtained for the Fenshanbao assemblage, these being 6800±80, 8220±250 and 8274±234 bp (He 1995a). Details of these dates are not available, but He (1995a) argues that the first date seems too late and the latter two dates too early (He 1995a). His argument is based upon cross-dating and pottery comparison with the Pengtoushan and Hujiawuchang assemblages. He places the Fenshanbao assemblage after Pengtoushan but prior to Hujiawuchang (below). The approximate date of Fenshanbao is suggested to be within the range of 8000 - 7800 BP (He 1995a), after the last phase of the Pengtoushan culture.

The Hujiawuchang assemblage is another important representative of the Neolithic culture in the middle Yangzi valley. The site is located less than 50 kms south of Pengtoushan [Fig. 7 (1)]. Discovered in 1984 and excavated in 1986, remains of post-holes, fired clay, pottery and stone tools have been discovered (IA, Hunan Province 1993). The lithic implements of Hujiawuchang are classified in Table 7(3).

Both pebble and small flakes of chert were major lithic implements in Hujiawuchang. Polished stone tools were still a small proportion of the toolkit; but compared to the polished tools found in Pengtoushan, both the typological variety and the workmanship in Hujiawuchang were advanced. In addition to axes, polished adzes, chisels and a knife were present in Hujiawuchang.

It is probable in Hujiawuchang that the flaked discoid 'chopper' first acquired a significant presence since its appearance in the terminal Pleistocene [Fig. 7(9)]. Although named a 'chopper', the function of this implement is not clear. Judged from the use-wear around the edge, Yuan (1995) suggests that it might have been used for cutting wood and bamboo (Yuan 1995). Another hypothesis is that it might have been used to cut rice panicles (He 1995b). But both arguments are virtually guess-works without evidence from use-wear analysis and/or experiment. The function of this tool, and its significance in the subsistence strategy from the end of the Pleistocene to the early Holocene in the middle Yangzi valley is still not clear. Further study is needed in this issue.

This discoid lithic implement has been widely discovered in other Neolithic sites in this area, such as at Zaoshi, Qili and Huangjiagang, dated to 7000 BP onwards (He 1995b; Yuan 1995). Yuan (1995) has pointed out that the discoid implement was first present at the end of the Pleistocene, became popular during the early Holocene (i.e. 7000 BP), and declined afterwards (Yuan 1995). Consequently, he proposes that this tool might have been a hallmark of the transition from the Palaeolithic to the Neolithic in the middle Yangzi valley (Yuan 1995).

No tools made of organic materials have been found in Hujiawuchang. But abundant animal remains were recovered, and the quantity is more greater in the earlier stratum than in the later ones (IA, Hunan Province 1993). According to the preliminary report, water buffalo, goat, pig, deer and rat, as well as birds, fish and shells have all been found (IA, Hunan Province 1993).

Remains of post-holes and one burial were also found in Hujiawuchang. Fired clay was found covering the surface of what would have been the floor of a house (IA, Hunan Province 1993). These remains were poorly preserved, and a reconstruction is not achievable (IA, Hunan Province 1993). However, these discoveries reveal that the life style of the prehistoric Hujiawuchang occupants was quite sedentary.

The pottery found in Hujiawuchang is advanced in terms of

technique, decoration and variety. Charcoal and sand were still the major tempers, but fine pottery was present in Hujiawuchang. The majority of the vessels were still built by slab-modelling, but it has been inferred that a slow wheel might have been used for trimming, as some vessels are quite symmetrical (IA, Hunan Province 1993). Black and red were prevailing colours. Vessels include various plates, pots, jars with two lugs, cooking *fu*, lids, bowls and stands [Fig. 7(7)]. Like the Fenshanbao pottery, vessels found in Hujiawuchang were also coated with a red slip. The decoration motifs include cord-marking, incision, stamping, punctuation, carving and perforating (IA, Hunan Province 1993) [Fig. 7(7)].

One rice grain was found in a pottery stand (IA, Hunan Province 1993), indicating that rice harvesting was one of the economic activities. But the rice has not been extracted from the pottery, so taxonomic identification is not feasible. However, the Hujiawuchang assemblage represents a prehistoric society that was sedentary, harvesting rice, but at the same time still hunted and gathered, as judged from the unearthed animal remains. Radiocarbon dates for Hujiawuchang cluster around 7600-7100 cal. BP [Table 7 (2)].

Numerous archaeological assemblages have been found as successors of the Hujiawuchang assemblage in the middle Yangzi valley, including the well-known Daxi culture in the sixth millennium BP (IA, CASS 1984). The cultural link between the Fenshanbao - Hujiawuchang assemblages and the Neolithic assemblages dated after 6000 BP in this region is indicated by their similarities in many aspects: rice cultivation (remains of domesticated rice have been widely discovered), painted and white pottery, the style of vessels, particularly the widespread of *dou*; house floors covered by fired clay, etc (IA, CASS 1984). All these suggest a clear cultural continuity in the region.

While both Fenshanbao and Hujiawuchang are in the southern alluvial plain of the middle Yangzi valley, the Chengbeixi culture is located along the Three Gorges of the Yangzi River. First discovered in 1978 at Chengbeixi, Hubei province, this culture is now represented by archaeological discoveries on the first and second terraces along the River. Several sites have been excavated since the early 1980s, but only very brief reports are published. Among these sites, the Chengbeixi and Liulinxi discoveries are representatives of this culture.

The Chengbeixi site is located on the western bank of the Yangzi River [Fig. 7(1)]. The site had been severely damaged by the river when it was discovered. It was excavated in 1983 and 1984 (IA, Hubei Province 1996), and surveyed twice after excavation (Honghuatao Archaeology Station and Zhicheng Museum 1988). Stone implements, bone tools, pottery, fish and animal bones have been discovered (IA, Hubei Province 1996). The lithic implements are summarised in Table 7(4).

Apparently stone axes are the major implements within the Chengbeixi toolkit. The majority of axes were flaked without polish [Fig. 7(10)]. Judged from the deep and abrupt scars on the surface, it has been suggested that these axes were made by flaking pebbles using direct percussion (IA, Hubei Province 1996). Some of these axes are irregularly shaped and the workmanship seems quite crude [Fig. 7(10)]. On the other hand, fully-polished axes were also found [Fig. 7(10):5]. Use-wear has been found on both flaked and polished stone axes, so both seem to have been final products and had been used.

The quantity of other stone tools is limited [Table 7(4)]. This may be partially due to the small size of the excavation (only 50 square meters were dug). Nevertheless, these tools were also made from flaked pebbles; some of them had been polished along the working edge, such as the adze and chisel; some had been almost fully polished, as the stone balls. The presence of net-sinker indicates that fishing was practised. As for the stone balls, they have been found in large numbers in various middle Palaeolithic assemblages in Central China, such as at Kehe, Dingcun and Xujiayao (Jia and Wei 1976). Jia and Wei (1976) have suggested that stone balls might have been used as bolas stone for hunting (Jia and Wei 1976). Similar function may be inferred for the stone balls in Chengbeixi. Bearing no scars of impact on their surface, they could not have been used as stone hammers (IA, Hebei Province 1996).

According to the brief report, animal remains of several species were found at Chengbeixi, including water buffalo, deer and cattle, turtle and mollusc shells, and fish (IA, Hubei Province 1996). The Latin name and full details of these animal remains are not reported, but the discovery further suggests that hunting and fishing were important economic activities at Chengbeixi .

In addition to stone tools, a spindle whorl of fired clay, one bone needle, one bone drill and one bone spade were found in Chengbeixi (IA, Hubei Province 1996). Spindle whorls and needles are specific tools for fibre weaving and sewing. Their presence indicates that weaving might have been practised by the Chengbeixi occupants.

Pottery is another important component of the Chengbeixi assemblage. Sand and charcoal were used as pottery tempers, and colours ranged from brownish red, greyish brown, black, yellow to white. Some vessels were coated with a red slip (IA, Hubei Province 1996). The vessels were made by slab-modelling, and many were asymmetric in shape (IA, Hubei Province 1996). Types of vessels include various cooking *fu*, pots, *dou*, stands, bowls, plates and globular jar with two lugs [Fig. 7(7)]. Decoration consists of cord-marking, stamping, incision and perforation, but cord-marking prevails.

It is notable that rice husks and straw have been found as tempers in the Chengbeixi pottery (Lu 1993). Again, the rice remains could not be extracted from the pottery, so a taxonomic identification is not possible. But even if these are the remains of wild rice, this discovery at least suggests that rice harvesting was practised by the Chengbeixi people.

Three radiocarbon dates have been obtained for the Chengbeixi assemblage. Two of these dates are based upon charcoal samples, and date to between 9000 and 8400 BP [Table 7(2)]. One date was obtained from an animal bone, dated to only about 7500 BP [Table 7(2)]. It has been argued that these samples were from different strata (Yang 1991). While the potsherds with charcoal were from lower strata in Chengbeixi, the bone sample was from a higher stratum, which might represent a later phase of the Chengbeixi culture (Yang 1991).

Another important site of the Chengbeixi culture is **Liulinxi**, which is located on the first terrace on the northern bank of the Yangzi River. The northern edge of the Liulinxi site is the mountain-side of the Three Gorges, and its southern edge the Yangzi River (IA, Hubei Province 1994) [Fig. 7(1)]. Excavated in 1991, substantial quantities of fish and animal bones, as well as stone implements and pottery, were discovered (IA, Hubei Province 1994).

Only a brief report of this excavation has been published. The quantities of both stone implements and pottery is not revealed, except for stone balls, grinding slabs and a roller. According to the report, the stone toolkit of Liulinxi consists of axes, adzes, chisels, scrapers, two stone balls,

four grinding slabs and one roller (IA, Hubei Province 1994). The stone implements were made from flakes detached from pebbles. The lithic implements of over 10 cm in length were seldom polished but flaked, but small adzes and axes (both under 10 cm in length) were partially or fully polished, although the flaked surfaces of some stone implements were still visible (IA, Hubei Province 1994) [Fig. 7(11)].

Beside the lithic implements, a substantial quantity of pottery vessels and potsherds were found in Liulinxi. The vessels were made by slab-modelling (IA, Hubei Province 1994). It has been claimed that red pottery with sand temper accounted for over 82.4% of the total potsherds in Liulinxi, grey pottery with sand accounted for 7.9%, and the rest were red fine pottery, or red pottery with charcoal, crushed shell or bone as tempers. A few painted potsherds have also been found (IA, Hubei Province 1994). Types of vessels include various pots, cooking *fu*, plates, bowls and stands [Fig. 7(7)]. Cord-marking is the dominating decoration motif, which was found on 83.9% of the total potsherds. Other motifs include incision (5.7%), stamping (2.8%), and painting (1.5%). The rest was plain pottery (6.1%) (IA, Hubei Province 1994). Rice husks and straws have been found in the pottery (IA, Hubei Province 1994), suggesting the practice of rice harvesting.

No absolute dates are available for Liulinxi. Based upon toolkit and pottery comparison, it has been suggested that the Liulinxi assemblage is contemporary with the Chengbeixi assemblage, but it might have been slightly later (Yang 1991).

In addition, a few other sites of the Chengbeixi culture have also been excavated since the mid 1980s. The excavation reports are very brief, providing only limited information (Chen 1991; Chen and Yang 1985a &b; The Archaeology Team of the Three Gorges 1989). In general, the characteristics of the Chengbeixi culture can be summarised as follows:

1). Archaeological assemblages of this culture have been only found in the Three Gorges area of the Yangzi River so far. The majority of these sites are located on the first or second terraces of the River, usually very close to the river bank (Lu 1993). Consequently, many sites have been partially or severely destroyed by the river floods. However, several sites have been found on the hills along the Yangzi River (Chen 1991).
2). Chipped stone tools made from river pebbles are major implements of the toolkit. Polishing has been applied to small lithic implements. Axes, adzes, chisels, stone balls, net sinkers, as well as grinding slabs and rollers are the prevalent implements. Tools made from bone have been also discovered. The toolkit of the Chengbeixi culture, as well as the substantial remains of bone and fish found in these sites, indicates that hunting and fishing were primary subsistence strategies. On the other hand, grinding tools and remains of rice husks found in pottery suggest that rice harvesting was also practised. Whether the rice was cultivated is not certain.
3). It is notable that no residential remains or burials have been reported for the Chengbeixi culture. Those sites found on the first terrace of the Yangzi River are always under threat from summer floods. According to geomorphologic reconstruction, this first terrace would have been close to the actual river bed in the early Holocene, and very likely to have been inundated by the Yangzi during summer; so they were probably not residential sites. But the function of these sites are not clear.

   As for the sites in the hills, it is impossible to judge their nature as no archaeological reports are available. Only one site called Qinglongshan is mentioned in an article on the Chengbeixi culture, where three refuse pits were found (Cheng and Yang 1985c), but no other residential remains have been reported. Yang (1991) hypothesises that the sites on the lowest terrace might represent places where the Chengbeixi occupants fished and hunted; and the hilly sites might have been their camps (Yang 1991). But more archaeological evidence is required to verify this hypothesis.

(4) Chengbeixi pottery was built by slab-modelling. Sand, rice husks and, less commonly, crushed shell were used as tempers. Red pottery dominates the collection. Round- and flat-based vessels dominate. Some vessels were coated with a red slip. Cord-marking, incision and stamping, as well as painting, are major decorative methods.

Thus, the archaeological data for the Chengbeixi culture are still scanty. Remains of fish and animals suggest that fishing and hunting were important subsistence strategies. Rice was harvested, but whether it was also cultivated is not clear. Very little is known about residential patterns, not to mention burial and social systems. Further study is necessary.

Major archaeological assemblages from the terminal Pleistocene to the early Holocene in the middle Yangzi basin have now been reviewed. It seems that the toolkit found in this region from the terminal Pleistocene into the early Holocene illustrates cultural continuity. Pebble tools prevailed in the lithic toolkits of the terminal Pleistocene, such as those from Yuchan and Yan'er. Pebble tools remained important in the early Holocene assemblages, particularly in the earlier ones such as Pengtoushan. On the other hand, small flakes of chert and quartzite first appeared in the terminal Pleistocene assemblages, and became more significant in the early Holocene.

After the first appearance of Neolithic assemblages in the middle Yangzi valley, as represented by Pengtoushan and Bashidang, they seem to have expanded to adjacent areas. The Fenshanbao assemblage represented the eastbound expansion, and the Zaoshi assemblage the southbound one [Fig.7(1)]. Expansion to the north is less clear.

Given the archaeological data, it seems that the Chengbeixi culture resembles the Pengtoushan culture in many ways, particularly in pottery. Both were built by slab-modelling. Sand and rice remains were used for tempers in both collections. The decoration methods were also similar. Red-slip, cord-marking and incision, as well as painting, were common in both. Given these resemblance, many scholars argue that the Chengbeixi assemblage is a development from the Pengtoushan culture in the Three Gorges area (He 1995b; Pei 1995; Yang 1991). This argument seems strong, but not without its problems.

Despite the pottery resemblance, too little can be said regarding the relationship between the Chengbeixi and Pengtoushan cultures. First, the lack of archaeological data makes it impossible to reconstruct the residential pattern of the Chengbeixi culture at this stage, whereas village sedentism in Pengtoushan and its successors seems quite certain. Second, the toolkit of Chengbeixi is different from that of Pengtoushan. It is notable that grinding slabs have never been found in any prehistoric assemblages in the Yangzi valley, except for Liulinxi of the Chengbeixi culture. The grinding slabs and stone balls of the Chengbeixi culture might signal cultural influence from the north, where both implements are quite common (above and chapter 4).

Both the Pengtoushan and Bashidang sites are located on

alluvial plains which today are still fertilised areas for rice cultivation. The rice remains found in Bashidang have been identified as in an initial stage of domestication (Zhang and Pei 1996). So there seems little doubt that rice cultivation was practised in the Pengtoushan culture. On the other hand, although rice remains have been found in the Chengbeixi culture, even the excavators of the aforementioned sites are not certain whether rice was cultivated (Chen 1991; Yang 1991).

In summary, there are similarities between the Pengtoushan and Chengbeixi cultures in terms of their pottery. But as discussed above, their toolkits seem different, and data for the mobility patterns and economic activities of the Chengbeixi assemblages are too vague to be compared with the Pengtoushan culture. The radiocarbon dates for these two cultures suggest that the earliest phase of Pengtoushan might have been slightly earlier than Chengbeixi [Table 7(2)]. As archaeological data for Chengbeixi are far from sufficient, it is premature to draw any conclusions regarding the relationship between the Chengbeixi and Pengtoushan cultures at this stage.

7B. Archaeological assemblages in the lower Yangzi valley.

Archaeological research in the lower Yangzi valley started in the 1950s, and many Neolithic assemblages have been found since then. The Neolithic cultures in the lower Yangzi valley from 7000 BP onwards are rich and diversified. Polished stone tools, jade ornaments, fine pottery vessels, the earliest surviving silk woven fabrics, remains of towns, burials, and cult platforms have all been found in this area, not to mention plentiful remains of rice (IA, CASS 1984).

However, the picture for the period before 7000 BP is much less clear. Since the discovery of the Hemudu culture in the 1970s, no further significant discoveries of the early Holocene have been made. Chinese archaeologists are all aware that the Hemudu assemblage is too advanced to be the oldest Neolithic in the lower Yangzi valley. But no new data have been uncovered despite continuous searching.

Another problem in this area is the lack of archaeological data for the terminal Pleistocene. To date, only two assemblages of this period have been found in the lower Yangzi valley. The best-known comes from Xianrendong (or Spirit Cave) in Jiangxi province, and a less-known one comes from Sanshan in Jiangsu province [Fig. 7(1)]. Apart from these two assemblages, animal bones were found in two caves in Jiangsu province (Jinhua and Lishui), but without archaeological remains.

The Sanshan assemblage was discovered in 1985 on a small island in Tai lake (Chen *et al.* 1987). According to the excavation report, 5263 lithic implements were found at Sanshan; of which 1019 have been described (Chen *et al.* 1987). These implements were unearthed from a sand and pebble layer adjacent to the entrance of a small cave (Chen *et al.* 1987). They are summarised in Table 7(5).

Apparently, flake implements prevail in this assemblage, accounting for over 98% of the total implements. Some of them are blade-like. The Sanshan assemblage is quite different from the lithic assemblages in other sites of the Yangzi valley and South China, the latter dominated by pebble tools and irregular flakes. Major raw materials were chert, chalcedony and agate in Sanshan (Chen *et al.* 1987). It is reported that direct percussion by hard hammer was the major method for flake detachment and retouch (Chen *et al.* 1987).

It is reported that various scrapers are the predominant implements of the Sanshan assemblage. Of these, the 'thumb-nail scrapers' (dimensions from 3 to 1. 5 cm) [Fig. 7(12):15-17] look similar to those found in Xiachuan, Central China, but it is claimed that the workmanship is cruder (Chen *et al.* 1987). A salient feature of the Sanshan assemblage is that 218 out of the 531 flakes were used without retouch (Chen *et al.* 1987).

No other artefacts were found at Sanshan. It is said that the majority of the assemblage was debitage (Chen *et al.* 1987). According to the excavators, chert and agate occur to the south of the limestone cave, and the raw materials used might have been quarried here. On the other hand, the lithic implements were found in a water-transported layer of pebbles, suggesting that the assemblage may not be entirely *in situ.* But the excavators make no statement about the nature of the Sanshan site, and the question of whether the Sanshan deposit was *in situ* remains unanswered.

Based upon lithic comparisons and geological study, the Sanshan lithic assemblage is suggested to be of terminal Pleistocene to early Holocene date (Chen *et al.* 1987). The flaked lithic implements of Sanshan look quite distinct from the Neolithic culture of the Tai Lake area, and no polished tools or potsherds have been found. Therefore, it is quite possible that the Sanshan assemblage is earlier than the Neolithic culture in the Tai Lake area. But how much earlier is not clear.

The second location is Xianrendong, a cave located in Dayuan basin, a limestone area south of the Yangzi River [Fig. 7(1)]. The cave was discovered in 1962, and excavated twice by the archaeological team of Jiangxi province in 1962 and 1964 (CPAM of Jiangxi Province 1963; Jiangxi Museum 1976). Shells, animal bones, pottery vessels, stone and bone tools, and remains of hearths were found during these excavations. Later in 1962, scholars from the Institute of Vertebrate Palaeontology and Palaeoanthropology investigated the cave, focusing on the stratigraphy and fauna (Huang and Ji 1963). The lower layers of the deposit were suggested to have been laid down in the terminal Pleistocene, and the upper layer in the early Holocene. Mammals, birds, molluscs and tortoises were found and identified during the survey, suggested to belong to the early Holocene (Huang and Ji 1963).

Since its discovery, the Xianrendong assemblage has been considered to be essentially Neolithic (IA, CASS 1984). Polished stone tools and potsherds were found in most layers in the two excavations in the 1960s. But the radiocarbon dates from these early excavations are problematic, as a bone sample from a lower stratum was dated to 8575 ± 235 bp, while an unidentified shell sample from an upper layer was dated to $10{,}870 \pm 240$ bp [Table 7(1)]. For a long time the Xianrendong assemblage was viewed as an important Neolithic discovery, but with inherent problems (IA, CASS 1984).

In 1993 a Sino-American archaeology team, led by professor Yan Wenming of Beijing University and Dr. Richard MacNeish of the Andover Foundation for Archaeological Research in USA, launched a multi-disciplinary investigation at Xianrendong. A small cave called Wangdong (or Diaotonghuan), about 800 meters west of Xianrendong was also excavated in 1993 (MacNeish and Libby 1995 eds.). Then, both caves were excavated again in 1995 by the same team (Liu 1996; Peng 1996). The major achievements of this project are twofold: first, a preceramic deposit was identified, thus giving a sequential cultural deposit from the terminal Pleistocene into the early Holocene; second, phytoliths of wild or domesticated rice were identified, thus suggesting that at least wild rice collection was probably practised during the terminal Pleistocene (Zhao 1998).

After four separate excavations the stratigraphic profile at Xianrendong has become very complicated. The 1993 excavation took place on the western side of the entrance, extending from the northern end of trench No. 3 which had been dug in 1962 [Fig. 7(13)]. Only the lower strata were revealed here in 1993, as the upper layers had already been removed. Thus, in order to obtain a complete stratigraphic profile, the team shifted to the east and extended the northern part of trenches No. 4 and 5 excavated in 1964 (Peng 1996). In 1995, digging continued in the northern end of trench No. 3 and the western end of trench No. 6 (Peng 1996) [Fig. 7(13)].

There are discrepancies between the reports of these excavations. As far as the research to 1993 is concerned (details of the 1995 excavation are not yet published), it is apparent that the Xianrendong cultural strata can be divided into two phases: a preceramic phase and a ceramic phase, with sub-divisions within the latter. The preceramic phase is represented by zones 3C2 to 4 in trench No. 3, and zones 3 to 4 in trench No. 6. The ceramic phase is represented by zones 2 to 3C1B in trench No. 3, zones 1B to 4 in trench No. 4, and zone 2 in trench No. 6 [Fig. 7(13)] [Table 7(6); 7(7)]. The surface zones 1A and 1B in trench No. 3, zone 1A in trench No. 4 of the 1964 excavation, and zone 1 in trench No. 6 all yielded mixed materials [Table 7(6)], some dating as late as the Shang dynasty (1600-1100 BC). The excavation report of 1993 defined these surface zones as a 'Neolithic 4' phase (MacNeish and Libby 1995 eds.), but I think it would be more prudent to regard them as disturbed.

The stratigraphy at Wangdong is less complicated. The deposit here is divided into 13 zones. The lower zones F to M represent cultural deposits claimed to be from approximately 25,000 to 14,000 BP; zone E dates from 14,000 to 11,800 BP; zones D and C2 from 11,800 to 9600 BP, and C1 from 9600 to 7700 BP (MacNeish and Libby 1995 eds.). There were mixed materials from the late Neolithic to the Shang dynasty in zones B to A (MacNeish and Libby 1995 eds.) [Table 7(7)].

As the preceramic deposit was not recognised in the 1962 excavation, and the identifying numbers assigned to the artefacts do not indicate sub-layers, it is difficult for other researchers to distinguish genuinely preceramic artefacts in the collections. During the 1964 excavation, artefacts were found in zones 3 and 4 in trench No. 6, which are now defined as preceramic. But exact quantities were not described. To judge from identifying numbers given in the report, at least one chert flake, one flake core and one pebble implement were found in the preceramic deposits at this time (data based upon Jiangxi Museum 1976). No lithic implements have been reported from the lower strata at Xianrendong in the 1993 excavation report. It is reported that cores and small flakes made of quartz and chert, together with scrapers, burins and points made from such small flakes were found in the lower zones in 1995 (Zhang and Liu 1996), but details are not yet published.

At Wangdong, lithic implements were found both in 1993 and 1995. The 1995 digging at Wangdong yielded artefacts such as 'microblades', flaked scrapers, points and burins (Liu 1996; Peng 1996). But details are not available. The lithic implements found in 1993 are summarised in Table 7(8).

No formal drawings are available, but according to the descriptions in the 1993 excavation report (MacNeish 1995) and my observations of the collection made in 1996, some characteristics of the Wangdong assemblage from the terminal Pleistocene can be summarised as follows:

1). Two lithic industries seem to be present within the assemblage: a pebble tool industry primarily of sandstone and mica schist, and a small flake industry of chert and quartz (MacNeish 1995). Pebbles were used for heavy-duty tools such as choppers and hammers, while small flakes were made into various "scrapers", "points" and "spokeshaves" (MacNeish 1995).
3) All the prismatic blades were produced by direct percussion from polyhedral cores (MacNeish 1995). Direct percussion seems to have been the major method of flake detachment, but MacNeish claims that pressure retouch was also used on a spokeshave (MacNeish 1995:28). Scrapers and points were retouched (MacNeish 1995).
4) The quantity of the Wangdong assemblage is limited, but most of the implements seem to have been tools used for multiple purposes. The precise tasks cannot be identified in the absence of use-wear analysis. It has been generally assumed that the pebble toolkit in the vast area south of the Yellow River valley was used by hunters and gatherers (IA, CASS 1984), but the tasks assigned to it are not clear.

It has also been claimed that microblades were discovered, in association with a "tongue-shaped core" (MacNeish 1995:29). However, no typical microcores have been reported in Xianrendong or Wangdong. During my fieldwork in 1996 I was allowed to see some of the samples from both Xianrendong and Wangdong. It seems that the 'microblades' found in Wangdong are different from those found in Central China. Judging from the samples I observed, and the descriptions given by MacNeish (1995:29), these small blades were produced by direct percussion, as characterised by a clear flake scar, an unprepared striking platform, and visible 'ridges paralleling the parallel lateral edges' (MacNeish 1995:29). As they do not fulfil the definition of a typical microblade: i.e. produced by indirect percussion from a prepared platform (Gai 1985); and no microcores were reported, whether 'microblades' were produced there is a question.

The small flakes found in Wangdong are different from those found in Central China, but similar to those found in the middle Yangzi valley. They were made from quartz or chert, and are quite small and thin. The specimens which I was allowed to observe in 1996 are all under 5 cm in maximum dimension; and less than 1 cm in thickness [Fig. 7(14)]. A few flakes made from quartz are less than 0.5 cm in thickness: they are so thin that they are transparent. Their shapes vary from round to oblong. Many were not retouched, but some seem to bear use-wear. Unlike the scrapers, points and burins made from flakes in Central China, the functions of these small flakes in the Yangzi River are not known. However, I noticed that there were a few small blades made from chert bearing a glossed edge. Whether this gloss is a result of cutting cereals remains to be investigated.

Apart from lithic implements, organic artefacts were also found in the preceramic strata at Xianrendong and Wangdong. Perforated shell implements were found in Xianrendong and Wangdong in the 1995 excavation (Liu 1996; Peng 1996), but details are not available. Judging from the identifying numbers of the artefacts, at least two bone arrowheads, one bone harpoon and one perforated shell implement should belong to the preceramic strata dug in 1964, but none were reported during the 1962 excavation.

Hearths were discovered in Xianrendong, but it is not clear whether they were found both in the lower and upper strata, or in the upper stratum only. Some human remains were also found during the excavations in the 1960s and 1990s. An isotopic study has been conducted on human bone in order to assess the diet of the prehistoric occupants of Xianrendong, and it is argued that rice, possibly wild rice, 'was consumed in late Palaeolithic times in Jiangxi

province' (MacNeish *et al.* 1995: 74). But the stratum of the bone sample of this isotopic study is not clear, and caution should be taken with this result.

In summary, the tools found in Xianrendong-Wangdong, particularly the arrowheads and harpoons, seem to indicate a subsistence strategy of hunting, gathering and fishing in the terminal Pleistocene. The presence of knives at Wangdong may indicate grass harvesting, but this hypothesis has to be checked by use-wear analysis. As the outcome of the 1990s excavation is still under analysis, and a full report is yet to be published, few conclusions can be drawn at this stage.

Several absolute dates have been obtained for the Xianrendong-Wangdong assemblage. Two were run on samples collected from the 1960s excavation, and six on samples from the 1993 excavation (MacNeish and Taylor 1995). The two dates from the preceramic zones G and H at Wangdong are in sequential order, dated to 17,040±270 and 19,770±360 bp respectively [Table 7(1)]. Another two absolute dates come from preceramic deposits at Xianrendong, one dated to 15,180±90 and another to 18,100±270 bp (MacNeish and Taylor 1995) [Table 7(1)]. It should be pointed out that the stratum of the sample for the 15,180±90 date is not very clear (MacNeish and Taylor 1995). Based upon these dates and cross-dating with other archaeological assemblages, MacNeish and Taylor estimate that the temporal range for the preceramic deposit in both Xianrendong and Wangdong would be from 25,000 to 15,200 BP (MacNeish and Taylor 1995).

But there is one problem which cannot be ignored. It has been long realized that radiocarbon dates based upon samples from limestone areas are likely to be older than their actual age, and research was conducted in South China in the early 1980s to deal with this problem. The outcome of this research suggests that C^{14} dates from shells and other aquatic materials can be one to two thousand years older than true ages, but the deviation is less significant in terrestrial samples such as animal bone and charcoal (The Radiocarbon Dating Laboratories of Beijing University and IA, CASS 1982). But is there still a deviation for samples of bone and charcoal which originated in limestone areas? No answer has been provided for this question. The samples collected from the 1993 excavation are either human bone or charcoal, so the dates from these samples should be more accurate than that from aquatic samples. But how accurate they are remains a question.

Cross-dating based on cultural comparison is another method which can be used to verify the date for an archaeological assemblage. Unfortunately, this method does not work well in the Xianrendong-Wangdong case. One of the lithic hallmarks of the preceramic deposit at Wangdong is the presence of small flakes made of chert, quartzite and quartz. This type of implement is also found in the middle Yangzi valley and South China. In the middle Yangzi valley, such small flakes have been found in the Jiangling and Yan'er assemblages, which are estimated to be terminal Pleistocene. But no absolute dates are reported for these assemblages. Similar small flakes and implements made from small flakes have been found in Bailian Cave, Liyuzui Cave, and a few other caves in Guangxi province, South China. The absolute dates for these assemblages range from over 28,000 BP (Bailian Cave) to 23,330 BP (Liyuzui) (Zhou 1994) [Table 7(2), additional table 1]. But these caves are all in limestone areas, and the reliability of their absolute dates is questionable.

However, pollen analysis may provide some clues for this issue. According to the pollen analysis, when zones G and H at Wangdong were deposited, the palaeoclimate was cold and dry, probably within the LGM (Wang *et al.* 1995). Zones 3C2 to 3B from Xianrendong may belong to a warm and moist period with fluctuations, which is in consistent with the improved palaeoclimate after 15,000 BP in the Yangzi basin, with sudden cool changes comparable to the Younger Dryas (Wang *et al.* 1995). Given these palaeoclimatic data, I suspect that the preceramic stage of Xianrendong/Wangdong would have been around 18,000 to 13,000 years BP [Table 7(7)]. The dates for the early ceramic deposit in Xianrendong are also quite controversial [Table 7(7)]. Again, pollen analysis also seem to suggest that Ceramic Phase I in Xianrendong might have been within the period of 13,000-8000 BP [Table 6(3)] (Wang *et al.* 1995). This phase will be discussed in the following section for the purpose of continuous description.

In the above section the archaeological assemblages of the terminal Pleistocene in the lower Yangzi valley have been reviewed. Only the Sanshan and Xianrendong assemblages can be assigned to this period, and they are quite different. While the Sanshan assemblage resembles the flake industries of Central China, the Xianrendong-Wangdong assemblage is similar to the assemblages found in the middle Yangzi valley (chapter 7). As both sites lie to the south of the Yangzi River, this difference may not be explained in terms of geographical location. Perhaps the Sanshan assemblage represents cultural diffusion from Central China. This issue awaits more data from further archaeological discoveries.

While there are many archaeological assemblages dated prior to 7000 BP in the middle Yangzi valley, in the lower Yangzi valley archaeological discoveries for this period is very scanty. The earliest archaeological assemblage of the Holocene in the lower Yangzi valley now seems to be that from the upper zones of Xianrendong and Wangdong. As discussed above, zones 2 to 3C1 in trench No. 3, zones 1B to 3B in trench No. 4 and zone 2 in trench No. 6 at Xianrendong, and zones C to E at Wangdong, all belong to the ceramic period (MacNeish and Libby 1995) [Table 7(7)].

These zones can be further divided. In the 1964 excavation at Xianrendong it was noted that pottery with tempers of white quartz sand occurred in lower strata, while shell-tempered pottery was found above (Jiangxi Museum 1976). In addition, in the Chinese context, if both pottery with coarse temper of various materials and fine pottery are present in an archaeological site, fine pottery often appears later than the coarse tempered. This is the case in Central China, the Yangzi River valley and South China. Using the respective presence of tempered and fine pottery at Xianrendong, two sub-divisions can be noted within the ceramic phase. Ceramic Phase I is represented by zones 3A to 3C1 in trench No. 3 and zones 2 to 4 in trench No. 4, where only pottery with coarse sand or quartz sand temper was found [Table 7(7)]. Ceramic Phase II is represented by zone 2 in trench No. 3, zone 1B in trench No. 4, and zones 2A and 2B in trench No. 6, where fine and shell-tempered pottery occurred [Table 7(7)]. In addition, zones 1A/1B in the 1993 excavation were present in a protected location under a huge rock (MacNeish and Libby 1995), and their pottery also belongs to Ceramic Phase II. These two phases roughly correspond to the lower and upper complexes described in the 1960s excavation reports (CPAM of Jiangxi Province 1963; Jiangxi Museum 1976).

The strata in Wangdong seem less clear. According to the 1993 excavation report, shell-tempered potsherds only occurred in zone B and feature 5 (the latter being a surface deposit) (MacNeish and Libby 1995), where they are mixed with pottery from the Shang period. While Ceramic Phase I is clearly defined, Ceramic Phase II was obviously disturbed.

The stratigraphic and cultural divisions at Xianrendong and the corresponding deposits at Wangdong, as well as the guesstimated dates, are summarised in Table 7(7). Apparently the absolute dates are much older than the dates I estimate. Given the concerns about samples from limestone areas, the estimated dates are based upon pollen analysis and cross-dating. For Ceramic Phase I, the pottery is similar to that found in early contexts in South China, being tempered with coarse sand, cord-marked and baked at a low firing temperature (Yang 1994). The earliest potsherds of this type found in Miaoyan cave in South China are claimed to be dated to about 13,000 to 12,000 BP (Yuan 1993). Chert small flakes and bone implements have been found in the cave in association with pottery, animal bones, and perforated pebble disks (Yang 1994).

Potsherds and rice husk remains, as well as pebble and shell tools, were also found in the Yuchan assemblage in Hunan province, tentatively dated to 12,000-8000 BP (IA, CASS 1997; Yuan 1996). It is quite possible that Xianrendong Ceramic Phase I and the above assemblages are about the same age. As for Ceramic Phase II, the pottery is comparable to that from Luojiajiao (below), with a beginning date of about 7200 BP [Table 7(2)].

During the 1993 excavation, lithic and organic implements were also found in the Ceramic Phase I strata at Xianrendong. The lithic artefacts are summarised in Table 7(9). Artefacts found in zones now classified as of Ceramic Phase I at Xianrendong in the excavations of 1960s also summarised in Table 7(10). Lithic implements in zones C to E at Wangdong in the 1993 excavation are listed in Table 7(11). It seems that the lithic implements of Ceramic Phase I found in both caves consist of three industries: flake and blade implements, flaked pebble implements, and polished tools. The proportions of each industry vary, as illustrated by Table 7(12).

Seemingly, pebble implements dominated the lithic assemblage from both caves in 1993, but a flake industry dominates the assemblages found at Xianrendong in the 1960s, particularly during the first excavation in 1962. These difference could reflect various reasons, one being differences in collecting policies by the excavators, and another being real differences in distribution patterns within different parts of Xianrendong. As only a small part of the Xianrendong deposit has been excavated so far, it is premature to draw any conclusion on this issue.

According to excavation reports, the raw materials used for flake implements were chert, quartz and quartzite (CPAM of Jiangxi Province 1963; Jiangxi Museum 1976; MacNeish 1995). Flakes seem to have been produced by direct percussion, as points of impact and fissures are clearly visible (Jiangxi Museum 1976; MacNeish 1995). Many flakes have not been secondarily retouched. Scrapers prevail amongst the flake implements, followed by "burins" and "spokeshaves".

While scrapers can be used for multiple purposes, the functions of the small flakes are not clear. Morphologically similar to counterparts in the preceramic strata of both Xianrendong and Wangdong caves, the small flakes from Ceramic Phase I are small and quite thin. Some of them are only 1.5 cm in greatest dimension and less than 0.3 cm in thickness (author's observation in 1996). The pebble implements found in Ceramic Phase I strata of the two caves are also morphologically similar to counterparts from the preceramic strata of Wangdong. The sizes of these pebble tools vary from 5.9 to 8.8 cm in maximum dimension (CPAM of Jiangxi Province 1963). Their working edges were flaked, but the major part of the pebble surface usually remained [Fig. 7(15)].

The perforated discs in these sites are similar to those found in South China (below latitude 25°N). This type of tool is widely found in areas south of the Yangzi River. Based upon ethnographic data from Yunnan province, southwest China, and comparison with similar implements found in South Africa, it has been suggested that these discs could have served as digging stick weights, or as a throwing weapon for hunting (Song and Zhou 1994).

Polished implements are the smallest proportion within Ceramic Phase I of Xianrendong assemblage, and are absent at Wangdong. According to the 1960s excavation reports, these implements were often polished only along their working edges (CPAM of Jiangxi Province 1963; Jiangxi Museum 1976). The polished implements include adzes, points/drills and chisels, and projectile points which might have been used for hunting. Putative spades and hoes (MacNeish 1995) could have been used for cultivation, but residue and use-wear analyses are required.

Apart from lithic implements, a considerable number of organic tools has been discovered in Ceramic Phase I of both Xianrendong and Wangdong [Table 7(13)]. Organic implements were also found in the 1990s excavations, but only briefly reported. Liu (1996) states that '318 bone implements, 26 perforated shells' were found in the 1995 excavation from the two caves, but no further details are given (Liu 1996).

The perforated shells merit special attention. They accounted for over 30% of the organic toolkit in the 1960s excavation. I observed some of these shells in 1996, and noticed that their edges were polished by usage. Similar shells, found in the Yuchan assemblage, middle Yangzi valley, also bear glossed edges (author's observation 1996). Could these shells have served as tools for rice harvesting? Given the presence of phytoliths of wild and possibly domesticated rice in Ceramic Phase I deposit at Xianrendong (Zhao *et al* 1995:52), and the rice husk found in Yuchan (Yuan 1995), this is possible. Again, use-wear and residue analysis is required before any conclusion can be reached.

In sum, although functions of many artefacts are not clear at this stage, the whole toolkit of Xianrendong and Wangdong, consisting of the lithic and organic implements, suggests a diversified subsistence strategy of hunting, gathering (including plant gathering), fishing, and possibly incipient cultivation.

Another important aspect of the Xianrendong-Wangdong assemblage is the pottery. Sherds of Ceramic Phase I are said to have been tempered with crushed white quartzite (Hill 1995; Jiangxi Museum 1976; CPAM of Jiangxi Province 1963). These tempers are quite coarse, measuring from 1-3 mm (Hill 1995). Pots were made by coiling (Hill 1995; CPAM of Jiangxi Province 1963; Jiangxi Museum 1976), and smoothed by hand or with a handful of grass (Hill 1995). Fabrics are of brownish-red colour, and mainly decorated by cord-marking (Hill 1995; CPAM of Jiangxi Province 1963; Jiangxi Museum 1976). Reconstructed forms include a cooking *fu*, which appear similar to the *fu* found in Pengtoushan in the middle Yangzi valley [Fig. 7(18)]. According to Hill (1995), such simple vessels lasted into Ceramic Phase II (Hill 1995: 42).

Animal bones have been found in substantial numbers in the two caves. Macaque, hare, leopard, wolf, raccoon dog, hog badger, polecat, civet, masked palm civet, wild pig, Chinese water deer, muntjac, spotted deer, sheep, and probably water buffalo were found in Xianrendong (Huang and Ji 1963; CPAM of Jiangxi Province 1963; Jiangxi Museum 1976) [Table 6(2)]. The faunal report for the 1995 excavation is not available, and the report of the 1993

excavation is too brief. But it is stated that animals found in the 1990s excavations are similar to those found in the 1960s (Redding 1995). Of these, spotted deer are the most common.

Tools made from deer antler were also found in both caves. As the growing season for deer antler is from August to the following March (Zhang *et al.* 1988), the remains of these antlers with skull fragments (Huang and Ji 1963) indicate that deer might have been hunted from late summer to early spring. This would also have overlapped with the time for wild rice collection, as contemporary wild rice ripens from late October to early November (Qi *et al.* 1984). Thus it seems that Xianrendong was occupied at least from late summer to the early spring. The continuous deposition in both caves suggests that they were occupied for a long period of time. But at this stage it is not clear whether there was year-round sedentism.

From the 1993 report it seems that the strata in Wangdong are not fully continuous and the Ceramic Phase II deposit is missing. The artefacts from Ceramic Phase II in Xianrendong are quite limited. Lithic implements from the three excavations in the 1960s and 1993 are summarised in Table 7(14).

The adzes could have been used for many purposes, but according to a use-wear analysis, they were likely to have been used for making wooden and organic implements (Tong 1989). As discussed above, the perforated pebble disks might have been used as digging stick weights. According to the phytolith report from the 1993 excavation, phytoliths of domesticated rice were present in the Ceramic Phase II deposits of both caves (Zhao *et al.* 1995), indicating the possibly practice of rice cultivation.

Organic tools were also found in Ceramic Phase II deposit in Xianrendong. Details for the 1990s excavations are not available. Those from the 1960s excavations are listed in Table 7(15). The common implements here are awls/drills and perforated shells. The function of the perforated shells is not clear. They were found in substantial numbers in deposits from Ceramic Phase I to Ceramic Phase II. Other types of organic tools - arrowheads, harpoons and points - indicate the practice of hunting and fishing [Fig. 7(16)].

Comparing the whole toolkit of Ceramic Phase I of Xianrendong with its Ceramic Phase II counterpart, one can see that changes are quite limited. This similarity strongly suggests cultural continuity. The most visible change from Ceramic Phase I to Ceramic Phase II is in the pottery tempers, with the switch to shell and crushed potsherds (CPAM of Jiangxi Province 1963; Hill 1995; Jiangxi Museum 1976). Pottery was still built by coiling, but paddles were used to thin the walls (Hill 1995). The decoration now includes cord-marking and basket and check impression (Hill 1995; Jiangxi Museum 1976; CPAM of Jiangxi Province 1963). But the reconstructable shapes seem to have changed little, and the *fu* and pot still dominated (Hill 1995).

It is notable that large quantities of mollusc shells and bone fragments were found in Ceramic Phase II deposit, in densities much greater than that in Ceramic Phase I and preceramic deposits (Liu 1996; MacNeish 1995). This suggests that fishing and mollusc collection remained primary economic activities. Phytolith analysis shows that domesticated rice (*O. sativa*) was present from zone E in Wangdong and zone 3B1 in Xianrendong upwards, suggesting that rice cultivation was practised by approximately 10,000 to 8000 BP (Zhao *et al.* 1995; Zhao 1998). But the quantity of phytolith for domesticated rice seems quite limited, ranging from 1 to 6 grains; and a substantial quantity (22 grains) was only present in zone B at Wangdong, which is guesstimated to be 4500 years BP by the excavators (Zhao *et al.* 1995; Zhao 1998). These prima facie archaeological evidence seem to suggest that mollusc collection, fishing and hunting were the main subsistence strategy of the Xianrendong occupants. Whether rice cultivation was primary in Xianrendong-Wangdong remains to be investigated.

To date, archaeological assemblages with solid evidence of extensive rice cultivation are found in the Yangzi delta. The best representative of these archaeological assemblages is still the well-known Hemudu culture, located on the alluvial plain of the Yangzi river mouth [Fig. 7(1)]. Discovered and excavated in the 1970s, Hemudu's extremely well preserved artefacts, pile-dwellings, animal and rice remains surprised the Chinese archaeological world, and challenged the traditional theory that the Yellow River was the centre of Chinese prehistory and civilization. Regarding the issue of agricultural origins, the dense rice remains found in Hemudu have been cited as evidence for an origin of rice agriculture in the lower Yangzi valley (i.e. Smith 1995; IA, CASS 1984). Unfortunately, although the report for the 1973 excavation has been published, a full report of the second excavation from 1977 to early 1978 is still not available. However, a summary report of the two excavations has recently been published (Liu and Yao 1993).

The richness of the Hemudu assemblage is due largely to the excellent preservation of organic materials. A major contributor to this is the location of the site on a flat alluvial plain, the current surface of which is only about 1.1 meters above sea level (Lang 1987). Most of the cultural strata are under the current sea level and are covered by marine sediments or peat deposits, which preserve organic materials very well. It is said that the rice husks were still golden yellow and the leaves green when they were unearthed (Liu and Yao 1993).

There are four layers in the Hemudu profile. The lowest, layer 4, represents the first phase of the Hemudu culture and is dated from 7100 to 6500 BP (Liu and Yao 1993) [Table 7(2)]. Layer 3 overlays layer 4, and dates from 6500 to 6000 BP (Liu and Yao 1993). Layer 2 and layer 1 date from 6000 to 5500, and 5500 to 5000 BP respectively (Liu and Yao 1993). The whole profile shows cultural continuity and development, but with some geomorphic interruptions such as the alluvial deposit between layers 2 and 3 (Liu and Yao 1993). Only the first phase (layer 4), which is also the richest cultural deposit, is discussed in this thesis.

Layer 4 was found overlying a light blue clay of marine origin (Liu and Yao 1993), indicating that the first occupation occurred when the site was closer to the sea than it is now. During the excavations in 1973 and 1977, abundant animal bones, remains of rice straw, husks and leaves, as well as foundations of pile-dwellings, pits, lithic and organic implements, pottery, and two baby skeletons were found in this layer (Liu and Yao 1993). Other plant remains included sawtooth oaks, Chinese jujubes, acorns and water chestnuts. It is claimed that 2838 artefacts were found during the two excavations, including 385 lithic artefacts, 1934 bone tools, 328 wooden tools, and 191 fired clay instruments (Liu and Yao 1993). But details of these implements are not fully available. Only 313 lithic items have been reported, summarised in Table 7(16).

Black chert was the major raw material for lithic implements (Liu and Yao 1993). The lithic tools were flaked, then polished, but most of them were only partially ground (Liu and Yao 1993). These lithic implements are not the most common section of the toolkit in Hemudu. The majority of tools were made from bone [Table 7(17)].
While lithic and bone implements were commonly found in

the archaeological assemblages of the early Holocene, wooden implements are very rare, mainly because they were seldom preserved in dryland sites. But it is claimed that 328 wooden implements were found in the waterlogged conditions of layer 4 of Hemudu (Liu and Yao 1993). Of these, 144 implements have been reported [Table 7(18)].

In addition to the these implements, 177 spindle whorls made of fired clay with an organic temper, and 14 pellets of fired clay were found in layer 4. There was also a unique class of implement, made of either stone or wood, and shaped like a butterfly (Liu and Yao 1993) [Fig. 7(17):14]. Their sizes vary. The stone ones measured 11 cm long and 8 cm wide, while the wooden ones measured around 22-23 cm in length and 13.4-13.5 cm in width (CPAM of Zhejiang Province and Zhejiang Museum 1978). The function of this tool type has been a topic for debate since its discovery. Yan (1989) suggests that it might have been used for preparing rice fields (Yan 1989). Liu and Yao (1993) think that they were building decorations (Liu and Yao 1993).

The whole toolkit found in layer 4 thus consists of various items for a wide range of tasks. While axes and adzes can be used for many purposes, from land clearing to wood working, particularly the construction of pile-dwellings, the bone spades are believed to be specifically for the preparation of rice fields (Liu and Yao 1993). Some of these spade blades were tied on wooden handles by rattan strings. Stone and wooden pestles, as well as some stone grinding slabs might have been used for rice processing (Liu and Yao 1993).

While tools for rice cultivation seem abundant, it is noticeable that cutting tools are quite scarce. According to Liu and Yao (1993), only one bone sickle was found in layer 4. Given the dense rice remains, the lack of a cutting tool for rice harvesting seems puzzling. Yan Wenming (1996, personal communication) suggests that harvesting might have been done by hand-pulling. This would have been possible. The Gaoshan minority in Taiwan was harvesting rice by hand pulling in the Ming dynasty (AD 1366-1644), according to an observation record written by a Chinese official during that time (cited by Chen and Lin 1988). However, during my visit to the Hemudu museum in 1996, I was allowed to visit the storeroom, and was told that bone knives were actually found in layer 4. One such bone knife was on display, and had a glossed edge. But details have not been published, and the function of such tools remains to be studied.

Apart from tools for cultivation, tools for hunting and gathering are abundant. Stone balls, arrowheads and spears made of wood and bone are all likely hunting tools. The numerous animal bones in the site further support this assumption. A total of 61 species of animals and birds was found in layer 4, including 8 species of birds, two species of macaques, pangolin, rat, porcupine, bear, raccoon dog, polecat, hog badger, two species of river otters, oriental civet and rasse, masked palm civet, tiger, Asiatic elephant, two species of rhinoceros, wild boar, four species of deer, two species of muntjac, water buffalo, as well as domesticated dog and pig (Liu and Yao 1993) [Table 6(2)]. It seems that the Hemudu people exploited their natural resources in full, and hunted almost everything, including dangerous animals like tiger and bear.

Fishing spears and net sinkers were apparently used for fishing, while bone shuttles might have been used for knitting fishing nets. The presence of wooden oars indicates a use of boats. In layer 4, bones of whale, shark, 8 species of other river or sea fish, river and sea turtles, and crocodiles, together with mollusc shells and crab were found (Liu and Yao 1993). The presence of these aquatic animal remains indicates that the Hemudu occupants also exploited their fresh water and sea territory extensively, and they must have been able to fish a certain distance out from the sea shore.

Besides economic activities, the Hemudu people also made pottery. A total of 952 vessels was found in layer 4, and they consist of items with different tempers. The black organic-tempered pottery, with rice husks and other plant materials, accounted for 77.58% of the potsherds (Liu and Yao 1993). Black sand-tempered pottery accounted for the other 22.42% (Liu and Yao 1993). Among the organic-tempered vessels, about 0.01% were covered by with a white slip and painted with brown and dark brown designs (Liu and Yao 1993). This is one of the earliest examples of painted pottery found in China.

The pottery from layer 4 was all made by coiling, and shapes are not regular. Types of vessels include 191 cooking *fu*, 114 pots, 325 bowls, 46 deep and large bowls, 193 plates, 25 stands, 44 lids, and 14 other small containers (Liu and Yao 1993) [Fig. 7(18)]. There is no doubt that the *fu* was used for rice cooking, as remains of over-cooked rice were found in one. Bowls and plates might have been used for serving food (Liu and Yao 1993). Apart from painting, vessels were decorated by cord-marking and incision with various motifs, including animals, plants, as well as patterns formed by small circles and lines (Liu and Yao 1993). These motifs were designed with liveliness and quite delicately made.

The toolkit found in layer 4 also indicates that some fibres might have been used to produce loom textiles (Liu and Yao 1993). Cloth would have been sewn with bone needles (it is a pity that no cloth was found). The Hemudu occupants were certainly able to knit reed mats and make string of rattan and straw, as remains of these materials were found in layer 4 (Liu and Yao 1994). Remains of ivory sculptures of birds, and various ornaments made of stone, bone and animal teeth, including beads, rings and hair-pins (Liu and Yao 1993), illustrate the artistic ability of the inhabitants.

In addition to the above achievements, the Hemudu people were also capable carpenters and builders. The skilfully-made wooden house components, such as piles, wall timbers and floor boards, particularly the various kinds of joinery, form another stunning discovery in Hemudu [Fig. 7(17)]. According to the remains found in layer 4, dwellings were long, rectangular houses with piles supporting them above the ground, and thatched roofs (Liu and Yao 1993). The house remains in layer 4 suggest that at least three large pile-dwellings were found during the excavations, and smaller rooms were delineated within each (Liu and Yao 1993). There was probably a corridor, or a long balcony, in the front part of each house. Such "long-house" pile-dwellings can prevent animal attacks, and reduce the dampness of living on the ground. They are an invention of humid subtropical and tropical areas, and identical structures can still be found in contemporary Southeast Asia (e.g. Bellwood 1978). The remains of such substantial houses suggest that the Hemudu people were sedentary.

Living in such a resourceful environment, and having access to such a wide range of food, the population of Hemudu must have been quite large. It is stated that the excavated part of layer 4 covered only 5% of the total deposit (Liu and Yao 1993). Yet the post-holes of three large houses were found, one being 23 m long and 7 m wide, with a front verandah 1.3 m in width. It is suggested that the sizes of the other two could have been similar (Liu and Yao 1993). According to similar houses found in contemporary Dulong and Pumi minorities in Southwest

China, such large houses could have been divided into small rooms for individual families (The Editing Committee of Yunnan Province 1981; 1990). Assuming each room was about 4 meters in width, then one large house in Hemudu could have been divided into six rooms, providing accommodation for six families, and 18 families in total within the three large houses. Further assuming that each family consisted of an average four members, then the total population would have been 72 people. And this would only have been part of the population of the whole of Hemudu village, as only 5% of the site was excavated.

Hemudu was not the only affluent village in the lower Yangzi valley in the early to middle Holocene. Similar discoveries have been made in another site, called Luojiajiao, north of Hemudu [Fig. 7(1)] (The Archaeology Team of Luojiajiao 1981). Four cultural deposits were found here on top of lake sediments, with the lowest (layer 4) dated from 7200 to 6780 BP [Table 7(2)], contemporary with layer 4 of the Hemudu assemblage. The cultural contents of layer 4 at Luojiajiao were not as rich as those of layer 4 at Hemudu, but to judge from available data, the majority, particularly the house timbers, bone spades and rice husk tempered pottery were similar to those found in Hemudu (The Archaeology Team of Luojiajiao 1981) [Fig. 7(19)]. Rice cultivation was also practised in Luojiajiao, and the cultural similarities may suggest contact between the two groups.

On the other hand, the Luojiajiao assemblage was distinguished by having shell-tempered pottery, which was not found in Hemudu. Although the Hemudu types of cooking *fu*, plates, bowls, and stands were commonly found in Luojiajiao, some vessels, such as 'tea pots', were unique to Luojiajiao [Fig. 7(18)].

The discoveries of Hemudu and Luojiajiao illustrate the existence of prosperous and sedentary societies with rice cultivation, supplemented by hunting, gathering and fishing. Chinese scholars unanimously agree that these cannot be the initial agricultural societies of their region, that there must have been antecedents (e.g. IA, CASS 1984). But what were the preceding assemblages, and where are they to be found? Chinese archaeologists have been searching for over 20 years, but still have not found a satisfactory answer. The new discoveries at Xianrendong and Wangdong in the 1990s may shed light on this question, but there are still many questions regarding the transition from hunting and gathering to rice cultivation in the lower Yangzi valley. On the other hand, it should be noted that there were some cultural similarities between the Hemudu/Luojiajiao assemblages and the Pengtoushan culture in the middle Yangzi valley. Was the former a result of cultural diffusion from the latter? This question will be discussed in more detail in chapter 11.

Table 7(1) Radiocarbon Dates of Archaeological Assemblages of the Terminal Pleistocene in the Yangzi valley.

Site	Location	Laboratory	Sample	Stratum of the sample	Radiocarbon dates (B.P.). H.L.5730	H.L.5568	Calibrated date	Referencs.
Zhangnaodong		No data.	?		13,490±15	3,080±15	15811-15290	Huang *et al.* 1987.
Xianrendong	117°13'E	ZK-39	Shell.	Bottom of zone 2, T3?	0,870±24	10563±240	12738-12176	MacNeish and Taylor 1995.
	28°44'N.	ZK-92	Animal bone.	Zone 3?	8825±24	8575±240	9867-9372	" "
		BA-93181	Charcoal.	Zone 3B1.	4,610±29	14197±290	17366-16670	" "
		AA-15008	Charcoal?	Floation sample from zone 3C1B.		17,420±13	20973-20447	" "
		UCR-3300	Human bone.	?		15,180±90	18238-17977	" "
		BA-93182	Charcoal.	Zone 3C2.	8,100±27	17588±270	21342-20524	" "
Wangdong.	117°13'E	BA-93180	Charcoal.	Zone G.	7,040±27	16558±270	19895-19155	" "
	28°44'N.	BA-93179.	Charcoal.	Zone H.	9,770±36	19211±360	n/a.	" "

Table 7(2) Radiocarbon Dates of Archaeological Assemblages of the Early Holocene in the Yangzi valleys.

Site	Location	Laboratory	Sample	Stratum of the sample	Radiocarbon dates (B.P.). H.L.5730	H.L.5568	Calibrated date	Referencs.
Pengtoushan	110°45'E 29°40'N.	BK89016	Carbonized rice straw.	T14(2)	7815±100	7594±100	8416-8213	Chen and Hedges 1994.
		QxA 2210	Carbonized rice straw	Potsherd II, T1(2)H2.	7775±90	7555±90	8405-8187	" "
		QxA 2211	Humic acid	Potsherd II, T1(2)H2.	7520±90	7307±90	8137-7960	" "
		QxA 2212	Charcoal in the potsherd.	Potsherd II, T1(2)H2.	8550±80	8308±80	9427-9058	" "
		QxA 2213	" "	Potsherd II, T1(2)H2.	9220±80	8959±80	9997-9897	" "
		QxA 2214	Carbonized rice straw.	Potsherd III, T14(3).	7250±140	7045±140	7944-7670	" "
		QxA 2215	Humic acid.	Potsherd III, T14(3).	7610±80	7395±80	8304-8019	" "
		QxA 2216	Charcoal.	" "	8290±80	8056±80	8995-8727	" "
		QxA 2217	Charcoal.	" "	8490±80	8250±80	9372-9000	" "
		BK 89017	soil with charcoal	T13(3)	7770±110	7550±110	8409-8172	" "
		BK 87050	charcoal	T11(4)	8200±120	7968±120	9973-9653	" "
		BK 89018	bamboo and charcoal	T14(6)	7945±170	7720±170	8599-8334	" "
		BK 87002	pottery	?	9100±120	8843±120	9972-9652	" "
		BK 89019	"	T 3D3	7700±110	7550±110	9442-9057	" "
		BK 89020	"	T1H1	7945±100	7945±100	8980-8557	" "
		BK 89021	"	T10D22	8380±115	8143±115	9244-8955	" "
		BK 89022	"	T1H5	8135±90	7905±90	8951-8547	Chen and Hedges 1994.
		QxA1274	Lipid.	Potsherd I.	7055±100	6856±100	7719-7549	" "
		QxA 1275	Humic acid.	" "	7930±100	7706±100	8548-8370	" "
		QxA 1277	Fulvic acid.	Potsherd I.	6252±100	6073±100	7145-6793	" "
		QxA 1280	Coarse charcoal grain	" "	9785±180	9508±180	10946-10301	" "
		QxA 1281	Fine charcoal grain.	" "	7890±90	7667±90	8494-8361	" " " "
		QxA 1282	" "	" "	8455±90	8216±90	9365-8992	" "
		QxA 1273	" "	" "	9065±300	8809±300	10036-9459	" "

Table 7(2) Radiocarbon Dates of Archaeological Assemblages of the Early Holocene in the Yangzi valleys (continued).

Site	Location	Laboratory	Sample	Stratum of the sample	Radiocarbon dates (B.P.). H.L.5730	H.L.5568	Calibrated date	Referencs.
Chengbeixi	110°27'E 30°28'N.	ZK-2643	Charcoal	Potsherd, H1.	8220±25	7988±250	9215-8435	IA, CASS 1993.
		ZK-2644	"	K1	8274±23	8040±234	9251-8546	" "
		BK 84028	Animal bone	T6(3).	6800±80	6610±80	7530-7390	" "
Hujiawuchang		BK87045	Charcoal.	T102K7.	7190±140	6987±140	7916-7633	Chen and Hedges 1991.
		BK87047	" "	T102(5).	6960+100	6763±100	7644-7478	" "
		QxA 2222	Carbonized straw.	Potsherd III, T3(5):15.	6500±100	6316±100	7281-7092	Chen and Hedges 1994
		QxA 2223	Acid.	Potsherd III, T3(5):14.	6715±80	6525±80	7471-7292	" "
		QxA 2224	Charcoal.	Potsherd III, T3(5):14.	7385±80	7176±80	8058-7904	" "
		QxA 2225	Charcoal.	Potsherd III, T3(5):14.	1,020±10	10708±100	12750-12520	" "
		BK87046	Charcoal.	T2(4).	7210±110	7006±110	7912-7664	IA, CASS 1991.
		QxA 2218.	Carbonized straw.	Potsherd II, T3(4):55.	6395±90	6214±90	7208-7002	Chen and Hedges 1994.
		QxA 2219	Acid.	Potsherd II, T3(4):55.	6695±80	6506±90	7431-7280	" "
		QxA 2220	Charcoal.	Potsherd II, T3(4):55.	7590±80	7375±80	8240-8008	" "
		QxA 2221	Charcoal.	Potsherd II, T3(4):55.	1,190±10	10874±100	12906-12689	" "
		QxA 2733	Charcoal.	Potsherd IV.	6540±170	6355±170	7389-7029	" "
Zaoshi	111°12' 29°36'N.	BK82081	Charcoal	T43(5).	6920±200	6724±200	7685-7389	IA, CASS 1991.
		QxA 2731	Carbonized straw.	Potsherd I, T102(2):33.	6775±90	6583±90	7526-7384	" "
Gaomiao		BK92034		T1015(4)B.	6040±120	5869±120	6851-6528	Chen et al. 1996.
		BK92037		T0914(4)A	6310±140	6132±140	7180-6806	" "
		BK92036		T115(6)	5420±80	5267±80	6176-5930	" "
		BK92031		T0914(9)	5460±80	5306±80	6191-5946	" "
		BK92038		T1115(9)	6790±90	6598±90	7529-7385	" "
		BK92035		T1015(10)	6360±80	6180±80	7176-6941	" "
		BK92033		T1115(12)	6740±80	6549±80	7516-7335	" "
		BK92032		T2005(19)	6530±100	6345±100	7368-7171	" "

Table 7(2) Radiocarbon Dates of Archaeological Assemblages of the Early Holocene in the Yangzi valleys (continued).

****

				Stratum of	Radiocarbon dates (B.P.).			
Site	Location	Laboratory	Sample	the sample	H.L.5730	H.L.5568	Calibrated date	Referencs.
Hemudu	121'22' E	BK75057	wood	T16(4):13	6310±100	6130±100	7167-6878	IA, CASS 1991.
(the first stag	29'58'N.	BK78101	wood	T211(4)	6060±100	5890±100	6852-6574	" "
only)		BK78102	"	T211(4)	6040±100	5810±100	6738-6484	" "
		BK78103	grass.	"	5910±90	5740±90	6666-6416	IA, CASS 1991.
		BK78104	oak seed.	"	6310±170	6130±170	7206-6787	" "
		BK78109	organic stuff	T212(4)	6260±200	6080±200	7197-6724	" "
		BK78114	husk	T232(4)	6240±100	6060±100	7018-6787	" "
		BK78116	wood.	T214	6200+85	6020+85	6986-6753	" "
		PV-0047	oak seed	T21(4)	6260±130	6080+130	7160-6782	" "
		WP 7701	wood	T16(4)	5975±100	5810±100	6738-6484	" "
		ZK-0263	oak seed	T21(4)	6060±120	5890±120	6859-6559	" "
		PV-0028	wood.	T16(4)	5320±100	5170±100	6024-5762	
		ZK-0263(2	husk of rice	pottery in layer 4.	6085±100	5910±100	6859-6646	IA, CASS 1991.
		ZK-0590	wood	T16(4)	6200±85	6020±85	6986-6753	" "
Luojiajiao	120'24'E.	BK 80004	charcoal	Layer (4)H16	6400±100	6200±100	7206-6942	IA, CASS 1991.
	30'36' N.	ZK-0860	"	"	6260±130	6080±130	7160-6782	" "

Additional table 1. Radiocarbon Dates of some Archaeological Assemblages in South China.

Site	Location	Laboratory	Sample	Stratum of the sample	Radiocarbon dates (B.P.). Half-life 5	Half-life 5	alibrated date(B	Referencs.
Dalongtan	109'24' E	GSU-87-3	shells	T1.west(4)B	1,217±61	20617±612	n./a.	IA, CASS 1991.
	24'18'N.	GSU-87-3	"	T1.west(4)A	8,923±33	18388±331	n/a.	" "
		BK 82091	"	lower layer	3,330±25	22670±250	n/a.	" "
		PV379(2)	"	"	1,430±45	20430±450	n/a.	" "
		PV379(1)	"	"	8,030±30	18030±300	n/a.	" "
		PV402	human bone	upper section of the lower layer.	1,440±15	1,450±15	13549-13198	" "
		PV401	"	"	1,210±15	0,893±15	12966-12663	" "
		BK82090	shells	upper layer.	2,880±22	12516±220	15042-14336	" "
Bailandong	109°36'E	BK 82096	calcium layer	East (7)	1,670±15	1,340±15	13424-13090	IA, CASS 1991.
	24°54'N.	BK 82097	"	West(2)	9,910±18	19347±180	n/a.	" "
		BK 82101	"	West(10).	7,000±200	5,950±20	n/a.	" "
		PV -0455	"	East (2)	3,895±25	3,500±25	16499-15803	" "
Huangyandon	111'43'E.	ZK-0676	shells	cave entrance.	1,930±20	1,590±20	13767-13291	IA, CASS 1991.
	23'32'N.	ZK-0677	"	inside the cave	0,950±30	10,640±30	12866-12205	" "
Zengpiyan	110'15'E.	BK78038	calcium layer	cover cultural deposit.	6600±150	6410±150	7395-7176	IA, CASS 1991.
	25'18'N.	BK79308	shells.	T5 upper calcium layer.	8970±100	8720±100	9872-9626	" "
		BK79314	animal bone.	DT5 upper calcium layer.	9100±150	8840±150	9978-9592	" "
		ZK-0910	charcoal.	0.3cm under DT6 upper cal.	7680±140	7460±140	8369-8024	" "
		BK 79301	shells.	0.6m under the upper cal. la	0,320±14	0,030±14	11924-11002	" "
		ZK-0906	shells.	0.6m under the upper cal. la	0,675±15	0,370±15	12457-11979	IA, CASS 1991.
		BK79309	shells.	0.8m under the upper cal.lay	0,300±10	0,010±10	11769-11005	" "
		ZK-0907	shells.	1.1m under the upper cal.lay	0,780±33	0,480±33	12743-11821	" "
		BK79310	shells.	1.3m under the upper cal. la	0,270±16	9980±160	11841-10993	" "
		BK79316	shells.	1.7m under the upper cal.lay	0,090±10	9800±105	11007-10948	" "
		ZK-0908	shells.	1.8m under the upper cal.lay	1,055±23	0,740±23	12895-12423	" "
		ZK-0909	shells.	2.2 m under the upper cal.la	0,990±33	0,680±33	12931-12220	" "
		ZK-0280	bones.	DT5(3)	7580±410	7370±410	8505-7722	" "
		Zk-0280	shells.	"	1,310±18	0,990±18	13086-12730	" "

Additional table 1. Radiocarbon Dates of some Archaeological Assemblages in South China.

Site	Location	Laboratory	Sample	Stratum of the sample	Radiocarbon dates (B.P.). Half-life 5	Half-life 5	alibrated date(B	Referencs.
Nanning	108'20'E.	ZK-0284	shells	T2(3)	0,720±26	0,420±26	12618-11846	IA, CASS 1991.
	22'48'N.	ZK--0839	"	0.5m under the ground.	9985±200	9700±200	11005-10481	" "
		ZK-0840	"	1.0m under the ground.	0,155±20	9870±200	11593-10919	" "
		ZK-0841	"	1.1m under the ground.	9625±120	9350±120	10532-10150	" "
		ZK-0842	"	1.9m under the ground.	0,560±20	0,270±20	12397-11269	" "
		ZK-0843	"	2m under the ground.	0,735±20	0,430±20	12570-12007	" "
		ZK-0856	"	T2(2).	5155±300	5010±300	4223-3380	" "

Table 7(3). Lithic implements of Hujiawuchang.

No.	Type of tools.	Quantity	% of the total lithic assemblage.
1	Polished axes	16	7.41%
2.	Polished adzes	2	0.93%
3.	Polished knife	1	0.46%
4.	Polished chisels	3	1.39%
5.	Flaked discoid	15	6.94%
6.	Flaked pebble choppers	8	3.70%
7.	Flakes detached from pebbles.	87	40.28%
8.	Small flakes of chert.	67	31.02%
9.	Flake cores (chert)	17	7.87%
Total		216	100%

(After IA, Hunan Province 1993).

Table 7(4). Lithic implements found in Chengbeixi.

No.	Type of lithic implements	Quantity	% of the total lithic assemblage
1	Flaked pebble axes	13	61.91%
2	Fully polished axes	3	14.29%
3	Flaked chisel	1	4.76%
4	Flaked and partially polished adze	1	4.76%
5	Stone balls	2	9.52%
6	Net sinker	1	4.76%
	Total	21	100%

(After IA, Hubei Province 1996).

Table 7(5) Lithic implements of Sanshan.

No..	Type of lithic implement.	Quantity.	% of the total lithic assemblage.
1	Core.	250	24.53%
2.	Flake.	531	52.11%
3.	Flake "scraper".	192	18.84%
4.	Flake "point".	22	2.16%
5.	Flake "drill".	10	0.98%
6.	Flake "burin".	1	0.10%
7.	Pebble chopper.	13	1.28%
	Total	1019	100%

(After Chen *et al*. 1987).

Table 7(6). Stratigraphic consolidation at Xianrendong.

Trench No. 3 (dug in 1962)		Trench No. 3 (dug in 1993)		Trench No. 4 (the northern part originally under rock. Dug in 1		Trench No. 4 (the southern part, dug in 1964).		Trench No. 6 (dug in 1964)	
Zones	Unearthed artefacts	Zones	Unearthed artefacts	Zones	Unearthed artefacts	Zon	Unearthed artefacts	Zon	Unearthed artefacts
1A	Potsherds from the late								
	Neolithic to Shang periods.								
1B	Similar to 1A.								
						1A	Potsherds from the late	1	Potsherds from the late
							Neolithic to Shang periods.		
									Neolithic to Shang periods.
				1A/1	Shell-tempered sherds;	1B	Red potsherds (fine pottery,	2A	Red potsherds (fine, sand
					some bone/shell artefacts.		sand and shell-tempered);		and shell-tempered). Stone
				1B	Similar to 1A.		stone and organic implements.		and organic implements.
2	Red/grey potsherds- fine,							2B	Similar to 2A.
	or sand-tempered. Polished								(Roof-fall divided
	adze, perforated disks, bone								zone 2 from zone 3)
	and shell implements.			2A	Sand-tempered sherd, b	2	Red potsherds (red-painted		(Peng 1996).
					and shell artefacts.		and sand-tempered). Stone		
3A	Sand-tempered red potsherds			2B	Artefacts similar to 2A		and organic implements.		
	with cord-marking. Pebble				but less in quantity.		Faunal and mollusc remains.		
	choppers and discoids.			2C	33 sherds.				
	Chipped scrapers; polished			3A	Artefacts further reduce	3A	Red pottery (coarse sand-	3	2 stone implements;
	drills, chisels. Bone drills,				(the lower strata of No.4		tempered), hearth. Others		50 bones.
	needles, knife and harpoon.				were not excavated in 1993.		similar to zone 2.		
3B1	Similar to 3A, plus 3	3B1	4 sherds, 8 stone tools,			3B	Similar to 3A.		
	human skulls.		many bones. Phytoliths of						
			Oryza sativa (?).						

****** ******

Table 7(6). Stratigraphic consolidation at Xianrendong (continued).

Trench No. 3 (dug in 1962)		Trench No. 3 (dug in 1993)		Trench No. 4 (the northern part originally under rock. Dug in 1		Trench No. 4 (the southern part, dug in 1964).		Trench No. 6 (dug in 1964)	
Zones	Unearthed artefacts	Zones	Unearthed artefacts	Zones	Unearthed artefacts	Zon	Unearthed artefacts	Zon	Unearthed artefacts
3B2	Fired clay. Artefacts rare.	3B2	23 sherds, red, tempered with white quartz sand. 3 stone implements, many. bones.						
3C1	Faunal and human remains, artefacts.	3C1A	20 sherds; 4 stone implements.						
		3C1B	18 sherds; 8 stone implements; many bones; limited carbon.			4	Bone needles, harpoons, drills, stone implements, red pottery with coarse sand. Dense bones and ashy deposit.		
3C2	Artefacts more limited.	3C2	A few bones. no sherds.			5A	?	4	Bone arrowhead, chert flakes and cores.
		4A	4 stone implements, some bones; no sherds; little carbon.			5B	Bone drill and chisel, faunal remains.		
		4B	Initial deposit?			6A	A few animal remains.		
		5A				6B	Bed rock?		

Sources: CPAM of Jiangxi Province 1963; Jiangxi Museum 1976; MacNeish and Libby 1995 eds.

Table 7(7) Stratigraphic and cultural divisions of Xianrendong and Wangdong.

Cultural phases	Xianrendong.			Wangdong	C^{14} dates(bp)	Stratum where the. sample yielded	Estimated date (BP)
	Trench No. 3	Trench No. 4.	Trench No 6.				
Disturbed deposit	Zones 1A and 1B.	Zone 1A in the 1964 digging (the southern part of the trench)	Zone 1.	Zones A and B.			
Ceramic Phase II	Zone 2.	Zone 1B in the 1964 digging, and zones 1A-1B in the 1993 digging (the northern part of the trench).	Zone 2A and 2B.	Missing?	10,870±240 (ZK-39).	Bottom of zone 2 at Xianrendong.	7700 - 6000?
					8825±240 (ZK-92).	From zone 2 or zone 3 at Xianrendong?	
Ceramic Phase I.	Zones 3A to 3C1.	Zones 2A to 4.	Missing?	Zones C1 to E.	14,610±290 (BA93181).	Xianrendong zone 3B1.	13,000 - 8,000? .
					17,420±130 (AA15008)	Xianrendong zone 3C1B.	
Preceramic	Zones 3C2 to 4B.	Zones 5 to 6.	Zones 3 to 4.	Zones F to M.	17,040±270 (BA93180).	Wangdong zone G.	18,000 - 14,000 ?
					18,110±270 (BA93182).	Xianrendong zone 3C2.	
					19,770±360 (BA93179).	Wandong zone H.	

Sources: MacNeish 1995; Jiangxi Museum 1976; CPAM of Jiangxi Province 1963.

Note: The stratum from which sample ZK-92 collected is not clear. The original report only states that it was from the lower strata (ie. zone 3. Jiangxi Museum 1976). But the then collectors suspect that the sample might have been from an intrusive pit from zone 2 (cited by MacNeish and Taylor 1995: 83).

Table 7(8) Lithic implements discovered in zones F-M at Wangdong in 1993.

No.	Type of implement	Quantity	% of total lithic assemblage
1	Laterally worked flake	18	34.61%
2	Crude prismatic blade	2	3.85%
3	Chipped unifacial spokeshave	3	5.77%
4.	Chipped unifacial graver	3	5.77%
5.	Scraper plane-core.	3	5.77%
6.	Chipped semilunar knife	1	1.92%
7.	Microblade	2	3.85%
8.	Tongue-shaped core	1	1.92%
9	Small flake scraper	2	3.85%
10	Pebble chopper	6	11.54%
11	Pebble hammer	8	15.39%
12	Pebble semilunar knife	1	1.92%
13.	Chipped point	1	1.92%
14.	Ground flat point	1	1.92%
	Total	52	100%

(Typological names after MacNeish 1995).

Table 7(9) Lithic implements of the Ceramic Phase I at Xianrendong (1993 excavated).

No.	Type of implement	Quantity	% of the total lithic assemblage
1.	Laterally worked flake	2	5.26%
2	Scraper plane-core	3	7.90%
3	Pebble chopper	7	18.42%
4	Pebble hammer	13	34.22%
5	Pebble semilunar knife	2	5.26%
6.	Pierced pebble disk	1	2.63%
7	Flat pebble pecker	4	10.53%
8	Flat pebble adze	2	5.26%
9	Chipped pebble netsinker	1	2.63%
10	Ground flat point	1	2.63%
11	Ground wedge	1	2.63%
12	Ground pebble lance	1	2.63%
	Total	38	100%

(Typological names after MacNeish 1995).

Table 7(10) Lithic implements of Ceramic Phase I discovered in the 1960s at Xianrendong.

No.	Type of implement	Quantity (1962)	% of the total lithic assemblage	Quantity (1964)	% of the total lithic assemblage
1	Flake	92	69.69%	12	29.27%
2.	Flake cores.	4	3.03%		
3	Flake "scrapers".	6	4.55%	5	12.20%
4	Pebble choppers.	11	8.33%	5	12.20%
5	Discoid pebble implements.	3	2.27%		
6.	Perforated pebble disk	1	0.76%	8	19.51%
7	Pebble grinding stones	6	4.55%	2	4.87%
8	Ground chisel	1	0.76%		
9	Ground point/drill	8	6.06%		
10	Ground pebble "lance"?			9	21.95%
	Total	132	100%	41	100%

(After CPAM of Jiangxi Province 1963 and Jiangxi Museum 1976).

Table 7 (11) Lithic implements discovered in zones C-E at Wangdong.

No.	Type of implement	Quantity	% of the total lithic assembla.
1.	Laterally worked flake	2	2.35%
2	Crude prismatic blade	2	2.35%
3	Scraper plane-core.	11	12.94%
4	Chipped semilunar knife	3	3.53%
5	Microblade	2	2.35%
6	Small flake scraper	1	1.18%
7	Pebble chopper	17	20.00%
8	Pebble hammer	20	23.53%
9	Pebble semilunar knife	11	12.94%
10	Small pebble pecker	11	12.94%
11	Perforated pebble disk	1	1.18%
12	Flat pebble adze	1	1.18%
13.	Flat pebble trianguloid hoe	3	3.53%
	Total	85	100%

(Typological names after MacNeish 1995).

Table 7(12) Proportions of lithic types in the Xianrendong-Wangdong Ceramic Phase I assemblage.

Excavations	% of flake implements	% of pebble implements	% of polished implements
Xianrendong 1993	13.16%	78.95%	7.89%
Xianrendong 1964	41.47%	36.58%	21.95%
Xianrendong 1962	77.27%	15.91%	6.82%
Wangdong 1993	24.70%	75.30%	--

Table 7(13) Organic implements of Ceramic Phase I at Xianrendong.

Items	Quantity (1962 digging)	% of the total organic assemblage	Quantity (1964 digging)	% of the total organic assemblage
Bone				
awl	27	40.30%	27	31.40%
knife	1	1.49%		
needle	7	10.45%	2	2.32%
harpoon	1	1.49%	5	5.81%
arrowhead			3	3.49%
chisel			4	4.65%
Antler				
chisel	3	4.48%	17	19.77%
Shell				
perforated	24	35.82%	27	31.40%
denticulated	3	4.48%		
Tooth				
"knife"	1	1.49%	1	1.16%
Total	67	100%	86	100%

(Based upon CPAM of Jiangxi Province 1963; Jiangxi Museum 1976).

Table 7(14) Lithic implements of Ceramic Phase II discovered in Xianrendong.

No	Type of implement	Quantity 1962	% of the total lithic assembla.	Quantity 1964	% of the total lithic assembla.	Quantity 1993	% of the total lithic assembla.
1	Flake			3	37.50%	1	7.14%
2.	Flaked "scraper".			1	12.50%	1	7.14%
3	Pebble hammer					4	28.57%
4	Pebble semilunar "knife"					2	14.29%
5	Pebble "pecker"					4	28.58%
6.	Pierced pebble disk	4	66.67%	1	12.50%	1	7.14%
7	Pebble adze	2	33.33%			1	7.14%
8	Ground chisel			1	12.50%		
9	Ground spade (?)			1	12.50%		
10	Ground netsinker			1	12.50%		
	Total	6	100%	8	100%	14	100%

(Typological terms after CPAM of Jiangxi Province 1963; Jiangxi Museum 1976 and MacNeish 1995 respectively. 1995 data not available).

Table 7(15) Organic implements of Ceramic Phase II at Xianrendong.

Items	Quantity (1962)	% of the total organic assemblage	Quantity (1964)	% of the total organic assemblage
Bone				
awl	1	4.76%	7	20.59%
spear			1	2.94%
harpoon			1	2.94%
arrowhead	2	9.52%		
chisel	1	4.76%		
Antler				
chisel			1	2.94%
Shell				
perforated	16	76.20%	24	70.59%
arrowhead	1	4.76%		
Total	21	100%	34	100%

(sources: CPAM of Jiangxi Province 1963; Jiangxi Museum 1976).

Table 7(16) Lithic implements found in layer 4, Hemudu site.

No.	Lithic implements	Quantity	% of the total lithic assemblage.
1	Polished axes	180	57.51%
2	Polished adzes	31	9.90%
3	Polished chisels	39	12.46%
4	Chipped stone balls	3	0.96%
5	Chipped netsinker	1	0.32%
6	Pebble pestles	2	0.64%
7	Grinding stones	57	18.21%
	Total	313	100%

(After Liu and Yao 1993).

Table 7(17) Bone implements found in layer 4, Hemudu.

No.	Bone implements	Quantity	% of the total bone assemblage
1	Spades	130	7.23%
2	Arrowheads	1079	60.01%
3	Whistles	139	7.73%
4	Chisels	138	7.67%
5	Drills	133	7.40%
6	Needles	102	5.67%
7	Daggers	69	3.84%
8	Net shuttles	5	0.28%
9	Fishing spears	2	0.11%
10	"Sickle"	1	0.06%
	Total	1798	100%

(After Liu and Yao 1993).

Table 7(18) Wooden implements found in layer 4, Hemudu.

No.	Wooden implements	Quantity	% of the total wooden assemblage
1	Spades	2	1.39%
2	Handles for axe, adze etc.	41	28.47%
3	Projectile points	42	29.17%
4	Oars	8	5.56%
5	Pestles	2	1.39%
6	Thread-spindles for looms?	35	24.30%
7	Spindle whorl	1	0.69%
8	"Daggers"	8	5.56%
9	Small sticks.	5	3.47%
	Total	144	100%

(After Liu and Yao 1993).

Chapter 8. Discussion on the transition from foraging to farming in the Yangzi River basin.

The archaeological assemblages from the terminal Pleistocene to the early Holocene in the middle and lower Yangzi valley have been reviewed in chapter 7. These data provide information for the transition from hunting and gathering to cultivation in the Yangzi River basin from different aspects.

8A. Toolkits as indicators of the transition from foraging to farming in the Yangzi valley.

A salient characteristic of the toolkits found in the Yangzi valley is the highly stability from the middle Pleistocene to about 7000 BP, after which polished stone and bone tools dominated the toolkit. The earliest stone tools found in the vast area south of the Yellow River basin are pebble and flake tools. They are made from river pebbles primarily of sandstone and mica schist, flaked along the working edges, and much of the pebble surface is retained. The tool types of this lithic industry are limited and changed little over time. Choppers and chopping tools prevailed from the middle to the terminal Pleistocene, and lasted into the early Holocene, such as those found in Pengtoushan (IA, Hunan Province and Lixian Museum 1990) and Hujiawuchang (IA, Hunan Province 1993).

However, some new lithic implements did appear during the transitional period from the terminal Pleistocene to the early Holocene, namely, the small flakes primarily of chert and quartz, and the pebble discoid implements and perforated disks. A quantitative analysis is not possible as quantities for many assemblages are not available, and there is too much ambiguities in the existing classifications. But it is clear that the small flakes and scrapers are found both in the middle and lower Yangzi valley, while the pebble discoids are found in the middle valley, and the perforated pebble disks seem to have been found mainly in Xianrendong and South China.

The significance of the small flakes found in the archaeological assemblages of both the terminal Pleistocene and early Holocene is twofold: first, they may indicate cultural continuity; second, their presence may indicate changes in subsistence activities during that time.

Because of the small sizes of these flakes, in the 1960s it was suggested that they might be debitage (Peng 1992). But these flakes cannot be debitage from the pebble tools, as the majority of the pebbles used to make tools in the Yangzi River basin are of sandstone, mica schist, and quartzite, whereas the small flakes are mainly of chert, followed by quartz and quartzite. If they are debitage there should be stone implements made from the same materials. In addition, many small flakes bear use-wear, indicating that they were used (author's observation 1996). The crucial task is to find out the functions of these small flakes.

According to my harvesting experiment in 1996, a chert flake about 1.5 cm in dimension can be used to cut grasses, although the harvesting was not as efficient as that derived from up-rooting (chapter 5). Could these small flakes have been used to cut wild rice? This question can only be answered by use-wear and experiments in the future. But the co-existence of wild or domesticated rice remains and these small flakes, as in Pengtoushan, Chengbeixi, and Xianrendong, may not be pure coincidence.

As discussed in chapter 7, the discoid implements are also found in archaeological assemblages from the terminal Pleistocene to the early Holocene in the middle Yangzi valley. Yuan (1995) has suggested that this tool is a hallmark of the transition from foraging to farming. But still, the function of these discoids remains unclear.

Another tool of the transitional period in the lower Yangzi Valley is the perforated pebble disk. Based upon ethnographic data from the Dulong minorities in Yunnan province, Southwest China, it is argued that this type of tool could have been used for various purposes, from adding weight to digging sticks, to netsinkers (Song and Zhao 1994). Although the actual function of this tool can only be solved by use-wear and experiment, its presence in the terminal Pleistocene should be noticed.

Apart from lithic implements, organic implements such as bone arrowheads, harpoons and spears, as well as shell implements (e.g. those found in the Yuchan assemblage), are another important tool category. They were present from the terminal Pleistocene onwards. Their presence may indicate an increased efficiency of hunting, since even after the emergence of rice cultivation, hunting and gathering clearly did not cease. The discovery of these tools, in association with animal bones found in Fenshanbao and Hujiawuchang in the middle Yangzi valley, in Xianrendong, Hemudu and Luojiajiao in the lower Yangzi valley, clearly illustrates the importance of hunting and gathering among prehistoric cultivators.

In summary, even though pebble tools still dominated the lithic toolkit during the transition from foraging to farming in the Yangzi valley, the small flakes, discoids and perforated disks were apparently new comers. In addition, bone and shell tools for hunting and fishing also appeared during this transitional period. The appearance of these new tools may indicate the emergence of a subsistence strategy of hunting and gathering, including fishing and plant harvesting.

8B. The cultural relevance of the botanical characteristics of wild rice during the transition from foraging to farming.

As discussed in chapter 6, many scholars agree that common wild rice (*Oryza rufipogon*) is the wild progenitor of domesticated rice (*Oryza sativa*). As in green foxtail (*Setaria viridis*), the ripening season of common wild rice is quite heterochronous, ranging from over two weeks to sixty days (Sun 1990). Like green foxtail, common wild rice also shatters their ripe seeds. In addition, because perennial common wild rice grows in a permanently watered habitat, its shattered seeds often fall into water and are almost impossible to gather (author's observation 1995). This means that seeds can only be gathered before they are fully ripe.

What would have been the harvesting methods used by prehistoric gatherers? According to my harvesting experiment conducted in Sanshui county, Guangdong province in 1995, beating rice with a wooden rod into a basket is not efficient, as the panicles still contain considerable moisture, and the grains won't come off until they are fully ripe. What is more, some of those grains fell to the swamp and could not be collected. But when the whole plant was pulled and left inside a house for one or two days, the panicles dried up and all the ripe seeds shattered.

Another method is to cut the whole plant. As no replicate lithic tools were available when I conducted my harvesting experiment, I had to use an iron sickle. After 10 minutes, 21 plants bearing about 80 grams of rice grains were harvested. It took another fifteen minutes to beat the seeds off the panicles. Unfortunately, these were all the stands of common wild rice I could find in the field site.

Nevertheless, based upon this preliminary result, it can be calculated that about 480 grams of common wild rice could have been harvested within one hour of labour, or 3840 grams for an eight-hour day. This is a much higher return than that for green foxtail (*S. viridis*). An iron sickle may be a more efficient tool than a lithic flake; but according to Harlan (1967), a flint sickle set in a wood handle is about as efficient as a modern steel sickle (Harlan 1967).

Judging from available data, it seems that common wild rice has quite a high return in harvesting. It is said that the starch quality of common wild rice is similar to that of the domesticated one, and the taste of wild rice is even better (Qin and Li 1990). What is more, rice seeds, like many other seeds of the grass family, can be stored for a long time. Yan (1995) has suggested that this advantage would have been particularly attractive to the prehistoric people who lived in a temperate area, such as the Yangzi valley, where the winter would have been quite cold, and other food would not have been easy to find (Yan 1995, personal communication). The archaeological discoveries in Xianrendong and Yuchan in the Yangzi basin suggest that common wild rice might have been harvested in the terminal Pleistocene; and rice cultivation began by no later than 8000 BP at Bashidang and Pengtoushan.

According to an observation conducted in Hunan province, the flowering season of modern perennial wild rice there was from 10th September to mid-November (Sun 1990), lasting for about two months. The precise ripening season is not reported, but would be about one week after flowering (information given by a local farmer in 1996). Assuming the botanical characteristics of wild rice have changed little in the last 8000 years, the seasonal availability of wild rice from the terminal Pleistocene to the early Holocene would have been from late September to late November in the Yangzi valley. However, if harvesting was conducted by cutting or pulling the whole plant, the seasonal availability of wild rice would have been much shortened (similar to green foxtail, chapter 3).

As discussed in chapter 6B, common wild rice might not have been found in substantial quantity during the cold LGM era in the Yangzi valley, due to its limited cold tolerance. The phytolith analysis conducted in Xianrendong seems to support the same argument (Zhao 1998). This scarce might have triggered the local wild rice collectors to make extra efforts in order to increase the availability of wild rice by cultivation.

8C. Pottery comparison.

The earliest pottery found in the middle and lower Yangzi valley occurs in Yuchan and Xianrendong, estimated to be from 13,000 to 8000 BP. Both assemblages are in the limestone area, and the accuracy of the early dates is questionable. The pottery found in Pengtoushan is dated to about 8400-8000 BP [Table 7(2)], but the Pengtoushan pottery does not seem to be at the initial stage of pottery making (Yuan 1996). Thus, it is possible that the earliest pottery appeared at about 10,000 BP, or slightly earlier.

The pottery vessels found in the Yangzi valley before 7000 BP were all made without use of a wheel. Slab-modelling prevailed in the middle Yangzi valley, but coiling appeared at about 7300 BP in Chengbeixi, and later in other areas (He 1995a). In the lower Yangzi valley, coiling dominated in the Xianrendong assemblage, but slab-modelling appeared in Hemudu. The walls of these early vessels are usually over 1 cm thick; vessel shapes are irregular, and firing temperatures low [Table 8(1)]. Cord-marking was the prevailing decoration on all pottery in the Yangzi valley until about 7000 BP. Various materials were used for tempers [Table 8(1)], and different temper materials may provide some information relating to the economic activities of the vessel makers. For example, rice remains in pottery temper must indicate the practice of rice harvesting, although they do not indicate clearly whether the rice was domesticated.

In the middle Yangzi valley, the development of pottery seems relatively clear. Plant remains and coarse sand were used as tempers in this area from the beginning at Yuchan (Yuan 1996). Both tempers were still used at 8000-7500 BP, but by this time the rice temper dominated in the Pengtoushan and Fenshanbao assemblages [Table 8(1)]. Shell-temper was used in the Chengbeixi culture, Ceramic Phase II at Xianrendong, and the Luojiajiao assemblage. This may indicate that shell collection was an important activity in these groups.

The pottery sequence seems much less clear in the lower Yangzi valley than in the middle. While the early pottery in Xianrendong is brownish-red in colour, and tempered by coarse crushed quartzite, the pottery found in Hemudu is black, with rice husks and sand tempers (Liu and Yao 1993). Hill (1995) states that the shell-tempered pottery found in Ceramic Phase II at Xianrendong is similar to that found in Hemudu (Hill 1995: 40). I have searched the excavation reports of Hemudu, but cannot find any shell-tempered pottery being mentioned. In fact, the pottery of Ceramic Phase II in Xianrendong may be more comparable to the Luojiajiao pottery, which was greyish red and black, with sand, shell and rice husk tempers (The Archaeology Team of Luojiajiao 1981).

In terms of vessel form, the dominant vessel in pottery assemblages in the Yangzi valley up to 7000 BP is the cooking *fu*, which has been discovered in the middle and the lower Yangzi valley. With its round bottom and deep belly, plus the burned rice found in one *fu* at Hemudu (Liu and Yao 1993), there seems little doubt that this vessel was used for rice-cooking. The earliest *fu* are found in Xianrendong and Pengtoushan. In Fenshanbao, Hemudu and Luojiajiao this vessel type still prevailed, and its shape and decoration further diversified [Fig. 7(7) and 7(18)]. Bowls and *dou* are also found in both the middle and lower Yangzi valley at around 7500-7000 BP, but those found in the middle Yangzi valley seem more elaborately decorated [Fig. 7(7), 7(18)]. Globular jar with two lugs is widely found in the middle Yangzi valley, but absent in the lower valley.

In general, the cord-marking, the cooking *fu*, bowl and *dou* are all common to the middle and lower Yangzi valley. There are also similarities in terms of pottery temper and colour between the Neolithic in the middle and lower Yangzi valley. Pile-dwellings seem to have been another common cultural component in the lowland areas of both the middle and the lower Yangzi valley (i.e. those found in Bashidang and Fenshanbao, Hemudu and Luojiajiao). These similarities may suggest cultural connections (more discussions in chapter 10 and 11). On the other hand, the Xianrendong assemblage seems different in many aspects from the other assemblages in the Yangzi valley, and its cultural relationships with the others remain to be studied.

Summary.

In chapters 6 and 7, the environmental and archaeological

data from the middle and lower Yangzi valley have been reviewed. Compared to Central China, it seems that the impact of the LGM was less significant here, probably due to the lower latitude. Although evergreen trees seem to have withdrawn from the Yangzi valley during the LGM, some deciduous still existed, and there is little evidence for any significant extinction of animals. According to pollen and faunal data, average temperature during the LGM might have been 4-6° C lower than at the present, with reduced moisture. But compared to the palaeoclimate in Central China, the situation in the Yangzi valley would have been less severe.

After the LGM the palaeoclimate seems to have improved quite significantly, although with fluctuations. Temperatures rose, precipitation increased, and sea levels became higher. The warm and moist early Holocene supported many animal and plant species in the Yangzi valley, thus providing food resources of great diversity. It is possible that after 10,000 BP common wild rice expanded into many areas of the Yangzi River valley, as its remains have been found in Hemudu at 7000 BP (Chen and Huang 1994).

The comparatively stable environment in the Yangzi valley might have contributed to the stability of the toolkit in this region. Pebble tools predominated this area until replaced by polished tools after approximately 7500 BP. The stability of the toolkit, as well as the animal species found in association, may indicate that hunting techniques and game for prehistoric people in the Yangzi valley changed little. But some new tools did appear in the terminal Pleistocene, namely the small flakes made of chert and quartzite, the discoids and perforated disks. If the latter were within the pebble tool tradition, the former were completely separate from the general characteristics of the pebble industry. Except for the perforated disks, the other two flourished into the early Holocene, particularly in the middle Yangzi valley. Although the functions of these tools are not clear, they appear to be marks of the transition from the terminal Pleistocene to the early Holocene.

All the three new tool categories are found in association with rice remains, either wild or domesticated, or both. The presence of rice remains suggests that rice harvesting might have begun in the terminal Pleistocene, and rice cultivation began by 8000 BP (Zhang and Peng 1996). The development from collection to cultivation in the middle Yangzi valley seems relatively clear, demonstrated by the wild rice found in Yuchan, and the domesticated rice remains from Bashidang and Pengtoushan. But in the lower Yangzi valley there are still questions.

Table 8 (1) Pottery Comparison

	Middle Yangzi valley					Lower Yangzi valley.			
Assemblage	Pottery colour and temper	Building method	Firing temperate.	Date	Assemblage	Pottery colour and temper	Building method	Firing Temperature.	Date
Chengbeixi	Red or black, with rice remains, sand, shell and crushed sherds.	Slab modelling. Coiling appeared.	n/a.	7500 to 7000 BP.	Luojiajiao- layer 4.	Greyish red and black, with sand, crushed shell and rice charcoal.	Coiling.	800-850°c	7170-6890 BP.
Fenshanbao	Black or grey, with rice husks. White and painted pottery appears.	Slab model-ling.	n/a	7500 to 7000 BP.	Hemudu-layer 4	Black, with rice charcoal and sand.	Coiling and slab-modelling.	800-850°C (charcoal) 880-930°C (sand).	7000-6500 BP.
Pengtoushan	Black or grey, with rice husks and other organic materials.	Pinching and slab modelling	n/a.	8000-8500 BP.	Xianrendong-middle ceramic	Brown/yellow-ish red, with shell and calcined quartzite.	Coiling, thinning by paddle and anvil.	920°C.	7700-6000 BP.
Yuchan	Brownish black with coarse sand and charcoal	Slab-modelling.	n/a.	15,000 to 12,000 BP?	Xianrendong-early ceramic	Brownish red with crushed quartzite.	Coiling, thinning by hand/ grass.	600-700°C	13,000 to 8000 BP?

Sources: Li *et al*. 1996. Hill 1995; Yuan 1996; Peng 1996.

Part IV. Comparison and Proposition: The Transition from Foraging to Farming in China.

The transition from foraging to farming was a development that occurred between about 10,000 and 5000 years ago in several regions of the Old and New Worlds (Golson 1989; Smith 1995) (chapter 1A). Apparently, the domestication of plants and animals in China was by no means an isolated process in prehistory, but part of a worldwide trend which occurred in the early Holocene.

The origins of agriculture involved the domestication of different plants and animals in different centres. The nature of the transition from foraging to farming was not the same in each of these centres, but there might still have been some common characteristics which they all shared. The possible existence of such common characteristics is assumed on the ground that, although different human groups have been living in a vast variety of environments, as a single species they might have shared biologically and socially common characteristics.

Given this possible commonness, the origins of agriculture in China can be better understood by comparing toolkits in China with those in other areas where agriculture independently emerged; by referring to contemporary and comparative ethnographic data on foragers; and by considering theoretical frameworks for the origins of agriculture based upon data from other agriculture centres. Through such comparisons, the research achievements in other areas may stimulate thought on this topic in the Chinese context.

The comparisons of archaeological assemblages undertaken here, however, will focus on the Middle East and China, for two reasons. First, the Middle East is the earliest and best-studied centre for agricultural origins, with the richest and most continuous archaeological data from the terminal Pleistocene to the Holocene, thus providing a sound database for comparison. Second, the date of agricultural emergence in the Middle East was similar to, but slightly earlier than that in China, being occurred in the late Natufian and PPNA period between about 13,000 and 11,000 BP. This temporal similarity may reflect the importance of postglacial palaeoclimate change.

On the other hand, ethnographic data on the subsistence strategies of hunters and gatherers are best-documented in Australia and the Americas. Further, it has been demonstrated that plant exploitation was practised in Australia by 19,000 BP (Furby *et al.* 1991). Therefore, ethnographic and archaeological data from Australia will be drawn upon, as well as relevant data from the Americas. A hypothesis for the transition from foraging to farming in China will be proposed later.

Chapter 9. Foragers of the terminal Pleistocene in China.

It is essential to understand the forager lifestyle prior to the emergence of the first farmers in China, in order to understand the transition from foraging to farming. As archaeological and archaeobotanical data in China are meagre, archaeological and ethnographic data from other continents are used here as a comparative database. It is realized that environments and material cultures differ significantly between China and other areas, and the risk of this broad-scale comparison has been considered. However, the major purpose of this comparison is to suggest and/or increase possible interpretation(s) for the functions of lithic implements and other archaeological remains found in China.

9A. The subsistence strategies of the foragers - a toolkit comparison.

It seems that the LGM strongly influenced the valleys of the two great rivers, particularly the Yellow River. During the LGM, temperatures dropped significantly, moisture was reduced, lakes and rivers shrank, broad-leafed trees withdrew, and some animal species became extinct. The LGM must have directly affected the floral and faunal resources which the foragers of that time would have relied upon. As deciduous trees retreated in favour of conifers, the quantity and variety of nuts and fruits from these trees would have been reduced or even have disappeared. The extinction or habitat shifting of various rodents, deer and small carnivores would have meant a decrease or disappearance of easier game [Table 3(1)]. During the LGM, coniferous trees and drought-resistant grasses dominated the landscape, while fast-running ungulates prevailed in the fauna.

The LGM caused significant changes in natural resources in Central China. Consequently, foragers would have had to change their strategies of food procurement. In theory, such change could be evidenced in faunal and floral remains in archaeological assemblages, and residues on tools. Unfortunately archaeobotanical and archaeozoologic data are scanty in Central China, and residue analysis is yet to be conducted in China. Thus the toolkit, which is really secondary evidence, becomes the primary database for studying changes in subsistence strategy.

This change is primarily illustrated by the new implements which make an appearance by about 18,000 BP, represented by the Xiachuan complex in the middle Yellow valley. The Xiachuan assemblage contained microblades, points, axes, and grinding tools (slabs and handstones), as well as numerous types of scrapers. As discussed in Chapter 5, it is inferred that these tools were used for extensive foraging activities. Unfortunately, the functions of these tools in China can only be inferred, based upon their morphological appearances; but archaeological and/or ethnographic parallels from the Middle East and Australia can be instructive.

In the Middle East, the LGM epoch is defined as from 22,000 to about 16,000 BP; there were climatic fluctuations from 16,000 to 10,500 BP, and the Holocene began with conditions close to those of the present at around 10,000 BP (Bar-Yosef 1990). The period between 20,000 and 10,000 years ago witnessed the spread of several archaeological complexes, in temporal sequence from the Kebaran, through the Geometric Kebaran and the Mushabian to the Natufian (Bar-Yosef and Belfer-Cohen 1989; Byrd 1994). The Kebaran complex dominated the period from around 20,000 to 14,000 BP; the Geometric Kebaran from 14,000 to 12,800 BP; the Mushabian probably from 14,000 to the

end of the 13th millennium, and the Natufian from 12,800/500 to 10,500 BP (Bar-Yosef and Belfer-Cohen 1989).

Generally speaking, archaeobotanical evidence from the Kebaran, Geometric Kebaran and Mushabian complexes is scanty (Bar-Yosef 1990), and inferences about the subsistence strategies of these complexes are mainly based upon toolkit interpretation. The major components of the lithic toolkits of these complexes are microliths in various geometric shapes, small bladelets detached from single platform cores, many of them retouched and backed, flaked scrapers, burins, truncations, notches, denticulates, and pounding/grinding tools (Bar-Yosef and Belfer-Cohen 1989; Edwards *et al.* 1996; Garrard *et al.* 1985, 1987, 1992, 1994; Ferring 1976; Gilead 1991; Marks 1976a, 1976b; Nadel and Hershkovitz 1991) [Fig. 9(1)].

It has been demonstrated by use-wear and hafting studies that the microliths and bladelets in the Levant were used for both hunting and cutting. Byrd (1989b) has proposed that some bladelets ('non-geometric microliths' as he calls them) from the Natufian assemblage at Beidha could have been used for cereal cutting, particularly when they exhibit 'sickle polish' (Byrd 1989b). In his harvesting experiment in 1967, Harlan (1967) constructed a composite sickle by inserting a few blades into a wooden haft. He claimed that such a lithic sickle was efficient for cereal cutting (Harlan 1967). According to a use-wear analysis and harvesting experiments, the sickle-blades found in the terminal Pleistocene in the Middle East were used for plant harvesting (Unger-Hamilton 1985).

The function of flaked scrapers is commonly interpreted as processing animal hides, while burins would have been used for wood, antler or bone working (Byrd 1989b). Regarding the grinding slabs, this tool was first present in the Upper Palaeolithic in Qafza Cave (Gilead 1991), dated to about 90,000 BP (Bar-Yosef and Kra 1994); and are still found in the PPNB (Pre-Pottery Neolithic B) at Beidha (Wright 1993). On the other hand, basalt mortars and pestles first appeared during the Kebaran and flourished in the Natufian period (Wright 1991). These grinding or pounding implements might have been used for multiple purposes, but it has been argued that one of the tasks associated with them would have been seed processing (Wright 1991).

The functions of grinding and pounding tools are further illustrated by a recent archaeological discovery. Remains of wild barley, wild wheat, pistachio and other nuts, together with another 30 species of plants, were discovered in a submerged Kebaran assemblage at Ohalo II, on the shore of the Sea of Galilee, Israel in 1989 and 1990 (Kislev *et al.* 1992). This discovery demonstrates that wild cereal collection began at as early as 19,000 BP (Kislev *et al.* 1992). It is also argued that wild cereals were collected in spring, and fruits were collected in autumn at Ohalo II, indicating that 'the site was occupied for more than six months at a time' (Nadel and Hershkovitz 1991: 633).

The toolkit of Ohalo II was dominated by bladelets from single-platform cores, accompanied by burins, end scrapers, awls, notches, retouched flakes and blades, broken basalt bowls and pestles (Nadel and Hershkovitz 1991). Faunal remains include fish, birds, reptiles, rodents, hare, fox, gazelle and deer (Nadel and Hershkovitz 1991). The presence of basalt bowl and pestle fragments in the Ohalo II, in association with the remains of wild barley and wheat, illustrates that there was a close link between the grinding/pounding tools and wild cereal exploitation.

The Ohalo II discovery clearly demonstrates the practice of a broad-spectrum strategy of hunting and gathering during the late Pleistocene in Israel. Taking the Ohalo II assemblage as a reference, it is now more convincing to argue that the exploitation of wild cereals was practised in the Middle East as early as in the LGM. Similar toolkits, particularly the specialised grinding/pounding tools, have been found in many Kebaran, Geometric Kebaran and Natufian places (eg. Bar-Yosef 1990; Gilead 1991). The discovery of wild barley, a variety of legumes and nuts from Wadi Hammeh 27 and Hayonim Cave, both early Natufian settlements (Byrd 1989a), documents the continuity of wild cereal exploitation from the Kebaran into the Natufian period.

Although the typology and morphology of Kebaran lithic implements are not exactly the same as those of the Xiachuan complex in China, there are many points of overlap. The Kebaran bladelets struck from single-platform cores are morphologically similar to the microblades in China, and both seem to have been used as the cutting edges of composite tools for hunting and gathering (chapter 4; Bar-Yosef 1990). Both the Kebaran and the Xiachuan assemblages contain grinding tools, presumably for seed processing. Both had a variety of flaked scrapers and burins [Figs. 4(3)-(5); Fig. 9(1)]. These similarities in terms of toolkits may indicate that the subsistence strategies of the two assemblages were comparable, although the species being hunted and gathered in the Middle East and in China were obviously different.

As discussed in chapter 3, various species of herbs dominated the vegetation during the LGM in the Yellow valley, including species with edible seeds from the Graminea and Chenopodiaceae families. The presence of grinding slabs, and some blades with glossed edges in the Xiachuan assemblages (author's observation 1996), suggests that wild cereals were exploited, at least in the Xiachuan basin. Animal remains were also found in association with the Xiachuan assemblage and other archaeological assemblages in the Shanxi Basin, indicating hunting activities focusing on ungulates (chapters 3 and 4).

The functions of grinding slabs are also demonstrated by ethnographic data. Grinding slabs are widely found in Australia in both archaeological and contemporary ethnographic assemblages. The Australian Aborigines still collect seeds of various wild grasses and other plants, and grind them before consumption (eg. Allen 1974; Cane 1984; Gould 1969; Jones and Meehan 1989; O'Connell *et al.* 1981; O'Connell *et al.* 1983; Tindale 1977). In contemporary Australia, grinding slabs are made of a large stone block, usually trimmed into an oval or oblong shape (Cane 1984; Smith 1985, 1986). Surfaces of the grinding slabs usually bear one or more shallow grooves, caused by wet-grinding (Cane 1984; Smith 1985, 1986) [Fig. 9 (2)]. But it has been reported that dry-grinding was also practised, and in these cases the use-wear is generally in the form of a circular depression (Cane 1984). As discussed in chapter 4, similar use-wear patterns of both grooved and circular depressions are found on the surfaces of grinding slabs in the Xiachuan assemblages in China [Figs. 4 (3) and 4(5)].

Seeds collected in Australia can be categorised into two groups in terms of their husks: the 'soft' seeds, which include seeds from the grass family [such as millet (*Panicum* spp.) and rice (*Oryza* spp.)], *Eucalyptus* sp., herbs and succulents; and the 'hard' seeds, such as various species of the Acacia family (O'Connell *et al.* 1983). The processing methods for the 'soft' and 'hard' seeds are different. The soft seeds were hulled either by rubbing between the hands (Cane 1984: 77), or by treading in a circular or square hole in the ground (Tindale 1977:346). The seeds then were winnowed, sometimes soaked; and finally ground with water on grinding slabs. On the other

hand, hard seeds, such as *Acacia coriacea*, *A. victoriae* and *Brachychiton gregorii*, were first pounded to crush their coats with a mortar and pestle, then dry-ground on grinding slabs (Cane 1984: 140; Smith 1985: 29).

According to archaeological data, the earliest grinding slabs found in Australia so far are from Cuddie Springs, dated from 28,310±200 to 19,270±320 bp (Furby *et al.* 1993). Although fragmentary, the use-wear on these grinding slabs is said to be in the form of a circular depression rather than shallow grooves (Furby *et al.* 1993). Starch and plant tissue have been found on the surfaces of grinding slabs from Cuddie Springs and Puntutjarpa Rockshelter, suggesting that these implements were tools for plant/seed processing (Furby *et al.* 1993; Veth *et al.* 1997). It is apparent that seeds were exploited during the terminal Pleistocene in Australia, and the use-wear pattern seems to suggest a dry-grinding technique.

Grinding tools have also been found in Africa from 48,900 BP; in Europe from 44,000 BP (Kraybill 1977); and in America from about 10,000 BP (Carter 1980). The emergence of grinding slabs during the late Pleistocene in several continents is an important issue. Although not exactly in the same form, and bearing different use-wear patterns, the presence of this tool seems to indicate world-wide technical progress in terms of plant exploitation. Further, archaeological data in the Middle East and Australia, and the ethnographic data from the America and Australia suggest that grinding slabs were often associated with intensive plant exploitation (including seeds from various grasses). Whether this observation is valid regarding the Xiachuan-like assemblages in China remains to be investigated, but as discussed in chapter 4, it is a very likely possibility.

Compared with the toolkit in the Yellow valley, the lithic implements found in the Yangzi sites bear little similarity to the tools found in the Middle East. As discussed in chapter 6, the LGM might have had less impact in the Yangzi valley than in the Yellow, particularly on the fauna. Judging from the animal remains found in Yangzi assemblages, the fauna in the terminal Pleistocene changed little (chapter 6). However, there were still changes, particularly in terms of flora, as some broad-leafed trees retreated from the region and drought-resistant grasses moved in during the LGM.

As discussed in chapter 8, a few new lithic forms appeared during the terminal Pleistocene in the Yangzi valley, especially discoids and small flakes. The functions of these new lithic implements are not clear, but they might reflect changes in some aspects of subsistence strategy.

Recent discoveries of wild rice remains suggest that wild rice was collected during the terminal Pleistocene in the Yangzi valley, but there appear to be no specialised tools for cereal processing. How can one interpret this? One possibility is that rice could have been processed by hand or foot. Rice is a member of the grass family, and its husk is soft. As mentioned above, soft-husked seeds were dehusked by hand or foot by the Aborigines in Australia (Cane 1984; Tindale 1977). It is possible that similar methods of dehusking were practised in the prehistoric Yangzi valley.

In Australia, the grain heads of wild rice (*Oryza* spp.) were 'collected into a wooden container, then winnowed so that the husks were blown away' by the Aborigines (Jones and Meehan 1989: 128). The dehusking method is not mentioned, but it is reported that the dehusked grains were ground into flour (Jones and Meehan 1989). However, rice in Neolithic China was processed differently. Cooked rice was found in a *fu* in Hemudu (Liu and Yao 1993), indicating that rice was cooked in whole grain form rather than ground into flour (rice is pulverised today but only for making desert and snacks). Therefore, grinding tools might not have been essential. If foragers of the terminal Pleistocene consumed rice in a similar manner, they probably did not need grinding tools at all.

Another possibility is that seed processing tools were used, but were made of organic materials in the Yangzi valley and did not survive. In Hemudu, two wooden pestles were found. According to Chen (1983), two circular holes in the ground were found in the Honghuatao site in Hubei province, middle Yangzi valley, associated with remains of a wooden pestle (Chen 1983). Details for this Honghuatao assemblage are not yet published, but it is reported that domesticated rice was found there, and the assemblage is suggested to be about 6000 BP or slightly earlier (IA, CASS 1984). These discoveries suggest that rice dehusking was conducted by ground holes and wooden pestles in Neolithic Yangzi valley.

Did foragers of the terminal Pleistocene also use wooden pestles? This possibility cannot be ruled out. Whatever the possibility, there is little doubt that wild rice was collected during the terminal Pleistocene. And according to an isotopic analysis, the diet of the Xianrendong occupant was very high in plant food (MacNeish *et al.* 1995). The discoveries of wild rice in the Yuchan and Xianrendong assemblages, as well as the isotopic analysis conducted, attests the importance of archaeobotanical search and other scientific approaches, as surviving archaeological toolkits are more likely to be an incomplete, sometimes biased indicator of the economic activities of prehistoric foragers.

Were the wild cereals ever staple foods in terminal Pleistocene China? The toolkits provide limited information for this question. Smith (1986) argues that if seed grinding was an important activity in an archaeological site, the quantity of grinding slabs should be considerable (Smith 1986: 34). It is not clear how to define 'considerable' in terms of quantity, but in the Xiachuan assemblage, a total of 34 grinding slabs and hand stones were found during the 1970s, only accounting for 0.67% of the total of lithic implements. No grinding tools were found in Xueguan. Grinding tools in the Shizitan assemblage only accounted for 0.16% of the total lithic implements (chapter 4). These percentages are very low. In the Yangzi valley, no grinding tools were found during the terminal Pleistocene at all.

Given these data, can one argue, prima facie, for only a minor importance of seed gathering in these Chinese assemblages? Caution must be taken here. The first factor to be considered regarding the low incidence of grinding tools in China is the availability of raw materials and the mobility of foragers. According to Australian ethnographic data, if raw materials for grinding tools are scarce in one region, the remains of these tools found in archaeological contexts can be rare (Cane 1984; Smith 1986). Furthermore, grinding tools are heavy implements, and Australian Aborigines tended to leave them in their base-camps (Cane 1984). In other words, if the discovered archaeological site is only a seasonal and/or temporary camp for tasks other than cereal processing, the chance of finding substantial grinding tools will be quite low. However, the brief reports on the Xiachuan assemblage do not provide information on either the availability of raw materials for grinding tools, or the activities carried out in each site, so these two issues await further investigation.

The second factor regarding incidence of grinding slabs is the differences in seed processing techniques. In Australia, Aborigines exploited numerous species of plants. In Western Desert Australia, Cane (1984) recorded that 'edible seeds were collected from 42 plants', and 102 plants were

used for different purposes (Cane 1984:70). In Cape York Peninsula, 37 plants with edible parts were exploited (Hynes and Chase 1982). At least 92 plants were exploited for different parts and among them, edible seeds were collected from 36 species by the Alyawara tribes in Central Australia (O'Connell *et al.* 1983). All these seeds were first dehusked, some were soaked or toasted, then milled. These complicated processes demanded long labour hours. Cane (1984) records that 67 minutes were spent in grinding 500 g millet (*Panicum* sp.) seeds (Cane 1984:80). The long hours spent on processing seeds, and the different grinding techniques for different seeds (i.e. wet-grinding for soft seeds and dry-grinding for hard seeds), together result in an intensive use of grinding and pounding tools. It is reported that in the Western Desert, every married Aboriginal woman has her own grinding tools (Cane 1984:138).

But the situation for prehistoric foragers in China might have been different. It is not clear how many species of herbs with edible seeds were exploited during the terminal Pleistocene. But if some of the seeds from species like green foxtail, wild broomcorn millet (*Panicum* sp.) and wild rice were only dehusked by hand or foot, or holes in the ground and wooden pestles (as those found in Honghuatao, above), then cooked without further grinding, few labour hours would have been spent on seed processing and fewer, or even no grinding tools would have been needed.

Neolithic discoveries in China also seem to support this assumption. According to Wu (1986), the proportions of grinding tools within stone toolkits in early Neolithic Central China ranged from 10.4% to 28%, but declined to 1.4% -7.7% in the Yangshao culture (Wu 1986:52). The incidence of mortars and pestles found in Yangshao assemblages is also quite low (Wu 1986). Yet agriculture was certainly well developed in these societies, and seed processing must have been important. This phenomenon has been interpreted as caused by possibly use of wooden pestles, and by the way of cereal consumption in China. It is proposed that millets were probably cooked in whole grains in mid-Neolithic Central China, as indicated by the widespread of pottery 'steamers' by 6000 BP (IA, CASS and Banpo Museum 1963).

In summary, the availability of raw materials for grinding tools, discard behaviour, and more importantly, different techniques of seed processing, are factors that would have affected the incidence of grinding tools. The existence and/or importance of cereal/plant exploitation within an archaeological assemblage is better defined by integrating data from use-wear and residue analysis of stone tools, pollen analysis and studies of any botanical remains, as well as grinding tools.

9B. The population density of prehistoric foragers in China.

On this issue, archaeological data are very meagre in China. Many archaeological assemblages of the terminal Pleistocene are not absolutely dated. And in the Yangzi valley, there is the concern about inaccurate radiocarbon dates affected by the limestone environment. Consequently, it is difficult to define the precise chronological sequence and contemporaneity of archaeological assemblages within any one area. Probably only in the western plateau and mountain area north of the Yellow River can archaeological discoveries provide some clues on population issues.

In the western plateau /mountain area covering about 170,100 square kms, over 40 sites and find places are suggested to have belonged to the late Palaeolithic (Chen 1993) [Fig. 9(3)]. The number of archaeological sites of the late Palaeolithic is greater than that of the middle Palaeolithic; the latter totaled only 14 sites (Chen 1993). This increased incidence of sites may suggest that population density during the late Palaeolithic (35,000 BP onwards in the Chinese context) had increased.

Given the archaeological data from this plateau and mountain area to date, there were at least two different lithic complexes during the terminal Pleistocene: one was dominated by microblades, the flake tradition and some heavy duty tools, represented by the Xiachuan assemblages; another was dominated mainly by flakes, represented by the Xibaimaying and Tashuihe assemblages (Chen 1993). These two lithic complexes differed significantly, as microblades and heavy duty tools, particularly grinding tools, were completely absent in the latter. Therefore, it is likely that the two complexes were created by different groups with different material cultures. The continuous co-existence and the closeness of the two lithic complexes (e.g. the distance between Xiachuan and Tashuihe is about 55 km) in the basin seems, prima facie, to indicate a dense population. But it is still difficult to estimate the number of forager groups in this plateau/mountain area at the terminal Pleistocene due to various uncertainties of chronology and site correlation.

Faunal remains have been found in many sites, and a few species are temporally diagnostic in a broader sense, particularly the woolly rhinoceros (*Coelodonta antiquitatis*), which is considered adaptive to a cold climate (Zhou 1978). Yet the time span of this animal is too long to be of precise value for archaeological cross-dating. For example, both the Tashuihe rockshelter and Beiyaowan cave in Shanxi province yielded remains of this animal, the former radiocarbon dated to 26,165 and 20, 000 bp, the latter to 28,730±1175 and 31,495±1570 bp (Chen 1993). If these dates are valid, woolly rhinoceros also existed in the area before the beginning of the LGM.

In summary, it can be said that the population density of the western plateau and mountain area in Central China during the terminal Pleistocene might have been denser than that of the previous period, but details of the demographic pattern are not known at this stage.

However, it is notable that probably after the LGM, similar lithic assemblages to Xiachuan have been found along the Yellow River, represented by Shayuan in the upper valley, Lingjing in the middle valley, and more than one hundred find places in the lower valley (chapter 4). Although none of these discoveries have absolute dates, it has been suggested that their approximate dates would have been later than the Xiachuan and Xueguan assemblages. This argument is based upon morphological comparison of their toolkits and associated animal remains (chapter 4). The archaeological data seem to suggest that the Xiachuan-Xueguan culture had dispersed to other regions along the Yellow River by the end of the Pleistocene. The wide spread of the microblade industry in the lower Yellow valley may also suggest an increase of population during the terminal Pleistocene.

As to the other areas concerned in this thesis, very few data are available. There is clear evidence of human occupation to the north and west of the North China Plain during the LGM, represented by the Tingsijian, Mengjiaquan and Xiaonanhai assemblages respectively. But no archaeological remains have been found in other parts of Central China [Fig. 4(1)]. There may be two possibilities: either there was human occupation during the terminal Pleistocene but remains have been destroyed or not found; or the other regions were only sparsely peopled.

The hunters and gatherer in the North China Plain during the LGM represented by the Xiaonanhai assemblage might

have had a different subsistence strategy from the Xiachuan groups. The Xiaonanhai toolkit differed significantly from the Xiachuan one, although both were roughly contemporary [Table 4(1)]. No microblades, axes or adzes, or grinding tools were found in Xiaonanhai (chapter 4). Although Xiaonanhai cave is located within the Cishan-Peiligang cultural region, and the late phase of the Xiaonanhai assemblage is dated to 11,000 bp, only about four thousand years prior to the Cishan-Peiligang assemblages, the lithic implements of Xiaonanhai and those of Cishan-Peiligang shared little in common. There is no evidence for close cultural links between the former and the latter. Given the toolkit and associated animal remains, the Xiaonanhai occupants might have been traditional hunters and gatherers. Whether this group of people disappeared after the emergence of agriculture in Central China, or their descendants also became agriculturists, remains a question.

The population issue in the Yangzi valley is even more problematic, caused by a lack of archaeological data with absolute dates, worsened by the inherent uncertainty of radiocarbon dating, and the homogenous and little-changing lithic implements. It is reported that many find-places occur along rivers in the middle Yangzi basin, but their chronological sequences cannot be clearly defined due to lack of absolute dates (Yuan 1995). The increased number of these find-places said to belong to the terminal Pleistocene in the middle Yangzi valley (Yuan 1995) may suggest an increased population, but much more study is certainly required before any assumption can be made.

9C. The mobility of foragers.

According to ethnographic data world-wide, contemporary hunters and gatherers move in pursuit of food, and their movement patterns vary. Some foragers move their residential camps several times a year, but make frequent short forays for special hunting and/or gathering tasks, such as the Anbarra in north Australia (Kelly 1995:120). Others may settle down in winter or dry-season camps for several months, but make seasonal journeys for special tasks, such as the Aborigines in the Western Desert in Australia and the Great Basin (Cane 1984; Steward 1938). Binford (1980) has proposed a dichotomic model of 'foragers' and 'collectors' to define the mobility patterns of foragers (Binford 1980), but actual patterns seem more diversified and overlapping, to judge from ethnographic data.

What would have been the mobility patterns of prehistoric foragers in China? Only limited data are available, but discoveries from a few site-clusters, namely the Xiachuan, Hutouliang and Xianrendong assemblages, may provide some clues.

As discussed in chapter 4, a total of 16 sites was found around the Xiachuan Basin, an area of about 93 square kms, all yielding Xiachuan lithic implements. The dates are mainly within the range from 23,900 to 13,500 bp (Wang *et al.* 1978) [Table 4(1)]. The locations of these sites are close to each other, some only 2 kms apart [Fig. 4(2)]. Did these sites belong to one forager group, or more than one? The identicalness of toolkits suggests the former possibility. This question can also be addressed from the aspect of carrying capacity.

According to world-wide ethnographic data from 205 foraging groups, population densities for hunter-gatherers living in various environments ranged from a lowest of 0.5 person/100 km^2 (Siberia), to a highest of 843 persons/100 km^2 (the Chumash in California) (Kelly 1995:223-226). However, the latter figure seems quite exceptional, as the average population density of these 205 groups is only 49.85 persons/100 km^2 (calculation based upon data from Kelly 1995). In other words, in a very broad sense, the average carrying capacity of an area of 100 km^2 is about 50 foragers.

Of all these forager groups, data from the Great Basin seem more comparable to the foragers in the Yellow valley. Both the Great Basin and the Yellow valley are within latitudes 32 to 42° N; both are temperate zones in inland areas. The vegetation of the Great Basin consisted of pine, spruce and various herbs in the 1930s (Steward 1938), which would have been similar to the palaeo-vegetation of the middle Yellow valley during the LGM, according to palynological reconstruction (chapter 3A). Fish resources were abundant along the rivers in the Great Basin and fishing was one of the major activities (Steward 1938). In the Chinese context, no fish remains have been reported in archaeological assemblages dated to the LGM in the western plateau/mountain area, but mollusc remains were found in the Xibaimaying and Shizitan assemblages of around 18,000 and 10,000 bp respectively [Table 3(1)], indicating an exploitation of fresh water food resources. The subsistence strategy of the foragers in the Great Basin included seed and nut gathering, fishing and game hunting in the 1930s (Steward 1938), which would have been comparable to that of the foragers in the Xiachuan area in China.

The population density in the Great Basin was quite low, ranging from 1 to 19 persons/100 km^2, with an average of 7.32 persons /100 km^2 (based upon data from Kelly 1995:224). In the temperate forests areas of Oregon Plateau, population densities range from 1.31 to 33.2 persons /100 km^2, with an average of about 12.2 persons /100 km^2 (Kelly 1995:224). All these data seem to suggest that the carrying capacity for foragers in a temperate steppe/plain or forest environments is around 10 to 20 persons /100 km^2, which, according to ethnographic data, roughly equals the minimum size of one foraging group (Kelly 1995:211).

Taking this figure as a reference, the natural resources of the Xiachuan area, of less than 100 km^2, might not have been able to support more than one foraging band of 10-20 people. Therefore, the 16 loci found in Xiachuan might have been left by one foraging group. As ten loci are in hilly areas and six on river terraces, it may be inferred that these could have been various camp sites or working stations, part of the annual round of one group.

According to surveys conducted in the 1990s, debitage was found in nine places at Xiachuan, two in the hilly areas and seven on terraces (Chen 1996). For example, in the Tanglishu terrace site, debitage pieces total 625, accounting for 36.61% of the total lithic remains; in the Fuyi River terrace site, debitage accounts for 41.22% of the total of 866 lithic pieces; in the Huashu terrace site, debitage accounts for 35.89% of a total of 599; and in the Yangling hill slope site, it accounts for 47.32% of the total of 317 implements (all calculations based upon data from Chen 1996). The rest of the lithic implements found in these sites are identical to those discovered in the 1970s. These data from the 1990s survey indicate that at least nine sites in the Xiachuan Basin served in part as working stations for tool manufacturing.

It seems that the Xiachuan foragers might have been moving around the terraces and hills within the area during the LGM. Current vegetation in the Xiachuan area is dominated by grass and cultivated crops, but forest still covers most of the mountains (Chen 1996). If, according to palynological studies, pine and spruce grew in the hills during the LGM, and herbs in the basin (chapter 3A; Yao 1989), then foragers could have collected seeds from grasses and other herbs from mid-summer to early autumn

in the basin, as this would have been the ripening season of the grasses (author's observation of current grasses and green foxtail 1996). They could also have hunted game in the nearby hills. Nuts from some pine species could have been another edible resource in autumn, but it is not known if they were collected.

The long time span of the Xiachuan culture is also notable. If the conventional radiocarbon dates are valid, then this assemblage existed in the Xiachuan basin from 23,900 to 13,900 bp, lasting for ten thousand years. Most of the dates, however, are within the range of 23,900 to 18,375 bp. The stratum containing the Xiachuan assemblage at one excavated site is reported to be 1-1.5 meters thick (Wang *et al.* 1978). Data for the other sites are not reported, but according to the 1990s survey report, the lithic implements in the Basin were extremely abundant. It is claimed that 'lithic implements were found on almost every hill slope. Lithic implements in the find-places were all collected during each survey, yet there were always new implements found in the next survey' (Chen 1996: 2).

This seeming abundance and long-lasting presence of the Xiachuan lithic assemblage in one small basin could seem quite exceptional. Although it is not clear whether the Xiachuan Basin really was continuously occupied for thousands of years, the widespread presence of the Xiachuan lithic implements in this Basin, and the absence of other lithic assemblages, indicates that foragers with a well-defined lithic technology were wandering the Basin for a long period of time.

Similar lithic assemblages were found at Xueguan and Shizitan. While the Xueguan assemblage was approximately contemporary with the last stage of Xiachuan, the Shizitan assemblage might be much younger [Table 4(1)]. It is not clear whether these sites represent groups closely related to those of Xiachuan, but archaeological data in the Xiachuan Basin suggest abandonment after 13,075 bp [Table 4(1)]. Were the Xueguan and Shizitan assemblages the results of a dispersal from the Xiachuan region? This possibility awaits further investigation.

A similar mobility pattern may be proposed for the Hutouliang assemblage, which consists of nine sites within an area of less than 10 km^2, the closest distance between any two sites being about 500 meters (Gai and Wei 1977). It is stated that the lithic implements from all the sites were identical (Gai and Wei 1977). Again, this suggests that the sites might have belonged to one foraging band. Further, ethnographic data seem to document that 10 km^2 is too small a territory for a single temperate foraging group (eg. Cane 1984; Kelly 1995; Steward 1938). The catchment area of the Hutouliang occupants might have been larger than the area where the nine sites were found.

The function of a few sites of the Hutouliang cluster have been suggested, based upon the archaeological discoveries. According to the excavators, one site was a camp, as hearths were found; another three sites were probably an observation spot, a butchering site, and a site for stone tool manufacturing respectively (Gai and Wei 1977; see details in chapter 4).

If this inference is valid, then the foragers at Hutouliang might have focused on a residential camp site, with other working stations for special tasks such as hunting ungulates and probably ostrich, butchering game, and making tools. This resembles a 'collector' pattern in Binford's terminology. There are no grinding tools reported from Hutouliang, and no archaeobotanical data are available. It is not clear whether the Hutouliang people collected any plants and/or seeds. Neither is it clear if the foragers stayed at Hutouliang year-round, or if they only visited on a seasonal basis. But the continuous strata profile from the terminal Pleistocene to the early Holocene recently found in Hutouliang (Xie 1996) seem to indicate a degree of continuity in foraging activities.

While the above data come from the Yellow River region, the re-investigation of the Xianrendong and Wangdong assemblages in the 1990s provides some information for forager mobility patterns in the lower Yangzi valley. Based upon faunal and lithic remains, MacNeish argues that Xianrendong might have been a camp/residential site, and Wangdong might have been a butchering site for the Xianrendong occupants (MacNeish 1995). The thick deposits found in both caves indicate that prehistoric foragers were active there over a long period, and the consistency in artefact typology suggests that the two caves might have been occupied by foragers from one material cultural complex. Although published data do not inform whether Xianrendong was occupied year-round, or only during a certain period of the year, remains of deer antler and wild rice seem to suggest that the caves were probably occupied at least from later summer to the next spring (chapter 7).

In the Middle East, the presence of human commensal species is often used as an indicator of human sedentism (Phillips 1979), but this issue is open for debate (Edwards 1989). In the Chinese context, a few rodent species are possibly human commensals, most commonly found is the black rat (*Rattus rattus*). If the identification reports are correct, black rat is reported in several archaeological assemblages in China, and its presence can be traced back to the middle Palaeolithic [eg. Locality 1 in Zhoukoudian (Qi 1989:319)]. Further, black rat is reported from the Upper Cave bone assemblage, which does not seem to relate to human activities. Therefore, the presence of "commensal" species as evidence of sedentism in the Chinese context should be treated with caution.

9D. Pottery and foraging

Pottery was present quite early in China. According to absolute dates from Hutouliang in the middle Yellow valley, and Yuchan and Xianrendong in the Yangzi basin, pottery was present by 11,800 BP in the former and from approximately 12,000 BP in the latter [Tables 4(1) and 7(1)]. If these dates are to be trusted, pottery was present well before the documented emergence of agriculture in both areas, thus suggesting an origin amongst forager communities, as in Japan. Pottery is also claimed in South China by 10,000 BP in the Miaoyan assemblage, Guangxi Province (Yuan 1993), without evidence of agriculture. Pottery dated to about 12,900 BP is also reported from the Amur River Basin in Russia (Tang 1997a). These dates seem quite controversial. In the Chinese context, the emergence of pottery might have related to wild cereal collection and preparation for eating, especially if grains were boiled or steamed whole in pottery as with rice and millets (foxtail millet and broomcorn millet) today in China.

It is notable that seed grinding tools were absent from the early Neolithic assemblages in the middle Yangzi valley. In the lower Yangzi valley, a total of 2373 lithic, bone and wood implements was found in Hemudu, but with only two wooden pestles (Liu and Yao 1993). On the other hand, rice remains found in a cooking *fu* from Hemudu demonstrate that the whole grain was cooked by boiling. This explains the extremely low incidence of grain processing tools: rice was probably hulled in wooden mortars, or in holes (functioned as mortars) in the ground (as those found in Honghuatao, above), winnowed, then cooked. Similar processes still practised in some countries in China today,

although dehusking has been performed by machinery.

As for foxtail millet, it is not clear how this cereal was consumed during the early Holocene. Grinding slabs and rollers are widely found during the early Neolithic in the Yellow valley, indicating the importance of grain processing. But whether grains were dehusked only, or milled as well, remains unclear. In addition, as discussed above, the incidence of grinding and pounding tools is very low in the middle Neolithic Central China. According to historical documents, the whole grain was cooked with water and then consumed during the Western Zhou dynasty, some 2800 years ago (Hu 1994). It is also notable that pottery steamers were popular from 6500 BP onwards in the Yellow valley (IA, CASS 1984). A document written in the third century BC describes that such steamers were used to steam-cook whole grains of millet (Hu 1994). A similar method was still used in the 1950s in villages in Central China (IA, CASS 1959) and in Taiwan (Chen 1968). It is possible that millet might have been consumed in whole grain form in prehistoric China .

In summary, grain processing in Neolithic and ancient China was probably simple as only dehusking was required, and milling was not always practised. This results in a much lower labour cost of seed processing, and consequently a relatively low occurrence or absence of grinding tools, but requires specialising cooking equipment. If the foragers of the terminal Pleistocene in China consumed wild cereals in the same manner, the need for cooking vessels would have arisen (Fujimoto 1983; Jin and Xu 1992).

According to ethnographic data, the Dai minority in Southwest China cook rice with water in sections of bamboo, sealed with clay and put on a fire (Cao 1988). This is a good method, but abundant bamboo must be available as a new bamboo section is needed each time, and the diameter needs to be quite large. Obviously, in temperate areas where bamboo is small or not available, this method cannot be used. It seems that fired clay was thus turned to by foragers in China for making cooking vessels, hence the invention of pottery.

If this proposition for the origins of pottery in China is valid, then it may be argued that the presence of pottery could be an indicator of intensive grain exploitation. This hypothesis is to be tested by archaeological and archaeobotanical studies in the future.

This hypothesis does not suggest that an absence of pottery indicates an absence of plant exploitation, as the cooking and consuming methods for cereal grains vary in different areas. For example, cereal seeds were extensively collected by Aborigines in Australia, yet most were ground into flour and baked as cakes, the rest eaten raw (Cane 1984). In the Great Basin and California likewise, seeds of wild cereals were collected and ground into flour, then cooked as soup in soapstone bowls or baskets (Kroeber 1925; Steward 1930). In the Middle East, where intensive cereal collection was practised in the late Pleistocene, pottery followed the invention of agriculture by almost 2000 years, and first occurred as standard equipment in sites only around 8500 BP (Bar-Yosef 1981; Rice 1987). This is perhaps because cereals grains in the Middle East were ground into flour, not generally cooked as whole grains (although the presence of pottery predecessors such as gypsum white ware in PPNB sites suggests that the situation might not have been as clear cut as in China).

This is not to say that cooking cereals was the only impetus for pottery invention. The need for seed storage could be another one. And the function of pottery is not restricted to grain cooking. Once invented, pottery vessels could have been used to cook and serve various foods, to store liquids etc. The origins of pottery might have been stimulated by different impetuses in different areas of the world, or by multiple needs in one area. The importance of cereal cooking as a causal factor for the pottery origination is tenable in the Chinese context, but may not be so in other areas.

A few points can be made regarding the forager populations in China during the terminal Pleistocene. First, foragers were active in both the Yellow and the Yangzi valleys during the LGM, documenting a constant human occupation of both regions. The lithic assemblages of the groups in the Yellow valley were more diversified, as some lithic assemblages were dominated by flakes, others by the combination of microblades, flakes and heavy duty tools (chapter 4). In the Yangzi valley, the lithic assemblages seem more homogeneous . This suggests that there were two separate cultural systems in the two valleys, but possibly in contact via the Huai River region (e.g. the Xiaokongshan lithic assemblage, chapter 4).

The second point is the occurrence of new heavy-duty stone implements - axes, adzes, grinding tools and discoid implements in the middle Yellow valley; discoids and perforated pebbles in the Yangzi valley, and the emergence of small size tools - microblades in the Yellow valley and small flakes in the Yangzi valley. The simultaneous occurrences of the discoid implements and the small size tools was an important phenomenon during the terminal Pleistocene, although the significance is not fully comprehended at this stage, mainly owing to problems in understanding the functions of the tools. Even so, the presence and spread of these lithic implements in their respective regions from the terminal Pleistocene to the early Holocene is very important.

The development of these new tools differed. In the Yellow River basin, the axes, adzes and grinding tools developed into polished items and flourished in the Neolithic; microblades diminished after the occurrence of polished stone tools in early Holocene; and the discoids remained little changed but were very popular in the middle Neolithic in the middle Yellow valley. In spite of these different developments, they all represent technical changes in subsistence strategies during the terminal Pleistocene to the early Holocene, and were hallmarks of the transitional period.

It has been argued that the diminution of microblade numbers in the Yellow River was contemporary with the emergence of agriculture (Lu 1998). Microblades and implements of the flake tradition are only found in very small quantities in the agricultural societies such as Cishan and Peiligang. In the lower Yellow valley, where over one hundred find places containing microblades have been found, not a single microblade has been reported from the Houli culture of the early Holocene in that region, although implements of the flake tradition are occasionally found (Wang 1994). The contemporaneity of the decline and disappearance of microblades and the appearance of agricultural societies in the Yellow valley suggests that the microblades were probably a lithic tradition primarily for hunting and gathering activities.

The situation in the Yangzi valley is different. Grinding tools are not popular in this region. The discovery of such tool in Liulinxi may indicate cultural influence from the north. The development of adze and axes is not clear, as no forerunners of these implements can be identified in the Yangzi basin in the terminal Pleistocene at this stage. The small flakes in the Yangzi basin were first present in the terminal Pleistocene, and remained quite common in the early Holocene. These flakes disappeared after

approximately 6500 BP when the Neolithic in the Yangzi River reached its zenith. Although the functions of these small flakes are not clear, their constant presence illustrates a degree of cultural continuity.

The cultural continuity is the third point that needs to be emphasised. In the Yellow River this continuity is strongly evidenced by the grinding tools, the adzes and axes, which were present in the terminal Pleistocene, and dominate the lithic toolkits of the early Holocene. The axes could have been used for many purposes, and the development of this tool from the terminal Pleistocene to the early Holocene was from chipped to polished. On the other hand, grinding tools altered slightly during this time span. They were associated with grinding handstones during the Pleistocene (eg. those found in Xiachuan), but with rollers in the early Holocene (eg. those found in Cishan, Peiligang and Houli). This difference suggests a change of seed processing technique from a circular to a rolling movement. It requires further study and experiment to investigate whether the rolling movement would have been more efficient, or whether there were any other causes for such a change.

Similar cultural continuity is also evident in the middle Yangzi valley, where the discoid choppers occurred in the terminal Pleistocene, and flourished in the early Holocene. The development of the perforated pebble in the lower Yangzi valley is less clear, as this lithic implement was present in some sites (e.g. Xianrendong) but absent in many prosperous agricultural societies such as Hemudu and Luojiajiao.

The last but not the least point is the possibility of cultural contact between the Yellow and the Yangzi valleys in the terminal Pleistocene. This is indicated by the discoid implements, which occurred earliest in Xiachuan north of the middle Yellow valley, then in Xiaokongshan south of the middle Yellow valley; then in the surface collections of the middle Yangzi valley. Although the lithic compositions of these three assemblages are significantly different, the distribution of this implement may indicate some diffusion or spread of cultural components from the Yellow valley to the Yangzi valley.

Based upon the lithic toolkits found in both river valleys, therefore, and comparison with archaeological and ethnographic records in the Middle East, the Great Basin and Australia, it can be argued that some foraging groups, such as those in the Xiachuan Basin in the Yellow valley, and Yuchan and Xianrendong in the Yangzi basin, practised extensive hunting and gathering strategies during the terminal Pleistocene, including wild plant exploitation. These groups seem to have been relatively sedentary within one given area (eg. Xiachuan and Xianrendong). Some might have adopted a logistic 'collector' mobility pattern (Hutouliang, probably Xiachuan). Pottery was apparently manufactured in both the middle and lower Yangzi valley by 12,000 BP, and in the middle Yellow valley by 11,800 BP. The presence of pottery might have been an outcome of intensive cereal collection and cooking.

The study of foragers in China so far has been concentrated on their economic perspective. Very little is known regarding their social structure, art and spiritual life. Ochre was found in the Shizitan assemblage, middle Yellow River, possibly in association with paintings (chapter 4). Although human remains have been found in several sites, such as Mengjiaquan in the Yellow valley, Yan'er and Xianrendong in the Yangzi valley, details of burial methods remain unknown.

Chapter 10. Farmers of the early Holocene in China.

The termination of the LGM marked the beginning of an improving palaeoclimate, with increased precipitation and moisture. As a result, more trees and animals became available in the valleys of the two rivers. The climatic improvement began at approximately 14,000 to 13,000 BP, with fluctuations comparable to the Younger Dryas in Europe occurring approximately between 12,000 and 10,500 BP. After 10,000 BP the warm Holocene began in the Yellow and Yangzi valleys , consistent with the global climate pattern.

Between 10,000 and 8500 BP few archaeological data have been found in both the Yellow and the Yangzi valleys. Although the grinding slabs and rollers found at Nanzhuangtou in the Yellow valley and the rice remains found at Yuchan and Xianrendong in the Yangzi valley indicate cereal collection, the cultural links between these assemblages and the Neolithic cultures of 8000 BP in each river basin remain to be investigated. Judging from archaeological data, unquestionable farming in the Yellow valley first occurred at about 8500 cal. BP, represented by the Jiahu assemblage; and 9000-8000 cal. BP in the middle Yangzi valley, represented by the Bashidang and Pengtoushan assemblages.

10A. The subsistence strategy and the cultivars of the early farmers.

The agricultural societies of 8500 to 7500 years ago in the Yellow valley were apparently quite advanced. The large quantity of stored foxtail millet found in Cishan indicates a high level of crop productivity. Remains of seeds of rape (*Brassica sp.*), foxtail millet (*Setaria italica*) and broomcorn millet (*Panicum miliaceum*) have been found in the Yellow River sites, as well as rice (*Oryza* spp.) (details in chapter 4).

Judging from botanical remains as well as lithic implements, particularly the specialised cultivation, harvesting and grain processing tools (spades, sickles, grinding slabs and rollers), there can be little doubt that agriculture was a primary economic activity by 8500 BP in the Yellow valley. On the other hand, hunting and gathering were still part of the subsistence strategy, especially in areas where rich faunal and floral resources were available. According to research on animal remains, wild animal bones accounted for about 50% of the Cishan bone assemblage, and about 40% in both the Bancun site of the Peiligang culture and Baijia of the Dadiwan culture (Yuan and An 1997). The organic tools such as arrowheads and harpoons found in Cishan and Jiahu for hunting and fishing, the rich faunal and floral remains from these two sites, as well as the stored hackberry seeds found in Cishan, document the importance of foraging activities.

Compared to the discoveries in Cishan and Jiahu, faunal remains and hunting tools were present in much lower frequencies in other Peiligang sites. Whether this indicates that foraging activities were less important in these Peiligang assemblages, or it was due to other reasons, remains to be studied.

According to pollen profiles of the early Holocene, the palaeoclimate was warm and moist at that time, and the quantity of trees had increased, particularly in mountain areas and in the Huai River delta (details in chapter 3A). However, it should be noticed that at the Nanzhuangtou site, dated from 12,000 to 10,000 cal. BP [Table 4(7)], there was a sharp decrease in tree cover shortly after 10,000 BP - after the first occurrence of archaeological remains [Fig. 3(9)]. This sudden decline of trees cannot be explained by climatic change, because pollen profiles from other contemporary boreholes all indicate that the palaeoclimate after 10,000 BP became warmer and moister. Thus the decline of trees might have been the result of human activity, i.e., land clearance for the purpose of either foraging (using fire), or cultivation.

In the Yangzi valley, the earliest farmers appeared at approximately 9000-8000 cal. BP, represented by the Pengtoushan and Bashidang assemblages. Rice (*Oryza* spp.) was the major cultivar. As a substantial quantity of rice has been found in Bashidang (Pei 1998b), it seems convincing that rice cultivation was a primary economic activity. Apart from rice cultivation, animal husbandry seems to have begun in Bashidang, as domesticated ox and pig have been reported (Pei 1998a). No animal remains are reported from Pengtoushan. As discussed in chapter 7, whether this absence is due to acid soil remains unclear. The incidence of faunal remains within archaeological assemblages increases after the Pengtoushan period, as illustrated by the discoveries at Fenshanbao and Hujiawuchang, although it is not clear whether there were domesticated animals.

In the lower Yangzi valley, the Hemudu assemblage is still the richest evidence of an agricultural society at 7000 BP. The abundance of rice, domesticated and wild animal remains, as well as the rich artefacts found in Hemudu illustrates the life of a prosperous prehistoric village. The Hemudu occupants seem to have been skilful rice cultivators. They were also fishers, gatherers, and hunters, experienced pile-dwelling builders, textile weavers and artists.

Regarding the cultivars in Neolithic China, it is now well known that millet was cultivated in the Yellow valley, and rice in the Yangzi (eg. Barnes 1992; Smith 1995). But it should be noted that rice was also cultivated in the Neolithic Yellow valley. The earliest cultivated rice to date occurs in the Jiahu assemblage, which is also the earliest Peiligang assemblage so far recovered. From 7000-5000 BP, rice remains are found in both the middle and the lower Yellow valley (Zhang *et al.* 1994) [Fig. 6(8)]. It has been argued that the expansion of rice cultivation might have been due to the warmer and moister climate from 7000 to 5000 BP, which allowed rice to be cultivated in northern area (Chen 1993). On the other hand, millets have not so far been found in the early Neolithic in the Yangzi valley, and the oldest dates for millet from this region come from the Qinglongquan site, Hubei province, dated to about 5000 BP (Jin 1990). The millet seeds were found on the bottom of a cooking pot (Jin 1984), indicates that millet was cooked in whole. Domesticated rice was also found in this site (Jin 1990).

Generally speaking, the subsistence strategy of the early farmers in China was a combination of cultivation, hunting and gathering (including fishing). The occurrence of faunal remains was lower at some sites than that at the others. But it is premature to draw any conclusions regarding these different incidences of animal remains.

10B. Population densities and degrees of sedentism of the early farmers.

The population during the early Holocene in both the Yellow and Yangzi valleys was denser than during the terminal Pleistocene. This density is particularly impressive in the middle Yellow valley, where over 100 sites of the Peiligang culture have been located in an area of about 135,530 km^2 in Henan province (Sun 1992), much greater than the less-than-20 late Palaeolithic find-places located in the same region (Anonymous 1991). The size and thickness of cultural deposit of 65 Peiligang sites are known, and

47.69% of these Peiligang sites are over 50,000 m^2 in scale [Table 10(1)].

However, not every site has been excavated. And for those excavated, only a portion of each site has been dug. The excavated sites contained both semisubterranean houses of square shape, and pits, some of which might have been used for residence. Because the quantity of residential pits is not disclosed in the archaeological reports, it is difficult to estimate the population of each site.

Another problem of estimating the population density in the early Holocene in Henan province is the uncertainty of the mobility pattern of the early farmers. It has been proposed that slash-and-burn would have been the technology for agriculture in the early Holocene in the Yellow valley (Chen 1991). This might have required residential movement, but the frequency of this movement would have depended upon some other factors, i.e. whether the soil was fertile, and how large the territory of the farming group was. If such movement did occur, more than one Peiligang sites might have been occupied by one group. However, durable dwellings, facilities such as kiln and storage pits and site longevity support the stability of Neolithic settlement. The greater numbers of Peiligang sites indicate that the overall population density would have been much greater in the early Holocene.

The occupation lengths of the Cishan and Peiligang sites are quite long. Two cultural phases are identified in Cishan, dated between 8000 and 7700 cal. BP, probably lasting for about 300 years [Table 4(7)]. Three cultural phases are identified in Jiahu of the Peiligang culture, with a much longer time span possibly from 8561 to 7529 cal. BP [Table 4(7)]. Some of the Peiligang sites, e.g. the Zhongshanzhai, Tanghu and Xiaobaizhaung sites, all in Henan province, were probably occupied from about 7000 to 4000 BP (Anonymous 1991).

Burials have also been found in the early Neolithic cultures in the Yellow valley. A total of 114 burials was found in Peiligang; 69 in Shigu, 68 in Er'gou and 120 in Shuiquan (Zheng 1993). These were all in well-planned cemeteries, and two temporal-phases were observed at Peiligang and Er'gou (Zheng 1993). The well-planned and long-lasting cemeteries suggest that the villagers were sedentary. In addition, the presence of burials and sets of tomb objects may also indicate concepts of territorial ownership (Underhill 1997).

Compared to the Yellow valley, the population in the Yangzi valley seems less dense. To date, less than 40 archaeological find-places of the early Holocene have been located in the middle Yangzi valley - three Pengtoushan sites, over 20 sites with Fenshanbao and Hujiawuchang assemblages, and less than 6 Chengbeixi sites (He 1995b). They are distributed in an area of about 356,000 km^2. This incidence is much lower than that in the Yellow valley (above). As for the lower Yangzi valley, only a few sites occur by 7000 BP, including Hemudu and Luojiajiao. This might be because there was a sea transgression between 8500 and 7000 BP, and many parts of the lower Yangzi valley were inundated and not inhabited (Wu 1983).

The sizes of the early Neolithic sites in the Yangzi valley are less clear [Table 10(2)]. Details for Pengtoushan are not available. Although several sites have been excavated, only a portion of each was dug, and many archaeological data have not been published. The geographical distances and the chronological differences among the above early Neolithic sites indicate that they were probably remains of different cultural groups, but it is difficult to estimate the population at each site. However, at Hemudu, the discovery of three big houses suggests a considerable population (see chapter 7 for details).

One important feature of population distribution in the middle Yangzi valley is the co-existence of groups of different material culture. Recently, a shellmidden assemblage was found at a site called Gaomiao, located on the southern edge of Dongting Lake (He 1994). Details of this assemblage are not available, but it is stated that the shellmidden deposit was 7 meters in thickness, containing large numbers of tortoise, wild and domesticated animal (pig) remains, yet it is reported that no cereal remains were found (He 1994). Stone net-sinkers, pebble flakes, as well as small flakes of black chert were the major lithic implements (He 1994). Seven radiocarbon dates have been obtained for the Gaomiao assemblage, within the range from 7529 to 5930 cal. BP (Chen *et al.* 1996) [Table 7(2)]. Sand-tempered pottery of red and brown colour was found at this site, some incised by geometric, sun and "phoenix" motifs (He 1994). The toolkit and pottery of this Gaomiao assemblage are significantly different from those of Pengtoushan and Fenshanbao, although both Gaomiao and Fenshanbao are located in the Dongting Lake area [Fig. 7(1)]. Judging from faunal remains and the toolkit, the excavator infers that the Gaomiao occupants were primarily hunters and gatherers, but animal domestication was probably practised (He 1994).

On the whole, the population density during the early Holocene in both the Yellow and the Yangzi valleys must have increased, indicated by the increased numbers and sizes of archaeological sites. The occupation lengths and the formal burials of the early Neolithic societies in China indicate that the early farmers were quite sedentary, and many sites were occupied for hundreds of years. The well-planned burials probably also indicate a close attachment to the land. However, there are still questions. Edwards has defined sedentism as 'a system in which the greater part of the population of a community resides perennially at one settlement' (Edwards 1989:9). In the Chinese context, the degree of sedentism might have been related to farming activities and techniques, which would have differed in dry-land and wet-land cultivation. Further research is required to investigate the settlement pattern for the farmers in China.

10C. Pottery of the early farmers.

The pottery of the early Holocene in the two great river valleys in China was extremely diversified. Yet some vessel types have been commonly found in both the Yellow and the Yangzi valleys. Such similarities may provide important messages regarding the cultural relationships among different archaeological assemblages.

The first commonly found vessel is the cooking *fu*. This vessel was a major item within the pottery of the Houli assemblage in the lower Yellow valley. In the Xihe site of the Houli assemblage, one *fu* was discovered sitting on three stone supports over a hearth (Tong and Wei 1994). This discovery clearly illustrates the function of such vessels. Combining the *fu* and the three supports together produced the *ding* in the later phase of the Houli assemblage [Fig. 10(1)].

As discussed in chapter 5, the Dadiwan, Cishan and Peiligang pottery assemblages in the middle Yellow valley differed significantly, but their tripod *bo* (defined in chapter 2) are almost exactly the same [Fig. 4 (10)], suggesting cultural contact. The cooking vessels in Peiligang and Dadiwan were deep pots, although the Dadiwan pots had tripods. The Cishan occupants used the *yu* (defined in chapter 2) and pottery stands to cook. The pottery assemblages seem to suggest that these were different farmer groups with some cultural contacts.

In the Yangzi basin, *fu* dominated the early Neolithic assemblages from the middle to the lower valley. *Ding* were present from approximately 6000 BP both in the middle (He 1995a) and the lower Yangzi valley (IA, CASS 1984). For a long period of time the farmers from Pengtoushan to Hemudu used pottery stands to support the cooking *fu*. This is quite different from the Cishan, Peiligang and Dadiwan cultures in the middle Yellow valley, but similar to the Houli culture in the lower Yellow valley.

It should be noticed that rice husks and other plant remains were used as pottery tempers in the Pengtoushan culture of the middle Yangzi valley, at 9000 to 8000 BP. The usage of such plant tempers progressively decreased in Fenshanbao and Hujiawuchang, from 8000 to 7000 BP (He 1995a). On the other hand, rice-husks and other organic tempering agents were used widely in the Hemudu and Luojiajiao assemblages in the lower Yangzi valley, at around 7000 BP (chapter 7). This similarity in tempering agents may suggest some cultural linkage between the middle and the lower Yangzi valley.

The pottery of Pengtoushan, Fenshanbao and Hujiawuchang clearly shows the development of vessels, particularly the cooking *fu*, which would have been an important item [Fig. 7(7)]. Geographically, the Fenshanbao assemblage is located to the east of Pengtoushan, and is probably only a few hundred years later than the latter. The similarity of pottery between Pengtoushan and Fenshanbao, as well as the remains of a possible pile-dwelling found in the lower layer of Fenshanbao (Zhang 1992) and Bashidang (Pei 1996 personal communication), suggest that the Fenshanbao assemblage may be an eastern expansion of the Pengtoushan culture.

It is difficult to demonstrate that Hemudu and Luojiajiao were descendant cultures of Fenshanbao, because the pottery of Fenshanbao was significantly different and much more complicated than the Hemudu and Luojiajiao pottery [Fig. 7(7), 7(18)]. However, given the similarities of pottery temper and the presence of *fu*, and the similar economic activity of rice cultivation, as well as the possible presence of pile-dwellings in Bashidang of the Pengtoushan culture, a dwelling style well developed in the Hemudu and Luojiajiao assemblages, there might have been some cultural connections between the Pengtoushan and the Hemudu/Luojiajiao assemblages. It is possible that early farmers in the middle Yangzi valley expanded from the Pengtoushan region towards the lower Yangzi valley, with their crude pottery; and they later settled in different areas, built pile-dwellings to suit the swampy environment, and developed different pottery vessels; but the tradition of using rice husk as temper and making *fu* remained. The Fenshanbao assemblage might have been one group dispatched from the Pengtoushan culture, while the Hemudu and Luojiajiao might have been other groups.

Pottery assemblages of the early farmers may also reflect cultural relationships between the Yellow and the Yangzi valleys, such as those between the Pengtoushan and Jiahu assemblages. The Pengtoushan pottery was tempered by rice husks and other plant remains. Plant-tempered pottery was also found in the second and third cultural phases at Jiahu, but has never been reported in any other Neolithic culture in the Yellow valley. In addition, a few vessels such as the globular jar with two lugs and the bowls found in these two sites are quite similar [Fig. 10(2)]. As the Pengtoushan assemblage is dated slightly earlier than the Jiahu one, and pottery of the early phase at Jiahu was sand-tempered, with little fine pottery (IA, Henan Province 1989), it seems that the Jiahu occupants received some influences from their southern neighbours. Worth of notice is that the globular jar and bowl were major forms in the Peiligang culture. As Jiahu is the earliest assemblage of the Peiligang culture to date, this cultural influence from Pengtoushan could have dispersed northbound.

On the whole, the Jiahu pottery is significantly different from Pengtoushan. For example, rice husks and plants were the only tempering agents in Pengtoushan, whereas sand was the only temper in the first phase of Jiahu; the cooking *fu* is an important vessel in Pengtoushan, but is absent in Jiahu, where a deep-bellied pot dominates (IA, Henan Province 1989; IA, Hunan Province and Lixian Museum 1990) [Fig. 10(2)]. The Jiahu pottery was probably slightly more advanced than the Pengtoushan, as fine pottery was present in a small quantity in the first phase of Jiahu, but absent in Pengtoushan (IA, Henan Province 1989; IA, Hunan Province and Lixian Museum 1990). Although rice was also cultivated in Jiahu, the lithic implements were completely different from those of Pengtoushan and its descendant cultures. The similarities between some pottery vessel from Pengtoushan and Jiahu may suggest a certain degree of cultural contact, but the difference of pottery assemblages and toolkits indicates that the Pengtoushan and Jiahu occupants were different groups.

A few points can be summarized about the pottery assemblages in the Yellow and the Yangzi valleys. First, pottery was diverse from the beginning. Sand was the major temper in the Yellow valley and in the early phase of Xianrendong in the lower Yangzi valley, but plant remains were used as tempers in the middle Yangzi valley from the beginning of pottery making (ie.Yuchan, Yuan 1996), and in the lower Yangzi valley at 7000 BP. However, plant remains were gradually replaced by sand in the Yangzi valley. This replacement is clearly demonstrated by the pottery sequences in Fenshanbao, Hujiawuchang, Hemudu and Luojiajiao. The diversity may suggest that pottery originated independently in both the Yellow and the Yangzi valleys, but whether the conformity of sand-tempering in later stages suggests cultural interaction or influence on a broad scale, or whether it was the result of independent technical change, remains to be investigated.

Second, tempered pottery vessels are older than fine pottery in both the great river valleys. The quantity of fine pottery increased progressively in the later cultural phases of the Neolithic. This development is evident in the pottery sequences found in Xianrendong, where only sand-tempered pottery was present in the early stage, and fine pottery occurred in the later stage (Hill 1995). In the Pengtoushan assemblage, only plant-tempered pottery was found. Yet small amounts of both sand-tempered and fine pottery occurred in the Fenshanbao and Hujiawuchang assemblages, although plant-tempered pottery still dominated.

According to pottery analyses conducted in China, tempered vessels are best for cooking, as they can stand the heat and remain intact; whereas fine pottery can only be used as containers or to serve food (Shanghai Silicate Institute CASS 1987). Prehistoric people might have already known this. The potsherds found in Yuchan and Xianrendong caves of the terminal Pleistocene show that rice husks and other plants, or crushed rock and white quartzite, were purposely mixed with clay as tempers (Yuan 1996; Hill 1995). The origins of tempered pottery in China seem to reinforce that it was initially made to meet the needs of cooking, which, as discussed above, would have been stimulated by grass exploitation.

The development of pottery from tempered-vessels-only to the presence of both tempered and fine vessels in the Yangzi valley seems to have been associated with the progress of rice cultivation and a more prosperous farming society, as demonstrated by the pottery sequence from

Pengtoushan to Fenshanbao. If a similar development occurred in the Yellow valley, then it may be inferred that the Jiahu-Peiligang and Cishan assemblages, and probably the Houli assemblage, are by no means the earliest Neolithic in the Yellow valley, as fine vessels were present in all these assemblages, and the vessels do not seem to be at the initial stage of pottery making. Logically, there should have been one or more preceding archaeological assemblages in the Yellow valley, with grinding tools, more primitive pottery, and sedentary or semi-sedentary life. The Nanzhuangtou discovery seems to represent such an assemblage, which contained fragments of a grinding slab and a roller. In addition, its potsherds were tempered by sand, crushed shell and mica (Jin and Xu 1992), all these tempers were found in the Cishan and Jiahu-Peiligang pottery. There might have been a cultural linkage between Nanzhuangtou and the Neolithic in the Yellow valley, but the unearthed materials from Nanzhuangtou are too scanty for further interpretation.

Third, the tempering agents of early pottery seem to have related to the economic activities of their makers, particularly for the rice cultivators in the Yangzi valley, as rice husks and straw are often found as tempers (from Pengtoushan, Fenshanbao to Hemudu). Crushed shell temper is found in Jiahu, Shigu and Zhongshanzhai of the Peiligang culture (Zhang 1989), Xianrendong and the Chengbeixi culture (Hill 1995; Lu 1993), often in association with shell remains in the archaeological assemblages.

Finally, the pottery assemblages in the Yellow and Yangzi valleys also indicate both cultural development and contemporary cultural contact, not only between adjacent districts, but also over large distances. The similarities in the pottery assemblages of Cishan, Dadiwan and Peiligang in the middle Yellow valley document contemporary cultural contacts. The pottery from Pengtoushan, Fenshanbao and Hujiawuchang illustrates both cultural development and contemporary cultural contacts. A possible cultural contact is also indicated between Pengtoushan and Jiahu. The similarities between the Pengtoushan, Hemudu and Luojiajiao pottery also suggests a possible cultural expansion.

The pottery assemblages of the early farmers in the Yellow and the Yangzi valleys show a communication network from 7000 BP, which was probably more complicated and diversified than previously thought. On one hand, it is clear that there was an independently cultural tradition in each valley; on the other hand, cultural contacts occurred between the two valleys, especially via the Huai River valley. The Neolithic farmers were not isolated, as they certainly accepted influences from each other; yet they still retained the basic elements of their own material culture, which distinguished one from the other.

10D. Cereal cultivation and animal husbandry.

In the Chinese context, animal husbandry seems to have occurred simultaneously with cereal cultivation in both the Yellow and the Yangzi valleys. Remains of domesticated dog and pig have been found in Cishan and Peiligang cultures, as well as in Bashidang and Hemudu. Water buffalo was probably domesticated in Hemudu. However, it should be noticed that remains of possibly domesticated pig were also reported from Gaomiao, the middle Yangzi valley, and Zengpinyan in South China.

Although animal husbandry is not the major topic of this thesis, it is worthwhile to point out the above phenomena. However, the identification of domesticated animals in China is an issue that is not fully solved, particularly for remains found in early Holocene. Consequently, many questions remain. For example, was pig domesticated by foragers as well as by farmers? What was the effect of animal domestication related to farming? Etc. These questions await further study.

10E. The social complexity of the early farmers.

The social complexity of the early farmers in China is much less well-known. It should be noticed that remains of a square house at about 33.6 m^2, built on the ground, was found in Pengtoushan, which is in contrast to the small semisubterranean houses of under 5 m^2 from the same site (IA, Hunan Province and Lixian Museum 1990). Whether this big house was used as a ritual centre for the tribe, or was occupied by the chief of the prehistoric society, remains a question. But the presence of these different dwellings may indicate that there were probably more significant differences in terms of the status of members of the early farmer societies, that the social structure was more complex than before.

This dwelling difference is not visible in the early Holocene in Central China, since only semisubterranean pit houses were found in both the Cishan and Peiligang sites, and the sizes are often under 6 m^2 (Henan Museum and CPAM of Mi County 1981; IA, Henan Province 1989). The semisubterranean houses in Dadiwan are bigger, measured to about 20 m^2 (Dadiwan Excavation Group 1982). However, large, on-the-ground square houses are found in the Yangshao culture (IA, CASS 1984), which was a direct descendant of the Peiligang and Dadiwan cultures in the Yellow basin.

The presence of well-planned cemeteries and sets of tomb objects in the Yellow basin may indicate different status and/or well-beings of the society members. Grave goods were widely found in both the Yellow and the Yangzi valleys, mostly being tools and pottery vessels. The grave goods found in Peiligang were assembled in a certain manner - grinding tools for females and digging tools for males (Zheng 1993). This indicates possible belief in an afterlife.

The tortoise shells found in Jiahu provide additional information on the spiritual life of the early farmers. These grave goods often buried in groups. Some contained small cobbles of various sizes and colours (IA, Henan Province 1989). This phenomenon has been found in succeeding Neolithic cultures in the Yellow valley, and the discovery in Jiahu seems to be the root of the tradition, although its meaning is not yet clear.

In ancient China, music was an essential part of ceremonies (Sun 1987). This was possibly also the case for the early farmers at Jiahu, where two bone flutes and eight pieces of tortoise shell were found in one tomb. The occupant of this tomb was a mature male, and the quantity and quality of tomb objects in this tomb was much greater than that in other tombs (IA, Henan Province 1989). This indicates that the tomb occupant was probably an important person. The flutes found in Jiahu are the earliest music implements found in China to date. After testing, it is said that seven consecutive notes can be produced (IA, Henan Province 1989).

The arts of the early farmers are represented by painting and engraving. Painting occurs primarily as pottery decoration in the early Neolithic, its earliest presence being at Fenshanbao. The motifs in Fenshanbao are squares and wide bands, painted on a white slip (Zhang 1992). Pottery painting further developed in the Neolithic cultures in the Yangzi valley, as represented by the pottery found at Datang and Gaomiao (He 1994, 1995a). Engraving, incision and perforation were also used as decoration on

pottery vessels from Fenshanbao and Hujiawuchang in the Yangzi valley, and the Peiligang culture in the Yellow valley. Sculptures have also been found in the early farming societies. Ivory bird sculptures, pottery fish and pig sculptures were found in the Hemudu assemblage (Liu and Yao 1993). Pottery animal sculptures were found at Jiahu (IA, Henan Province 1989).

In summary, early farmers in China emerged by no later than 8500 BP. After the initial emergence, farming societies rapidly spread through both the Yellow and the Yangzi valleys by 7000 BP. These societies shared many common characteristics. They all seem sedentary, as house remains have been found at each site; polished lithic implements were major tools for most of them, probably except the Pengtoushan assemblage where only limited numbers of tools were found. Pottery was found at each site, and the types of vessels varied. Millets were cultivated in the Yellow valley, and rice in the Huai and Yangzi valleys (chapter 4, 7). Dog, pig and probably chicken were domesticated in the Yellow valley, as indicated by bones found in the Cishan assemblage (Zhou 1981); dog and pig were domesticated in Bashidang and Hemudu in the lower Yangzi valley (Liu and Yao 1993; Pei 1998a). Cemeteries were widely found in both river valleys, and remains of closed kilns were first found at Jiahu of the Peiligang culture, and in the Houli assemblage (Wang 1994). It is notable that a small amount of small flake tools, and/or microblades are often found within the toolkits of these early agricultural societies, indicating strong cultural links between the farmers of the early Holocene and the foragers of the terminal Pleistocene in each region.

Dwelling remains, cemeteries and pottery kilns found in the early Neolithic sites in the Yellow valley illustrate the sedentary life. Burials, the earliest bone flutes and tortoise shells bearing inscriptions and drills provide information for art and religion in the Huai valley. The discoveries suggest that the Neolithic cultures dated at around 8500-7500 BP in the Yellow and Huai River valleys should not have been the earliest Neolithic, but should have been descendants of some undiscovered early Neolithic societies. The recently discovered Nanzhuangtou material is probably a representative of such societies. In the Yangzi valley, the earliest farmers seem to have been represented by the Bashidang/Pengtoushan assemblages, which were probably descended from the foragers of the terminal Pleistocene.

Archaeological data also suggest a close cultural contact between the Neolithic groups in the two great river valleys, most obviously the distribution of cultivated rice, which quickly expanded northwards from the Yangzi to the Yellow valley by 5000 BP [Fig. 6(8)]. The toolkits of these prehistoric rice-cultivators in the Yellow valley were different from those in the Yangzi valley, so this seems to represent a diffusion of cultural idea from the south to the north.

Table 10(1). Size and thickness of 65 Peiligang sites.

Groups	Known size (, 000 m^2)	Thickness of cultural deposit (meters)	Number of sites	Representative site	% of total sites.
Group 1.	10 - 20.	0.5 - 2.0.	16	Peiligang.	24.62%.
Group 2.	20 - 50	0.8 - 2.0.	18	Malianggou	27.69%
Group 3.	50 - 100	1.0 - 4.0.	24	Jiahu.	36.92%
Group 4.	100 - 240	1.0 - 4.0 (or more).	7	Zhongshan-zhai.	10.77%
Total			65		100%

(Source: Anonymous 1991).

Table 10(2). Size and thickness of some early Neolithic sites in the Yangzi valley.

Site	Known size (,000 m^2)	Thickness of cultural deposit	Occupation length
Bashidang	8	0.3 - 1 meter.	Not yet reported.
Fenshanbao	over 40.	0.6 - 1.45 meter.	From early to mid-Holocene.
Hujiawuchang	7.	?	From early to late Holocene (Zhou dynasty).
Dingjiagang	about 10.	0.6 - 1 meter.	Early to middle Holocene.
Xianrendong	0.1	1. 6 -2.9 meter.	Late Pleistocene to mid-Holocene.
Luojiajiao	120.	0.2 - 5.5 meter.	Early to middle Holocene.
Hemudu	About 40.	1.95 - 4 meter.	Same as above.

(Sources: He 1995a; Jiangxi Museum 1976; Liu and Yao 1993; Luojiajiao Archaeology Team 1981; Pei 1995; The Archaeology Team of Yueyang City 1994).

Chapter 11. Hypotheses and propositions: the origins of agriculture in China.

As discussed in chapter 1, various hypotheses and models have been proposed for the origins of agriculture in the Middle East and other continents. Are any of these hypotheses and models relevant to the origins of agriculture in China? Can the Chinese case be explained by any of these theories? Given the current database, is it possible to address the question of when, where, why and how agriculture originated in China? And, was there one centre, or two centres for indigenous agricultural origination? These are the questions to be discussed.

11A. When did agriculture originate in China? - the chronology of early farmers in the Yellow and the Yangzi valleys.

The chronology of the early farming societies in both the Yellow and the Yangzi valleys is primarily based upon conventional and AMS C^{14} dating [Tables 4(1), 4(7), 7(1), 7(2)]. However, some of these dates seem problematic, and only a tentative chronology can be proposed at present.

In the Yellow valley, the date of 9988 cal. BP for the Peiligang site [Table 4(7), Lab. No. ZK572] is an outlier which should probably be disregarded. The earliest farmers are then represented at Jiahu, a site of the Peiligang culture. Six dates were obtained for this assemblage, ranging from 8561 to 7529 cal. BP [Table 4(7)], representing three phases. The dates are in sequential order, and consistent with the sequence of strata. The Jiahu site is situated on an alluvial plain and thus without the concerns about dating accuracy in limestone environments, so these dates seem reliable.

However, the Jiahu assemblage shows a cultural complexity that is more advanced than other Peiligang assemblages which are dated slightly later [Table 4(7)] This cultural complexity suggests that the Jiahu occupants should not have been the earliest farmers in that region. There must have been a preceding culture as its ancestor.

But all the other Neolithic assemblages in the Yellow valley are dated later than the early phase of Jiahu, or roughly to the same period. These assemblages cannot be ancestors. Presently, it can only be said that agriculture emerged no later than 8500 BP in the middle Yellow basin.

The Houli culture in the lower Yellow valley raises another question. Dated from 8501 to 7667 cal. BP (the Xihe assemblage of the Houli culture) [Table 4(7)], these assemblages contain a toolkit that seems identical to that of the other farming societies of the middle Yellow valley. Remains of houses, kilns, burials and pits in several sites suggest that the Houli and Xihe occupants were sedentary. It is only the absence of domesticated cereal remains from the Houli assemblage that casts doubt over the assumption that they were farmers. This issue awaits further archaeological investigation.

In the Yangzi valley, the earliest farmers may be represented by the Pengtoushan assemblage. A total of 24 absolute dates have been obtained for Pengtoushan, but some seem problematic. The charcoal extracted from potsherds is dated much older than the organic acid and rice straw from the same sherds [Table 7(2)]. This problem occurred repeatedly in three separate sherds [potsherd I, II, III, see Table 7(2)], and may reflect the inclusion of old organic materials in the clay source (Chen and Hedges 1994). Most other dates from the site range around 8000 to 8995 BP [Table 7(2)]. This is probably the temporal range for the Pengtoushan assemblage.

The problem for the Pengtoushan assemblage is the taxonomic identification of the rice remains. As the rice husks found in Pengtoushan cannot be identified to species, the question of whether it was a wild or domesticated species remains. The earliest identified cultivated rice in the Yangzi valley to date is found in Bashidang. The Bashidang assemblage is said to be slightly later than Pengtoushan, approximately at around 8000 BP (Pei 1998a). Until further data become available, it would be prudent to suggest that rice cultivation commenced by 8000 BP in the middle Yangzi valley.

For the lower Yangzi valley, it is not certain whether agriculture is represented in the early phase of Xianrendong assemblage. A phytolith analysis suggests that both wild and domesticated rice were present in zone D and E of Daotonghuan (the site has been assigned to the middle Yangzi valley by Zhao) (Zhao 1998), which is estimated to be around 10,000-8000 BP (Zhao 1998). But as no macro-remains of domesticated rice have been found there, and it has been argued that the identification of rice species based upon phytoliths is very difficult (Lu *et al.* 1997), this proposition is skeptical.

It is certain that by 7000 BP several agriculture societies occupied the lower Yangzi valley. As discussed in chapter 8, the pottery of these agricultural societies seems quite different from that found in Xianrendong, but shows similarities to the Pengtoushan pottery. Whether rice cultivation in the lower Yangzi valley originated independently from a Xianrendong foundation, or it was a result of cultural diffusion from the middle Yangzi valley, is still a question. Current archaeological data seem to suggest the latter (below).

In summary, agricultural societies existed in both the Yellow and the Yangzi valleys by 8000 BP. Did agriculture originate independently in both river valleys? Or did it emerge in one centre and expand ? WHERE did agriculture first originate in China?

11B. The WHERE question - one, or two centres, or multiple centres?

At present, archaeological data seem to suggest that agriculture emerged in both the Yellow and the Yangzi valleys by 8000 BP. In the Yellow valley, the wide dispersal and contemporaneity of the early agricultural societies are two prominent characteristics. The Dadiwan culture in the Wei valley is dated to 7884-7385 BP; the Cishan assemblage in the Hebei Plain is dated from 8052 to 7550 BP, the Peiligang culture in the middle Yellow valley dates from about 8500 to 7500 BP [Table 4(7)]. The Houli culture, a possible agriculture society in the lower Yellow valley, dates from around 8500 to 7660 BP [Table 4(7)].

Spreading over the vast area of Central China, these assemblages all consist of similar house forms and similar toolkits, particularly the polished axes and spades, grinding slabs and rollers, and a small number of microblades and/or implements of the flake tradition. But their pottery differed in some aspects [Fig. 4(10)]. The toolkit similarities suggest two points: first, these Neolithic assemblages might have come from a common lithic tradition; second, their subsistence strategies would have been similar - an economy of cereal cultivation supplemented by hunting and gathering. On the other hand, the differences of pottery indicate that they were different Neolithic groups and had been developing in their own regions for some time, although with cultural contact (chapter 10).

The microblades and the flake tradition, as well as the forerunners of the Neolithic axes, grinding slabs and rollers, have been found in the Xiachuan assemblages in Shanxi province (also called the Shanxi Basin), at about 20,000 to 13,000 BP [Table 4(1)]. At about 10,000 BP or slightly earlier there was the Shizitan assemblage in the western mountain/plateau area; the Shayuan assemblage in the Wei River valley, with Lingjing and Dagang in the middle Yellow valley, and various surface archaeological remains in the lower Yellow valley. All these assemblages contain microblades, the flake tradition, and various heavy-duty stone tools. At about 8500 to 7500 BP numerous Neolithic groups occurred from the middle to the lower Yellow valley, with different pottery vessels, but similar lithic tools. All these discoveries seem to suggest that the Xiachuan culture, or a direct descendant, was a common ancestor for the early Neolithic assemblages in the middle and the lower Yellow valley. This common ancestor seems to have dispersed along the Yellow River by 10,000 BP towards the west and the east, represented by the Shayuan, Lingjing and Qingfengling assemblages [Fig. 4(1)].

The toolkit of Xiachuan indicates that extensive foraging activities were practised, very possibly with cereal exploitation. It was based upon this subsistence strategy that various agricultural societies developed by 8500-8000 BP, represented by the Dadiwan, Cishan, Peiligang, and probably Houli and Xiaoshankou cultures. These early Neolithic groups still retained some of their lithic inheritance - the flake tradition and microblades, while the axes and grinding tools were further refined and developed for cereal exploitation and cultivation, and became predominant tools in the early farming societies.

Presently, it is difficult to pinpoint the exact place for the origin of agriculture in the Yellow basin due to lack of data. There are two possibilities: the first one is that agriculture originated in the Shanxi Basin amongst the descendants of the Xiachuan people, and dispersed to the middle and lower Yellow valley at the beginning of the Holocene. The second possibility is that the Xiachuan people first dispersed along the Yellow valley in the late Pleistocene, and agriculture originated from one or more of these descendant groups. Only further archaeological discoveries and research can solve this question. At this stage, the whole middle Yellow valley should be viewed as one centre.

As no pottery was found at Dali, Lingjing, Dagang or in the surface collections in the lower Yellow valley, pottery manufacturing must have occurred after the dispersal. The potsherds found in Hutouliang and Nanzhuangtou probably represent the early stage of pottery manufacturing in Central China to date. The microblades found in Hutouliang, and the fragment of a grinding slab found in Nanzhuangtou, indicate some cultural links between these two and the Xiachuan assemblages. Archaeological discoveries seem to suggest that the Xiachuan complex dispersed its cultural influences towards several directions after 18,000 BP [Fig. 11(1)]. This dispersal may be with or without human migrations.

The close relationship among the various early Neolithic assemblages in the Yellow valley can also be understood by physical anthropology. Only one population of the Peiligang culture from Shigu in Henan province has been subjected to analysis, but several middle Neolithic groups have been analysed, and different propositions have been submitted (e.g. Wang 1986; Zhang *et al.* 1982). Chinese scholars all agree that the populations from the middle to the lower Yellow valley at around 7000 to 6000 BP belonged to one regional group (Zhang 1989). Further, a cluster analysis shows that there was close physical resemblance between groups in the western Fen-Wei river valley, and the eastern lower Yellow valley (Zhang 1989) [Fig. 11(2)] [Table 11(1)].

Given the result of this analysis, two inferences can be proposed: first, various prehistoric groups had dispersed to different areas along the Yellow River long before 7000 BP, as indicated by certain degree of differences in terms of their physical features; second, these groups might have shared a common ancestor; otherwise it cannot be explained why the prehistoric peoples in the western Fen-Wei River and in the eastern lower Yellow valley would have had a close physical resemblance.

This dispersal might have been stimulated by the increasing population during the improving palaeoclimate after the LGM. As discussed in chapter 9, the population during the terminal Pleistocene in the western plateau/mountain of the middle Yellow valley seems to have been quite dense, while the North China Plain seems to have been only sparsely inhabited. Yet after the LGM the incidence of archaeological sites increased significantly in the North China Plain [Fig. 4(1)]

On the other hand, it is puzzling that so few remains of the early Neolithic have been found in the western plateau/mountain area. Whether the descendants of the Xiachuan-Xueguan-Shizitan groups abandoned this area by the end of the Pleistocene and moved to the alluvial plains more suitable for cultivation, or whether there are early Neolithic remains in the western plateau/mountain region yet to be discovered, is a subject which still awaits further investigation.

This 'out of the plateau/mountain' hypothesis will have to be tested in future by archaeological and physical anthropological investigations. At present, human remains are so insufficient that a detailed comparison of physical features between the Palaeolithic and the Neolithic inhabitants of the Yellow valley is not feasible.

In the middle Yangzi valley, it is clear that the material culture during the early Holocene differed from that of the Yellow valley. It also seems quite certain that wild rice was collected in this region then. The analysis of rice found in Bashidang indicates that rice cultivation began at around 8000 BP (Zhang and Pei 1997). Whether this analysis can support the argument that rice was also cultivated in Pengtoushan is another question.

Bashidang was succeeded by the Fenshanbao and Hujiawuchang assemblages (chapter 7). The Chengbeixi assemblage seems to have been influenced by the Pengtoushan assemblage in terms of rice collection or cultivation, but it might have been a different group whose subsistence strategy was primarily based upon fishing, hunting and gathering along the Three Gorges in the early Holocene. Rice cultivation later became more important in the Daxi culture, which is the descendent of the Chengbeixi culture (Chen 1991).

It has been proposed that the middle Yangzi valley is the incipient centre of agriculture origination, and that the Jiahu assemblage of the Peiligang culture was a descendant of the Pengtoushan culture. This argument is based upon pottery analysis, the presence of rice cultivation at Jiahu, and the earlier dates of the Pengtoushan assemblage (Liu 1994). As the early stage of Jiahu is dated earlier than all the other Peiligang, Dadiwan and Cishan assemblages, one question arises: is it possible that agriculture first originated in the middle Yangzi valley, then expanded northwards, and because the environment in the Yellow valley was not suitable for rice cultivation, millet was cultivated instead?

However, a crucial question on this proposition is, again,

the uncertainty of whether the Pengtoushan occupants were rice farmers, or only collectors. If they were only collectors, and rice was only cultivated in Bashidang, then as the date of Bashidang is later than the date of the early stage of Jiahu, a counter-argument becomes possible. As cultivated rice has been reported from the early phase in Jiahu, is it possible that the millet cultivators in Jiahu from the Yellow valley took advantage of the warm and moist palaeoclimate in the Huai valley, and added rice to their crops? Is it possible that the collectors in the middle Yangzi valley received cultural influences from the Jiahu group, and developed their rice cultivation?

In addition, there is little evidence to suggest that Pengtoushan and Jiahu belong to one archaeological culture. The Pengtoushan and Jiahu lithic toolkits are dissimilar. On the other hand, the Jiahu and Peiligang tools are highly identical [Table 11(2)]. The lithic toolkit of Pengtoushan forms part of the pebble industry in the Yangzi valley, and the Jiahu assemblage belongs to the lithic tradition in the Yellow valley.

Regarding the pottery similarities between Pengtoushan and Jiahu, it is true that wood charcoal and other plant remains were used as tempers in both assemblages, and the outcome was a black pottery with carbonaceous temper. But it should be noticed that, first, charcoal and plant tempers were absent in the first phase of Jiahu, and only present from the second to the third phases; second, while black pottery was the only type in Pengtoushan, the dominant colour in Jiahu pottery was red instead of black (IA, Henan Province 1989). On the other hand, red pottery with sand temper was the predominant type in the Peiligang culture. Apparently, Jiahu pottery belonged to the Peiligang culture, but received some influence from the Pengtoushan culture.

These comparisons show that the Jiahu occupants were not descendants of the Pengtoushan people. However, the possibility of cultural contact and interactions cannot be ruled out. At this stage, it is difficult to conclude whether the idea of rice cultivation was diffused from the Yangzi valley to the Huai valley, or vice versa, or even occurred independently in the two areas.

Another question regarding the centre of the origin of agriculture is the uncertainty of subsistence strategy of the Nanzhuangtou and Houli sites in the Yellow valley, which are dated to around 10, 000 and 8500 BP respectively [Table 4(7)]. If it can be confirmed that cereal cultivation was practised either in Nanzhuangtou and/or in Houli, then clearly there was an independent centre for millet cultivation in the Yellow valley before 8500 BP.

In the Yangzi valley, recent research on the Xianrendong assemblage stimulates the proposition that rice cultivation might have originated in the lower Yangzi valley between 9600 and 12,000 years ago (MacNeish and Libby 1995 eds.). This proposition, however, is arguable. While there is evidence for rice gathering in Xianrendong, no solid evidence for intensive cultivation has been found in strata dated to that period. And the increased shell midden deposit found in the early Holocene stratum at Xianrendong (MacNeish 1995) does not support the assumption that rice cultivation was a primary economic activity there (see chapter 7 for details).

The difference of stone toolkits of Hemudu and Xianrendong should be noted. While flakes and pebble tools are major implements at Xianrendong, polished tools dominate the Hemudu assemblage (Liu and Yao 1993). The polished axe is a major tool in the Hemudu assemblage, yet it is completely absent at Xianrendong even in the late phase (MacNeish 1995), which should be contemporary with Hemudu. Pebble tools such as perforated discs and 'lances' become abundant in the early Holocene phase in Xianrendong, but are absent in Hemudu, where pebbles are only found as abraders (Liu and Yao 1993). In sum, the dissimilarity of lithic implements between Xianrendong and Hemudu makes it difficult to suggest a close cultural link between the two.

The pottery vessels of Xianrendong and Hemudu also differ. Most of the pottery vessels in Xianrendong are of red to yellow colour with a sand temper (MacNeish 1995), but the Hemudu pottery is dominated by a black colour with carbonised rice husks and other plant remains as tempers (Liu and Yao 1993). While *fu* have been found at Xianrendong, and a variety of *fu* dominated the Hemudu pottery, this cooking vessel has been widely found in the middle and lower Yangzi valley, as well as in South China from the early to the mid Holocene. Therefore, the presence of *fu* alone cannot confirm a close cultural link between the Xianrendong and Hemudu assemblages.

If the Hemudu assemblage was not a direct descendant of the Xianrendong assemblage, was there another ancestor? The first phase of the Hemudu assemblage (layer 4) was accumulated on a marine-sediment (Liu and Yao 1993), indicating that the earliest occupants must have come from somewhere else. According to a physical anthropological analysis of two skulls found in layer 3 at Hemudu, the occupants were anthropologically close to the Zengpiyan group in Guangxi administrative district, but differed from those in the Yellow valley (Han and Pan 1983).

As discussed in chapter 10, the Pengtoushan/Bashidang pottery in the middle Yangzi valley is also black, filled with carbonised rice husks and other plant remains (Pei 1995). The lithic implements of these assemblages consist of pebbles and polished tools, including polished axes (Pei 1995, 1998a). The Pengtoushan/Bashidang assemblages are dated from 9000 to 8000 BP onwards, which are one to two thousand years earlier than the Hemudu assemblage. It is possible that the Hemudu and Luojiajiao assemblages represent cultural dispersal from the middle Yangzi valley.

In summary, current archaeological data suggest that agriculture occurred in both the Yellow and the Yangzi valleys by 8500- 8000 years ago, characterised by different cultivars, and created by different groups. However, because there were similar cultural components shared between the Jiahu and Pengtoushan assemblages, a possibility of 'cultural idea dispersal' between the Yangzi and the Yellow valleys remains open.

11C. How did agriculture originate in China?

Studies on the question of how agriculture originated have been conducted by many scholars in the Middle East, Europe and America, and various techniques have been applied, including harvesting and cultivation experiments using wild cereals, use-wear analysis, tool-residue analysis, and ethnographic observations (eg. Anderson 1992 ed.; Harlan 1967; Unger-Hamilton 1985). The outcome of these studies seems to indicate one general proposition: agriculture was developed from wild cereal collection, through cultivation, to cereal domestication.

This proposition seems also tenable in the Yangzi valley, as remains of wild rice were found in the Yuchan and Xianrendong caves, whereas cultivated rice has been uncovered only in open sites. In the Yellow basin, although no botanical remains of wild cereals have yet been found, the presence of grinding tools and blades with glossed edges also seem to suggest the collection of cereals prior to the emergence of agriculture.

Based on cultivation experiments with wild wheat and

barley, Hillman and Davies (1992) suggest that two husbandry methods would have been essential for the development of domesticated crops:
"1. particular harvesting methods were used.
2. the crops were harvested when partially ripe or <<near ripe>>"
(Hillman and Davies 1992:124). They also suggest that "annual extensions or shifts in the areas under cultivation" were important in order to maintain the mutant phenotypes (Hillman and Davies 1992:124).

In the Chinese context, the initial husbandry methods used by early farmers are little known, and one can only get a glimpse of what would have happened when agriculture first emerged. In the Yellow valley, regular rotation of cultivated crops would have to be practised for foxtail millet. Even today, cultivation of this crop is on a rotating basis, or field shifting, otherwise there will be a substantial increase of weeds, insects and diseases in the second year (Anonymous 1976). It is clear that regular rotation was crucial for the success of millet cultivation.

Ethnographic data from Taiwan document that millet is cultivated by a slash and burn method (Fogg 1983). In Southwest China during the Ching Dynasty (1644-1911), some minorities cultivated corn and millet by the slash-and-burn method, and they did not attend to the crop until ripening time (Chen 1991). No weeding or fertilising was carried out. It is recorded that a site cultivated for one year was abandoned for another eight to ten years before being selected for cultivation again (Chen 1991). This period was to allow the growth of trees and grasses for the next burning as fertilisers (Chen 1991). When all the land near the residential site was used, people moved to another area and started the same process again (Chen 1991).

The early farming techniques for millets in the Yellow valley are not clear at this stage, although the 'slash and burn' method has been proposed as the initial technique (Zhang 1984). This is only a hypothesis without evidence from archaeological discoveries or use-wear analysis. Thus the question remains to be further investigated.

As discussed above, the Cishan and Peiligang cultures do not seem to have been the earliest farming societies in the Yellow valley. But some of their specialised tools may still provide a clue for the husbandry methods used by their ancestors. For example, stone sickles are widely found in the Cishan and Peiligang cultures. This tool is quite standard and well polished [Figs. 4(18) and 4(22)]. It may be inferred that sickles were also used by the earliest farmers in the Yellow valley for harvesting, which would have produced a favourable selection for a tough rachis and synchronous ripening, thus leading to the domestication of millet (chapter 5; Hillman and Davies 1992).

Regarding the initial husbandry method for rice in the Yangzi valley, very little is known. The Pengtoushan toolkit contains nothing specialised for soil digging, or harvesting. It is not know if rice was cut, or pulled, or if only the panicles were collected. Nor is it known whether field shifting would have been required. Bone and wooden tools are reported from Bashidang, and it is proposed that some of the wooden tools could have been used for soil digging (Pei 1998a). Again, this proposition awaits further testing.

Although cultivation experiments on wild rice were carried out in the 1950s and 1960s (Oka and Chang 1959; Oka and Morishima 1971), the experimental method was to soak, and then transplant the seedlings in the plot, like modern cultivation. On the other hand, White (1995) proposes that 'initial cultivation of an area with wetland rice need not require extensive or thorough clearance'. She observed in northern Thailand that "rice was planted in natural depressions where water tends to collect from runoff and seepage, again without the need for a dike", and "rice seed can be broadcast or even dibbled into the soil without transplanting" (White 1995: 51).These seem more likely to be initial methods for rice cultivation.

It is generally agreed that annual wild rice of Asia (*Oryza spontanea*) was derived from the perennial wild rice (*O. rufipogon*), and the latter is the ultimate progenitor of domesticated rice (Oka 1988). Ding (1933) has argued that perennial wild rice is an aquatic plant which can only survive in swamps (Ding 1933). On the other hand, annual wild rice is able to survive in seasonally wet and disturbed habitats (Oka 1975). As annual wild rice is derived from the perennial type, these botanical characteristics of annual wild rice must be a later adaptive mechanisms. The unclear point here is whether these later adaptations were due to human interference, or to the need for survival and expansion of this species in different environments.

However, even the annual rice needs water during most parts of its life cycle. But it is not clear whether rice was initially cultivated in the annual, or the perennial form. Nor is it known whether initial cultivation was in a permanent swamp, or in a disturbed habitat. The earliest rice fields excavated to date are those found in Chengtoushan, middle Yangzi valley, located in the same Li county where the Pengtoushan assemblage was found; and in Caoxieshan, Jiangsu province, in the lower Yangzi valley. Both are dated to about 6000 BP (Jiang 1997b; Zhou 1996 personnel communication). It is said that the Chengtoushan rice field is similar to those of the present (Jiang 1997b), and the Caoxieshan rice field combines one large water pool connecting several plots with small water channels (Jiang 1997b). Abundant domesticated rice phytoliths and/or grains have been found in both Chengtoushan and Caoxieshan; the quantity of rice phytoliths found in Chengtoushan is said to be similar to that of a modern rice field (Jiang 1997b). In Caoxieshan, it is proposed that rice was probably grown in the plots, and the water supply was controlled by the pool and channels. A similar system of cultivation is still found in that area for cultivating some aquatic vegetables (Zhou 1996, personal communication). Was initial rice cultivation similar to this method? This question awaits further investigation and experiments.

11D. "Why should I farm when there are so many mongongo nuts?" - the impetus for the origins of agriculture in China.

As discussed in chapter 1, numerous hypothesis and models have been created for the impetus of the origins of agriculture. But as Harlan (1995) points out, these models are created by modern, university-educated scholars, and may be too far from reality (Harlan 1995:25). The above quoted comment from a Bushman (cited from Harlan 1995:25) may be more close to the way of thinking of foragers.

In fact, why should one bother to put in all the effort required to farm, when there was plenty of food just waiting for one to gather? A similar comment is also reported from Australian Aborigines (Harlan 1995: 26). The way of life of foragers, in the 1960s and 1970s, was described as 'affluent' (Kelly 1995). It was reported then that foragers spent limited time in food procurement, had more leisure time, did not suffer starvation and were in harmony with their environments (for a review, see Kelly 1995). In addition, the Australian Aborigines had their feasts and ceremonies based upon a foraging economy (eg. Cane 1984). If foragers were not in need of food, nor in need of exotic items for ceremonies, neither the proposition of food crisis nor that of social complexity can explain the

origins of agriculture. If the prehistoric foragers lived the same way as their present counterparts, what would have been the impetus for the origins of agriculture?

But, two questions arise from the above observations. First, was the foraging way of life really as rosy as described? Second, did prehistoric foragers live in a way similar to their present counterparts? For the first question, the answer is 'not quite'. It is now clear that some foragers do work long hours, and suffer malnutrition due to seasonal fluctuations and failure of natural resources (for a comprehensive review, see Kelly 1995). The ethnographic data seem to indicate that, in a foraging economy which completely relies on natural resources, seasonality and the unpredictable availability of resources are two major problems, particularly the latter, which is determined by the natural abilities of reproduction of floral and faunal species, also climatic changes and other natural disasters. These two problems are vital for the survival of foragers.

While climatic deterioration and other natural disasters may not happen every year, the reproductive ability (RA for short) is a vital factor for the annually continuous availability of natural resources. RA of herbs and grasses is often very high, probably over a hundred times (eg. one seed of green foxtail can produce a whole plant with hundreds of seeds in the next season), and the reproduction is often on annual basis. The RA for trees is lower than herbs due to two reasons: first, trees can not produce nuts or fruits until they are several to tens of years old, depending on their species (Allaby 1985 ed.). Second, the abundance of nuts or fruits fluctuates. One year there are many, but the next year there can be many less. This has been reported for foragers collecting pine cones in the Great Basin (Steward 1938).

Regarding animals, the RA of fish and molluscs is also high (probably over a hundred offsprings per adult), and often is annual. But for mammals, particularly ungulates which are major game, the RA is very low. Only one to several young can be produced each time, and the reproduction may not be annual (Nowak 1991). In addition, the young are usually targeted by human hunters and carnivores, thus further reducing the availability of these animals.

Contemporary foragers overcome (or attempt to overcome) the problems of seasonality and availability by placing their food supplies on a very broad base, in order to ensure that first, there are always food supplies during each season of the year; and second, if one or more foods fail, there are always many supplements (eg. Cane 1984; Kroeber 1925). The Australian Aborigines gathered their food from over 70 plants in the desert area, among which 42 were seed-producing species (Cane 1984). They also hunted and gathered over forty species of small game, birds and insects with high RA (Gould 1980). Steward (1938) reported that 81 species of plants were gathered by the Gosiute in the Great Basin, and 54 species were exploited in Owens Valley, in addition to game hunting and fishing (Steward 1938:19).

Even so, food shortages still occur. Steward (1938) observed that the Indians in the Great Basin had problems of food supply due to the unreliability and unpredictability of plant harvests (Steward 1938). In Australia, foragers also have problems with getting food in the dry season (eg. Allen 1974; Cane 1984). It seems that the seasonality and unpredictability of natural food supplies are not easily overcome by simply practising a broader spectrum foraging strategy.

Therefore, methods to combat the two problems developed. The first is storage, practised by many foraging groups in Australia (Allen 1974; Cane 1984) and America (Steward 1938). The second method, or methods, is to maintain or even to increase the productivity of some staple food. The present-day foragers have developed various methods to maintain food resources. Some groups of Indians in the Great Basin sowed grass seeds and/or irrigated plots in order to maintain or increase productivity (Asch 1994; Keeley 1995; Steward 1938). In Australia, it is reported that when Aborigines in the Cape York Peninsula collect a tuber, part of the plant will be replaced into the soil to assure future supply (Hynes and Chase 1982). In addition, fire is frequently used to increase the growth of seed-bearing grasses (Nicholson 1981), and barriers are erected to protect plants and seedlings which sprouted near camps (Hynes and Chase 1982). These behaviours may not necessarily lead to cultivation and agriculture, but they clearly demonstrate a desire to maintain and increase food supplies.

It should be noted that not every forager group can have their food supplies as diverse as they like. The diversity of food resources depends on the environment. If there were still food shortages for people living on diversified food resources, like the Indians in the Great Basin whose food composition included floral and faunal resources of both high and low RA, the life for poorer-habitat foragers living on less diversified resources must have been even harder.

It probably will never be certain whether prehistoric foragers 10,000 years ago lived the same way of life as any contemporary foragers do, as so many changes in environment, climate as well as human culture and society have happened since then. However, archaeological discoveries illustrate that some foragers of the terminal Pleistocene also based their food supplies on a broad resource base, as revealed by discoveries in the Middle East (see chapter 9). It seems logical to infer that the problems confronted by contemporary foragers would have occurred in prehistory as well, and there would also have been the desire of maintaining and/or increasing food productivity. Although this desire alone might not have led to the emergence of agriculture, it would have been the fundamental impetus for agricultural origination.

The palaeoclimatic changes during the terminal Pleistocene and their impact on natural resources might have played an important role for the transition from foraging to farming. The LGM was a global event. It is noticeable that immediately prior to, and/or during the LGM, microliths or microblades occurred widely in the Middle East (eg. Bar-Yosef and Belfer-Cohen 1989), Europe (Champion *et al.* 1984), the Nile valley (Clark 1985), India (Misra 1985) and East Asia (Chen 1984). In the same period, grinding and/or pounding tools are found in Africa (Kraybill 1978), the Middle East (Wright 1991), Australia (Veth *et al.* 1997), America (Carter 1980) and China. This is probably not a coincidence. The presence of grinding tools and microliths may not only indicate technical progress in terms of manufacturing specialised tools for efficient exploitation of lithic raw materials. They also suggest a worldwide change of subsistence strategy, probably with an inclination towards plant exploitation, particularly seeds from wild grasses. In some areas this exploitation became the prelude to agriculture, as in the Middle East, China and America. In other areas it developed into an even broader spectrum of foraging strategies, such as those in ethnographic Australia.

Grass seeds are smaller than fruits and tubers, which means that more are needed to produce a given amount of calories. The harvesting return for some grass species is high (eg. Cane 1984; Harlan 1967; author's work on rice in 1996), but is quite low for others such as green foxtail (chapter 5). Further, ethnographic data demonstrate that seed-processing is often a laborious and time-consuming task (eg. Cane 1984; Horne and Aiston 1924). No wonder the

Australian Aborigines stated clearly that they would always prefer fruits and tubers to grass seeds (Cane 1984).

So, what forced prehistoric foragers to collect seeds? Contemporary foragers collect seeds because they are abundant; and because they can be stored for the lean seasons (Cane 1984; Kelly 1995), thus seeds are often insurance for the hard time (Keeley 1995). It is legitimate to infer that similar causes were the impetus for prehistoric foragers in the late Pleistocene.

During the LGM, the palaeoclimatic changes would have dramatically reduced natural resources. This would have forced foragers to have shifted their attentions to grass seeds with high RA, abundant in a dry and cool climate. Foragers would have gradually come to know the benefits of seeds under drying conditions, and some species probably became staple or secondary staple foods, similar to the situation in Australia (Veth and Walsh 1988). Ethnographic data suggest that present foragers cultivate some grass species to insure the food supply during the hard time (Keeley 1995). A similar impetus can be proposed for the prehistoric foragers. Another possibility would be that when the productivity of these staple foods could not meet the demands of an increasing population, and there were not sufficient supplements, cultivation would have been stimulated.

This hypothesis seems tenable in China. According to pollen reports, it is clear that grasses and herbs dominated the vegetation both in the Yellow and the Yangzi valleys during the LGM, while fruits and nuts decreased or were absent. Faunal resources also reduced, and ungulates with low RA dominated the animals. In this circumstance the importance of grasses would have progressively increased.

As discussed in chapter 3B, green foxtail is a highly adaptive plant in the Yellow valley, and its expansion and survival ability are excellent. This plant today is still widely distributed in the Eurasian continent. In the Yangzi valley, wild rice was another plant with a large grain, good taste and a high return on harvesting. Seeds from both species are storable. While the RA for perennial wild rice is not very high (below), each plant of green foxtail can produce hundreds of seeds every year. These values would have been recognized by the prehistoric foragers.

Was the palaeoclimatic change of the LGM the only cause for the emergence of extensive seed collection? Pollen analysis indicates that herbs and grasses dominated in Central China during the glacial periods of the Pleistocene (Tong *et al.* 1992). However, archaeological evidence of extensive seed collection is only present during the terminal Pleistocene in China. So other causes might have contributed to the emergence of extensive seed collection.

According to geological studies, the continuous uplift of the Qin Mountains since the middle Pleistocene blocked the warm and moist summer monsoon from the southeast, and Central China became much drier and cooler during the terminal Pleistocene (Qi 1989). Consequently, grasses and herbs would have become more dominant in this region during the LGM. Meanwhile, it became harder for sub-tropical plants to survive (Qi 1989:279). It seems that the availability of high-return floral resources, such as nuts and fruits, was further reduced during the terminal Pleistocene .

On the other hand, the distribution of archaeological sites indicates that the population during the terminal Pleistocene in China, particularly in the middle Yellow basin, was denser than that in previous periods. Consequently, the territory and the terrestrial resources available for each prehistoric forager group would have been reduced. When there were insufficient foods of high rank (i.e. high return), people would have had to shift to grass seeds.

Grass exploitation not only enabled people to survive, it might have also reduced the mobility of the foragers. The Australian Aborigines and the Great Basin Indians both resided in a certain area, and usually moved among seasonal camps in their own territory, unless ceremonies or harsh seasons forced them to move further (eg. Cane 1984; Kroeber 1925; Steward 1938). In a broad sense, these groups were quite sedentary within a given area. This is probably because seeds and other plants were gathered each year in the same areas and the same season, so the foragers were relatively tied to a certain region. Such a relatively sedentary way of life would have increased the population, which would have led to further exploitation of natural resources, and would have accelerated the possibility of food shortage. The desire of maintaining or increasing food supplies would have been a strong impetus for cultivation to begin.

In the Chinese context, the toolkit of the Xiachuan assemblage indicates that grasses were exploited; and the long-lasting presence of the same archaeological assemblages in the Xiachuan Basin probably indicates continuing exploitation. On the other hand, the low incidence of animal bones found in Xiachuan may indicate a lower availability of faunal resources.

A persistence of human exploitation in a given area would have led to a twofold result: first, the natural resources, particularly the faunal resources in that region would have decreased. This would have forced foragers into more intensive exploitation of wild cereals. Second, their increasingly sedentary life - partially due to their reliance on annual grasses and probably also restricted by the limited territory - would have led to an increase of population, as illustrated by ethnographic data and analysis of contemporary foragers (Kelly 1995). Depletion of natural resources and increase of population would have led to a break of balance between the foragers and their natural resource base.

Initially, this problem could have been solved by moving to another area, which is what contemporary foragers often do (Cane 1984; Steward 1938), and was probably what prehistoric foragers often did. But if there was limited possibility of moving out; if the population was quite dense in the whole area and migration was not permitted, then foragers would have had to opt for an attempt to maintain the availability of their staple food by cultivation.

As discussed in chapter 9, the population in the western plateau/mountain area (also called the 'Shanxi Basin') was quite dense during the terminal Pleistocene. Again, environmental change during the terminal Pleistocene might have played a role here. It seems that the population in the Shanxi Basin was denser than in other areas during the LGM. Although the present effective temperature in the Shanxi Basin is lower than that in the North China Plain [Table 11(3)], and the situation during the LGM would probably have been similar, the forest in the western and eastern mountains with its higher moisture might have supported more useful fauna and flora than the eastern plains, the latter mainly being steppe and pebble desert then (Wang *et al.* 1995). It is possible that the harsh LGM environment forced foragers to remain in a basin where the natural resources might have been relatively abundant. Consequently, the population density in that area increased, and the territory for each group was probably quite limited. This would have accelerated the depletion of natural resources, particularly animal resources. Consequently, plant exploitation would have been more intensified, and eventually caused the need for increasing plant availability

by cultivation.

In summary, the combined effect of reduced natural resources, a harsher environment from 18,000 to 13,000 BP in China, an increase of population and a decrease of wild food diversity would have forced foragers to focus on grass exploitation, and the demand for more grass seeds would have been the impetus for the emergence of agriculture in Central China. Interestingly, a similar explanation has been recorded in ancient Chinese documents (see chapter 1 for details).

For the Yangzi valley, the situation was probably slightly different. While the LGM impact seems less severe there, local pollen profiles indicate that the climate was still colder than the present during the terminal Pleistocene (Wang *et al.* 1990), and vegetable resources decreased. Unlike South China, yams and tubers were probably much less abundant during the LGM in the Yangzi valley, as these were mainly sub-tropical and tropical plants, whereas the Yangzi valley during the LGM would have been a temperate to cool-temperate zone [Table 11(3)]. On the other hand, pollen reports show that herbs dominated the vegetation in that area from 18,000 to 13,000 BP. Again, foragers would have been forced to turn their attention to the grass seeds, which also eventually led to the emergence of cereal domestication.

There is little evidence to indicate that population in the Yangzi valley was dense during the LGM. Presently, the population pressure in the Yangzi valley seems less than in the Yellow valley. In this case, why would foragers bother to cultivate rice? To address this question it is necessary to investigate the availability of wild rice in the Yangzi valley, and the cultural relationships between the Yangzi valley and South China.

As discussed in chapter 6B, wild rice is a sub-tropical to tropical plant, with scanty stands in the middle and lower Yangzi valley today (The National Survey Group of Wild Rice 1984) [Fig. 6(8)]. Of course, the present distribution of this wild plant may not represent its distribution some 8000 to 10,000 years ago, but according to the palaeoclimate reconstruction, the temperature during the LGM in the Yangzi valley then would have been cooler than that of the present [Table 11(3)], with less rainfall (chapter 6). Therefore, wild rice is not likely to have been abundant during the LGM in the Yangzi valley.

On the other hand, pollen reports document that South China was still a sub-tropical zone, the palaeoclimate being temperate to warm, and relative moist even during the LGM (Wang *et al.* 1990). In South China, the terminal Pleistocene still witnessed the dominance of evergreen and broad-leafed trees with some coniferous species; while grasses and tubers were abundant (Wang *et al.* 1990). It is estimated that the average annual temperature during the LGM was 15-18°C, about 4-5°C lower than the present (Wang *et al.* 1990) [Table 11(3)]. Given the cold tolerance of wild rice (Sun 1990), this species would have survived well in South China from the terminal Pleistocene to the early Holocene.

The prehistory of the terminal Pleistocene to the early Holocene in areas south of Central China is characterised by the dominance of a pebble tool industry, distributed from the Yangzi valley southwards. It is notable that the lithic assemblages found in Yuchan and Xianrendong strongly resemble the pebble tools in South China (Peng 1996; Yuan 1996), clearly indicating a close cultural connection between foragers in the Yangzi basin and South China. In fact, similar tools have been found in many caves in South China, namely, Miaoyan, Zengpiyan and Bailandong in Guangxi district, Dushizi and Huangyandong in Guangdong province (Jiao 1994) [Fig. 7(1)]. Small flakes, discoids and perforated pebbles are found in South China from the terminal Pleistocene to the early Holocene (Jiao 1994), and pottery is first present in Dalongtan and Miaoyan, the former dated to a controversial 23000 to 18000 bp (An 1994), and the latter to around 10,000 bp (Yuan 1993).

If a similar toolkit indicates a similar subsistence strategy, it can be inferred that the economic activities in South China during the terminal Pleistocene would have been similar to those of the Yuchan and Xianrendong occupants, probably on a more diversified resource base, as pollen reports document that plant resources were richer in South China from the terminal Pleistocene to the early Holocene (Wang *et al.* 1990). If the presence of pottery can be used as indirect evidence of grass exploitation, it is also possible that wild rice and other plants were collected by foragers in South China in the terminal Pleistocene.

But why is there no evidence for early agriculture based upon rice cultivation in South China? This is probably due to the substantial natural resources, including wild rice in that region. According to a survey, even today, wild rice is extremely abundant in some areas of South China. In the 1978-1979 survey in Guangxi, wild rice was found at 758 places; among which, 13 patches were each measured to over 2.6 hectares, the biggest stand covering an area of 31.1 hectares (Wu 1990). It can be inferred that a few thousand years ago, with much less population and none of the modern human disturbances such as fields, buildings, highways and roads, the stands of wild rice would have been much denser.

Such a rich availability of wild rice, as well as numerous other resources including animals and species of tubers, yams and nuts, many of them with a high RA, would have provided abundant food for foragers in South China. In addition, South China is a sub-tropical area, and foods can be found year-round. The problems of seasonality and availability would have been less severe there. In such an affluent environment, why should they bother to farm? In fact, evidence of rice cultivation does not occur in South China until 4500 BP, in the Shixia assemblage (IA, CASS 1984). And, judging from the jade ornaments and pottery found in Shixia, this assemblage is likely to have been deposited by a group moving from the lower Yangzi valley.

But for the foragers in the Yangzi valley, the situation would have been different. It is not clear whether the rice collectors of Yuchan and Xianrendong were migrants from South China, or were local people receiving the idea of rice collection from South China, or even if they found the value of wild rice themselves. Whatever the case, there must have been fewer natural resources for these foragers from the LGM to the beginning of the Holocene, since nuts and fruits were hard to find, wild rice would have been either absent or scarce, as the Yangzi valley was much colder [Table 11(3)]. In addition, the Yangzi valley was a temperate to cool-temperate zone in the LGM, so the problems of both seasonality and availability would have been more serious there. If the Yangzi foragers collected wild rice and, knew the value of this species, then the scarcity of wild rice at that latitude might have caused a desire to increase its availability by cultivation (Bellwood 1996) (Yan 1992a).

It is worthy of notice that perennial wild rice produces a limited quantity of seeds per year, as this plant is primarily regenerated by vegetative reproduction (Oka 1975). In addition, Oka (1975) has observed that most of the seeds are consumed by birds and animals, and very few are able to bury themselves in the soil (Oka 1975). Perennial wild rice also has a lower survival rate in disturbed habitats, and

a considerate mortality rate in the seedling stage (Oka 1975). All these botanical characteristics suggest that the natural abilities of reproduction and expansion of perennial wild rice were limited. If this plant was exploited by foragers, the desire to increase its productivity would have also caused a desire for cultivation.

One important feature in terms of grain processing in China might have also played an important role in the intensification of grass exploitation. Grass seeds in China seem to have been only dehusked and cooked whole, instead of being milled into flour. The consequence of this simple processing is twofold: on one hand, the labour and time required for seed processing were significantly reduced, thus making grass exploitation a much less formidable task; on the other hand, the type of cooking required the emergence of cooking equipment, thus stimulating the manufacturing of pottery. Once pottery was invented, seed-cooking would have become a much easier task, and the cooked seeds were a tasteful, nutritive and easily digested food. Consequently, grass exploitation could have become more attractive to foragers, causing continuous and intensive exploitation. When grass resources were not sufficient to meet the increased demand (caused either by the decline of other resources, or the increase of population, or the combination of both), it would only be natural that methods of maintaining and increasing its availability were tried, one of them being cultivation, which eventually led to the domestication.

In summary, the changes of palaeoclimate and palaeoenvironment during and after the LGM in China caused changes in natural resources; thus encouraging prehistoric foragers to exploit wild grasses. In the middle Yellow valley, this environmental change also forced people to crowd into smaller areas, thus the population density increased while natural resources decreased. This would have forced people to intensively exploit the natural resources in their limited districts, and to develop methods to maintain resources in their own territory, which eventually led to cereal cultivation. In the Yangzi valley, the storability and nutritional value of rice were probably the main elements behind exploitation, and the scarcity of this plant in local areas might have stimulated cultivation.

Summary and Remaining questions

Generally speaking, the transition from foraging to farming in China occurred in the valleys of the two greatest rivers, a temperate zone between latitudes 25 to 35° N, with relatively fertile soil, easy access to water, and moderate temperature. Agriculture probably emerged around 9000 to 8000 BP in the middle Yangzi valley, and by no later than 8500 years in the middle Yellow valley. Once emerged, agriculture quickly expanded to adjacent areas in both river valleys.

Archaeology and physical anthropology suggest that there were independent culture centres in these two valleys. The two river valleys are likely to have been sub-centres for the origins of agriculture, but idea diffusion between the middle Yangzi and the middle Yellow valleys cannot be excluded at present.

The ancestors of the early farmers in China seem to have been groups of foragers who not only hunted and gathered during the LGM, but also collected wild grasses. After the LGM it seems that the economic strategy of grass seed collection spread to other part of the middle Yellow River, while in the Yangzi valley, wild rice was collected during the LGM. Intensified grass exploitation was a necessary prelude to agriculture, although it did not automatically lead to agriculture. The agricultural societies which appeared by 8000 BP in both river valleys show a strong cultural link between them and the preceding assemblages in the same region, indicating a local cultural continuity. Wild cereal exploitation during the LGM in China was at least partially due to the environment and natural resource changes at that time, as the colder and drier climate caused the reduction of natural resources, and forced foragers towards grass exploitation. This exploitation probably caused a reduced mobility of the foragers and an increase of population in some areas. It is notable that natural resources for foragers in the Yellow valley during the LGM were mainly ungulates and herbs, the former with a very low RA and vulnerable to human exploitation. Thus, herbs of high RA would have become more important to foragers, and the desire to increase their production would have led to cereal cultivation, and eventually, domestication.

The improvement of human intelligence would have also played an important role in the transition from foraging to farming. Relevant parts of this intelligence would have been the accumulation of knowledge and experience by many generations about the characteristics of wild grasses, the economic values of these plants, their response to human interference, the feasibility of cultivation, the habitats for the plants, and tools specialised for cultivation activities. The origin of agriculture itself might have occurred within a few hundred years, but in an evolutionary sense, this process was a result of thousands of years of human evolution in terms of material culture, as well as physical and mental progress.

It seems that the transition from foraging to farming in China largely related to the natural environment. The current effective temperature in China declines from north to south. In addition, the precipitation increases southbound. As precipitation and solar radiation are both higher in the south, the primary production of terrestrial plants must also be higher in the south. According to Binford's theory, in areas where the effective temperature (ET for short) is high, the resources are rich, hunter-gatherers tend to be 'foragers', there is no need for storage, and a mobile subsistence strategy is selected for. In lower ET areas with fewer resources, hunter-gatherers tend to be 'collectors', with a more complicated mobility pattern and subsistence strategy (Binford 1980). The ET in both Central China and the Yangzi valley is quite low; shortfalls of food would have occurred during the winter in these regions. In this aspect, being storable food with a high RA, cereal seeds would have been more important for people in these areas than in South China.

But the environmental changes are not the only cause for the origin of agriculture in China. Population pressure and an imbalance between human and natural resources might also have played a role in the Yellow valley, whereas in the Yangzi valley the scarcity of wild rice might have been another important cause. Hypotheses of environmental changes combined with demographic change seem tenable in China (eg. Xu 1994), and the hypothesis of cultivation occurring in a marginal area of wild cereal distribution (Yan 1992a) is useful in explaining the situation in the Yangzi valley. However, no single hypothesis alone can satisfactorily explain the transition. It seems that the transition from foraging to farming was an process of multiple causes, and even in China the impetus for this transition in two different areas is not the same.

Such is a sketch on the transition from foraging to farming in China. Obviously there are many problems remaining. The first one is the lack of archaeobotanical data. The majority of the archaeological assemblages in Central China were discovered prior to the 1990s, and virtually no pollen or phytolith analysis was applied during the excavations. As a result, the presence of cereals, wild or domesticated, can only be recognized from macro-forms.

The lack of data on microscopic remains of cereals has proved to be a crucial drawback: without these we cannot tell when the domestication of millets and other cereals in the Yellow River basin began. The remains of domesticated millet (*Setaria italica*) found in Cishan were presumably not the earliest ones, as the Cishan assemblage does not seem to have been the earliest Neolithic culture. Judging from the advanced Jiahu assemblage and other early Neolithic assemblages in Central China, the transition from foraging to farming might have occurred between 10,000 to 9000 BP. But more archaeobotanical evidence is needed to explain this assumption. Similar problem also exists in the Yangzi valley and South China, although recently this issue has been addressed in the excavations at Yuchan and Xianrendong, but not in South China.

The second problem is the insufficient study on prehistoric foragers, particularly in terms of their mobility patterns, seasonal camps, population densities and subsistence strategies, all still inferred based upon toolkit interpretation. Residue analysis and use-wear analysis are yet to be applied to this research. This insufficiency makes the reconstruction of the way of life for the prehistoric foragers very difficult, and consequently, the transition from foraging to farming is poorly documented.

Absolute dating is the third problem. Only a few archaeological assemblages of the terminal Pleistocene have been absolutely dated. And quite often, only one or a few dates are available from each assemblage, thus the dates are less convincing. In addition, limestone areas in the Yangzi valley and South China have proved to be problematic. Because of the insufficiency and inaccuracy of absolute dates, a proper chronology for the assemblages from the terminal Pleistocene to the early Holocene cannot be set up.

The lack of experiments is another problem. Prior to my field work, no one had conducted either observation, or cultivation and/or harvesting experiments on wild cereals from an archaeological viewpoint in China. And my field work was only very preliminary. Further observation, harvesting and cultivation experiments will enable a much better understanding of the botanical characteristics of wild cereals, their responses to human cultivation and selection, domesticate rates, as well as the possible methods used by the prehistoric cultivators.

Finally, prehistoric cultural development in China might have been more complicated than we realized. Before the 1960s, the Yellow valley was perceived as the primary centre of Chinese prehistoric culture. As numerous archaeological discoveries were made from the 1960s onwards in the Yangzi valley, it was gradually accepted that the Yangzi valley might have been another centre (IA, CASS 1984). Since the later 1980s, remains of several villages have been found in both Inner Mongolia and the southern alluvial plain of Northeast China, represented by two sites - Chahai in Inner Mongolia, and Xinglongwa in Liaoning province [Fig. 4(1)]. Both dated from approximately 8000 to 7000 BP (Neimeng Archaeology Team 1985, 1997; Xin 1995). The discovery of these sites, and the succeeding Neolithic cultures in Northeast China, seems to raise new questions in terms of the cultural centres.

Remains of houses, burials and refuse pits, as well as pottery and implements of stone and organic materials were found in these sites (Neimeng Archaeology Team CASS 1985, 1997; IA, Liaoning province 1994). The houses found were well-planned, built in different phases. The largest one of over $120m^2$ floor was found at Chahai, surrounded by other small houses (Xin 1995). A protective ditch surrounding the houses was also found at Chahai (Xin 1995). A cobble 'dragon' over 19 meters long was found at Chahai, indicating the existence of ceremony. Jade implements were found in these sites, being the earliest ones in China (Neimeng Archaeology Team CASS 1985, 1997; IA, Liaoning province 1994; Xin 1995). All these discoveries suggest that these groups were sedentary, and their social structure might have been quite advanced.

The lithic toolkits from Chahai and Xinglongwa contain notched spades, knives, grinding slabs and rollers [Fig. 11(3)] (Neimeng Archaeology Team CASS 1985, 1997; IA, Liaoning province 1994). Prima facie, this toolkit seems to indicate that cereal cultivation would have been the primary economic activity. The remains of large sedentary villages (over 100 house remains were found at Chahai, and over 30 at Xinglongwa), burials, and the high level of cultural complexity would further support the proposition that the occupants of these sites were farmers. But because substantial animal bones were uncovered at both sites, yet no cultivated cereals were found (only wild walnuts), even with flotation (Liu 1995, personal communication), the primary subsistence strategy of these groups is still under debate (Yan 1995, personal communication).

The discoveries of the Chahai and Xinglongwa assemblages document the existence of quite advanced societies on the northern edge of Central China in the early Holocene. It is not clear whether these groups were affluent foragers, or farmers, or a combination of both. However, the toolkit of this group contains microblades and grinding tools, which probably signal cultural contacts or connections with the microblades and grinding slabs found in the middle Yellow valley during the terminal Pleistocene. Were these people also farmers? If so, were they from the Yellow River, or from somewhere else? Or did they develop farming independently? None of these questions are clear, and further studies are needed.

It is clear that studies on the origins of agriculture in China are only at an initial stage, and many questions remain. Hopefully, these questions can be addressed in future, not only by new archaeological discoveries, but more importantly, by the application of new methods of analysis which have been developed in recent years, and by well-designed cultivation experiments.

Table 11(1) Cranial Measurement of the Early to Middle Neolithic Inhabitants in the Yellow Valleys.

Item	Sex	Baoji	Banpo	Huaxian	Hengzhen	Shigu	Dawenkou	Xixiahou	Yedian
Maximum cranial length	M.	180.22	180.84	178.84	180.40	166.50	181.11	180.30	181.40
(g-op)	F.	175.30	178.12	174.25	n/a	160.30	160.45	169.31	160.22
Maximum cranial breadth	M.	143.25	138.93	140.69	144.80	129.00	145.70	140.90	146.00
(eu-eu)	F.	138.56	141.13	140.67	n/a	145.82	145.21	144.25	153.33
Basi-bregmatic height (ba-b).	M.	141.55	138.80	144.30	141.40	136.33	142.89	148.30	141.70
	F.	135.94	n/a.	136.17	n/a	138.25	142.96	142.60	135.50
Mininum frontal breadth	M.	93.29	93.10	94.25	93.10	93.25	91.64	93.94	94.30
(ft-ft)	F.	91.96	89.33	94.00	n/a	92.20	90.17	91.33	92.90
Maximum bizygonatic	M.	137.13	130.50	133.86	138.70	137.00	140.56	139.43	137.3
breadth (zy-zy)	F.	128.25	105.11	133.84	n/a	133.67	129.06	133.40	144.19
Upper facial height (n-pr).	M.	72.66	75.96	75.23	69.50	79.00	74.84	72.03	73.30
	F.	68.26	69.00	70.50	n/a	64.75	69.02	67.20	68.00
Orbital height	M.	33.96	33.94	33.28	32.90	34.50	35.14	34.27	34.00
	F.	33.13	33.55	33.30	n/a	33.62	34.41	32.98	34.15
Orbital breadth (mf-ek).	M.	43.55	42.57	42.90	43.40	43.00	42.95	44.13	41.10
	F.	41.49	40.00	41.71	n/a	41.52	42.13	43.20	41.50
Nasal breadth	M.	27.29	27.05	28.52	27.50	26.17	27.45	27.66	26.10
	F.	25.86	26.10	23.25	n/a	24.53	25.71	25.58	26.55
Nasal height (n-ns).	M.	52.13	55.50	53.51	53.60	53.67	54.72	57.12	55.15
	F.	49.12	51.10	51.00	n/a	47.50	51.68	53.87	49.00
Total prognathism	M.	82.35	81.00	83.61	80.40	87.00	83.61	84.38	85.50
(n-nr<MH)	F.	83.22	n/a	87.00	n/a	82.75	84.40	84.63	83.00

Source: Zhang 1989.

Table 11(2). Comparison of lithic implements between Pengtoushan, Jiahu and Peiligang assemblages.

Pengtoushan	**Jiahu**	**Peiligang**
Polished axe.	Polished axe.	Polished axe.
Pebble chopper.	Polished spade.	Polished spade.
Stone hammer	Sickle.	Sickle.
Large flaked scraper (sandstone)	Quern and roller.	Quern and roller.
Small flakes made of black chert.	Flaked scraper of quartzite.	Flaked scraper of quartzite and chert.
Drill (black chert)		Knife
Burin (black chert)		

Sources: CPAM of Hebei Province and Handan City 1981; CPAM of Handan District *et al.* 1977; IA, Henan Province 1989; IA, Hunan Province and Lixian Museum 1990.

Table 11 (3) Effective Temperatures in China.

Areas.	Present W (°C)	Present C (°C)	Present ET(°C)	(18,000-13,000 years BP) W (°C)	C (°C)	ET(°C)	(13,000-10,000 years BP) W (°C)	C (°C)	ET(°C)
West plateau	22	-12	12.3	14 (-8)	-20 (-8)	10.8	18 (-4)	-16 (-4)	11.5
/mountains	"	-8	12.5	"	-16 (-8)	10.8	"	-12 (-4)	11.7
North China	26	-6	13.2	18.5*(-7.5)	-13.5 (-7.5)	11.7	24 (-2.0)	-8 (-2.0)	12.8
Plain & Yellow	"	-4	13.4	"	-11.5 (-7.5)	11.8	"	-6 (-2)	13.0
Valley	"	-2	13.6	"	-9.5 (-7.5)	11.9	"	-4 (-2)	13.1
North Huai	26	0	13.8	20 (-6)	-6 (-6)	12.4	25 (-1)	-1 (-1)	13.5
valley	28	0	14.0	22 (-6)	-6 (-6)	12.7	27 (-1)	-1 (-1)	13.5
Yangzi valley	28	2	14.2	23 (-5)	-3 (-5)	13.1	27 (-1)	1 (-1)	14.0
	"	4	14.5	"	-1 (-5)	13.3	"	3 (-1)	14.3
South China	28	12	16	24 (-4)	8 (-4)	14.7	27 (-1)	11 (-1)	15.7
	"	14	17.2	"	10 (-10)	15.1	"	13 (-1)	16.2

Formula (Kelly 1995): ET= 18W-10C/(W-C)+8

W. average temperature of the warmest month.

C. average temperature of the coldest month (after Kelly1995)

* Temperature deduction is based upon the reconstruction of palaeoclimate in chapters 3 and 6. Minimum values are applied here. For example, the mean annual temperature in the North China Plain is estimated to be 7.5 to 8.5°C lower than the present during the LGM; so 7.5°C is applied in the calculation. Further, it is assumed that the average temperatures of both the warmest and the coldest month were evenly decreased by 7.5°C. The values in brackets indicate estimated minimum decrease of palaeo-temperature below the present level.

Temperature source: The Editing Committee of the Physiography of China 1984.

Text Bibliography

A large proportion of the references comes from Chinese literature. They are cited here following the seemingly international standards: the titles of the Chinese articles/books on journal are first Romanized, then the meaning of the titles are translated into English.

Many Chinese articles are published as written by archaeological teams or institutes, most commonly the federal and provincial institutes of archaeology, local managing committee of antiquities and regional museums. Abbreviations will be used for the first two organisations: IA for the Institute of Archaeology and CPAM for the managing committee of antiquities (CPAM is the English term commonly used by the latter in Chinese literature). IA CASS is preserved for the Institute of Archaeology, Chinese Academy of Social Science (CASS), and names of each province are added after IA for each provincial institute of archaeology. However, some books were complied by an editing committee, and the authors are not disclosed. These reference are cited under 'anonymous'.

The name (s) of individual authors are all Romanized by Pingyin. The name of a team or a group as author(s) are translated into English. For the title of the articles and books, if there are original English translations, then the original ones are retained. Others are translated into English by myself.

Some major Chinese journals are translated and abbreviated as follows:

H. K. Huaxia Kaogu (Archaeology in Central China). Edited by the Institute of Archaeology, Henan Province, China.

G. D. X. Gujizhui Dongwu Xuebao (Vertebrata PalAsiatica). Edited by the Institute of Vertebrate palaeontology and Palaeoanthropology, Academia Sinica, China.

J.K. Jianghan Kaogu (Archaeology in the Middle Yangzi Valley). Edited by the Institute of Archaeology, Hubei Province, China.

Kg Kaogu (Archaeology journal). Edited by the Institute of Archaeology CASS, China.

K.Y.W. Kaogu yu Wenwu (Archaeology and Relics). Edited by the Institute of Archaeology, Shaanxi Province.

K. X. Kaogu Xuebao (Archaeologica Sinica.) Edited by the Institute of Archaeology. CASS, China.

N.K. Nongye Kaogu (Agricultural Archaeology). Edited by Chinese Agriculture Museum and the Sociology Institute of Jiangxi Province, China.

R. X. Renleixue Xuebao. ACTA Anthropologica Sinica. Edited by the Institute of Vertebrate Palaeontology and Palaeoanthropology, Academia Sinica, China.

S. Y. Shiqian Yanjiu (Prehistory Research). Edited by the Institute of Archaeology, Shaanxi Province, China.

Ww. Wenwu (The Relics). Edited by the National Bureau for Antiquity Management, China.

W.Y.K. Wenwu yu Kaogu (Relics and Archaeology). Edited by the National Bureau for Antiquity Management, China.

Ww.C. Wenwu Chunqiu (Discussion on the Relics). Edited by the Relics Bureau of Hebei Province, Shijiazhuang, Hebei, China.

Z. N. Zhongguo Nongshi (The History of Chinese Agriculture). Edited by ACTA The Institute of Agriculture, Beijing, China.

Z. W. B. Zhongguo Wenwu Bao (News of Chinese Relics). Published by the National Bureau for Antiquity Management, China.

Z. W. Zhongyuan Wenwu (Antiquity in Central China). Edited by the Institute of Archaeology, Henan Province, China.

Allaby, M. (ed.), 1985. *The Oxford Dictionary of Natural History*. Oxford: Oxford University Press.

Allen, H., 1974. The Bagundji of the Darling Basin: cereal gatherers in an uncertain environment. *World Archaeology* Volume 5:309-322.

Ahn, S.M., 1993. *Origin and Differentiation of Domestication Rice in Asia - Review of Archaeological and Botanical Evidences.* Unpublished PhD thesis. University of London.

An, Z.M., 1965. Henan Anyang Xiaonanhai Jiushiqi Shidai Doxue Duiji de Shijue (The Trial Excavation of a Palaeolithic Cave in Xiaonanhai of Anyang County, Henan Province). *K.X.* 1: 1-28.

An, Z.M., 1979. Peiligang, Cishan he Yangshao (Peiligang, Cishan and Yangshao). *Kg.* 4: 335-346.

An, Z.M., 1983. Zhongguo Wanqi Jiushiqi de Tanshisi Duandai he Wenti (Radiocarbon dating and problems of the Late Palaeolithic in China). *R. X.* Volume 4: 342-351.

An, Z.M., 1984. Luelun Huabei de Zaoqi Xinshiqi Wenhua (A discussion on the Neolithic in North China). *Kg.* 10: 936-944.

An, Z.M., 1994. Tanshisi Duandai he Zhongguo Shiqian Kaoguxue (Radiocarbon dating and the prehistoric archaeology in China). *Ww.* 3:83-87.

An, Z.M. and R.Z. Wu, 1957. Shaanxi Zhaoyong Dali Shayuan Diqu de Shiqi Shidai Yichun (Remains of Stone Age in Shayuan Area Between Dali and Zhaoyang Counties, Shaanxi Province. *K.X.*3: 1-12.

An, Z.S., X. H. Wu, Y.C. Lu, D.E. Zhang, X.J. Sun and G.R. Dong, 1990. Zujin Er'wannian Zhongguo Guhuanjing Bianqian de Chubu Yanjiu (A study on palaeoenvironmental changes in China since of the last 20,000 years). In *Huangtu, Disiji Dizhi - Quanqiu Bianhua* (Loess, Quaternary and Global Change) *Volume 2*, ed. T.S. Liu. Beijing: Science Press, pp. 1-29.

An, Z.S., X.H. Wu, P.X. Wang and another 6 authors, 1991. Changes in the monsoon and associated environmental changes in China since the last interglacial. In *Loess, Environment and Global Change- the Series of The XIII INQUA Congress*, ed. T.S. Liu. Science Press. Beijing, pp. 1-29.

Anderson, J.G., 1929. *Children of the Yellow Earth.* London: Kegan Paul, Trench, Trubner & Co. Ltd.

Anderson, P.C. (ed.), 1992. *Prehistorie de L'Agriculture.* Paris: CNRS press.

Anonymous, 1976. *Guzi* (Domesticated Millet). Beijing: Science Press.

Anonymous, 1991. *Zhongguo Wenwu Dituji - Henan Fence* (The Antiquity Atlas of China - Henan province). Beijing: The Map Press.

Anonymous, 1994. *Xinshiqi Shidai* (The Neolithic). In *Forty Years of Archaeology in Shanxi Province*, ed.

The Archaeology Institute of Shanxi Province. Taiyuan: The People's Press of Shanxi, pp. 50-65.

Anonymous, 1995. Caoxieshan Yizhi Shouci Faixian Woguo Liuqiannian Qian de Shuidaotian Yizhi (The discovery of rice field of 6000 years ago at Caoxieshan) *Z. W. B.* 18th, June. the front page.

Asch, D.L., 1994. Aboriginal speciality-plant cultivation in Eastern North America: Illinois prehistory and a post-contact perspective. In *Agricultural Origins and Development in the Midcontinent*, ed. W. Green. Iowa: Office of the State Archaeologists, pp. 25-86.

Ban-po Museum and CPAM of Dali County, 1983. Shaanxi Dali Shayuan Diqu Kaogu Diaocha Baogao (A survey report on Shayuan, Dali county, Shaanxi Province). *S. Y.* 1:101-113.

Bar-Yosef, O., 1981. The Pre-pottery Neolithic period in the Southern Levant. In *Prehistoire du Levant*, ed. J. Cauvin and P. Sanlaville. Paris: CNRS Press, pp. 551-602.

Bar-Yosef, O., 1990. The last glacial maximum in the Mediterranean Levant. In *The World at 18,000 BP*, eds. C. Gamble and O. Soffer. London: Unwin Hyman, pp. 58-77.

Bar-Yosef, O. and A. Belfer-Cohen, 1989. The origins of sedentism and farming communities in the Levant. *Journal of World Prehistory*, Volume 3(4): 447-498.

Bar-Yosef, O. and M.E. Kislev, 1989. Early farming communities in the Jordan valley. In *Foraging and Farming: the Evolution of Plant Domestication,* ed. D.R. Harris and G.C. Hillman. London: Unwin Hyman, pp. 632-642.

Bar-Yosef, O. and R.S. Kra, 1994. Dating eastern Mediterranean sequences: introductory remarks. In *Late Quaternary Chronology and Palaeoclimate of the Eastern Mediterranean*, eds. Bar-Yosef, O. and R.S. Kra. Tuscon: University of Arizona Press, pp. 1-12.

Barnes, G., 1992. China. In *People of the Stone Age-Hunter and gatherers and Early Farmers: Volume 2 of the Landmark Series from the America Museum of the National History*, ed. G. Burenhult. Australia: University of Queensland Press, pp. 134-137.

Bellwood, P., 1978. *Man's Conquest of the Pacific*. Hong Kong: Dai Nippon Printing Co.

Bellwood, P., 1995. Early agriculture, language history and the archaeological record in China and Southeast Asia. Paper prepared for the conference on Archaeology in Southeast Asia, University Museum and Art Gallery, University of Hong Kong.

Bellwood, P., 1996. The origins and spread of agriculture in the Indo-Pacific region: gradualism and diffusion or revolution and colonization? In *The Origins and Spread of Agriculture and Pastoralism in Eurasia*, ed. D. R. Harris. London: UCL Press, pp. 465-498.

Bender, B., 1978. Gather-hunter to farmer: a social perspective. *World Archaeology* Volume 10 (2): 204-222.

Bettinger, R.L., 1991. *Hunter-gatherers - Archaeological and Evolutionary Theory*. New York: Plenum Press.

Binford, L.R., 1968. Post-Pleistocene adaptations. In *New Perspectives in Archaeology*, eds. S.R. Binford and L.R. Binford. Chicago: Aldine Publishing Company, pp. 313-342.

Binford, L.R., 1980. Willow smoke and dog's tails: hunter-gatherer settlement systems and archaeological site formation. *American Antiquity* Volume 45: 4-21.

Braidwood, R.J, 1975. *Prehistoric Men.* Eighth edition. Illinois: Scott, Foresman and Company.

Byrd, B. F., 1989a. The Natufian: settlement variability and economic adaptations in the Levant at the end of the Pleistocene. *Journal of World Prehistory* Volume 3(2): 159-197.

Byrd, B.F., 1989b. Subsistence patterns and settlement organization. In *The Excavation at Beidha: the Natufian Encampment at Beidha,* ed. B.F. Byrd. Hojbjerg: Jysk arkaeologisk selskab, 75-85.

Byrd, B.F., 1994. Late quaternary hunter-gatherer complexes in the Levant between 20,000 and 10,000 BP. In *Late Quaternary chronology and Palaeoclimates of the Eastern Mediterranean*, eds. O. Bar-Yosef and R. S. Kra. Tuscon: The University of Arizona, pp. 205-226.

Byrne, R., 1987 Climate change and the origins of agriculture. In *Studies in the Neolithic and Urban Revolutions*, ed. L. Manzanilla. BAR International Series 349, pp. 21-35.

C^{14} Laboratory, Dep. of Archaeology, Beijing University., 1996. Tanshisi Niandai Ceding Baogao (A radiocarbon dating report). *Ww.* 6:91-95.

Caldwell, J.R., 1977. Cultural evolution in the Old World and the New, leading to the beginning and spread of Agriculture. In *Origins of Agriculture,* ed. C. A. Reed. The Hague: Mouton Press, pp. 77-88.

Cane, S., 1984. *Desert Camps - A Case Study of Stone Artefacts and Aboriginal Behaviour in the Western Desert*. unpublished Ph.D. thesis. ANU.

Cao, C.Z., 1988. *Daizu Shehui Diaocha* (A Social Survey of the Dai Ethnic). Kunming:Yunan People's Press.

Carter, G.F., 1980. The metate: an early grain-grinding implement in the New World. In *Early Native Americans*, ed. D.L. Browman. UK: Mouton Publishers, pp. 21-40.

Champion, T., C. Gamble, S. Shennan & A. Whittle, 1984. *Prehistoric Europe*. London: Academic Press.

Chang, T.T., 1989. Domestication and spread of the cultivated rices. In *Foraging and Farming - the Evolution of Plant Exploitation*, eds. D.R. Harris and G.C. Hillman. London: Unwin Hyman Ltd., pp. 408-416.

Chang, Y.B and J.L. Liu, 1997. 11,000 nian Yilai Taihu de Xingcheng he Yanbian (The formation and development of the Tai Lake since the last 11,000 years B.P.). *Gushengwu Xuebao* (ACTA Palaeontologica Sinica). Volume 35(2): 129-135.

Chen, B.Z., J. Zhang, and H. Lu, 1995. Shuidao Yingke Zhiguishi yu Daozhong Qiyuan Yanjiu (A study of rice phytoliths and the origin of cultivated species). *Kexue Tongbao* (Chinese Science Bulletin) 4: 339-342.

Chen, C.L., 1968. *Material Culture of the Formosan Aborigines*. Taiwan: Taiwan Museum.

Chen, C., 1984. The microlithic in China. *Journal of Anthropological Archaeology* 3(2): 79-115.

Chen, C. and X.Q. Wang, 1989. Upper Palaeolithic microblade industries in North China and their relationships with Northeast Asia and North America. *Arctic Anthropology Volume* 26(2): 127-156.

Chen, C., Z. Zhang, M. Wang, W. Gu, and Q. Yao, 1987. Sanshan Wenhua (The Sanshan Culture). *Nanjiing Bowuyuan Jikan* (Bulletin of Nanjing Museum) Volume 9: 2-20.

Chen, G.Q. and J.H. Lin, 1988. *Gaoshanzu Wenhua* (The Gaoshan Culture). Shanghai: Academy Press.

Chen, T.M., & R.E. Hedges, 1994. Pengtoushan deng Yizhi Taopian he Woguo Zuizao Shidao Yichun de Jiasuqi Zhipu C^{14} Cenian (AMS radiocarbon dating of pottery from Pengtoushan and Hujiawuchang sites and the earliest rice remains in China). *Ww.* 3: 88-94.

Chen, T.M., S.X. Yuan, S.J. Gao, L.X. Wang, & G.Y. Zhao, 1982. Xujiayao Yizhi Puru Dongwu Huashi de Youxi Cenian (Uranium series dating of the Xujiayao assemblage). *R. X.* Volume 1(1): 91-95.

Chen, T.M., R.E. M. Hedges, & Z.X. Yuan, 1992 Shandindong Yizhi de Di'erpi Jiasuqi Zhipu Tanshisi Nianling Shuju yu Taolun (The second batch of

accelerator radiocarbon dates for Upper Cave site of Zhoukoudian). *R. X.* Volume 11(2): 112-116.
Chen, W.Y., 1991. Shanxi Dingcunren Shengho Shiqi de Guqihou (The palaeoclimate for the Dingcun occupants). *Shanxi Wenshi Ziliao* (Historic Data of the Shanxi Province) Volume 78 (6): 117-129.
Chen, W.H., 1989. Guanyu Zhongguo Nongye Qiyuan de Jige Wenti (A few problems on the origins of agriculture in China). *N. K.* 2: 84-98.
Chen, W.H., 1991. *Zhongguo Nongye Kejishi Tuji* (Atlas of the History of Agricultural Technology in China). Beijing: Agriculture Press.
Chen, W.H., 1993. *Zhongguo Nongye Kaogu Tulu* (Atlas of The Agricultural Archaeology in China). Nancheng: Jiangxi Science and Technology Press.
Chen, X.B., 1986. Hunansheng Disiji Puru Dongwu Huashi Xingcailiao (A new discovery of mammals in Hunan Province). *G. D. X.* Volume 24(3): 242-244.
Chen, X.Q., & Huang M.M., 1994. Hemudu Wenhua Guoji Taolunhui Zhaiyao (A summary of the international conference on Hemudu culture). *Ww.* 10: 84-86.
Chen, Z.Y., 1993. Shanxi Jiushiqi Kaogu Gaixu (A summary on the Palaeolithic archaeology in Shanxi province). In *Shanxi Jiushiqi Shidai Kaogu Lunwenji* (The Palaeolithic of Shanxi Province- Selected Treaties), ed. Z.Y. Chen. Taiyuan: The Shanxi Economic Press, pp. 1-14.
Chen, Z.Y., 1996. Xiachuan Yizhi de Xincailiao (New collections in Xiachuan). *Z. W.* 4: 1-22.
Chen, Z.Y., 1983. Hubei Nongye Kaogu Gaisu (A summary on the archaeological agriculture in Hubei province). *N. K.* 1:91-102.
Chen, Z.Y., 1991. Chengbeixi Wenhua (The Chengbeixi culture). *Dongnan Wenhua* (Southeast Culture) 5: 206-212.
Chen, Z.Y. & Q.X. Yang, 1985a. Yidou Zhichengbei Jiaoqi Xinshiqi Shida Yizhi (The early Neolithic site at Zhichengbei, Yidou city). *Zhongguo Kaoguxue Nianjian* (Chinese Archaeology Annual), ed. the Institute of Archaeology CASS. Beijing: The Relics Publishing House, pp. 177-178.
Chen, Z.Y. & Q.X. Yang, 1985b. Yidouxian Huamiaoti deng Sichu Xinshiqi Shidai Yizhi (Four early Neolithic sites in Yidou county). In *Zhongguo Kaoguxue Nianjian* (Chinese Archaeology Annual), ed. the Institute of Archaeology CASS. Beijing: The Relics Publishing House, pp. 178-179.
Chen, Z.Y. & Q.X. Yang, 1985c. Qinglongshan Jiaoqi Xinshiqi Shidai Yizhi (The Qinglongshan early Neolithic site). In *Zhongguo Kaoguxue Nianjian* (Chinese Archaeology Annual), ed. the Institute of Archaeology CASS. Beijing: The Relics Publishing House, pp. 179.
Childe, V.G., 1951 *The Neolithic Revolution.* Reprinted in 1971. New York: The Natural History Press.
Clark, J. D., 1985. The microlithic industries of Africa: their antiquity and possible implications. In *Recent Advance in Indo-Pacific Prehistory*, eds. V.N. Misra and P. Bellwood. New Delhi: Oxford and IBH Publishing Co., pp. 95-104.
Cohen, M. N., 1977. *The Food Crisis in Prehistory: Overpopulation and the Origins of Agriculture.* New Haven: Yale University Press.
CPAM of Anyang District & CPAM of Qi County, 1981. Henan Jixian Huawo Yizhi Shijue (A trial-excavation of the Huawo site in Ji county, Henan province). *Kg.* 2:279-281.
CPAM of Baoding District, CPAM of Xushui County, Department of Archaeology, Beijing University, & Department of History, Hebei University, 1992. Hebei Xushui Nanzhuangtou Yizhi Fajue Jianbao (A preliminary report on the excavation of Nanzhuangtou, Xiushui, Hebei province). *Kg.* 11: 961-970.
CPAM of Handan District & Students of the Archaeological Training Class of Handan District, 1977. Hebei Cishan Xinshiqi Yizhi Shijue (A trial excavation at Cishan). *Kg.* 6: 361-372.
CPAM of Hebei Province & CPAM of Handan City, 1981. Hebei Wu'an Cishan Yizhi (The Cishan site in Wu'an county, Hebei province). *K.X.* 3: 303-338.
CPAM of Jiangxi Province, 1963. Jiangxi Wannian Dayuan Xianrendong Yizhi Shijue (A trial excavation at Xianrendong, Wannian, Jiangxi province). *K. X.* 1: 1-17.
CPAM of Kaifeng District & CPAM of Xinzheng County, 1978. Henan Xinzheng Peiligang Zaoqi Xinshiqi Shidai Yizhi (The Peiligang early Neolithic site in Xinzheng county, Henan province). *Kg.* 2: 73-79.
CPAM of Kaifeng District, CPAM of Xinzheng County, & Department of History, Zhengzhou University, 1979. Peiligang Yizhi 1978 nian Fajue Jianbao (A preliminary report of the 1978 excavation of Peiligang). *Kg.* 3: 197-205.
CPAM of Linyi District, 1983. Shandong Linyixian Fenghuangling Faxian Xishiqi (Microblades were discovered in Fenghuangling, Linyi county, Shandong province). *Kg.* 5:385-388.
CPAM of Nanyang District, 1982. Nanzhaoxian Xiaokongshan Faxian Jiushiqi Shidai Wenhua (The discovery of Palaeolithic in Xiaokongshan, Nanzhao county). *Z.W.* 1:31-36.
CPAM of Zhejiang Province & Museum of Zhejiang Province, 1978. Hemudu Yizhi Diyiqi Fajue Baogao (The report of the first excavation season at Hemudu). *K. X.* 1: 39-93.
Cui, Z.J. & Y.Y. Xie, 1984. Lun Woguo Dongbei, Huabei Wangengxinshi Wanqi Duonian Dongtu Nanjie yu Bingyuan Huanjing (On the South Border of Frozen-ground and Periglacial of the Northeast and North of China in Late Pleistocene). *Dizhi Xuebao* (ACTA Geologica Sinica) (2): 165-174.
Cui, Z.J. & Y.Y. Xie, 1985. Lun Wogo Beifang Wangengxinshi Bingyuan Huanjing (On the Periglacial Environment of the Late Pleistocene in North China). *Disiji Yanjiu* (Quaternary Research) Volume 6(2): 115-120.
Dadiwan Excavation Group, 1981. Gansu Qin'an Dadiwan Xinshiqi Shidai Jiaoqi Yizhi (An early Neolithic remain in Dadiwan, Qin'an county, Gansu province). *Ww.* 4: 1-7.
Dadiwan Excavation Group, 1982. Qin'an Dadiwan 1980 nian Fajue Jianbao (A brief report on the excavation of Qin'an Dadiwan in the 1980 season). *K. Y. W.* 2: 1-8.
Dai, G.H., 1985. Huanan Diqu Xinshiqi Shidai Zaoqi Wenhua de Dongwu Kaoguxue Yanjiu (The zoological - archaeological study on the environment of the Neolithic in South China). *S. Q. Y. J.* 2:95-99.
Ding, Y., 1933. Guangdong Yeshengdao de Fenbu (The distribution of wild rice in Guangdong). In *Dingying Daozuo Yanjiu Lunwenji* (Collected Works on the Origin of Rice Cultivation by Ding Ying), complied by the Agronomy University of South China. Beijing: Science Press, pp. 421-426.
Ding, Y., 1957. Zhongguo Zaipei Daozhong de Qiyuan Jiqi Yanbian (The origin and evolution of cultivated rice in China). In *Dingying Daozuo Yanjiu Lunwenji* (Collected Works on the Origin of Rice Cultivation by Ding Ying), complied by the Agronomy University of South China. Beijing: Science Press, pp. 25-39.
Doebley, J.F., 1984. "Seeds" of wild grasses: a major food of Southwest Indians. *Economic Botany* Volume 38(1): 52-64.
Editors of the Archaeology Journal, 1989. Guanyu Zhongguo Wenming Qiyuan de Taolun (Discussions

on the origin of Chinese civilisation). *Kg.* 12: 1110-1120.

Edwards, P.C., 1989. Problems of recognizing earliest sedentism: the Natufian example. *Journal of Mediterranean Archaeology* Volume 2(1): 5-48.

Edwards, P.C., 1990. Kebaran occupation at the Last Glacial Maximum in Wadi al-Hammerh, Jordan Valley. In *The World at 18,000 years BP*, eds. O. Soffer & C. Gramble. London: Unwin Hyman, pp. 97-118.

Edwards, P.C., P. G. Macumber, & M. J. Head, 1996. The early Epipalaeolithic of Wadi al-Hammeh. *Levant* Volume 28: 115-130.

Fan, Z.M., 1993. Shilun Zhongguo Zhongshiqi Shida de Yuanshi Nongye Mengya (The seeds of primitive agriculture in the Mesolithic period in China). *Z. N. Volume* 12 (2): 1-9.

Feng, M.Q., 1992. Luotan cong Kaogu Faxian Panduan Nongye Fasheng de Ruogan Biaozhun Wenti (A discussion on indicators of the origins of agriculture from archaeological discoveries). *N. K.* 1:38-43.

Ferring, C. R., 1976. SDE Division: an upper Palaeolithic site on the Divshon Plain. In *Prehistory and Palaeoenvironments in the Central Negev, Israel. Volume I*, ed. A. E. Marks. Dallas: SMU Press, pp. 199-226.

Flannery, K.V., 1973. The origins of agriculture. *Annual Review of Anthropology* Volume 2: 271-310.

Fogg, W. H., 1983. Swidden cultivation of foxtail millet by Taiwan Aborigines: a cultural analogue of the domestication of *Setaria italica*. In *The Origins of Chinese Civilisation*, ed. D.N. Keightley. Berkeley: University of California Press, pp. 95-115.

Fu, X.G., 1989. Lun Youjian Shifu he Youduan Shibeng (On shouldered axe and carinated adze). *K.X.* 1:1-35.

Fujimoto, T., 1983. Grinding-slabs, hand-stones, mortars, pestles, and saddle querns. *Bulletin of the Department of Archaeology, Faculties of Letters, University of Tokyo*. 2: 73-75.

Furby, J.H., R. Fullagar, J. R. Dodson, & I. Prosser, 1993. The Cuddie Springs bone bed revisited, 1991. In *Sahul in Review*, eds. M. Spriggs and B. Fankhauser. Canberra: ANU, pp. 204-210.

Gai, P., 1985. Microlithic industries in China. In *Palaeoanthropology and Palaeolithic Archaeology in the People's Republic of China*, eds. R.K. Wu and J.W. Olsen. Florida: Academic Press Inc., pp. 225-241.

Gai, P. & Q. Wei, 1977. Hutouliang Jiushiqi Shidai Wanqi Yizhi (The discovery of the Upper Palaeolithic site at Hutouliang). *G. D. X.* Volume 15(4): 286-300.

Gao, F.Q. & H.D. Li, 1986. Disiji Nitan Chengtanqi de Huafen yu Qihou Bodong (Periodicity of quaternary peat formation and climatic fluctuation). *Haiyang Dizhi yu Disiji Dizhi* (Marine Geology and Quaternary Geology) Volume 6(2): 95-104.

Gao G.M. & X.Y. Dou, 1981. *Hunan Ziran Dili (Physiography of Hunan Province)*. Changsha: Hunan People's Press.

Gao, M.J. & J.J. Chen, 1988. Zaipeisu Qiyuan de Tonggongmei Yanjiu (Isozymic studies on the origin of cultivated foxtail millet). *Zuowu Xuebao* (ACTA Agronomica Sinica) Volume 14(2): 131-136.

Gao, X., Y.Z. You, & Z.J. Wu, 1991. Shanxi Yuci Dafa Jiushiqi Shidai Yizhi (The Palaeolithic Site at Dafa Village, Yuci district of Shanxi Province). *R. X.* Volume 10(2): 147-154.

Garrard, A.N. & B.F. Byrd,. 1992. New dimensions to the Epipalaeolithic of the Wadi el-Jilat in Central Jordan. *Paleorient Volume* 18(1): 47-62.

Garrard, A., B. Byrd, P. Harvey, & F. Hivernel, 1985. Prehistoric environment and settlement in the Azraq Basin. A report on the 1982 survey season. *Levant Volume* 17: 1-28.

Garrard, A., B. Byrd, P. Harvey, & F. Hivernel, 1987. Prehistoric environment and settlement in the Azraq Basin: an interim report on the 1985 excavation season. *Levant Volume* 19: 5-25.

Garrard, A., B. Byrd, P. Harvey, & F. Hivernel, 1992. Prehistoric environment and settlement in the Azraq Basin: an interim report on the 1989 excavation season. *Levant Volume* 24: 1-31.

Garrard, A., B. Byrd, P. Harvey, & F. Hivernel, 1994. Prehistoric environment and settlement in the Azraq Basin: an interim report on the 1987 and 1988 excavation seasons. *Levant Volume* 26: 73-109.

Gebauer, A.B. & T.D. Price, 1992. Foragers to farmers: an introduction. In *Transition to Agriculture in Prehistory*, eds. A.B. Gebauer & T.D. Price. Madison, Wisconsin: Prehistory Press, pp. 1-10.

Geng, X.S., 1981. Zhongguo Dongbu Wan Gengxinshi Yilai de Haishui Jingtui (Marine transgression and regression in East China since the late Pleistocene). *Haiyang Xuebao* (ACTA Oceanologia Sinica). *Volume* 3 (1): 117-130.

Gilead, I., 1991. The upper Palaeolithic period in the Levant. *Journal of World Prehistory*. Volume 5 (2): 105-154.

Golson, J., 1989. The origins and development of New Guinea agriculture. In *Foraging and Farming: the Evolution of Plant Domestication*, eds. D.R. Harris & G.C. Hillman. London: Unwin Hyman, pp. 678-687.

Gould R.A., 1969. Subsistence behaviour among the Western Desert Aborigines of Australia. *Oceania Volume* 39: 253-274.

Gould R.A., 1980. *Living Archaeology*. UK: Cambridge University Press.

Gu, H.B., 1991. Dongtinghu Diqu Disiji Guhuanjing Yanbian Jiqi dui Renlei Huodong Yingxiang de Chutan (Quaternary palaeoenvironment of Dongting Lake and its impacts upon human activities). *Huanjing Kaogu* (Environmental Archaeology). Volume 1: 173-178.

Gu, T.M., 1991. *Huabei Pingyuan de Qihou (Climate of the North China Plain)*. Beijing: Climatology Press.

Han, D.F. & C.H. Xu, 1989. Zhongguo Nanfang Disiji Puru Dongwuqun Jianlun Yuanshi Renlei de Shenghuo Huanjing (Quaternary mammalian faunas and environment of fossil humans in South China). In *Zhongguo Yuangu Renlei (*Early Humankind in China), eds. R.K. Wu, X.Z. Wu & S.S. Zhang. Beijing: Science Press, pp. 338-391.

Han, K.X. & Q.F. Pan, 1983. Zhejiang Yuyao Hemudu Xinshiqi Shidai Renlei Tougu (Early Neolithic human skulls from Hemudu, Yuyao, Zhejiang). *R. X.* Volume 2(2): 124-131.

Han, R., 1985a. Tancheng Heilongtan Jiushiqi Shidai Yichun (Tancheng Heilongtan Palaeolithic remains). In *Zhongguo Kaoguxue Nianjian* (Chinese Archaeology Annual), ed. The Institute of Archaeology CASS. Beijing: The Cultural Relics Publishing House, pp. 154-155.

Han, R., 1985b. Linyi Qingfengling Faxian de Xishiqi (The microblade remains found in Qingfengling, Linyi city). In *Zhongguo Kaoguxue Nianjian* (Chinese Archaeology Annual), ed. The Institute of Archaeology CASS. Beijing: The Cultural Relics Publishing House, pp. 155.

Harlan, J.R., 1967. A wild wheat harvest in Turkey. *Archaeology* Volume 20(1): 197-201.

Harlan, J.R., 1995. *The Living Fields - Our Agriculture Heritage*. Cambridge: Cambridge University Press.

Harrold, R.L. & J.D. Nalewaja, 1977. Proximate, mineral and amino acid composition of 15 weed seeds. *Journal of Animal Science*. Volume 44 (3): 389-394.

Harrold, R.L., D.L. Craig, J.D. Nalewaja, & B.B. North, 1980. Nutritive value of green or yellow foxtail, wild

oats, wild buckwheat or redroot pigweed seed as determined with the rat. *Journal of Animal Science.* Volume 51 (1): 127-131.

He, J.J., 1994. Huanzhujiangkou de Shiqian Caitao yu Daxi Wenhua (Prehistoric painted pottery and the Daxi culture at the Circum-Pearl Delta). In *Ancient Cultures of South China and Neighbouring Regions*, ed. Centre for Chinese Archaeology and Art, ICS The Chinese University of Hong Kong. Hong Kong: the Chinese University Press, pp. 321-330.

He, J.J., 1995a. Hunan Zaoqi Xinshiqi Shidai Wenhua Yichun (Early Neolithic relics in Hunan). In *Conference Papers on Archaeology in Southeast Asia*, eds. C. T. Yeung & W.L. Li. Hong Kong: Oak Promotion & Marketing Company, pp. 71-78.

He, J.J., 1995b. Dongtinghuqu Xinshiqi Shidai Zaoqi Wenhua Tansuo (A discussion on the early Neolithic culture in the Dongting Lake District). In *Hunan Shiqian Kaogu Lunwen Xuanpian* (Treaties of the Prehistoric Archaeology in Hunan Province), ed. and printed by The Archaeology Institute of Hunan Province, pp. 52-64.

Henan Museum & CPAM of Mi County, 1981. Henan Mixian Er'gou Beigang Xinshiqi Shidai Yizhi (The Er'gou Neolithic site in Mi county, Henan province). In *Kaoguxue Jikan* (Collection of Archaeology Papers), ed. The Institute of Archaeology CASS. Beijing: Sociology Press, pp. 1-26.

Henry, D.O., 1976. Rosh Zin: A Natufian settlement near Ein Avdat. In *Prehistory and Palaeoenvironments in the Central Negev, Israel.* Volume I, ed. A.E. Marks. SMU Press. Dallas, pp. 317-347.

Henry, D.O., 1989. *From Foraging to Agriculture.* Philadephia: University of Pennsylvania Press.

Higham, C. F.W., 1995. Lingnan and Southeast Asia in Prehistory. In *Conference Papers on Archaeology in Southeast Asia*, ed. Yang C.T. & Li H.L. Hong Kong: The University Museum and Art Gallery, the University of Hong Kong, pp. 23-38.

Hill, D. V., 1995. Ceramic analysis. In *Origins of Rice Agriculture*, eds. R.S. MacNeish & J.G. Libby. Elpaso: The University of Texas, pp. 35-45.

Hillman, G.C. & M.S. Davies, 1992. Domesticated rate in wild wheat and barley under primitive cultivation: preliminary results and archaeological implications of field measurements of selection coefficient. In *Prehistoire de L'agriculture*, ed. P.C. Anderson. Paris: CNRS Press, pp. 113-148.

Ho, P.T., 1969. *Huangtu he Zhongguo Nongye de Qiyuan* (The Loess and the origin of Chinese agriculture). Hong Kong: The University of Hong Kong Press.

Honghuatao Archaeology Station & Zhicheng Museum, 1988 Chengbeixi Yizhi Fucha Ji (A report on the survey of Chengbeixi site). *J K.*4: 7-14.

Horne, G. & G. Aiston, 1924. *Savage Life in Central Australia.* London: MacMillan and Co. Ltd.

Hu, Z.X., 1994. Handai Yiqian de Yingshi Xiguan (The custom of eating and drinking prior to the Qin Dynasty). *N. K.*1:214-218.

Huang, Q.X., 1982. Huanghe Liuyu Xinshiqi Nonggeng Wenhua zhong de Nongzuowu (Cultivars in the Neolithic Yellow Valley). *N. K.* 2: 55-61.

Huang, Q.X., 1983. Peiligang, Jericho yu Tehuacan (Peiligang, Jericho and Tehuacan). *N. K.*1: 182-192.

Huang, W.B. & H.X. Ji, 1963. Jiangxi Wannian Xianrendong Quanxinshi Dongxue Duiji (Note on Holocene Xianren Cave deposit of Wannian county, Jiangxi). *G. D. X.* Volume 7 (3): 263-272.

Huang, W.B. & Y.J. Tang, 1989. Nihewang Pendi Wanxinshengdai Jige Diceng Poumian de Guanca (A survey on several stratigraphic cross-sections in the Nihewan basin). In *Nihewan Yanju Lunwen Xuanbian* (Selected Treaties on Nihewan), eds. Q. Wei & F. Xie. Beijing: Cultural Relics Publishing House, pp. 353-364.

Huang, W.B., X. Xu, & T. Li, 1987. Hubei Fangxian Zhangnaodong Jiushiqi Shidai Yizhi Fajue Baogao (An excavation report on the Palaeolithic site of Zhangnao Cave, Fang county, Hubei province). *R. X. Volume* 6(4): 298-304.

Huang, W.W., 1989. Zhongguo Jiushiqi Shidai Wanqi Wenhua (The Later Palaeolithic of China). In *Zhongguo de Yuangu Renlei* (Early Humankind in China), eds. R.K. Wu, X.Z. Wu & S.S. Zhang. Beijing: Science Press, pp. 220-243.

Hunan Museum, 1986. Hunan Shimengxian Zaoshi Xingshiqi Yichun (The early Neolithic remains in the lower strata of Zaoshi in Shimeng County, Hunan Province). *Kg.* 1:1-11.

Hynes, R.A & A.K. Chase, 1982. Plants, sites and domiculture: Aboriginal influence upon plant communities in Cape York Peninsula. *Archaeology in Oceania* Volume 17:38-50.

IA CASS, 1959. *Miaodigou he Sanliqiao* (Miaodigou and Sanliqiao). Beijing: Cultural Relics Publishing House.

IA CASS, 1984. *Xinzhongguo de Kaogu Faxian he Yanjiu* (Archaeological Excavation and Research in New China). Beijing: Cultural Relics Publishing House.

IA CASS, 1991. *Kaoguxue Tanshisi Niandai Shujuji 1965-1991* (Archaeological C^{14} dates from 1965-1991). Beijing: Cultural Relics Publishing House.

IA CASS, 1993. Tanshisi Niandai Ceding Baogao (A radiocarbon dating report). *Kg.* 7: 645-648.

IA CASS & Banpo Museum, 1963. *Xi'an Banpo.* Beijing: Cultural Relics Publishing House.

IA, CASS & CPAM of Shanxi Province, 1978. *Xiachuan-Yuanshi Gongshe Yizhi Fajue Baogao* (The Xiachuan Report). Unpublished draft.

IA, Hebei Province, 1989. Hebei Yangyuan Xibaimaying Wanqi Jiushiqi Yanjiu (Research on the Late Palaeolithic Assemblage of Xibaimaying of Yangyuan County, Hebei Province). *Ww. C.* (3): 13-26.

IA, Hebei Province, 1993. Jingjitan Jiushiqi Shidai Wanqi Xishiqi Shidai Yizhi (A Palaeolithic Microblade site in Jingjitan). *Ww. C.* 2: 1-22.

IA, Hebei Province, CPAM of Tangshan City & CPAM of Yutian County, 1991. Hebei Yutianxian Mengjiaquan Jiushiqi Shidai Yizhi Fajue Jianbao (A preliminary report of the Palaeolithic site at Mengjiaquan of Yutan county, Hebei province). *Ww. C.* 1: 1-13.

IA, Henan Province, 1989. Henan Wuyang Jiahu Xinshiqi Shidai Yizhi Di'er zi Diliu ci Fajue Jianbao (The preliminary report of the 2nd to 6th excavations at Jiahu, Wuyang county, Henan province). *Ww.* 1:1-14.

IA, Hubei Province, 1994. 1982. Nian Ziguixian Liulinxi Fajue de Xinshiqi Zaoqi Yichun (The discovery of an early Neolithic site at Liulinxi, Zigui county). *J. K.* 1:1-11.

IA, Hubei Province, 1996. 1983 nian Hubei Yidou Chengbeixi Yizhi Fajue Jianbao (A brief report on the excavation of Chengbeixi site in 1983). *J. K.*4: 1-17.

IA, Hunan Province, 1993. Hunan Linlixian Hujiawuchang Xinshiqi Shidai Yizhi (The Hujiawuchang Neolithic site in Linli county, Hunan province). *K.X.* 2: 171-203.

IA, Hunan Province & Lixian Museum, 1990. Hunan Lixian Pengtoushan Xinshiqi Shidai Zaoqi Yizhi Fajue Baogao (A preliminary excavation report of early Neolithic site Pengtoushan, Lixian county, Hunan province). *Ww.* 8: 17-29.

IA, Hunan Province & Shimen Museum, 1994. Shimengxian Yan'erdong Jiushiqi Yizhi Shijue (A

trial excavation at Yan'er Cave, Shimeng county). *Hunan Kaogu Jikan* (Archaeological Collections of Hunan Province) Volume 6:1-7.

IA, Liaolin province, 1994. Liaolin Fuxinxian Chahai Yizhi 1987-1990 nian Sanci Fajue (Three seasons of excavation at the Chahai site, Fuxin county, Liaolin province). *Ww.* 11:4-19.

IA, Shanxi Province, 1994. *Shanxi Kaogu Sishi Nian* (Forty Years of Archaeology in Shanxi Province). Taiyuan: People's Press of Shanxi Province.

Ji, H.X., 1974. Shaanxi Lantian Laochihe Wangengxinshi Puru Dongwu Huashi (Late Pleistocene Mammals from Lantian, Shaanxi). *G. D. X.* Volume 12(3): 222-227.

Jia, L.P., 1958. *Beijingren Zijia* (The Home of Peking Man). Beijing: Beijing Press.

Jia, L.P., 1978. Zhongguo Xishiqi de Tezheng he Tade Chuantong, Qiyuan yu Fenbu (On the phase, origin, and tradition of the microliths in China). *G. D. X. Volume* 16 (2): 137-144.

Jia, L.P., 1980. *Early Man in China*. Beijing: Foreign Languages Press.

Jia, L.P. & W.W. Huang, 1985. The late Palaeolithic of China. In *Palaeoanthropology and Palaeolithic Archaeology in the People's Republic of China*, eds. R.K. Wu & J.W Olsen. Florida: Academic Press, pp. 211-224.

Jia, L.P. & Q.Wei, 1976. Yanggao Xujiayao Jiushiqi Shidai Wanqi Yizhi (The Xujiayao Palaeolithic site in Yanggao county). *K. X*. 2: 97-114.

Jia, L.P., P. Gai & Y.Z. You, 1972. Shanxi Shiyu Jiushiqi Shidai Yizhi Fajue Baogao (The Excavation Report of A Palaeolithic Site at Shiyu, Shanxi Province). *K.X.* 1: 39-58.

Jiang, T.Y. & S.L. Peng, 1990. Guangxi Chuantong Dazhi Jiushiqi (The traditional style of chipped stone tools in Guangxi province). *K. Y. W.* 3: 8-18.

Jiang, W.Z., Y. Tu, Z. Ding & another 5 authors, 1990. Dongxiang Yeshengdao Qingkuang Chubao (A study on the wild rice in Dongxiang). In *Yeshengdao Ziyuan Lunwenji* (Treaties of Wild Rice Resources), ed. M.S. Wu. Beijing: China's Science and Technology, pp. 51-55.

Jiang, Y.C., 1997a. 1996 nian Zhongguo Shida Kaogu Xinfaxian (Ten prominent archaeological discoveries in China in 1996). *Lishi* (History) 4:4-11.

Jiang, Y.C., 1997b. Chengtoushan wei Zhongguo Yizhi Shidai Zuizao Guchengzhi (Chengtoushan is the earliest city remain in China today). *Z. W. B.* 10th, August 1997, front page.

Jiangxi Museum, 1976. Jiangxi Wannian Xianrendong Di'erqi Fajue Baogao (The report of the second excavation at Xianrendong, Wannian, Jiangxi province). *Ww.* 12:23-30.

Jiao, T.L., 1994. Gengxinshi Mo zi Quanxinshi Chuqi Lingnan Diqu de Shiqian Wenhua (The prehistoric culture from the terminal Pleistocene to the early Holocene in South China). *K.X.* 1: 1-24.

Jin, J.G., 1995. Cishan Wanqi 'Zuhewu' Yiji Chutan (A discussion on the 'combined implements' of Cishan). *Kg.* 3: 231-238.

Jin, J. & H. Xu, 1992. Guanyu Nanzhuangtou Zaoqi Xinshiqi Shidai Yizhi de Jidian Kanfa (Opinions about the early Neolithic site of Nanzhuangtou at Xushui). *Kg.*11: 1018-1022.

Jin, Q., P. Wang & S.G. Wang, 1990. *Anhui Huabei Pingyuan Disixi* (The Quaternary in the North Huai Delta in Anhui Province). Beijing: Geology Press.

Jin, X.F., 1984. Bohai Xibu ji Yan'an Diqu Wangengxinshi Yilai Baofeng Zuhe ji Diceng, Gudili Yanjiu (Research on the Pollen Profile, Stratigraphy and Palaeo-geography of the Late Pleistocene on the Coastal Region of Bo-hai). *Haiyang Kexue* (Marine Science) 3: 16-24.

Jin, X.S., 1990. Erxibei Shanqu Shiqianqi zi Nongye (The prehistoric agriculture in the northwest mountains of Hubei province). *N. K.* 2: 108-110.

Jiqing Archaeology Team, 1992. Shandong Linzi Houli Yizhi Diyi, Er'ci Fajue Jianbao (A preliminary report of the first and second excavations of the Houli site in Linzi county, Shandong province). *Kg.* 11: 987-995.

Jiqing Archaeology Team, 1994. Shandong Linzi Houli Yizhi Disan, Sici Fajue Jianbao (A preliminary report of the third and forth excavations of the Houli site in Lingzi county, Shandong province).*Kg.* 2: 97-112.

Jones, R. & B. Meehan, 1989. Plant foods of the Gidjingali: ethnographic and archaeological perspectives from northern Australia on tuber and seed exploitation. In *Foraging and Farming: the Evolution of Plant Exploitation*, eds. D.R. Harris and G.C. Hillman. London: Unwin Hyman, pp. 120-135.

Keeley, L.H., 1992. The use of plant foods among hunter-gatherers: a cross-cultural survey. In *Prehistorire de L'Agriculture*, ed. P.C. Anderson. Paris: CNRS Press, pp. 29-38.

Keeley, L.H., 1995. Protoagricultural practice among hunter-gatherers. In Last Hunters-First Farmers, ed. T.D. Price and A.B. Gabauer. Sanata Fe: SAR Press, pp. 243-272.

Kelly, R.L., 1995. *The Foraging Spectrum - Diversity in Hunter-Gatherer Lifeways*. Washington: Smithsonian Institution Press.

Kislev, M.E., D. Nadel, & I. Carmi, 1992. Epipalaeolithic (19,000 BP) cereal and fruit diet at Ohalo II, Sea of Galilee, Israel. *Review of Palaeobotany and Palynology*. Volume 73:161-166.

Kraybill, N., 1977. Pre-agricultural tools for the preparation of foods in the Old World. In *Origins of Agriculture*, ed. C.A. Reed. Mouton Publishers: The Hague, pp. 485-522.

Kroeber, A.L., 1925. *Handbook of the Indians of California*. Washington: Smithsonian Institution Bureau of American Ethnology Bulletin No. 78.

Lang, H.R., 1987. Zhejiang Yuyao Hemudu Xinshiqi Shidai Yizhi yu Quanxinshi Haipingmian Banhua (The Neolithic Hemudu site and the sea level change of the Holocene). *Zhejiang Dizhi* (Geology of Zhejiang) 1: 5-13.

Layton, R., R. Foley & E. William, 1991. The transition between the specialised hunting and gathering and husbandry of resources - a social-ecological approach. *Current Anthropology* 32(3): 255-264.

Li, F.Q., 1990. Shilun Huanan Diqu Yuanshi Nongye de Qiyuan (A discussion on the remains of prehistoric agriculture in South China). *N. K.* 2: 84-95.

Li, H. W., H. Li, & W. K. Pao, 1945. Cytological and genetical studies of the interspecific cross of the cultivated foxtail millet (*Setaria italica*) and the green foxtail (*S. viridis*). *Journal of the American Society of Agronomy*. 9: 32-54.

Li, J.Z., Z. Zhang, Q. Ze & B. Liang, 1996. Xinshiqi Shidai Zaoqi Taoqi de Qiyuan - Jianlun Zhongguo Taoqi Qiyuan (A study on early Neolithic pottery - also a discussion on the origin of Chinese pottery). *Kg.* 5: 83-91.

Li, P., 1982. Zhongguo Zaipei Zhiwu de Qiyuan yu Fazhan (The origins and distribution of cultivated cereals in China). *Shijie Nongye* (World Agriculture) 2: 40-56.

Li, P. & X. Lu, 1981. Nongye Qiyuan yu Shandi Kao (On the Origins of Agriculture in Hilly Area). *N. K.*1:21-31.

Li, W.J., S. Lang, & J. Zhao, 1996. Gansu Qin'an Dadiwan Yiqi Zhitao Gongyi Yanjiu (Studying the manufacturing technique of the Dadiwan pottery in Qin'an, Gansu province). *K.Y.W.* 2: 22-34.

Li, W.M., Z. Zhang, Y. Gu, Y. Lin, & F. Yan, 1982. Jiangsu Dantu Lianhuadong Dongwuqun (The fauna in Lianhuadong, Dantu county, Jiangsu Province). *R. X.* Volume 1(2): 169-179.

Li, W.Y., 1987. Lun Zhongguo Dongbu Disiji Lengqi Zhibei yu Huanjing (Vegetation and environment of Eastern China in Quaternary cold stage). *Dili Xuebao* (ACTA Geographica Sinica) 4: 299-307.

Li, W.Y., 1991. Vegetation and Environment of the Quaternary in China. In *Quaternary Geology and Environment in China*, ed. T.S. Liu. Beijing: Science Press, pp. 151-157.

Li, Y.X., 1993. Zhongguo Jiushiqi Wanqi Wenhua de Huafen (On the division of the Upper Palaeolithic industries of China). *R. X. Volume* 12(3): 214-223.

Li, Y.X. & C.Y. Lei, 1980. Jiangsu Lishui Shengxiandong Faxian de Dongwu Huashi (The faunal fossils found in the Spirits Cave, Lishui county, Jiangsu province). *G. D. X.* Volume 18(1): 59-63.

Li, Y.H. & D.F. Han, 1978. Guangxi Guilin Zengpiyan Yizhi Dongwuqun (The fauna in Zengpiyan, Guilin City, Guangxi Province). *G. D. X* Volume 16(4): 244-254.

Li, Z.W., 1984. Shanxi Jiushiqi Shidai Yizhi Maicang Gulu (The Distribution of the Palaeolithic Sites in Shanxi Province). In *Shanxi Jiushiqi Shidai Kaogu Wenji* (The Palaeolithic of Shanxi Province: Selected Treaties), ed. The Institute of Archaeology of Shanxi Province. Taiyuan: Shanxi Economic Press, pp. 548-554.

Li, Z.W. & Z.G. Wang, 1992. Shanxi Yuci Dafa Didian de Jiushiqi (Palaeolithic from Dafa, Yuci County of Shanxi Province). *R. X.* Volume 11(4): 345-353.

Liang, G.S., 1984. Zhongguo Zaipei Daozhong de Dili Qiyuan yu Shengtai Fenhua (The geographical origin and deviation of cultivated rice in China). *Zuowu Pinzhong Ziyuan Yanjiu* (Resources of Plant Species) Volume 1: 10-17.

Liang, G.S. & J.W. Qi, 1980. Zhongguo Zaipeidao de Qiyuan (Study on the origins of cultivated rice in China). *Kejishi Wenji* (Collected Works on Technological History) 4: 165-174.

Lin, D.H., 1990. Yeshengdao Rengong Qihouxiang Nailengxing Jianding Yanjiu Chubao (A preliminary report on cold tolerance of common wild rice by experiment in human-made climatic box). In *Yeshengdao Ziyuan Yanjiu Lunwen Xuanpian* (Treaties on Wild Rice Resources), ed. M.S. Wu. Beijing: Chinese Science and Technology Press, pp. 63-68.

Lin, H.D., 1992. *Hemudu Wenhua Chutian* (A Study on the Hemudu Culture). Hangzhou: Zhejiang People's Press.

Lin, S.L., 1993. Guanyu Jianzhuangqi de Dingyi - Zhongxifang de Bijiao (On the definition of point - a comparison between China and the West). R. X. Volume 12 (1): 8-22.

Linfen Culture Bureau, 1989. Shanxi Jixian Shizitan Zhongshiqi Wenhua Yizhi (The Mesolithic site of Shizitan, Ji County, Shanxi Province). *K.X.* 3: 305-323.

Liu, D.Y., 1993. Woguo Jiushiqi Shidai Kaogu de Zhongda Tupo (An important breakthrough of Palaeolithic archaeology in our country). *Z.W.B.* 2nd, May, 1993. Third page.

Liu, G.X., 1991. Jianghan Pingyuan Wanbingqi ji Bingqihou de Zhibei yu Huanjing (Late-glacial and postglacial vegetation and associated environment in Jianghan Plain). *Zhiwu Xuebao* (ACTA Botanica Sinica). *Volume* 33(8): 581-588.

Liu, J.L. & W. Chang, 1996. Gengju Baofeng Ziliao Tulun Changjiang Sanjiaozhou Diqu 12,000 nian Yilai de Huanjing Bianqian (A reconstruction of the palaeoenvironment of the Yangzi delta in the last 12,000 years based upon pollen data). *Gushengwu Xuebao* (ACTA Palaeontologica Sinica Volume 35(2): 136-153.

Liu, J. & Z.Y. Yao, 1993. *Zhongguo Hemudu Wenhua* (The Hemudu Culture of China). Zhejiang: Zhejiang People's Press (with English translation).

Liu, K.B., 1988. Quaternary history of the temperate forests of China. *Quaternary Science Reviews*. Volume 7: 1-20.

Liu, K.B., S. Sun, & X. Jiang, 1992. Environmental changes in the Yangzi River delta since 12,000 years BP. *Quaternary Research* Volume 38: 32-45.

Liu, L.S., 1995. Changjiang Zhongxiayou Diqu Gengxinshi Yilai Guhuanjing Yanbian de Chubu Yanjiu (A preliminary study on palaeoenvironmental changes in the Middle and Lower Yangzi River basin since the Pleistocene). *Disiji Bingchuan yu Disiji Dizhi Lunwenji* (Quaternary Glaciology and Quaternary Geology). *Volume* 8: 30-36. Beijing: Geology Press.

Liu, S.Z., 1996. Jiangxi Wannian Xianrendong he Diaotonghuan Fajue Huo Zhongyao Chengguo (Important achievements in the excavation of Xianrendong and Diaotonghuan). *Z. W. B.* 28th, January, the front page.

Liu, X.L., 1991. Zhongguo Shiqian Nongye Fasheng Yuangyin Shishuo (On the origins of prehistoric agriculture in China). *Zhongguo Nongshi* (Chinese Agriculture History) 3: 1-9.

Liu, Z.Y., 1994. Guanyu Daozuo Qiyuan de Tongxin (Letters about the origins of rice cultivation). *N. K.* 3: 54-70.

Lixian Museum, 1992. Hunan Lixian Zhangjiatan, Xiangongtan Jiushiqi Didian Diaocha Baogao (A survey on several Palaeolithic sites in Lixian, Hunan province).*H. K.* 4: 1-8.

Lu, D.P., 1993, Yicheng Diqu Yuanshi Wenhua Zongsu (A summary of the prehistoric culture in Yicheng area). *J. K.*2:42-49.

Lu, H.Y, N.Q. Wu and B.Z. Liu, 1997. Recognition of rice phytoliths, in *First European Meeting on Phytolith Research*. Eds. Ascension Pinilla and Jordi Juan-Tresserras. Spain: Imprime Graficas Fersan. S.L.: 159-174.

Lu, L.D., 1998a. The microblade tradition in China: regional chronological and significance in the transition to agriculture. *Asian Perspectives* Volume 37(1):84-112.

Luojiajiao Archaeology Team, 1981. Tongxiang Luojiajiao Yizhi Fajue Baogao (The excavation report of Luojiajiao site). *Zhejiangsheng Wenwu Kaogusuo Xuekan* (Journal of the Institute of Archaeology, Zhejiang Province) 1:1-36.

Ma, A.C. & H.L. Tang, 1992. Zhejiang Jinghua Quanxinshi Daxiongmao-Jianchixiang Dongwuqun de Faxian Jiqi Yiyi (On discovery and significance of a Holocene *Ailuropoda-Stegodon* fauna from Jinhua, Zhejiang). *G. D. X. Volume* 30(4): 295-312.

MacNeish, R.S. & J. G. Libby (eds), 1995. *Origins of Rice Agriculture*. Elpaso: The University of Texas.

MacNeish, R.S., 1995. Lithic typology. In *Origins of Rice Agriculture*, eds. R.S. MacNeish & J.G. Libby. Elpaso: The University of Texas, pp. 23-34.

MacNeish, R.S. & R.E. Taylor, 1995. Chronometric. In *Origins of Rice Agriculture*, eds. R.S. MacNeish & J. G. Libby. Elpaso: The University of Texas, pp. 77-86.

MacNeish, R.S., B. Marino, G. Lin, L. Cai, & S. Qiu, 1995. Isotopic analysis to determine ancient diet in the Neolithic-Epipalaeolithic of China. In *Origins of Rice Agriculture*, eds. R.S. MacNeish & J.G. Libby. Elpaso: The University of Texas, pp.69-76.

Marks, A.E., 1976a. Ein Aqev: a late Levantine Upper Palaeolithic site in the Nahal Aqev. In *Prehistory and Palaeoenvironments in the Central Negev,*

Isreal. Volume I, ed. A.E. Marks. Dallas: SMU Press, pp. 227-291.

Marks, A.E., 1976b. Site D5: A geometric Kebara 'A' occupation in the Nahal Zin. In *Prehistory and Palaeoenvironments in the Central Negev, Isreal. Volume I,* ed. A. E. Marks. Dallas: SMU Press, pp. 293-316.

Miller, N. F., 1992. The origins of plant cultivation in the Near East. In *The Origins of Agriculture - An International Perspective*, eds. C. Cowan & P.J. Watson. Washington: Smithsonian Institutes Press, pp. 39-58.

Minze, S. & P.J. Reimer, 1993. *Calib. 3* (a computer program). Quaternary Research Centre AK-60, University of Washington, Seattle, USA.

Misra, V.N., 1985. Microlithic industries in India. In *Recent Advance in Indo-Pacific Prehistory*, eds. V.N. Misra & P. Bellwood. New Delhi: Oxford & IBH Publishing Co., pp. 111-122.

Nadel, D. & I. Hershkovitz, 1991. New subsistence data and human remains from the earliest Levantine Epipalaeolithic. *Current Anthropology*. Volume 32(5): 631- 635.

Neimeng Archaeology Team CASS, 1985. Neimenggu Ao'hanqi Xinglongwa Yizhi Fajue Jianbao (A brief report on the excavation at Xinglongwa, Ao'hanqi, Inner Mongolia). *Kg.* 10: 865-867.

Neimeng Archaeology Team CASS, 1997. Neimenggu Ao'hanqi Xinglongwa Yizhi 1992 nian Fajue Jianbao (A brief report of the 1992 excavation season at Xinglongwa, Ao'hanqi, Inner Mongolia). *Kg.* 1: 1-27.

Nelson, S.M., 1993. *The Archaeology of Korea.* UK: Cambridge University Press.

Nicholson, P.H., 1981. Fire and the Australian Aborigines - an enigma. In *Fire and The Australian Biota*, eds. A.M. Gill *et al.* Canberra: Australian Academy of Science, pp. 55-77.

Nie, G.Z. & D.M. He, 1991. A new explanation regarding the provenance of materials of China's Xiashu loess. In *Loess, Environment and Global Change,* ed. T.S. Liu. Beijing: Science Press, pp. 213-227.

Nowak, R.M., 1991. *Walker's Mammals of the World.* fifth edition. Baltimore and London: The Johns Hopkins University Press.

Oka, H.I., 1975. Mortality and adaptive mechanisms of *Oryza perennis* strains. *Evolution* Volume 30:380-392.

Oka, H.I., 1988. *Origin of Cultivated Rice*. Tokyo: Japan Scientific Societies Press.

Oka, H-I & W.T. Chang, 1959. The impact of cultivation on populations of wild rice, *Oryza sativa f. spontanea. Pyton* Volume 13(2): 105-117.

Oka, H.I. & H. Morishima, 1971. The dynamics of plant domestication: cultivation experiments with *Oryza* perennial and its hybrid with O. sativa. *Evolution* Volume 25:356-364.

O'Connell, J.F. & K. Hawkes, 1981. Alyawara plant use and optimal foraging theory. In *Hunter-Gatherer Foraging Strategies: Ethnographic and Archaeological Analyses*, eds. B.Winterhalder and E. Smith. Chicago: University of Chicago Press, pp. 99-125.

O'Connell, J.F., P. K. Latz, & P. Barnett, 1983. Traditional and modern plant use among the Alyawara of Central Australia. *Economic Botany* Volume 37(1): 80-109.

Pan, X.G. & X.Z. Rao, 1990. Jiangxi Dongxiang Yeshengdao Kaocha ji Texing Jianding Baogao (A survey on the wild rice found in Dongxiang, Jiangxi province). In *Yeshengdao Ziyuan Yanjiu Lunwen Xuanpian* (Treaties on Wild Rice Resources), ed. M.S. Wu. Beijing: Chinese Science and Technology Press, pp. 11-16.

Pang, H.H. & X.Q. Wang, 1996. Zhongguo Putong Yeshengdao Ziyuanzhong Yiniansheng Leixing de Yanjiu (A study on the annual type of common wild rice in China). *Zuowu Pingzhong Ziyuan* (Resources of Plant Species) (3):8-11.

Pang, H.H., H.W. Cai & X.Q. Wang, 1995. Zhongguo Putong Yeshengdao de Xingtai Fenlei Yanjiu (The classification of the common wild rice *(Oryza rufipogon)* in China). *Zuowu Xuebao* (ACTA Agronomica Sinica) Volume 21(1): 17-23.

Pei, A.P., 1994. Lixian Faxian Wo Zuizao Julo Weihao yu Weiqiang (The earliest protective ditch found in Xi county). *Z. W. B.* 4th, December. the front page.

Pei, A.P., 1995. Pengtoushan Wenhua Chulun (A preliminary discussion on the Pengtoushan culture). Paper presented to the *Conference on the Prehistoric Culture in the Middle Yangzi Valley*. Unpublished paper.

Pei, A.P., 1998a. Pengtoushan Wenhua de Daozuo Yichun yu Zhongguo Shiqian Daozuo Nongye Zailun (Remains of rice cultivation found in the Pengtoushan assemblages and a re-discussion on the origin of rice cultivation in prehistoric China). *N.K.* 1:193-203.

Pei, A.P., 1998b. Notes on new advancements and revelations in the agricultural archaeology of early rice domestication in the Dongting Lake region. *Antiquity* Volume 72: 878-885.

Peng, S.F., 1992. Wannian Xianrendong Xinshiqi Shidai Zaoqi Wenhua de Jige Wenti (A few problems of the early Neolithic in Xianrendong culture). In *Jiangxi Xianqin Kaogu* (Prehistoric Archaeology in Jiangxi province), ed. S.F. Peng. Jiangxi: Jiangxi Tertiary Education Press, pp. 2-14.

Peng, S.F., 1996. Jiangxi Shiqian Kaogu de Zhongda Tupo (An important breakthrough in Jiangxi archaeology). Unpublished paper.

Phillips, J.L. 1979. Origins of agriculture in Israel. In *The Quaternary of Israel*, ed. A. Horowitz. New York: Academic Press, pp. 317-319.

Prausnitz, M.W., 1970. *From Hunter to Farmer and Trader - Studies in the Lithic Industries of Israel and Adjacent Countries.* Jerusalem: Sivan Press Ltd.

Price, T.D. & A.B. Gebauer (eds), 1995. *Last Hunters, First Farmers.* Santa Fe, New Mexico. USA.

Price, T. D. & A.B. Gebauer, 1995. New Perspectives on the Transition to Agriculture. In *Last Hunters, First Farmers,* eds. T.D. Price & A.B. Gebauer. Santa Fe, New Mexico, pp. 3-20.

Qi, G.Q., 1989. Zhongguo Beifang Disiji Puru Dongwuqun Jianlun Yuanshi Renlei Shenghuo Huanjing (Quaternary Fauna of North China and the Palaeoenvironment). In *Zhongguo Yuangu Renlei* (*Early Humankind in China*), eds. R.K. Wu, X.Z. Wu & S.S. Zhang. Beijing: Science Press, pp. 277-337.

Qi, J.W., Z.B. Li & Y.F. Chen, 1984. Guangdong Dalu he Hainandao Putong Yeshengdao de Bijiao Guanca (A comparative observation on common wild rice of Guangdong continent vs. Hainan island). In *Zuowu Pingzhong Ziyuan Yanjiu* (Research on Plant Species), ed. The Institute of Plant Species. Beijing: Agriculture Press, pp. 35-44.

Qin, X.Y. & R.P. Li, 1990. Guangxi Putong Yeshengdao Pingzhi Jianding Chubao (A preliminary report of the quality of common wild rice in Guangxi province). In *Yeshengdao Ziyuan Yanjiu Lunwen Xuanpian* (Treaties On Wild Rice Resources), ed. M.S. Wu. Beijing: Chinese Science and Technology Press, pp. 6-10.

Redding, R.W., 1995. Preliminary report on faunal remains recovered from the 1993 excavation. In *Origins of Rice Agriculture*, eds. R.S. MacNeish & J.G. Libby. Elpaso: The University of Texas, pp. 53-58.

Ren, M.E., 1985. *Zhongguo Ziran Dili Gangyao* (The Physiography of China). Beijing: Commercial Press.

Rice, P.M., 1987. *Pottery Analysis - A Source Book.* Chicago: The University of Chicago Press.

Rindos, D., 1980. Symbiosis, instability, and the origins and spread of agriculture - a new model. *Current Anthropology.* Volume 21:751-767.

Rindos, D., 1984. *The origins of Agriculture - An Evolutionary Perspective.* Orlando: Academic Press.

Rozoy, J-G., 1985. The revolution of the bowmen in Europe. In *The Mesolithic in Europe*, ed. C. Bonsall. Edinburgh: J. Donald Published Ltd., pp. 13-28.

Sato, H.Y., 1995. Transitional meanings of Araya microlithic industry at Kanto region in Japan. In *Treaties of the People and Lithic Culture of 20,000 years B.P. after the Explosion of Aira Volcano in Kyushu*, ed. and printed. the International Exchange Centre of Kumamoto City, pp. 149-158.

Shanghai Silicate Institute, CASS, 1987. *Zhongguo Gutaoci Yanjiu* (Study the ancient ceramic in China). Beijing: Science Press.

Shen, Y.L., 1994. Quanxinshi Huanjing yu Pengtoushan Daozuo Yichun (The Holocene environment and rice remains in Pengtoushan). *N. K.* 3: 84-87.

Shi, S.H., 1993. Zhongguo Quanxinshi Gaowenqizhong de Qihou Tubian Shijian Jiqi dui Renlei de Yingxiang (Climatic abrupt change events and their impact on human civilisation during the Holocene Megathermal in China). *Haiyang Dizhi Yu Disiji Dizhi* (Marine Geology and Quaternary Geology) Volume 13(4): 65-73.

Shi, X.B., 1986. Qian-Yangshao Wenhua de Faxian Jiqi Yiyi (The discovery of the pre-Yangshao culture and its significance). In *Zhongguo Kaoguxue Yanjiu - Xianai Xiansheng Kaogu Wushinian Jinian Lunwenji* (Archaeological Researches in China - A Collection of Papers in Commemoration of the Fiftieth Years of Prof. Xia Nai's Work in Archaeology) *Volume 2*, complied by the Institute of Archaeology CASS. Beijing: Cultural Relics Publishing House, pp. 1-23.

Shi, X.B., 1989. Xiachuan Wenhua Yanjiu (Study of the Xiachuan culture). In *Qingzhu Subingqi Kaogu Gongzuo Wushi Wunian Lunwenji* (Celebration of 55 years of archaeology work of Su Bingqi), ed. Editing group of the Articles. Beijing: The Cultural Relics Publishing House, pp. 144-151.

Shi, Y.F. & J.T. Wang, 1982. Zhongguo Wan Disiji de Qihou, Bingchuan he Haipingmian Bianhua (The change of palaeoclimate, glacier and sea-level in quaternary China). In *Disanjie Quanguo Disiji Xueshu Huiyi Lunwenji* (Treaties of the Third Conference of the National Quaternary Research), ed. the Research Committee of the Quaternary Research in China. Beijing: Science Press, pp. 111-121.

Smith, B.D., 1995. *The Emergence of Agriculture.* New York: Scientific American Library.

Smith, M.A., 1985. A morphological comparison of central Australian seedgrinding implements and Australian Pleistocene-Age grindstones. *The Beagle Volume* 2(1): 23-38.

Smith, M.A., 1986. The antiquity of seedgrinding in arid Australia. *Archaeology in Oceania* Volume 21:29-39.

Society of Crop Genetic Resources, 1994. *Zhongguo Zuowu Pinzhong Ziyuan* (Crop Genetic Resources in China). Beijing: Agriculture Press.

Song, E.C., 1986. *Yunan Shaoshu Mingzu Yanjiu Lunwenji* (Papers on studying the Minorities in Yunan Province). Kuming: Yunan People's Press.

Song, Z.L. & G.X. Zhou, 1994. Yuanshi Juetubeng shang de Chuankong Zhongshi (The perforated weighting stones). In *Zhongri Gurenlei yu Shiqian Wenhua Yuanyuan Guanxi Guoji Xueshu Yantaohui Lunwenji* (International Symposium Relationship between Chinese and Japanese Ancient Men and Prehistoric Cultures), eds. G.X. Zhou, Y. Zhang & X. Zhang. Beijing: China International Radio Press, pp. 254-262.

Steward, J.H., 1938. *Basin-Plateau Aboriginal Sociopolitical Groups.* Smithsonian Institution, Bureau of American Ethnology. Bulletin 120.

Stuiver, M. & P.J. Reimer, 1993. CALIB Rev. 3.0.3. (A computer program). Quaternary Research Center, University of Washington. Seattle, USA.

Sun, C.Q., X.Q. Wang & Z.C. Li, 1998. Cong Putong Yeshendao DNA de Xiangeng Fenhuai kan Yazhou Zaipeidao de Qiyuan yu Yanhuai (Studying the origin and evolution of domesticated rice in Asia from the DNA differentiation of wild rice). *N.K.* 1:21-29.

Sun, G.Q., 1992. Henan Peiligang Wenhua de Fengbu he Diyu Leixing (The distribution and types of the Peiligang culture in Henan province). *H. K.* 4: 47-62.

Sun, G.Z., 1990. Hunan Yeshengdao de Shengjing Jiqi Tezheng, Texing Yanjiu (The ecology and characteristics of wild rice in Hunan province). In *Yeshengdao Ziyuan Yanjiu Lunwen Xuanpian* (Treaties on Wild Rice Resources), ed. M.S. Wu. Beijing: Chinese Science and Technology Press, pp. 31-34.

Sun, M., 1987. *Xia-Shang Shigao* (The History of Xia and Shang Dynasties). Beijing: The Cultural Relics Publishing House.

Sun X.J., C. Q. Song, B. Y. Wang & M. R. Sun, 1995. Huangtu Gaoyuan Nanyuan Shiwannian Yilai de Zhibei (Vegetation of the last 100,000 years in the south edge of the loess plateau). *Kexue Tongbao* (Chinese Science Bulletin). Volume 40(13): 1222-1224.

Tang, C., 1997. Notes on nail-impressed ceramics along with Microblades in Northeast Asia. In *International Research on the Origin of Pottery in Eastern Asia*, ed. and printed by Kokugakuin University in Japan, pp. 5-8.

Tao, L.H., 1994. Lun Nongye Qiyuan de Dili Huanjing (The natural environment of the origins of agriculture). *N. K.*1: 36-40.

The Anhui Archaeology Team CASS, 1993. Xiaoshankou he Gutaishi Yizhi Shijue (A trial excavation of Xiaoshankou and Gutaishi sites). *Kg.* 12:1062-1075

The Archaeology Team of the Three Gorges, 1989. Hubei Zigui Chaotianzui Yizhi Fajue Jianbao (A brief report on the excavation of Chaotianzui, Zigui county, Hubei province). *Ww.* 2: 41-51.

The Archaeology Team of Yueyang City & CPAM of Qianlianghu Farm, 1994. Qianlianghu Fenshanbao Xinshiqi Shidai Yizhi Shijue Baogao (A brief report on the trial excavation of the Neolithic site at Fenshanbao). *Hunan Kaogu Jikan* (Archaeological Collections of Hunan Province) Volume 6: 17-33.

The Editing Committee of the Agriculture Encyclopedia of China, 1991. *Zhongguo Nongye Dabaike - Nongzuowu Juan* (Agriculture Encyclopedia of China - the Cultivars. Volume II). Beijing: Agriculture Press.

The Editing Committee of the Physiography of China, 1981. *Zhongguo Ziran Dili - Shui Ziyuan* (Physiography of China-Water Resources). Beijing: Science Press.

The Editing Committee of the Physiography of China, 1983. *Zhongguo Ziran Dili -Zhiwu Dili* (Physiography of China - Botanical geography). Beijing: Science Press.

The Editing Committee of the Physiography of China, 1984. *Zhongguo Ziran Dili -Qihou* (Physiography of China -The Climate). Beijing: Science Press.

The Editing Committee of Yunnan Province, 1981. *Bulangzu Shehui Lishi Diaocha* (A Survey on the Social History of Bulang Ethnic). Kungming: Yunnan People's Press.

The Editing Committee of Yunnan Province, 1990. *Jino he Pumi zu Shehui Lishi Diaocha* (A survey on the Social History of Jino and Pumi Minorities). Beijing: Ethnic Press.

The First Henan Archaeology Team CASS, 1983. Henan Xinzheng Shawoli Xinshiqi Shidai Yizhi (The Shawoli Neolithic site in Xinzheng county, Henan province). *Kg.* 12: 1057-1065.

The First Henan Archaeology Team CASS, 1984. 1979 nian Peiligang Yizhi Fajue Baogao (A report of the 1979 excavation at Peiligang). *K.X.* 1: 23-51.

The First Henan Archaeology Team CASS, 1991. Henan Ruzhou Zhongshanzhai Yizhi (The Zhongshanzhai site in Ruzhou city, Henan province). *K.X.* 1: 57-89.

The First Henan Archaeology Team CASS, 1992. Henan Jiaxian Shuiquan Xinshiqi Shidai Yizhi Fajue Jianbao (A preliminary report of the excavation of Shuiquan site in Jia county, Henan province). *Kg.* 10: 865-874.

The Guangxi Survey Group of Wild Rice, 1990. Guangxi Yeshengdao de Dili Fenbu Jiqi Texing Tezheng (The distribution and Characteristics of wild rice in Guangxi). In *Yeshengdao Ziyuan Lunwenji* (Treaties On Wild Rice Resources), ed. M.S. Wu. Beijing: Chinese Science and Technology Press, pp. 26-34.

The Institute of Agronomy, CAS, 1986. *Zhongguo Daozuoxue* (Rice Cultivation in China). Beijing: Agriculture Press.

The Institute of Botany, CAS, 1972. Zhongguo Gaodeng Zhiwu Tujian (A Handbook of High Plants in China). Beijing: Science Press

The Pollen Laboratory of the Institute of Archaeology, Hunan Province, 1990. Hunan Lixian Pengtoushan Yizhi Baofeng Fenxi yu Guhuanjing Tantao (Pollen analysis and reconstruction of the palaeoenvironment of Pengtoushan). *Ww.* (8): 30-32.

The Radiocarbon Dating Laboratories of Beijing University & IA CASS, 1982. Shihuiyan Diqu Tanshisi Yangping Niandai de Kekaoxing yu Zengpiyan deng Yizhi de Niandai Wenti (The reliability of radiocarbon dating in limestone region and the problem of the date of Zengpiyan). *K. X.* 2: 243-250.

The Sixth Shaanxi Archaeological Team, 1984. Shaanxi Lintong Baijiacun Xinshiqi Shidai Yizhi Fajue Jianbao (Excavation of the Baijiacun Neolithic site at Lintong, Shaanxi). *Kg.* 11: 961-973.

The National Survey Group of Wild Rice, 1984. Zhongguo Yeshengdao Ziyuan Daocha (The survey of wild rice resources in China). *Zhongguo Nongye Kexue* (Chinese Agriculture Science) (6): 27-34.

Tindale, N. B., 1977. Adaptive significance of the Panara or grass seed culture of Australia. In *Stone Tools as Cultural Makers: Change, Evolution and Complexity*, ed. R.V.S. Wright. Canberra: Australian Institute of Aboriginal Studies, pp. 340-349.

Tong, E.Z., 1989. Luosu Dongnanya ji Zhongguo Nanbu Nongye Qiyuan de Ruogan Wenti (The origins of agriculture in South China and the characteristics of this agriculture). *N. K.* 2: 57-71.

Tong, G.B., M.H. Ke & S.F. Yu, 1983. Hebei Pingyuan Disiji Baofeng Zuhe Jiqi Dizhi Yiyi (The pollen profile of the North China Plain and its geological significance). *Haiyang Dizhi Yu Disiji Dizhi* (*Marine Geology and Quaternary Geology*) Volume 3(4): 91-101.

Tong, G.B., J. Zhang, & S. Fan, 1992. Zhongguo Disiji Baofeng Zhiwuqun de Fenbu (Distribution of Quaternary Palynoflora in China). *Haiyang Dizhi yu Disiji Dizhi (Marine Geology and Quaternary Geology) Volume* 12 (3): 45-56.

Tong, P.H. & C.M. Wei, 1994. Zhangqiu Xihe Xinshiqi Shidai Yizhi (The Xihe Neolithic site in Zhangqiu). *Z.W.B.* 20, February, the third page.

Tong, Z.C., 1989. Yangshao, Longshan Wenhua Shizhi Gongju de Gongyi Yanjiu (Research on the lithic tools of Yangshao and Longshan culture). In *Zhongguo Dongbei Diqu he Xinshiqi Shidai Kaogu Lunji* (Selected Works of the Archaeology in Northeast China and the Neolithic Archaeology of China), Complied by Z.C. Tong. Beijing: Cultural Relics Publishing House, pp. 127-147.

Tsunoda S. & N. Takahashi, 1984. *Biology of Rice.* Tokyo: Japanese Scientific Societies Press.

Underhill, A.P., 1997. Current Issues in Chinese Neolithic. *Journal of World Prehistory* Volume 11(2):103-160.

Unger-Hamilton, R., 1985. Microscopic striations on flint sickle-blades as an indication of plant cultivation: preliminary results. *World Archaeology Volume* 17(1): 121-126.

Vavilov, N.I., 1926. Centers of origin of cultivated plants. In *Origin and Geography of Cultivated Plants* : 22-135. English translation by Doris Love. UK: Cambridge University Press. Published in 1992.

Veth, P., R. Fullagar & R. Gould, 1997. Residue and use-wear analysis of grinding implements from Puntutiarpa Rockshelter in the Western Desert: current and proposed research. *Australian Archaeology* 44:23-25.

Veth, P.M. & F. J. Walsh, 1988. The concept of 'staple' plant food in the Western Desert region of Western Australia. *Australian Aborigines Studies* 2: 19-24.

Wang, J., 1986. Guanyu Xiachuan Yizhi he Dingcun Yizhiqun 7701 Didian de Shidai, Xingzhi Wenti (Problems of the date and cultural features of Xiachuan and Loc. 7701 of Dingcun site). *R. X.*vol. 5(2): 172-177.

Wang, J. & F.H. Tao, 1991. Dingcun Yizhiqun Jiushiqi Gaisu (A summary of the Palaeolithic site-cluster in Dingcun). *Shanxi Wenshi Ziliao* (Historical Documents in Shanxi) Volume 78(6): 73-90.

Wang, J., X.Q. Wang & Z. Y. Chen, 1978. Xiachuan Wenhua (Xiachuan culture). *K.X.* 3: 259-288.

Wang, J.F., K.Y. Yang, Y. F. Ma & another 18 authors, 1995. *Rendi Guanxi Yanjing Jiqi Tiaokong* (The Evolution and Adjustment of the Relationship between Mankind and the Environment). Beijing: Science Press.

Wang, K.F., H. Jiang, & Y. Zhang, 1990. *Nanhai ji Yan'an Diqu Baofeng Zaolei yu Huanjing* (The Pollen Profile and Quaternary Environment in the South China Sea and Adjacent Areas). Shanghai: Tongji University Press.

Wang, L.H., 1986. Zhongguo Xinshiqi Shidai he Xiandai Juming de Shidai Bianhua he Dili Bianyi-Lugu Xingzhuang Celiang de Tongji Fengxi (Secular change and geographical variation in Chinese Neolithic and modern inhabitants - a statistical study of cranial metric traits). *R. X.* Volume 5(3):243-258.

Wang, X.Q., J. P. Ding, & F. H. Tao, 1983. Shanxi Puxian Xueguan Xishiqi (The Microlithic found in Xueguan, Pu county, Shanxi province). *R. X.* Volume 2(2): 162-171.

Wang, X.Q., 1994. Guanyu Zhongguo Yeshengdao Yanjiu de Jige Wenti de Taolun [A preliminary discussion on a few questions regarding the common wild rice (*Oryza rufipogon*) in China]. *N. K.* (1): 48-51.

Wang, X.Q., Sun C.Q. & J.Z. Zhang, 1998. Zhongguo Zaipeidao Qiyuan Yanjiu de Xianzhuang yu Zhanwang (Current studies and future development on the origin of rice cultivation in China). *N. K.* 1:11-20.

Wang, X.Z., Q. H. Jiang, & R. S. MacNeish, 1995. Palynology and Palaeoclimate in the Dayuan Basin. In *Origins of Rice Agriculture*, eds. R.S. MacNeish

& J.G. Libby. Elpaso: The University of Texas, pp. 59-68.

Wang, Y.B., 1993. Houli Wenhua yu Beixin Wenhua de Guanxi (The relationship between Houli and Beixin cultures). *Z.W.B.* 18, April, the third page.

Wang, Y.B., 1994. Zailun Houli Wenhua yu Beixin Wenhua de Guanxi (More discussion on the relationship between Houli and Beixin cultures). *Gugong Bowuyuan Yuankan* (Academic Journal of the Forbidden City Museum). 2: 81-106.

Wang, Y.P., 1988. Xiaokongshan he Zhangnaodong (Xiaokongshan and Zhangnaodong). *H. K.* 4:38-42.

Wei, Q., H. Meng, & S. Cheng, 1985. Nihewan Cengzhong Xinfaxian Yichu Jiushiqi Didian (New Palaeolithic site from the Nihewan (Nihowan) beds). *R. X. Volume* 4(3): 223-232.

Weidenreich, F., 1939. On the earliest representatives of modern mankind recovered on the soil of east Asia. In *Anthropological Papers of Franz Weidenreich 1939-1948 - A Memorial Volume,* compiled. S.L. Washburn & W. Davida. New York: the Viking Fund Inc., pp. 205-224.

de Wet, J.M. J., L.L. Oestry-Stidd & J.I. Cubero, 1979. Origins and evolution of foxtail millet. *Journ. d'Agric. Trad. et de Bota.* Appl. 1: 53-64.

White, J.C., 1995. Modeling the development of early rice agriculture: ethnoecological perspectives from Northeast Thailand. *Asian Perspectives.* Volume 34(1): 37-68.

Whittle, A., 1985. *Neolithic Europe - A Survey.* UK: Cambridge University Press.

Wright, H.E., 1993. Environmental determinism in Near East prehistory. *Current Anthropology Volume* 34(4): 458-469.

Wright, K., 1991. The origins and development of ground stone assemblages in Late Pleistocene Southwest Asia. *Paleorient.* Volume 17/1:19-46.

Wright, K., 1993. Early Holocene ground stone assemblages in the Levant. *Levant Volume* 25:93-111.

Wu, B.P., M.R. Chen, Z.X. Chen, N.X. Mai & K.W. Liu, 1991. *Zhongguo de Huangtu Gaoyuan* (The Loess Plateau of China.). Xi'an: People's Press of Shaanxi Xi Province.

Wu, J.A., 1986. Shiqi Shidai de Shimopan (The stone querns of the Stone Age). *S. Y.* (1-2): 46-54.

Wu, J.A., 1989. Luelun Huanghe Liuyu Qian-Yangshao Wenhua Shiqi de Nongye (A discussion on agriculture of the proto-Yangshao culture in the Yellow Valley). *N. K.*2: 118-125.

Wu, M.S. & C.B. Chen, 1986. Guangxi Yeshengdao de Tonggongmei Yanjiu (Research report on isozymes of wild rice in Guangxi Province). *Zuowu Xuebao* (ACTA Agronomica Sinica) Volume 12 (2): 87-94.

Wu, M.S., 1990. Guangxi Yeshengdao Ziyuan Daocha (A survey on wild rice in Guangxi. In *Yeshengdao Ziyuan Yanjiu Lunwen Xuanpian* (Treaties on Wild Rice Resources), ed. M.S. Wu. Beijing: Chinese Science and Technology Press, pp. 17-21.

Wu, N.Q., G. Gu & Y. Song, 1994. Zhiwu Guisuanti-Qihou Yingzi Zhuanhuan Hanshu Jiqi zai Weinan Wanbingqi Yilai Guhuanjing Yanjiu zhong de Yiyong (The conversion function of phytoliths - climate factors and its application on study of palaeoenvironment in the southern Wei since the Last Glacial Maximum). *Disiji Yanjiu* (Quaternary Research) (8): 270 - 277.

Wu, R.Z., 1983. Lun Lijiacun-Laoguantai Wenhua de Xingzhi (A discussion on the characteristics of Lijiacun and Laoguantai cultures). *K. Y. W.* 2: 52-59.

Wu, S.G., R. Chen, & D. Yang, 1992. Jiangsu Zhenjiang Xiashutu de Chubu Yanjiu (A preliminary study on the Xiashu loess in Zhenjiang, Jiangsu province). In *Huangtu, Disiji, Quanqiu Bianhua* (Loess, Quaternary Geology and Global Change) *Volume* 3, ed. T.S. Liu *et al.* Beijing: Science Press, pp. 122-127.

Wu, W.T., 1983. Cong Xinshiqi Shidai Wenhua Yizhi Kan Hangzhouwan Liang'an de Quanxinshi Gudili (A reconstruction of the palaeo-geographic environment of the Hangzhou Bay based upon Neolithic sites). *Dili Xuebao (*ACTA Geographica Sinica) Volume 38 (2): 113-127.

Wu, Y.L., 1994. Huanghe Liuyu de Daozuo Nongye (The rice cultivation in the valleys of the Yellow River). *N. K.*(1): 78-83.

Wu, Z.Q., B. Sun, & Z. Wu, 1990. Yuci Dafa Xishiqi Yizhi Jiqi zai Dicengxue ji Guqihouxue shang de Yiyi (The Microlithic site of Yuci Dafa and its stratigraphical and palaeoenvironmental significance). *K.Y.W.* 5:63-69.

Wymer, J., 1991. *Mesolithic Britain.* U.K: Shire Publication Ltd.

Xiang, A.Q., 1991. Lun Changjiang Zhongyou Xinshiqi Shidai Jiaoqi Yichun de Nongye (The agriculture of the early Neolithic in the middle Yangzi valley). *N. K.* 1: 121-135.

Xiang, A.Q., 1992. Hunan Lixian Jiushiqi Didian Diaocha (A survey on the Palaeolithic site in Lixian, Hunan province). *Dongnan Wenhua* (Southeast Culture) 1: 113-123.

Xiang, A.Q., 1993. Dongtinghu Shiqian Nongye Chutan (A preliminary discussion on the prehistoric agriculture of the Dongting Lake). *N.K.* 1: 31-35.

Xiaokongshan Excavation Team, 1988. 1987 nian Henan Nanzhao Xiaokongshan Jiushiqi Shidai Yizhi Fajue Baogao (An excavation report of the Xiaokongshan Palaeolithic site in Nanzhao, Henan province). *H. K.* 4:1-15.

Xie, F., 1991. Nihewan Pendi Jiushiqi Wenhua Yanjiu Xinjinzhan (New progress of studies on Palaeolithic culture in Nihewan basin). *R. X.* Volume 10 (4): 324-332.

Xie, F., 1996. Yangyuan Jiangjialiang Faxian Xinshiqi Shidai Mudi (The Neolithic cemetery found in Jiangjialiang, Yangyuan). *Z.W.B.* 7th, April. The front page.

Xie, F. & S.Q. Cheng, 1989. Hebei Yangyuan Youfang Xishiqi Fajue Baogao (A report on the excavation at Youfang, Yangyuan county, Hebei province). *R. X.* Volume 8(1): 59-68.

Xin, Y., 1995. Chahai Yizhi Fajue Zahuo Zhongda Chengguo (A significant discovery at Chahai site). *Z. W. B.* 19th, March , front page.

Xiong, Y. & C.P. Xi, 1965. *Huabei Pingyuan de Turang* (The Soil of the North China Plain). Beijing: Science Press.

Xu, R., Z.C. Kong & N.Q. Du, 1989. Zhongguo Gengxinshi de Yushan-Lengshan Zhiwuqun Jiqi zai Disiji Yanjiu shang de Yiyi (The *Abies-Picea* flora in Pleistocene in China and its significance in quaternary research). *Disiji Yanjiu* (Quaternary Research). *Volume* 5(1): 48-55.

Xu, S.B., 1994. Lunan Yi-Shuhe Liuyu Zhongshiqi Shidai Wenhua Yanjiu (A study on the Mesolithic culture along the Yi and Shu river valleys, south of Shandong province). In *Zhongri Gurenlei yu Shiqian Wenhua Yuanyuan Guanxi Guoji Xueshu Yantaohui Lunwenji* (International Symposium Relationship between Chinese and Japanese Ancient Men and Prehistoric Cultures), eds. G.X. Zhou, Y. Zhang & X Zhang. Beijing: China International Radio Press, pp. 35-57.

Xu, W.S., 1994. Zhongguo Nongye Bengtu Qiyuan Xinlun (A new opinion on the origin of agriculture in China). *Zhongguo Nongyeshi* (Chinese Agriculture History) Volume 13(1): 24-32.

Xu, W.S., 1998. Cong Nonggeng Qiyuan de Jiaodu kan Zhongguo Daozuo de Qiyuan (Discussing the origins of rice cultivation from the viewpoint of the origin of agriculture). *N.K.* 1:70-77.

Xu, X., 1985. Disiji Bingqi zhongde Woguo Zhiwuqun (Vegetation of the quaternary glaciation in China). In *Zhongguo Disiji Bingchuan Bingyuan Xueshu Taolunhui Lunwenji* (Symposium of a National Conference on the Quaternary Glacier and Periglacial), ed. The Geology Society of China etc. Beijing: Science Press, pp. 125-128.

Xu, X. & M.L. Zhu, 1984. Zhenjiang Diqu 15,000 nian Yilai Guzhibei yu Guqihou Bianhua (Changes on palaeo-vegetation and palaeoclimate in Zhengjiang area in the last 15,000 years BP). *Dili Xuebao* (ACTA Geographica Sinica) Volume 39(3): 277-283.

Xu, X., S. W. Zhang, & S. Zhou, 1987. Wuhu-Jiangyin Diqu 30,000 nianlai de Zhibei, Qihou yu Huanjing de Chubu Yanjiu (A preliminary research on the palaeoclimate, vegetation and palaeoenvironment of Wuhu-Jiangyin area since the last 30,000 years BP). *Nanjing Daxue Xuebao (Ziran Kexue Ban)* [Journal of Nanjing University (Natural Sciences Edition)]. Volume 23 (3): 556-574.

Xu, X.M., W. Chang & J. Liu, 1996. 11,000 nian Yilai Taihu Diqu de Zhibei yu Qihou Bianhua (The floral and climatic changes in the Tai Lake area since 11,000 years ago). *Gushengwu Xuebao* (ACTA Palaeontologica Sinica) Volume 35(2): 175-185.

Yan, F.H., Y.Y. Ye, X.S. Mai, & Y.X. Liu, 1979. Ju Huafeng Fengxi Lun Xujiayao Yizhi de Shidai he Guhuanjing (Discussion on the age and palaeoenvironment of the Xujiayao site based upon pollen analysis). *Dizhen Dizhi* (Seismology and Geology) *Volume* 1 (4): 72-78.

Yan, F.H., Y.Y. Ye, X.S. Mai, & Y.X. Liu, 1984. Ju Huafeng Fengxi Shilun Beijing Diqu Quanxintong de Huafeng (On the division of Holocene in Beijing area based upon pollen analysis). In *Disiji Baofeng Fengxi Yu Guhuanjing* (*Quaternary Pollen Analysis and Palaeoenvironment*), ed. Anonymous. Beijing: Science Press, pp. 86-97.

Yan, Q.S. & S. Huang, 1987. Hangjiahu Pingyuan Quanxinshi Chenji Huanjing Yanbian (Evolution of Holocene sedimentary environment in the Hangzhou - Jiaxing Huzhou plain). *Dili Xuebao* (ACTA Geographica Sinica) Volume 42 (1): 1-16.

Yan, W.M., 1989. Zhongguo Nongye he Jiazhu Xunyang de Qiyuan (The origins of agriculture and pig husbandry in China). *Liaohai Wenwu Xuekan* (The Antiquity in Liaohai Area) 2: 1-11.

Yan, W.M., 1990. Zhongguo Shiqian Daozuo de Xinfaxian (A new discovery of the prehistoric rice cultivation in China). *J. K.* 3: 10-17.

Yan, W.M., 1992a. The origins of agriculture and animal husbandry in China. In *Pacific Northeast Asia in Prehistory*, ed. C.M. Aikens & N.R. Song. Washington: Washington University Press, pp. 113-123.

Yan, W.M., 1992b. Luelun Zhongguo Wenming de Qiyuan (A discussion on the origins of Chinese civilisation). *Ww.* 1: 40-49.

Yan, W.M. 1997. Woguo Daozuo Qiyuan Yanjiu de Xinjinzhan (New achievements of the origin of rice cultivation in China). *Kg.* 9: 71-76.

Yang, D.Y., 1986. Wangengxinshi Bingqi Zuishengshi Changjiang Zhongxiayou Diqu de Guhuanjing (The Palaeoenvironment of the middle and lower valley of the Yangzi River during the LGM). *Dili Xuebao* (ACTA Geographica Sinica). *Volume* 41(4): 302-310.

Yang, H.R. & X.Q. Chen, 1988. Zhongguo Moci Bingqi de Guqihou (The Palaeoclimates during the Last Glacial Maximum in China), *Disiji Bingchuan yu Disiji Yanjiu Lunwenji* (The Quartenary Glacial and Quaternary Research) Volume 5: 22-42.

Yang, H.R. & Z.R. Xie, 1984. Zhongguo Dongbu Jin 20,000 nianlai de Qihou Bodong yu Haimian Shengjiang Yudong (The palaeoclimatic fluctuation and sea level changes of the last 20,000 years BP in East China). *Haiyang yu Huzhao* (Oceanologia et Limnologia Sinica). Volume 15(1): 1-12.

Yang, H.R., X. Xu, & G.C. Li, 1989. Disiji Zhongguo Ziran Huanjing Bianqian de Yuanyin Jizhi (The mechanism of the quaternary environmental change in China). *Disiji Yanjiu* (Quaternary Research) (2): 97-110.

Yang, J.C., 1994. Guilin Xinshiqi Shidai Dongxue Yizhi ji Youguan Wenti (The Neolithic caves in Guilin and some related problems). In *Zhongri Gurenlei yu Shiqian Wenhua Yuanyuan Guanxi Guoji Xueshu Yantaohui Lunwenji* (International Symposium Relationship between Chinese and Japanese Ancient Men and Prehistoric Cultures), eds. G.X. Zhou, Y. Zhang & X. Zhang. Beijing: China International Radio Press, pp. 94-102.

Yang, Q.X., 1991. Shilun Chengbeixi Wenhua (A discussion on the Chengbeixi culture). *Dongnan Wenhua* (Southeast Culture) 5: 206-212.

Yang, X.D., S.M. Wang & G.B. Tong, 1996. Jiangsu Gucheng Huqu Yiwan Duonianlai de Baofeng Zhiwuqun ji Gujifeng Qihou Bianqian (Character of palynology and changes of monsoon climate over the last 10,000 years in Gucheng lake, Jiangsu province). Zhiwu Xuebao (ACTA Botanica Sinica) Volume 38(7): 576-581.

Yang, Z.G. & H.M. Lin, 1993. *Zhongguo Dongbu Disiji Jincheng yu Guoji Duibi* (The Quaternary Progress of East China and International Comparison). Beijing: Geology Press.

Yang, Z.G., Y.J. Li, Q.L. Ding & B.C. He, 1979. Lun Hebei Pingyuan Dongbu Disiji Dizhi Jige Jiben Wenti (Some fundamental problems of the quaternary geology in the East Hebei Plain). *Dizhi Xuebao* (*ACTA Geologica Sinica*) (4): 262 - 279.

Yao, Z.J., 1989. Shanxi Zhongtiaoshan Diqu Biaotu Huafeng Fengxi (Surface pollen analysis in Zhongtiao Mountain, Shanxi Province). Dili Xuebao (ACTA Geographica Sinica) Volume 44 (4): 469-478.

You, Y.Z. & Q.Q. Xu, 1981. Zhongguo Beifang Wangengxinshi Dongwuqun yu Shenhai Chenjiwu de Duibi (A comparison between fauna of the late Pleistocene in North China and the deep-sea deposit). *G. D. X. Volume* 19(1): 77-86.

You, R.J., 1983. Cong Yuyanxue Jiaodu Shilun Yazhou Zaipeidao de Qiyuan yu Chuanbo (A discussion on the origins and expansion of cultivated rice in Asia from a linguistic viewpoint). *Nongshi Yanjiu* (Study on the Agriculture History) 3: 131-141.

You, X.L., 1995. *Zhongguo Daozuo Shi* (The History of Rice Cultivation in China). Beijing: Chinese Agriculture Press.

You, Y.Z, X.F. Xu, X.F. Yuan, S.B. Xu, S.F. Yang & Y. Hu, 1989. Shandong Rizhao Yanhai Faxian de Jiushiqi Jiqi Yiyi (Palaeolithic implements from the coastal area in Rizhao, Shandong province). *R. X.* Volume 8 (2): 101-107.

Yu, Z. J. & D.C. Huang, 1993. Anhuisheng Huaibei Pingyuan Disiji Baofeng Xulie (Quaternary pollen sequence of the North Huai Plain in Anhui province). *Haiyang Dizhi Yu Disiji Dizhi* (Marine Geology and Quaternary Geology) *Volume* 13 (1): 21-32.

Yuan, J.R., 1995. Luetan Hunan Jiushiqi de Jige Wenti (A few questions regarding the Palaeolithic in Hunan province). In *Hunan Shiqian Kaogu Lunwen*

Xuanpian (Treaties of the Prehistoric Archaeology in Hunan Province), ed. and printed. The Archaeology Institute of Hunan Province, pp. 1-12.

Yuan, J.R., 1996. Yuchanyan Huo Shuidao Qiyuan Zhongyao Xinwuzheng (An important evidence of the origin of rice cultivation found from discovery in Yuchan cave). *Z.W.B.* 3rd, March. front page.

Yuan, J. & J.Y. An, 1997. Zhongguo Dongwu Kaoguxue Yanjiu de Liangge Wenti (Two questions on the archaeozoological study in China). *Z. W. B.* 27th, April 1997, the third page.

Yuan, S.X., 1993. Huanan Zaoqi Xinshiqi Tanshisi Niandai Shuju Yiqi de Kunhuo yu Zhenshi Niandai (The radiocarbon dating problem of the early Neolithic in south China). *Kg.* 4: 367-375.

Yuan, S.X., T.M. Chen, & Q.S. Zhou, 1991. Nanzhuangtou Yizhi Tanshisi Niandai Ceding yu Wenhuaceng Baofeng Fengxi (Pollen analysis and radiocarbon dating of the cultural deposit in Nanzhuangtou site). *Huanjing Kaogu* (Environmental Archaeology) 1: 136 -139.

Zhang, C. & S.Z. Liu, 1996. Jiangxi Wannian Xianrendong yu Diaotonghuan Yizhi (Remains of Xianrendong and Diaotonghuan, Jiangxi province). *Lishi* (History). June: 47-49.

Zhang, C.L., 1992. Dongtinghu Diqu Xingshiqi Kaogu Xinshouhuo (The new Neolithic discovery in Dongting Lake area). *Z.W.B.* 14th, June, the third page.

Zhang, J.E., 1985. Changjiang Xiayou Wanbingqi Baofeng Zuhe he Qihou Huinuan Wenti (Palynological evidences of the climatic warming trend during the late glacial period in the lower reaches of the Yangzi River). In *Zhongguo Disiji Bingchuan Bingyuan Xueshu Taolunhui Lunwenji* (Symposium of National Conference on Quaternary Glacier and Periclacial), ed. The Geology Society of China etc. Beijing: Science Press, pp. 175-179.

Zhang, J.Z., 1989. Shilun Jiahu Leixing de Tezheng ji yu Zhouwei Wenhuai de Guanxi (A preliminary discussion on the characteristics of Jiahu and its relationships with other cultures in adjacent areas). *Ww.* 1:18-20.

Zhang, J.Z., 1991. Huanjing yu Peiligang Wenhua (Environments and the Peiligang culture). *Huanjing Kaogu* (Environmental Archaeology) Volume 1:122-129.

Zhang, J.Z., 1994. Henan Shiqian Shuidao Zaipei Chuyi (A discussion on the prehistoric rice cultivation in Henan province). *Wenwu Tiendi* (The World of Antiquity). 3: 24-26.

Zhang, J.Z. & Z.Y. Li, 1996. Henan Wuyang Dagang Xishiqi Didian Fajue Baogao (The excavation report of the microblade at Wuyang, Henan province). *R. X.* Volume 15(2): 105-112.

Zhang, J.Z., Z.C. Kong, & C.J. Liu, 1994. Wuyang Shiqian Daozuo Yichun yu Huanghuai Diqu Shiqian Nongye (The prehistoric rice remains in Wuyang and the prehistoric agriculture in the Yellow and Huai river delta). *N. K.* 1: 68-73.

Zhang, L.Y., J.J. Li, & S.Z. Zhou, 1985. Guanyu Woguo Yushan, Lengshan Baofeng Zuhe de Guqihou Yiyi zi Sangque (Palaeoclimatic significance of the spore-pollen composition of *Picea* and *Abies* in China). In *Zhongguo Disiji Bingchuan Bingyuan Xueshu Taolunhui Lunwenji* (Symposium of a National Conference on the Quaternary Glacier and Periglacial), ed. The Geology Society of China. Beijing: Science Press, pp. 129-137.

Zhang, S.S., 1989. Zhongguo Beifang Jiushiqi Shidai Zaoqi Wenhua (The early Palaeolithic of north China). In *Zhongguo Yuangu Renlei* (Early Humankind in China), eds R.K. Wu, X.Z. Wu & S.S. Zhang. Beijing: Science Press. 97-158.

Zhang, S.S., 1990. Zhongguo Beifang Jiushiqi Gongye de Quyu Jianjin yu Wenhua Jiaoliu (Regional industrial gradual advance and cultural exchange of Palaeolithic in North China). *R. X.* Volume 9 (4): 322-334.

Zhang, S.Q., 1985. Hainandao Lizu Nonggeng Zhishushi Tantao (A discussion on agriculture technology of the Li ethnic in Hainan Island). *Nongshi Yanjiu* (Study on the Agriculture History) *Volume* 6:15-19.

Zhang, W.X. & A.P. Pei, 1997. Lixian Mengxi Bashidang Chutu Daogu de Yanjiu (A study on the ancient rice from Bashidang, Mengxi in Lixian County). *Ww.* (1): 36-41.

Zhang, Z.H., F. Song, & L. Wang, 1988. Luokeng Dongwuqun de Chubu Yanjiu (A preliminary research on the Luokeng fauna). In *Jinian Mabaren Huashi Faxian Sanshi Zhounian Lunwenji* (Treaties in Commemoration of the 30th Anniversary of the Discovery of Maba Human Cranium), ed. Guangdong Museum. Beijing: the Cultural Relics Publishing House, pp. 54-64.

Zhang, Z.B., 1989. Zhongguo Xinshiqi Shidai Renlei Yihai (Neolithic Human Skeleton in China). In *Zhongguo Yuangu Renlei* (Early Humankind in China), ed. R.K. Wu, X.Z. Wu & S.S. Zhang. Beijing: Science Press, pp. 62-80.

Zhang, Z.B., L. Wang, & L. Ouyang, 1982. Zhongguo Xinshiqi Shidai Juming Tizheng Leixing Chutan (Physical patterns of Neolithic Skulls in China in view of Cluster Analysis). *G. D. X.* Volume 20(1): 72-80.

Zhang, Z.B., D. Wang, & J.X. Ding, 1981. Beijing Diqu Yiwan Sanqian Nianlai Ziran Huanjing de Yanbian (The evolution of natural environment of Beijing area for the last 13,000 Years). *Dizhi Kexue (Scientia Geologica Sinica)* (3): 259-268.

Zhang, Z.H., 1984. Zhongguo Yuanshi Nongye de Chansheng he Fazhan (The origins and development of indigenous agriculture in China). *N. K.* 2:36-42.

Zhang, Z.F. *et al.*, 1987. Taihu Sanshandao de Puru Dongwu Huashi (The faunal fossils in Sanshan island, Tai lake). *Nanjing Bowuyuan Yuankan* (Journal of Nanjing Museum). Volume 9: 21-24.

Zhao, Y.S., 1987. Huabei Pingyuan Shi'er Wannian Yilai de Guqihou Bianhua (Palaeoclimate changes of the North China Plain since the last 120,000 Years). *Dili Yanjiu (*Geographical Research). Volume 6 (4): 54-61.

Zhao, Z.J., 1998. The middle Yangtze region in China is one place where rice was domesticated: phytolith evidence from the Diaotonghuan Cave, northern Jiangxi. *Antiquity* Volume 72:885-897.

Zhao, Z.J., D.M. Pearsall, & Q.H. Jiang, 1995. Analysis of phytoliths from Xian Ren Dong and Wang Dong. In *Origins of Rice Agriculture,* eds. R.S. MacNeish and J.G. Libby. Elpaso: The University of Texas, pp. 47-52.

Zheng, N.W., 1993. Luelun Peiligang Wenhua de Maizang Zhidu (Briefly on the burial custom of the Peiligang culture). In *Zhongguo Kaoguxue Luncong - Jinian Kaogusuo Chengli Sishi Zhounian* (Treatises on Chinese Archaeology - to the 40th Anniversary of the Founding of the Institute of Archaeology, CASS), ed. The Institute of Archaeology CASS. Beijing; Science Press, pp. 29-40.

Zhou, B.X., 1965. Henan Anyang Xiaonanhai Jiushiqi Shidai Dongxue Yizhi Jizhui Dongwu Huashi de Yanjiu (A study on the fauna of the Xiaonanhai cave in Anyang county, Henan province). *K.X.* (1): 29-48.

Zhou, B.X., 1978. Pimaoxi he Mengmaxiang de Dili Fengbu, Gushengtai yu Youguan de Guqihou Wenti (Geographic distribution of *Ceolodonta antiquitatis* and *Mammuthus primigenius* and the problems of palaeoclimate). *G. D. X.* Volume 16(1): 47-58.

Zhou, B.X., 1981. Hebei Wu'an Cishan Yizhi de Dongwu Guhai (The animal remains discovered at Cishan village, Wu'an, Hebei province). *K.X.* 3:339-347.

Zhou, B.X., 1992. Hebeisheng Xushuixian Nanzhuangtou Yizhi de Dongwu Yihai (Animal remains from Nanzhuangtou, Xushui county, Hebei Province). *Kg.* 11:966-967.

Zhou, G.X., 1974. Henan Xucheng Lingjing de Shiqi Shidai Yichun (The lithic implements found in Lingjing, Henan Province). *Kg.* 2: 91-98.

Zhou, G.X., 1994. Zailun Bailiandong Wenhua (A further study on the Bailian Cave culture). In *Zhongri Gurenlei yu Shiqian Wenhua Yuanyuan Guanxi Guoji Xueshu Yantaohui Lunwenji* (International Symposium Relationship between Chinese and Japanese Ancient Men and Prehistoric Cultures), eds. G.X. Zhou, Y. Zhang & X. Zhang. Beijing: China International Radio Press, pp. 203-254.

Zhou, J., 1988. Shilun Xiaokongshan Xiadong Jiushiqi Yizhi de Niandai (A discussion on the age of the Xiaokongshan site). *H. K.* (4): 43-49.

Zhou, K.S., F.H. Yan, X.L. Liang, & Y.Y Ye, 1978. Beijing Pingyuan Disiji anqi Huafeng Fengxi Jiqi Yiyi (Late quaternary pollen analysis of Beijing Plain and its significance). *Dizhi Kexue* (*Scientia Geologica Sinica*) (1):57-63.

Zhou, K.S., X.L. Liang, F.H. Yan, and Y.Y Ye, 1983. Cong Nihewanceng Baofeng Fengxi Tan Nangou Lengqi Wenti (Some remarks on the cold period based on pollen analysis from Nihewan stratum). *Dizhi Kexue* (Scientia Geologica Sinica) 1: 82-92.

Zhou, K.S., S.M. Chen, C.W. Chen, Y.Y Ye, & X.L. Liang, 1984. Huabeiqu Disiji Zhibei Yanti yu Qihou Bianhua (Quaternary succession of vegetation and climatic evolution in North China). *Dizhi Kexue (Scientia Geologica Sinica)* 2: 165-172.

Zhou, M.Z., 1955. *Zhongguo Shiqian Renleixue de Faxian yu Yanjiu* (The Discovery and Research of Palaeoanthropology In China). Beijing: Science Press.

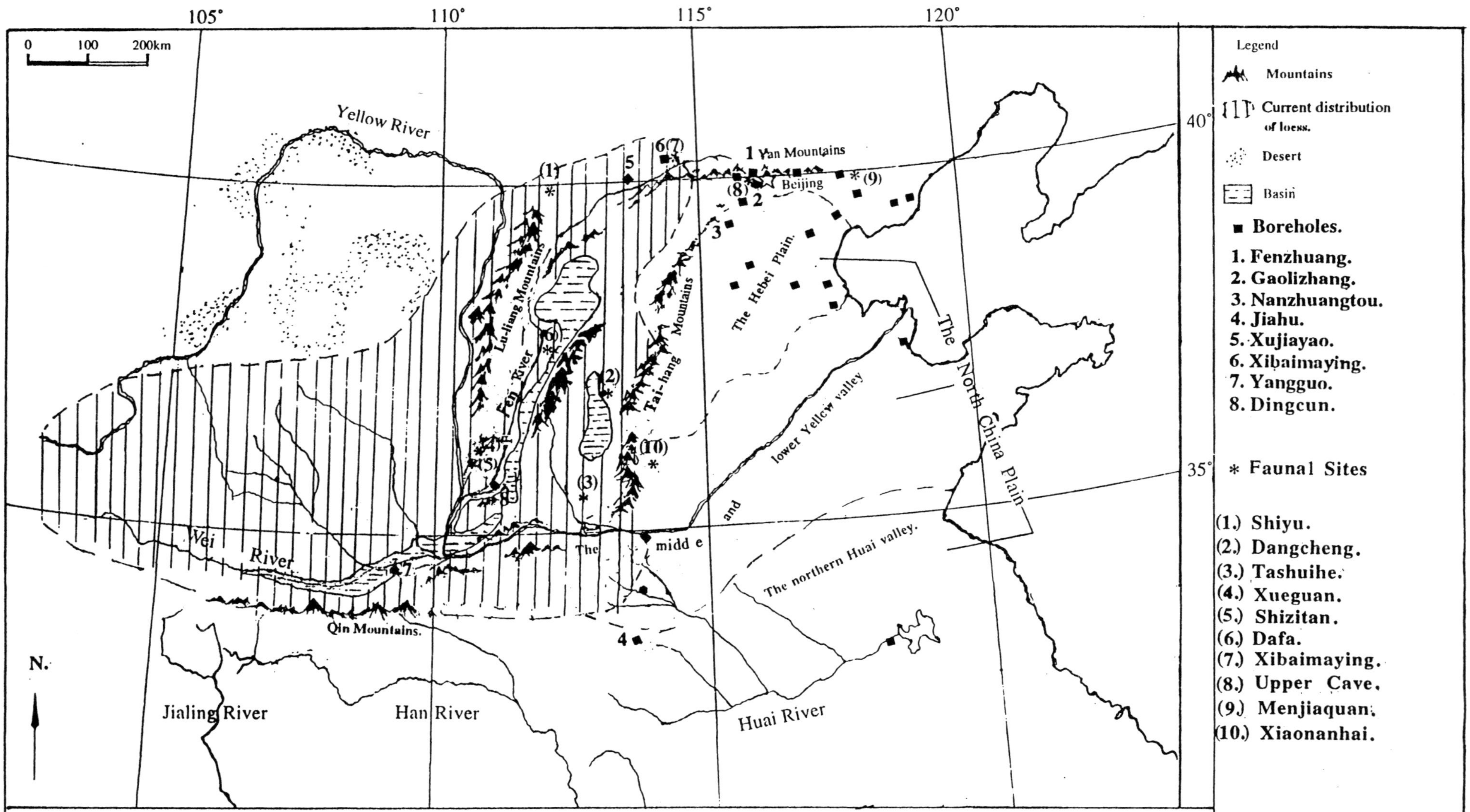

Fig. 3 (1) The distribution of pollen borehole and faunal sites in Central China.

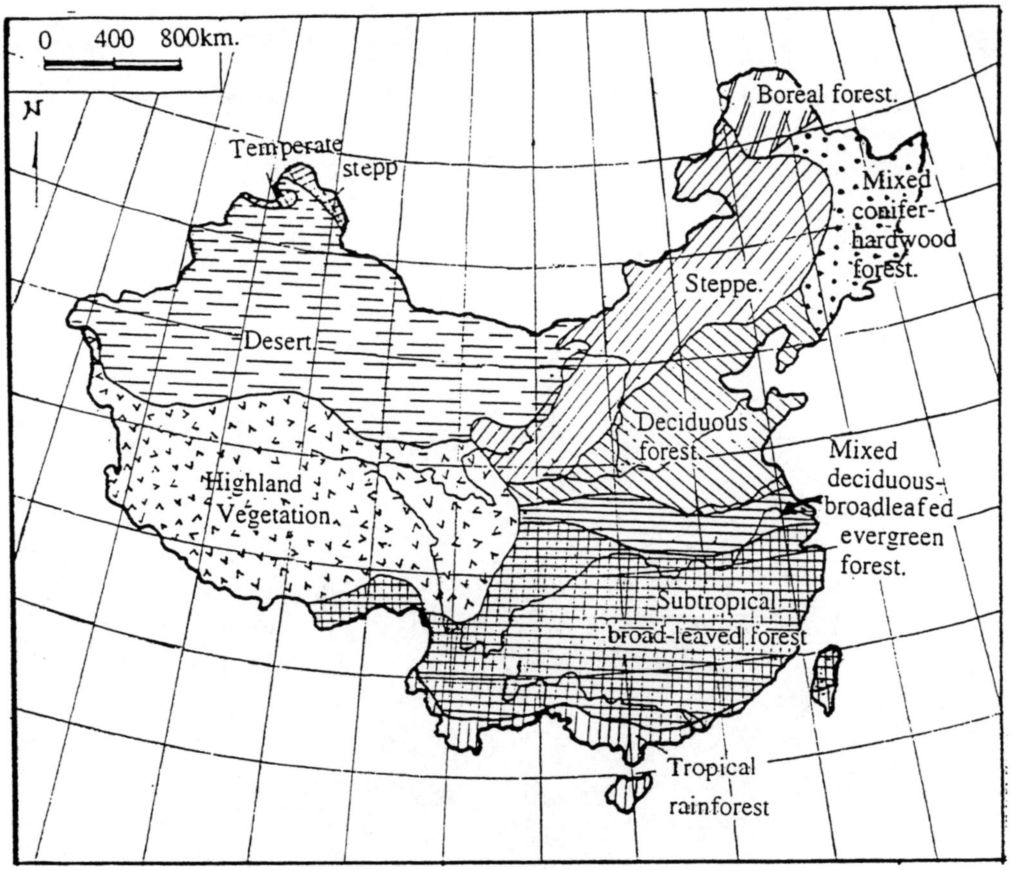

Fig. 3 (2) Modern vegetation zones in China. (after Liu 1988).

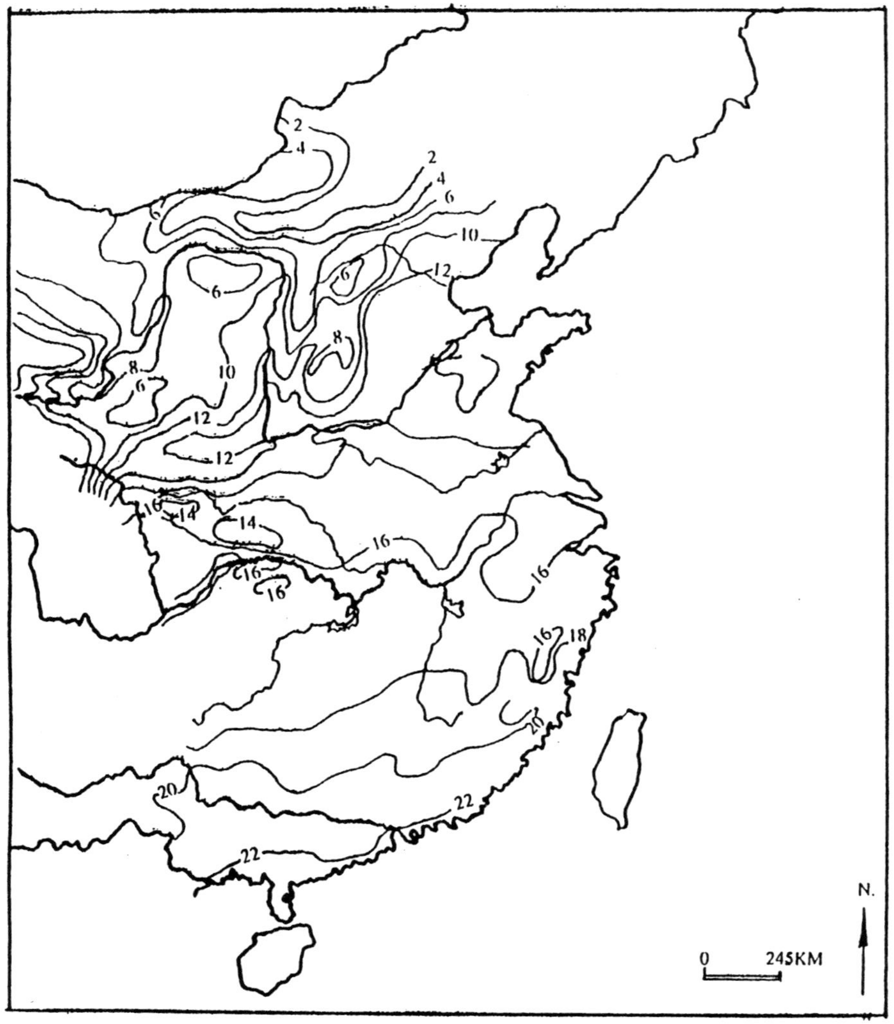

Fig. 3 (3) Average yearly temperatures (Celsius) in East China.
(after The Editing Committee of Physiography of China 1984).

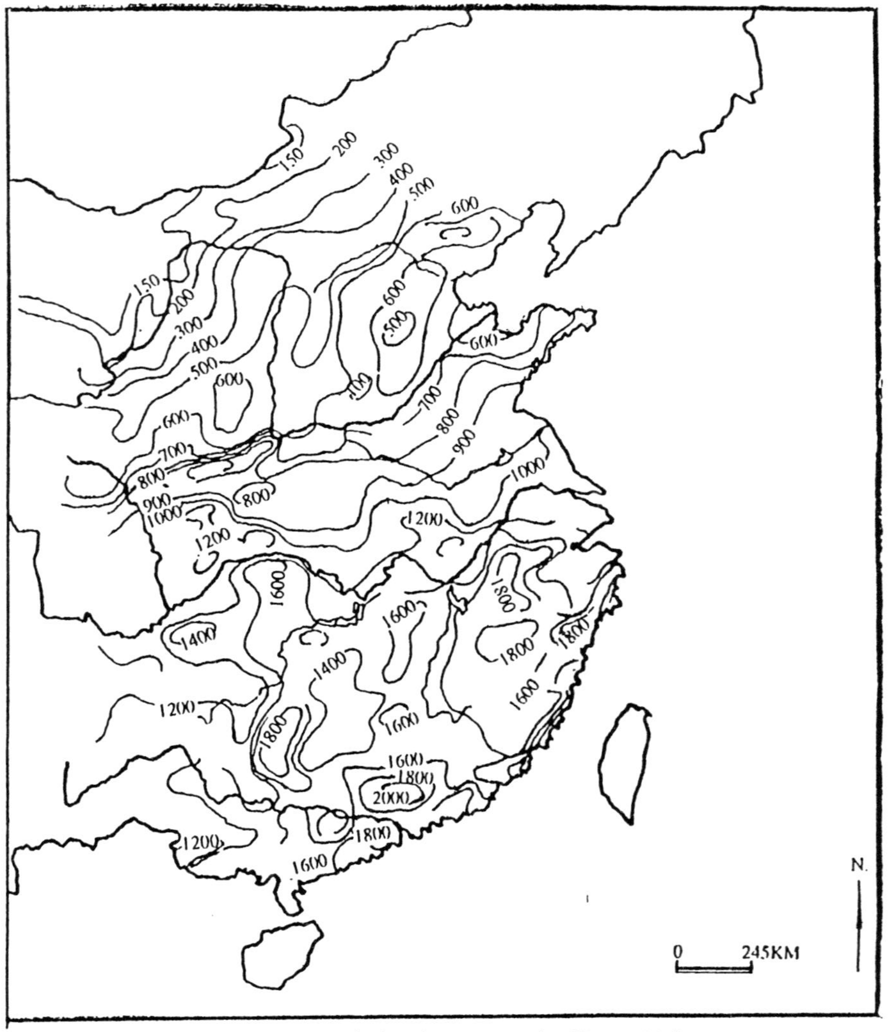

Fig. 3 (4) Average precipitation (mm) in East China.
(after The Editing Committee of Physiography of China 1984).

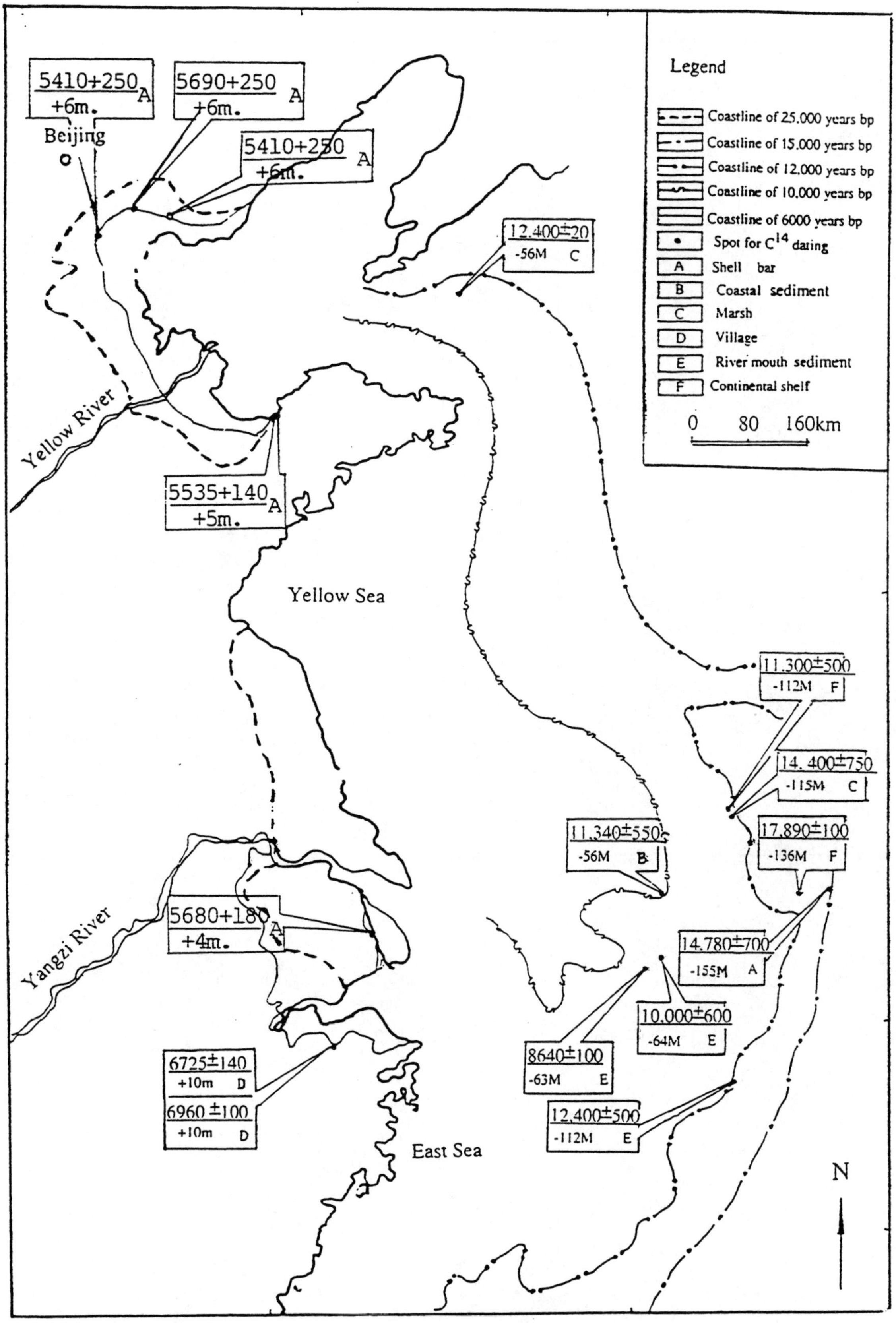

Fig.3(5) Coastline changes in East China since 25,000 years BP (after Shi and Wang 1982).

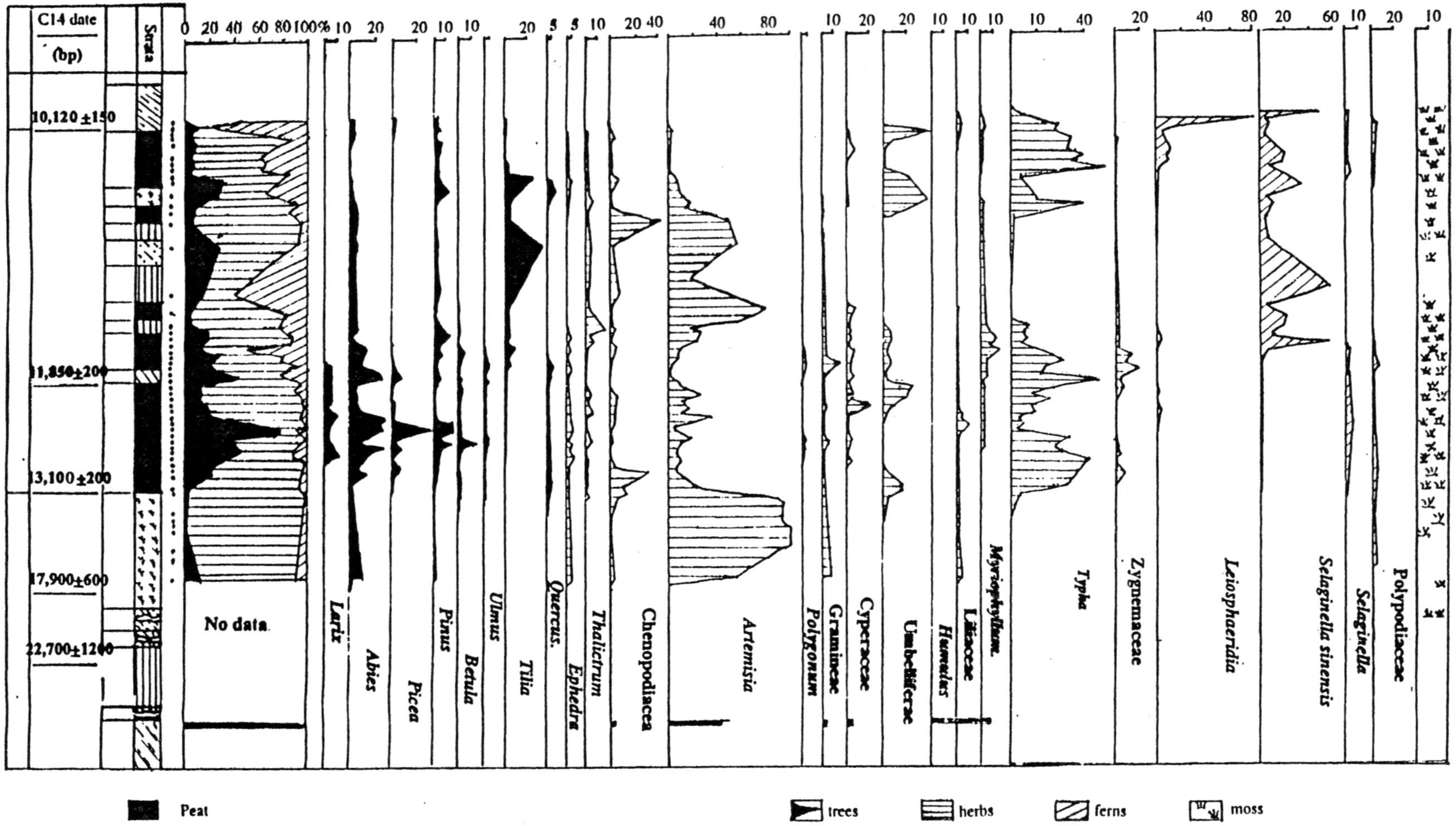

Fig. 3(6) Pollen profile from the terminal Pleistocene to the early Holocene at Fenzhuang (site No. 1). (After Zhang *et al.* 1981).

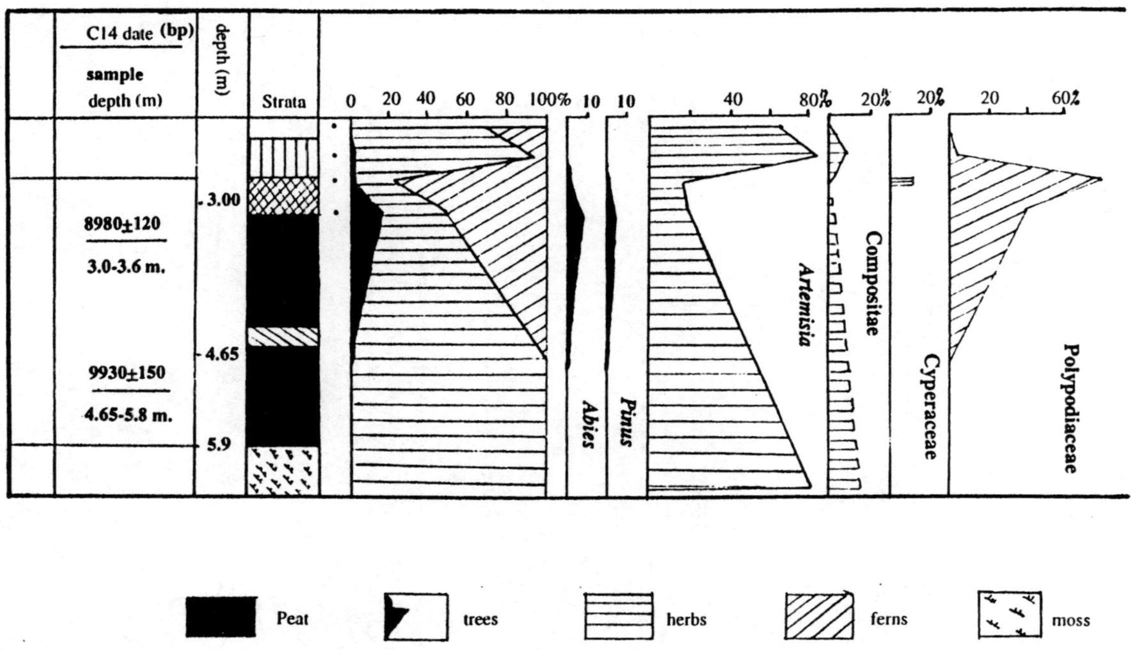

Fig. 3 (7) Pollen profile at GaoliZhang (Site No. 2) (after Zhang *et al.* 1981)

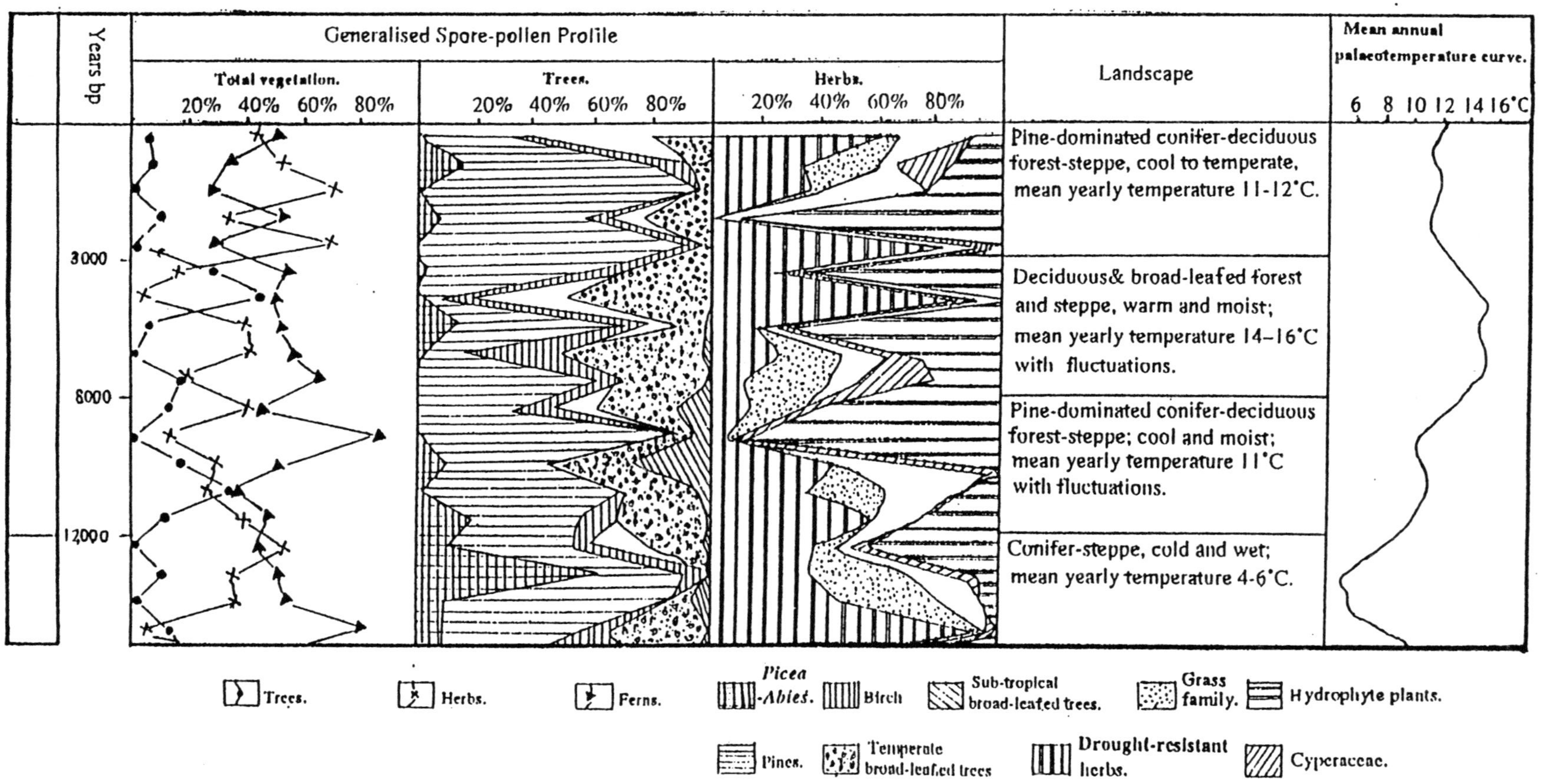

Fig. 3 (8) Generalised spore-pollen scheme and palaeotemperature curve since the late Pleistocene in the North China Plain. (after. Zhao 1987).

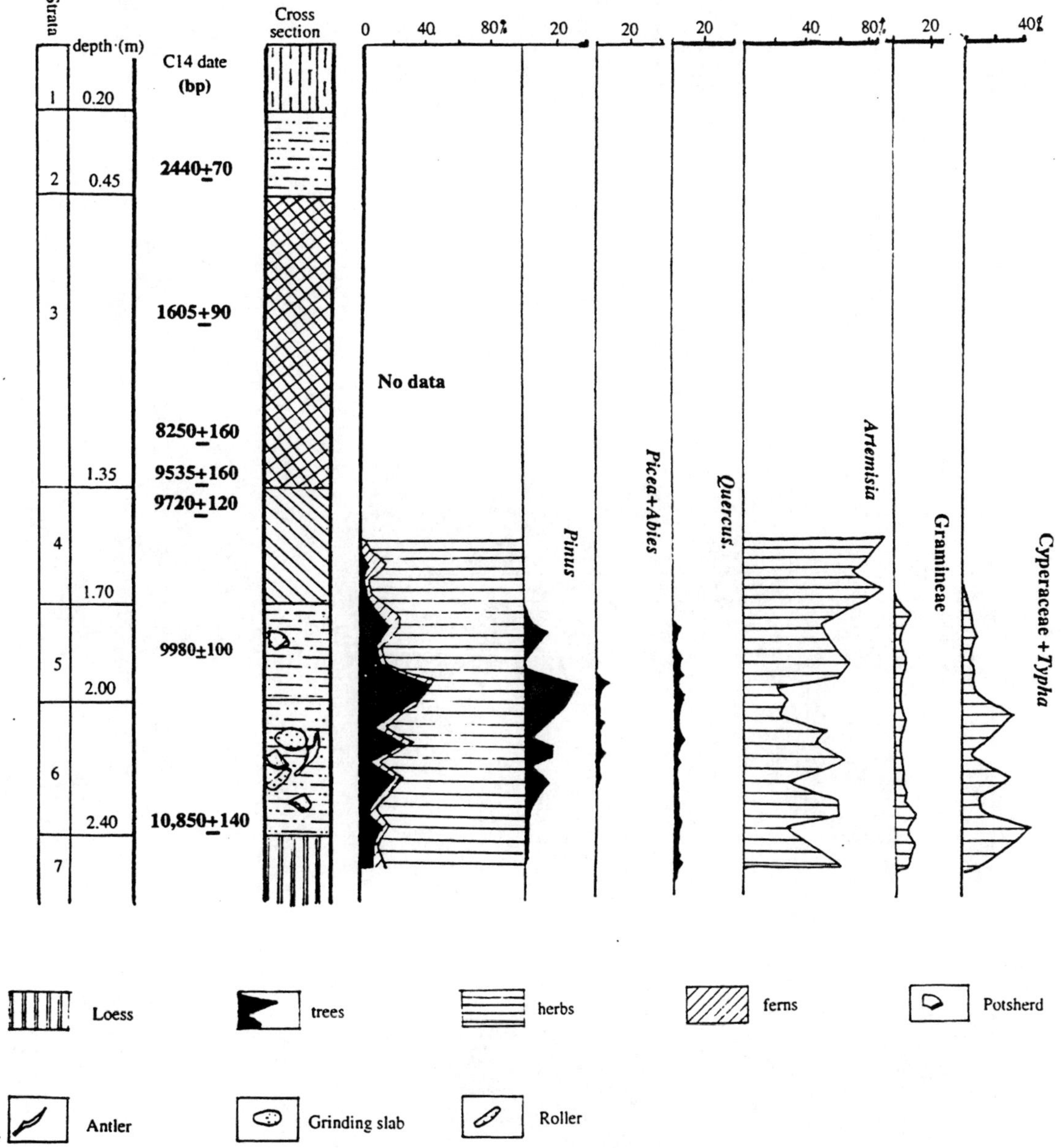

Fig. 3 (9) Pollen profile at Nanzhuangtou (site No. 3) (after Yuan *et al.* 1991).

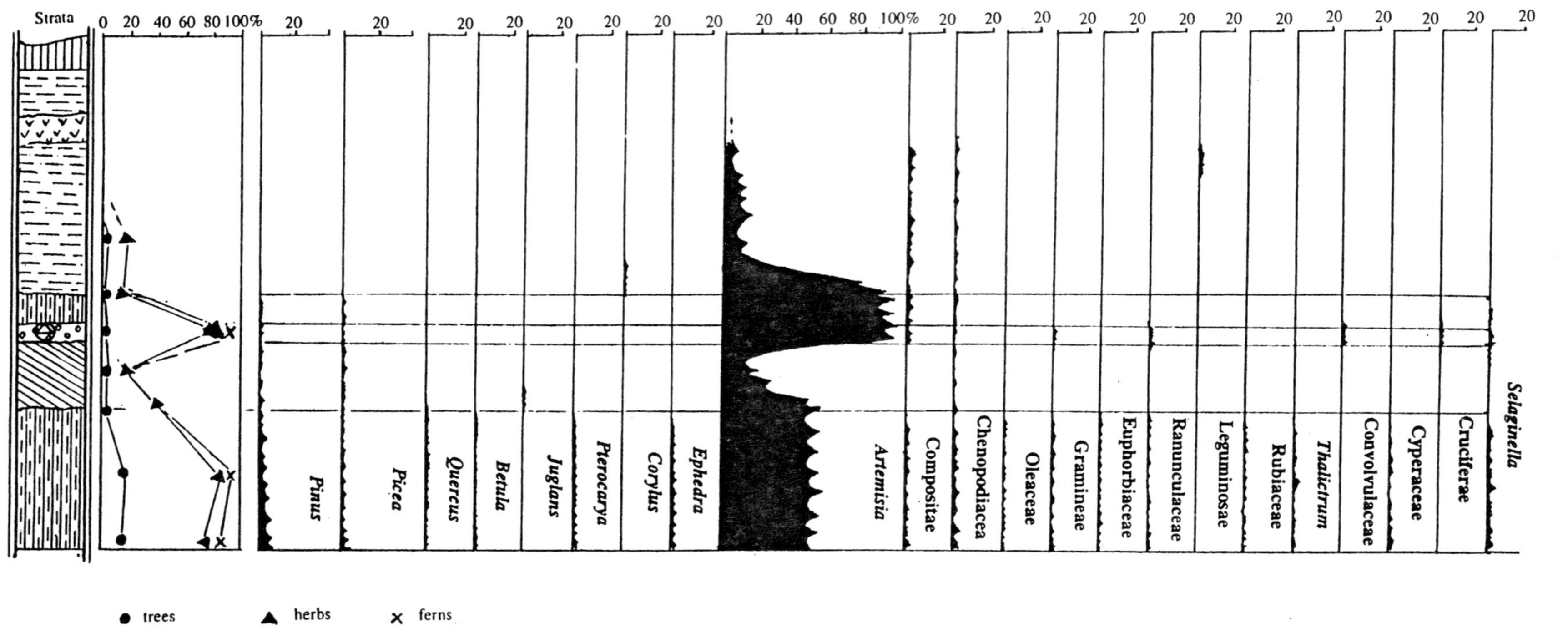

Fig. 3 (10) Pollen profile at Xibaimaying (Site No. 6) (based upon data from IA, Hebei Province 1989).

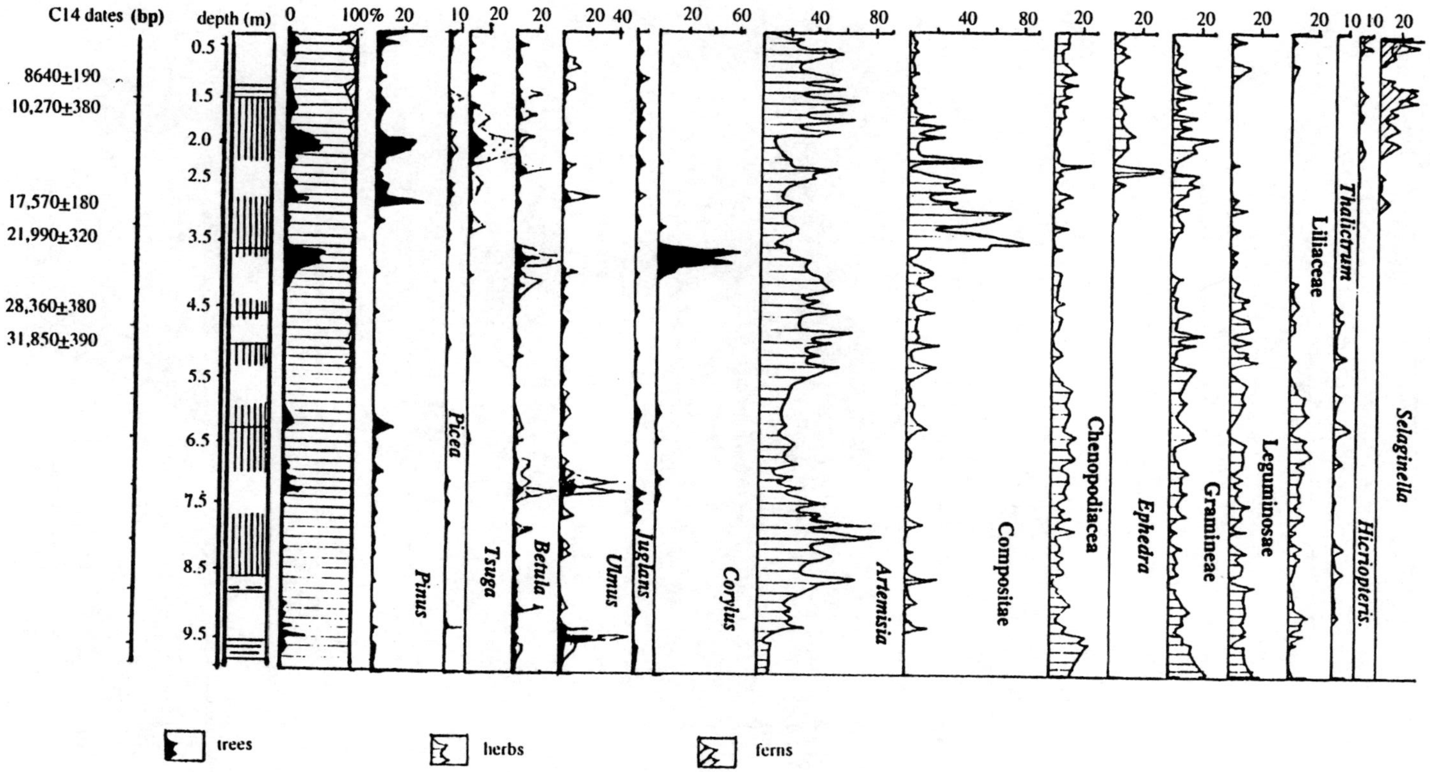

Fig. 3 (11) Pollen profile at Yangguo (Site No. 7) (After Sun *et al.* 1995).

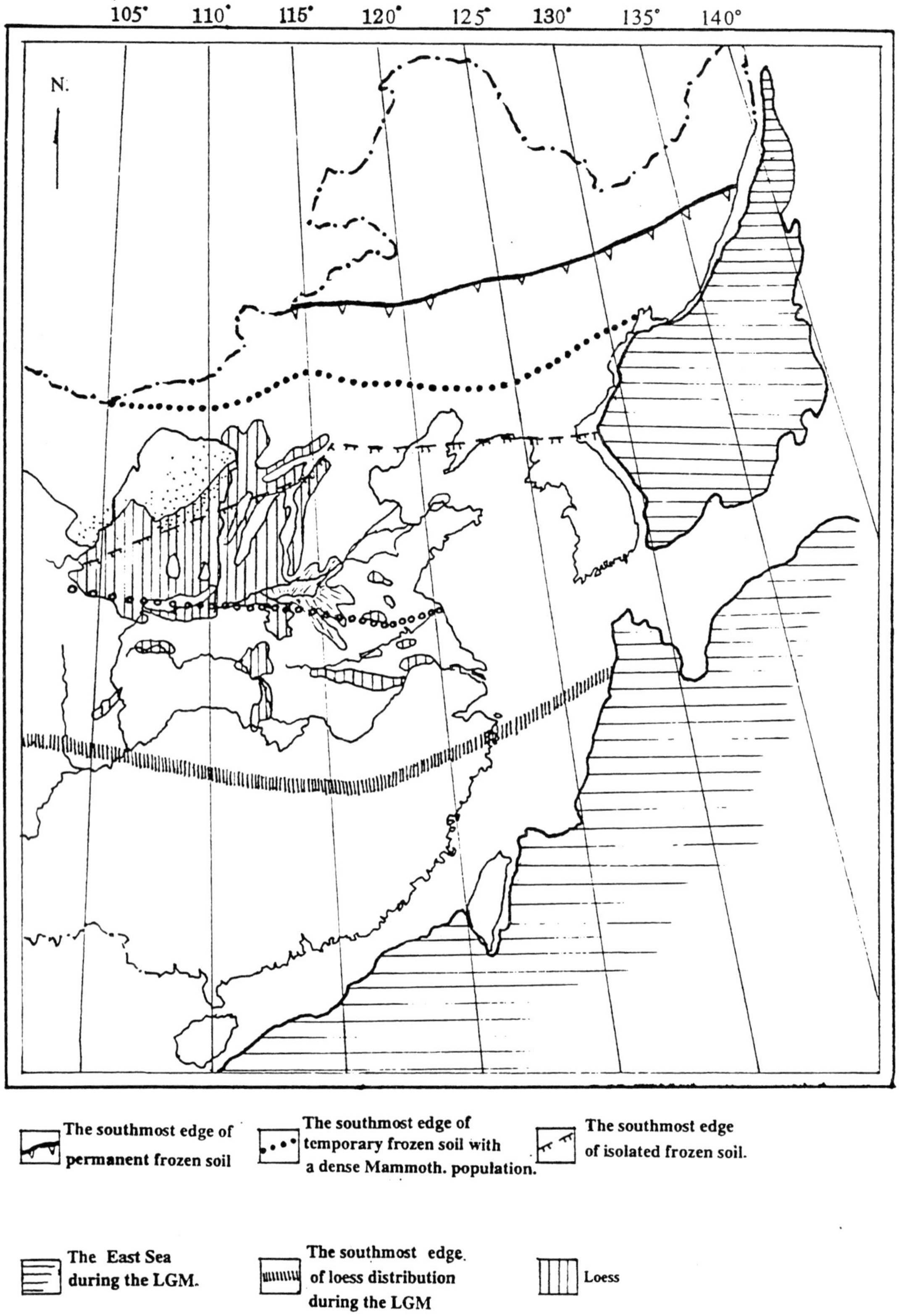

Fig. 3(12) The Palaeoenvironment during the LGM and the distribution of Mammuth us in Central China (Data based upon An *et al.* 1990).

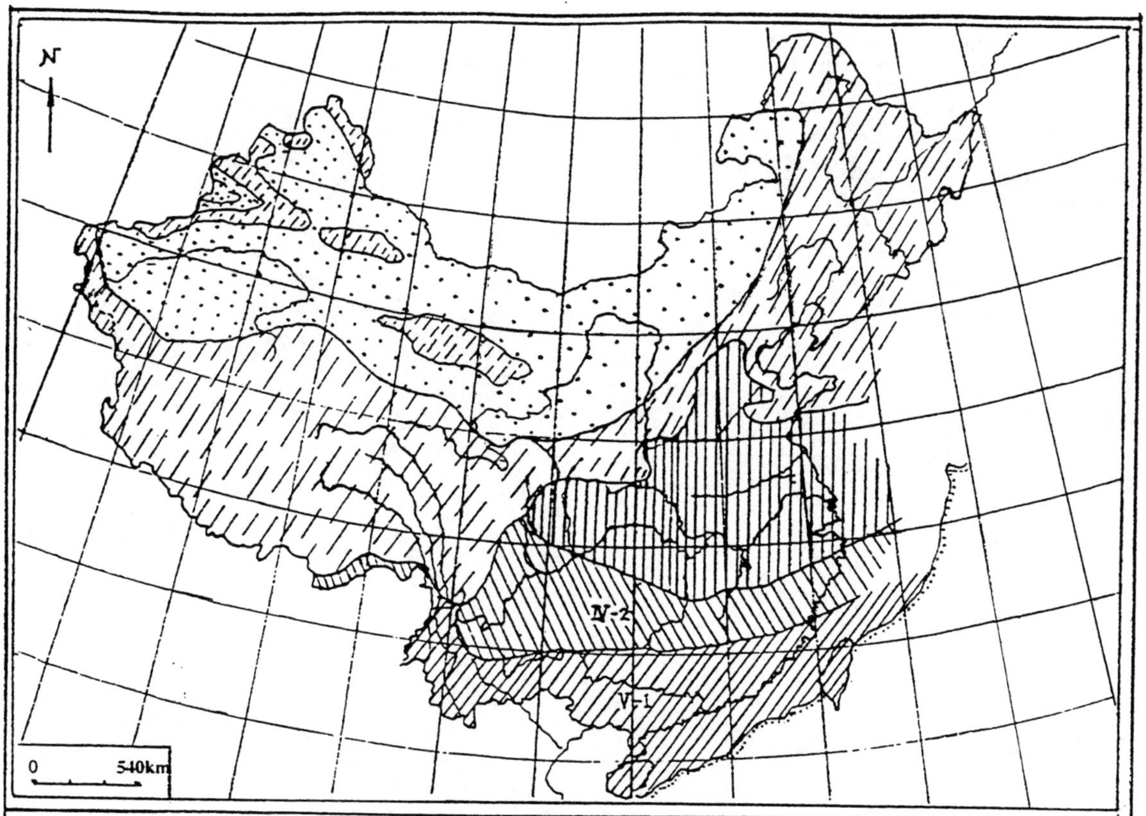

Fig. 3 (13) Palaeoenvironment of Central China from 18,000 to 13,000 BP (After An *et al.* 1990).

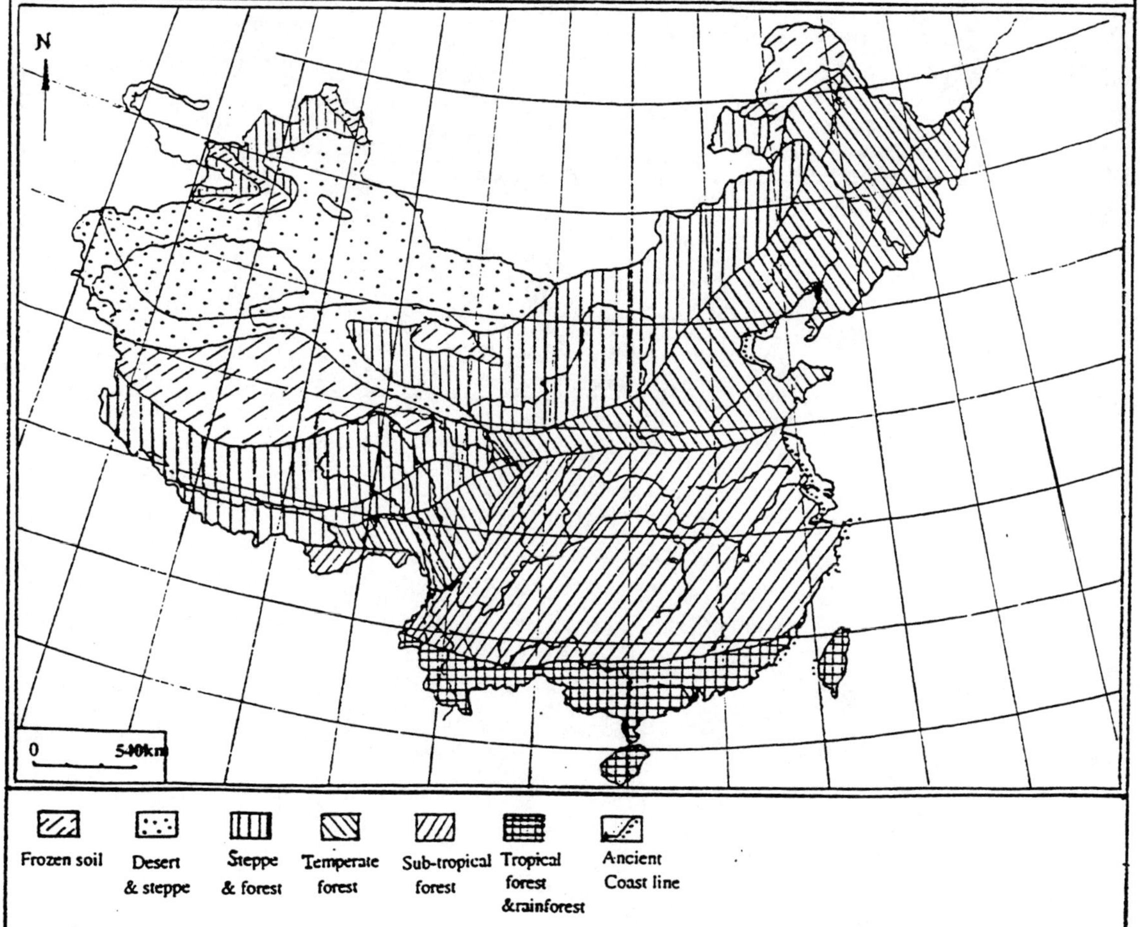

Fig. 3 (14) Palaeoenvironment of Central China from 9000 to 5000 BP (After An *et al.* 1990).

Fig. 3(15) The habitat of green foxtail (*Setaria viridis*). (photo taken by the author on 24th, July 1996 in Qufu, Shandong province, China).

Fig. 3(16). Simultaneous flowering, ripening and shattering on one plant of green foxtail (*S. viridis*). Noted that the hairs remain on the spikelet after the grain has shattered.

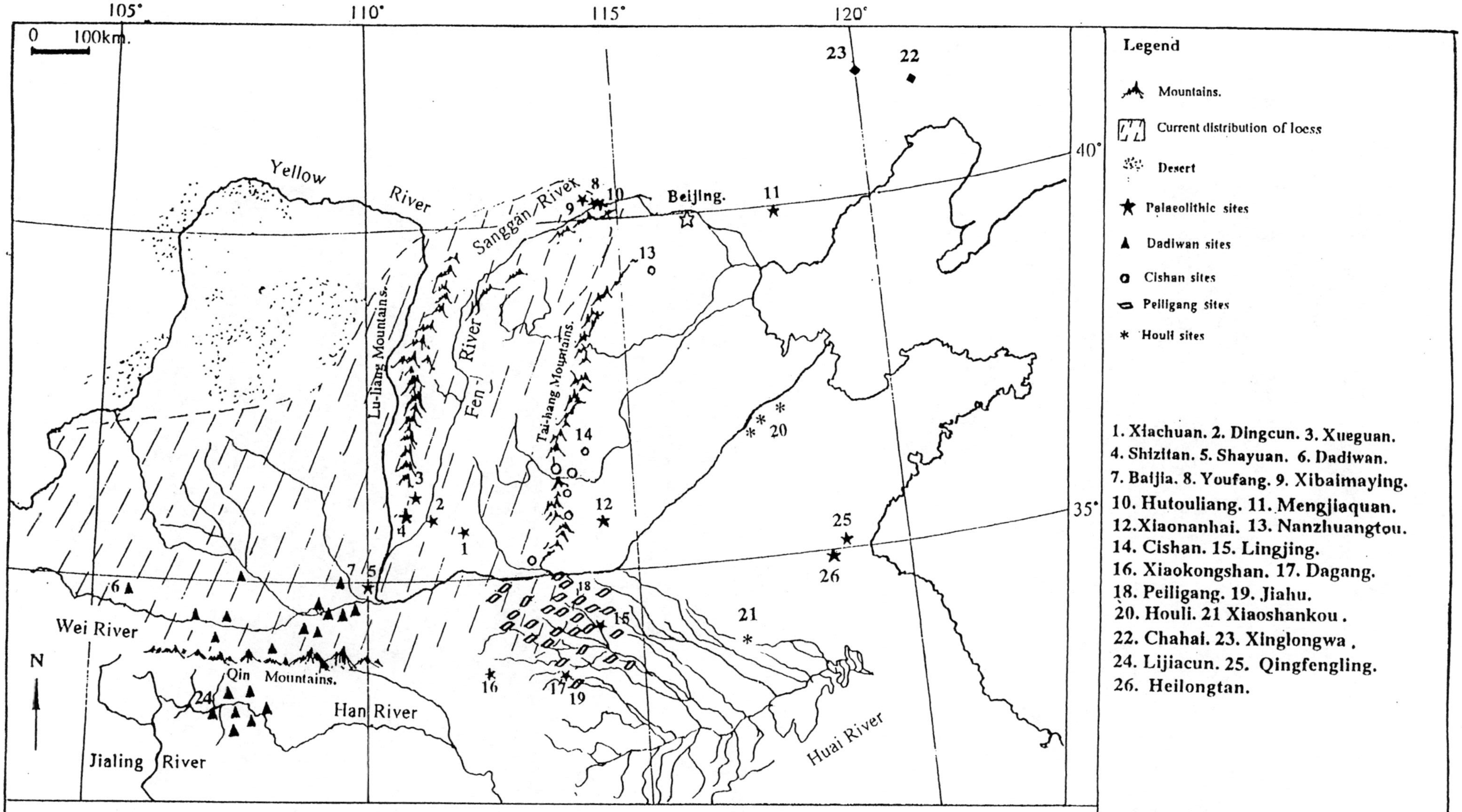

Fig. 4 (1) Locations of archaeological sites mentioned in the text.

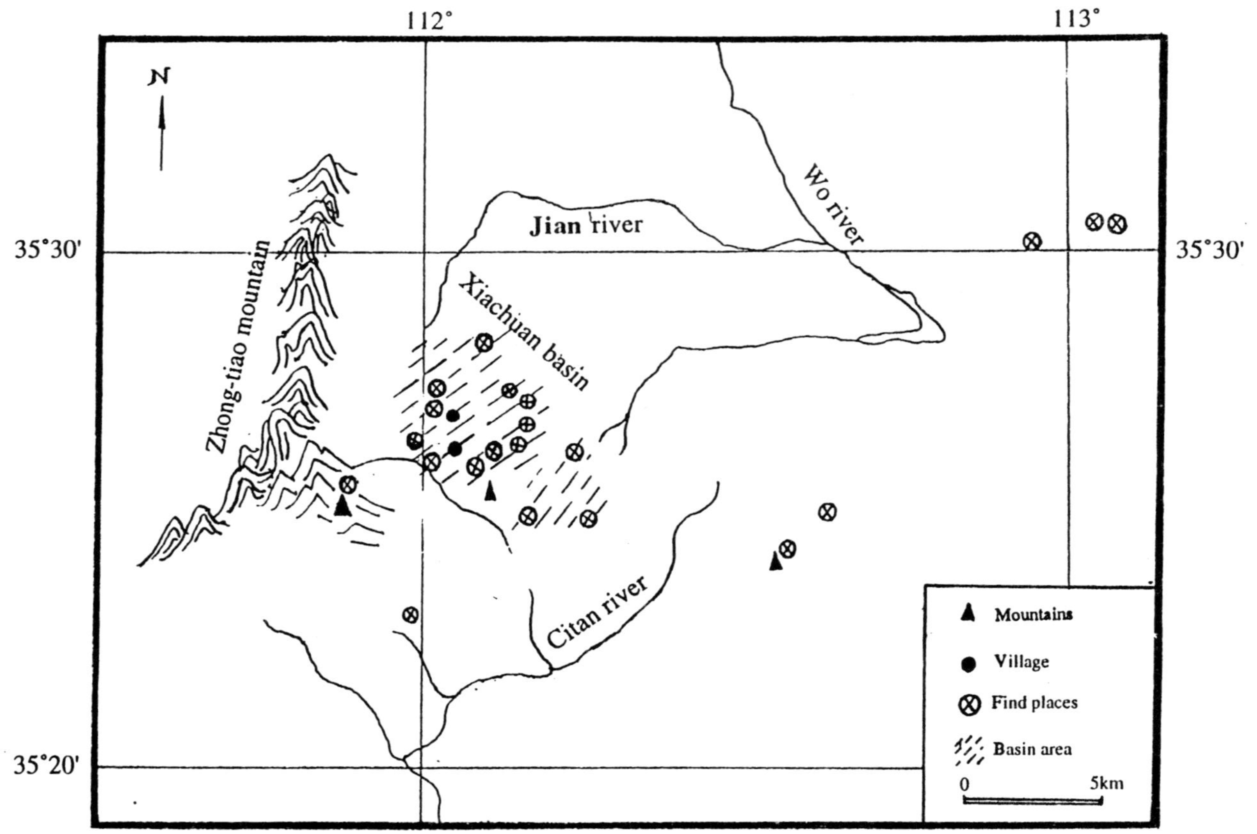

Fig. 4(2) The location of find-places in the Xiachuan Basin
(After Wang *et al.* 1978)
(The Fuyi river was not marked on the original map)

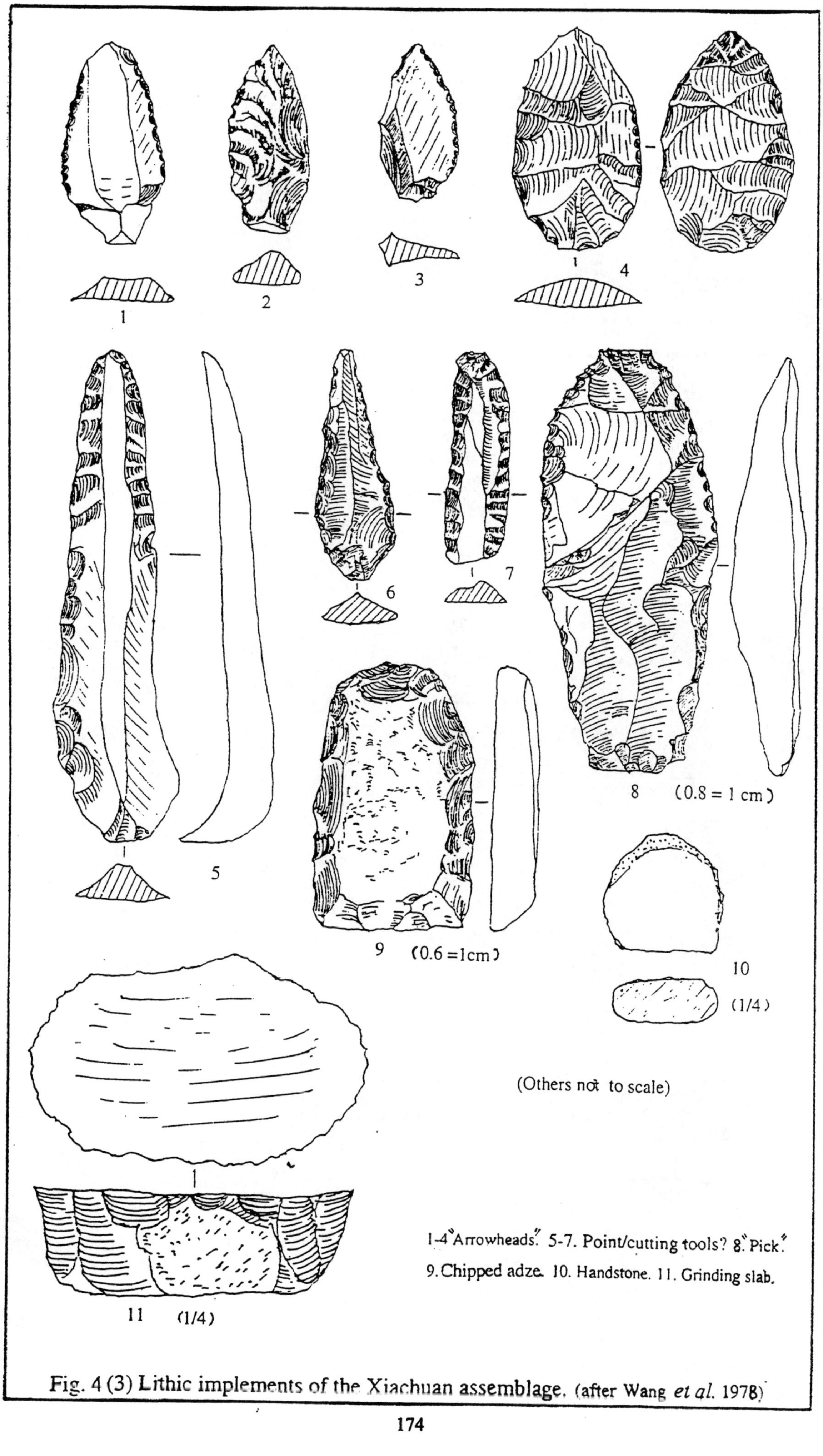

Fig. 4 (3) Lithic implements of the Xiachuan assemblage. (after Wang *et al.* 1978)

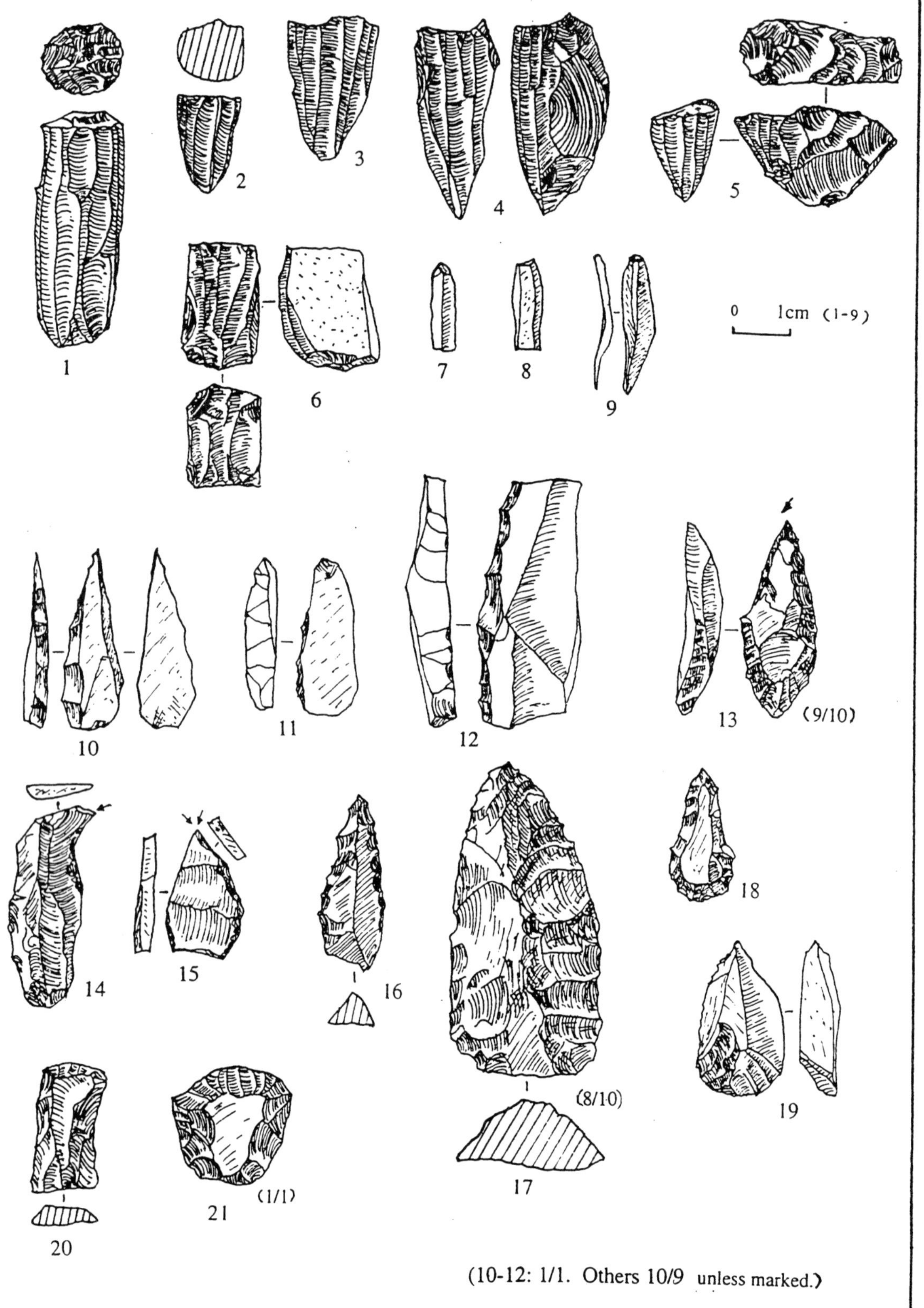

1-6. Microcores. 7-9. Microblades. 10-12. Backed blades. 13-15."Burins." 16-18."Points". 19."Point/scraper" 20-21."Scrapers".

Fig. 4 (4) Lithic implements of the Xiachuan assemblage. (after Wang *et al.* 1978).

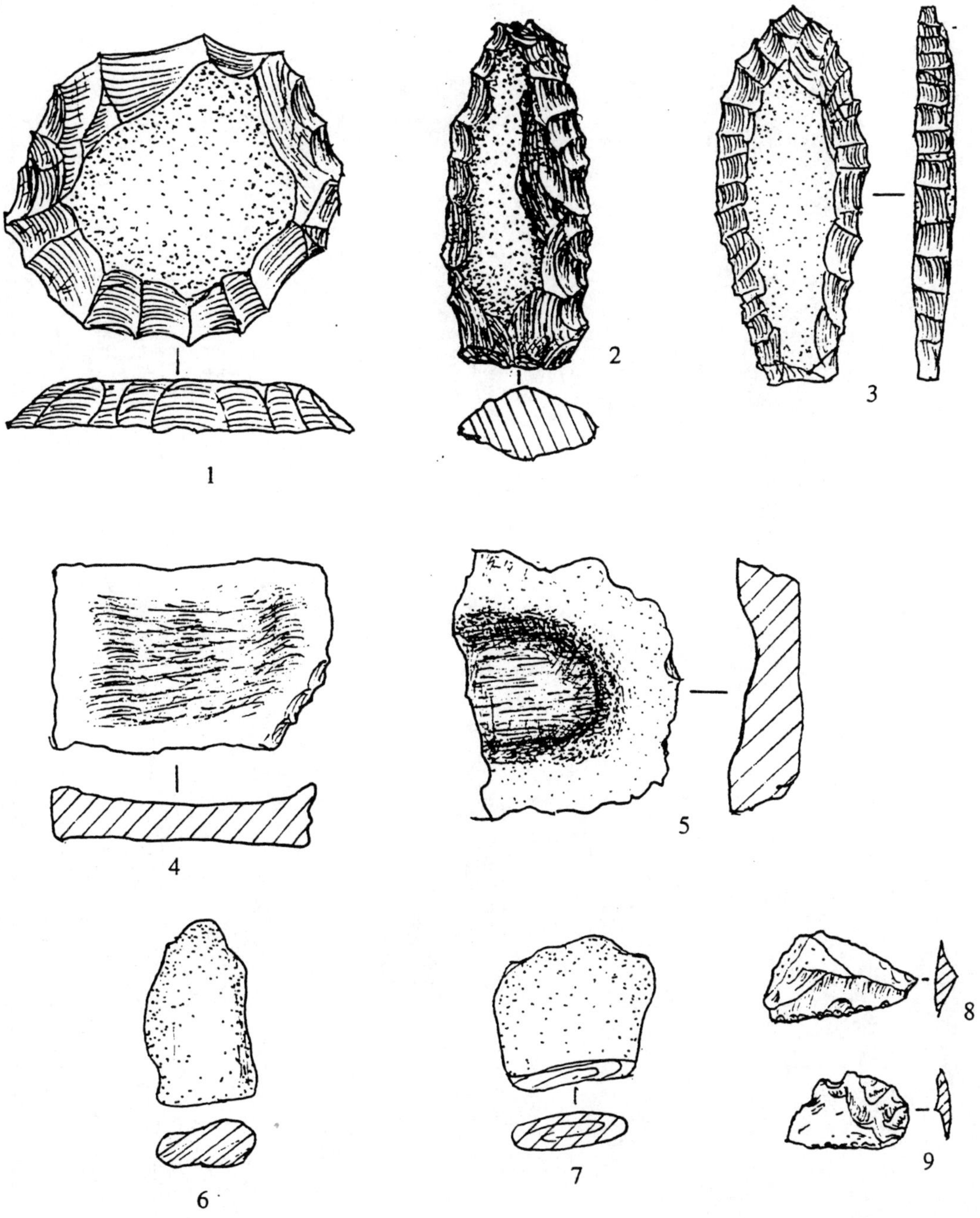

(Nos 1-3: 1/3; others not to scale)

1."Discoid." 2. Chipped Axe. 3."Pick." 4, 5. Grinding slabs.
6, 7. Handstones. 8-9. Flake "sickles."

Fig. 4 (5) Lithic implements of the Xiachuan assemblage. (Nos. 8-9 after Shi 1989; others after IA, CASS and CPAM of Shanxi Province 1978).

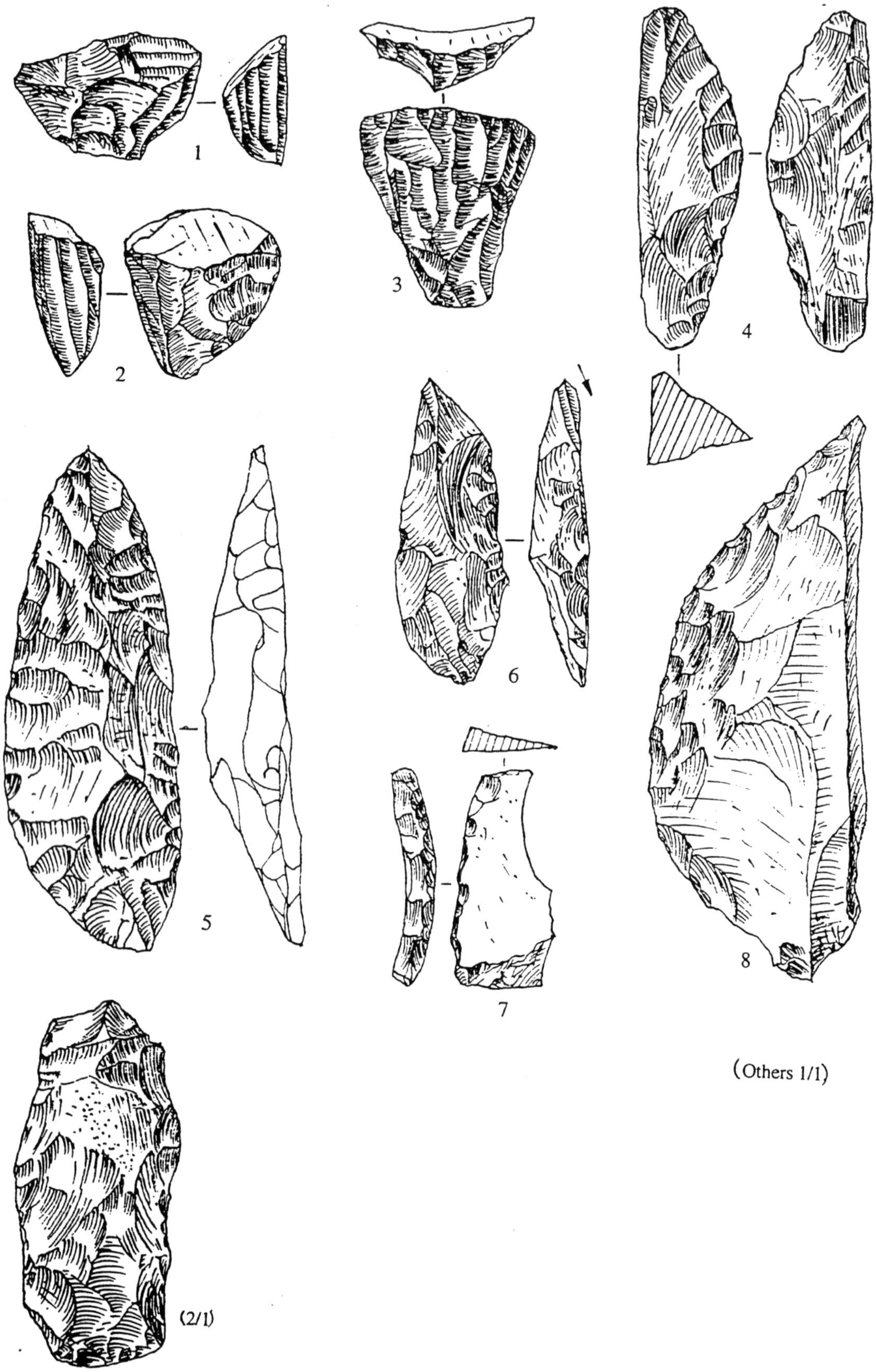

1-3. Microcores. 4,8. Scrapers. 5. Point. 6,7."Burins." 9. Chipped axe.

Fig. 4 (6) Lithic implements of the Xueguan assemblage. (after Wang *et al.* 1983.)

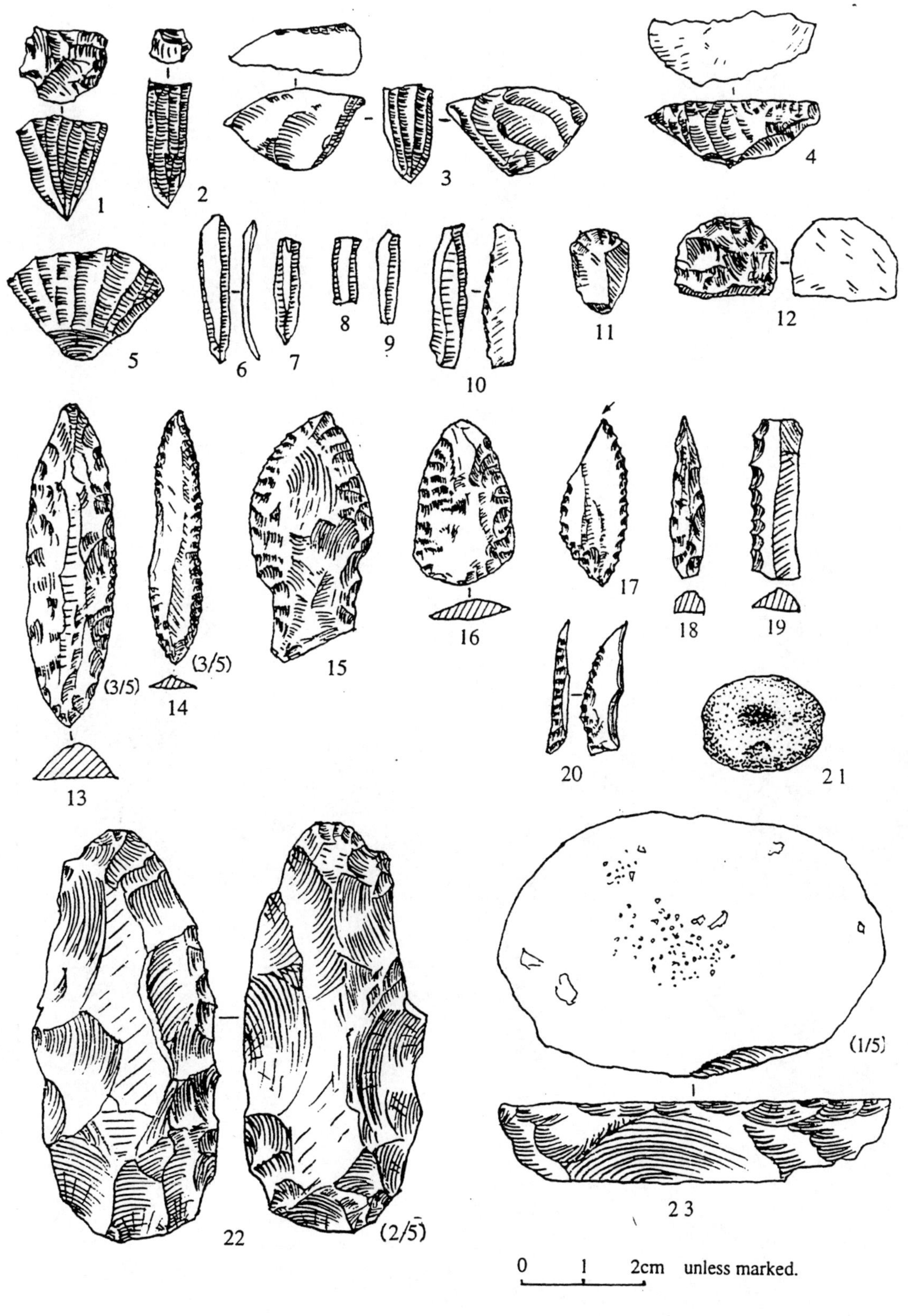

1-5 . Microcores. 6-9. Microblades. 10. Backed blade. 11,12. Scrapers. 13, 14. Points. 15, 16. Arrowheads. 17. Burin 18. Drill 19, 20. Backed blades . 21. Handstone. 22. Chipped axe. 23. Grinding slab.

Fig. 4 (7) Lithic implements of the Shizitan assemblage.
(after Linfen Culture Bureau 1989).

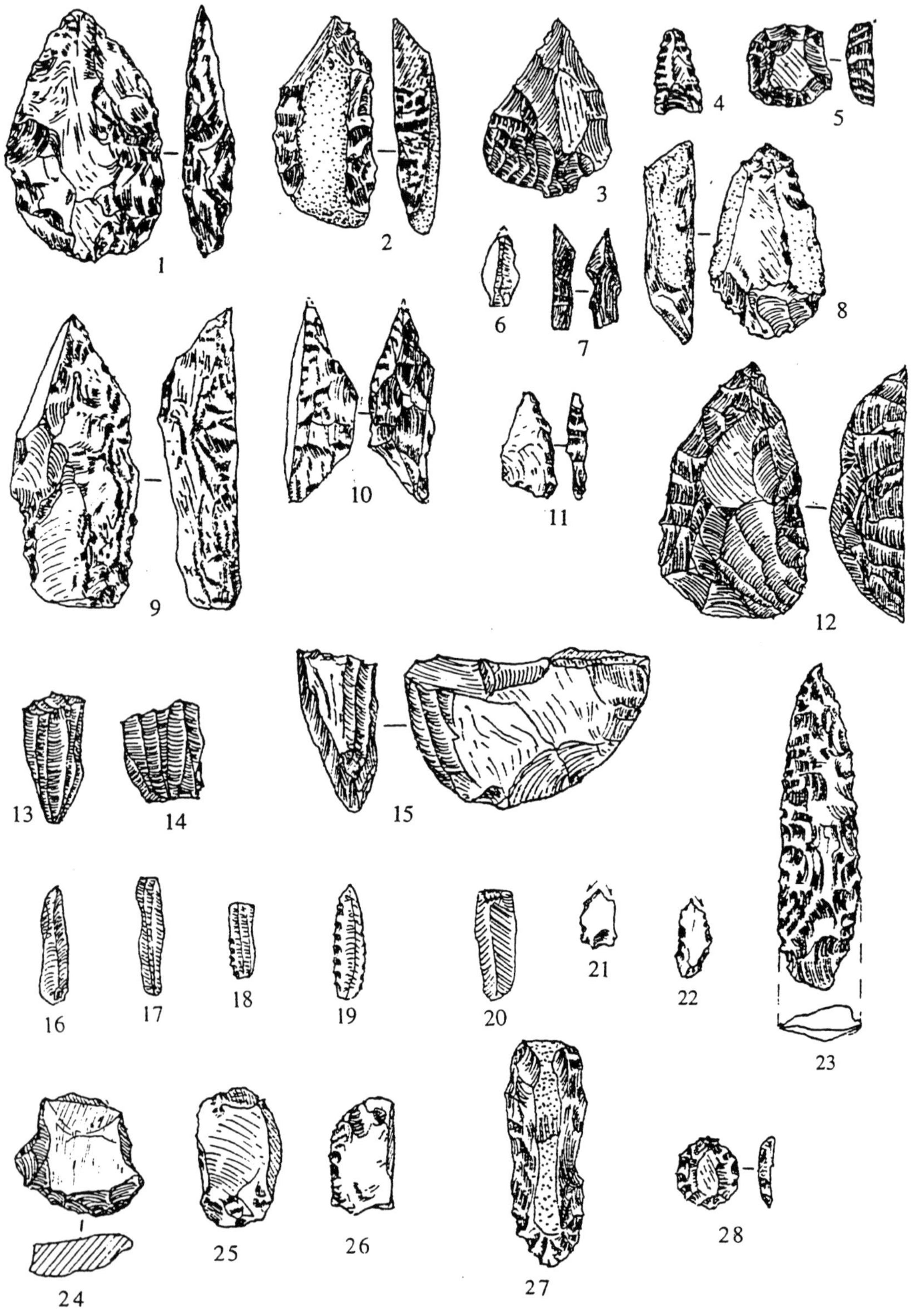

Scale: No. 1, 2.. 6-10, 25, 26: 1/2. Others 1:1.

1, 2, 9, 10. Points. 3, 4, 6, 7. Arrowheads. 5, 8, 25-28. Scrapers. 11. Backed blades. 12. Point/scraper. 13-15. Microcores. 16-20. Microblades. 21, 22. Burins. 23. Long flake. 24. Chipped Axe.

Fig. 4 (8) Lithic implements of the Shayuan assemblage. (after An and Wu 1957).

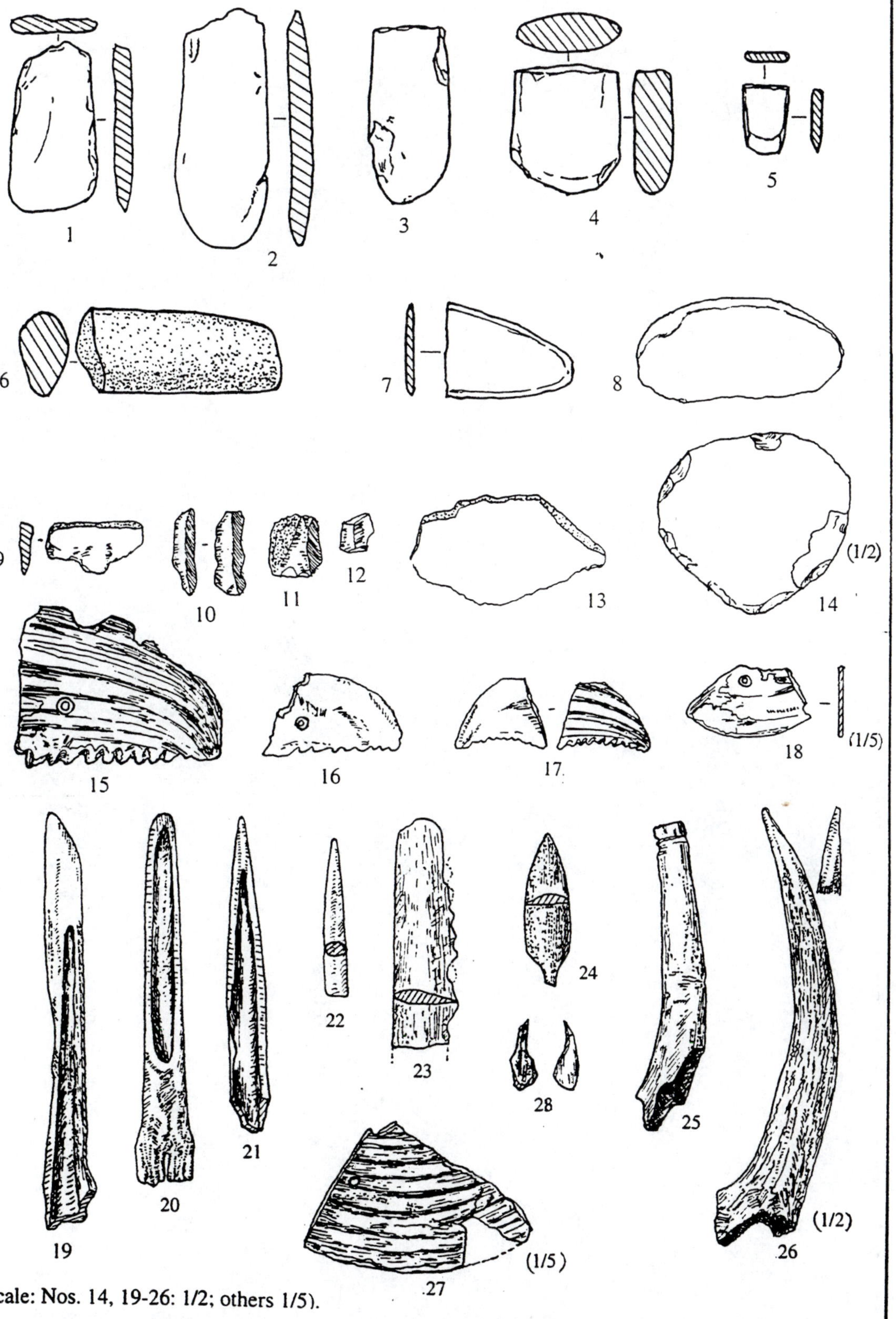

(Scale: Nos. 14, 19-26: 1/2; others 1/5).

1-3. Stone spades. 4. Stone axe. 5. Stone chisel. 6. Stone roller. 7-8. Stone knives. 9-12. Stone flakes. 13. Stone scrapers. 14. Stone chopper . 15 -17. Shell sickles. 18, 27. Shell knives. 19. Bone knife. 20. Bone spearhead. 21, 22. Bone awls. 23. Bone saw. 24. Bone arrowhead. 25. Carved antler. 26. Antler point. 28. Tooth awl.

(Nos. 6, 8, 9-14 from Dadiwan; others from Baijia).

Fig. 4 (9) Tools of the Dadiwan Culture (after Dadiwan Excavation Group 1981, 1982)

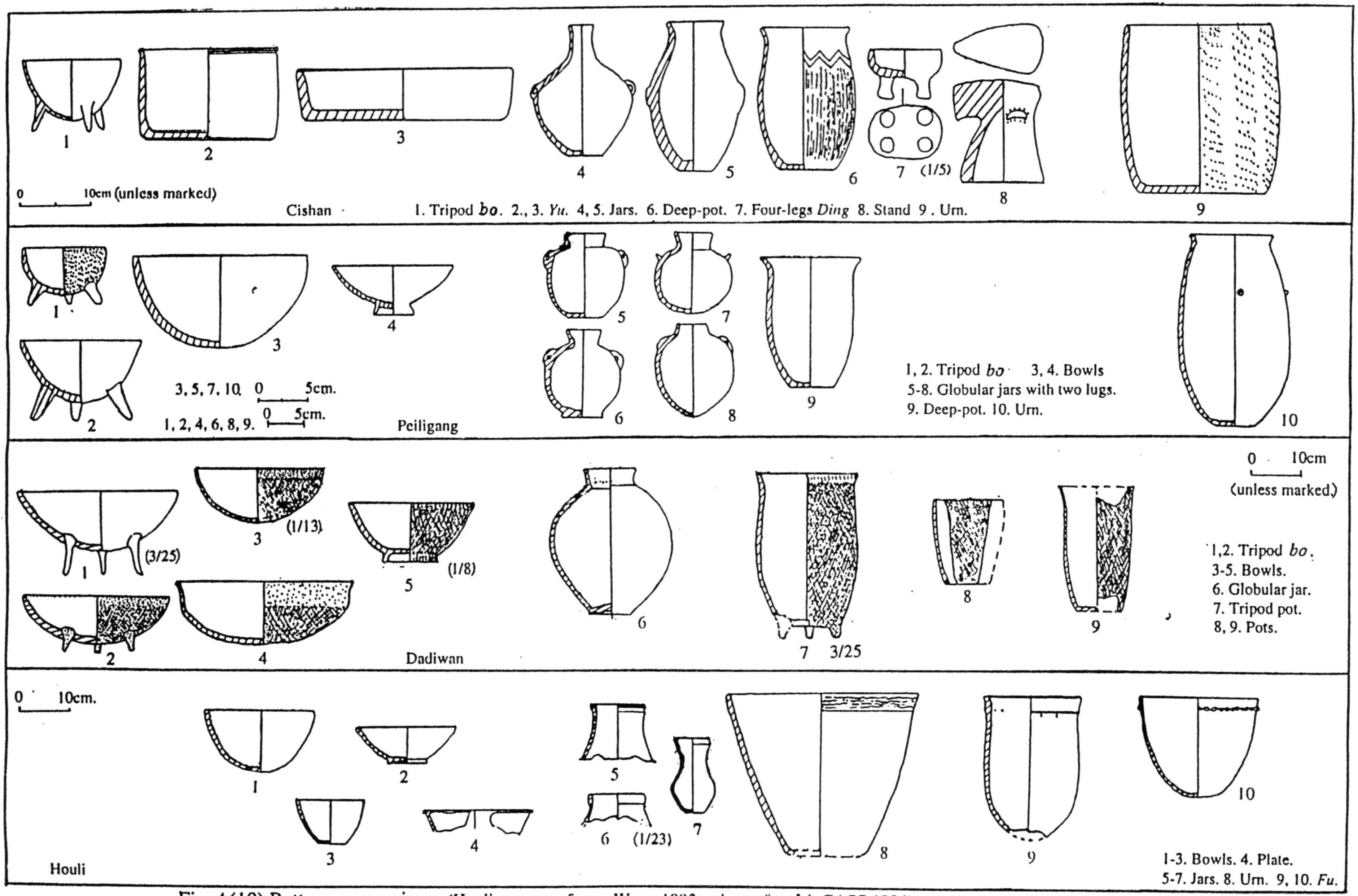

Fig. 4 (10) Pottery comparison. (Houli pottery: after Wang 1993; others after IA CASS 1984)

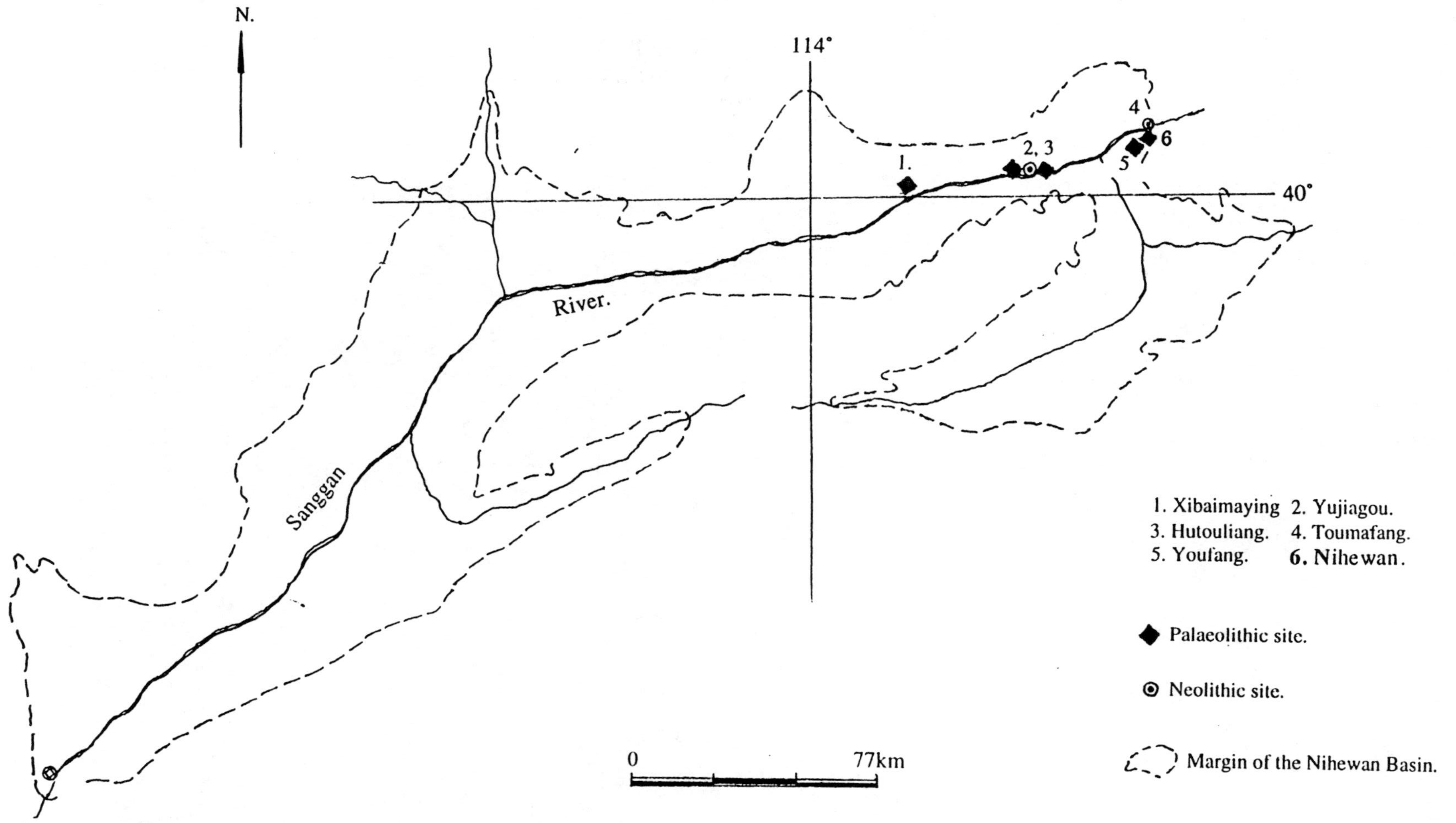

Fig. 4(11) Archaeological sites in the Sanggan River area mentioned in the text (based upon Huang and Tang 1989 Updated).

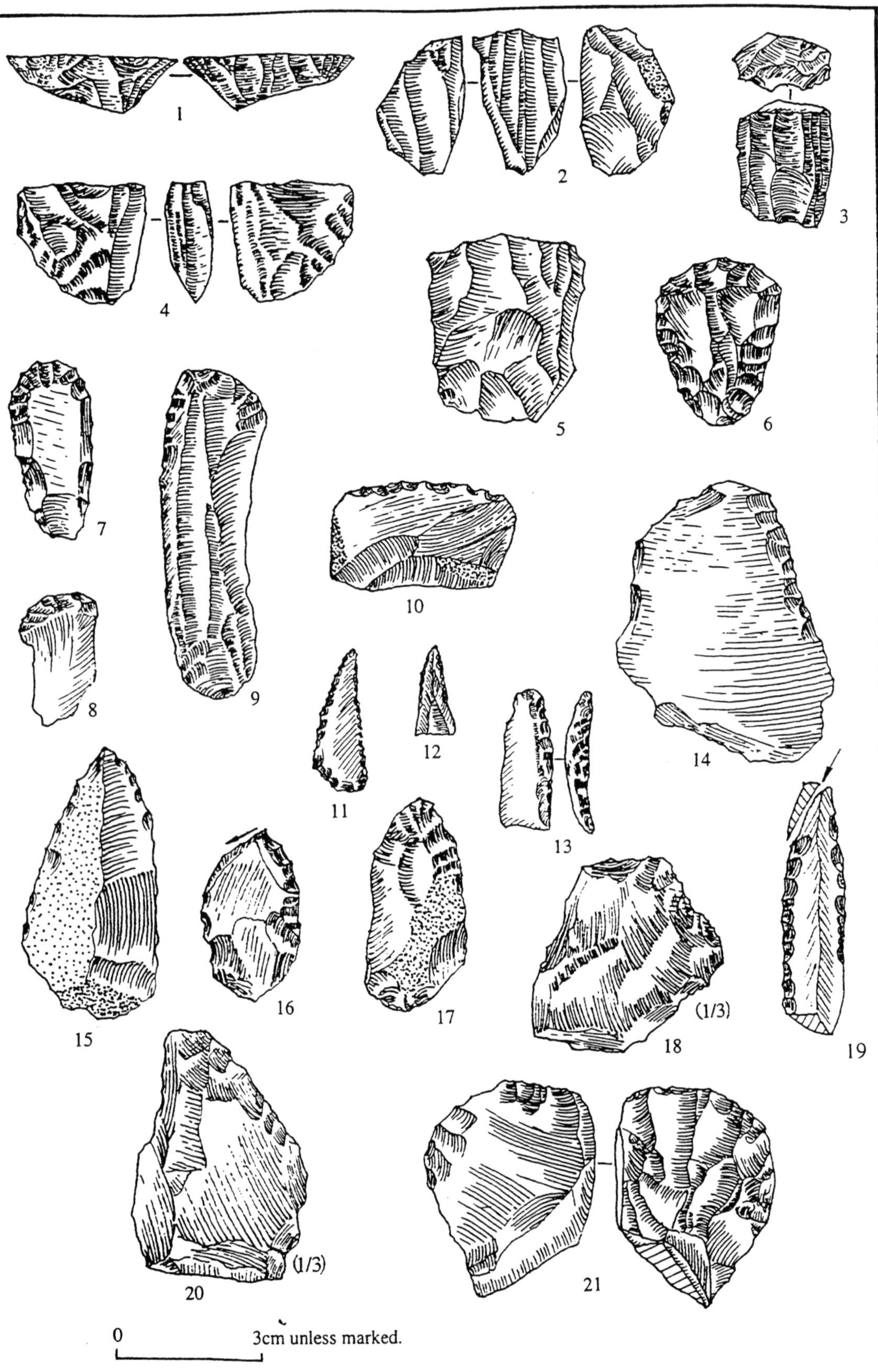

1-5. Microcores. 6-10, 14. Scrapers. 11,12, 15. Points. 13. Backed blade. 16, 19. Burins.
17. Point/scraper. 18, 20. Choppers. 21. Core scraper.

Fig. 4 (12) Lithic implements of the Youfang assemblage. (after Xie and Cheng 1989)

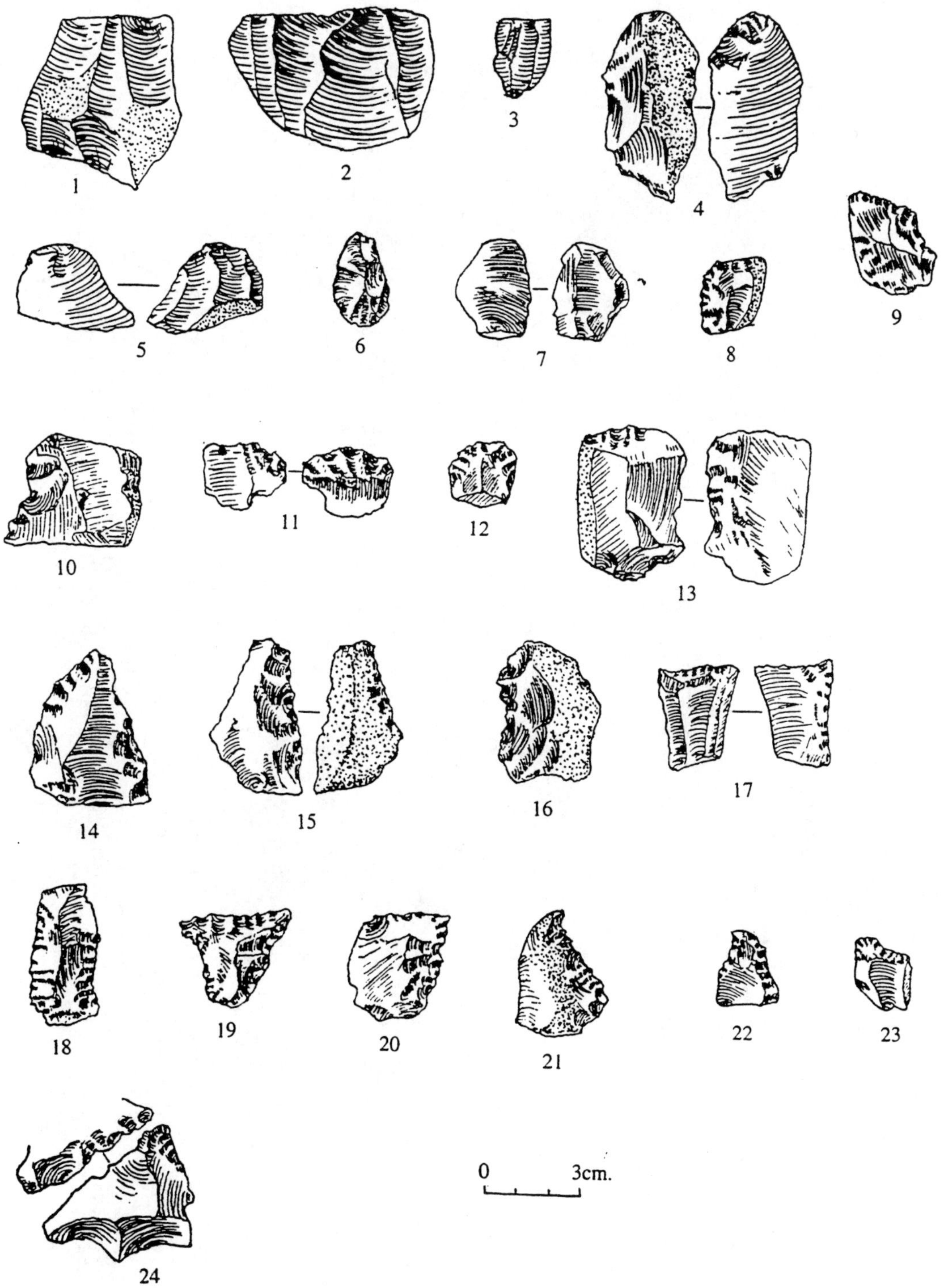

1-3. Cores. 4, 5. Flakes. 6-18."Scrapers." 19-24. Retouched flakes.

Fig. 4 (13) Lithic implements of the Xibaimaying assemblage.
(after IA, Hebei Province 1989)

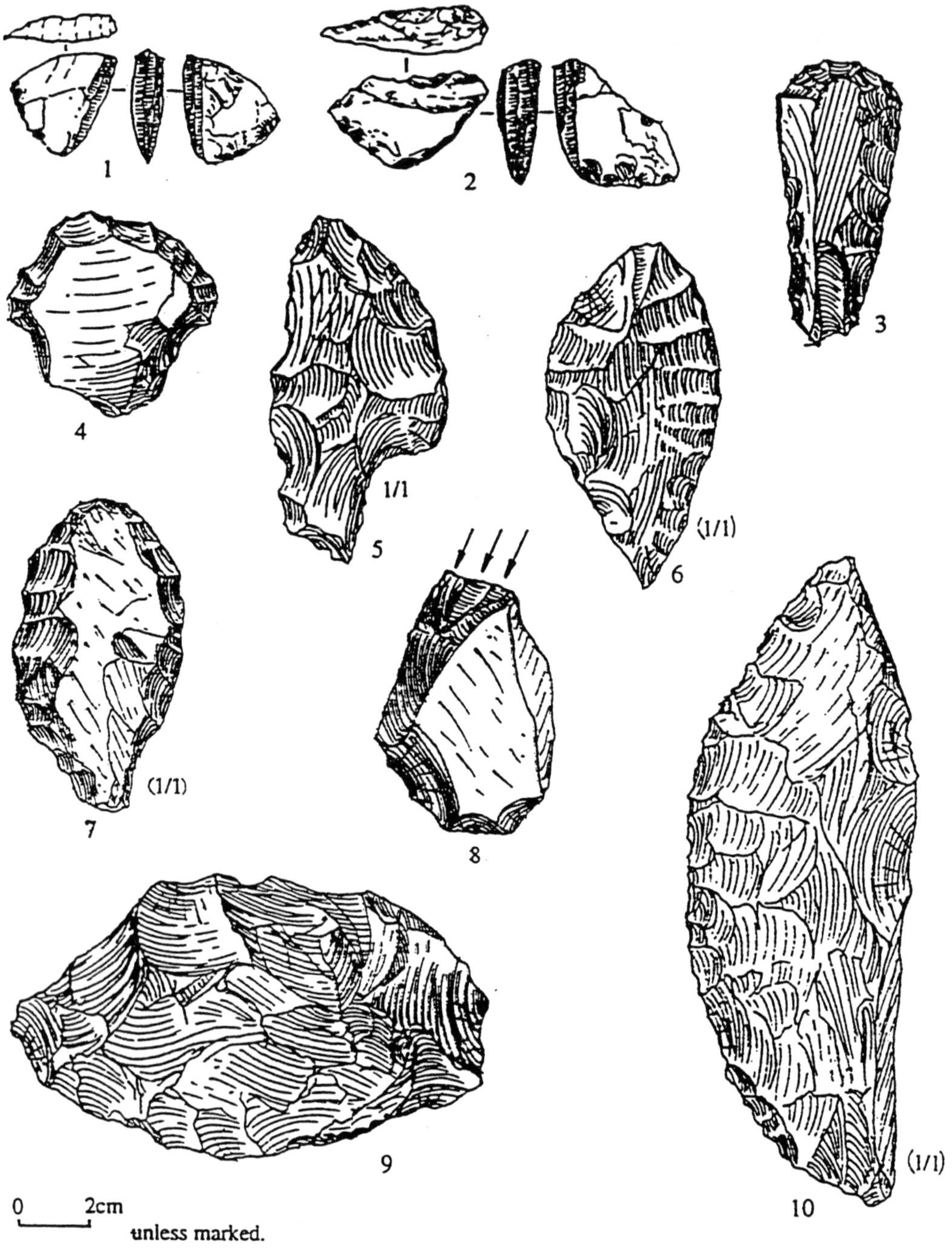

1-2. Microcores. 3, 4. "Scrapers." 5-7, 10. "Points." 8. "Burin." 9. Chopper.

Fig. 4(14) Lithic implements of the Hutouliang assemblage. (after Gai and Wei 1977)

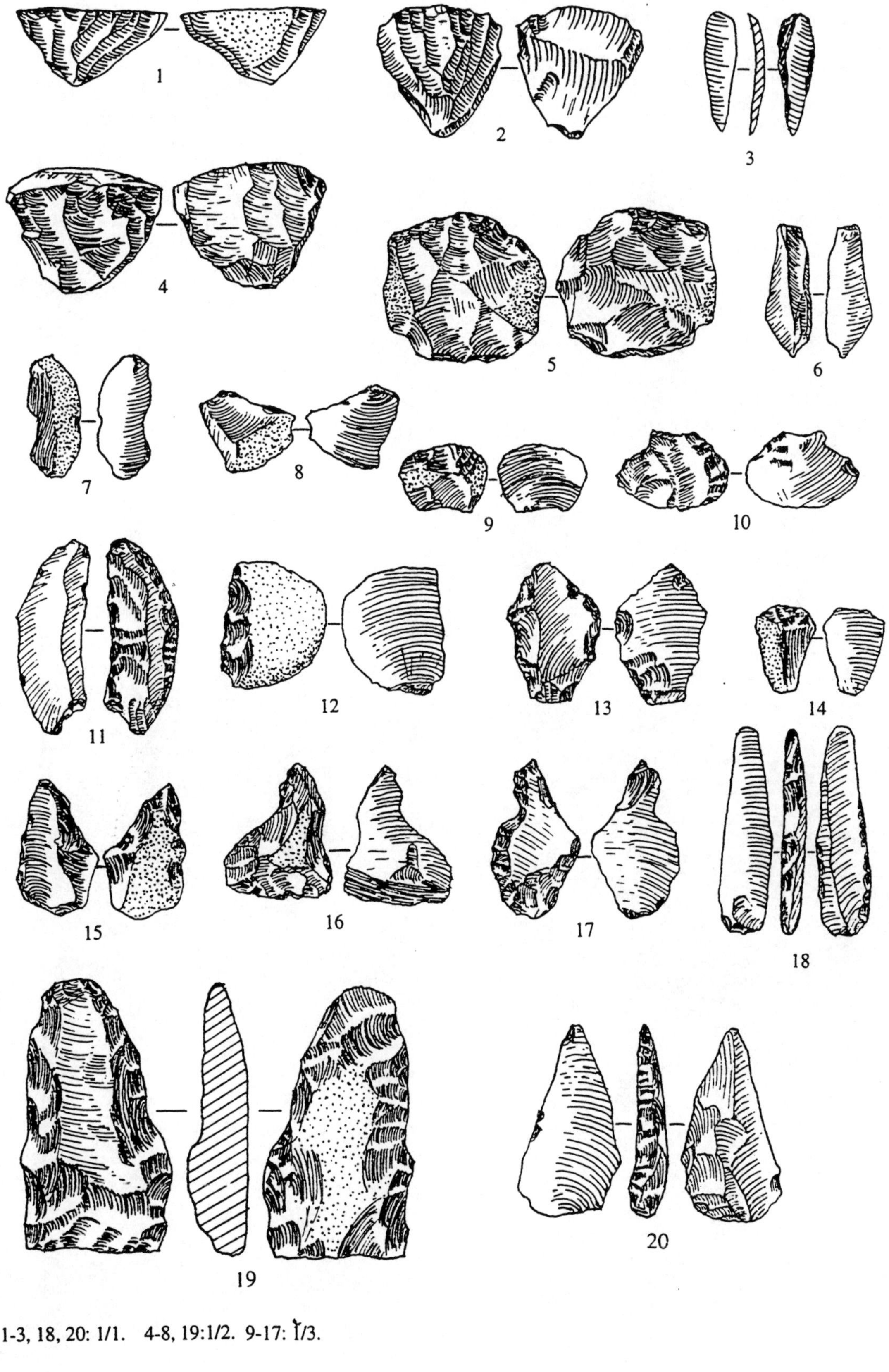

1-3, 18, 20: 1/1. 4-8, 19:1/2. 9-17: 1/3.

1-2. Microcores. 3. Microblade. 4, 5. Flake cores. 6-8. Flakes. 9-14."Scrapers." 15."Point." 16."Point/scraper." 17."Drill." 18, 20. Backed blades. 19. Flaked adze.

Fig. 4 (15) Lithic implements of the Mengjiaquan assemblage.
(after IA, Hebei Province *et al*. 1991).

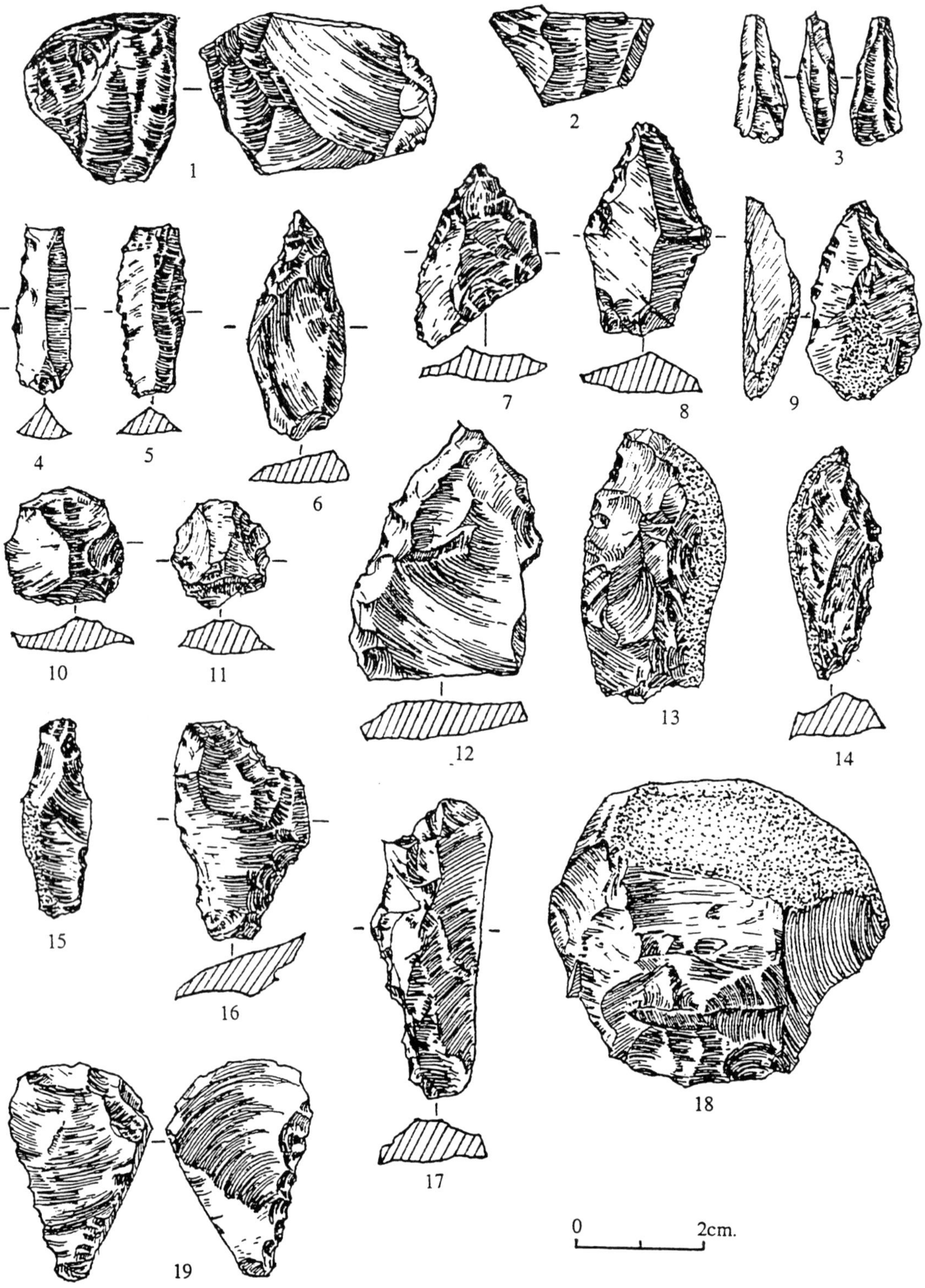

1-3. Flake cores. 4, 5. Blade-like flakes. 6-9."Points."10-17, 19."Scrapers."18. Chopper.

Fig. 4 (16) Lithic implements of the Xiaonanhai assemblage (after An 1965).

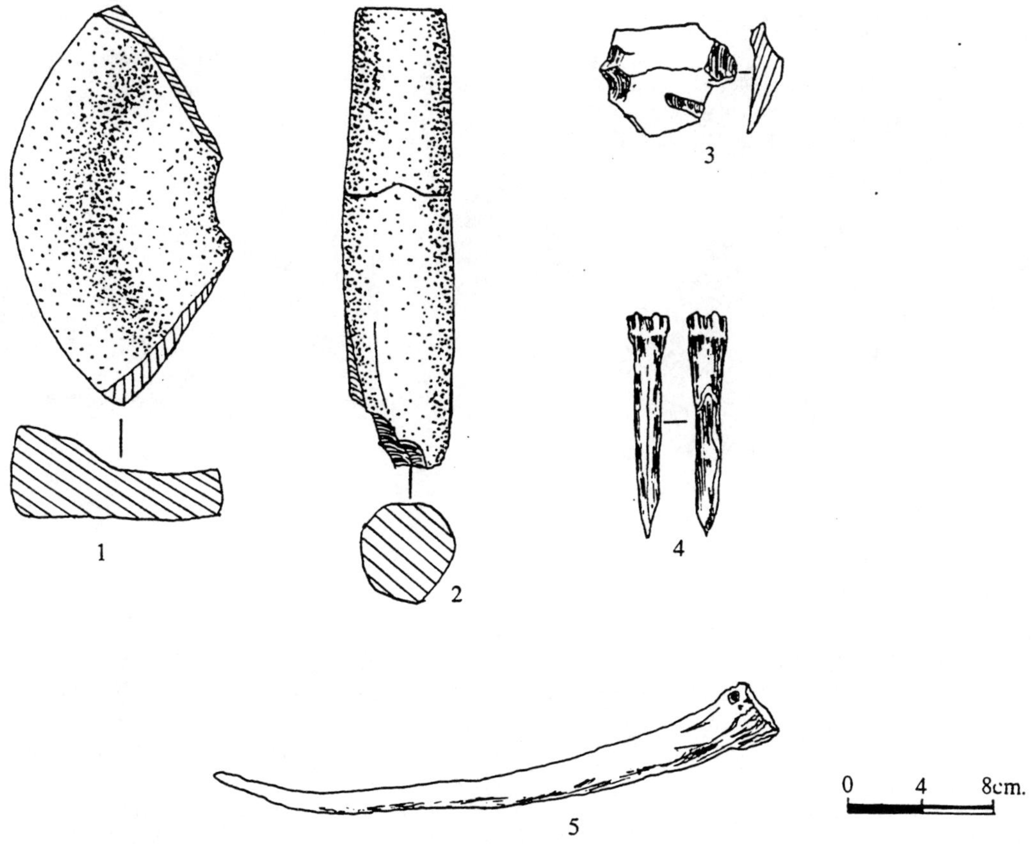

1. Fragment of a grinding slab. 2. Stone roller. 3. Flake. 4. Bone awl. 5. Antler point.

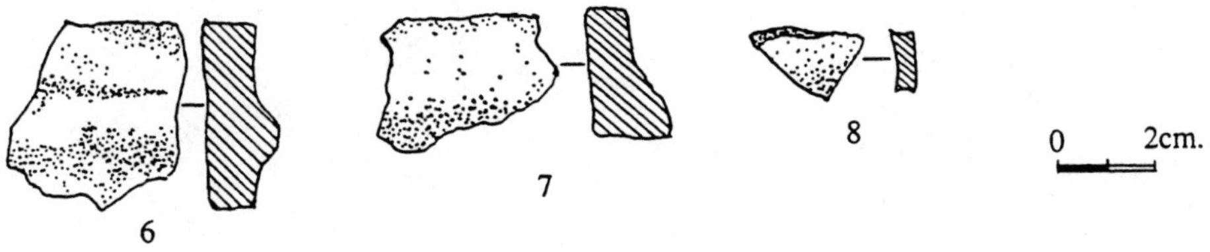

6-8. Potsherds.

Fig. 4 (17) Artefacts found at Nanzhuangtou
(after CPAM of Baoding District *et al.* 1992.).

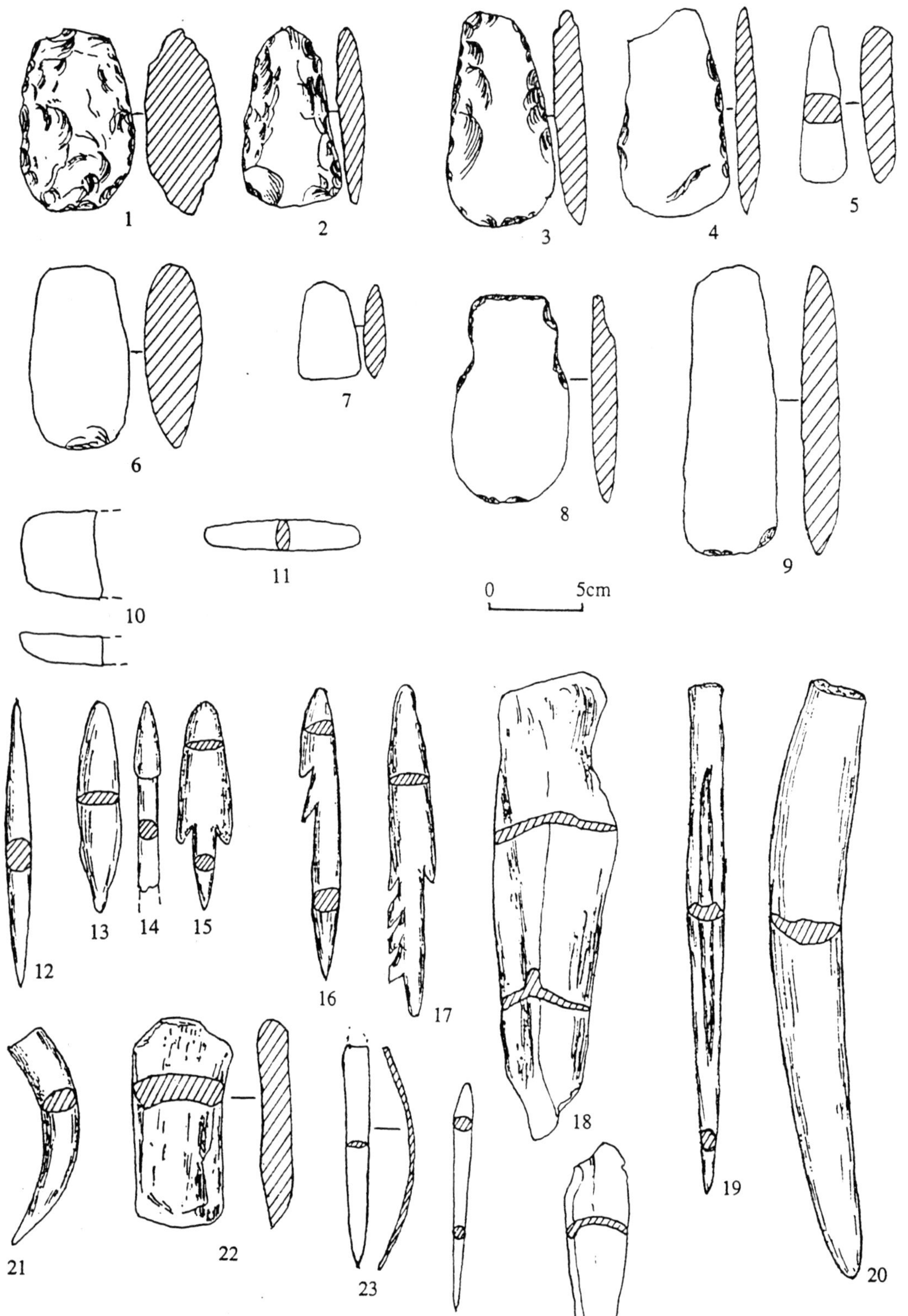

1, 2, 6. Stone axes. 3,4, 8,9. Stone spades. 5. Stone chisel. 10. Fragment of a stone grinding slab. 11. Stone roller. 12-15. Bone arrowheads. 16, 17. Bone harpoons. 18. Bone knife. 19. Bone drill. 20. Antler drill. 21. Tooth drill. 22. Bone chisel. 23. Bone shuttle. 24. Bone needle. 25. Shell knife.

Fig. 4 (18) Tools of the Cishan assemblage (layer 3). (after CPAM of Hebei Province and CPAM of Handan City 1981).

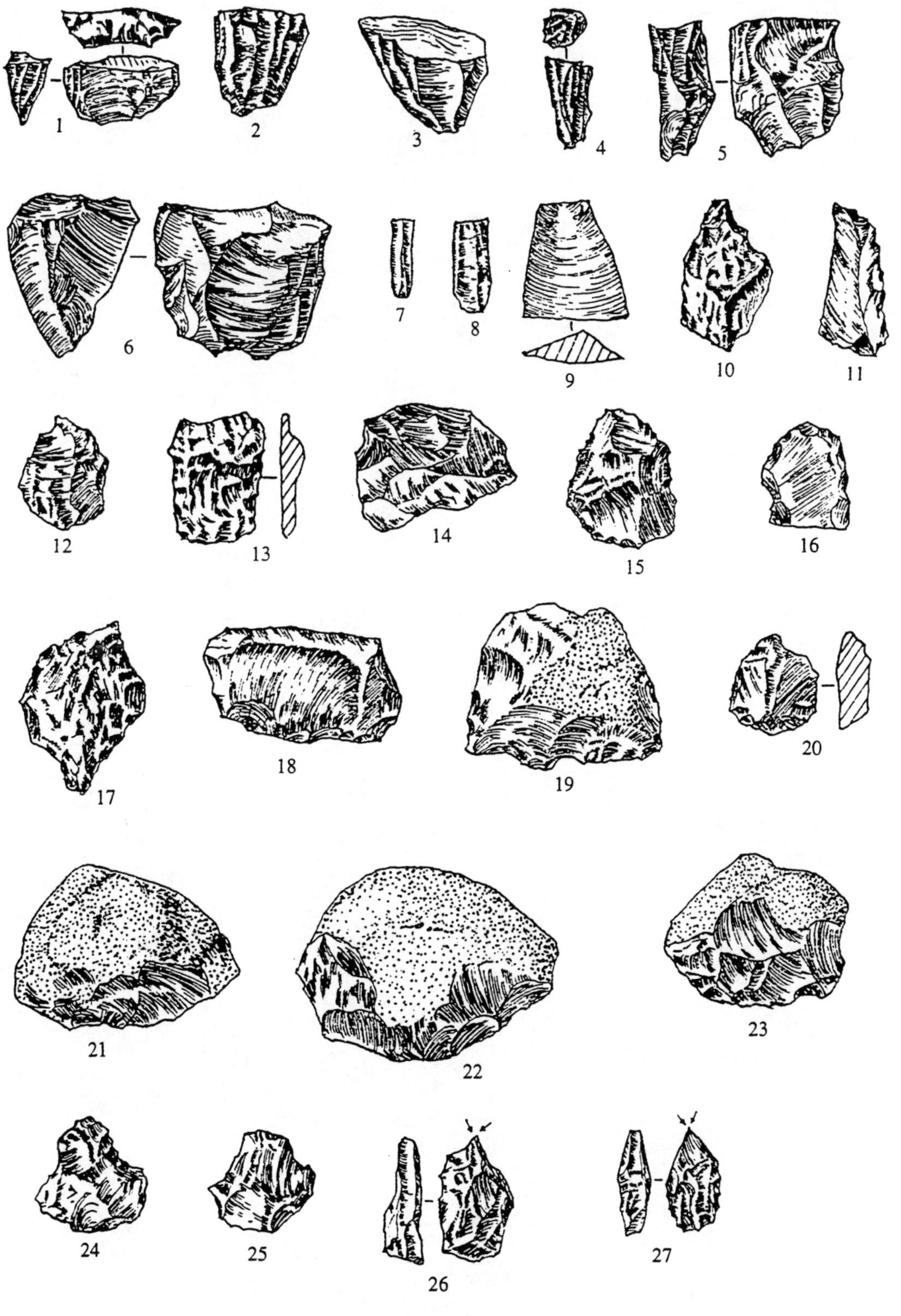

Scale: 7, 8, 26, 27:1/1. 17, 22, 23: 1/3. The rest 5/6.

1-4. Microcores. 5,6. Cores. 7. Microblade. 8,9. Flakes. 10-12, 17, 20. Points. 13-16, 18. Scrapers. 19, 21-23. Choppers & chopping tools. 24, 25. Axes. 26,27. Burins.

Fig. 4 (19) Lithic implements of the Lingjing assemblage (after Zhou 1974).

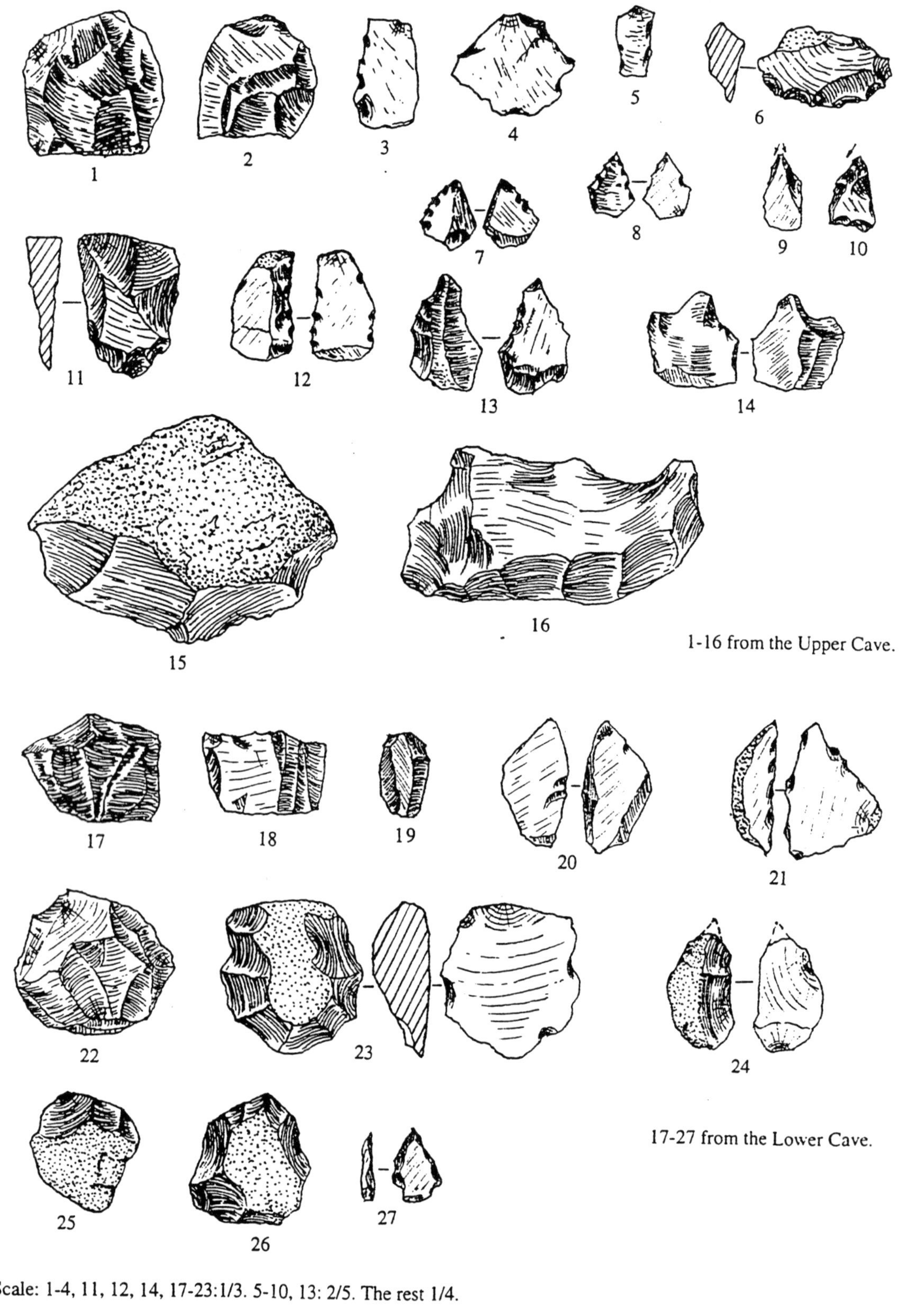

Scale: 1-4, 11, 12, 14, 17-23:1/3. 5-10, 13: 2/5. The rest 1/4.

1, 2, 17-19. Cores.3-5, 20: Flakes. 6, 11, 12, 21: Scrapers. 7. Point/scraper. 8, 13, 14, 24, 27. Points. 9, 10. Burins. 15, 16, 25. Choppers. 22. Stone ball. 23, 26. Discoid implements.

Fig. 4 (20) Lithic implements of the Xiaokongshan assemblage (after Xiaokongshan Excavation Team 1988).

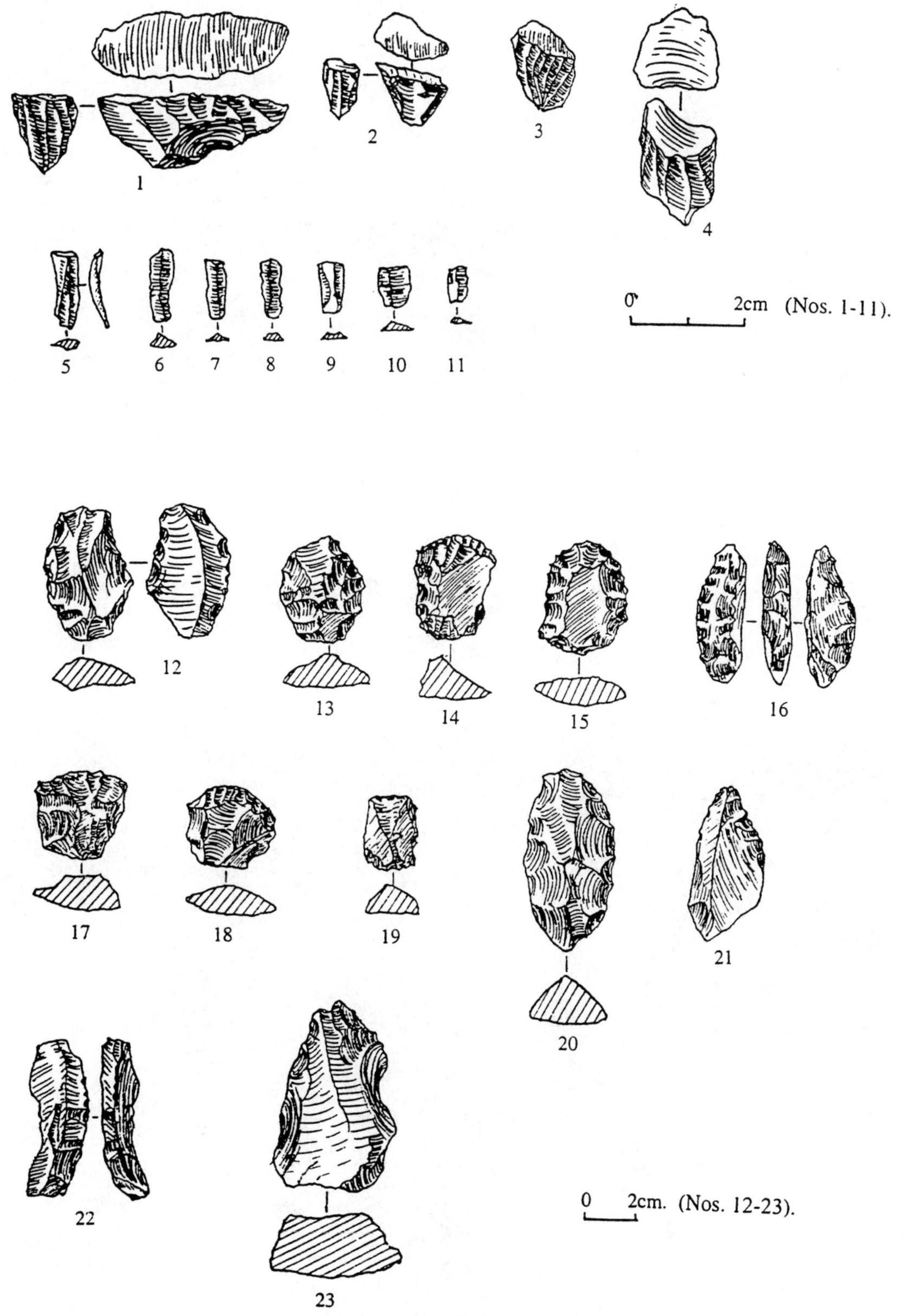

1-4. Microcores. 5-11. Microblades. 12-19, 22, 23. Scrapers. 20, 21. Points.

Fig. 4 (21) Lithic implements of the Dagang assemblage. (after Zhang and Li 1996).

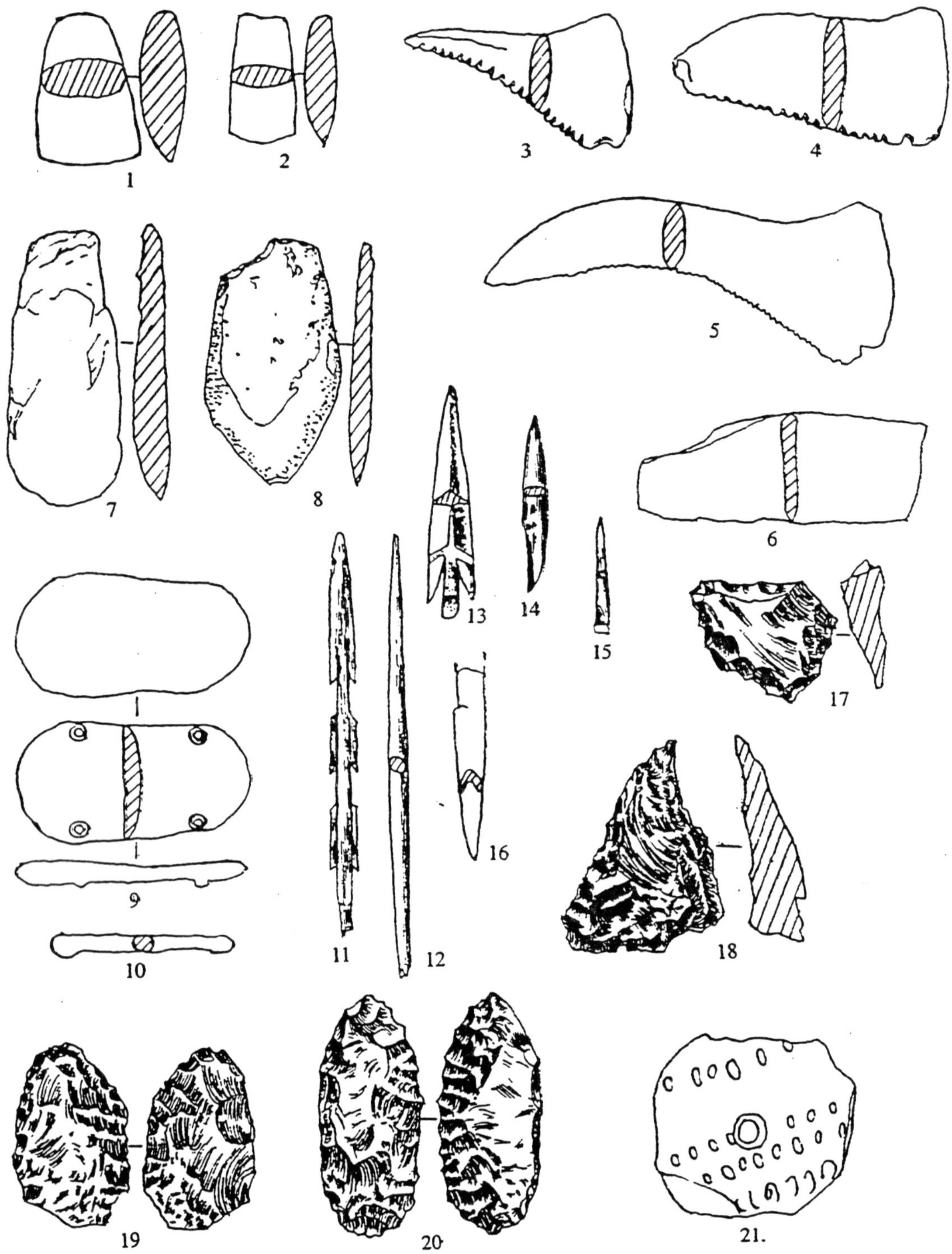

Scale: 1-6: 2/5. 7, 8: 1/5. 9, 10: 1/15. 11-15: 1/2. 16, 21: 4/5. 17-20: 1/1.

1, 2. Polished stone axes. 3-5. Polished stone sickles. 6. Stone knife. 7, 8. Polished stone spades.
9, 10. Stone quern and roller. 11, 12. Bone harpoons. 13, 14. Bone arrowheads. 15. Bone needle. 16. Bone drill.
17-20. Stone scrapers. 21. Spindle whorl made from a potsherd.

(No. 11-15 from the Jiahu site. Others are from the Peiligang site.)

Fig. 4 (22) Tools of the Peiligang assemblages (after CPAM of Kaifeng District and CPAM of Xinzheng County 1978; IA, Henan Province 1989).

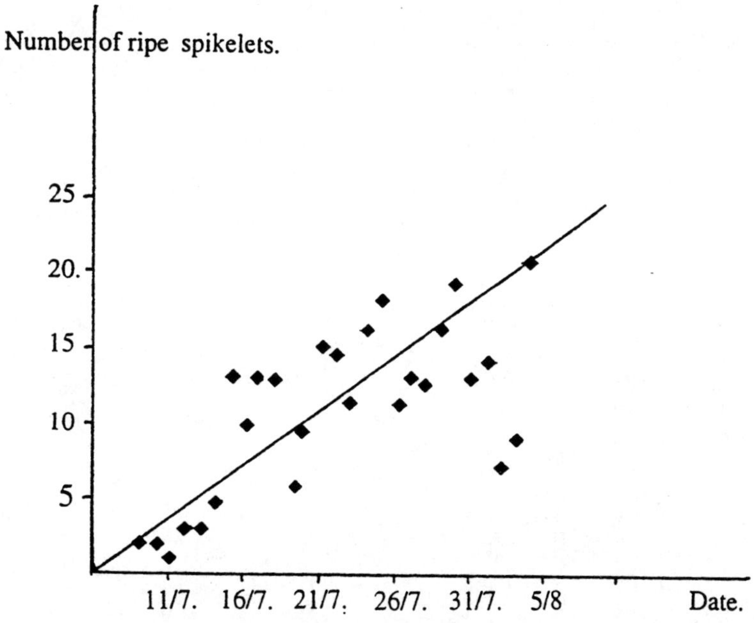

Fig.5(1) Daily availability of ripe spikelets of green foxtail (*Setaria viridis*).

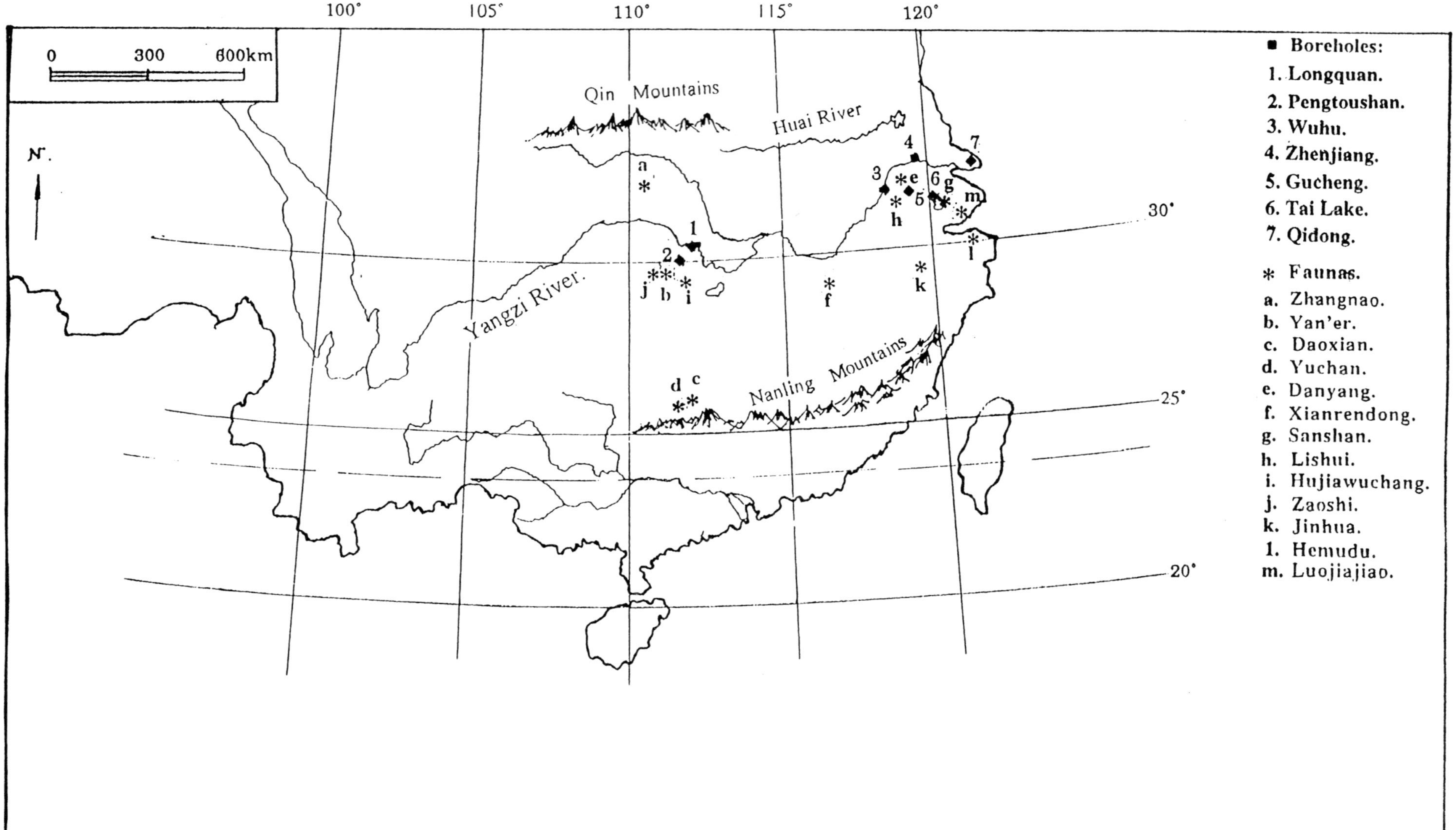

Fig. 6 (1). Locations of boreholes and faunas mentioned in the text.

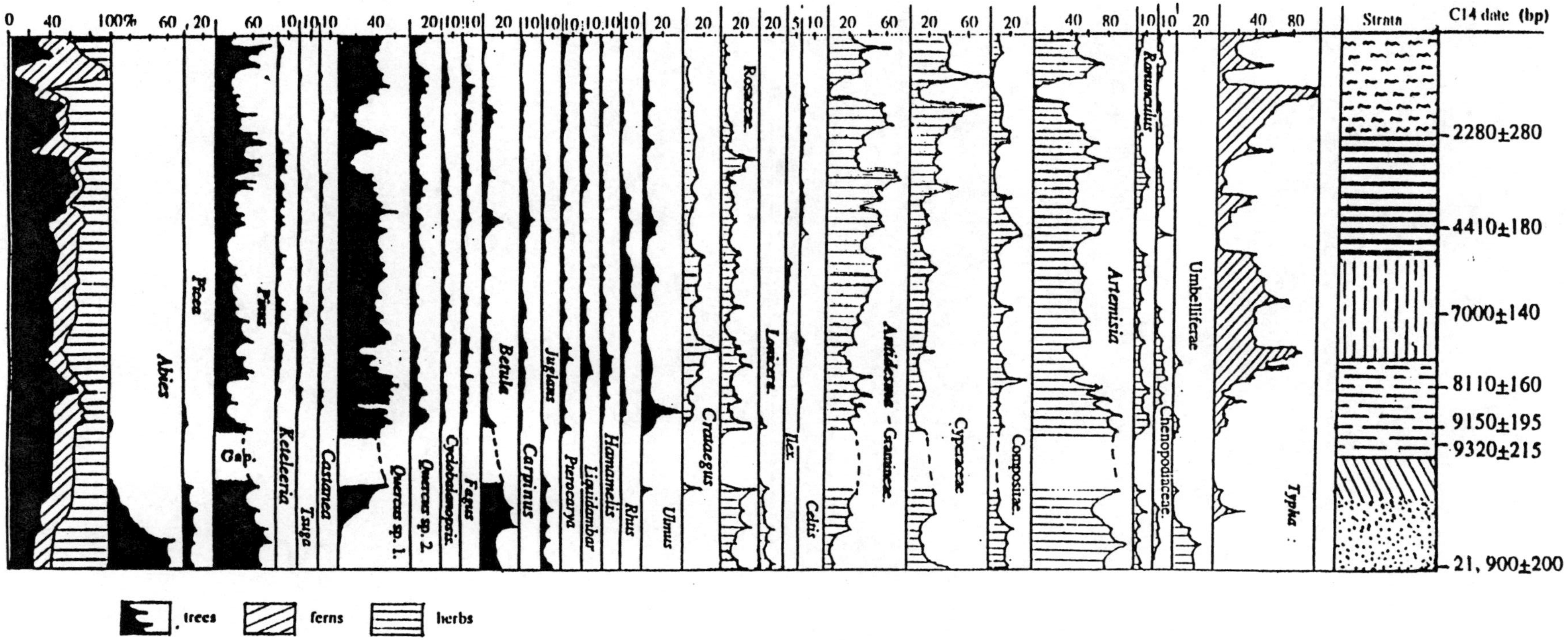

Fig. 6 (2). Pollen profile at Longquan Lake (site No. 1). (after Liu 1991).

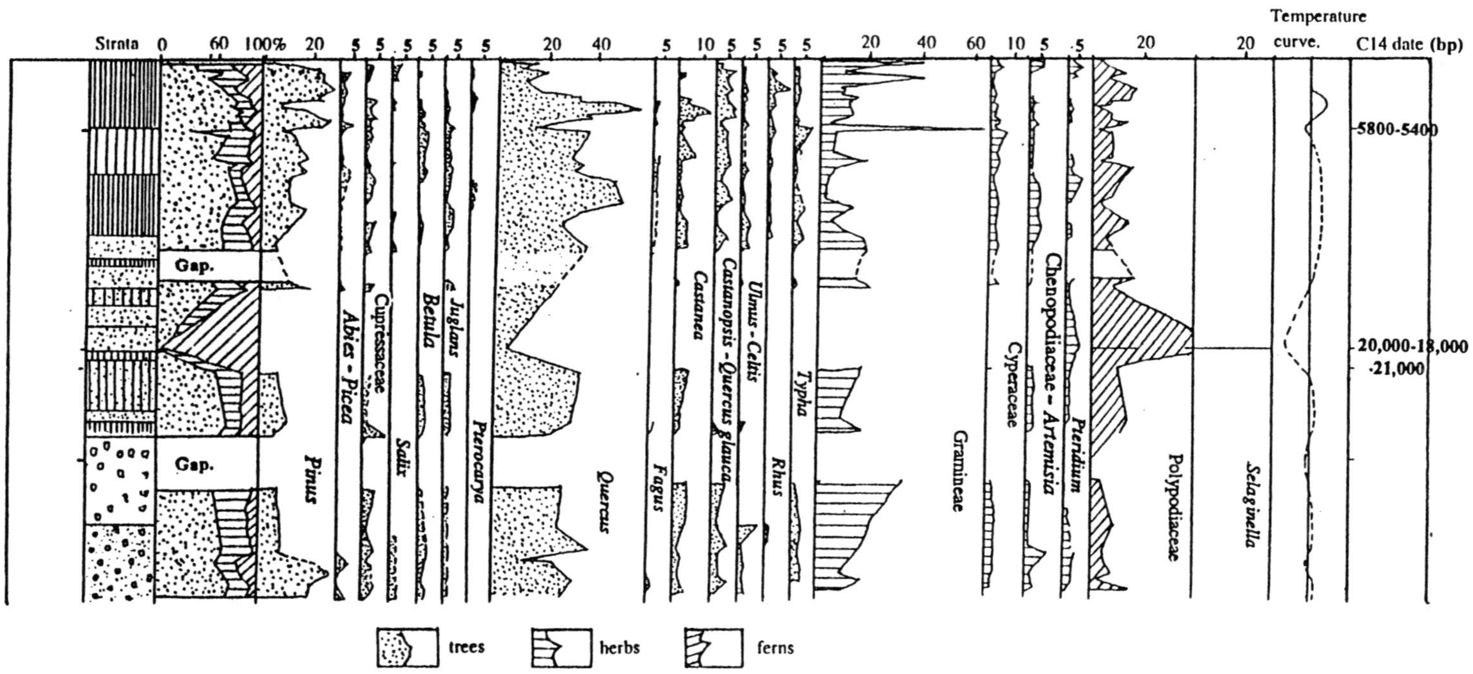

Fig. 6 (3). Pollen profile at Wuhu borehole (site No. 3). (after Xu *et al.* 1987).

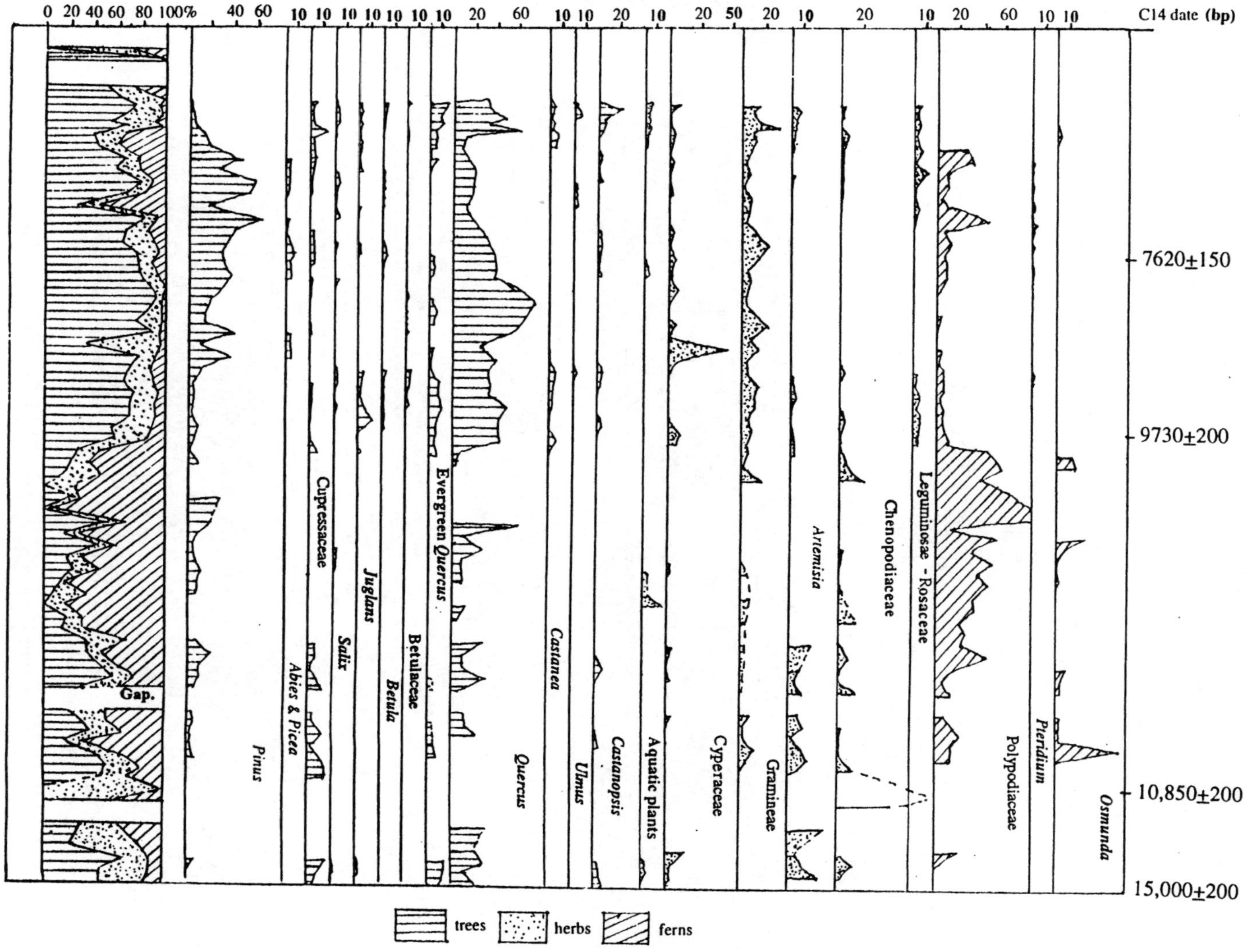

Fig. 6 (4). Pollen profile at Zhenjiang borehole (site No. 4). (after Xu *et al.* 1987).

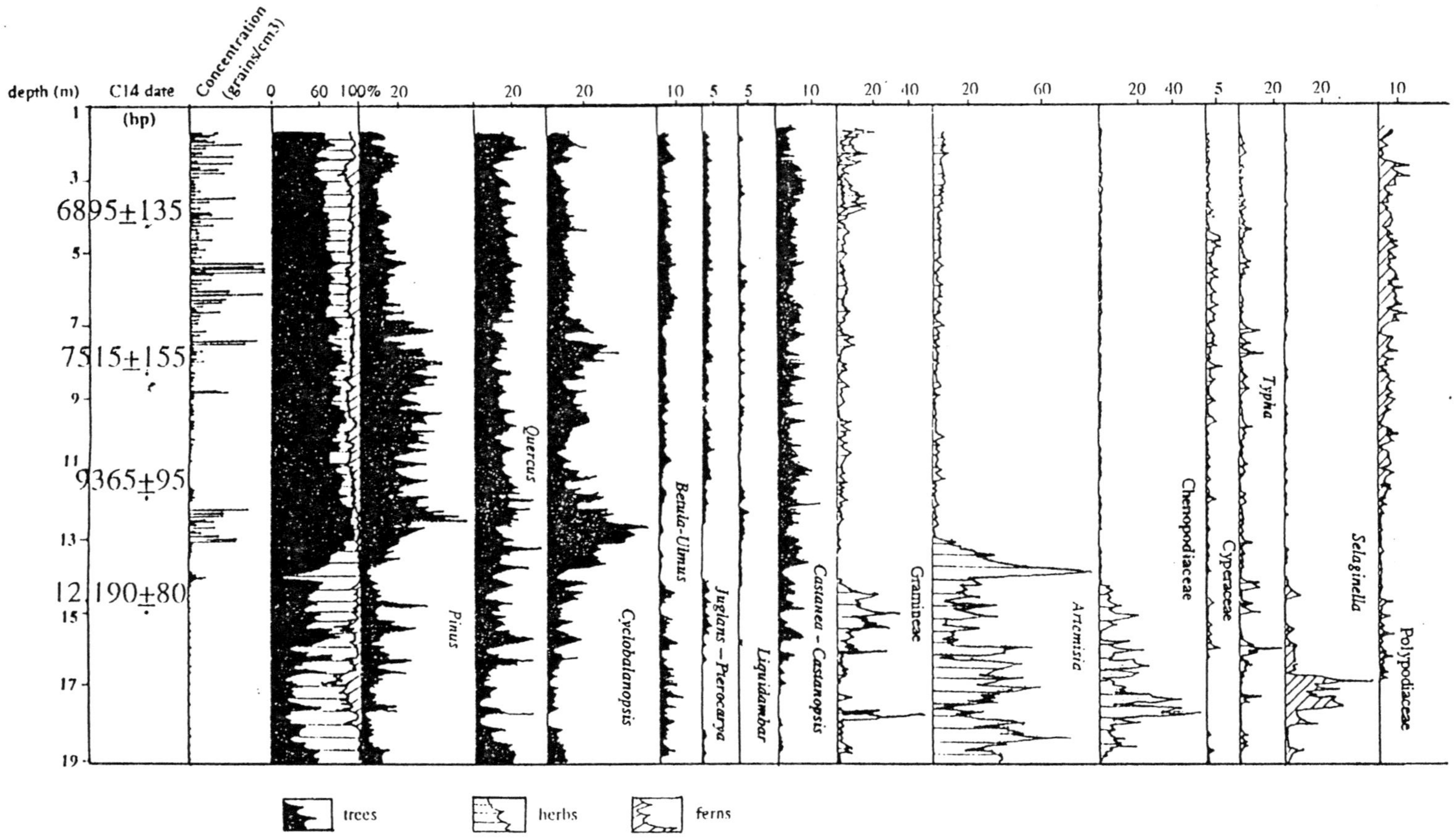

Fig. 6 (5). Pollen profile at Gucheng borehole (site No. 5). (after Yang *et al.* 1996).

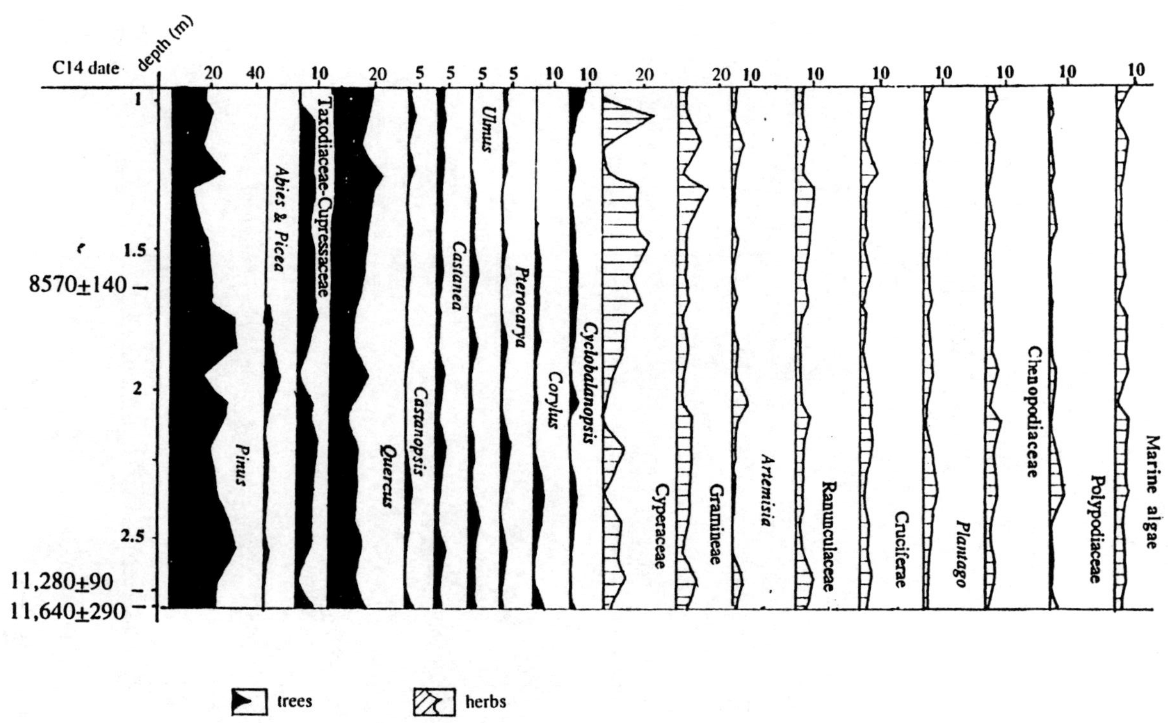

Fig. 6 (6). Pollen profile at Tai Lake borehole (site No. 6). (after Xu *et al.* 1996).

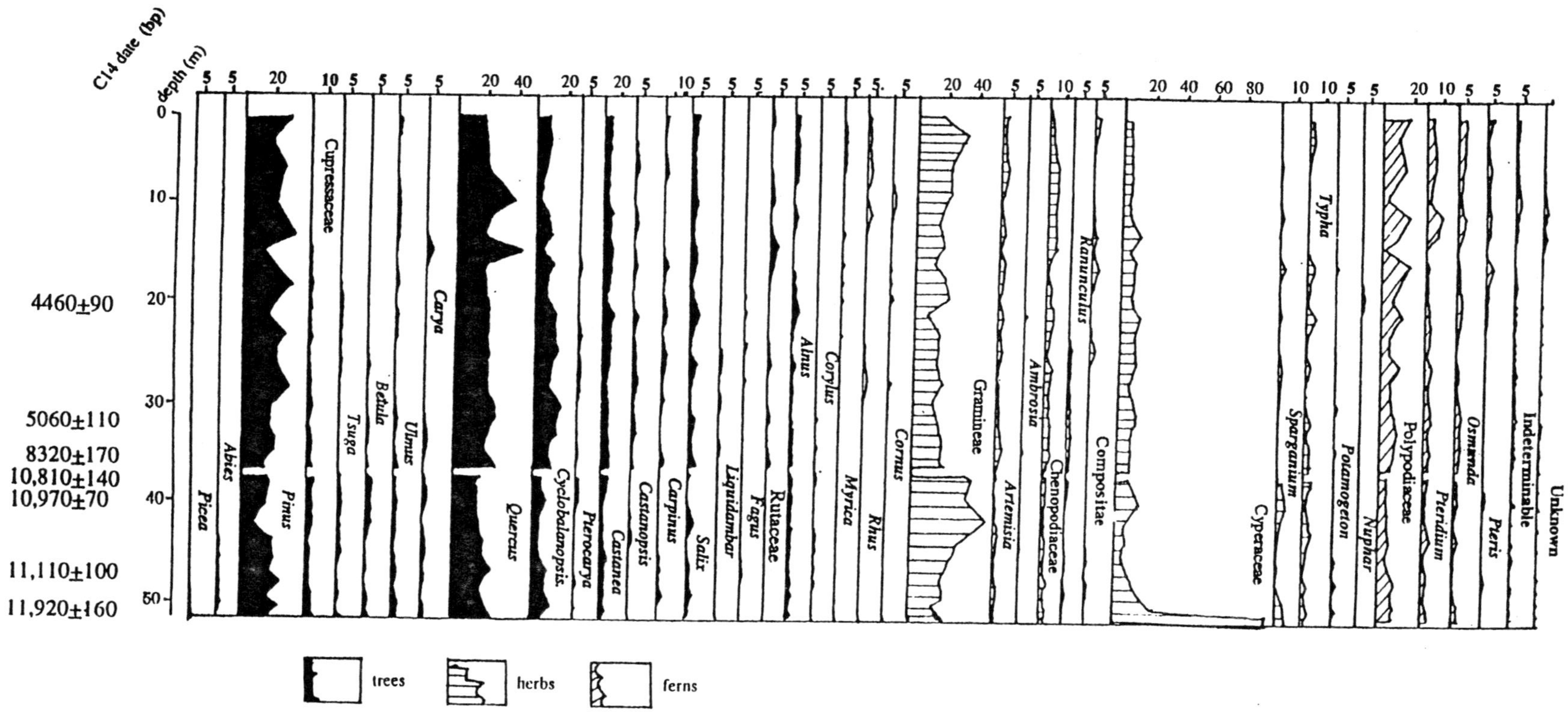

Fig. 6 (7). Pollen profile at Qidong borehole (site No. 7). (after Liu *et al.* 1992).

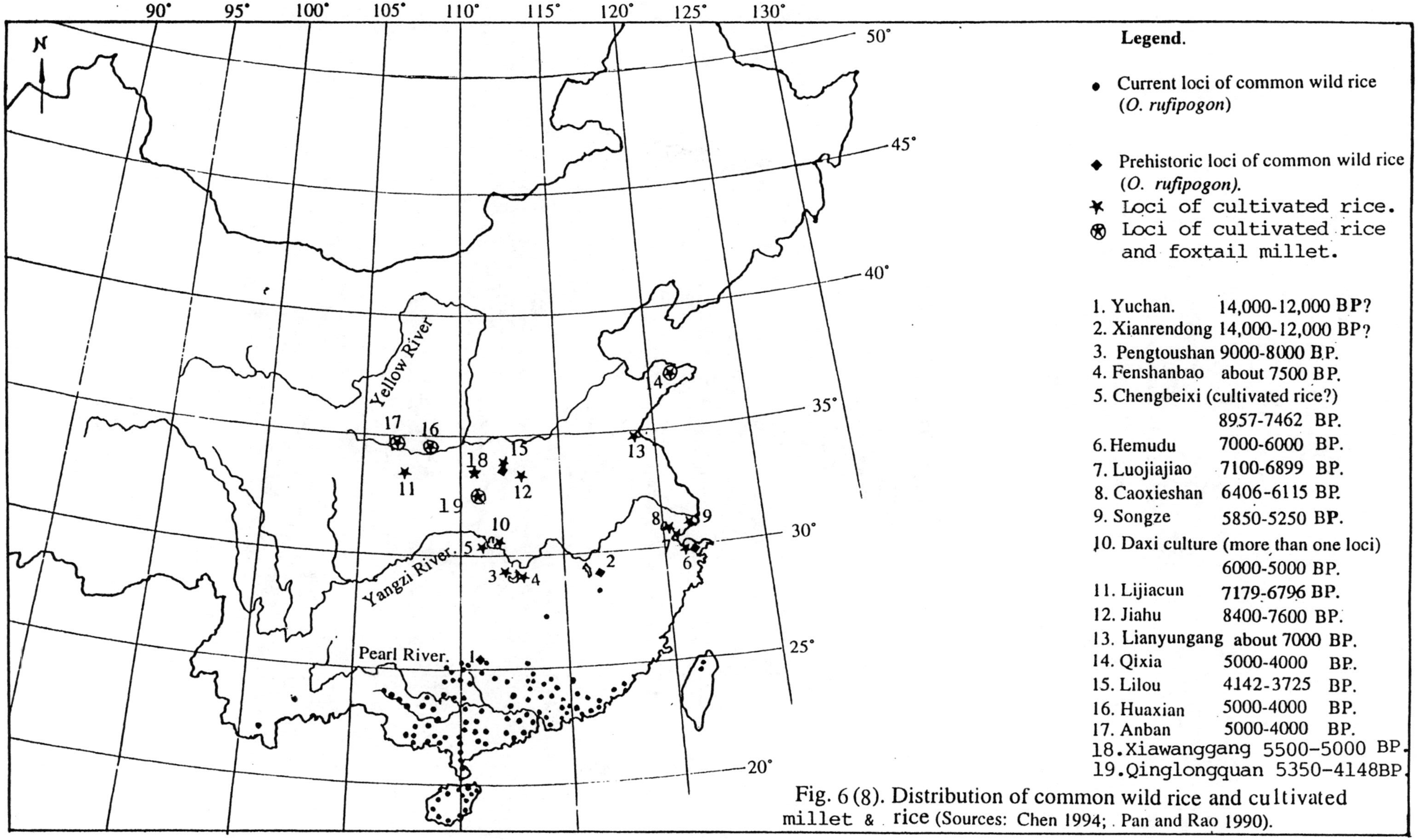

Fig. 6 (8). Distribution of common wild rice and cultivated millet & rice (Sources: Chen 1994; Pan and Rao 1990).

Fig. 6(9). Habitat of common wild rice (*Oryza rufipogon*).

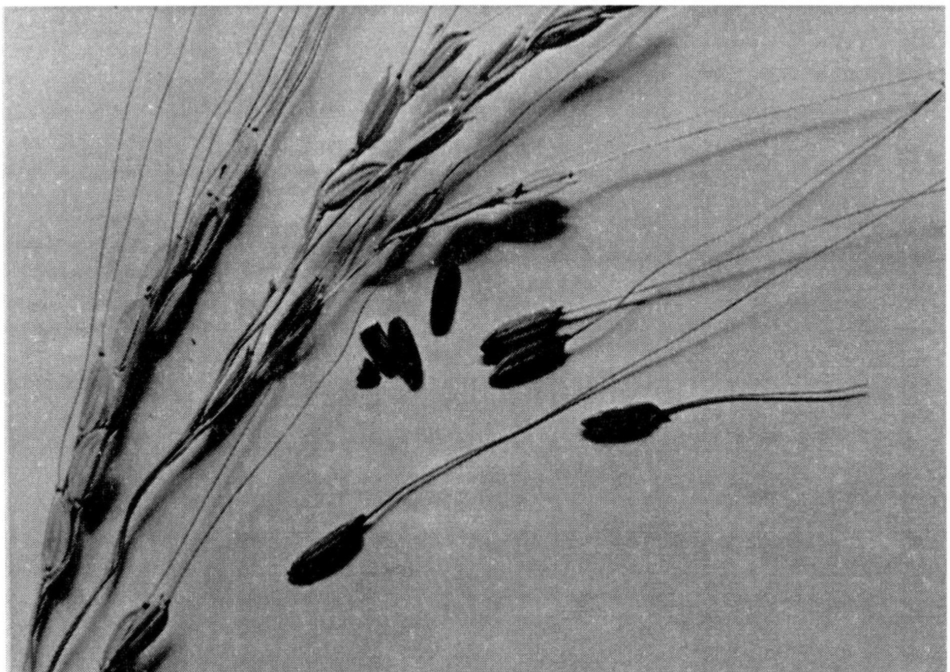

Fig. 6(10). Simultaneous flowering, ripening and shattering of common wild rice (*O. rufipogon*) on one plant.
(photo taken on 28th, October 1995 in Sanshui County, Guangdong province, China).

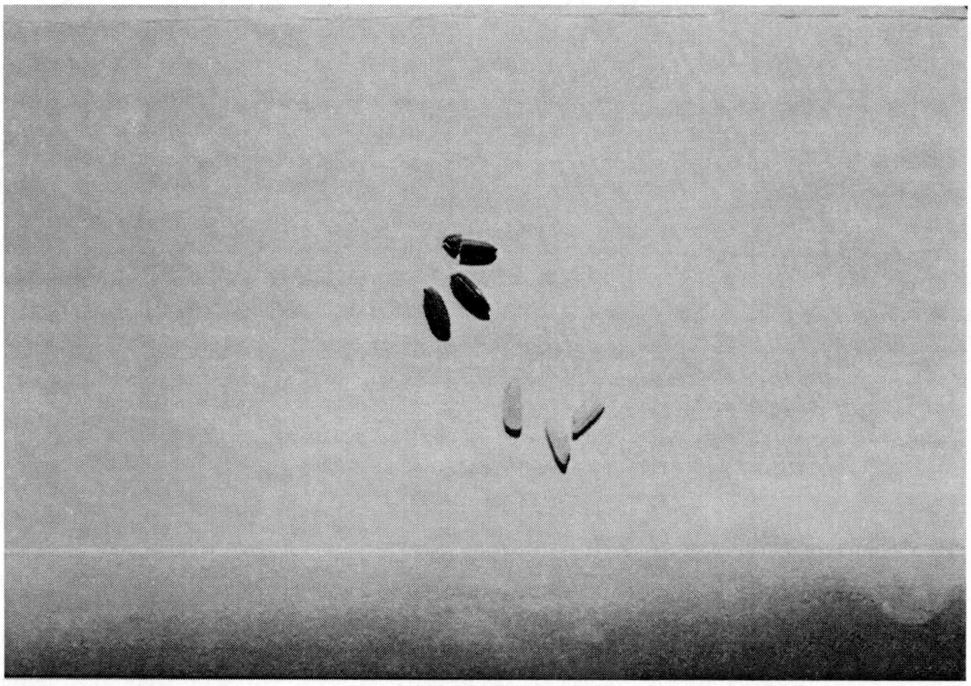

Fig. 6(11) Unhusked grains of common wild *rice (O. rufipogon)* (above) and domesticated rice (*O. sativa*) (below).

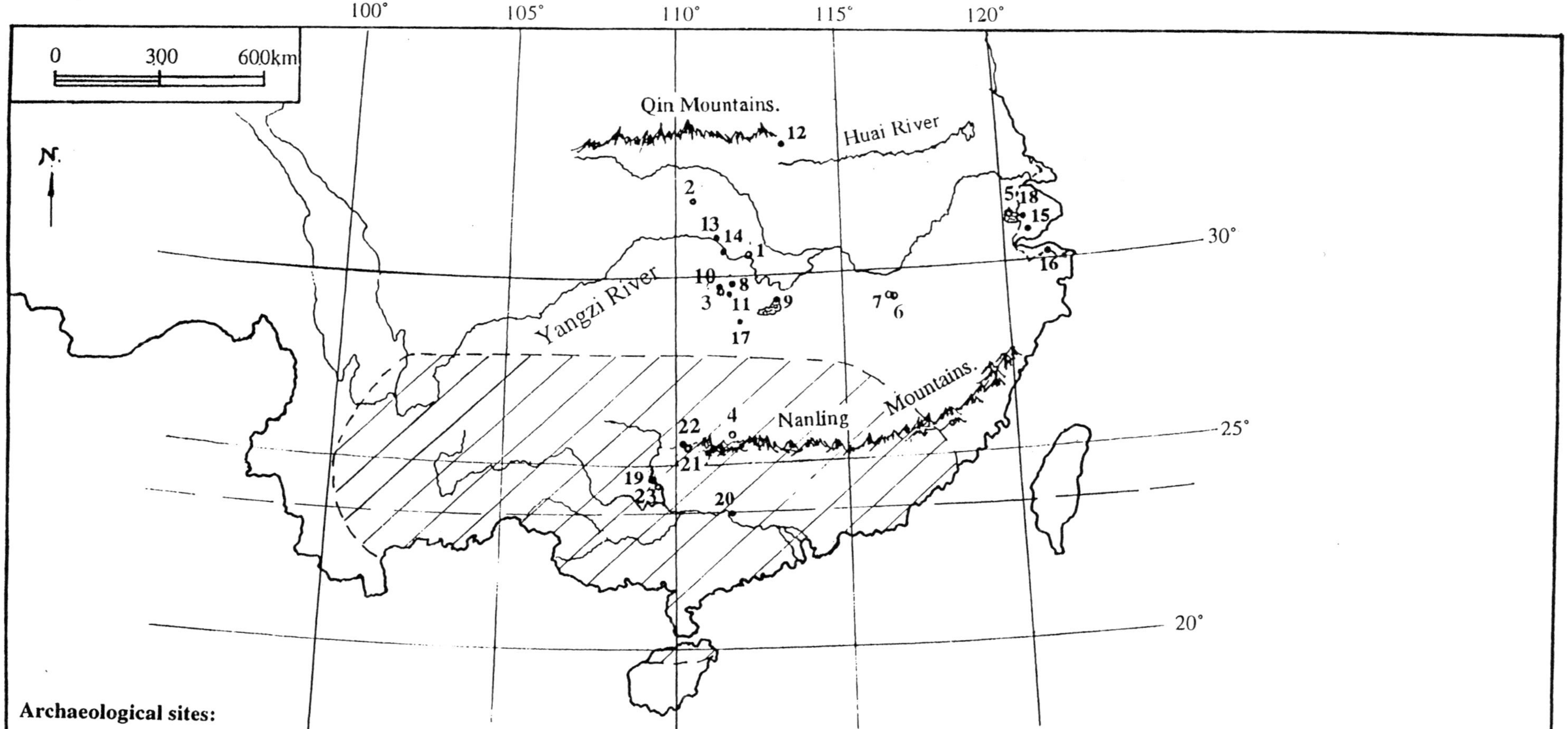

Archaeological sites:

1. Jiangling. 2. Zhangnao. 3. Yan'er. 4. Yuchan. 5. Sanshan. 6. Xianrendong. 7. Wangdong. 8. Pengtoushan/Bashidang 9. Fenshanbao. 10. Zaoshi. 11. Hujiawuchang. 12. Jiahu. 13. Chengbeixi. 14. Liulinxi. 15. Luojiajiao. 16. Hemudu. 17. Gaomiao. 18. Caoxieshan. 19. Dalongtan. 20. Huangyandong. 21. Miaoyan. 22. Zengpinyan. 23. Bailiandong.

Legends.

- o **Pleistocene sites.**
- • **Holocene sites.**
- Current distribution of wild rice.
- **The mid-Holocene coastline (6000 BP).**

Fig. 7 (I). Archaeological sites mentioned in the text.

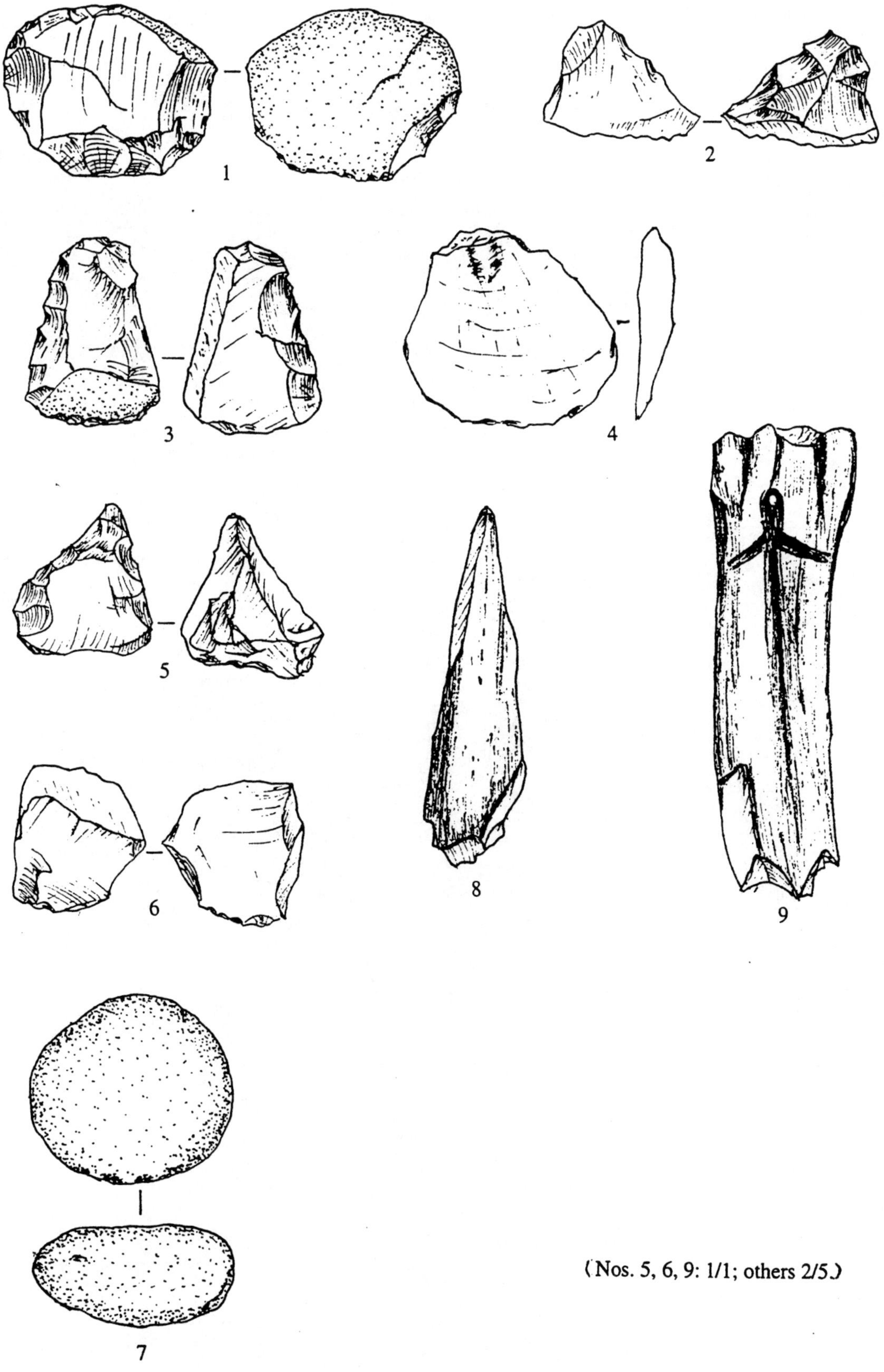

1. Chopper. 2.-6. Scrapers. 7. Hand stone. 8. Bone awl. 9. Bone handle.

Fig. 7(2) Artefacts of the Yan'er assemblage (after IA, Hunan Province and Shimeng Museum 1994).

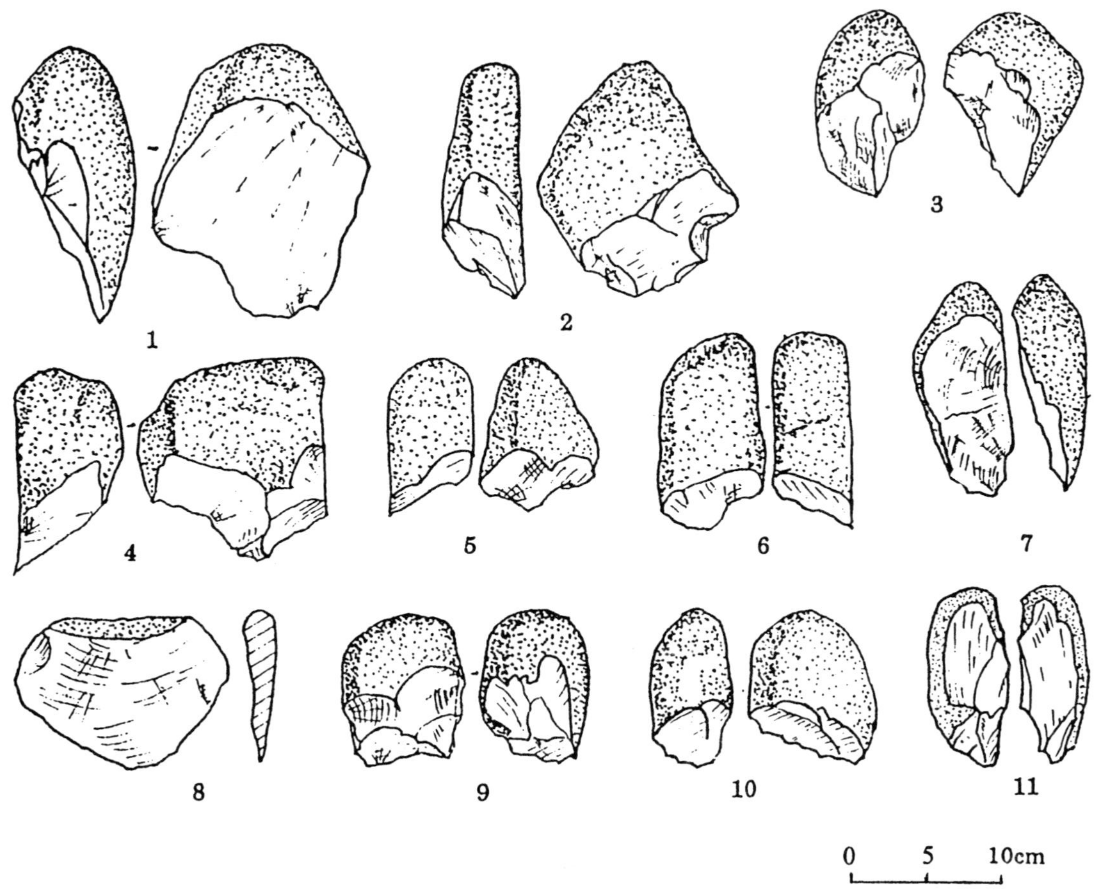

Fig. 7(3) Lithic implements from surface collections in Lixian county in the middle Yangzi valley. (after Xiang 1992).

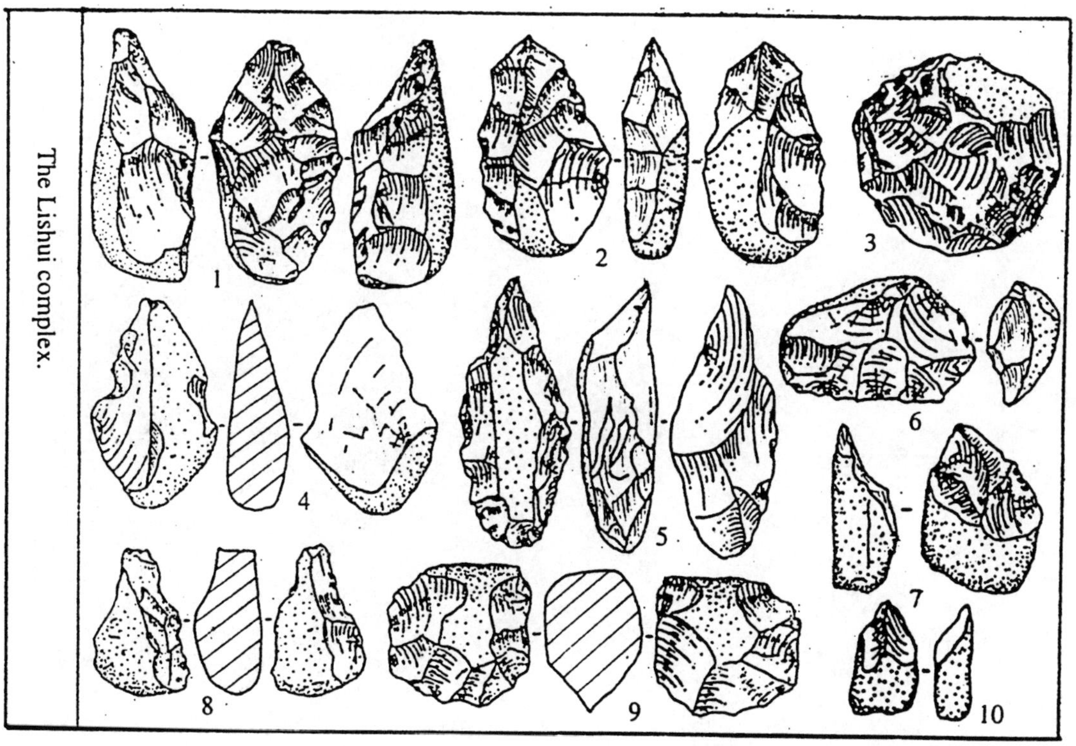

Fig. 7(4). The Lishui lithic complex found in the middle Yangzi Valley (after Yuan 1995).

Nos. 1,2,4,5,7,8,10: 'Triangle points'. Nos. 3 and 6: Choppers.
(not to scale. Typological terms after Yuan 1995).

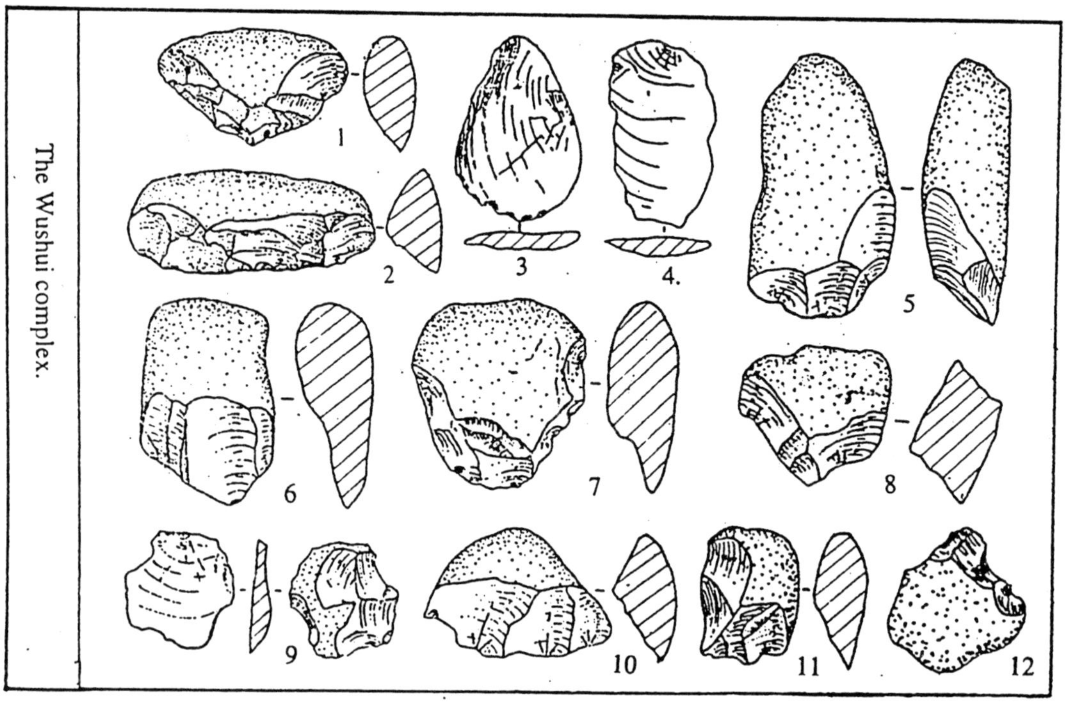

Fig. 7(5). The Wushui lithic complex found in the middle Yangzi Valley (after Yuan 1995).

Nos. 1, 5-8, 10-12. Chopping tools. No. 2. Knife? Nos. 3,4,9. Flake 'scrapers'.
(not to scale. Typological terms after Yuan 1995).

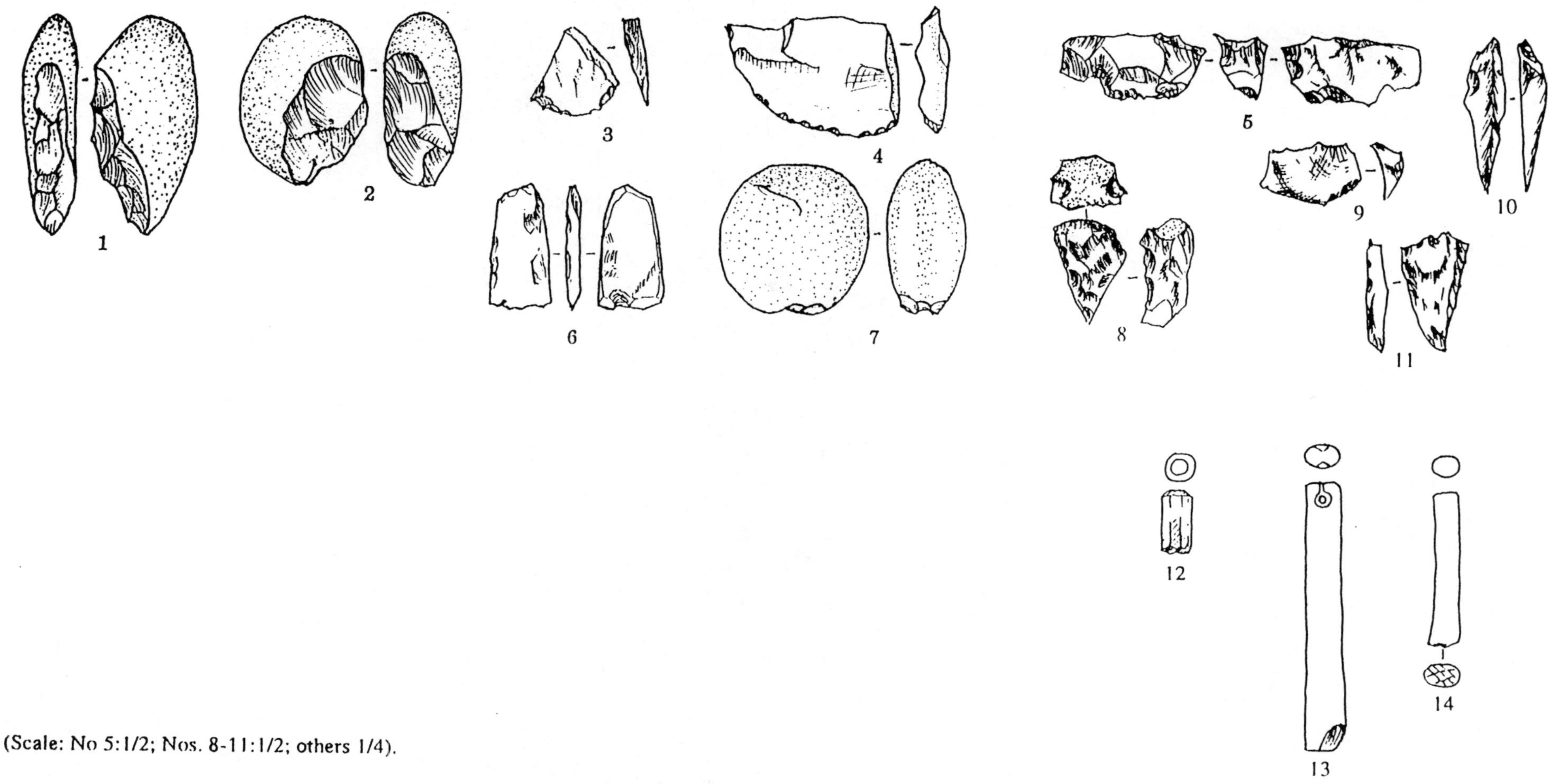

(Scale: No 5:1/2; Nos. 8-11:1/2; others 1/4).

1,2. Pebble tools. 6. 'Axe'? 7. Handstone. 12. Stone bead. 13,14. Stone pendants. The rest flakes.

Fig. 7 (6) Lithic implements of Pengtoushan (after IA, Hunan Province and Lixian Museum 1990)

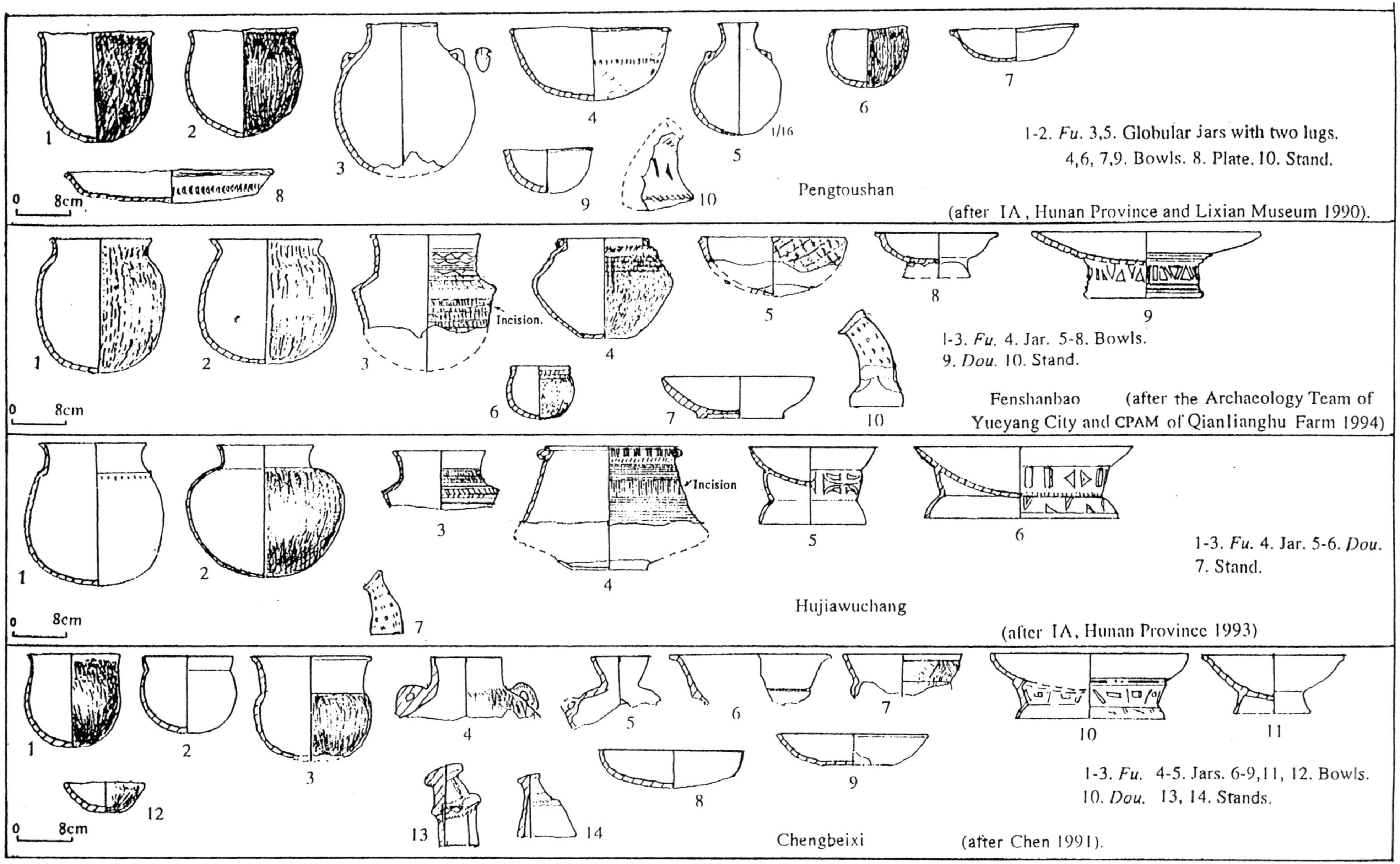

Fig. 7(7). Pottery comparison - the middle Yangzi valley.

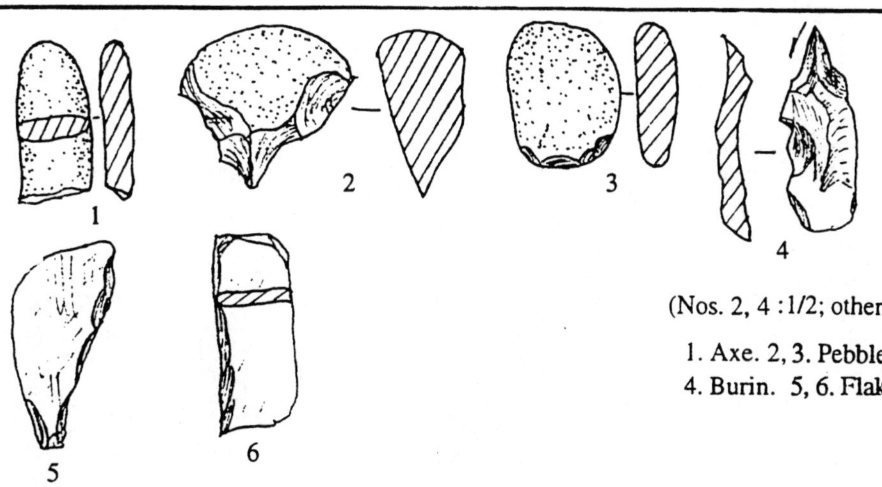

(Nos. 2, 4 :1/2; others 1/4).

1. Axe. 2, 3. Pebble tools.
4. Burin. 5, 6. Flakes.

Fig. 7 (8). Lithic implements of the Fenshanbao assemblage (after The Archaeology Team of Yueyang City and CPAM of Qianlianghu Farm 1994).

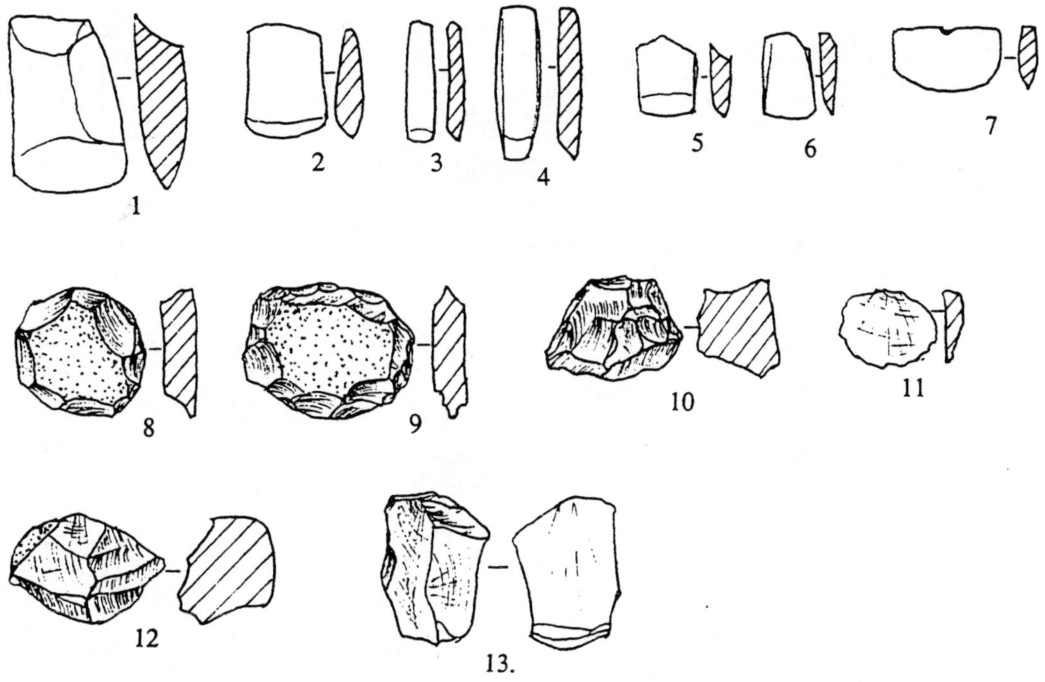

(Scale: Nos. 1-7:1/6; Nos. 8-11:1/4; Nos. 12, 13:1/2)

1, 2. Axes. 3,4. Chisels. 5. 6. Adzes. 7. Knife. 8-10. Discoids 11, 13. Flakes. 12. Flake core.

Fig. 7 (9). Lithic implements of the Hujiawuchang assemblage (after IA, Hunan province 1993).

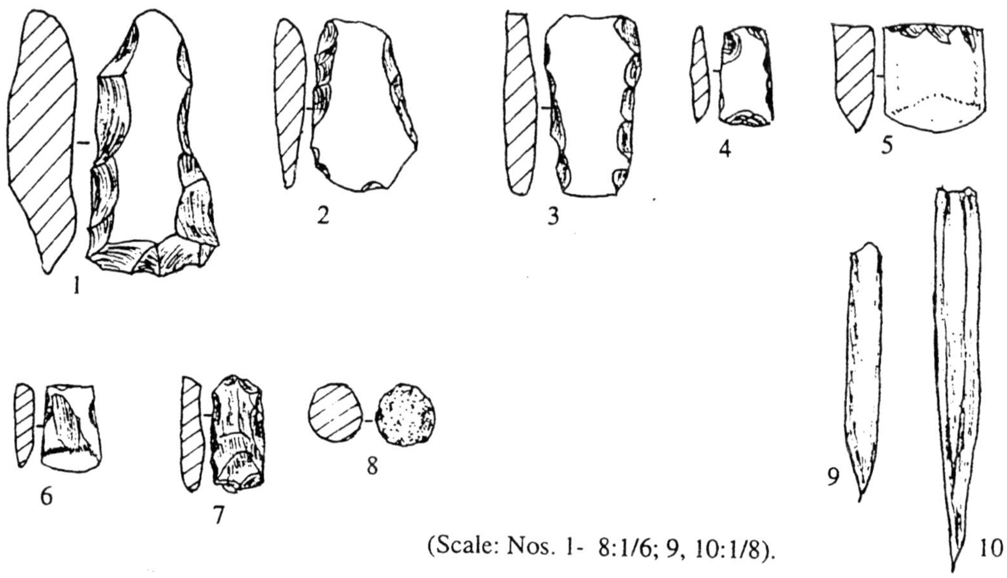

(Scale: Nos. 1- 8:1/6; 9, 10:1/8).

1-5. Axes. 6. Adze. 7. Chisel. 8. Stone ball. 9, 10. Bone awls. (1-8: Stone implements).

Fig. 7(10). Artefacts of the Chengbeixi assemblage (after Chen 1991).

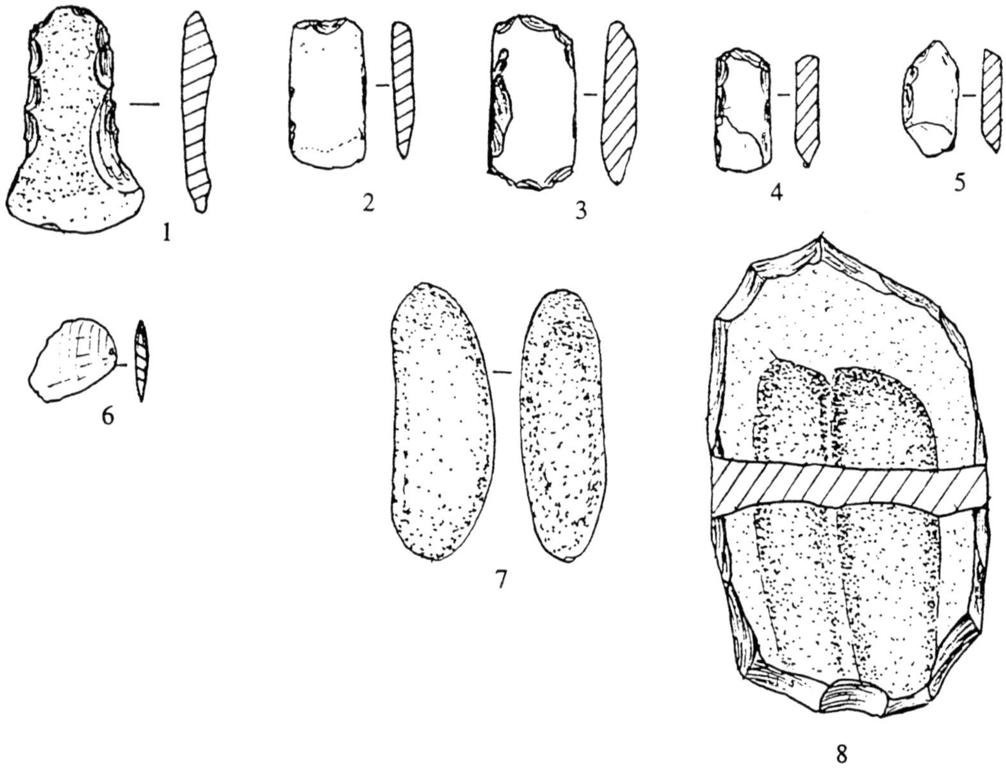

(Scale: all 1/5).

1-3. Axes. 4. Adze. 5. Chisel. 6. Flake. 7. Roller. 8. Grinding slab.

Fig. 7(11). Lithic implements of the Liulinxi assemblage
(after IA, Hubei Province 1994)

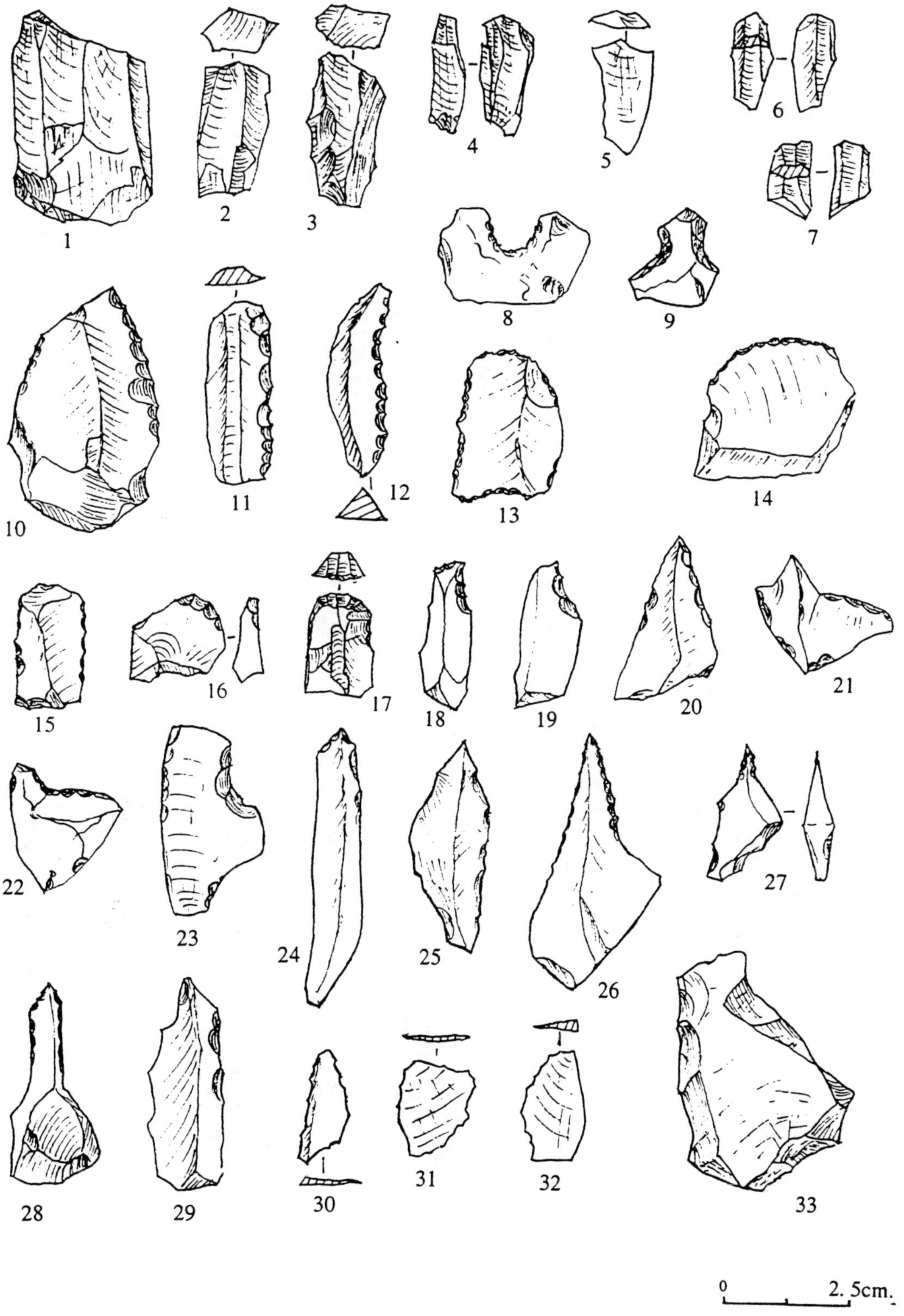

1-4. Flake-blade cores. 5-7. Flake & blades. 8-19."Scrapers." 20-25."Points."
26-29."Awls." 30-32. Used flakes. 33."Chopper."

Fig. 7(12). Lithic implements of the Sanshan assemblage (after Chen *et al.* 1987)

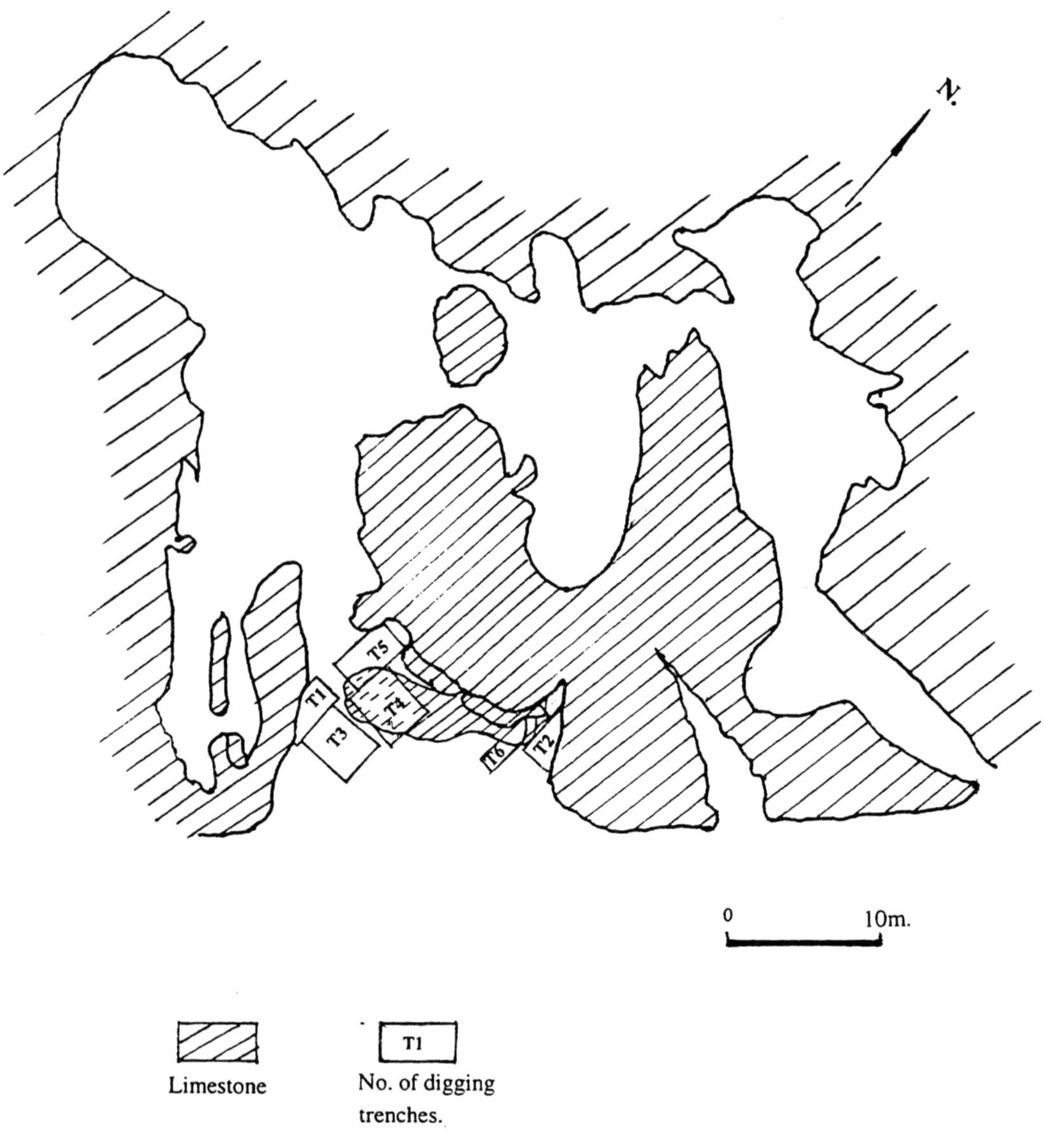

Fig. 7(13). Xianrendong cave (after Jiangxi Museum 1976).

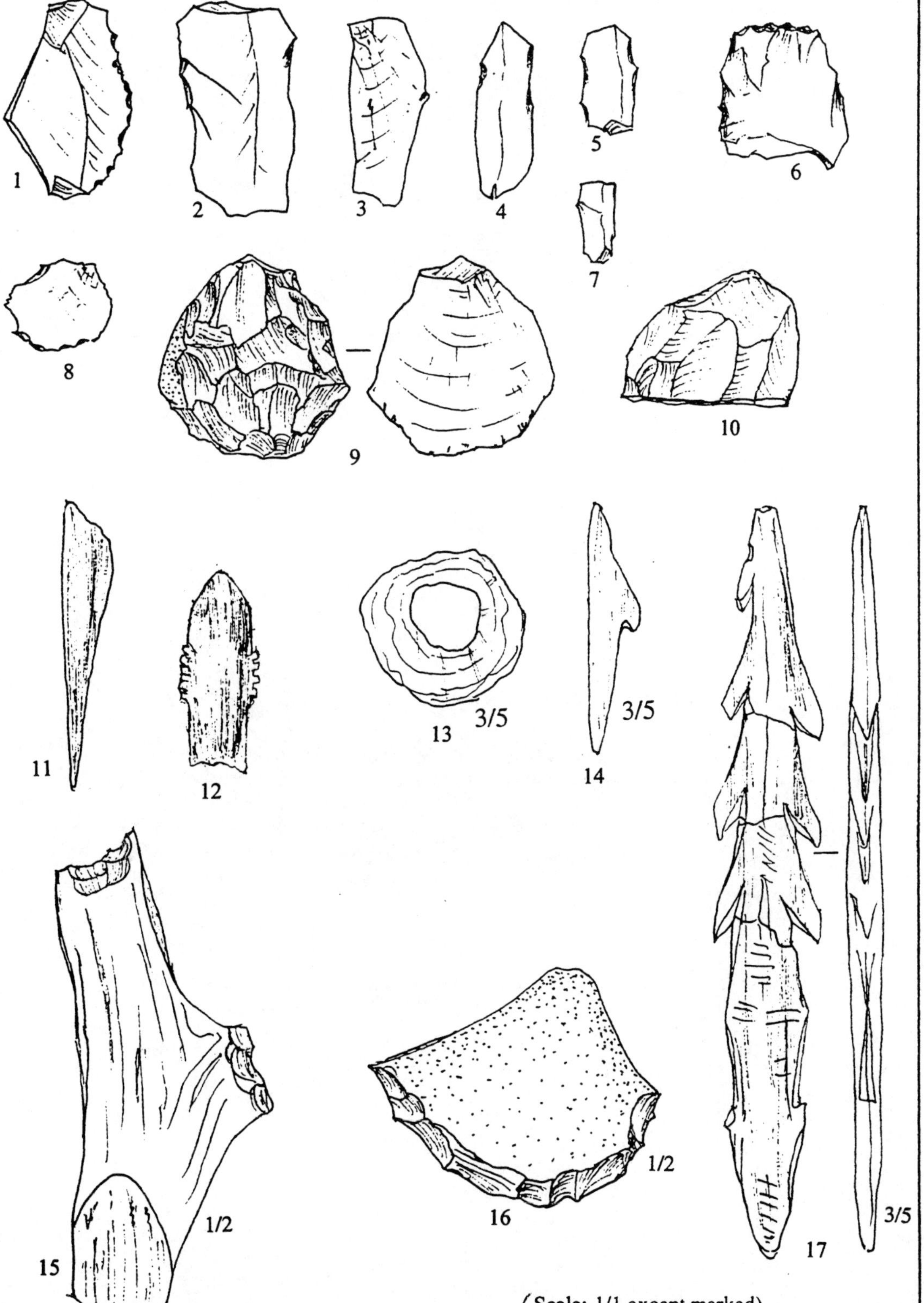

(Scale: 1/1 except marked)
(Nos. 11-15,17: are organic tools; others stone implements)

1-8: Flakes and blade-like flakes. 9. "Scraper". 10. Core. 11. Bone awl. 12. Bone arrowhead. 13. Perforated shell. 14, 17: Antler harpoons. 15. Antler axe. 16. Pebble chopper.

(Nos. 9, 14-16: from Wangdong; others from Xianrendong).

Fig. 7 (14). Pre-ceramic artefacts from Xianrendong and Wangdong.
(Nos. 1-8: author's drawings; 9, 14-16 after Tang 1997;
others after Jiangxi Museum 1976).

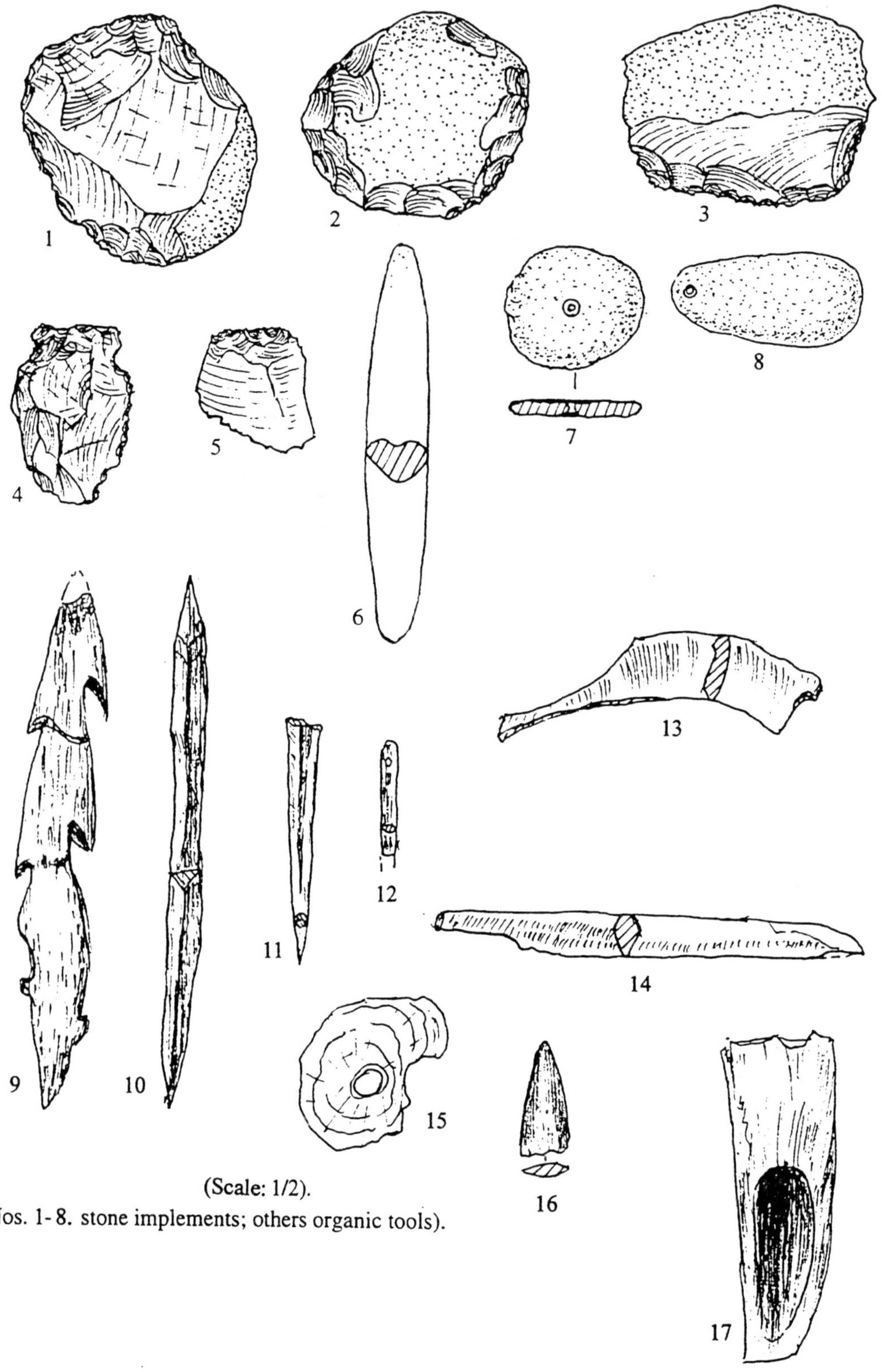

1-3. Pebble choppers. 4, 5. Flake "scrapers." 6. Polished "awl." 7, 8. Perforated pebbles. 9. Bone harpoon. 10, 11: Bone awls. 12. Bone needle. 13. Tooth knife. 14. Bone knife. 15. Perforated shell. 16. Bone arrowhead. 17. Antler chisel.

Fig. 7(15) Artefacts of Ceramic Phase I from Xianrendong.
(Nos. 1-7, 9-14: after CPAM of Jiangxi Province 1963; others after Jiangxi Museum 1976)

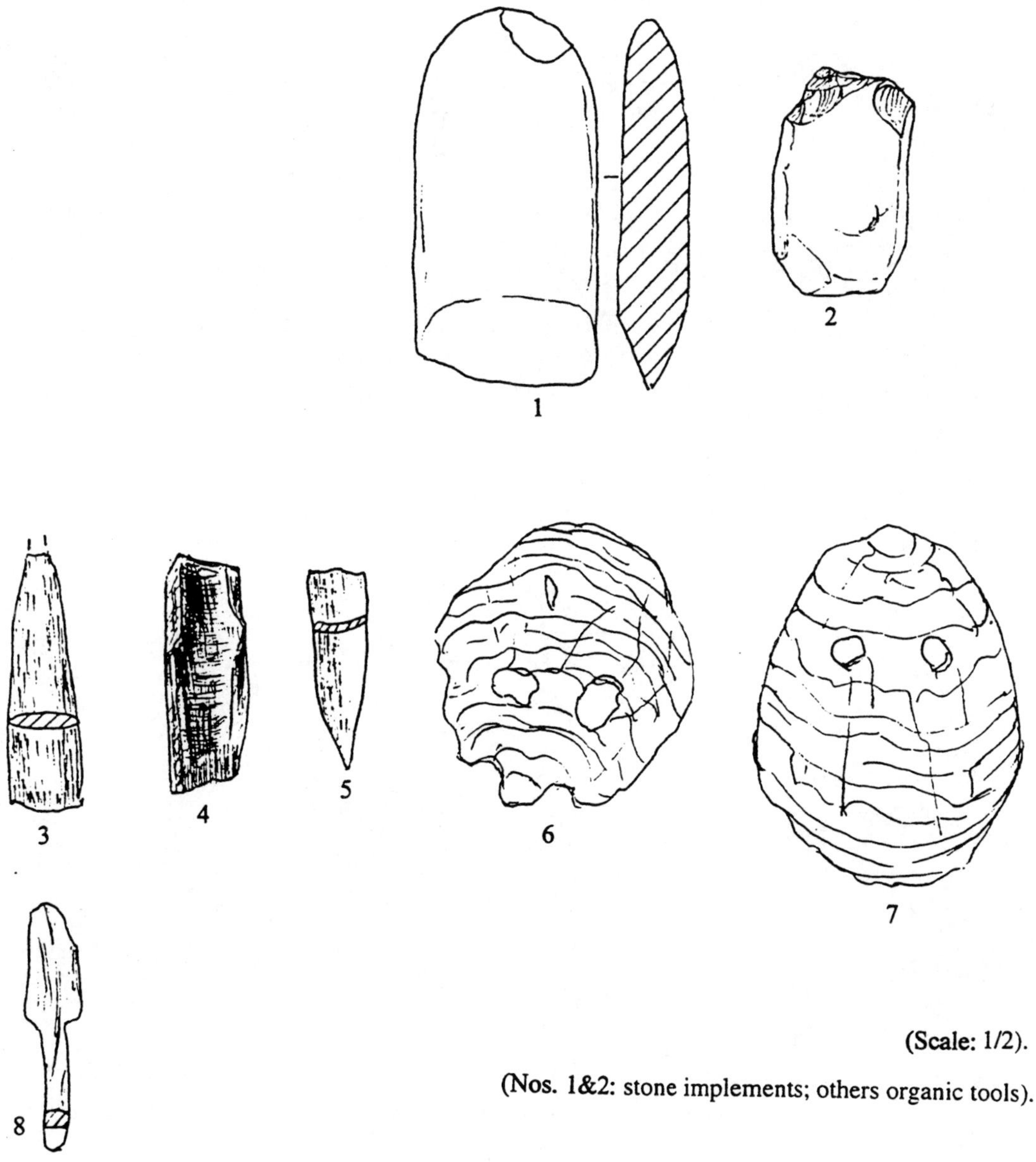

(Scale: 1/2).

(Nos. 1&2: stone implements; others organic tools).

(Nos. 1&2: Polished adzes.
3. Bone arrowhead. 4. Bone chisel. 5. Bone awl. 6, 7. Perforated shells. 8. Shell arrowhead.

Fig. 7 (16) Artefacts of Ceramic Phase II from Xianrendong.
(Nos. 1&8: after CPAM of Jiangxi Province 1963; others after Jiangxi Museum 1976)

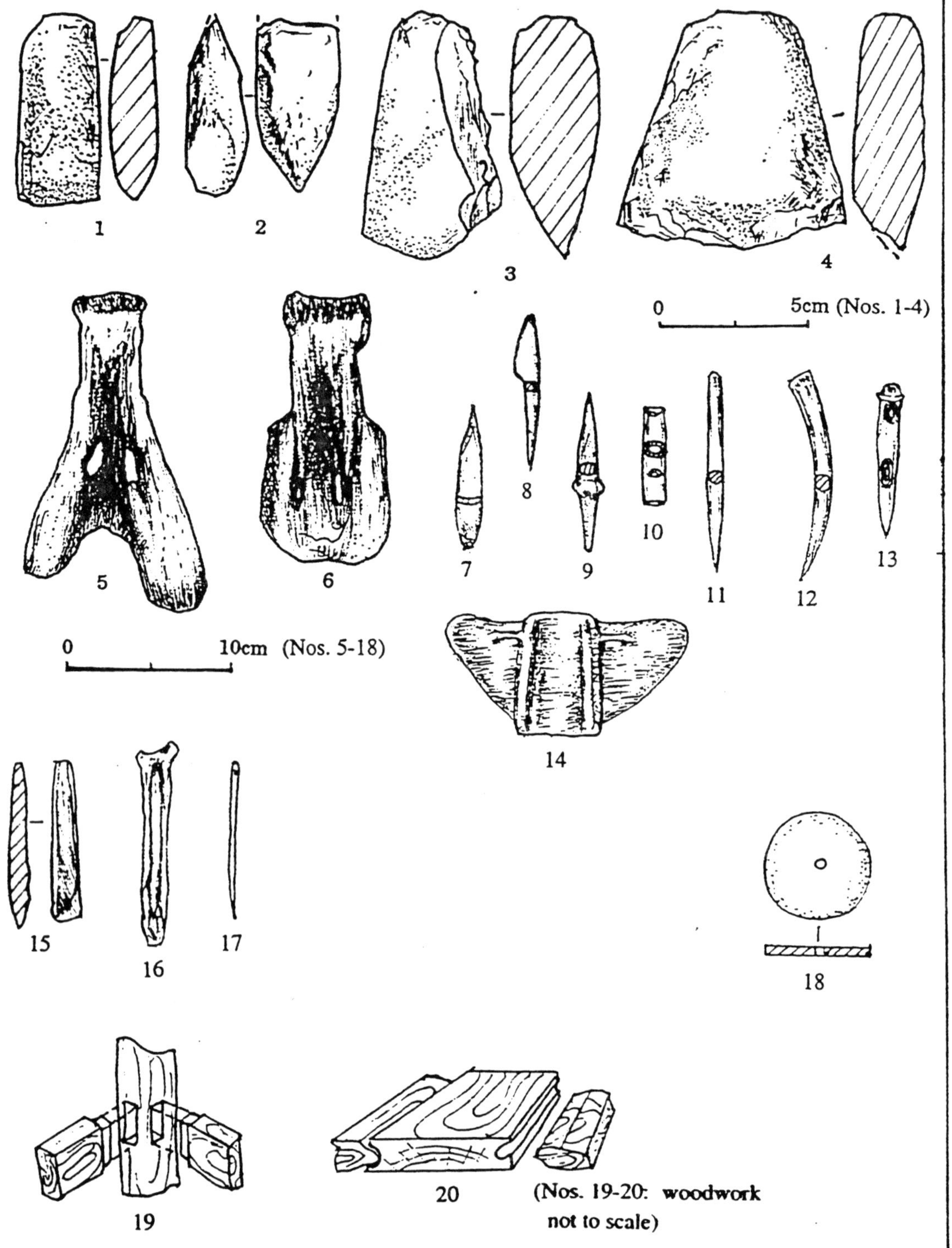

1, 2. Stone chisels. 3, 4. Polished stone axes. 5, 6. Bone spades. 7-9. Bone arrowheads. 10. Bone whistle. 11, 12. Bone awls. 13, Bone shuttle. 14. Wooden implement ('butterfly'). 15, 16. Bone chisels. 17. Bone needle. 18. Spindle whorl of fired clay.

Fig. 7 (17). Tools and woodwork of the Hemudu assemblage (layer 4)
(after CPAM, Zhejiang Province and Zhejiang Museum 1978)

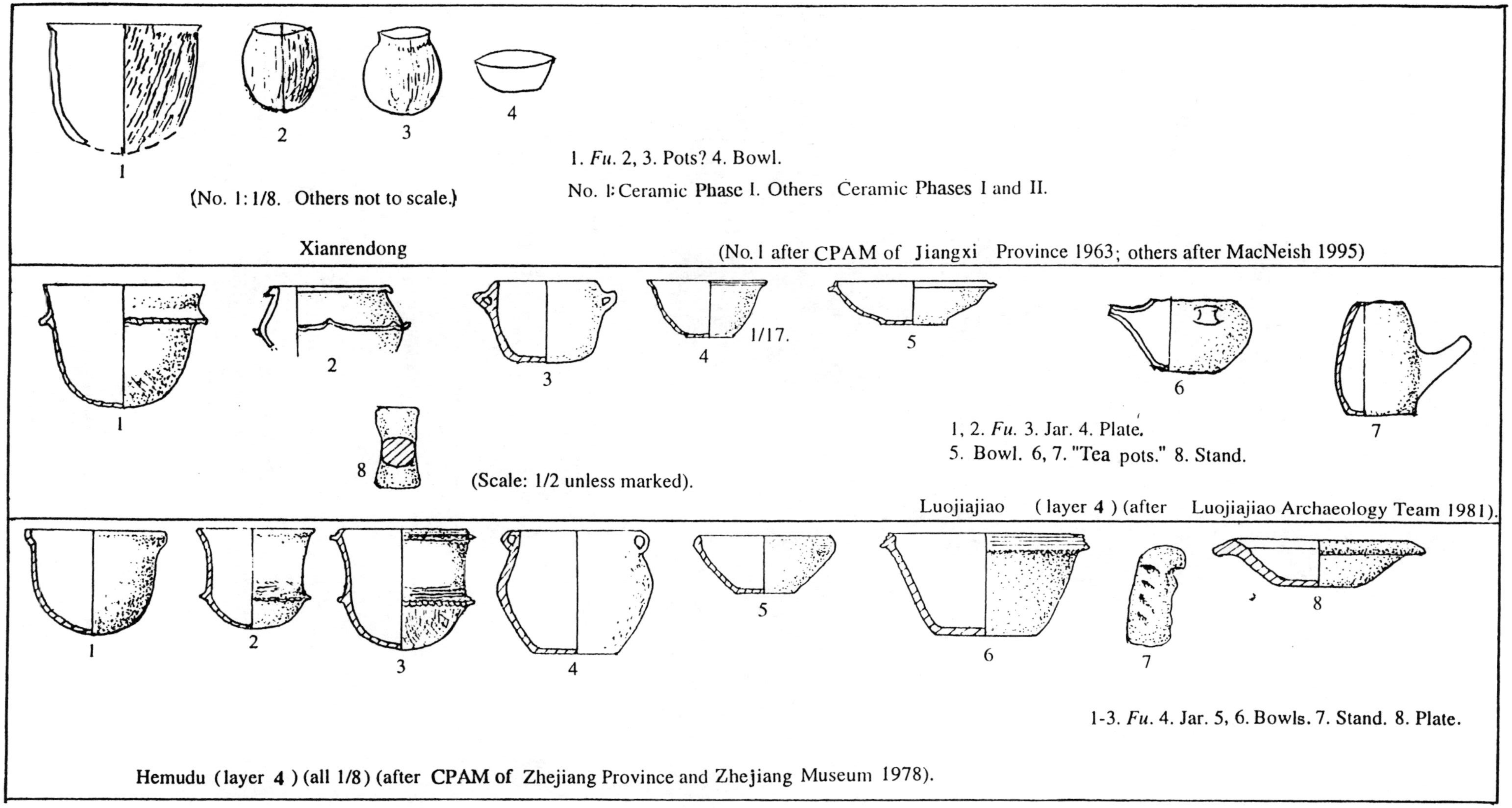

Fig. 7 (18). Pottery of the Lower Yangzi valley.

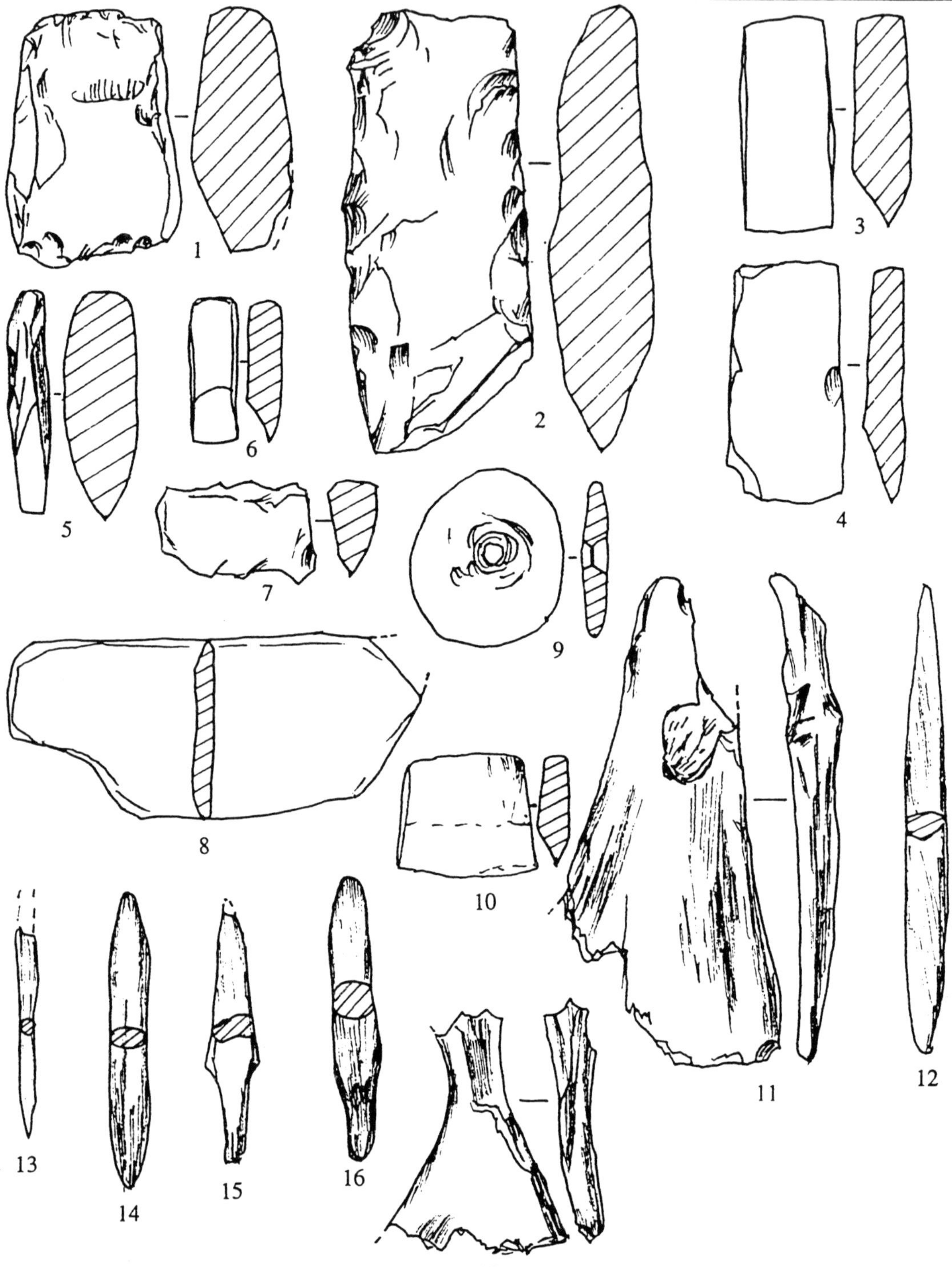

(Scale: not reported).

(Nos. 1-10: stone implements; others organic tools)

1, 2. Axes. 3, 4. 10. Adzes. 5, 6. Chisels. 7. Flake. 8. Knife. 9. Spindle whorl.
11, 17. Bone spades. 12.-14. Bone drills. 15, 16. Bone arrowheads.

Fig. 7(19). Toolkit of the Luojiajiao assemblage (layer 4)
(after Luojiajiao Archaeology Team 1981)

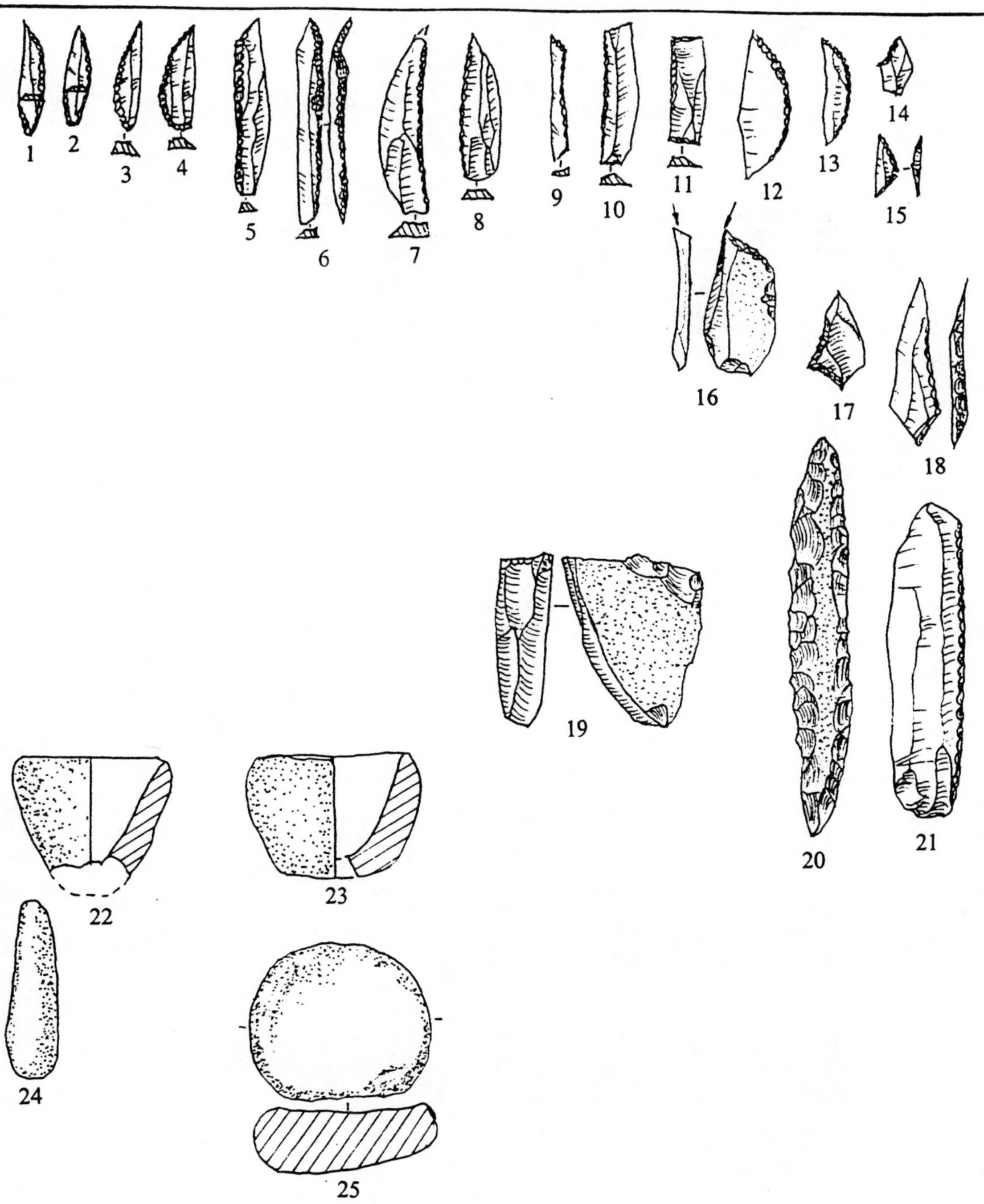

(Scale: No. 19. 1/3; No. 25. 1/4; others 2/3).
(Nos. 12-15, 18-20, 23: Natufian implements; others Kebaran and Geometric Kebaran implements)

Microliths. 16,17: Burins. 18: Point. 19: Single platform blade cores. 20: Natufian pick. 21: Sickle blade. 22-23: Mortar. 24: Pestle. 25: Grinding slab.

Fig. 9 (1). Lithic implements of the Epipalaeolithic in the Middle East. (No. 19. after Garrard *et al.* 1985; others after Bar-Yosef and Belfer-Cohen 1989).

1

2

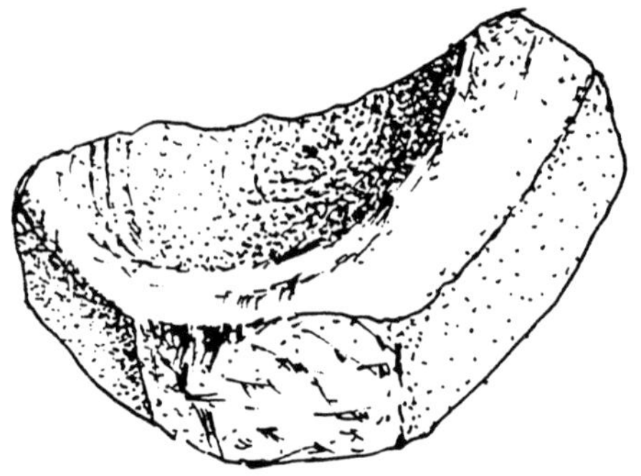

3

(Scale: 1/5)

1. Contemporary grinding slab with grooved use-wear in West Australia.
2. Grinding slab with grooved use-wear, dated to 700 years BP from Central Australia.
3. Contemporary mortar from Central Australia.

Fig. 9 (2). Grinding slabs found in Australia (all fragments only).

(No. 1 after Cane 1984; No. 2 after Smith 1986; No. 3. after Smith 1985).

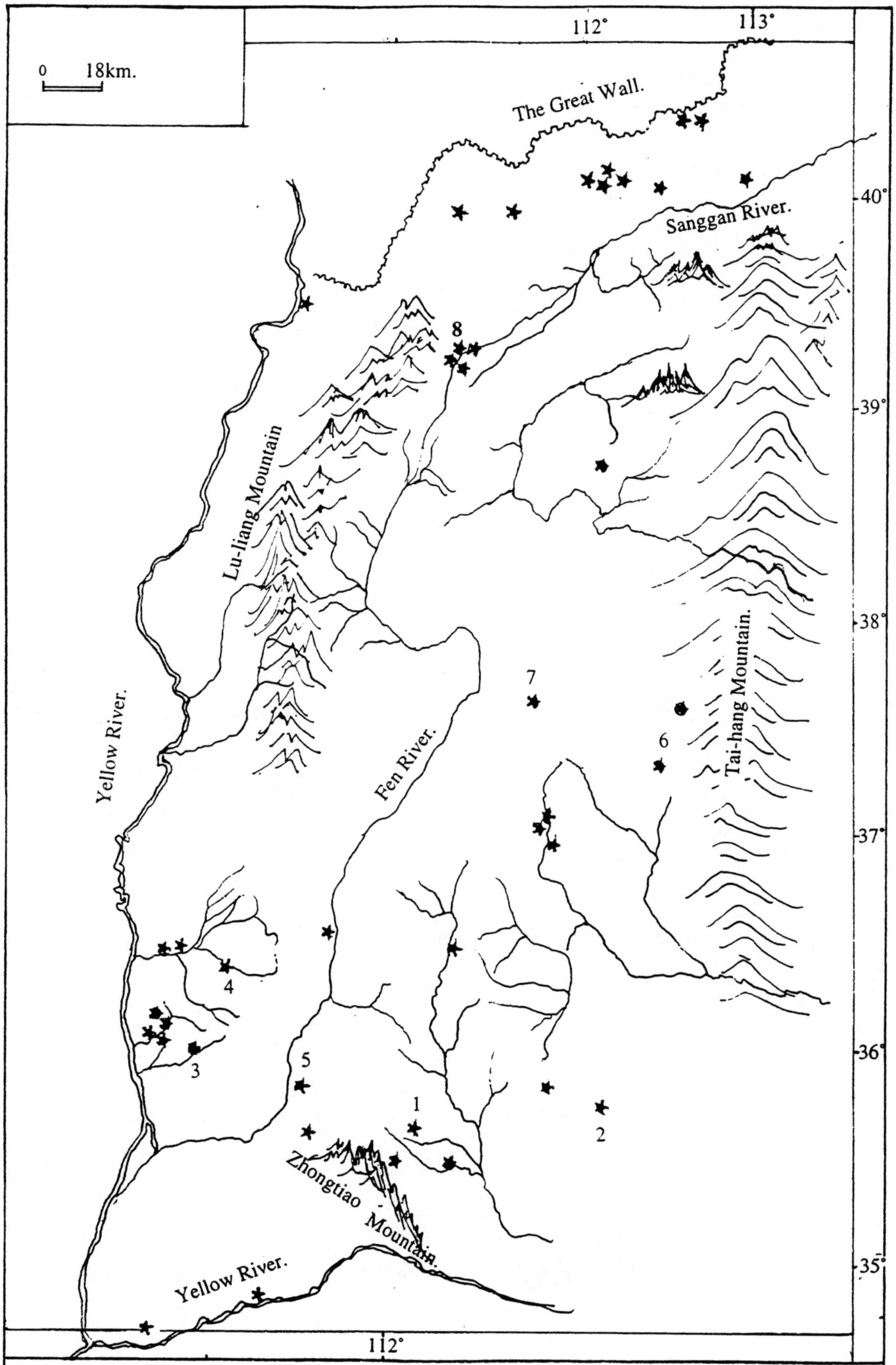

★ Upper Palaeolithic sites.

1. Xiachuan. 2. Tashuihe. 3. Xueguan. 4. Shizitan. 5. Dingcun. 6. Dangcheng. 7. Dafa. 8. Shiyu.

Fig. 9(3). Upper Palaeolithic sites found in the Shanxi Basin (after Chen 1993).

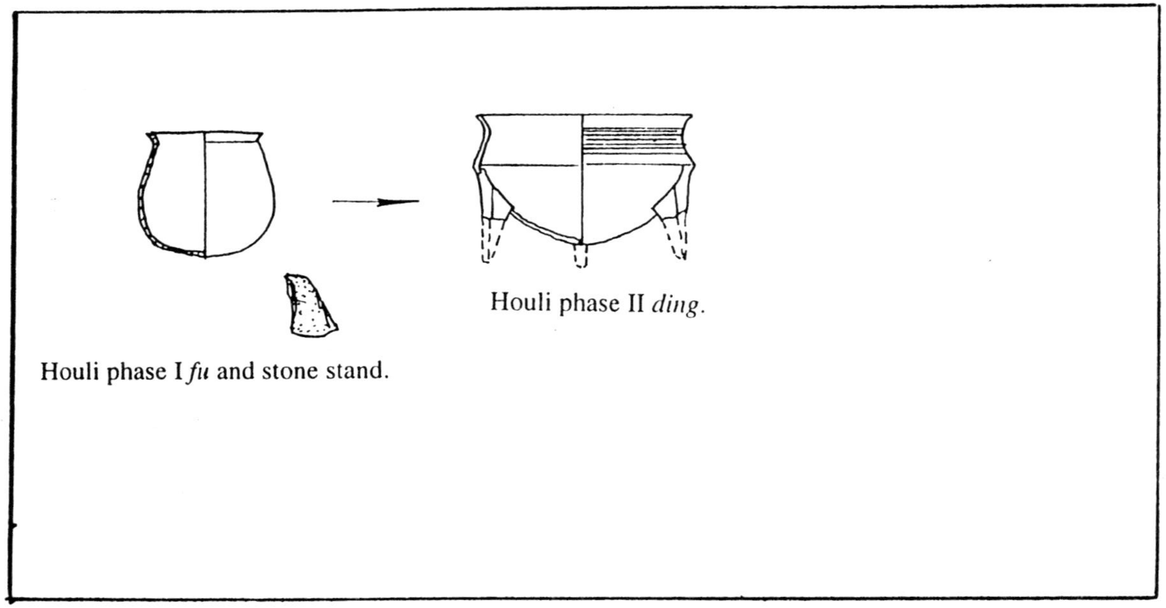

Fig. 10 (1) The development from *fu* to *ding* (not to scale).

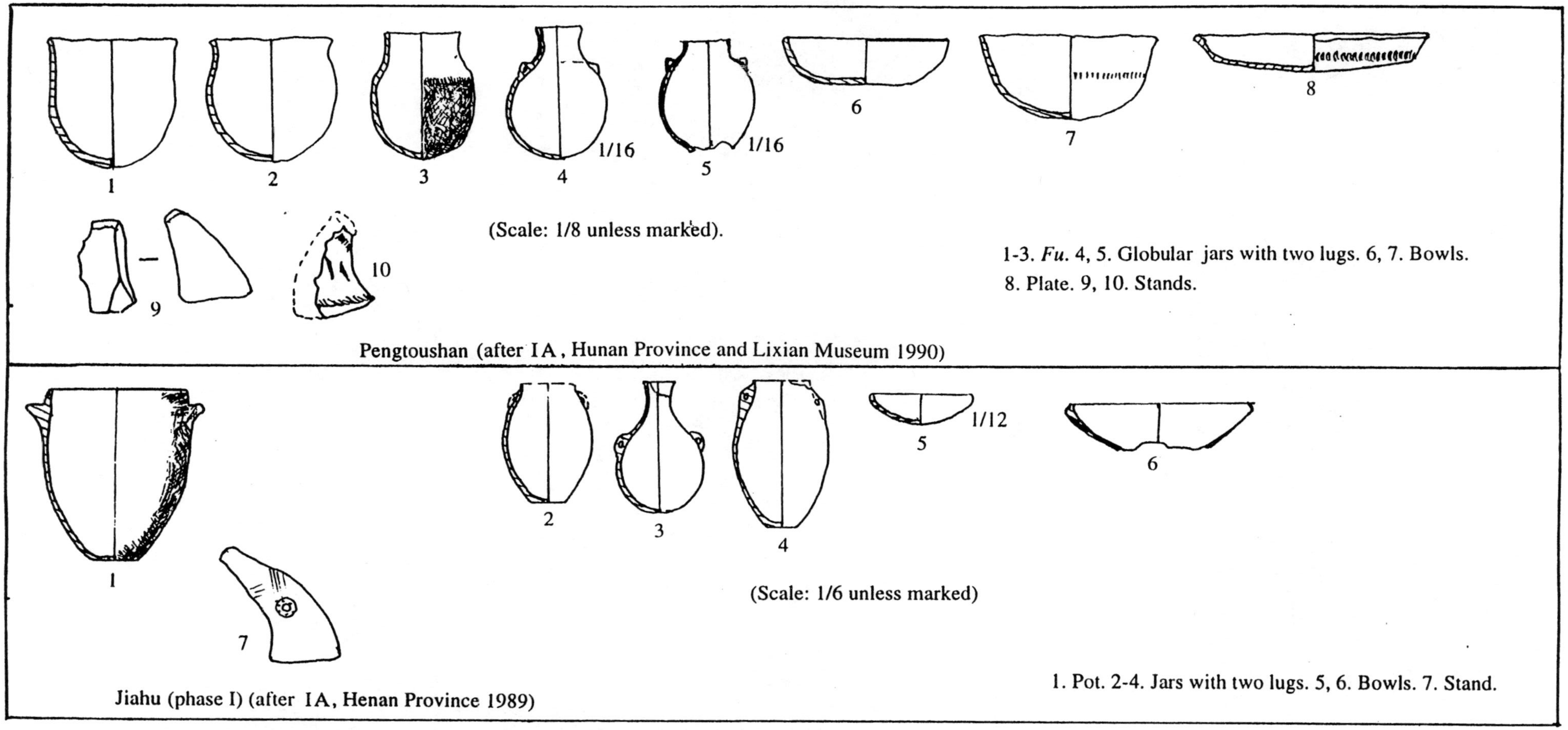

Fig. 10 (2) Pottery comparison - Pengtoushan and Jiahu (the lowest layer).

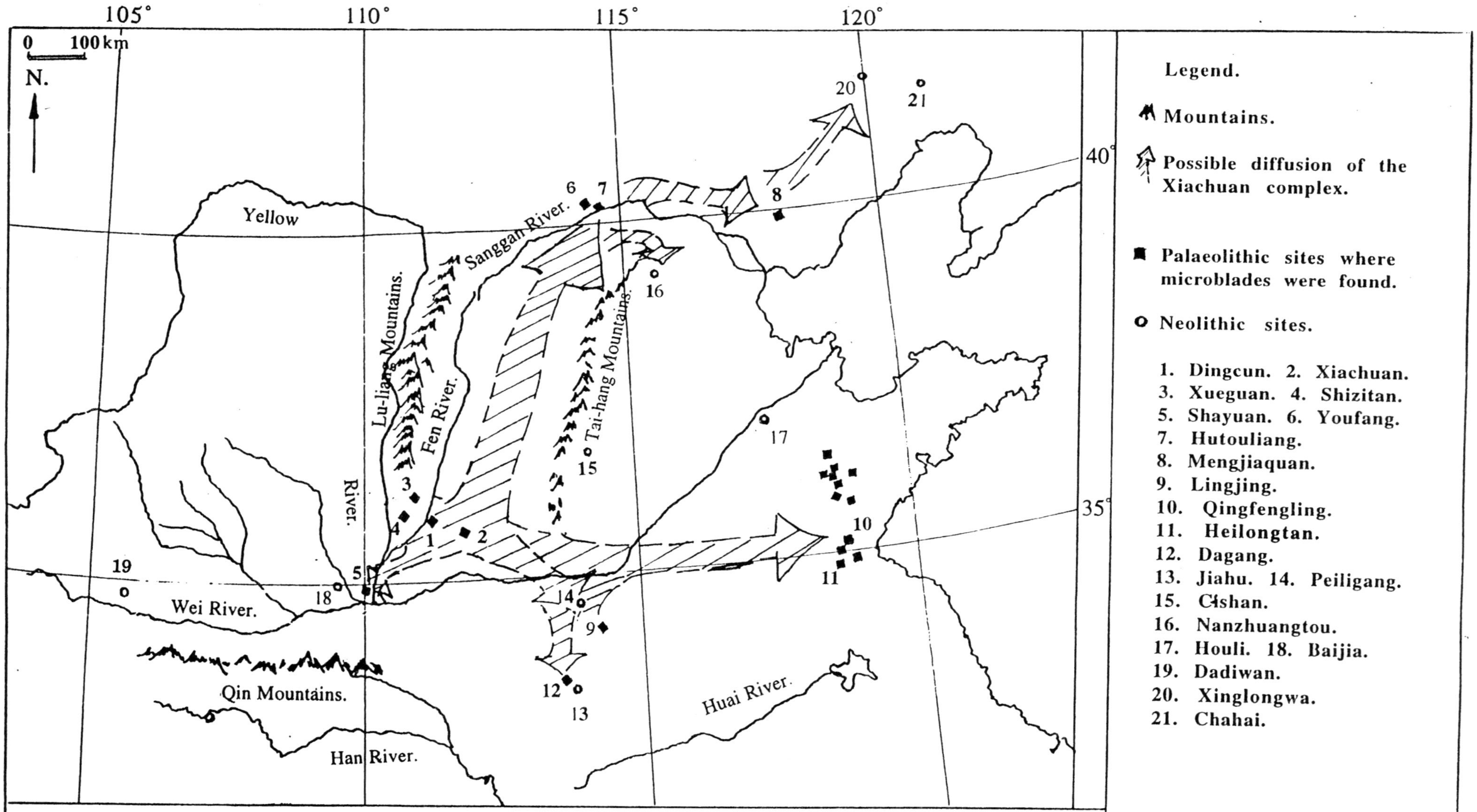

Fig. 11(1). Possible diffusion of the Xiachuan complex.

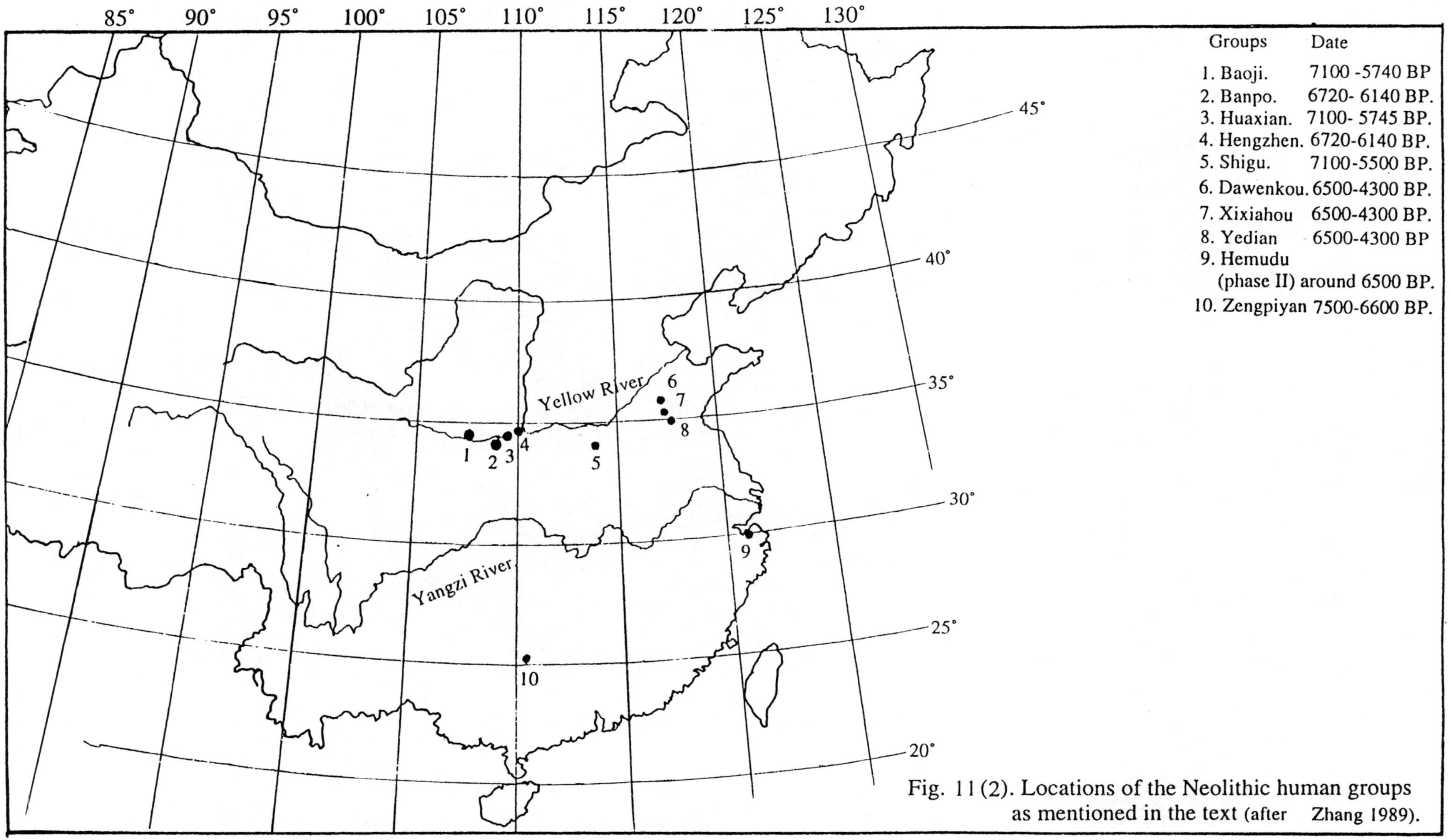

Fig. 11 (2). Locations of the Neolithic human groups as mentioned in the text (after Zhang 1989).

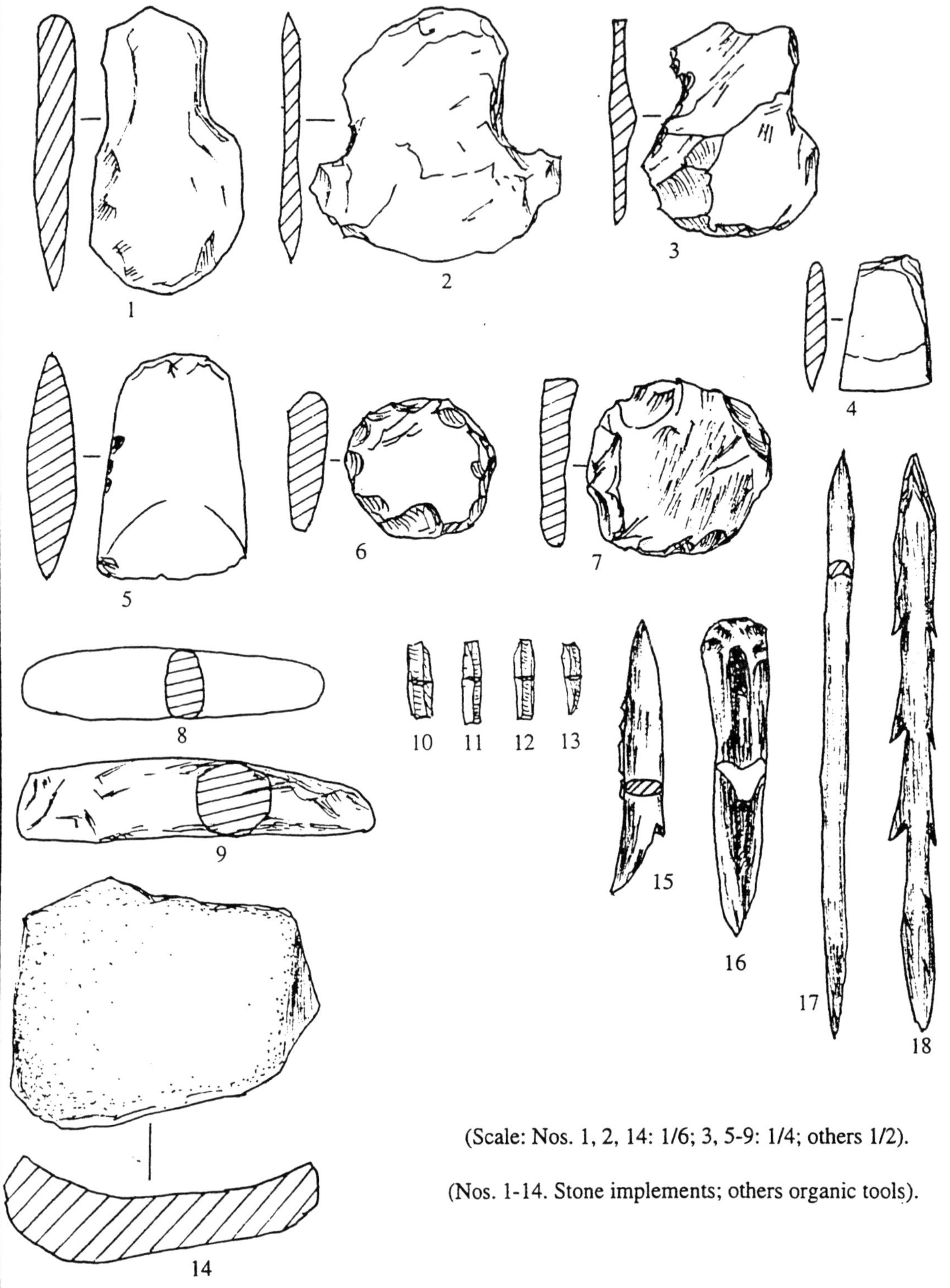

1. Spade. 2, 3. Notched hoes. 4. Adze. 5. Axe. 6, 7. Discoid choppers. 8, 9: Rollers.
10-13. Microblades. 14. Grinding slab. 15. Bone harpoon inserted with microblades as cutting edge.
16. Bone awl. 17. Bone point. 18. Bone harpoon.

Fig. 11 (3). Toolkit from the Xinglongwa assemblage
(after Neimeng Archaeology Team CASS 1997)

Appendix I.

A list of Latin and English names of plants commonly found in China.

Abies sp. Fir tree.
Acer sp. Maples.
Adiantum sp. Maidenhair fern.
Alnus sp. Alder. Genus of deciduous trees which are common in wet places. Often found in temperate zone.
Ambrosia sp. Genus of weedy herbs.
Antidesma sp. Genus of shrubs or small trees in the Old World (sub)tropics, esp. in Asia.
Apocynaceae Family mostly of trees, climbers or shrubs, mainly found in tropics with a few temperate outliers.
Artemisia sp. Genus of perennial (or, rarely, annual) herbs or low shrubs.
Betula sp. Birches. Genus of fast-growing trees. The fruit is a small, winged nut.
Carpinus sp. Hornbeams. Genus of monoecious, catkin-bearing trees in northern temperate zones and Central America.
Carya sp. Hickory. Genus of trees. The fruit is a drupe with an edible seed (pecan). 25 species, mainly found in E. Asia and N. America.
Castanea sp. Genus of trees which yield useful timber and fruit. *C. sativa* is the sweet chestnut. 12 species. Mainly found in northern temperate regions.
Castanopsis sp. Genus of medium trees, some fruits edible. 120 species, occurring in tropical and subtropical Asia.
Cedrus sp. Cedars. Genus of evergreen conifers.
Celtis sp. Hackberries, sugar berries and nettle-trees. Genus of trees, fruits are drupes. 80 species distribute in northern hemisphere to Old World tropics.
Chenopodiacea Goose feet. Family most of herbs but some shrubs. The fruits are one-seed nuts. *Beta* (sugar beet) and *Spinachia* (spinach) are important vegetables or sources of sugar. About 75 genera, 500 species, distribution mainly in arid regions or saline habitats, also as weed in enriched cultivated or waste ground.
Cinnamomum sp. Genus of trees. The fruit is a berry. 250 species, occurring in E. and SE. Asia.
Compositae The Daisy family. The largest family of flowering plants. The family include may cultivated plants, eg. *Lactuca sativa* (lettuce), oil-producing plants [*Helianthus annus* (sunflower)], etc. with cosmopolitan distribution.
Concentricystes sp. Genus mainly of herbs? Pleistocene fossil.
Convolvulaceae Family mostly of twinning herbs, but with some trees and shrubs. 1000 species occur mainly in tropical but also in temperate regions.
Cornus sp. Genus of trees mostly temperate in distribution.
Corylus sp. Genus of small trees or shrubs. Some have edible nuts: the hazel, filbert, and cobnut. 15 species occur in northern hemisphere forests.
Crataegus sp. Genus of shrubs and trees.
Cruciferae Family of mostly annual or perennial herbs. 2000 species, mainly in N. hemisphere but cosmopolitan.
Cunninghamia sp. Broad-leafed China fir. Genus of tall, coniferous trees. 2 species, one in SW China and another in Taiwan.
Cupressus sp. Cypress. Genus of trees, 12 species in N. temperate zone.
Cyatheaceae Family of ferns, most are massive tree ferns. 8 genera, occurring mainly in the tropics and moist, warm, temperate southern hemisphere.

Cyclobalanopsis. = *Quercus* sp.

Cyperaceae — Family of monocotyledonous herbs of grass-like habit, occurring mostly in wet places or on poor soils. The family contains some garden plants, but of little economic importance.

Ephedra sp. — Genus of shrubs. 400 species, found in the warm-temperate regions of each hemisphere.

Euphorbiaceae — Huge family mostly of trees and shrubs but a few herbs. 5000 species cosmopolitan distribution.

Fagus sp. — Beeches. Genus of trees, 10 species occurring in N. temperate regions. Some with edible mast, but rarely produce until the tree is 40 years old.

Gramineae — The Grass family.

Hamamelis sp. — Genus of witch hazel.

Humulus sp. — Genus of perennial climbing herbs. 4 species mainly found in N. temperate zones.

Ilex sp. — Genus of trees and shrubs. The fruit is a berry. 400 species of cosmopolitan distribution except for N. America.

Juglans sp. — Walnut. Genus of deciduous, nut-bearing trees. 15 species mainly found in N. temperate regions, but extending to the tropics and native to Asia and America.

Keteleeria sp. — Genus of conifers. Native to the mountains of China.

Larix sp. — Larches. Genus of coniferous trees with deciduous leaves in tufts. 12 species found in the cold temperate parts of the N. hemisphere.

Lauraceae — Family of trees and shrubs. 2000-2500 species occurring in the tropics and sub-tropics, and centred in Amazonia and SE Asia.

Leguminosae — The pea, lentil or pulse family, contains trees, climbers, shrubs and herbs, some of which are aquatic. 170,000 species with a cosmopolitan distribution.

Liliaceae — Large family of monocotyledons, mostly herbs. 2500 species with a cosmopolitan distribution.

Liquidambar sp. — Small genus of trees and shrubs occur in temperate areas of E. Asia, N. America etc.

Lonicera sp. — Honeysuckles. Genus of shrubs or woody climbers, most of them deciduous.

Myricaceae — Family of shrubs. One genus, with 60 species, found through most of the world. The fruit is a nut or a drupe.

Myriophyllum sp. — Water milfoils. Genus of submerged fresh water aquatic herbs with a few terrestrial species. The genus forms winter buds, which survive in the bottom of a lake or river, and from which the plant can develop in the spring. There are about 45 species, with a cosmopolitan distribution.

Nuphar sp. — Genus of aquatic herbs. Found throughout tropical and temperate areas of the world.

Oleaceae — Family of trees and shrubs. 21 genera, 400 species, widely distributed but centred on Asia.

Osmunda sp. — Genus of large ferns with massive, erect rhizomes, and bipinnate leaves. 12 species, cosmopolitan in distribution but absent from Australia.

Palmae — The palms family. Family of trees, many species yielding economic products, including thatch, starch and oil. 212 genera and 2780 species, chiefly tropical, but some subtropical and a few occurring as temperate outliers.

Picea sp. — Spruce.

Plantago sp. — Genus of herbs.

Polygonaceae — Buckwheats, docks, rhubarb and sorrels. Family of herbs, with a few shrubs and trees. 30 genera, 700 species, distributed throughout the world but with a concentration in the northern temperate regions.

Polypodiaceae — Family of ferns. Cosmopolitan distribution.

Polypodium sp. — Polypodies. Genus of ferns. Cosmopolitan distribution.

Potamogetonaceae — Pondweeds. Cosmopolitan family of aquatic herbs, found mostly in fresh water.

Potamogeton sp. — Genus of water plants, found throughout the world in a variety of aquatic habitats.

Pteridium sp. — Bracken. Monospecific genus of fern.

Pteris sp. — Genus of ferns with creeping rhizomes and long pinnate. The genus is largely tropical.

Pterocarya sp. — Wing nuts. Genus of deciduous trees, 12 species occurring from the Caucasus to Japan.

Quercus sp. — Oak. Genus of deciduous and evergreen trees and shrubs of the beech family. 500 species, found in N. hemisphere temperate zones, sub-tropical and tropical Asia, and the Andes.

Ranunculaceae — Family of dicotyledons, mostly herbs but including a few shrubs. 1300 species, occurring mostly in the N. hemisphere, arctic and alpine zones.

Rhus sp.	Sumach. Genus of shrubs, small trees or climbers. 250 species occurring in tropical and warm-temperate regions.
Rosaceae	The Rose family. Family of trees, shrubs and herbs. Many are cultivated for their edible fruits or flowers. Some 90 genera, 2000 species, mostly in temperate zones, but cosmopolitan overall.
Rubiaceae	One of the biggest family of flowering plants, comprising trees, shrubs and a few herbs. Various species are the sources of quinine, coffee, and many ornamentals. 500 genera, 7000 species, cosmopolitan distribution but strongly centred in the tropics.
Rutaceae	Family of trees and shrubs, occur in tropical and warm-temperate regions.
Sagittaria sp.	Arrow- root.
Salix sp.	Willows. Genus of trees. 300 species, most of them northern temperate or arctic, but with a few tropical and southern temperate outliers.
Selaginella sp.	Genus of trees and ferns.
Sparganium sp.	Genus of aquatic plants.
Taxodiaceae	Family of conifers, but leaves in a few genera are deciduous. Mainly found in temperate zones.
Thalictrum sp.	Genus of herbs.
Tilia sp.	Lime tree, Basswood, Linden.
Trapa sp.	Water chestnuts.
Tsuga sp.	Hemlock firs. Genus of evergreen conifers. 10 species, some native to temperate N. America, others to E. Asia.
Typha sp.	Reedmace. Genus of tall, stout, rhizomatous herbs growing in shallow water. 10-20 species found throughout shallow, fresh-water habitats of temperate and tropical regions.
Ulmus sp.	Elms. Genus of mostly large trees. 30 species in N. temperate zone and mountains of the Asian tropics.
Umbelliferae	Family of dicotyledonous herbs with a few shrubs. 200 genera, 2700 species found throughout most of the world, but mainly in N. temperate regions.
Zizyphus jujuba	Jujube (like date). Genus of small trees. 100 species occurring in the tropics and subtropics.

Sources:

Allaby, Michael ed.
1985 *The Oxford Dictionary of Natural History.* Oxford: Oxford University Press.
Mabberley, D. J.
1993 *The Plant Book.* Cambridge: Cambridge University Press.
Smith, William T.
1972 *A Gardener's Dictionary of Plant Names.* London: The Camelot Press Ltd.